THE SEA

Ideas and Observations on Progress in the Study of the Seas

Volume 1	HILL/PHYSICAL OCEANOGRAPHY
Volume 2	HILL/COMPOSITION OF SEA WATER
Volume 3	HILL/THE EARTH BENEATH THE SEA
Volume 4	MAXWELL/NEW CONCEPTS OF SEA FLOOR EVOLUTION (IN TWO PARTS)
Volume 5	GOLDBERG/MARINE CHEMISTRY
Volume 6	GOLDBERG/MARINE MODELING
Volume 7	EMILIANI/OCEANIC LITHOSPHERE
Volume 8	ROWE/DEEP-SEA BIOLOGY
Volume 9	Le MÉHAUTÉ/OCEAN ENGINEERING SCIENCE
Volume 10	BRINK AND ROBINSON/THE GLOBAL COASTAL OCEAN: PROCESSES AND METHODS
Volume 11	ROBINSON AND BRINK/THE GLOBAL COASTAL OCEAN: REGIONAL STUDIES AND SYNTHESES
Volume 12	ROBINSON, McCARTHY, AND ROTHSCHILD/BIOLOGICAL-PHYSICAL INTERACTIONS IN THE SEA
Volume 13	ROBINSON AND BRINK/THE GLOBAL COASTAL OCEAN: MULTISCALE INTERDISCIPLINARY PROCESSES
Volume 14	ROBINSON AND BRINK/THE GLOBAL COASTAL OCEAN: INTERDISCIPLINARY REGIONAL STUDIES AND SYNTHESES (IN TWO PARTS)

THE SEA

Ideas and Observations on Progress in the Study of the Seas

THE GLOBAL COASTAL OCEAN

INTERDISCIPLINARY REGIONAL STUDIES AND SYNTHESES

PART A: PANREGIONAL SYNTHESES AND THE COASTS OF NORTH AND SOUTH AMERICA AND ASIA

PART B: THE COASTS OF AFRICA, EUROPE, MIDDLE EAST, OCEANIA AND POLAR REGIONS

Edited by

ALLAN R. ROBINSON and **KENNETH H. BRINK**
Harvard University *Woods Hole Oceanographic Institution*

THE SEA

Ideas and Observations on Progress in the Study of the Seas

Volume 14
Part B

Harvard University Press
Cambridge, MA

Printed in the United States of America

Published with the financial assistance of UNESCO/IOC.

Library of Congress Cataloging-in-Publication Data

Robinson, Allan R.

The Sea: Ideas and Observations on Progress in the Study of Seas
Volume 14. Interdisciplinary Regional Studies and Syntheses
Part A. Panregional Syntheses and the Coasts of North and South America and Asia
Part B. The Coasts of Africa, Europe, Middle East, Oceania, and Polar Regions
Allan R. Robinson and Kenneth H. Brink
Includes bibliographic information and index

1. Oceanography   2. Submarine geology
Part A. ISBN-13: 978-0674-1527-2
         ISBN-10: 0-674-01527-4

Part B. ISBN-13: 978-0-674--2117-4
         ISBN-10: 0-674-02117-7
Library of Congress catalog card number: 62018366

CONTENTS

THE GLOBAL COASTAL OCEAN INTERDISCIPLINARY REGIONAL STUDIES AND SYNTHESES

**The number and letter in parentheses following each title refer respectively to the coastal segment indicated on the map following the contents and to the panregional overview in Part 1 which includes that segment. Note that the region numbers differ from the chapter numbers.*

PART B

FOREWORD

Coastal areas are vital to the life-support system of our planet. They represent 20 percent of the Earth's surface, yet serve as the home for over 50 percent of the entire human population. Coastal populations are expected to account for 75 percent of the total world population by the year 2025. A significant number of the world's megacities, with more than 8 million inhabitants, are located in coastal areas. Coastal ecosystems produce about 25 percent of global biological productivity and yield 90 percent of global fisheries. As a result of these dense human populations, the coasts suffer the effects of intense coastal development that threatens the integrity of coastal ecosystems.

The effective management and protection of coastal ecosystems must be science-based. With this general purpose in mind, the COASTS Programme, sponsored by the Intergovernmental Oceanographic Commission of UNESCO and the Scientific Committee on Oceanic Research, was established *to promote and facilitate research and applications in interdisciplinary coastal and shelf ocean sciences and technology on a global basis to increase scientific understanding of coastal ocean processes.*

The production line of science functions constantly and without pause. It works through a myriad of individual contributions and the publication of scientific "papers," often written in isolation from other work. Each project answers the next question down a logical chain of analysis of a given phenomenon. Each answer reflects and acknowledges the particular perspective of the authors. Apparently, there is no "a priori" plan. But collectively, all these elemental contributions build the edifice of science, and push back the boundary of the unknown.

For this collective knowledge to be used, a major effort of synthesis must be organized on a regular basis. This is the purpose for COASTS for the management of coastal oceans. Through this programme, the IOC is facilitating the synthesis of oceanographic knowledge and communicating those results to a wide audience.

We are fortunate in that the recent advances in interdisciplinary ocean sciences allow us to identify research issues better and to make feasible definitive studies on several topics. These advances need to be understood by all at national, regional and global levels. This need is very effectively served by the monographic volumes of *The Sea*, which are a reference point in the advance of oceanography, helping to train future generations of oceanographers around the world.

The results emerging from the present volumes are expected to lead to the sustainable use of the resources of the coastal oceans, producing a knowledge base on the science of the coastal oceans for developed and developing nations alike. The Intergovernmental Oceanographic Commission of UNESCO is proud to be associated with this very meaningful and high quality academic activity.

IOC is grateful to Professor Allan Robinson for his leadership of the COASTS Programme, and to Dr. Kenneth Brink, along with the authors of the introductory chapters and the external reviewers, together with a large international ocean science community, for their contributions towards the realization of Volumes 13 and 14 of *The Sea*.

PATRICIO A. BERNAL
ASSISTANT DIRECTOR GENERAL, UNESCO
EXECUTIVE SECRETARY, IOC

July, 2004

PREFACE
THE INTERDISCIPLINARY GLOBAL COASTAL OCEAN

Companion Volumes 13 and 14 of THE SEA deal respectively with fundamental multiscale interdisciplinary processes that occur generally throughout the coastal seas of our planet, and with the specific combination of processes which govern the dynamics of regions of the coastal seas which together constitute the whole. The term *coastal ocean* here is used for the combined shelf and slope seas and the term *interdisciplinary processes* implies coupled physical-biological-chemical-sedimentological processes and interactions. These volumes complement and supplement Volumes 10 and 11 of THE SEA, which respectively deal generally and regionally with the physical oceanography of the global coastal oceans. The intent of these volumes is to review and overview the status and prospectus of fundamental coastal ocean science in the context of advanced applications and as input to enhanced rational management of coastal seas.

Volumes 13 and 14 are comprised of six parts. Part 1 (13) provides general perspective, and Chapter 1 on multiscale interdisciplinary processes, which has been prepared as a general introduction, refers throughout to all the other chapters of the volume. Part 2 (13) on sediment, biogeochemical and ecosystem dynamics and Part 3 (13) on episodic and long time scale dynamics together constitute the dynamical core of the volume. Part 4 (13) presents scientific issues for applications. The chapters of Part 1 (14) are panregional syntheses for western ocean boundaries (W), eastern ocean boundaries (E), polar ocean boundaries (P), and semi-enclosed seas, islands and Australia (S). These four chapters serve to summarize and introduce the material in the chapters on regional interdisciplinary oceanography, Part 2 (14). The regional segments of the global coastal ocean are indicated on the map following the table of contents of Volume 14 and their region numbers and panregional attributions are indicated following the chapter titles of Part 2 (14) in the table of contents. (Note that the region numbers differ from the chapter numbers.) In order to facilitate coordinated study of material presented in Volumes 11 and 14, the numbering of regional segments is the same. However, here it was necessary to subdivide segment 10 off the coast of eastern Asia into two segments (one coastward of the Oyashio current and the other coastward of the Kuroshio current), and to extend segment 18 northward into the Bay of Biscay, which was not treated in Volume 11. Thus the map in Volume 14 indicates four segments as 10 a, b and 18 a, b.

Note that the coastal segments on the map are numbered geographically, i.e., consecutively moving along a global coastline rather than by panregional groupings. Because Volume 14 comprises 1568 pages of text, it was necessary to publish it in two parts with the geographically numbered coastal segments split between segments 14 and 15. Thus Volume 14, Part A is subtitled "Panregional Syntheses and the Coasts of North and South America and Asia," and Volume 14, Part B is subtitled "The Coasts of Africa, Europe, Middle East, Oceania and Polar Regions." Both Parts A and B contain all introductory material and a complete index.

In the preface to Volume 1 of THE SEA in 1962, Dr. Maurice N. Hill states, "In collecting the material for these volumes we have had most helpful cooperation from our contributors. In some topics, however, ... we must admit ... to omissions, some of which are conspicuous and important." In Volume 13 we had intended to include a chapter on coral reefs. All thirty-two regional segments of the global coastal ocean are treated in Part 2 (14) except for region 25 (northern North America and West Greenland) because of the withdrawal of the potential author. However, aspects of that region's oceanography are discussed in the chapter on the Northwest Passage (region 26). Additionally, region 17 (the Gulf of Guinea) is treated only by a note, and region 9 (the Northeastern Pacific shelves) only by a bibliography.

The in-depth and comprehensive synthesis of interdisciplinary coastal ocean science presented in these volumes was possible only through the dedicated and scholarly collaboration of the international community of coastal ocean scientists. Chapters were reviewed both by authors of other chapters in the volumes and by external reviewers, whose expertise has contributed essentially to this study. Following the list of contributors is a list of external reviewers in alphabetical order.

This pair of Volumes is a contribution of the COASTS (Coastal Ocean Advanced Science and Technology Studies) program of the Intergovernmental Oceanographic Commission (IOC) of UNESCO and the Scientific Committee on Oceanic Research (SCOR). The scope and structure of the volumes were developed and most chapter topics were presented and discussed at the second COASTS international workshop attended by over sixty scientists, held in Paris in August 2001 under the support of IOC/UNESCO, SCOR, and the U.S. ONR (Office of Naval Research). Subsequent support for the preparation and production of the volumes was provided by IOC and ONR. We sincerely appreciate the support of these agencies which has enabled the substantial international cooperation necessary to accomplish this study.

It is a pleasure to thank Dr. Patricio A. Bernal, Assistant Director General, UNESCO/ Executive Secretary, IOC, for his encouragement and stimulation and for preparing a foreword to these volumes, and Dr. Umit Unluata, Head of the Ocean Science Section of IOC, for his guidance and advice. We are most pleased that Harvard University Press (HUP) is now publisher of *The Sea* series starting with these volumes, and thank Mr. Michael G. Fisher, Executive Editor in Science and Medicine, for his interest and help. The challenging task of preparing for publication this comprehensive study of 60 chapters authored by 170 international scientists was only made possible by the dedicated scientific expository skills of Mr. Wayne G. Leslie and the dedicated administrative editorial expertise of Ms. Gioia L. Sweetland of Harvard University. We thank M. Julian Barbiere and Dr. Maria Hood (IOC), Ms. Sara Davis (HUP), and Mr. Oleg Logoutov and Ms. Margaret S. Zaldivar (Harvard) for their valued contributions to this study and its publication. ARR acknowledges ONR grant N00014-02-1-0989 and KHB acknowledges grant NSF-OCE-0227679 for support throughout this study.

ALLAN R. ROBINSON
KENNETH H. BRINK

July, 2004

CONTRIBUTORS

PABLO ABAUNZA
Centro Oceanográfico de Santander
Instituto Español de Oceanografía
Apdo. 240
San Martín s/n
39080 Santander
Spain

AHMAD ABU-HILAL
Department of Earth and Environmental Sciences
Yarmouk University
Irbid
Jordan

XOSÉ A. ÁLVAREZ-SALGADO
Departamento de Oceanografía
Instituto de Investigacións Mariñas
Consejo Superior de Investigaciones Científicas (CSIC)
Eduardo Cabello 6
36208 Vigo
Spain

JAVIER ARISTEGUI
Facultad de Ciencias del Mar
Universidad Las Palmas de Gran Canaria
Campus Universitario de Tafira
35017 Las Palmas de Gran Canaria
Canary Islands
Spain

ENRICO ARNERI
Istituto di Scienze Marine-ISMAR
Consiglio Nazionale delle Ricerche
Sezione Pesca Marittima di Ancona
Largo Fiera della Pesca
60125 Ancona
Italy

ERIC D. BARTON
Departamento de Oceanografía
Instituto de Investigacións Mariñas
Consejo Superior de Investigaciones Científicas (CSIC)
Eduardo Cabello 6
36208 Vigo
Spain

TIM BAUMGARTNER
Departmento de Oceanografia Biologica
Centro de Investigacion Cientifica y Educacion Superior de Ensenada
Km. 107, Carretera Tijuana-Ensenada, C.P. 22860
Ensenada, Baja California
Mexico

JEAN BOUCHER
IFREMER
Technopole Brest-Iroise
B.P. 70
Fr-29280 Plouzané
France

FREDERICO P. BRANDINI
Av. Beiramar, s/n°
83255-000 - Pontal do Paraná – PR
Brazil

EDWARD C. CARMACK
Climate Research Oceanographer
Institute of Ocean Sciences
9860 W. Saanich Road
P. O. Box 6000
Sidney, B.C. V8L 4B2
Canada

BELMIRO M. CASTRO
University of São Paulo, Oceanographic Institute
191 Praca do Oceanographico
05508-900, São Paulo
Brazil

SHENN-YU CHAO
Horn Point Laboratory
University of Maryland
Center for Environmental Science
2020 Horns Point Road
P. O. Box 775
Cambridge, MD 21613-0775

FRANCISCO CHAVEZ
MBARI
Monterey Bay Aquarium Research Institute
7700 Sandholt Rd.
Moss Landing, CA 95039-9644

NATALIA V. CHERNOVA
Laboratory for Ichthyology
Zoological Institute
1, Universitetskaja nab.
199034, St.Petersburg
Russia

SCOTT A. CONDIE
CSIRO Marine Research
GPO Box 1538
Hobart, Tasmania 7001
Australia

ALESSANDRO CRISE
Dipartimento di Oceanografia
Istituto Nazionale di Oceanografia e di Geofisica Sperimentale -OGS
Borgo Grotta Gigante 42/c
34010 Sgonico, Trieste
Italy

EHRLICH DESA
National Institute of Oceanography (CISR)
Dona Paula, Goa, 403 004
India

VEDAT EDIGER
Institute of Marine Sciences
P.O Box 28, 33731
Erdemli, Icel
Turkey

ELVA ESCOBAR-BRIONES
Universidad Nacional Autonoma de México
Instituto de Ciencias del Mar y Limnologia
Unidad Academica Sistemas Oceanográficos y Costeros
A.P. 70-305 Ciudad Universitaria 04510
México

JOHN G. FIELD
Marine Biology Research Institute
Zoology Department
University of Cape Town
7701 Rondebosch
South Africa

FRANCISCO G. FIGUEIRAS
Departamento de Oceanografía
Instituto de Investigacións Mariñas
Consejo Superior de Investigaciones Científicas (CSIC)
Eduardo Cabello 6
36208 Vigo
Spain

ANDRÉ FOREST
IFREMER
Laboratoire Maerha
Rue de l'Ile de Yeu
B.P. 21105
FR-44311 Nantes cedex 3
France

EILIF GAARD
Faroese Fisheries Laboratory
Box 3051, FO-110 Torshavn
Faroe Islands

ASTTHOR GISLASON
Marine Research Institute
Skulagata 4, P.O. Box 1390
121 Reykjavik
Iceland

JANET M. BRADFORD GRIEVE
NIWA Wellington
301 Evans Bay Parade
Private Bag 14901
Kilbirnie, Wellington
New Zealand

PETER T. HARRIS
Petroleum & Marine Division
Geoscience Australia
GPO Box 378
Canberra ACT 2601
Australia

JUAN CARLOS HERGUERA
Centro de Investigación Científica y de Educación Superior
Ensenada, Baja California, 22800
México

SANTIAGO HERNANDEZ-LEON
Facultad de Ciencias del Mar
Universidad Las Palmas de Gran Canaria
Campus Universitario de Tafira
35017 Las Palmas de Gran Canaria
Canary Islands
Spain

JENS A. HOLEMANN
Alfred-Wegener-Institut für Pola
r-und Meeresforschung (AWI)
Columbusstrasse
D-27568 Bremerhaven
Germany

PATRICK M. HOLLIGAN
Southampton Oceanography Centre
Empress Dock
Southampton SO14 3ZH
United Kingdom

GEORGE L. HUNT, JR.
Department of Ecology and Evolutionary Biology
465 Steinhaus Hall
University of California
Irvine, CA 92697-2525

SANGMIN HYUN
Marine Geology and Geochemical Lab
Korea Ocean Research and Development Institute
391 Jangmok-ri, Jangmok-myon
Geoje 656.830, Korea

R. GRANT INGRAM
Department of Earth and Ocean Sciences
University of British Columbia
6339 Stores Road
Vancouver, B.C., V6T 1Z4
Canada

OSAMU ISOGUCHI
Center for Atmospheric and Oceanic Studies
Graduate School of Science
Tohoku University
Sendai, Miyagi 980-8578
Japan

ANNE-MARIE JEGOU
IFREMER/DEL/AO
BP 70
29280 Plouzane
France

HEIDEMARIE KASSENS
Forschungzentrum für marine Geowissenschaften (GEOMAR)
Wischhofstrasse 1-3, Geb.4
24148 Kiel
Germany

HIROSHI KAWAMURA
Center for Atmospheric and Oceanic Studies
Graduate School of Science
Tohoku University
Sendai, Miyagi 980-8578
Japan

A. ERKAN KIDEYS
Institute of Marine Sciences
P.O Box 28, 33731
Erdemli, Icel
Turkey

MICHIO J. KISHI
Professor in Marine Environmental Science
Graduate School of Fisheries Sciences
Hokkaido University
Hakodate, Hokkaido, 041-8611
Japan

OLGA A. KIYKO
Department of System Geoecological Study
All-Russia Research Institute for Geology and Mineral Resources of the World Ocean
(VNII Okeangeologia)
1, Angliysky ave.
190121, St. Petersburg
Russia

NILGUN KUBILAY
Institute of Marine Sciences
P.O Box 28
33731 Erdemli, Icel
Turkey

MIKHAIL YU. KULAKOV
Department of Oceanography
Arctic and Antarctic Research Institute
38, Bering str.
St Petersburg, 199397
Russia

ALICIA M. LAVIN
Centro Oceanografico de Santander
Instituto Español de Oceanografia
Apdo 240
Pr. San Martin s/n
39080 Santander
Spain

PASCAL LAZURE
IFREMER
B.P. 70
Fr-29280 Brest
France

KEITH B. LEWIS
NIWA
235 The Terrace
Wellington
New Zealand

KON-KEE LIU
Institute of Hydrological Sciences
National Central University
Jungli, Taiwan 320
Republic of China

STEVEN E. LOHRENZ
University of Southern Mississippi
Department of Marine Science
1020 Balch Boulevard
Stennis Space Center, MS 39529-5005

JOHANN R. E. LUTJEHARMS
Department of Oceanography
University of Cape Town
7700 Rondebosch
South Africa

DAVID L. MACKAS
Institute of Ocean Sciences, Fisheries and Oceans
Sidney, BC Canada V8L 4B2
1000 Pope Road, MSB 505
Honolulu, HI 96822

JORGE MARCOVECCHIO
Instituto Argentino de Oceanografia
CC 804
Florida 8000 Complejo CRIBABB
Edificio El
8000 Bahia Blanca
Argentina

JOHN MARRA
Lamont-Doherty Earth Observatory
Columbia University
61 RT 9W
Palisades, NY 10964-8000

LAWRENCE M. MAYER
School of Marine Sciences
University of Maine
Darling Maine Center
193 Clark's Cove Road
Walpole, ME 04573

FIONA A. MCLAUGHLIN
Institute of Ocean Sciences
9860 W. Saanich Road
Sidney, B.C., V8L 4B2
Canada

WEBJØRN MELLE
Institute of Marine Research
P.O. Box 1870 Nordnes
NO-5817 Bergen
Norway

CHRISTINE MICHEL
Freshwater Institute
501 University Crescent
Winnipeg, Manitoba
Canada

LUIZ B. MIRANDA
Praça do Oceanográfico, 191
05508-900 - São Paulo - SP
Brazil

VIVIAN MONTECINO
Facultad de Ciencas
Universitad de Chile
Casilla 653
Las Palmeras 3425
Santiago, Chile

JEFFREY M. NAPP
Alaska Fisheries Science Center
Bldg: 4
7600 Sand Point Way, NE
Seattle, WA 98115-6349

S. WAJIH A. NAQVI
National Institute of Oceanography
Chemical Oceanography
Dona Paula Goa, 403 004
India

PRADIP V. NARVEKAR
National Institute of Oceanography
Dona Paula, Goa 403 004
India

VRADIM V. NAVROTSKY
V. I. Ilyichev Pacific Oceanological Institute
43 Baltiyskaya Street
Vladivosotk 690041
Russia

STEPHEN NICOL
Department of the Environmental and Heritage
Australian Antarctic Division and the Antarctic Climate and Ecosystems, CRC
Channel Highway
Kingston, Tasmania, 7050
Australia

TEMEL OGUZ
Institute of Marine Sciences
Middle East Technical University
P.O.Box 28
33731, Erdemli, Icel
Turkey

ALAN R. ORPIN
NIWA Wellington
301 Evans Bay Parade
Private Bag 14901
Kilbirnie, Wellington
New Zealand

GERARDO M.E. PERILLO
Instituto Argentino de Oceanografia
CC 804 - Florida 8000 Complejo CRIBABB Edificio E1
8000 Bahía Blanca
Argentina

M. CINTIA PICCOLO
Instituto Argentino de Oceanografie
CC 804
Florida 8000 Complejo CRIBABB Edificio El
8000 Bahia Blanca
Argentina

DIETER PIEPENBURG
Akademie der Wissenschaften und der Literatur Mainz.
Institut für Polarökologie der Universität Kiel
Wischhofstrasse 1-3, Geb.12
24148 Kiel
Germany

ANA MARIA S. PIRES-VANIN
Praça do Oceanográfico, 191
05508-900 - São Paulo - SP
Brazil

NADIA PINARDI
Laboratorio SINCEM
Corso di Laurea in Scienze Ambientali
University of Bologna
Via S. Alberto 163
I-48100 – Ravenna
Italy

SERGEY V. PIVOVAROV
Senior Scientist, Department of Oceanography
Arctic and Antarctic Research Institute
38, Bering Str.
St. Petersburg, 199397
Russia

VLADIMIR B. POGREBOV
Laboratory for Ecology and Nature Conservation
Arctic and Antarctic Research Institute
38, Bering str.
St. Petersburg, 199397
Russia

P. KEITH PROBERT
Department of Marine Science
University of Otago
P. O. Box 56
Dunedin
New Zealand

JOHN A. QUINLAN
Institute of Marine and Coastal Science
Rutgers University
71 Dudley Road
New Brunswick, NJ 08901-8521

VLADIMIR I. RADCHENKO
Sakhalin Research Institute of Fisheries and Oceanography (SakhNIRO)
196 Komsomolskaya Street
Yuzhno-Sakhalinsk, 693023
Russia

MARIANGELA RAVAIOLI
Istituto di Scienze Marine-ISMAR
Consiglio Nazionale delle Ricerche
Sezione di Geologia Marina
Via Gobetti, 101
Bologna
Italy

CLAUDIO RICHTER
Center for Tropical Marine Ecology (ZMT)
Fahrenheitstr. 6
D-28359 Bremen
Germany

JOHAN RODHE
Department of Earth Sciences
Oceanography
Göteborg University, Box 460
SE-405 30 Göteborg
Sweden

CLAUDE ROY
Centre IRD de Bretagne
BP 70
29280 Plouzané
France

THOMAS C. ROYER
Center for Coastal Physical Oceanography
Department of Ocean, Earth and Atmospheric Sciences
Old Dominion University
768 W. 52nd Street
Norfolk, VA 23529

FRANCISCO SANCHEZ
Centro Oceanográfico de Santander
Instituto Español de Oceanografía
Apdo. 240
San Martín s/n
39080 Santander
Spain

ANTONIO M.P. SANTOS
IPIMAR
Instituto Nacional de Investigação Agrária e das Pescas (INIAP)
Av. Brasilia s/n
1449-006 Lisboa
Portugal

MICHAEL K. SCHMID
Alfred-Wegener-Intitute für Polar- und Meeresforschung (AWI)
Institut für Polarökologie der Universität Kiel
Wischhofstrasse 1-3, Geb.12
24148 Kiel
Germany

JAMES D. SCHUMACHER
288 Ivan Road
Friday Harbor, WA 98250

JONATHAN SHARPLES
Proudman Oceanographic Institute
Bidston Observatory
Birkenhead CH43 7RA
United Kingdom

FRANK A. SHILLINGTON
Oceanography Department
University of Cape Town
7701 Rondebosch
South Africa

ANOND SNIDVONGS
Marine Science Department
Sea Start RC
Chulalongkorn University
Bangkok 10330
Thailand

PHYLLIS J. STABENO
Pacific Marine Environmental Laboratory
Bldg: 3
7600 Sand Point Way, NE
Seattle WA 98115-6349

P. TED STRUB
College of Oceanic & Atmospheric Sciences
Oregon State University
104 Ocean Admin Bldg
Corvallis, OR 97331-5503

PETER G. STRUTTON
Marine Sciences Research Center
State University of New York
Stony Brook, NY 11794-5000

SU JILAN
Second Institute of Oceanography
State Oceanic Administration
P.O. Box 1207, 36 Bao-Chu-Bei-Lu
Hangzhou, Zhejiang 310012
China

TAKASHIGE SUGIMOTO
Ocean Research Institute
University of Tokyo
1-15-1, Minamidai
Nakanoku, Tokyo 164-8639
Japan

PHILIP SUTTON
NIWA Wellington
301 Evans Bay Parade
Private Bag 14901
Kilbirnie, Wellington
New Zealand

JUAN TARAZONA
Facultad de Ciencias Biologicas
Universidad Nacioinal Mayor de San Marcos
P. O. Box 1898
Lima 100
Peru

PAUL TETT
School of Life Sciences
Napier University
10 Colinton Rd.
Edinburgh EH10 5DT
Scotland

ANDREW C. THOMAS
School of Marine Sciences
5741 Libby Hall
University of Maine
Orono, ME 04469-5741

MAURA A. THOMAS
School of Marine Sciences
5706 Aubert Hall
University of Maine
Orono, ME 04469-5706

SERGEY F. TIMOFEYEV
Laboratory for Plankton
Murmansk Marine Biological Institute
17, Vladimirskaia str.
183010, Murmansk
Russia

DAVID W. TOWNSEND
School of Marine Sciences
5706 Aubert Hall
University of Maine
Orono, ME 04469-5706

THOMAS W. TRULL
Antarctic Climate and Ecosystem Cooperative Research Centre
University of Tasmania
Private Bag 80
Hobart, 7001, Tasmania
Australia

SULEYMAN TUGRUL
Institute of Marine Sciences
P.O Box 28
33731, Erdemli, Icel
Turkey

LUIS VALDES
Centro Oceanográfico de Gijón
Instituto Español de Oceanografía
Avda. Principe de Asturias 70 bis
33212 Gijón
Spain

PETER G. VERITY
Skidaway Institute of Oceanography
10 Ocean Science Circle
Savannah, GA 31411-1011

WILLIAM J. WILLIAMS
Institute of Ocean Sciences
9860 W. Saanich Road
Sidney, B.C., V8L 4B2
Canada

ANTHONY P. WORBY
Department of the Environment and Heritage
Australian Antarctic Division and the Antarctic Climate and Ecosystems
CRC
Channel Highway
Kingston, 7050, Tasmania
Australia

FREDRIK WULFF
Department of Systems Ecology
Stockholm University
SE-10691 Stockholm
Sweden

MARCO ZAVATARELLI
Laboratorio SINCEM
Corso di Laurea in Scienze Ambientali
University of Bologna
Via S. Alberto 163
I-48100 – Ravenna
Italy

JOHN ZELDIS
NIWA Christchurch
10 Kyle Street
P. O. Box 8602
Riccarton, Christchurch
New Zealand

CHANG-IK ZHANG
Department of Marine Production Management
Pukyong National University
Pusan 608.737
Korea

JING ZHANG
State Key Laboratory of Estuarine and Coastal Research
East China Normal University
3663 Zhongshan Road North
Shanghai, 200062
China

EXTERNAL REVIEWERS

KNUT AAGAARD
University of Washington
Seattle, Washington

ICARUS ALLEN
Plymouth Marine Laboratory
Plymouth, United Kingdom

CARIN J. ASHJIAN
Wood Hole Oceanographic Institution
Woods Hole, Massachusetts

PHILIP BOYD
University of Otago
Dunedin, New Zealand

GREGG J. BRUNSKILL
Australian Institute of Marine Science
Queensland, Australia

C. T. ARTHUR CHEN
National Sun Yat-Sen University
TAIWAN, R.o.C.

PETER CRAIG
CSIRO
Tasmania, Australia

PAUL C. FIEDLER
National Marine Fisheries Service
La Jolla, California

EILEEN HOFMANN
Old Dominion University
Norfolk, Virginia

ROBERT W. HOUGHTON
Lamont-Doherty Earth Observatory at Columbia University
Palisades, New York

IAN JOINT
Plymouth Marine Laboratory
Plymouth, United Kingdom

STEIN KAARTVEDT
University of Oslo
Oslo, Norway

BJÖRN KJERFVE
Texas A&M University
College Station, Texas

STEVEN J. LENTZ
Woods Hole Oceanographic Institution
Woods Hole, Massachusetts

ARNOLDO VALLE LEVINSON
Old Dominion University
Norfolk, Virginia

SKIP MCKINNELL
Institute of Ocean Sciences
Sidney, BC, Canada

CHRISTOPHER MOOERS
University of Miami
Miami, Florida

JAMES W. MURRAY
University of Washington
Seattle, Washington

DONALD OLSON
University of Miami
Miami, Florida

WILLIAM PETERSON
National Marine Fisheries Service
Newport, Oregon

ALBERT J. PLUEDDEMANN
Woods Hole Oceanographic Institution
Woods Hole, Massachusetts

LAWRENCE R. POMEROY
University of Georgia
Athens, Georgia

RAYMOND N. SAMBROTTO
Lamont-Doherty Earth Observatory
at Columbia University
Palisades, New York

CHARLES SHEPPARD
University of Warwick
Coventry, United Kingdom

JOHN SIMPSON
University of Wales Bangor
United Kingdom

EINAR SVENDSEN
Institute of Marine Research
Bergen, Norway

HIDETAKA TAKEOKA
Center for Marine Environmental
Studies
Matsuyama City, Japan

JOAQUIN TINTORE
IMEDEA (CSIC-UIB)
Esporles (Mallorca), Spain

MICHELLE WOOD
University of Oregon
Eugene, Oregon

ICHIRO YASUDA
The University of Tokyo
Tokyo, Japan

Chapter 20. THE COASTAL OCEANS OF SOUTH-EASTERN AFRICA (15,W)

JOHANN R.E. LUTJEHARMS

University of Cape Town

Contents

1. Introduction to the region

The coastal ocean off south-eastern Africa is characterised by at least one common, coherent aspect: it forms part of what may be considered to be the greater Agulhas Current circulation and all its components are largely dominated by this current system. In other respects it is very diverse (Schumann, 1998). It extends from the tropics to a region adjacent to the Subantarctic. The shelf regions are very narrow in some distinct parts and quite wide in others (Fig. 20.1). Certain parts of the shelf regions have been studied fairly intensively, while for other regions there is no data or information to speak of. The current extent of knowledge on this coastal ocean region is therefore, relying on the existing database, very inhomogeneous.

Knowledge on the bathymetry and geology of the region has not changed significantly since a previous review of this kind (Schumann, 1998), so will only be dealt with here where it affects other aspects of the nature of the coastal oceans. The equatorward border of the system lies at the northern mouth to the Mozambique Channel (Fig. 20.1). This is a useful and generic choice and not just one brought about by geographical pragmatism. The major influence of the Indian monsoon system on the dominant ocean currents lies entirely to the north of this border (e.g., Ridderinkhof and de Ruijter, 2003). Numerical model studies (e.g., Maltrud et al., 1998) suggest a residual monsoonal effect on currents in the Mo-

The Sea, Volume 14, edited by Allan R. Robinson and Kenneth H. Brink
ISBN 0–674–01527–4 ©2005 by the President and Fellows of Harvard College

zambique Channel, but direct current observations to date do not support this. By contrast, the poleward border of the region coincides with the termination of the Agulhas Current at the Agulhas Current Retroflection south-west of the southern tip of Africa. The eastern border of the coastal oceans of south-eastern Africa has to include the waters to the east of the island of Madagascar. As is described below, the coastal waters of the east coast of Madagascar are not a distinctly separate system (Cooke et al., 2004) but in many ways form a coherent part of the greater Agulhas Current system. Nevertheless, for organisational purposes the coastal oceans of this system can be separated into four distinctive parts: the shelf regions in the Mozambique Channel, the shelf regions east and south of Madagascar, the shelf inshore of the northern Agulhas Current and, last, inshore of the southern Agulhas Current. As will be seen, making this geographic distinction between the southern and northern parts of the Agulhas Current is important since the nature of the waters and the circulation on their respective shelves are dissimilar (viz. Fig. 20.1). This comes about as a result of the different behaviour of the edge of the Agulhas Current in these two regions. For an understanding of the characteristics of waters on the continental shelves it is therefore necessary briefly to describe the existing knowledge on the offshore ocean currents of the greater Agulhas Current system.

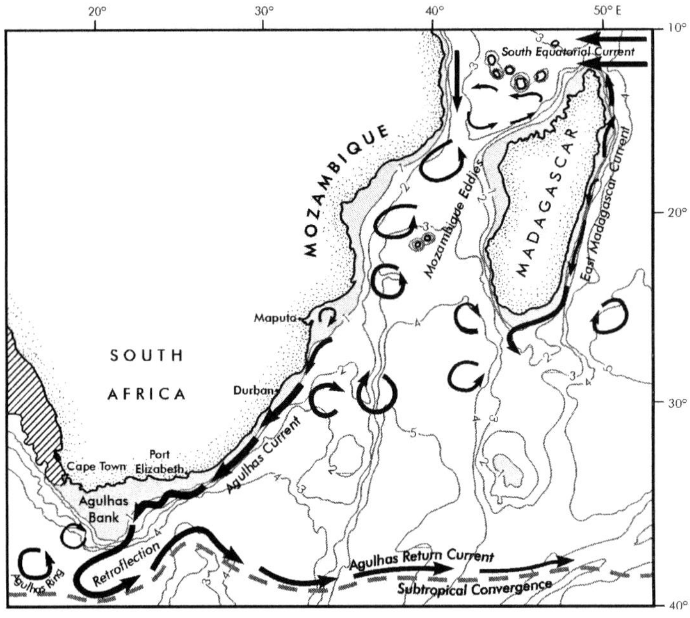

Figure 20.1 Bathymetry of the South West Indian Ocean in km (after Dingle et al., 1987a and Simpson, 1974) and the major circulation features. Shelf regions shallower than 1 km are shaded; hatching indicates upwelling. Some major place and feature names used in the text are given.

The Agulhas Current is supplied with water from essentially two different sources: the South Equatorial Current and recirculation in a South-West Indian Ocean subgyre (Stramma and Lutjeharms, 1997; Fig. 20.2). The greater part (40 Sv out of a total 65 Sv in the upper 1 000 m) comes from the subgyre. This configuration of the basin-wide circulation has been thought to be reflected in the behaviour and natural history of marine animals (Heydorn et al., 1978), for example the migrations of leatherback sea turtles (Hughes et al., 1998). The manner in which the South Equatorial Current acts as a source for the Agulhas Current is not entirely clear. It was previously thought that this current bifurcates on reaching the east coast of Madagascar—forming the southern and the northern branches of the East Madagascar Current. The northern branch of this current and the remainder of the South Equatorial Current would then pass the northern tip of Madagascar and move onwards to the east coast of the African continent. Here a similar split would occur with some of the water passing northwards into the Somali Current and the rest southward into the Mozambique Channel as the Mozambique Current.

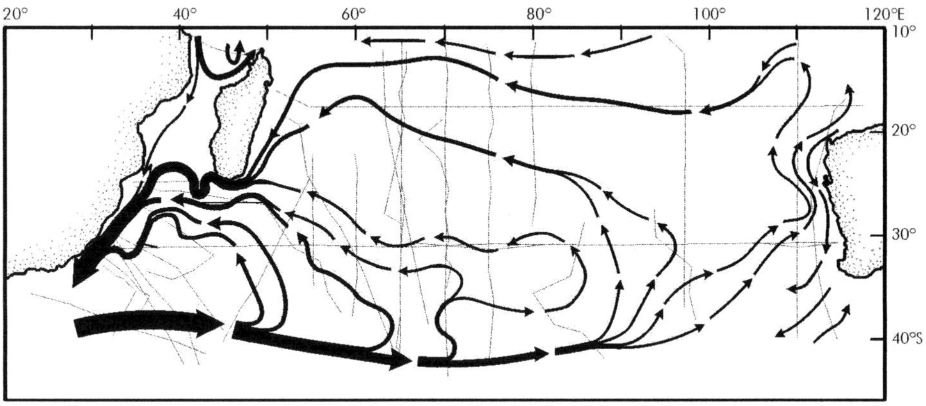

Figure 20.2 Circulation patterns of the South West Indian Ocean. (After Stramma and Lutjeharms, 1997.) The thickness of the lines denotes the volume flux in the upper 1 500 m. Thin lines indicate lines of historical hydrographic stations on which the portrayal is based. The Agulhas Current along south-eastern Africa is largely supported by recirculation in a South West Indian Ocean subgyre.

These two western boundary currents—the southern limb of the East Madagascar Current and the Mozambique Current—would then have a confluence somewhere off South Africa to form the Agulhas Current. This is the classical portrayal of flow in this ocean region (Michaelis, 1923) still found in many textbooks and most atlases and is largely based on analyses of ships' drift (e.g., Sætre, 1985; Lutjeharms et al., 2000b). However, modern observations have shown this depiction to be largely incorrect. The new findings on these currents are of substantial importance for understanding the flow patterns on the adjacent continental shelves.

It has been demonstrated that no coherent, unbroken western boundary current exists in the Mozambique Channel. First adumbrated by using the full set of non-synoptic hydrographic observations (Sætre and Jorge da Silva, 1984) as well as by modelling (Biastoch and Krauß, 1999), modern synoptic observations (De Ruijter et al., 2002) have unequivocally shown that mesoscale eddies are formed at the narrows of the channel and progress from here southward. No continuous current exists. On averaging the motions of the drifting eddies (as happens by plotting the means of many ships' drift observations; e.g., Lutjeharms et al., 2000b) there is the suggestion of a consistent and continuous boundary current, leading to the classical misinterpretation.

These Mozambique eddies drift southward along the shelf edge at speeds of about 5 cm/s and may eventually reach the Agulhas Current (Schouten et al., 2002). They form the major elements of the circulation in the Mozambique Channel. The movement in the rest of the channel seems to be sluggish and quite variable, but there are few observations to support any firm conclusion in this regard, particularly on the eastern side of the channel. Here there is a dearth of modern hydrographic or current meter observations (Lutjeharms, 1977). In contrast to classical portrayals, the Mozambique Channel is therefore now seen as a minor source of water for the Agulhas Current proper (Fig. 20.2), and only in an intermittent manner. The same holds for the East Madagascar Current.

Instead of being a direct tributary to the Agulhas Current, the southern branch of the East Madagascar Current has been noted to retroflect south of Madagascar both in hydrographic data (Lutjeharms et al., 1981a) as well as in satellite remote sensing (Lutjeharms, 1988). This implies that this current, as in the case of the purported Mozambique Current, contributes very little to the Agulhas Current, and if so, only spasmodically. However, the retroflection of the East Madagascar Current has been shown to be a region that generates both cyclonic as well as anticyclonic eddies (De Ruijter et al., 2003), carrying away some of the shelf waters in the process. Although these two currents systems—East Madagascar Current and Mozambique eddies—therefore do not form a continuum with the Agulhas Current itself, they do influence its behaviour indirectly and can therefore be considered to constitute an inherent part of the greater Agulhas system.

The Agulhas Current proper is fully constituted somewhere near 28° S, along the east coast of the province of KwaZulu-Natal of South Africa, between Maputo and Durban (Fig. 20.1). Here the continental shelf is narrow and the course of the current very stable. This is true for the whole of what might be considered the northern Agulhas Current (Gründlingh, 1983) extending downstream as far as Port Elizabeth, at the eastern end of the broad shelf region south of Africa known as the Agulhas Bank (Fig. 20.1). The coincidence of a narrow shelf and a very stable juxtapositioned current is not fortuitous; De Ruijter et al. (1999a) have demonstrated that the continental slope plays an important role in stabilising the trajectory of the current. The only part of the shelf that does not comply with these stabilising requirements lies just upstream of Durban, the Natal Bight. At this coastal offset the shelf is anomalously wide and the continental slope considerably gentler than at other locations adjacent to the north Agulhas Current. Perturbations on the current path that occur here will grow, move downstream and have considerable effects on the subsequent behaviour of the current (Van Leeuwen et al., 2000). This intermittent meander on the Agulhas Current (Lutjeharms and Roberts, 1988) is called the Natal Pulse and has a considerable effect on the circulation of the adjoining shelf region. When the waters in the Agulhas Current reach the Agulhas Bank, the nature of the current changes dramatically.

From the latitude of Port Elizabeth it is know as the southern Agulhas Current and, in contrast to the northern Agulhas Current, a range of meanders are formed on its shoreward side (Lutjeharms et al., 1989a). These meanders cause shear edge eddies and attendant plumes that move with the current and influence the upper waters of the shelf in this region. The average trajectory of the current follows the shelf edge, thus lying farther and farther from the coastline (viz. Fig. 20.1), until the tip of the Agulhas Bank is passed. In the western lee of the shelf this major current generates an intense cyclonic eddy (Penven et al., 2001a) that eventually drifts off into the South Atlantic Ocean. The current itself continues southwestward into the South Atlantic until it retroflects.

The Agulhas Retroflection is a region exhibiting some of the highest levels of mesoscale variability in the world ocean (Lutjeharms and van Ballegooyen, 1988; Garzoli et al., 1996). This is due to the manner in which the retroflection loop occludes, intermittently forming large Agulhas rings (Lutjeharms and Gordon, 1987) that then drift off into the South Atlantic (Duncombe Rae, 1991; Duncombe Rae et al., 1996; Schouten et al., 2000). It has been suggested (Duncombe Rae et al., 1992) that these rings may interact with the coastal upwelling on South Africa's west coast, but analyses of the tracks of such rings (e.g., Schouten et al., 2000) indicates that such interaction must be a rare exception. Most rings spend some time in the vicinity of where they have originally been shed (Boebel et al., 2003), losing a substantial part of their characteristic heat and salt (Arhan et al., 1999) and having their water masses exchanged with other rings (Fine et al., 1988) and with Agulhas cyclones (Lutjeharms et al., 2003b). There is some evidence that Agulhas rings may often be found next to the western edge of the Agulhas Bank (Lutjeharms and Valentine, 1988) and may play a role in carrying surface water directly from the Agulhas Current northward past the western edge of the bank (Lutjeharms and Cooper, 1996).

On moving away from their source region, Agulhas rings tend to move further away from the African coast (Byrne et al., 1995; Schouten et al., 2000). Comprehensive reviews of what is currently known about the inter-ocean exchange that takes place here can be found in Lutjeharms (1996) and in De Ruijter et al. (1999b). That part of the flow not involved in ring formation, the Agulhas Return Current (Lutjeharms and Ansorge, 2001), moves water eastward along the Subtropical Convergence. These deep-sea components of the greater Agulhas system therefore no longer have a direct influence on the shelf regions. They, as well as the Agulhas Current proper, may have an indirect effect via their influence on the atmosphere and on biota.

The Agulhas Current has a marked, visible effect on the overlying atmosphere through the creation of cumulus clouds (Lutjeharms et al., 1986b), especially along its southern part and at the Agulhas Retroflection where the heat and moisture loss to the atmosphere is substantial (Mey et al., 1990). In these regions, under the right conditions, the thermodynamic considerations adequately account for the formation of cumulus clouds (Lee-Thorp et al., 1998, 1999). The uptake of moisture in the marine boundary layer (e.g., Jury and Walker, 1988) can have a marked effect on the moisture of the adjacent coastal zone and in consequence on the intensities of local storms and thus on rainfall (Rouault et al., 2002). Increasing rainfall may in turn be felt in the salinity, water column stratification and colour of the adjacent shelf waters. Jury et al. (1993) have in fact shown that the distance of the Agulhas Current offshore has a noticeable effect on the coastal rainfall. Under different climate change scenarios, the behaviour of the Agulhas Current may change significantly and the effect on coastal and shelf rainfall may change in concert (Lutjeharms and de Ruijter, 1996).

The Agulhas Current not only has a direct effect on rainfall over the adjacent coast and shelf, it also has an effect on regional atmospheric circulation patterns (Reason, 2001) and thus on rainfall over much wider areas. An increase in sea surface temperature of 2°C at the Agulhas Retroflection (Crimp et al., 1998) can substantially affect the atmospheric circulation over the whole southern African subcontinent. An increase in sea surface temperature over the South Indian Ocean will statistically lead to an increase in rainfall over regions that form the drainage regions for some of the main rivers of the South African east coast (Walker, 1990). The effect of increased runoff on different shelf seas may be very uneven. An investigation of the shelf waters in the Natal Bight shortly after a major rainfall event (Lutjeharms et al., 2000c) suggests that it might be relatively insignificant here and restricted to a small area just offshore of the mouth of the river. Along the Mozambican coast it may, by contrast, affect the surface salinities of substantial parts of the shelf (Sætre and de Paula e Silva, 1979).

To summarise: the south-western Indian Ocean may be considered to be that part of the South Indian Ocean that is not directly influenced by the monsoonally driven ocean currents. The shelf regions of this particular region are largely dominated by what may be considered to be the greater Agulhas Current system. Along the east coast of Madagascar and the east coast of South Africa this consists of well-developed western boundary currents. Along the east coast of Mozambique it consists of a series of eddies drifting poleward and along the west coast of Madagascar it may be a sluggish flow with no distinct pattern or temporal behaviour. Too few observations are available to characterise the latter unambiguously. This is also true of the behaviour of shelf waters in many other parts of the Mozambique Channel.

2. Mozambique Channel

A proper understanding of the nature of the waters over the shelves of the Mozambique Channel is constrained by the large degree of ignorance on the water movement in the adjacent deep sea. As noted above, historically the flow here has been visualised as consisting of an intense western boundary current along the east coast of Mozambique. Although this has now been shown to be incorrect (e.g., Ridderinkhof et al., 2001), the average drift patterns at the sea surface (e.g., Sætre, 1985) nevertheless indicate a strong movement poleward along the eastern shelf of Mozambique (Fig. 20.4) whereas elsewhere in the channel the average flow is small and its direction indistinct. The variability of the flow is very high in the western side of the channel, but low in the eastern side. This variability is evident in analyses of ships' drift (Lutjeharms et al., 2000b), altimetric observations (e.g., Lutjeharms et al., 2000d) and in modelling of the region (e.g., Biastoch and Krauß, 1999). These results—both strong currents at the shelf edge and high variability—of course all agree with the concept of a train of eddies moving poleward through the channel. Knowledge on the deep-sea circulation that may affect the shelves is even more lacking on the eastern side of the channel.

Ships' drift observations, altimetry and even the few hydrographic observations give no unambiguous indication of the movement of water here. In general speeds are low and to the north in the ships' drift observations (Sætre, 1985; Lutjeharms et al., 2000b) whereas modelling results (e.g., Biastoch et al., 1999) indicate an average flow to the south. Interpretations from the very few hydrographic data (e.g., Menaché, 1963; Sætre and Jorge da Silva, 1984; Donguy and Piton, 1991) have included a northward flow as well as a southward flow. The only known direct observations indicated a weak northward (Martin et al., 1965) and a weak southward flow (Piton and Poulain, 1974). Satellite observations have indicated the presence of cyclonic eddies off the shelf edge at the south-western coast of Madagascar that draw coastal water, rich in chlorophyll-*a*, seaward (Quartly and Srokosz, 2004). What does stand out in all appropriate data is that the intensity of current variability on the eastern side of the channel is considerably lower than on the western side.

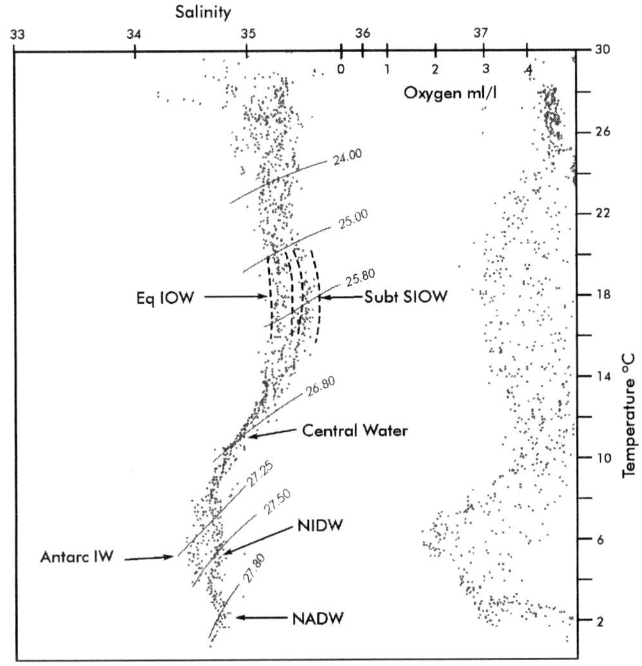

Figure 20.3 The temperature-salinity and temperature-oxygen characteristics of the water masses in the Mozambique Channel. (After Lutjeharms, 1991.) Sigma-t isolines have been added on the T-S scattergram. Water types that are evident are Equatorial Indian Ocean Water (Eq IOW), Subtropical South Indian Ocean Water (Subt SIOW), Central Water, Antarctic Intermediate Water (Antarc IW), North Indian Deep Water (NIDW) and North Atlantic Deep Water (NADW).

2.1 General water mass characteristics in the channel

The temperature-salinity characteristics of the waters in the Mozambique Channel are shown in Fig. 20.3. This figure is based on relatively old hydrographic data, but homogeneously covers a greater part of the channel than more modern data. It demonstrates that the salinities of waters in the upper layers lie in a band between 35.00 and 35.40, with a few outliers, mostly as fresher water. These particular outliers may be due to river runoff. At 18° C there are two different water masses, identifiable by distinct salinities. They are Equatorial and Subtropical Indian Ocean Water respectively and are found in different parts of the channel on different occasions. No durable delimitation for the distribution of either is to be found (Sætre and Jorge da Silva, 1984), although Equatorial Indian Ocean Water (also called Tropical Surface Water) is found largely in the northern part of the channel.

Since waters to a depth of 900 m are know to upwell onto the Mozambican shelf (e.g., Lutjeharms and Jorge da Silva, 1988) one can expect to find water types down to the less saline water of the Central Water (viz. Fig. 20.3) in some coastal regions, but most of the waters on shallow shelves would come from waters in more superficial layers.

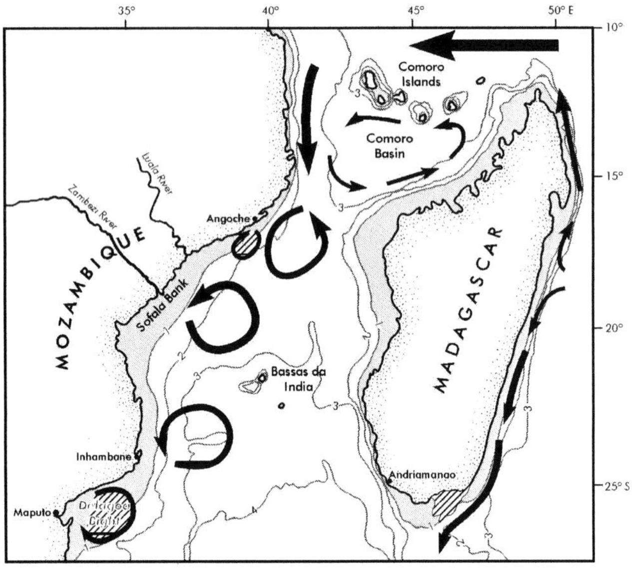

Figure 20.4 Bathymetry of the Mozambique Channel and the continental shelf off Madagascar in km (after Simpson, 1974) with the major circulatory features. Shaded areas are shallower than 1 km; hatched areas denote upwelling. Names of places and features mentioned in the text are given.

2.2 *Effects of shelf morphology*

The general bottom morphology of the shelves in the Mozambique Channel is shown in Fig. 20.4. The shelf is narrow on the western side of the narrows of the channel at about 16° S, but wide on the eastern side. They are fairly narrow on both sides of the channel at its southern mouth. The rest of the Mozambican shelf is wide, while that off Madagascar is narrower. Just south of the mouth of the channel there is an extensive offset in the coastline just off the Mozambican capital of Maputo, the Delagoa Bight. There is also an offset south of Angoche (Fig. 20.4). The shelves around the numerous islands, particularly the Comores in the northern mount of the channel, are narrow. These shapes of the shelf edge have a decided effect on the coastal water movement.

One of the consequences of the changes in direction of the coastline is in the formation of coastal lee eddies. An example of such a feature can be seen off the town of Angoche (Fig. 20.5). Here the flow along the greater part of the shelf edge—and probably on the shelf as well—is strongly poleward. At the time it was assumed that this formed part of a continuous Mozambique Current (Nehring, 1984); currently the consensus is that this most probably was part of the edge of an anti-cyclonic eddy drifting southward (e.g., De Ruijter et al., 2002). Notwithstanding the ignorance on its source, the important thing to note is the fact that there was a strong current poleward on this occasion, even though it might have been intermittent. This current overshot the offset in the coast at about 16° S, forming a distinct lee eddy to the south (Fig. 20.5). This eddy had a diameter of about 100 km and deeper water was upwelled in its core (Schemainda and Hagen, 1983). This is shown by an enhanced nutrient content of more than 12 μmol/ι nitrate-nitrogen at 75 m depth compared to 2–4 μmol/ι in the offshore current at the same depth. A resultant peak in chlorophyll-*a* concentration was also observed in this lee eddy (Nehring et al., 1987). Although no hydrographic stations were carried out on the shelf itself, the implication clearly is that the motion on the adjacent shelf would be equatorward.

The shelf configuration at this presumed lee-eddy is similar to that of the St Lucia and the Port Alfred upwelling cells (Lutjeharms et al., 1989b; Lutjeharms et al., 2000a; Figs 20.1 and 20.16). At all three of these locations a strong current along the shelf edge moves past an offset in the coastline, from a narrow shelf and then past a wider shelf. According to the theory put forward by Gill and Schumann (1979), this should lead to upwelling inshore of the strong current. In the former two cases it does, therefore perhaps also off Angoche at about 16° S in the Mozambique Channel. In his analysis of seasonal sea surface temperatures and the depths of isotherms along this coastline, Jorge da Silva (1984a) has shown the prevalence of upwelling at this spot, but only for certain periods of the year. His database for such seasonal analysis was not very large, so that this result can only be considered tentative. If the hypothesis is correct and this upwelling is driven by poleward currents as part of passing eddies, it would be sporadic and not a continuous feature. Nevertheless, this spot represents the highest observed chlorophyll-*a* values along this coastline (Nehring et al., 1987) and may therefore have a decided influence on the ecosystem of this whole shelf region.

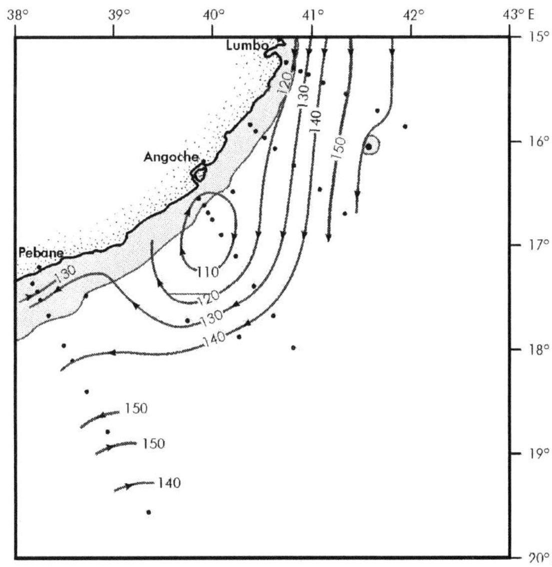

Figure 20.5 Lee eddy at Angoche along the Mozambican coast (viz. Fig. 20.4). The dynamic topography of the sea surface relative to the 600 dbar level is given in dynamic centimetre, based on a cruise undertaken in 1980. Dots represent station positions. The continental margin shallower than 1 km is shaded. (After Nehring et al., 1987.)

If the offshore eddies that form at the narrows (De Ruijter et al., 2002; Ridderinkhof and de Ruijter, 2003; Fig. 20.4) consistently move along the shelf edge it is likely that the waters on the adjacent shelf would experience alternating poleward and equatorward setting currents. This situation is also found on the shelf adjacent to the northern Agulhas Current where the shelf waters move largely in harmony with the current, but on the intermittent passing of a Natal Pulse, reverse and set strongly against the direction of the Agulhas Current (Lutjeharms and Connell, 1989). Quartly and Srokosz (2004) have used satellite observations of ocean colour to demonstrate that passing Mozambique eddies extract water from the neighbouring Mozambican shelf and inject it into the mid-channel region. In this way the exchange of waters between the shelf and the deeper part of the channel will consist of rather frequent episodes, driven from afar. They (Quartly and Srokosz, 2004) have also shown that the passage of eddies past the Delagoa Bight appears to affect the circulation there.

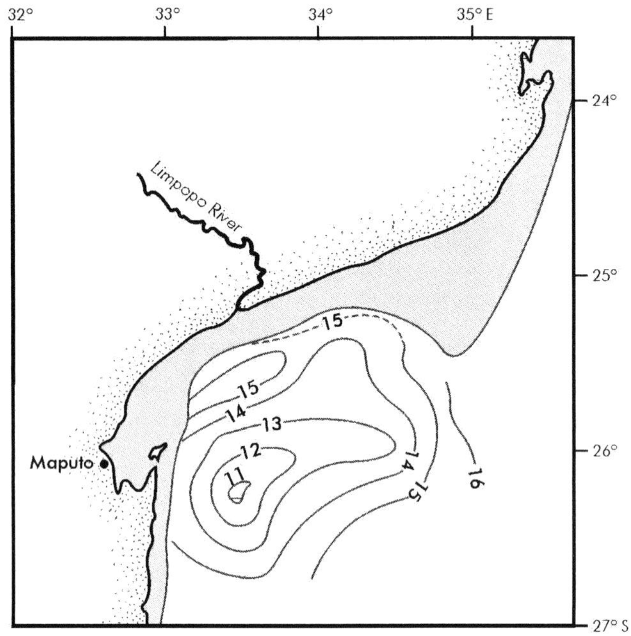

Figure 20.6 The Delagoa Bight eddy off the city of Maputo in southern Mozambique (viz. Fig. 20.4). Isotherms at 200 m depth show the cold water upwelled in the centre of the eddy, based on a cruise undertaken in 1982. (After Lutjeharms and Jorge da Silva, 1988.) The offshore line shows the intersection of the 200 m isobath with the shelf edge; the shelf itself is shaded.

The Delagoa Bight is a large offset in the coastline at the latitude of Maputo (see Fig. 20.4). The flow past here, either as the start of the Agulhas Current or as the landward end of passing, anti-cyclonic Mozambique eddies, is poleward. This passing water generates a cyclonic flow in the bight (Lutjeharms and Jorge da Silva, 1988) and the resultant Delagoa Bight eddy dominates the flow at the shelf here throughout most of the year (Fig. 20.6). Over a period of 23 years it has been observed 11 times in the hydrographic results of research cruises. It has been hypothesised that this recurrent eddy has largely determined the distribution of the sediment base of the Delagoa Bight (Martin, 1981). There is evidence (Gründlingh, 1992a) that these eddies may on occasion escape from the bight and drift into the deep ocean. On this part of the Mozambican shelf the water movement will most probably be totally dominated by this lee-eddy. If the same effects are found here as in lee-eddies in the other offsets in this coastline (e.g., Nehring et al., 1987), it can be assumed that there is considerable vertical movement of water in the lee eddy (Schemainda and Hagen, 1983). Water masses in this eddy have temperature-salinity characteristics implying substantial upwelling in the core of the eddy from depths of at least 900 m. Hence there should be nutrient enrichment of the surface layers and thus increases in the chlorophyll-*a* content. To date this has not been observed, except intermittently at the north-eastern point of the Delagoa Bight (Quartly and Srokosz, 2004). This sporadic increase in primary productivity may be the result of current-induced upwelling as predicted by Gill and Schumann (1979). If driven by passing eddies, such intermittent upwelling would be expected.

The shelf morphology of the eastern part of the Mozambique Channel is different to that of the western part (Fig. 20.4). For the most part the shelf is narrow except at the narrowest part of the channel—at about 17° S—where the eastern shelf is widest. As mentioned before, the flow along this eastern shelf edge is quiescent compared to the western side of the channel (Sætre, 1985; Lutjeharms et al., 2000b) with low speeds and low eddy kinetic energy. The average current direction seems to be equatorward. Nevertheless, there are tentative indications from the distribution of chlorophyll-*a* (Quartly and Srokosz, 2004), as observed from satellite, that cyclonic eddies form along the southern tip of Madagascar and drift into the channel. They seem to draw water off the shelf and inject it into the waters of the central channel.

In summary, the shelf edges of the western part of the Mozambique Channel seem to be influenced mainly by passing Mozambique eddies that have their origin in the narrows of the channel and by secondary effects due to these eddies, namely upwelling at coastal offsets, the generation of lee eddies at shelf offsets and the extraction of shelf waters to mid-channel. It is possible that parts of the eastern shelf of the Mozambique Channel are also influenced by eddies, but the data at hand are inadequate to show this unambiguously. It is evident that such eddies would be considerably more infrequent than and not as intense as those that affect the western shelf of the channel. Besides the direct influence of the currents at the shelf edge, the waters over the shallower parts of the shelf may be substantially influenced by the reigning winds.

2.3 Effect of winds and tides

The meteorological conditions for the Mozambique Channel have been summarised by van Heerden and Taljaard (1998). In austral summer the mean wind direction is uniformly from a south-easterly direction and weak. In winter the average wind direction differs for the southern and the northern part of the channel. The southern part experiences south-easterly winds; the northern part north-westerlies. The northern part thus may be considered to form part of the monsoonal wind system of the Indian Ocean up to 15° S (Sætre and Jorge da Silva, 1982) whereas the southern part does not. The border between these two systems is the Inter Tropical Convergence Zone. Donguy and Piton (1991) have attempted to relate the currents in the channel to the monsoonal winds, but with only a few cruises and short sea level records at their disposal a conclusion of monsoonal seasonality in the currents is probably premature. As mentioned above, the large-scale current systems show no evidence of monsoonal influence.

The winds over the shelf regions agree in part with the current direction hypothesised by Sætre and Jorge da Silva (1984). Except for the most southern part of the western shelf – the Delagoa Bight – the winds along this coastline all have a strong equatorward component year round. Sætre and Jorge da Silva (1984) have therefore concluded that the shelf currents here also are in that direction. These inferred current directions are in contrast to ships' drift observations (Sætre, 1985; Lutjeharms et al., 2000b) that show consistent movement poleward over the western shelf region. The winds over the eastern shelf are also largely northward, but vary with season.

During summer the shelf waters of the Mozambique Channel may also be affected by passing tropical cyclones that, as a rule, move poleward through the channel (Van Heerden and Taljaard, 1998). Some, however, make landfall somewhere along the coast of Mozambique. The effect of passing cyclones on the water movement over the shelves is dramatic. In situ current observations have shown (H. Ridderinkhof, personal communication) a reversal of current direction to a depth of at least 200 m at the passing of a severe cyclone.

Tides are an important component of water motion in the Mozambique Channel, especially when compared to other parts of the South-West Indian Ocean where tidal ranges are small. There is a gradual increase in tidal range from less than 2 m to the north and to the south of the channel to more than 5 m in its central part. This is the product of a double standing wave system, driven from either end, which develops in the channel (Pugh, 1987). Low lying coastal regions and estuaries, particularly on the Mozambican side, contain extensive areas of salt marshes and mangrove swamps as a result of the tidal motion (G. B. Brundrit, personal communication). Over the wide, shallow parts of the Sofala Bank (viz. Fig. 20.7) strong tidal currents lead to the continuous movement of sand banks and other mobile sedimentary seabed features.

In summary, tidal currents are important in the shallow parts of the shelves, but the usually weak winds have a limited affect on the main water movement over the shelves of the Mozambique Channel, with the exception of passing cyclones whose influence might be short lived.

Apart from the wind regimes, in a region of high rainfall (Van Heerden and Taljaard, 1998) the influence of river runoff on the shelf waters may be important.

2.4 *Effect of river runoff*

The runoff from the Mozambican land mass varies seasonally, but there also are considerable interannual differences (Jorge da Silva, 1984c). The total runoff from the Zambezi River (viz. Fig. 20.4, 20.7), for example, was 168.9 km^3 for 1977–78; whereas for 1982–83 it was a mere 50.3 km^3. Intermittent tropical cyclones bring event-scale rainfall episodes that may totally dominate the annual rainfall distribution. This naturally also holds for the Madagascar land mass, but runoff records for that country are difficult to obtain. One of the few coastal regions for which accurate observations of the effect of the river runoff on the shelf waters have been made on a number of occasions (Jorge da Silva, 1984c; 1984b) is along the Sofala Bank. It lies between 16° and 21° S on the western shelf of the channel (Fig. 20.4).

The salinity of surface waters close to the Zambezi River mouth on the Sofala Bank may drop as low as 20.00 (Sætre and de Paula e Silva, 1979) at a time when water at the shelf edge may be 35.4 (viz. Fig. 20.7). This river water may be severely discoloured, giving a Secchi disc depth of less than 2 m (Jorge da Silva, 1984b). The plume of muddy water can be quite extensive (Sidorn et al., 2001) and has been thought to play a major role in the natural history of local biota, such as shrimp. The fresher water may extend over the shelf to a distance of 50 km offshore. It may be confined to the top 15 m of the water column, or it may extend to the full depth of the shelf (Fig. 20.7), presumably dependent on the density differences between the river and shelf waters as well as the concurrent wind action. Plumes of fresh river water have been observed to extend both equatorward as well as poleward from most rivers here. No clear pattern of movement is therefore immediately evident. Sætre and de Paula e Silva (1979) have shown that the greater part of the Sofala Bank is affected by fresher surface water.

The chlorophyll-*a* distribution as well as the primary productivity on the Sofala Bank seems to be largely controlled by the effluent from rivers. These carry loads of nutrients (Jorge da Silva, 1984c) that, as mentioned above, may penetrate halfway across the shelf. The chlorophyll-*a* distributions exhibit very analogous patterns. That these distributions occur with a high frequency can be seen by the organic matter content of the surface sediments on this part of the shelf. It is highest directly off the Zambezi River mouth and in a narrow strip adjacent to the coast to either side of this river mouth. Some small pelagic fish seem to concentrate in these waters (Jorge da Silva, 1984c). Otherwise pelagic fish on the Sofala Bank seem chiefly to frequent strips parallel to the coast, most extending along the whole coastline (Brinca et al., 1981).

One may quite reasonably expect the observed effects of fresh water runoff on the Sofala Bank to hold for the rest of the shelf regions of the Mozambique Channel as well, where there are fewer measurements of this kind. The salinity of shelf waters along most of Mozambique varies seasonally with the river outflow, with the lowest salinities found in February (Sætre and de Paula e Silva, 1979). Apart from the expected seasonality, results for the Sofala Bank show great inter-annual variability and one could assume that this holds for all the other Mozambican shelf regions as well. The freshening of the shelf waters off Mozambique by river run-off seems ubiquitous.

There is at least one clear exception. At the southern extremity of the Sofala Bank, between 20° to 21° 30' S (viz. Fig. 20.7), there is a large expanse of coastline that is subject to inundation by seawater. The subsequent runoff from this region can be extremely salty. Salinity values in excess of 36 are not uncommon (e.g., Jorge da Silva, 1984c). During September of 1982 surface values of 37.2 were observed near the coast (Brinca et al., 1983). The influence of this saline water has been observed a distance of 100 km offshore and throughout the shallow water column of 50 m.

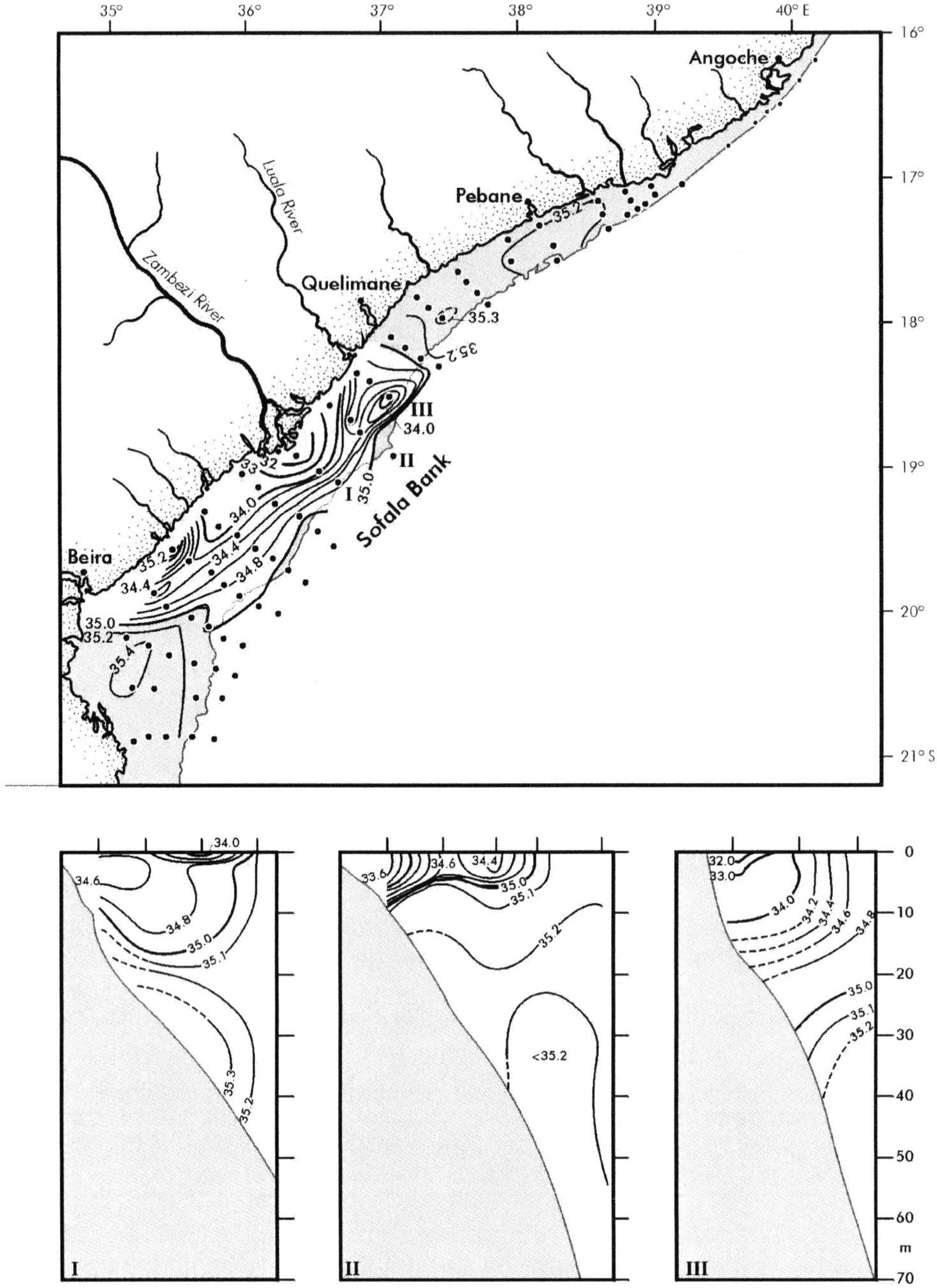

Figure 20.7 Surface salinities on the Sofala Bank—the wide, shallow shelf off central Mozambique—based on a cruise of 1982. (After Jorge da Silva, 1984c.) For general location see Fig. 20.4. Dots show station positions. Locations of station sections in the bottom panels are shown by roman numerals in the upper panel. The shelf shallower than 50 m is shaded. Fresher water (< 35.0) extended to a depth of 15 m directly off the Zambezi River mouth; to 30 m off the Luala River mouth. Note the high salinity values south of the city of Beira, due to marsh runoff.

2.5 Characteristics of the western shelf

To recapitulate, the water mass characteristics of the shelf waters on the western shelf of the Mozambique Channel have been presented by Jorge da Silva et al. (1981), Jorge da Silva (1984a), Lutjeharms and Jorge da Silva (1988) and Sætre and Jorge da Silva (1982). Except for regions and times where river run-off plays a major role, the water masses over the shelf are identical to those offshore (viz. Fig. 20.3) at the same depths. This means that the subsurface salinity maximum of Subtropical Surface Water (at depths of 150 – 300 m) is much more pronounced in the southern part of the channel than farther north. In the Delagoa Bight eddy, at the southernmost extremity of the shelf (Lutjeharms and Jorge da Silva, 1988), there is hardly any evidence of Tropical Surface Water left. The exchange of water masses between the shelf and the deep ocean seems to vary considerably. The passage of Mozambique eddies may play a key role here. At parts of the western shelf where the shelf is very narrow (viz. Fig. 20.4) and the current at the shelf edge strong, such as at the narrows of the channel (Angoche; Sætre, 1985) and just north of the Delagoa Bight (Inhambane; viz. Fig. 20.4; Lutjeharms et al., 2000b), one may expect that the shelf edge currents may have a more decided effect, whereas along the Sofala Bank where the shelf is widest (Fig. 20.4), this effect would be substantially less.

Sætre and Jorge da Silva (1982; 1984) have carried out the most detailed analyses of water masses on this shelf to date. They have claimed that the circulation patterns of the shelf waters can be visualised by the temperature distribution at 150 m depth. This leads to a different pattern for the hydrographic results of each research cruise for the region. A set of cyclonic eddies of various shapes and sizes are evident. Based on these data one may therefore safely assume that the waters and the circulation on this shelf region are very variable. What effect does this have on the ecosystem of this shelf region?

Surveys of organisms and in particular of fish resources have been made on the western shelf region of the Mozambique Channel (Nehring, 1984; Nehring et al., 1987; Sætre and de Paula e Silva, 1979; Jorge da Silva, 1984a), but particularly on the Sofala Bank (Brinca et al., 1981; Jorge da Silva, 1984c). The values of column chlorophyll-*a* concentration are relatively low over most of the outer parts of the shelf, somewhat higher at the shelf edge (Sætre and de Paula e Silva, 1979). Over inner parts it can rise to 98 mg/m^2 (Nehring et al., 1987). The exception, mentioned above, are the observations in the upwelling cell off Angoche where values of 600 mg/m^2 have been observed. The distribution of zooplankton biomass in the upper 30 m of the water column shows a similar general distribution with low values of 20–40 mg/m^3 over most outer parts of the shelf with higher values, up to 160 mg/m^3 at inner stations. The exception is for the region directly poleward of the Angoche upwelling cell where observations of 320 mg/m^3 have been made. No evidence for such major increases in either phytoplankton or zooplankton have to date been found in the Delagoa Bight eddy. Even though Nehring et al. (1987) have shown that the primary productivity at inshore stations on the shelf was double that of stations farther offshore, there were considerable differences between stations close to each other on the shelf. Large degrees of spatial and temporal variability can therefore be assumed.

An analysis of the seasonal distribution of the depth of the 20° and the 23° C isotherms has shown that large parts of the outer shelf would in principle be suitable for yellowfin tuna fisheries, whereas the skipjack tuna are most likely to be found off Angoche and Maputo, i.e. at the sites of upwelling induced by lee eddies (Fig. 20.8). The distribution of fish; demersal, small pelagic, larger pelagic, mesopelagic as well as that of crustaceans has been summarised by Sætre and de Paula e Silva (1979). In most cases where more than one survey cruise was carried out there was a considerable difference between the cruise observations and this was usually considered to be due to seasonality in the distribution of organisms. It may have been due to irregular temporal changes of shorter duration.

In short, the circulation on the Mozambican shelf is very variable in both space and time and may be influenced by offshore currents only where the shelf is narrow. Run-off from land plays a key role, but exhibits both seasonal and interannual variations. The biogeography exhibits the same variability. Although the interpretation of the circulation patterns as well as the biogeography of this western shelf of the Mozambique Channel is constrained by limited data, this is even more the case for the eastern shelf.

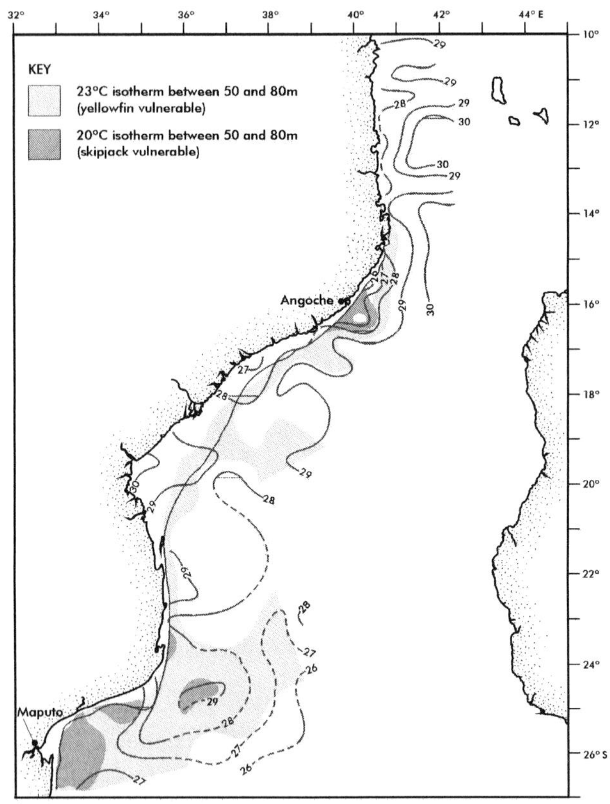

Figure 20.8 Tuna vulnerability to catching by surface gear on the Mozambican shelf, superimposed on average sea surface temperatures during the period January to March. (After Jorge da Silva, 1984a.) The 50 m isobath is shown. Note the increased concentration in the lee eddies off Angoche and in the Delagoa Bight off Maputo.

2.6 Characteristics of the eastern shelf

The eastern shelf of the Mozambique Channel can—using information currently available—be divided into three specific regions: the very south where the shelf is more or less zonal, the very north, where the shelf forms part of the Comoro Basin (viz. Fig. 20.4) and, third, the meridional shelf in between.

The Comoro Basin lies directly south of the South Equatorial Current (Piton and Poulain, 1974). Between it and the narrows of the channel an anti-cyclonic gyre is formed that seems relatively stable (Donguy and Piton, 1991). The surface currents of which the gyre consists are not strong (Sætre, 1985) except near the African coast where mean speeds in excess of 0.5 m/s are to be found (Lutjeharms et al., 2000b). The variability on the African side is high. The currents over the eastern shelf in the Comoro Basin are weak but in general in an equatorward direction. This rather inadequate portrayal is nonetheless supported by direct measurements (Piton and Poulain, 1974; Martin et al., 1965) as well as by a variety of models (e.g., Biastoch and Krauß, 1999; Asplin et al., 2004; Sætre, 1985).

The water types on the shelf are dominated here by Tropical Surface Water with a salinity range from 34.3 to 35.2 (Donguy and Piton, 1969). There is no evidence to date that Subtropical Surface Water with greater salinity is found on the shelf, although in principle it is entirely possible since this water mass is found at about 200 m depth in the region. Surface temperatures exceed 26° C throughout the year, except in the months of August and September. Highest temperatures (> 29°) are found in April; lowest (25° to 26° C) in August. Surface salinities are lowest (34.4) in March; highest in the period September to November (>35.10). Recent observations on this shelf region (Roberts, personal communication) shows that the water column consists of a warm mixed layer to a depth of about 80 m with the seasonal thermocline extending to 250 m. Salinities in the centre of the Comoro Gyre are slightly elevated above those found on the adjacent shelves. An oxygen minimum lies at 200—300 m depth and is less strongly developed in the centre of the gyre. The hydrography of the middle part of the western shelf of Madagascar is not much different.

Here, according to the only observations to be found (Roberts, personal communication) the warm mixed layer extends to 150 m, with a uniform salinity between 35 and 35.25 (Fig. 20.9). Particularly noteworthy is the oxygen minimum layer found between 150 and 250 m that is strongly developed here, more so than anywhere else on the shelves of the region. How representative these observations are remains unknown. The little that is known about the currents along the shelf edge (e.g., Sætre, 1985; Lutjeharms et al., 2000b) here, suggest that they are very weak and will most probably have very little influence on the movement of the waters on the shelf itself. As mentioned before, satellite imagery, particularly of ocean colour (Quartly and Srokosz, 2004), shows that cyclonic eddies along this coastline may on occasion draw off substantial amounts of shelf water and with it chlorophyll-*a* into the deeper parts of the channel. The origin of these eddies, between 200 to 300 km in diameter, is uncertain. De Ruijter et al. (2003) have shown that cyclones can be formed at the termination of the southern East Madagascar Current and these could conceivably drift into the Mozambique Channel along the shelf edge. Although a number of hydrographic interpretations (e.g., Sætre and Jorge da Silva, 1984; Donguy and Piton, 1991) imply a southward setting current along this shelf, direct measurements (Martin et al., 1965) and ships' drift observations (Lutjeharms et al., 2000b) indicate otherwise. Cyclonic eddies could therefore be advected equatorward along this coast.

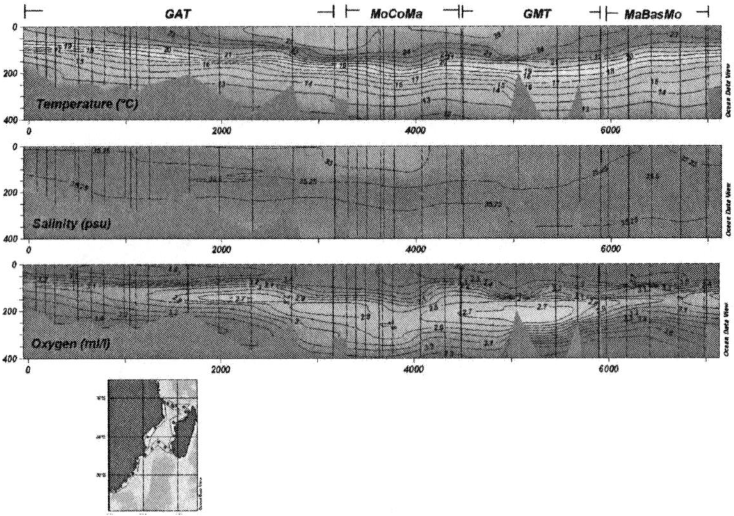

Figure 20.9 Hydrographic characteristics of shelf waters along the east coast of South Africa and the continental shelves in the Mozambique Channel. (With special permission of M. Roberts.) GAT is the section from Port Elizabeth in the south along the east coast of South Africa and of Mozambique; MoCoMa represents the zonal section from the coast of Mozambique to the west coast of Madagascar via the Comores; GMT a poleward section along the west coast of Madagascar and MaBasMo a zigzag section from the southern point of Madagascar to the island of Bassas da India and on to Maputo on the east coast of Mozambique.

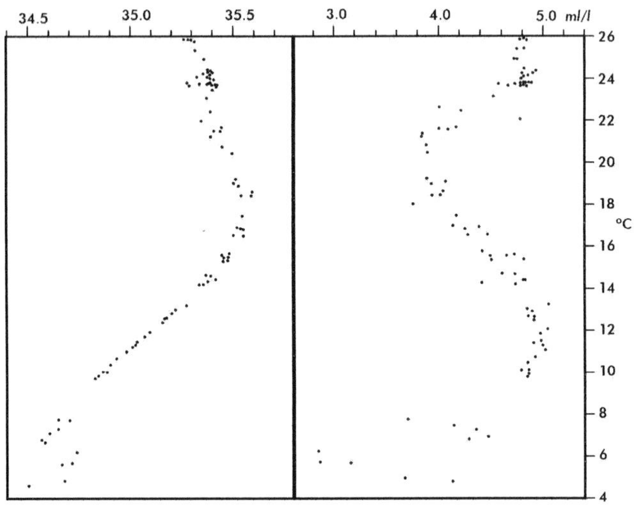

Figure 20.10 The temperature/salinity (left panel) and the temperature/dissolved oxygen characteristics for the shelf waters south of Madagascar. (After Anonymous, 1983.) The line of stations on which this was based was taken to the south-west of Andriamanao (viz. Fig. 20.4).

The third and last component of the eastern shelf of Madagascar is the southern part. The movement of the shelf waters here may be more dynamic, since it could conceivably be affected by the southern limb of the East Madagascar Current (viz. Fig. 20.4). This would most likely be true largely for the eastern part of this shelf region (Lutjeharms et al., 2000b). Hydrographic observations on the shelf show the normal temperature/salinity characteristics to be expected (Fig. 20.10). Surface temperatures (in June) were 24 °C and surface salinities 35.3. (The higher temperatures shown in the scatter diagram of Fig. 20.10 represent temperatures farther offshore.) Both Tropical and Subtropical Surface Waters are evident in the temperature/salinity values. The Tropical Surface Water extended from the surface to about 100 m depth. The Subtropical Surface Water below that extended to about 250 m. A well-developed subsurface oxygen minimum was found at a depth of 80 to 200 m.

The surface salinities and temperatures all suggest an upwelling regime on this southern shelf. During the first extensive cruise over this shelf (June 1983; Anonymous, 1983) the temperatures at the coast were 2 °C lower than further offshore. The salinities were up to 35.6, indicating upwelled Subtropical Surface Water (see Fig. 20.10). Ocean colour also shows signs of enhanced chlorophyll-*a* values along this coastal segment (Quartly and Srokosz, 2004) and that these may be drawn off the shelf in plumes. The presumed upwelling is concentrated in an upwelling cell on the south-eastern corner of Madagascar (Lutjeharms and Machu, 2000; DiMarco et al., 2000) where enhanced concentrations of chlorophyll-*a* and lower temperatures are found most frequently. The question remains if these remotely sensed suggestions of upwelling may not be due to runoff from land.

Hydrographic evidence specifically collected to ascertain the origin of these elevated values of chlorophyll-*a* (Machu et al., 2002) has recently shown unequivocally that there is indeed upwelling at this location. The upwelling does not seem to be strictly related to the wind patterns and it has therefore been hypothesised (Lutjeharms and Machu, 2000) that the driving force for the upwelling is the juxtapositioned East Madagascar Current. This would be comparable to the upwelling at the eastern extremity of the Agulhas Bank (Lutjeharms et al., 2000a; viz. Fig. 20.16) and at the northern extremity of the Natal Bight (Lutjeharms et al., 1989b; viz. Fig. 20.11). However, there is some evidence (Quartly and Srokosz, personal communication) that the chlorophyll-*a* concentration at this location has a distinct seasonal pattern with the highest concentrations found in the austral winter months of July and August, the lowest in December. Major winds at this location are from the east in winter, from the north-east in summer (Sætre, 1985) suggesting winds more favourable for upwelling along the full south coast of Madagascar in winter, at the south-eastern corner in summer. The importance of the wind compared to the current in driving this upwelling therefore remains unresolved.

The biological implications of this upwelling are intriguing, but to date not properly quantified. Apart from the remotely sensed chlorophyll-*a*, surveys of fish stocks (Anonymous, 1983) have shown slightly higher concentrations of demersal fish on the southern shelf than off the adjacent, eastern shelf. Mackerel numbers were higher on the southern shelf, but scad lower. In general fish were found with such a very scattered distribution on this shelf that no firm conclusions can be reached on their biogeography.

In summary, the waters over the shelves of western Madagascar seem fairly unusual, based on the current—very limited—data. Water masses are those found offshore, except for an intensification of the subsurface oxygen minimum, and the currents are most probably weak and variable. The wider, southern shelf exhibits characteristics of upwelling, but this does not seem to have a marked effect on higher trophic levels.

An outline of the characteristics of the shelf waters of the Mozambique Channel as a whole is, as can be seen from the above descriptions, largely a function of the amount of available data and their quality. The broad Mozambican shelf is characterised by substantial terrestrial influence from runoff both from rivers and salt marshes. Lee eddies at Angoche and in the Delagoa Bight may play an important, but local role. The effect of Mozambique eddies intermittently passing by the shelf edge is not known. Water masses over the northern shelves are predominantly Tropical Surface Water, those to the south Subtropical Surface Water. This seems true for the western as well as the eastern shelves of the Mozambique Channel. About the latter very little is known, except that it is relatively broad.

The shelf to the east of Madagascar is by contrast much narrower and the offshore circulation totally different.

3. Region east of Madagascar

Not only is the shelf narrow here, but the continental slope is precipitous (viz. Fig. 20.4). Along the shelf flows a small, but intense western boundary current, the East Madagascar Current (e.g., Lutjeharms et al., 1981a). Swallow et al. (1988) have observed a speed in the southern limb of this current of 0.66 m/s, at a latitude of 23 °S, about 50 km from the coast, with a standard deviation of only 12 cm/s. The speeds in the northern branch of this current (Schott et al., 1988) are not much different. The separation point between the northern part, flowing equatorward, and the southern part, flowing poleward, is estimated to lie between 17 and 18 °S (e.g., Lutjeharms et al., 2000b; viz. Fig. 20.4). This may vary with season as the wind patterns shift northward along this coastline in the austral winter (Van Heerden and Taljaard, 1998). No direct observations have been made to date, but one may assume that the waters over the shelf move in concert with the strong offshore currents.

As can be expected, the temperature/salinity relationships of the waters on this part of the shelf are indistinguishable (Anonymous, 1983) from those found on the shelf south of Madagascar (Fig. 20.10).

Little is known about the biological productivity of the region. Fish distributions are scattered with a decrease in pelagic as well as demersal fish as one moves equatorward. Catches of sharks and rays increase on going northward along this coast (Anonymous, 1983).

In summary, all that is known with a certain degree of certainty about the shelf seas east and south of Madagascar is the presence of the East Madagascar Current at the shelf edge and the likelihood of persistent upwelling south-east of the island. The interaction between these or the influence of the current on the shelf circulation remains unknown due to an extreme paucity of observations.

Compared to the lack of data and knowledge about this particular part of the shelf seas of the South-West Indian Ocean, the region inshore of the northern Agulhas Current is very much better studied and understood.

4. Northern Agulhas regime

The northern Agulhas Current is defined as that part of the current extending from the southern mouth of the Mozambique Channel downstream to the eastern edge of the Agulhas Bank (viz. Fig. 20.1). This component of the current flows past a shelf that may be considered to consist of two categories. For the greater part the shelf is narrow and the continental slope has a steep gradient. The only exception is a part of the shelf along the province of KwaZulu-Natal known as the Natal Bight. Here the shelf is considerably wider and the slope much broader and with a gentler gradient (Fig. 20.11). As will be seen below, this shelf morphology has some remarkable effects on the offshore currents.

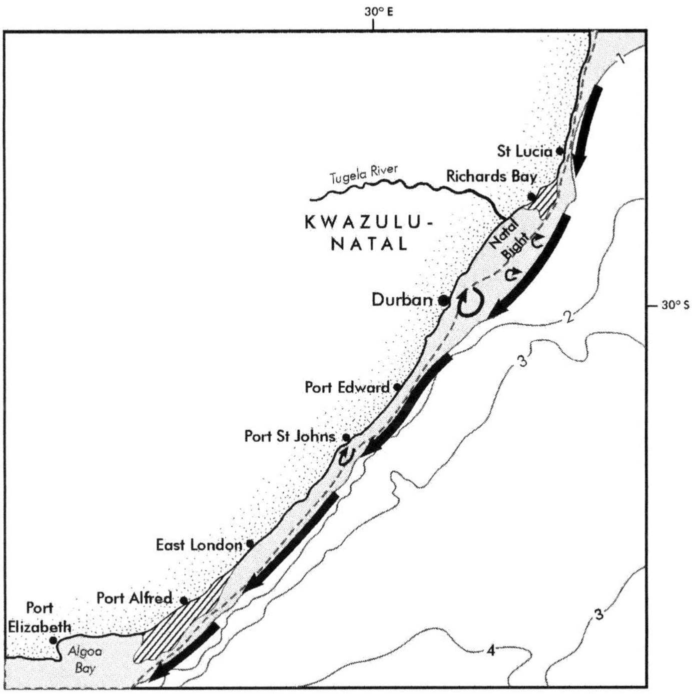

Figure 20.11 The bathymetry of the continental shelf along the northern Agulhas Current. (After Simpson, 1974.) The continental shelf area is shaded. The 200 m isobath is shown as a broken line. Hatched areas denote upwelling. Place names and circulatory features are mentioned in the text.

The core of the northern Agulhas Current follows the shelf edge very closely almost all of the time (Tripp, 1967) meandering less than 15 km to either side (Gründlingh, 1983). For a western boundary current this is quite unusual, but it has important consequences for the circulation on the adjacent shelf. As can be expected, in its very surface layers the behaviour of the current is not as stable (Pearce, 1977a) with changes in speed occurring from day to day and the penetration of surface water of the current onto the shelf, taking place at irregular intervals. In some cases short-term current reversals at the edge of the current have been observed (Gründlingh, 1974; Schumann, 1981; Pearce et al., 1978), possibly due to shear edge eddies or to the effect of the wind. Surface speeds of the inshore edge of the current may exceed 1.5 m/s, salinities may lie between 35.00 and 35.50 and temperatures in summer may exceed 28. In winter these sea surface temperatures drop to less than 21°C (Pearce, 1978). Shallow water near the shelf edge is usually Tropical Surface Water. The characteristic salinity maximum of Subtropical Surface Water is found at a depth of 150 to 250 m at least 60 km offshore (Pearce, 1977a) although this distance may vary on a near-daily basis. One would therefore expect the waters over the adjacent shelves to consist largely of modified Tropical Surface Water, but as it turns out, this is not the case.

This established current disposition is not entirely stable, as mentioned above. During about 15% of the time—and at irregular intervals—the current moves offshore in a sudden, single meander (Gründlingh, 1979). This Natal Pulse (Lutjeharms and Roberts, 1988) moves downstream with the current at a rate of about 20 km/day. Features of this kind have also been observed north of the Natal Bight (Gründlingh and Pearce, 1984; Gründlingh, 1992a), however, all information currently available suggests that it is only meanders that originate at the Natal Bight that consistently progress downstream with the current. Theoretical studies (De Ruijter et al., 1999a) have shown that it is the weak gradient of the shelf at the Natal Bight that will allow baroclinic instability in the Agulhas Current to occur here and to grow once the core of the current has been detached from the sharp slope gradient. The trigger for this meander has been thought to be the adsorption of offshore eddies, the tell-tale signs of which have been seen in many satellite images in the thermal infrared (e.g., Gründlingh, 1986; Lutjeharms and Roberts, 1988). This has recently been proven to be the case (Schouten et al., 2002). It is interesting to note that some marine animals such as leatherback sea turtles carefully use all these circulation features to move about in the ocean (Hughes et al., 1998; Luschi et al., 2003). The question remains what, if any, effects these unusual meanders have on the shelf waters. This will be discussed in a section to follow. It is necessary first to describe the wind regimes over this shelf region.

4.1 Winds along the shelf off south-eastern Africa

Tropical cyclones hardly ever reach this coastline, in contrast to that of Madagascar and Mozambique (Jury and Pathack, 1991). On the infrequent occasions when they do arrive at the coast (e.g., Poolman and Terblanche, 1984), one would expect the shallow waters of the shelves over which they move to be thoroughly mixed. Otherwise the shallow waters, as measured by moored current meters (upper ~20 m), follow the reigning winds closely (Pearce et al., 1978).

Coastal winds for the region have been analysed in detail by Schumann (1989). He has shown that the main wind axis is parallel to the coast. At Durban the wind is 5 times more likely to blow along the shelf than across it, whereas at East London (viz. Fig. 20.11) it is three times more likely. Average wind speeds are about 2.5 m/s at Durban; 3.2 m/s at East London (for 1984). The average wind speeds along the coast and across it were not very different. The north-easterly wind and the south-westerly winds both occur about 50% of the time (Schumann and Martin, 1991), with both showing seasonality in wind speed, the north-easterly winds having slightly greater seasonality. During summer the alongshore component of the wind is considerably higher (Hunter, 1988), particularly farther downstream.

An important additional wind process for the coastal waters is diurnal land and sea breezes. These can exhibit speeds of the same magnitude as those brought about by normal synoptic systems (Hunter, 1988). Hunter (1981) has used offshore wind observations to show that land breezes can here extend at least 60 km seaward. The direction of these winds may have a decided influence on cloud formation, precipitation over the shelf as well as coastal runoff.

As mentioned before, it has been demonstrated that cumulus cloud lines frequently form over the northern Agulhas Current (Lutjeharms et al., 1986b; Lee-Thorp et al., 1998) but mostly when the winds are along-current, from the northeast. During such along-current air motion there is an enormous uptake of moisture from the current (Lee-Thorp et al., 1999; Rouault et al., 2000). About 5 times as much water vapour is transferred to the atmosphere above the current itself than from ambient waters. During on-shore winds this moisture is advected inland and may contribute significantly to moisture convergence and rainfall over the interior of South Africa. In fact, it has been shown that this leads to local intensification of storm systems and the concurrent flood events (Rouault et al., 2002). To what extent this leads to measurable dilutions of coastal waters by river runoff is not known. What is known is that the presence of the Agulhas Current has a consistent effect on coastal rainfall all along its northern part (Jury et al., 1993; Fig. 20.12). Wherever the current is close to the coast, such as between Durban and Port Elizabeth, the rainfall is enhanced; wherever the current axis diverges from the coastline, such as at the Natal Bight and at the southern part of the Agulhas Current, coastal rainfall is significantly reduced. This is not the only process that makes the Natal Bight an unusual shelf region.

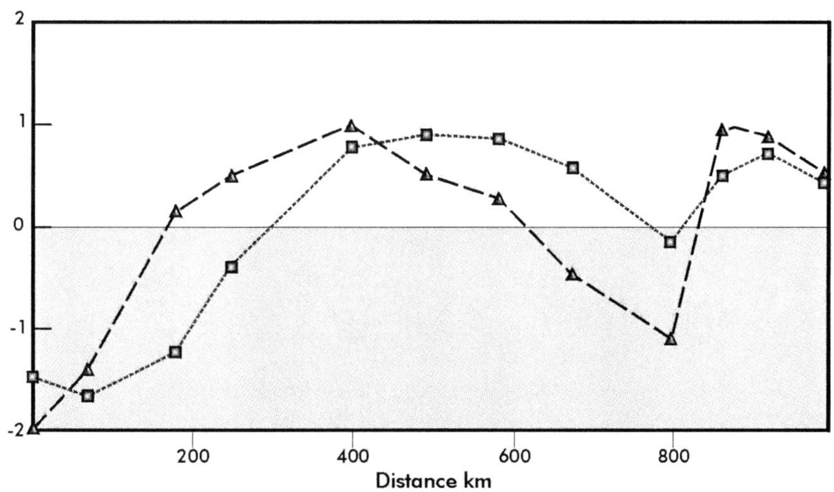

Figure 20.12 The influence on coastal rainfall of the distance of the core of the Agulhas Current from the coastline. (After Jury et al., 1993.) The abscissa gives the distance upstream from Port Elizabeth (see Fig. 20.11). The solid curve gives the distance from the coast to the core of the Agulhas Current as expressed by sea surface temperatures. Note that the distance is greater off the Agulhas Bank (0 km) and at the Natal Bight (800 km). The broken line shows the coastal rainfall. Both curves are expressed as standardized departures; that of the distance from the coast having been inverted for comparison.

4.2 The Natal Bight

The Natal Bight is formed by a landward offset between Richard's Bay and Durban in an otherwise rather linear coastline (Fig. 20.11). The northern part of the bight is shallower than 50 m; the southern part deeper. There are some well-developed canyons in the bathymetry of the continental slope, but these do not extend onto the shelf, where they have been filled in by sediment (Martin and Flemming, 1988). The major depocentre of the region is the offshelf Tugela Cone (viz. Fig. 20.13), evidence that the Tugela River is the major source of sediment for this shelf region. Sediments over the shelf itself consist largely of sand. The percentage is in excess of 75% over all parts of the Natal Bight shelf except seaward of the Tugela River (Flemming and Hay, 1988) where mud is the dominant sediment type. Gravel patches are found largely, but not exclusively, at the shelf edge where scouring from the Agulhas Current is to be expected. The distribution of sediments (Flemming and Hay, 1988) is particularly instructive here since it gives a clear indication of the integrated movement of the bottom waters where other data may not be available.

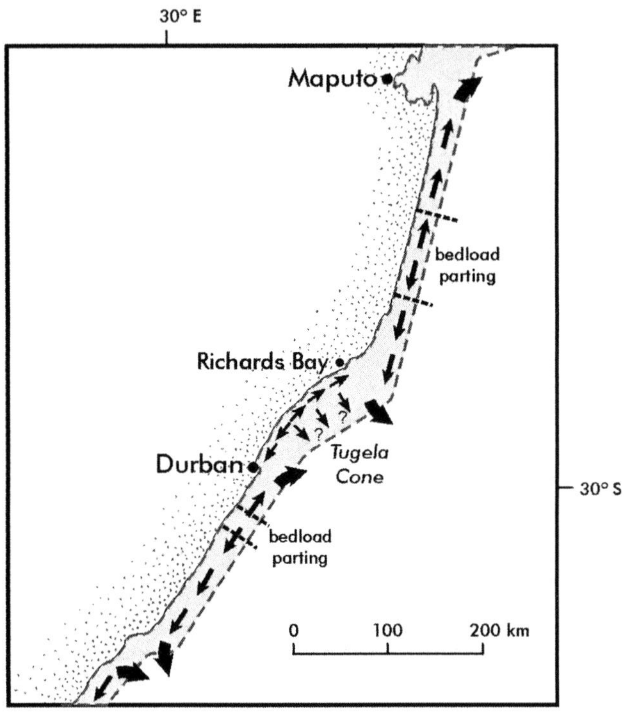

Figure. 20.13 A conceptual model of the bedload movement on the continental shelves adjacent to the northern Agulhas Current. (After Flemming and Hay, 1988.)

The shelf directly equatorward of the Natal Bight has, for instance, seen hardly any hydrographic investigations. However, the general bedload dispersal model suggests that the shelf waters move equatorward here, in clear disagreement with the concept of a straightforward Mozambique-Agulhas Current continuity. A bedload parting is found at about 28°S (Fig. 20.13). Analyses of ships' drift (Harris, 1978) indicate that poleward of this point the currents over the shelf follow the Agulhas Current 75% of the time. This implied movement is consistent along the whole coastline except just downstream of Durban where it again is equatorward. This latter discrepancy may be due to an embedded lee eddy that is a recurring feature of the circulation at this location just south of the Natal Bight (e.g., Anderson et al., 1988; Meyer et al., 2002) and will be discussed in greater detail below.

The submarine bedform distribution is even more instructive. Active submarine dune fields, moving with the current, are found north of the Natal Bight and along the shelf break of the northernmost half of the bight. Along the southern half of the shelf break there is no evidence of the influence of the current, suggesting that it overshoots here (viz. Fig. 20.11), maintaining some distance from the shelf edge and thus not affecting the sediments. It is intriguing that off Durban the movement of the mobile dune field is, by contrast, northward, substantiating the persistence of the lee eddy surmised to occur here (Pearce et al., 1978; Meyer et al., 2002). Flemming and Hay (1988) have inferred a complex shelf circulation from the sediment distributions and the bedforms, consisting of a cyclonic movement over the northern, shallower part of the bight and a dipolar structure of an inner anticyclonic and an outer cyclonic eddy over the southern, deeper parts. What is in fact known about the circulation here?

First, the presence of a persistent upwelling cell at the upstream end of the Natal Bight is the most prominent part of the hydrodynamics of the shelf waters of the Natal Bight and a fundamental key to understanding the ecosystem of this shelf sea. From all other perspectives it may be considered to be a semi-enclosed system. The strong and ever-present Agulhas Current at the shelf edge forms a formidable barrier to exchanges of water and biota with the open ocean. At the northern end of the bight, between Richard's Bay and Cape St Lucia, the shelf widens as the current sweeps poleward. This bathymetric arrangement is believed to lead to topographically induced upwelling (Gill and Schumann, 1979), as it does elsewhere along the trajectory of the Agulhas Current.

In this general region sea surface temperatures are about 26°C in the summer months, peaking in February (Pearce, 1978) and dropping to about 21°C in August. Observations of sea surface temperature in the region (Gründlingh, 1974; Gründlingh and Pearce, 1990) have shown that off Richard's Bay the temperatures are always a few degrees lower. As could be expected, the water here is largely Tropical Surface Water with only the occasional presence of Subtropical Surface Water (Pearce, 1978). Others (Lutjeharms et al., 2000c) have shown that the purest Subtropical Surface Waters is found on the shelf edge off Richard's Bay and St Lucia. This sporadic presence of Subtropical Surface Water on the shelf, otherwise found at depths of 150 m or more offshore (Pearce, 1977b), is highly suggestive. Subsequent investigations using satellite images (Lutjeharms et al., 1989b) and a dedicated hydrographic cruise (Lutjeharms et al., 2000c; Meyer et al., 2002) have demonstrated unequivocally that this is indeed a persistent upwelling cell.

Lower temperatures are observed here more or less continuously, although the areal extent of the surface expression may vary considerably. This surface expression seems to have no clear seasonal pattern, neither is it clearly related to potential upwelling inducing winds (Lutjeharms et al., 1989b). Pearce (1978) has shown that evidence of 16°C water (Subtropical Surface Water, viz. 20.3) at depths of 125 m, sometimes less than 100 m depth, on the shelf is intermittent, with no clear pattern. It can therefore be accepted that this upwelling is not wind-driven. The effect of this upwelling can be observed at the sea surface along the inner edge of the Agulhas Current as far downstream as Durban as the colder surface water is dragged southward as a cool filament (Lutjeharms et al., 1989b). Further evidence for the nature of this upwelling cell comes from nutrient distributions (Carter and d'Aubrey, 1988; Meyer et al., 2002; Fig. 20.14). It shows that the influence of the upwelling cell extends over a sizeable part of the bight. Carter and d'Aubrey (1988) have given historical nutrient values all over the bight and state that there is no clear seasonal pattern in the occurrence of nutrients. Vertical sections show clearly (Lutjeharms et al., 2000c) that this nutrient-rich water is upwelled at the St Lucia upwelling cell and from there moves over the floor of the bight southwards. Further evidence for the effect of this upwelling cell comes from biological observations.

The distribution of chlorophyll-*a* exhibits a very similar pattern to that of the nutrients (e.g., Meyer et al., 2002), with the enhanced values slightly lower very close to the coast, implying an active upwelling process taking place during the observations. Oliff (1973; as quoted by Carter and Schleyer, 1988) has shown how the phytoplankton production reacts almost instantaneously to an upwelling event at Richard's Bay. Reviews of the plankton, zooplankton as well as the benthic species found in the Natal Bight have been given by Carter and Schleyer (1988) and by McClurg (1988) respectively. These studies were based on information that was geographically very inhomogeneous, since they had to depend on an eclectic set of previous collections not designed uniformly to cover the shelf as a whole. From these scattered observations it is impossible to infer the extent of the biological influence of the St Lucia upwelling cell over the Natal Bight shelf, particularly over the southern part.

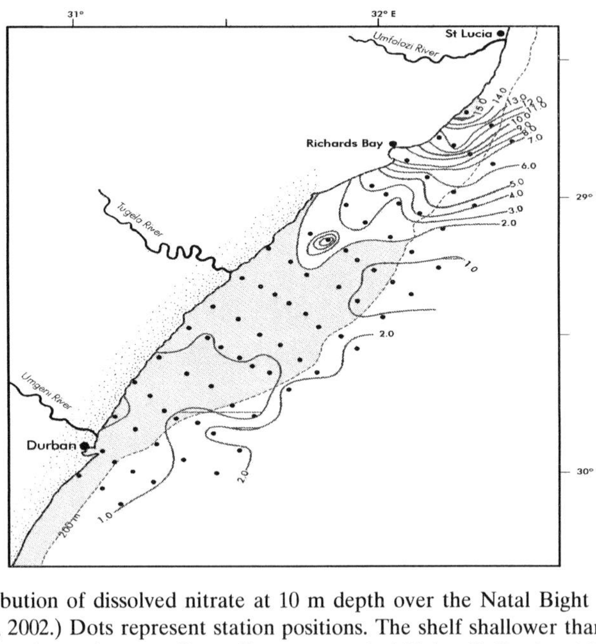

Figure 20.14 Distribution of dissolved nitrate at 10 m depth over the Natal Bight during July 1989. (After Meyer et al., 2002.) Dots represent station positions. The shelf shallower than 200 m has been shaded. The presence of an active upwelling cell equatorward of Richard's Bay is evident.

The waters of the poleward part of the Natal Bight are by contrast unspectacular and fit well into the ranges, both physical and biological, to be expected at these latitudes. The nutrients are low (Carter and d'Aubrey, 1988; Meyer et al., 2002), coming from Tropical Surface Waters. The nutrient concentrations are 1.01 – 1.86 μmol/ι (nitrate), 0.48 – 0.72 μmol/ι (phosphate) and 3.50 – 4.69 μmol/ι (silicate) at 10 m depth. The values on this part of the shelf do not exhibit great differences from those found in the surface waters of the Agulhas Current. The exception is to be found close to the Tugela River mouth (viz. Fig. 20.11) where values of all nutrients are higher during floods and chlorophyll-*a* values are enhanced (Carter and Schleyer, 1988). This is reflected in greater densities of fish larvae at such times (Beckley and van Ballegooyen, 1992). When this outflow was directly observed during such a flood event, the indications were that most of the outflow occurred at a depth of 30 m and extended at least 25 km offshore. Salinities and temperatures over the shelf otherwise closely follow those of the Agulhas Current itself, also its seasonal cycle.

The circulation in the southern part of the bight is much harder to establish. Remote sensing has suggested a cyclonic eddy (Malan and Schumann, 1979) and this has been considered the main element in many conceptual portrayals ever since (e.g., Pearce, 1977b, Gründlingh and Pearce, 1990; Schumann, 1987; Harris, 1978). Observations show (Pearce, 1977b) that close to the coast the currents only follow the Agulhas Current 50% of the time. Ship's drift close inshore (1.6 km) is also about equally divided between poleward and equatorward drift (Harris, 1964; Pearce et al., 1978), agreeing with wind frequencies. The closer to the current, the greater the tendency is to follow its direction closely (Harris, 1978). The only quasi-synoptic hydrographic survey of the bight as a whole (Lutjeharms et al., 2000c) gives no indication of a consistent circulation. There are indications that the location of the edge of the Agulhas Current may show greater shifts in location along the edge of this particular part of the shelf than farther up- or downstream (Gründlingh and Pearce, 1990) and that shear edge eddies may play an important role on occasions (e.g., Lutjeharms and Roberts, 1988; their Fig. 10b). Putting it all together, Pearce et al. (1978) have concluded, on the basis of an eclectic set of observation, that "at any one time a succession of eddies of a variety of scales (are) generated by shear processes or meteorological forcing probably exists in the (Natal Bight)" and this is as good a summary of what is currently known as data will allow. Directly south of the bight, adjacent to the city of Durban (Fig. 20.11), the situation seems much simpler.

As mentioned above, a number of investigators (e.g., Pearce et al., 1978; Schumann, 1982; Anderson et al., 1988; Lutjeharms et al., 2000c and Meyer et al., 2002) have pointed out the presence of a cyclonic eddy directly off Durban, in the lee of the broader shelf that forms the Natal Bight. Currents measured off Durban do show a dominant north-eastward component (Fig. 20.15). Ship's drift close inshore (1.6 km) is—similar to further north on the shelf—about equally divided between poleward and equatorward drift (Harris, 1964; Pearce et al., 1978), which agrees with wind frequencies. Drifters have shown the same bi-polar tendency. Results presented by Tripp (1967) are also in agreement, showing that the current sets north-eastwards only about 50% of the time, with speeds of between 0.25 to 0.51 m/s. Current observations for a period of a month (Schumann, 1988b), as well as over shorter periods (Gründlingh and Pearce, 1984), indicate frequent current reversals here. There is therefore abundant evidence that this eddy does not seem to be present all the time. Current reversals at its location are measured over the full depth, indicating the barotropicity of this eddy. The main driving force of the shelf circulation at this location is most probably the Agulhas Current and not the wind (Schumann, 1981). It is clear from hydrographic observations that there is considerable upwelling in this eddy, bringing nutrient rich water closer to the surface, but this also occurs in a very spasmodic way.

When there is no eddy present there may be a mixed layer to a depth of 50 m with a maximum gradient in nitrate-nitrogen lying below 100 m. With an eddy present this nutricline can be lifted to about 40 m (Carter and d'Aubrey, 1988) with a concurrent drop in surface temperatures. This surface cooling has also been observed in satellite thermal infrared observations (Lutjeharms and Connell, 1989). Average values of nitrates in the Durban eddy have been measured around 3.33 μmol/ι, but maximum values have reached 16.79 μmol/ι, demonstrating the effect of the eddy's presence at times. These higher values in nutrients do not seem to be reflected in higher chlorophyll-*a* values. There is evidence of higher zooplankton biomass here, but only intermittently. Phytoplankton production rates are below 1 gC/m^2/ day (Burchall, 1968). There has been a greater density of fish larvae off Durban on occasion (Beckley and van Ballegooyen, 1992), but not consistently (Beckley and Hewitson, 1994). A hydrographic cruise covering the whole Natal Bight has clearly demonstrated (Lutjeharms et al., 2000c; Meyer et al., 2002) the vertical structure of this eddy compared to the waters over the rest of the shelf. The upwelling of nutrient-rich water did not enhance chlorophyll-*a* values on that occasion either.

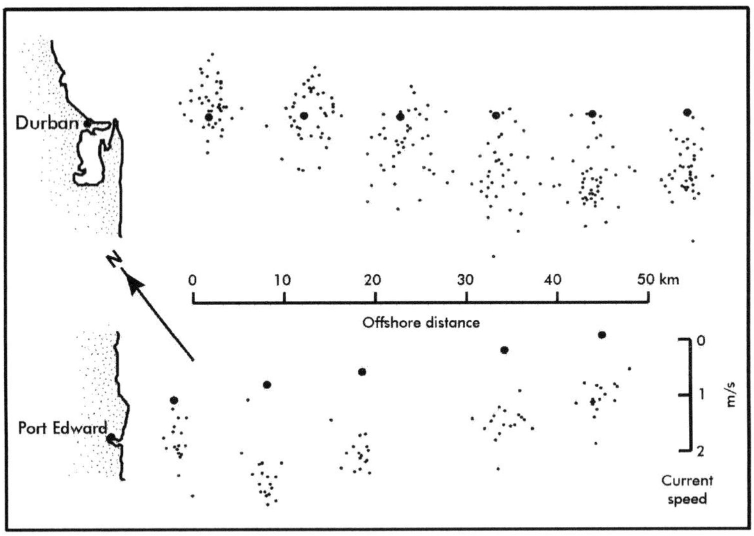

Figure 20.15 Current observations off Durban and Port Edward. (According to Schumann, 1982.) For the location of these two places, see Fig. 20.11. The large dot indicates the location of the observations; the small dots the tips of current arrows from the station location. These arrow points show speed and direction integrated over the top 100 m of the water column. Each dot represents one spot measurement.

In summary, a cyclonic lee eddy is a recurrent, but not ever-present part of the circulation on the shelf directly off Durban, but seems to have hardly any local biological impact, perhaps because it is not enduring. There is evidence that this cyclonic eddy and its nutrient contents form the core to incipient Natal Pulses (Lutjeharms and Roberts, 1988).

When a Natal Pulse forms, due to an offshore eddy or otherwise, it seems to carry the Durban lee eddy with it (Lutjeharms et al., 2003b) all the way to the southern tip of the continent, intensifying the cyclonic motion in the eddy in its downstream journey. This would imply that inshore currents from Durban downstream would experience a sudden, but short-lived, reversal. As mentioned before, such a reversal has actually been observed (Lutjeharms and Connell, 1989) and directly related to a passing Natal Pulse. This incorporation of the Durban eddy in Natal Pulses also means that a part of the shelf fauna of the Natal Bight will intermittently be carried away. The effect of such a mechanism on the shelf biota is not known. Most local fish seem to spawn in Cape waters downstream and migrate upstream in July and November (Garratt, 1988). Only geelbek (*Atractoscion aequidens*) and seventy-four (*Polysteganus undulosis*) appear to spawn off the east coast of southern Africa. A small proportion of the fish caught off Natal are present throughout the year, most are distinctly seasonal, being either summer or winter species (Van der Elst, 1988). The most dramatic of the upstream migrations is that of the pilchard *Sardinops ocellata,* that takes place close inshore and is therefore very noticeable. No correlation with the passing of Natal Pulses in any of these fish migrations has to date been established.

4.3 Between Durban and the Agulhas Bank

Downstream of Durban the bathymetry changes dramatically (Fig. 20.11). Here the shelf is narrow and without any indentations to speak of. This configuration of the continental border has a controlling influence on the behaviour of the Agulhas Current. The current closely follows the edge of the shelf (Gründlingh, 1983) and the currents on the shelf move largely in sympathy (Fig. 20.15). Current speeds up to 1 m/s have been measured to 10 km off the coast. The currents at the shelf edge are, on average, about 0.5 m/s just downstream of Durban (based on ships' drift; Tripp, 1967), but increase steadily till they are 2.16 m/s off Port Elizabeth. Current speeds inshore are much lower and vary in a seemingly random manner from 0.38 to 0.77 m/s. At Port St Johns there is a minor coastal offset (viz. Fig. 20.11) and this seems to cause a higher frequency of counter currents (Harris, 1964), up to 40% of the time (Tripp, 1967). Observed currents at Port Edward (Fig. 20.15) are with the Agulhas Current, but somewhat reduced in speed. The strong currents are reflected in the sand transport and bedform patterns along this whole shelf edge. Substantial underwater sand dunes are formed (Flemming, 1978, 1980, 1981) and this whole dune field is mobilised to move downstream with the current. It has been surmised that at certain locations where canyons cut into the shelf they act as sediment traps, carrying sediment off the shelf.

Surface drifts over this shelf region, as measured by drifters (Anderson et al., 1988), are parallel to the current, but in both alongshore directions, suggesting that they are driven largely by the wind. Movement at greater depths is with the current (Schumann, 1982; 1987) except during very strong south-westerly winds when the shelf currents may briefly change direction. The latter is also reflected in observations of ships' drift (Pearce et al., 1978), although the incidence of current reversal is much reduced here compared to the movement in the Natal Bight, demonstrating the greater influence of the juxtapositioned Agulhas Current. From a direct comparison of current meter data and concurrent wind records it has been shown (Schumann, 1981) that the correlation between these signals is smaller than expected. Many current reversals therefore seem likely to be due to the passing of a Natal Pulse. However, there are other possible low frequency fluctuations in the currents on the shelf on this coast.

Schumann (1981; with Perrins, 1982) has analysed current records for signs of tidal and inertial signals. He found limited inertial motion; the energy spectrum being dominated by much longer periods that were assumed to be associated with variations in the Agulhas Current. Schumann (1986), in an investigation of the bottom boundary layer along this coastline, has furthermore demonstrated that in 50 m of shelf water Ekman veering took place over the lower 35 m. This would cause some upwelling of deeper water, possibly enhancing biological productivity.

Observations of the vertical distribution of nutrients do indeed show higher nutrient levels at depth (Carter and d'Aubrey, 1988) and over the deeper parts of the shelf. In general shelf waters were poor in nutrients (nitrate: 2.48 μmol/ι mean; 15.80 μmol/ι maximum) as is to be expected where surface waters of the Agulhas Current dominate. No observations have to date been made during the passage of a Natal Pulse and its imbedded Durban eddy. Conceivably such an event could cause a sudden, but short-lived elevation of nutrient concentrations. The effect of such pulses on the productivity along this stretch of shelf remains unknown. Observations of zooplankton (Carter, 1977) indicate higher biomass at the edge of the Agulhas Current here than upstream. Ichthyoplankton observations (Beckley and van Ballegooyen, 1992) reveal the same, but with great variability in space and time.

To recapitulate: the influence of the northern Agulhas Current on the shelf circulation seems to start only at about 28°S. The circulation in the wider shelf region of the Natal Bight is principally influenced by the passing current in the creation of an upwelling cell at St Lucia. By inserting nutrients into shelf waters this cell may have a controlling influence on the ecology of the whole Natal Bight. At the southern end of the bight the current drives a persistent lee eddy off Durban. The movement of waters on the narrow shelf south of Durban are for the most part parallel to the Agulhas Current. With the possible exception of what happens during the passage of a Natal Pulse, the circulation on this particular shelf region consequently seems uncomplicated, with low productivity and biological activity. The exception seems to be at Port Alfred, at the eastern tip of the Agulhas Bank.

5. Southern Agulhas regime

The Agulhas Bank forms the triangular continental shelf south of Africa (Fig. 20.16). It lies between Port Elizabeth in the east and Cape Town in the west. It is about 250 km at its widest. A shallow part, the Alphard Rise, where the shelf is widest (viz. Fig. 20.16), constitutes the border between the western and the eastern Agulhas Bank that, as a result of this partition, have distinctly different characteristics. This is exemplified by the contrasting sedimentary nature of the eastern and western Agulhas Bank.

The sediments on the eastern bank are dominated by shelly fragments whereas there are mainly foraminiferal oozes on the western bank (Rogers and Bremner, 1991). The organic matter content of the eastern Agulhas Bank sediments lies between 0.0 and 3.9 % (per unit mass) compared to 4.0 to 11.9 % on the western Agulhas Bank This places the western bank squarely in the province of the Benguela upwelling system (Dingle et al., 1987b) A further reason for this difference is the contrasting characteristics of the offshelf currents on either side.

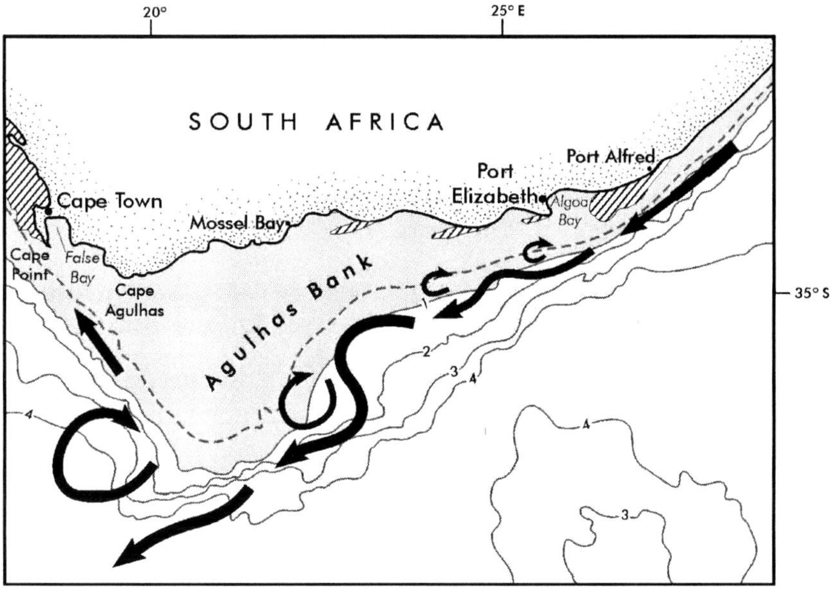

Figure 20.16 Bathymetry of the continental shelf off the south coast of South Africa in km (after Simpson, 1974; Dingle et al., 1987a) and its major circulatory elements. This covers the southern Agulhas Current regime. Areas shallower than 1 km are shaded. The broken line denotes the edge of the continental shelf as defined by the 200 m isobath. Upwelling is shown by hatching. Place and feature names used in the text are given here.

On the eastern side the warm waters of the intense southern Agulhas Current follow the shelf edge (Fig. 20.16), but much less closely than upstream. The current starts to meander near Port Elizabeth and these meanders grow in amplitude downstream (Lutjeharms et al., 1989a). The meanders come with cyclonic shear edge eddies and attendant warm plumes. These plumes in general are shallow features, but may eventually spread their warm water over large parts of the adjacent shelf. By contrast, on the western side of the Agulhas Bank the sluggish Benguela Current carries cold water equatorward, but this is complicated by the presence of a lee eddy directly to the west of the Agulhas Bank (Penven et al., 2001a), the recurrent passage of Agulhas rings (Lutjeharms and Valentine, 1988; Boebel et al., 2003) and the intermittent advection of warm Agulhas surface water in the form of Agulhas filaments (Lutjeharms and Cooper, 1996). However, this shelf region is wide and the inner parts may be less affected by the offshore currents than by winds.

5.1 Wind regimes south of Africa

As is the case along the east coast, the winds along the south coast are mainly parallel to the coastline, more so at Port Elizabeth than in the central part of the south coast (Schumann, 1989). The ratio of offshore/onshore to coast-parallel winds is also lower along the south coast to that of Port Elizabeth where the winds are more strongly oriented parallel to the coast. North-easterly winds show the greatest seasonality with a greater than 40% occurrence frequency in austral summer, dropping to 25% in winter (Schumann and Martin, 1991). Average speeds for north-easterly winds at Port Elizabeth are greater than 4 m/s in summer, but only 1.5 m/s in winter. South-westerly winds show an inverse seasonal occurrence compared to north-easterly winds, but the wind speeds always lie around 4 m/s. Based on ships' reports, Jury (1994) has shown that on the eastern Agulhas Bank the most frequent and the strongest winds are from the east in summer and from the west in winter. Winds are significantly weaker over the southern parts of the shelf compared to closer to the coast. These winds and their directions naturally are part and parcel of the global ENSO and other perturbations (e.g., Schumann, 1992). Little work has to date been carried out to pin down these relationships.

A significant proportion of these winds come about as a consequence of coastal lows. These are formed as a result of the topography of southern Africa (Gill, 1977), most often associated with cold fronts that move from west to east (Hunter, 1987). The coastal lows move along the south coast with periods of 2 to 5 days and propagation speeds of 14 to 20 m/s. Analyses of the climatological variability of the major axis winds along the coast of the Agulhas Bank (Schumann et al., 1991) has therefore shown a distinct spectral peak at 6 days.

Over the western Agulhas Bank the wind patterns are significantly different to those on the eastern bank. At Cape Town the most seasonal wind is the south-easterly wind with an 80% occurrence frequency in summer and a 40% frequency in winter. Average speeds in summer are in excess of 5 m/s; only 1.5 m/s in winter. By contrast the north-westerly winds are only slightly more prevalent in winter and maintain average speeds of about 3 m/s.

Wind strength is increased over the Agulhas Current itself. The loss of heat from the Agulhas Current to the atmosphere is about 200 W/m^2 higher than that of ambient water masses (Rouault et al., 2000). During along-current winds an atmospheric moisture and thermal front develops at the inshore edge of the current off Port Alfred, where the current passes a distinct upwelling cell.

5.2 The Port Alfred upwelling cell

As mentioned before, if the theoretical portrayal of Gill and Schumann (1979) holds, wherever a western boundary current moves from a narrow shelf past a wider shelf a degree of upwelling should be experienced. This has been observed south of Madagascar, at the northern corner of the Natal Bight and even in the Kuroshio system (Lutjeharms et al., 1993). It would therefore be expected to occur also where the Agulhas Current starts flowing along the Agulhas Bank and there it has indeed been observed (Lutjeharms et al., 2000a).

Thermal infrared imagery from satellite shows that this upwelling has its centre off Port Alfred (viz. Fig. 20.16), but that cold water inshore of the Agulhas Current may extend up to a maximum of 300 km upstream. On average it is about 30 km wide and extends 180 km along the edge of the current (Fig. 20.17). Water masses in the upper 200 m of the Agulhas Current are Subtropical Surface Water with the occasional pulse of Tropical Surface Water, mostly at the landward side (Gordon et al., 1987). During those times that the upwelling is evident at the sea surface the water in the upwelling cell is all South Indian Central Water (Lutjeharms et al., 2000a). This implies that it is upwelled onto the shelf from offshore waters deeper than 400 m and this has been confirmed by hydrographic observations in the region (Goshen and Schumann, 1988; their Fig. 8). Nutrient values in the upwelling cell can exceed, for example, 20 μmol/ι at 100 m, whereas they are less than 5 μmol/ι on the adjacent shelf. This upwelling cell may furthermore have considerable implications for the Agulhas Bank as a whole.

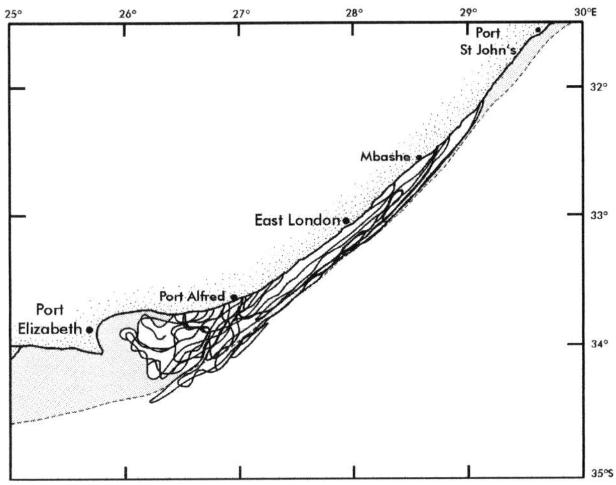

Figure 20.17 An ensemble of outlines of cold water of the Port Alfred upwelling cell inshore of the Agulhas Current, using the 17°C isotherm as an indicator. (As given by Lutjeharms et al., 2000a.) The 200 m isobath is shown as a dotted line and the shelf inshore of this isobath shaded.

First, it forms one of what may be considered to be the three main physical/chemical provinces of the bank. The others are the greater part of the shelf and the coastal upwelling regime of the western Agulhas Bank (Lutjeharms et al., 1996). These provinces are well-defined, considering the poor and inhomogeneous distribution of hydrographic stations available. Second, from this upwelling cell water colder than 10°C moves over the bottom of the Agulhas Bank (Lutjeharms and Meyer, 2004). The seasonal thermoclines over the bank are unusually intense. It is believed that this is partially due to the continual input of cold water along the bottom. It has previously been surmised that this cold water is upwelled inshore of the Agulhas Current along the eastern edge of the bank (Chapman and Largier, 1989). To date little direct evidence for this has been found. In fact, almost all hydrographic sections across the bank show the water colder than 10°C near the shelf edge only in the vicinity of Port Alfred. If this upwelling cell plays such a crucial role in the stratification of the whole Agulhas Bank, how permanent is it?

Regrettably, there are no data to address this important question. Surface observations show that it is present almost 50% of the time (Lutjeharms et al., 2000a), but this surface outcropping may be largely wind dependent, as has been observed at sea (Rouault et al., 1995). With the limited hydrographic data available, the presence of upwelled water at depth seems considerably more enduring. If this upwelling cell is so persistent, how does it affect the distribution of primary productivity and biota at higher trophic levels?

A chlorophyll-*a* maximum zone extends from the vicinity of Port Alfred roughly along the 100 m isobath across the eastern Agulhas Bank (Probyn et al., 1994). This "upwelling ridge" has primary production rates of 104 mg/m^2/h, whereas the shelf near Port Alfred boasts values of 888 mg/m^2/h. Values at the shelf break are 231 mg/m^2/h. In this respect it is interesting that ocean colour observations from satellite (Lutjeharms and Walters, 1985) show slightly enhanced values along the whole landward edge of the Agulhas Current. By contrast the zooplankton biomass is low near Port Alfred (Verheye et al., 1994), increasing westward. Spawning of chokka squid, as detected by eggs trawled, show a distinct concentration at Port Alfred, with a downstream decrease across the shelf, downstream (Augustyn et al., 1994). About predators much less is known (Smale et al., 1994). The size distribution of fish varies noticeably across the Agulhas Bank (Japp et al., 1994), the larger fish being found on the eastern Agulhas Bank, some species densities higher at the horizontal thermal gradients associated with the shelf edge (Barange, 1994). The biological influence of the Port Alfred upwelling cell therefore seems indirect, particularly via the inflow of bottom water over the eastern Agulhas Bank.

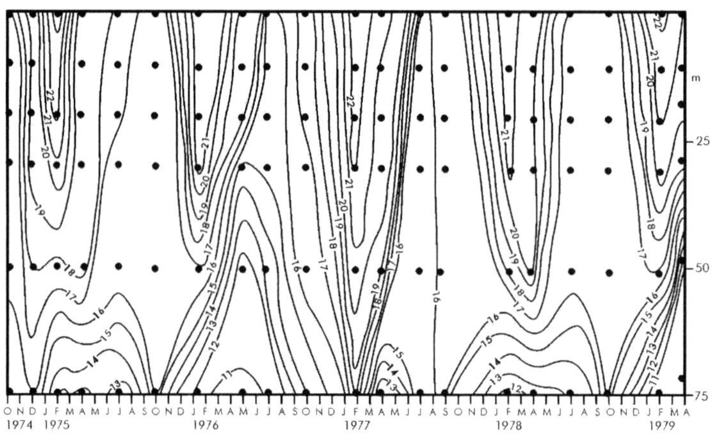

Figure 20.18 Time series of the temperature profile for a hydrographic station on the central Agulhas Bank for the period 1974 to 1979. Dots indicate standard depths at which nutrient samples were taken. (After Lutjeharms et al., 1996.)

5.3 The eastern Agulhas Bank

Much has been made of the ridge of cold water (< 10 °C) that overlies the 100 m isobath over much of the eastern Agulhas Bank (e.g., Boyd and Shillington, 1994; Verheye et al., 1994; Probyn et al., 1994). This ridge is evident in temperature sections across the bank; at times it crops out, producing areas of cold water at the sea surface (e.g., Walker, 1986; Swart and Largier, 1987) over the bank. The location of this ridge corresponds with the distribution of a number of copepod development stages (Largier and Swart, 1987), medium size chokka squid (Augustyn et al., 1994) and a subsurface chlorophyll-*a* maximum at the nutricline (Probyn et al., 1994). It has been proposed (Lutjeharms and Meyer, 2004) that this ridge of cold, nutrient-rich water originates in the Port Alfred upwelling cell. It enhances the thermal stratification of the water column from below over most of the bank and brings nutrients onto the full extent of the shelf. These nutrients may not be available to the phytoplankton throughout the year, but follow an annual cycle. Observations show (Carter et al., 1987) that the subsurface chlorophyll-*a* peak is closely associated with the behaviour of the seasonal thermocline. This stratification on the Agulhas Bank follows a rather peculiar cycle (Eagle and Orren, 1985; Lutjeharms et al., 1996; Fig. 20.18).

During the winter months storm winds and convection mix the upper water column to depths of 75 m or more; in summer fewer extreme wind events and increased insolation establish a seasonal thermocline that may be at 50 m or deeper. Summer winds have been shown to be too weak (Largier and Swart, 1987) to break down the seasonal thermocline. The winter mixing of the water column raises the temperature of the bottom water in winter, being warmest in July (Swart and Largier, 1987). All this is normal for shelf regimes in the subtropics. On the Agulhas Bank this seasonal stratification is however enhanced by cold water from *below* (Fig. 20.18) thus establishing a seasonal thermocline and nutricline about twice as strong as would normally be expected. It is believed that the inflow of all this cold, nutrient-rich water is from the Port Alfred upwelling cell.

The result of this seasonal behaviour of the water column is also reflected in the temperature/salinity characteristics of the Agulhas Bank. In spring the temperature range is restricted to 14 – 18°C, salinity to 35.0 to 35.5; by contrast, in winter this ranges from 10 – 23°C, salinity 34.6 to 35.8. The nutrients follow suite. Average sea surface temperatures are between 21 – 26°C over the eastern Agulhas Bank, representing the horizontal offshore gradient towards the Agulhas Current (Schumann and Beekman, 1984). In winter this is reduced to 16 – 19°C. In summer the average temperature difference between the water at the surface and at 80 m depth is as high as 11°C; in winter the highest gradient is 5°C. With the seasonal collapse of the nutricline the subsurface chlorophyll-*a* peak also disappears (McMurray et al., 1993). Nevertheless the seasonal thermocline on the eastern bank is usually shallower and better developed compared to that of the western part of the bank.

The movement of the waters of the eastern Agulhas Bank can most probably be placed into three categories: the very surface layer, the bottom layer and the rest of the water column. The very surface layer moves largely with the wind (Lutjeharms et al., 1986a). Comparison of the movement of a drifter with progressive wind stress vectors rotated 35° to the left has demonstrated this quite admirably (Harris, 1978). Analyses of ships' drift, corrected for windage on ships' superstructure, show (Tripp, 1967) no distinct patterns or seasonal cycles. Since the dominant wind directions are parallel to the coast, so will the top water layer therefore move. This is particularly important from a pollution point of view since tar balls (Shannon and Chapman, 1983; Shannon et al., 1983) will be beached only under onshore wind conditions. The bottom waters, where they have been measured (Swart and Largier, 1987), move largely parallel to the isobaths and in the direction with the Agulhas Current. The movement of the rest of the water column is more complex. Schumann and Perrins (1982) have shown that the highest energies on the shelf are to be found at inertial periods. The bottom motion described earlier consists of the residual drift after the inertial current signals have been removed. The influence of the Agulhas Current is not always clear.

As mentioned above, the major characteristic of the Agulhas Current along the shelf edge of the eastern Agulhas Bank is the presence of shear edge eddies of various sizes (Lutjeharms et al., 1989a). These eddies move downstream with the current and are most prevalent in the concave part of the shelf edge (Lutjeharms et al., 2003a). There is evidence (Hutchings, 1994) that the spawning of anchovy takes place preferentially just inshore of the shelf edge where these eddies are common. The passage (Schumann and van Heerden, 1988) of such eddies inverts the shelf edge current that is usually dominated by the Agulhas Current. Cold water is upwelled in the core of these cyclonic features and is has therefore been surmised that they are instrumental in bringing this cold water onto the shelf (Chapman and Largier, 1989). This has been observed very infrequently (e.g., Schumann and Beekman, 1984). The warm water plumes associated with the shear edge features bring warm Agulhas Current water onto the shelf and this may enhance the shelf thermocline. In general these plumes may represent a relatively thin layer of warm water (e.g., Goschen and Schumann, 1990) and their subsequent movement over the shelf may be almost entirely driven by winds (Goschen and Schumann, 1994).

Once past the southern tip of the Agulhas Bank, the Agulhas Current proper moves offshore, but some of the plumes may be drawn along the edge of the western Agulhas Bank as Agulhas filaments.

5.4 *The western Agulhas Bank*

Agulhas filaments are relatively shallow features (~ 50 m) and may carry warm surface water as far as the coastal upwelling zone off south-western Africa (Lutjeharms and Cooper, 1996). To date there has been only occasional evidence (e.g., Mitchell-Innes et al., 1999) that warm surface water from these filaments may intrude over the shelf in an analogous fashion to that of plumes adjacent to the eastern Agulhas Bank. Why Agulhas filaments are drawn so rapidly equatorward is also not entirely clear. This has been assumed (Lutjeharms and Valentine, 1988) to be due to passing Agulhas rings. Such ring prevalence may be the origin of the hypothesised shelf edge jet (e.g., Hutchings, 1994) that has been put forward as a mechanism to move anchovy larvae northward towards the coastal upwelling region of the west coast (e.g., Boyd et al., 1992). Then again, whenever direct current observations of the currents over the shelf edge have been made (e.g., Fowler and Boyd, 1998) no evidence for any persistent north-westerly setting current could be found.

However, there is strong evidence that there is a tendency for the circulation off the shelf of the western Agulhas Bank to be cyclonic, which would lead to currents at the shelf edge to move poleward instead. Modelling (Lutjeharms et al., 2003a) and hydrographic observations (Penven et al., 2001a) now suggest that this is due to a cyclonic eddy driven by the Agulhas Current in the lee of the Agulhas Bank. The discrepancy between an equatorward shelf edge jet and a poleward movement driven by an Agulhas Bank lee eddy had not yet been resolved. Apart from the difference in current behaviour at the shelf edge, the western Agulhas Bank also differs from the eastern bank by extensive coastal upwelling.

The main, wind-driven upwelling along the west coast normally starts at Cape Point (south of Cape Town, viz. Fig. 20.16), but during about 10% of the time there also is contiguous upwelling all the way between Cape Agulhas and Cape Point (Lutjeharms and Meeuwis, 1987; Lutjeharms and Stockton, 1991). This latter upwelling, as expressed as a biologically productive region, may extend from less than 20 to 100 km offshore (Mitchell-Innes et al., 1999). At Hermanus and Gansbaai, between Cape Point and Cape Agulhas, annual surface temperatures range (Largier et al., 1992) from 17°C in January to 15°C in August with a small maximum in May. With a higher incidence of south-westerly winds in summer, the upwelling here is most intense in that season (Boyd et al., 1985), as can easily be modelled (Penven et al., 2001b). Constant westerly or south-westerly winds may induce downwelling, but upwelling can recur within a matter of days under the influence of winds of the correct direction and sufficient speed (Jury, 1988). The vertical temperature gradients in this upwelling are strongest in summer, weakest in winter (Schumann and Beekman, 1984). Cool, low-salinity Central Water was upwelled to within 20–40 m of the sea surface from January to April during a survey in 1975 (Boyd et al., 1985) and from September to November when south-easterly winds are dominant. Only between February and April was it observed to break the surface. Climatologically it is also present in July and December (Lutjeharms and Stockton, 1991). Vertical stratification was best developed in summer, due to the uplift of cooler Central Water and the onshore advection of warm surface water from the Agulhas Current and the Agulhas Bank. This is reminiscent of the cycle of stratification over the central Agulhas Bank (Eagle and Orren, 1985; Fig. 20.18). Nutrients moved closer to the surface in spring and summer, causing a modest spring bloom (Mitchell-Innes et al., 1999), but generally did not penetrate the thermocline until vigorous mixing to a depth of 60–80 m in winter.

Over the rest of the western Agulhas Bank the stratification seems much weaker than over the eastern bank. The uplift of the isopycnals towards the coast over this shelf region is naturally evident in the distribution of nutrients with high values covering the greater part of the shelf (Lutjeharms et al., 1996) than over the eastern Agulhas Bank.

The prevailing currents on the western Agulhas Bank are in a north-westerly direction in summer; in winter those along the southern parts of this coastline may be south-eastward (Harris, 1978). Water characteristics are the same above the thermocline than on the eastern Agulhas Bank, but below the thermocline the waters are slightly less salty and have lower nutrient loads (Chapman and Largier, 1989). This has been taken as evidence that this water has an Atlantic rather than an Indian Ocean origin and that the bottom water here on average moves in a south-easterly direction. In general the water movement here is parallel to the isobaths (Largier et al., 1992), but in both directions. Nevertheless, inertial currents may still be the dominant movements on shorter time scales (Schumann and Perrins, 1982).

Evidently, the coastal upwelling on the western Agulhas Bank will have a dramatic effect on the ecosystem of this part of the bank. In contrast to the eastern bank, chlorophyll-*a* values are considerably higher (Largier et al., 1992). Phytoplankton concentrations are lower (De Decker, 1973) than in the upwelling region along the west coast, both measured as numbers of cells and as settled volume. Remarkably, zooplankton settled volumes are almost consistently higher. Biota in higher trophic levels (e.g., Roel et al., 1994) do not show any patterns clearly related to the upwelling, including euphausids.

5.5 *Coastal processes*

The wind-driven upwelling along the western part of the Agulhas Bank is a specific coastal process, but there are a number of other such locations and processes on the Agulhas Bank that are largely divorced from what happens over the broader expanses of the shelf. Perhaps the most important of these is intermittent upwelling at the promontories along the coast of the eastern Agulhas Bank (Schumann et al., 1982; viz. Fig. 20.16).

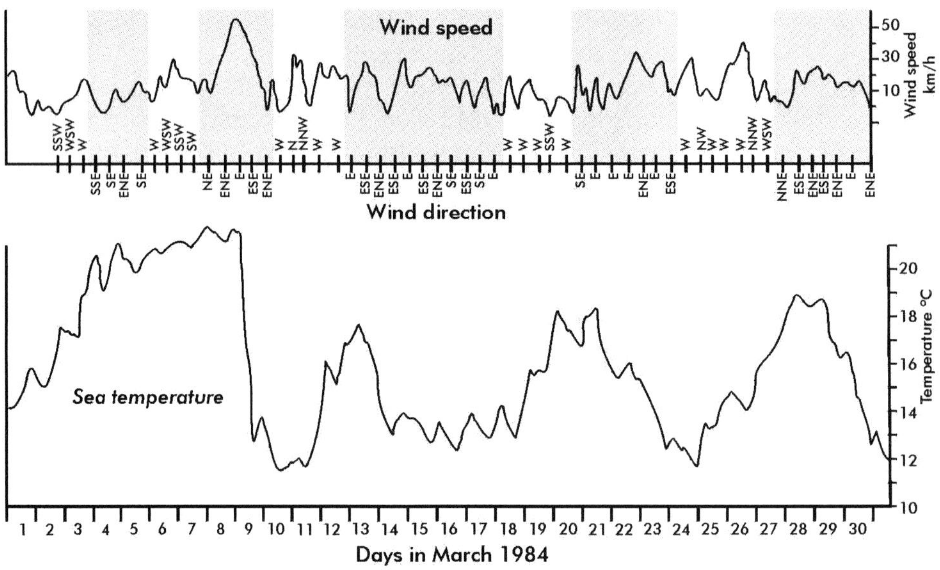

Figure 20.19 Sea surface temperatures and wind speeds at Gouriqua (near Mosselbaai, viz. Fig. 20.16) on the south coast of South Africa during March 1984. Wind speed is shown as a curve; wind direction by letters on the abscissa of the wind speed curve. (According to Lutjeharms, 1998.) Temperatures rise and fall at the coastline in direct response to the direction and speed of the local wind.

On the whole the seasonal cycle of sea surface temperatures at this coast is considerably better developed than that along the northern Agulhas Current. Average temperatures at Port Elizabeth, for example, range from 21°C in January to about 15°C in July/August (Lutjeharms, 1998) whereas those at East London vary only from 19°C to 17°C. Intermittent coastal upwelling along the south coast is superimposed on this seasonal sequence. It occurs only at prominent capes (Schumann et al., 1982) under suitable wind directions (e.g., Goschen and Schumann, 1995) and with winds above a certain threshold (Schumann et al., 1995).

Upwelling is seen less than 10% of the time, the highest incidence being in summer. The local effects can be quite dramatic (Fig. 20.19). Beckley (1983, 1988) has reported temperatures dropping by 8°C over a period of 24 h at a cape site compared to one less than 40 km away. Goschen and Schumann (1995) have observed reaction times of a few hours. These upwelling events may last from one to 5 days (Schumann, 1999). Wind events in along-coast winds have a frequency peak between 4 and 8.5 days, with a maximum at 7 days (Schumann et al., 1991) indicating the driving role of the winds.

Nevertheless an analysis of chlorophyll-*a* intensity at this coastline (De Villiers, 1998) suggests that the correlations between primary productivity and sea surface temperature is much better than between primary productivity and upwelling inducing winds. Along the very farthest eastern side of the Agulhas Bank it may be difficult to distinguish between local, wind-induced upwelling and the advection into the region of cold water that has been upwelled in the Port Alfred upwelling cell (e.g., Schumann et al., 1988). Upwelling events of this kind have in fact been observed to attract large flocks of sea birds. This phenomenon remains unexplained, since the short duration of most upwelling events would not result in greatly enhanced primary productivity. It has been hypothesised (Schumann et al., 1988) that the convergence associated with an upwelling front may have caused concentration of food items for birds. The counterpart of coastal upwelling of cold water is advection of warm water into coastal embayments.

Along the far eastern side of the Agulhas Bank the likelihood of an intrusion of warm surface water from the adjacent Agulhas Current right to the coast may be high. Under southerly wind conditions water from plumes may reach the coastline and cause a temperature rise of 3°C or more in surface water. They may also cause a distinct layering in offshore water. To some extent the coastal morphology plays a role in the manner in which offshore events may influence the water masses or circulation at the coast. At certain locations, such as Algoa Bay at Port Elizabeth (viz. Fig. 20.16), the local winds may create circulations restricted to a bay (Goschen and Schumann, 1994). A prominent case of localised circulation is found in False Bay, the large, rectangular Bay on the most western tip of the western Agulhas Bank, just south-east of Cape Town (viz. Fig. 20.16).

The surface circulation of this bay is complex (Gründlingh and Largier, 1991). On the eastern side there is intermittent, wind-driven upwelling (Cram, 1970) that brings nutrient rich water to the surface (Taljaard, 1991). The border of this upwelling may on occasion form a zonal front across the mouth of the bay (Lutjeharms et al., 1991). Only one hydrographic survey of the bay as a whole has been made (Gründlingh, 1992b, Taljaard, 1991) and as part of that investigation two sections across the bay were repeated shortly after. These repeat measurements have dramatically demonstrated the variability to be expected.

During the observations in 25–27 April 1989, cold, fresher upwelled water from farther east along the coastline was advected into the western side of the bay, creating a meridional front, evident throughout the water column. During the repeat survey between 27 and 29 April 1989, the situation had changed completely. Now there again was a meridional front near the centre of the bay, but with the cold water on the *east,* brought about by substantial upwelling on that side. A shallow thermocline (30 m) remained well-developed in the northern, shallower part, but more variable in the southern part of the bay. Surface currents are also very variable and show a range of different patterns (Atkins, 1970), probably wind generated (Jury, 1991). The bay is surrounded by mountains, creating a strong orographic effect on the air flow. At depth, by contrast, the currents seem much more steady and consistent (Gründlingh et al., 1989). Water enters on the western side and leaves on the eastern side. As might be expected the vertical stratification is very seasonal, particularly in the southern, deeper parts (Gründlingh, 1992b). Huge demographic shifts to the shores of the bay have placed enormous pressure on its self-cleaning ability. As in many other parts of the shelf seas discussed above, ignorance of the flushing rate and many other aspects of the circulation and the ecosystem make its proper management hazardous.

To sum up: the shelf waters of the Agulhas Bank may be placed into three physical/chemical provinces according to the influence of the Agulhas Current, or lack thereof. These are: the Port Alfred upwelling cell, the coastal upwelling of the western Agulhas Bank and, third, the wider shelf outside the immediate influence of the Agulhas Current. The Port Alfred upwelling cell may be permanent at depth, with a surface outcropping driven by local winds. Cold, nutrient-rich water is brought onto the shelf here and may spread over the greater part of the Agulhas Bank. There is a high likelihood that this inflow of bottom water is the driving force for the very intense seasonal thermoclines over the wider shelf region. The northward advection of warm Agulhas plumes is assumed to play a minor role. The unusual intensity of the stratification over this shelf may be crucial to the local ecosystem and, in particular, to the successful spawning of economically important fish species. The upwelling on the western Agulhas Bank is a seasonal phenomena and forms part and parcel of the greater Benguela upwelling system.

6. Future directions

From the above brief review it should be abundantly clear that the multidisciplinary characteristics of this part of the world ocean are very imperfectly understood. This is particularly true for the coastal oceans. This ignorance is easily explained: very few observations have been made on these continental shelves as a whole; in large parts, none. In some respects this makes it easy to suggest future research directions.

In shelf regions where hardly any data whatsoever are available, any new measurements are welcome. In other respects it makes it very difficult. To identify the key elements that need to be addressed as a matter of urgency—particularly for contiguous Third World countries with severely limited funds and research capability—is near to impossible. With these caveats, let us proceed by examining the research needs for the coastal oceans in the sequence in which they were discussed in the preceding sections, starting with the shelf regions of the Mozambique Channel.

The wide shelf on the eastern seaboard of Mozambique plays a recognised important role in local fisheries and thus in the economy of the adjacent country (e.g., Sætre and de Paula e Silva, 1979; Brinca et al., 1981). Its waters are likely to be predominantly influenced by the reigning winds and in particular the occasional incidence of severe cyclones. The effects of such winds on the hydrographic characteristics, especially vertical stratification, the primary productivity and the distribution of biota need urgent research attention. It has now been shown conclusively (De Ruijter et al., 2002) that there is no continuous, intense Mozambique Current that borders this shelf region, but that instead a series of eddies are formed in the narrows of the channel that subsequently progress poleward. The influence of these drifting eddies on the adjacent shelf is not known yet. Quartly and Srokosz (2004) have shown that they extract water, rich in phytoplankton, from the surface layers of the shelf. These potentially important processes need to be studied with *in situ* measurements. These recent research developments also point to the need for more, extensive deep-sea investigations.

The lack of hydrographic stations to describe and demarcate the water masses and the circulation in the deeper parts of the Mozambique Channel (Lutjeharms, 1977) also has an effect on what is known on the influence of the deep-sea on shelf waters. Although recent work (e.g., De Ruijter et al., 2002; Ridderinkhof and De Ruijter, 2003) has started to fill the most obvious gaps, much remains to be done. The hydrography and kinematics of the eastern side of the Mozambique Channel, for instance, remain cloaked in mystery. The influence of this particular deep-sea region on the western shelf of Madagascar therefore also remains unknown.

For the whole South West Indian Ocean this eastern side of the Mozambique Channel is probably the coastal ocean about which least is known. At the northern tip of Madagascar some work was done on shelf and adjacent coastal processes during the 1960s (e.g., Angot and Ménaché, 1963; Gerard, 1964; Angot and Gerard, 1966; Donguy and Piton, 1969), but along the central and southern parts there is – to my knowledge – no adequate information whatsoever. The most basic investigations on water masses, current patterns, biota and their respective variabilities need to be undertaken here. This is also true for the eastern shelf of Madagascar.

As was seen in the preceding sections, only the most basic facts are know about the two branches of the East Madagascar Current along this shelf of Madagascar. Since the shelf here is very narrow (Figure 20.1) the fast moving waters of this current are more than likely to have a decisive influence on the shelf waters. However, as long as no detailed observations are made and no careful monitoring is done, this reasonable hypothesis has to remain in the realm of speculation. Of particular importance is a better understanding of the upwelling cell at the south-eastern corner of Madagascar. Only one set of dedicated observations have been made here to date (Machu et al., 2002) and much more needs to be learnt about the possible influence of this upwelling on the biological productivity of the region. Anecdotal information from local subsistence fishermen suggests (M. Rouault, personal communication) that it may be a rich region for fisheries. Its investigation therefore also may hold considerable economic and fisheries management consequences. In comparison to these shelf regions off Madagascar and Mozambique, comparatively much more is known about the shelf regions off South Africa.

For the shelf region inshore of the northern Agulhas Current a number of things stand out that need further investigation if we are to understand the physico-chemical and biological behaviour of these particular shelf waters. It has been hypothesised that the upwelling cell at the northern tip of the Natal Bight, at St Lucia, may control the nutrient supply to the whole bight and thus the ecosystem of this shelf province. The intensity of this upwelling cell and the progress of water from there onto the shelf of the rest of the bight need to be monitored carefully if this hypothesis is to be verified. The shelf downstream from the bight is much narrower (Figure 20.1) and the water movement on it therefore conceivably much simpler. The one exception to this simple movement in sympathy with that of the Agulhas Current may come about due to the passage of Natal Pulses.

This irregularly occurring meander on the trajectory of the Agulhas Current has been shown to have a dramatic influence on the currents over the adjacent shelf (Gründlingh, 1979; Lutjeharms and Connell, 1989) during its passage. To date this has been observed only fortuitously, although it is evident in some historical current meter records (e.g., Schumann, 1982). This shelf is the known conduit for the migration of whales and the annual sardine run. It is unknown what effect the passage of a Natal Pulse has on these migrations or if the unusual, but potentially beneficial, currents associated with these current meanders are in fact purposefully used by these animals. It would seem clear that this needs investigation. For the shelf off the southern Agulhas Current the research questions are different.

It has been surmised (Lutjeharms and Meyer, 2004) that the upwelling cell at the eastern end of the Agulhas Bank, adjacent to Port Alfred, may dominate the flux of bottom water onto this shelf region. This proposition has considerable implications for a proper understanding of the vertical stratification of the waters over the shelf, the nutrient supply to the shelf and the ecology of this large coastal region. Observations on this upwelling cell to date have all been inadvertent. The need for a dedicated observational programme seems essential, particularly since the Agulhas Bank is the spawning region for anchovy that sustain the prime pelagic fisheries of South Africa. Such a programme should at a minimum study the nature and driving forces of the Port Alfred upwelling cell, the movement of upwelled water from here and the variability of the system of bottom water supply. It is not only the bottom water that is of importance; the whole physico-chemical nature of the waters over the Agulhas Bank is—bearing in mind its probable economic importance—amazingly poorly known.

To date only two quasi-synoptic research cruises have been undertaken to establish the hydrographic structure of the Agulhas Bank as a whole (Lutjeharms et al., 1981b; 1983). During these cruises only the thermal structure of the water masses was measured. Notwithstanding the importance of this shelf region for the local fisheries, no dedicated hydrographic cruise or set of cruises has covered the full extent of the bank since the aforementioned cruises, nor are there immediate plans to rectify this situation. This seems a particularly serious gap in the knowledge of these coastal oceans that needs urgent attention.

The waters on the Agulhas Bank are influenced not only by normal solar and air-sea interaction processes, nor only by the imbedded upwelling cell, but also by interaction with the juxtapositioned Agulhas Current. As is evident from the above review, plumes of warm surface water from the Agulhas Current move onto the bank and may spread over extensive parts of it. It has consequently been surmised (Swart and Largier, 1987) that this input of advected surface water may play an important, if not decisive, role in maintaining the very strong vertical temperature gradients over the bank. Attempts at quantifying this process and comparing it with the role of insolation would be valuable and contribute to a better understanding of the factors playing a role in the hydrography of this important part of the South African shelf and its role in the ecology of the region, particularly the spawning of anchovy.

In short, even an entirely subjective listing of research priorities for the coastal seas of the South West Indian Ocean demonstrates unequivocally the urgent need for research in order to understand even the most fundamental, descriptive aspects of many of the shelf regions here.

Acknowledgements

This review was undertaken during a personally difficult time. I therefore thank the editors for their considerable patience and understanding. I wish to thank in particular Dr Mike Roberts for kindly making data from the Mozambique Channel available to me before publication. The volume on coastal ocean studies off Natal (Schumann, 1988a) is a veritable gold mine of useful information for a shelf region where investigations have virtually stopped since the commercialisation of the South African CSIR (Lutjeharms and Thompson, 1993). Prof. Geoff Brundrit and Dr John Rogers are thanked for a critical reading of the manuscript and for valuable comments that made the final product better. Financial support came from an award by the National Research Foundation of South Africa and the IDYLE programme of the French *Institut de Recherche pour le Développement.*

Bibliography

Anderson, F. P., M. L. Gründlingh and C. C. Stavropoulos, 1988. Kinematics of the southern Natal coastal circulation: some historic measurements 1962–63. *S. Afr. J. Sci.,* **84,** 857–860.

Angot, M. and R. Gerard, 1966. Charactères hydrologiques de l'eau de surface au Centre ORSTOM de Nosy-Bé de 1962 à 1965. *Cah. ORSTOM Océanogr.,* **4,** 37–54.

Angot, M. and M. Ménaché, 1963. Premiéres données hydrologiques sur la région voisine de Nosy Bé (nord-ouest de Madagascar). *Cah. ORSTOM Océanogr.,* **3,** 7–15.

Anonymous, 1983. Cruise report R/V "Dr. Fridtjof Nansen", Fisheries Resources Survey, Madagascar, 16–28 June 1983. *Reports on Surveys with the R/V Dr Fridtjof Nansen,* Institute of Marine Research, Bergen, 9 pp.

Arhan, M., H. Mercier and J. R. E. Lutjeharms, 1999. The disparate evolution of three Agulhas rings in the South Atlantic Ocean. *J. Geophys. Res.,* **104,** 20,987–21,005.

Asplin, L., M. D. Skogen, W. P. Budgell, V. Dove, E. Andre, T. Gammelsrød, A. M. Hoguane, 2004. Numerical modelling of currents and hydrography of the Mozambique Channel. In preparation.

Atkins, G. R., 1970. Wind and current patterns in False Bay. *Trans. Roy. Soc. S. Afr.,* **39,** 139- 148.

Augustyn, C. J., M. R. Lipiński, W. H. H. Sauer, M. J. Roberts and B. A. Mitchell-Innes, 1994. Chokka squid on the Agulhas Bank: life history and ecology. *S. Afr. J. Sci.,* **90,** 143–154.

Barange, M., 1994. Acoustic identification, classification and structure of biological patchiness on the edge of the Agulhas Bank and its relation to frontal features. *S. Afr. J. Mar. Sci.,* **14,** 333–347.

Beckley, L. E., 1983. Sea-surface temperature variability around Cape Recife, South Africa. *S. Afr. J. Sci.,* **79,** 436–438.

Beckley, L. E., 1988. Spatial and temporal variability in sea temperature in Algoa Bay, South Africa. *S. Afr. J. Sci.,* **84,** 67–69.

Beckley, L. E. and R. C. van Ballegooyen, 1992. Oceanographic conditions during three ichthyoplankton surveys of the Agulhas Current in 1990/91. *S. Afr. J. Mar. Sci.,* **12,** 83–93.

Beckley, L. E. and J. D. Hewitson, 1994. Distribution and abundance of clupeoid larvae along the east coast of South Africa in 1990/91. *S. Afr. J. Mar. Sci.,* **14,** 205–212.

Biastoch, A. and W. Krauß, 1999. The role of mesoscale eddies in the source regions of the Agulhas Current. *J. Phys. Oceanogr.,* **29,** 2303–2317.

Biastoch, A., C. J. C. Reason, J. R. E. Lutjeharms and O. Boebel, 1999. The importance of flow in the Mozambique Channel to seasonality in the greater Agulhas Current system. *Geophys. Res. Lett.,* **26,** 3321–3324.

Boebel, O., J. Lutjeharms, C. Schmid, W. Zenk, T. Rossby and C. Barron, 2003. The Cape Cauldron: a regime of turbulent inter-ocean exchange. *Deep-Sea Res. II,* **50,** 57–86.

Boyd, A. J., B. B. S. Tromp and D. A. Horstman, 1985. The hydrology off the South African south-western coast between Cape Point and Danger Point in 1975. *S. Afr. J. Mar. Sci.,* **3,** 145–168.

Boyd, A. J., J. Tauton-Clark and G. P. J. Oberholster, 1992. Spatial features of the near-surface and midwater circulation patterns off western and southern South Africa and their role in the life histories of various commercially fished species. *S. Afr. J. Mar. Sci.,* **12,** 189–206.

Boyd, A. J., and F. A. Shillington, 1994. Physical forcing and circulation patterns on the Agulhas Bank. *S. Afr. J. Sci.,* **90,** 114–122.

Brinca, L., A. Jorge da Silva, L. Sousa, I. M. Sousa and R. Sætre, 1983. A survey of the fish resources at Sofala Bank, Mozambique. *Reports on Surveys with the R/V Dr Fridtjof Nansen,* Serviço de Investigações Pesqueiras, Maputo, Institute of Marine Research, Bergen, 70 + 15 pp.

Brinca, L., F. Rey, C. Silva and R. Sætre, 1981. A survey on the marine fish resources of Mozambique. *Reports on Surveys with the R/V Dr Fridtjof Nansen,* Instituto de Desenvolvimento Pesqueiro, Maputo, Institute of Marine Research, Bergen, 58 pp.

Burchall, J., 1968. An evaluation of primary productivity studies in the continental shelf region of the Agulhas Current near Durban (1961–1966). *Investigatl Rep., Oceanogr. Res. Inst., Durban,* **21,** 44 pp.

Byrne, D. A., A. L. Gordon and W. F. Haxby, 1995. Agulhas eddies: a synoptic view using Geosat ERM data. *J. Phys. Oceanogr.,* **25,** 902–917.

Carter, R. A., 1977. The distribution of calando copepods in the Agulhas Current system off Natal, South Africa. Unpublished MSc thesis, University of Natal, 165 pp.

Carter, R. A., H. F. McMurray and J. L. Largier, 1987. Thermocline characteristics and phytoplankton dynamics in Agulhas Bank waters. *S. Afr. J. Mar. Sci.,* **5,** 327–336.

Carter, R. and J. d'Aubrey, 1988. Inorganic nutrients in Natal continental shelf waters. In *Coastal Ocean Studies off Natal, South Africa,* E. H. Schumann, ed. Lecture Notes on Coastal and Estuarine Studies 26, Springer-Verlag, Berlin, pp. 131–151.

Carter, R. and M. H. Schleyer, 1988. Plankton distribution in Natal coastal waters. In *Coastal Ocean Studies off Natal, South Africa,* E. H. Schumann, ed. Lecture Notes on Coastal and Estuarine Studies 26, Springer-Verlag, Berlin, pp. 152–177.

Chapman, P. and J. L. Largier, 1989. On the origin of Agulhas Bank bottom water. *S. Afr. J. Sci.,* **85,** 515–519.

Cooke, A., J. R. E. Lutjeharms and P. Vasseur, 2004. Marine and coastal natural history and conservation in Madagascar. In *The Natural History of Madagascar,* S. M. Goodman and J. P. Benstead, eds, The University of Chicago Press, Chicago, pp. 179–209.

Cram, D. L., 1970. A suggested origin for the cold surface water in central False Bay. *Trans. Roy. Soc. S. Afr.,* **39,** 129–137.

Crimp, S. J., J. R. E. Lutjeharms and S. J. Mason, 1998. Sensitivity of a tropical-temperate trough to sea-surface temperature anomalies in the Agulhas retroflection region. *Water S. A.,* **24,** 93–101.

De Decker, A. H. B., 1973. Agulhas Bank plankton. In *The Biology of the Indian Ocean,* B. Zeitzschel, ed., Springer-Verlag, Berlin, pp. 189–219.

De Ruijter, W. P. M., P. J. van Leeuwen and J. R. E. Lutjeharms, 1999a. Generation and evolution of Natal Pulses: Solitary meanders in the Agulhas Current. *J. Phys. Oceanogr.,* **29,** 3043–3055.

De Ruijter, W. P. M., A. Biastoch, S. S. Drijfhout, J. R. E. Lutjeharms, R. P. Matano, T. Pichevin, P. J. van Leeuwen and W. Weijer, 1999b. Indian-Atlantic inter-ocean exchange: dynamics, estimation and impact. *J. Geophys. Res.,* **104,** 20,885–20,911.

De Ruijter, W. P. M., H. Ridderinkhof, J. R. E. Lutjeharms, M. W. Schouten and C. Veth, 2002. Observations of the flow in the Mozambique Channel. *Geophys. Res. Lett.,* **29,** 10.1029/2001GL013714.

De Ruijter, W. P. M., H. M. van Aken, E. Beier, J. R. E. Lutjeharms, R. P. Matano and M. W. Schouten, 2004. Eddies and dipoles around South Madagascar: formation, pathways and large-scale impact. *Deep-Sea Res. I,* **51,** 383–400.

De Villiers, S., 1998. Seasonal and interannual variability in phytoplankton biomass on the southern African continental shelf: evidence from satellite-derived pigment concentrations. *S. Afr. J. Mar. Sci.,* **19,** 169–179.

DiMarco, S. F., P. Chapman and W. D. Nowlin, 2000. Satellite observations of upwelling on the continental shelf south of Madagascar. *Geophys. Res. Lett.,* **27,** 3965–3968.

Dingle, R. V., G. V. Birch, J. M. Bremner, R. H. de Decker, A. du Plessis, J. A. Engelbrecht, M. J. Fincham, T. Fitton, B. W. Flemming, R. I. Gentle, S. H. Goodlad, A. K. Martin, E. G. Mills, G. J. Moir, R. J. Parker, S. H. Robson, J. Rogers, D. A. Salmon, W. G. Sieser, E. S. W. Simpson, C. P. Summerhayes, F. Westall, A. Winter and M. W. Woodborne, 1987a. Bathymetry around Southern Africa (SE Atlantic & SW Indian Oceans). *Ann. S. Afr. Mus.,* **98,** 1–27 (separate map 1).

Dingle, R. V., G. V. Birch, J. M. Bremner, R. H. de Decker, A. du Plessis, J. A. Engelbrecht, M. J. Fincham, T. Fitton, B. W. Flemming, R. I. Gentle, S. H. Goodlad, A. K. Martin, E. G. Mills, G. J. Moir, R. J. Parker, S. H. Robson, J. Rogers, D. A. Salmon, W. G. Sieser, E. S. W. Simpson, C. P. Summerhayes, F. Westall, A. Winter and M. W. Woodborne, 1987b. Deep-sea sedimentary environments around southern Africa (South-East Atlantic and South-West Indian Oceans). *Ann. S. Afr. - Mus.,* **98,** 1–27 (separate map 2).

Donguy, J.-R. and B. Piton, 1969. A perçu des conditions hydrologiques de la partie nord du canal de Mozambique. *Cah. ORSTOM Océanogr.,* **7,** 3–26.

Donguy, J.-R. and B. Piton, 1991. The Mozambique Channel revisited. *Oceanologica Acta,* **14,** 549–558.

Duncombe Rae, C. M., 1991. Agulhas retroflection rings in the South Atlantic Ocean; an overview. *S. Afr. J. Mar. Sci.,* **11,** 327–344.

Duncombe Rae, C. M., F. A. Shillington, J. J. Agenbag, J. Taunton-Clark and M. L. Gründlingh, 1992. An Agulhas Ring in the South East Atlantic Ocean and its interaction with the Benguela upwelling frontal system. *Deep-Sea Res.,* **39,** 2009–2027.

Duncombe Rae, C. M., S. L. Garzoli and A. L. Gordon, 1996. The eddy field of the southeast Atlantic Ocean: a statistical census from the Benguela Sources and Transports Project. *J. Geophys. Res.,* **101,** 11,949–11,964.

Eagle, G. A. and M. J. Orren, 1985. A seasonal investigation of the nutrients and dissolved oxygen in the water column along two lines of stations south and west of South Africa. *Nat. Res. Inst. Oceanol., CSIR. Res. Rep.,* **567:** 52 pp.

Fine, R. A., M. J. Warner and R. F. Weiss, 1988. Water mass modification of the Agulhas retroflection: chlorofluoromethane studies. *Deep-Sea Res.,* **35,** 311–332.

Flemming, B. W., 1978. Underwater sand dunes along the southeast African continental margin - observations and implications. *Mar. Geol.,* **26,** 177–198.

Flemming, B. W., 1980. Sand transport and bedform patterns on the continental shelf between Durban and Port Elizabeth (Southeast African Continental Margin). *Sediment. Geol.,* **26,** 179–205.

Flemming, B. W., 1981. Factors controlling shelf sediment dispersal along the southeast African continental margin. *Mar. Geol.* **42,** 259–277.

Flemming, B. and R. Hay, 1988. Sediment distribution and dynamics of the Natal continental shelf. In *Coastal Ocean Studies off Natal, South Africa,* E. H. Schumann, ed., Lecture Notes on Coastal and Estuarine Studies 26, Springer-Verlag, Berlin, pp. 47–80.

Fowler, J. L. and A. J. Boyd, 1998. Transport of anchovy and sardine eggs and larvae from the western Agulhas Bank to the west coast during the 1993/94 and 1994/95 spawning seasons. *S. Afr. J. Mar. Sci.,* **19,** 181–195.

Garratt, P. A., 1988. Notes on seasonal abundance and spawning of some important offshore linefish in Natal and Transkei waters, southern Africa. *S. Afr. J. Mar. Sci.,* **7,** 1–8.

Garzoli, S. L., A. L. Gordon, V. Kamenkovich, D. Pillsbury and C. Duncombe-Rae (sic), 1996. Variability and sources of the southeastern Atlantic circulation. *J. Mar. Res.*, **54,** 1039–1071.

Gerard, R., 1964. Étude de l'eau de mer de surface dans une baie de Nosy-Bé. *Cah. ORSTOM Océanogr.*, **2,** 5–26.

Gill, A. E., 1977. Coastally trapped waves in the atmosphere. *Quart.J. Roy.Meteorol. Soc.*, **103,** 431–440.

Gill, A. E. and E. H. Schumann, 1979. Topographically induced changes in the structure of an inertial coastal jet: application to the Agulhas Current. *J. Phys. Oceanogr.*, **9,** 975–991.

Gordon, A. L., J. R. E. Lutjeharms and M. L. Gründlingh, 1987. Stratification and circulation at the Agulhas Retroflection. *Deep-Sea Res.*, **34,** 565–599.

Goschen, W. S. and E. H. Schumann, 1988. Ocean current and temperature structures in Algoa Bay and beyond in November 1986. *S. Afr. J. Mar. Sci.*, **7,** 101–116.

Goschen, W. S. and E. H. Schumann, 1990. Agulhas Current variability and inshore structures off the Cape Province, South Africa. *J. Geophys. Res.*, **95,** 667–678.

Goschen, W. S. and E. H. Schumann, 1994. An Agulhas Current intrusion into Algoa Bay during August 1988. *S. Afr. J. Mar. Sci.*, **14,** 47–57.

Goschen, W. S. and E. H. Schumann, 1995. Upwelling and the occurrence of cold water around Cape Recife, Algoa Bay, South Africa. *S. Afr. J. Mar. Sci.*, **16,** 57–67.

Gründlingh, M. L., 1974. A description of inshore current reversals off Richard's Bay based on airborne radiation thermometry. *Deep-Sea Res.*, **21,** 47–55.

Gründlingh, M. L., 1979. Observation of a large meander in the Agulhas Current. *J. Geophys. Res.*, **84,** 3776–3778.

Gründlingh, M. L., 1983. On the course of the Agulhas Current. *S. Afr. Geograph. J.*, **65,** 49–57.

Gründlingh, M. L. and A. F. Pearce, 1984. Large vortices in the northern Agulhas Current. *Deep-Sea Res.*, **31,** 1149–1156.

Gründlingh, M. L., 1986. Features of the Northern Agulhas Current in spring, 1983. *S. Afr. J. Sci.*, **82,** 18–20.

Gründlingh, M. L., I. T. Hunter and E. Potgieter, 1989. Bottom currents at the entrance to False Bay, South Africa. *Cont. Shelf Res.*, **9,** 1029–1048.

Gründlingh, M. L. and A. F. Pearce, 1990. Frontal features of the Agulhas Current in the Natal Bight. *S. Afr. Geograph. J.*, **72,** 11–14.

Gründlingh, M. L. and J. L. Largier, 1991. Physical oceanography of False Bay: a review. *Trans. Roy. Soc. S. Afr.*, **47,** 387–400.

Gründlingh, M. L., 1992a. Agulhas Current meanders: review and a case study. *S. Afr. Geogr. J.*, **74,** 19–28.

Gründlingh, M. L., 1992b. Quasi-synoptic survey of the thermohaline properties of False Bay. *S. Afr. J. Sci.*, **88,** 325–334.

Harris, T. F. W., 1964. Notes on Natal coastal waters. *S. Afr. J. Sci.*, **60,** 237–241.

Harris, T. F. W., 1978. Review of coastal currents in Southern African waters. *South African National Science Programmes Report, CSIR Rep.* 30, vii + 103 pp.

Heydorn, A. E. F., N. D. Bang, A. F. Pearce, B. W. Flemming, R. A. Carter, M. H. Schleyer, P. F. Berry, G. R. Hughes, A. J. Bass, J. H. Wallace, R. P. van der Elst, R. J. M. Crawford and P. A. Shelton, 1978. Ecology of the Agulhas Current region: an assessment of biological responses to environmental parameters in the South-West Indian Ocean. *Trans. Roy. Soc. S. Afr.*, **43,** 151–190.

Hughes, G. R., P. Luschi, R. Mencacci and F. Papi, 1998. The 7000-km oceanic journey of a leatherback turtle tracked by satellite. *J. Exp. Mar. Biol. Ecol.*, **229,** 209–217.

Hunter, I. T., 1981. On the land breeze circulation of the Natal coast. *S. Afr. J. Sci.*, **77,** 376–378.

Hunter, I. T., 1987. The weather of the Agulhas Bank and Cape south coast. *CSIR Res. Rep.*, **634,** 184 pp.

Hunter, I. T., 1988. Climate and weather off Natal. In *Coastal Ocean Studies off Natal, South Africa,* E. H. Schumann, ed., Lecture Notes on Coastal and Estuarine Studies 26, Springer-Verlag, Berlin, pp. 81- 100.

Hutchings, L., 1994. The Agulhas Bank: a synthesis of available information and a brief comparison with other east-coast shelf regions. *S. Afr. J. Sci.,* **90,** 179–185.

Japp, D. W., P. Sims and M. J. Smale, 1994. A review of the fish resources of the Agulhas Bank. *S. Afr. J. Sci.,* **90,** 123–134.

Jorge da Silva, A., A. Mubango and R. Sætre, 1981. Information on oceanographic cruises in the Mozambique Channel. *Revista de Investigação Pesquira,* **2,** Instituto de Desenvolvimento Pesqueiro, Maputo, República Popular de Moçambique, 89 pp.

Jorge da Silva, A., 1984a. Circulation system and areas of potentially successful tuna fishing with surface methods off Mozambique. *Revista de Investigação Pesquira,* **11,** Instituto de Investigação Pesqueiro, Moçambique, pp. 5–40.

Jorge da Silva, A., 1984b. Report on the oceanographic investigations carried out at the Sofala Bank by the Soviet trawler "Sevastopolsky Rybak" in September-December 1982. *Revista de Investigação Pesquira,* **10,** Instituto de Investigação Pesqueiro, Maputo, Moçambique, pp. 5–35.

Jorge da Silva, A., 1984c. Hydrology and fish distribution at the Sofala Bank (Mozambique). *Revista de Investigação Pesquira,* **12,** Instituto de Investigação Pesqueiro, Moçambique, pp. 5–36.

Jury, M. R., 1988. A climatological mechanism for wind-driven upwelling near Walker Bay and Danger Point, South Africa. *S. Afr. J. Mar. Res.,* **6,** 175–181.

Jury, M. (R.) and N. Walker, 1988. Marine boundary layer modification across the edge of the Agulhas Current. *J. Geophys. Res.,* **93,** 647–654.

Jury, M. (R.), 1991. The weather of False Bay. *Trans. Roy. Soc. S. Afr.,* **47,** 401–417.

Jury, M. R. and B. Pathack, 1991. A study of climate and weather variability over the tropical southwestern Indian Ocean. *Meteor. Atmos. Phys.,* **47,** 37–48.

Jury, M. R., H. R. Valentine and J. R.E. Lutjeharms, 1993. Influence of the Agulhas Current on summer rainfall on the southeast coast of South Africa. *J. Appl. Meteorol.,* **32,** 1282–1287.

Jury, M. R., 1994. A review of the meteorology of the eastern Agulhas Bank. *S. Afr. J. Sci.,* **90,** 109–113.

Largier, J. L. and V. P. Swart, 1987. East-west variation in thermocline breakdown on the Agulhas Bank. *S. Afr. J. Mar. Sci.,* **5,** 263–272.

Largier, J. L., P. Chapman, W. T. Peterson and V. P. Swart, 1992. The western Agulhas Bank: circulation, stratification and ecology. *S. Afr. J. Mar. Sci.,* **12,** 319–339.

Lee-Thorp, A. M., M. Rouault and J. R. E. Lutjeharms, 1998. Cumulus cloud formation above the Agulhas Current. *S. Afr. J. Sci.,* **94,** 351–354.

Lee-Thorp, A. M., M. Rouault and J. R. E. Lutjeharms, 1999. Moisture uptake in the boundary layer above the Agulhas Current: a case study. *J. Geophys. Res.,* **104,** 1423–1430.

Luschi, P., A. Sale, R. Mencacci, G. R. Hughes, J. R. E. Lutjeharms and F. Papi, 2003. Current transport of leatherback sea turtles (*Dermochelys coriacea*) in the ocean. *Proc. Royal Soc. London, Ser. B – Biol. Scis,* **270** (Supplement): S129-S132.

Lutjeharms, J. R. E., 1977. The need for oceanologic research in the South-West Indian Ocean. *S. Afr. J. Sci.,* **73,** 40–43.

Lutjeharms, J. R. E., N. D. Bang and C. P. Duncan, 1981a. Characteristics of the currents east and south of Madagascar. *Deep-Sea Res.,* **28,** 879–899.

Lutjeharms, J. R. E., N. D. Bang and H. R. Valentine, 1981b. Die fisiese oseanologie van die Agulhasbank. Deel I: Vaart 170 van die N.S. Thomas B. Davie. WNNR Navorsingsverslag 386, 38 pp.

Lutjeharms, J. R. E. and H. R. Valentine, 1983. Die fisiese oseanologie van die Agulhasbank. Deel 2: Vaart 185 van die N.S. *Thomas B. Davie.* WNNR Navorsingsverslag 557, 15 pp.

Lutjeharms, J. R. E. and N. M. Walters, 1985. Ocean colour and thermal fronts south of Africa. In *South African Ocean Colour and Upwelling Experiment,* L. V. Shannon, ed., Sea Fisheries Research Institute, Cape Town, pp. 227–237.

Lutjeharms, J. R. E., D. Baird and I. T. Hunter, 1986a. Seeoppervlak dryfgedrag aan die Suid-Afrikaanse suidkus in 1979. *S. Afr. J. Sci.,* **82,** 324–326.

Lutjeharms, J. R. E., R. D. Mey and I. T. Hunter, 1986b. Cloud lines over the Agulhas Current. *S. Afr. J. Sci.,* **82,** 635–640.

Lutjeharms, J. R. E. and A. L. Gordon, 1987. Shedding of an Agulhas Ring observed at sea. *Nature,* **325,** 138–140.

Lutjeharms, J. R. E. and J. M. Meeuwis, 1987. The extent and variability of South-East Atlantic upwelling. *S. Afr. J. Mar. Sci.,* **5,** 51–62.

Lutjeharms, J. R. E., 1988. Remote sensing corroboration of retroflection of the East Madagascar Current. *Deep-Sea Res.,* **35,** 2045–2050.

Lutjeharms, J. R. E. and H. R. Roberts, 1988. The Natal Pulse; an extreme transient on the Agulhas Current. *J. Geophys. Res.,* **93,** 631–645.

Lutjeharms, J. R. E. and H. R. Valentine, 1988. Evidence for persistent Agulhas rings southwest of Cape Town. *S. Afr. J. Sci.,* **84,** 781–783.

Lutjeharms, J. R. E. and R. C. van Ballegooyen, 1988. The retroflection of the Agulhas Current. *J. Phys. Oceanogr.,* **18,** 1570–1583.

Lutjeharms, J. R. E. and A. Jorge da Silva, 1988. The Delagoa Bight eddy. *Deep-Sea Res.,* **35,** 619–634.

Lutjeharms, J. R. E. and A. D. Connell, 1989. The Natal Pulse and inshore counter currents off the South African east coast. *S. Afr. J. Sci.,* **85,** 533–535.

Lutjeharms, J. R. E., R. Catzel and H. R. Valentine, 1989a. Eddies and other border phenomena of the Agulhas Current. *Cont. Shelf Res.,* **9,** 597–616.

Lutjeharms, J. R. E., M. L. Gründlingh and R. A. Carter, 1989b. Topographically induced upwelling in the Natal Bight. *S. Afr. J. Sci.,* **85,** 310–316.

Lutjeharms, J. R. E., 1991. The temperature/salinity relationships of the South West Indian Ocean. *S. Afr. Geographer,* **18,** 15–31.

Lutjeharms, J. R. E. and P. L. Stockton, 1991. Aspects of the upwelling regime between Cape Point and Cape Agulhas. *S. Afr. J. Mar. Sci.,* **10,** 91–102.

Lutjeharms, J. R. E., J. Olivier and E. Lourens, 1991. Surface fronts of False Bay and vicinity. *Trans. Roy. Soc. S. Afr.,* **47,** 433–445.

Lutjeharms, J. R. E. and J. A. Thomson, 1993. Commercializing the CSIR and the death of science. *S. Afr. J. Sci.,* **89,** 8–14.

Lutjeharms, J. R. E., C.-T. Liu, W.-S. Chuan and C.-Z. Shyu, 1993. On some similarities between the oceanic circulations off Southern Africa and off Taiwan. *S. Afr. J. Sci.,* **89,** 367–371.

Lutjeharms, J. R. E., 1996. The exchange of water between the South Indian and the South Atlantic. In *The South Atlantic: Present and Past Circulation,* G. Wefer, W. H. Berger, G. Siedler and D. Webb, eds., Springer-Verlag, Berlin, pp. 125–162.

Lutjeharms, J. R. E. and J. Cooper, 1996. Interbasin leakage through Agulhas Current filaments. *Deep-Sea Res. I,* **43,** 213–238.

Lutjeharms, J. R. E. and W. P. M. de Ruijter, 1996. The influence of the Agulhas Current on the adjacent coastal zone: possible impacts of climate change. *J. Mar. Syst.,* **7,** 321–336.

Lutjeharms, J. R. E., A. A. Meyer, I. J. Ansorge, G. A. Eagle and M. J. Orren, 1996. The nutrient characteristics of the Agulhas Bank. *S. Afr. J. Mar. Sci.,* **17,** 253–274.

Lutjeharms, J. R. E., 1998. Coastal hydrography. In *A Field Guide to the Eastern and Southern Cape Coast,* R. Lubke and I. de Moor, eds, The Wildlife and Environment Society of Southern Africa, Grahamstown, University of Cape Town Press, Rondebosch, pp. 50–61.

Lutjeharms, J. R. E. and E. Machu, 2000. An upwelling cell inshore of the East Madagascar Current. *Deep-Sea Res. I,* **47,** 2405–2411.

Lutjeharms, J. R. E., J. Cooper and M. Roberts, 2000a. Upwelling at the inshore edge of the Agulhas Current. *Cont. Shelf Res.,* **20,** 737–761.

Lutjeharms, J. R. E., P. M. Wedepohl and J. M. Meeuwis, 2000b. On the surface drift of the East Madagascar and the Mozambique Currents. *S. Afr. J. Sci.,* **96,** 141–147.

Lutjeharms, J. R. E., H. R. Valentine and R. C. van Ballegooyen, 2000c. The hydrography and water masses of the Natal Bight, South Africa. *Cont. Shelf Res.,* **20,** 1907–1939.

Lutjeharms, J. R. E., W. P. M. de Ruijter, H. Ridderinkhof, H. van Aken, C. Veth, P. J. van Leeuwen, S. S. Drijfhout, J. H. F. Jansen and G.-J. A. Brummer, 2000d. MARE and ACSEX: new research programmes on the Agulhas Current system. *S. Afr. J. Sci.,* **96,** 105–110.

Lutjeharms, J. R. E. and I. Ansorge, 2001. The Agulhas Return Current. *J. Mar. Syst.,* **30,** 115–138.

Lutjeharms, J. R. E., P. Penven and C. Roy, 2003a. Modelling the shear edge eddies of the southern Agulhas Current. *Cont. Shelf Res.,* **23,** 1099–1115.

Lutjeharms, J. R. E., O. Boebel and H. T. Rossby, 2003b. Agulhas cyclones. *Deep-Sea Research II,* **50,** (1): 35–56.

Lutjeharms, J. R. E. and A. A. Meyer, 2004. The origin and circulation of bottom water on the Agulhas Bank, South Africa. *Cont. Shelf Res.,* in preparation.

Machu, E., J. R. E. Lutjeharms, A. M. Webb and H. M. van Aken, 2002. First hydrographic evidence of the south-east Madagascar upwelling cell. *Geophys. Res. Lett.,* **29,** doi:10.1029/2002GL015381.

Malan, O. G. and E. H. Schumann,1979. Natal shelf circulation revealed by Landsat imagery. *S.Afr. J. Sci.,* **75,** 136–137.

Maltrud, M. E., R. D. Smith, A. J. Semtner and R. C. Malone, 1998. Global eddy-resolving ocean simulations driven by1985–1995 atmospheric winds. *J. Geophys. Res.,* **103,** 30,825–30,853.

Martin, A. K., 1981.The influence of the Agulhas Current on the physiographic development of the northernmost Natal Valley (SW Indian Ocean). *Mar. Geol,* **39,** 259–276.

Martin, A. K., and B. W. Flemming, 1988. Physiography, structure and geological evolution of the Natal continental shelf. In *Coastal Ocean Studies off Natal, South Africa,* E. H. Schumann, ed., Lecture Notes on Coastal and Estuarine Studies 26, Springer-Verlag, Berlin, pp. 11–46.

Martin, J., P. Guibout, M. Crepon and J.-C. Lizaray, 1965. Circulation superficielle dans l'ocean Indien. Résultats de mesures faites à électrodes remorguées G.E.K. entre 1955–1963. *Cah. Oceanogr.,* **17,** suppl. 3, 221–241.

McClurg, T. M., 1988. Benthos of the Natal continental shelf. In *Coastal Ocean Studies off Natal, South Africa,* E. H. Schumann, ed. Lecture Notes on Coastal and Estuarine Studies 26, Springer-Verlag, Berlin, pp. 178–208.

McMurray, H. F., R. A. Carter and M. I. Lucas, 1993. Size-fractionated phytoplankton production in western Agulhas Bank continental shelf waters. *Cont. Shelf Res.,* **13,** 307–329.

Menaché, M., 1963. Première compagne océanographique du *Commandant Robert Giraud* dans le canal de Mozambique, 11 octobre au 28 novembre 1957. *Cah. Oceanogr.,* **15,** 224–235.

Mey, R. D., N. D. Walker and M. R. Jury, 1990. Surface heat fluxes and marine boundary layer modification in the Agulhas retroflection region. *J. Geophys. Res.,* **95,** 15,997–16,015.

Meyer, A. A., J. R. E. Lutjeharms and S. de Villiers, 2002. The nutrient characteristics of the Natal Bight, South Africa. *J. Mar. Res.,* **35,** 11–37.

Michaelis, G., 1923. Die Wasserbewegung an der Oberfläche des Indischen Ozeans im Januar und Juli. *Veröffentl. Inst. Meeresk. Uni. Berlin,* n.f. **A8,** 32 pp.

Mitchell-Innes, B. A., A. J. Richardson and S. J. Painting, 1999. Seasonal changes in phytoplankton biomass on the western Agulhas bank, South Africa. *S. Afr. J. Mar. Sci.,* **21,** 217–233.

Nehring, D. ed., 1984. The oceanological conditions in the western part of the Mozambique Channel in February-March 1980. *Geodät. geophys. Veröffentl.,* **4,** 163 pp.

Nehring, D., E. Hagen, A. Jorge da Silva, R. Schemainda, G. Wolf, N. Michelchen, W. Kaiser, L. Postel, F. Gosselck, U. Brenning, E. Kühner, G. Arlt, H. Siegel, L. Gohs and G. Bublitz, 1987. Results of oceanological studies in the Mozambique Channel in February – March 1980. *Beitr. Meereskd.,* **56,** 51–63.

Oliff, W. D., 1973. Chemistry and productivity at Richard's Bay. *NPRL Oceanogr. Div. Contract Report,* **CFIS 37B,** Durban, South Africa.

Pearce, A. F., 1977a. Some features of the upper 500 m of the Agulhas Current. *J. Mar. Res.,* **35,** 731–753.

Pearce, A. F., 1977b. The shelf circulation off the east coast of South Africa. *Council for Scientific and Industrial Research, CSIR Res. Rep.,* **361,** 220 pp.

Pearce, A. F., 1978. Seasonal variations of temperature and salinity on the northern Natal continental shelf. *S. Afr. Geogr. J.,* **60,** 135–143.

Pearce, A. F., E. H. Schumann and G. S. H. Lundie, 1978. Features of the shelf circulation off the Natal coast. *S. Afr. J. Sci.,* **74,** 328–331.

Penven, P., J. R. E. Lutjeharms, P. Marchesiello, S. J. Weeks and C. Roy, 2001a. Generation of cyclonic eddies by the Agulhas Current in the lee of the Agulhas Bank. *Geophys. Res. Lett.,* **26,** 1055–1058.

Penven, P., C. Roy, G. B. Brundrit, A. Colin de Verdière, P. Fréon, A. S. Johnson, J. R. E. Lutjeharms and F. A. Shillington, 2001b. A regional hydrodynamic model of the Southern Benguela. *S. Afr. J. Sci.,* **97,** 472 - 475.

Piton, B., and J. F. Poulain, 1974. Résultats de mesures de courants superficiels au G.E.K. effectuées avec N.O. Vauban dans le sud-ouest de l'océan Indien (1973–1974). Office de la Recherche Scientifique et Technique Outre-mer, *Documents Scientifique de la Mission de Nosy-Bé,* **47,** 14 pp.

Poolman, E. and D. Terblanche, 1984. Tropical cyclones Domoina and Imboa. *S. Afr.Weather Bureau Newsl.,* **420,** 37–46.

Probyn, T. A., B. A. Mitchell-Innes, P. C. Brown, L. Hutchings and R. A. Carter, 1994. A review of primary production and related processes on the Agulhas Bank. *S. Afr. J. Sci.,* **90,** 166–173.

Pugh, D. T., 1987. *Tides, surges and Mean Sea Level.* John Wiley and Sons, New York, 472 pp.

Quartly, G. D. and M. A. Srokosz, 2004. Eddies in the southern Mozambique Channel. *Deep-Sea Res. II,* **51,** 69–83.

Reason, C. J. C., 2001. Evidence for the influence of the Agulhas Current on regional atmospheric circulation patterns. *J. Climate,* **14,** 2769–2778.

Ridderinkhof, H., J. R. E. Lutjeharms and W. P. M. de Ruijter, 2001. A research cruise to investigate the Mozambique Current. *S. Afr. J. Sci.,* **97,** 461 - 464.

Ridderinkhof, H., and W. P. M. de Ruijter, 2003. Moored current observations in the Mozambique Channel. *Deep-Sea Res. II,* **50,** 1933–1955.

Roel, B. A., J. Hewitson, S. Kerstan and I. Hampton, 1994. The role of the Agulhas Bank in the life cycle of pelagic fish. *S. Afr. J. Sci.,* **90,** 185–196.

Rogers, J. and J. M. Bremner, 1991. The Benguela Ecosystem. Part VII. Marine-geological aspects. *Oceanogr. Mar. Biol. Annu. Rev.,* **29,** 1–85.

Rouault, M., A. M. Lee-Thorp, I. Ansorge and J. R. E. Lutjeharms, 1995. Agulhas Current Air-Sea Exchange Experiment. *S. Afr. J. Sci.,* **91,** 493–496.

Rouault, M., A. M. Lee-Thorp and J. R. E. Lutjeharms, 2000. The atmospheric boundary layer above the Agulhas Current during alongcurrent winds. *J. Phys. Oceanogr.,* **30,** 40–50.

Rouault, M., S. A. White, C. J. C. Reason and J. R. E. Lutjeharms and I. Jobard, 2002. Ocean-atmosphere interaction in the Agulhas Current and a South African extreme weather event. *Weather Forecast.,* **17,** 655–669.

Sætre, R. and R. de Paula e Silva, 1979. The marine fish resources of Mozambique. *Reports on surveys with the R/V Dr Fridtjof Nansen,* Serviço de Investigações Pesqueiras, Maputo, Institute of Marine Research, Bergen, 179 pp.

Sætre, R. and A. Jorge da Silva, 1982. Water masses and circulation of the Mozambique Channel. *Revista de Investigação Pesquira,* **3,** Instituto de Desenvolvimento Pesqueiro, Maputo, República Popular de Moçambique, 83 pp.

Sætre, R. and A. Jorge da Silva, 1984. The circulation of the Mozambique Channel. *Deep-Sea Res.,* **31,** 485–508.

Sætre, R., 1985. Surface currents in the Mozambique Channel. *Deep-Sea Res.,* **32,** 1457–1467.

Schemainda, R. and E. Hagen, 1983. On steady state intermediate vertical currents induced by the Mozambique Current. *Océanogr. Trop.,* **18,** 81–88.

Schott, F., M. Fieux, J. Kindle, J. Swallow and R. Zantopp, 1988. The boundary currents east and north of Madagascar. Part II. Direct measurements and model comparisons. *J. Geophys. Res.,* **93,** 4963–4974.

Schouten, M. W., W. P. M. de Ruijter, P. J. van Leeuwen and J. R. E. Lutjeharms, 2000. Translation, decay and splitting of Agulhas rings in the south-eastern Atlantic ocean. *J. Geophys. Res.,* **105,** 21,913–21,925.

Schouten, M. W., W. P. M. de Ruijter and P. J. van Leeuwen, 2002. Upstream control of Agulhas ring shedding. *J. Geophys. Res.,* doi: 10.1029/2001JC000804.

Schumann, E. H., 1981. Low frequency fluctuations off the Natal coast. *J. Geophys. Res.,* **86,** 6499–6508.

Schumann, E. H., 1982. Inshore circulation of the Agulhas Current off Natal. *J. Mar. Res.,* **40,** 43–55.

Schumann, E. H. and L.-A. Perrins, 1982. Tidal and inertial currents around South Africa. In *Proceedings of the Eighteenth International Coastal Engineering Conference,* American Society of Civil Engineers, Cape Town, South Africa, Nov. 14–19, 1982, pp. 2562–2580.

Schumann, E. H., L.-A. Perrins and I. T. Hunter, 1982. Upwelling along the south coast of the Cape Province, South Africa. *S. Afr. J. Sci.,* **78,** 238–242.

Schumann, E. H. and L. J. Beekman, 1984. Ocean temperature structures on the Agulhas Bank. *Trans. Roy. Soc. S. Afr.,* **34,** 191–203.

Schumann, E. H., 1986. The bottom boundary layer inshore of the Agulhas Current off Natal in August 1975. *S. Afr. J. Mar. Sci.,* **4,** 93–102.

Schumann, E. H., 1987. The coastal ocean off the east coast of South Africa. *Trans. Roy. Soc. S. Afr.,* **46,** 215–229.

Schumann, E. H. (editor), 1988a. *Coastal Ocean Studies off Natal, South Africa,* Lecture Notes on Coastal and Estuarine Studies 26, Springer-Verlag, Berlin, 271 pp.

Schumann, E. H., 1988b. Physical oceanography off Natal. In *Coastal Ocean Studies off Natal, South Africa,* E. H. Schumann, ed., Lecture Notes on Coastal and Estuarine Studies 26, Springer-Verlag, Berlin, pp. 101–130.

Schumann, E. H. and I. Li (sic) van Heerden, 1988. Observations of Agulhas Current frontal features south of Africa. *Deep-Sea Res.,* **35,** 1355–1362.

Schumann, E. H., G. J. B. Ross and W. S. Goschen, 1988. Cold water events in Algoa Bay and along the Cape south coast, South Africa, in March/April 1987. *S. Afr. J. Sci.,* **84,** 579–584.

Schumann, E. H., 1989. The propagation of air pressure and wind systems along the South African coast. *S. Afr. J. Sci.,* **85,** 382–385.

Schumann, E. H. and J. A. Martin, 1991. Climatological aspects of the coastal wind field at Cape Town, Port Elizabeth and Durban. *S. Afr. Geogr. J.,* **73,** 48–51.

Schumann, E. H., W. K. Illenberger and W. S. Goschen, 1991. Surface winds over Algoa Bay. *S. Afr. J. Sci.,* **87,** 202–207.

Schumann, E. H., 1992. Interannual wind variability on the south and east coasts of South Africa. *J. Geophys. Res.*, **97,** 20,397–20,403.

Schumann, E. H., A. L. Cohen and M. R. Jury, 1995. Coastal sea surface temperature variability along the south coast of South Africa and the relationship to regional and global climate. *J. Mar. Res.*, **53,** 231–248.

Schumann, E. H., 1998. The coastal ocean off southeast Africa, including Madagascar. In *The Sea,* Volume 11, Chapter 19, A. R. Robinson and K. H. Brink, eds, John Wiley & Sons, pp. 557–581.

Schumann, E. H., 1999. Wind-driven mixed layer and coastal upwelling processes off the south coast of South Africa. *J. Mar. Res.*, **57,** 671–691.

Shannon, L. V. and P. Chapman, 1983. Suggested mechanism for the chronic pollution by oil of beaches east of Cape Agulhas, South Africa. *S. Afr. J. Mar. Sci.,* **1,** 231–244.

Shannon, L. V., P. Chapman, G. A. Eagle and T. P. McClurg, 1983. A comparative study of tar ball distribution and movement in two boundary current regimes. *Oil and Petrochemical Pollution,* **1,** 243–259.

Sidorn, J. R., D. G. Bowers and A. M. Hoguane, 2001. Detecting the Zambezi river plume using observed optical properties. *Mar. Pollut. Bull.,* **42,** 942–950.

Simpson, E. S. W., 1974. *Southeast Atlantic and Southwest Indian Oceans.* Chart 125A, bathymetry.

Smale, M. J., N. T. Klages, J. H. M. David and V. G. Cockroft, 1994. Predators of the Agulhas Bank. *S. Afr. J. Sci.,* **90,** 135–142.

Stramma, L. and J. R. E. Lutjeharms, 1997. The flow field of the subtropical gyre of the South Indian Ocean. *J. Geophys. Res.,* **102,** 5513–5530.

Swallow, J. C., M. Fieux and F. Schott, 1988. The boundary currents east and north of Madagascar. Part I. Geostrophic currents and transports. *J. Geophys. Res.,* **93,** 4951–4962.

Swart, V. P. and J. L. Largier, 1987. Thermal structure of Agulhas Bank water. *S. Afr. J. Mar. Sci.,* **5,** 243–253.

Taljaard, S., 1991. The origin and distribution of dissolved nutrients in False Bay. *Trans. Roy. Soc. S. Afr.,* **47,** 483- 493.

Tripp, R. T., (1967). *An Atlas of Coastal Surface Drifts; Cape Town to Durban.* South African Oceanographic Data Centre, Department of Oceanography, University of Cape Town, Rondebosch, South Africa, 12 pp.

Van der Elst, R. P., 1988. Shelf ichthyofauna off Natal. In *Coastal Ocean Studies off Natal, South Africa,* E. H. Schumann, ed., Lecture Notes of Coastal and Estuarine Studies, 26, Springer-Verlag, Berlin, pp. 209–225.

Van Heerden, J. and J. J. Taljaard, 1998. Africa and surrounding waters. In *Meteorology and the Southern Hemisphere,* D. J. Karoly and D. G. Vincent, eds, Meteorological Monographs, **27**(49): 141–174.

Van Leeuwen, P. J., W. P. M. de Ruijter and J. R. E. Lutjeharms, 2000. Natal Pulses and the formation of Agulhas rings. *J. Geophys. Res.,* **105,** 6425–6436.

Verheye, H. M., L. Hutchings. J. A. Huggett, R. A. Carter, W. T. Peterson and S. J. Painting, 1994. Community structure, distribution and tropic ecology of zooplankton on the Agulhas Bank with special reference to copepods. *S. Afr. J. Sci.,* **90,** 154–165.

Walker, N. D., 1986. Satellite observations of the Agulhas Current and episodic upwelling south of Africa. *Deep-Sea Res.,* **33,** 1083–1106.

Walker, N. D., 1990. Links between South African summer rainfall and temperature variability of the Agulhas and Benguela Current Systems. *J. Geophys. Res.,* **95,** 3297–3319.

Chapter 21. VARIABILITY OF THE BENGUELA CURRENT SYSTEM (16,E)

J.G. FIELD

Marine Biology Research Institute, University of Cape Town

F.A. SHILLINGTON

Department of Oceanography, University of Cape Town

Contents

1. Introduction and Summary

The Benguela Current System (BCS) is one of four major eastern boundary upwelling systems of the world oceans (Hill *et al.*, 1998). It spans three countries on the west coast of Africa from about 14°S in Angola, through the entire coast of Namibia to about 37°S off the southern tip of Africa (Shannon and Nelson, 1996; Shillington, 1998). It extends along the south coast of South Africa to the eastern edge of the Agulhas Bank at about 27°E. The BCS has similar characteristics to some of the other eastern boundary upwelling areas; there is a region with tropical ocean characteristics off Angola (from the equator to 15°S); there is a region of year-round upwelling (approximately 15°-30°S; Boyer *et al.*, 2000); a region of seasonal upwelling (approximately 30°-34°S). The surface currents are generally equatorward, with vigorous upwelling cells, strong and narrow equatorward shelf edge jets (near Cape Town), and a poleward undercurrent along the shelf bottom. Coastal trapped shelf waves travel poleward at regular synoptic time scales (~10 day periods) from Walvis Bay in Namibia (20°S) to Port Elizabeth on the east coast of South Africa (Brundrit *et al.*, 1987; Schumann and Brink, 1990). Along the

The Sea, Volume 14, edited by Allan R. Robinson and Kenneth H. Brink
ISBN 0–674–

southern coast of Africa, the Agulhas Bank is a wide shelf region that is highly stratified in the west in summer, and well mixed in the winter. Closer to the coast, there is summer upwelling along the major embayments of the Agulhas Bank. On the Agulhas Bank, there is a seasonal cyclonic cool "ridge" feature in the near-surface waters (Boyd and Shillington, 1994). The Agulhas Bank region is very important for pelagic fish spawning from September to March (Hutchings *et al.*, 2002). After spawning, the eggs and larvae drift northwards, until juvenile fish recruitment occurs north of St Helena Bay. Adult fish then make their way back to the Agulhas Bank to spawn in the following austral spring-summer. The BCS is the only Eastern Boundary Current (EBC) upwelling system bounded by warm water at both ends, the warm Angola Current in the north and the warm Agulhas Current in the south and east. These features are summarised in Fig. 21.1.

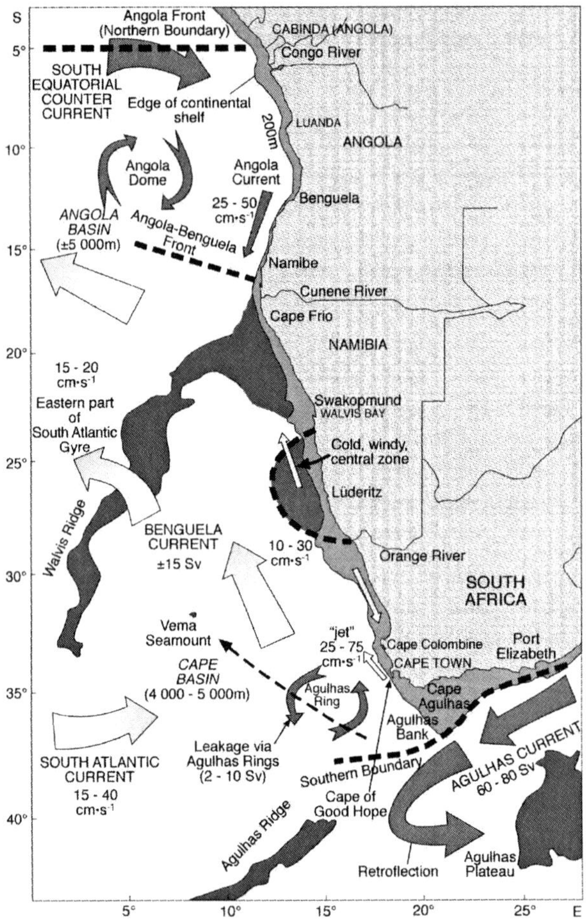

Figure 21.1 Cartoon of the main features of the Benguela Current System (Courtesy Benguela Current Large Marine Ecosystem Study, BCLME of the Global Environmental Facility, GEF)

The BCS is physically forced at a variety of space and time scales, with cascades to smaller scales and biological responses and feedbacks at different scales. The remainder of this chapter is organised according to time-scale, going from shortest to longest. At the daily time scale, the diel cycle of photosynthesis and respiration occurs as everywhere. It is also the scale at which zooplankton and fish migrate vertically. Vertical migration of zooplankton has been shown to vary with the stage of development in the southern Benguela system, and provides a mechanism to maintain position within the system. On the outer shelf, the general movement of the upper layers of water is offshore and northwards, whereas the lower layers tend to move onshore and southwards, particularly with the passage of coastal trapped waves. Thus regular vertical migration between the upper and lower layers may provide a mechanism for maintaining position over the shelf and may also provide a mechanism for the recruitment of fish larvae to the inshore nursery grounds of the Benguela system.

At the synoptic time scale of about a week, the passage of atmospheric coastal low systems causes reduction in the southerly (equatorward) wind component and relaxation of upwelling. With relaxation of wind forcing, the upwelling front tends to move back towards the coast (Bakun, 1996). This occurs particularly in the southern subsystem. The pulses of upwelling result in blooms of phytoplankton that take some 4–8 days to develop fully. These are often out of synchrony with the longer cycles of zooplankton development, resulting in diatoms and other phytoplankton sinking out of the euphotic zone to form organically rich mud on the shelf and slope. These are sites of nutrient regeneration and enrichment of the shelf/slope bottom water. The lack of synchrony also results in inefficient ecological transfer of energy and matter from primary producers to fish. The general position of the upwelling front is usually approximately at the shelf edge; the shelf varies greatly in width along the BCS, from being very wide over the Agulhas Bank in the south to narrow off the Cape Peninsula, and wide off the Orange River and again off Walvis Bay (Strub *et al.,* 1998). The process of the front moving back towards the coast brings with it the fish eggs and larvae associated with the frontal jet current, and may provide a mechanism for retention of these in nourishing coastal waters.

At a time scale of about 1–2 months, the upper atmosphere jet stream wavers north and south, guiding the eastward passage of cold fronts every 4–8 days. In the northerly position in summer, the cyclonic wind systems and their associated cold fronts just brush the southern tip of Africa, causing reduction in upwelling in the southern BCS. At quasi-monthly periods, anticyclonic Agulhas rings may break off the Agulhas retroflection and carry warm water from the Indian Ocean into the Atlantic, as part of the thermohaline circulation system. If these Agulhas Rings move close to the west coast, they have a profound effect, both on the weather, and also in interacting with the upwelling system and carrying plankton, fish eggs and larvae offshore in jets and squirts. These may or may not return into coastal waters.

On a seasonal scale upwelling is weakest in austral summer (November–February) at the northernmost cell (17°S). Upwelling is strongly seasonal in the southern Benguela system being strongest in summer and weakest in winter. The northern boundary of the Benguela system, the Angola-Benguela Front, moves seasonally from about 14°-17°S, being furthest south in March and September

(Veitch *et al.,* 2005). The frontal region appears to form the northern limit of sardine spawning and is associated with horse mackerel catches. There are periodic incursions of hypoxic water from the north onto the shelf of the Benguela system. The main oxygen minimum forms off Angola and is transported from the Angola Dome (10°-11°S) southwards onto the shelf by the poleward undercurrent. Further hypoxic water is formed by decomposition of plankton blooms over the Namibian shelf, and the formation of sulphurous mud-islands of H_2S is a late-summer seasonal phenomenon off Walvis Bay in central Namibia (Weeks *et al.,* 2002; Weeks *et al.,* 2004). Hypoxic water also forms further south off the Orange River and St Helena Bay, causing mass strandings of rock-lobster on occasions. This is a seasonal phenomenon, often coinciding with the occurrence of red tides in late-summer—autumn in the southern Benguela region. There is also considerable inter-annual variability in its severity.

At the decadal scale, warm Angola Current water penetrates large distances southwards into central Namibia with a periodicity of some 10 years as Benguela Niños (Shannon *et al.,* 1986), analogous to the Humboldt Current El Niños. Benguela Niños cause disruption of fish spawning and recruitment patterns in Namibian waters, with dramatic effects on pelagic fish: sardine, anchovy and horse mackerel and also the valuable semi-pelagic hake stocks. In turn, seal pup survival is greatly reduced by the displacement of their fish prey. These effects are confined to Namibian waters and do not penetrate as far south as the Orange River border with South Africa. Benguela Niños are not directly linked to ENSO, but ENSO does influence the Benguela system through its effect on the whole southern hemisphere atmospheric pressure system, which in turn affects the wind field. During El Niño, southerly winds tend to be less prevalent in the southern Benguela system, driving less upwelling, whereas in La Niña conditions the upwelling tends to be stronger and more prevalent. The BCS upwelling is driven by prevailing southerly winds caused by the South Atlantic semi-permanent atmosphere high-pressure cell. Wind stress is strongest at 27°S, creating a perennial upwelling cell near Lüderitz. This appears to act as a barrier to the migration of small pelagic fish such as sardine and anchovy and divides the BCS into northern and southern sub-systems.

2. Overview

This section provides a geographical overview and details of the forcing of the BCS, including remote forcing from the north (the Angola Current) and the south (effects of the Agulhas Current Retroflection and rings), the general SST and the circulation. For further details of the physical processes, see the description by Shillington (1998), and the eastern boundary current comparison by Hill *et al.* (1998).

The continental shelf off the west coast of Africa is very narrow off northern Angola, but widens off southern Angola and Namibia, to its greatest offshore width at the Orange River cone (29°S). Moving southwards, the largest embayment is St Helena Bay (near Cape Columbine), with a narrowing of the shelf at the Cape Peninsula (near Cape Town). South of Africa, the Agulhas Bank has its widest extent east off Cape Agulhas at 21°E (about 250 km), and then narrows again to Port Elizabeth (Fig. 21.1).

There is a major freshwater input at the Congo River (7°S) with an average annual flux of 45,000 $m^3 s^{-1}$, and then minor inputs at the Cunene (northern border of Namibia at 17°S), Orange River (northern border of South Africa at 27.5°S) with a maximum flux of about 8,000 $m^3 s^{-1}$ in flood, and Berg River. Major offshore diamond mining takes place between Lüderitz and the Berg Rivers in three different concessions: the beach mining by scuba divers down to 20 m depth, the mid zone from 20–60 m depth and the deep zone greater than 60 m.

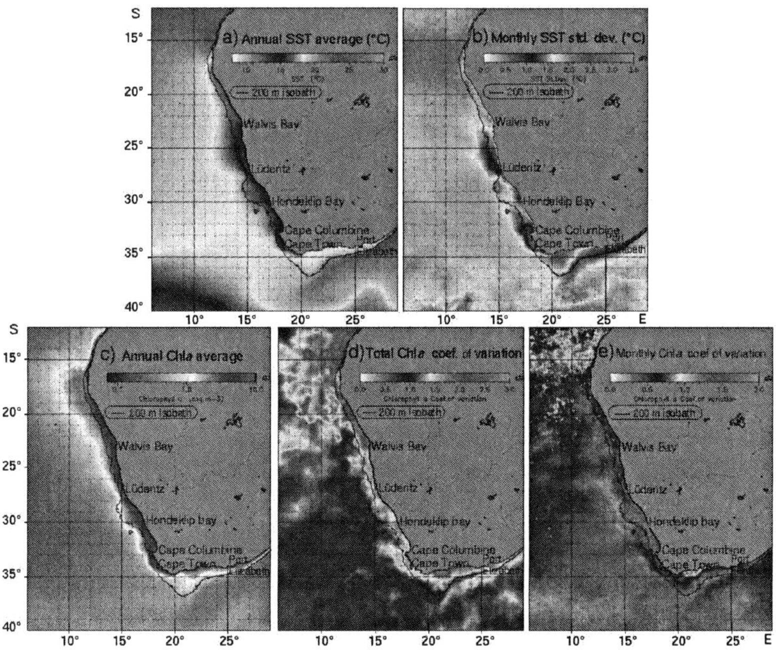

Figure 21.2 a) Annual average SST and b) mean monthly SD for SST computed from monthly composites for the period 1982 – 1999; c) annual average Chlorophyll and d) total CV of chlorophyll computed from 5-day weighted composites from Sept. 1997 – April 2002; e) mean monthly CV of chlorophyll computed from monthly composites (From Demarcq *et al.*, 2003).

The Angola-Benguela Frontal Zone (ABFZ) has a sharp signal in SST and Chl a between 14–17°S, separating the south flowing warm, nutrient poor Angola Current, from the northern extremity of the cool Benguela Upwelling System (Fig. 21.2). The ABFZ is at its southern position in March and September, and at its most northerly position in December and June (Fidel 2001; Hocutt and Verheye 2001). A multidisciplinary survey of the ABFZ carried out in April 1999 showed that there were several mesoscale cyclonic eddies embedded in the Angola Gyre (Mohrholz *et al.,* 2001). A special feature of the area north of the ABFZ is the Angola Dome, which is a seasonal uplift of the thermocline at depths of 20–150 m during January-May (Wacongne and Piton, 1992). This area also has lowered salinity (by 0.3–0.5 psu) and reduced oxygen (by 2–3 ml l^{-1}) (Mohrholz *et al.,* 2001).

In the Pacific Ocean, El Niños have striking impacts on the local ecosystem and on rainfall. Similar warm, poleward flowing intrusions from the Angolan Current sporadically disrupt the ABFZ in the South Atlantic off the coasts of Angola and Namibia and have been named Benguela Niños (Shannon *et al.*, 1986, Gammelsrød *et al.*, 1998). The most striking differences between Benguela and Pacific Niños are their different frequency of occurrence and duration (see Section 7). Benguela Niños do not necessarily occur in unison with their Pacific counterpart and may thus have different forcing mechanisms. Using a general ocean circulation model, forced by real wind measurements and verified with satellite remote sensing, Florenchie *et al.* (2003) have shown that the major Benguela Niños in 1984 and 1995 during the past 20 years, were generated by specific wind stress events in the west-central equatorial Atlantic, and progressed from there as subsurface temperature anomalies that eventually outcropped at the south-west African coast, near the ABFZ. Florenchie *et al.* (2004) extended the initial analysis to include cold episodes as well as warm episodes, which both created identifiable SST anomalies near 15°S. The elucidation of the remote equatorial wind forcing mechanism suggests that it may be possible to predict the occurrence of these disruptive events with a lead-time of 2 months.

Carr (2002) provided a satellite-based estimate of potential primary production in the four Eastern Boundary Currents, (EBC) i.e. the California, Humboldt, Canary, and Benguela systems from the first 24 months of the Sea-Viewing Wide Field of View Sensor, SeaWiFS. Within each EBC, production was estimated for the area of high chlorophyll concentration (>1mg m^{-3}) or active area, which is likely to determine the production that can be utilized by higher trophic levels. Primary production decreased with latitude within each EBC while the extent of the active area was related to the magnitude of offshore transport. The most productive EBC was the Benguela Current (0.37 Gt C yr^{-1}).

3. Fine-scale and diel-scale processes

There has been little work on physical processes at a very fine scale in the BCS. However, size-based modelling of the microbial loop in the BCS by Moloney *et al.*, (1991) has shown short-term predator-prey oscillations of picoplankton, bacteria and bactivorous micro-zooplankton of a few hours periodicity. This has been supported by the laboratory experiments of Painting *et al.* (1993) who found similar short-term periodicity. Touratier *et al.* (2003) have developed a detailed ecosystem model of biogeochemical processes in the BCS during relaxation of upwelling, based on the St Helena Bay anchor station data (see Section 4 below). They conclude that sediments on the shelf play a crucial role in increasing the level of phytoplankton production. Another conclusion is that exudation of dissolved organic matter is the main fate of carbon fixed by phytoplankton, with micro-heterotrophic pathways playing an important role in recycling nutrients. These are processes that occur at the diel or sub-diel scale.

On the diel scale, Shelton and Hutchings (1982) show that anchovy form a subsurface acoustic scattering layer at night, coinciding with the appearance of newly spawned anchovy eggs. This suggests strongly that anchovy spawn at night. Acoustic surveys in the BCS have shown clear day-night patterns in the vertical distribution of both zooplankton and fish, including gobies off Namibia (Hampton, 1987). However, one of the difficulties in studying fish recruitment is the confounding of diel vertical migration (DVM) with daytime net-avoidance by larger larvae and juvenile fish. It appears that organisms that can swim strongly enough to migrate vertically can also swim strongly enough to avoid nets they can see approaching.

Gibbons *et al.* (1991), Verheye and Field (1992), and Stuart and Verheye (1991) have demonstrated DVM in euphausiids, calanoid copepods, and chaetognath zooplankton, respectively. Both the euphausiids and copepods showed increasing DVM with increasing age and size, the younger stages and males not migrating as much as the older ones. *Calanoides carinatus* was found to migrate vertically in response to food availability, and did not migrate up to feed at night when cells of unsuitable size or quality dominated the phytoplankton community (see Fig. 21.5). DVM in upwelling systems is believed to serve more than one purpose, one being to avoid predation in the upper layers, another to maintain position in the upwelling system by moving shorewards and poleward at depth and being carried offshore and equatorward in the surface Ekman transport of the upwelling system. Fig. 21.3 depicts the 3-D upwelling process in cartoon form. It can be seen that vertically migrating organisms would be carried offshore and northwards in the upper layers (at night) and onshore and southwards in deeper layers if they migrated down by day.

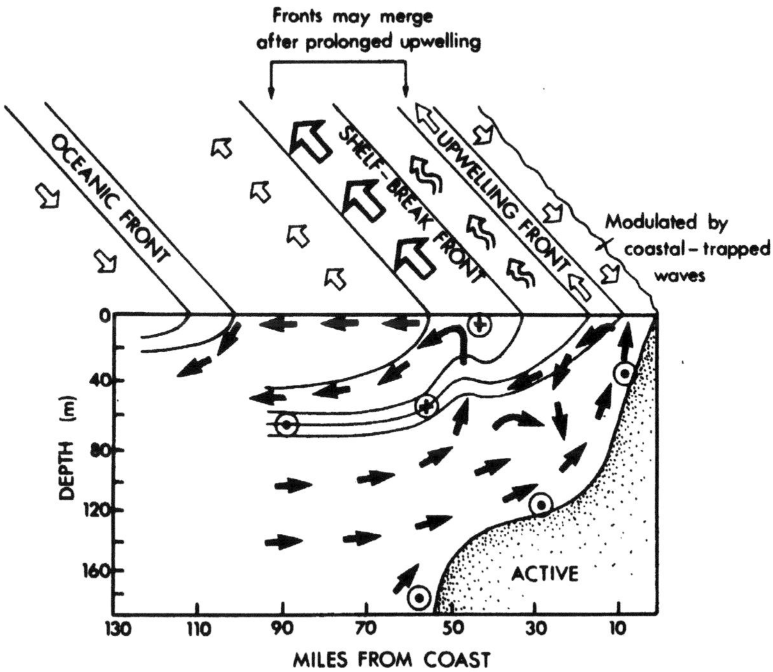

Figure 21.3 Three-dimensional cartoon, depicting water movements in the upwelling system in the northern BCS where the shelf is wide (from Barange *et al.,* 1992). In the southern BCS, the shelf is narrower and the upwelling and shelf-break fronts tend to merge. + signs indicate northerly equatorward movement, closed circles indicate poleward movement. Note the upwelling front and thermocline, with onshore movement at depth and offshore movement near the surface.

Downward migration would also allow onshore transport of recruiting juvenile fish. Recent confirmation of this is provided by individual based models (IBM's) of the early life-history of anchovy coupled to a 3-D physical model "PLUME" (Penven, 2000). The IBM models show that the anchovy larvae are only advected from the spawning area on the Agulhas Bank to the general region of the nursery area on the west coast, and do not reach the inshore heart of the nursery area in numbers unless vertical migratory behaviour of the late larvae is incorporated in the model (Parada *et al.,* 2003). This suggests that downward DVM of late anchovy larvae (and probably other similar fish) out of the surface Ekman layer is necessary for full recruitment to the inshore nursery area. Pre-recruit surveys are being planned to validate the model findings. Similarly, in the northern BCS, Stenevik *et al.* (2001) have shown that sardine (*Sardinops sagax*) spawn inshore just below the surface layers affected by offshore Ekman transport, the buoyant eggs float upwards but tend to be mixed downwards by wind-mixing. Later, the growing larvae tend to migrate down away from the Ekman layer, and this would tend to carry them onshore and southward.

4. Intra-seasonal and synoptic event scale processes

The main driving force of the Benguela upwelling system results from the winds induced by the semi permanent high-pressure system over the subtropical South Atlantic, together with the continental low pressure atmospheric system that develops over southern Africa during the summer months. The prevailing summer winds in the region blow from a south to south-easterly direction. The dominant south-easterly wind regime is modulated, south of the Orange River (28.5°S), by the east-moving passage of mid-latitude cyclones south of the continent, and their associated frontal systems. During the austral autumn months, the effect of this is usually seen in a weakening of the South Atlantic Anticyclone and an abatement of south-easterly winds along the coast. During the winter months, however, westerly disturbances can, at times, result in gale force north-westerly and south-westerly winds. Preston-Whyte and Tyson (2000) suggest that this modulation occurs in cycles of 2–8 days. Risien *et al.* (2004) confirm this using a wavelet decomposition of Quikscat satellite derived winds with a two-day time resolution; to show that there was significant energy (wind speed variability) in the 4–12 day period band.

Moving to the biological effects of physical events, a series of drogue studies followed the fate of a patch of upwelled water at different stages after upwelling events (Brown and Hutchings, 1987). When combined, these studies show that phytoplankton blooms take 3–5 days to build up to maximum biomass and then 1–3 days to decline, if uninterrupted by another upwelling event. These results are consistent with the studies on the development of red tides in the southern BCS by Pitcher *et al.* (1998). In their study, the wind pulsed at a 3–5 day periodicity, with chlorophyll building up at 15–20m depth as the water column stratified during the relaxation phase after upwelling. With relaxation, the upwelling front moved closer inshore concentrating surface phytoplankton in the process. At the same time, dinoflagellates became dominant over the initial diatom bloom, forming the seasonal red tides characteristic of late summer and autumn, but also found at other times of year.

Armstrong *et al.* (1987) conducted repeated transects down the length of an upwelling plume off Cape Columbine north of Cape Town during a quiescent phase after upwelling, studying its physical, chemical and biological changes. The spatial changes correspond to temporal ones in an upwelling cycle: a nearshore zone influenced by recent upwelling, a frontal zone through which the longshore jet current passes, and an oligotrophic offshore zone. Evidence of denser concentrations of both fish larvae and copepod larvae was found in the jet current at the front, suggesting that the frontal jet current provides the transport mechanism from the spawning grounds in the south to nursery grounds in the north.

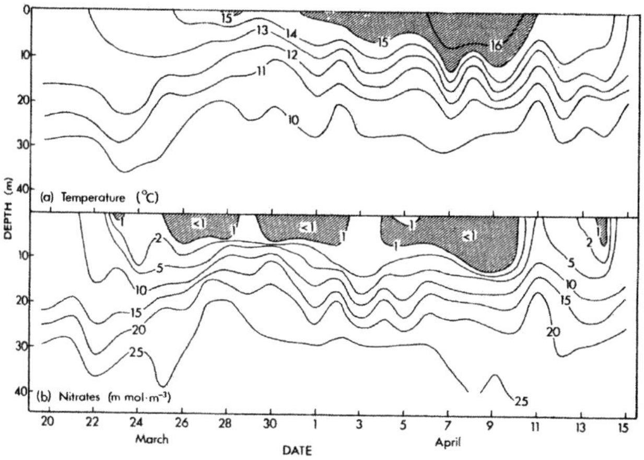

Figure 21.4 Time series sections of temperature and nitrate nitrogen over the full period of the anchor station, showing several pulses of upwelling over the month-long period. Note the pulses of upwelling on April 10 and again on April 14 (from Mitchell-Innes and Walker, 1991).

An anchor station study in St Helena Bay lasting a month (Chapman and Bailey, 1991) has demonstrated the short-term temporal fluctuations that occur in the BCS. Southerly upwelling-inducing winds tend to fluctuate with a 4–8 day periodicity associated with the poleward passage of coastal-trapped low-pressure atmospheric systems. The coastal lows cause relaxation of prevailing southerly winds during spring and summer. As the coastal low pressure system passes eastwards past Cape Agulhas, the Atlantic High pressure system ridges in behind it and southerly winds freshen, driving a new bout of upwelling. In this study, profiles of the water column were sampled every few hours day and night, giving the opportunity of measuring chemical, bacterial, phytoplankton and zooplankton variations. There were several upwelling events during the study (Mitchell-Innes and Walker, 1991). Fig. 21.4 shows the increase in surface nitrate after upwelling on April 10, followed by its decline as primary production and chlorophyll biomass increased over 4 days as the wind relaxed and the water column became stratified. This was interrupted by another wind change and another upwelling event on 14 April. An interesting feature of this study is the change shown in phytoplankton community composition (Fig. 21.5), and how this affected copepod egg production and diel vertical migration. Fig. 21.6 shows that when the phytoplankton cells were too large (*Coscinodiscus gigas,* April 6–9), *Calanoides* could not handle them, and remained above 35 m depth by day and night. When phytoplankton consisted of suitably sized cells (April 10–15), *Calanoides* females migrate vertically, feed at shallow depths at night, and reproduce. It can also be seen that the smallest stages migrate the least and the largest ones the farthest (Verheye and Field, 1992). Thus the diel cycles of primary production and vertical migration are strongly influenced by events on a longer time scale and by the stage of development and size of the organisms concerned.

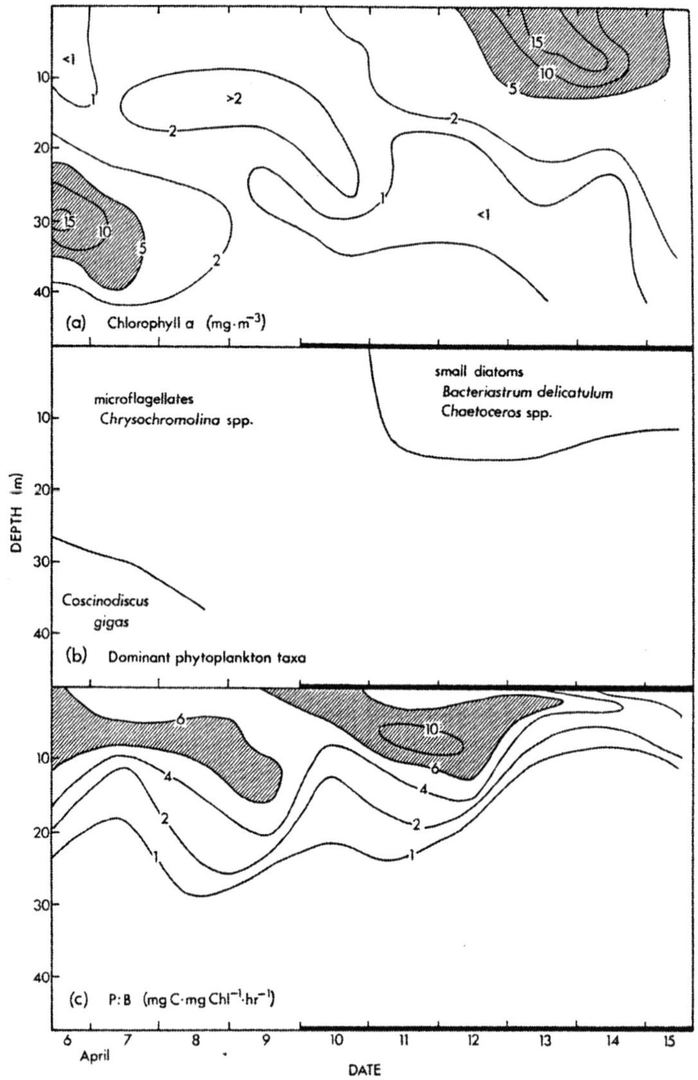

Figure 21.5 Time series section during the second half of the anchor station from 6 - 15 April, showing chlorophyll concentration, major phytoplankton groups, and production to biomass ratios (P/B). Note the deep diatoms from 6 – 9 April and the new bloom of small diatoms that developed after upwelling on 10 April and was ended by upwelling on 14 April. (From Verheye and Field, 1992).

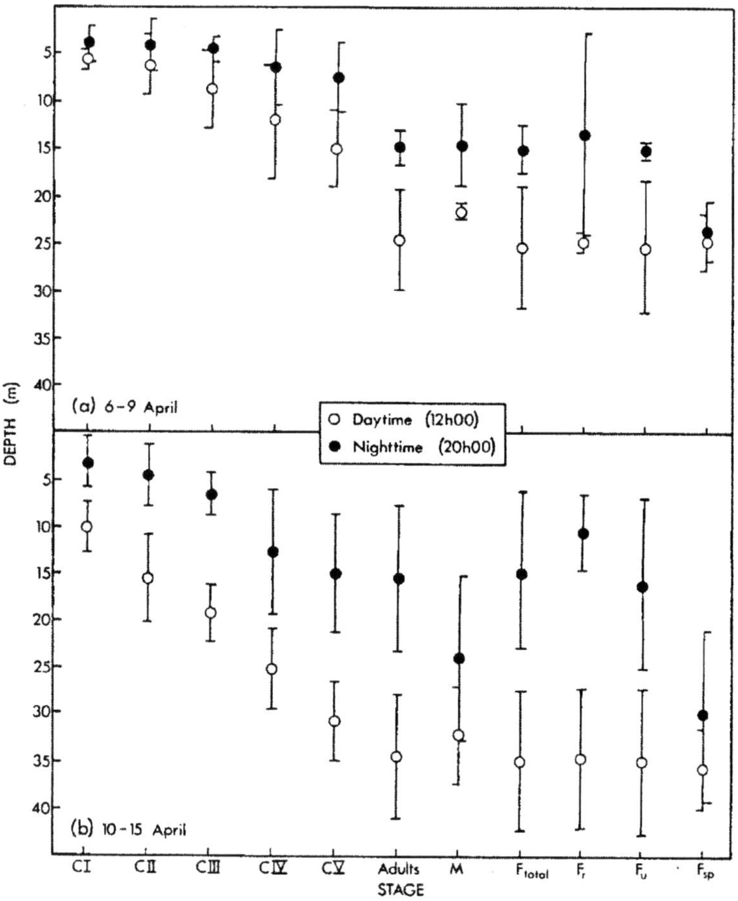

Figure 21.6 Weighted mean depth of various developmental stages of the copepod, *Calanoides carinatus,* showing diel vertical migration during day and night in St Helena Bay (after Verheye and Field, 1992). The upper panel shows weighted mean depths during the period 6–10 April with a deep mature bloom of large diatoms, the lower panel the same soon after upwelling from 10–14 April when small diatoms occurred in the upper layers (see Fig. 21.5).

Hutchings (1992) shows the mismatch in temporal scales between phytoplankton blooms as shown by the drogue and anchor station studies (5–8 days) and zooplankton life cycles, exemplified by the dominant copepod *C. carinatus* (ca. 30 days). This mismatch allows much primary production to go ungrazed, sedimenting to the shelf and slope floor and contributing to the mud belts characteristic of upwelling systems. It is believed that similar spatial and temporal mismatches occur between zooplankton and small pelagic fish, also reducing the efficiency of energy transfer up the food web. These mismatches in temporal and spatial scales may be a reason for the paradox that the BCS is highly productive at the level of primary production (Carr, 2002, see Section 5, below), yet the fish catch is lower than in other comparable upwelling regions (Hutchings, 1992).

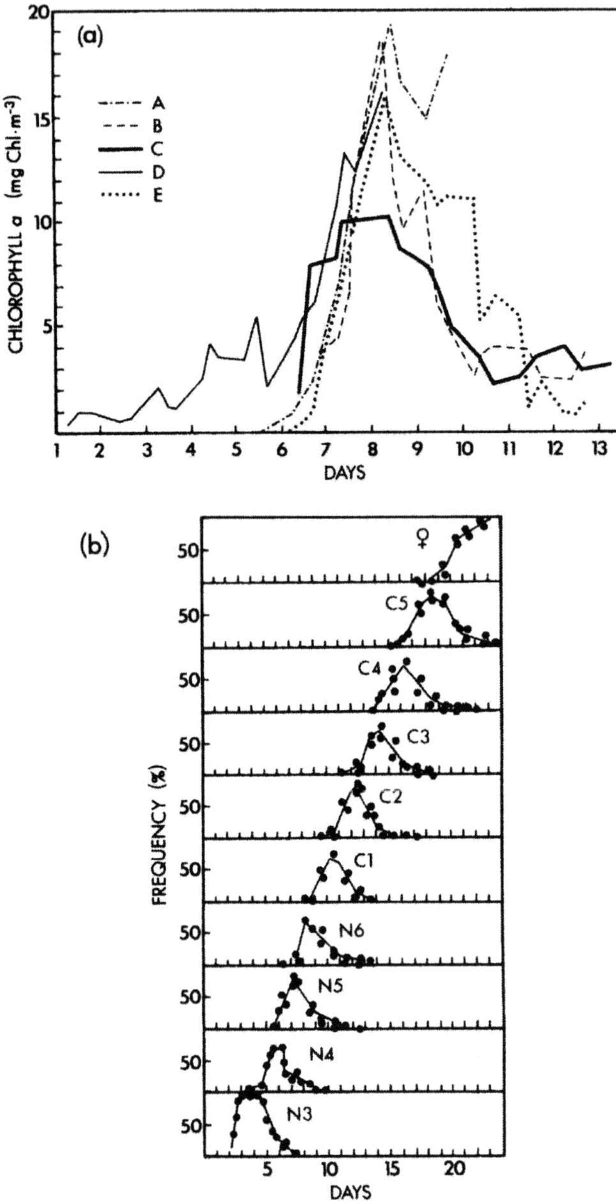

Figure 21.7 Mismatch between the timescales for the development and decay of phytoplankton blooms (6–10 days) after pulses of upwelling on the one hand and the development of the copepod *Calanoides carinatus* (some 30 days) on the other (after Hutchings, 1992).

One of the puzzles of upwelling systems is how fish recruit to nursery grounds (usually inshore) in the face of offshore Ekman transport of the surface waters, when fish eggs and larvae usually inhabit the upper layers. In the southern BCS, one possibility is that fish larvae and juveniles, being carried northward in the frontal jet current, are moved shorewards with the front as upwelling relaxes with the passage of coastal lows. If this were coupled with a short downward migration at night, they would not have far to travel to escape the jet current and be carried inshore by the upwelling water. This is an example of the interaction of different scales: the front moving at the event scale and the DVM of fish larvae to take advantage of the inshore position of the front at the diel scale (see Section 3 above).

5. The Seasonal scale

Bakun and Nelson (1991) used monthly composites of shipboard wind measurements, wind stress and wind stress curl in the Benguela ecosystem, to reveal the main seasonal characteristics. The cyclonic wind stress curl appears as a wedge from 20°S, narrowing in offshore extent to the south. The wedge has its most limited extent in austral winter (June-July), when it reaches just south of Lüderitz. By early spring the wedge has extended to Cape Columbine and by late spring south of Cape Town to Cape Agulhas. The pattern of wind stress curl then retreats north again in winter. (The shipping lanes diverge from the coast north of Cape Town, so that caution has to be applied to the interpretation of the sparse wind field data close to the coast in the north near Cape Frio.) This type of study will be repeated using much higher resolution Quikscat satellite derived winds (C. Risien *pers. comm.*).

Demarcq *et al.* (2003) have analysed the climatology and variability of SST and Surface Chlorophyll Concentration (SCC) in the BCS using weekly mean SST and SCC satellite images (Fig. 21.8). The dominant pattern in the annual SST is the cold upwelled water on the western continental shelf of southwest Africa and Namibia. High SST's are evident at the northern (Angola/Benguela Front 15–17°S) and southern (Agulhas Current) extremities of the upwelling system. Upwelling regions north of Cape Town and north of Lüderitz display the lowest seasonal variability. To the south of Africa, the SST on the western Agulhas Bank displays a clear seasonal pattern of warm surface water in summer, and cool surface water in winter. The monthly variability is always the strongest source of variability for the SST in the region. The annual average SCC is given in Fig. 21.2, showing a band of high values (>5 to 10 mg m^{-3}) close to the coast from the Angola/Benguela Frontal Zone in the north to Cape Town in the south, with a noticeable relative minimum at Lüderitz. Along the south coast of Africa, the highest SCC values occur between Cape Agulhas and Port Elizabeth (>3 mg m^{-3}), in the form of a plume moving offshore. In contrast to the SST variability, the intra-monthly SCC variability is the major source of variability and shows patchy distributions of high values north of the Angola/Benguela Front in summer and in the upwelling regions in winter. The Agulhas bank and its surroundings also show high short-term variability.

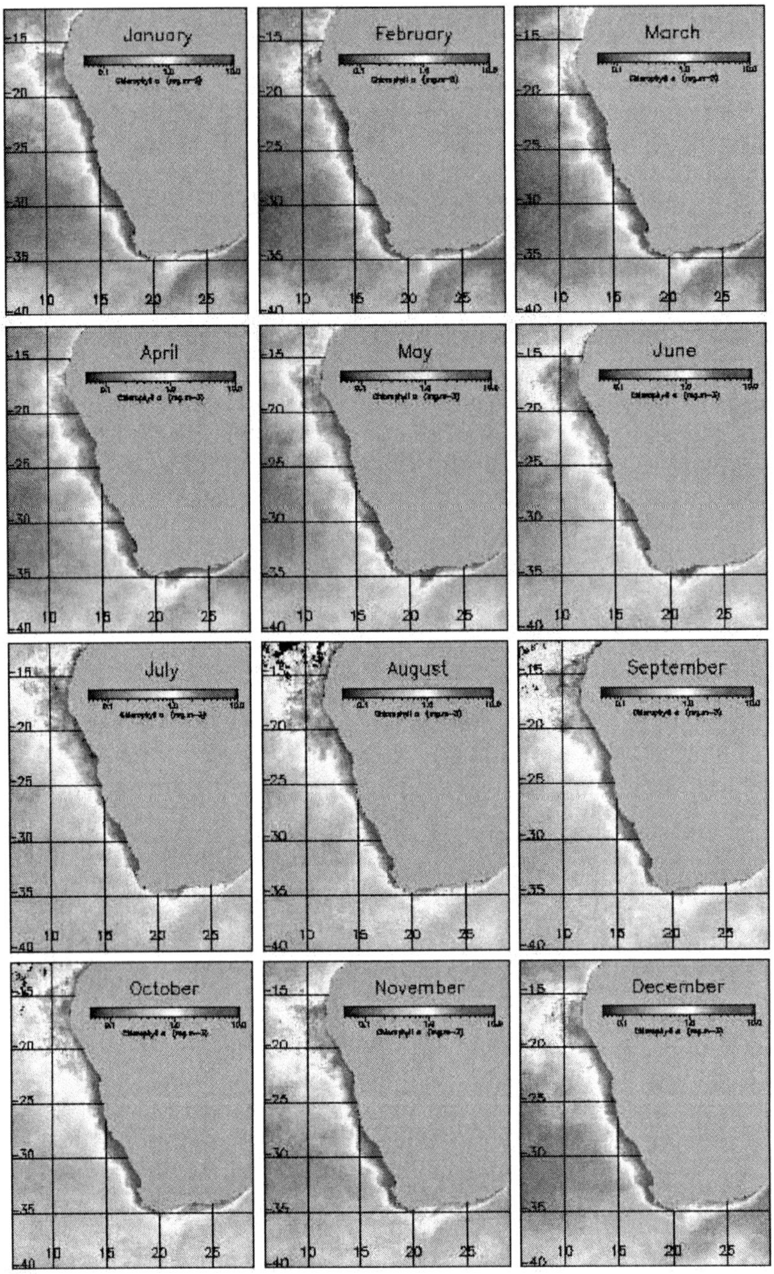

Figure 21.8 Monthly climatology of BCS chlorophyll from SeaWiFS GAC data for the period Sept 1997 – April 2002 (Demarcq *et al.,* 2003). Note the southward movement of the chlorophyll-rich upwelled water south of the Angola-Benguela front near 15 deg. S. during the austral summer months (Nov. – April). There are also seasonal changes on the Agulhas Bank.

Seasonality in upwelling is reflected by the spawning patterns of the main fish stocks, anchovy, sardine, and hake. Anchovy spawn in spring and summer with a peak in November (Shelton 1986, 1987), whereas sardine have a slightly longer spawning season with peaks in October and March (Shelton, 1986). Batch spawning in sardine and anchovy appears to continue at roughly weekly intervals for 6–8 months to take advantage of the upwelling season, but tends to be spread out because of the variability in pulses of upwelling, and in their frequency and duration (Shelton, 1987). Hake have a spawning peak from October to December in the Northern BCS (Assorov and Berenbeim 1983; Sundby *et al.,* 2001), with a secondary peak in February-March in the southern BCS (Botha, 1986), again this appears to be linked with the upwelling season.

Zooplankton biomass seems to vary seasonally in parts of the southern BCS (Andrews and Hutchings, 1980; Pillar 1986; Verheye *et al.,* 1991; Verheye *et al.,* 1998) generally decreasing from summer to winter. One hypothesis for the seasonal decline is that the zooplankton are grazed down by recruiting anchovy in this region of the BCS during the early austral winter (Verheye *et al.,* 1998).

Andrews and Hutchings (1980) found higher biomasses of phytoplankton in the upwelling season of the southern BCS than during the slack period from April - August, with greatest variability in summer, attributed to the very low biomass found in newly upwelled water followed by very high biomasses as the blooms mature. The variability of phytoplankton in the summer upwelling season was confirmed by Brown and Cochrane (1991). Pitcher *et al.* (1992) found seasonal changes in chlorophyll distribution from ocean colour satellite imagery, the highest near-surface concentrations of Chl a being found in the southern BCS inshore of the upwelling front in summer and autumn when seasonal upwelling is strongest. During winter the front is more diffuse, with lower near-surface concentrations. The productive area also varies from summer to winter. The areas of fairly low biomass (>2 and >4 mg. m^{-2}) are greater in winter, but areas of mean Chl a concentrations >6mg m^{-2} are only found in summer. North of Cape Columbine, there is more persistent upwelling, with a broader chlorophyll-rich belt and less seasonal variation (Shannon *et al.,* 1986, Pitcher *et al.,* 1992).

6. The Inter-annual scale

Hagen *et al.* (2001) developed an upwelling index based on the area of 13°C monthly SST for the BCS for the period 1982–1999. They detected both a decreasing trend in the index and strong interannual variability. The years of smallest area of the 13°C SST were 1984, 1993, 1996, 1997 and 1999. From the ENVIFISH project (Hardman-Mountford *et al.,* 2003), SST for the period 1982 -1999 can be divided into three different periods: 1982–1986 that was cooler than normal; 1987–1994 that was intermediate; and, 1995–1999 that was warmer than average. Superimposed on these trends were notable anomalous warm and cool events. In the equatorial and tropical region the extreme warm events of 1984 and 1995 were clearly observable and other warm events occurred in 1991, 1993, 1996 and 1998–99, while cool events happened in 1982, 1983, 1991–92 and 1997. In the Northern Benguela, some of these events were also observed: the extreme warm years of 1984 and 1995 (Gammelsrød *et al.,* 1998), other warm events in 1999 and cool events in 1982, 1983, and 1997. These warm events do not appear to be interannual in the true sense of an anomaly building over more than one year, but appear to be a seasonal amplification of the summer values which only last a month or two.

Warm events in the equatorial and southeastern Atlantic have been noted by several authors. The extreme warm event of 1984 was well documented in the eastern equatorial Atlantic and many parallels were drawn with Pacific El Niño events (Philander, 1986), partly prompted by the proximity of the very strong 1983 El Niño event in the Pacific and possible teleconnections between the two basins. The extreme Benguela warm year of 1984 was termed a "Benguela Niño" by Shannon *et al.* (1986), and other Niño-type years were identified retrospectively as 1934 and 1963, with the most recently described Benguela Niño occurring in 1995 (Gammelsrød *et al.,* 1998). From a time series of coastal sea levels from 1959 to 1985, Brundrit *et al.* (1987) observed high sea level events in 1963, 1974 and 1984 along the entire South Atlantic coast of Africa, from the equatorial area to Cape Town suggesting a ten-year periodicity in these events.

In the southern BCS, including the Agulhas Bank, SST variability is not dominated by interannual trends, as in the equatorial and Angolan coastal areas. Instead, high frequency variability appears more important. Occasional strong intrusions of warm Agulhas water into the southern BCS have been detected in the in 1986, 1992 SST (Hardman-Mountford *et al.,* 2003). During the period 1997–1999, there was a strong warming trend in SST. In the austral summer 2000, abnormally warm water around the Cape Peninsula may have contributed to the high anchovy fish recruitment (Roy *et al.,* 2001).

Carr (2002) demonstrated that the interannual differences of potential primary production between 1997, 1998,and 1999 were largest for 1997 (measured by the Ocean Colour Temperature Scanner, OCTS on the Japanese ADEOS satellite), but cautioned that this might be due primarily to the different sensor and algorithm used. Actual pelagic fish catch data in the Benguela region were 20 times smaller than in the Humboldt Current. The most likely explanations for the differences in potential and observed fish catch are related to differing trophic structure and spatial and temporal accessibility in different EBCs (see Section 4 above). If the estimated yield is an upper bound that will be decreased to 10% or 20% by environmental accessibility, the small pelagic fishery in all four EBCs is likely to be food-limited.

The inter-annual variation of pelagic fish stocks in the BCS has been well documented (Hampton, 1992; Crawford, 1998; Hutchings *et al.,* 1998; Schwartzlose *et al.,* 1999; Barange *et al.,* 1999; Boyer *et al.,* 2001; Payne and Bannister, 2003) and a series of expert systems and rule-based models has been built in attempts to forecast approximate recruitment success of anchovy in the southern BCS using environmental data (Cochrane and Starfield, 1992; Bloomer *et al.,* 1994; Cochrane and Hutchings, 1995; Korrubel *et al.,* 1998; Painting and Korrubel, 1998; Miller and Field, 2002). Almost every time an expert system has been built, new data in the time series give poor forecasts for future years. This suggests that the patterns are not regular or that different factors are called into play as conditions change.

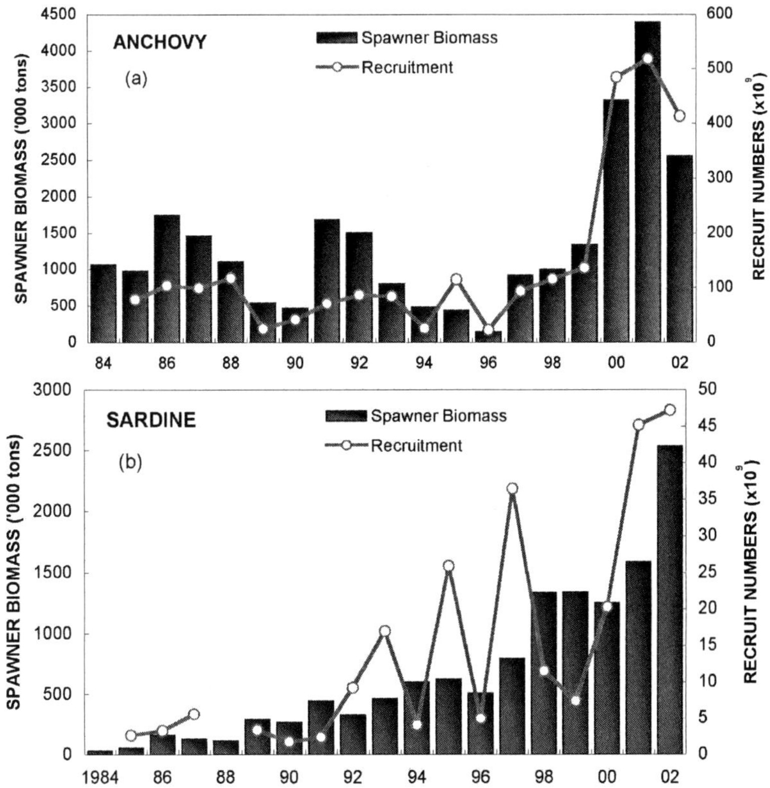

Figure 21.9 Spawner biomass (10^3tons, bars) and recruit numbers (x 10^9, circles) in the southern BCS from 1984–2003 of: a) anchovy and, b)sardine.
Note the different scales on both axes and the gradual build-up of longer-lived sardine in spite of fluctuating recruitment, compared to the variable biomass of short-lived anchovy. (Janet Coetzee, Marine and Coastal Management, DEAT, South Africa, *pers. comm.*).

Fig. 21.9a shows anchovy spawner biomass and recruitment numbers in the southern BCS. The steady build up of the sardine biomass since1984 is notable, in spite of large variations in sardine recruitment from year to year (Fig. 21.9b). This is attributed partly to careful management measures, designed to rebuild the stock of this longer-lived species. On the other hand, the huge increase in anchovy recruitment in 2000, followed by the increased spawner biomass later that year (anchovy start spawning as 1 year-olds), led to a doubling in the resource for at least three years. It seems that two phenomena contributed to the strong anchovy recruitment of 2000: an above average spawning stock the previous year and very unusual environmental conditions in December 1999 (reduced upwelling and warm SST) followed by very strong upwelling and a negative SST anomaly in March-April 2000 (Fig 8, Roy *et al.,* 2001). These are interpreted as warm conditions and reduced upwelling favouring the development and retention of eggs and larvae fairly close to the coast, followed by upwelling and the associated increased plankton productivity once the larvae and juvenile recruits are large enough to migrate vertically and avoid being swept offshore in the surface Ekman layer (See Section 3 above). An expert system has been built to crudely forecast anchovy recruitment, based on these concepts (Miller and Field, 2002). This has very good success (89%) over the period modelled (1985–2001), but it remains to be seen whether this can be maintained in the future.

7. The Decadal scale

Taunton-Clark and Shannon (1988) deduced that, over an 80-year period, Benguela Niños had occurred with a near-decadal periodicity. This was confirmed by Gammelsrød *et al.* (1998) in describing the 1995 Benguela Niño. The apparent decadal-scale period of Benguela Niño-type events occurring in the Angola/Benguela region is much longer than is observed in the Pacific, whereas one would expect a shorter period if Benguela Niños were directly analogous to ENSO. However, they are really an exaggeration of the seasonal movement southwards of the Angola-Benguela front. Florenchie *et al.* (2003) have applied a model of the Equatorial Atlantic Ocean that explains Benguela Niños on the basis of anomalies in the tropical windfield off Brazil in the western tropical Atlantic. If the anomalies occur at a particular time of year and under particular conditions, then a tropical Kelvin wave propagates eastwards across the Atlantic and manifests itself at the surface along the Angolan coast, pushing the Angola-Benguela front further southwards than usual in the austral late summer-autumn. Thus the mechanism of ocean propagation of Benguela Niños appears to be similar to Pacific Niños, but the longer frequency is unexpected in the narrower South Atlantic Ocean basin. Fig. 21.10 shows the good agreement between the model and measured sea surface temperatures off southern Angola/northern Namibia, together with the variability of monthly SST anomalies from 1982–1999. Clearly the greatest SST anomaly variability occurs at 12° – 16° S latitude, the region of the Angola-Benguela front.

Voges *et al.* (2002) have shown that good recruitment of Namibian hake in the northern BCS appears to be associated with Benguela Niños at a decadal scale, over a 14-year period, which is hardly sufficient to demonstrate periodicity. They related the success of hake recruitment to a Benguela Nino index (southward penetration of the Angola-Benguela front in the January-March period) using multinomial logistic regression. The analysis also uses two upwelling indices. Hake recruit to the fishery at 2 years of age and the model is based on the premise that reduced upwelling in the main spawning season (Sept.–March) will create warm conditions favourable for larval (age 0) development and retention in the coastal area. On the other hand, increased upwelling at age 1 will increase productivity when the hake are old enough to migrate vertically to avoid transport offshore in the upper Ekman layer (see Section 3), and reduced intrusion of the warm Angola –Benguela front will increase the area suitable for larval development at age 0 compared to Benguela Niño years. The model predicts the three years of poor recruitment very well and shows an overall skill of correctly predicting into three recruitment categories in 10 out of 14 years.

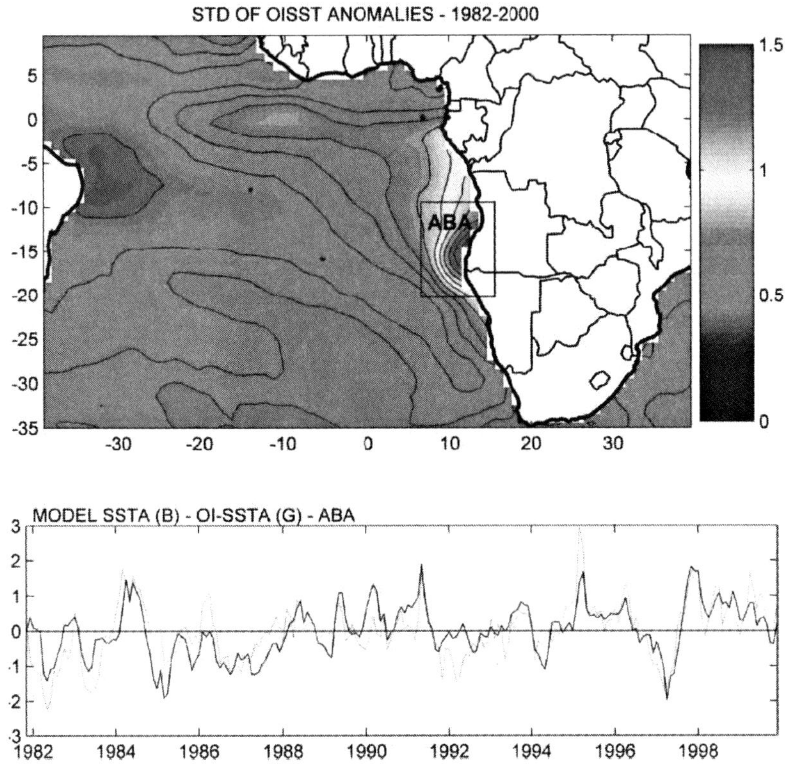

Figure 21.10 The top panel shows the standard deviation of Sea Surface Temperature anomalies from 1982 to 1999 (SST °C). Lowest variability is off the Brazilian coast, highest in the area labeled ABA. Area ABA corresponds to the location of sea-surface temperature anomalies associated with Benguela Ninos. Bottom panel shows temporal history of SST anomalies (°C), spatially averaged from 10 – 20 °S and from 8 °E to the coast of south-western Africa (area ABA on top panel), according to thermal infra-red imagery (grey line) and the model (black line). (From Florenchie *et al.*, 2004).

Schwartzlose *et al.* (1999) have reviewed regime shifts in pelagic fish in the BCS and other upwelling regions. Records of seabird guano yields from coastal islands, mainly reflecting the availability of sardine and anchovy, suggest fluctuations of 20–30 year periodicity since the 1890s, long before pelagic fish were harvested commercially. The quasi-simultaneous shifts of regime in the Pacific and Atlantic Oceans suggest strongly that regime shifts are caused by remote physical forcing, although the mechanisms are still unclear. The semi-independent stocks of sardine and anchovy in the northern and southern BCS show some similarities. In both sub-regions, sardine catches peaked in the 1960s followed by steep declines. Sardine catches have remained at a very low level in the northern BCS (Boyer *et al.*, 2001), but recovered in the south during the mid-late 1990s under a cautious management policy of rebuilding the stock. The anchovy fishery started in the mid-1960s as sardine catches dwindled. Anchovy catches remained high in both sub-regions during the 1970s and peaked in the late 1980s, followed by a sharp decline that has not recovered in Namibia but in the southern BCS both sardine and anchovy have been at record high levels since 2000 (see Fig. 21.9). Thus while there was a regime shift from dominance by sardine to anchovy in the 1960s, and sardine appeared to be dominating the southern BCS again in the mid-late 1990s, there now appears to be a period in which both have remained at high levels for several years. Is this a third regime of joint dominance of both sardine and anchovy?

Verheye and Richardson (1998) and Verheye *et al.* (1998) have assembled long-term data on zooplankton abundance in St Helena Bay in the southern BCS from 1950 – 1996. These data were collected in the period March - June when recruiting anchovy migrate through the area, although there are unfortunate gaps in the 1970s and early 1980s. The data show a clear increase in the biomass and abundance of meso-zooplankton groups, from copepods and euphausiids to chaetognaths by two orders of magnitude (Fig. 21.11). The suggestion is that this was partly forced by increasing annual average wind stress and increased upwelling from 1950 until the mid-1980s when wind stress started to decline (Verheye and Richardson, 1998). Another component may be that heavy fishing on anchovy in St Helena Bay reduced their impact on larger meso-zooplankton over the period 1965–1995. Since the mid-1990s, the abundance of meso-zooplankton has declined in the area (H. Verheye, *pers. comm.*) as the combined biomasses of sardine and anchovy have reached record levels. Thus the evidence of time series observations of fish catches and zooplankton, and also simulation models suggests both bottom up and top down controls on zooplankton, resulting in decade-long changes that may be referred to as regime shifts. The combination of top-down and bottom up controls has been referred to as "wasp-waist" control (Cury *et al.*, 2000).

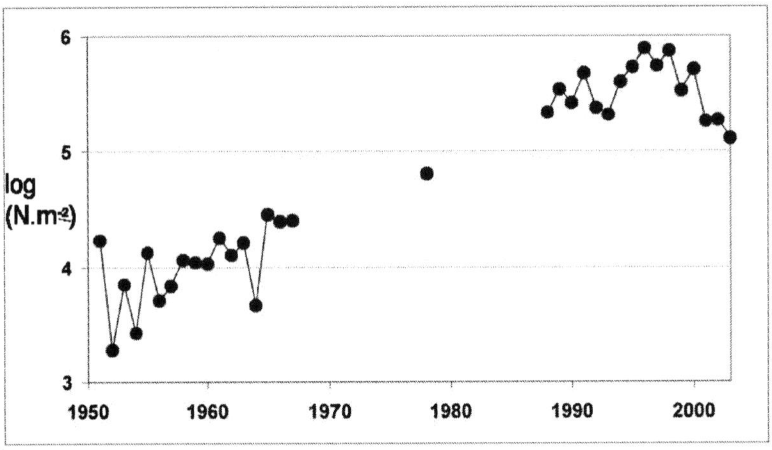

Figure 21.11 Time series of total crustacean copepod abundance (log numbers m^{-2}) from 1951 – 2003 in St Helena Bay in the southern BCS. Annual means of samples taken in March – May each year when recruiting pelagic fish are common in St Helena Bay. Note the hundred-fold increase from 1950s to mid- 1990s (H. Verheye, *pers comm.*)

Shannon *et al.* (2004) have developed a trophic simulation model of the southern BCS to test hypotheses about the causes of regime shifts. Several scenarios of increased fishing mortality suggest that "top-down" changes in fishing pressure are unlikely to have caused the shifts in anchovy and sardine biomass observed in the

1980s and 1990s. On the other hand, when forced from the "bottom up" to simulate a warm period with reduced upwelling, small-celled phytoplankton, fewer macro-zooplankton, then anchovy biomass decreases and sardine biomass increases. The model simulates the generalised filter-feeding behaviour of sardine compared to the more specialist biting behaviour on large meso-zooplankton and macro-zooplankton preferred by anchovy (van der Lingen, 2002). This scenario gives a plausible 10-year regime shift from anchovy to sardine forced by a 2–4 year shift in anchovy and sardine diet (Fig 21.12).

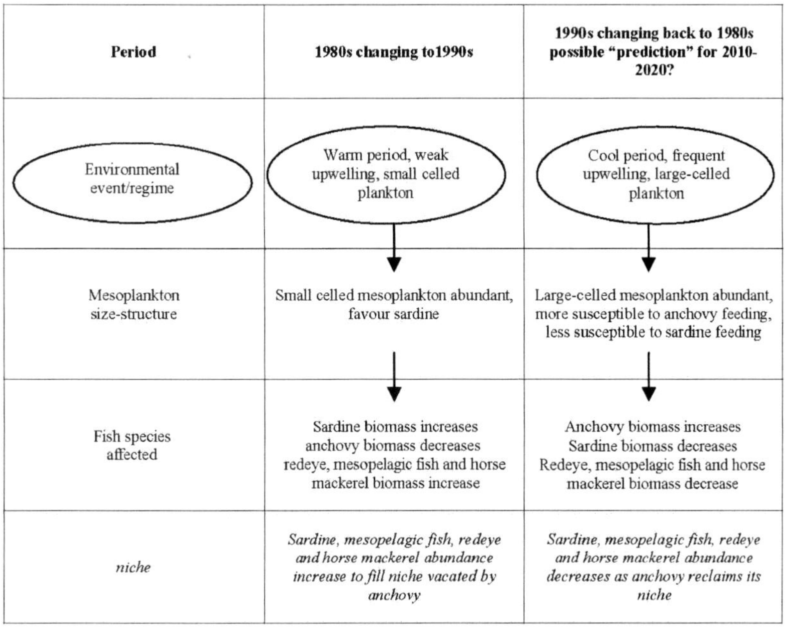

Figure 21.12 Cartoon showing hypothesis of links between periods of less (warm) and more (cool) upwelling, plankton abundance, and sardine/anchovy regime shifts in the southern BCS. (From Shannon *et al.*, 2003).

8. Multiscale Considerations

Satellite remote sensing in the BCS has revealed a number of novel unsampled aspects and continues to be an important tool for observing the coastal ocean at a variety of space and time scales. Routine SeaWiFS imagery has revealed the nature of the chl signal (Demarcq *et al.*, 2003), and in combination with suitable numerical models will be used to calculate the primary productivity of the region, to compare with the global estimates of other eastern boundary systems (e.g. Carr, 2002). Widespread patches of highly reflective surface features have been seen intermittently both near the coast (Weeks *et al.*, 2002), and at greater distances offshore, and have been interpreted as outbreaks of sulphur (Weeks *et al.*, 2004). There is clearly a need for a combined simultaneous ship borne/remote sensing series of measurements to confirm various assumptions that have been made during these conjectures, particularly for the cases at large distances from the shore.

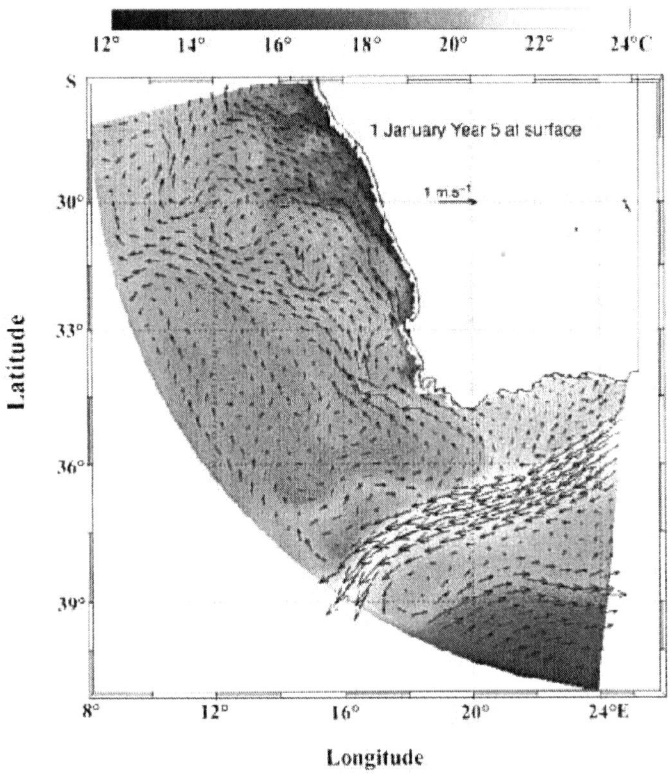

Figure 21.13 January Snapshot of SST and currents (arrows, only every third vector drawn) in the southern BCS simulated by the PLUME ROMS model. (From Penven *et al.*, 2001b)

In the past five years, there has been a sharp increase in the hydrodynamic modelling of the southern Benguela Upwelling ecosystem by implementing the 3-D ROMS numerical code (Penven 2000; Penven *et al.*, 2001a; Penven *et al.*, 2001b). A ten-year model run with a time resolution of two days, and variable horizontal grid spacing from 9–18 km has provided the community with output for use of a number of individual based model (IBM) configurations (Fig. 21.13). These have tested a variety of hypotheses related to anchovy life histories in the southern Benguela (Mullon *et al.*, 2002; Mullon *et al.*, 2003; Parada, 2003; Parada *et al.*, 2003; Huggett *et al.*, 2003). The three open boundary conditions for the hydrodynamic model were supplied by the basin scale AGAPE model Biastoch and Kraus (1999). Climatological wind forcing was used initially (Penven, 2000), and then Blanke *et al.* (2002) reran the model with a more realistic ERS1/2 satellite derived weekly windfields. A comparison of the two outputs (climatology forced and realistic satellite derived wind) from the model, made it possible to delineate the areas which were directly forced by the local wind, and those that appeared to respond more to the remote forcing. The cascade of scales involved in anchovy recruitment appears to be captured in the Individual Based Models (IBM). Thus changes in the wind fields driving the 3-d PLUME model, result in changes in inter-annual, seasonal and event scale model output. Realistic mesoscale phenomena appear as emergent properties (Fig. 21.13). These are used as input for the IBM model, which reflects inter-annual, seasonal and event scale effects on anchovy larvae, which only recruit successfully to the inshore area in autumn-winter if they exhibit diel vertical migration (DVM) behaviour (See Sections 3, 4, above). The IBM also reflects the trade off between different factors affecting fish eggs and larvae, in this example (Fig. 21.14) the trade off is between on the one hand, unfavourable offshore Ekman transport in the upper 20 metres but with water temperature favourable for fast development, and on the other hand, cool temperatures below 50 m depth, unfavourable for development, but more favourable for transport. The optimum trade-off between these two factors appears to be at about 30–50 m depth. (Fig. 21.15)

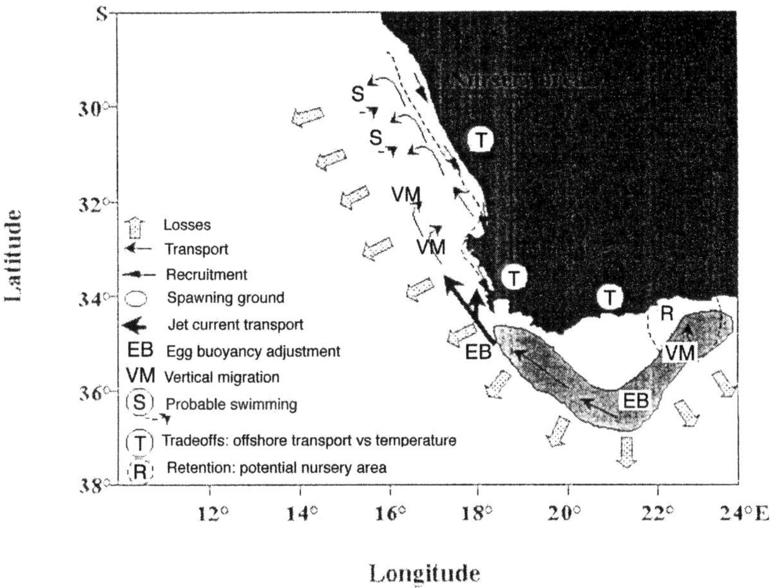

Figure 21.14 Cartoon depicting the processes modeled in the recruitment of young anchovy to the fishery in the southern BCS (from Parada, 2003). The spawning ground on the Agulhas Bank in the south and fish recruit on the west coast of South Africa. Processes include spawning at optimal depths, transport by currents, egg buoyancy adjustment, trade-offs between optimal development temperatures and advection offshore, vertical migration of larvae and swimming of juveniles.

Fig 21.14 shows the factors that have emerged as important in the early life history of anchovy from IBM modelling (Parada, 2003). Spawning occurs over the Agulhas Bank in the south during the summer upwelling season (Section 5). Transport towards the nursery grounds on the west coast is by the Benguela frontal jet current which varies in strength and position on a 4–12 day scale according to the passage of atmospheric low-pressure systems (Section 4). Eggs need to adjust their buoyancy to remain at the optimal depth for development and longshore transport during this phase, followed by the later larvae, which start diel vertical migration to escape predation and get transported inshore in the upwelling water (Section 3). Similarly, anchor-station studies (Section 4) have shown how the diel vertical migration behaviour of zooplankton (Section 3) may be modified by the development of phytoplankton communities in the upwelled water on the scale of days to weeks. Thus the cascade of scales of physical phenomena is mirrored by a suite of biological processes that have evolved to take advantage of the physical system.

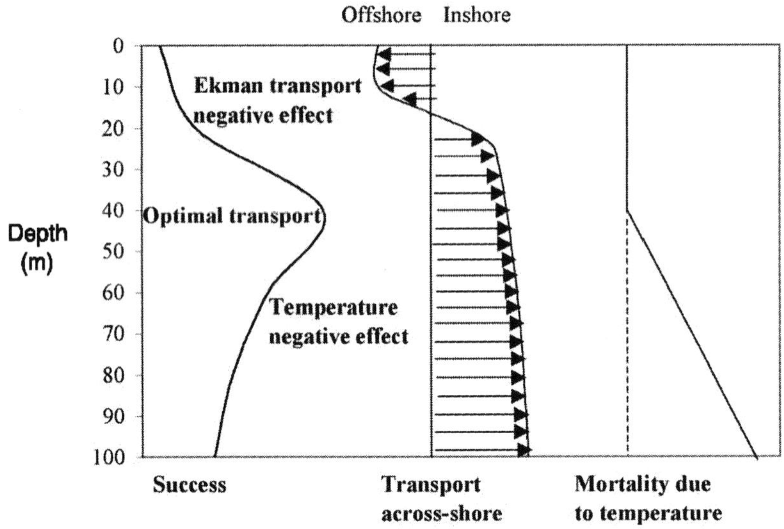

Figure 21.15 Hypothetical vertical section showing "optimal transport success" and the tradeoff between Ekman transport and mortality of larvae due to cool temperatures during upwelling (From Parada, 2003)

9. Regional research programmes in the BCS

Regional and international funding (BENEFIT *http://www.benguela.org.za* and IDYLE *http://sea.uct.ac.za/idyle*) has made it possible to increase research in all aspects in the BCS over the past 5–10 years. More recently, the Benguela Current Large Marine Ecosystem (BCLME) Programme funded by the World Bank and GEF has entered a five-year phase (2002–2006). The BCLME (*http://www.bclme.org*) is a multi-sectoral regional initiative by Angola, Namibia and South Africa whose objective is to facilitate the integrated management, sustainable development and protection of this unique eastern boundary upwelling ecosystem. It is funded by the Global Environmental Facility (GEF) under its International Water Portfolio and is implemented by the United Nations Development Programme (UNDP) with the United Nations Office for Project Services (UNOPS) as executing agency. The three member countries provide further financial and in-kind contributions. The Eur-Oceans network of excellence *(http://www.univ-brest.fr/IUEM/EUR-OCEANS/eur-oceans_uk.htm)* includes a four-year study of comparative aspects of major EBC systems, including the Benguela Current System, with a view to incorporate these in an Institute of European Research on Ocean Ecosystems under Anthropogenic and Natural forcings.

Acknowledgements

Thanks are expressed to Ken Brink and Allan Robinson for the invitation to attend the COASTS meeting in Paris and for help in suggesting this review. We thank Ted Strub for helpful comments on an early draft of the MS and Paula de Coito for help in preparing the MS. Funding from the South African National Foundation for Research and the South African Department of Environmental Affairs: Marine and Coastal Management helped support the research reviewed here. Collaboration with the French Institut Reserche pour le Developpement and the French-South African IDYLE project is gratefully acknowledged. JGF acknowledges support from the Ray Lankester Fund of the Marine Biological Association of the UK where much of this review was completed.

Bibliography

Armstrong, D. A., B. A. Mitchell-Innes, F. Verheye-Dua, H. Waldron and L. Hutchings, 1987. Physical and biological features across an upwelling front in the southern Benguela. *S. Afr. J. mar. Sci.,* **5,** 171–190.

Assorov, B.B. and D.Y. Berenbeim, 1983. Spawning grounds and cycles of Cape hakes in the southeast Atlantic. *Colln scient. Pap. Int. Commn SE Atl. Fish.,* **10** (1), 27–30.

Andrews, W.R.H. and L. Hutchings, 1980. Upwelling in the southern Benguela Current. *Progr. Oceanogr.,* **9,** (1), 81pp.

Bakun, A., and C. S. Nelson, 1991. Wind stress curl in subtropical eastern boundary current regions. *J. Phys. Oceanogr.,* **21,** 1815–1834.

Bakun, A.,1996 - *Patterns in the Ocean: ocean processes and marine population dynamics.* University of California Sea Grant, San Diego, Calif., in cooperation with Centro de Investigaciones Biolgicas de Noroeste, La Paz, Baja California Sur, Mexico.

Barange, M., S.C. Pillar, and L. Hutchings, 1992. Major pelagic borders of the Benguela upwelling system according to Euphausiid species distribution. *S. Afr. J. Mar. Sci.,* **12:** 3 – 17.

Barange, M., I. Hampton and B. A. Roel, 1999. Trends in the abundance and distribution of anchovy and sardine on the South African continental shelf in the 1990's, deduced from acoustic surveys. *S. Afr. J. mar. Sci.,* **21,** 367–392.

Biastoch, A. and W. Krauß, 1999. The Role of Mesoscale Eddies in the Source Regions of the Agulhas Current, *J. Phys. Oceanogr.,* **29,** 2303–2317.

Blanke, B., C. Roy, P. Penven, S. Speich, J. McWilliams, and G. Nelson, 2002. Linking wind and interannual upwelling variability in a regional model of the southern Benguela. *Geophys. Res. Lett.,* **29,**(24), 2188, doi:10.1029/2002GL015718.

Bloomer, S.F., K.L. Cochrane, J.G. Field, 1994. Towards predicting recruitment success of anchovy *Engraulis capensis* Gilchrist in the southern Benguela system using environmental variables: a rule-based model. *S. Afr. J. mar. Sci.,* **14,** 107 – 119.

Botha, L. 1986. Reproduction, sex ratio and rate of natural mortality of Cape hakes *Merluccius capensis* Cast. and *M. paradoxus* Franca in the Cape of Good Hope area. *S. Afr. J. mar. Sci.,* **4:** 23–35.

Boyd, A.J. and F.A. Shillington 1994. The Agulhas Bank: A review of the physical processes. *S. Afr. J. Sci.,* **90,** 114–122.

Boyer, D.C., H.J. Boyer, I. Fossen, A. Kreiner, 2001. Changes in abundance of the northern Benguela sardine stock during the decade 1990–2000, with comments on the relative importance of fishing and the environment. *S. Afr. J. mar. Sci.*, **23,** 76–84.

Boyer, D., J. Cole , and C. Bartholomae, 2000. Southwestern Africa: Northern Benguela Current Region, in *Seas at the Millenium: an environmental evaluation,* Ed. C.R.C. Sheppard, Vol 1, Elsevier Science Ltd., 821–840.

Brown, P.C. and K.L. Cochrane, 1991. Chlorophyll *a* distribution in the southern Benguela, possible effects of global warming on phytoplankton and its implications for pelagic fish. *S. Afr. J. Sci.,* **87,** (6), 233 - 242.

Brown, P. C. and L. Hutchings 1987. The development and decline of phytoplankton blooms in the southern Benguela upwelling system. 1. Drogue movements, hydrography and bloom development. *S. Afr. J. mar. Sci.,* **5,** 357–392.

Brundrit, G.B., B.A. De Cuevas, and A.M. Shipley, 1987. Long-term variability in the eastern South Atlantic and a comparison with that in the eastern Pacific. Payne, A.I.L., Gulland, J.A., Brink, K.H. (Eds), The Benguela and comparable ecosystems, *S. Afr. J. mar. Sci.,* **5,** 73–78.

Carr, M.-E., 2002. Estimation of potential productivity in Eastern Boundary Currents using remote sensing. *Deep-Sea Res., II,* **49,**59 -80.

Chapman, P. and G. W. Bailey, 1991. Short-Term Variability During an Anchor Station Study in the Southern Benguela Upwelling System - Introduction. *Progr. Oceanogr.,* **28,**(1–2), 1–7.

Cochrane, K.L. and L. Hutchings, 1995. A structured approach to using biological and environmental parameters to forecast anchovy recruitment. *Fish. Oceanogr.,* **42,** 102 - 107.

Cochrane, K. L. and A. M. Starfield, 1992. The potential use of predictions of recruitment success in the management of the South African anchovy resource. *S. Afr. J. mar. Sci.,* **12,** 891–902.

Crawford, R.J.M., 1998. Responses of African penguins to regime changes of sardine and anchovy in the Benguela system. *S. Afr. J. mar. Sci.,* **19,** 355 - 364.

Cury, P.M., A. Bakun, R.J.M. Crawford, A. Jarre-Teichmann, R. Quinones, L.J. Shannon, H.M. Verheye, 2000. Small pelagics in upwelling systems: patterns of interaction and structural changes in "wasp-waist" ecosystems. *ICES J. mar. Sci.,* **57,** 603 -618.

Demarcq, H., R. G. Barlow and F. A. Shillington, 2003. Climatology and variability of SST and surface chlorophyll for the Benguela and Agulhas region. *Afr. J. mar. Sci.,* **25,** 363–376.

Fidel, Q., 2001. Unpublished MSc Thesis, Department of Oceanography, University of Cape Town.

Florenchie, P., J. R. E. Lutjeharms, C. J. C. Reason, S. Masson, and M. Rouault, 2003. The source of Benguela Niños in the South Atlantic Ocean, *Geophys. Res. Lett.,* **30,** (10), 1505, doi:10.1029/2003GL017172.

Florenchie, P., C.J.C. Reason, J.R.E. Lutjeharms, M. Rouault, C. Roy, and S. Masson, 2004. Evolution of Interannual Warm and Cold Events in the Southeast Atlantic Ocean. *J. Climate,* **17,** 2318–2334.

Gammelsrød, T., C. H. Batholomae, D.C. Boyer, V.L.L. Filipe, and M.J. O'Toole, 1998. Intrusion of warm surface water along the Angolan-Namibian coast in February-March 1995: the 1995 Benguela Niño. In Benguela dynamics. Edited by S.C. Pillar, C.L. Moloney, A.I.L. Payne, and F.A. Shillington. *S. Afr. J. Mar. Sci.,* **19,** 51–56.

Gibbons, M.J., M. Barange and S.C. Pillar, 1991. Vertical migration and feeding of *Euphausia lucens* (Euphausiacea) in the southern Benguela. J. Plankt. Res., 13, 473 - 486.

Hagen, E., R. Feistel, J.J. Agenbag and T.Ohde, 2001. Seasonal and interannual changes in Intense Benguela Upwelling (1982–1999). *Oceanol. Acta,* **24,** (6), 557–568.

Hardman-Mountford, N.J, A.J. Richardson, J.J. Agenbag, E. Hagen, L. Nykjaer, F.A. Shillington, and C. Villacastin, 2003. Climatic variability of the South East Atlantic Ocean observed from satellite. *Progr. Oceanogr.,* **59,** 181–221.

Hampton, I., 1987. Acoustic study on the abundance and distribution of anchovy spawners and recruits in South African waters. *S. Afr. J. mar. Sci.,* **5,** 901–918.

Hampton, I. 1992. The role of acoustic surveys in the assessment of pelagic fish resources on the South African continental shelf. *S. Afr. J. mar. Sci.*, **12**,1031–1050.

Hill, A.E., B.M Hickey, F.A. Shillington, P.T. Strub, K.H. Brink, E.D. Barton and A.C. Thomas, 1998. Eastern ocean boundaries: coastal segment (E). In Robinson A.R. and Brink K.H. Eds, *The Sea* 11, pp. 29–68. John Wiley & Sons.

Hocutt, C.H. and H.M. Verheye, 2001. BENEFIT marine science in the Benguela Current region during 1990s: Introduction. *S. Afr. J. Sci.*, **97**, 195–198.

Huggett, J., P. Fréon , C. Mullon , and P. Penven, 2003. Modelling the transport success of anchovy *Engraulis encrasicolus* eggs and larvae in the southern Benguela: the effect of spatio-temporal spawning patterns. *Mar. Ecol. Prog. Ser.*, **250**, 247–262.

Hutchings, L. 1992. Fish harvesting in a variable productive environment - searching for rules or searching for exceptions? *S. Afr. J. mar. Sci.*, **12**, 297–318.

Hutchings, L., M. Barange, S. F. Bloomer, A. J. Boyd, R. J. M. Crawford, J. A. Huggett, M. Kerstan, J. L. Korrubel, J. A. A. De Oliveira, S. J. Painting, A. J. Richardson, L. J. Shannon, F. H. Schulein, C. D. van der Lingen and H. M. Verheye, 1998. Multiple factors affecting South African anchovy recruitment in the spawning, transport and nursery areas. *S. Afr. J. mar. Sci.*, **19**, 211–226.

Hutchings, L., L. E. Beckley, M. H. Griffiths, M. J. Roberts, S. Sundby and C. van der Lingen, 2002. Spawning on the edge: spawning grounds and nursery areas around the southern African coastline. *Mar. Freshw. Res.*, **53**, 307–318.

Korrubel, J. L., S. F. Bloomer, K. L. Cochrane, L. Hutchings and J. G. Field, 1998. Forecasting in South African pelagic fisheries management: the use of expert and decision support systems. *S. Afr. J. mar. Sci.*, **19**, 415–424.

Miller, D. C. M., and J.G. Field, 2002. Predicting anchovy recruitment in the southern Benguela ecosystem: developing an expert system using classification trees. *S. Afr. J. Sci.*, **98**, 465–472.

Mitchell-Innes, B.A. and D.R.Walker, 1991. Short-term variability during an anchor station study in the southern Benguela upwelling system: phytoplankton production and biomass in relation to species changes. *Prog. Oceanogr.*, **28**, 65–89.

Moloney, C. L., J. G. Field, and M.I. Lucas, 1991. The size-based dynamics of plankton food webs .2. Simulations of 3 contrasting Southern Benguela food webs. *J. Plankton Res.*, **13**, 1039–1092.

Mohrholz, V., M. Schmidt, and J. R. E. Lutjeharms, 2001. The Hydrography and dynamics of the Angola-Benguela Frontal Zone and environment in April 1999. *S. Afr. J. Sci.*, **97**, 199–208.

Mullon, C., P. Cury and P. Penven, 2002. Evolutionary individual-based model for the recruitment of anchovy *(Engraulis capensis*) in the southern Benguela. *Can. J. Fish. Aquat. Sci.*, **59**, 910–922.

Mullon, C., P. Freon, C. Parada, C. van der Lingen and J. Huggett, 2003. From particles to individuals: modelling the early stages of anchovy (*Engraulis capensis/encrasicolus*) in the southern Benguela. *Fish. Oceanogr.*, **12**, 396–406.

Painting, S. J. and J. L. Korrubel, 1998. Forecasts in recruitment in South African anchovy from SARP field data using a simple deterministic expert system. *S. Afr. J. mar. Sci.*, **19**, 245–262.

Painting, S. J., C. L. Moloney, and M.I. Lucas, 1993. Simulation and Field-Measurements of Phytoplankton-Bacteria- Zooplankton Interactions in the Southern Benguela Upwelling Region. *Mar. Ecol. Prog. Ser.*, **100**, 55–69.

Parada, C. 2003. Modelling the effects of environmental and ecological processes on the transport, mortality, growth and distribution of early stages of Cape anchovy (*Engraulis encrasicolus)* in the Benguela system. Unpublished PhD Thesis, Department of Oceanography, University of Cape Town, pp 123.

Parada, C., C.D. van der Lingen, C. Mullon and P. Penven, 2003. Modelling the effect of buoyancy on the transport of anchovy (*Engraulis capensis*) eggs from spawning to nursery grounds in the southern Benguela: an IBM approach. *Fish. Oceanogr.*, **12**, (3), 170–184.

Payne, A.I.L. and R.C.A. Bannister, 2003. Science and fisheries management in southern Africa and Europe. *Afr. J. mar. Sci.*, **25**, 1–24.

Penven, P. 2000. *A numerical study of the Southern circulation with an application to fish recruitment.* Ph.D. thesis, Université de Bretagne Occidentale, Brest, France.

Penven, P., J.R.E. Lutjeharms, P. Marchesiello, C. Roy and S.J.Weeks, 2001a. Generation of cyclonic eddies by the Agulhas Current in the lee of the Agulhas Bank. *Geophys. Res. Lett.,* **28,** 1055–1058.

Penven, P., C. Roy, G.B. Brundrit, A. Colin de Verdière, P. Fréon, A.S. Johnson, J.R.E Lutjeharms and F.A. Shillington, 2001b. A regional hydrodynamic model of upwelling in the Southern Benguela. *S. Afr. J. Sci.,* **97,** 472–475.

Philander, S. G. H., 1986. Unusual conditions in the tropical Atlantic Ocean in 1984. *Nature,* **322,** (6076), 236–238.

Pillar, S.C. 1986. Temporal and spatial variations in copepod and euphausiid biomass off the southern and south-western coasts of South Africa 1977/78. *S. Afr. J. mar. Sci.,* **4,** 219 - 229.

Pitcher, G. C., P. C. Brown and B. A. Mitchell-Innes, 1992. Spatio-temporal variability of phytoplankton in the southern Benguela upwelling system. *S. Afr. J. mar. Sci.,* **12,** 439–456.

Pitcher, G. C., A. J. Boyd, D.A. Horstman and B.A. Mitchell-Innes,1998. Subsurface dinoflagellate populations, frontal blooms and the formation of red tide in the southern Benguela upwelling system. *Mar. Ecol. Prog. Ser.,* **172,** 253–264.

Preston-Whyte, R. A. and P. D. Tyson 2000. *The Weather and Climate of Southern Africa,* Oxford University Press, Cape Town.

Risien, C.M., C.J.C. Reason, F.A. Shillington and D. B. Chelton, 2004. Variability in satellite winds over the Benguela Upwelling System. *J. Geophys. Res.,* **109,** C03010, doi:10.1029/2003JC001880.

Roy, C., S. Weeks, M. Rouault, G. Nelson, R. Barlow and C. van der Lingen, 2001. Extreme oceanographic events recorded in the southern Benguela during the 1999–2000 summer season. *S. Afr. J. Sci.,* **97,** 465–471.

Schumann, E.H. and K.H. Brink, 1990. Coastal-trapped waves off the coast of South Africa: generation, propogation and current structures. *J. Phys. Oceanogr.,* **20,** 1206–1218.

Schwarzlose, R. A., J. Alheit, A. Bakun, T. R. Baumgartner, R. Cloete, R. J. M. Crawford, W. J. Fletcher, Y. Green-Ruiz, E. Hagen, T. Kawasaki, D. Lluch-Belda, S. E. Lluch-Cota, A. D. MacCall, Y. Matsuura, M. O. Nevarez-Martinez, R. H. Parrish, C. Roy, R. Serra, K. V. Shust, M. N. Ward and J. Z. Zuzunaga, 1999. Worldwide large-scale fluctuations of sardine and anchovy populations. *S. Afr. J. mar. Sci.,* **21,** 289–348.

Shannon, L. J., J.G. Field, and C.L.Moloney, 2004. Simulating anchovy-sardine regime shifts in the southern Benguela ecosystem. *Ecol. Model.,* **172,** 269–281.

Shannon, L. J., C.L. Moloney, A. Jarre, and J.G. Field, 2003. Trophic flows in the Southern Benguela region during the 1980's and 1990's. *J. Mar. Systems,* **39,** 83–116.

Shannon, L.V., A. J. Boyd, G. B. Brundrit and J. Taunton-Clark, 1986. On the existence of an El Niño-type phenomenon in the Benguela system, *J. mar. Res.,* **44,** 495–520.

Shannon, L.V., and G. Nelson, 1996. *The South Atlantic: present and past circulation,* edited by G. Wefer, W. H. Berger, G. Siedler, D. J. Webb, pp. 163–210, Springer-Verlag, Berlin.

Shelton, P.A. 1986. Fish Spawning strategies in the variable southern Benguela Current region. PhD thesis, University of Cape Town, 327 pp.

Shelton, P.A. 1987. Life history traits displayed by neritic fish in the Benguela current ecosystem. *S. Afr. J. mar. Sci.,* **5,** 235 - 242.

Shelton, P.A. and L. Hutchings, 1982. Transport of anchovy *Engraulis capensis* Gilchrist, eggs and early larvae by a f rontal jet current. *J. Cons. perm. int. Explor. Mer.,* **40** (2), 185 - 198.

Shillington, F.A. 1998. The Benguela upwelling system off southwestern Africa. Coastal segment (16,E). In: Robinson A.R. and Brink K.H. Eds, *The Sea* 11, pp. 583–604. John Wiley & Sons.

Stuart, V. and H.M. Verheye, 1991. Diel migration and feeding patterns of the chaetognath, *Sagitta friderici* off the west coast of South Africa. *J. Mar. Res.,* **49,**493–515.

Stenevik, E. K., S. Sundby and R. Cloete, 2001. Influence of buoyancy and vertical distribution of sardine Sardinops sagax eggs and larvae on their transport in the northern Benguela ecosystem. *S. Afr. J. mar. Sci.*, **23,** 85–97.

Strub, P. T., F. A. Shillington, C. James, and S. J. Weeks, 1998. Satellite comparison of the seasonal circulation in the Benguela and California current systems, *S. Afr. J. mar. Sci.*, **19,** 99–112.

Sundby, S., A. J. Boyd, L.Hutchings, M.J. O'Toole, K. Thorisson, A.Thorsen, 2001. Interaction between Cape hake spawning and the circulation in the northern Benguela Upwelling ecosystem. *S. Afr. J. mar. Sci.*, **23,** 317–336.

Taunton-Clark, J., and L.V. Shannon, 1988. Annual and interannual variability in the South-East Atlantic during the 20th century. *S. Afr. J. mar. Sci.*, **6,** 97–106.

Touratier, F., J. G. Field, and C.L. Moloney, 2003. Simulated carbon and nitrogen flows of the planktonic food web during an upwelling relaxation period in St. Helena Bay (southern Benguela ecosystem). *Prog. Oceanogr.*, **58,** 1–41.

van der Lingen, C. D. 2002. Diet of sardine *Sardinops sagax* in the southern Benguela upwelling ecosystem. *S. Afr. J. mar. Sci.*, **24,** 301–316.

Veitch, J. A., P. Florenchie and F. A. Shillington, 2005. Seasonal and interannual fluctuations of the Angola Benguela frontal zone using high resolution satellite imagery from 1982 to 1999. (Submitted to *Int. J. Remote Sensing*), in press.

Verheye, H. M. and J.G. Field, 1992. Vertical distribution and diel vertical migration of *Calanoides carinatus* (Kroyer, 1849) developmental stages in the southern Benguela upwelling region. *J. exp. Mar. Biol. Ecol.*, **158,** 123–140.

Verheye, H. M., L. Hutchings, and W.T. Petersen, 1991. Life history and population maintenance strategies of *Calanoides carinatus* (Copepoda: Calanoida) in the southern Benguela ecosystem. *S. Afr. J. mar. Sci.*, **11,** 179 - 191.

Verheye, H. M. and A. J. Richardson 1998. Long-term increase in crustacean zooplankton abundance in the southern Benguela upwelling region (1951–1996): bottom-up or top-down control? *ICES J. mar. Sci.*, **55,**(4), 803–807.

Verheye, H.M., A.J. Richardson, L. Hutchings, G. Marska, and D. Gianakouras, 1998. Long-term trends in the abundance and community structure of coastal zooplankton in the southern Benguela system, 1951–1996. *In* Benguela Dynamics. *Edited by* S.C. Pillar, C.L. Moloney, A.I.L. Payne, and F.A. Shillington. *S. Afr. J. mar. Sci.*, **19,** 317 - 332.

Voges, E., A. Gordoa, C.H. Bartholomae and J.G. Field, 2002. Estimating the probability of different levels of recruitment for Cape hakes *Merluccius capensis* off Namibia, using environmental indices. *Fisheries Res.*, **1350,** 1–8.

Wacongne, S., and B. Piton, 1992. The near-surface circulation in the northeastern corner of the South Atlantic Ocean. *Deep-Sea Res.*, **39,** 1273–1298.

Weeks, S.J., B. Currie, and A. Bakun, 2002. Satellite Imaging: massive emissions of toxic gas in the Atlantic. *Nature,* **415,** 493–494.

Weeks, S.J., B. Currie, A. Bakun and K.R. Peard, 2004. Hydrogen sulphide eruptions in the Atlantic Ocean off southern Africa: implications of a new view based on SeaWiFS satellite imagery. *Deep-Sea Res., I,* **51,** 153–172.

Chapter 22. A NOTE ON COASTAL UPWELLINGS AND FISHERIES IN THE GULF OF GUINEA (17,E)

C. Roy

Centre IRD de Bretagne, BP 70, 29280 Plouzané, France

Contents

1. Introduction

The Equatorial West Africa coastal region lies within the Guinea Current Coastal Province defined by Longhurst (1998). It is bounded to the north by the Canary Current coastal upwelling region and to the south by the Benguela Current system. The waters of the Equatorial West Africa region include the Exclusive Economic Zones (EEZ) of sixteen countries between Guinea Bissau in the northwest and Angola in the southwest. The east-west orientation of the coast, from Côte d'Ivoire to the eastern border of Nigeria, is a unique characteristic of this tropical region (Fig. 1). Most of the northern boundary of the region is located at about 5°N while the eastern boundary lies between 7°E and 12°E. The oceanic domain delineated by these zonal and meridional boundaries is known as the Gulf of Guinea.

The continental shelf (defined by the 200m isobath) is quite narrow with a width ranging between 15 and 105 km and being widest on the northwestern part of the region. The shelf generally breaks at depths of between 100 and 120m. The coastline of the sub-region is generally low lying and interspersed with marshes, lagoons and mangrove swamps (Hardman-Mountford et al., 2000). A number of estuaries interrupt the barrier beaches that separate mangrove swamps from the sea. These estuaries and coastal lagoons constitute an important component of the inshore ecosystem. Two major rivers flow into the region: the Niger through its wide delta in the North East and the Congo in the South East. The Congo River is by far the

The Sea, Volume 14, edited by Allan R. Robinson and Kenneth H. Brink
ISBN 0–674–01527–4

major contributor with an annual runoff representing almost 50% of the total freshwater input in the region (Mahé, 1998).

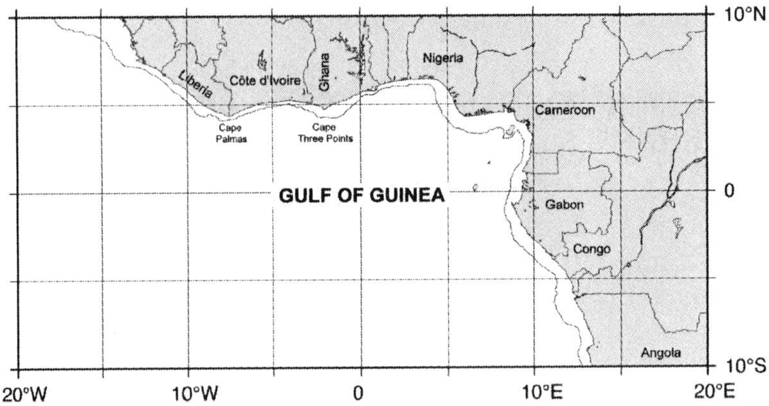

Figure 22.1 Map of the Equatorial West Africa region with the 200m isobath.

Due to the asymmetry in the distribution of water and land in this part of the Atlantic, there is a northward shift of the meteorological equator. As a result, the Gulf of Guinea is located mostly south of the Inter-Tropical-Convergence-Zone (ITCZ) and under the influence of the moderate southeast trades (Ajao and Houghton, 1998). North of the geographic equator, the oceanic southeast trades are deflected and become south-westerlies when they reach the shore of the northern part of the region.

In the late 1960's, the first modern synthesis on the oceanography and fisheries resources in the tropical Atlantic was published (Anon., 1966). It presented a series of review papers and contributions summarizing the results of the International Cooperative Investigation of the Tropical Atlantic (ICITA) and the Guinean Trawling Survey (GTS) projects. Most of the recently published scientific literature on the Gulf of Guinea addresses topics related to offshore oceanography and equatorial climate variability. The oceanography of the coastal domain is much less documented and the major part of the available data and original material date back from the colonial and post-colonial era, most of it concentrated along the coast off Côte d'Ivoire and Ghana (see for example Longhurst and Pauly (1987) or Le Loeuf et al. (1993)). For the eastern coast, an overview of the ecology of the continental shelf of the Republic of Congo is available in Fontana (1981). The coastal system of the eastern part of the Gulf of Guinea, from Cameroon to Angola, today remains far from being fully documented and understood. Fisheries related issues represent the dominant topics of the regional syntheses or reviews published during the last two decades (Fontana, 1981; Cury and Roy, 1991; Le Loeuf et al., 1993; McGlade et al., 2002). The Longhurst and Pauly (1987) contribution is unique by providing a comprehensive synthesis of the ecology of the coastal ecosystems of the Gulf of Guinea that covers topics ranging from circulation, benthos and plankton production, species assemblages and dynamics of fish populations.

Because of the importance of this region to interdisciplinary coastal ocean science, we have prepared this brief introduction in the absence of a complete review so as to bring to the reader's attention some of the most important issues and concerns. The material here is most useful when considered together with the review focused on the physical oceanography of the region by Ajao and Houghton in Volume 11 of The Sea (1998) as well as with the other reviews mentioned in the previous paragraph.

2. Coastal upwelling in the Gulf of Guinea

The coastal upwellings of the Gulf of Guinea are the most well known oceanographic feature of the region. They have attracted great scientific interest during the past decades not only because of their importance in the functioning of the coastal ecosystems but also because of the unique driving mechanism, which involves a combination of both local and remote forcing factors (Picaut, 1984; Ajao and Houghton, 1998). The Gulf of Guinea coastal upwellings have a major impact on the productivity of the coastal domain and consequently on the coastal fisheries off the northern (Côte d'Ivoire and Ghana) and eastern (Gabon, Congo and northern Angola) coasts. Fish abundance as well as fish availability to the fisheries are affected by the seasonal and interannual variability of the upwelling process (Bakun, 1978; Binet, 1982; Cury and Roy, 1987; Mendelssohn and Cury, 1987; Koranteng, 1998). Data collected at coastal stations as well as a limited set of oceanographic cruises over the shelf provide most of the available scientific material to investigate the oceanography of the coastal domain. Detailed studies of the coastal upwelling variability along the northern coast have been made possible using temperature and salinity time series from the coastal stations (Arfi et al., 1991; Koranteng and Pezennec, 1998). Surface temperature data from satellite observations have recently been used to investigate environmental variability over the shelf as well as for looking at the broad spatial structure of the Gulf of Guinea coastal system (Hardman-Mountford and McGlade, 2002a, 2002b).

Coastal upwelling occurs seasonally along the northern coast (from the eastern part of Liberia to the western part of Nigeria) and along the eastern coast (off Gabon, Congo and Angola). As a result of the shoaling of the thermocline, Sea Surface Temperature (SST) can reach relatively low values (sometimes al low as 20°C along the northern coast) for an equatorial region, with a maximum annual amplitude reaching 10°C (Colin et al., 1993). A unique characteristic of the region is the occurrence of two upwelling seasons (Berrit, 1962). The major upwelling season occurs during the boreal summer from late June to August (Berrit, 1958; Morlière, 1970). The summer upwelling occurs as a series of events of variable intensity and duration ranging from a few days to several weeks (Berrit, 1962; Picaut and Verstraete, 1976; Colin et al., 1993). It is most developed from Côte d'Ivoire to Ghana and it reaches its maximum intensity in the vicinity of cape Three Points (Ghana) where the width of the shelf presents a local maximum. The minor upwelling that occurs during the boreal winter has received much less attention than the summer upwelling. The minor upwelling season is characterised by sporadic upwelling events that eventually lead to significant drops in SST lasting a few days. Along the eastern coast, the surface signature of the minor upwelling has been recorded as early as October (Gallardo, 1981). Along the northern coast, it occurs mainly in January and February (Morlière, 1970). SST data from coastal station shows that its intensity reaches a maximum in the vicinity of Cape Palmas (Côte d'Ivoire) and sharply decreases toward the east to become almost unnoticeable on the eastern part of Ghana (Arfi et al., 1991). Little is know about the mechanism responsible for the minor upwelling. The upward movement of the thermocline resulting from the intensification of the Guinea current in January and February has been mentioned as a potential contributor (Morlière, 1970; Binet and Servain, 1993). SST interannual variability during the minor upwelling off Côte d'Ivoire-Ghana has also been related to local wind forcing (Roy, 1995).

The most remarkable characteristic of the Gulf of Guinea boreal summer upwelling is the absence of correlation between wind stress and coastal temperature. This has attracted much scientific interest during the past decades (Ajao and Houghton, 1998). Along the Côte d'Ivoire and Ghana coasts, Ekman transport induced by south-west winds is directed offshore and should contribute to the coastal upwelling but all attempts to correlate the intensity and duration of the upwelling with the coastal winds have failed (Houghton, 1976). Bakun (1978) also pointed out that maximum values of offshore Ekman transport occur to the east and downstream (in reference to the Guinea current) of the location of the summer minimum temperature. The low values of the coastal wind and the absence of a marked seasonal cycle seem also to question the contribution of the local wind to the upwelling process (Verstraete et al., 1980) but Colin (1988), using data collected a few miles offshore with moored buoys, showed that the offshore wind is stronger than the wind measured at coastal stations and has a pronounced seasonal cycle. Ingham (1970) proposed the idea of a current induced upwelling along the north coast of the Gulf of Guinea. The Guinea Current, in geostrophic balance, is associated with an upward slope of the thermocline toward the coast. The acceleration in summer of the eastward flowing Guinea current increases the slope and thus could lead to a coastal upwelling. An attempt by Colin (1998) to test this hypothesis by correlating coastal current and temperature was not conclusive. Remote forcing was proposed as a key process contributing to the dynamic of the Côte d'Ivoire and Ghana summer upwelling (Moore et al., 1978). An increase of the easterly wind in the western equatorial Atlantic could create an internal Kelvin wave that would propagate eastward along the equator; when reaching the African continent, this wave would reflect as coastal Kelvin waves and Rossby waves. The coastal Kelvin waves are trapped along the coast and propagate poleward. The Rossby waves propagate westward on both sides of the equator. Picaut (1983) showed that the phase lag observed in the SST signal from several coastal stations along the coast is in agreement with the idea of a westward propagation of the upwelling signal; vertical propagation was also observed. Servain et al. (1982) found a correlation between the zonal wind stress in the western Atlantic and the SST in the Gulf of Guinea. However, results from the SEQUAL-FOCAL experiment showed that there is no apparent relationship between the onset of the equatorial upwelling and the onset of the coastal upwelling (Houghton and Colin, 1986). Gaps in the existing surface and subsurface measurement as well as the lack of data in the eastern part of the Gulf of Guinea preclude a definitive test of the various hypotheses (Ajao and Houghton, 1998). It seems most likely that a combination of processes contribute to the coastal upwelling in the Gulf of Guinea. One should also expect to find differences in the processes responsible for upwelling on the northern and the eastern coasts.

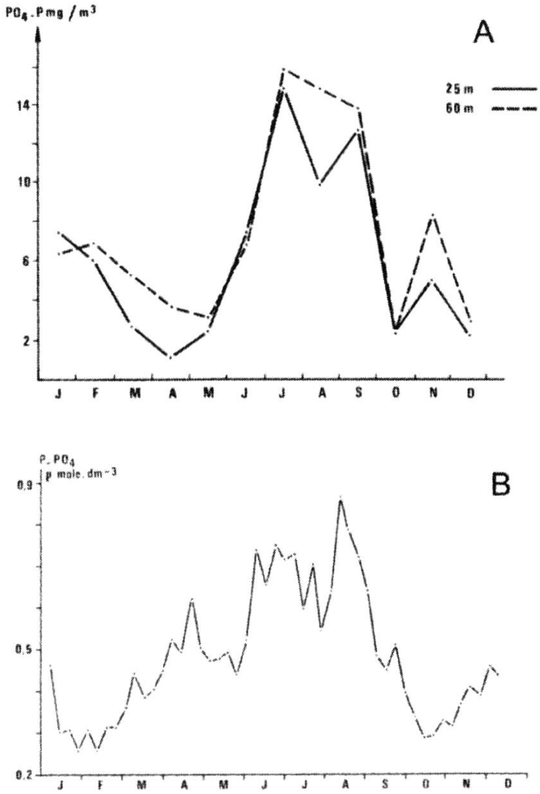

Figure 22.2 Seasonal cycle of phosphate in the surface layers of the coastal upwellings of the Gulf of Guinea . A)off Côte d'Ivoire: surface concentration over the isobath 25m and 60m (Reyssac, 1970). B) off Congo: concentration at 5m depth at the Pointe Noire coastal station (Dessier, 1981). Figure redrawn from Binet (1983).

Coastal upwelling and river discharges are the major sources of enrichment within the region. From Guinea to Liberia, productivity mostly relies upon river flow and land drainage (Longhurst, 1998; Hardman-Mountford et al., 2000). Further east, upwelling is the major contributor to the enrichment of the continental shelf off Côte d'Ivoire and Ghana while river discharges contribute to enhance productivity during the rainy season (Binet, 1983a). Nutrients concentration in the surface layers reach maximum values during the major upwelling season while a secondary maximum is often observed during the minor upwelling (Fig. 2). Off the eastern coast, the plume of the Congo river is a major feature of almost every satellite ocean colour image, however these low salinity waters seems to have a negative impact on productivity in the coastal domain off Congo and Gabon (Dessier, 1981; Binet, 1983b). The seasonality of the chlorophyll signal is pronounced with the highest values (up to 10 mg/m^3) recorded during the austral winter upwelling season (Binet, 1983a). Primary production in the coastal area is estimated to be in the range of 150 gC.m^{-2}.y^{-1}, well below the values observed in the major upwelling systems (Binet, 1983a). The impact of the minor upwelling season on the enrichment process is mainly restricted to the inshore area and lasts for only a few weeks. The species composition of the phytoplankton changes seasonally in response to the occurrence of the upwelling. Dinoflagelates are dominant during the warm seasons while diatoms become the major species during the upwelling seasons (Binet, 1983b). Zooplankton biomass peaks after the two upwelling seasons (Fig. 3). Vertical migration of zooplankton contributes to maintain zooplankton in the main upwelling cells off Côte d'Ivoire-Ghana by making use of the strongly stratified vertical circulation structure (Binet, 1979).

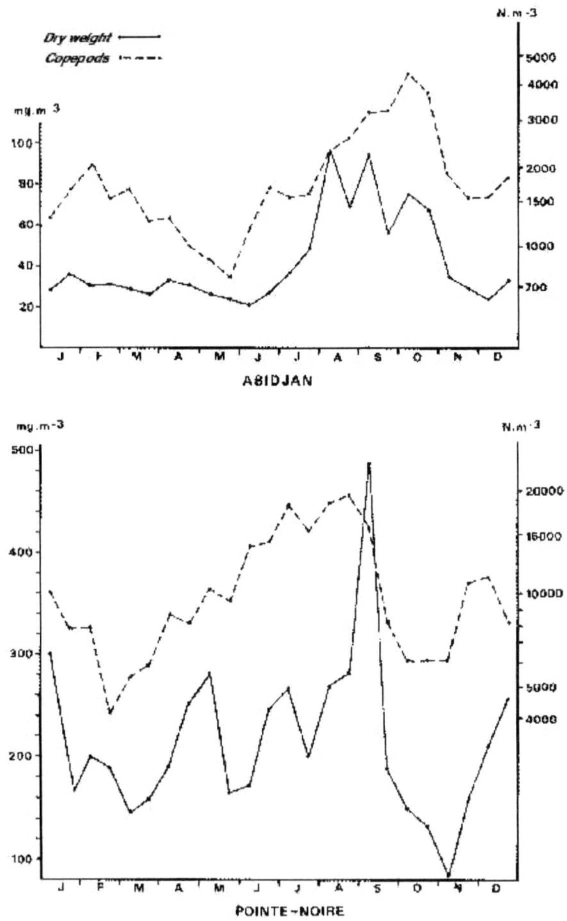

Figure 22.3 Seasonal cycle of zooplankton in the coastal upwelling of the Gulf of Guinea (solid line = dry weight, dashed line = total copepods (N.m^{-3}). Upper panel: at the Abidjan coastal station (Côte d'Ivoire) ; lower panel: at the Pointe Noire coastal station (Congo). From Binet (1983).

3. Coastal Upwelling Variability

Monitoring of the coastal upwelling of Côte d'Ivoire and Ghana since the late 1960s has been possible through a set of coastal stations where temperature and salinity have been recorded on a regular basis (Arfi et al., 1991; Koranteng and Pezennec; 1998, Koranteng and McGlade, 2001). One of the major changes that have been noticed is an intensification of the minor upwelling season, starting in the mid 1970s (Fig. 4). This intensification results in cooler surface temperature being recorded at the coastal stations off Ghana and Côte d'Ivoire in January and February (Pezennec and Bard, 1992; Koranteng and McGlade, 2001). It has also been recorded in SST time series extracted from the COADS dataset (Cury and Roy, 2002). An intensification of the upwelling favourable winds during the first quarter as well as a strengthening of the Guinea Current have been proposed as mechanisms to explain the observed changes (Binet and Servain, 1993; Cury and Roy, 2002). Changes in SST during the major upwelling season (July to September) appear to be opposite to those observed earlier in the year (Koranteng and Pezennec, 1998). A slight shift in the seasonality of the major upwelling has also been observed: while the maximum cooling remains in August, a sharp intensification of the cooling has been observed in July during the 1990s (Cury and Roy, 2002).

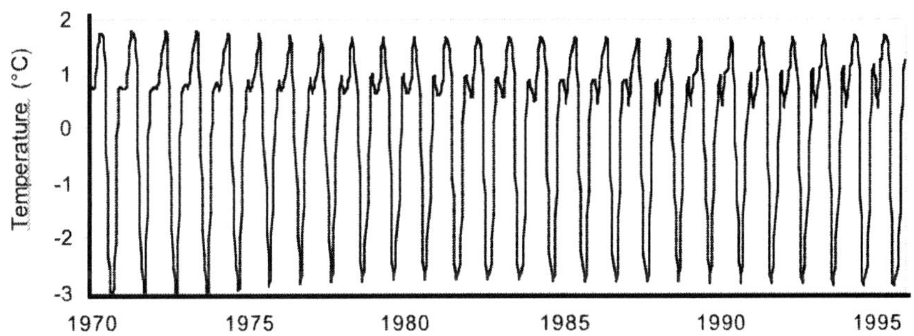

Figure 22.4 Seasonal component of SST from 1963 to 1995 given a decomposition of the monthly SST time series off Côte d'Ivoire and Ghana (from Cury and Roy, 2002).

Low frequency variability has been documented in the Ghana-Côte d'Ivoire coastal region. From Koranteng and McGlade (2001) analysis of coastal temperature and salinity time-series, it is possible to identify two major turning points from the late 1960s to the early 1990s, namely 1977 and 1989. The first one marked the peak of a cooling period characterised by low temperature and high salinity water. The second one marked the end of a sustained warming period. As expected, there are similarities between the timing of those turning points in the coastal regions of the Gulf of Guinea and major climatic events in the Pacific Ocean (Enfield and Mayers, 1997).

4. Fish populations and fisheries.

Fisheries landings off Ghana and Côte d'Ivoire have been expanding during the last four decades: the total marine fish catch has been increasing from less than fifty thousand tonnes in the fifties to around four hundred thousand tonnes in the mid-nineties (Cury and Roy, 2002). Data from FAO show that from 1950 to 2001, total catches reached a maximum in 1996 with 412 444 t. Ghanaian catches represent the dominant component of the catch with a contribution of more than 70% during the last decade. Pelagic species catches are increasing since the 1950s and the total catch fluctuated around 300 000 tonnes in the late 1990s (Fig. 5). In contrast landings of the demersal fisheries in Côte d'Ivoire and Ghana increased up to the early-1980s when they reached almost 70000t. After the mid 1980s, the total landings of demersal species show a negative trend (Fig. 5).

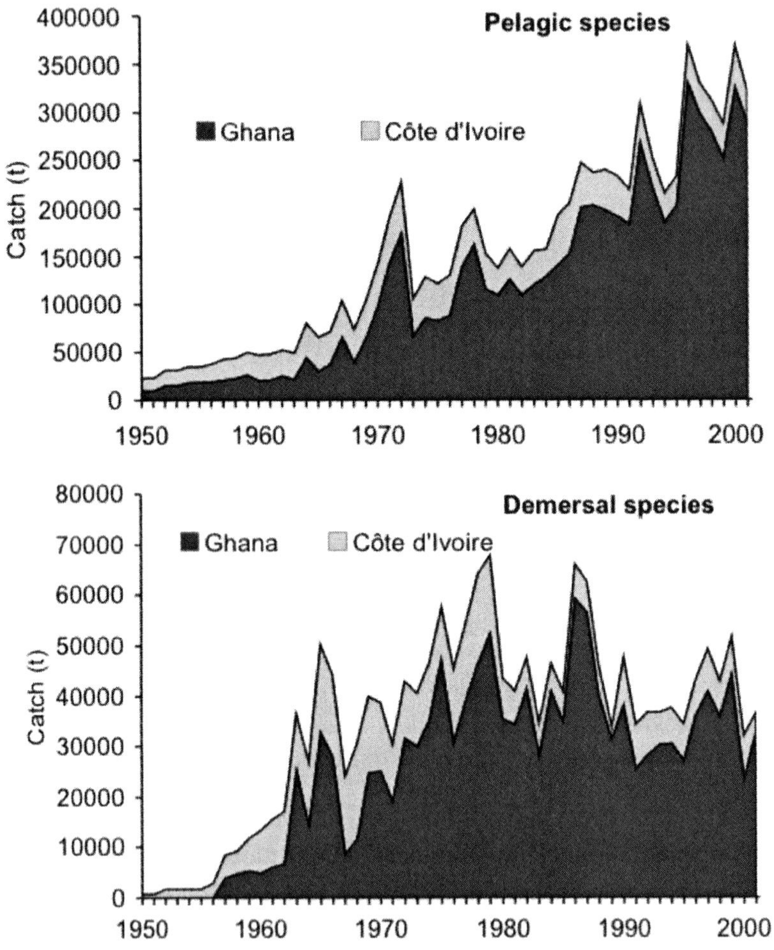

Figure 22.5 Total pelagics (upper) and demersal (lower) fish landings in Ghana and Côte d'Ivoire from 1950 to 2001 (source FAO).

The Guinean Trawling Survey (GTS) performed during the early 1960's provided detailed information on demersal resources in the Gulf of Guinea (Williams, 1966). Using the GTS data, Longhurst (1969) presents an in depth analysis of the fish communities identified in the region. A recent analysis by Koranteng (1998, 2002) using data off Ghana shows that minor changes occurred in the demersal species assemblage since the 1960s. This analysis also shows that the dynamics of the assemblage, including the seasonal movements, are influenced by the physical and chemical properties of the water masses. This indicated that interannual and decadal environmental fluctuations could affect the dynamics of species assemblage over the continental shelf (Koranteng, 2002). The synchrony observed between total demersal production off Côte d'Ivoire and SST supports this assumption (Cury and Roy, 2002). Cold temperature in the 1970s seems to have favoured the abundance demersal fish communities.

The massive outbreak of triggerfish (*Balistes capriscus*) in the Gulf of Guinea during the 1970s has been described as one of the most phenomenal population outbursts in the history of fish population dynamics (Fig. 6) (Bakun, 1995). Before the 1970s, *Balistes capriscus* was rather rare in the eastern Atlantic, but by 1981 it was estimated that it represented up to 83% of the pelagic biomass off Côte d'Ivoire and Ghana (Ansa-Emmin, 1979; Koranteng, 1998). This increase in abundance was associated with a wide geographical spreading that reached the southern part of the Canary Current, south of 20°N (Gulland and Garcia, 1984). In the latter part of the 1980s the population declined and triggerfish no longer represents a dominant component of the catches. While several authors believe that the fisheries induced reduction of the sparid communities facilitated the outburst of the Balistes (Gulland and Garcia, 1984), environmental changes (reduced freshwater inputs to the region and increased salinity of continental shelf waters) have also been proposed as a potential trigger for the outburst of *Balistes capriscus* (Caverivière 1991). Bakun (Vol 13 Chapter 24) provides further details on the ecological implication of the triggerfish outbreak.

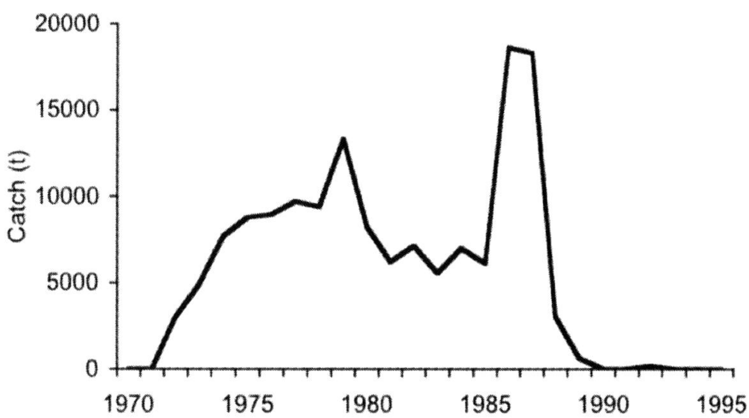

Figure 22.6 Total landings of trigger fish (*Balistes capriscus*) in the Gulf of Guinea from 1970 to 1995 (source FAO).

Another remarkable population change is the collapse of *Sardinella aurita* in the early 1970s that followed unusually high catches by the Côte d'Ivoire and Ghana fisheries in 1972 (Binet, 1982; Koranteng, 1991). Overfishing and adverse environmental conditions were thought to have initiated the collapse, but by the mid-1980's the stock has recovered and unprecedented sustained landings, far greater than the 1972 catches, were recorded from the mid-1980s up to early 1990s. The recovery of *S. aurita* fisheries was associated with important changes in the fishing areas and seasons. While in the earlier period, the bulk of the catch occurred off Ghana during the major upwelling season, sustained catches were recorded all year round over the Côte d'Ivoire shelf during the 1990s. The intensification of the minor upwelling seemed to have played a major role in the recovery of *Sardinella aurita* after the 1973 collapse (Pezennec and Bard, 1992).

5. Conclusion

The diversity of the coastal regions of the Gulf of Guinea is large and among those, the coastal upwelling systems of Côte d'Ivoire and Ghana and the associated fisheries have received much attention during the last two decades. Most of the research published in the recent years is based on historical time-series that have been rescued and analysed (Le Loeuff et al., 1993; McGlade et al., 2002). Several hypotheses have been proposed to explain the functioning of the systems and its variability but a full understanding of the processes involved remains elusive. While the analysis of recent remote sensing data corroborated previous observations (Hardman-Mountford and McGlade, 2002a, 2003), there is an urgent need for new in situ data to be collected in a coherent way at the regional scale. One needs also to stress the importance of large scale ocean forcing and its influence on the coastal environment in the Gulf of Guinea.

Bibliography

Ajao E.A. and R. W. Houghton. 1998. Coastal ocean of equatorial West Africa from 10°N to 10°S. Coastal segment (17,E). In: Global Coastal Ocean (eds. A.R. Robinson and K.H. Brink). The Sea, Vol. 11, John Wiley & Sons, NY, p605–631.

Anon. 1966. Proceedings of the Symposium on the Oceanography and Fisheries of the tropical Atlantic. UNESCO, Paris. 430p.

Arfi R., O. Pezennec, S. Cissoko and M. Mensah. 1991. Variations spatiales et temporelles de la résurgence ivoiro-ghanéenne. In : Variabilité, instabilité et changement dans les pêcheries ouest africaines, P. Cury et C. Roy eds. ORSTOM, Paris, p162–172.

Ansa-Emmin, M.1979. Occurrence of the Triggerfish, *Balistes capriscus* (Gmel), on the continental shelf of Ghana. In: report of the special working group on the evaluation of demersal stock of the Ivory Coast-Zaire sector, CECAF/ECAF SERIES/79/14(EN) FAO, Rome, p20–27.

Bakun, A. 1978. Guinea current upwelling. *Nature,* 271: 147–150.

Bakun, A. 1995. Global climate variations and potential impacts on the Gulf of Guinea sardinella fishery. In : Dynamique et usage des ressources en sardinelles de l'upwelling côtier du Ghana et de la Côte d'Ivoire. F.X. Bard et K.A. Koranteng eds., ORSTOM éditions Paris. p60–84.

Berrit, G.R. 1958. Les saisons marines à Pointe-Noire. *Bull. Inf. C.O.E.C.,* 6, 335–360.

Berrit, G.R. 1962. Contribution à la connaissance des variations saisonnières dans le golfe de Guinée. Observations de surface le long des lignes de navigation. II Etude régionale. Cah. Océanogr. COEC 14(9), 633–643.

Binet D. 1979. Le zooplancton du plateau continental ivoirien. Essai de synthèse écologique. Oceanologica Acta, 2,4, p397–410.

Binet, D. 1982. Influence des variations climatiques sur la pêcherie des *S. aurita* ivoiro-ghanéennes: relation sécheresse-surpêche. *Oceanologica Acta,* 5, 4, p443–452.

Binet, D. 1983a. Phytoplancton et production primaire des régions côtières à upwelling saisonniers dans le Golfe de Guinée. *Océanographie tropicale,* 18, 2, p331–335.

Binet D. 1983b. Zooplancton des régions côtières à upwelling saisonniers dans le Golfe de Guinée. *Océanographie tropicale,* 18, 2, p357–380.

Binet D. and J. Servain, 1993 Have the recent hydrological changes in the northern Gulf of Guinea induced the *Sardinella aurita* outburst. *Oceanologica Acta,* 16,3, p247–260.

Caverivière A. 1991. L'explosion démographique du baliste (*Balistes carolinensis*) en Afrique de l'Ouest et son évolution en relation avec les tendances climatiques. In Pêcheries ouest-africaines: variabilité, instabilité et changement. Cury P. and C. Roy (eds.). Orstom Editions, Paris, p354–367.

Colin, C. 1988. Coastal upwelling events in front of the Ivory Coast during the FOCAL program. *Oceanologica Acta,* 11, 125–138.

Colin C., Y. Gallardo, R. Chuchla and S. Cissoko. 1993. Environnements climatique et océanographique sur le plateau continental de Côte d'Ivoire. In : Environnement et resources aquatiques de Côte d'Ivoire. Le Loeuff P., Marchal E. and J.B.A. Kothias, (eds.). Tome 1: le milieu marin. ORSTOM, Paris. 75–110.

Cury, P. and C. Roy, 1987. Upwelling et pêche des espèces pélagiques côtières de Côte-d'Ivoire : une approche globale. *Oceanologica Acta.* 10, 3, 347–357.

Cury, P. and C. Roy, (eds). 1991. Pêcheries Ouest-africaines: variabilité, instabilité et changement. Orstom éditions, Paris, 525p.

Cury P. and C. Roy. 2002. Environmental forcing and fisheries resources in Côte d'Ivoire and Ghana: did something happen ? In: The Gulf of Guinea Large Marine Ecosystem. J. McGlade, P. Cury, K. Koranteng and N. Hardman-Mountford (Eds.). Large Marine Ecosystems series, Elsevier Science B.V., 241–260.

Dessier, A. 1981. La production planctonique: phytoplancton et zooplancton. *In* : Milieu marin et ressources halieutiques de la République Populaire du Congo, Fontana A. Ed., *Trav. et Doc. de l'ORSTOM.* Paris, 138, 75–150.

Enfield, D. B. and D. A. Mayer, 1997: Tropical Atlantic SST variability and its relation to El Niño-Southern Oscillation. *J. Geophys. Res.,* 102, 929–945.

Fontana, A (ed). 1981. Milieu marin et ressources halieutiques de la République Populaire du Congo. Trav. Doc. ORSTOM 138, éditions de l'ORSTOM, Paris, 339pp.

Gallardo, Y. 1981. Océanographie physique. *In* : Milieu marin et ressources halieutiques de la République Populaire du Congo, Fontana A. Ed., *Trav. et Doc. ORSTOM.* Paris, 138, 47–73

Gulland J.A. & S. Garcia. 1984. Observed patterns in multispecies fisheries. In : Exploitation of marine communities, R.M. May (ed.). Dahlem Konferenzen. Berlin, Heidelberg, New-York, Tokyo : Springler- Verlag: 155–190.

Hardman-Mountford N.J., K.A. Koranteng and A.R.G. Price. 2000. The Gulf of Guinea large marine ecosystem. In: Seas at the Millennium: an environmental evaluation, Volume I: Regional Chapters: Europe, The Americas and West Africa. (ed. C. Shepperd), Elsevier Science Ltd; Oxford, England , p773–796.

Hardman-Mountford N. J. and J. M. McGlade. 2002a. Variability of physical environmental processes in the Gulf of Guinea and implications for fisheries recruitment. An investigation using remotely sensed SST. In: The Gulf of Guinea Large Marine Ecosystem. J. McGlade, P. Cury, K. Koranteng and N. Hardman-Mountford (Eds.). Large Marine Ecosystems series, Elsevier Science B.V., 49–66.

Hardman-Mountford N. J. and J. M. McGlade. 2002b. Defining ecosystem structure from natural variability: application of principal components analysis to remote sened SST. In: The Gulf of Guinea Large Marine Ecosystem. J. McGlade, P. Cury, K. Koranteng and N. Hardman-Mountford (Eds.). Large Marine Ecosystems series, Elsevier Science B.V., 67–82.

Hardman-Mountford N. J. and J. M. McGlade. 2003.Seasonal and interannual variability of oceanographic processes in the Gulf of Guinea: an investigation using AVHRR sea surface temperature data. *International Journal of Remote sensing.* 24, (16): 3247–3268.

Houghton, R. W. 1976. Circulation and hydrographic structure over the Ghana continental shelf during the 1974 upwelling. *J. Phys. Oceanogr.,* 6, 909–924.

Houghton R.W. and C. Colin, 1986. Thermal structures along 4°W in teh Gulf of Guinea during teh 1983–1984. *J. Geophys. Res,* 91, 11727–11739.

Ingham, M. C. 1970. Coastal upwelling in the northwestern of Gulf of Guinea. *Bull. Mar. Sci.,* 20, 2–34.

Koranteng, K.A. 1991. Some aspects of the Sardinella fishery in Ghana. In : Ph. Cury and C. Roy eds. Variabilité, instabilité et changement dans les pêcheries ouest africaines, Editions ORSTOM, Paris, p269–277.

Koranteng K. A. 1998. The impacts of environmental forcing on the dynamics of demersal fishery resources of Ghana. PhD Thesis, University of Warwick, 377p.

Koranteng, K.A. 2002. Fish species assemblage on the continental shelf and upper slope off Ghana. In: The Gulf of Guinea Large Marine Ecosystem. J. McGlade, P. Cury, K. Koranteng and N. Hardman-Mountford (Eds.). Large Marine Ecosystems series, Elsevier Science B.V., 173–188.

Koranteng K. A. and O. Pezennec. 1998. Variability and trends in some environmental time series along the Ivoirian and the Ghanaian coasts. In : Durand M.H., P. Cury, R. Mendelssohn, C. Roy, A. Bakun et D. Pauly (eds). Global versus local changes in upwelling systems. Editions ORSTOM Paris. p167–178.

Koranteng K. A. and J.M. McGlade. 2001. Climatic trends in continental shelf waters off Ghana and in the Gulf of Guinea 1963–1992. *Oceanologica Acta,* 24, 2, p187–198

Le Loeuff P., Marchal E. and J.B.A. Kothias, (eds.). 1993. Environnement et resources aquatiques de Côte d'Ivoire. Tome 1: le milieu marin. ORSTOM, Paris, 589 p.

Longhurst A. R. 1969. Species assemblages in tropical (Atlantic) demersal fisheries. *FAO Fisheries Rep.* 51, 147–167.

Longhurst, A.R. 1998. Ecological Geography of the Sea. Academic Press, San Diego. 398p.

Longhurst A.R. and D. Pauly. 1987. Ecology of tropical oceans. Academic Press. San Diego. 407p.

Mahé, G. 1998. Freshwater yields to the Atlantic Ocean: local and regional variations from Sénégal to Angola. In: Durand M-H, P. Cury, R. Mendelssohn, C. Roy, A. Bakun et D. Pauly (eds). Global versus local changes in upwelling systems. Editions ORSTOM Paris. P127–138

McGlade J., P. Cury, K. Koranteng and N. Hardman-Mountford (Eds.). 2002. The Gulf of Guinea Large Marine Ecosystem. Environmental forcing and sustainable development of marine resources. Large Marine Ecosystems series, Elsevier Science B.V., 392p.

Mendelssohn, R. and P. Cury. 1987. Fluctuations of a fortnightly abundance index of the Ivoirian coastal pelagic species and associated environmental conditions. *Can. J. Fish. Aquat. Sci.* 44: 408–428.

Morlière A., 1970. Les saisons marines devant Abidjan. *Doc. Sci. Centre Rech. Oceanogr.* Abidjan, 1, 1–15.

Moore D. W., P. Hisard, J.P. McCreary, J. Merle, J.J. O'Brien, J. Picaut, J.M. Verstraete et C. Wunsch, 1978. Equatorial adjustment in the eastern Atlantic ocean. *Geophys. Res. Lett.,* 5, 637–640.

Pezennec O. and F.X. Bard, 1992. Importance écologique de la petite saison d'upwelling ivoiro-ghnanéenne et changements dans la pêcherie de *Sardinella aurita. Aquating Living Ressources,* 5, 249–259.

Picaut, J. and J.M. Verstraete. 1976. Mise en évidence d'une onde de 40 - 50 jours de période sur les côtes du Golfe de Guinée. *Cah. O.R.S.T.O.M. sér. Océanogr.* 14 (1): 3–14.

Picaut, J. 1983. Propagation of the seasonal upwelling in the eastern equatorial Atlantic. *J. Phys. Oceanogr.,* 13, 1, 18–37.

Picaut, J. 1984. On the dynamics of thermal variations in the Gulf of Guinea. *Océanographie tropicale.* 19, 2, p127–154.

Reyssac J. 1970. Phytoplancton et production primaire au large de la Côte d'Ivoire. *Bull. I.F.A.N.,* 32, 4, 869–981

Roy C. 1995. The Cote d'Ivoire and Ghana coastal upwellings : dynamics and Change. In : Dynamiqueet usage des ressources en sardinelles de l'upwelling côtier du Ghana et de la Côte d'Ivoire. F.X. Bard et K.A. Koranteng (eds.), ORSTOM éditions, Paris. p346–361.

Servain, J., J. Picaut and J. Merle. 1982. Evidence of remote forcing in the equatorial Atlantic ocean. *J. Phys Oceanogr.,* 12, 457–463.

Verstraete J.M., J. Picaut and A. Morlière, 1980. Atmospheric and tidal observations along the shelf of the Guinea Gulf. *Deep Sea Res.,* 26(suppl. II), 343–356.

Williams, F. (1966) Review of the principal results of the GTS. Proc. Symp. Oceanog. Fish. Trop. Atlantic. UNESCO, Paris 139–146.

Chapter 23. OCEANOGRAPHY AND FISHERIES OF THE CANARY CURRENT/IBERIAN REGION OF THE EASTERN NORTH ATLANTIC (18a,E)

JAVIER ARÍSTEGUI

University of Las Palmas de Gran Canaria

XOSÉ A. ÁLVAREZ-SALGADO

CSIC, Instituto de Investigacións Mariñas, Vigo

ERIC D. BARTON

CSIC, Instituto de Investigacións Mariñas, Vigo

FRANCISCO G. FIGUEIRAS

CSIC, Instituto de Investigacións Mariñas, Vigo

SANTIAGO HERNÁNDEZ-LEÓN

University of Las Palmas de Gran Canaria

CLAUDE ROY

Centre IRD de Bretagne, Plouzané

ANTONIO M.P. SANTOS

Instituto de Investigaçao das Pescas e do Mar, Lisboa

Contents

The Sea, Volume 14, edited by Allan R. Robinson and Kenneth H. Brink
ISBN 0–674–01527–4

1. Introduction

The eastern boundary of the North Atlantic subtropical gyre extends from the northern tip of the Iberian Peninsula at 43°N to south of Senegal at about 10°N, approximately the range of displacement of the Trade wind band (Fig. 23.1). It is one of the four major eastern boundary upwelling systems of the world ocean, and thus an area of intensive fisheries activity. The meridional shift of the Trade wind system causes seasonal upwelling in the extremes of the band, while in the central region upwelling is relatively continuous all year round (Wooster et al., 1976). Superimposed on the seasonal variation, short-term variability in wind direction and intensity may induce or suppress upwelling, affecting the dynamics of the ecosystem. At long-term scale, decadal fluctuations in fisheries landings - particularly north of 20°N- have been related to environmental changes due to the North Atlantic Oscillation (NAO) (Borges et al., 2003). In the southern part of the region, the influence of the El Niño Southern Oscillation (ENSO) may be also responsible of some of the recorded inter-annual variability in fisheries landings (Roy and Reason, 2001).

The upwelling region is separated into two distinct areas - the Iberian coast and the Northwest African coast- with apparently little continuity in the flow between them. This is caused by the interruption of the coastline at the Strait of Gibraltar, which allows the exchange of water between the Mediterranean Sea and the Atlantic Ocean. The continental shelf of the whole region is the most extensive of any eastern boundary current and persistent hydrographical features are associated with the topography of the shelf. Several submarine canyons are distributed along the Iberian margin, acting as sites for coastal sediment deposition, the Nazaré canyon being the most significant depocenter. At the northern part of the Iberian coast, the Rías Baixas represent a singular ecosystem, which transforms the 3D-variability of continental shelf waters into 2D-variability, causing an amplification of the biogeochemical signals. Large filaments of coastal upwelled water stretch offshore from the numerous capes and promontories, exchanging water and biological properties with the ocean boundary. The exchange is particularly noticeable along the giant filaments of Cape Guir and Cape Blanc, which stretch up to several hundred kilometres into the open ocean, transporting rich-organic matter waters into the impoverished oligotrophic waters of the subtropical gyre. Another singularity of this eastern boundary system is the presence of the Canary Archipelago, close to the NW African coast, which interrupts the main flow of the Canary Current and introduces large mesoscale variability, mainly in the form of vortex streets downstream of the islands (Arístegui et al., 1994). Island eddies and upwelling filaments interact to exchange water properties, acting as an efficient route for transporting organic matter to the open ocean.

Research effort has been unevenly distributed through the whole region. The western and northern coasts of the Iberian Peninsula have been extensively studied from the hydrographic and dynamic point of view (Barton et al., 1998). However, most of the field studies on carbon and nutrient biogeochemistry in the Iberian margin concentrate in the Galician and Cantabrian coast (42°–44°N, 5–10 °W), at the boundary between the temperate and subpolar regimes of the North Atlantic. There, intensive research was supported during the past decade by the European scientific community in the frame of several research projects, like "The

Control of Phytoplankton Dominance" (Figueiras et al., 1994; Moncoiffé et al., 2000), MORENA (Multidisciplinary Oceanographic Research in the Eastern Boundary of the North Atlantic; Fiuza et al., 1998; Pérez et al., 1999; 2001), or OMEX II (Ocean Margin Exchange; Joint and Wassmann, 2001; Huthnance et al., 2002; Joint et al., 2002; van Weering and McCave, 2002).

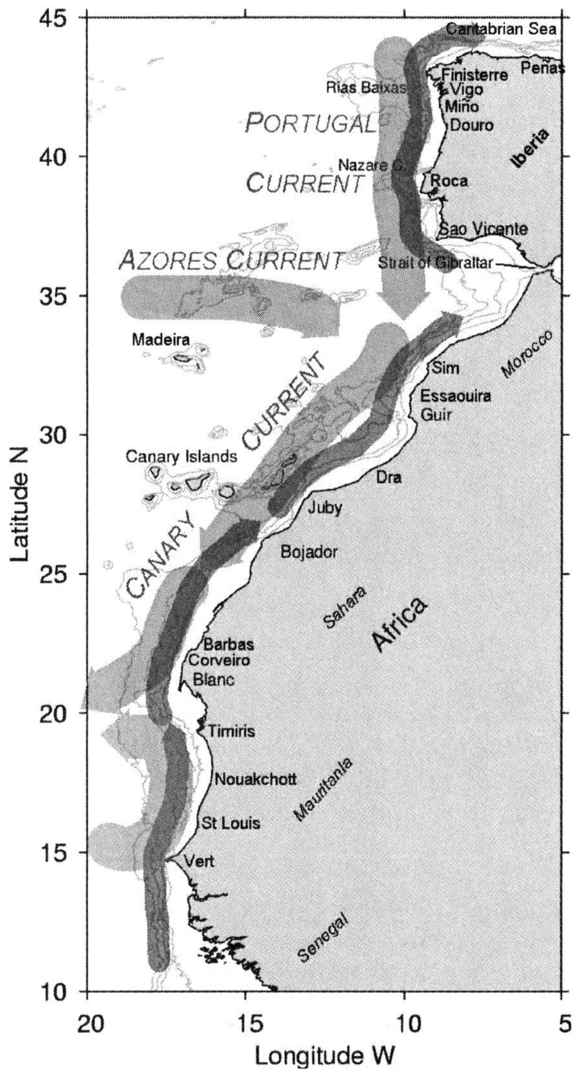

Figure 23.1 Fig. 23.1. Chart of the Eastern North Atlantic Iberian and NW African margin indicating locations referred to in the text. The 200, 1000, 2000 and 3000 m isobaths are shown. A schematic summer current regime is shown by broad, light arrows representing surface flows and narrower, dark arrows representing the poleward slope undercurrent. Seasonal variation of the currents is discussed in the text.

The upwelling region of NW Africa was intensively studied in the 1970s, during the International Decade of Ocean Exploration (IDOE), under the international programme of Cooperative Investigation of the Northern Part of Eastern Central Atlantic (CINECA; Hempel, 1982), but almost ignored during the 1980s. Some interdisciplinary studies, focused on the export of organic matter from the NW Africa upwelling into the open ocean, were however carried out during the last decade in the Canary Current region. These were the cases of the French EUMELI (Eutrophic, Mesotrophic, Oligotrophic) project (Morel, 1996), and the European Canaries-Coastal Transition Zone (Barton et al., 1998) and CANIGO (Canary Islands, Azores, Gibraltar Observations; Parrilla et al., 2002) projects.

In this chapter, we present a review bringing together for the first time up-to-date knowledge on inter-disciplinary aspects of the oceanography and fisheries of the eastern boundary of the North Atlantic. Our synthesis provides a global view of the trophic functioning of these coastal ecosystems. Aspects covered range from nutrient dynamics, through production and respiration, to fish populations and fisheries, all of which are affected by processes that occur on distances from mesoscale to large scale and vary over periods from days to decades.

2. The coastal upwelling system in Iberia and NW Africa

2.1 Temporal and spatial variability

The surface circulation off Iberia (Fig. 23.1), far away from the main currents of the North Atlantic Ocean, has been systematised by Pelíz and Fiúza (1999). They coined it the 'Portugal Current System' (although Iberian would be more inclusive) because of similarities with the 'California Current System'. The oceanic side of the Iberian basin is occupied by the weak 'Portugal Current' (PC), which flows southwards year round from 45°–50°N and 10°–20°W, at the interface between the areas of influence of the North Atlantic Current and the Azores Current (Krauss, 1986). The circulation pattern is more complex at the ocean margin, showing a marked seasonal variability defined by the coastal wind regime of the area either in the western (Huthnance et al., 2002) and northern coast (van Aken, 2002). During spring and summer (from March–April to September–October) north-easterly winds (Fig. 23.2a) predominate in the Iberian basin (Wooster et al., 1976; Bakun and Nelson, 1991), producing the southward flowing 'Portugal Coastal Current' at the surface (<100m) and the northward flowing 'Portugal Coastal Under Current' (PCUC) at the slope. In contrast, during the rest of the year, south westerly winds are predominant, provoking a reversal of the surface circulation to form the 'Portugal Coastal Counter Current (PCCC)' which flows northward from the surface to 1500m depth, including the propagation of the Mediterranean Overflow Water along the western and northern Iberian slope. The existence of a poleward surface flow during the autumn and winter months in the Iberian slope was described first by Wooster et al. (1976) and subsequently found by Frouin et al. (1990) and Haynes and Barton (1990) in the western Iberian coast and by Pingree and Le Cann (1990) in the Cantabrian coast.

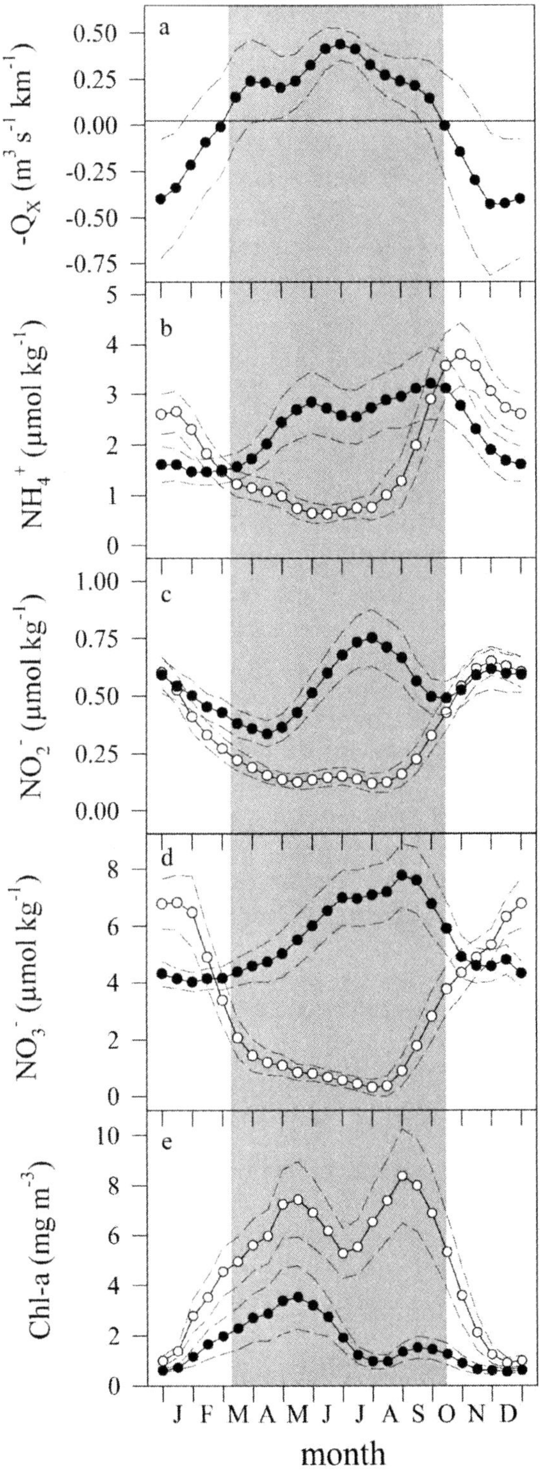

Figure 23.2 Seasonal cycles of (a) offshore Ekman transport ($-Q_X$, $m^3 s^{-1} km^{-1}$) in a 2°×2° geostrophic cell centred at 43°N 11°W; (b) surface and bottom ammonium (NH_4^+, µmol kg^{-1}); (c) nitrite (NO_2^-, µmol kg^{-1}); (d) nitrate (NO_3^-, µmol kg^{-1}); and (e) Chlorophyll (mg m^{-3}) in the coastal upwelling ecosystem of the Ría de Vigo. Shaded area: spring-summer upwelling season. Black circles in panel a: 1987–96 fortnightly average; black circles in panels b–e: 1987–96 fortnightly bottom average; white circles in panels b–e: 1987–96 fortnightly surface average; dashed lines: average ± std.

The Portugal Coastal Current is associated with seasonal coastal upwelling at the Iberian margin, and the export of coastal surface waters to the open ocean, especially at the recurrent upwelling centres and filaments along the western Iberian coast (e.g. Fiúza, 1983; McClain et al., 1986; Sousa and Bricaud, 1992; Haynes et al., 1993; Pelíz and Fiúza, 1999). On the contrary, the PCCC is associated with downwelling on the coast, and the piling of coastal waters onto the Portuguese (Pelíz and Fiúza, 1999), Galician (e.g Castro et al., 1997; Álvarez–Salgado et al., 2003) and Cantabrian (e.g Bode et al., 1990; Fernández et al., 1993) coasts. Anticyclonic eddies, named SWODDIES ('Slope Water Oceanic Eddies') by Pingree and Le Cann (1992), detach from the PCCC at certain points along the Cantabrian coast and inject slope water into the southern Bay of Biscay, affecting its kinematics (Pingree, 1994) and biogeochemistry.

The extension and intensity of the seasonal upwelling and downwelling favourable periods varies strongly from year to year (Fig. 23.2a), describing a decadal cycle linked to the NAO. Whereas the onset of the upwelling season can occur within two months, from the beginning of April to the end of May, its cessation falls within a period of just one month, from middle September to middle October. However, this seasonal cycle only explains about 10% of the variability of the wind regime off NW Spain, whereas >70% of the variability arises from periods <30 days (Álvarez–Salgado et al., 2002; 2003). In fact, the upwelling season off NW Spain is described well as a succession of upwelling/relaxation events of period 1–3 wk (Blanton et al., 1987; Silva, 1992; Álvarez–Salgado et al., 1993). It is also remarkable that due to the different orientation of the western and northern coasts (Fig. 23.1), northerly winds produce upwelling off the western coast whereas easterly winds do it off the northern coast. The orientation of the coast changes abruptly north of Cape Finisterre, in such a way that both northerly and easterly winds are upwelling favourable there (McClain et al., 1986; Torres et al., 2003). Similar considerations apply to Cape Sao Vicente and the coast of southern Portugal (Fiúza 1983; Relvas and Barton, 2002).

The coastal upwelling region from Gibraltar to Cape Blanc is maintained by the presence of favourable northeasterly winds throughout the year, although winds and upwelling are more intense during the summer months. Between Cape Blanc and Cape Vert, the upwelling has a marked seasonal periodicity, reaching its peak of intensity during winter. In contrast with the Iberian coast, the NW African coast is largely influenced by the general circulation of the North Atlantic subtropical gyre, particularly by its eastern branch, termed the "Canary Current". The Canary Current flows equatorward while interacting with the coastal upwelling waters. It detaches from the coast near Cape Blanc (21° N), flowing westward at the latitude of Cape Vert (15° N). South of Cape Blanc a large permanent cyclonic recirculation develops as a consequence of the offshore displacement of the Canary Current (Fig. 23.1). Nearshore in winter, a narrow equatorward flow develops over the shelf in response to upwelling forced by the southward migration of the trade wind band (e.g. Hughes and Barton, 1974).

Several studies (e.g. Stramma and Siedler, 1988; Siedler and Onken, 1996) have described the seasonal variability of the Canary Current, confirming the existence of water inflow from the open ocean into the coastal upwelling region north of the Canary Islands. The coupling between the coastal and open ocean waters at the Canary- Coastal Transition Zone (Canary-CTZ) region, has been addressed in more recent works (Pelegrí et al., 1997; Barton, 1998; Barton et al., 1998; Hernández Guerra et al., 2001; 2002; Pelegrí et al., 2005), in which it is shown how the water recirculates south along the continental slope, where quasi-permanent filaments stretch offshore and exchange water properties with island eddies (Arístegui et al., 1997; Barton et al., 1998).

Flow reversals in the main flow have been observed close to the upwelling-Canary CTZ, during late-fall and winter (Navarro-Pérez and Barton 2001; Hernández Guerra et al. 2002). These flow diversions, probably caused by a weakening of the trade winds south of Cape Guir (Pelegrí et al. 2005), allow the presence of a northward flow from Cape Blanc to Cape Juby, and consequently an offshore spread of organic matter produced in upwelling waters near the Canary Islands region (Arístegui et al. 1997). Pelegrí et al. (2005) suggested that the Canary region is characterised by the presence of two cells transporting upwelled waters into the open ocean. The first one would be the standard vertical cell, present in all upwelling systems, with Ekman offshore transport responding to the wind variations. The second one would be the horizontal circulation cell originated by the impinging of open ocean water north of Cape Guir, which is closed by the offshore export of water through several upwelling filaments and the flow diversion at Cape Guir. The joint action of both cells would cause this upwelling region to be a key region for export of organic matter and nutrients to the open ocean.

2.2. *Water masses and nutrients*

The distribution of water masses in the region has been summarized by Barton (1998). Most of the region, from Cape Finisterre to Cape Blanc, is dominated by North Atlantic Central Water (NACW), responsible for the fertility of the coast during upwelling processes, although there is considerable variation in this water mass with latitude. Eastern North Atlantic Central Water (ENACW) of subtropical (13°C–15°C) and subpolar (11°C–13°C) origin (Fiúza, 1984; Ríos et al., 1992; Fiúza et al., 1998) extends from 50–100m (marked by a salinity maximum, >35.9) to 450–750 m depth (marked by a salinity minimum, <35.4), depending on latitude. Subtropical (σ_θ <27.1) ENACW lays above subpolar (σ_θ >27.1) off the western Iberian coast and north of the Canary region, so nutrient–poorer (0–6 μM of nitrate) subtropical waters upwell first, and nutrient–richer (6–10 μM of nitrate) subpolar ENACW enters the shelf only during strong upwelling events. Additionally, a variety of subpolar ENACW, the so–called 'Bay of Biscay Central Water' (BBCW), characterized by 12°C and 6 μM of nitrate, dominates the central waters domain off the Cantabrian coast (Treguer et al., 1979; Fraga et al., 1982; Botas et al., 1989). A subsurface front between subtropical ENACW and BBCW forms in the surrounding of Cape Finisterre (Fraga et al., 1982), although it experiences a seasonal displacement from south of the River Miño in early spring to west of Cape Peñas in late autumn (Fig. 23.1; Castro, 1997). There is also evidence that BBCW can intermittently upwell off the western Iberian coast during the summer (Álvarez–Salgado et al., 1993).

Subtropical ENACW is transported northward year round, by the PCUC during the upwelling season and by the PCCC during the downwelling season. A southward displacement of the origin of the subtropical ENACW which arrives to the Galician coast has been inferred from a progressive increase of salinity and temperature and a decrease of nutrients from the early spring to the winter mixing period (Ríos et al., 1992; Álvarez–Salgado et al., 1993; 2003). On the other hand, subpolar ENACW is conveyed southwards by the oceanic PC and it mixes laterally with the northward flowing PCUC and PCCC being eventually conveyed northwards at the slope (Pérez et al., 2001).

During the spring–summer period, surface (0–100 m) eutrophic coastal waters (affected by intermittent coastal upwelling) and surface oligotrophic surrounding ocean waters are separated by a slope front in the northern Iberian coast (Marañón and Fernández, 1995) but connected throughout upwelling centres and filaments at the wider coastal transition zone (CTZ) of the western coast (Haynes et al., 1993). In contrast, during the autumn–winter period, eutrophic coastal and oceanic waters are separated by the oligotrophic subtropical waters carried by the PCCC, which occupies the CTZ.

The variability (from weekly to interannual) of nutrient levels in the coastal domain has been studied in detail by Nogueira et al. (1997) in the Ría de Vigo (Fig. 23.1), a large (2.5 km^3) coastal embayment that behaves as an extension of the shelf (Figueiras et al., 2002). The seasonal evolution of inorganic nitrogen (Fig. 23.2b–d) allows clear definition of the spring–summer upwelling period, when nitrate levels are the highest in the cold bottom layer, because of intermittent coastal upwelling, and the lowest in the warm surface layer, because of efficient utilisation of upwelled nitrate by coastal phytoplankton populations. At the short–time–scale of an upwelling episode (1–3 wk), the sequence consists of upwelling of nitrate rich ENACW followed by nutrient consumption during the subsequent upwelling relaxation (Pérez et al., 2000a). The autumn–winter downwelling period is characterised by high nitrate levels throughout the water column, being maximum in the surface layer in association with continental runoff and intense regeneration processes. The importance of regeneration during the autumn–winter period is indicated by the succession of ammonium, nitrite and nitrate maxima in the surface layer, separated by ~1 month, which is characteristic of nitrification processes (Wada and Hattori, 1991). The succession of ammonium, nitrite and nitrate maxima also occurs in the bottom layer during the spring–summer period, indicating that regeneration processes are also important in the bottom layer, as will be shown below.

In the Iberian CTZ, during the spring–summer period, surface nutrient levels are detectable only where upwelling centres and filaments develop, being specially intensified in the surroundings of Cape Finisterre, where surface nitrate can reach up to 6 μM (Fraga, 1981; McClain et al., 1986; Castro et al., 1994). On the contrary, during the autumn–winter period, the CTZ is occupied by the PCCC that transports northwards warm (>15°C), salty (>35.8) and nutrient poor (<2 μM nitrate, >0.2 μM nitrite) subtropical waters (Bode et al., 1990; Castro et al., 1997; Álvarez–Salgado et al., 2003). This has important implications for the development of the spring bloom in the CTZ, because reduced nutrient levels transported by the PCCC do not allow massive chlorophyll accumulation as in the coastal and ocean domains.

The nutrient regime in coastal waters of NW Africa is conditioned by the presence of two different water masses. A marked front at 21°N (Cape Blanc) separates the NACW from the slightly cooler, less saline, and richer in nutrients South Atlantic Central Water (SACW). The boundary between these two water masses is convoluted, variable in position and characterised by intense mixing and interleaving processes (Fraga, 1974; Barton and Hughes, 1982; Minas et al., 1982; Barton, 1987; Hagen and Schemainda, 1987). South of Cape Blanc, the SACW is advected northward along the inshore side of the cyclonic recirculation to meet the equatorward flow of the Canary Current. Beyond this front, the northward propagation of SACW depends mainly on the poleward undercurrent, which is associated with the coastal upwelling. Hughes and Barton (1974) and Gardener (1977) traced this poleward flow between 16° and 28°N, and at depths of 200 to 400 m over the continental slope. Since the SACW is richer in nutrients than the NACW, a meridionally decreasing nutrient gradient is apparent in the northward flowing waters (Table 1). Off Mauritania (~18°N), high nutrient concentrations are observed at surface, probably related to the doming in the centre of the cyclonic recirculation. However, according to Minas et al. (1982), this is a region of anomalously low productivity and chlorophyll concentrations in relation to the observed nutrient concentrations.

TABLE 1.
Nutrient concentrations (in μmol kg^{-1}) observed in waters upwelled in the NW Africa and NW Iberia upwelling system.

	T (°C)	NO_3^-	PO_4^{3-}	SiO_2	Reference
NW AFRICA					
CapeTimiris-Nouakchott	14	20	1.5	10	Minas et al .. 1982
Cape Blanc - Cape Corveiro	15.5	14–15.5	0.9–1.0	6.5–7.5	Minas et al., 1982
Cape Sim, Cape Guir (summer)	14.5	8–9	0.6–0.7	4–5	Minas et al., 1982
NW IBERIA					
Galicia (May 91)	11.8±0.3	8±1	0.5±0.1	3.5±0.9	Castro et al., 2000
Galicia (July 84)	11.9–12.8	7±2	0.5±0.1	3.0±0.7	Castro et al., 2000
42°18'N, 8°57'W	12–13	9–12	0.5–0.8	5–10	Álvarez-Salgado et al., 1993

Compared to other eastern boundary regions in which upwelling takes place, the source waters off NW Africa are somewhat poorer in nutrients but richer in oxygen, largely as result of the global scale circulation (Codispoti et al., 1982). This could explain some of the differences in productivity and regeneration rates found, for instance, between the NW Africa and the Peruvian upwelling systems (Minas et al. 1982). However, differences in local topography, together with dissimilar seasonal cycles in the nutrient regimes, a larger interannual variability in Peru, a higher wind stress and turbidity in NW Africa, are all factors that must be considered in understanding the spatial-temporal variability in the productivity regimes of these regions.

Nutrient assimilation and regeneration ratios in the NW African coast are in general high, as in other coastal upwelling regions, although there are significant regional differences, particularly in silicon supply. Minas et al. (1982) have hypothesized that silicon limitation is a function of the geomorphology and dynamics of the coast. In the Moroccan upwelling, silicon, which should in theory be limiting because it is present in low concentrations in the source waters, is frequently in excess, due to high regeneration processes. In the Mauritanian upwelling, between Cape Timiris and Nouakchott, Herbland and Voituriez (1974) have shown that silicon is the first nutrient to be exhausted. Apparently, regeneration on this narrow shelf cannot compensate for the high demand. This region differs from more northern zones with wider shelves, such as Capes Dra and Corveiro, where regeneration compensates for the silicon deficiencies of the upwelling source waters. Indeed, Friederich and Codispoti (1982) made a detailed study of dissolved silicon regeneration over the shelf off Cape Corveiro, and observed that silicon regeneration in upwelling waters exceeded the inorganic nitrogen regeneration rate in spite of low (<1) Si/NO_3 ratios in the source water. This finding agrees with similar results obtained from the NW Iberian upwelling (Álvarez-Salgado et al. 1997)

High ammonia concentrations and high regenerations rates are frequently found in the inner shelf of the African coast (Codispoti and Friederich 1978; Minas et al 1982; Head et al. 1996). As a result of high wind stress, which produces mixing in the whole water column, ammonia is homogeneously distributed in the turbid inshore waters. This distribution contrasts with other coastal upwelling systems, which are less affected by strong wind regimes and present more stratified water columns. The turbidity, which may result also from aeolian dust deposition blown into the sea from the Sahara desert, produces a poor light regime for phytoplankton, which may inhibit ammonia uptake, in spite of high concentrations (Huntsman and Barber, 1977; Codispoti et al., 1982). Head et al. (1996), however, did not find any correlation between integrated primary production rates and light intensity in the inshore waters off the coast of Morocco, despite a two-fold variation in the latter.

2.3. *Primary production*

The seasonal and event–scale variability of coastal winds and upwelling in the Iberian and NW African coasts determines 1) the limiting nutrient flux to the photic layer; 2) the efficiency of utilisation of the limiting nutrient; 3) the fraction of the primary production required to maintain the metabolism of the pelagic ecosystem; and 4) the fate of the excess 'new' production: net transference to higher trophic levels, off–shelf export or deposition on shelf sediments, with subsequent benthic mineralization processes, reinjection of nutrients into the water column and eventual promotion to the photic layer.

The effect of spring–summer upwelling and autumn–winter downwelling periods in the productivity of the Iberian margin is illustrated by the 1987–1996 seasonal cycle of chlorophyll–*a* (Chl–*a*) in the Ría de Vigo (Fig. 23.2e; Nogueira et al., 1997). Apart from the spring and autumn Chl–*a* maxima, characteristic of any temperate ecosystem, Chl–*a* levels remain relatively high throughout the summer because of nutrient fertilisation by coastal upwelling. In addition, the spring and autumn Chl–*a* maxima occur within the transitional periods of onset and cessation of the upwelling season (Fig. 23.2a), respectively, determining the fate of the accumulated Chl–*a:* off–shelf export *versus in situ* mineralization (Álvarez–Salgado et al., 2003).

Although dramatic changes occur over space and time scales of a few kilometres and days, a seasonal pattern of gross primary production (P) rates in the Iberian margin can be drawn from field observations (Bode et al., 1994; 1996; Tenore et al., 1995; Bode and Varela, 1998; Teira et al., 2001; Joint et al., 2002; Álvarez-Salgado et al., 2003, Tilstone et al., 2003). The highest, but very variable, P rates are recorded during the spring and summer (1.0–8.0 g C m^{-2} d^{-1}) and the lowest in winter (~0.2 g C m^{-2} d^{-1}), as expected in any temperate coastal upwelling system. However, the succession of 1–3 wk coastal wind stress/relaxation cycles produces a remarkable short–time–scale variability in P rates during the seven months of the upwelling season (Fig. 23.3). The expected decrease of P rates from the inner shelf to the open ocean is also observed in the Iberian margin yearound (Tenore et al., 1995; Joint et al., 2002; Álvarez–Salgado et al., 2003; Tilstone et al., 2003), except when the PCCC occupies the CTZ between the spring bloom and the onset of the upwelling season (April–May), when P rates can be higher in the PCCC than in coastal waters (Álvarez–Salgado et al., 2003; Tilstone et al., 2003). These authors argue that this is because downwelling on the coast, generated by the PCCC, forces the sinking of coastal phytoplankton, mainly diatoms, into the aphotic layer.

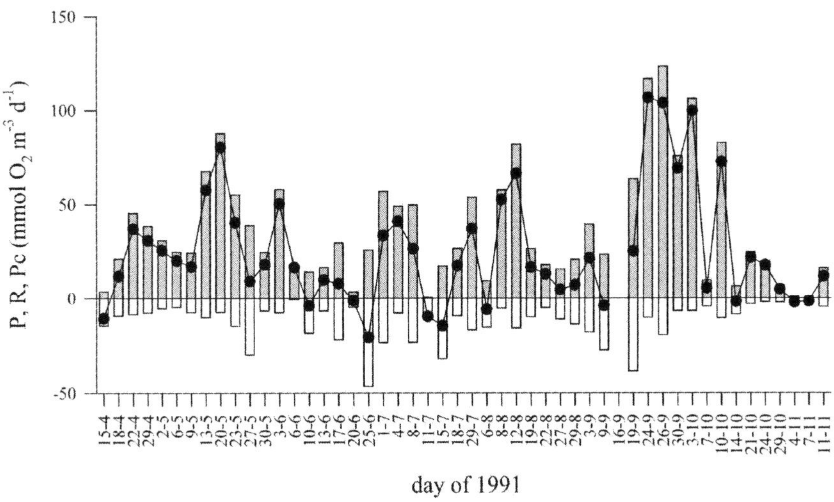

Figure 23.3 Short–time scale (3–4 days) evolution of community respiration (*R*, white bars), gross primary production (*P*, black bars) and net community production (*Pc*, black circles), in surface waters of the coastal upwelling ecosystem of the Ría de Vigo, from April to November 1991 (all rates in mmol O_2 m^{-3} d^{-1}).

Recently, the seasonal cycle of primary production in the Iberian margin (from 42° to 44°N) has been addressed by Joint et al. (2002), who produced monthly averages of net production (Pn), derived from satellite chlorophyll and 24 hours *in situ* ^{14}C incubations. They concluded that the average Pn for shelf waters during the 1998–2000 upwelling season was 1.5 g C m^{-2} d^{-1}. This value coincides with average net community production (Pc), estimated from 24 h in vitro oxygen experiments, over the upwelling season for the area of the Rías Baixas (Moncoiffé et al., 2000). Since the mean ratio of oxygen net community production/ gross production (Pc/P) for the upwelling season is ~0.6 (Moncoiffé et al., 2000), P in the Iberian margin should be ~2.5 g C m^{-2} d^{-1} for the upwelling season. Assuming a P of 2.5 g C m^{-2} d^{-1} during 225 days of upwelling period and a P of 0.2 g C m^{-2} d^{-1} during 140 days of downwelling period (see above, and Fig. 23.2) the average annual P value in the Iberian upwelling system would be 590 g C m^{-2} y^{-1}.

Average microbial respiration in the photic layer represents ~40% of P (Fig. 23.4), although it ranges from <10% in spring and during summer upwelling events to >100% (net heterotrophy) during upwelling relaxation periods (Serret et al., 1999; Moncoiffé et al., 2000; Teira et al., 2001; Barbosa et al., 2001). The coupling between small phytoplankton (<20μm) and microzooplankton activities causes these relatively high microbial respiration rates. Small phytoplankton (<20 μm) usually accounts for >50% of P (Bode et al., 1994, Tilstone et al., 1999; 2003; Joint et al., 2001a) and microzooplankton, which can reach a biomass equivalent to 60% of phytoplankton biomass during upwelling relaxation on the shelf, is able to graze 40–80% of the chlorophyll standing stocks and ~60% of the daily P (Fileman and Burkill, 2001; Tilstone et al., 2003). Dominance of large phytoplankton (>20 μm), particularly diatoms, is restricted to the spring bloom and during strong upwelling events at the inner shelf (Varela et al., 1991; Bode et al., 1994; Abrantes and Moita, 1999; Tilstone et al., 1999, Cachâo and Moita, 2000). Upwelling relaxation and downwelling favours diatom sinking at these sites (Figueiras et al., 1994, Fermín et al., 1996, Castro et al., 1997; Tilstone et al., 2000), contributing to enhanced nutrient mineralization in bottom shelf waters (Álvarez–Salgado et al., 1993; 1997; Prego and Bao, 1997). It has been argued that microzooplankton herbivory is an important source of dissolved organic matter (DOM) in the sea (Strom et al., 1997; Nagata, 2000), and the correlation found between a hetrotrophic/autotrophic microplankton assemblage and the concentration of DOM in the Iberian margin supports this view (Joint et al., 2001b). The possible role of microzooplankton grazing in DOM production is further reinforced by the fact that bacteria do not remove all DOM produced during an upwelling–relaxation cycle (Barbosa et al., 2001), while DOM released by phytoplankton, which is a small fraction (~6%) of P (Teira et al., 2001; Morán et al., 2002a), is not enough to satisfy the bacterial demand of labile carbon (Morán et al., 2002a; 2002b). Zooplankton grazing impact on phytoplankton seems to be rather moderate, ~5% of phytoplankton stock and ~10% of P (Barquero et al., 1998; Halvorsen et al., 2001), even though previous studies in the Rías Baixas indicated that zooplankton could consume the entire daily phytoplankton production (Hanson et al., 1986). Results of model simulations (Slagstad and Wassmann, 2001) also produce P values >2 g C m^{-2} d^{-1} under upwelling conditions and emphasise the predominant role played by small flagellates in P and by microzooplankton in grazing.

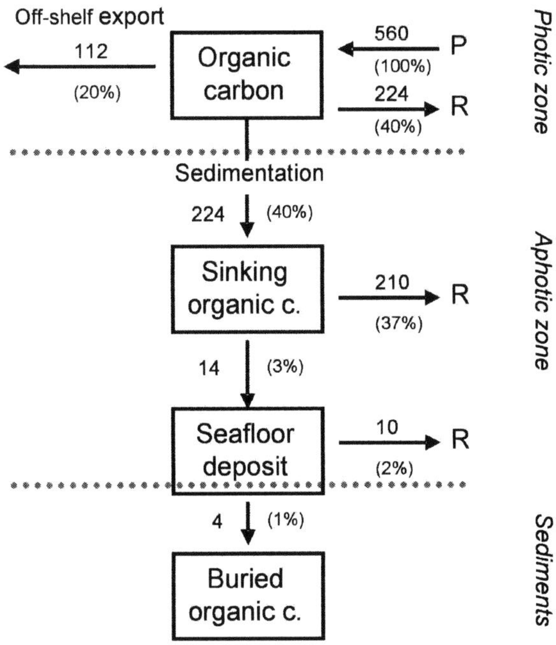

Figure 23.4 Tentative organic carbon budget of the NW Iberian margin, from 42° to 44°N, during the upwelling season. All fluxes are expressed in gC m^{-2} and are referred to an average P of 560 gC m^{-2} over the upwelling season. This value results from multiplying 2.5 gC m^{-2} d^{-1} (the average daily P) times 225 days (average duration of the upwelling season).

New production (New P) has been also estimated in shelf waters between 42° and 44°N combining upwelling rates, bottom shelf temperatures and ENACW nutrient–temperature relationships (Álvarez-Salgado et al., 2002). Shelf New P, fuelled exclusively by upwelled ENACW, was relatively low (0.5±0.15 g C m^{-2} d^{-1} for the 1982–1999 period) and, consequently, it represents ~20% of P. These low New P rates compared with other coastal upwelling systems are attributable to reduced continental inputs during the upwelling season, low nutrient levels in upwelled ENACW, and low average coastal winds because of the 1–3 wk wind stress–relaxation cycles. Therefore, although P rates, ~2.5 g C m^{-2} d^{-1}, are among the expected values for an upwelling system (e.g. Barber and Smith, 1981; Brown and Field, 1986; Minas et al., 1986; Pilskan et al., 1996) the average *f*–ratio (=New P/P) of ~0.2 over the upwelling season seems to be very low.

The succession of periodic 1–3 wk wind stress–relaxation cycles throughout the upwelling season is the reason behind both the high productivity and the low *f*–ratio of the Iberian margin. A well-described spatial sequence in coastal upwelling systems is initiated with upwelling of nutrient–rich subsurface waters near coast and continues with progressive nutrient consumption during seaward displacement of the upwelled water parcel (e.g. MacIsaac et al., 1985; Dugdale and Wilkerson, 1989). In the Iberian upwelling, this 'spatial sequence' is also a periodic 'time sequence' (Pérez et al., 2000a), which determines that the Iberian margin periodically switches from net autotrophy to net heterotrophy (high microbial respiration) producing a low average *f*–ratio.

Compared to the Iberian region, recent studies on primary production in the NW African region are scarce, and annual integrated rates have to be estimated from local studies on individual cruises. Compiled P values in the NW Africa upwelling region (Table 2) yields an average annual estimate of 2.4 g C m^{-2} d^{-1}, which coincides with P estimates from the NW African coast, computed from satellite chlrorophyll and averaged photosynthetic parameters (Longhurst et al., 1995; Morel et al., 1996). On a yearly basis, this estimate (874 g C m^{-2} y^{-1}) is about 50% higher than the average annual P value in the Iberian upwelling system (590 g C m^{-2} y^{-1}). Although the NW Africa estimate may be biased in time (since many of the studies correspond to the spring and summer periods), the difference in magnitudes is mainly caused by the strong seasonality in the Iberian upwelling, compared to the more continuous upwelling off NW Africa. Indeed, the average P values for the upwelling season in the Iberian margin (spring-summer) coincide with the annual average values from NW Africa.

TABLE 2.
Compiled daily primary production values in coastal waters from the Northwest Africa upwelling system (a), and average annual primary production for the Iberia and NW Africa coastal upwelling regions (b).

(a)Region	Latitude (N)	Season	Daily Primary Production g C m^{-2} d^{-1} Average	Min.	Max.	Reference
C.Sim-C.Ghir	31.5–30.5	Summer	2.4	0	4.2	Minas et al., 1982
C.Sim-C.Ghir	31.5–30.5	Winter	1.3	0.2	2.5	Minas et al., 1982
C.Sim-C.Ghir	31.5–30.5	Summer	1.3	0.6	2.8	Grall et al., 1982
C.Sim-C.Ghir	31.5–30.5	Autumn	1.5	1	2.5	Head et al., 1996
C.Jubi-C.Bojador	28.5–26.5	Summer	3.1	1.3	5.3	Basterretxea and Arístegui, 2000
C.Juby-C.Bojador	28.5–26.5	Summer	2.4			Minas et al., 1982
C.Corveiro-C.Blanc	22.0–21.0	Spring	2.4	0.8	5.0	Huntsman and Barber, 1977
C.Corveiro-C.Blanc	22.0–21.0	Spring		1.1	3.4	Lloyd, 1971
C. Blanc	21.0	Spring	1.15	0.3	2.6	Schulz, 1982
C. Blanc	21.0	Spring	0.96	0.2	1.8	Schulz, 1982
C. Blanc	21.0	Summer	1.62	0.4	3.6	Schulz, 1982
C.Timiris-Nouakchott	19.5–18.0	Spring	3.9	1.6	4.3	Minas et al., 1982
NW Africa upwelling	31.5–18		2.4 ± 1.5			

(b)Region	Latitude (N)	Annual Prim. Production g C m^{-2} y^{-1}	Data	Reference
Iberia upwelling	43.0–41.0	590	In situ	This work
NW Africa upwelling	31.5–18.0	874	In situ	Several authors
Iberia and NW Africa	43.0–15.0	732	Satellite based	Longhurst et al., 1995

Regional variability in P in the NW Africa region probably results from differences in nutrient regimes, although the lack of detailed studies in the region preclude identification of clear spatial-temporal patterns as in other eastern-boundary uwelling regions. The highest annual P rates are presumably produced between Cape Barbas and Cape Blanc, where the more nutrient-rich SACW is available as source water, and upwelling is produced year round (Minas et al., 1982). Yet, local topographic and environmental factors may be responsible of short-term changes in the nutrient regime and primary production. Codispoti (1981) concluded from a study comparing different coastal upwelling regions, that, in contrast with other upwelling regions, much of the nutrient variability off NW Africa occurs on the scale of several days. In agreement with this, Grall et al. (1992) described large spatial variations in P (from 0.8 to 2.8 g C m^{-2} d^{-1}) off Morocco, between Cape Sim and Cape Guir, which were correlated to short (~3 days) but intense wind pulses. He also observed that the centre of the upwelled water migrated from the inner shelf to the shelf break as the upwelling progressed, probably reflecting the offshore intensification of the Cape Guir filament. In a more recent study, Arístegui and Harrison (2002) observed also a similar large daily variability in P (1.1 to 2.8 g C m^{-2} d^{-1}) and chlorophyll a (25–94 mg m^{-2}), following a drifting buoy along the Cape Guir filament, during 7 days of consecutive measurements.

2.4. *Shelf-ocean exchange of organic matter*

The excess production of any coastal upwelling system can be 1) transferred to higher trophic levels (section 4); 2) transported downwards to the coastal sediments, where it can experience mineralization in the dark water column and sediments or be buried (section 2.6); and 3) exported to the adjacent open ocean.

The shelf–ocean exchange of organic matter is enhanced during the productive upwelling season, especially at sites where large filaments develop. Several recurrent filaments are distributed all along the Iberian coast (Sousa and Bricaud, 1992; Haynes et al., 1993; Pelíz and Fiúza, 1999), although only the filament off the Rías Baixas has been studied from a biogeochemical point of view (Joint and Wassmann, 2001). One of the most relevant conclusions of that process–orientated study is that fresh dissolved and particulate organic materials are exported in equal amounts to the ocean by the upwelling filament (Álvarez–Salgado et al., 2001). Total organic carbon export by the filament off the Rías Baixas was estimated to be ~100 g C m^{-2} during the upwelling season, a number that is very close to the New P estimate of 0.5±15 g C m^{-2} d^{-1} (=110±30 g C m^{-2}) by Álvarez–Salgado et al. (2002), suggesting that, at the time scale of the upwelling season, a significant fraction of the New P is exported to the adjacent ocean (Fig. 23.4). On the contrary, during the unproductive downwelling season (winter), the PCCC acts as an insulator between shelf and ocean waters, confining coastal primary production on the shelf and enhancing sedimentation and *in situ* mineralization (Castro et al., 1997; Álvarez–Salgado et al., 2003). Recent estimates of particle residence times (average: from 26 days in the early summer upwelling season to 113 days in winter) in surface waters of the NW Iberian margin support this view (Schmidt et al., 2002a).

Transitional periods, from the upwelling to the downwelling season, are especially important for the fate of coastal P because they occur at the time of the spring and autumn phytoplankton blooms (Fig. 23.2a,e). In this manner, a coastal bloom under upwelling conditions might be exported to the adjacent ocean, while a bloom under downwelling conditions would sediment on the shelf, being eventually processed by the benthic communities (Álvarez–Salgado et al., 2003).

Since ~40% of P is recycled in the photic layer by the microbial component and ~20% is exported (Fig. 23.4), the remaining 40% of P would be respired in the aphotic layer and the sediments. Estimates of vertical fluxes of organic matter from the photic layer (Fernández et al., 1995; Bode et al., 1998; Hall et al., 2000) provide values between 19 and 50% of P, although extremely low values of 4–9% may occur during strong upwelling events and prolonged relaxation periods (Bode et al., 1998; Olli et al., 2001). Strong upwelling precludes the development of phytoplankton on the shelf because off–shelf export of upwelled water is enhanced. Prolonged relaxation favours small phytoplankton (Joint et al., 2001a, Tilstone et al., 2003), microbial recycling (Fileman and Burkill, 2001) and high retention of particulate material in the upper 200m (Riser et al., 2001) at the expense of export and sinking (Olli et al., 2001).

Off NW Africa, numerous upwelling filaments are distributed along the coastal-offshore upwelling boundary (Fig. 23.5). These filaments may arise for one or a combination of several factors: baroclinic instability of the coastal current, irregularities in coastline and bottom topography, coastal convergence caused by wind stress, and the interaction of the coastal region with offshore eddies (Brink and Cowles, 1991; Strub et al., 1991; Barton, 1998). The latter process is very common south of the Canary Archipelago, where island eddies are sequentially spun off downstream the islands. However, only two of the filaments (the Cape Guir and Cape Blanc filaments) remain as major permanent features, even during non-favourable upwelling winds, and thus represent key sites for the export of organic matter to the open ocean waters of the subtropical gyre.

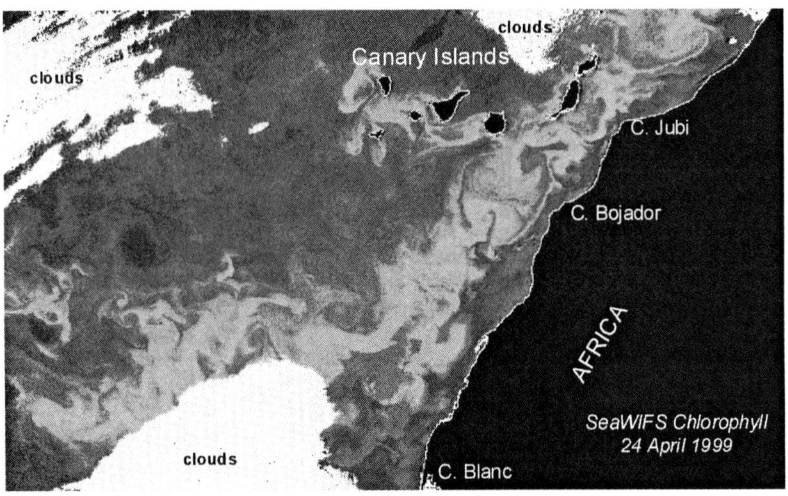

Figure 23.5 SeaWIFS chlorophyll image from the Canary Islands region (24 April, 1999), showing the presence of numerous cyclonic and anticyclonic eddies downstream the islands. The eddy field extends southward down to the latitude of Cape Blanc, where the Canary Current detaches from the coast and flows westward. As a reference, green colour represents chlorophyll values >1 mg m^{-3} and red colour corresponds to values >3 mg m^{-3}

The Cape Guir filament has been documented in several studies (Mittlestaedt, 1991; Hernández Guerra and Nykjaer, 1997; Pelegrí et al., 2005) as a recurrent feature, which may extend several hundred kms offshore (Fig. 23.5), transporting particulate organic carbon (POC). The impact of POC export by the Cape Guir filament has been observed as far as 700 km from the filament origin, well into the subtropical gyre province, by means of deep-water sediment traps deployed northwest of La Palma, the easternmost island of the Canaries (Neuer et al., 2002). The filament presumably represents also a significant oceanic route for the transport of excess dissolved organic and inorganic nutrients produced in the coastal upwelling waters. García-Muñoz et al. (2005) calculated a net offshore export of nitrate of 2.6 kmol s^{-1} by this filament, during non-favourable upwelling conditions. This value is much larger than the offshore Ekman transport (0.4 kmol s^{-1} of nitrate) that would take place in an upwelling system along a shore distance of 10 km, equivalent to the filament width. However, although the offshore transport of filaments is usually significantly larger than Ekman transport (e.g. Kostianoy and Zatsepin, 1996) the return flow of filament water into the coastal upwelling jet may reduce the impact of the overall offshore transport. This was observed in an upwelling filament stretching between Cape Juby and Cape Bojador during August 1993 (Navarro Pérez and Barton, 1998). The filament extended approximately 150 km offshore, wrapping around a cyclonic eddy of 100 km diameter, before returning shoreward. Most of the particulate organic matter exported by this filament was returned by the recirculation path of the filament (Basterretxea and Arístegui, 2000).

Off Mauritania, the giant Cape Blanc filament appears as a persistent low-temperature and high-chlorophyll feature, extending several hundred kms offshore. The filament is linked to the convergence of NACW and SACW at the Cape Vert frontal zone. At this site, the large-scale circulation pattern induces the upwelling centre to shift towards the shelf break area, allowing phytoplankton to achieve maximal growth rates in oceanic waters. According to Gabric et al. (1993) this filament exports more than 50% of New P to oceanic waters. Deep-water sediment traps deployed during the EUMELI project at a mesotrophic station located >500 km off Cape Blanc (Bory et al. 2001) recorded biogenic fluxes higher than other stations in the Atlantic Ocean (Jickells et al., 1996; Wefer and Fischer, 1993). The temporal variability of these fluxes seemed to be linked to westward surface currents, which are likely to transport seaward carbon-rich waters from the coastal upwelling. Nevertheless, a large interannual variability is appreciable in the downward particle flux, presumably caused by climatic changes (Bory et al., 2001). Similar interannual variability was appreciated in particle fluxes recorded off Cape Guir (Neuer et al., 2002).

2.5. *The coastal upwelling as a source–sink of CO_2*

The behaviour of coastal upwelling systems as sources or sinks of CO_2 to the atmosphere depends on the balance of two opposing tendencies. On one side, the higher the nutrient concentration of the source water, the higher is its partial CO_2 pressure (pCO_2). On the other, the higher the P rates (enhanced by the nutrient input), the higher must be the reduction of pCO_2 in the surface layer (e.g. Watson, 1995; Borges and Frankignoulle, 2002a).

With these simple considerations in mind, the water masses and the dynamics of the Iberian margin would favour its behaviour as a CO_2 sink. First, nutrients and, consequently, pCO_2 levels of upwelled ENACW are relatively low (400–500 μatm) compared with the aged central waters of the South Atlantic, the Indian and the Pacific Ocean. Second, the intermittency of coastal upwelling in the Iberian margin allows efficient utilisation of upwelled nutrients, leading to P rates comparable with other coastal upwelling systems, where the nutrient loads of upwelled waters are much higher.

Surface pCO_2 measurements in the western Iberian margin (Pérez et al., 1999; Borges and Frankignoulle, 2001; 2002a,b) confirm this general view. Despite seasonal upwelling, surface pCO_2 undersaturation occurs throughout the spring and summer, except at the Cape Finisterre upwelling centre throughout the upwelling season (Borges and Frankignoulle, 2002b) and along the northern Portuguese coast during strong upwelling events (Pérez et al., 1999). Shelf waters off the Rías Baixas are usually undersaturated because of the combination of outwelling of pCO_2 equilibrated/undersaturated waters from the rías and the considerable width of the continental shelf compared with the Cape Finisterre area (Borges and Frankignoulle, 2002a). It should be highlighted that the 'outwelled' water from the rías is just upwelled ENACW that experiences pCO_2 reduction during seaward displacement from the inner to the outer part of these embayment. However, Álvarez et al. (1999) followed the variability of surface pCO_2 levels along a transect from the central Ría de Vigo to the middle shelf, from May 1994 to September 1995, observing pCO_2 supersaturation in the inner shelf and the ría under strong upwelling conditions. Moreover, bottom regeneration associated with the intensive culture of mussels on hanging ropes in the Ría de Arousa, which increases the pCO_2 of upwelled ENACW, together with enhanced flushing rates in the rias, do not allow P to reduce pCO_2 below the atmospheric level during the upwelling season of 1989 (Rosón et al., 1999).

During the autumn and winter period, pCO_2 undersaturation is associated with low salinity continental waters, which allow a sequence of stratification, Chl–*a* accumulation and pCO_2 decrease. On the contrary, pCO_2 supersaturation occurs anywhere continental runoff is reduced, because the aged shelf bottom waters enter in contact with the atmosphere as a result of irradiative loss and strong vertical mixing (Fiúza et al., 1998; Vitorino et al., 2002a). Equilibrium with the atmosphere or slight undersaturation is usually found in the high salinity subtropical waters occupying the CTZ and the surrounding ocean year-round (Pérez et al., 1999; Borges and Frankignoulle, 2002a).

Borges and Frankignoulle (2002b) computed air–sea exchange fluxes of CO_2 on the western Iberian shelf from 42° to 44°N yielding a net influx in the range of –2.3 to –4.7 mmol C m^{-2} d^{-1} during the upwelling season and of –3.5 to –7.0 mmol C m^{-2} d^{-1} on an annual basis, using different formulations of the CO_2 exchange coefficient. A similar calculation made by Álvarez et al. (1999) off the Ría de Vigo, using the lower flux estimate, yielded a net influx from about –0.1 mmol C m^{-2} d^{-1} in the middle ría to –2.0 mmol C m^{-2} d^{-1} in the middle shelf during the upwelling season. CO_2 uptake is maximum at the time of the spring and autumn blooms, when influxes up to –4.3 mmol C m^{-2} d^{-1} were recorded in the middle shelf. On the contrary, the area acted as a CO_2 source during the winter, with maximum fluxes of 2.3 mmol C m^{-2} d^{-1} again in the middle shelf. In a more recent study, Gago et al. (2003)

found that the inner Ría de Vigo is usually a CO_2 source to the atmosphere, except in December.

Recent studies in the NW Africa upwelling near the Canary Islands region, have identified the coastal upwelling as a weak CO_2 source, with average carbon fluxes of 0.5 mmol $m^{-2} d^{-1}$ (Santana-Casiano et al., 2001; Pelegrí et al., 2005). Yet, most of these studies were performed during autumn and winter, when winds are low to moderate and upwelling is weaker than the rest of the year. It is therefore plausible that during strong upwelling events and higher productivity, the system behaves as a carbon sink, as off the north Iberian coast.

Upwelling filaments can support an important export of excess inorganic carbon in the upwelling region to the open ocean, providing carbon uptake by phytoplankton is not large enough to decrease significantly the pCO_2 along the filament extension. Pelegrí et al. (2005) observed a net surface flux of CO_2 from the coast to the open ocean through the Cape Guir filament, during October 1999. The calculated biological consumption of CO_2 along the filament was low enough to allow supersaturation of CO_2 in the warmer open ocean waters, increasing the net flux of CO_2 to the atmosphere.

2.6. *Benthic production and sediment processes*

The fate of phytogenic organic materials exported from shelf surface waters to the sea floor depends on the supply of organic matter, the preservation conditions, the dilution with terrigenous sediments and the near bed hydrodynamics (van Weering et al., 2002). In this sense, the role of the Galician Rías Baixas in the north Iberian margin (42–43°N), is significant, because they trap the terrestrial input from their 6800 km^2 drainage basin (Araújo et al., 2002; Dias et al., 2002a) and export endogenous detritus to the adjacent shelf (Prego, 1994; Rosón et al., 1999; Pérez et al., 2000a). In the northern Portuguese coast, the river Douro (Fig. 23.1), with the largest drainage basin of the Iberian Peninsula, constitutes the main sediment source for both the coarse material deposited close to the coast and the active mud belts of fine-grained materials in the middle shelf off northern Portugal, especially during winter high floods (Drago et al. 1998). Further south, several submarine canyons, and particularly the Nazaré canyon, which intersects the entire continental shelf (Fig. 23.1), are sites where export of shelf sediments to the Iberian Abyssal Plain is likely enhanced (van Weering et al., 2002). Regarding the near bed hydrodynamics, the predominantly along-shore flow of the 'Portugal Current System' at the western Iberian margin, favours summer offshore export in the surface layer, with onshore compensation below, and a winter onshore component through the water column with some offshore export being confined to the bottom layer (Drago et al., 1998; Vitorino et al., 2002a). Consequently, resuspension and transport of shelf fine sediments in bottom nepheloid layers (BNLs), occur mainly in the alongshore direction (Dias et al., 2002a; Oliveira et al., 2002). Any export to the adjacent ocean is confined to the shelf edge during the summer but it is enhanced during the winter, when resuspension is maximum (Vitorino et al., 2002b). Well-developed BNLs are observed at the shelf break during winter storms, likely making a significant contribution to the off-shelf export (McCave and Hall, 2002) when they detach from the slope in the form of intermediate nepheloid layers (Oliveira et al., 2002). In the case of the middle-shelf Douro mud patch, fine sediments are resuspended and transported northwards during the winter to meet the outer shelf mud patch off the Rías Baixas, because the plateau west of the Douro acts as a barrier to off–shelf export (Dias et al., 2002b; Jouanneau et al., 2002).

Intense biogeochemical processes occur at the sediment–water interface of the 42°–43°N region during the upwelling season, as a consequence of the enhanced productivity of the Rías Baixas (Tilstone et al., 1999; Moncoiffé et al., 2000; Pérez et al., 2000a), the outwelling of detritus, and the onshore compensating flow in bottom shelf waters, which favours particle retention. Upwelled ENACW experiences a remarkable nutrient enrichment, which is maximum for water parcels in direct contact with shelf sediments (Fraga, 1981; Álvarez–Salgado et al., 1993; Prego et al., 1999). The agreement between the spatial distributions of organic carbon and silica debris in sediments and the corresponding nutrient salts in the overlying water (López–Jamar et al., 1992; Prego and Bao, 1997) confirm the benthic origin of this enrichment. It has been observed that nutrient enrichment increases progressively throughout the upwelling season (Álvarez–Salgado et al., 1993) and from the outer to the inner shelf (Álvarez–Salgado et al., 1997), being maximum inside the Rías Baixas (Prego et al., 1999). This nutrient enrichment increases P rates by 'secondary recycling' and dampens the temporal and spatial variability in nutrient conditions seen when driven exclusively by upwelling.

Schmidt et al. (2002b) observed that whereas the Iberian margin between 43° and 44°N is not an efficient depocenter, rapid sedimentation appears to be more important on the shelf and slope of the 41°–43°N area at a time scale of 100 days. In this sense, van Weering et al. (2002) obtained organic carbon burial rates of 1.0–34.3 g C m^{-2} y^{-1} for the shelf and 0.01–0.69 g C m^{-2} y^{-1} for the slope off the Rías Baixas. In contrast, carbon burial rates up to 182 g C m^{-2} y^{-1} were observed at the Nazaré canyon, mainly on the upper and middle canyon, from where it is episodically released to the deep sea (van Weering et al., 2002).

Estimates of organic carbon deposition rates at the sediment–water interface (~14 g C m^{-2} y^{-1}; Epping et al., 2002) are low compared with the relatively high P rates (~560 g C m^{-2} y^{-1}; Fig. 23.4) of the NW Iberian shelf during the upwelling season. Since ~40% of P is deposited on the shelf and the near bed hydrodynamics favours particle retention -organic carbon deposition on the deep ocean is only ~20% of that on the shelf (Epping et al., 2002)- most of the organic carbon delivered to the sea floor (~87%) has to be mineralised at the sediment–water interface to produce the observed nutrient enrichment in bottom shelf waters. Finally, ~30% of organic carbon deposited in the sediments is buried and ~70% experiences mainly aerobic oxidation, although denitrification is important (up to 50%) on the inner shelf (Epping et al., 2002), in the rías (Dale and Prego, 2002) and in the canyons (Epping et al,. 2002), where higher sedimentation rates occur. Considering that these deposition rates are representative of the upwelling season and assuming that continental inputs (Álvarez–Salgado et al., 2002) and off–shelf export of sediments is negligible, the tentative carbon balance for the NW Iberian shelf off the Rías Baixas presented in Figure 23.4 can be completed. This situation contrasts with the preponderant role previously assigned to continental shelves as organic carbon sinks but agrees with the most recent assessments which point out shelves primarily as mineralization sites (de Haas et al., 2002).

Along the NW African coast, the high wind stress and strong equatorward and cross-shelf currents may prevent accumulation of organic matter in sediments, where aerobic respiration dominates. In fact, concentrations of particulate organic matter nearshore are in the lower range of coastal upwelling regions, with a predominance of particles <50 μm (Lenz, 1982). Nevertheless, there is a trend towards increasing accumulation rates of organic material from north to south, which partly reflects a southward increase of productivity, but also an input of clay minerals by the Senegal River and better preservation of organic matter from plankton or land plants in these fine grained sediments (Seibold, 1982).

Relexans et al (1996) studied the biological and chemical characteristics of sediments sampled at three different sites along a trophic gradient from the eutrophic waters of the Mauritanian upwelling to the oligotrophic open ocean waters of the subtropical gyre. As expected, they observed a general decrease, from eutrophic to oligotrophic waters in POC, macromolecular contents, biomass and metabolic activities in microbenthos, reflecting the fertility of the water column above the sediments. However, while POC fluxes at the sediment water interface ranged from 1 g C $m^{-2} y^{-1}$ (oligotrophic) to >20 g C $m^{-2} y^{-1}$ (eutrophic), carbon mineralization rates varied only 3 fold: from 1.7 gC $m^{-2} y^{-1}$ in the oligotrophic station, to an average 5 g C $m^{-2} y^{-1}$ in the eutrohic region. These results led Relexans and coworkers (1996) to suggest a more efficient utilization of the vertical carbon supply in deep oceanic waters than in the shallower shelf of the upwelling system. Yet, organic carbon flux rates near the shelf bottom may be higher than fluxes computed from surface productivity and empirical relationships of decrease of POC with depth, indicating that a large fraction of the nearshore material is resuspended (Bory et al., 2001; Neuer et al., 2002). This material may be lost out from the system by cross-shelf advective transport. Particularly, high offshore carbon export rates have been observed in the region associated with the Cape Blanc filament (e.g. Gabric et al., 1993).

The sediments off Cape Guir represent an important depocenter site (Henderiks et al., 2002). Nevertheless sedimentation rates in the Moroccan coast may actually be low compared with the offshore transport. Head et al., (1996) carried out a carbon flux study in this region, following drifting arrays deployed in the core of the filament during 7 days. These authors estimated from in situ P measurements and shallow sediment traps, that the *f*-ratio (export flux/total P; Eppley and Peterson 1979) ranged from 0.16 to 0.24. These ratios are significantly lower than those seen in other coastal regions during periods of active upwelling (Codispoti et al., 1982; Dugdale et al., 1990), indicating that remineralization rates are very high in the upper water column, or the offshore export is higher than in other regions, or both. The high offshore export hypothesis is strongly supported by recent studies, which ascribe a key role in the off-shelf transport of organic matter to the Cape Guir filament (Neuer et al., 2002: García-Muñoz et al., 2005; Pelegrí et al., 2005). The distribution of surface-sediment analysis underlying the Cape Guir filament mirrors the surface gradients in temperature and chlorophyll observed from satellite images. Total organic carbon, benthic foraminifera and *Globigerina* shells decrease from coastal sediments to open ocean sediments (Meggers et al., 2002). Moreover, the occurrence of coccolithophorids (Sprengel et al., 2002), diatom and foraminifera species (Abrantes et al., 2002), characteristic of high productivity coastal areas, in deep sediment traps deployed several hundred of kms offshore, reflects the export capacity of the Cape Guir filament.

3. The Canary Islands Coastal Transition Zone

South of Cape Guir, the Canaries archipelago (28°N) spans the transition zone between the NW Africa coastal upwelling waters and the open ocean waters of the subtropical gyre. The sharpest changes in primary production (Basterretxea and Arístegui, 2000; Fig. 23.6), as well as in the distributions of particulate and dissolved organic carbon (Arístegui et al. 2003), are found in the eastern sector of the region, close to the upwelling region. The zonal gradient extends however for the width of the Archipelago's area of influence (>300 km) because of the recurrence of mesoscale eddies and fronts, induced by island disturbance of wind and currents (Barton et al. , 1998). Eddies are spun off in the form of vortex streets, extending southward to about 22° N, where the Canary Current detaches from the coast and flows westwards. While drifting southward, eddies may interact with the coastal upwelling waters and act as a route to export organic matter to the oligotrophic open ocean (Fig. 23.5).

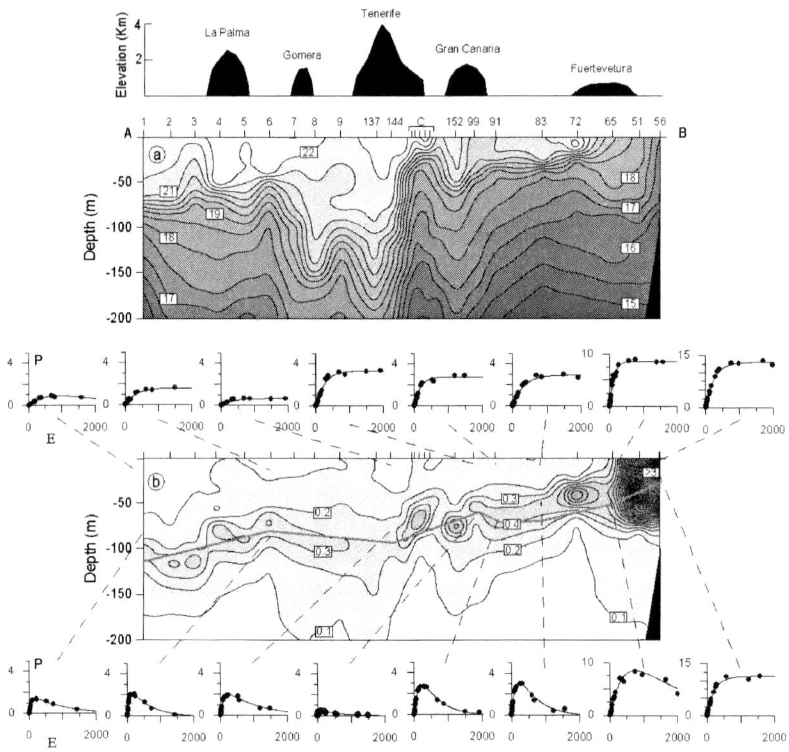

Figure 23.6 Contour plots of temperature (°C; middle panel, a) and chlorophyll (mg m^{-3}; lower panel, b) along a transect extending from the African coast to the open ocean subtropical gyre, south of the Canary Islands (upper panel). Dashed lines indicate the location and depth of photosynthesis (P; mgC mg^{-1} chla h^{-1}) - irradiance (E; µmol m^{-2} s^{-1}) experiments; grey line marks depth of 1% surface irradiance. Notice the high range of variation in the photosynthetic parameters along the coastal-open ocean gradient, comparable to latitudinal ranges observed in basin-scale studies. The chlorophyll maximum coincides with the depth of 1% surface irradiance along the whole section. (Adapted from Basterretxea and Arístegui, 2000)

3.1. Interaction of island eddies and upwelling filaments

The temporal and spatial variability in plankton biomass, community structure, and metabolic activities in the Canaries-Coastal Transition Zone (CTZ) region, largely results from the mesoscale variability generated by the perturbation of the main flow by the islands (Fig. 23.7). Wind and current shear at the flanks of the islands enhance plankton productivity and respiratory activity, by increasing vertical mixing and nutrient availability in surface waters (Hernández- León, 1988; Arístegui et al., 1989; Hernández-León, 1991; Arístegui and Montero, 2005). Additionally, Ekman pumping on the wind shear boundaries of the islands produces convergence and divergence fronts (Barton et al., 2000), which affect plankton distribution and productivity. Divergence fronts induce upwelling of deep nutrient-rich water, increasing primary production and chlorophyll (Basterretxea et al., 2002).

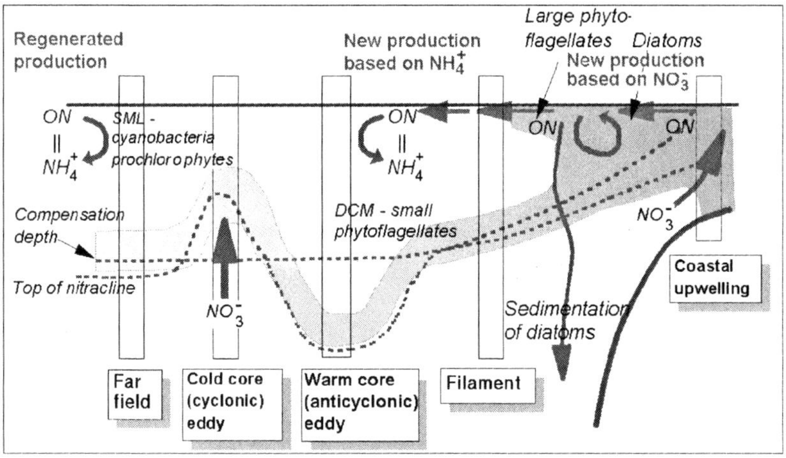

Figure 23.7 Schematic of vertical processes along the NW Africa-Canaries CTZ region. Vertical rectangles indicate typical situations. The *Far-field* nitracline lies below the compensation depth for phytoplankton growth, and so production is low, carried out by pico- or nanoplankton, and likely sustained by recycled ammonium. *Cyclonic eddies* lift isopycnals and nitracline relative to compensation depth, so locally stimulating new production. *Anticyclonic eddies* depress isopycnals and nitracline, deepening the deep chlorophyll maximum (*DCM*) below the compensation depth. Counter paired cyclonic and anticyclonic eddies act as a two-way "biological pump" enhancing the formation and vertical transport of organic matter in the water column. The DCM intensifies toward the coast and higher nitrate concentrations are exposed above the compensation depth. In the *Upwelling* region the nitracline reaches the surface mixed layer, resulting in high production of diatoms, converting nitrate into Organic Nitrogen (*ON*). These diatoms sink or are eaten as upwelled water moves offshore in the *Filament,* leaving large phytoflagellates as the dominant producers, supported by recycled ammonia from the organic matter transported by the filament. (Modified from Barton et al., 1998)

Downstream of the islands, cyclonic and anticyclonic eddies are generated, by a combined mechanism of flow perturbation and Ekman pumping (Arístegui et al., 1994; 1997; Barton et al., 2000). These eddies exchange water properties between themselves and with the coastal upwelled waters at its offshore boundary or in upwelling filaments (Fig. 23.8). In their early digenetic stages, island eddies present strong vertical transport of water in their cores to compensate for the ageostrophic movement. Cyclonic eddies enhance primary production by upwelling nutrient-rich thermocline waters into the euphotic zone. Primary production in the core of cyclonic eddies may increase several fold in magnitude compared with the ambient waters (Basterretxea and Arístegui, 2000). As in coastal upwelling, high nutrient concentrations favour the growth of larger cells, like diatoms, increasing considerably the chlorophyll content inside cyclonic eddies, associated with a shift from small to large phytoplankton cells (Barton et al. 1998; van Lenning 2000). Conversely, anticyclonic eddies collect and downwell surface water, deepening the mixed layer and the chlorophyll maximum to depths well below the euphotic zone (>100 m) (Arístegui et al., 1997). These eddies may act as effective organic carbon pumps, sequestering dissolved and particulate organic carbon from the nearby waters and sinking it into the dark ocean. Overall, counter-paired eddies behave as a two-way biological pump, accelerating the production and transport of organic matter in the water column (Arístegui and Montero, 2005).

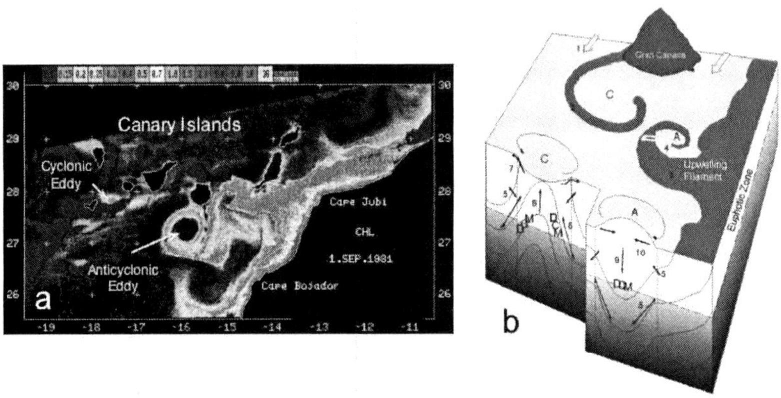

Figure 23.8 (a) CZCS chlorophyll image from the Canary Islands region (1 September 1981), showing the entrainment of high-chlorophyll water from an upwelling filament by an anticyclonic island eddy. (b) Schematic diagram showing vertical and horizontal fluxes associated with island-generated cyclonic (C) and anticyclonic (A) eddies downstream Gran Canaria Island. (1) Mean surface flow and wind direction; (2) advection of high-chlorophyll coastal water during cyclonic eddy formation; (3) entrainment of high-chlorophyll water upwelled from the African coast by anticyclonic eddy; (4) lateral exchange of chlorophyll between the periphery and the eddy centre; (5) diapycnal mixing; (6) upwelling or uplifting of the thermocline; (7) outward radial advection; (8) isopycnal mixing; (9) downwelling or deepening of the thermocline; (10) inward radial advection. DCM = Deep Chlorophyll Maximum. (Panel b, adapted from Arístegui et al., 1997)

Anticyclonic eddies are revealed by SST or chlorophyll like-pigments images thanks to the entrainment of cold, chlorophyll-rich waters when interacting with cyclonic eddies or upwelling filaments stretching from the African coast (Hernández Guerra et al., 1993; Arístegui et al., 1997; Barton et al., 1998; Pacheco and Hernández Guerra, 1999). Satellite images show that anticyclonic eddies detached from the eastern islands of the Archipelago, occasionally interact with the offshore boundary of the coastal upwelling, entraining upwelled water and extending the filament further offshore (Fig. 23.8). The exchange of water and organic matter between filaments and eddies may be extended to the open ocean waters of the subtropical gyre, by successive interactions between cyclonic and anticyclonic eddies all along the Canaries-CTZ region. This has been confirmed by a drifting lagrangian buoy deployed in the core of a filament near Cape Juby, which followed a westward trajectory to the open ocean, crossing the Canaries-CTZ while spinning around several island eddies (Barton et al., unpubl.). Nevertheless, the water mass characteristics and their associated biological properties must be drastically modified along the way from the coastal upwelling to the open ocean, by mixing of cyclonic and anticyclonic eddy surface waters with different bio-chemical properties, and by vertical displacements of isopycnals and downwelling/ upwelling processes, associated with eddies.

The biological effects of upwelling filaments on open ocean waters depend in large part on the source coastal water that is advected offshore. Arístegui and Montero (2005) observed that the Cape Juby filament could either transport water with large phytoplankton cells and low community respiration, or water with high respiratory rates associated with smaller cells. These contrasting differences in the composition and metabolism of the transported organic matter may be explained by a different origin of the filament waters, and the time that these waters take to be advected offshore. Significant differences in nutrients and chlorophyll concentrations, and phytoplankton composition, have been described also for an upwelling filament in the California Current upwelling system (Chavez et al., 1991; Jones et al., 1991). These authors attributed a different origin of the upwelled water along the coastal jet to interpret the chemical and biological differences in the upwelling filament sampled during two different cruises.

3.2. *Carbon imbalance in the Canary region*

The advection of upwelling waters into the Canary region has important consequences not only by prompting changes in plankton community structure, but also by altering the metabolic balance of the region. The average seasonal P in the Canary region varies more than one order of magnitude, from ~0.1 g C m^{-2} d^{-1} in autumn to >1 g C m^{-2} d^{-1} in spring (Basterretxea, 1994; Basterretxea and Arístegui, 2000). Comparatively, the average integrated community respiration (R) rates are higher and the range of variation (2 fold) lower: from 1.3 g C m^{-2} d^{-1} in summer to 2.3 g C m^{-2} d^{-1} in spring (Arístegui and Montero, 2005). The annual average P/R ratio for the region is <0.5, and only approximates equilibrium during late winter/ early spring, when phytoplankton blooms (Arístegui and Montero, 2005). These results agree with most of the studies on plankton metabolism in the subtropical Northeast Atlantic, where net hetetotrophy (P<R) seems to be the dominant state for the surface planktonic community (Duarte et al., 2001; González et al., 2001, Arístegui and Harrison, 2002). Nevertheless, heterotrophy is considerably larger in the Canary region, where R may at times be more than ten-fold higher than P (Arístegui and Montero, 2005).

Although cyclonic eddies contribute to increase P and R in the region, frequent loadings of organic matter from the coastal upwelling would be necessary to balance the high R rates measured in the Canary region. Satellite images show the yearly recurrence of high-chlorophyll filaments invading the Canary region (Arístegui et al., 1997; Pacheco and Hernández Guerra, 1999), and thus spreading organic matter into the open ocean waters. Neuer et al (2002) observed a marked seasonal correlation between carbon fluxes collected in 3 moored sediment traps deployed along the Canaries CTZ region (up to 250 km offshore) and the variability in the intensity of the NW Africa upwelling system. They concluded that particle fluxes in all the traps were affected by particle dispersion from the upwelling filaments of Cape Juby and Cape Guir. Apart from the particulate carbon export, upwelling filaments may export dissolved organic carbon too, as demonstrated by García-Muñoz et al. (2005). Most of the exported particulate and dissolved organic matter is thought to be respired in the Canary region, although some must be transported, by means of eddy exchange, to the oligotrophic waters of the subtropical gyre, where R is consistently higher than P (Duarte et al., 2001; González et al., 2001; Serret et al., 2002).

3.3. *Upwelling filaments and fish larvae survival*

Filaments in the Canary region are also known to transport neritic fish larvae to the open ocean acting as a strong tracer of the movement of upwelled water (Rodríguez et al., 1999). Considerable interannual variability is observed in the predominance of fish larvae groups, presumably reflecting different environmental conditions in the upwelling system or different source waters. In a 1993 survey immediately south of the Canaries archipelago, 94% of the neritic larvae were captured in the offshore extension of a filament, the two most common species being sardine (*Sardina pilchardus;* 27.6%) and anchovy (*Engraulis encrasicolus;* 7.5%). In contrast, in 1999, the dominant clupeid species in the area was *Engraulis encrasicolus* (J.M. Rodríguez com. pers.). In 2001 the situation had reversed to its earlier state of dominance by *Sardina pilchardus* (Hernández-León et al., unpubl.). In 1993, the mean length of sardine larvae increased along the filament suggesting that the high zooplankton biomass inside the filament could maintain their growth.

As with phytoplankton, filaments transport zooplankton species originating on the shelf to the open ocean. However, while chlorophyll decreases sharply beyond the shelf break, due to decreasing nutrient concentration and grazing (Basterretxea and Arístegui, 2000), zooplankton show high biomass in the oceanic domain as result of their longer generation times and longevity (Hernández-León et al., 2002). Indices of feeding, metabolism and growth progressively decrease along the offshore extension of the filament, suggesting that advection, rather than local enrichment processes inside the filament, are mostly responsible for the high biomass. Additionally, a change in feeding behaviour takes place between coastal waters and the open ocean. Over the shelf edge zooplankton are mostly herbivorous but they gradually change their diet to fulfil their metabolic demands with non-pigmented food (e.g., microzooplankton) as they move to the ocean (Hernández-León et al., 2002).

In the sense of Bakun's (1996) triad of enrichment, concentration and retention, filaments can act as either retentive or dispersive agents. Coastal upwelling provides the basic enrichment necessary for a favourable reproductive habitat while fronts between upwelled and oceanic water provide a means of concentration. Filaments, by providing a transport mechanism to remove neritic fish eggs and larvae from the area of their continental shelf spawning to the open ocean, may act dispersively. On the other hand, the Canary Islands filament near Cape Juby has been shown to be entrained around a quasi-permanent cyclonic eddy (Barton et al., 1998; Navarro-Pérez and Barton, 2001). Filament waters may so be returned to the continental shelf on a time scale of 7–10 days by the cyclonic circulation (Navarro-Pérez and Barton, 1998). The filament in this case acts retentively and provides a nursery within which the larvae may grow before returning to their shelf environment (Rodríguez et al., 1999).

Moreover, both cyclonic and anticyclonic eddies spun off the islands may interact with the filament to entrain upwelled water carrying larvae (Rodríguez et al., 2001), in some cases facilitating their transport from the African coast to near shore of the islands. Since zooplankton accumulates near the islands, (Hernández-León, 1991), it has been suggested that the interaction between filaments and eddies enhances recruitment there. Recent studies (Hernández-León et al., unpubl.) suggest a connection between the dominant species in the African upwelling area, their larvae transported in the filaments, and the presence of their juveniles in the fishing catch around the islands. This genetic refreshing from the African coast is thought to promote the high fish abundances sometimes observed around the Canaries Archipelago.

4. Fish and fisheries

4.1. Link with the environment

Off the Iberian Peninsula, the long-term changes in alongshore winds during recent decades are related to the NAO, and lead to variations in the patterns of upwelling in the region and decadal fluctuations in the annual catch of sardine (Borges et al., 2003). Both sustained and intermittent northerly winds, and therefore upwelling, during the winter spawning season have a negative impact on sardine (*Sardina pilchardus*) and horse mackerel (*Trachurus trachurus*) recruitment and catches the following year, even if adequate upwelling conditions occur later during the summer upwelling (feeding) season (Santos et al., 2001).

South of 36°N, variability of recruitment is poorly understood due to the lack of adequate data. However, several patterns relating the environment with fish abundance or reproductive strategies have been documented (Cury and Roy, 1991; Durand et al., 1998). A comparative analysis of the spawning pattern of the major small pelagic species in the Canary Current shows that there is no correspondence between the spawning seasons of sardine and sardinellas (*Sardinella aurita* and *S. maderensis*) and the occurrence of upwelling (Roy et al., 1989; 1992). In some areas spawning occurs during the upwelling season (Senegal) and in other areas outside the upwelling season (Morocco) or when upwelling activity reaches a seasonal minimum (Sahara). Rather than being associated with the maximum upwelling intensity, it appears that the timing of spawning is associated with the occurrence of wind speed of about 5–6 m s^{-1} (Roy et al., 1992). This wind speed range corresponds to the optimal wind conditions for recruitment success, as defined by Cury and Roy (1989). This correspondence between the spawning peaks and the optimal wind value for recruitment success illustrates the long-term adaptation of small pelagic fish reproductive strategies to the environment.

The dynamics of the Morrocan sardine fisheries was studied in great detail by Belvèze and Erzini (1983) and Belvèze (1991). Three main populations of sardine are found along the northwest African coastline. An analysis of the catches over the last 40 years showed that these three populations have distinct patterns of abundance variability (Kifani, 1998). The boundaries between the three populations are located approximately at 27°N and 34°N, but these stocks migrate latitudinally in relation to the seasonal dynamics of upwelling in the region. These migrations have a minor effect on the fisheries occurring in the southern part of the region. However, the migration of the central sardine stock to the upwelling area that develops in summer north of Cape Guir (29°N) is a major component of the dynamics of the traditional Moroccan fishery that operates north of 30°N (Belvèze and Erzini, 1983; Belvèze, 1991). The intensity of this migration appears to be linked to the strength of the upwelling, with higher catches being recorded during years of enhanced upwelling. Sustained upwelling is thought to enhance primary production and sardine feeding condition in summer north of Cape Guir, resulting in a strengthening of the migration flux and consequently higher catches by the fishery. The traditional Moroccan fishery collapsed in the early 1990s following a constant decline of the catches since the mid-1970s. The reason for this collapse is still an unresolved issue and several hypotheses have been proposed (Do-Chi and Kiefer, 1996). Changes of the migration pattern, unfavourable environment in the traditional fishery zones affecting fish migration, a southward shift in the centre of gravity of the sardine population due to climatic changes were the different mechanisms that were put forward. However, in the early eighties, Belvèze and Erzini (1983) expressed concern about the persistent decline of the catches that they observed in the traditional fishery areas. They linked this decline to a reduced sardine summer migration related to a decrease of the upwelling favourable wind recorded at a coastal station (Essaouira, 31°30' N). They predicted that if the observed tendencies persisted, a geographical concentration of the central stock between 24°N and 30°N could likely occur and severely impact the fisheries operating at the geographical boundaries of the stock. It is worth noting that this scenario, proposed in the early eighties, adequately anticipated what happened during the following decade.

The southernmost sardine population is mainly exploited by foreign fishing fleets, most of them from Eastern Europe. The catch record showed a pattern of drastic booms and bursts with two major peaks in the late seventies and late eighties during which annual catches reach 600000 t. It is recognised that these periods of enhanced catches were associated with a dramatic increase of the abundance and with a significant southward extension of the southern population (Binet, 1997; Kifani, 1998). During the two periods of high abundance, significant sardine catches were recorded as far south as Cape Vert (14°N). There is an apparent synchrony between the two outbursts of the southern sardine population and periods of enhanced upwelling in the central Canary Current region, around 24°N (Binet, 1997; Roy and Reason, 2001).

The highly seasonal activity of the upwelling off Mauritania and Senegal (between 22°N and 14°N) induces profound changes in the biological community structure: the ecosystem shifts from tropical to a sub-tropical influence within a few months. Abrupt seasonal shifts in the environment induce a pronounced seasonal migration of the major pelagic and demersal fish stocks between Senegal,

Mauritania and the Sahara region. The following example illustrates how the seasonal dynamic of the upwelling impacts on the latitudinal distribution of the fish, on the fishing activities, as well as on the fish landings. The appearance of a migrant population of thiof (*Epinephelus aeneus*) along the north coast of Senegal is related to the onset of the Senegalese upwelling, which takes place between November and May (Cury and Roy, 1988). Anomalies of SST data collected at coastal stations were used to characterize the upwelling intensity. Using CPUE in the two main landing ports (Saint-Louis and Kayar, 15°-16°N) a mean lag of about one month was found between the occurrence of the upwelling and the arrival of the thiof off Kayar (Fig. 23.9a). The migration of the thiof from Mauritania to Senegal appears to be not only related to the onset of the Senegalese upwelling but is also linked to the relaxation of the upwelling off northern Mauritania (Fig. 23.9b).

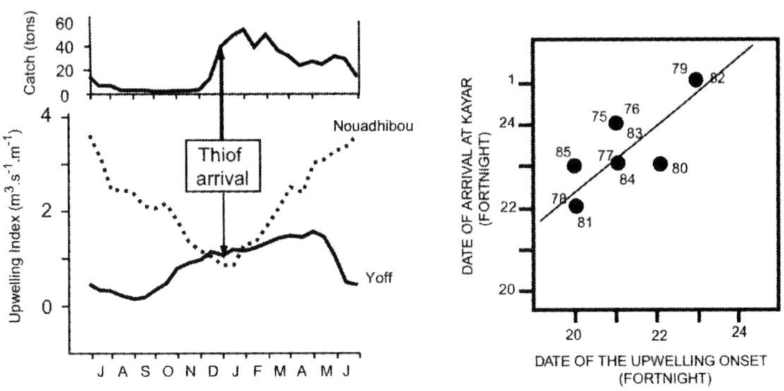

Figure 23.9 (a) The relationship between the seasonal arrival of Thiof at Kayar and the onset of the upwelling (1975–1985). (b) Mean monthly upwelling (m^3 s^{-1} m^{-1}) indices at Yoff (Senegal) and Nouadhibou (Mauritania) and mean monthly thiof catches (tons) at Kayar. (Adapted from Cury and Roy, 1988).

The effect of the environment on the Canary Current fish populations and fisheries has been investigated using production models incorporating an environmental variable in addition to fishing effort (Fréon et al., 1993). These models have been successfully applied to several pelagic stocks in the region (Fréon, 1983; 1988). In the case of Senegal for instance, the abundance of the sardinella stock appears to be related to the interanual variations of a wind derived upwelling index. Off Morocco, the variability of the upwelling has a significant impact on the catchability of sardine in the northern region.

4.2. *Major patterns of changes off the Iberian Peninsula*

In the northern part of the Canary Current, there is evidence of changes in fish distribution and community structure related to an increase of water temperature. Air and sea temperatures over most of the NE Atlantic increased since the late 1980s by at least 0.4° C per decade (Dias et al., 1992; Brander et al., 2003). Over the same period extensive northward shifts in distribution of commercial and non-commercial fish species occurred from southern Portugal to northern Norway (Brander et al., 2003). Off the Iberian Peninsula, there is evidence that warm water species have been extending their range to southern Portugal with, during the last decade, an increase of species whose distribution was previously limited to the Mediterranean and/or NW Africa (Brander et al., 2003). Quero et al. (1998) observed also that a variety of tropical species have systematically extended their ranges northward along the European continental slope since the early 1960s.

Decadal fluctuations in the catches of small pelagic fishes species off the western Iberian Peninsula were observed in the last century (Borges et al., 2003; Cendrero, 2002; ICES, 2002). Sardine landings in Portugal peaked in the 1930s and especially in the 1960s, when landing reached an historical maximum of 158000t. Landings in Spain also presented high values in the 1960s. At the end of the 1960s and beginning of 1970s there was an important decline in the landings of the two countries. However, the major decline was recorded in the main sardine fishing grounds north of 41° N, while in the southern areas there was a slight increase (ICES, 2002). After the abrupt decline at the end of the 1960s, the Portuguese landings in the northern region remained stable until 1994 with values around 45000t, followed by a slight decrease of the catches during the following period (ICES, 2002). Since 1985, sardine catches off Galicia (Spain) have been declining continuously with an historical low recorded in 1999–2000 (ICES, 2002).

Using information from eggs and larvae surveys during the period 1985–2000, Stratoudakis et al. (2003) found significant changes in the distribution of sardine eggs and larvae off Portugal. They showed that the area where fish eggs are encountered significantly decreased from 11800 km^2 in 1988 to around 7200 km^2 in late 1990s. This decline is related to a marked reduction in egg abundance in the northern Portuguese spawning grounds, and corresponds to changes observed in the distribution of larvae. The pattern of change is coherent with changes observed in sardine catches off northwestern Iberia, and is also corroborated by changes in the area of distribution and abundance of sardine estimated by Portuguese and Spanish acoustic surveys (Stratoudakis et al., 2002; 2003; ICES, 2002). Other documented changes in northern Portugal during the 1980s and 1990s, include a reduction of the abundance of sardine larger than 16 cm and a decrease in the length at first maturation (Stratoudakis et al., 2002).

4.3. Fisheries in the NW African coast

Total landings in the NW Africa region fluctuated between 1.3Mt and 2.6Mt over the last 30 years. As in the other three major eastern boundary currents, fisheries landings in the region are based on pelagic fish species which represent 70% of the total catches on average (Fig. 23.10). Catches of pelagic species are dominated by the sardine. Its contribution to the total pelagic catches peaked in 1977, sharply declined in the late seventies and showed a constant positive trend up to the mid 1990s when it suddenly dropped to pre mid-1970s levels (Fig. 23.11). These fluctuations are thought to be the result of environmental fluctuations as well as changes in the exploitation (Binet et al., 1998; Roy and Reason, 2001). Demersal catches peaked in the early 1970s with a maximum of 440000t in 1974 (Fig. 23.10). The persistent negative trend that characterises the demersal catches since the early

1980s is an indication that most of the demersal fish populations are fully exploited in the region (FAO 1997).

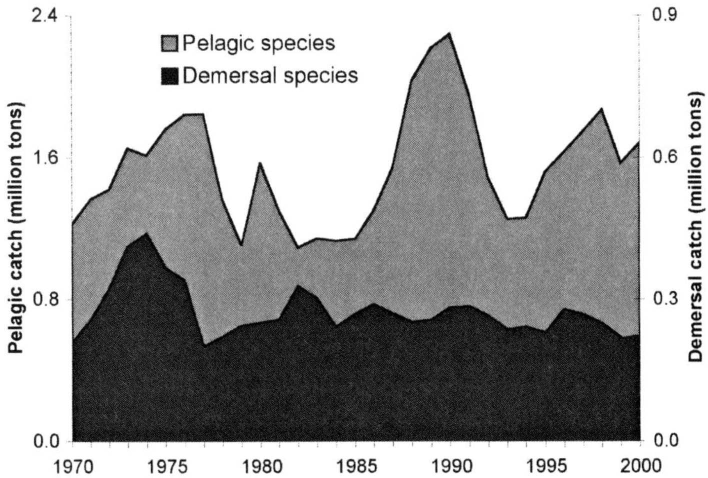

Figure 23.10 Annual catches of pelagic and demersal fish species in the Canary Current region from 1970 to 2000. Source: CECAF (Eastern Central Atlantic) capture production 1970–2000 from FAO.

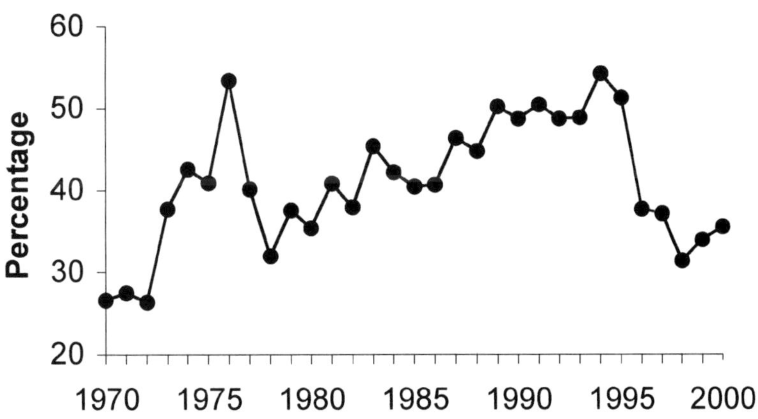

Figure 23.11 Contribution of the European sardine (*Sardina pilchardus*) to the total pelagic catches in the Canary Current from 1970 to 2000. Source : CECAF (Eastern Central Atlantic) capture production 1970–2000 from FAO.

Drastic changes in the exploitation occurred over the last 50 years. Both small scale and industrial-type fisheries have drastically expanded over the whole region by taking advantage of technological changes. Artisanal fisheries in West Africa have been in operation for centuries. Since the early 1960s, they have considerably improved their efficiency by increasing their size, integrating new technologies such as outboard engines and new fishing gears (Chauveau, 1991). The regional importance of the artisanal sector is best illustrated by the Senegalese small-scale fisheries which drastically "mutated" over the last 30 years to become as socially and economically efficient as the industrial fishery sector (Kébé, 1994). With a fleet of about 5000 motorized canoes, this fleet provides employment to about 35000 fishermen. Total catches reached 270000t of pelagic fish and 46000t of demersal fish in the early 1990s (Ferraris et al., 1998). During the 1970s and 1980s, exploitation by foreign nations was particularly active in the Canary Current region with a total 25 long-range fishing fleets operating mainly in the central part of the region (FAO, 1997; Maus, 1997). This pattern changed drastically in the 1990s following the political transformation in the Eastern European countries. By targeting mostly pelagic species, these fleets contributed up to 50% of the total regional marine catches, but their contribution has been declining rapidly since the early 1990s, with the partial withdrawal of Eastern European and former USSR fleets from West Africa (FAO 1997).

4.4. Fish population outburst in the Canary Current

In the Canary Current several species that were known to be rare, developed huge biomass during several years then suddenly vanished in a relatively short time period. Off Morocco an outburst of blue whiting (*Micromesistius poutassou*) was recorded in the 1960s, which later disappeared. In the 1970s snipefishes (*Macrorhamphosus scolopax* and *M. gracilis*), which have very different life-history traits, were found all along the Moroccan coast. In 1976, a biomass of about 1 Mt was estimated during an acoustic survey. In the 1980s, the abundance was drastically reduced and today the two species are rarely encountered.

In the late 1960s, trigger fish (*Balistes carolinensis*) was scarce in the Central Atlantic. Between 1972 and 1980, the biomass of this species drastically increased in the Gulf of Guinea to reach more than 1 Mt. This increase in abundance was associated with a wide geographical spreading that reached the southern part of the Canary Current, south of 20°N (Gulland and Garcia, 1984). Off Senegal, the peak in abundance was recorded in the early 1980s. A sharp decline in abundance followed the outburst of the population and at the end of the 1980s this species almost disappeared from the different ecosystems. While several authors believe that the fisheries-induced reduction of the sparid communities facilitated the outburst of the trigger fish (Gulland and Garcia, 1984), environmental changes may also have affected the trigger fish population dynamics (Caverivière, 1991).

The octopus (*Octopus vulgaris*) population increased significantly in abundance in the mid-1960s in the Canary Current (Caddy and Rodhouse, 1998). Three stocks of octopus are presently exploited by both industrial and artisanal fisheries off southern Morocco, Mauritania and Senegal. The rapid emergence of the octopus in Mauritania in the late 1960s led to historical yields of 52900t and 45600t recorded in 1976 and 1987, respectively. During recent years, the production has decreased to about 20000 t (Faure et al, 2000). A rapid expansion of the stock was observed off Morocco with catches reaching 100000t at the beginning of the 1980s. In the mid 1980s, octopus abundance started to increase off Senegal and both the local small-scale and industrial fisheries shifted a significant part of their activity to target this highly economically valuable resource. A peak catch of 17000t was recorded in 1986 whereas catches represented only a few hundreds tonnes before.

Altogether the combined catches in the Canary Current fisheries ranged between 40000t and 90000t from 1985 to 1995. During recent years, catches off southern Morocco and Mauritania have been drastically reduced (Inejih, 2000). Like in the case of the trigger fish, the outburst of the octopus in the Canary Current appears to be related to a lesser abundance of the sparid community caused by strong fishing pressure (Gulland and Garcia 1984), which is supposed to lower larval mortality and competition for food in the octopus (Caddy and Rodhouse 1998).

5. Decadal environmental changes

Wooster et al. (1976), using a compilation of merchant ship SST and wind data from 43°N to 7.5°N, summarised the seasonal patterns of the Canary Current coastal upwelling. In the mid 1970s, intensive process oriented studies were carried along the coast to study the Canary Current upwelling with the main focus being off the Sahara coast between 20°N and 26°N (see Hempel, 1982 for a collection of papers presenting detailed results of the CINECA program). After the intensive process oriented studies carried during the 1970s, most of the research and data collection effort was directed toward stock assessment and fisheries related studies. Except from some intensive surveys by the former GDR and USSR research groups off Mauritania, few oceanographic cruises have been carried out by the National Fisheries Research Centres. To our knowledge, there is no long-term time series of subsurface data available. A network of coastal stations, where daily SST and sometimes nutrient data are collected, is maintained by several fisheries research centres along the coast (see Cury and Roy, 1991 and Durand et al., 1998 for details) but the accessibility of these data remains limited.

There is an abundant literature on the decadal climatic variability in the subtropical and Northern Atlantic basin (Enfield and Mayer, 1997; Seager et al., 2000; Eden and Jung, 2001; Marshall et al., 2001). However, most of these analyses are performed on relatively low spatial resolution dataset and one can wonder how much of the coastal signal is diluted within the broad oceanic signal. Several studies presented detailed information on the variability in a given area. Arfi (1985), using wind and sea surface temperature measurements at a coastal station, studied the variability of the wind-driven upwelling off Cape Blanc (Mauritania 21°N) from 1955 to 1982. The link between wind forcing and temperature fluctuations at 21°N was further explored by Ould-Dedah et al. (1999). An analysis of the interannual variability of the Senegalese upwelling (14°N) from 1963 to 1986 was presented by Roy (1989) using wind data from a coastal station. High-resolution SST data from satellite, available since the early 1980s, have also been used to investigate the SST structure and the variability over the shelf of the Canary Current (Nykjaer and Van Camp, 1994; Hernandez Guerra and Nykjaer, 1997; Demarcq and Faure, 2000).

In the northern part of the Canary Current, a significant change on the monthly wind distribution pattern on the Portuguese west coast occurred during the period of 1947–1991, namely an increase in the frequency and intensity of upwelling-favourable winds in winter (Borges et al., 2003). During the spring-summer season, several authors noticed an increasing trend in the alongshore wind time series off the Iberian Peninsula (Dickson et al., 1988; Bakun, 1990; 1992). However, contrary results were reported from the analysis of wind and sea surface temperature observations (1947 to 1992) at the meteorological station of Cape Carvoeiro (Dias et al., 1996) and from daily atmospheric pressure data (1966 to 1999) off Galicia (Lavín et al., 2000). Pérez et al. (1995; 2000b) found a good correlation between the thermohaline properties of ENACW and the wind stress, the cumulative river discharge and the NAO index, illustrating the close coupling between water mass formation and climate change in the North Atlantic.

The interannual and long-term variability in the shelf domain of the Canary Current has not yet been summarised. To get an overview of the main pattern of variability at the surface, we extract SST data collected by merchant ships from the COADS database (Woodruff et al., 1987) from 1950 to 1995 within sixteen boxes along the Canary Current coast from 10°N up to 43°N (Fig. 23.12). The shipping line from Europe to the Indian Ocean follows the shape of the West African coast south of 28°N, and so in these regions data density is high along the coast. Between 28°N and 32°N, the shipping line moves offshore and data density sharply decreases. Further north, along the coast of Morocco, data density increases again and reaches a maximum off Spain and Portugal. To avoid potential biases due to seasonal changes in the data distribution, observations from former USSR fishing fleet which was quite active in the 1980s and early 1990s in the central region were excluded. For each of the sixteen boxes, monthly SST time series from 1950 to 1995 are constructed. Monthly SST anomalies are computed by subtracting the monthly climatology from the 1950–1995 monthly time series.

SST anomalies can be used to track changes in the intensity of the upwelling. Intensification (relaxation) of the upwelling process enhances (reduces) the upward flux of cold water along the coast; the offshore extension of the cold upwelled water is also enhanced (reduced) during intensification (relaxation) of the upwelling process. As a result, negative (positive) SST anomalies are expected during an intensified (relaxed) phase of the upwelling.

To get a synthetic view of the variability, monthly anomalies are then averaged by quarter. The data are presented on a time/latitude diagram and smoothed to get the long-term pattern of the SST variability (Fig. 23.13). In the early and mid-1970s, an intensive cooling (negative SST anomalies), characterised the variability of the SST anomalies during the first quarter (Fig. 23.13). It had a large latitudinal extension, affecting the whole Canary Current coastal domain from Spain to Senegal. South of 16°N, the low frequency variability was characterised by a succession of warm and cold periods. Between 16°N and 23°N, the variability appeared to increase after the mid-1970s with a succession of warm and cold periods, similar to further south. North of 23°N the variability decreases, the cooling of the 1970s being the major climatic event.

The variability of SST anomalies during the second quarter shares some common characteristics with that during the first quarter. There was an extensive cooling affecting the whole region in the 1970s but, although less pronounced, this cooling appeared to start in the mid 1960s and extended to the late 1970s. South of 16°N, the variability was also characterised by a succession of warm and cold periods, but with a slightly different pattern than during the first quarter. Further north, a pronounced warm event developed in the early 1960s between 25°N and 43°N. Following the cooling of the 1970s, the variability north of 25°N remains weak with a slight warming.

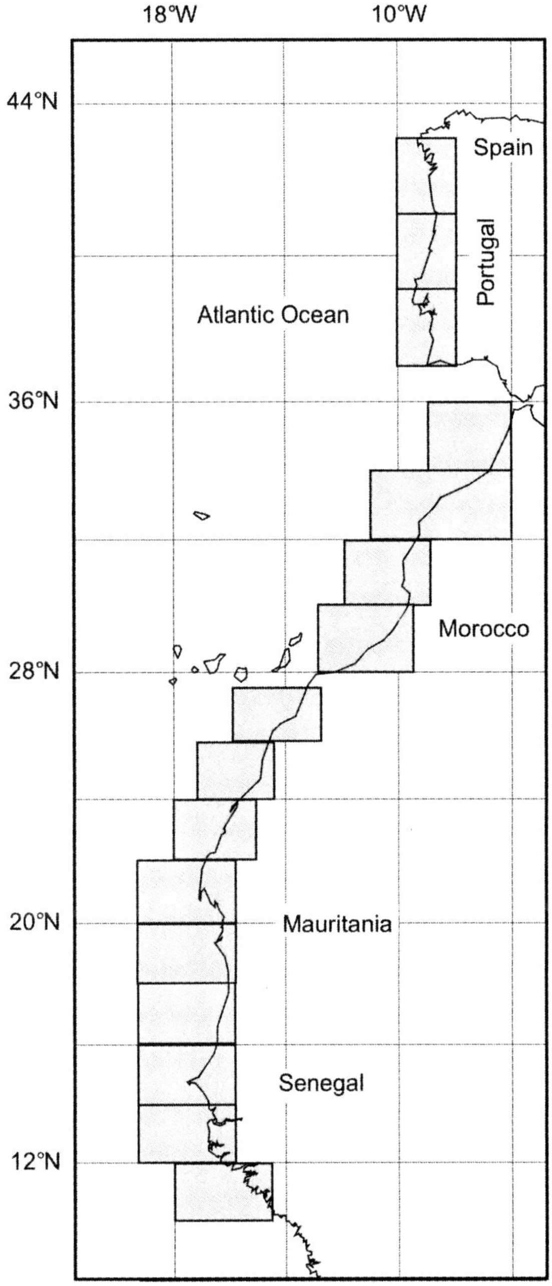

Figure 23.12 Location of the 16 coastal boxes where SST time-series have been built using data extracted from COADS.

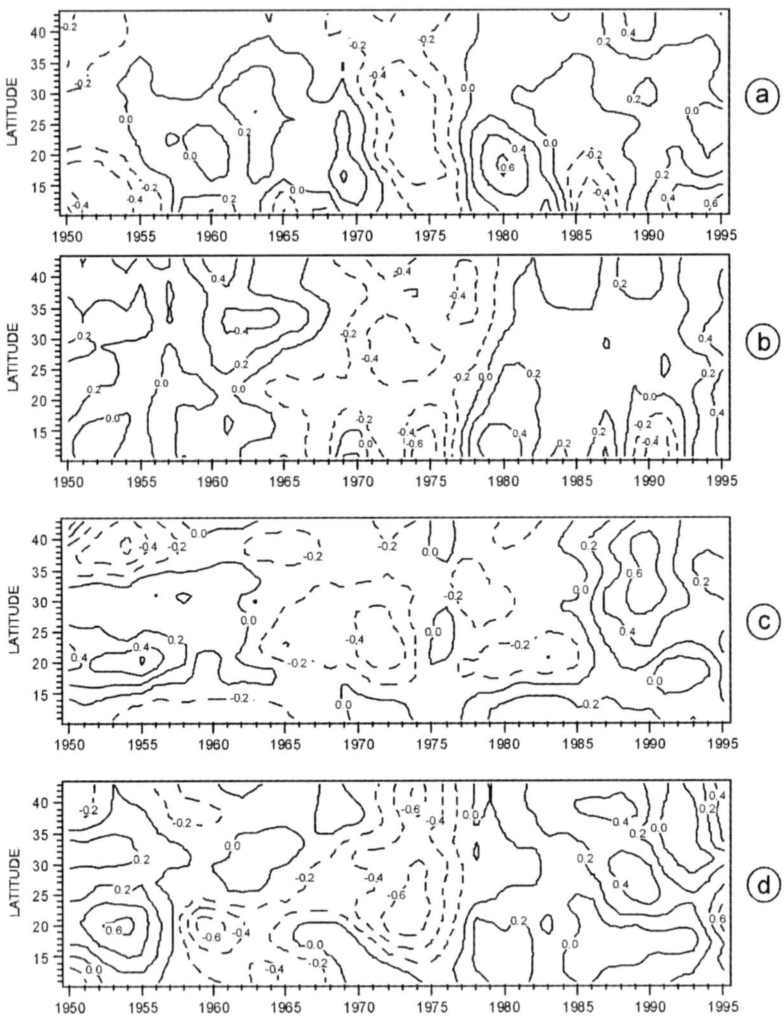

Figure 23.13 Time/latitude plot of SST anomalies (°C) in the Canary Current/Iberia from 43°N to 11°N and from 1950 to 1995 for the first (a), second (b), third (c) and fourth quarter (d).

During the third quarter, SST anomalies were characterised by a pronounced cooling from the mid-1960s to the mid-1980s. It affected the region north of 18°N up to 35°N. Another emergent pattern with large latitudinal extent was the strong warming developing north of 20°N during the late 1980s and early 1990s. From 1950 to 1965, the latitudinal variability contrasted with warm anomalies in the central region (maximum intensity at 20°N) and cold anomalies at both high and low latitudes. During the fourth quarter, the global 1970s cooling pattern over the region was again one of the salient features of the SST anomalies. It reached maximum intensity between 15°N and 30°N from 1972 to 1975. A warming affecting all the Canary Current since the 1980s followed this cooling.

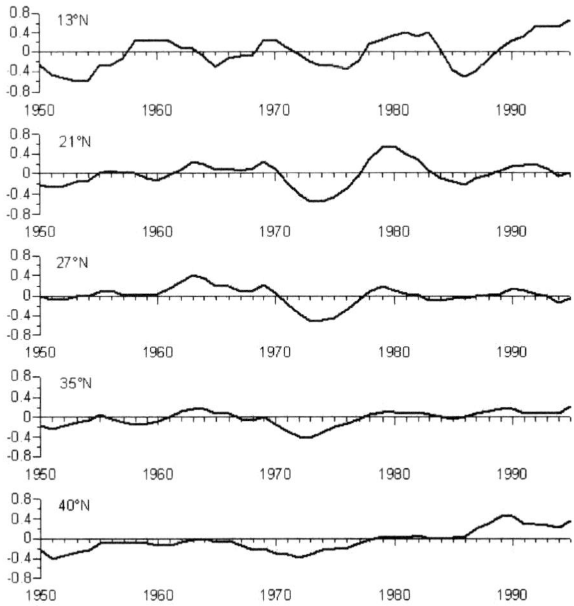

Figure 23.14a Trend of the SST anomalies (°C) at selected latitudes in the Canary Current/Iberia during the first quarter, from 1950 to 1995

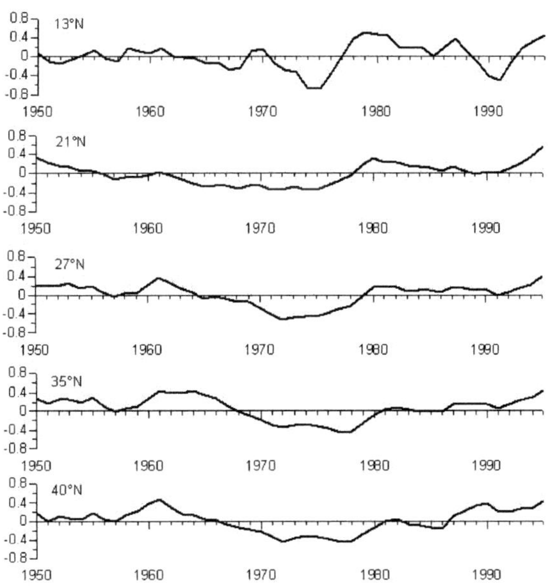

Figure 23.14b Trend of the SST anomalies (°C) at selected latitudes in the Canary Current/Iberia during the second quarter, from 1950 to 1995

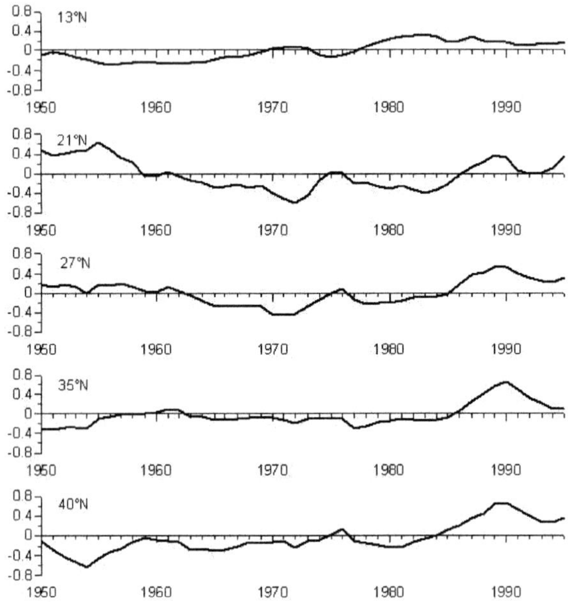

Figure 23.14c Trend of the SST anomalies (°C) at selected latitudes in the Canary Current/Iberia during the third quarter, from 1950 to 1995

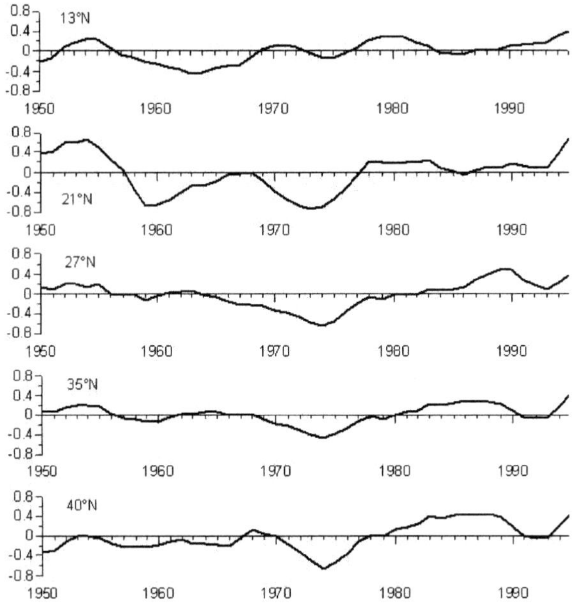

Figure 23.14d Trend of the SST anomalies (°C) at selected latitudes in the Canary Current/Iberia during the fourth quarter, from 1950 to 1995

In summary, time series in selected regions are presented to highlight the dominant patterns of variability of SST anomalies by quarter from 1950 to 1995 (Fig. 23.14a to 23.14d). The cooling in the 1970s is a major feature. It affected the whole region during and outside the upwelling seasons (Binet, 1997). The southern part of the region presents a quite dynamic pattern of variability unique to the area. This is confirmed by a comparative analysis of the variability of SST anomaly time series using the standard deviation as an index of the variability (Fig. 23.15). It shows that the SST variability reaches a maximum during the first and second quarter in the southern part of the region. During the third and fourth quarters, the interannual variability is maximum around 20°N. This suggests that the variability of the upwelling strongly enhances the interannual variability of SST anomalies in the southern part of the Canary Current.

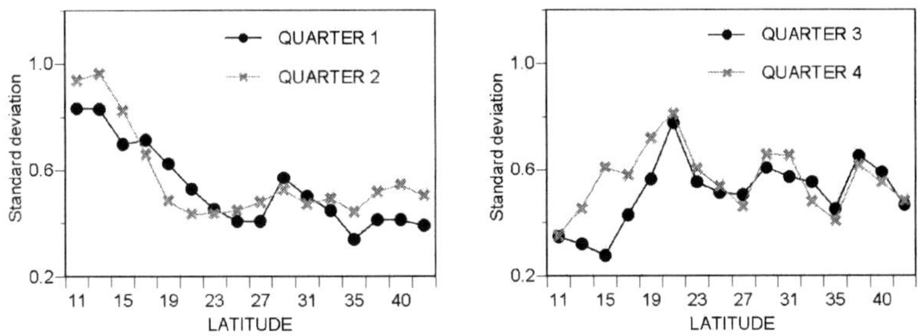

Figure 23.15 Trend of the SST anomalies (°C) at selected latitudes in the Canary Current/Iberia during the fourth quarter, from 1950 to 1995

5.1. Link with the North Atlantic Oscillation (NAO)

The North Atlantic Oscillation (NAO) characterises a meridional oscillation in atmospheric mass with centres of action being the Icelandic low and the Azores high. It is most pronounced in amplitude and areal coverage during winter. NAO is an important contributor to the North Atlantic climatic variability (Hurrel, 1995, Marshall et al., 2001) and has a measurable impact on the North Atlantic ecosystem (Fromentin and Planque, 1996).

The link between the NAO and the Canary Current upwelling is investigated by looking at the correlation between quarterly averaged values of the NAO index and the corresponding SST anomalies over the regions (Table 3). The NAO index is based on the difference of normalized sea level pressures between Ponta Delgada, Azores and Stykkisholmur, Iceland from 1865 through 1995. This index is slightly different from the winter version of the index that uses data from Lisbon. The Ponta Delgada station is chosen instead of Lisbon to adequately capture the NAO during the four quarters. A positive NAO index indicates stronger than average westerlies and anomalously high pressures across the sub-tropical Atlantic.

TABLE 3.
Correlation between quarterly averaged SST anomalies in the Canary Current and the corresponding NAO anomalies, from 1957 to 1995 (* = p<0.05, ** = p<0.01).

Latitude	1st quarter	2nd quarter	3rd quarter	4th quarter
11°N	0.02	-0.05	0.03	-0.19
13°N	-0.01	-0.10	0.04	-0.26
15°N	-0.16	-0.16	0.07	-0.38*
17°N	-0.37 *	-0.22	0.13	-0.35 *
19°N	-0.41 **	-0.31	0.04	-0.44 **
21°N	-0.30	-0.39 *	-0.01	-0.44 **
23°N	-0.48 **	-0.44 **	-0.06	-0.49 **
25°N	-0.60 **	-0.40 **	-0.22	-0.41 **
27°N	-0.56 **	-0.34 *	-0.29	-0.54 **
29°N	-0.01	-0.44 **	-0.20	-0.58 **
31°N	-0.15	-0.29	-0.17	-0.40 **
33°N	-0.09	-0.13	-0.04	-0.44 **
35°N	-0.20	-0.14	-0.05	-0.50 **
38°N	0.13	-0.31	-0.19	-0.57 **
40°N	0.26	-0.29	-0.24	-0.50 **
42°N	0.26	-0.24	-0.25	-0.50 **

Negative and statistically significant (p<0.01) correlations between NAO and SST anomalies off the Canary Current occur during the first and second quarters in the central region (between 18°N and 30°N), as well as during the fourth quarter north of 20°N. The inverse relationship indicates that negative (positive) SST anomalies are related to positive (negative) NAO anomalies. In the central region, this correlation suggests that an intensification of the westerlies (positive NAO index) across the sub-tropical Atlantic induces an intensification of the upwelling favourable wind and enhances the upwelling process (negative SST anomalies). In the case of a relaxed meridional oscillation (negative NAO index), the weaker than average atmospheric circulation contributes to a relaxation of the Canary Current upwelling (positive SST anomalies). In the northern region (north of 30°N), upwelling is not a dominant oceanographic process during the 4th quarter. An alternative mechanism accounting for the correlation between NAO and SST involves the intensification of the westerlies in early winter leading to a premature erosion of the thermocline and to a deepening of the surface mixed layer. Both the erosion of the thermocline and the deepening of the surface mixed layer may result in negative SST anomalies.

5.2 *ENSO Teleconnection*

Pacific El-Nino/Southern-Oscillation (ENSO) events have a major impact on the world climate. In the tropical Atlantic, the SST and wind fields are regularly affected by Pacific equatorial variability (Hastenrath et al., 1987, Nobre and Shukla, 1996; Enfield and Mayer, 1997). Large-scale analyses have shown that the North Atlantic warms in response to the Pacific ENSO with a lag of about 3 to 6 months, the effect being apparently stronger in the north-western part of the basin and during the boreal spring and early summer (Enfield and Mayer, 1997). The origin of this lag is not clear, but Klein et al. (1999) argue that the El Nino warming signal is communicated to the tropical Atlantic via a reduction in surface latent heat flux associated with reduced trade winds. In the Canary Current, the wind is the driving force of the upwelling and an alteration of the trade wind activity has a pronounced effect on the ecosystem, thus one can expect to observe a strong connection between ENSO and the coastal upwelling. This is investigated by looking at the correlation between quarterly SST anomalies during the first semester and the Southern Oscillation Index (SOI) during the fourth quarter of the preceding year. In the southern part of the region (south of 19°N), the correlation is statistically significant ($p<0.01$) during the first and second quarter of the year (Table 4). As expected from the weakening of the Atlantic trade wind that is associated with ENSO events, there is a negative correlation between SOI and SST anomalies: warm events in the Pacific (negative SOI) lead to the development of positive SST anomalies in the southern part of the Canary Current during late winter and early spring. Moreover, it appears that the correlation holds also for cold events in the Pacific (positive SOI), suggesting that there is a strong teleconnection between the coastal upwelling activity in the Atlantic and the state of the Pacific Ocean. This link between ENSO and the Canary Currrent upwelling has been explored in details by Roy and Reason (2001). It is thought that the mechanism responsible for this remote forcing involves a tropospheric connection along 10°N-20°N between the Atlantic and the Pacific resulting in an alteration of the Atlantic trade winds by the conditions over the Pacific Ocean (Enfield and Mayer, 1997).

TABLE 4.
Correlation between quarterly averaged SST anomalies in the Canary Current and the SOI anomalies during the fourth quarter of the preceding year, from 1957 to 1995 (* = $p<0.05$, ** = $p<0.01$).

Latitude	1st quarter	2nd quarter
11°N	-0.41 **	-0.46 **
13°N	-0.48 **	-0.47 **
15°N	-0.48 **	-0.54 **
17°N	-0.40 **	-0.51 **
19°N	-0.38 *	-0.48 **
21°N	-0.42 **	-0.38 *
23°N	-0.36 *	-0.21
25°N	-0.28	-0.15
27°N	-0.17	-0.19
29°N	-0.11	-0.25
31°N	-0.26	-0.28
33°N	-0.28	-0.24
35°N	-0.18	-0.16
38°N	-0.16	-0.09
40°N	-0.15	-0.03
42°N	-0.29	-0.04

6. Synthesis and future research

Considerable information has been gathered on biogeochemical cycles and fisheries of the eastern boundary of the subtropical gyre of the North Atlantic, but knowledge of some space and time scales, of some areas and of certain seasons is lacking. For example the northwestern Iberian margin has been intensively studied whereas southern Iberia is much less well known. The dynamics and biogeochemistry of much of the African coast have been little sampled. Some basic questions about seasonal variation remain. The winter downwelling regime off Iberia is poorly known. Recent evidence suggests the development of a previously unrecognized surface poleward counterflow in the Canaries region during winter, despite favourable, albeit weaker, Trade winds than in summer. It has been hypothesized that this is linked to a horizontal re-circulation associated with the Cape Guir filament, which would have major consequences for carbon fluxes in the area. Large scale, high resolution, multidisciplinary surveys, such as have been made in the California Current, would throw much light on these issues and their implications.

Both the NW African coast and the Canary Islands region are considered important sources of primary productivity. Although most of the organic matter produced is thought to be respired in the surface waters of the eastern boundary of the Canary Current, the interplay between island eddies and upwelling filaments may enhance the export of coastal enriched water to the oligotrophic open ocean. This would explain the strong imbalance between phytoplankton production (P) and community respiration (R) observed in the subtropical gyre, where the P/R ratio is normally <1. However, little is still known about the relative importance of vertical sedimentation of organic carbon in the coast vs. offshore transport, the nature of the organic matter exported (e.g. dissolved vs. particulate), or the different mechanisms of filament-eddy exchange through the Canaries Coastal Transition Zone. Multi-disciplinary oceanographic studies should be strongly supported in order to achieve the basic knowledge of the functioning of these complex coastal ecosystems in order to test and validate trophic and biogeochemical models.

South of Gibraltar much knowledge of the coastal upwelling ecosystem was obtained during the intensive studies of the 1970s, but recent research has been largely (though not entirely) confined to coastal regions in relation to fisheries and dynamics of fish populations. Major changes like the redistribution of Moroccan sardine populations and outbursts of octopus and other species underline profound changes in the ecosystem over the last 30 years. In a context of global climate change, escalating fisheries exploitation and increasing pressure on coastal resources, the need for further intensive study and greater understanding of the system is clear if it is to be managed in a sustainable way. The requirement is for a dual approach of more systematic and detailed observational programs that take advantage of advances in remote sensing and instrumentation, coupled with the development of linked numerical modeling that combines knowledge of ocean dynamics, the ecosystem, fisheries and proper management of the ecosystem.

Acknowledgements

This work was partially supported by the Spanish Plan Nacional de I+D (Projects COCA and MESOPELAGIC), the French IRD IDYLE project, the Portuguese Science Foundation (Projects SURVIVAL and PO-SPACC, and the programme PELAGICOS) and by the European Union (Projects "The Control of Phytoplankton Dominance", "European Canaries-Coastal Transition Zone", OMEX II, MORENA, CANIGO and OASIS). This is a contribution to the GLOBEC-SPACC program. EDB was partially funded by Mexican Conacyt Catedra Patrimonial EX-010009A. Hernández-Guerra kindly provided the satellite images for Figures 23.5 and 23.8.

Bibliography

Abrantes, F. and M.T. Moita, 1999. Water column and recent sediment data on diatoms and coccolithophorids off Portugal confirm sediment record of upwelling events. *Oceanol. Acta,* **22,** 319–336

Abrantes, F., H. Meggers, S. Nave, J. Bollman, S. Palma, C. Sprengel, J. Henderiks, A. Spies, E. Salgueiro, M.T. Moita and S. Neuer, 2002. Fluxes of micro-organisms along a productivity gradient in the Canary Islands region (29°N): implications for paleoconstructions. *Deep Sea Res. II,* **49,** 3599–3629.

Álvarez, M., E. Fernández and F.F. Pérez, 1999. Air–sea CO_2 fluxes in a coastal embayment affected by upwelling: physical versus biological control. *Oceanol. Acta,* **22,** 499–515.

Álvarez–Salgado, X.A., G. Rosón, F.F. Pérez and Y. Pazos, 1993. Hydrographic variability off the Rías Baixas (NW Spain) during the upwelling season. *J. Geophys. Res.,* **98,** 14447–14455.

Álvarez–Salgado, X.A., C. G. Castro, F.F. Pérez and F. Fraga, 1997. Nutrient mineralization patterns in shelf waters of the Western Iberian upwelling. *Cont. Shelf Res.,* **17,** 1247–1270.

Álvarez–Salgado, X.A., M.D. Doval, A.V. Borges, I. Joint, M. Frankinouglle, E.M.S. Woodward and F.G. Figueiras, 2001. Off–shelf fluxes of labile materials by an upwelling filament off the NW Iberian upwelling system. *Prog. Oceanogr.,* **51,** 321–339.

Álvarez–Salgado, X.A., S. Beloso, I. Joint, E. Nogueira, L. Chou, F.F. Pérez, S. Groom, J.M. Cabanas, A.P. Rees and M. Elskens, 2002. New Production of the NW Iberian shelf during the upwelling season over the period 1982–1999. *Deep Sea Res. I,* **49,** 1725–1739.

Álvarez–Salgado, X.A., F.G. Figueiras, F.F. Pérez, S. Groom, E. Nogueira, A. Borges, L. Chou, C.G. Castro, G. Moncoiffé, A.F. Ríos, A.E.J. Miller, M. Frankignoulle, G. Savidge and R. Wollast, 2003. The Portugal Coastal Counter Current off NW Spain: new insight on its biogeochemical variability. *Prog. Oceanogr.,* **56,** 281–321.

Araujo, M., J. Jouanneau,, P. Valerio, T. Barbosa, A. Gouveia,, O. Weber, A. Oliveira, A. Rodrigues and J. Dias, 2002. Goechemical tracers of northern Portuguese estuarine sediments on the shelf. *Prog. Oceanogr.,* **52,** 277–297.

Arfi, R. 1985. Variabilité inter-annuelle d'un indice d'intensité des remontées d'eaux dans le secteur du Cap-Blanc (Mauritanie). *Canad. J. Fish. Aquat. Sci.,* **42,** 1969–1978.

Arístegui, J. and W.G. Harrison, 2002. Decoupling of primary production and community respiration in the ocean: implications for regional carbon studies. *Aquat. Microb. Ecol.,* **29,** 199–209.

Arístegui, J. and M. F. Montero, 2005. Temporal and spatial changes in microplankton respiration and biomass in the Canary Islands: the effect of mesoscale variability. *J. Mar. Sys.* 54, 65–82

Arístegui, J., S. Hernández-León, M. Gómez, L. Medina, A. Ojeda and S. Torres, 1989. Influence of the wind on neritic plankton around Gran Canaria. In *Topics in Marine Biology.* Ros, J.D., ed., *Sci. Mar.,* **53,** 223–229.

Arístegui, J, P. Sangrà, S. Hernández-León, M. Cantón, A. Hernández_Guerra and J. L. Kerling, 1994. Island-induced eddies in the Canary Islands. *Deep-Sea Res. I,* **41,** 1509–1525.

Arístegui, J., P. Tett, A. Hernández-Guerra, G. Basterretxea, M.F. Montero, K. Wild, P. Sangrà, S. Hernández-León, M. Cantón, J.A. García-Braun, M. Pacheco and E.D. Barton, 1997. The influence of island-generated eddies on chloropyll distribution: a study of mesoscale variation around Gran Canaria. *Deep-Sea Res. I,* **44,** 71–96.

Arístegui, J, E.D. Barton, M.F. Montero, M. García-Muñoz and J. Escánez. 2003. Organic carbon distribution and water-column respiration in the NW Africa-Canaries Coastal Transition Zone. *Aquat. Microb. Ecol.* 33, 289–301

Bakun, A., 1990. Global climate change and intensification of coastal ocean upwelling. *Science,* **247,** 198–201.

Bakun, A., 1992. Global greenhouse effects, multi-decadal wind trends, and potential imapcts on coastal pelagic fish populations. *ICES Marine Science Symposia,* **195,** 316–325.

Bakun, A., 1996. Patterns in the ocean: ocean processes and marine population dynamics. Calif. Sea Grant College Syst. Univ. Calif., La Jolla, 323p.

Bakun, A., C.S. Nelson, 1991. The seasonal cycle of wind–stress curl in subtropical eastern boundary current regions. *J. Phys. Oceanogr.,* **21,** 1815–1834.

Barber, R.T. and R.L. Smith, 1981. Coastal Upwelling Ecosystems. In: *Analysis of marine ecosystems,* Longhurst, A.R. ed., Academic press, New York, pp. 31–68.

Barbosa, A.B., H.M. Galvao, P.A. Mendes, X.A. Álvarez–Salgado, F.G. Figueiras and I. Joint, 2001. Short–term variability of heterotrophic bacterioplankton during upwelling off the NW Iberian margin. *Prog. Oceanogr.,* **51,** 339–359.

Barquero, S., J. A. Cabal, R. Anadón, E. Fernández, M. Varela and A. Bode, 1998. Ingestion rates of phytoplankton by copepod size fractions on a bloom associated with an off–shelf front off NW Spain. *J. Plankton Res.,* **20,** 957–972.

Barton, E. D., 1987. Meanders, eddies and intrusions in the thermohaline front off northwest Africa. *Deep Sea Res.,* **24,** 1–6.

Barton, E. D., 1998. Eastern boundary of the North Atlantic: Northwest Africa and Iberia. Coastal segment (18,E). In *The Sea,* Vol. 11, Robinson, A. R. and K. H. Brink, eds, John Wiley and Sons, Inc., New York, pp. 633–657.

Barton, E. D. and P. Hughes, 1982. Variability of water mass interleaving off NW Africa. *J. Mar. Res.,* **40,** 963–984.

Barton, E. D., J. Arístegui, P. Tett, M. Cantón, J. García-Braun, S. Hernández-León, L. Nykjaer, C. Almeida, J. Almunia, S. Ballesteros, G. Basterretxea, J. Escánez, L. García-Weil, A. Hernández-Guerra, F. Ópez-Laatzen, R. Molina, M. F. Montero, E. Navarro-Pérez, J. M. Rodríguez, K. van Lenning, H. Vélez and K. Wild, 1998. The transition zone of the Canary Current upwelling region. *Prog. Oceanogr.,* **41,** 455–504.

Barton, E.D., G. Basterretxea, P. Flament, E.G. Mitchelson-Jacob, B. Jones, J. Arístegui and H. Felix, 2000. Lee region of Gran Canaria. *J. Geophys. Res.,* **105,** 17173–17193.

Basterretxea, G., 1994. Influencia de las estructuras oceanográficas mesoescalares sobre la producción primaria en la región Canaria. PhD. Thesis. University of Las Palmas de Gran Canaria, 113 pp.

Basterretxea, G. and J. Aristegui, 2000. Mesoscale variability in phytoplankton biomass distribution and photosynthetic parameters in the Canary-NW African coastal transition zone. *Mar. Ecol. Prog. Ser.,* **197,** 27–40.

Basterretxea, G., E.D. Barton, P. Tett, P. Sangrà, E. Navarro-Pérez and J. Aristegui, 2002. Eddy and DCM response to wind-shear in the lee of Gran Canaria. *Deep-Sea Research I,* **49,** 1087–1101

Belvèze, H., 1991. Influence des facteurs hydroclimatiques sur la pêcherie marocaine de petits pélagiques côtiers. In *Les pêcheries ouest-africaines: variabilité, instabilité et changement.* Cury, P. and C. Roy, eds., ORSTOM, Paris.

Belvèze, H. and K. Erzini, 1983. The influence of hydroclimatic factors on the availability of the sardine (Sardina pilchardus, Walbaum) in the Moroccan Atlantic fishery. In *Proceedings of the expert consul-*

tation to examine changes in abundance and species composition of neritic fish resources G.D. Sharp and J. Csirke, eds., FAO, Fish. Rep. 291, pp. 285–327.

Binet, D., 1997. Climate and pelagic fisheries in the Canary and Guinea currents 1964–1993: the role of trade winds and the southern oscillation. *Oceanol. Acta,* **20,** 177–190.

Binet, D., B. Samb, M.T. Sidi, J.J. Levenez and J. Servain, 1998. Sardine and other pelagic fisheries changes associated with multi-year trade wind increases in the southern Canary current. In *Global versus local changes in upwelling systems,* Durand, M.H., P. Cury, R. Mendelssohn, C. Roy, A. Bakun and D. Pauly, eds., ORSTOM Editions, Paris, pp. 211–233.

Blanton, J.O., K.R. Tenore, F.F. Castillejo, L.P. Atkinson, F.B. Schwing and A. Lavín, 1987. The relationship of upwelling to mussel production in the rías on the Western coast of Spain. *J. Mar. Res.,* **45,** 497–511.

Bode, A. and M. Varela, 1998. Mesoscale estimations of primary production in shelf waters: a case study in the Golfo Artrabo (NW Spain). *J. Exp. Mar. Biol. Ecol.,* **229,** 111–131.

Bode, A., E. Fernández, J.A. Botas and R. Anadón, 1990. Distribution and composition of suspended particulate matter related to a shelf–break saline intrusion in the Cantabrian Sea (Bay of Biscay). *Oceanol. Acta,* **13,** 219–228.

Bode, A., B. Casas and M. Varela, 1994. Size–fractioned primary productivity and biomass in the Galician shelf (NW Spain): Netplankton versus nanoplankton dominance. *Sci. Mar.,* **58,** 131–141.

Bode, A., B. Casas, E. Fernández, E. Marañón, P. Serret and M. Varela, 1996. Phytoplankton biomass and production in the shelf waters off NW Spain: spatial and seasonal variability in relation to upwelling. *Hidrobiologia,* **341,** 225–234.

Bode, A., M. Varela, S. Barquero, M.T. Álvarez–Osorio and N. González, 1998. Preliminary studies on the export of organic matter during phytoplankton blooms off La Coruña (North–western Spain). *J. Mar. Biol. Ass. U.K.,* **78,** 1–15.

Borges, A.V. and M. Frankignoulle, 2001. Short–term variations of the partial pressure of CO_2 in surface waters of the Galician upwelling system. *Prog. Oceanogr.,* **51,** 283–302.

Borges, A.V. and M. Frankignoulle, 2002a. Aspects of dissolved inorganic carbon dynamics in the upwelling system off the Galician coast. *J. Mar. Sys.,* **32,** 181–198.

Borges, A.V. and M. Frankignoulle, 2002b. Distribution of surface carbon dioxide and air-sea exchange in the upwelling system off the Galician coast. *Global Biogeochem. Cycles,* **16,** 10.1029/2000GB001385.

Borges, M. F., A. M. P. Santos, N. Crato, H. Mendes and B. Mota, 2003. Sardine regime shifts off Portugal: a time series analysis of catches and wind conditions. *Sci. Mar.,* **67**Suppl.1, 235–244.

Bory, A., C. Jeandel, N. Leblond, A. Vangriesheim, A. Khripounoff, L. Beaufort, C. Rabouille, E. Nicolas, K. Tachikawa, H. Etcheber and P. Buat-Ménard, 2001. Downward particle fluxes within different productivity regimes off the Mauritanian upwelling zone (EUMELI program). *Deep Sea Res. I,* **48,** 2251–2282.

Botas, J.A., E. Fernández, A. Bode and R. Anadón (1989). Water masses off the Central Cantabrian coast. *Sci. Mar.,* **53,** 755–761.

Brander, K., G. Blom, M. F. Borges, K. Erzini, G. Henderson, B.R. MacKenzie, H. Mendes, J. Ribeiro, A.M. P. Santos and R. Toresen, 2003. Changes in fish distribution in the eastern North Atlantic; are we seeing a coherent response to changing temperature?. *ICES Marine Science Symposia,* **219,** 261–270.

Brink, K.H. and T.J. Cowles, 1991. The Coastal Transition Zone Experiment. *J. Geophys. Res.,* **96,** 14637–14647.

Brown, P.C. and J.C. Field, 1986. Factors limiting phytoplankton production in a neashore upwelling area. *J. Plankton Res.,* **8,** 55–68.

Cachâo, M. and M.T. Moita, 2000. *Coccolithus pelagicus* a productivity proxy related to moderate fronts off Western Iberia. *Mar. Micropaleontol.,* **39,** 13–155

Caddy, J.F. and P.G. Rodhouse, 1998. Cephalopod and groudfish landings: evidence for ecological change in global fisheries. *Rev. Fish Biol. Fish.,* **8,** 431–444.

Castro, C.G., 1997. Caracterización química del agua subsuperficial del Atlántico Nororiental y su modificación por procesos biogeoquímicos. Ph.D. Thesis, University of Santiago, 244 pp.

Castro, C.G., F.F. Pérez, X.A. Álvarez–Salgado, G. Rosón and A.F. Ríos, 1994. Hydrographic conditions associated with the relaxation of an upwelling event off the Galician Coast (NW Spain). *J. Geophys. Res.,* **99,** 5135–5147.

Castro, C.G., X.A. Álvarez–Salgado, F.G. Figueiras, F.F. Pérez and F.Fraga, 1997. Transient hydrographic and chemical conditions affecting microplankton populations in the coastal transition zone of the Iberian upwelling system (NW Spain) in September 1986. *J. Mar. Res.,* **55,** 321–352.

Castro, C.G., F.F. Pérez, X.A. Álvarez-Salgado and F. Fraga, 2000. Coupling between thermohaline and chemical fields during two contrasting upwelling events off the NW Iberian Peninsula. *Cont. Shelf Res,,* **20,** 189–210.

Caverivière, A., 1991. L'explosion démographique du baliste (Balistes carolinensis) en Afrique de l'Ouest et son évolution en relation avec les tendances climatiques. In *Pêcheries ouest-africaines: variabilité, instabilité et changement,* Cury P. and C. Roy, eds., Orstom Editions, Paris, pp. 354–367.

Cendrero, O., 2002. Sardine and anchovy crises in northern Spain: natural variations or an effect of human activities?. *ICES Marine Science Symposia,* 215:279–285.

Chauveau, J.P., 1991. Les variations spatiales et temporelles de l'environnement soci-économique et l'évolution de la pêche maritime artisanale sur les côtes ouest-africaines. Essai d'analyse en longue période: XVè-XXè siècle. In *Pêcheries ouest-africaines: variabilité, instabilité et changement.* Cury P. and C. Roy, eds., ORSTOM Editions, Paris, pp. 14–25.

Chavez, F.P., R.T. Barber, P.M. Kosro, A. Huyer, S.R. Ramp, T.P. Stanton and B. Rojas de Mendiola, 1991. Horizontal transport and the distribution of nutrients in the coastal transition zone off Northern California: effects on primary production, phytoplankton biomass and species composition. *J. Geophys. Res.,* **96 (C8),** 14833–14848.

Codispoti, L.A., 1981. Temporal nutrient variability in three different upwelling regions. In *Coastal upwelling,* Richards, F.A., ed., Coastal and Estuarine Sciences 1, Washington D.C., pp. 209–220.

Codispoti, L.A. and G.E. Friederich, 1978. Local and mesoscale influences on nutrient variability in the northwest African upwelling region near Cabo Corbeiro. *Deep Sea Res.,* **25,** 751–770.

Codispoti, L.A., R. C. Dugdale and H. J. Minas, 1982. A comparison of the nutrient regimes off Northwest Africa, Peru and Baja California. *Rapp. Proc. Reun. Cons. Int. Expl. Mer.,* **180,** 184–201.

Cury, P. and C. Roy, 1988. Migration saisonnière du thiof (Epinephelus aeneus) au Sénégal : influence des upwellings sénégalais et mauritaniens. *Oceanol. Acta,* **11,** 1, 25–36.

Cury, P. and C. Roy, 1989. Optimal environmental window and pelagic fish recruitment success in upwelling areas. *Canad. J. Fish. Aquat. Sci.,* **46,** 670–680.

Cury, P. and C. Roy (eds), 1991. Pêcheries ouest-africaines: variabilité, instabilité et changement. ORSTOM Editions, Paris, 525p.

Dale, A.W. and R. Prego, 2002. Physico–biogeochemical controls on benthic–pelagic coupling of nutrient fluxes and recycling in a costal upwelling system. *Mar. Ecol. Prog. Ser.,* **235,** 15–28.

Demarcq, H. and V. Faure, 2000. Coastal upwelling and associated retention indices from satellite SST. Application to Octopus vulgaris recruitment. *Oceanol. Acta,* **23,** 391–408.

De Hass, H., T.C.E. van Weering and H. de Stigter, 2002. Organic carbon in shelf seas: sinks or sources, processes and products. *Cont. Shelf Res.,* **22,** 691–717.

Dias, C. A., A. Amorim and M. das Dores Vacas, 1992. Sea-surface temperature: seasonal variation between the Iberian coast and the Madeira Islands, 1981–1987. *ICES Marine Science Symposia,* **195,** 177–186.

Dias, C. A., G. Pestana, E. Soares and V. Marques, 1996. Present state of sardine stock in ICES Divisions VIIIc and IXa. Working Document to the Working Group on the Assessment of Mackerel, Horse Mackerel, Sardine and Anchovy, 17 pp.

Dias, J., R. Gonzalez, C. Garcia and V. Diaz del Río, 2002a. Sediment distribution patterns on the Galicia-Minho continental shelf. *Prog. Oceanogr.* **52,** 215–231.

Dias, J., J. Jouanneau, R. Gonzalez, M. Araujo, T. Drago, T., C. Garcia, A. Oliveira, A. Rodrigues, J. Vitorino and O. Weber, 2002b. Present day sedimentary processes on the northern Iberian shelf. *Prog. Oceanogr,* **52,** 249–259.

Dickson, R. R., P. M. Kelly, J. M. Colebrook, W. S. Wooster and D. H. Cushing, 1988. North winds and production in the eastern North Atlantic. *J. Plankton Res.,* **10,** 151–169.

DoChi, T. and D.A. Kiefer, 1996. Workshop on the coastal pelagic resources of the upwelling ecosystem of Northwest Africa: research and predictions, Casablanca 15–17 April 1996. FAO:TCP/MOR/4556(A)

Drago, T., A. Oliveira, F. Magalhaes, J. Cascalho, J. Jouanneau and J. Vitorino, 1998. Some evidences of northward fine sediment transport in the northern Portuguese continental shelf. *Oceanol. Acta,* **21,** 223–231.

Duarte, C.M., S. Agustí, J. Arístegui, N. González and R. Anadón, 2001. Evidence for a heterotrophic subtropical northeast Atlantic. *Limnol. Oceanogr.,* **46,** 425–428.

Dugdale, R.C. and F.P. Wilkerson, 1989. New production in the upwelling center at Point Conception, California: Temporal and spatial patterns. *Deep Sea Res.,* **36,** 985–1007.

Dugdale, R.C., F.P. Wilkerson and A. Morel, 1990. Realization of new production in coastal upwelling areas: a means to compare relative performance. *Limnol. Oceanogr.,* **35,** 822–829.

Durand, M-H., P. Cury, R. Mendelssohn, C. Roy, A. Bakun and D. Pauly (eds), 1998. Global versus local changes in upwelling systems. Editions ORSTOM Paris. 594pp.

Eden, C., and T. Jung, 2001. North Atlantic interdecadal variability: Oceanic response to the North Atlantic Oscillation (1865–1997). *J. Climate,* **14,** 676–691.

Enfield, D. B., and D. A. Mayer, 1997. Tropical Atlantic sea surface temperature variability and its relation to El Nino-Southern Oscillation. *J. Geophys. Res.,* **102,** 929–945.

Epping, E., C. Van der Zee, K. Soetaert and W. Helder, 2002. On the oxidation and burial of organic carbon in sediments of the Iberian margin and Nazare Canyon (NE Atlantic). *Prog. Oceanogr.,* **52,** 399–431.

Eppley, R.W. and B.J. Peterson, 1979. Particulate organic matter flux and plankton new production in the deep ocean. *Nature,* **282,** 677–680.

FAO, 1997. Review of the state of world fishery resources: marine fisheries. FIRM/C920, Fisheries Circular N°920.

Faure, V., C.A. Inejih, H. Demarcq and P. Cury, 2000. The importance of retention processes in upwelling areas for recruitment of Octopus vulgaris: the example of the Arguin Bank (Mauritania). *Fish. Oceanogr.,* **9,** 343–355.

Fermín, E. G., F.G. Figueiras, B.Arbones and M.L. Villarino, 1996. Short-time evolution of a Gymnodinium catenatum population in the Ría de Vigo. *J. Phycol.,* **32,** 212–221.

Fernández, E., J. Cabal, J.L. Acuña, A. Bode, J.A. Botas and C. García-Soto, 1993. Plankton distribution across a slope current-induced front in the southern Bay of Biscay. *J. Plankton Res.,* 15, 619–641.

Fernández, E., E. Marañón, J. Cabal, F. Álvarez and R. Anadón, 1995. Vertical particle flux in outer shelf waters of the southern Bay of Biscay in summer 1993. *Oceanol. Acta,* **18,** 379–384.

Ferraris J., K. Koranteng and A. Samba, 1998. Comparative study of the dynamics of small-scale marine fisheries in Senegal and Ghana. In *Global versus local changes in upwelling systems:* Durand M.H., P. Cury, R. Mendelssohn, C. Roy, A. Bakun and D. Pauly, eds., Editions ORSTOM, Paris, pp. 447–464.

Figueiras, F.G., K. Jones, A.M. Mosquera, X. A. Álvarez-Salgado, A. Edwards and N. MacDougall., 1994. Red tide assemblage formation in an estuarine upwelling ecosystem: Ría de Vigo. *J. Plank Res,* **16,** 857–878.

Figueiras, F.G., U. Labarta and M. J. Fernández Reiriz, 2002. Coastal upwelling, primary production and mussel growth in the Rías Baixas of Galicia. *Hydrobiologia,* **484,** 121–131.

Fileman, E. and P. Burkill, 2001. The herbivorous impact of microzooplankton during two short-term Lagrangian experiments off the NW coast of Galicia in summer 1998. *Prog. Oceanogr.*, **51,** 361–383.

Fiúza, A.F.G., 1983. Upwelling patterns off Portugal. In *Coastal upwelling.* Suess, E. and J. Thiede, eds., pp. 85–98

Fiúza, A.F.G., 1984 Hidrología e dinámica das augas costeiras de Portugal. Ph.D. Thesis, University of Lisbon, 294 pp.

Fiúza, A.F.G., M. Hamann, I. Ambar, G. Díaz del Río, N. González and J.M. Cabanas, 1998. Water masses and their circulation off western Iberia during May 1993. *Deep Sea Res.,* **45,** 1127–1160.

Fraga, F., 1974. Distribution des masses d'eau ans l'upwelling de Mauritanie. *Téthys,* **6,** 5–10

Fraga, F., 1981. Upwelling off the Galician coast, northwest Spain. In *Coastal upwelling.* Richards, F.A., ed., Washington, DC, pp. 176–182.

Fraga, F., C. Mouriño and M. Manríquez, 1982. Las masas de agua en las costas de Galicia: junio–octubre. Resultados Expediciones Científicas, **10** ,51–77.

Fréon, P., 1983. Production models as applied to sub-stocks depending on upwelling fluctuations. In *Proceedings of the expert consultation to examine changes in abundance and species composition of neritic fish resources.* Sharp, G. D. and J. Csirke, eds., FAO, Fish. Rep. 291 (3), 1047–1064.

Fréon, P. 1988. Réponses et adaptations des stocks de clupéidés d'Afrique de l'ouest à la variabilité du milieu et de l'exploitation : Analyse et réflexion à partir de l'exemple du Sénégal. Etudes et Thèses, ORSTOM, 287 pp.

Fréon, P. Mullon, C. and Pichon, G. 1993. CLIMPROD: Experimental interactive software for choosing and fitting surplus production models including environmental variables. *FAO Computerized information series,* 5, 76 pp.

Friederich, G.E. and L.A. Codispoti, 1982. Some factors influencing dissolved silicon distribution over the Northwest African shelf. *Rapp. Proc. Reun. Cons. Int. Expl. Mer.,* **180,** 205.

Fromentin, J.-M., and B. Planque, 1996. Calanus and environment in the eastern North Atlantic. II. Influence of the North Atlantic Oscillation on C. Finmarchicus and C. Helgolandicus. *Mar. Ecol. Prog. Ser.,* **134,** 111–118.

Frouin, R., A.F.G. Fiúza, I. Ambar and T.J. Boyd, 1990. Observations of a poleward surface current off the coasts of Portugal and Spain during winter. *J. Geophys. Res.,* **95:** 679–691.

Gabric, A.J., L. García, L. van Camp, L. Nykjaer, W. Eifler and W. Schrimpf, 1993. Offshore export of shelf production in the Cape Blanc giant filament as derived from CZSC imagery. *J. Geophys. Res.,* **98,** 4697–4712.

Gago, J., M. Gilcoto, F.F. Pérez and A.F. Ríos, 2003. Short-term varibility of $f\mathrm{CO}_2$ in seawater and air-sea CO_2 fluxes in a coastal upwelling system (Ría de Vigo, NW Spain). *Mar. Chem.,* **80,** 247–264.

García-Muñoz, M., J. Arístegui, J.L. Pelegrí, A. Antoranz, A. Ojeda and M. Torres, 2005. Exchange of carbon and nutrients by an upwelling filament off Cape Guir (NW Africa). *J. Mar. Sys.* 54, 83–95

Gardener, D., 1977. Nutrients as tracer of water mass structure in the coastal upwelling of Northwest Africa. In *A voyage of Discovery,* Angel, M., ed., Pergamon Press, Oxford and N.Y., 712 pp.

González, N., R. Anadón, B. Mouriño, E. Fernández, B. Sinha, J. Escánez and D. de Armas, 2001. The metabolic balance of plankton community in the N Atlantic Subtropical Gyre: the role of mesoscale instabilities. *Limnol. Oceanogr.,* **46,** 946–952.

Grall, J.R., P. Le Corre, P. Treguer, 1982. Short-term variability of primary production in coastal upwelling of Morocco. *Rapp. Proc. Reun. Cons. Int. Expl. Mer.,* **180,** 221–227

Gulland, J.A. and S. Garcia, 1984. Observed patterns in multispecies fisheries. In *Exploitation of marine communities.* R.M. May, ed., Dahlem Konferenzen. Berlin, Heidelberg, New-York, Tokyo, Springler-Verlag, pp. 155–190.

Hagen, E. and R. Schemainda, 1987. On the zonal distribution of South Atlantic Central Water (SACW) along a section off Cape Blanc, Northwest Africa. *Oceanol. Acta,* vol. special, **6,** 61–70

Hall, I.R., S. Schmidt, I.N. McCave and J.L. Reyss, 2000. Particulate matter distribution and 234Th/238U disequilibrium along the Northern Iberian Margin: implications for particulate organic carbon export. *Deep Sea Res.*, **47,** 557–582.

Halvorsen, E., A.G. Hirst, S.D. Batten, K.S. Tande and R.S. Lampitt, 2001. Diet and community grazing by copepods in an upwelled filament off the NW coast of Spain. *Prog. Oceanogr.*, **51,** 399–421.

Hanson, R.B., M. T. Álvarez–Osorio, R. Cal, M.J. Campos, M. Roman, G. Santiago, M. Varela and J. A. Yoder, 1986. Plankton response following a spring upwelling event in the Ría de Arosa, Spain. *Mar. Ecol. Prog. Ser.*, **32,** 101–113.

Hastenrath, S., L. C. de Castro and P. Aceituno, 1987: The Southern Oscillation in the Atlantic sector. *Contrib. Atmos. Phys.*, **60,** 447–463

Haynes, R. and E.D. Barton, 1990. A poleward flow along the Atlantic coast of the Iberian Peninsula. *J. Geophys. Res.*, **95,** 11425–11441.

Haynes, R., E.D. Barton and I. Pilling, 1993. Development, persistence and variability of upwelling filaments off the Atlantic coast of Iberian Peninsula. *J. Geophys. Res.*, **98,** 22681–22692.

Head, E.J.H., W.G. Harrison, B.D. Irwin, E.P.W. Horne and W.K.W. Li, 1996. Plankton dynamics and carbon flux in an area of upwelling off the coast of Morocco. *Deep Sea Res. I,* **43,** 1713–1738

Hempel G., 1982, The Canary current: studies of an upwelling system. Introduction. *Rapp. Proc. Reun. Cons. Int. Expl. Mer.*, **180,** 7–8.

Hendericks, J., T. Freudenthal, H. Meggers, S. Nave, F. Abrantes, J. Bollmann and H.R. Thierstein, 2002. Glacial-interglacial variability of particle accumulation in the Canary Basin: a time-slice approach. *Deep Sea Res. II,* **49,** 3675–3705.

Herbland, A. and B. Voitiurez, 1974. La production primaire dans l'upwelling mauritanien enmars 1973. *Cah. ORSTOM, sér. Océanogr.*, **12,** 187–201.

Hernandez Guerra, A. and L. Nykjaer, 1997. Sea surface temperature variability off North-West Africa. *Int. J. Remote Sens.*, **18,** 2539–2558.

Hernández-Guerra, A., J. Arístegui, M.Cantón and L. Nykjaer, 1993. Phytoplankton pigments patterns in the Canary Islands as determined using Coastal Zone Colour Scanner Data. *Int. J. Remote Sens.*, **14,** 1431–1437.

Hernández-Guerra, A., F. López-Laatzen, F. Machín, D. de Armas and J. L. Pelegrí, 2001. Water masses, circulation and transport in the eastern boundary current of the North Atlantic subtropical gyre. *Sci. Mar.*, **65 (S1),** 177–186.

Hernández-Guerra, A., F. Machín, A. Antoranz, J. Cisneros-Aguirre, C. Gordo, A. Marrero-Díaz, A. Martínez, A. Ratsimandresy, A. Rodríguez-Santana, P Sangrá, F. López-Laatzen, G. Parrilla and J. L. Pelegrí, 2002. Temporal variability of mass transport in the Canary Current. *Deep-Sea Res. II* 49, 3415–3426.

Hernández-León, S. 1988. Gradients of mesozooplankton biomass and ETS activity in the wind-shear area as evidence o fan island mass effect in te Canary Islands waters. *J. Plankton Res.,* **10,** 1141–1154

Hernández-León, S. 1991. Accumulation of mesozooplankton in a wake area as a causative mechanism of the "Island-mass effect". *Mar. Biol.,* **109,** 141–147.

Hernández-León, S., C. Almeida, L. Yebra, J. Arístegui, 2002. Lunar cycle of zooplankton biomass in subtropical waters: biogeochemical implications. *J. Plankton Res.,* **24,** 935–939.

Hughes, P. and E. D. Barton, 1974. Stratification and water mass structure in the upwelling area off northwest Africa. *Mém. Soc. Roy. Sciences Liège,* 6 sér., **X,** 31–42.

Huntsman, S.A. and R.T. Barber, 1977. Primary production off northwest Africa: the relationship to wind and nutrient conditions. *Deep-Sea Res. I,* **24,** 25–33.

Hurrell, J. W., 1995. Decadal trends in the North Atlantic Oscillation: Regional temperatures and precipitation. *Science,* **269,** 676–679.

Huthnance, J.M., H.M.V.Aken, M. White, E.D. Barton, B. Le Cann, B., E.F. Coelho, E., E.A. Fanjul, P. Miller and J. Vitorino, 2002. Ocean margin exchange–water flux estimates. *J. Mar. Sys.*, **32,** 107–137.

ICES, 2002. Report of the Working Group on the Assessment of Mackerel, Horse Mackerel, Sardine and Anchovy. ICES CM 2002/ACFM:6, 474 pp.

Inejih, C. A., 2000. Dynamique spatio-temporelle et biologie du poulpe (Octopus vulgaris) dans les eaux mauritaniennes: modélisation de l'abondance et aménagement des pêches. Thèse de doctorat. Université de Bretagne Occidentale, Brest, 155pp.

Jickells, T.D., P.P. Newton, P. King, R.S. Lampitt and C. Boutle, 1996. A comparison of sediment traps record of particle fluxes from 19° to 48° in the Northeast Atlantic and their relation to surface water productivity. *Deep Sea Res. I,* **43,** 971–986.

Joint, I. and P. Wassmann, 2001. Lagrangian studies of the Iberian upwelling system–an introduction. A study of the temporal evolution of surface production and fate of organic matter during upwelling an and off the NW Spanish continental margin. *Prog. Oceanogr.,* **51,** 217–221.

Joint, I., A.P. Rees and E.M.S. Woodward 2001a. Primary production and nutrient assimilation in the Iberian upwelling in August 1998. *Prog. Oceanogr.,* **51,** 303–320.

Joint, I., M. Inall, R. Torres, F.G. Figueiras, X.A. Álvarez–Salgado, A.P Rees and E.M.S. Woodward, 2001b. Two lagrangian experiments in the Iberian upwelling system: tracking an upwelling event and an off–shore filament. *Prog. Oceanogr.,* **51,** 221–248.

Joint, I., S.B. Groom, R. Wollast, L. Chou, G.H. Tilstone, F.G. Figueiras, M. Loijens and T.J. Smyth., 2002. The response of phytoplankton production to periodic upwelling and relaxation events at the Iberian shelf break: estimates by the 14C method and by satellite remote sensing. *J. Mar. Sys.,* **32,** 219–238.

Jones, B.H., C.N.K. Mooers, M.M. Rienecker, T. Stanton and L. Washburn, 1991. Chemical and biological structure and transport of a cool filament associated with a jet-eddy system off northern California in July 1986 (OPTOMA21). *J. Geophys. Res.,* **96 (C12),** 22207–22225

Jouanneau, J., O. Weber, T. Drago, A. Rodrigues, A. Oliveira, J. Dias, C. Garcia, S. Schmidt and J. Reyss, 2002. Recent sedimentation and sedimentary budgets on the western Iberian shelf. *Prog. Oceanogr.,* **52,** 261–275.

Kébé, M., 1994. Principales mutations de la pêche artisanale maritime sénégalaise. In *L'évaluation des resources exploitables par la pieche artisanale sénégalaise.* Barry-Gérard M.B., T. Diouf and A. Fonteneau., tome 2. ORSTOM Editions, Paris, 43:58.

Kifani S., 1998. Climate dependant fluctuatiosn of the Moroccan sardine and their impact on fisheries. In *Global versus local changes in upwelling systems.* Durand M.H., P. Cury, R. Mendelssohn, C. Roy, A. Bakun and D. Pauly, eds., Editions ORSTOM Paris, 235–248.

Klein, S. A., B. J. Soden and N.-C. Lau., 1999. Remote Sea Surface Temperature Variations during ENSO: Evidence for a Tropical Atmospheric Bridge. *J. Climate,* **12,** 917–932.

Kostianoy, A.G. and A.G. Zatsepin, 1996. The west African coastal upwelling filaments and cross-frontal water exchange conditioned by them. In *The coastal ocean in a global change perspective,* Djenidi, S., ed., *J. Mar Sys.,* **7,** 349–359

Krauss, W., 1986. The North Atlantic Current. *J. Geophys. Res.* **91,** 5061–5074.

Lavín, A., G. Díaz del Rio, G. Casas and J. M. Cabanas, 2000. Afloramiento en el noroeste de la Península Ibérica. Índices de afloramiento para el punto 43° N, 11° O, Periodo 1990–1999 [Upwelling in the NW of the Iberian Peninsula. Upwelling indexes at 43° N, 11° W, between 1990–1999]. Datos y Resúmenes Instituto Español de Oceanografia (15), 25 pp.

Lenz, J., 1982. Particulate organic matter off Northwest Africa: size distribution and composition in relation to shore distance and depth. *Rapp. Proc. Reun. Cons. Int. Expl. Mer.,* **180,** 234–238.

Lloyd, I.J., 1971. Primary production off the coast of north-west Africa. *J. Cons. Int. Explor. Mer,* **33,** 312–323.

Longhurst, A., S. Sathyendranath, T. Platt and C. Caverhill, 1995. An estimate of global primary production in the ocean from satellite radiometer data. *J. Plankton Res.*, **17,** 1245–1272.

López-Jamar, E., R.M. Cal, G. González, R.B. Hanson, J. Rey, G. Santiago and K.R. Tenore, 1992. Upwelling and outwelling effects on the benthic regime of the continental shelf off Galicia, NW Spain. *J. Mar. Res.* **50,** 465–488.

MacIsaac, J.J., R.C. Dugdale, R.T. Barber, D. Blasco, and T.T. Packard, 1985. Primary production cycle in an upwelling center. *Deep Sea Res,* **32,** 503–529.

Marañón, E. and E. Fernández, 1995. Changes in phytoplankton ecophysiology across a coastal upwelling front. *J. Plankton Res.*, **10,** 1999–2008.

Marshall, J., Y. Kushnir, D. Battisti, P. Chang, A. Czaja, R. Dickson, J. Hurrell, M. McCartney, R. Saravanan and M. Visbeck, 2001. North Atlantic Climate variability: phenomena, impacts and mechanisms. *Int. J. Climat.*, **21,** 1863–1898.

Maus, J., 1997. Sustainable fisheries information management in Mauritania. PhD thesis. Ecosystem analysis and Management Group. University if Warwick, Coventry, 269pp.

McCave, I. and I. Hall, 2002. Turbidity of waters over the Northwest Iberian continental margin. *Prog. Oceanogr.*, **52,** 299–313.

McClain, C.R., S.-Y. Chao, L.P. Atkinson, J.O. Blanton and F.F. de Castillejo, 1986. Wind-driven upwelling in the vicinity of Cape Finisterre, Spain. *J. Geophys. Res.*, **91,** 8470–8486.

Meggers, H., T. Freudenthal, S. Nave, J. Targarona, F. Abrantes and P. Helmke, 2002. Assessment of geochemical and micopaleontological sedimentary parameters as proxies of surface water properties in the Canary Islands region. *Deep Sea Res. II,* **49,** 3631–3654.

Minas, H.J., L.A. Codispoti and R.C. Dugdale, 1982. Nutrients and primary production in the upwelling region off Northwest Africa. *Rapp. Proc. Reun. Cons. Int. Expl. Mer.*, **180,** 148–183.

Minas, H.J., M. Minas and T.T. Packard, 1986. Productivity in upwelling areas deduced from hydrographic and chemical fields. *Limnol. Oceanogr.*, **31,** 1182–1206.

Mittlelstaedt, E., 1991. The ocean boundary along the northwest African coast: circulation and oceanographic properties at the sea surface. *Prog. Oceanogr.*, **26,** 307–355.

Moncoiffé, G., X.A. Álvarez-Salgado, F.G. Figueiras and G. Savidge, 2000. Seasonal and short time-scale dynamics of microplankton community production and respiration in an inshore upwelling system. *Mar. Ecol. Prog. Ser.*, **196,** 111–126.

Morán, X.A.G., J.M. Gasol, C. Pedrós-Alió and M. Estrada, 2002a. Partitioning of phytoplankton organic carbon production and bacterial production along a coastal-offshore gradient in the NE Atlantic during different hydrographic regimes. *Aquat. Microb. Ecol.*, **29,** 239–252.

Morán, X.A.G., M. Estrada, J.M. Gasol and C. Pedrós-Alió, 2002b. Dissolved primary production and the strength of phytoplankton-bacterioplankton coupling in contrasting marine regions. *Microb. Ecol.*, **44,** 217–223.

Morel, A., 1996. An Ocean flux study in eutrophic, mesotrophic and oligotrophic situations: the EUMELI program. *Deep-Sea Res. I,* **43,** 1185–1190.

Morel, A., D. Antoine, M. Babin and Y. Dandonneau, 1996. Measured and modeled primary production in the northeast Atlantic (EUMELI JGOFS program): the impact of natural variations in photosynthetic parameters on model predictive skills. *Deep-Sea Res. I,* **43,** 1273–1304.

Nagata, T. 2000. Production mechanims of dissolved organic matter. In *Microbial ecology of the oceans* Kirchman, D.L., ed., Wiley and Sons, New York, pp. 121–152.

Navarro-Pérez, E. and E. D. Barton, 1998. The physical structure of an upwelling filament off the north-west African coast during August 1993. In *Benguela dynamics: impacts of variability of shelf – sea environments and their living resources,* Pillar, S.C., C.L. Moloney, A.I.L. Payne and F.A. Shillington, eds., *S. Afr. J. Mar. Sci.*, **19,** 61–74

Navarro-Pérez, E. and E. D. Barton, 2001. Seasonal and interannual variability of the Canary Current. *Sci. Mar.*, **65 (S1),** 205–213.

Neuer, S., T. Freudenthal, R. Davenport, O. Llinás, and M.J. Rueda, 2002. Seasonality of surface water properties and particle flux along a productivity gradient off N.W. Africa. *Deep Sea Res. II,* **49,** 3561–3576.

Nobre, P. and J. Shukla, 1996. Variations of sea surface temperature, wind stress, and rainfall over the tropical Atlantic and South America. *J. Climate,* **9,** 2464–2479.

Nogueira, E., F.F. Pérez and A.F. Ríos, 1997. Modelling thermohaline properties in an estuarine upwelling ecosystem (Ría de Vigo: NW Spain) using box–Jenkins transfer function models. *Est. Coast. Shelf Sci.,* **44,** 685–702.

Nykjaer L. and L. Van Camp, 1994. Seasonal and interannual variability of coastal upwelling along Norhwest Africa and Portugal from 1981 to 1991. *J. Geophys. Res.,* **99,** C7, 14197–14207.

Oliveira, A., J. Vitorino, A. Rodrigues, J. Jouanneau, J. Dias and O. Weber, 2002. Nepheloid layer dynamics in the northern Portuguese shelf. *Prog. Oceanogr.* , **52,** 195–213.

Olli, K., C.W. Riser, P. Wassmann, T. Ratkova, E. Arashkevich and A. Pasternak, 2001. Vertical flux of biogenic matter during a Lagrangian study off the NW Spanish continental margin. *Prog. Oceanogr,* **51,** 443–466.

Ould-Dedah, S., W.J. Wiseman Jr. and R.F. Shaw, 1999. Spatial and temporal trends of sea surface temperature in the northwest African region. *Oceanol. Acta,* **22,** 265–279.

Pacheco, M. and A. Hernández-Guerra, 1999. Seasonal variability of recurrent phytoplankton pigment patterns in the Canary Islands area .. *Int. J. Remote Sens.,* **29,** 1405–1418.

Parrilla, G., S. Neuer, P.-Y. Le Traon and E. Fernández-Suárez, 2002. Topical studies in oceanography : Canary Islands Azores Gibraltar Observations (CANIGO). Volume 1: Studies in the northern Canary Islands basin. *Deep Sea Res. II,* **49,** 3409–3413

Pelegrí, J. L., P. Sangrà and A. Hernández-Guerra, 1997. Heat gain in the eastern North Atlantic subtropical gyre. In *The mathematics of models for climatology and environment,* Díaz J. I., ed., NATO ASI Series, Vol. I 48, Springer-Verlag, Berlin, pp. 419–436.

Pelegrí, J. L., J. Arístegui, L. Cana, M. González-Dávila, A. Hernández-Guerra, S. Hernández-León, A. Marrero-Díaz, M. F. Montero, P. Sangrà and M. Santana-Casiano, 2005. Coupling between the open ocean and the coastal upwelling region off Northwest Africa: Water recirculation and offshore pumping of organic matter. *J. Mar. Sys.* 54, 3–37

Pelíz, A.J. and A.F.G. Fiúza , 1999. Temporal and spatial variability of CZCS–derived phytoplankton pigment concentrations off the western Iberian Peninsula. *Int. J. Remote Sens.,* **20,** 1363–1403.

Pérez, F.F., A.F. Ríos, B.A. King and R.T. Pollard, 1995. Decadal changes of the O-S relationship of the Eastern North Atlantic Central Water. *Deep-Sea Res.,* **42,** 1849–1864.

Pérez, F.F., A.F. Ríos and G. Rosón, 1999. Sea surface carbon dioxide off the Iberian Peninsula (NE Atlantic Ocean). *J. Mar. Sys.,* **19,** 27–46.

Pérez, F.F., X.A. Álvarez–Salgado and G. Rosón, 2000a. Stochiometry of nutrients (C, N, P and Si) consumption and organic matter production in a coastal inlet affected by upwelling. *Mar. Chem.,* **69,** 217–236.

Pérez, F.F., R.T. Pollard, J.F. Read, V. Valencia, J.M. Cabanas and A.F. Ríos, 2000b. Climatological coupling of the thermohaline decadal changes in Central Water of the Eastern North Atlantic. *Sci. Mar.,* **64,** 347–353

Pérez, F.F., C.G. Castro, X.A. Álvarez–Salgado and A.F. Ríos, 2001. Coupling between the Iberian basin–scale circulation and the Portugal boundary current system. A chemical study. *Deep Sea Res I,* **48,** 1519–1533.

Pilskan, C.H., J.B. Paduan, F.P. Chavez, R.Y. Anderson and W.M. Berelson, 1996. Carbon export and regeneration in the coastal upwelling system of Monterey Bay, central California. *J. Mar. Res.,* **54,** 1149–1178.

Pingree, R.D., 1994. Winter warming in the southern bay of Biscay and Lagrangian Eddy kinematics from a deep–drogued argos buoy. *J. Mar. Biol. Ass. U.K.,* **74,** 107–128.

Pingree, R.D. and B. Le Cann, 1990. Structure, strength and seasonality of the slope currents in the bay of Biscay region. *J. Mar. Biol. Ass.U.K.,* **70,** 857–885.

Pingree, R.D. and B. Le Cann, 1992. Three anticyclonic Slope Water oceanic eDDIES (SWODDIES) in the Southern Bay of Biscay in 1990. *Deep Sea Research II,* **39,** 1147–1175.

Prego, R., 1994. Nitrogen interchanges generated by biogeochemical processes in a Galician Ria. *Mar. Chem.,* **45,** 167–176.

Prego, R. and R. Bao, 1997. Upwelling influence on the Galician coast: silicate in shelf water and underlying surface sediments. *Cont. Shelf Res.,* **17,** 307–318.

Prego, R., M.C. Barciela and M. Varela, 1999. Nutrient dynamics in the Galician coastal area (Northwestern Iberian Peninsula): Do the Rias Bajas receive more nutrient salts than the Rías Altas?. *Cont. Shelf Res,* **19,** 317–334.

Quero, J-C., M-H. Du Buit and J-J. Vayne, 1998. Les observations de poissons tropicaux et le réchauffement des eaux dans l'Atlantique européen. *Oceanol. Acta,* **21,** 345–351.

Relexans, J.-C., J. Deming, A. Donet, J.-F. Gaillard and M. Sibuet, 1996. Sedimentary organic matter and micro-meiobenthos with relation to trophic conditions in the tropical northeast Atlantic. *Deep Sea Res. I,* **43,** 1343–1368.

Relvas, P. and E. D. Barton, 2002. Mesoscale patterns in the Cape São Vicente (Iberian Peninsula) upwelling region. *J. Geophys. Res.,* 107(C10), 3164, doi:10.1029/2000JC000456, 2002.

Ríos, A.F., F.F. Pérez and F. Fraga, 1992. Water masses in upper and middle North Atlantic Ocean east of Azores. *Deep Sea Res. I,* **39,** 645–658.

Riser, C.W., P. Wassmann, K. Olli and E. Arashkevich, 2001. Production, retention and export of zooplankton faecal pellets on and off the Iberian shelf, north-west Spain. *Prog. Oceanogr,* **51,** 423–441.

Rodríguez, J.M., S. Hernández-León and E.D. Barton, 1999. Mesoscale distribution of fish larvae in relation to an upwelling filament off Northwest Africa. *Deep Sea Res. I* , **46,** 1969–1984

Rodríguez, J.M., E.D. Barton, L. Eve and S. Hernández-León, 2001. Mesozooplankton and ichtyoplankton distribution around Gran Canaria, an oceanic island in the NE Atlantic. *Deep Sea Res. I* , **48,** 2161–2183

Rosón, G., X.A. Álvarez–Salgado and F.F. Pérez, 1999. Carbon cycling in a large coastal embayment affected by wind–driven upwelling. Short–time–scale variability and spatial differences. *Mar. Ecol. Prog. Ser.,* **176,** 215–230.

Roy, C., 1989. Fluctuations des vents et variabilité de l'upwelling devant les côtes du Sénégal. *Oceanol. Acta,* **12,** 361–369.

Roy, C. and C. Reason. 2001. ENSO related modulation of coastal upwelling in the eastern Atlantic. *Prog. Oceanogr,* **49,** 245–255.

Roy, C., P. Cury, A. Fontana and H. Belvèze. 1989. Stratégies spatio-temporelles de la reproduction des clupéidés des zones d'upwelling d'Afrique de l'Ouest. *Aquat. Liv. Resources,* **2,** 21–29.

Roy, C., P. Cury and S. Kifani, 1992. Pelagic fish recruitment success and reproductive strategy in upwelling areas : environmental compromises. In *Benguela Trophic Functioning.* Payne, A.I.L., Brink, K.H., Mann, K.H. and R. Hilborn, eds., *S. Afr. J. Mar. Sci.,* **12,** 135–146.

Santana-Casiano, J.M., M. González-Dávila, L.M. Laglera-Baquer and M.J. Rodriguez-Somoza, 2001. Carbon dioxide system in the Canary region during October 1995. *Sci. Mar.,* **65,** 41–49.

Santos, A. M. P., M. F. Borges and S. Groom, 2001. Sardine and horse mackerel recruitment and upwelling off Portugal. *ICES J. Mar. Sci.,* **58,** 589–596.

Schmidt, S., L. Chou and I.R. Hal, 2002a. Particle residence times in surface waters over the north–western Iberian Margin: comparison of pre–upwelling and winter periods. *J. Mar. Sys.* **32,** 3–11.

Schmidt, S., T.C. van Weering, J.L. Reyss and P. van Beek, 2002b. Seasonal deposition and reworking at the sediment–water interface on the northwestern Iberian margin. *Prog. Oceanogr,* **52,** 331–348.

Schulz, S., 1982. A comparison of primary production in upwelling regions off Northwest and Southwest Africa. *Rapp. P. -v. Reún. Cons. Int. Explor .Mer,* **180,** 202–204.

Seager, R., Y. Kushnir, M. Visbeck, N. Naik, J. Miller, G. Krahmann, and H. Cullen, 2000. Causes of Atlantic Ocean climate variability between 1958 and 1998. *J. Climate,* **13,** 2845–2862.

Seibold, E., 1982. Sediments in upwelling areas, particularly off Northwest Africa. *Rapp. Proc. Reun. Cons. Int. Expl. Mer.,* **180,** 315–322.

Serret, P., E. Fernández, F.A. Sostres, and R. Anadón, 1999. Seasonal compensation of microbial production and respiration in a temperate sea. *Mar. Ecol. Prog. Ser.,* **187,** 43–57.

Serret, P., E. Fernández and C. Robinson, 2002. Biogeographic differences in the net ecosystem metabolism of the open ocean. *Ecology,* **83,** 3225–3234.

Siedler, G. and R. Onken, 1996. Eastern recirculation. In *The warmwatersphere of the North Atlantic Ocean,* W. Krauss, ed., Gebrüder Borntraeger, Berlin, pp. 339–364.

Silva, A.J., 1992. Dependence of upwelling related circulation on wind forcing and stratification over the Portuguese northern shelf. International Council for the Exploitation of the Sea, Council Meeting, Hydrography Committee C17, 1–12.

Slagstad, D. and P. Wassmann, 2001. Modelling the 3–D carbon flux across the Iberian margin during the upwelling season in 1998. *Prog. Oceanogr,* **51,** 467–497.

Sousa, F.M. and A. Bricaud, 1992. Satellite-derived phytoplankton pigment structures in the Protuguese upwelling area. *J. Geophys. Res.,* **97,** 11343–11356.

Sprengel, C., K-H. Baumann, J. Henderiks, R. Henrich and S. Neuer, 2002. Modern coccolitophore carbonate sedimentation along a productivity gradient in the Canary Islands region: seasonal export production and surface accumulation rates. *Deep Sea Res. II,* **49,** 3577–3598.

Stramma, L. and G. Siedler, 1988. Seasonal changes in the North Atlantic subtropical gyre. *J. Geophys. Res.,* **93,** 8111–8118.

Stratoudakis, Y., A. Morais, A. Silva, V. Marques and C. A. Dias, 2002. Sardine distribution, abundance and population structure off Portugal: acoustic surveys in 1984–2000. Working Document to the Working Group On the Assessment of Mackerel, Horse Mackerel, Sardine and Anchovy. ICES CM 2002/ACFM:6, p. 473

Stratoudakis, Y., M. Bernal, D. Borchers and M. F. Borges, 2003. Changes in the distribution of sardine eggs and larvae off Portugal, 1985–2000. Fisheries Oceanography, **12,** 49–60

Strom, L.S., R. Benner, S. Ziegler and M.J. Gagg, 1997. Planktonic grazers are a potentially important source of marine dissolved organic carbon. *Limnol. Oceanogr.,* **42,** 1364–1374.

Strub, P.T., P.M. Kostro and A. Huyer, 1991. The nature of cold filaments in the California Current System. *J. Geophys. Res.,* **96 (C8),** 14743–14768.

Teira, E., P. Serret and E. Fernández, 2001. Phytoplankton size–structure, particulate and dissolved organic carbon production and oxygen fluxes through microbial communities in the NW Iberian coastal transition zone. *Mar. Ecol. Prog. Ser.,* **219,** 65–83.

Tenore, K.R., M. Alonso–Noval, M. Álvarez–Osorio, L.P. Atkinson, J.M. Cabanas, R.M. Cabal, H.J. Campos, F. Castillejo, E.J. Chesney, N. González, R. B. Hanson, C.R. McClain, A. Miranda, M.R. Roman, J. Sanchez, G. Santiago, L. Valdes, M. Varela and J. Yoder, 1995. Fisheries and oceanography off Galicia, NW Spain: Mesoscale spatial and temporal changes in physical processes and resultant patterns of biological productivity. *J. Geophys. Res.,* **100,** 10943–10966.

Tilstone, G.H., F. G. Figueiras, E.G. Fermín and B. Arbones, 1999. Significance of nanophytoplankton photosynthesis and primary production in a coastal upwelling system (Ría de Vigo, NW Spain). *Mar. Ecol. Prog. Ser.,* **183,** 13–27.

Tilstone, G.H., B.M. Míguez, F.G. Figueiras and E.G. Fermín, 2000. Diatom dynamics in a coastal ecosystem affected by upwelling: coupling between species succession, estuarine circulation and biogeochemical processes. *Mar. Ecol. Prog. Ser.,* **205,** 23–41.

Tilstone, G.H., F.G. Figueiras, L.M. Lorenzo and B. Arbones, 2003. Phytoplankton composition, photosynthesis and primary production during different hydrographic conditions at the NW Iberian upwelling system. *Mar. Ecol. Prog. Ser.,* **252,** 89–104.

Torres, R., E.D. Barton, P. Miller and R. Fanjul, 2003. Spatial patterns of wind and sea surface temperature in the Galician upwelling region, *J. Geophys. Res.,* 108(C4), 3130, doi: 10.1029/2002JC001361, 2003.

Tréguer, P., P. Le Corre and J.R. Grall, 1979. The seasonal variations of nutrients in the upper waters of the Bay of Biscay region and their relation to phytoplankton growth. *Deep–Sea Res.,* **26A,** 1121–1152.

Van Aken, H.W., 2002. Surface currents in the bay of Biscay as observed with drifters between 1995 and 1999. *Deep–Sea Res. I,* **49,** 1071–1086.

Van Lenning, K., 2000. Variability in biomass and structure of phytoplankton populations in the Canary Islands waters, as determined by HPLC analyses of pigments. PhD. Thesis. University of Las Palmas de Gran Canaria, 270 pp.

Van Weering, T.C. and I.N. McCave, 2002. Benthic processes and dynamics at the NW Iberian margin: an introduction. *Progr. Oceanogr., 52,* 123–128

Van Weering, T.C., H. de Stigter, W. Boer and H. de Haas, 2002. Recent sediment transport and accumulation on the NW Iberian margin. *Prog. Oceanogr.,* **52,** 349–371.

Varela, M., G. Diaz del Río, M. T. Álvarez–Osorio and E. Costas, 1991. Factors controlling phytoplankton size class distribution in the upwelling area of the Galician continental shelf (NW Spain). *Sci. Mar.,* **55,** 505–518.

Vitorino, J.O., J.M. Jouanneau and T. Drago, 2002a. Winter dynamics on the northern Portuguese shelf. Part I: physical processes. *Prog. Oceanogr.,* **52,** 129–153.

Vitorino, J.O., J.M. Jouanneau and T. Drago, 2002b. Winter dynamics on the northern Portuguese shelf. Part 2: bottom boundary layers and sediment dispersal. *Prog. Oceanogr.,* **52,** 155–170.

Wada, E. and A. Hattori, 1991. Nitrogen in the sea: forms, abundances and rate processes. CRC Press, Boca Raton.

Watson, A.J., 1995. Are upwelling zones sources or sinks of CO_2?,. In *Upwelling in the ocean. Modern processes and ancient records.* C.P. Summerhayes, K.–C. Emeis, M. V. Angel, R. L. Smith and B. Zeitzschel, eds, Wiley Sons, pp. 321–336.

Wefer, G. and G. Fisher, 1993. Seasonal patterns of vertical particle flux in equatorial and coastal upwelling areas of the eastern Atlantic. *Deep Sea Res. I,* **40,** 1613–1645.

Woodruff, S.D., R.J. Slutz, R. L. Jenne and P. M. Steurer, 1987. A Comprehensive Ocean-Atmosphere Data-Set. *Bull. Amer. Meteor. Soc.,* **68,** 1239–1250.

Wooster, W.S., A. Bakun and D.R. McLain, 1976. The seasonal upwelling cycle along the eastern boundary of the North Atlantic. *J. Mar. Res.,* **34,** 131–141.

Chapter 24. THE BAY OF BISCAY: THE ENCOUNTERING OF THE OCEAN AND THE SHELF (18b,E)

ALICIA LAVIN, LUIS VALDES, FRANCISCO SANCHEZ, PABLO ABAUNZA

Instituto Español de Oceanografía (IEO)

ANDRE FOREST, JEAN BOUCHER, PASCAL LAZURE, ANNE-MARIE JEGOU

Institut Français de Recherche pour l'Exploitation de la MER (IFREMER)

Contents

1. Introduction

The Bay of Biscay is an arm of the Atlantic Ocean, indenting the coast of W Europe from NW France (Offshore of Brittany) to NW Spain (Galicia). Traditionally the southern limit is considered to be Cape Ortegal in NW Spain, but in this contribution we follow the criterion of other authors (i.e. Sánchez and Olaso, 2004) that extends the southern limit up to Cape Finisterre, at 43° N latitude, in order to get a more consistent analysis of oceanographic, geomorphological and biological characteristics observed in the bay. The Bay of Biscay forms a fairly regular curve, broken on the French coast by the estuaries of the rivers (i.e. Loire and Gironde). The southeastern shore is straight and sandy whereas the Spanish coast is rugged and its northwest part is characterized by many large V-shaped coastal inlets (rias) (Evans and Prego, 2003). The area has been identified as a unit since Roman times, when it was called Sinus Aquitanicus, Sinus Cantabricus or Cantaber Oceanus. The coast has been inhabited since prehistoric times and nowadays the region supports an important population (Valdés and Lavín, 2002) with various noteworthy commercial and fishing ports (i.e. Brest, Nantes, Lorient, Saint-

The Sea, Volume 14, edited by Allan R. Robinson and Kenneth H. Brink
ISBN 0-674-01527-4

Nazaire, La Rochelle and Bordeaux in France; Ondarroa, Bilbao, Santander, Gijón, Avilés and A Coruña in Spain).

Marine studies have been performed extensively in the Bay of Biscay, where some of the oldest marine laboratories in the world are located (e.g. the French Laboratoire de Zoologie et de Physiologie Maritime of Concarneau and Station Marine de la Société Scientifique d'Arcachon funded in 1857 and 1867 respectively, and the Spanish Estación de Biología Marina de Santander funded in 1886 by Dr. A. G. de Linares, and the Sociedad Oceanográfica de Guipúzcoa funded in 1908 in San Sebastián). Despite having these laboratories on the fringes, oceanography has developed slowly and in the older maps of the North Atlantic circulation a gap each time appears in the Bay of Biscay showing the lack of knowledge of the area. Activities were always restricted to some PhD studies inside 'rades' or bays. In the 1970s activity began to develop all over the bay, not only in the shelf (such as the POLYGAS and PHYGAS cruises) and a large increase occurred in the 1980s. In the physical oceanographic studies on the Bay of Biscay there is a remarkable figure from Britain - Dr. Robin Pingree. Dr. Pingree developed a great deal of work either alone or together with some colleagues and has a large number of publications on the area, 16 of which are cited in references in the present chapter. He has strongly improved the knowledge of the oceanography of the Bay of Biscay and we would like to honor him in this chapter.

The Bay of Biscay is a well-differentiated geomorphological unit in the northeast Atlantic. The abyssal basin has a mean depth of 4,800 m. The shelf in the south of the bay (Spanish coast) is quite narrow whereas on the French coast it is much wider, increasing with latitude. In the Bay of Biscay there are various deep-sea canyons that have generally narrow, steep-sided, linear and sinuous channels. The deep-sea valleys allow continental sediments to be transported to oceanic basins. Most of the water masses occupying the bay have a North Atlantic origin or are the result of interaction between waters formed in the Atlantic with water of Mediterranean origin. The hydrodynamics of the bay are dominated by: a) a weak anticyclonic circulation in the oceanic part, b) a poleward-flowing slope current, c) coastal upwelling, d) the northward flow of Mediterranean water, e) the shelf circulation and f) the cross-shelf transport along the axes of submarine canyons (OSPAR, 2000). Most of these features show a marked seasonality (Koutsikopoulos and Le Cann, 1996). The Bay of Biscay is a region of large tidal amplitudes and strong thermohaline forcing (Piraud et al., 2003). It is well known for its energetic internal tides, caused by the combination of summer stratification, steep shelf-edge topography, and strong (cross-slope) tidal currents, especially at spring tides (Lam et al., 2003).

In a temperate sea like the Bay of Biscay the annual phytoplankton and zooplankton seasonal cycle is governed by the alternation between mixing and stratification of the water column (Fernández and Bode, 1991; Valdés et al., 1991). The highest planktonic biomass appears in spring followed by a secondary peak in autumn. This cycle can be modified by the development of shelf-edge fronts and coastal upwelling (García-Soto and Pingree, 1998; Díez et al., 2000). Mesozooplankton biomass is highly variable, following the production of phytoplankton, which is mainly determined by the seasonally changing environmental conditions. The primary production levels and the topographic characteristics of the shelf basins have allowed the development of important fisheries and mariculture areas of oysters and mussels. Living marine resources exploited in the Bay of Biscay include a wide range of organisms, from seaweeds to shellfishes (i.e. Norway lobster, cephalopods), fishes (i.e. mackerel, hake, anchovy, sole) and whales. Indeed, it has been reported that molluscs were exploited in the bay as early as the Palaeolithic period and whales since the Middle Ages (Valdés and Lavín, 2002). The study of the exploitation of some pelagic fishes in the Bay of Biscay started in the middle of the last century under the auspices of ICES (International Council for the Exploration of the Sea) (Cendrero, 2002). The economic and social importance of these and other fisheries and the objective of obtaining appropriate yields led to the development of studies to explain the variability of fish abundance and to predict the strength of incoming recruitment. In this way, important developments have been made in research fields related to marine primary production, ichthyoplankton, zooplankton and applied oceanography. Many of the exploited resources in the Bay of Biscay are managed as a particular unit or stock. In spite of the amount of information about the biology of these species, there is much to work on species interactions, including man, and in their relationships with the variable environment. Thus, the study of ecosystems should be one of the milestones for future research in the Bay of Biscay.

There are many descriptive studies on different aspects of the Bay of Biscay as we will show throughout this contribution and recently a few reviews were published integrating all the information available. The main contributions are Quality Status Report from OSPAR (2000) and the work of Valdés and Lavín (2002), which considers the Bay of Biscay as a large marine ecosystem (LME). Díez et al. (2000) reviewed the information on the southern part of the Bay of Biscay (the Cantabrian Sea).

The ocean is an inhomogeneous space. Therefore, the regionalization of the oceans arises to explain the discontinuities observed in the distribution of oceanic properties, substances and movements (Vanney, 2002). The aim of this chapter is to bring out the common characteristics from geography, geology, oceanography and ecology that are met with in the Bay of Biscay and that allow it to be defined as a particular marine region.

2. Geography of the Bay of Biscay

The North Atlantic Ocean has a general circulation made up of two enormous gyres: the anticyclonic subtropical gyre, and the cyclonic subpolar gyre. The Bay of Biscay is found at mid-latitudes between the limits of both gyres - the Azores current from the subtropical gyre and North Atlantic current from the subpolar gyre.

2.1 Topography and geology

The Bay of Biscay is bounded zonally by the parallels 48°N and 43°N. The northern boundary roughly corresponds to the separation between the Armorican shelf and the Celtic Sea. The slope of the bay is formed by three main areas with different orientation, the Armorican slope NW-SE, the Aquitaine slope N-S and the Cantabrian slope with an E-W orientation. The southern boundary is composed of Cape Finisterre and the Galician Bank (Fig. 24.1).

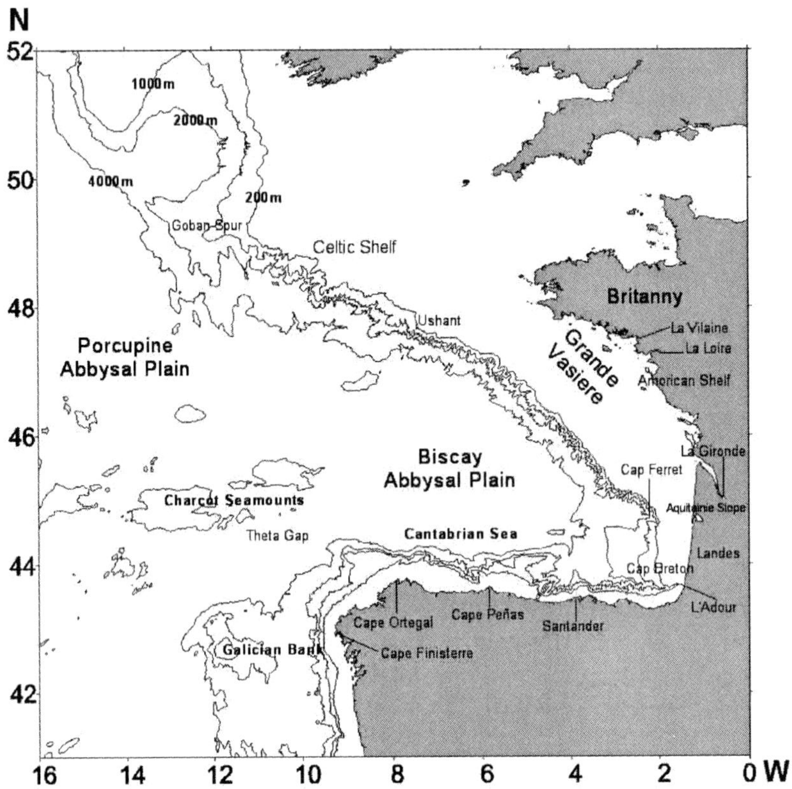

Figure 24.1 Bay of Biscay topography map based on the GEBCO97 dataset.

The Biscay abyssal plain is adjacent to the Porcupine abyssal plain in the northern part but separated from the Iberian Abyssal Basin and the West Iberian Margin by the Charcot Seamounts and the Galician Bank (Fig. 24.1).

The French Central Massif, the northern and western Pyrenees and the Iberian mountain systems correspond to the most prominent continental limits of a river catchment area, which is limited to the north by the hills of Normandy and Brittany. Four French rivers are the main pathways for freshwater into the bay: Vilaine, Loire, Gironde and Adour.

The morphological diversity of the sea bottom is a very good example of the interaction between the internal and external geological processes that have affected the continental margin since the opening of the Atlantic Ocean. One may easily recognize the erosional and accumulation features associated with the major geological units of the sub-bottom as well as with the structural alignments of the European plate. On the other hand, the bottom topography provides excellent clues to understand the present sediment dynamics (erosion and accumulation areas, pathways of sediment flow). Terrigenous sediments thus dominate the sedimentary input into the shelf and upper continental slope. These sediments are mostly made up of sandstone (OSPAR, 2000).

2.2 *Main features (rivers and canyons)*

The margin is divided into units by the presence of seamounts and banks as well as of pronounced submarine canyons (Fig. 24.1). Some are very prominent, such as the Cap Breton Canyon, where the 1000 m isobath is found at just 3 km from the coast. At shallower depths, the continental shelf is an area of a more gentle slope affected by smaller scale rock outcrops.

The French continental shelf is about 150 km wide in the northern part with a gentle slope of 0.12%. In contrast, the width of the shelf falls below 30 km off the Basque Country. Along the Cantabrian coast the shelf may be as narrow as 12 km, widening westwards. The continental slope, an area of transition between the shelf and the deep sea is very pronounced (slope of the order of 10–12%) and cut by numerous canyons.

River catchments represent the principal sources of freshwater that drain into the Atlantic along the French coast from north to south. The Loire and the Gironde are the two main rivers with an annual mean outflow of about 900 m^3 s^{-1}. They contribute 80% to the freshwater discharges onto the French shelf. Each of them has peak runoff in winter or spring that exceeds 3000 m^3 s^{-1} and a minimum in summer of about 200 m^3 s^{-1}. Rivers in the Cantabrian shelf are small due to the proximity of the mountains to the sea and total run-off represents a third of the sum of the Loire and Gironde.

During the last decade of the twentieth century a minimum of freshwater discharges occurred in 1992 with total mean runoffs from the Loire and Gironde of less than 1000 $m^3 s^{-1}$ from January to March. The highest runoffs were recorded in 1994, with 5000 $m^3 s^{-1}$ for the same period (Fig. 24.2).

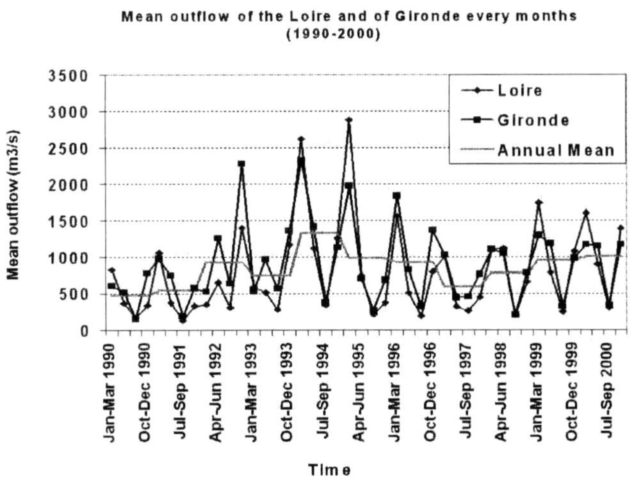

Figure 24.2 Quarterly mean outflow of the Loire and Gironde (1990–2000).

2.3 *Meteorology*

Atmospheric circulation at the middle latitudes of the North Atlantic and over western Europe is governed by the existence of two main centres of activity: an anticyclonic zone south of 40°N centred near the Azores (the Azores High), and a low pressure area centred around 60°N near Iceland (the Iceland Low). Between these two areas, the prevailing winds are from west to southwest, strongest in winter and lighter and less regular in summer.

In a common situation in spring and summer, the anticyclone extends in a ridge towards Iceland, and a low-pressure area is present to the east of the British Isles. Moderate to strong northwesterly winds occur over the Bay of Biscay, with a weak circulation over most of the Iberian Peninsula. A relatively common winter situation corresponds to an anticyclone over the continent and a fairly broad, deep low-pressure area in the North Atlantic, causing a southwesterly flow to prevail over most of the bay.

In the French shelf, prevailing winds are southwesterly and downwelling favorable in autumn and winter, while they are northwesterly and upwelling favorable in spring and summer. Nevertheless, average patterns of individual years reveal a great inter-annual variability, mainly in spring (March-April). The Progressive vector diagram (PDV) of the annual mean of winds on the French coast is shown in Fig. 24.3A. In the Cantabrian Sea, prevailing winds are southwesterly in winter and autumn and northeasterly, upwelling favorable, in summer and spring. The monthly mean of the north-south component of the winds calculated as Ekman transport at 43°N, 11°W in the southwestern Bay of Biscay for the period 1967–2000 (from Lavín et al., 1991, 2000) presents a southern component from October to March and a northern component from April to September, although variability is high as shown in the Fig. 24.3B.

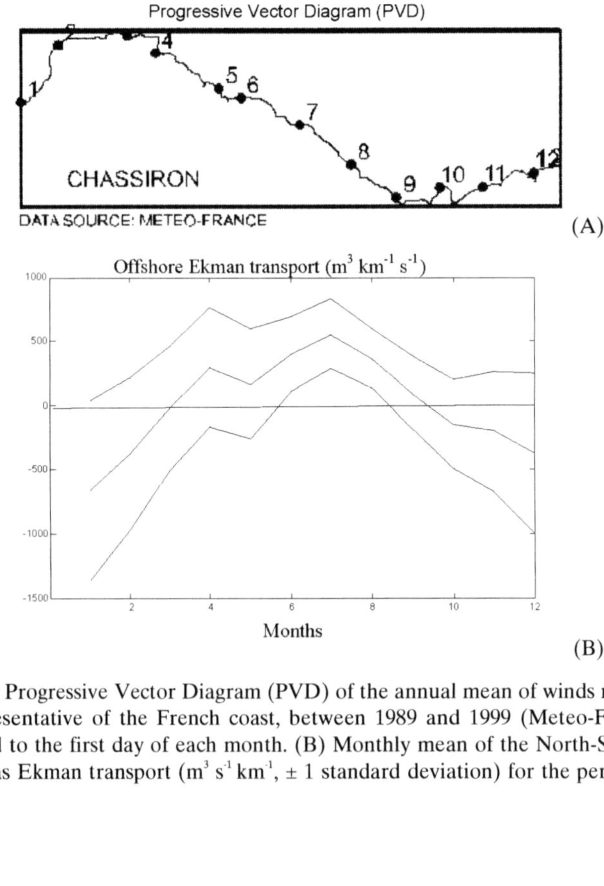

Figure 24.3a (A) Progressive Vector Diagram (PVD) of the annual mean of winds recorded at Chassiron (46°N), representative of the French coast, between 1989 and 1999 (Meteo-France). Numbered points correspond to the first day of each month. (B) Monthly mean of the North-South wind components calculated as Ekman transport (m^3 s^{-1} km^{-1}, ± 1 standard deviation) for the period (1967–2000) at 43°N 11°W.

The annual mean wind stress near the European ocean margin of about 43°N is directed to the east, while south of that latitude the wind stress turns to a more southward direction (Isemer and Hasse, 1987). This implies a southward Ekman transport north of 43°N, The annual mean curl of the surface wind stress along the European ocean margin east of 20°W and south of 50°N derived from Isemer and Hasse (1987) is negative, inducing a typical downward Ekman velocity $w_{EK} \approx -30$ m yr^{-1} (van Aken, 2001). This indicates that Porcupine and the Biscay area form part of the anti-cyclonic wind-driven Atlantic Ocean gyre.

3. Hydrography

In the last two decades a large number of projects have been developed in the Bay of Biscay to gain a better understanding and description of circulation processes, mixing and long-term transport of water masses, but more effort is needed. Hydrodynamic models, satellite imagery and operational oceanography systems have helped traditional hydrography to that purpose. Some projects have been funded by the EU, such as SEFOS and SEAMAR, others have been funded by France, such as ARCANE, PNOC and PNEC, and others by Spain, such as the deep standard sections (up to 4800 m depth) off Cape Finisterre, Cape Ortegal and Santander, the coastal standard sections of A Coruña, Cudillero, Gijón and Santander, both of which were funded by the IEO and the stations off San Sebastián operated by AZTI. Most of them have been sampled for over a decade in the southern Bay of Biscay.

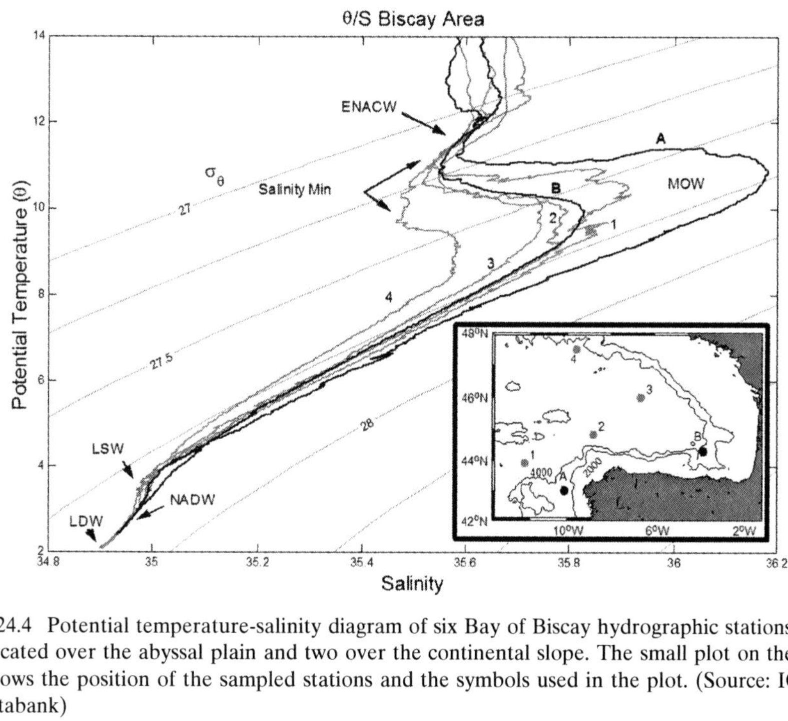

Figure 24.4 Potential temperature-salinity diagram of six Bay of Biscay hydrographic stations, four of them located over the abyssal plain and two over the continental slope. The small plot on the bottom right shows the position of the sampled stations and the symbols used in the plot. (Source: ICES and IEO databank)

3.1 Water Masses and related circulation

Most of the water masses of the Bay of Biscay are of North Atlantic origin, including those that have been transformed after mixing with the Mediterranean water that outflows through Gibraltar. The northern part of the Bay of Biscay is also an area of winter water masses formation due to large vertical convection, with a strong inter-annual variability (Pollard et al., 1996). The water mass structure of the different areas is shown in the Φ-S diagram of Fig. 24.4.

3.1.1 Bottom and deep waters

The influence of Antarctic Bottom Water (AABW) at the deepest levels of the eastern North Atlantic is a well known feature. It flows at 11°N through the Vema Fracture zone (McCartney et al., 1991) and spreads northwards through the Discovery Gap to the Theta Gap, the only deep passage (>5000 m) to the Biscay Abyssal Plain. This lower deep water (with AABW influence) is represented by the deepest segment of the Φ-S diagram in Fig. 24.4. Arhan et al., (1994) indicated a northeastward flow of bottom water and a cyclonic flow around the Charcot and Biscay seamounts associated with horizontal convergence as the basin is closed to the north.

Between approximately 2500 and 3000 m depth a core of northeast Atlantic Deep Water (NEADW) is found, which originates in the overflow from the Nordic Seas into the Iceland Basin. In the continental slope of the Bay of Biscay the amount of Iceland Scotland Overflow Water (ISOW) in the NEADW core is reduced to about 10%, due to additional mixing with the underlying LDW and the overlying cores of Labrador Sea Water (LSW) and Mediterranean Outflow Water (MOW) (van Aken, 2000a). Paillet and Mercier (1997) found a cyclonic recirculation cell over the abyssal Biscay plain with a characteristic poleward velocity near the continental margin of 1.2 ±1.0 cm s^{-1} (Dickson et al., 1985).

3.1.2 Intermediate waters

At a depth of about 1800 m, a core of LSW is characterized by a deep salinity minimum more visible in the northern and western part of the bay (see Fig. 24.4). Pingree (1973) traced the salinity minimum along a neutral surface from the Labrador Sea eastward to the Bay of Biscay east of 10°W. Paillet et al. (1998) reported low salinity intrusions in the northern bay with a possibly northwestward return flow in the saline boundary over the Celtic continental slope. The presence of isolated salinity extreme in isopycnal surfaces evidences intense diapycnal mixing over the continental slope at intermediate levels in the eastern Bay of Biscay (van Aken, 2000b). The energy source for such intensified diapycnal mixing probably has to be sought in the highly energetic internal tidal waves (Pingree et al., 1986; Pingree and New, 1995). Slope enhanced diapycnal mixing apparently makes the eastern Bay of Biscay a focal point for the local transformation of intermediate water masses (van Aken, 2000b).

Above the LSW layer, at 1000 db, a core of northward moving saline Mediterranean Overflow Water (MOW) is observed hugging the continental slope of Portugal and enters the Bay of Biscay. The salinity of this core decreases poleward along the continental slope due to isopycnal and diapycnal mixing with less saline water types (Díaz del Río et al., 1998; van Aken, 2000b). Over the abyssal plane salinity values are lower (Fig. 24.4).

Iorga and Lozier (1999) show the eastern turn of the Mediterranean Water into the Bay of Biscay as two branches around the Galician Bank and a cyclonic recirculation in the bay. In the NW corner of the Iberian Peninsula the maximum of the MOW is found offshore, but from Cape Ortegal the vein is trapped again near the upper slope flowing easterly along the northern Iberian coast with maximum salinities of 35.9 at 7° 55'W and 35.8 at 5° 10'W. North of 46°N, the maximum salinity is found near the shelf at 6°W and offshore at 8°W and at 11°W. The northern edge of the cyclonic recirculation of MOW in the Bay of Biscay is identified offshore at 47°N. Further north, the salinity maximum can be traced as 35.7 at 47°N and 35.65 at 48°30'W in the slope. At northern latitudes the MOW cannot be well distinguished from the surrounding water. Lately, values higher than the climatological mean have been found in the southern bay, saltier than 35.8 in the slope off Santander (González-Pola et al., 2003) and in the Cap Breton Canyon (Valencia et al., 2004).

Perhaps the narrowing and the restriction of the near slope area in the southwestern Bay of Biscay is due to the increasing influence of mixing with the LSW core (van Aken, 2000b). Low salinity values observed in the Armorican continental slope (eastern Bay of Biscay) may reflect a depletion of the MOW core by diapycnal mixing over the slope with overlaying and underlying fresher water masses (van Aken, 2000b).

Paillet et al. (2002) reported observations of a meddy (lenses of warm, salty Mediterranean Water that rotate anticyclonically) in the western Bay of Biscay (near 45°N 11° 30′ W). Their results suggest that it was generated in the Cape Finisterre-Cape Ortegal area, since this meddy and others also found in their study had lower temperature and salinity than the meddies formed further south. In the same paper they report that "northern meddies" (meddies north of 40°N) "are indeed common features".

3.1.3 Upper Waters

The upper 1000 db in the North Atlantic is the domain of mode waters, in this case, the Eastern North Atlantic Central Water (ENACW) variety (Harvey, 1982). According to their formation area, two main water masses are identifiable: a subpolar mode water formed in an area south of the North Atlantic Current (NAC) spreading southwards (Pollard et al., 1996) and a subtropical branch formed in the northern margin of the Azores current, which moves north towards the Iberian Coast (Pingree, 1997). They meet in the southwest corner of the Bay of Biscay where the subpolar branch subsides to spread southwards under the subtropical branch (Fraga et al., 1982; Ríos et al., 1992). The part of the subpolar mode waters formed in the northern Bay of Biscay was called Bay of Biscay Central Water (BBCW) (Tréger et al., 1979; Fraga et al., 1982).

The wind stress condition described in Chapter 2 causes a downward Ekman pumping transport between Porcupine Sea Bight and the Biscay Abyssal Plain, which allows the assumption that there the ocean boundary is part of the anticyclonic wind-driven ocean gyre, with a mean southward flow component in the upper layers. The subpolar branch presents a southward decreasing maximum winter mixed layer thickness at 20°W of around 600 m at 48°N and 350 m at 43°N (van Aken 2001). In the western Biscay region (40 to 48°N) van Aken (2001) inferred a mean southward velocity of around 1 cm s^{-1} along the European ocean margin that could maintain a subduction velocity of -125 m yr^{-1}, mainly supported by the advective term.

Other authors (Saunders, 1982; Maillard, 1986; Koutsikopoulos and Le Cann, 1996) have confirmed that the oceanic part of the Bay of Biscay is characterized by a weak (1–2 cm s^{-1}) and variable anticyclonic circulation and the presence of cyclonic and anticyclonic eddies shed by the slope current (Pingree and Le Cann, 1992a). From studies of Lagrangian drifters, Pingree (1993) showed that ENACW penetrates southwards into the bay following an anticyclonic circulation over the abyssal plain at 400 m depth. Surface currents derived from shallow drogued drifters (≈15 m) (van Aken, 2002) confirmed the southward flow tendency with a seasonal trend affected by an eastward component in autumn and a southward component in spring and summer. The presence of internal tide waves in the centre of the Bay of Biscay may enhance the amplitude of surface currents at tidal frequencies to values well over 5 cm s^{-1} (Pingree and New, 1991). The relatively large tidal motions over the abyssal plain, compared with the barotropic tides, confirm the importance of internal tides in the Bay of Biscay.

Above the MOW core, the Θ–S curves depict an inverted "S" shape (Fig. 24.4) with a salinity minimum around 450 m. At shallower levels, both temperature and salinity increase upwards from the salinity minimum (indicated in Figure 24.4) to the lower level of the seasonal thermocline, where a slight salinity maximum is generally observed at about 100 m. In summer, strong temperature gradients are observed in the upper 100 m, with only limited salinity gradients.

Off Galicia and in the Bay of Biscay, the highest salinity in the thermocline isopycnals is also found over the upper slope. The generic salt enrichment over the continental slope indicates that diapycnal mixing is quite important near the shelf. Salination of thermocline waters near the continental slope is due to local downward mixing from higher surface values in the poleward slope current (van Aken, 2001). Salinity in the upper seasonal thermocline varies seasonally because of the variable slope current; the highest salinities are found at the end of winter.

McCartney and Mauritzen (2001) deduce from the climatological database that the subpolar mode water is not only an extensive pool in the subtropical interior off the Bay of Biscay and northward to off Porcupine Bank, but also blankets the upper continental slope (the Armorican and Celtic shelf/slope sectors) of the Bay of Biscay. LSW, MW and ENACWp are indistinguishable on the basis of their phosphate, silicate and nitrate contents, all three having very similar nutrient levels. Hovewer, they can be distinguished on basis of their oxygen concentrations. In spite of their remineralization processes that have occurred in LSW from the time of its formation until it arrives in the study area, it still contains higher oxygen levels, than ENACWp and MOW (Castro et al, 1998).

3.1.4 Variability

Inter-annual variations observed in Eastern North Atlantic Central Water (ENACW; Fig. 24.4) are attributable to changes in the salinity and/or temperature of the subducting Central Water. Winter vertical convection in the upper western Bay of Biscay is a process subject to significant inter-annual variability (Pollard et al., 1996). The characteristic salinity of the ENACW envelope at 12°C shows an inter-annual variation (Fig. 24.5) with a salinity maximum in 1992. In the preceding period, the characteristic salinities were about 0.1 lower, while in 1997–1999, characteristic salinities at the 12°C isotherm were 0.05 lower. The inter-annual salinity variations of the thermocline near the Armorican slope in the Bay of Biscay are well correlated with the variations observed west of the Iberian Peninsula, so lateral salinity gradients seem to be more or less maintained on inter-annual time scales (Huthnance et al., 2002). Extending the time series to the southern Bay of Biscay, inter-annual salinity variations at 11.75°C are also well correlated with the Armorican and west Iberian time series for the 1992–2001 period. A maximum at the beginning of the 1990s, a minimum around 1995, an increase towards 1998 and a decrease in 2001 is common to the eastern North Atlantic from western Iberia towards Rockall Trough, including the North Sea entrance (Hudges and Lavín, 2003). This correlation indicates a strong connection of upper water masses in this part of the eastern North Atlantic including the Bay of Biscay.

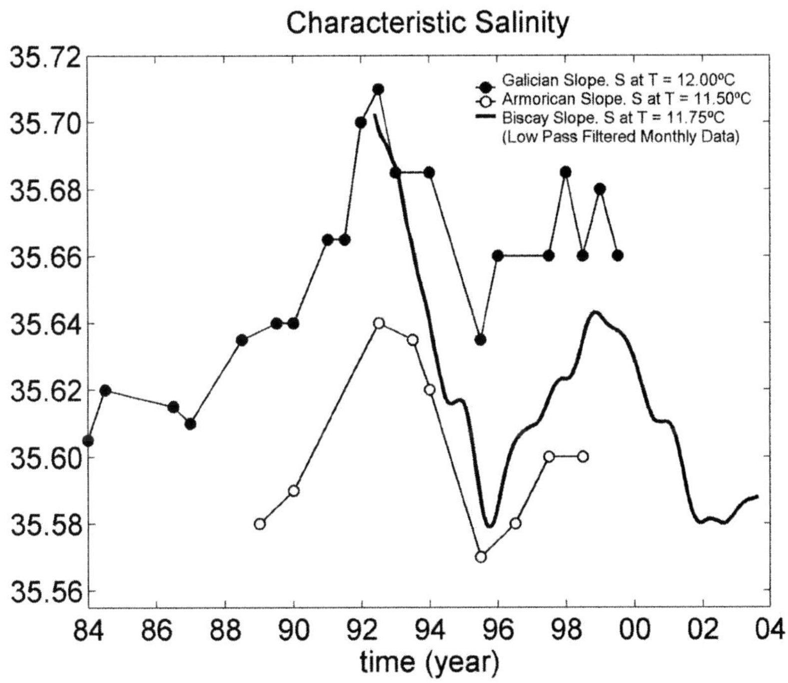

Figure 24.5 Characteristic salinity of the Eastern North Atlantic Central Water over the Galician slope (black dots), the Armorican slope (open circles) (Huthnance et al., 2002), and southern Bay of Biscay (Santander section).

Pérez et al. (1995), showed a significant year to year variability in the T/S characteristic of the upper 800 m of water column from data collected at 43°N 10°W. This variability correlates well (Pérez et al., 2000) with the wind stress at 43°N 11°W, with the cumulative river discharge and with the North Atlantic Oscillation (NAO).

3.1.5 Seasonality

Seasonal atmospheric cycle and meteorological conditions have a strong influence in temperate seas such as the Bay of Biscay. Pingree et al. (1999) have analysed the annual response reflecting a cycle related to September-October and March-April (with some variability of ± 1 month) periods in the slope current. This effect has been termed the "SOMA" response. They present a summary of responses in the slope region of the Bay of Biscay of SOMA-type behavior. The main changes are found in March-April and reflect the change of current direction in the northern Spanish slope, Ushant and Goban Spur. They also noted the importance of the spur or abrupt changes of topography that represent a loss in the slope current continuity (transport) with the subsequent inertial overshoot from the slope flow moving towards the spur.

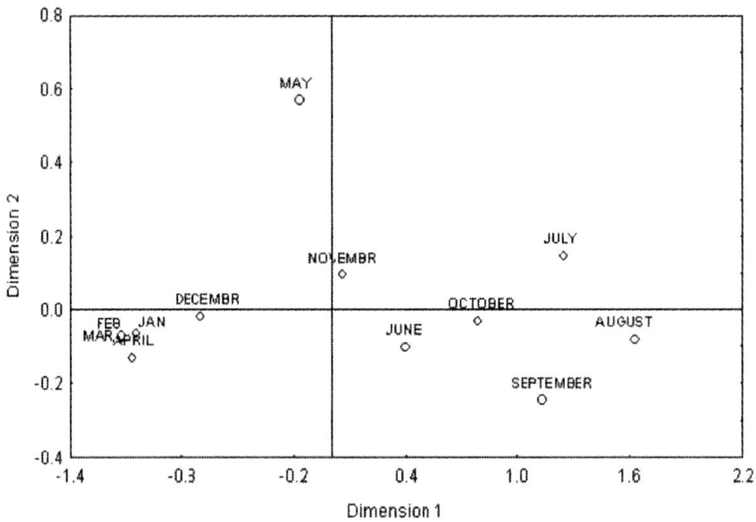

Figure 24.6 Statistical ordination of the monthly mean SST images obtained by the NOAA-AVHRR satellite from January 1994 to December 1996 in the Bay of Biscay (Moreno-Ventas et al., 1997). The annual cycle can be divided into two main periods: the first, from December to April, and the second, from June to October. The transitions between the two periods occur in May and November.

Using a collection of Sea Surface Temperature (SST) images obtained by the NOAA-AVHRR satellite from January 1994 to December 1996 in the Bay of Biscay and based on a statistical ordination of the monthly mean SST images, Moreno-Ventas et al. (1997) showed that the annual cycle can be divided into two main periods. In the first, December to April, the isotherms in the Bay of Biscay follow the general eastern North Atlantic pattern, characterised by a meridional temperature gradient. In the second, from June to October, the Bay of Biscay is isolated from its oceanic surroundings and develops an independent pattern in the distribution of isotherms. The transitions between both periods occur in May and November (Fig. 24.6).

Koutsikopoulos and Le Cann (1996) have summarized the hydrographic features in the Bay of Biscay for each season. Inspired by their schematic representation, a sketch has been produced with the main features of the hydrography and circulation in the area, with emphasis on the shelf and upper ocean layers (Fig. 24.7). At the surface and particularly over the shelf, wind forcing, heating, rainfall and river runoff modify the water characteristics and impose high variability (spatial, seasonal and inter-annual).

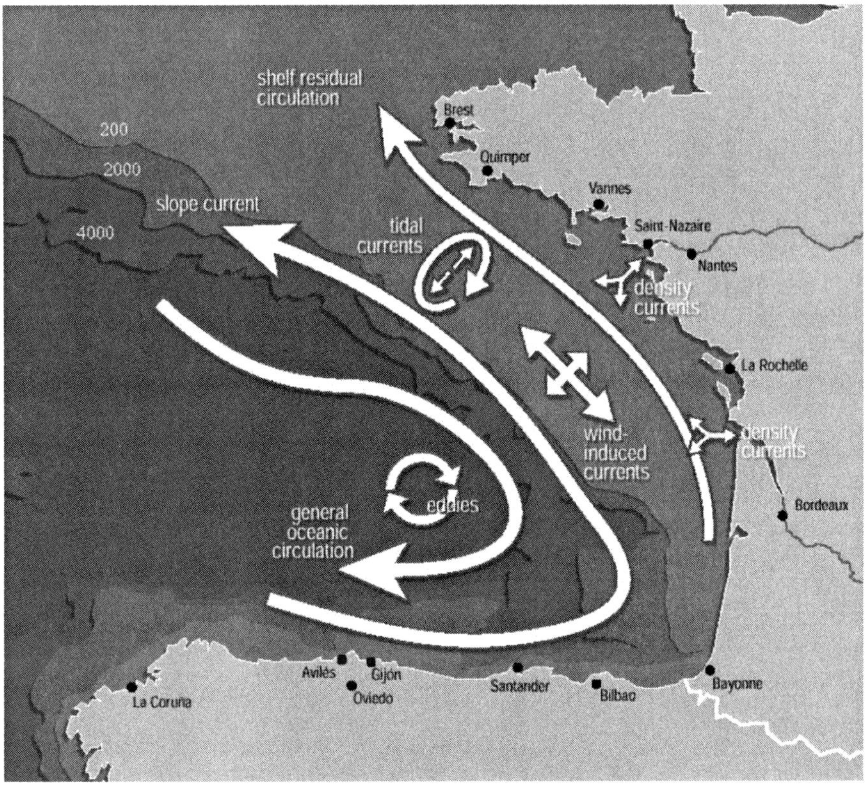

Figure 24.7 Schematic illustration of Bay of Biscay circulation summarized by Koutsikopoulos and Le Cann (1996) modified by OSPAR 2000.

In winter, (Fig. 24.8A) the continental shelf off the main rivers - Loire, Gironde and Adour- is marked by cold, low salinity waters. Thermal inversions are often observed over the shelf in the vicinity of the estuaries sustained by the haline stratification. This is also the season when the warm, high salinity, poleward slope current presents its clearest surface signal and its northernmost extent. The contribution of river runoff will be diluted in the whole water column, lowering salinity and temperature over the inner shelf and helping to sustain a cross-shore gradient between shelf water and the slope current.

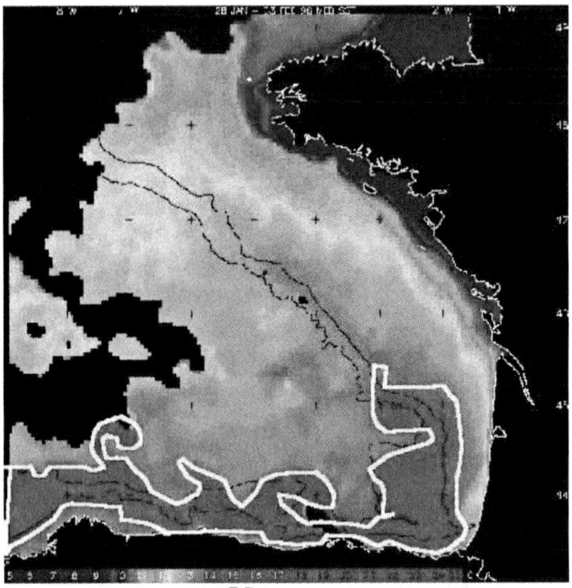

PC (A)

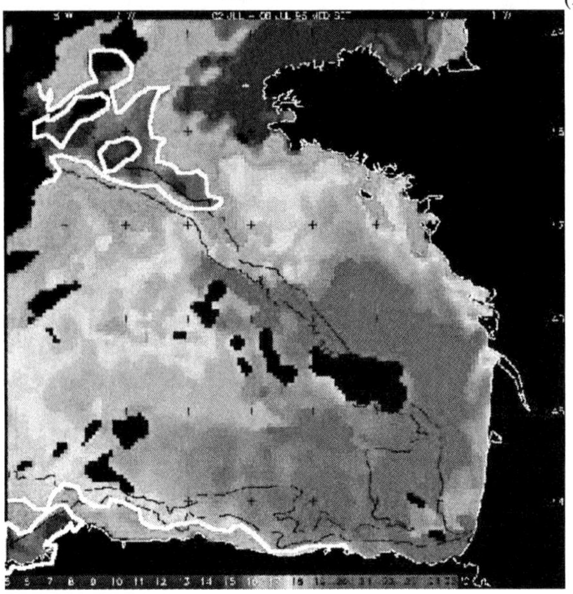

UPW (B)

Figure 24.8 Mean features of Bay of Biscay SST in winter (A) and summer (B). Strong Poleward Current (warm feature marked with white line) and cold water over the inner slope on the Armorican shelf (North France) in winter and strong warming in the inner Bay of Biscay and cold signal due to upwelling in the western Cantabrian Shelf and in the north part of the Bay of Biscay Ushant front in summer are presented. Mesoscale activity is noted in both periods. Satellite images were received by the NERC Dundee Satellite Receiving Station and processed by Peter Miller at the Plymouth Marine Laboratory Remote Sensing Group (www.npm.ac.uk/rsdas/).

During the winter months, the flow over the deep basin is generally towards the east with a mean velocity between 1.5 and 2 cm s^{-1} converging into a poleward flow over the continental slope with a mean velocity of 5 cm s^{-1} (van Aken, 2002). Most of the drifters crossing the shelf break from the slope region towards the continental shelf do so between the middle of November and the beginning of April (van Aken, 2002). He also reported that the most energetic eddies were also found in winter, confirming that eddies are preferentially generated in the slope region in winter and dissipate later in the year in the bay interior as proposed by Pingree and Le Cann (1990, 1992a and b) and Paillet et al. (1999).

In spring the low salinity waters cover a great part of the continental shelf depending on the river runoff and on the wind regime. The seasonal thermocline appears in April in the outer shelf and reaches the coastal area in May. Below the thermocline, a cold pool (11°C) appears off the French coast, extending from the southern Brittany area to the latitude of the Gironde estuary, centred over the 100 m depth contour. The pool is observed throughout the year and shows weak temperature variations (less than 1°C). Although river runoff adds to the thermal stratification, its importance is reduced throughout spring and summer.

In summer and early autumn (Fig. 24.8B) coastal upwelling appears off the Iberian Peninsula in late spring and reaches its maximum expression in summer, when it also occurs in the southeastern Bay of Biscay over the shelf off the Landes coast. Interactions of tidal currents with bottom topography are responsible for the formation of seasonal thermal fronts in the Bay of Biscay, such as the Ushant front off western Brittany. Several other mixed areas occur along the French coast, generally in the vicinity of islands. Along the Armorican and Celtic slopes, frontal zones are induced by internal waves.

The surface currents over the deep basin vary between southward and southeastward, with a mean scalar velocity of 1.7 cm s^{-1}, while over the continental slope the flow is equatorward with a mean velocity of 1.5 cm s^{-1} (van Aken, 2002).

With the changes in atmospheric forcing, upwelling ceases and late autumn is marked by the appearance at the surface of the poleward slope current signal off west Iberia.

3.2 Slope current

As has been mentioned when talking about the seasonal pattern, a warm, saline intrusion has been recurrently observed during winter off the western Iberian coast, trapped within 50 km of the shelf edge, flowing poleward with speeds of 20–30 cm s^{-1}, the transport increasing in the flow direction (Frouin et al., 1990; Haynes and Barton, 1990). Intermittently, it reaches the northern Spanish coast margin (Pingree and Le Cann, 1990) around the Christmas period, which led Pingree and Le Cann, (1992b) to name it "Navidad " (Fig. 24.8A).

According to Huthnance (1984), the slope current is primarily non-wind driven. The meridional density gradient, observed in the upper 200–300 m of the NE Atlantic (Pollard and Pu, 1985) is the result of the poleward cooling of the ocean surface. It is balanced by an eastward flow of light water, which in turn causes a height difference between shelf waters and deep sea. A cross-slope density gradient develops and a geostrophically balanced poleward flow is thus generated. In the absence of any additional forcing, the poleward flow can only decay off those boundaries, as is the case on the northern coast of the Iberian Peninsula. Winter westerlies pile up water at the Iberian margin and reinforce the slope current at the southern Bay of Biscay. In that case, the slope current develops a surface expression and achieves its greatest velocities. It is present over the upper slope and eventually invades the outer shelf. During this period the slope current tends to act as a boundary preventing upper layer exchanges between the shelf and the deep sea. This exchange, however, is not totally blocked, since current instabilities caused by interactions with topography may result in the generation of eddies (OSPAR, 2000).

Serpette et al. (in press) described the slope current at the upper and lower part of the ENACW and found an alternation between poleward and equatorward directions: at the lower part, the current is poleward over the northern slopes of the Bay of Biscay and tends to be equatorward over the Armorican and northern Spanish slopes. They link this fact with the eastward penetration of waters over the abyssal plain near 46°N (Pingree, 1993; van Aken, 2002; Reid, 1994; McCartney and Mauritzen, 2001). This westward flow over the northern Spanish Slopes was postulated by Schopp (1993) and attributed to positive windstress curl forcing (Isemer and Hasse, 1987), which drives a southward Sverdrup flow component over the Bay of Biscay abyssal plain producing a westward flow over the Spanish slopes. Current meter measurements obtained during the European SEFOS project in the central part of the southern Bay of Biscay slope (Díaz del Rio et al., 1996) show this alternation of flows with more a southward component in winter 1995 and an intensification of the eastward slope current in autumn 1995 and winter 1996, mainly in the upper part of the ENACW (Fig. 24.9).

Surface drifters released near the shelf edge on the western Iberian coast during the French ARCANE project tended to remain trapped in this region in winter. Nevertheless, the flow may have a branch around Galicia Bank as suggested by the float tracks (Huthnance et al., 2002). The surface temperature field, obtained from remote-sensed images, often shows a much broader (hundreds of kilometres) northward extension of warm water, probably affected by winds (García-Soto et al., 2002).

Winter temperature along the Spanish northern slope can be used to examine the inter-annual variability of the Poleward Current (Pingree, 1994). Some attempts to relate it to climate forces as NAO and Ekman transport have been made. Garcia-Soto et al. (2002) did not find a clear relationship between winter NAO (previous November-December) and SST for the time series they analysed (1967–2000), but when they limited the years to those of a marked satellite-observed Poleward Current, they found that SST values at 4° and 8°W were tightly associated with negative values of the November-December NAO index. They attributed this relationship to the decreasing wind speeds in the negative phase of NAO and associated decrease in latent and sensible heat fluxes (Cayan, 1992a, b). Cabanas et al. (2003) found a correlation between the November-December Ekman transport at 43°N, 11°W and the December SST in NW Iberia (42°N, 9°W). Correlation reduces over time (November or January) and distance into the Bay of Biscay (7°W using December or 4°W using January SST values). The previous December maximum Ekman transport coincides with years of strong Poleward Current during the 1990s (Garcia-Soto et al., 2002), but in previous decades the coincidence is only occasional. Planque et al. (2003) also show that there is no significant relationship between the winter NAO index and SST, wind or river runoff during the last century and a half.

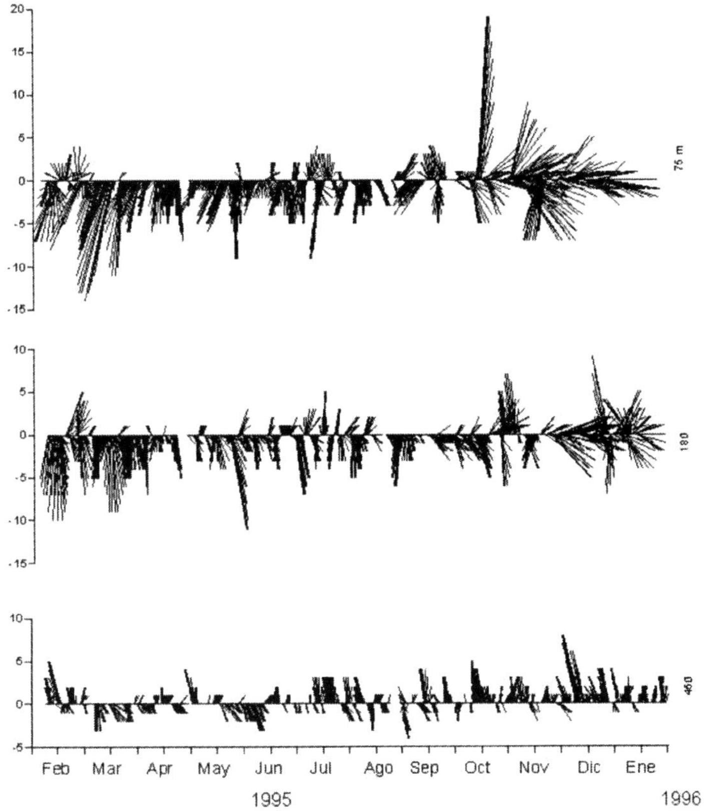

Figure 24.9 Velocity vectors (cm 5′) at 75, 180 and 450 m depth in the Central Cantabrian Sea (Díaz del Río et al., 1996). The data series have been smoothed by a Godin filter to eliminate high frequency noise. Low frequency oscillations have been obtained by applying a Godin A24A24A25 filter.

Garcia-Soto et al. (2002) showed pulses of warm water in the winter of 1990 extending northwards along the shelfbreak/slopes from the Subtropical front (35°N) to near the polar front. The surface pulses had a more pronounced northwards advance along the shelfbreak in the Armorican and Celtic shelves. In the Cantabrian region, where the shelf is oriented E-W, water spreads over it.

3.3 Mesoscale features (off-shore)

As the winter Poleward Current enters the Bay of Biscay around Cape Finisterre, warm water moves eastward along the Cantabrian continental shelf and slope. Some flow continues its poleward advance across the Landes Plateau and the continental slope of Cape Ferret canyon (see Fig. 24.8A). Where topography changes abruptly, as in Cape Ortegal, Cape Ferret canyon and Goban Spur, the slope water is injected into the oceanic region to form anticyclonic eddies that contain a core of slope water (Pingree and Le Cann, 1989, 1990). These slope water oceanic eddies (or "swoddies") contain surface warm water and their movements have been followed using infrared and color satellite images (Pingree and Le Cann, 1992a; Bardey et al., 1999 Garcia-Soto et al., 2002;) and floats (van Aken, 2002; Huthnance et al., 2002).

Serpette et al. (in press) described slope current and eddies at two levels of the NACW. They suggest that another possible mechanism for the westward current in the southern Bay of Biscay may be the collective effects of anticyclonic eddies, which are more persistent in this area. In the vicinity of the slope they may drive the observed equatorial flow. The change in the flow direction will cause slope-ocean exchange, a key process for biogeochemical studies (Huthnance et al., 2002).

Van Aken (2002), on the mesoscale motion over the abyssal plain, reported a 2:1 dominance of the number of anticyclonic eddies over cyclonic ones, characteristic of the European ocean margin. Garcia-Soto et al. (2002) presented a statistical report of summer swoddy-like eddies in the southern Bay of Biscay (south of 46°N) and found them during either years of strong or weak Poleward Current, but eddies at 4°W appear to be confined to the first case and eddies west of 6°W to years without a clear slope current, perhaps indicating swoddy production only from near Cape Ortegal.

3.4 Shelf Seas

3.4.1 Cantabrian Sea shelf

The Cantabrian shelf has marked thermal zoning, from a cold western part where oceanic influence is greater, towards the inner part where continental influence is stronger. The intensity and frequency of upwelling events decreases eastward as temperature and stratification increase. Hydrographic variables follow the seasonality previously described. The main features are:

Seasonal cycle. Temperature follows the expected seasonal warming and cooling pattern, which determines a seasonal process of stratification and mixing of the water column. The stratification period occurs between May and October in a layer of about 50 m depth from the neritic part to beyond the shelf-break. Between November and April the water column remains mixed. During spring and summer low salinity values are found in the surface due to continental runoff and advection from eastern low salinity waters. In late autumn and winter, the salinity pattern is governed by the influx of salty water associated with the Poleward Current. As in other temperate latitudes nitrates show the highest values in winter throughout the whole water column and the lowest values at the surface during the stratified period (Lavín et al., 1998).

The seasonal cycle in sea surface temperatures varies from the western Cantabrian Sea towards the east. In the west, near A Coruña, the cycle has a small temperature range from summer to winter, as upwelling cools water in summer. The range between summer and winter SST was reported as being around 4°C in A coruña (Valdés et al., 1991), increasing in the inner part of the Bay where there is more influence from the land (Valencia et al., 2004) and where temperatures are higher in summer and lower in winter than in the surrounding area. Near the surface (10 m), monthly mean temperatures at the shelf break off Santander (43° 42'N, 3° 47'W) during the 1990s indicate a typical summer (August) maximum and a late winter (February-March) minimum. The amplitude of the cycle, based on an annual and semi-annual harmonic, is around 8.5°C (between 12.8 and 21.3°C). Deeper in the water column, the maximum occurs later in the year and has a reduced amplitude (Cabanas et al., 2003). The time series of temperature, salinity and nitrates from 1992 to 2003 in the whole water column at the central shelf of the Santander standard section (43° 34.5'N, 3° 47'W) are shown in Fig. 24.10.

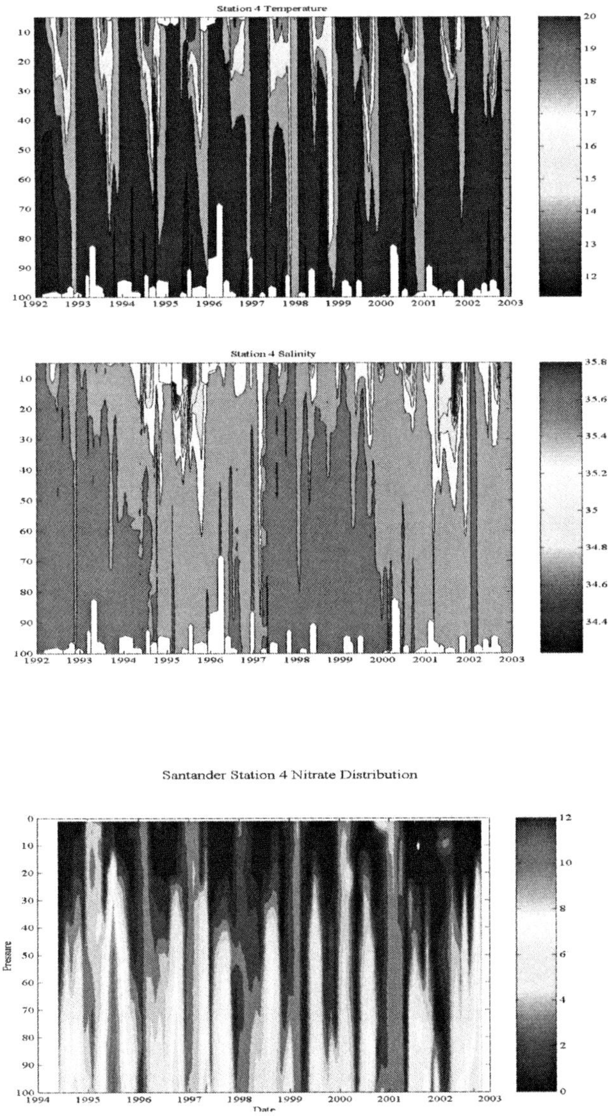

Figure 24.10 Seasonal and interannual variations in temperature from 1991 to 2003 (A), salinity from 1991 to 2003 (B) and nitrates from 1993 to 2003 (C) over the shelf off Santander (43° 34.5'N, 3° 47'W, southern Bay of Biscay).

Western incursions. The hydrographic properties of the southern Bay of Biscay upper water layer are eventually influenced by intrusions of Atlantic water advected from the west of the Iberian Peninsula. These intrusions of warm and saltier water of subtropical origin are more intense in winter, when wind conditions are more favorable, but during other seasons spring events have also previously been reported over the western part of the Cantabrian shelf with relatively high nitrate concentrations (Bode et al., 1990; Lavín et al., 2001). These features influence the planktonic community as well as the fish larval distribution during spring, a key period for eggs and larval development and survival in the western Cantabrian Sea, an important nursery area.

Fresh water advections. The large runoff from the French rivers not only conditions the saline distribution on the French shelf, but also affects the eastern Cantabrian Sea (Lavín et al., 1998; Gil et al., 2002; Valencia et al., 2003). These fresh water advections produce significant variations in the seasonal cycle along the Cantabrian Sea in late spring and summer. They have been described as areas of accumulation of eggs and larvae of some pelagic and demersal fish in the inner Bay of Biscay (Sánchez and Gil, 2000).

Upwelling. Northeasterly winds between Cape Finisterre and Ortegal and easterly winds in the Cantabrian shelf produce upwelling in the Cantabrian shelf mainly in spring and summer. Topography enhances the effect. Upwelling intensity decreases from Cape Finisterre where the more intense effect is found, towards Cape Ortegal and Cape Peñas where it is detected to a lesser extent. As the upwelled water is cooler than the surrounding surface water, it is easily detectable in images of sea surface temperature (see UPW in Fig. 24.8B).

Upwelling events have been reported in the western part of the southern Bay of Biscay (Molina, 1972; Fraga 1981, Botas et al., 1990; Lavín et al., 1998; Moreno-Ventas et al., 1997) and small effects have also been reported in the inner part (Borja et al., 1996). This process is considered to be an important mechanism of fertilization of the photic layer (Fernández and Bode, 1991; Gil et al., 2002).

Upwelling is the main nutrient supply during the period of water column stratification. The intensity and duration of the upwelling period is known to vary from year to year but there is poor information on other sources of variability, such as the nutrient load of the upwelled waters. Tréger et al. (1979) gave a late winter nutrient distribution for the Bay of Biscay sub-surface water masses and found that subtropical ENACW has a relatively low nutrient concentration and in contrast subpolar mode waters were richer. ENACWt has the lowest nutrient levels reflecting its formation in the nutrient-poor waters of the subtropical gyre (Pollard and Pu, 1985, Perez et al, 1993, Castro et al, 1998). González et al. (2003) reported upwelling during 1991 and 1992 from water of southern origin with high salinity (>35.80) and low nutrients (5 μM NO_3) associated with a strong poleward current, and during 1995 and 1996 when salinity was low and nitrate values were as high as 12 μM NO_3. In this latter case water was of northern origin. These values are in the same range as those that Valdés et al. (1991) reported for an annual cycle in A Coruña during 1988 and Lavín et al. (1998) at a neritic station (St-4) off Santander. Fig. 24.10C shows nitrate concentrations detected at St-4 off Santander (43° 34.5'N, 3° 47'W) during the 1994–2002 time series (Lavín et al., 2003).

Mesoscale and submesoscale filaments and eddies. Gil et al. (2002) reported a hydrographic condition in spring over the Cantabrian shelf. An incipient wind forced a coastal upwelling event that exhibited offshore filaments, rings and eventually quasi-geostrophically isolated, semi-closed mesoscale eddies. Some of these structures concentrated phyto-, zoo- and ichthyoplanktonic material within them. Residual currents have been pointed out as a major source of the westward displacement of the mesoscale eddies, producing a net westward transport of planktonic patches through the southern Bay of Biscay in spring and summer. Formation of eddies over the continental shelf produces ageostrophic vertical forcing, which provides nutrients to the upper layers and seems to have a positive effect on the abundance and distribution of recruits (Sánchez and Gil, 2000; Sánchez et al., 2001). They also reported the favorable effect of the anticyclonic eddies in westward transport of larvae inside the nuclei, close to or over the continental shelf.

3.4.2 French Continental shelf

Tides. The tides in the Bay of Biscay are semidiurnal. The amplitudes increase from the shelf (1.3 m) to the shore (1.6–1.7 m). Instantaneous tidal current strength is partly related to the topography and width of the continental shelf. Thus, in the southern part of the bay (south of 45°N) these currents are quite weak, less than ~15 cm s^{-1} (Le Cann, 1990). Over the Armorican shelf, instantaneous tidal currents are stronger (~30 cm s^{-1}) with several maximum local values in the vicinity of islands (>50 cm s^{-1}).

At the shelf break, when the water column is vertically stratified, tides generate internal waves that propagate both on- and off-shelf from about 5° to 9°W. They appear to be responsible for significant mixing and upwelling of nutrients over the shelf break.

Wind-induced currents and upwelling. The NW wind induces SW-S surface currents over the French shelf. The SW winds invert the circulation towards the NW and produce local cross shelf set-ups with little vertical integrated flow. Over the Armorican shelf and the northern part of the Aquitaine shelf, the currents generated are typically ~10 cm s^{-1} and locally 20–30 cm s^{-1}. In the southeastern part of the bay, the situation is more complex: the circulation over the slope should affect this area because of the narrowness of the continental shelf.

In spring and summer the prevailing NW winds (see Fig. 24.2) induce an offshore transport in the surface Ekman layer and upwelling along the coasts is evenly observed (Pingree, 1984; Froidefond et al., 1996; Puillat et al., 2003). Upwelling is not generated under the same wind direction on either side of the Loire estuary because of the coastline orientations: northerly winds in the southern part and west to northwesterlies along the Brittany coast. Jégou and Lazure (1995) showed that vertical movements induced by upwelling-favorable winds are stronger when tidal currents are weak.

Seasonal cycle. In winter the water column is generally homogeneous. The inner shelf is marked by a cold strip, primarily linked to the presence of freshwater from the Loire, Gironde and Adour estuaries in the surface layer. Downwelling-favorable winds and high turbulence levels confine low salinity waters (< 35) into a strip about 50 km wide on the inner shelf and prevent stratification. Near the bottom, low-salinity water spreading hardly varies, reaching about the same extent each year.

Thermal stratification occurs from May until mid-September in a layer ~30–50 m thick. Maximum temperatures are reached in August, the mean SST is roughly ~2°C higher near 45°N than at 47°N (along the 200 m isobath) and the thermocline appears less marked in the north (Vincent and Kurc, 1969; Koutsikopoulos and Le Cann, 1996). This north-south difference can be partially explained by the occurrence of greater tidal mixing in the north, but also by less solar radiation and the influence of stronger winds on an extended area of shallow water. Thermal fronts can be observed over the continental shelf off western Brittany and along the French coast in the vicinity of islands. They are due to interactions between the topography and tidal currents (Pingree et al., 1982; Mariette and Le Cann, 1985) whereas along the northern Armorican slope they mainly result from mixing induced by internal tides propagating across the shelf-break (e.g. Pingree et al., 1982; Mazé et al., 1986; Le Tareau and Mazé, 1993; Pingree and New, 1995; Druon et al., 2001).

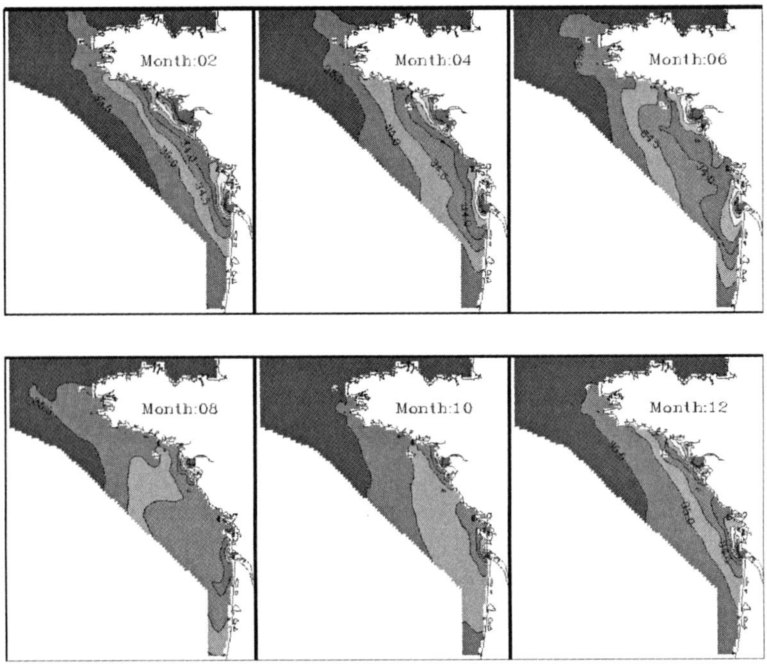

Figure 24.11 Average monthly surface salinity calculated by Mars 3D model for 15 years.

From late winter to summer, the haline stratification is strong due to high river runoff and relatively low vertical mixing. In May and June, low salinity water (< 35) can extend offshore beyond the 100 m isobath due to prevailing upwelling-favorable winds which may drive river plumes offshore.

In autumn, the haline stratification decreases because of low river discharge and wind-induced vertical mixing. North of 46°N, surface salinity increases by up to 1 inshore of the 100 m-isobath from spring to the end of summer (Puillat et al., 2004). This seasonal evolution agrees with salinity measurements made on the inner shelf (Lazure et al., in press). They show a general decrease from north to south along the coast in summer, which may be linked to northwesterly wind conditions, favorable for driving surface water to the SW of the bay. Average monthly surface salinity calculated by Mars 3D model for 15 years is shown in Fig. 24.11.

Cold pool. Below the thermocline, a cold water mass called the "Bourrelet froid" (Vincent and Kurc, 1969) is isolated on the middle part of the shelf. This cold pool is nearly homogeneous (11–12°C) from ~40–50 m to the bottom. It is present throughout a large part of the year when water is stratified with weak temperature variations (less than 1°C).

Puillat et al. (2004) have shown the inter-annual variability of this structure, both in extent and temperature. For instance, in 1999 the cold pool was larger and colder (< 0.4°C) compared with the 1998 situation. In 1999, its extent had more than doubled on a NW-SE axis and the vertical thermal gradient was also greater than in 1998. This variability may be linked to the wind effect: September 1998's wind event induced significant upwelling and probably increased vertical mixing, thus reducing thermal stratification, whereas in 1999 the wind was not upwelling-favorable.

River plumes fluctuations and lower salinity lenses. In winter and spring, the main mesoscale structures are river plumes. The behavior of Loire and Gironde plumes has been described by a 3-D model of the Atlantic shelf, first for winter and spring, then over a period of several years (1990 to 1996) under realistic forcing (Lazure and Jegou, 1998). Maximum runoff and SW/downwelling-favorable winds usually occur from winter to spring and decreased runoff and NW/upwelling-favorable winds from spring to autumn. Thus in winter, plumes usually spread northwards and along the shore and vertical stratification is weak and almost non-existent on the shelf.

In spring, when river discharges are reduced and when prevailing winds are from the NW, the northward spreading of plumes may be stopped. In that case, plumes may be driven offshore or southwards in a 30 m surface layer and a strong stratification may develop on the shelf. This path change usually occurs in late March or early April. The low-salinity strip along the shore seldom builds up again, and the shelf circulation of water masses becomes mainly wind-driven.

The greatest offshore spreading of low-salinity waters on the shelf occurred in years of very large river discharges and strong upwelling-favorable winds, such as in spring 1994 and 1995. In those years, the Loire and Gironde plumes were connected and low-salinity waters influenced a large part of the shelf. On the other hand, the plumes still spread northwards along the coast in spring 1993 and low-salinity waters reached the western part of Brittany. The mid and outer shelf remained unaffected by the Loire and Gironde influences in spring that year. This situation corresponds to runoffs close to the average and long periods of weak wind stress. Thus, low-salinity waters are mainly driven by density gradients.

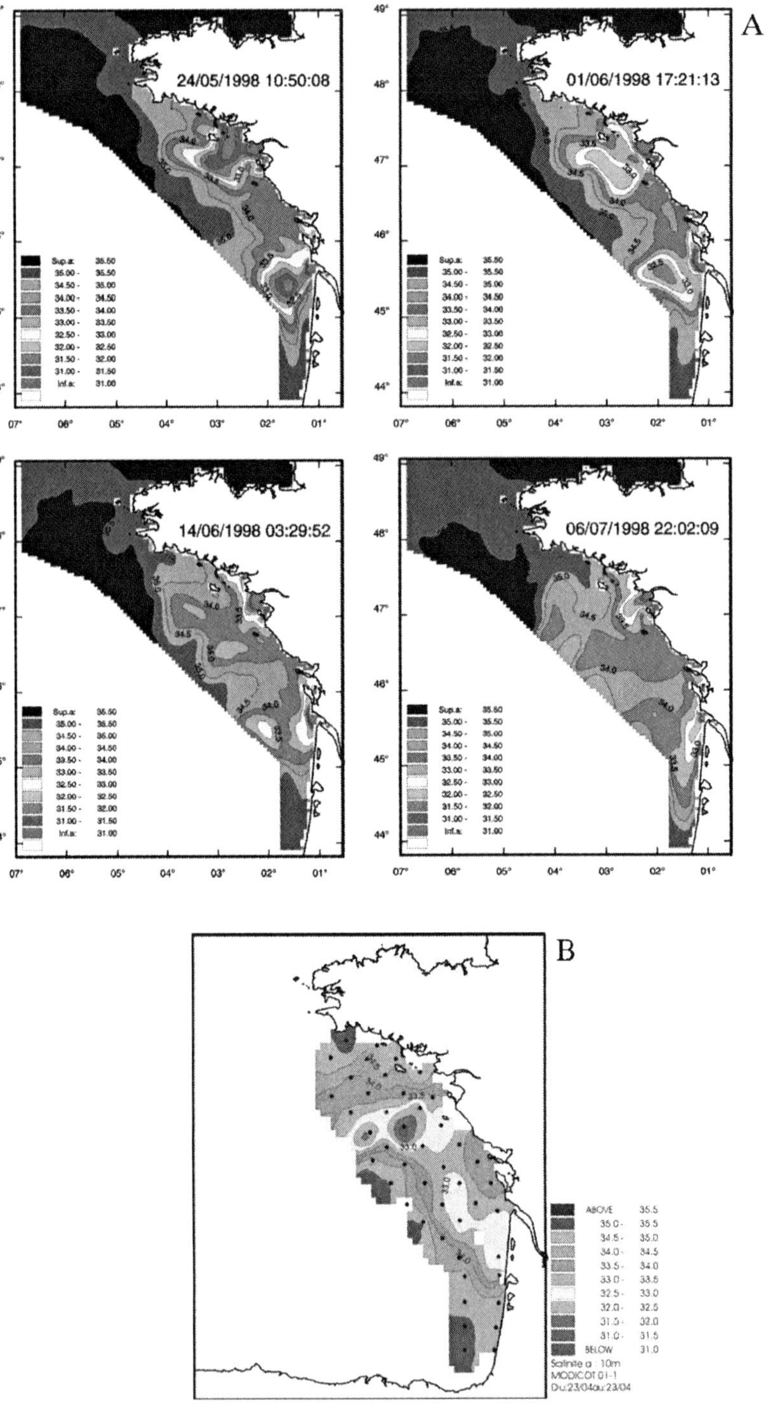

Figure 24.12 (A) Formation and fate of low salinity lenses from Loire and Gironde in spring 1998 from Mars 3D model and (B) map of salinity distribution during the Modycot cruise (April 2001) with an example of low salinity.

Offshore, parts of river plumes can sometimes be detached and isolated from the main plume structure to become independent mesoscale low salinity lenses with lower salinity in their core compared to the surrounding water. Such patterns have been revealed in spring and summer in model results and observed in hydrological measurements (Froidefond et al., 1996; Uriarte et al., 2001; Puillat et al., 2004) on the shelf. These lenses are 50–80 km wide and ~30 m thick and occur during westerly to northerly wind events. The low-salinity lenses observed during five cruises in 1998 and 1999 by Puillat et al. (2004) show variability in diameter, shape, mean depth and intensity. The formation and fate of low-salinity lenses from the Loire and Gironde in spring 1998 from the Mars 3D model are shown in Fig. 24.12A and a map of salinity distribution during the Modycot cruise (April 2001) with an example of low-salinity lenses is shown in Fig. 24.12B. Lenses seen in spring have a lower minimum salinity of 33–34 in the core and a higher gradient.

Low salinity lenses are expected to behave like retention systems for larvae and eggs. Thus, high numbers of anchovy eggs have sometimes been found in isolated lower-salinity lenses far from the coast in the southern Bay of Biscay and along the Cantabrian coast (Motos et al., 1996; Sánchez and Gil, 2000; Sánchez et al., 2001). These low water patterns have a French shelf origin and affect the configuration of water masses by locally increasing the stratification of the upper layers.

Fresh water inputs can induce significant density currents of ~10 cm. s^{-1} in model simulations (Lazure and Jégou, 1998). These density currents generally flow northwards in the surface layer because of the Earth's rotation but they are frequently affected by wind-induced circulation (Jégou and Lazure, 1995; Lazure and Jégou, 1998; Froidefond et al., 1998; Hermida et al., 1998).

4. Biology of the pelagic ecosystem

Biological cycles, population dynamics and coupling of ecological responses of plankton species and communities to physical forcing have been studied in the Bay of Biscay in the frame of wider interdisciplinary research programmes directed at understanding classical topics such as the conditions for the recruitment of living marine resources, the balances of matter and biogeochemical cycles, and the adaptation of marine ecosystems to global change. These programmes usually involve plankton-dedicated surveys which include a grid of stations covering the spatial occurrence of a given process (e.g. the spawning area of a given species). The grid is then sampled at such time intervals (hours, day or days) as to allow the observation of plankton dynamics (e.g. changes in abundance and biomass) and the identification of patterns in the plankton response under defined physical and biological conditions. Nevertheless, these studies are not homogeneous in terms of sampling strategy and seasonality, thus limiting comparisons among different regions and surveys.

Recently, monitoring programmes based on routine observations of ocean properties and biological communities were established in the Bay of Biscay and its environs. These include the transects off Santander, Gijón, Cudillero and A Coruña (N. Spain), Plymouth (S. England), the CPR routes (which cross the entire Bay of Biscay), and ships of opportunity (e.g. the Ferrybox project, which started a pilot experience in the Bay of Biscay). The temporal resolution of these projects allows us to study and resolve plankton dynamics, seasonal cycles, inter-annual variability and mid-term trends, and they also allow for making easy comparisons among areas. The main variables measured in time series programmes include structural parameters such as abundance, biomass, species composition, body size and also physiological rates such as production, respiration and consumption.

Not all the regions in the Bay of Biscay were sampled and studied equally. Balances of matter and biogeochemical cycles were investigated in detail in the area of Galicia where upwelling processes are stronger and govern pelagic production (Bode and Fernández, 1992; Casas et al.,1997; Serret et al., 1999; Teira et al., 2003; Varela et al., 2003; Bode et al., 2004, etc.), whereas in the central Cantabrian Sea and inner Bay of Biscay most of the fisheries and marine research institutions prioritized their research aims towards understanding the plankton dynamics in relation to living marine resources, recruitment processes and interannual variability. The results presented in this review are focussed on these latter topics.

4.1. Phytoplankton dynamics and regional variability

In general we have gained a good knowledge of the main biological processes and driving forces that govern the cycles and variability of plankton populations in the Bay of Biscay (Varela, 1996; Poulet et al., 1996; Valdés et al., 2000; Valdés and Lavín 2002).

Like the entire northeastern Atlantic, this region undergoes a seasonal climatic cycle, which strongly affects the pelagic ecosystem through three interrelated forcing factors over the year: sunlight exposure, heat exchange with the atmosphere input and mechanical forcing on the surface due to wind. The effect of these forces produces a regular pattern in hydrographic conditions characterized by winter mixing of waters, followed by summer stratification. Phytoplankton blooms occur during the transition between the two periods. In addition, productivity and plankton growth in the shelf seas of the Bay of Biscay are influenced by river discharges and freshwater advection and other mesoscale and submesoscale features.

Monthly composites of SeaWIFS satellite images (Fig 24.13) reveal the main spatial and temporal patterns of fluorescence (as a proxy for chlorophyll, see for example Gohin, 2002). Even after correction for turbidity, high values of chlorophyll can be observed over the wide French continental shelf most of the year round (particularly in late winter) and even over the narrower Spanish continental shelf we can observe several peaks of plankton growth during the year. In contrast, the oceanic waters in the Bay of Biscay basin show an oligotrophic state with a single peak of biomass in spring.

The onset of the spring bloom occurs with remarkable regularity in March on the southern coast when seasonal re-stratification starts and consequently the mixed layer becomes shallower than the critical depth in which phytoplankton growth exceeds mortality and respiration according to Sverdrup´s classical mechanism; by March-early April the spring bloom covers the entire Bay of Biscay (Fig. 24.14). From May onwards, chlorophyll drops sharply, and the lowest values are observed in summer when water column stratification prevents nutrient supply to photic layers. In this season new production is often related to mesoscale processes associated with topographical features. The autumn bloom is quite variable in timing and intensity, and restricted to coastal areas generally inwards of the 75 m isobath. During winter months and in the coastal areas inwards of the 100 m isobath, chlorophyll estimates remain relatively high (Fig. 24.13), presumably due to turbid plumes related to runoff and river discharges (Gohin et al., 2003). A succession of phytoplankton assemblages in relation to hydrography in the southern Bay of Biscay was studied by Valdés et al. (1991), Fernandez and Bode (1994), Varela (1996), Casas et al (1997) and others, who noted the prevalence of diatoms in spring, whereas flagellates were dominant during summer stratification.

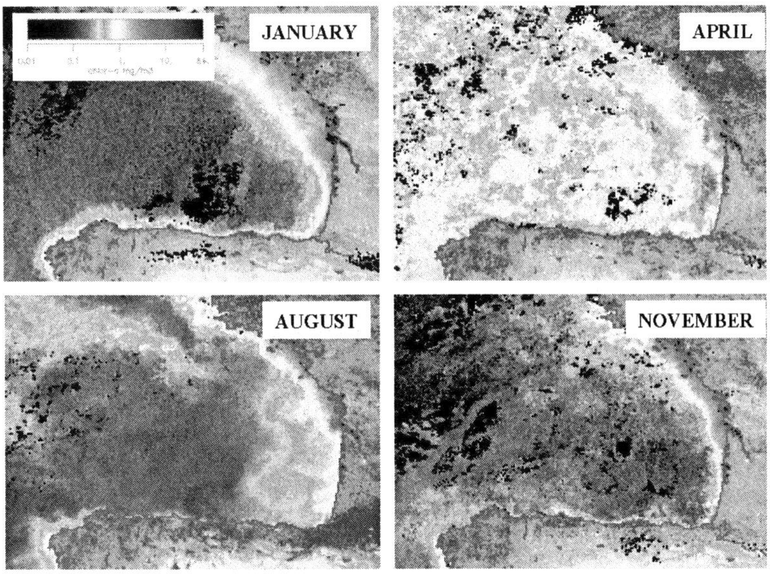

Figure 24.13 SeaWIFS composite images of the Bay of Biscay: January, April, August and November 1999.

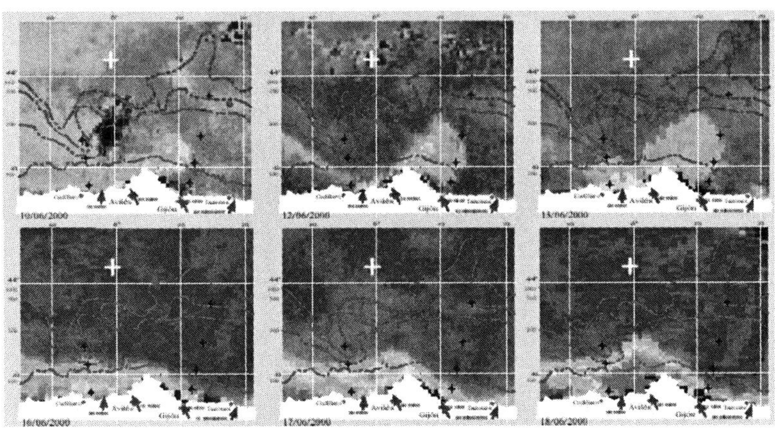

Figure 24.14 Upwelling signature in Cape Peñas observed by remote sensing (SeaWIFS, 7–18 June 2000). Color scale as in Fig. 24.13.

Nevertheless, this main scheme is far from being unchanging and uniform. As explained in the previous section, episodic mesoscale and submesocale events are common in the Bay of Biscay, introducing a high variability in the ecosystem's productivity from year to year. Most important features enhancing primary production are coastal upwelling, coastal runoff and river plumes, seasonal currents and internal waves and tidal fronts.

Coastal upwelling events occur mainly on the Spanish continental margin. These are produced by NE winds prevailing from late May to September. Upwelling events are highly variable in intensity and frequency from year to year, but in general they are more common and intense to the west of Cape Peñas and act as a mechanism generating spatial variability between the western and eastern parts of the Cantabrian Sea and between the coastal mixed waters and the neighboring oceanic stratified areas. When upwelling is particularly intense, surface signs of upwelling are observed first on the northern coast of Galicia, then off Asturias, and later to the east off Cantabria, and can be observed and studied by remote sensing (Revilla et al., 2002). Fig. 24.14 shows a SeaWIFS sequence of an upwelling event centred on Cape Peñas. This was a typical short-lived event (June 7–18, 2000) and covered ~1500 km^2. Upwelled waters are usually advected offshore forming filaments that extend westward following the 200 m isobaths and transporting biological material towards oceanic waters. The appearance of upwelling pulses during the summer is important in the formation of a permanent and deep stratified surface layer, enhancing phytoplankton growth (Estrada, 1984; Botas et al., 1990; Tenore et al., 1995). Under conditions of moderate upwelling, the innermost coastal 25 km are about 10 times more productive than offshore waters and the upwelling centres about 20 times more productive than offshore waters (Bode et al., 1996; Teira et al., 2003; Varela et al., 2003). On the French continental margin, mainly in summer, weak upwelling events are induced along South Brittany and the Landes coastline by northerly winds and their intensity seems to be closely related to anchovy recruitment (Borja et al., 1996).

Filaments and fronts associated with high salinity water of subtropical origin (Eastern North Atlantic Central Water, ENACWt, also called Navidad current) are important along the coast of the southern Bay of Biscay. During winter and spring, this current gives rise to a convergence front, which is localized in the boundary zone between the coastal water and the oceanic water. The effects on phytoplankton (distribution, standing stock, growth rates, species composition and functioning) were described by Botas et al. (1988), Fernández et al. (1993), Varela (1996), Varela et al. (1995, 1998). When saline intrusion is weak, the development of fronts and the formation of a seasonal thermocline are enhanced and phytoplankton blooms occur. Intense saline intrusion causes strong vertical mixing, preventing phytoplankton growth during spring.

The Gironde, Loire and, to a lesser extent, the Vilaine and Adour rivers provide large volumes of fresh water (which are turbid but rich in nutrients from the mainland) at the end of the winter. This results in the formation of dilution plumes at the surface of coastal waters, which drive significant northward currents over the inner Armorican shelf. Because the tidal currents along the Atlantic coast are much weaker than in the English Channel, these plumes can extend over several hundred kilometres in length, becoming progressively diffuse. Because of the continual input of nutrients they provide, these river plumes maintain a very large amount of new phytoplankton production along the coastal fringe, and sometimes even to the edge of the continental shelf. Due to the haline stratification that maintains the phytoplankton cells in a very thin layer of water, phytoplankton production can start very early in the year. An anticyclone regime, frequent in the region during winter (January to March), may be favorable to triggering this production. Studies have shown that these winter blooms linked to plumes from large rivers are relatively short, because they are quickly limited by lack of phosphorus due to the high N/P ratios of river waters, which are highly unbalanced in favor of nitrogen. A lower volume of river runoff and a much narrower shelf off the Iberian Peninsula act in tandem to make buoyant plumes much less persistent over the Cantabrian coasts than over the Armorican shelf (Prego and Vergara, 1998).

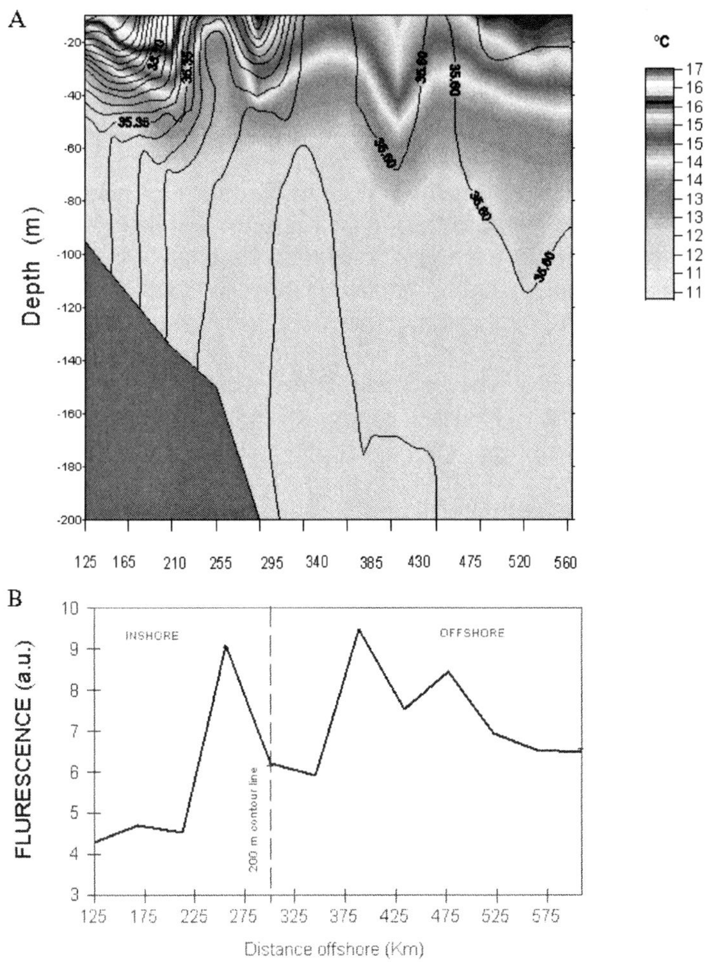

Figure 24.15 (A) Temperature and salinity profiles in a transect over the Armorican shelf in June 2000. (B) Fluorescence signal (arbitrary units) in the same transect.

At the Armorican shelf break when the water column is vertically stratified in summer, tides generate internal waves that propagate both on- and off-shelf from about 5° W to 9° W. Such internal waves appear to be responsible for significant mixing and upwelling of nutrients and occur with remarkable regularity, starting in June or July at the shelf-break, where they have their maximum intensity, and thus their greatest impact on primary production. Biological signs of this tidal front include an increase in fluorescence values (Fig. 24.15), zooplanktonic biomass, and eggs and larvae of pelagic fishes, e.g. sardine, mackerel and horse-mackerel (Lago de Lanzós et al., 1997). The abundant, sustainable new production makes this 100-km wide fringe overhanging the continental shelf break, an oasis in the open sea and it is used by pelagic species as spawning grounds to nourish future larvae (Arbault and Boutin, 1968; Lago de Lanzós et al., 1997).

4.2. Zooplankton dynamics and regional variability

The high variability in time and space of zooplankton populations in the Bay of Biscay and their complex relationship with environmental factors severely limits our understanding of ecosystem functioning. Time series programmes established in the late 1980s (Valdés et al., 2002) together with the implementation of new technologies have allowed us a deeper insight into zooplankton dynamics. New findings are promising and we are now starting to understand how the zooplankton adapt to environmental factors, how species respond to local or to broad-scale perturbations and how to distinguish regional patterns from local anomalies.

4.2.1 Main patterns and regional dynamics of zooplankton

Zooplankton in the Bay of Biscay is very rich in terms of taxonomic groups and species. The most important group by specific richness, persistence, abundance and ecological significance is that of copepods, which account for 60% and 85% of total zooplankton abundance in coastal and oceanic areas respectively off the north coast of Spain. Copepods are present all the year round whereas other holoplankton and meroplankton groups have a marked seasonal distribution (D'Elbee and Castel, 1991; Poulet, 1996; Valdés and Alvarez-Ossorio, 1996; Valdés and Moral, 1998).

Time series studies carried out on the north coast of Spain show that the annual cycle of abundance and biomass of zooplankton presents one main peak in spring corresponding to, but lagging behind, the pulse of phytoplanktonic production. Other less regular peaks may be observed in summer and autumn depending on the oceanographic conditions (upwelling, wind turbulence, etc.) (Bode et al., 1998). Winter is well defined with the lowest values. The oceanic areas present a pattern of oligotrophic areas with slight variations in values of abundance and biomass throughout the annual cycle and a single period, which generally coincides with April, when communities develop and reach annual peaks.

An interesting example of a regional pattern of plankton variability in response to environmental conditions was observed in the southern Bay of Biscay. Fig. 24.16 shows the regional pattern of surface (10 m depth) temperature and the zooplankton biomass at shelf sampling stations off Santander, Cudillero and A Coruña from 1991–2002. Sea surface temperature in Santander follows a regular seasonal cycle of heating and cooling, with maximum values around 21°C in August. This pattern is also observed in Cudillero, but what was also noted here was the effect of the upwelling events that happened in the neighboring Cape Peñas, causing small decreases in temperature during the summer months. Maximum values were reached in August with temperatures close to 20°C. In A Coruña the seasonal cycle of sea surface temperature is amply modified by the strong effect of the upwelling in the west Iberia, which starts in late spring and extends until late summer. Maximum values rarely surpass 17°C.

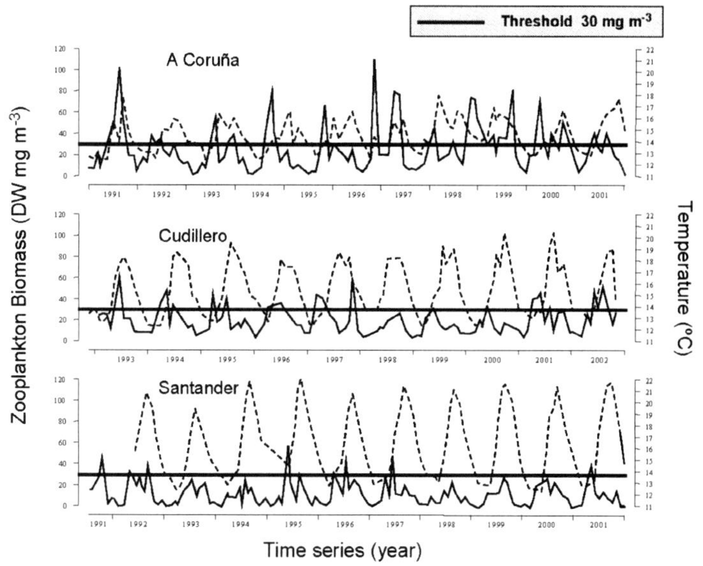

Figure 24.16 Zooplankton biomass patterns in the time series 1991–2002 at the mid-shelf sampling stations in the transects off Santander, Cudillero and A Coruña. The continuous line represents zooplankton biomass and the dotted line stands for temperature. (Llope et al., 2003, modified by Valdés).

Zooplankton biomass also shows a cyclical seasonal pattern with maximum values usually in spring when the heating of the water column starts. The regional pattern of the zooplankton biomass shows higher production on the Galician coast where for several months it surpasses 30 mg DW m^{-3} and on many occasions the biomass is doubled (Fig. 24.16). A significant portion of this production is linked to the frequency and intensity of summer upwelling (Bode et al., 1998). In Cudillero the biomass is lower than in A Coruña and although in some months it accounts for 30 mg DW m^{-3}, it never reaches 60 mgDW m^{-3}. Santander has the lowest values of the three locations and only on some occasions does biomass reach 30 mgDW m^{-3}. In Santander the thermal regime determines a strong thermohaline stratification that limits the supply of nutrients to the photic layers and once the nutrients are consumed, production drops. It is clear that the thermal regime of the water column (heating of surface water, upwelling events, stratification of water column) strongly determines the abundance of zooplankton in the Cantabrian Sea and in Galicia and creates a well-defined regional pattern.

Regarding the whole Bay of Biscay, since 1992 mesozooplankton abundance and biomass has been studied in large surveys covering the whole north Spanish coast and Bay of Biscay in spring (mid March-mid May) during the spawning season of the main pelagic commercial species. In general terms, biomass distribution of mesozooplankton (200–2000 μm) shows the same patterns as those described for phytoplankton. Zooplankton reach maximum abundances and biomass (values of ~70 mgDW m^{-3}, Fig. 24.17) soon after the phytoplankton spring bloom, when high values are regularly observed over the whole continental shelf of both margins of the Bay of Biscay. Once the spring bloom relaxes, zooplankton also decrease in number and biomass showing a more patchy distribution with some hot spots coinciding with upwelling regions and freshwater plumes from big rivers (Fig. 24.17). Oceanic and oligotrophic waters of the Bay of Biscay basin have very low abundances most of the time. In consequence these poor waters in the middle of the Bay of Biscay do not support spawning or nurseries of the main pelagic species. Fig. 24.18 shows the total abundances of eggs and larvae of sardine, mackerel and horse mackerel along a N-S Bay of Biscay transect from the coast of Brest (N Bay of Biscay) to A Coruña (S Bay of Biscay) during the spawning season of the above species. It can be seen that significant numbers were obtained only at the sampling stations located over the shelf at both extremes of the transect, whereas in the deep waters in the middle of the transect abundances were ten times lower.

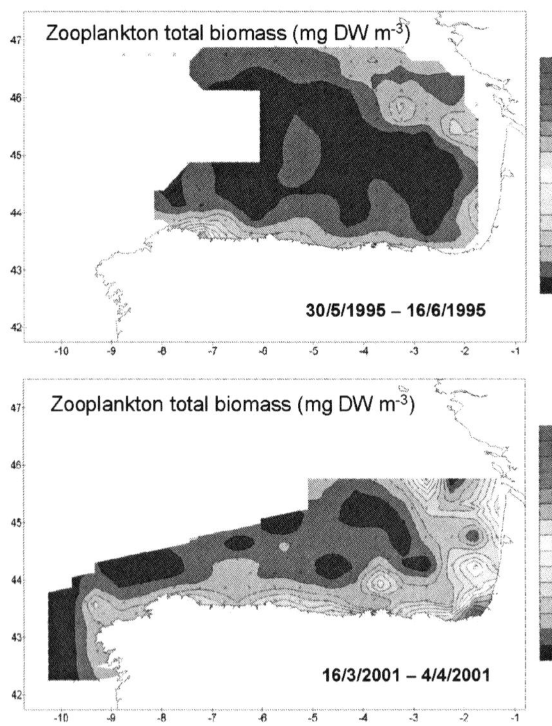

Figure 24.17 Spatial distribution of mesozooplankton biomass in the Bay of Biscay in June (1995) and March (2001).

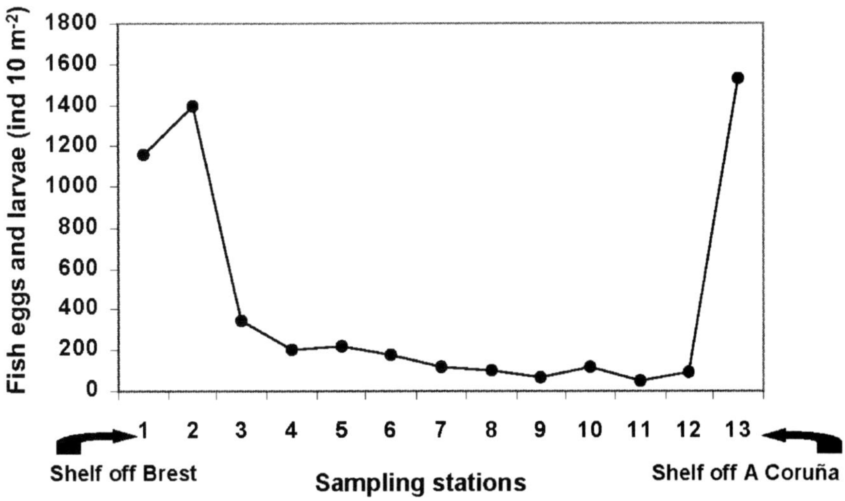

Figure 24.18 Sardine, mackerel and horse mackerel eggs and larvae abundance along a transect crossing the entire Bay of Biscay form Brest (N) to A Coruña (S). Sampling stations regularly spaced; st-1= continental shelf off Brest, st-13= continental shelf off A Coruña.

4.2.2 Seasonal cycles and year-to-year variability of key species

Seasonal cycle and year-to-year variability of *Acartia spp.* and *Calanus helgolandicus* in the Bay of Biscay and Celtic Sea was studied by Valdés et al. (2001) using data from two sampling stations off Santander (N. Spain), L4 station off Plymouth (S. England) and the CPR routes in Biscay and Celtic Sea. Mean monthly abundances (1992–1999) of *Acartia spp.* (Fig. 24.19) show a gradual latitudinal pattern in seasonality of the *Acartia* population. The growth season starts earlier in the southern regions (February-March on the shelf and oceanic waters off Santander) than in northern regions (May on the shelf off Plymouth). The annual cycle observed from CPR data shows that seasonality in the Bay of Biscay is similar to that observed at the oceanic station off Santander, and that the annual cycle at the Celtic Sea follows the temporal pattern observed at L4 off Plymouth. The growth season off Santander extends from February to July, and population decreases after August in the Bay of Biscay. In the Celtic Sea and Plymouth, *Acartia* shows a first peak in May and a second one in August-September, then population decreases in November. The magnitude of absolute abundances is similar in Santander and Plymouth, CPR values, however, are much lower.

The growth season of *C. helgolandicus* also shows a latitudinal pattern in its seasonality (Fig. 24.19). A spring peak occurs earlier (March-April) in Santander and the Bay of Biscay than in the Celtic Sea and Plymouth (May). At the inner station off Santander the growth season lasts for 3 months (March-May), whereas at the outer station and in the Bay of Biscay the growth season is more limited in time. The Celtic Sea and Plymouth show two different peaks, one in spring (May), and a second in late summer-autumn (August-September). The highest abundances were found over the shelf off Santander. Oceanic waters off Santander and Plymouth show similar values and CPR values are much lower. These findings suggest a synchronicity at a regional scale in the biological cycles of these species.

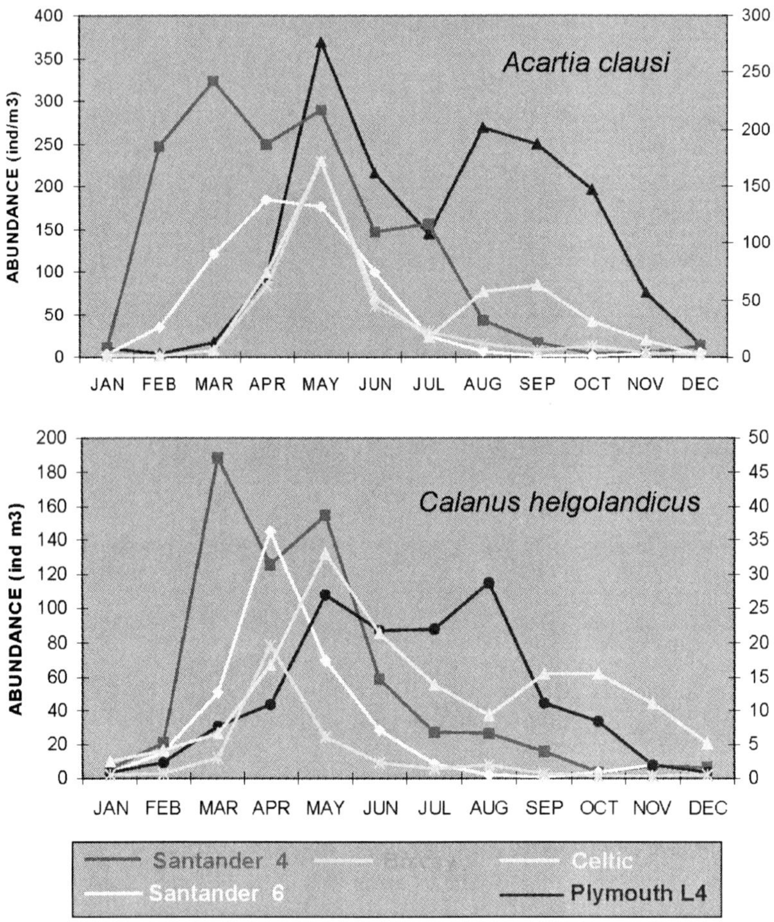

Figure 24.19 Mean monthly abundance (1992–1999) of *Acartia spp.* (upper panel) and *Calanus helgolandicus* (lower panel) over the shelf off Santander (dark grey ■), an oceanic station off Santander (white ◆), open ocean in the Bay of Biscay (light grey ✱, from CPR), Celtic Sea (light grey ▲, from CPR) and over the shelf off Plymouth (black). The left axis represents abundance off Santander and Plymouth, the right axis abundance from CPR results in the Bay of Biscay and Celtic Sea (Valdés et al., 2001).

Good records on the spatial distribution (coastal, shelf and oceanic stations off Santander) and year-to-year variability (Jun-1991/Nov-2000) of mesozooplankton groups and species are available from the time series project carried out on the north coast of Spain. As explained previously, *A. clausi* populations in the southern Bay of Biscay develop in the first half of the year. They exhibit coastal preferences and reach maximum abundances at the station located in the inner shelf (Fig. 24.20), showing a remarkable year-to-year variability, with years 1994, 1997 and 2000 above the mean expected values and 1995, 1998 and 1999 with values lower than expected. Statistical trends on cumulative anomalies show an interannual cycle of a three year period: every third year abundance peaks to a maximum followed by a decay and a recovery in the next two years before peaking again in the third.

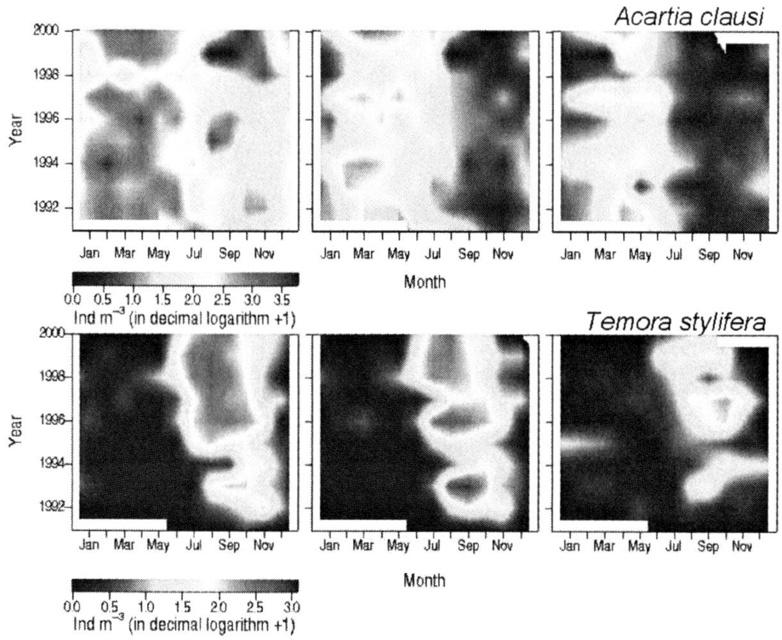

Figure 24.20 Year-to-year variability of *A. clausi* (upper panel) and *Temora stylifera* (lower panel) at the coastal, shelf and oceanic stations off Santander during the time series 1992–2000.

Temora stylifera populations reach maximum values in August and the growth season lasts for almost the whole of the second half of the year (Fig. 24.20). Its presence in the Bay of Biscay has been reported recently and related to the increase in temperature due to climatic variations. The year-to-year variability of this species shows a shift in its seasonality as the annual peak now occurs one month earlier than it used to ten years ago. This pattern was observed in the three stations studied (Fig. 24.20). The progression of this species towards northern waters was reported up to Helgoland (Halsband-Lenk et al., 2003). Both facts, shifts in seasonality and northern displacement, reveal a regional pattern that makes of *T. stylifera* a target species for understanding the effects of climatic variability on plankton populations (Villate et al., 1997).

4.3. Carbon uptake, grazing and the pelagic food web

Photosynthetic activity of natural assemblages of phytoplankton in the three fractions of pico, nano and microphytoplankton (0.2–2, 2–20 and >20 μm respectively) have been studied in surface waters of the Spanish continental shelf by ^{14}C uptake measurements in *in situ* incubation periods of several hours. The results show that maximum rates of incorporation of ^{14}C are obtained during the spring and winter blooms when nutrients are available in the photic layer (Fig. 24.21A). Production during the spring blooms is driven by nano and microphytoplankton, whereas during the oligotrophic state in summer and in the winter blooms, the incorporation of ^{14}C is mainly supported by picophytoplankton (not shown). This seasonal alternation in production rates from nano-microphytoplankton in spring to picophytoplankton in summer, when nutrients are depleted in surface waters and microbial processes promote regenerated production through the smaller cells, have also been described in other temperate seas (Smetacek et al., 1984; Legendre and Gosselin, 1989; Legendre and Le Févre, 1991; Legendre and Rassoulzadegan, 1995) and it is probably true for other regions of the Bay of Biscay.

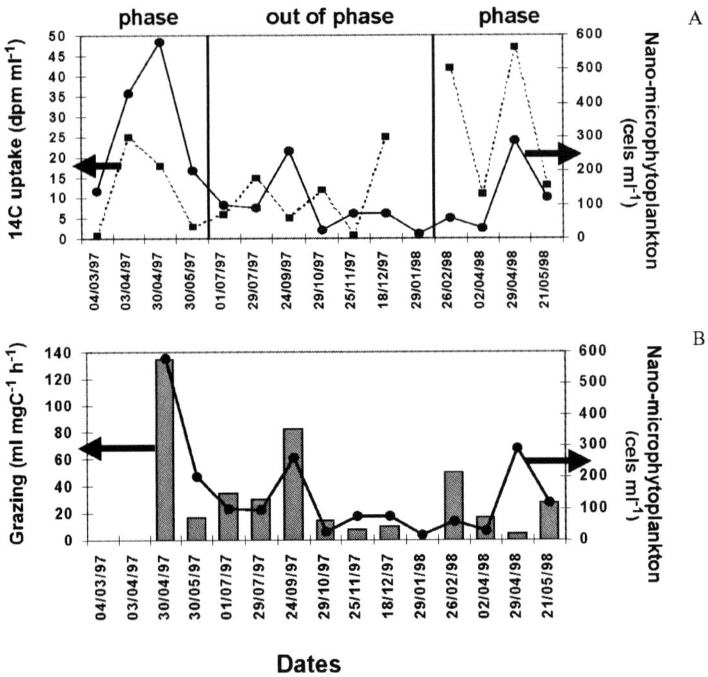

Figure 24.21 (A) Uptake of ^{14}C by the natural community of phytoplankton off Santander and abundance of nano and microphytoplankton. (B) Grazing of zooplankton (fraction 250–500 μm) and abundance of nano and microphytoplankton.

The study of the trophic relationships in pelagic ecosystems is complicated by the large variability in diet of most species, which leads to unstructured food webs (Isaacs, 1973). Preferences in mesozooplankton feeding habits were studied in shelf waters off Santander using natural assemblages of mesozooplankton and natural assemblages of phytoplankton. The results suggest that the most abundant fractions of mesozooplankton (200–500 and 500–1000 μm) graze mainly on nano and microphytoplankton cells (Fig. 24.21B). Grazing values of the larger fraction of mesozooplankton (>1000 μm) showed neither a clear relationship with nano-microphytoplankton nor with picophytoplankton, suggesting feeding habits are not based exclusively on primary producers.

A different approach to the study of plankton feeding habits was followed by Bode et al. (2003a, b and in press) in shelf waters off Galicia (NW boundary of the Bay of Biscay). These authors studied the gains of stable isotopes of C (^{13}C) and N (^{15}N) through the trophic chain from plankton to sardines and dolphins. Fig. 24.22A shows isotopic enrichment as the individual size of organisms increases from plankton (four size classes from 20 to 2000 μm) to dolphins. The gain in ^{15}N by the different classes of plankton suggests that as plankton size increases, organisms feed on more complex diets composed of a mixture of phytoplankton, detritus and microplankton, which is in agreement with the above findings.

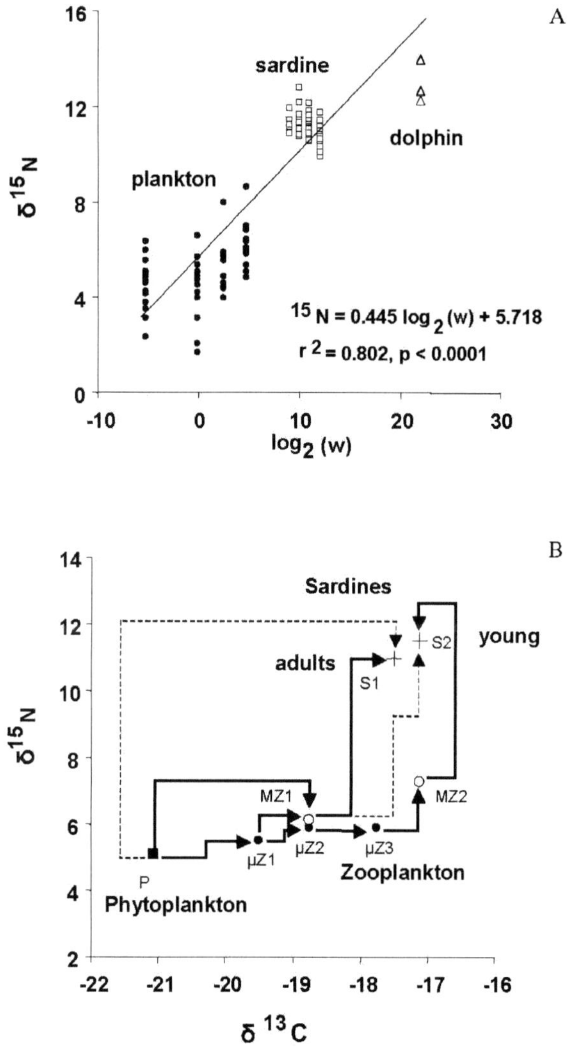

Figure 24.22 (A) Variation of natural abundance of ^{15}N with the individual size of plankton (four size-classes from 20 to 2000 μm), sardines and common dolphin. (B) Reconstruction of trophic pathways linking plankton and sardine components in the northern Iberian shelf ecosystem from ^{13}C and ^{15}N abundance values. Each component is located according to average values of isotopic abundance. Thick arrows indicate the most probable links, dashed arrows indicate less probable links (from Bode et al., 2003b).

But the determination of natural abundance of stable C and N isotopes also provided a way to quantify the main exploitation patterns of plankton by sardines and to reconstruct the pathways of the pelagic food web (Bode et al., 2003a, b). A change in the diet of sardines after the first year of life was inferred from the decrease in ^{15}N (and also ^{13}C) abundance in sardines ≥ 18 cm in length (Fig. 24.22A). Such a change implies the increasing consumption of plankton of low trophic position in the food web as sardines grow and as their filtering apparatus develops, suggesting an opportunistic feeding behavior as adults.

The reconstructed food web taking into account isotopic signatures included at least three trophic steps between phytoplankton and sardines, while direct consumption of large amounts of phytoplankton seemed unlikely (Fig. 24.22B). Such a result was also found in other clupeids from upwelling areas (Monteiro et al., 1991). Nevertheless, differences in ^{13}C and ^{15}N enrichment between plankton and sardine suggest that adult sardines obtain most of their muscle protein from zooplankton while phytoplankton would provide reserve materials.

In addition to the pathways it is also important to quantify the transfer of matter at each trophic level. According to data from grazing experiments, the copepods and other mesozooplankters (200–2000 μm) off Santander remove less than 1% of phytoplankton standing stock per day. These values are in agreement with other data from the Bay of Biscay (Barquero, 1995) and from other temperate seas (Calbet, 2001), indicating that most of the production in the shelf waters of the Bay of Biscay is not transferred directly into the pelagic food web but moved towards the sediment where the organic matter is incorporated in the benthic food webs or remineralized (Fortier et al., 1994). This has strong implications for supporting the main demersal fisheries in the Bay of Biscay.

5. Biology of fishes and main fisheries

5.1. Pelagic fishes

Pelagic fishes belong to Nekton, i.e. those animals capable of swimming to such a degree that they can overcome many ocean currents. This allows them to move great distances within a day and they are adapted to the demands of continual swimming, with fusiform shapes and an efficient circulatory system (Helfman et al., 1997). The term 'pelagic fishes' is used here as a synonym of epipelagic fishes, i.e. those fishes that swim in the upper 200 m of coastal and open-sea areas.

The geographical location of the Bay of Biscay favors a diversity of pelagic ichthyofauna where species characteristic of cold North Atlantic waters such as herring (*Clupea harengus*) share the area with those from more temperate subtropical waters such as chub mackerel (*Scomber japonicus*). The phenomenon of global warming seems to have led to an increase in the presence of temperate water fish species in the Bay of Biscay (e.g., among pelagic fishes *Megalops atlanticus, Seriola rivoliana*) over the last twenty years (Quéro et al., 1998; Stebbing et al., 2002). From an ecological point of view and also in relation to fishing activity, pelagic fishes can be divided into three large groups: small-sized pelagics, middle-sized pelagics and large migrators (Bas, 1995).

5.1.1 Small-sized pelagic fishes

Small pelagic fishes are distinguished by their low trophic level, feeding on phytoplankton and zooplankton typically in upwelling and surrounding areas. Growth is fast, reproduction early and lifespan short, giving rise to the formation of very large populations (Bas, 1995). The population dynamics of these species are dominated by the strength of the generation born each year (recruitment). The most representative species of this group in the Bay of Biscay are: anchovy (*Engraulis encrasicolus*) and sardine (*Sardina pilchardus*). Another characteristic species is sprat (*Sprattus sprattus*).

Anchovy: Anchovy in the Bay of Biscay may grow to >20 cm and rarely lives beyond three years of age. It forms large schools located at between 5 and 15 metres above the bottom during the day (Massé, 1996). It is a serial spawner (several spawns per year) and reproduces in spring. The spawning area stretches to the south of 47°N latitude and to the east of 5°W longitude. Most spawning takes place over the continental shelf in areas under the influence of the river plumes of the Gironde, Adour and Cantabrian rivers (Fig. 24.23) (Motos et al., 1996). As spring and summer progress the anchovy migrates from the interior of the Bay of Biscay towards the north along the French coast and towards the east through the Cantabrian Sea. It spends the autumn in these areas and in winter migrates in the opposite direction towards the southeast of the Bay of Biscay (Prouzet et al., 1994). It has a high and very variable natural mortality. Mesoscale processes in relation to the vertical structure of the water column (stratification, upwelling and river plume extent) appear to have a great influence on the survival of larvae (Allain et al., 2001). In the northeast Atlantic the largest and most stable anchovy population seems to be that of the Bay of Biscay (Uriarte et al., 1996).

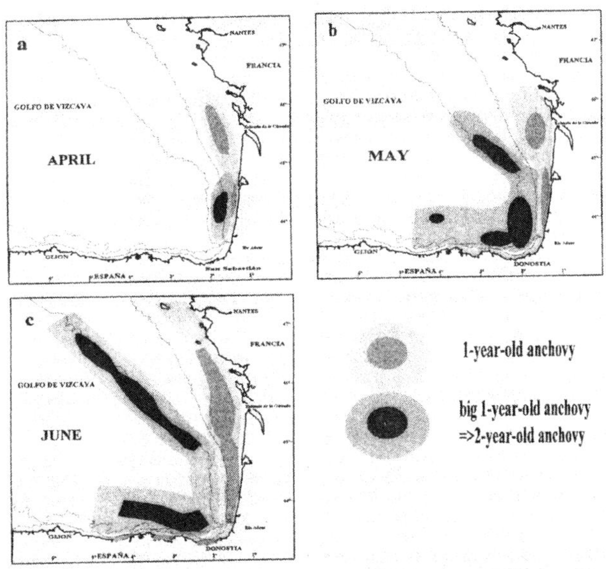

Figure 24.23 Spawning grounds of anchovy in the Bay of Biscay (from Motos et al., 1996).

Sardine: Sardine in the Bay of Biscay can reach up to 25 cm in length and more than ten years of age. It forms large schools usually close to the coast and up to around 50 m depth from the sea surface. It is a serial spawner. In general, the spawning peak appears in spring, although there is a second peak in autumn (Solá et al., 1992). These two peaks may correspond to the existence of sympatric (or parapatric) spring and autumn sardines (Wyatt and Porteiro, 2002), although no conclusive evidence has been found. The concept of metapopulation for the sardine resident in the Bay of Biscay may help to explain the significant variations in abundance which have taken place, especially in the southeast of the region in the last decade (Carrera and Porteiro, 2003). The decrease in sardine recruitment has been related to global warming (Lavín et al., 1997; Cabanas and Porteiro, 1998; Guisande et al., 2001; Valdés and Lavín, 2002; Wyatt and Porteiro, 2002) and this hypothesis is currently under investigation.

5.1.2 Middle-sized pelagic fishes

Middle-sized pelagic fishes are characterized by a greater plasticity in the food spectrum than that of the small pelagic fishes. They are closely tied to areas of high productivity but the relationship is looser and less direct than that of the small pelagic fishes (Bas, 1995). The diet is mainly made up of large copepods and mesozooplankton. They have greater mobility and make longer migrations, both horizontally and vertically, than the small pelagic fishes. They also have a longer lifespan and populations are made up of several age groups. All of these characteristics favor stability in the abundance of these species (Bas, 1995).

This category mainly includes species from the Scombridae and Carangidae families. In the Bay of Biscay, the most important are mackerel (*Scomber scombrus*) and horse mackerel (*Trachurus trachurus*). Also characteristic are other species more common to temperate and subtropical waters, such as chub mackerel (*S. japonicus*), Mediterranean horse mackerel (*T. mediterraneus*) and blue jack mackerel (*T. picturatus*). Other families with species in this category are Mugilidae and Belonidae.

Mackerel: Distributed throughout the northeast Atlantic-Mediterranean Sea and in the northwest Atlantic, mackerel is an active migrator in schools, which are sometimes of very great density. It is a serial spawner whose spawning area extends throughout the west of the British Isles, the Bay of Biscay and the North Sea. Mackerel spawns at the beginning of spring, and once spawning is over at the end of spring, it makes a food migration along the west of the British Isles to the north of the North Sea. From September to December mackerel is found in the Norwegian Sea and the north of the North Sea. During winter, the mackerel migrates southward once more towards spawning grounds down through waters to the west of the British Isles (Uriarte et al., 2001). This migratory behavior seems to be associated with waters of the slope current (SEC) (Reid, 2001). Migration is the cause of the seasonality seen in the mackerel fishery of the Bay of Biscay (Villamor et al., 1997). Northeast Atlantic mackerel is considered to be a single management unit or stock, in which three spawning components are distinguished (south, west and North Sea) (ICES, 2004a).

Horse mackerel: It is a carangid with a distribution from the coasts of Cape Verde up to the north of the North Sea, as well as the Mediterranean. Its behavior is more demersal than that of mackerel or other pelagic species. It is a serial spawner whose spawning area stretches from the British Isles towards the south throughout its area of distribution. In the Bay of Biscay spawning takes place mainly in spring and the beginning of summer. It is a long-lived fish, which can reach up to 40 years of age (Abaunza et al., 2003). It makes spawning and feeding migrations, but these are less evident than in the case of mackerel. Three stocks are currently recognized in the northeast Atlantic: Southern stock, North Sea stock and Western stock. Recently it has been considered that the Bay of Biscay horse mackerel should be included within the Western stock, which extends from Galicia northwards throughout the western European coast (ICES, 2004a).

5.1.3 Large migratory pelagic fishes

These are large in size and strong swimmers, which enables them to perform long migrations. In general, the small and middle-sized pelagic fishes make up their primary food source, which places them at the highest levels of the trophic chain. Some families of the sub-order Scombridae (tuna-like fishes) and sharks from the Carchariniforms and Lamniforms typically belong to this group. Tuna-like fishes are serial spawners whose spawning area is usually located in tropical and subtropical waters. In tropical areas food is relatively scarce and so tuna fishes have to actively search for food patches. This means that their life is nomadic, based on continuous long displacements (Helfman et al., 1997). In the Bay of Biscay the most characteristic species are albacore (*Thunnus alalunga*) and bluefin tuna (*Thunnus thynnus*). Other tuna and tuna-like fishes such as bigeye (*Thunnus obesus*), Atlantic bonito (*Sarda sarda*), skipjack tuna (*Euthynnus pelamis*) and swordfish (*Xiphias gladius*) may also be present.

Tuna: The presence of bluefin and albacore in the Bay of Biscay is seasonal. They normally appear at the beginning of summer and disappear at the beginning of autumn, following a trophic migration in search of food. In the case of bluefin, they are usually young (Cort, 1990). Abiotic variables such as surface temperature and overall climatic and oceanographic indices play an important role in their distribution and migratory behavior (ICCAT, 2003). Three albacore stocks are currently considered, and the albacore from the Bay of Biscay belongs to the North Atlantic stock. Bluefin is regarded as having two stocks and the Bay of Biscay belongs to the East Atlantic stock (ICCAT, 2003).

Sharks: These large predatory fishes have internal fertilization and females either lay eggs or nourish embryos internally for several months before giving birth (Helfman et al., 1997). The type of reproduction, in which the offspring is limited, makes their populations very vulnerable to fishing pressure. In the Bay of Biscay the epipelagic sharks: blue shark (*Prionace glauca*), shortfin mako (*Isurus oxyrrinchus*) and porbeagle (*Lamna nasus*) are common. They predate on a wide range of pelagic and demersal fishes. The largest shark in the Bay of Biscay is the basking shark (*Cetorhinus maximus*), with a length of more than 9 m. It is also characterized by its planktonic feeding (Quéro, 1984).

5.1.4 The main pelagic fisheries

The pelagic species dealt with here form the basis of important fisheries in the Bay of Biscay (see Fig. 24.24), which represent an important source of income for local economies and sustain a considerable number of jobs. Small pelagic fishes are generally caught by purse seiners and pelagic trawlers. A wide variety of gears are used to catch middle-sized pelagic fishes, including the two previously described, the hand-line and the bottom trawl. Large pelagic fishes are also caught using surface hooks, currican and live-bait. Stocks of the pelagic species related to the Bay of Biscay are close to over-exploitation if not over-exploited (ICCAT, 2003; ICES, 2004a). Most of them are subject to active management policies to limit fishing effort and total catches by means of TACs and quotas. The variability inherent in the abundance of these species, migratory movements and seasonality make their management difficult. A typical case is anchovy, for which it is difficult to make reliable predictions of abundance in the short and medium term due to its short lifespan and high natural mortality. Spain and France are the main countries involved in the pelagic fisheries of the Bay of Biscay, although recently other European Union countries, such us The Netherlands and Denmark have also developed interests in resources such as mackerel and horse mackerel.

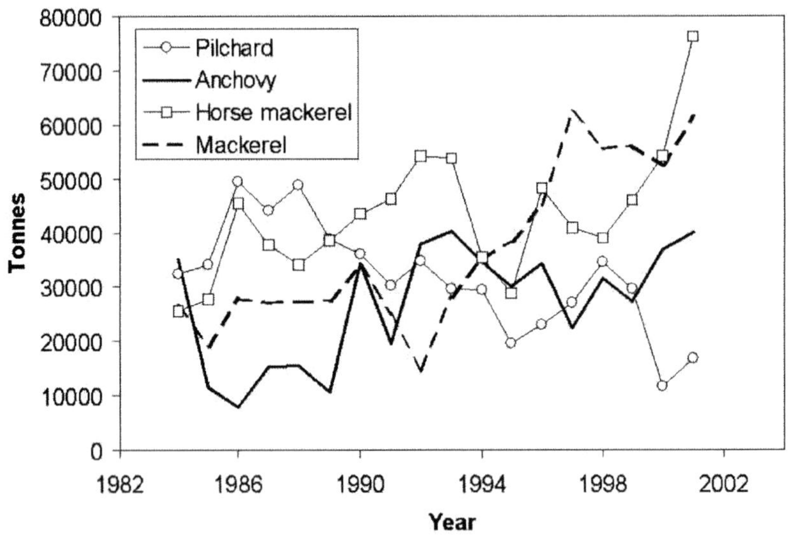

Figure 24.24 Catches of pelagic fishes in the Bay of Biscay (ICES Subarea VIII) from 1984 to 2001 (Source: ICES, 2004). (Mackerel: dashed line; Horse Mackerel: thin line with squares; Anchovy: thick line; Sardine: thin line with circles).

5.2. *Demersal and benthic fishes*

Demersal and benthic fishes live on or near the bottom. Most of them are long-lived species (more than 10 years) with a lower growth rate compared to small-sized pelagic fishes. They are mainly suprabenthic plankton feeders or predators mostly feeding on benthos and fishes. They are characterized by their diversity with nearly 200 demersal and benthic fish species recorded in this area, 100 of which are of commercial interest.

In the Bay of Biscay, many species reach their limits of distribution. Some species characteristic of cold waters such as whiting (*Merlangius merlangius*) or dab (*Limanda limanda*) reach their southern limits and other species from temperate waters, such as meagre (*Argyrosomus regius*), various sargo breams (*Diplodus spp*), Senegalese sole (*Solea senegalensis*) or wedge sole (*Dicologoglossa cuneata*) reach their northern limits.

5.2.1 Main key species

Hake: European hake (*Merluccius merluccius*) is both commercially and ecologically one of the most important species in the Bay of Biscay. Hake spawns in winter, with the adults concentrating in canyons and rocky grounds of the shelf break area. The drift of larvae from spawning sites to the nursery areas has been associated with the physical characteristics of the region, and particularly with the current regime during the spring. Hake recruitment processes lead to well-defined patches of juveniles in autumn (SESITS, 2000; Sánchez et al., 2001). The size and location of the patches show that the Bay of Biscay appears to be the main nursery area of hake (Fig. 24.25). Areas of high concentration of hake recruits have been located between 80 to 200 m depth and over predominantly muddy bottoms. In the Cantabrian Sea these concentrations vary in density according to the strength of the year class, although they remain relatively stable in size and spatial location and depend on larval retention and transport from the spawning areas in anticyclonic eddies (Sánchez and Gil, 2000). Consequently, the strength of recruitment is dependent on an optimal environmental window that controls this feature (Sánchez et al., 2003). Over the Grande Vasière area, the recruitment processes depend on the readjustment of fluxes from the north (cold waters) and warm and fresh waters from river discharges (Sánchez et al., 2001). After recruitment, hakes juveniles are dispersed on the shelf and there is a tendency for adults to live in the deep waters of the shelf break (Sánchez and Gil, 2000; Poulard, 2001), where they find their main preys, blue whiting, horse-mackerel and crustaceans (Guichet, 1995; Velasco and Olaso, 1998). Population limits are uncertain but two hake stocks (Northern and Southern) are recognized in the East Atlantic area with the boundary in the Bay of Biscay inner corner.

Megrim: Two species of megrims, four-spot megrim (*Lepidorhombus boscii*) and megrim (*L. whiffiagonis*) live in the Bay of Biscay. The four-spot megrim shows a preference for southern regions while the megrim is mainly associated with northern waters, the Bay of Biscay being the overlapping area of the distribution of both species (Fig. 24.26). A degree of specialization can be observed in the two species' habitats (Aubin-Ottenheimer, 1986; Sánchez et al., 1998) probably due to their different feeding systems, with *L. whiffiagonis* (more ichthyophagous) occupying shallower waters (100 to 300) and *L. boscii* (crustacean feeder) living in muddy and deeper grounds (200 to 600 m). Both species of megrims disappear at the mouth of the most important rivers, probably associated with the occurrence of continental run-off, which acts mainly by modifying the composition of the grounds on which megrim depend for food, and creating grounds which are more appropriate for other flatfish, such as *Solea spp.* or *Dicologoglossa cuneata,* adapted to estuarine conditions (Sánchez et al., 2001).

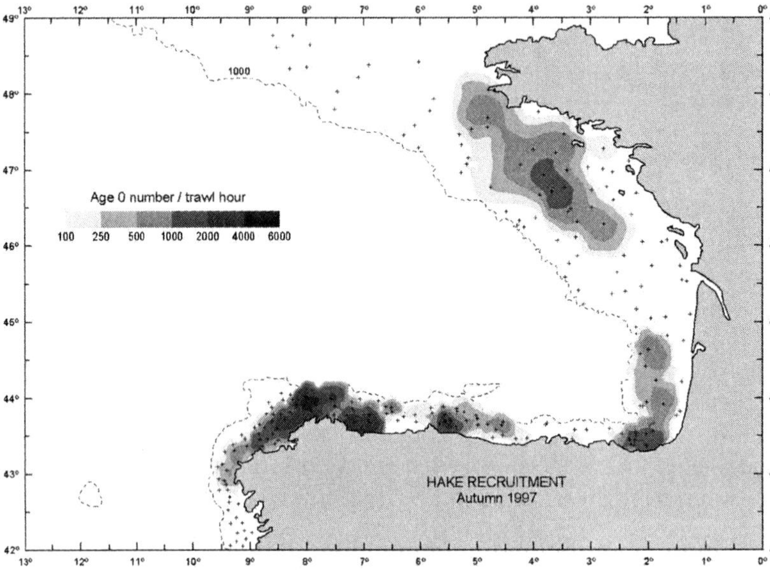

Figure 24.25 Main nurseries of European hake in the Bay of Biscay in autumn 1997. Data from standardized bottom trawl surveys carried out during the SESITS international project (SESITS, 2000).

The more abundant species in the French area of the Bay of Biscay is *L. whiffiagonis* but its distribution is more restricted to the external shelf that those observed in the Celtic Sea (Fig. 24.26). No evidence of geographical migrations was found in either species of megrims, although a bathymetric expansion occurs with age: juveniles are more stenobathic and live in deeper waters than adults (Sánchez et al., 1998).

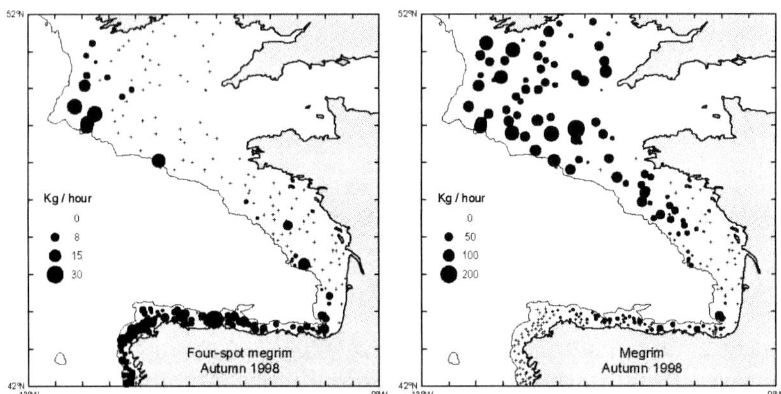

Figure 24.26 Spatial distribution of biomass of four-spot megrim (Lepidorhombus boscii) and megrim (L. whiffiagonis) in the Bay of Biscay in autumn 1998. Data from standardized bottom trawl surveys carried out during the SESITS international project (SESITS, 2000).

Anglerfish: Two species of Anglerfish (*Lophius piscatorius* and *L. budegassa*) are widely distributed along the European coast and are recorded from the Barents Sea to the Mediterranean and into the Black Sea. They are bottom-living species, morphologically similar but distinguished by the color of the peritoneum (*L. piscatorius* white, *L. budegassa* black). The two species are now of major commercial importance, but our understanding of their basic biology is far from complete. In the Bay of Biscay they have been recorded from shallow inshore waters to depths of 800 m (*L. piscatorius*) and even more than 1000 m (*L. budegassa*) (Dardignac, 1988; Azevedo and Pereda, 1994). Large specimens of both species are mostly found in deep waters. In contrast, juvenile anglerfish are distributed both in deep and shallow waters and discrete nursery areas have not been identified. Compared to many other groundfish, anglerfish are particularly late to mature with sexual maturity being attained from around 10 years old for black anglerfish and 7 years old for white anglerfish (Quincoces et al, 1998; Duarte et al., 2001). The ovaries are peculiar. Eggs are released in a buoyant gelatinous ribbon, which floats in the surface layers and can reach more than 10 metres. Localization of spawning grounds are not known in the Bay of Biscay but some indications suggest that anglerfish may spawn in deep waters (Duarte et al., 2001; Hislop et al., 2001). Anglerfish are mostly ichthyophagous, with bib (*Trisopterus luscus*), horse mackerel, flatfish, blue whiting and cephalopods as main preys.

Norway lobster: This species *(Nephrops norvegicus)* is distributed from Iceland to Portugal and the Mediterranean between 5 and 800 m depth. Its general biology has been reviewed by Chapman (1980). *Nephrops* are sedentary and rather common on muddy grounds, in which they dig their burrows where they spend most of their time. *Nephrops* distribution, therefore, is limited by the extent of suitable muddy ground. In the Bay of Biscay, three populations are distinguished, one on the French shelf and two in the Cantabrian Sea. In common with other crustaceans, the growth of the Norway lobster consists of a succession of moults continuing throughout its life (Farmer, 1973). Females spawn from April to August and carry eggs under their tails ("berried" females) until they hatch about 7 months later. The larvae develop in the plankton for one month before settling to the seabed. When berried, females rarely come out of the burrow, and are therefore naturally protected from trawlers. *Nephrops* are mainly nocturnal and feed on detritus, crustaceans and worms.

Sole: Sole (*Solea solea*) is a common benthic species in the central part of the Bay of Biscay between 0 and 100 m depth on sandy and muddy grounds. They mainly hunt for food at night and feed on small bivalves, annelids and small crustaceans. Results from tagging and genetic studies confirm that sole in the northern Bay of Biscay must be considered an homogeneous population (Koutsikopoulos et al., 1995). Its life history is marked by transport and migration processes. Spawning takes place in winter in two main areas located between 30 and 70 m depth but juveniles are concentrated in shallow and muddy estuarine areas. The distance between spawning grounds and the nursery areas varies from 40 to 80 km. There is also an inshore migration in spring and an offshore one in early winter, in phase with the seasonal change in water temperature in the coastal zone (Koutsikopoulos et al., 1995). Different factors influence recruitment but river flow seems to be important in governing the abundance of young sole: in spring fluvial discharges have a positive effect on the water of the area, which is covered by high densities of young juveniles and the recruitment of the Bay of Biscay sole stock depends partly on the influence of river plumes on nursery grounds. (Le Pape et al., 2003).

5.3 Diversity and distribution

The Bay of Biscay area forms the subtropical/boreal transition zone of the eastern Atlantic, where typical temperate-water species from the south occur together with those of northern origin and, consequently, high biodiversity indices exist in comparison with adjacent areas (Quéro et al., 1989; Sánchez et al., 2002). In addition, the topographical complexity and wide range of substrates on its continental shelf result in many different types of habitat. It is also the winter and spring spawning area for some species, such as hake, megrims, red-sea bream, mackerel, horse mackerel and anchovy, and the feeding area for others, e.g. tuna.

The bathymetric differences in species distribution show a progressive decrease in mean fish species richness with depth, as described by Sánchez (1993) in the Cantabrian Sea. This pattern is reasonably stable in the Bay of Biscay. The decline in the number of fish species with depth is probably brought about by the higher productivity of the coastal waters, in contrast with the inverse phenomena appearing in invertebrates (Olaso, 1990), which prefer deeper water and muddy substrates due to their predominantly detritivorous feeding habits. In general terms, fish diversity in the Bay of Biscay increases towards the extreme depth strata (shallowest and deepest). This is due to the major homogeneity of grounds located in the medium and external shelf (in relation to the coastal or break shelf) and the impact of the low selective trawl fishery (Sánchez and Serrano, 2003).

The groundfish species assemblages on the Bay of Biscay continental shelf are spatially organized mainly according to depth, latitude, longitude and substrate type (Sánchez, 1993; Souissi et al., 2001; Sánchez and Serrano, 2003). The relative stability of the demersal fish communities throughout the last decade contrasts with the strong variability of some mesoscale hydrodynamic features (upwellings, lower salinity water lenses and cold pool, Puillat et al., 2003) encountered on the French continental shelf (Poulard et al., 2003). Also, the narrowest surface of the Cantabrian Sea shelf produces strong environmental gradients over a short distance. From Canonical Correspondence Analysis (CCA), depth is the most influential and stable factor determining the assemblages observed (Sánchez, 1993; Sánchez and Serrano, 2003) and five main groups were described: coastal, inner-shelf, middle-shelf, outer-shelf and shelf-break (Fig. 24.27). The most discriminatory were the coastal group and the deepest strata groups (outer-shelf and shelf-break), situated at the extremes of the environmental gradients analysed. On the other hand, in the middle segments of these gradients, the inner-shelf and middle-shelf groups displayed reduced dispersion, with a position closer to the centroid of the biplot. This centroid is occupied by ubiquitous species of a wide optimal environmental range that corresponds to the top predators (hake, anglerfish and conger eel) that live in the area. Although the spatial pattern of fish communities remains stable, decadal changes in the structure of these communities have been described and related to climate change and fishing effects (Blanchard et al., 2002). Within these communities in particular, the relative abundances of species that are at their latitudinal limit of geographical distribution exhibit changes concommitant with a mean temperature increase of 1.5°C: the tropical species show increasing trends in abundance, whereas the boreal ones are decreasing (Poulard et al., 2003). Fishing on these communities alters species interactions, leading to increasing variability of the total production of the ground fish communities (Blanchard, 2001).

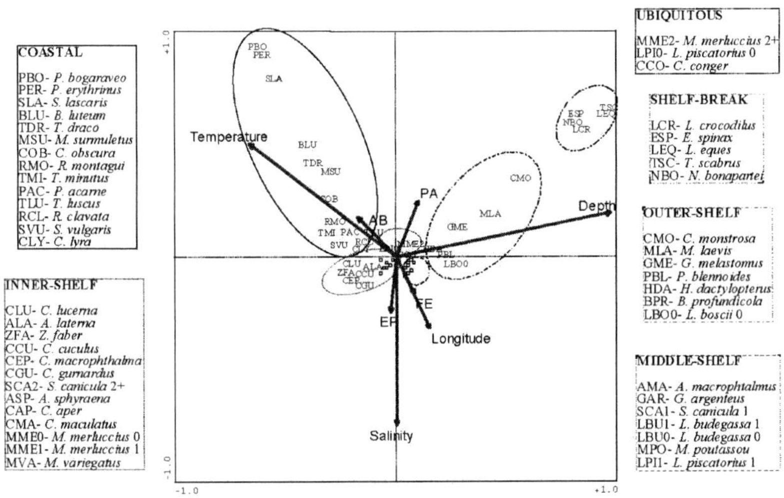

Figure 24.27 Canonical Correspondence Analysis (CCA) biplot of species vs. variables of the Cantabrian Sea from data of the 1993 bottom trawl survey. Environmental variables analysed are depth, bottom temperature, bottom salinity, longitude and geographical sectors (FE: Finisterre-Estaca; EP: Estaca-Peñas; PA: Peñas-Ajo and AB: Ajo-Bidasoa). Adapted from Sánchez and Serrano, 2003.

The influence of continental waters in the French shelf of the Bay of Biscay benefits species with estuarine dependencies, which are absent in the Spanish area (e.g. *Dicologoglossa cuneata*). Species whose recruitment is highly dependent on larval drift processes, such as hake or megrims, show strong links with hydrographic structures of retention and transport, independently of whether they have cyclonic (cold and fresh) or anticyclonic (warm and saltier) characteristics in relation to the general pattern of the surrounding waters (Sánchez et al., 2001).

5.3.1 The main demersal fisheries in the Bay of Biscay

Demersal resources of the Bay of Biscay are characterized by their diversity and contain more than 100 demersal species, including fish, cephalopods and crustaceans. Many of these are target species landed each year by a very large variety of fleets from France, Portugal and Spain, but also from some other nations (e.g. Belgium and The Netherlands). The French fleet is composed of about 2 500 vessels, two thirds of them being less than 12 m long. An important part of these vessels (around 70%) prosecute their activity inside the 12 nautical mile limit. In the coastal area, demersal resources are exploited using a large range of fishing gears, including trawls and dredges, gillnets and trammels, line, traps, etc. Trawling is predominant in the offshore zone (Léauté, 1998). The northern Spanish demersal fleet is composed of about 5 000 vessels, of which 187 are trawlers, 256 longliners 184 gillnetters and the rest (more than 4 000) make up small artisanal fisheries.

The fishery of large crustaceans involves a relatively small number of vessels (predominantly French) dedicated to working with crustacean traps. The finfish and *Nephrops* fisheries involve the largest number of vessels, which may include some of those vessels participating in the pelagic fisheries.

About 230 French trawlers are currently involved in the *Nephrops* fishery of the Bay of Biscay. Small hake are an important by-catch of this fishery. A mixed trawl fishery, mostly made up of Spanish vessels, targets hake with anglerfish and megrim as by-catches. In shallower waters, another trawl fishery (mostly French) exploits sole, hake, anglerfish, cuttlefish (*Sepia officinalis*) and squid (*Loligo spp.*) with other demersal species as by-catch: *Nephrops,* megrims, gurnards, red-mullets, bib, pollack, etc. The distribution of hake nurseries in the northern part of the Bay of Biscay coincides with the *Nephrops* fishing grounds and a major conflict of interest exists over the effect of the trawl fishery for *Nephrops* on juvenile hake and subsequently on the abundance of adults.

Also used extensively in the Bay of Biscay are fixed gears. Spanish offshore long liners target hake, while in coastal waters French vessels exploit different species such as whiting or sea bass. A French gillnet and trammel fishery for sole started in the mid 1980s and then expanded dramatically. Hake is also a target species for gillnetters, most of which originate from Spain. In the Cantabrian Sea, hake is taken in mixed trawl fisheries, with blue whiting (*Micromesistius poutassou*), mackerel, horse mackerel, megrims, anglerfishes and *Nephrops* as important components of catches.

Some Belgian and Dutch beam trawlers fish for sole in the French Bay of Biscay and form a readily identifiable group with characteristics different from other trawlers.

Single-species assessment approaches have been used historically for their management and a TAC system has been implemented by EU for some of them: hake, sole, *Nephrops,* megrim, anglerfish. These species account for a large part of the total landings (about 50% of the total French landings). Each of those species that are not under TAC regulation represent only a few hundred tonnes annually, except for some whose captures exceed 1 000 t (ling, sea bass, skates, crabs, conger, cuttlefish, pollack, bib, etc.). As in the case of pelagic species the majority of demersal and benthic resources in the Bay of Biscay are either fully exploited or overexploited and are subject to active management regulations (ICES 2004).

5.4. *Ecosystem trophodynamics on the shelf*

The high species diversity of the Bay of Biscay and the strong dependencies among them due to feeding habits implies the necessity to define the main existing interactions. In particularly, the presence of numerous multispecific fisheries focussing on the benthic and demersal species (trawlers, gillnetters, etc.) means that the present management scheme, based on monospecific stock units, is very unrealistic and hardly applicable. Trophodynamic models (mass-balance) attempt to parameterize the energy flows that link the different trophic groups characterising each ecosystem. This new approach can be a valuable tool for understanding ecosystem functioning, and for the design of ecosystem-scale adaptive management experiments. It may also be possible to simulate the consequences of certain management measures, such as effort reduction or closed areas, on the ecosystem.

5.4.1 Foodweb interactions

In the Bay of Biscay ecosystem, most of the biomass and production are contained within the pelagic domain. The main flow is determined by the interaction between phytoplankton, mesozooplankton, planktophagous fish and tuna. Nevertheless, feeding pressure on phytoplankton is low, which means that a large percentage of the primary production passes to detritus. Studies in the area indicate that a high percentage of the primary production is exported to the bottom as particulate organic matter (OSPAR, 2000). The impact of copepods (main zooplankton group in biomass) on the phytoplankton bloom is negligible (Barquero et al., 1998) and the detritus constitutes one of the main energy flow inputs in the system. Consequently, detritivorous species are an important component of the ecosystem and suspension feeders (e.g. suprabenthic zooplankton, shrimps) and deposit feeders (polychaetes, echinoderms, gastropods, hermit crabs) constitute a high percentage of the biomass to the detriment of pelagic plankton (Sánchez and Olaso, 2004). In particular, the high availability of suprabenthic zooplankton in the system, accessible to both middle-sized pelagic fish (mackerel and horse mackerel) and to small demersal fish (blue whiting, *Gadiculus argenteus, Capros aper*), means that this whole group of fish has high biomass and consumption values compared to the small pelagic fish at a lower trophic level (anchovy and sardine), which only have access to the pelagic plankton.

In order to compare the relative role of the pelagic, demersal and benthic subsystems, Fig. 24.28 presents the major biomass flows for the southern Bay of Biscay ecosystem (Cantabrian Sea). The groups represented by small plankton and benthic invertebrate filter feeders and detritivores are in trophic level II. Part of their production passes to the large plankton, benthic and suprabenthic invertebrates, and small pelagic fish (level III). The planktophagous fish of medium size, together with the elasmobranch, benthic cephalopods and benthic fish are in level IV. The highest level, close to level V, corresponds to apex pelagic fish (albacore and bluefin), squid, and large demersal (hake) and benthic fish (anglerfish).

The number of links among all the trophic groups in the area is important and governs the high complexity of internal flows. Strong relationships exist among the three domains due to key groups, which transfer the flow from primary production to the upper trophic levels. Groups linking the pelagic and demersal domains through vertical migration are the suprabenthic zooplankton, horse mackerel and squid. The demersal domain connects with the mesozooplankton at low levels through the suprabenthic zooplankton eaten by a multitude of small demersal fish and blue whiting, and constitutes the main demersal flow. Species linking the benthic and demersal domains are the large demersal fish and dogfish group (mainly *Scyliorhinus canicula*), since their diets include a large quantity of benthic organisms.

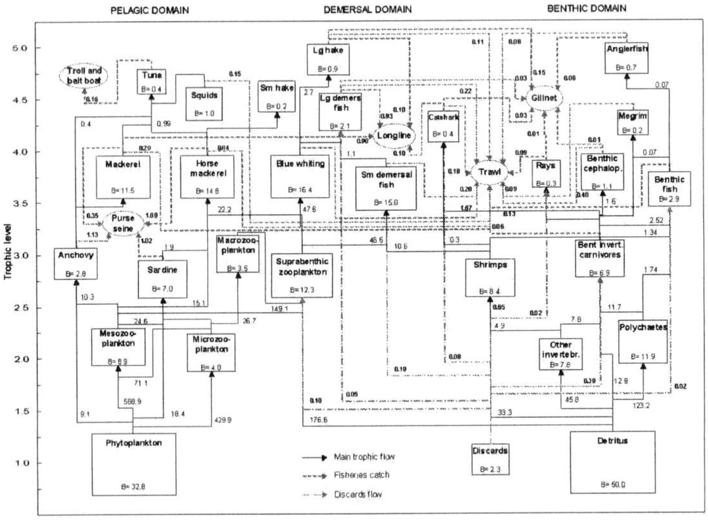

Figure 24.28 Main trophic interactions in the Cantabrian Sea, 1994. The boxes (size roughly proportional to biomass) are arranged on the y-axis by trophic level, and to some degree on a pelagic to benthic scale on the x-axis. Main flows are expressed in t $km^{-2} yr^{-1}$ and the biomass of each trophic group in t km^{-2}. Minor flows, respiration, catch and all backflows to the detritus are omitted. Adapted from Sánchez and Olaso, 2004.

High values of niche overlapping (competition for the same preys) appear among fish from the low trophic level both in the pelagic (sardine and anchovy - 98%) and in the demersal domain (blue whiting and small demersal fish - 96%). The two top predators from the demersal and benthic domain, large hake and anglerfish, also have a high dietary overlap (82%) mainly because both show a high consumption of blue whiting, which constitutes one of the main species in biomass close to the bottom (Sánchez and Olaso, 2004). There is also, a low level of niche overlap between large hake and small hake (31%). This is due to their different diets, large hake mainly base their diet on blue whiting, whereas small hake prey on horse mackerel, small demersal fish and shrimps. Blue whiting inhabit deeper waters than small hake (Sánchez, 1993), and therefore the co-existence of hakes with large size differences is rare on the bottom and explains the low level of cannibalism observed in the hake of the Bay of Biscay (Olaso, 1990; Velasco and Olaso 1998). In the benthic domain, the marked presence of invertebrate detritivores and shrimps in the lower trophic levels produces a strong overlapping of diets among the benthic cephalopods, rays and benthic fish.

6. Changes and risks to the Bay of Biscay Marine Ecosystem

6.1 Hydrography trends, climate indices and biological processes

Cabanas et al. (2003) showed that a notable shift in the winds has occurred during the last two decades, resulting in a reduction in the spring-summer upwelling off the northwest of the Iberian Peninsula. The annual mean of the summer (April-

September) upwelling index decreased from approximately 400 $m^3 s^{-1} km^{-1}$ in the 1970s to around 200 $m^3 s^{-1} km^{-1}$ through the 1980s and 1990s.

On a large scale, it has been demonstrated that the North Atlantic Oscillation (NAO), the dominant mode of atmospheric inter-annual variability in the North Atlantic sector, can be related to inter-annual variability in wind, precipitation and SST fields in the eastern North Atlantic (Hurrel, 1995). Planque et al. (2003), in a time series going back 150 years, found a significant but weak relationship between NAO and SST in the Bay of Biscay, although during the 1990s there is no significant correlation. This may be due to the fact that the Bay of Biscay lies between two regions with different wind and precipitation responses to NAO (Pérez et al., 2000).

Koutsikopoulos et al. (1998) determined from Meteo-France SST data a mean increase of 1.4°C in the surface waters of the southeast Bay of Biscay for the period 1972–1993 (0.6°C per decade), which was slightly higher in winter than in summer. This trend was also reported for shorter periods. Over the last century, COADS records analysed by Planque et al. (2003) show an increase in the mean annual SST of 1.03°C. The analysis of the monthly increase from COADS data indicates that this warming is more pronounced during the winter season (December through to March), with 1.21°C in 100 years, than in summer, which is in agreement with Koutsikopoulos et al. (1998). Planque et al (2003) also reported changes in the wind pattern and rain, i.e. an increase in wind speed in all seasons, but most significantly in the winter period. On average, annual mean wind speed has decreased in the southern Bay of Biscay but has increased in the northern part. The decrease in wind mixing is in part responsible for the warming of the southeastern part of the bay.

Variations in the flow of ENACW towards the southeast of the bay may be responsible for this change. Observations carried out during the 1990s in a section off Santander by González-Pola and Lavín (2003) has suggested that an increase in heat content stored in the water column was greater in the 200–300 m layer. This water mass changes around 0.035°C y^{-1} mostly due to the isopycnal deepening (González-Pola et al., 2003). In this layer ENACW responds quickly to climatological forcing (Pérez et al., 2000). González-Pola et al. (2003) also reported increases in temperature of 0.02±0.01°C y^{-1} in the MOW from 1994 to 2003 in the southern Bay of Biscay, a change which occurs along the isopycnal surfaces.

Cabanas et al. (2003) reported a long-term trend in sea level, varying from 2.02 mm yr^{-1} in Santander (southeastern bay) to 1.45 mm yr^{-1} in A Coruña (southwestern), but estimates over recent decades are greater than for the whole period (1945–1999).

The time series of environmental variables that characterises atmospheric and oceanic variability, such as the North Atlantic Oscillation (NAO) index, the oceanic transport given by differences in Potential Energy Anomaly (PEA) (Curry and McCartney, 2001), the position of the Gulf Stream Index (GSI) (Taylor and Stephens, 1998) and local variables, such as air and sea surface temperature, precipitation, wind, mean sea level and Ekman transport in the southwestern and southeastern Bay of Biscay were analysed by Lavín et al. (2002) using Principal Component Analysis. Periods of different hydro-climatic conditions have been characterized as well as strong shifts. The first component, the thermal one, presents low values at the beginning and some shifts may be found during the 1966–

2000 period, in agreement with the results presented by Planque et al (2003) and Southward et al. (1995).

PCA has also been related to variations in catches of two pelagic fishes: horse mackerel (coastal) and albacore (oceanic). In this context, horse mackerel recruitment is correlated with the thermal/upwelling first component (r=0.78) (see Fig 24.29). Finally, a Multiple Regression analysis was carried out for both pelagic fishes, which showed that Sea Surface Temperature (SST), Upwelling index and 2-year lag NAO index explained 67% of horse mackerel recruitment variance and the PEA and the NAO indices explained 47% of the variance in age 3 albacore catches.

Figure 24.29 Time series of annual mean Sea Surface Temperature in the southwestern (43°N 11°W) and southeastern (45°N 3°W) Bay of Biscay from COADS, and horse mackerel recruitment since 1985 (from Lavín et al., 2002).

These changes related to global warming tend to operate slowly but have severe long-term consequences for the ecology of the ecosystem. They can affect: i) the behavior of species (e.g. changes in migratory routes), ii) their recruitment (due to changes in the environmental conditions in the spawning and/or recruitment areas) and iii) the spatial distribution of species (since more meridional species can expand their area of distribution). In fact, this increase in temperature is likely to be responsible for the appearance of tropical fish species in the southeast shelf of the Bay of Biscay. Several tropical species have been caught throughout the Bay of Biscay and its vicinities (Quéro et al., 1997, 1998; Stebbing et al., 2002), where a trend of temperature increase has been noticed. The first records of tropical fish species in this area were made in the south of Portugal in 1963 and in the Bay of Biscay in 1968 for *Cyttopsis roseus,* and 1970 for *Zenopsis conchifer* and 1980 for

Sphoeroides pachygaster (Figure 24.30) (Quéro et al., 1997, 1998) among other species.

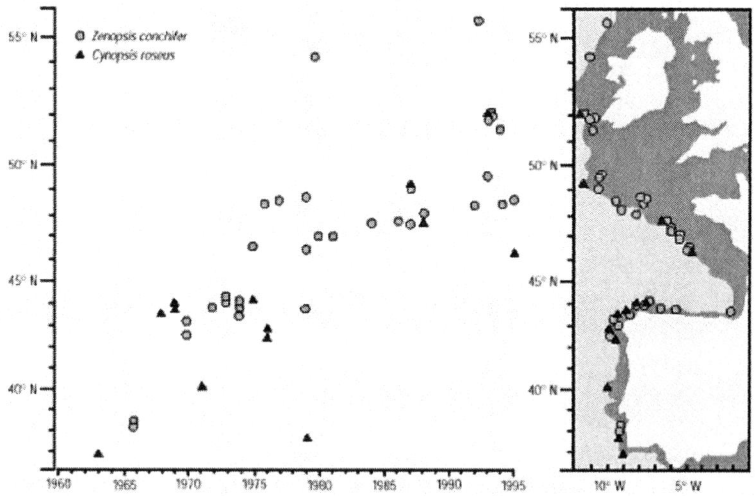

Figure 24.30 Distribution of two tropical fish species (*Zenopsis conchifer* and *Cynopsis roseus*) along the Atlantic coast (based on commercial catches, Quéro et al., 1998)

Southward and Boalch (1994) and Southward et al. (1995) suggest that if the predicted global warming results in increases in temperature of the order of 2°C, then there will be considerable changes in marine communities producing latitudinal shifts in species distribution of the order of 200–400 miles, especially for mobile animals such as fish. The subtropical copepod *Temora stylifera* has also shown a significant increasing presence during the last 15 years in the Bay of Biscay (Villate et al., 1997). Sessile intertidal organisms such as limpets and barnacles have also been reported to show changes in their spatial distribution towards northern areas (Southward et al., 1995).

Other biological effects caused by global warming are related to water column stratification, which will develop earlier in the year, experience a higher degree of stratification and remain stratified for a longer period of time. This effect has been noted in the Bay of Biscay. Lavín et al. (1998), and Valdés and Moral (1998) have related this higher degree of water-column stratification to a less intense exchange of nutrients with the surface layers, which in the end means a reinforcement of the microbial loop, a decrease in mesozooplankton biomass, a drop in the number of species per unit of volume, and a shift of the classic trophic chain to a less efficient one.

6.2 Impact of fisheries on the ecosystem

The fisheries, which have been operating for centuries in the Bay of Biscay, have become more industrialized over the last 50 years, with catches reaching about 90 000 and 200 000 tonnes per year for France and Spain respectively. Trawlers operate on the muddy bottoms of the shelf, whereas long-liners operate mainly at the bottom of the shelf-break and gill nets are used on rocky grounds near the coast and shelf-break. There are also seasonal pelagic fisheries for anchovy (purse seine), mackerel (hand line, trawlers, purse seine) and tunas (troll and bait boats) during their migrations in spring and summer.

Bay of Biscay fisheries have had a strong impact on the bottom communities and have induced changes in their structure. This impact has been mainly direct (fishing mortality on target species and bycatch) and indirect by means of modifications to the habitat through erosion of the sediment and damage to the benthos by different elements of the gears.

The discards that constitute more than the 20% of the total of catches in the Bay of Biscay also impact indirectly causing changes to the biological interactions. One of the main consequences of the fishing pressure on the structure of the communities is that it benefits species with high growth rates and wide distribution, whereas the more specialized and those with low growth rates are eliminated, resulting in a simplification of the communities. Associated with it comes an increase in the values of dominance and a fall in the biodiversity, both of which have been reported in the fish communities of the Cantabrian Sea (Sánchez, 1993).

In the pelagic domain, planktivorous fish of small and medium size are captured by the purse seine, and bait and troll (surface hook) boats catch tuna exclusively. In the demersal domain, large piscivorous fish, such as hake and other large demersal fish, are captured by the selective bottom longline. In the benthic domain, the less selective bottom trawl catch a great variety of organisms and produce a high level of discards. Gillnets exploit certain predators from the demersal as well as the benthic domain. Using the Leontief matrix routine and following the subsequent development described by Ulanowicz and Puccia (1990), the negative trophic impact of trawling on the different groups of the system is high and much stronger than for other gears (Sánchez and Olaso, 2004). Longlining produces the least negative trophic impact on the ecosystem because it only catches predators. All fishing gears, except the purse seine, impact negatively on piscivorous fish and elasmobranchs.

Scavenger fauna and sea birds have found a source of food from discards. It is estimated that 20% of the food of sea birds in the North Sea is obtained from discards (Camphuysen et al., 1995). On the other hand, Olaso et al. (2002) demonstrated that the abundance of sparids and the portunid crabs increases significantly after baiting the sea bed of a trawling beach with discards, as well as increasing the feeding intensity of skates, dogfish and *Trachinus draco.* These are groups at medium trophic levels, since species belonging to higher trophic levels require moving living prey and the species from lower levels are basically filter feeders or detritivores. The importance of discards as food in the ecosystem is low with respect to detritus, primary producers or other low trophic levels, and makes up 0.07% of the total food intake in the Cantabrian Sea (Sánchez and Olaso, 2004).

The level of fisheries impact in the Cantabrian Sea is comparable to the most intensively exploited temperate shelf ecosystems in the world. The high PPR (Primary Production Required) value corroborates the fact that most commercially important shelf stocks in the area are either fully exploited or overexploited, and landings are expected to decrease with the current fishing pressure. From 1983 to 1993, the mean trophic level of Cantabrian Sea demersal fisheries declined from

4.1 to 3.7. This is reflected in a gradual transition of landings from long-lived, high trophic level piscivorous groundfish (hake, anglerfish, megrim) towards lower trophic level planktivorous fish (blue whiting, horse mackerel). Since 1993 there has been no further decrease and the fisheries may have reached their lowest historical trophic level limit. The depressed abundance and productivity of top predators impedes the recovery of trophic levels in the catch (Sánchez and Olaso, 2004).

6.3. Risks derived from shipping and oil transport activities

Other main human activities that cause environmental degradation in the Bay of Biscay are shipping and oil transport. The Bay of Biscay is located on the main route of supertankers transporting oil from the Middle East and Africa to EU harbors. More than 70% of the total oil consumed in the EU is moved by shipping through the Finisterre pass directly towards the English Channel and then to the final destination in different European harbors. In recent years several oil spills have occurred in the Bay of Biscay (Fig. 24.31), for example 5 supertankers carrying more than 50 000 t have wrecked since 1976, and the last 3 in an interval of just a decade (1992, Aegean Sea; 1999, Erika; 2002, Prestige), which has made this region the most severely affected by this kind of accident in the world.

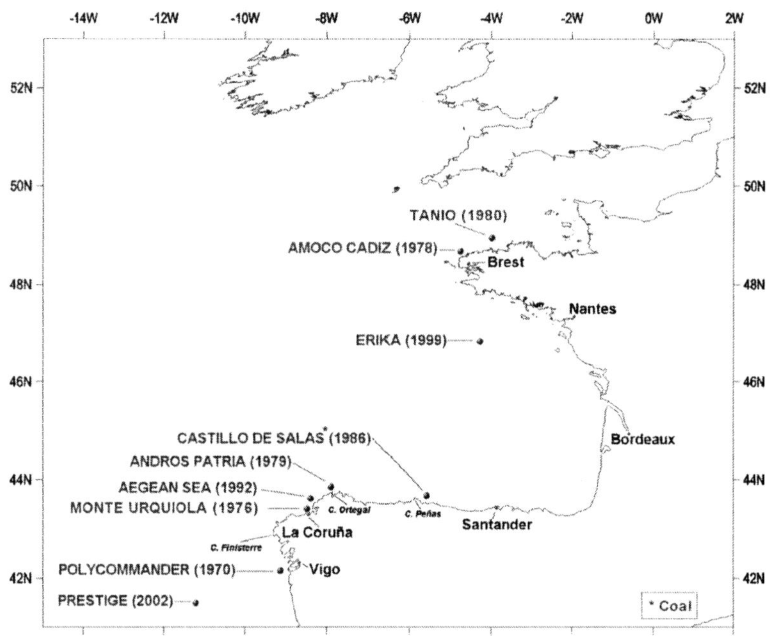

Figure 24.31 Main oil spill events in the Bay of Biscay and its vicinities since 1970 with indication of the tanker's name and year of the spill.

Two carriers were wrecked in the tidal estuary of A Coruña when approaching the harbor, the Urkiola in 1976 and the Aegean Sea in 1992. These carriers spilt 30000 and 80 000 t of oil respectively in an enclosed bay of less than 50 m depth. Several research institutions monitored the effects of the oil spills on the ecosystem. A temporal reduction of the infauna (turbellaria and harpacticoida copepods) and macroinfauna (amphipods, echinoderms, molluscs) and a simultaneous dramatic increase of opportunist nematodes and polychaetes was observed (Giere, 1979; Parra and López-Jamar, 1997) which coincided with the period of higher concentrations of hydrocarbons in the sediments. The black tide extended out towards the NE carried by the dominant current, but the extension of the tide was nevertheless quite limited spatially. The beaches most highly exposed to hydrodynamism suffered the greatest impact. The interstitial system of the beach was damaged by the spill, accumulating hydrocarbons up to very deep levels in areas of coarse sand (140 cm). Although the immediate effects of the contamination were very serious, the abundance and diversity of the meiofauna recovered after three years.

Evaluation of ecosystem effects and monitoring of ecosystem recovery were more complex in the case of the tankers Erika (South of Brittany in 1999) and Prestige (west of Galicia in 2002). Both were wrecked offshore and the fuel (50 000 and ~60 000 t respectively) was dispersed by the dominant currents in a wide region inside the Bay of Biscay, affecting beaches, cliffs, fishing grounds, marine protected areas, etc.

More than 40 000 t of fuel were removed from the sea and beaches after the Prestige spill, but another 10 000 t were dispersed along the coast or deposited on the bottom. Experts and resources were mobilized in order to do a surveillance of the entire north Spanish coast, waters and marine sediments on the shelf. Investigations on the effects of this oil spill are still ongoing and unfortunately we cannot present conclusive results here. Shellfish and demersal fisheries were closed in Galicia and some other regions in the north of Spain for several months and economic losses were enormous. One year after the spill, some traces of fuel still remain in sediments and a few of the rocky and pebble beaches most highly exposed to hydrodynamism that suffered the greatest impact remain patchy with fuel. The mortality of marine birds was high but fortunately communities seem to be recovering well. We did not find any impact of fuel on plankton abundance, seasonality or species composition and no evidence of oil transfer through the food web was observed. Catches of pelagic fishes during 2003 in the Bay of Biscay were close to average (with the exception of anchovy in spring), and the same occurred with tuna and albacore. These last data suggest that the impact of this event on the marine pelagial was less than presumed at the beginning of the crisis.

Oil spills are a real risk in the Bay of Biscay. The events monitored during recent years can be used as a basis for comparison and for a better understanding of ecosystem response to these perturbations, as well as for preparing adequate action plans (Carballeira, 2003). Combining and coordinating the efforts of both Spanish and French experts and resources to preserve the quality of our environment is a challenge for both oceanographic communities.

7. Concluding remarks

As shown in the present review, descriptive aspects of the oceanography and ecology of the Bay of Biscay, both on the French and Spanish continental margins, have been extensively studied and it is possible to recognize general patterns in annual cycles of hydrography, planktonic communities, and in the distribution of demersal and pelagic fish populations. In spite of the effort devoted to the description of the marine communities, difficulties in explaining why some species extend and contract their spatial distribution and how their abundances increase and decrease are generally recognized. This is due to the complexity of relationships among abiotic and biotic properties of the ecosystem, which are subjected to different sources of variability (both natural and anthropogenic).

It is also evident that all the pressing environmental problems caused by man's activities and the complexity of ecosystem interactions add a high degree of uncertainty to the proper management of living marine resources and uses of coastal areas, which makes it imperative that the future of their management has a more solid basis of scientific knowledge. Accordingly, OSPAR (2000) has recommended to the appropriate authorities a list of actions including the foundation of Marine Protected Areas for the protection and conservation of biological biodiversity in this region. Regarding this, it must be noted that no marine reserve exists in the Bay of Biscay (except in coastal and estuarine areas) whereas on the French and Spanish Mediterranean coasts there are more than 15 different marine protected areas. This is a simple comparison but quite thought-provoking.

Research possibilities of oceanographic and ecological development in the Bay of Biscay are high. There is a large number of laboratories in France and Spain working in these areas - Brest, Lorient, Nantes, La Rochelle, Arcachon, Biarritz, Guipuzcoa, Vizcaya, Santander, Gijón and A Coruña belonging to IFREMER, IEO, AZTI and also some Universities, Maritime Museums and Associations. There is an international colloquium of the Bay of Biscay every two years, in which new research developments are presented. A good number of collaborations between France and Spain have been developed over the last two decades, some funded by the European Union, others funded by local or national institutions and under international institutions such as ICES, ICCAT, and GOOS. In spite of these facilities, more integration of disciplines is needed to provide comprehensive answers to the central issues of holistic ecosystem management. The existing lack of knowledge reveals the great need for more studies on ecosystem functioning in the Bay of Biscay. Ocean observation programmes on adequate spatial and time scales supported by governmental agencies should be encouraged. Fortunately, some international proposals have recently been developed to forward these aims. There is also a need for improved forecasting of the human impact on the ecosystem. Searching for changing trends based on key species, and monitoring the preservation state of selected areas should also be major priorities.

Acknowledgements

We wish to thank G. Parrilla, M. Rúiz, C. González-Pola , J. C. Sorbe, P. Barquín and R. Herrera for their help, R. Pingree for his interest in the development of this chapter and for sending us a good number of papers. We would also like to thank our families for the amount of free time we have needed to complete the chapter.

Bibliography

Abaunza, P., L. Gordo, C. Karlou-Riga, A. Murta, A.T.G.W. Eltink, M.T. García Santamaría, C. Zimmermann, C. Hammer, P. Lucio, S.A. Iversen, J. Molloy and E. Gallo, 2003. Growth and reproduction of horse mackerel, *Trachurus trachurus* (Carangidae). *Rev. Fish Biol. Fish.*, **13 (1),** 27–61.

Allain, G., P. Petitgas and P. Lazure, 2001. The influence of mesoscale ocean processes on anchovy (*Engraulis encrasicolus*) recruitment in the Bay of Biscay estimated with a three-dimensional hydrodinamic model. *Fish. Oceanogr.*, **10 (2),** 151–163.

Arbault, S. and N. Boutin, 1968. Ichtyoplancton. Oeufs et larves de poissons téléostéens dans le Golfe de Gascogne en 1964. *Rev. Trav. Inst. Pêches marit.*, **32 (4),** 413–476.

Arhan, M., A. Colin de Verdière and L. Mémery, 1994. The eastern boundary of the subtropical North Atlantic. *J. Phys. Oceanogr.*, **24,** 1295–1316.

Aubin-Ottenheimer, G., 1986. Bilan des connaissances sur la cardine (*Lepidorhombus whiffiagonis*). Etude du stock de mer Celtique. *Rev. Trav. Inst. Tech. Pêches mar.*, **49 (3–4),** 205–214.

Azevedo, M. and P. Pereda, 1994. Comparing monkfish (*Lophius piscatorius* and *L. budegassa*) abundance in ICES Division VIIIc by year and depth strata. ICES CM 1994/G:22, 7 pp.

Bardey, P., P. Garnesson, G. Moussu and L. Wald, 1999. Joint analysis of temperature and ocean colour satellite images for mesoscale activities in the Gulf of Biscay. *Int. J. Remote sensing,* **7,** 1329–1341.

Barquero, S.B., 1995. Consumo de los copépodos herbívoros sobre una floracion primaveral de fitoplancton (costa NO de Galicia, mayo 1994). Seminario de investigación, Univ. Oviedo, 26 pp.

Barquero, S., J.A. Cabal, R. Anadon, E. Fernandez, M. Varela and A. Bode, 1998. Ingestion rates of phytoplankton by copepod size fractions on a bloom associated with an off-shelf front off NW Spain. *J. Plankton Res.*, **20** (5), 957–972.

Bas, C., 1995. Ecological structures: expansion and replacement. *Sci. Mar.*, **59,** 373–380.

Blanchard, F., 2001. The effects of fishing on demersal fish community dynamics: an hypothesis. *ICES J. Mar. Sci.*, **58,** 711–718.

Blanchard, F., J. Boucher and J.C. Poulard, 2002. General trends in the fish community of the Bay of Biscay from 1973 to nowadays: ecosystem effects of fishing or climate?. 8th International Symposium on Oceanography of the Bay of Biscay, Gijón (April 2002).

Bode, A. and E. Fernandez, 1992. Variability of biochemical composition and size distributions of seston in the eurphotic zone of the Bay of Biscay: implications for microplankton trophic structure. *Mar. Biol.*, **114:** 147- 155.

Bode, A., E. Fernández, A. Botas and R. Anadón, 1990. Distribution and composition of suspended particulate matter related to a shelf-break saline intrusion in the Cantabrian Sea (Bay of Biscay). *Oceanol. Acta,* **13 (2):** 219–228.

Bode, A., B. Casas, E. Fernández, E. Marañon, P. Serret and M. Varela, 1996. Phytoplankton biomass and production in shelf waters off NW Spain: spatial and seasonal variability in relation to upwelling. *Hydrobiologia,* **241:** 225–234

Bode, A., M.T. Alvarez-Ossorio and N. González, 1998. Estimations of mesozooplankton biomass in a coastal upwelling area off NW Spain. *J. Plankton Res.*, **20:** 1005–1014.

Bode, A., P. Carrera and S. Lens, 2003a. The pelagic foodweb in the upwelling ecosystem of Galicia (NW Spain) during spring: natural abundance of stable carbon and nitrogen isotopes. *ICES J. Mar. Sci.*, **60,** 11–22.

Bode, A., P. Carrera and C. Porteiro, 2003b. Pelagic foodwebs and sardine in the north Iberian shelf. *GLOBEC Int. Newsletters April 2003,* 27–29.

Bode, A., M.T. Alvarez-Ossorio, P. Carrera and J. Lorenzo, (in press). Reconstruction of trophic pathways between plankton and the North Iberian sardine (*Sardina pilchardus*) using stable isotopes. *Sci. Mar.*

Bode, A., S. Barquero, N. González, M.T. Alvarez-Osorio and M. Varela, 2004. Contribution of heterotrophic plankton to nitrogen regeneration in the upwellin ecosystem of A Coruña (NW Spain). *J. Plankton Res.*, **26 (1):** 11–28.

Borja, A., A. Uriarte, V. Valencia, L. Motos and A. Uriarte, 1996. Relationships between anchovy (*Engraulis encrasicholus*) recruitment and environment in the Bay of Biscay. *Sci. Mar.*, **60 (2),** 179–192.

Botas, J.A., E. Fernández and R. Anadón, 1990. A persistent upwelling off the central Cantabrian coast (Bay of Biscay). *Estuar. coast. shelf Sci.*, **30,** 185–199.

Botas, J.A., A. Bode, E. Fernández and R. Anadón, 1988. Descripción de una intrusión de agua de elevada salinidad en el Cantábrico Central: distribución de los nutrientes inorgánicos y su relación con el fitoplancton. *Inv. Pesq.*, **52 (4):** 559–572.

Cabanas, J.M. and C. Porteiro, 1998. Links between the North Atlantic Sardine recruitment and their environment. *ICES CM 1998/*R: 23.

Cabanas, J.M., A. Lavín, M.J. García, C. González-Pola and E. Tel Pérez, 2003. Oceanographic variability in the northern shelf of the Iberian Peninsula 1990–1999. *ICES mar. Sci. Symp,* **219,** 71–79.

Calbet, A, 2001. Mesozooplankton grazing impact on primary production: a global comparative analysis in marine ecosystems. *Limnol. Oceanogr.*, **46,** 1824–1830.

Camphuysen, C.J., B. Calvo, J. Dunrik, K. Ensor, A. Follestad, R.W. Furness, S. Garthe, G. Leaper, H. Kov, M.L. Tasker and C.J.N. Winter, 1995. Consumption of discards by seabirds in the North Sea. NSRPDU, Den Burg, Texel. NIOZ-Rapport 1995–5, 202 pp.

Carballeira, A., 2003. Considerations in the design of a monitoring program of the biological effects of the prestige oil spill. *Cienc. Mar.*, **29 (1),** 123–139.

Carrera, P. and C. Porteiro, 2003. Stock dynamic of Iberian sardine (*Sardina pilchardus,* W.) and its implication on the fishery off Galicia (NW Spain). *Sci. Mar.*, **67 (1),** 245–258.

Casas, B., M. Varela, M. Canle, N. González and A. Bode, 1997. Seasonal variations of nutrients, seston and phytoplankton, and upwelling intensity off La Coruña (NW Spain). *Estuar. coast. shelf Sci.*, **44:** 767–778.

Castro, C., F. Perez, S. Holley and A. Rios, 1998. Chemical characterisation and modelling of water masses in the Northeast Atlantic. *Prog. Oceanogr.*, **41,** 249–279.

Cayan, D.R., 1992a. Latent and sensible heat flux anomalies over the northern oceans: The connection to monthly atmospheric circulation. *J. Clim.*, **5,** 354–369.

Cayan, D.R., 1992b. Latent and sensible heat flux anomalies over the northern oceans: Driving the sea surface temperature. *J. Phys. Oceanogr.*, **22,** 859–881.

Cendrero, O., 2002. Sardine and anchovy crisis in northern Spain: natural variations or an effect of human activities? *ICES mar. Sci. Symp.*, **215,** 279–285.

Chapman, C.J., 1980. Ecology of juvenile and adult Nephrops. In *The biology and management of lobsters.* Cobb S. and B. Phillips (eds), Vol II. Academic Press, New York, 143–178.

Cort, J.L., 1990. Biología y pesca del atún rojo, *Thunnus thynnus* (L.), del Mar Cantábrico. *Pub. Esp. Inst. Esp. Oceanogr.*, **4,** 272 pp.

Curry, R. G. and M.S. McCartney, 2001. Ocean gyre circulation changes associated with the North Atlantic Oscillation. *J. Phys. Oceanogr.*, **31,** 3374–3400.

Dardignac, J., 1988. Les pêcheries du golfe de Gascogne. Bilan des connaissances. *Rapp. Scient. Techn. IFREMER,* **9,** 204 pp.

D'Elbée, J. and J. Castel, 1991. Zooplankton from the continental shelf of the southern Bay of Biscay exchange with Arcachon basin, France. *Ann. Inst. Oceanogr.*, **67 (1),** 35–48.

Díaz del Río, G., A. Lavín, J. Alonso, J.M. Cabanas and X. Moreno-Ventas, 1996. Hydrographic variability in Bay of Biscay shelf and slope waters in spring 1994, 1995, 1996 and relation to biological drifting material. ICES CM 1996/S:18, 8 pp.

Díaz del Río, G., N. Gonzalez and D. Marcote, 1998. The intermediate Mediterranean water inflow along the northern slope of the Iberian Peninsula. *Oceanol. Acta,* **21 (2):** 157–163.

Dickson, R.R., W.J. Gould, T.J. Muller and C. Maillard, 1985. Estimates of the mean circulation in the deep (>2000 m) layer of the eastern North Atlantic. *Prog. Oceanogr.,* **14,** 103–127.

Díez, I., A. Secilla, A. Santolaria and J.M. Gorostiaga, 2000. The north coast of Spain. In *Seas at the Millennium. An environmental evaluation.* C. Sheppard (Ed.). Pergamon, Amsterdam, Volume I, 135–150.

Druon, J.N., G. Langlois and J. Le Fèvre, 2001. Simulating vertical mixing in a shelf-break region: addition of a shear instability model, accounting for the overall effect of internal tides, on top of a one-dimensional turbulence closure mixed layer model. *Cont. Shelf Res.,* **21,** 423–454.

Duarte, R., M. Azevedo, J. Landa and P. Pereda, 2001. Reproduction of anglerfish (*Lophius budegassa* Spinola and *Lophius piscatorius* Linnaeus) from the Atlantic Iberian coast. *Fish. Res.,* **51,** 349–361.

Estrada, M., 1984. Phytoplankton distribution and composition off the coast of Galicia (northwest of Spain). *J. Plankton Res.,* **6,** 417–434.

Evans, G.T. and R. Prego, 2003. Rias, estuaries and incised valleys: is a ria an estuary?. *Mar. Geol.,* **196:** 171–175.

Farmer, A.D., 1973. Age and growth in *Nephrops norvegicus* (Decapoda: Nephropidae). *Mar. Biol.,* **23,** 315–325.

Fernández, E. and A. Bode, 1991. Seasonal patterns of primary production in the Central Cantabrian Sea (Bay of Biscay). *Sci. Mar.,* **55 (4),** 629–636.

Fernández, E., J. Cabal, J.L. Acuña, A. Bode, A. Botas and C. García-Soto, 1993. Plankton distribution across a slope current-induced front in the southern Bay of Biscay. *J. Plankton Res.* **15 (6)**: 619–641.

Fernández, E. and A. Bode, 1994. Succession of phytoplankton assemblages relation to the hydrography in the southern Bay of Biscay: A multivariate approach. *Sci. Mar.,* **58(3):** 191–205.

Fortier, L., J. Le Fevre and L. Legendre, 1994. Export of biogenic carbon to fish and to the deep ocean: the role of large planktonicmicrophages. *J. Plankton Res,* **16 (7),** 809–839.

Fraga, F., 1981. Upwelling off the Galician coast, Northwest Spain. In: Coastal Upwelling. Richards F.A. (ed.) American Geophysical Union, Washington DC, 176–182.

Fraga, F., C. Mourino and M. Manriquez, 1982. Las masas de agua en la costa de Galicia: junio-octubre. *Res. Exp. Cient.,* **10,** 51–77.

Froidefond, J.M., P. Castaing and J.M. Jouanneau, 1996. Distribution of suspended matter in a coastal upwelling area. Satellite data and *in situ* measurements. *J. Mar. Sys.,* **8,** 91–105.

Froidefond, J.M., A.M. Jégou, J. Hermida, P. Lazure and P. Castaing, 1998. Variabilité du panache turbide de la Gironde par télédétection. Effets des facteurs climatiques. *Oceanol. Acta,* **21 (2),** 191–207.

Frouin, R., A.F.G. Fiúza, I. Âmbar and T.J. Boyd, 1990. Observations of a poleward surface current off the coasts of Portugal and Spain during winter. *J. Geophys. Res.,* **95 (C1),** 679–691.

García-Soto, C. and R.D. Pingree, 1998. Late autumn distribution and seasonality of chlorophyll-a at the shelf-break/slope region of the Armorican and Celtic shelf. *J. Mar. Biol. Ass. U.K.,* **78,** 17–33.

Garcia-Soto, C., R.D. Pingree and L. Valdés, 2002. Navidad development in the southern Bay of Biscay: Climate change and swoddy structure from remote sensing and in situ measurements. *J. Geophys. Res.,* **107 (C8),** 10.1029/2001JC00101.

Gil, J., L. Valdés, M. Moral, R. Sánchez and C. García-Soto, 2002. Mesoscale variability in a high-resolution grid in the Cantabrian Sea (Southern Bay of Biscay), May 1995, *Deep Sea Res. I,* **49,** 1591–1607.

Gohin, F., J.N. Druon and L. Lampert, 2002. A five channel chlorophyll concentration algorithm applied to SeaWIFS data processed by SeaDAS in coastal waters. *Int. J. Remote Sensing,* **23,** 1639–1661.

Gohin, F., L. Lampert, J.F. Guillaud, A. Herbland and E. Nezan, 2003. Satellite and in situ observations of a late winter phytoplankton bloom in the northern Bay of Biscay. *Cont. Shelf Res.,* **23,** 1117–1141.

González, N., A. Bode, M. Varela and R. Carballo, 2003. Interannual variability in hydrobiological variables in the coast of A Coruña (NW Spain) from 1991 to 1999. *ICES mar. Sci. Symp,* **219,** 382–383.

González-Pola, C. and A. Lavín, 2003. Seasonal cycle and interanual variability of the heat content on a hydrographic section off Santandert (Southern Bay of Biscay), 1991–2000. *ICES mar. Sci. Symp.,* **219,** 343–345.

González-Pola, C., A. Lavín and M. Vargas-Yañez, 2003. Thermohaline variability on the intermediate waters of the Southern Bay of Biscay from 1992. ICES CM 2003/T:10.

Giere, O. 1979. The impact of oil pollution on the intertida meiofauna. Field studies after the La Coruña-spill, May 1976. *Cah. Biol. Mar.,* **20,** 231–251.

Guichet, R., 1995. The diet of European hake (*Merluccius merluccius*) in the northern part of the Bay of Biscay. *ICES J. mar. Sci.,* **52,** 21–31.

Guisande, C., J.M. Cabanas, A.R. Vergara and I. Riveiro, 2001. Effect of climate on recruiment success of Atlantic Iberian sardine Sardina pilchardus. *Mar. Ecol. Prog. Ser.,* **223:** 243–250

Halsband-Lenk, C., F. Carlotti and W. Greve, 2003. Life cycle strategies of calanoid copepod congeners: Significance for climate change studies. 3rd International Zooplankton Production Symposium, Gijón (May 2003).

Harvey, J., 1982. Theta-S relationships and water masses in the eastern North Atlantic. *Deep Sea Res.,* **29 (8A),** 1021–1033.

Haynes, R. and E.D. Barton, 1990. A poleward flow along the Atlantic coast of the Iberian Peninsula. *J. Geophys. Res.,* **95 (C7),** 11425–11441.

Helfman, G.S., B.B. Collette and D.E. Facey, 1997. *The diversity of fishes.* Blackwell Science. Oxford, 528 pp.

Hermida, J., P. Lazure, J.M. Froidefond, A.M Jégou and P. Castaing, 1998. La dispersion des apports de la Gironde sur le plateau continental. Données *in situ,* satellitales et numériques. *Oceanol. Acta,* **21 (2)**, 209–221.

Hislop, J.R.G, A. Gallego, M.R. Heath, F.M. Kennedy, S.A. Reeves and P.J. Wright, 2001. A synthesis of the early life history of the anglerfish, *Lophius piscatorius* (Linneaus, 1758) in northern British waters. *ICES J. Mar. Sci.,* **58,** 70–86.

Hurrell, J.W., 1995. Decadal trends in the North Atlantic Oscillation: Regional Temperatures and Precipitation. *Science,* **269,** 676–679.

Huthnance, J.M., 1984. Slope currents and JEBAR. *J. Phys. Oceanogr.,* **14,** 795–810.

Huthnance, J.M., H.M. van Aken, M. White, E.D. Barton, B. Le Cann, E.F. Coelho, E. Alvarez-Fanjul, P. Miller and J. Vitorino, 2002. Ocean margin exchange-water flux estimates. *J. Mar. Sys.,* **32,** 107–137.

ICCAT, 2003. Informe del periodo bienal 2002/03. **2(1),** 223 pp.

Hudges, S. and A. Lavín, 2003. The 2002/2003 ICES Annual Ocean Climate Status summary. *ICES Coop. Res. Rep.,* **259,** 30 pp.

ICES, 2004a. Report of the Working Group on the Assessment of Mackerel, Horse Mackerel, Sardine and Anchovy. ICES CM 2004/ACFM:08.

ICES, 2004b. Report of the Working Group on the Assessment of Southern Shelf Stocks of Hake, Monk and Megrim. ICES CM 2004/ACFM:02.

Iorga, M.C. and M.S. Lozier, 1999. Signatures of the Mediterranean outflow from a North Atlantic climatology 1. Salinity and density fields. *J. Geophys. Res.,* **104 (C11),** 25985–26009.

Isaacs, J.D., 1973. Potential trophic biomasses and trace-substance concentration in unstructured marine food webs. *Mar. Biol.,* **22,** 97–104.

Isemer, H.J. and L. Hasse, 1987. *The Bunker climate Atlas of the North Atlantic Ocean, Vol. 2: Air-Sea interactions.* Springer Berlin, 252 pp.

Jégou, A.M. and P. Lazure, 1995. Quelques aspects de la circulation sur le plateau atlantique. Acta del IV Coloquio Internacional Sobre Oceanografica del Golfo de Vizcaya, 96–106.

Koutsikopoulos, C. and B. Le Cann, 1996. Physical processes and hydrological structures related to the Bay of Biscay Anchovy. *Sci. Mar.,* **60,** 9–19.

Koutsikopoulos, C., D. Dorel and Y. Désaunay, 1995. Movement of sole (*Solea solea*) in the Bay of Biscay: coastal environment and spawning migration. *J. mar. biol. Ass. U.K.,* **75,** 109–126.

Koutsikopoulos, C., P. Beillois, C. Leroy and F. Taillefer, 1998. Temporal trends and spatial structures of the sea surface temperature in the Bay of Biscay. *Oceanol. Acta,* **21 (2),** 335–344.

Lago de Lanzós, A., L. Valdés, C. Franco, A. Solá and X. Moreno-Ventas, 1997. Bay of Biscay environmental scenary in June and distribution of early fish stages. Annual Symposium of Fisheries Society of the British Iles: Ichthyoplankton Ecology. Galway (July 1997).

Lam, F.P.A., T. Gerkema and L.R.M. Maas, 2003. Preliminary results from observations of internal tides and solitary waves in the Bay of Biscay. *http://www.whoi.edu/science/AOPE/people/tduda/isww/text/lam/*

Lavín, A., G. Díaz del Río, J. M. Cabanas and G. Casas, 1991. Afloramiento en el noroeste de la península Ibérica. Indices de afloramiento para el punto 43° N 11° W. Periodo 1966–1989. *Inf. Téc. Inst. Esp. Oceanogr.,* **91,** 40 pp.

Lavín, A., L. Valdés, J. Gil and M. Moral, 1998. Seasonal and interannual variability in properties of surface water off Santander (Bay of Biscay) (1991–1995). *Oceanol. Acta,* **21 (2),** 179–190.

Lavín, A., L. Valdés, X. Moreno-Ventas, V. Ortiz de Zárate and C. Porteiro, 1997. Common signals between physical, atmospheric variables, North Iberian sardine recruitment and North Atlantic albacore recruitment. *ICES/GLOBEC Workshop on Prediction and decadal-scale fluctuations on the North Atlantic.* Copenhaguen, Sep 1997.

Lavín, A., G. Díaz del Río, J. M. Cabanas and G. Casas, 2000. Afloramiento en el noroeste de la península Ibérica. Indices de afloramiento para el punto 43°N 11°W. Periodo 1990–1999. *Datos y resúmenes Inst. Esp. Oceanogr.,* **15,** 25 pp.

Lavín, A., C. González-Pola and J.M. Cabanas, 2001. Upper water hydrographic conditions in the SEAMAR 0400 and SEAMAR 0500 cruises. In *2nd year Report of the E.U. SEAMAR Project.*

Lavín, A., X. Moreno-Ventas, V. Ortiz de Zárate, P. Abaunza and J.M. Cabanas, 2002. Environmental variability in the North Atlantic and Iberian waters and its influence on pelagic fish: the cases of horse mackerel and albacore. ICES CM 2002/O:26.

Lavín, A., C. González-Pola and J.M. Cabanas, 2003. Spanish Standard Sections (area 4). Annex K of Report of the Working Group on Oceanic Hydrography. ICES CM 2003/C:07.

Lazure, P. and A.M Jégou, 1998. 3D modelling of seasonal evolution of Loire and Gironde plumes on Biscay Bay continental shelf. *Oceanol. Acta,* **21 (2),** 165–177.

Lazure P., A.M. Jégou and M. Kerdreux, (in press). Analysis of salinity measurements near islands on the French continental shelf of the Bay of Biscay. *Sci. mar.*

Léauté, J.P., 1998. Les flottilles de pêche de l'Union Européenne dans le golfe de Gascogne vues du ciel. *Oceanol. Acta,* **21 (2),** 371–381.

Legendre, L. and M. Gosselin, 1989. New production and export of organic matter to the deep ocean: consequences of some recent discoveries. *Limnol. Oceanogr.,* **34 (7),** 1374–1380.

Legendre, L. and J. Le Févre, 1991. From indivudual plankton cells to pelagic marine ecosystems and to global biogeochemical cycles. In *Particle analysis in Oceanography.* Ed S. Demers. *NATO ASI Series,* **G 27,** 261–300.

Legendre, L. and F. Rassoulzadegan, 1995. Plankton and nutrient dynamics in marine waters. *Ophelia,* **41,** 153–172.

Le Cann, B., 1990. Barotropic tidal dynamics of the Bay of Biscay shelf: observations, numerical modelling and physical interpretation. *Cont. Shelf Res.,* **10 (8),** 723–758.

Le Pape, O., F. Chauvet, S. Mahevas, P. Lazure, D. Guérault and Y. Desaunay, 2003. Quantitative description of habitat suitability for the juvenile common sole (*Solea solea,* L.) in the Bay of Biscay (France) and the contribution of different habitats to the adult population. *J. Sea Res.,* **50 (2–3),** 139–149.

Le Tareau, J.Y. and R. Mazé, 1993. Storm effects on the baroclinic tidal field in the Bay of Biscay. *J. Mar. Sys.,* **4,** 327–347.

Llope, M., R. Anadón, M. Alvarez-Ossorio, L. Valdés and M. Varela, 2003. Zooplankton biomass timing with temperature in South Bay of Biscay. 3rd International Zooplankton Production Symposium, Gijón (May 2003).

Maillard, C. 1986. *Atlas Hydrologique de l'Atlantique Nord-Est.* IFREMER, Brest, 260 pp.

Mariette, V. and B. Le Cann, 1985. Simulation of the formation of the Ushant thermal front. *Cont. Shelf Res.,* **4 (6),** 637–660.

Massé, J., 1996. Acoustic observations in the Bay of Biscay: Schooling, vertical distribution, species assemblages and behaviour. *Sci. Mar.,* **60 (2),** 227–234.

Mazé, R., Y. Camus and J.Y Le Tareau, 1986. Formation de gradients thermiques à la surface de l'océan, au-dessus d'un talus, par interaction entre les ondes internes et le mélange dû au vent. *J. Cons. Int. Explor. Mer,* **42,** 221–240.

McCartney, M.S. and C. Mauritzen, 2001. On the origin of the warm inflow to the Nordic Seas. *Prog. Oceanogr.,* **51,** 125–214.

McCartney, M.S., S.L. Bennet and M.E. Woodgate-Jones, 1991. Eastward Flow through the Mid-Atlantic Ridge at 11°N and its influence on the Abyss of the Eastern Basin. *J. Physic. Oceanogr.,* **21,** 1089–1121.

Molina, R., 1972. Contribución al estudio del upwelling frente a la costa nordoccidental de la Península Ibérica. *Bol. Inst. Esp. Oceanogr.,* **152,** 39 pp.

Monteiro, P.M.S., A.G. James, A.D. Sholto-Douglas and J.G. Field, 1991. The $\delta^{13}C$ trophic position isotope spectrum as a tool to define and quantify carbon pathways in marine food webs. *Mar. Ecol. Prog. Ser.,* **78,** 33–40.

Moreno-Ventas, X., A. Lavín and L. Valdés, 1997. Hydrodynamic Singularities Observed by Satellite Imagery in the Continental Margin of the Bay of Biscay. 2nd International Symposium on the Ibero-Atlantic Continental Margin, (Cádiz, 1997).

Motos, L., A. Uriarte and V. Valencia, 1996. The spawning environment of the Bay of Biscay anchovy (*Engraulis encrasicolus* L.). *Sci. Mar.,* **60 (2),** 140–177.

Olaso, I., 1990. Distribución y abundancia del megabentos invertebrado en fondos de la plataforma cantábrica. *Publ. Esp. Inst. Esp. Oceanogr.,* **5,** 128 pp.

Olaso, I., F. Sánchez, C. Rodríguez-Cabello and F. Velasco, 2002. The feeding behaviour of some demersal fish species in response to artificial discarding. *Sci. Mar,* **66 (3):** 301–311.

OSPAR Commission, 2000. Quality Status Report 2000: Region IV - Bay of Biscay and Iberian Coast. Ospar Commission. London. 134 + xiii pp.

Paillet, J. and H. Mercier, 1997. An inverse model of the eastern North Atlantic general circulation and thermocline ventilation. *Deep Sea Res. I,* **44,** 1293–1328.

Paillet, J., M. Arhan and M. McCartney, 1998. The spreading of Labrador Sea Water in the eastern North Atlantic. *J. Geophys. Res.,* **103,** 10223–10239.

Paillet, J., B. Le Cann, A. Serpette, Y. Morel and X. Carton, 1999. Real-time tracking of a Galician Meddy. *Geophys. Res.,* **26 (13),** 1877–1880.

Paillet, J., B. Le Cann, X. Carton, Y. Morel and A. Serpette, 2002. Dynamics and evolution of a northern Meddy. *J. Phys. Oceanogr.*, **32,** 55–79.

Parra, S. and López-Jamar, E., 1997. Cambios en el ciclo temporal de algunas especies infaunales como consecuencia del vertido del petrolero "Aegean Sea". *Publ. Espec. Inst. Esp. Oceanogr.*, **23,** 71–82.

Perez F. F. C. Mourinio, F. Fraga and A. F. Rios, 1993. Displacement of water masses and remineralization rates off the Iberian Peninsula by nutrient anomalies. J. Mar. Res. **51,** 869–892.

Pérez, F.F., A.F. Ríos, B.A. King and R.T. Pollard, 1995. Decadal changes of Θ-S relationships of the Eastern North Atlantic Central Water. *Deep Sea Res. I,* **42(11/12):** 1849–1864.

Pérez, F.F., R.T. Pollard, J.F. Read, V. Valencia, J.M. Cabanas and A.F. Rios, 2000. Climatological coupling of the thermohaline decadal changes in the Central Water of the Eastern North Atlantic. *Sci. Mar.*, **64,** 347–353.

Pingree, R.D., 1973 A component of Labrador Sea Water in the Bay of Biscay. *Limnol. Oceanogr.*, **18,** 711–718.

Pingree, R.D., 1984. Some applications of remote sensing to studies in the Bay of Biscay, Celtic sea and English Channel. Remote sensing of the shelf sea hydrodynamics, Proceeding of the 15th International Liege Collocium on Ocean Hydrodynamics. *Elsevier Oceanography series* **38,** 285–315.

Pingree, R.D., 1993. Flow of surface waters to the west of the British Isles and in the Bay of Biscay. *Deep Sea Res. II,* **40 (12).** 369–388.

Pingree, R.D., 1994. Winter warming in the Southern Bay of Biscay and Lagrangian eddy kinematic from a deep-drogued Argos buoy. *J. Mar. Biol. Ass. U. K.*, **74,** 106–128.

Pingree, R.D., 1997. The eastern Subtropical Gyre (North Atlantic): Flow Rings Recirculations Structure and Subduction. *J. Mar. Biol. Ass. U.K.*, **78,** 351–376.

Pingree, R.D. and B. Le Cann, 1989. Celtic and Armorican slope and shelf residual currents. *Prog. Oceanogr.*, **23,** 303–338.

Pingree, R.D. and B. Le Cann, 1990. Structure, strength, and seasonality of the slope current in the Bay of Biscay region. *J. Mar. Biol. Ass. U.K.*, **70,** 857–885.

Pingree, R.D., B. Le Cann, 1992a. Three anticyclonic Slope Water Oceanic eDDIES (SWODDIES) in the southern Bay of Biscay in 1990. *Deep Sea Res.*, **39,** 1147–1175.

Pingree, R.D. and B. Le Cann, 1992b. Anticyclonic eddy X91 in the southern Bay of Biscay, May 1991 to February 1992. *J. Geophys. Res.*, **97 (C9),** 14353–14367.

Pingree, R.D. and A.L. New, 1991. Abyssal penetration and bottom reflection of internal tidal energy in the Bay of Biscay. *J. Phys. Oceanogr.*, **21,** 28–39.

Pingree, R.D. and A.L. New, 1995. Structure, seasonal development and sunglint spatial coherence of the internal tide on the Celtic and Armorican shelves in the Bay of Biscay. *Deep Sea Res. I,* **42,** 245–284.

Pingree, R.D., G.D. Mardell, P.M. Holligan, D.K. Griffiths and J. Smithers, 1982. Celtic Sea and Armorican current structure and the vertical distributions of temperature and chlorophyll. *Cont. Shelf Res.*, **1 (1),** 99–116.

Pingree, R.D., G.D. Mardell and A.L. New, 1986. Propagation of internal tides from the upper slopes of the Bay of Biscay. *Nature,* **321,** 154–158.

Pingree, R.D., B. Sinha and C.R. Griffiths, 1999. Seasonality of the European slope current (Goban Spur) and ocean margin exchange. *C. Shelf Res.*, **19,** 929–975.

Piraud, I., Marseleix, P. and F. Auclair, 2003. Tidal and thermohaline circulation in the Bay of Biscay. *Geophys. Res. Abstracts,* **5,** 07058.

Planque B., P. Beillois, A.M. Jegou, P. Lazure, P. Petitgas and I. Puillat, 2003. Large scale hydrodynamic variability. 1990s in the context of the interdecadal changes. *ICES mar. Sci. Symp.*, **219:** 61–70.

Pollard, R.T. and S. Pu, 1985. Structure and circulation of the upper Atlantic Ocean northeast of the Azores. *Prog. Oceanogr.*, **14,** 443–462.

Pollard, R.T., M.J. Griffiths, S.A. Cunningham, J.F. Reid, F.F. Perez and A. Rios, 1996. Vivaldi-1991, A study of the formation, circulation and ventilation of eastern North Atlantic Central Water. *Prog. Oceanogr.,* **37,** 167–192.

Poulard, J.Ch., 2001. Distribution of hake (*Merluccius merluccius,* Linnaeus, 1758) in the Bay of Biscay and the Celtic sea from the analysis of French commercial data. *Fish. Res.,* **50,** 173–187.

Poulard, J.Ch., F. Blanchard, J. Boucher and S. Souissi, 2003. Variability in the demersal fish assemblages of the Bay of Biscay during the 1990s. *ICES mar. Sci. Symp.,* **219,** 411–414.

Poulet, S.A., M. Laabir and Y. Chaudron, 1996. Characteristic features of secondary production in the Gulf of Biscay. *Sci. Mar.,* **60 (2),** 79–95.

Prego, R. and J. Vergara, 1998. Nutrients fluxes to the Bay of Biscay from Cantabrian rivers (Spain). *Oceanol. Acta,* **21 (2),** 271–278.

Prouzet, P., K. Metuzals and C. Caboche, 1994. L'anchois du golfe de Gascogne. Caractéristiques biologiques et campagne de pêche française en 1992. Rapport CNPM-IMA-IFREMER.

Puillat I., P. Lazure, A.M. Jégou, B. Planque and L. Lampert, 2003. Mesoscale, interannual and seasonal hydrological variability over the French continental shelf of the Bay of Biscay during the 1990s. *ICES mar. Sci. Symp.,* **219,** 333–336.

Puillat I., P. Lazure, A.M. Jégou, L. Lampert and P.I. Miller, 2004. Hydrographical variability on the French continental shelf in the Bay of Biscay, during the 1990's. *Cont. Shelf Res.* **24:**1143–1164

Quéro, J.C., 1984. Cetorhinidae. In *Fishes of the North-eastern Atlantic and the Mediterranean.* Volume I. Ed. P.J.P. Whitehead, M.L Bauchot, J.C. Hureau, J. Nielsen and E. Tortonese, 1984. Unesco, 89–90.

Quéro J.C., J. Dardignac and J.J. Vayne, 1989. *Les poissons du golfe de Gascogne.* IFREMER. Brest: 229 pp.

Quéro, J.C., M.H. Du Buit and J.J. Vayne, 1997. Les captures de poissons à affinités tropicales le long des côtes atlantiques européennes. *Ann. Soc. Sci. Nat. Charente maritime,* **8 (6):** 651–673.

Quéro, J.C., M.H. Du Buit and J.J. Vayne, 1998. Les observations de poissons tropicaux et le réchauffement des eaux dans l'Atlantique européen. *Oceanol. Acta,* **21 (2),** 345–351.

Quincoces, I., M. Santurtún and P. Lucio, 1998. Biological aspects of white anglerfish (*Lophius piscatorius*) in the Bay of Biscay (ICES Divisions VIIIa, b and d) in 1996–1997. ICES CM 1998/O:48, 29 pp.

Reid, D.G., 2001. SEFOS: Shelf Edge Fisheries and Oceanography Studies: an overview. *Fish. Res.,* **50,** 1–15.

Reid, J.L., 1994. On the total geostrophic circulation of the North Atlantic Ocean: Flow pattern, tracers and transports. *Prog. Oceanogr.,* **33,** 1–92.

Revilla, R., E. Nogueira, X.A.G. Morán, L. Valdés, C. González-Pola, J.M. Rodríguez, T. Smyth and J. Cabal, 2002. Surface chlorophyll-a estimates in the central Cantabrian sea (Cape Peñas area) from SeaWIFS data 1998–2000. 8th International Symposium on Oceanography of the Bay of Biscay, Gijón (April 2002).

Ríos, A.F., F.F. Pérez and F. Fraga, 1992. Water masses in the upper and middle North Atlantic Ocean east of the Azores. *Deep Sea Res.,* **39,** 645–658.

Sánchez, F. 1993. Las comunidades de peces de la plataforma del Cantábrico. *Publ. Espec. Inst. Esp. Oceanogr.,* **13,** 137 pp.

Sánchez, F. and J. Gil, 2000. Hydrographic mesoscale structures and Poleward Current as a determinant of hake (*Merluccius merluccius*) recruitment in southern Bay of Biscay. *ICES J. Mar. Sci.,* **57,** 152–170.

Sánchez, F. and I. Olaso, 2004. Effects of fisheries on the Cantabrian Sea shelf ecosystem. *Ecological Modelling* **172** (2004), 151–174.

Sánchez, F. and A. Serrano, 2003. Variability of groundfish communities of the Cantabrian Sea during the 1990s. *ICES mar. Sci. Symp.,* **219,** 249–260.

Sánchez, F., N. Pérez and J. Landa, 1998. Distribution and abundance of megrim (*L. boscii* and *L. whiffiagonis*) on the northern Spanish shelf. *ICES J. Mar. Sci.* **55,** 494–514.

Sánchez, F., J. Gil, R. Sánchez, J.C. Mahé and Ph. Moguedet, 2001. Links between demersal species distribution pattern and hydrographic structures in the Bay of Biscay and Celtic Sea. In *Océanographie du golfe de Gascogne. VII° Colloq. Int., Biarritz, 4–6 avril 2000. Ed. IFREMER, Actes Colloq.,* **31,** 173–180.

Sánchez, F., M. Blanco and R. Gancedo, 2002. *Atlas de los peces demersales y de los invertebrados de interés comercial de Galicia y el Cantábrico, Otoño 1997–1999.* Ed. CYAN, Madrid, 158 pp.

Sánchez, R., F. Sánchez and J. Gil, 2003. The optimal environmental window that controls hake (*Merluccius merluccius*) recruitment in the Cantabrian Sea. *ICES mar. Sci. Symp.,* **219,** 415–417.

Saunders, P.M., 1982. Circulation in the Eastern North Atlantic. *J. Mar. Res.,* **40,** 641–657.

Schopp, R., 1993. Effets des frontières, structures frontales et circulation générale dans l'Atlantique Nord Est. Rapport Tech, LPO. Etude théorique, rapport scientifique Dyane, Volet 2.

Serpette, A., B. Le Cann and F. Colas, (in press). North Atlantic Central Water Lagrangian Circulation over the abyssal plain and continental slopes of the Bay of Biscay: description of selected mesoscale features. *Sci. Mar.*

Serret, P., E. Fernández, J.A. Sostres and R. Anadón, 1999. Seasonal compensation of microbial production and respiration in a temperate sea. *Mar. Ecol. Prog. Ser.,* **187:**43–57.

SESITS, 2000. Evaluation of demersal resources of Southwesthern Europe from standardised groundfish surveys. *Final Report to the Commission of European Communities,* 195 pp.

Smetacek, V., B. von Bodungen, B. Knoppers, R. Peinert, F. Pollehne, P. Stegmann and B. Zeitzschel, 1984. Seasonal stages characterizing the annual cycle of an inshore pelagic system. *Rapp. P.-v. Réun. Cons. int. Explor. Mer,* **183,** 126–135.

Solá, A., C. Franco, A. Lago de Lanzós and L. Motos, 1992. Temporal evolution of *Sardina pilchardus* (Walb.) spawning on the N-NW coast of the Iberian Peninsula. *Bol. Inst. Esp. Oceanogr.,* **8 (1),** 123–138.

Souissi, S., F. Ibanez, R. Ben Hamadou J. Boucher, A.C Cathelineau, F. Blanchard, and J.C. Poulard, 2001. A new multivariate mapping method for studying species assemblages and their habitats: example using bottom trawl surveys in the Bay of Biscay (France). *Sarsia,* **86,** 527–542.

Southward, A. J. and G.T. Boalch, 1994. The effect of changing climate on marine life: Past events and future predictions. *Exeter Maritime Studies,* **9,** 101–143.

Southward, A. J., S. H. Hawkins and M.T. Burrows, 1995. Seventy years' observations of changes in distribution and abundance of zooplankton and intertidal organisms in the western English Channel in relation to rising sea temperature. *J. thermal Biol.,* **20 (1/2),** 127–155.

Stebbing, A. R. D., S. M. T. Turk, A. Wheeler and K. R. Clarke, 2002. Immigration of southern fish species to south-west England linked to warming of the North Atlantic (1960–2000). *J. Mar. Biol. Ass. U.K.,* **82,** 177–180.

Taylor A. H. and J. A. Stephens, 1998. The North Atlantic Oscillation and the latitude of the Gulf Stream. *Tellus,* **50A,** 134–142.

Teira, E., J. Abalde, M.T. Alvarez- Osorio, A. Bode, C. Cariño, A. Cid, E. Fernández, N. González, J. Lorenzo, J. Valencia and M. Varela, 2003. Plankton carbon budget in a coastal wind driven upwelling station off A Coruña (NW Iberian Peninsula). *Mar. Ecol. Prog. Ser.,* **265:** 31–43.

Tenore, K.R., M .T. Alvarez-Ossorio, L.P. Atkinson, J.M. Cabanas, R.M. Cal, M.J. Campos, F. Castillejo, E.J. Chesney, N. González, R.B. Hanson, C.R. McLain, A. Miranda, M. Noval, M.R. Roman, J. Sánchez, G. Santiago, L. Valdés, M. Varela and J. Yoder, 1995. Fisheries and Oceanography off Galicia. NW Spain (FOG): Mesoscale spatial and temporal changes in physical processes and resultant patterns of biological productivity. *J. Geophys. Res.,* **100**(C6): 10943–10966.

Tréger, P., P. Le Corre and J.R. Grall, 1979. The seasonal variations of nutrients in the upper waters of the Bay of Biscay region and the relation to phytoplankton growth. *Deep Sea Res. II,* **26a:** 1121–1152.

Ulanowicz, R.E. and C.J. Puccia, 1990. Mixed trophic impacts in ecosystems. *Coenoses,* **5 (1),** 7–16.

Uriarte, A., P. Prouzet and B. Villamor, 1996. Bay of Biscay and Ibero Atlantic anchovy populations and their fisheries. *Sci. Mar.*, **60 (2),** 237–255.

Uriarte, A., P. Alvarez, S. Iversen, J. Molloy, B. Villamor, M.M. Martins and S. Myklevoll, 2001. Spatial pattern of migration and recruitment of north-east Atlantic mackerel. ICES CM 2001/O:17.

Valdés, L. and M.T. Alvarez-Ossorio, 1996. Tipificación de los copépodos pelágicos de la costa de Lugo. *Thalassas,* **12,** 9–17.

Valdés, L. and M. Moral, 1998. Time series analysis of copepod diversity and species richness in the Southern Bay of Biscay (Santander, Spain) and their relationships with environmental conditions. *ICES J. Mar. Sci.,* **55,** 783–792.

Valdés, L. and A. Lavín, 2002. Dynamics and human impact in the Bay of Biscay: An ecological perspective. In *Large Marine Ecosystems of the North Atlantic: Changing States and Sustainability.* K. Shermann and H.R. Skjoldal (ed.). Elsevier Science B.V., Amsterdam, 293–320.

Valdés, L., M.T. Alvarez-Ossorio, A. Lavín, M. Varela and R. Carballo, 1991. Ciclo anual de parámetros hidrográficos, nutrientes y plancton en la plataforma continental de La Coruña (NO, España). *Bol. Inst. Esp. Oceanog.,* **7 (1),** 91–138.

Valdés, L. (chapter co-ordinator) et 72 co-authors, 2000. Biology. In *Quality Status Repot 2000: Bay of Biscay and Iberian Coast.* OSPAR Commission 2000, Redhouse L. Communications Ltd, London, Vol IV, 83–112.

Valdés, L., G. Beaugrand, R. Harris, X. Irigoen, A. López-Urrutia, M. Moral and B. Planque, 2001. Seasonal dynamics and year-to-year variability (1991–1999) of *Calanus helgolandicus* and *Acartia clausi* in the Bay of Biscay and Celtic Sea. ICES Symposium on Hydrobiological variability in the ICES Area, 1990–1999. Edinburg (UK), (August 2001).

Valdés, L., A. Lavín, M.L. Fernández de Puelles, M. Varela, R. Anadón, A. Miranda, J. Camiñas and J. Mas, 2002. Spanish Ocean Observation System. IEO Core Project: Studies on time series of oceanographic data. In *Operational Oceanography: Implementation at the European and Regional Scales.* N.C. Fleming et al. (eds.), Elsevier Science B.V, Amsterdam, 99–105.

Valencia, V., A. Borja, A. Fontán, F.F. Pérez and A. F. Rios, 2003. Temperature and salinity fluctuations in the Basque Coast (Southeastern Bay of Biscay), from 1986 to 2000, related to the climatic factors. *ICES mar. Sci. Symp.,* **219,** 340–342.

Valencia, V., J. Franco, A. Borja and A. Fontán, 2004. Hydrography of the southeastern Bay of Biscay. In *Oceanography and Marine Environment of the Basque Country.* A. Borja and M. Collins (Eds.), *Elsevier Oceanography Series,* Elsevier, Amsterdam, 159–194.

van Aken, H.M., 2000a. The hydrography of the mid-latitude northeast Atlantic Ocean: I. The deep water masses. *Deep Sea Res. I,* **47,** 757–788.

van Aken, H.M., 2000b. The hydrography of the mid-latitude northeast Atlantic Ocean: II. The intermediate water masses. *Deep Sea Res. I,* **47,** 789–824.

van Aken, H.M., 2001. The hydrography of the mid-latitude northeast Atlantic Ocean: III. The thermocline water mass. *Deep Sea Res. I,* **48,** 237–267.

van Aken, H.M., 2002. Surface currents in the Bay of Biscay as observed with drifters between 1995 and 1999. *Deep Sea Res. I,* **49,** 1071–1086.

Vanney, J.R., 2002. *Géographie de l'océan global.* Collection Géosciences. Editions Scientifiques GB, CPI, Paris, 335 pp + VIII planches.

Varela, M., 1996. Phytoplankton ecology in the Bay of Biscay. *Sci. Mar.,* **60 (2),** 45–53.

Varela, M., S. Barquero, B. Casas, M.T. Alvarez-Ossorio and A. Bode, 1995. Are variations in salinity of north atlantic central waters affecting plankton biomass and production in NW Spain. *First JGOFS International Scientific Symposium.* Villefranche-sur-Mer, France.

Varela, M., A. Bode and N. González, 1998. Intrusiones de agua Nortatlántica de origen subtropical y fitoplancton durante la primavera en el Mar Cantábrico. *VI Coloquio Internacional de Oceanografía del Golfo de Vizcaya.* San Sebastián.

Varela, M., S. Barquero, A. Bode, E. Fernández, N. González, E. Teira and M. Varela, 2003. Microplanktonic regeneration of ammonium and disolved organig nitrogen in the upwelling area of the NW of Spain: relationships with dissolved organic carbon production and phytoplankton size-structure. *J. Plankton Res.,* **25 (7):**719–736.

Velasco, F. and I. Olaso, 1998. European hake *Merluccius merluccius* (L., 1758) feeding in the Cantabrian Sea: seasonal, bathymetric and length variations. *Fish. Res.,* **38,** 33–34.

Vincent, A. and G. Kurc, 1969. Hydrologie, variations saisonnières de la situation thermique du Golfe de Gascogne en 1967. Rev. *Trav. Inst. Pêches marit.,* **33 (1),** 79–96.

Villamor, B., P. Abaunza, P. Lucio and C. Porteiro, 1997. Distribution and age structure of mackerel (*Scomber scombrus,* L.) and horse mackerel (*Trachurus trachurus,* L.) in the northern coast of Spain, 1989–1994. *Sci. Mar.,* **61 (3),** 345–366.

Villate, F., M. Moral and V. Valencia, 1997. Mesozooplankton community indicates climate changes in a shelf area of the inner Bay of Biscay throughout 1988 to 1990. *J. Plankton Res.,* **19 (11),** 1617–1636.

Wyatt, T. and C. Porteiro, 2002. Iberian sardine fisheries: trends and crises. In *Large Marine Ecosystems of the North Atlantic: Changing States and Sustainability.* K. Sherman and H.R. Skjoldal (ed.). Elsevier Science, B.V., Amsterdam, 321–338.

Chapter 25. INTERDISCIPLINARY STUDIES IN THE CELTIC SEAS (19,E)

JONATHAN SHARPLES

Proudman Oceanographic Laboratory

PATRICK M. HOLLIGAN

Southampton Oceanography Centre

Contents

1. Introduction—the physical and sediment environment.

The Celtic seas comprise the English Channel, the Celtic Sea and the Irish Sea, the shelf west of Ireland and north to the Malin shelf and the Hebrides shelf west of Scotland (Fig. 25.1a). The region is bounded to the southwest and west by a shelf edge at a typical depth of 200 metres, running northward from the northern Bay of Biscay, to the southwest approaches of the English Channel, west of Ireland, along the Northwest Scottish coast, and turning to the east into the northern North Sea. Coastal regions open to the shelf edge occur off the west coast of Scotland (the Malin and Hebrides shelves), west of Ireland, and the Celtic Sea southwest of the English Channel. Along with the adjacent North Sea, this part of the world's coastal oceans have been studied extensively, and used to develop much of our basic understanding of the physics of shelf sea environments, and the links between the physical environment and the biochemistry and sediment dynamics. Much of the early work in this area was conducted in the context of a rapidly developing fisheries industry through the 20th Century. Latterly oceanographic research has been carried out in response to the demands placed on the Celtic Seas by the high population densities of Northwest Europe, with often incompatible pressures on the environment arising from fishing, the use of the coastal seas to carry away and dilute industrial and urban wastes, offshore hydrocarbon extraction

The Sea, Volume 14, edited by Allan R. Robinson and Kenneth H. Brink
ISBN 0–674–

and (more recently) the development of offshore renewable energy sources, and the leisure activities of the large population. The research questions associated with these conflicting demands have often led to an inherent need for oceanographic work to be cross-disciplinary.

Mean flows within the Celtic Sea are typically 0.005–0.015 m s^{-1} (Thompson and Pugh, 1986), while a density-driven jet drives water northward through St. Georges Channel at an average of 0.13 m s^{-1}, with maximum speeds reaching almost 0.3 m s^{-1} (Horsburgh et al., 1998). Flow through the Irish Sea and the North Channel has complex depth and lateral variability, but results in an observed net northward flux through the North Channel of 0.08×10^6 m^3 s^{-1} (Knight and Howarth, 1999). A residence time for water in the Irish Sea has been calculated to be about 1 year (Knight and Howarth, 1999). Closer to the coasts mean flows become dominated by the circulation associated with the freshwater-driven horizontal density gradients. A south-westward coastal buoyancy current is thought to flow along the east Irish coast during autumn through to early spring, as a result of a narrow coastal band of low salinity water. The mean flow becomes affected by a cyclonic gyre circulation during late spring and summer (Horsburgh et al., 2000), as thermal stratification develops over a deep region of the western Irish Sea. A freshwater-driven coastal current also flows along the coast of NW Scotland with speeds of about 5 km day^{-1} (Hill et al., 1997). Mean flows at the shelf edge are dominated by a continuous northward flow of slope water, with speeds of typically 0.2 m s^{-1} (Souza et al., 2001).

Over most of the region currents are dominated by the tides (Fig. 25.1b). On the Malin shelf the dominant M_2 barotropic tidal amplitude is typically 0.1–0.2 m s^{-1}, with internal tidal currents of similar size. In the Celtic Sea the M_2 tidal amplitude ranges from 0.4 m s^{-1} near the shelf edge, again with significant internal tidal currents superimposed, to 1 m s^{-1} within St. Georges Channel and around the Ushant peninsular of NW France. Tidal wave co-oscillation within the English Channel and Irish Sea results in M_2 current amplitudes of 1–1.5 m s^{-1}. The dissipation of tidal energy within the English Channel and Irish Sea is very high, with both damping the energy of the tidal wave so much that a degenerate amphidromic system is set up.

The pattern of tidal currents is a key factor in determining the seasonal structure of the overlying water column, to the extent that much of the physical oceanographic structure of the region is dominated by the effects of tidal stirring (Simpson, 1998). In the deeper regions with weaker tidal currents, such as the Celtic Sea, tidal stirring in summer is not able to redistribute the surface heat input. As a result the water column becomes stratified, with a warm, wind-mixed surface layer separated from a cold, tidally-mixed bottom layer by a thermocline (e.g. Simpson and Bowers, 1984). Seasonally these areas start the year vertically mixed, as a result of convective heat losses. Around the vernal equinox, with solar elevation increasing, the net heat flux switches to warm the sea surface. Initially the tidal mixing is still able to keep redistributing the heat. Eventually the increasing solar elevation results in a net heat supply to the sea surface that overcomes the ability of the tidal stirring to maintain vertical homogeneity, and the water column stratifies. Vertical homogeneity returns in autumn via a combination of convective overturning, and tidal and wind mixing. Within regions of stronger tidal currents, and/or shallower depth (e.g. the Irish Sea), tidal stirring is always able to overcome the rate of solar heating and the water column remains mixed throughout the year, though with a seasonally-varying temperature.

A summer satellite image of sea surface temperature illustrates this physical partitioning of the shelf seas in response to the competition between tidal stirring and surface heating (Fig. 25.1d). The shelf sea tidal mixing fronts clearly separate seasonally stratified (warm) and permanently mixed (cool) regions. An indication of the strength of the control that tidal mixing has on the regional oceanography is that the persistent structures seen in the satellite image can be described solely by a consideration of the energetics of tidal mixing and surface heating, leading to the successful prediction that the shelf sea fronts should follow contours of a critical value of depth/(tidal current amplitude)^3 (Simpson and Hunter, 1974; Simpson and Sharples, 1994). Both observational and modelling work has shown that the frontal positions oscillate, typically by about 2 – 4 km, in response to the spring-neap modulation of tidal currents (Simpson and Bowers, 1981; Sharples and Simpson, 1996). Similar frontal features are formed around islands within stratified regions of shelf seas, as the tidal flow is accelerated around the flanks (Simpson et al., 1982).

The marked control exerted by tidal stirring on much of the environment of the Celtic Seas is almost unique in the world's coastal oceans. Similar analyses have been carried out, for instance, off Georges Bank (Loder et al., 1992), in Long Island Sound (Bowman et al., 1981), and in Cook Strait, New Zealand (Bowman et al., 1983). Of all these examples, the Celtic Seas probably demonstrate the greatest influence of tidal mixing, with processes such as wind-driven mixing and mean flows modifying to only a small extent the predictions of the simple heating-stirring model of Simpson and Hunter, 1974. Arguably the seasonally consistent, largely predictable patterns of shelf sea environments in the Celtic Seas makes them ideal regions for the design of cross-disciplinary process studies.

In many areas close to the coast the tidal currents compete with the lateral input of freshwater buoyancy to determine the water column structure. On the Malin shelf this leads to a departure of the position of the Islay front away from the prediction of Simpson and Hunter, 1974, (Hill and Simpson, 1989). In the eastern Irish Sea (Liverpool Bay), the semi-diurnal tidal cycle causes stratification of the water column during each ebb tide, while the spring-neap modulation of tidal mixing leads to the development of more lasting mean stratification at neap tides (Simpson et al., 1990; Sharples and Simpson, 1995).

Water column structure at the Malin and Celtic Sea shelf edges is significantly affected by strong mixing associated with internal tidal wave dissipation (Sherwin, 1981; Pingree et al., 1982). Regions of cooler water lying along the shelf edge of the Celtic Sea are often clearly visible in satellite SST imagery, caused by the breaking of the internal tidal wave and energetic mixing by internal solitons.

Away from regions of coastal erosion and fluvial inputs, the seabed sediments of the Celtic seas have their origins during the last glacial period. Sediment type reflects the distribution of tidal stress (Fig. 25.1c), with the coarsest sediments (gravel/sand) associated with the strongest tidal currents in St. Georges Channel, the central Irish Sea, the English Channel, and off northwest France. In the weaker tidal regions of the Celtic Sea and on the Malin shelf the seabed is made up of mud/sand, and mud over the shelf edge. Reworking of these sediments, and the generation/movement of large bedforms, occurs in response to tidal flows (particularly during spring tides) and the superposition of storm waves. A large depositional centre lies in the western Irish Sea, where the deeper water and weaker tidal currents have allowed the accumulation of about 33 metres of holocene mud. Sea-bed bedforms have been observed at the shelf edge and slope consistent with the peaks in observed slope current speeds (Kenyon, 1986). Also at the shelf edge, stresses due to the internal tide and internal waves have also been observed to play a role in sediment resuspension and transport (Heathershaw et al., 1987), suggesting that the shelf edge is a region of bedload parting.

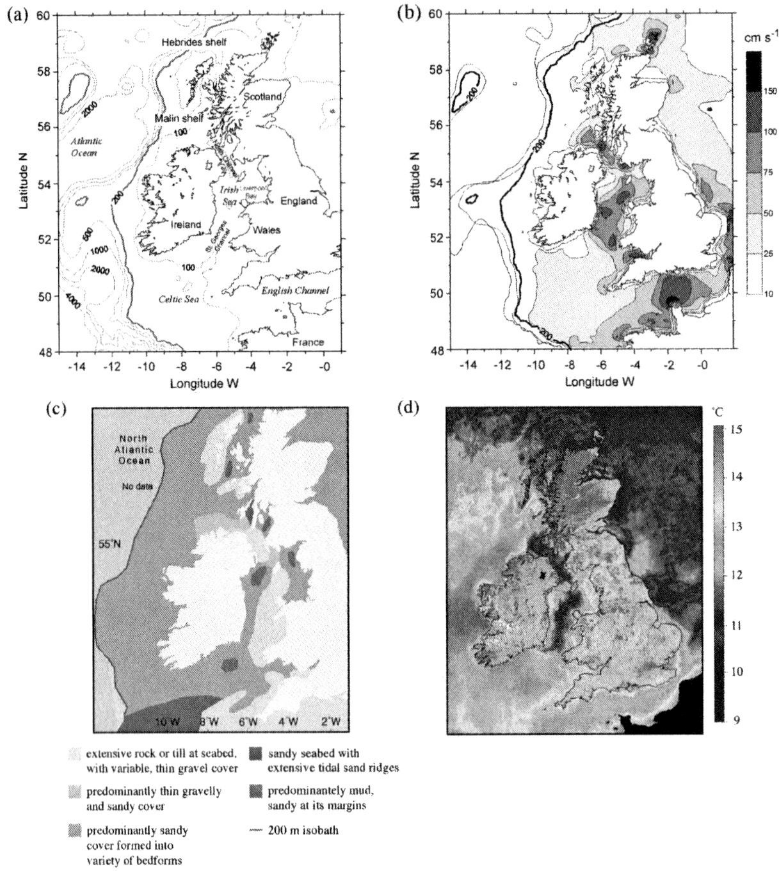

Figure 25.1
(a) Bathymetry of the Celtic seas.
(b) Numerical model prediction of the M_2 tidal current amplitude.
(c) Bottom sediment type.
(d) A composite sea surface temperature image for June 1997.
Data in (a) and (b) were supplied by Sarah Wakelin, Proudman Oceanographic Laboratory, UK. Sediment distribution in (c) after OSPAR Commission 2000. Image (d) courtesy of Peter Miller, Remote Sensing Group, Plymouth Marine Laboratory, UK.

Resuspension and shoreward transport of coastal sediments by wave action has received particular attention in the Irish Sea during investigations of transport pathways for radioisotopes released from the Sellafield nuclear power and reprocessing plant in Northwest England (e.g. Cook et al., 1997; MacKenzie et al., 1998).

Remote sensing using the AVHRR sensor has been used successfully to investigate the time scales of sediment resuspension variability over the Irish Sea, indicating a strong seasonal cycle (maximum suspended sediments in January/February) and a small but significant spring-neap cycle (maximum concentrations during spring tides) (Bowers et al., 1998).

2. The Open Shelf.

In the open shelf seas autotrophic phytoplanton represent the base of the food chain for both pelagic and benthic heterotrophs. Mixing processes in the water column, driven largely by bottom tidal flow and by surface wind and convection, determine the light environment for photosynthetic organisms (Fig. 25.2). Apart from seasonal changes in solar irradiance, the key parameters are depth of the surface mixed layer, h, which for well mixed water is also the depth of the water column, and the diffuse attenuation for photosynthetically active radiation, k. Values for k are proportional to loads of particulate (sediment, coccoliths and organic matter including phytoplankton) and dissolved (mainly organic matter) materials in the water. In general light availability is proportional to $(kh)^{-1}$ (Pingree *et al.,* 1986; Gowen *et al.,* 1995), and when a critical value of $(kh)^{-1}$ is reached there is net growth of the phytoplankton. For waters that become stratified in the summer (Fig. 25.2a), the onset of the spring diatom bloom generally corresponds to the time of the establishment of the seasonal thermocline when both the depth of the mixed layer and the load of particulate material derived from tidal suspension of bottom sediments are significantly reduced. Conversely, in the autumn, phytoplankton growth suddenly becomes light limited when the seasonal thermocline is eroded (Pingree *et al.,* 1976). A small autumn bloom of phytoplankton is sometimes seen just prior to the vertical homogenisation of the water column (Fig. 25.2a), as the deepening thermocline entrains bottom water nutrients into the surface layer. For mixed waters (Fig. 25.2b) the timing of the onset of the spring bloom depends largely on the depth and turbidity of the water, and is generally between late February and mid-May.

The nutrient environment for phytoplankton is largely determined by the late winter stocks of dissolved nitrate, phosphate and silicate, and by regenerative processes in both the water column (e.g. Maguer *et al.,* 1999) and bottom sediments (e.g. Trimmer *et al.,* 1999) which return essential elements in dissolved organic or inorganic forms to surface waters. The winter nutrient levels reflect net inputs from the ocean, the land (e.g. Gowen *et al.,* 2000) and (to a minor extent) the atmosphere. They are generally highest in outer shelf waters, especially in the north, as a result of advection from the N Atlantic, and in inshore waters close to sources of nutrient-rich fresh water. In summer the surface layer of stratified waters tends to become nutrient depleted (e.g. Fig. 25.2a), while nutrient concentrations in mixed waters depend on the balance between the assimilation by phytoplankton and the rates of replenishment.

As nutrients are utilised in the spring and early summer, the nitrate-to-phosphate ratio falls to values much lower than the Redfield ratio, suggesting that phytoplankton become nitrogen limited at this time of year (Pingree *et al.,* 1977, Gibson *et al.,* 1997; Jordan and Joint, 1998). This pattern of nutrient depletion may be explained by some combination of slower recycling of nitrogen compared to phosphorus, net loss of nitrogen by denitrification (Gibson *et al.*, 1997), and/or net gain of phosphorus through mechanisms yet to be identified (Jordan and Joint, 1998). Forms of regenerated nitrogen utilised by phytoplankton include ammonium and urea derived from animal excretion and the microbial foodweb (Holligan *et al.*, 1984b; Le Corre *et al.*, 1993; Maguer *et al.*, 1998). A better understanding of the nature and control of regeneration processes in shelf seas is needed for predicting how future changes in external nutrient supply might affect pelagic productivity.

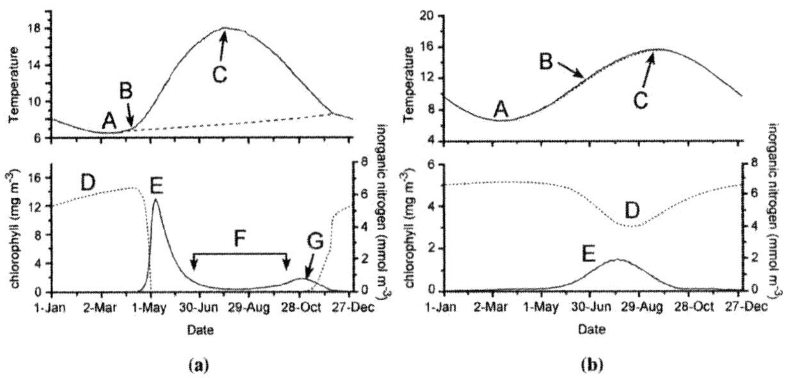

Figure 25.2 Model representation of the seasonal cycles of phytoplankton biomass in stratifying and permanently mixed water columns (Sharples, 1999). In each case the upper panel shows surface (solid line) and bottom (dashed line) water temperature, the lower panel shows surface chlorophyll concentration (solid line) and surface concentration of dissolved nitrate (dashed line).

(a) A station representative of the central Celtic Sea, with a depth of 120 metres and daily mean tidal current speed of 0.2 m s^{-1}. $kh^{-1} = 0.07$ during mixed conditions, and approximately 0.3 when the thermocline is developed. A: increasing solar irradiance begins to heat the water column in spring. B: stratification begins as tidal and wind mixing are unable to prevent the stabilising effect of the solar heating. C: maximum stratification reached in mid summer. D: high surface inorganic nitrogen in the mixed water column in winter and early spring. E: the spring bloom of surface biomass begins as soon as the water column stratifies, eventually limited by a lack of inorganic nitrogen as the thermocline prevents re-supply from the deeper water. F: a lack of nitrogen prevents significant growth in the surface water during summer, phytoplankton biomass at this time is concentrated deeper in the thermocline. G: convective deepening of the thermocline in autumn can bring deep nitrogen into the surface layer, allowing an autumn bloom of surface biomass.

(b) A station representative of the Irish Sea, with a depth of 80 metres and a daily mean tidal current speed of 0.65 m s^{-1}. $kh^{-1} = 0.1$. A: increasing solar irradiance begins to heat the water column in spring. B: the water column remains vertically homogeneous throughout summer, as the tidal and wind mixing prevent stratification developing. C: maximum surface temperature reached in late summer. D: surface nitrate remains high throughout the year, as long as re-supply from regeneration in the sediments and water column is equivalent to assimilation by phytoplankton. E: growth of phytoplankton is possible in summer during high solar irradiance if the phytoplankton critical depth is greater than the depth of the water column.

In stratified waters changes in nutrient levels below the thermocline during summer reflect the balance between gains from regeneration and losses due to upward mixing across the thermocline. In areas of relatively high vertical stability corresponding to seasonal gyres (Hill, 1993), bottom nutrient concentrations in late summer and autumn can exceed by a considerable amount water column values for late winter (Pingree *et al,* 1976), as a result of the oxidation of organic material that sinks through the seasonal thermocline. This sequestration process leads to a corresponding drop in dissolved oxygen below the thermocline (Morin *et al.,* 1985, Raine and McMahon, 1998). When affected by extreme or episodic mixing events (storms, spring tides etc.) such nutrient reservoirs are potential significant sources of nutrients to the surface waters, especially during late summer and autumn as the seasonal thermocline weakens. Attempts to derive such fluxes from satellite measurements of sea surface temperature (Morin *et al.,* 1993) are valid only for relatively deep water regimes in which consistent relationships between water temperature and nutrient concentration are maintained.

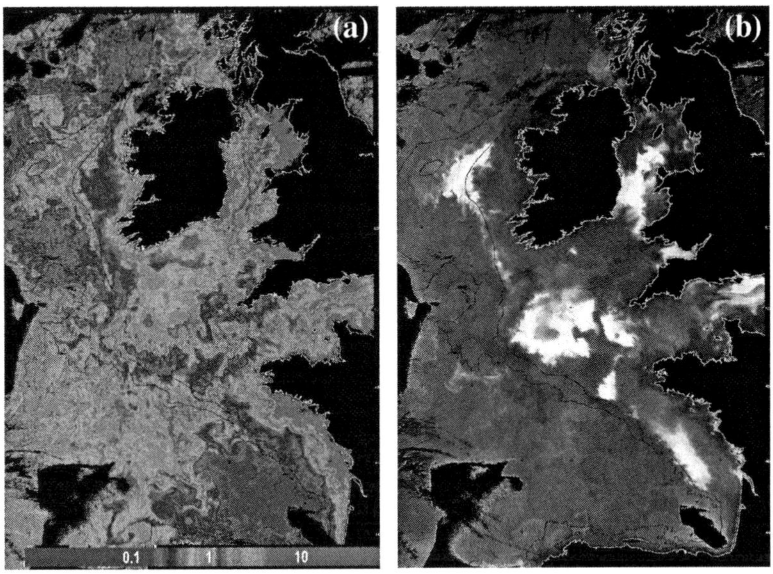

Figure 25.3 SeaWiFS images for the Celtic Seas, 18 May, 1998, supplied by S. Groom (Plymouth Marine Laboratory).

a) Chlorophyll image showing the distribution of surface phytoplankton in late spring about one month after the onset of stratification in the Celtic Sea.

b) Colour composite image showing strong reflectance due to extensive blooms of the coccolithophore *E. huxleyi* in the western Celtic Sea and to the west of Ireland, and due to suspended sediments in the regions of strong tidal mixing in the Irish Sea and the English Channel. Under both conditions algorithms for the determination of chlorophyll tend to break down so that the estimates of phytoplankton abundance in panel (a) for the high reflectance areas are likely to be in error.

Once the physical processes that determine the development of the seasonal thermocline across the northwest European shelf had been identified and parameterised (Pingree and Griffiths, 1978), it became possible to recognise clear patterns in the distribution and seasonal species succession of phytoplankton (Pingree *et al.,* 1976; Pingree *et al,* 1978, Richardson *et al.,* 1985). Small surface-to-bottom differences of water temperature (~0.5°C) in spring were shown to give sufficient surface stability for blooms of diatoms to develop, with concentrations of chlorophyll *a* up to 10 mg m^{-3} (Fasham *et al.,* 1983). During the summer months persistent chlorophyll maxima, with values locally higher than 50 mg m^{-3}, can be found within the seasonal thermocline and along tidal fronts at the boundary between nutrient limited surface stratified waters and light limited mixed waters (Pingree *et al.*, 1982). These blooms are dominated by dinoflagellates and, occasionally, coccolithophores. In the surface stratified waters of the open shelf, populations of the coccolithophore *Emiliania huxleyi* are commonly observed in early summer (Fig. 25.3) before levels of chlorophyll become low (<0.5 mg m^{-3} chlorophyll *a*) and dominated by picophytoplankton (Joint *et al.*, 1986). By contrast, mixed waters lying between the coast and offshore tidal fronts are characterised by persistent diatom populations, with the levels of chlorophyll generally in the range 1–2 mg m^{-3}, together with variable proportions of other types of phytoplankton dependent on the depth, clarity and nutrient status of the water (Pingree *et al.*, 1978; Holligan *et al.*, 1984a).

Investigations in the western Irish Sea have shown that annual primary production and the length of the productive season depends on the stratification regime, water turbidity and nutrient inputs from land (Gowen *et al.*, 1995). Similarly, in regions that remain well mixed through the year, such as the eastern English Channel, the development of phytoplankton populations in spring can be clearly linked to hydrodynamic conditions, in particular variations in water column irradiance and nutrient levels (Pingree *et al.*, 1986; Brunet *et al.*, 1996).

Growth of phytoplankton along surface tidal fronts and within the seasonal thermocline of the Celtic seas (Fig. 25.4) is affected by variations in isopycnal and diapycnal mixing and exchange processes related to the strength of tides and wind (Loder and Platt, 1985) and in the ability of phytoplankton to adjust physiologically to varying environmental conditions (Jordan and Joint, 1984; Holligan *et al.* 1984b; Videau, 1987; Moore *et al.*, 2003). Cell motility and buoyancy may be critical in maintaining the position of phytoplankton cells with respect to nutrient and light gradients, but their ecological significance remains poorly understood. The dynamic nature of such boundary environments has been investigated using numerical models (Sharples and Tett, 1994) and is illustrated by the observations of Sharples *et al.* (2001) who showed that subsurface chlorophyll maxima are eroded from below at times of strong tidal mixing. These phytoplankton populations are thought to be maintained by growth and accumulation of cells during neap tides, making use of nutrients that were mixed upwards during the preceding period of stronger mixing at spring tides.

One fundamental problem in understanding how the distributions of phytoplankton are controlled is a lack of knowledge of their physiology in relation to variations in irradiance and nutrient supply. Although this issue has been explored with numerical models (Mills *et al.*, 1997; Lizon *et al.*, 1998) it is only with new *in situ* sensors such as the Fast Repetition Rate Fluorometer (FRRF) that fully compatible data on cell physiology and water column turbulence can be gathered (Moore *et al.*, 2003). In general it appears hydrodynamic gradients are characterised by gradients in both species composition and physiological properties.

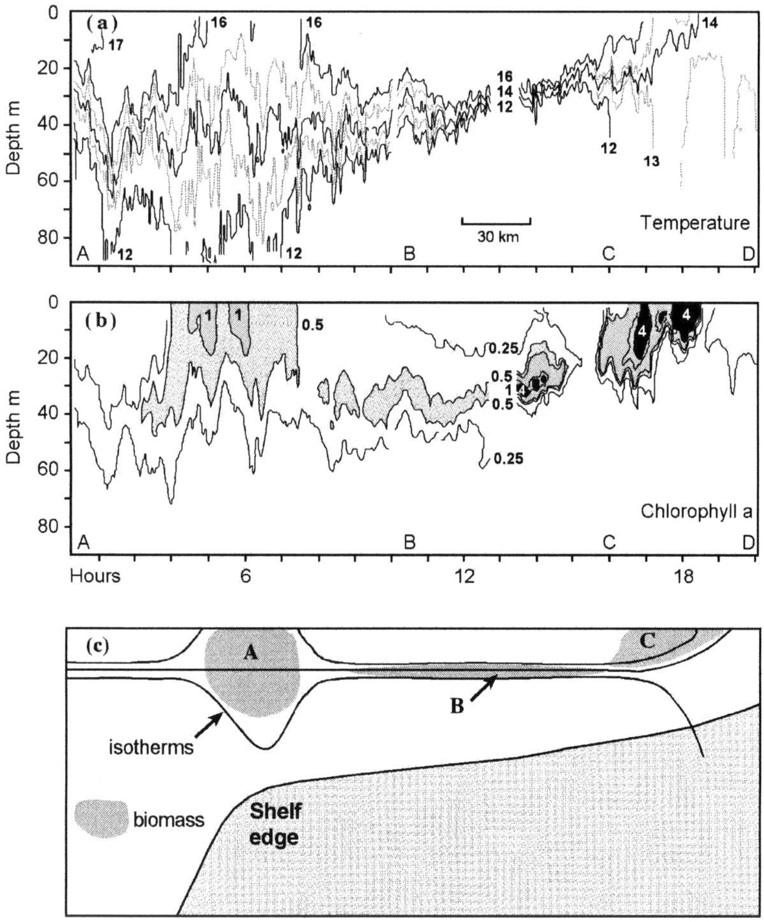

Figure 25.4 (a) temperature (°C) and (b) chlorophyll *a* (mg m^{-3}) sections made on 6 August, 1980 between mixed waters off France and across the edge of the continental shelf in the southwest Celtic Sea (from Pingree *et al,* 1982 courtesy of Elsevier, Ltd.). A is the open ocean offshore of the shelf edge, B is the open shelf sea, stratified during summer, C is the shelf sea tidal mixing front where stratification is broken down by mixing, and D is the permanently mixed, tidally energetic shelf water.
(c) Schematic illustration of the environments sampled during the section with A: Internal tidal mixing at the shelf edge supplying surface phytoplankton with nutrients. The surface signature of cool, nutrient-rich water and elevated chlorophyll concentrations is often seen in satellite imagery.
B: A strongly-stratified shelf sea, with phytoplankton biomass within the thermocline fuelled by weak mixing of nutrients from below.
C: The shelf sea tidal mixing front, with high phytoplankton biomass at the boundary between the light-limited (mixed) water and the nutrient-limited (stratified surface layer) water.

Comparative estimates of primary production in NW European shelf seas, based on standard methods of ^{14}C assimilation or oxygen production, range between ~100 and >250 g C m^{-2} y^{-1}. The lower values are representative of offshore areas with low tidal mixing, and the higher ones of inshore areas that receive significant inputs from land. However, a lack of consistency in experimental procedures and poor resolution of measurements means that the relationship between physical forcing and rate of primary production cannot be well defined. The application of ocean colour remote sensing for determining productivity is still in its infancy, largely as a result of the problems of establishing suitable algorithms for optically diverse waters (CASE I and II), of difficulties in detecting the subsurface chlorophyll maximum, and of limitations of sampling due to clouds. Initial studies for the Celtic Sea (Joint and Groom, 2000) are encouraging, and demonstrate how synoptic patterns in the estimated distributions of chlorophyll and primary productivity are strongly related to known physical structures and to occurrences of blooms. Similarly physiologically-based *in situ* methods, such as fast repetition rate fluorometry, are providing new information on the growth of sub-surface populations of phytoplankton (Moore *et al.,* 2003) which cannot be detected by remote sensing and are difficult to sample routinely.

Relationships between the physicochemical properties of shelf seas and the distributions of heterotrophs are generally more complex and less well known than for autotrophs. Distinct differences are observed in the trophic structure of plankton communities from mixed, frontal and stratified waters (Fig. 25.5) (Holligan *et al.,* 1984a; Le Fevre-Lehoerff *et al.,* 1993; Richardson *et al.,* 1998). In general, mixed waters are characterised by a relatively simple food chain based on microphytoplankton and herbivorous zooplankton, whereas picophytoplankton and an active microbial foodweb are prominent features of summer stratified shelf waters, especially where the oceanic influence is strong. Seasonal changes in the composition of heterotroph communities (Rodriguez *et al.,* 2000) are consistent with prevailing hydrographic conditions, which influence their growth and reproduction both directly (water temperature) and indirectly (food quantity and quality, predation).

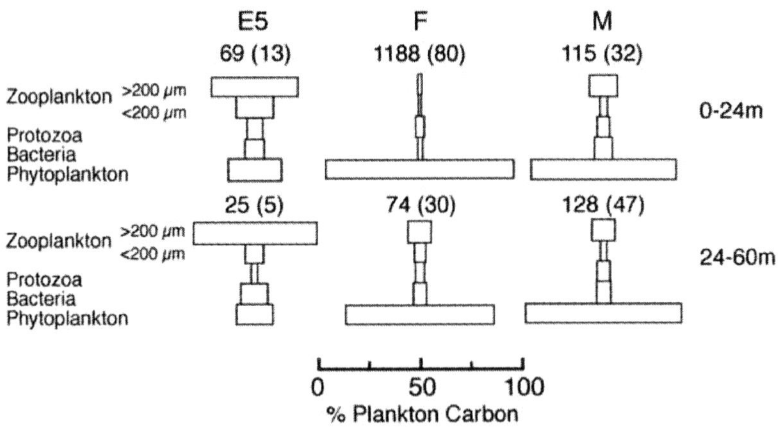

Figure 25.5 Distributions of organic carbon between the main autotrophic and heterotrophic components of plankton populations in mixed, frontal and stratified waters of the western English Channel (from Holligan *et al.,* 1984a courtesy of Inter-Research)

Populations of mesozooplankton and larval fish are also often described as typical of offshore (oceanic) or inshore (coastal) waters (e.g. Gowen *et al.,* 1998b).

This distinction is indicative of the importance of advection, as well as *in situ* reproduction and growth, for controlling the distributional patterns for organisms with relatively long generation times (weeks to months). Advection determines how organisms are dispersed away from their breeding and spawning areas, or how they might be entrained within waters favourable for their subsequent development. For example, gyral circulation associated with bodies of summer stratified water in the western Irish Sea has been shown to retain the planktonic larvae of *Nephrops* above the sediments where the juveniles will later need to settle to become breeding adults. Larvae that enter the water column before this seasonal gyral circulation is established risk being transported away from the adult breeding grounds by the coastal buoyancy current (White *et al.*, 1988; Dickey-Collas *et al,* 1996a; Hill *et al.*, 1996). Interactions between the biota and their environment are particularly complex for organisms that migrate vertically (Williams, 1985; Williams *et al.*, 1994; Dauvin *et al.*, 1998; Tarling *et al.*, 2002) especially in stratified waters. Numerical experiments by Hill (1994, 1998) show how different components of tidal motion lead to different types of spatial distribution depending on the diel rhythm and depth range of migration.

The most detailed studies of the relationships between hydrographic parameters, planktonic organisms (including copepods, larval fish, *Nephrops* larvae) (Dickey-Collas *et al.*, 1996a, 1996b, 1997; Gowen *et al.*, 1997, 1998a), and even top predators such as seabirds (Durazo *et al.*, 1998), have been carried out in the Irish Sea. In a region that includes stratified, mixed and coastal regimes, parameters that affect secondary production include length of the productive season, chlorophyll standing stock, and surface-to-bottom temperature difference, as well as advection of organisms into and out of favourable feeding areas.

Comparable studies in the Celtic Sea and English Channel have also shown that that the distribution, feeding, trophic interactions, and vertical migration of zooplankton (Williams, 1985; Holligan *et al.*, 1984a; Williams *et al.*, 1994; Laabir *et al.*, 1998) and larval fish (Grioche and Koubbi, 1997; Acevedo *et al.*, 2002) are related to hydrographic conditions and to positions of frontal boundaries. Most observations have been limited in space or time so that an overall picture of such relationships remains hard to put together (Rodriguez *et al.*, 2000). In particular, there is still insufficient information about processes occurring close to frontal boundaries (Le Fevre, 1986) and within the seasonal thermocline (Sharples *et al,* 2001) to allow hypotheses about the ecological and biogeochemical significance of persistently high chlorophyll concentrations in these situations during summer to be properly tested. For example, little is known about the importance of such boundary conditions for the survival of larval fish, as indicated by the observations of Kiorboe *et al.*, (1988) in the North Sea.

Relationships between hydrography and the distributions of benthic organisms are well established from benthic surveys (e.g. Sanvincente-Anorve *et al.*, 1996; Rees *et al.*, 1999; Ellis *et al.*, 2000), and are based in part on the role of tidal mixing in determining sediment type, and in part on the way food is supplied to the benthos (Vallet and Dauvin, 2001; Trimmer *et al.*, 1999). In tidally well-mixed waters the sediments are generally coarse, and inputs of organic matter maintained through the whole productive season. By contrast, in areas of seasonally stratified waters (relatively weak tides), the sediments are fine and the food supply appears to be relatively low after the spring diatom bloom due to a combination of surface nutrient depletion and active recycling of organic matter within the water column through the microbial foodweb. The benthic ecosystem contributes to the regeneration of nutrients, but the overall importance and timescale of this process remains uncertain. In this context it is noteworthy that bottom nutrient concentrations in the strongly stratified parts of the Celtic Sea are maximal in late autumn (Pingree *et al.*, 1976).

Various links between hydrography and adult fish have also been noted, including the effects of temporal variability in water temperature on recruitment and spawning in the Irish Sea (Planque and Fox, 1998; Nash and Geffen, 1999) and of spatial variability in plankton production on mackerel spawning in the Celtic Sea (Coombs *et al.*, 1990). Overall impacts on fish production are difficult to determine but, in a comparative study of the Irish and North Seas, Brander and Dickson (1984) concluded that the short production season for the deeper waters of the Irish Sea was the main cause of relatively low fish yields.

Over timescales longer than the annual cycle plankton populations are also influenced by climate and hydrography. The strength of northward moving currents in the eastern Atlantic Ocean affects the advection of oceanic waters and species onto the shelf (Reid *et al.*, 2001). Also correlations between the strength of the North Atlantic Oscillation (NAO) and the distributions and abundance of plankton have also been demonstrated (Beaugrand *et al.*, 2000, Irgoien *et al.*, 2000). The precise causes of such relationships, and of cyclic changes in plankton abundance as have been described for the western English Channel, remain uncertain, and the form that they take depends on the degree of oceanic influence and on more general climatic trends (Southward *et al.*, 1995).

3. Coastal Waters.

In the Celtic Seas three distinct types of coastal hydrodynamic regime can be recognised. In each one inputs of freshwater from land largely determine the position and strength of density gradients, which, in turn, affect both the productivity of the water column and the distributions of organisms within it. At one extreme in the fjords and sea lochs of western Scotland, where the tides are relatively weak and the water relatively deep, freshwater provides a source of buoyancy, as well as nutrients, which maintains a stratified water column for much of the year. At the other, in the eastern English Channel and eastern Irish Sea, strong tidal mixing leads to the formation of bands of fresher water of variable width along the coast but there is little or no vertical stratification of the water column. Intermediate conditions are found around most of the coast of Ireland.

Most of the fresh water entering the English Channel comes from French rivers, including the Loire on the Atlantic coast, which can influence the distributions of salinity and phytoplankton in the western English Channel during periods of strong flow in winter and early spring (Pingree *et al.*, 1986). In the eastern English Channel, water from the rivers Seine and Somme results in the formation of a zone of relatively fresh, turbid water, which flows northeast along the coast towards the North Sea. It is bounded offshore by a coastal front marking the transition to saltier water in the central Channel (Dupont *et al.*, 1991; Brylinski *et al.*, 1991, 1996). Phytoplankton abundance and primary production are generally highest in shallow inshore waters, which are fertilised by rivers, and just offshore from the coastal front where there is a transition to less turbid water (Brunet *et al.*, 1992; Brylinski *et al.*, 1996; Gentilhomme and Lizon, 1998). Along the frontal boundary itself physical accumulation of phytoplankton has been observed. Small-scale observations show that the relationship between variability in physical and biological parameters is complex (Seuront and Lagadeuc, 1998). However, at larger scales the abundance of different groups of plankton and the efficiency at which energy is transferred from one trophic level to another does appear to be linked to hydrodynamic conditions (Le Fevre-Lehoerff *et al.*, 1993).

Around the south and west coasts of Ireland the clockwise flow of lower salinity water forms on its outer boundary the Irish Shelf Front (ISF) (Raine *et al.*, 1990; McMahon *et al.*, 1995; Raine and McMahon, 1998). Tides are generally relatively small so that the coastal water is often weakly stratified and, in summer, lateral and vertical temperature gradients form in response to surface heating. The position and structure of the ISF is affected by spatial and temporal variations both in the strength of tidal mixing and in wind-induced upwelling (Raine and McMahon, 1998). The action of different mixing processes on thermal and haline sources of buoyancy produces complex physical structures which are readily apparent on satellite images of sea surface temperature (Huang *et al.*, 1991) and ocean colour (Fig. 25.3). The responses of phytoplankton to ambient physical conditions range from variations in the timing of seasonal succession (Roden and Raine, 1994) to the formation of transitory blooms of dinoflagellates and other species (Edwards *et al*, 1996; Raine *et al.*, 1993). The appearance and disappearance of such blooms is linked with changes in the depth of the seasonal thermocline and associated subsurface chlorophyll maximum, and with patterns of wind-driven advection (Edwards *et al.*, 1996). The inshore blooms have significant impacts on water chemistry and on other organisms that come into contact with them. Their prediction, however, requires better understanding of the complex interactions between mean hydrographic conditions and mixing due to both tide and wind (Raine and McMahon, 1998).

To the west of Scotland surface stratification has a strong influence on phytoplankton ecology (Jones *et al.*, 1984; Tett *et al.*, 1986) and, together with knowledge about light penetration, can be used to predict the occurrence of the main groups of phytoplankton (Jones and Gowen, 1990). The region of freshwater influence that has been most intensively studied is the Clyde Sea (Jones *et al.* 1995). The dynamics of water movement in the basin is controlled by freshwater inputs and by exchanges with the offshore North Channel. In summer the waters of the Clyde Sea become isolated below sill depth by a strong seasonal thermocline and at the surface by a well-developed thermohaline front. Exchanges of water and of dissolved and particulate matter occur largely in winter. Factors that control the seasonal cycles of phytoplankton and nutrients have been explored with a one dimensional box model (Rippeth and Jones, 1997), which shows that nitrogen stored in the Clyde during the summer is transferred offshore during winter largely as nitrate. Summer isolation of the deep water of the Clyde and other sea lochs leads to oxygen depletion of the deep water as organic matter sinks from the surface and becomes oxidised.

Further south, along the coast of NW England, interaction between freshwater input and the tides results in periodic stratification on semi-diurnal and fortnightly time scales (Sharples and Simpson, 1995). However, while such tidal modulation clearly affects the stability of the water, and may also affect residual transport mechanisms, there has to date been no concerted effort to link these well-defined physical cycles to the biochemistry or sedimentology of the region. An intriguing observation of the possible response of primary production to short-term stability cycles in this region has been identified (Floodgate et al., 1981).

In all coastal regions influenced by freshwater, strong relationships between hydrography, plankton growth and water chemistry can be discerned. The general nature of biological responses to variations in freshwater inputs varies with the relative importance of different physical forcing parameters, and is most predictable for semi-enclosed systems such as the Clyde Sea. In areas of strong tidal mixing hydrographic conditions are largely related to the input of freshwater and the strength and direction of the wind. By contrast off the south and west coasts of Ireland, where tides are variable, there are complex interactions between offshore and inshore waters with different thermohaline properties, leading to biological responses that are difficult to predict and also likely to be sensitive to climate change

4. The Shelf Edge.

The edge of the shelf of the Celtic seas is an important boundary, both in terms of the physical environment and its control of advection and shelf-ocean transfer. It is also a key determinant of the biochemical environment of the shelf edge and much of the shelf seas. A moderate slope current and internal tide-driven mixing dominate the physical environment. Also, at these latitudes important physical variability is driven by the generally west-to-east passage of weather systems, in particular driving strong variations in the surface wind stress. Cross-shelf edge transfers have a large contribution from variability in the along-shelf wind stress, typically resulting in annual mean flux rates of 0.9 $m^2 s^{-1}$ at the Celtic Sea shelf edge, and 1.2 $m^2 s^{-1}$ west of Ireland and Scotland (Huthnance, submitted). Wind-driven variability is likely to be a key controller of nutrient and carbon fluxes between the shelf sea and the NE Atlantic.

Two large, detailed, interdisciplinary observational programmes have been carried out with a particular focus on physical and biogeochemical processes at the shelf edge of the Celtic seas. The Land-Ocean-Interaction-Study Shelf Edge Study (LOIS-SES) took place in 1995/96 off the west coast of Scotland. The Ocean Margin Exchange programmme (OMEX) focused on the southwest approaches to the Celtic Sea, including the shelf edge and slope, and took place between 1993 and 1997. Most of the recent interdiscipinary work associated with these shelf edge regions was conducted under one or other of these programmes. The work at the Celtic Sea shelf edge and slope conducted during OMEX can be found in a detailed special edition of Deep Sea Research (numbers 14–15, volume 48, 2001).

The slope current flowing from Biscay northward to northwest Scotland is observed to be associated with a core of warm, high salinity water, and is thought to be driven by a latitudinal density (temperature) gradient in the upper ocean (Huthnance, 1986). In the region studied during LOIS-SES current speeds have been observed to average at around 0.1–0.2 m s^{-1} west of Ireland (White and Bowyer, 1997), and 0.2 m s^{-1} west of Scotland (Souza et al., 2001). Off the Malin shelf the flow tends to be more energetic and variable, with more influence further on the shelf, during winter. There is evidence of weak down-slope near-bed flows generated within the bottom Ekman layer of the current (Souza et al., 2001).

At the shelf edge west of Ireland, associated with the front between the shelf water and the North Atlantic Water of the slope current, high concentrations of surface water DMSP have been observed with elevated chlorophyll concentrations (Locarnini et al., 1998). The shelf edge was also a site of elevated DMS flux into the atmosphere. One reason for the enhanced chlorophyll concentration is likely to be mixing by the breaking of internal waves at the shelf edge. The vertical turbulent diffusivity of this mixing off the Malin shelf has been observed to be of 5 x 10^{-4} m^2 s^{-1} (neap tide) and 1.2 x 10^{-3} m^2 s^{-1} (spring tide) (Inall et al., 2000).

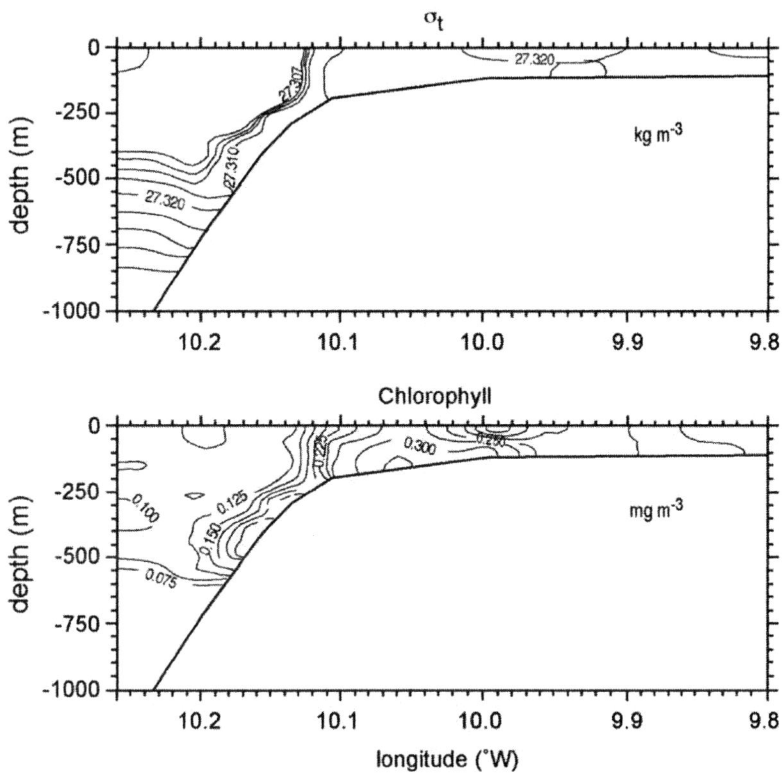

Figure 25.6 Observations of a cascade of dense water off the Malin Shelf in winter 1996. *Top:* σ_{tt} (kg m^{-3}); *Bottom:* Chlorophyll concentration (mg m^{-3}). Adapted from Hill et al., 1998.

During the LOIS-SES work observations of shelf edge sediments have shown that, despite the seasonality of shelf edge primary production, the total organic and inorganic carbon content in the sediments does not change through the year (Mitchell et al., 1997). This result is thought to indicate that most phytodetritus is consumed by organisms on the sediment surface before the material is buried, and/or that the material is further transported down the slope prior to burial. The isotopic ratio does have some correspondence with the increase in the supply of organic detritus during spring/summer, probably reflecting changes in the origins of the material (i.e. local phytoplankton growth versus land plant detritus). These results of Mitchell et al. (1997) suggested a small, rapid burial flux, though the area is not seen as a significant depocentre of organic carbon.

There are now several indications of off-shelf transport and carbon export in the region of the Malin shelf. In summer non-linear internal waves associated with the internal tide have been quantified in terms of the mass transport between shelf and ocean, with sporadic packets of the waves at the Malin shelf edge observed to transport an average of 0.3 m^2 s^{-1} offshore in the lower layer (Inall et al., 2001). During winter, surface cooling of the shallower shelf water has been seen to lead to dense water cascades of bottom water at the shelf edge to depths of at least 500 metres on the slope (Hill et al., 1998). These observations clearly showed the export of chlorophyll to the slope (Fig. 25.6). Combining the observations of Hill et al., 1998, Inall et al., 2001, and Souza et al., 2001, it could be suggested that there are 3 main processes that can export bottom shelf water to the upper slope. In summer both the bottom Ekman layer of the slope current and the transport driven by the non-linear internal waves drive offshore flow in the bottom water. In winter dense water cascades, along with (and possibly triggered by) the bottom Ekman layer of the slope flow provide the off-shelf transport, with internal waves absent because of the convective mixing and homogenisation of the water column. A 2-dimensional slice model of the physical and biogeochemical environment of the LOIS-SES region (Proctor *et al.,* 2003) has shown a near-surface flux of nitrate to the shelf from the ocean (65 kmol M m^{-1} yr^{-1}), and an off-shelf near-bottom flux of carbon at the shelf edge (23 kmol C m^{-1} yr^{-1}).

In the OMEX region there is evidence of maximum slope flows occurring in autumn/winter (Pingree *et al.,* 1999). Flow speeds tend to be significantly lower than in the LOIS-SES region (O(0.05 m s^{-1})), as a result of the complex shelf edge topography and the non-meridional alignment (Huthnance *et al.,* 2001). This along-slope flow acts as both a limiter of cross-slope transfer between the shelf and the NE Atlantic Ocean, and as a south-to-north advector of water properties (Pingree *et al.,* 1999). Cross-shelf edge exchange is dominated by wind-, tide-, and wave-forced currents, with an annual mean transport O(1 m^2 s^{-1}) (Huthnance *et al.,* 2001). Southwest of Ireland the mixing is provided mainly by surface wind and waves, while to the northwest of France tidally-driven internal waves become increasingly important (Huthnance *et al.,* 2001). As a result of the internal wave mixing cool bands of surface water are often clearly visible in satellite sea surface temperature images. Earlier work has shown that this mixing can supply useful quantities of deeper nutrients to the surface waters, leading to locally enhanced chlorophyll biomass at the shelf edge of the Celtic Sea (Pingree and Mardell, 1981) (Fig. 25.4 and Fig. 25.7). With the advent of frequent satellite images of sea surface chlorophyll concentration, these cool bands of water are now clearly related to shelf edge bands of elevated chlorophyll concentration, often observed both at the Celtic Sea shelf edge and off the Malin/Hebrides shelf edge.

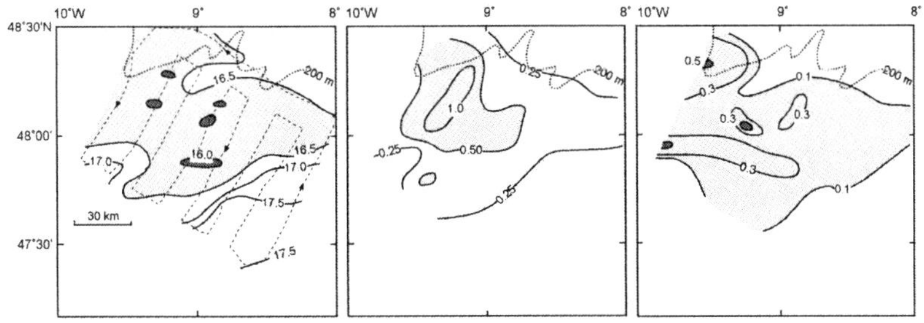

Figure 25.7 From left to right: surface temperature (°C), surface chlorophyll (mg m^{-3}), and surface inorganic nitrate (mmol m^{-3}) at the shelf edge of the Celtic Sea. The 200 metre isobath marks the position of the shelf edge. After Pingree & Mardell, 1981. Courtesy of the Royal Society, London.

Nutrient supply and demand in the OMEX study has been described by Hydes et al., 2001. Underway surface nitrate and salinity data collected during winter showed three important regimes between the shelf edge and the coastal zone (Fig. 25.8). Marked discontinuities in the salinity-nitrate relationship indicate the high nitrate concentrations at the shelf edge and in waters influenced by coastal run-off are not seen in open shelf waters. This has important consequences for the fuelling of primary production on the shelf, where winter nutrient concentrations are dependent on local remineralisation processes within the sediments. This highlights the difficulty in defining "new" production in systems where apparently new growth is fuelled by longer-term regeneration of organic material (e.g. Wollast, 2003). Pelagic primary production at the Celtic Sea shelf edge has been observed as 160 g C m^{-2} yr^{-1}, with about half of that being new production (Rees et al., 1999; Joint et al., 2001). Most of the new production occurs during the spring bloom (f~0.7–0.8). Estimates of production rates and f-ratios made by applying temperature-nitrate relations to satellite temperature imagery compared well with the direct measurements. Joint et al., 2001, estimated that 38% of the pelagic production became available to the midwater heterotrophs and the benthos.

The fate of organic material reaching the seabed (De Wilde et al., 1998) was extensively studied during OMEX. Re-suspension of material, and subsequent apparently rapid transport off the slope, is thought to transfer 6 – 9% of shelf edge primary production to below the depth of winter mixing, with final deposition occurring at the foot of the slope and abyssal ocean (Antia et al., 2001). Well-developed bottom nepheloid layers have been observed (McCave et al., 2001), with some indication of intermediate layers transferring resuspended material offshore. Re-suspension of these high-density bottom layers was seen to be forced by strong, semi-diurnal tidal currents at the seabed (van Weering et al., 2001). A deposition rate of > 44 g m^{-2} yr^{-1}, comprising about 98% organic carbon, was measured at the shelf edge. About half of this deposition was mineralised at the sediment-water interface and below the seabed, driven by oxygen (van Weering et al., 2001). Heip et al., 2001, observed >50% of the organic material that reached the shelf edge seabed to be respired by macrofauna, with intensive reworking of the shelf edge sediments. In a model of the Goban Spur shelf break, respiration of organic nitrogen was found to occur predominantly in the euphotic zone (42%) and in the deeper water (51%), with 7% taking place in the sediments (Soetaert et al., 2001). Over one year the model indicated that half of the required nitrogen was supplied by mixing from below the euphotic zone. The results also indicated important dependence on the meteorological forcing, with real meteorological data leading to a 40% increase in carbon production compared to a model forced by smooth, climatalogical data.

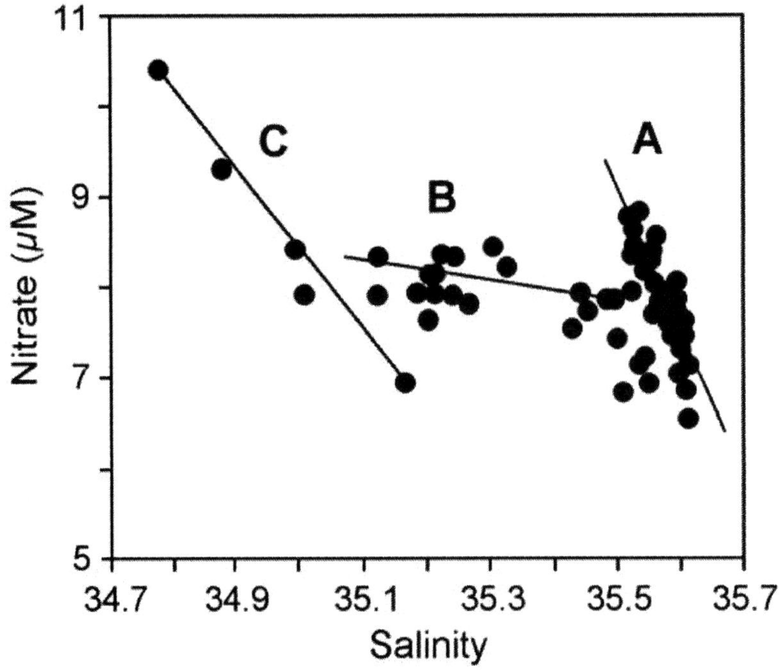

Figure 25.8 Surface nitrate and salinity measured between the Celtic Sea shelf edge and the UK coast during winter of 1994. Line A marks the mixing environment at the shelf edge of the Celtic Sea; line B is data collected from the broad shelf region within the Celtic Sea; line C marks data collected close to the UK coast where conditions were influenced by the freshwater input to the Bristol Channel. After Hydes et al., 2001 Courtesy of Elsevier Ltd.

The carbon and nitrogen cycles observed during the OMEX study have been summarised by Wollast and Chou, 2001 (Fig. 25.9). About half of the shelf edge annual production of 200 g C m^{-2} was exported from the euphotic zone, with about 10% reaching the sediment water interface to be respired. They estimated that 30 g C m^{-2} yr^{-1} is exported to the slope and deep ocean, suggesting a flux of 1500 tonnes of organic matter deposited in the slope area per km of shelf break. The role of vertical mixing in fuelling the shelf break production was highlighted (e.g. Holligan et al., 1985).

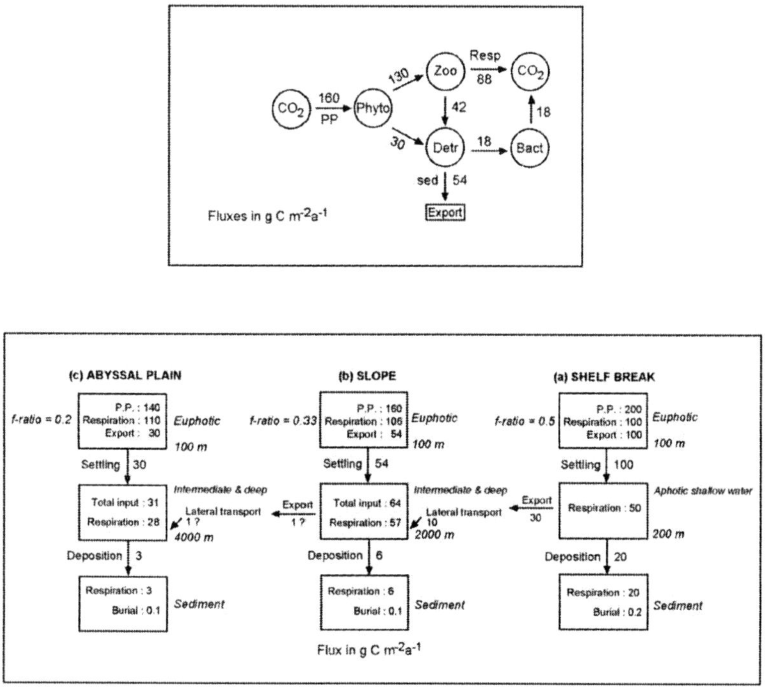

Figure 25.9 Summary flux estimates for the OMEX I study of the Goban Spur shelf edge region, after Wollast & Chou, 2001 Courtesy of Elsevier Ltd. *Top:* Organic carbon cycle in the euphotic zone. *Bottom:* Organic carbon cycle for the northern Gulf of Biscay.

The entire shelf edge of the Celtic seas, and reaching further south to the Iberian peninsula and north to the Norwegian and North Seas, has been studied in the context of oceanographic controls on fisheries. The Shelf Edge Fisheries and Oceanography Studies (SEFOS) was a major European research effort focussing on the life histories, distribution, and migration patterns of several species in relation to the oceanography of the shelf edge (Reid, 2001). As an example, Western mackerel are known to migrate southward along the shelf edge to their spawning grounds southwest of Ireland, and then return northward along the same route to the northern North Sea (Reid et al., 1997). The trigger for the start of the spawning migration is believed to be the winter decrease in water temperature. Schools of mackerel actively migrate along the shelf edge in this cold water, while the episodic introduction of the warmer slope current water onto the shelf temporarily halts the migration. Hotspots of mackerel spawning were often associated with the more mixed water at the shelf edge, indicating the possibly important efect of the enhanced primary production in this region for the feeding of adults and larvae (Bez and Rivoirard, 2001).

5. Discussion and Conclusions

The environment of the Celtic seas is controlled largely by tidal mixing and advection, and by seasonal heating and cooling of the water column, with inputs of freshwater also locally important in some coastal areas, and internal tidal mixing having increased importance near the shelf edge. Over much of the Celtic Seas the physical environment is dominated by a simple competition between tidal mixing and surface heating, which in turn controls the seasonality of biochemical cycling. The seasonal stratification in the low tidal energy, or deeper water, regimes leads to a broadly predictable pattern of spring bloom, followed by a summer sub-surface biomass maximum, and often ending with a small autumn bloom (Fig. 25.2a). Separating these seasonally-stratified waters from the permanently (tidally-energetic) mixed waters, the tidal mixing fronts provide a predictable spatial gradient in physical and associated biochemical properties. This underlying, relatively simple, spatial and temporal variability in water column structure has led to the Celtic Seas providing important practical models for the parallel understanding of primary production, zooplankton distributions, and other higher trophic interactions. It is recognised that many of the observed physical-biological interactions are similar to those known to occur in other temperate shelf areas. The North Sea (Richardson et al., 1998), Gulf of Maine (Townsend et al., 2001; Townsend and Thomas, 2002) and Patagonian shelf (Sabatini et al., 2000) are all areas with strong tidal currents that show many similar features in the basic physical environment and the corresponding cycles in the biochemistry.

The wide range of physical conditions in the Celtic seas leads to great heterogeneity in biological and chemical properties. Rates of primary production, nutrient regeneration, exchange of materials across the seasonal thermocline and across fronts are spatially and temporally variable, and generally not well constrained. Many important questions are only just becoming tractable to observation as developments in technology open up new areas for exploration. An important requirement here is to utilise new developments in biological and chemical instrumentation to gather well-resolved spatial (e.g. satellites and towed-vehicles) and temporal (i.e. moorings) series of compatible physical-biological-chemical information, allowing the patchiness of the biochemistry to be seen within the context of the underlying physics. As annual budgets for elements such as carbon and nitrogen are still difficult to estimate directly, coupled physical-biological models will increasingly be the mainstay of attempts to predict the effects of environmental variability and change on ecosystem dynamics and biogeochemical fluxes. However, the outputs of such models will still require validation, which is only possible through well-coordinated observational programmes bringing together scientists from different disciplines. An example is the EU-supported OMEX programme at the Celtic Sea shelf edge (see Wollast and Chou, 2001), but comparable studies using modern technologies have yet to be undertaken for the open shelf seas and tidal fronts.

A central focus for any new interdisciplinary programme must be the determination of nutrient fluxes within the water column and sediments, and across interfaces such as the water-sediment boundary, the seasonal thermocline, and the shelf edge. The existence of inshore, mid-shelf and oceanic nutrient regimes, as illustrated in Fig. 25.8, also highlights the importance of vertical processes for determining nutrient distributions over most of the shelf. In the spring and early summer, stocks of winter nutrients are depleted by actively growing phytoplankton so that nutrient regeneration becomes increasingly important for fuelling primary production. This change is reflected in estimates of the *f*-ratio for nitrogen assimilation, which range from ~0.8 in spring to typically <0.25 in mid-summer (Joint *et al.*, 2001). Inorganic nutrient distributions in summer are largely determined by vertical processes, including the sinking of particulate organic matter and nutrient exchange across the sediment-water interface and the seasonal thermocline. Without physically-constrained numerical models (e.g. Pingree and Pennycuick, 1975) that accurately simulate vertical gradients in nutrient concentrations (e.g. for the water column, Holligan *et al.*, 1984b), realistic nutrient budgets for the range of hydrographic regimes across the shelf still cannot be constructed. Another uncertainty for such models is the effect of irradiance on the balance between nitrate and regenerated ammonium assimilation by autotrophs (Gentilhomme and Lizon, 1998; Maguer *et al.*, 1998) especially under conditions of a deep (>40m) surface mixed layer.

Other important problems for future study include the dynamics of the summer chlorophyll maxima associated with the thermocline and surface fronts (Figs. 25.4 and 25.7), the role of cascading in shelf-ocean exchange (Fig. 25.6), and the impacts of hydrographic conditions (especially in coastal waters) on the occurrence of phytoplankton blooms (Fig. 25.3b) and on recruitment within populations of economically-important fish and shellfish. Studies of the thermocline in the North Sea have indicated that the subsurface chlorophyll maximum fuels pelagic production through the summer months (Richardson *et al.*, 2000) and that surface fronts are important areas for the growth of larval fish (Kiorboe at al., 1988). Understanding how phytoplankton populations within density gradients are maintained requires a detailed knowledge of physical turbulence and phytoplankton cell physiology in order to quantify both the supply of nutrients, and the behaviour and dispersal of cells. The problem is particularly difficult for subsurface populations, which cannot be detected by remote sensing and are exposed to internal mixing processes. Another major uncertainty is how circulation patterns associated with different types of frontal boundary redistribute biological and chemical properties (Bo Pedersen, 1994).

At the shelf edge new work is needed to establish the frequency and scale of cascading events. It is not known whether or not previous observations (Fig. 25.8) are representative of the whole shelf edge or what their overall contribution is the shelf-ocean exchange. Within inshore waters physical-biological interactions are a function of local tide, weather and stratification, and the implications that this hydrographic variability may have for important biological processes, such as the development of red tides and larval survival, are only just beginning to be understood.

Interdisciplinary studies are already providing a broad framework for the development of new management policies for the living resources of shelf seas (e.g. Hill *et al.*, 1997). In the context of climate change, increasing attention is being given to the role of the ocean margins in global biogeochemical cycles. Within both areas of work progress in understanding the dynamics of coastal and shelf systems, and how they might respond to climate change, will depend largely on improving our understanding of physical-biological interactions. In the same way that the Celtic Seas provided a robust physical framework for the practical and conceptual models of the fundamentals of primary production, this knowledge of the physical and biological environments now provides the context for addressing more complex multi-disciplinary questions.

Acknowledgements

This work was carried out as part of a United Kingdom Natural Environment Research Council grant NER/A/S/2001/00449. We would like to thank Steve Groom and Peter Miller at the Remote Sensing Data Analysis Service at the Plymouth Marine Laboratory, UK, for help with the satellite imagery used. Also Sarah Wakelin and Roger Proctor, Proudman Oceanographic Laboratory UK, for data from the POLCOMS numerical model of the NW European shelf.

Bibliography.

Acevedo, S., O. Dwane and J. M. Fives. 2002. The community structure of larval fish populations in an area of the Celtic Sea in 1998. *J. Mar. Biol. Ass. U. K.,* **82,** 641–648.

Antia, A. N., J. Maaβen, P. Herman, M. Voβ, J. Scholten, S. Groom, and P. Miller. 2001. Spatial and temporal variability of particle flux at the NW European continental margin. *Deep-Sea Res. II,* **48,** 3083–3106.

Beaugrand, G., F. Ibanez and P. C. Reid. 2000. Spatial, seasonal and long-term fluctuations of plankton in relation to hydroclimatic features in the English Channel, Celtic Sea and Bay of Biscay. *Mar. Ecol. Prog. Ser.,* **200,** 93–102.

Bez, N., and J. Rivoirard. 2001. Transitative geostatistics to characterise spatial aggregations with diffuse limits: an application on mackerel ichthyoplankton. *Fish. Res.* **50,** 41–58.

Bo Pedersen, F. 1994. The oceanographic and tidal cycle succession in shallow sea fronts in the North Sea and English Channel. *Estua. Coast. Shelf Sci.,* **38,** 249–269.

Bowers, D.G., S. Boudjelas, G.E.L. Harker. 1998. The distribution of fine suspended sediments in the surface waters of the Irish Sea and its relation to tidal stirring. *Int. J. Remote Sens.,* **19,** 2789–2805.

Bowman, M. J., W. E. Esaias, and M. B. Schnitzer. 1981. Tidal stirring and the distribution of phytoplankton in Long Island and Block Island Sounds. *J. Mar. Res.,* **39,** 587–603.

Bowman, M. J., A. C. Kibblewhite, S. M. Chiswell, and R. A. Murtagh. 1983. Shelf fronts and tidal stirring in Greater Cook Strait, New Zealand. *Oceanol. Acta* **6,** 119–129.

Brander, K.M. and R.R. Dickson. 1984. An investigation of the low level of fish production in the Irish Sea. *Rapp. P.-V. Reun. Cons. Int. Explor. Mer,* **183,** 234–242.

Brunet, C., J.M. Brylinski, L. Bodineau, G. Thoumelin, D. Bentley and D. Hilde. 1996. Phytoplankton dynamics during the spring bloom in the south-eastern English Channel. *Estuar. Coast. Shelf Sci.,* **43,** 469–483.

Brunet, C., J.M. Brylinski and S. Frontier. 1992. Productivity, photosynthetic pigments and hydrology in the coastal front of the Eastern English Channel. *J. Plankt. Res.,* **11,** 1541–1552.

Brylinski, J.M., C. Brunet, D. Bentley, G. Thoumelin and D. Hilde. 1996. Hydrography and phytoplankton biomass in the eastern English Channel in spring 1992. *Estuar. Coast. Shelf Sci.,* **43,** 507–519.

Brylinski, J.M., Y. Lagadeuc, V. Gentilhomme, J-P. Dupont, R. Lafite, P-A. Dupeuble, M-F. Huault, Y. Auger, E. Puskaric, M. Wartel and L. Cabioch. 1991. Le 'fleuve cotiere': un phenomene hydrologique important en Manche orientale. Exemple du Pas-de-Calais. *Oceanol. Acta, Spec. Issue* **11,** 197–203.

Cook, G.T., A.B. MacKenzie, P. McDonald, S.R. Jones. 1997. Remobilization of Sellafield-derived radionuclides and transport from the north-east Irish Sea. *J. Environ. Radioact.,* **35,** 227–241.

Coombs, S. H., J. Aiken and T.D. Griffin. 1990. The aetiology of mackerel spawning to the west of the British Isles. *Meeresforsch.,* **33,** 52–75.

Dauvin, J-C., E. Thiebaut and Z. Wang. 1998. Short-term changes in the mesozooplanktonic community in the Seine ROFI (Region of Freshwater Influence) (eastern English Channel). *J. Plankt. Res.,* **20,** 1145–1167.

De Wilde, P.A.W.J., G.C.A. Duinveld, E.M. Berghuis, M.S.S. Lavaleye and A. Kok. 1998. Late-summer mass deposition of gelatinous phytodetritus along the slope of the N.W. European continental margin. *Prog. Oceanogr.,* **42,** 165–187.

Dickey-Collas, M., J. Brown, L. Fernand, A.E. Hill, K.J. Horsburgh and R.W. Garvine. 1997. Does the western Irish Sea gyre influence the distribution of juvenile fish? *J. Fish Biol.,* **51,** 206–229.

Dickey-Collas, M., R.J. Gowen and C.J. Fox. 1996a. Distribution of larval and juvenile fish in the western Irish Sea: Relationship to phytoplankton, zooplankton biomass and recurrent physical features. *Mar. Freshwat. Res.,* **47,** 169–181.

Dickey-Collas, M., B.M. Stewart and R.J.Gowen. 1996b. The role of thermal stratification on the population dynamics of *Sagitta elegans* Verrill in the western Irish Sea. *J. Plankt. Res.,* **18,** 1659–1674.

Dupont, J-P., R. Lafite, M-F. Huault, P-A. Dupeuble, J-M. Brylinski, P. Gueguenjat, M. Lamboy and L. Cabioch. 1991. La dynamique des masses d'eaux et des matieres en suspension en Manche orientale. *Oceanol. Acta, Spec. Issue,* **11,** 177–186.

Durazo, R., N.M. Harrison and A.E. Hill. 1998. Seabird observations at a tidal mixing front in the Irish Sea. *Estuar. Cstl. Shelf Sci.,* **47,** 153–164.

Edwards A., K. Jones, J.M. Graham, C.R. Griffiths, N. MacDougall, J. Patching, J.M. Richard and R. Raine. 1996. Transient coastal upwelling and water circulation in Bantry Bay, a ria on the south-west coast of Ireland. *Estuar. Coast. Shelf Sci.,* **42,** 213–230.

Ellis, J.R., S.I. Rogers and S.M. Freeman. 2000. Demersal assemblages in the Irish Sea, St George's Channel and Bristol Channel. *Estuar. Coast. Shelf Sci.,* **51,** 299–315.

Fasham, M.J.R., P.M. Holligan and P.R. Pugh. 1983. The spatial and temporal development of the spring phytoplankton bloom in the Celtic Sea. *Prog. Oceanogr.,* **12,** 87–145.

Floodgate, G.D., G.E. Fogg, D.A. Jones, K. Lochte and C.M. Turley. 1981. Microbiological and zooplankton activity at a front in Liverpool Bay. *Nature,* **290,** 133–136.

Gentilhomme, V. and F. Lizon. 1998. Seasonal cycle of nitrogen and phytoplankton biomass in a well-mixed coastal system (eastern English Channel). *Hydrobiol.,* **361,** 191–199.

Gibson, C.E., B.M. Stewart and R.J. Gowen. 1997. A synoptic study of nutrients in the north-west Irish Sea. *Estuar. Coast. Shelf Sci.,* **45,** 27–38.

Gowen, R.J., M. Dickey-Collas and G. McCullough. 1997. The occurrence of *Calanus finmarchicus* (Gunnerus) and *Calanus helgolandicus* (Claus) in the western Irish Sea. *J. Plankt. Res.,* **19,** 1175–1182.

Gowen, R.J., G. McCullough, M. Dickey-Collas and G.S. Kleppel. 1998a. Copepod abundance in the western Irish Sea: relationship to physical regime, phytoplankton production and standing stock. *J. Plankt. Res.,* **20,** 315–330.

Gowen, R.J., D.K. Mills, M. Trimmer and D.B. Nedwell. 2000. Production and its fate in two coastal regions of the Irish Sea: the influence of anthropogenic nutrients. *Mar Ecol. Prog, Ser.,* **208,** 51–64.

Gowen, R.J., R. Raine, M. Dickey-Collas and M. White. 1998b. Plankton distributions in relation to physical oceanographic features on the southern Malin Shelf, August 1996. *ICES J. Mar. Sci.,* **55,** 1095–1111.

Gowen, R.J., B.M. Stewart, D.K. Mills and P. Elliott. 1995. Regional differences in stratification and its effect on phytoplankton production and biomass in the northwestern Irish Sea. *J. Plankt. Res.,* **17,** 753–769.

Grioche, A. and P. Koubbi. 1997. A preliminary study of the influence of a coastal frontal structure on ichthyoplankton assemblages in the English Channel. *ICES J. Mar. Sci.,* **54,** 93–104.

Heathershaw, A.D., A.L. New and P.D. Edwards. 1987. Internal tides and sediment transport at the shelf break in the Celtic Sea. *Cont. Shelf Res.,* **7,** 485–517.

Heip, C. H. R., G. Duineveld, E. Flach, G. Graf, W. Helder, P. M. J. Herman, M. Lavaleye, J. Middelburg, O. Pfannkuche, K. Soetaert, T. Soltwedel, H. de Stigter, L. Thomsen, J. Vanaverbeke, and P. de Wilde. 2001. The role of the benthic biota in sedimentary metabolism and sediment-water exchange processes in the Goban Spur area (NE Atlantic). *Deep-Sea Res. II,* **48,** 3223–3243.

Hill, A.E. 1993. Seasonal gyres in shelf seas. *Ann. Geophys.*, **11,** 1130–1137.

Hill, A.E. 1994. Horizontal zooplankton dispersal by diel vertical migration in S_2 tidal currents on the northwestern European continental shelf. *Cont. Shelf Res.*, **14,** 491–506.

Hill, A.E. 1998. Diel migration in stratified tidal flows: Implications for plankton dispersal. *J. Mar. Res.*, **56,** 1069–1096.

Hill, A.E., J. Brown and L. Fernand. 1996. The western Irish Sea gyre: a retention system for Norway lobster (*Nephrops norvegicus*)? *Oceanol. Acta,* **19,** 357–368.

Hill, A.E., J. Brown and L. Fernand. 1997. The summer gyre in the western Irish Sea: Shelf sea paradigms and management implications. *Estuar. Coast. Shelf Sci.,* **44** (Suppl. A), 83–95.

Hill, A.E., K.J. Horsburgh, R.W. Garvine, P.A. Gillibrand, G. Slesser, W.R. Turrell, and R.D. Adams. 1997. Observations of a density-driven recirculation of the Scottish coastal current in the Minch. *Estuar. Coast. Shelf Sci.,* **45,** 473–484.

Hill, A.E., and J.H. Simpson. 1989. On the interaction of thermal and haline fronts: The Islay front revisited. *Estuar. Coast. Shelf Sci.,* **28,** 495–505.

Hill A. E., A. J. Souza, K. Jones, J. H. Simpson, G. I. Shapiro, R. McCandliss, H. Wilson, and J. Leftley. 1998. The Malin cascade in winter 1996. *J. Mar. Res.* **56,** 87–106.

Holligan, P.M., R.P. Harris, R.C. Newell, D.S. Harbour, R.N. Head, E.A.S. Linley, M.I. Lucas, P.R.G. Tranter and C.M. Weekley. 1984a. Vertical distribution and partitioning of organic carbon in mixed, frontal and stratified waters of the English Channel. *Mar Ecol. Prog. Ser.,* **14,** 111–127.

Holligan, P.M., R.D. Pingree, and G.T. Mardell. 1985. Oceanic solitons, nutrient pulses and phytoplankton growth. *Nature,* **314,** 348–350.

Holligan, P.M., P.J.LeB. Williams, D. Purdie and R.P. Harris. 1984b. Photosynthesis, respiration and nitrogen supply of plankton populations in stratified, frontal and tidally mixed waters. *Mar. Ecol. Prog. Ser.,* **17,** 201–213.

Horsburgh, K.J., A.E. Hill, and J. Brown. 1998. A summer jet in the St George's Channel of the Irish Sea. *Estuar. Coast. Shelf Sci.,* **47,** 285–294.

Horsburgh, K.J., A.E. Hill, J. Brown, L. Fernand, R.W. Garvine, and M. M. P. Angelico. 2000. Seasonal evolution of the cold pool gyre in the western Irish Sea. *Prog. Oceanogr.,* **46,** 1–58.

Huang, W.G., A.P. Cracknell, R.A. Vaughan and P.A. Davies. 1991. A satellite and field view of the Irish Shelf Front. *Cont. Shelf Res.,* **11,** 543–562.

Huthnance, J.M .. 1986. The Rockall slope current and shelf-edge processes. *Proc. R. Soc. Edinb.,* **B88,** 83–101.

Huthnance, J. M .. The Northeast Atlantic continental margin. In: *Carbon and nutrient fluxes in continental margins: a global synthesis.* (Eds: L. Atkinson, K.K. Liu, R. Quinones, L. Talaue-McManus), submitted.

Huthnance, J.M., H. Coelho, C.R. Griffiths, P.J. Knight, A.P. Rees, B. Sinha, A. Vangriesheim, M. White, and P.G. Chatwin. 2001. Physical structure, advection and mixing in the region of the Goban Spur. *Deep-Sea Res. II,* **48,** 2979–3021.

Hydes, D.J., A.C. Le Gall, A.E.J. Miller, U. Brockmann, T. Raabe, S. Holley, X. Alvarez-Salgado, A. Antia, W. Balzer, L. Chou, M. Elskens, W. Helder, I. Joint, and M. Orren. 2001. Supply and demand of nutrients and organic matter at and across the NW European shelf break in relation to hydrography and biogeochemical activity. *Deep-Sea Res. II,* **48,** 3023–3047.

Inall, M.E., T.P. Rippeth, and T.J. Sherwin. 2000. The impact of non-linear waves on the dissipation of internal tidal energy at a shelf break. *J. Geophys. Res.,* **105,** 8687–8706.

Inall, M.E., G.I. Shapiro, and T.J. Sherwin. 2001. Mass transport by non-linear internal waves on the Malin Shelf. *Cont. Shelf Res.,* **21,** 1449–1472.

Irgoien, X., R.P. Harris, R.N. Head and D. Harbour. 2000. North Atlantic Oscillation and spring bloom phytoplankton composition in the English Channel. *J. Plankt. Res.,* **22,** 2367–2371.

Joint, I. and S.B. Groom. 2000. Estimation of phytoplankton production from space: current status and future potential of satellite remote sensing. *J. Exp. Mar. Biol. Ecol.,* **250,** 233–255.

Joint, I.R., N.J.P. Owens and A.J. Pomroy. 1986. Seasonal production of photosynthetic picoplankton and nanoplankton in the Celtic Sea. *Mar. Ecol. Prog. Ser.,* **28,** 251–258.

Joint, I.,R. Wollast, L. Chou, S. Batten, M. Elskens, E. Edwards, A. Hirst, P. Burkill, S. Groom, S. Gibb, A. Miller, D. Hydes, F. Dehairs, A. Antia, R. Barlow, A. Rees, A. Pomroy, U. Brockmann, D. Cummings, R.S. Lampitt, M. Loijens, F. Mantoura, P. Miller, T. Raabe, X. Alvarez-Salgado, C. Stelfox, and J. Woolfenden. 2001. Pelagic production at the Celtic Sea shelf break. *Deep-Sea Res. II,* **48,** 3049–3081.

Jones, K.J. and R.J. Gowen. 1990. Influence of stratification and irradiance regime on summer phytoplankton composition in coastal and shelf seas of the British Isles. *Estuar. Coast. Shelf Sci.,* **30,** 557–567.

Jones, K.J., R.J. Gowen and P. Tett. 1984. Water column structure and phytoplankton distribution in the Sound of Jura, Scotland. *J. Exp. Mar. Biol. Ecol.* **78,** 269–289.

Jones, K.J., B. Grantham, I. Ezzi, T. Rippeth and J. Simpson. 1995. Physical controls on phytoplankton and nutrient cycles in the Clyde Sea, a fjordic system on the west coast of Scotland. In *Ecology of Fjords and Coastal Waters,* Eds. Skjoldal, H.R., C. Hopkins, K.E. Erikstad and H.P. Leinaas. Elsevier.

Jordan, M.B. and I. Joint. 1984. Studies on phytoplankton distribution and primary production in the western English Channel in 1980 and 1981. *Cont. Shelf Res.* **3,** 24–34 ..

Jordan, M.B. and I. Joint. 1998. Seasonal variation in nitrate:phosphate ratios in the English Channel 1923–1987. *Estuar. Coast. Shelf Sci.,* **46,** 157–164.

Kenyon, N.H .. 1986. Evidence from bedforms for a strong poleward current along the upper continental slope of northwest Europe. *Mar. Geol.,* **72,** 187–198.

Kiorboe, T., P. Munk, K. Richardson, V. Christensen and H. Paulsen. 1988. Plankton dynamics and larval herring growth, drift and survival in a frontal area. *Mar. Ecol. Prog. Ser.,* **44,** 205–219.

Knight P.J., and M. J. Howarth. 1999. The flow through the north channel of the Irish Sea. *Cont. Shelf Res.* **19,** 693–716.

Laabir, M., S.A. Poulet, R.P. Harris, D.W. Pond, A. Cueff, R.N. Head and A. Ianora. 1998. Comparative study of the reproduction of *Calanus helgolandicus* in well-mixed and seasonally stratified coastal waters of the western English Channel. *J. Plankt. Res.,* **20,** 407–421.

Le Corre, P., S. L'Helguen and M. Wafar. 1993. Nitrogen source for uptake by *Gyrodinium* cf. *aureolum* in a tidal front. *Limnol. Oceanogr.,* **38,** 446–451.

Le Fèvre, J. 1986. Aspects of the biology of frontal systems. *Adv. Mar. Biol,.* **23,** 163–299.

Le Fevre-Lehoerff, G., E. Erard-Le Denn and G. Arzul. 1993. Planktonic ecosystems in the Channel. Trophic relations. *Oceanol. Acta,* **16,** 661–670.

Lizon, F., L. Seuront and Y. Lagadeuc. 1998. Photoadaptaion and primary production study in tidally mixed coastal waters using a Lagrangian model. *Mar. Ecol. Prog. Ser.,* **169,** 43–54.

Locarnini, S.J.P., S.M. Turner, and P.S. Liss. 1998. The distribution of dimethylsulphide, DMS, and dimethylsulphoniopropionate, DMSP, in waters off the western coast of Ireland. *Cont. Shelf Res.,* **18,** 1455–1473.

Loder, J. W., D. Brickman, and E. P. W. Horne. 1992. Detailed structure of currents and hydrography on the northern side of Georges Bank. *J. Geophys. Res.,* **97,** 14331–14351.

Loder J.W. and T. Platt. 1985. Physical controls on phytoplankton production at tidal fronts. In*: Proc. 19th Eur. Mar. Biol. Symp.* Ed. Gibbs, P.E. pp 3–21. Cambridge Univ. Press.

MacKenzie, A.B., G.T. Cook, P. McDonald, S.R. Jones. 1998. The influence of mixing timescales and re-dissolution processes on the distribution of radionuclides in northeast Irish Sea sediments. *J. Environ. Radioact.,* **39,** 35–53.

Maguer, J-F., S. L'Helguen, C. Madec and P. Le Corre. 1998. Uptake and regeneration of nitrogen in the well-mixed waters of the English Channel: new and regenerated productions. *Oceanol. Acta,* **21,** 861–870.

Maguer, J-F., S. L'Helguen, C. Madec and P. Le Corre. 1999. Seasonal patterns of ammonium generation from size-fractionated microheterotrophs. *Cont. Shelf Res.,* **19,** 1755–1770.

McCave, I.N., I.R. Hall, A.N. Antia, L. Chou, F. Dehairs, R.S. Lampitt, L. Thomsen, T.C.E. van Weering, and R. Wollast. 2001. Distribution, composition and flux of particulate material over the European margin at 47°-50° N. *Deep-Sea Res. II,* **48,** 3107–3139.

McMahon, T., R. Raine, O. Titov and S. Boychuk. 1995. Some oceanographic features of northeastern Atlantic waters west of Ireland. *ICES J.Mar. Sci.,* **52,** 221–232.

Mills, D.K., R.J. Gowen and E.A. Woods. 1997. Observation and simulation of the spring bloom in the north-western Irish Sea. *J. Plankt. Res.,* **19,** 63–77.

Mitchell, L., S.M. Harvey, J.D. Gage, and A.E. Fallick. 1997. Organic carbon dynamics in shelf edge sediments off the Hebrides: a seasonal perspective. *Int. Revue ges. Hydrobiol.,* **3,** 424–435.

Moore, C. M., D. Suggett, P.M. Holligan, J. Sharples, E.R. Abraham, M.I. Lucas, T.P. Rippeth, N.R. Fisher, J.H. Simpson and D.J. Hydes. 2003. Physical controls on phytoplankton physiology and production at a shelf sea front: an FRRF based field study. *Mar. Ecol. Prog. Ser.,* **259,** 29–45.

Morin, P., P. Le Corre and J. le Fevre. 1985. Assimilation and regeneration of nutrients off the west coast of Brittany. *J. Mar. Biol. Ass. U.K.,* **65,** 677–695.

Morin, P., M.V.M. Wafar and P. Le Corre. 1993. Estimation of nitrate flux in a tidal front from satellite-derived temperature data. *J. Geophys. Res.,* **98,** 4689–4695.

Nash, R.D.M. and A.J. Geffen. 1999. Variability in stage I egg production of plaice *Pleuronectes platessa* on the west side of the Isle of Man, Irish Sea. *Mar. Ecol. Prog. Ser.,* **188,** 241–250.

OSPAR Commission 2000. Quality Status Report 2000, Region III – Celtic Seas. OSPAR Commission, London. 116 + xii pp.

Pingree, R.D. and D.K. Griffiths. 1978. Tidal fronts on the shelf seas around the British Isles. *J. Geophys. Res.,* 83, 4615–4622.

Pingree, R.D., P.M. Holligan and G.T. Mardell. 1978. The effects of vertical stability on phytoplankton distributions in the summer on the northwest European shelf. *Deep-Sea Res.,* **25,** 1011–1028.

Pingree, R.D., P.M. Holligan, G.T. Mardell and R.N. Head. 1976. The influence of physical stability on spring, summer and autumn phytoplankton blooms in the Celtic Sea. *J. Mar. Biol. Ass. U.K.,* **56,** 845–873.

Pingree, R.D., L. Maddock and E.I. Butler. 1977. The influence of biological activity and physical stability in determining the chemical distributions of inorganic phosphate, silicate and nitrate. *J. Mar. Biol. Ass. U.K.,* **57,** 1065–1073.

Pingree, R.D., and G.T. Mardell. 1981. Slope turbulence, internal waves and phytoplankton growth at the Celtic Sea shelf-break. *Phil. Trans. R. Soc. Lond.* **A302,** 663–682.

Pingree, R.D., G.T. Mardell, P.M. Holligan, D.K. Griffiths, and J. Smithers. 1982. Celtic Sea and Armorican Current structure and the vertical distributions of temperature and chlorophyll. *Cont. Shelf Res.,* **1,** 99–116.

Pingree, R.D., G.T. Mardell, P.C. Reid and A.W.G. John. 1986. The influence of tidal mixing on the timing of the spring phytoplankton development in the southern part of the North Sea, the English Channel, and on the northern Armorican shelf. In: *Tidal mixing and plankton dynamics.* Ed. Bowman, M.J., C.M. Yentsch and W.T. Patterson. Pp 164–192. Springer Verlag.

Pingree, R.D. and L. Pennycuick. 1975. Transfer of heat, fresh water and nutrients through the seasonal thermocline. *J. Mar. Biol. Ass. U.K.,* **55,** 261–275.

Pingree, R.D., B. Sinha, and C.R. Griffiths. 1999. Seasonality of the European slope current (Goban Spur) and ocean margin exchange. *Cont. Shelf Res.,* **19,** 929–975.

Planque, B. and C.J. Fox. 1998. Interannual variability in temperature and the recruitment of Irish Sea cod. *Mar. Ecol. Prog. Ser.*, **172,** 101–105.

Proctor, R., F. Chen, and P.B. Tett. 2003. Carbon and nitrogen fluxes across the Hebridean shelf break, estimated by a 2D couples physical-microbiological model. *Sci. Total Environ.* **314,** 787–800.

Raine, R., B. Joyce, J. Richard, Y. Pazos, M. Moloney, K.J. Jones and J.W. Patching. 1993. The development of a bloom of the dinoflagellate *Gyrodinium aureolum* (Hulbert) on the south-west Irish coast. *ICES J. Mar. Sci.*, **50,** 461–469.

Raine, R., J. O'Mahony, T. McMahon and C. Roden. 1990. Hydrography and phytoplankton of waters off south-west Ireland. *Estuar. Coast. Shelf Sci.*, **30,** 579–592.

Raine, R. and T. McMahon. 1998. Physical dynamics on the continental shelf off southwestern Ireland and their influence on coastal phytoplankton blooms. *Cont. Shelf Res.*, **18,** 883–914.

Rees, A.P., I. Joint and K. M. Donald. 1999. Early spring bloom phytoplankton-nutrient dynamics at the Celtic Sea shelf edge. *Deep-Sea Res. I,* **46,** 483–510.

Rees, H.L., M.A. Pendle, R. Waldock, D.S. Limpenny and S.E Boyd. 1999. A comparison of the benthic biodiversity in the North Sea, English Channel, and Celtic Seas. *ICES J. Mar. Sci.*, **56,** 228–246.

Reid, D.G .. 2001. SEFOS – shelf edge fisheries and oceanography studies: an overview. *Fish. Res.*, **50,** 1–15.

Reid, D.G., W.R. Turrell, M. Walsh, and A. Corten.1997. Cross-shelf processes north of Scotland in relation to the southerly migration of Western mackeral. *ICES J. Mar. Sci.*, **54,** 168–178.

Reid, P.C., N. P. Holliday and T.J. Smyth. 2001. Pulses in the eastern margin current and warmer water off the north west European shelf linked to North Sea ecosystem changes. *Mar. Ecol. Prog. Ser.*, **215,** 283–287.

Richardson, K., M.F. Lavin-Peregrina, E.G. Mitchelson and J.H. Simpson. 1985. Seasonal distribution of chlorophyll *a* in relation to physical structure in the western Irish Sea. *Oceanol. Acta,* **8,** 77–86.

Richardson, K., T.G. Nielsen, F.B. Pedersen, J.P. Heilmann, B. Lokkegaard and H. Kaas. 1998. Spatial heterogeneity in the structure of the planktonic food web in the North Sea. *Mar. Ecol. Prog. Ser.*, **168,** 197–211.

Richardson, K., A.W. Visser and F.B. Pedersen. 2000. Subsurface phytoplankton blooms fuel pelagic production in the North Sea. *J. Plankt. Res.*, **22,** 1663–1671.

Rippeth, T.P. and K.J. Jones. 1997. The seasonal cycle of nitrate in the Clyde Sea. *J. Mar. Syst.*, **12,** 299–310.

Roden, C.M. and R. Raine. 1994. Phtoplankton blooms and a coastal thermocline boundary along the west coast of Ireland. *Estuar. Coast. Shelf Sci.*, **39,** 511–526.

Rodriguez, F., E. Fernandez, R.N. Head, D.S. Harbour, G. Bratbak, M. Heldal and R.P. Harris. 2000. Temporal variability of viruses, bacteria, phytoplankton and zooplankton in the western English Channel off Plymouth. *J. Mar. Biol. Ass. U.K.* **80,** 575–586.

Sabatini, M.E., F.C. Ramirez and P. Martos. 2000. Distribution pattern and population structure of *Calanus australis* Brodsky, 1959 over the southern Patagonian shelf off Argentina in summer. *ICES J. Mar. Sci.*, **57,** 1856–1866.

Sanvincente-Anorve, L., A. Lepretre and D. Davoult. 1996. Large-scale spatial patterns of the macrobenthic diversity in the eastern English Channel. *J. Mar. Biol. Ass. U.K.* **76,** 153–160.

Seuront, L. and Y. Lagadeuc. 1998. Spatio-temporal structure of tidally mixed coastal waters: variability and heterogeneity. *J. Plankt. Res.*, **20,** 1387–1401.

Sharples, J. 1999. Investigating the seasonal vertical structure of phytoplankton in shelf seas. *Prog. Oceanogr.*, Suppl. S, 3–38.

Sharples, J., C.M. Moore, T.P. Rippeth, P.M. Holligan, D.J. Hydes, N.R. Fisher and J.H. Simpson. 2001. Phytoplankton distribution and survival in the thermocline. *Limnol. Oceanogr.*, **46,** 486–496.

Sharples, J. and J.H. Simpson. 1995. Semi-diurnal and longer period stability cycles in the Liverpool Bay region of freshwater influence. *Cont. Shelf Res.*, **15,** 295–313.

Sharples, J. and J.H.Simpson. 1996. The influence of the springs-neaps cycle on the position of shelf sea fronts. In: *Buoyancy Effects on Coastal Dynamics,* D.G.Aubrey and C.T.Friedrichs (Eds). Coastal and Estuarine Studies **53,** AGU, 71–82.

Sharples, J. and P. Tett. 1994. Modelling the effect of physical variability on the midwater chlorophyll maximum. *J. Mar. Res.,* **52,** 219–238.

Sherwin, T.J. 1981. Analysis of an internal tide observed on the Malin shelf, north of Ireland. *J. Phys. Oceanogr.,* **18,** 1035–1050.

Simpson, J.H .. 1998. The Celtic Seas. In: *The Sea* vol 11, A.R. Robinson and K.H. Brink, eds. John Wiley and Sons Inc.

Simpson, J.H., and D.G. Bowers. 1981. Models of Stratification and Frontal Movement in Shelf Seas. *Deep-Sea Res.,* **28,** 727–738.

Simpson, J.H., and D.G. Bowers. 1984. The role of tidal stirring in controlling the seasonal heat cycle in shelf seas. *Ann. Geophys.,* **2,** 411–416.

Simpson, J.H., J. Brown, J. Matthews, and G. Allen. 1990. Tidal straining, density currents, and stirring in the control of estuarine stratification. *Estuaries,* **13,** 124–132.

Simpson, J. H., and J. R. Hunter, 1974. Fronts in the Irish Sea. *Nature,* **250,** 404–406.

Simpson, J.H. and J. Sharples. 1994. Does the earth's rotation influence the position of the shelf sea fronts? *J. Geophys. Res.,* **99,** 3315–3319.

Simpson, J.H., P.B. Tett, M.L. Argote-Espinoza, A. Edwards, K.J. Jones, and G. Savidge. 1982. Mixing and phytoplankton growth around an island in a stratified sea. *Cont. Shelf Res.,* **1,** 15–31.

Soetaert, K., P.M.J. Herman, J.J. Middleberg, C. Heip, C.L. Smith, P. Tett, and K. Wild-Allen. 2001. Numerical modelling of the shelf break ecosystem: reproducing benthic and pelagic measurements. *Deep-Sea Res. II,* **48,** 3141–3177.

Southward, A.J., S.J. Hawkins and M.T. Burrows. 1995. Seventy years' observations of changes in distribution and abundance of zooplankton and intertidal organisms in the western English Channel in relation to rising sea temperature. *J. Therm. Biol.,* **20,** 127–155.

Souza, A.J., J.H. Simpson, M. Harikkrishnan, J. Malarkey. 2001. Flow structure and seasonality in the Hebridean slope current. *Oceanol. Acta.,* **24S,** S63-S76.

Tarling, G.A., T. Jarvis, S.M. Emsley and J.B.L. Matthews. 2002. Midnight sinking behaviour in *Calanus finmarchicus:* a response to satiation or krill predation. *Mar. Ecol. Prog. Ser.,* **240,** 183–194.

Tett, P., R. Gowen, B. Grantham, K. Jones and B.S. Miller. 1986. The phytoplankton ecology of the Firth of Clyde sea-lochs Striven and Fyne. *Proc. R. Soc. Edinb.,* **90B,** 223–228.

Thompson, K.R., and D.T. Pugh. 1986. The subtidal behaviour of the Celtic Sea – 2 Currents. *Cont. Shelf Res.,* **5,** 321–346.

Townsend, D.W., Pettigrew, N.R. and A.C. Thomas. 2001. Offshore blooms of the red tide dinoflagellate, *Alexandrium* sp., in the Gulf of Maine. *Cont. Shelf Res.,* **21,** 347–369.

Townsend, D.W. and M. Thomas. 2002. Springtime nutrient and phytoplankton dynamics on Georges Bank. *Mar. Ecol. Prog. Ser.,* **228,** 57–74.

Trimmer, M., R.J. Gowen, B.M. Stewart and D.B. Nedwell. 1999. The spring bloom and its impact on benthic mineralisation rates in western Irish Sea sediments. *Mar Ecol. Prog. Ser.,* **185,** 47–46.

Vallet, C. and J-C. Dauvin. 2001. Biomass changes and benthic-pelagic transfers throughout the benthic boundary layer in the English Channel. *J. Plankt. Res.,* **23,** 903–922.

van Weering, T.C.E., H.C. de Stigter, W. Balzer, E.H.G. Epping, G. Graf, I.R. Hall, W. Helder, A. Khripounoff, L. Lohse, I.N. McCave, L. Thomsen, and A. Vangriesheim. 2001. Benthic dynamics and carbon fluxes on the NW European continental margin. *Deep-Sea Res. II,* **48,** 3191–3221.

Videau, C. 1987. Primary production and physiological state of phytoplankton at the Ushant tidal front (west coast of Brittany, France). *Mar. Ecol. Prog. Ser.,* **35,** 141–151.

White, M., and P. Bowyer. 1997. The shelf-edge current north-west of Ireland. *Ann. Geophys.*, **15,** 1076–1083.

White, R.G., A.E. Hill and D.A. Jones. 1988. Distribution of *Nephrops norvegicus* (L.) larvae in the western Irish Sea: an example of advective control on recruitment. *J. Plankt. Res.*, **10,** 735–747.

Williams, R. 1985. Vertical distribution of *Calanus finmarchicus and Calanus helgolandicus* in relation to the development of the seasonal thermocline in the Celtic Sea. *Mar. Biol.* **86,** 145–149.

Williams, R., D.V.P. Conway and H.G. Hunt. 1994. The role of copepods in the planktonic ecosystems of mixed and stratified waters of the European shelf seas. *Hydrobiol.*, **292/293,** 521–530.

Wollast, R .. 2003. Continental margins: review of geochemical settings. In: *Ocean Margin Systems* (eds. G. Wefer, D. Billet, D. Hebbeln, B. B. Jörgensen, M. Schlüter, T. C. E. van Weering), Springer-Verlag, Berlin-Heidelberg-New York.

Wollast, R., and L. Chou. 2001. The carbon cycle at the ocean margin in the northern Gulf of Biscay. *Deep-Sea Res. II,* **48,** 3265–3293.

Chapter 26. THE BALTIC AND NORTH SEAS: A REGIONAL REVIEW OF SOME IMPORTANT PHYSICAL-CHEMICAL-BIOLOGICAL INTERACTION PROCESSES (20,S)

JOHAN RODHE

Göteborg University

PAUL TETT

Napier University

FREDRIK WULFF

Stockholm University

Contents

1. Introduction

The goal of the present review is to indicate how physical-chemical-biological interactions control ecosystem dynamics in waters that range from almost fresh in the innermost part of the semi-enclosed Baltic Sea to oceanic salinities in the northern part of the North Sea. Along the rim of these seas are found some of the most densely populated areas of the world. A total of about 85 million people live in the Baltic Sea catchment area, and 184 in that of the North Sea. It is therefore no surprise that effects of eutrophication, toxic substances, oil spills and overfishing have significantly perturbed, or even damaged, parts of these coastal seas. There are comprehensive assessments of the environmental status of these seas, by the Helsinki Commission, Baltic Marine Environment Protection Commission

The Sea, Volume 14, edited by Allan R. Robinson and Kenneth H. Brink
ISBN 0–674–01527–4

(HELCOM, 2002) and by the OSPAR Commission for the Protection of the Marine Environment of the North-East Atlantic (OSPAR Commission, 2000). The reader is referred to these for thorough reviews on this subject. In the present review of physical-chemical-biological interaction processes we will only give a short overview of the state of the seas.

Sections 2, 3 and 4 give large-scale overviews of the physical oceanography, the sediment dynamics and the nutrient fluxes and the geo-chemical environment. This is in Section 5 followed by descriptions of the interactive processes of the different ecosystem regions found in the Baltic and North Seas. Some environmental problems are shed light on in Section 6 and concluding remarks are given in Section 7, including a short comment on climate variation effects.

2. The large-scale physical environment

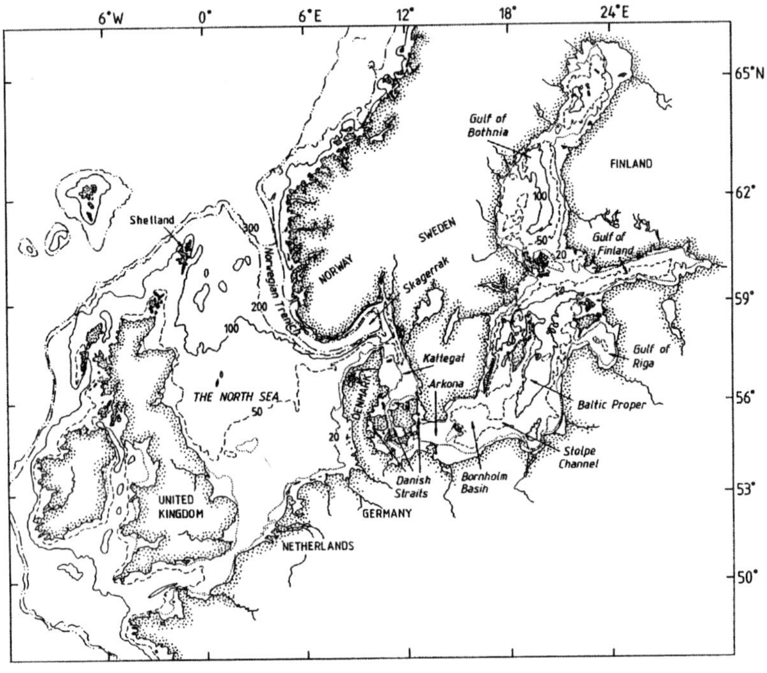

Figure 26.1 Topography of the Baltic and North Seas

Although the North and Baltic Seas make up an interconnected system (Fig. 26.1), they are often treated as 3 units: the North Sea proper; the Skagerak-Kattegat-Belt Sea region; and the Baltic Sea. When the words are used without qualification, the 'North Sea' comprises waters between Britain (including the Orkney and Shetland islands), continental Europe, and Norway, from the Dover Strait (51°N) to the continental shelf-break at 61–62°N. The remaining boundary of this natural North Sea lies at the interface between the Baltic outflow in the Kattegat and the more saline waters of the Skagerrak. The 'ERSEM North Sea' (Baretta-Bekker and

Baretta,1997) has its eastern limit at roughly 8°E at the entrance to the Skagerrak, and extends only to 61°N. The 'Greater North Sea' of the Quality Status Reviews (North Sea Task Force, 1993; OSPAR Commission, 2000) includes the (English) Channel, the entire Skagerak-Kattegat region, and all waters north of Scotland to 62°N. Including estuaries and fjords, it has a surface area of about 750 000 km^2 and a volume of about 94 000 km^3. Rivers draining about 850 000 km^2 discharge about 300 km^3 of freshwater annually into this region. The Baltic Sea is commonly considered as the region inside the Danish Straits but encompass 'officially' also Kattegat in the Helsinki Commission. Although the Baltic Sea is smaller, at 377,000 km^2 (415,000 km^2 including Kattegat), its rivers drain about 1 650 000 (1 720,000) km^2 and discharge about 440 (480) km^3 yr^{-1}. The volume, including Kattegat, is 22,000 km^3.

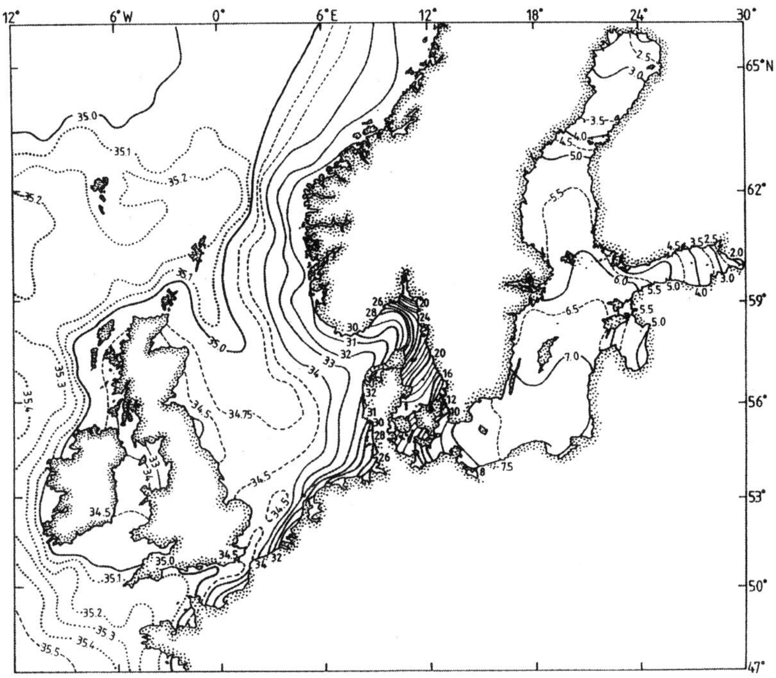

Figure 26.2 Baltic and North Seas surface salinity (August).

The net freshwater input results in a lower-than-ocean salinity (Fig. 26.2) in the region. The gradient of salinity, together with variations in topography (Fig. 26.1) and tidal stirring, creates a variety of physical environments in the Baltic and North Seas. The reader is referred to Chapter 24 of The Sea, Volume 11 (Rodhe, 1998) for a thorough discussion of the large-scale physical environment of this region. Here follows a summary.

2.1. The North Sea

The North Sea is a shelf sea, widely open towards the North Atlantic to the north, and having a comparatively narrow connection to the North Atlantic through the English Channel in south-west. The average depth is about 100 m. An outstanding feature is the Norwegian Trench with a fjord-like topography. The trench cuts into the shelf along the Norwegian coast. A maximum depth of 700 m is found in the innermost part of this trench, in the Skagerrak, which leads by way of the Kattegat and shallow Belt Sea to the Baltic Sea.

A general cyclonic circulation, on the average, renews the water in the North Sea on a timescale of the order of one year. Most of the water enters from the north and follows the Norwegian Trench in a cyclonic direction. Significant amounts of water also enter the North Sea just to the east of Shetland and between the Shetland and the Orkeneys (OSPAR Commission, 2000). The average transports involved are some $2 \cdot 10^6$ m^3s^{-1}. This is large compared to the freshwater input (including that to the Baltic Sea), which is only about $2.5 \cdot 10^4$ m^3s^{-1}. In the south relatively high-salinity water from the Channel enters the North Sea with an average (but highly variable) transport of about $0.1 \cdot 10^6$ m^3s^{-1}. The salinity of the North Sea is close to 35 PSU, except close to river outlets in the south-eastern North Sea, and along the coasts in the northeast, where the outflows from the Baltic Sea and large rivers lower the surface salinity to about 25–30 PSU.

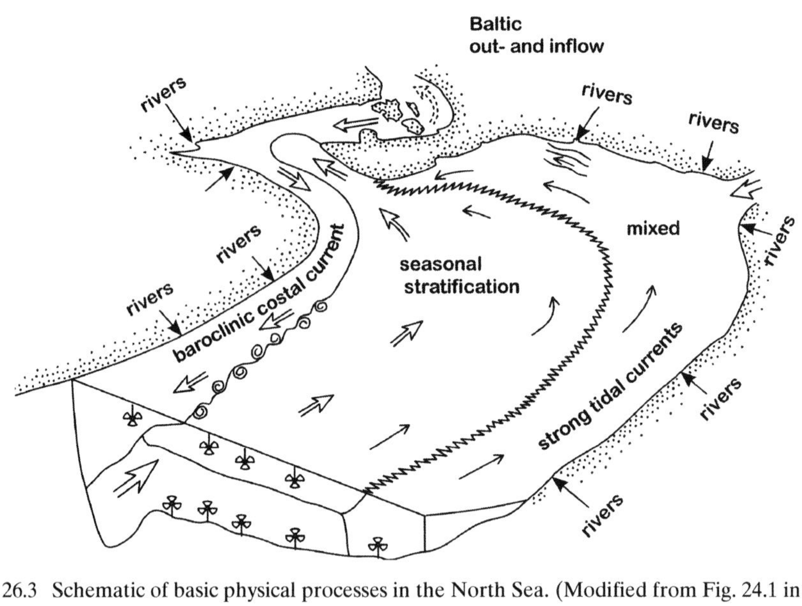

Figure 26.3 Schematic of basic physical processes in the North Sea. (Modified from Fig. 24.1 in Rodhe, 1998).

Fig. 26.3 illustrates how the North Sea can be divided into regions classified according to which processes dominate to determine the physical environment. The central part of the North Sea becomes density stratified due to heating in summer and mixes vertically during winter. Along the coasts in the southern and western parts of the North Sea, however, the tides are strong enough to keep the water vertically mixed throughout the year except in some river outflow regions.

The Norwegian Trench region is stratified by salinity all around the year due to the outflow from the Baltic Sea and local rivers. An inflow of high-saline Atlantic water along the western slope of the trench renews the underlying water more or less continuously in this region. On the contrary, the water in the deepest part of the trench (in the Skagerrak) is renewed intermittently with stagnation periods lasting for some years in between. A more ore less continuous upwelling and entrainment of subsurface water makes the salinity of the outflowing water of Baltic origin increase along its path from eastern Skagerrak and outwards along the Norwegian coast.

Approaching the Baltic-Sea entrance through the Kattegat the surface salinity drops to about 15 PSU. Kattegat is strongly stratified. A deep layer, having almost oceanic salinity, underlies the low-saline surface layer. The thickness of the surface layer is about 15 m. The internal circulation in the Kattegat is of estuarine type. This includes entrainment of deep-water into the surface layer implying an increasing surface-layer outflow and a decreasing deep-layer inflow. Being a comparatively narrow and shallow connection between the Baltic Sea and the North Sea the hydrographic conditions in the Kattegat undergo large changes in concert with the 'breathing' of the Baltic.

2.2. The Baltic Sea

The Baltic Sea has shallow and narrow connections with the North Sea through the Danish Straits. The deepest connection is through the Great Belt. It has a sill depth of about 18 m. A second connection through the Öresund has a sill depth of about 8 m. Both sills are on the Baltic Sea side of the straits. The Baltic Sea consists of a number of basins connected by relatively deep channels. A main division within the Baltic Sea separates the Baltic proper in the south from the Gulf of Bothnia in the north. Another sub region separated from the Baltic proper is the Gulf of Riga. The average depth of the entire Baltic Sea is about 60 m, and typical basin depths are in the range 100 to 250 m.

The restricted water exchange through the Danish straits implies a long residence time for the water in the Baltic Sea—about 30 years. This, and a comparatively large fresh water supply (about $1.6 \cdot 10^4$ $m^3 s^{-1}$) give the Baltic-Sea a very low mean salinity.

The exchange of water through the Danish Straits is mainly forced by an alternating atmospheric pressure- and wind-induced sea-level difference between the Kattegat and the Baltic Sea. The flow is essentially barotropic at the shallow sills, where low-saline outflows alternate with high-saline inflows. Typical flow rates are several times the net outflow, which equals the average fresh water supply to the Baltic Sea.

The Baltic proper is strongly stratified by salinity. A 50 to 70 m deep surface layer with no apparent vertical salinity gradient is separated from the weakly stratified deep water by a halocline. The surface salinity ranges from 8–9 PSU in the south, to 6–7 PSU in the northern part of Baltic proper. This surface layer becomes thermally stratified in summer leaving surface layer winter water between the thermocline and the halocline. Surface layer cooling opens up for deep-reaching wind mixing during winter whereby the upper part of the halocline becomes eroded and deep-water mixes into the surface layer as one step in the slow estuarine circulation found in the Baltic proper. The inflows of high-saline water over the sills form the first step of the estuarine circulation. However, outflowing Baltic Sea surface water dilutes the inflowing water to a comparatively low salinity before passing the sills. In the first basin inside the sills the deep water has a salinity of about 17 PSU, as an average. Descending into the deeper basins of the Baltic proper the salinity decreases even more due to entrainment of surface water along the flow. Most inflowing water interleaves just below the surface layer. In that way it renews the halocline water and the upper part of the deep water. The water in the deepest parts of the basins is renewed by long-lasting extreme inflows, which occur with several years interval. Between these inflows the deep water is stagnant and eventually becomes anoxic. The frequency of these deep-water inflows has decreased in recent decades. This has partly been attributed to the North Atlantic Oscillation (NAO) and the related variations in winds and precipitation in the region. The low inflow frequency during the last decades seems to be correlated with a pronounced positive NAO index during the same period.

In the Gulf of Bothnia the surface salinity drops from 6 PSU in the south to 2–3 PSU in the north. The deep-water salinity is a few PSU higher than that of the surface water. Convection processes renew the deep water every year and the comparatively good exchange with the Baltic Sea surface water to the south implies a residence time of about 5 years for the water in the Gulf of Bothnia.

Sea ice is formed in the Baltic Sea every year. However, the ice extent varies. During mild winters, only the northernmost basin and parts of the coastal archipelagos in the more southerly regions are covered by ice. During severe winters almost the entire Baltic Sea is covered by ice. During these conditions most salinity-stratified areas outside the Baltic also becomes ice covered. These include the Danish straits, the Kattegat and coastal regions in the Skagerrak and along the southeastern corner of the North Sea.

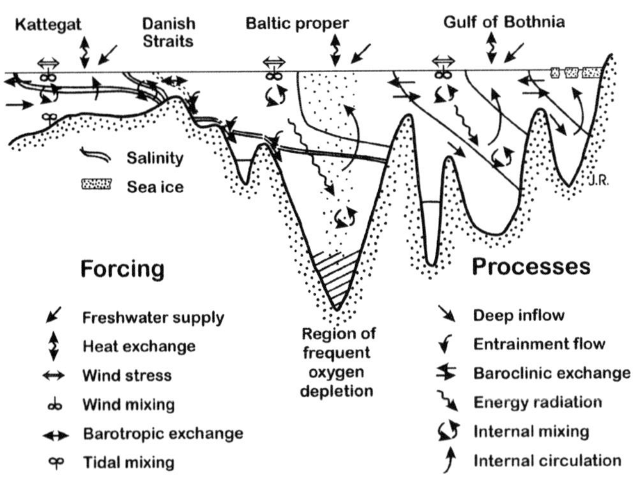

Figure 26.4 Schematic of stratification, forcing and basic physical processes in the Baltic Sea. (Modified from Winsor et al., 2001.)

3. Large-scale sediment dynamics

From a geological perspective, the Baltic Sea is a very young and dynamic region. The crustal rebound, currently ranging from nil in the south to 80–90 cm per 100 years in the northernmost region, is the result of a retreat of a glacier that covered the whole region about 15000 years ago (Flodén and Winterhalter, 1981). Due to this postglacial uplift, glacial and postglacial clays deposited hundreds to thousands of years ago are eroded from shallow areas (< 80 m). This eroded material is then deposited in the deepest parts of the basins together with recently produced organic matter from the overlying water column. This mixture of materials with very different ages, make it difficult to use traditional dating techniques of sediment cores in mass balance calculations (Jonsson et al., 1990; Blomqvist and Larsson, 1994). However, the deeper basins in the Baltic proper as well as many enclosed coastal basins have been anoxic and devoid of benthic macrofauna during many decades in the last century. Without perturbation, annually laminated sediments occur and these can be dated more easily. From such studies, Jonsson and Carman (1984) calculated a 1.7 fold increase in sediment organic matter, which in turn is related to an enhanced primary production, due to eutrophication during the last decades. The drastic expansion of the area covered by laminated sediments since the 1940's, that well mimics the expansion of anoxia, is illustrated in Fig. 26.5

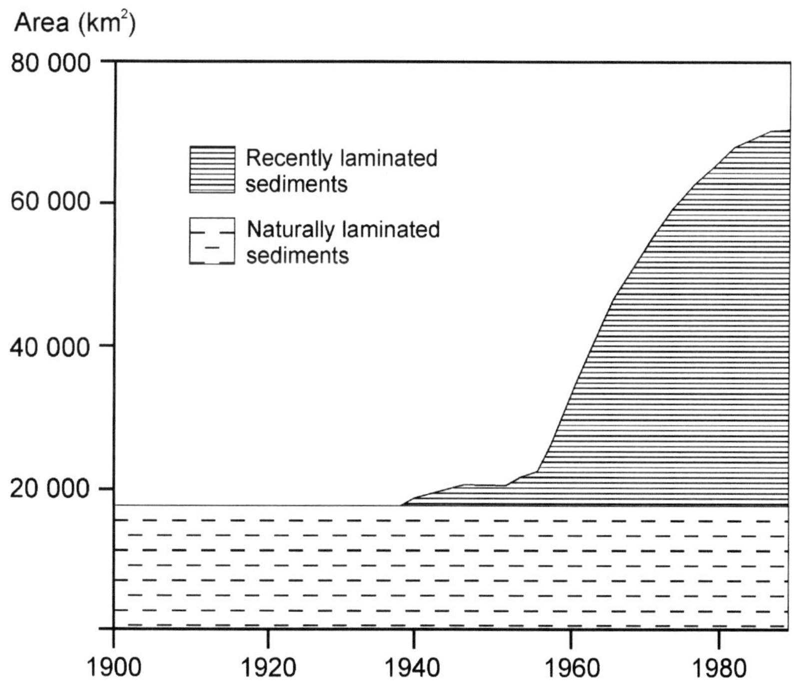

Figure 26.5 The expansion of laminated sediments in the Baltic Sea during the 20th century (Modified from Jonsson et al., 1990).

During the recent Ice Age, Scandinavian and Scottish mountain glaciers spread over the North Sea, sculpturing the topography and leaving elevations such as the

Dogger-Fisher Bank and depressions like the Oyster Ground, the submerged part of the Elbe valley, Devil's Hole, Fladen Ground and the Norwegian Trench. After the retreat of the ice there has been a subsidence of the crust in the southern part of the North Sea. This still continuing subsidence is today a threat to the vast lowland region bordering to the North Sea in the south.

The glaciers left much sediment. Cliff erosion, especially severe in eastern England, and river inputs continue to add material to the North Sea. All this is continuously reworked by wind waves and tidal currents, resulting in a predominantly sandy bed over much of the North Sea, with gravel banks and shifting sandbanks in shallower waters, and finer-grained sediments accumulating in the depressions. The general cyclonic water movements implies that most of the fine-grained material supplied to the North Sea finds its final deposition in the deep parts of the Norwegian Trench, particularly in the Skagerrak (Eisma and Kalf, 1987). With this transport of fine-grained material also follows different particle-bound substances, metals and other, indicated by high concentrations in the sediments.

4. Large-scale nutrient fluxes and geo-chemical environment

4.1. The North Sea

The primary production of the North Sea is supported by a northern inflow of nutrient-rich water from the Atlantic Ocean and by anthropogenic nutrients added mainly to the southern part. Because most parts of the North Sea are shallow and generally well oxygenated, relatively little nutrient is lost through burial in sediments or exported in deep water leaving the region with increased nutrient concentration. Winter convection of the seasonally stratified parts of the North Sea returns to the upper waters much of those nutrients remineralized in or near the seabed (excepting losses due to denitrification). In contrast, winter convection is hampered by the salinity stratification in the Norwegian-Trench region. Here upwelling and entrainment of North Atlantic water feeds the upper layer with nutrients. This is especially important in the Skagerrak where the upwelling is of the order of one meter per day.

The nutrients considered limiting organic primary production are nitrogen, phosphorus and silicon. Although the availability of iron and cobalt may sometimes also be important (see Tett et al., 2003), these metals will not be further discussed here.

Budgets for nutrient elements should take account of all relevant compounds and fluxes, but there may be difficulties in deciding how much refractory particulate and dissolved organic nitrogen (PON and DON), and how much particle-bound phosphorus to include. Such budgets for the North Sea have been constructed by several authors, and recently reviewed by Sundermann et al. (2001). The situation is unsatisfactory, in that there are large differences between authorities in their budgets, the result of differences in original data and interpretation.

For example, in the budget of the OSPAR Commission (2000), the losses of nutrient nitrogen in the Norwegian coastal current outflow (5587 kt N yr^{-1}) and to the Baltic (136 kt N yr^{-1}) were calculated so as to balance a budget, which included a large value (2700 kt N yr^{-1}) for nitrogen fixation in the North Sea. This value, based on an old study, now seems implausible. Instead, assuming zero fixation, implies a net input of nutrients into the North Sea from the Atlantic, and perhaps also into the North Sea from the Baltic. It is suggested below that the Baltic exports about 140 kt N yr^{-1} and 11 kt P yr^{-1} into the Kattegat (see Fig. 26.7). The average denitrification flux, of between 0.2 and 0.3 mol NO_3^- m^{-2} yr^{-1} required by the overall loss of 1600 kt N yr^{-1} in the North Sea, is compatible with modern measurements of sediment fluxes (Tett et al., 2003). Thus, nitrogen inputs from rivers, waste pipes, and atmosphere (totalling 1150 kt N yr^{-1}: OSPAR Commission, 2000) are insufficient to counterbalance North Sea denitrification, which is about 4 times nitrogen fixation in the Baltic (see below).

As will be discussed in Section 6, the river inputs to the North Sea are largely of anthropogenic nutrients. However, despite systematic monitoring of river nutrient loads, the removal of nutrients in estuaries - often places of rapid denitrification and complex interactions between phosphate and particles - brings about a somewhat uncertain reduction in discharge. For example, the OSPAR Commission (2000) estimate that only about 700 kt N yr^{-1} and 26 kt P yr^{-1} reaches the 'Greater North Sea' through estuaries into which rivers bring 1400 kt N yr^{-1} and 75 kt P yr^{-1}.

Table 26.1 proposes a balanced budget for N and P in the North Sea that assumes zero denitrification and includes estimates of Baltic exports given in this chapter. With the possible exception of direct inputs, which may contain labile DOM (dissolved organic mater) and POM (particulate organic matter), the inputs are assumed to be predominantly DAIN (dissolved available inorganic nitrogen) and DIP (dissolved inorganic phosphorus). Losses to the sediment, resulting in denitrification, are mainly PON, particulate organic nitrogen, (on the grounds that denitrification largely uses nitrate produced in sediment pore water by oxidation of ammonia, itself resulting from mineralization). The deduced exports into the depths of the Norwegian Trench, and in the Norwegian Coastal Current, may include POM and refractory DON. Loss of particle-bound phosphate may also be included.

The values given in Table 26.1 should be considered highly approximate. In addition to the sources of inaccuracy already mentioned, the estimates for ocean boundary fluxes take little account of what may be large interannual variations in water flows, or may be biased by observations made in particular years. Nevertheless, the table does suggest that (given a typical water residence time of 1 year in the North Sea) each atom of nutrient-N and -P cycles 2–3 times through biomass on their way from sources to sinks. Local nitrogen inputs, mainly anthropogenic, are about a quarter of the total (and a much larger proportion in the southern parts of the North Sea. Local phosphorus inputs are less important, partly because of the role of estuaries in limiting exported concentrations.

Atomic ratios of N:P are close to the Redfield 16:1 over much of the North Sea in winter (Radach and Gekeler, 1996) and in the external inputs. These winter and imported nutrients are largely inorganic. Although plankton optimally construct their biomass at Redfield, cell-quota models of micro-organism growth suggest that N rather than P is limiting while supply ratios of N:P are less than 30:1 (Tett et al., 2003) and this confirms the opinion of those authors who assume or conclude that primary production in the North Sea is more likely to be limited by nitrogen than by any other nutrient. In some parts of the southern North Sea, nitrogen enrichment has brought about winter or supply ratios of N:P exceeding 30:1, and arguments that P are limiting here are more convincing. The silica content of water from the northern Atlantic is relatively low, and the winter (atomic) ratio of N:Si in the northern North Sea exceeds 2:1. Silicon exhaustion by the spring bloom may play a part in controlling the annual succession of phytoplankters from diatoms to flagellates and dinoflagellates. Lower winter DAIN in the central North Sea brings the ratio below 1.5:1, but it is increased again in coastal waters by the anthropogenic enrichment with nitrogen (whereas silicon discharge tends to be decreased in human-influenced rivers - see Billen et al., 1991). In these waters, the depletion of silica relative to nitrogen by spring diatom growth may encourage growth of the colonial alga *Phaeocystis*.

TABLE 26.1:
Proposed N and P budgets for the North Sea. Production is that for the 'ERSEM North Sea'; some of the fluxes include inputs to or losses from the Skagerrak-Kattegat region. Note considerable uncertainties in most values.

	kt N yr^{-1}	kt P yr^{-1}	Source of values (notes)
internal turnover	11530	1596	based on production, at C:N:P of Redfield
GAINS (from)			
Atlantic inflow	3179	528	net water flow of 37 000 km^3 yr^{-1} at 6 _M DAIN, 0.45 _M DIP
Baltic	140	11	from this chapter
Channel	300	33	Brockmann et al., 1990
External inputs total	3619	572	(atomic N:P = 14:1)
as percent turnover	*31%*	*36%*	
Estuaries net	700	26	OSPAR Commission, 2000
Direct	100	13	OSPAR Commission, 2000
Atmosphere	350		OSPAR Commission, 2000
Local inputs total	1150	39	(atomic N:P = 65:1)
as percent turnover	*10%*	*2%*	
LOSSES (to)			
Denitrification	-1600		Seitzinger and Giblin, 1996
Norwegian Trench and sediment burial	-3169	-611	to balance budget
Losses total	4769	611	(atomic N:P = 17:1)
as percent turnover	*41%*	*38%*	
TOTALS	0	0	

According to North Sea Task Force (1993), "the average annual primary production of the North Sea is probably in the range 150–250 g C m^{-2} yr^{-1}." However, the estimate is imprecise "because of difficulties with methods and definitions, and because of variability between areas and seasons." There have been no surveys of primary production in the whole North Sea during an annual cycle, but the southern and central North Sea was systematically sampled during 1988–89 (Joint and Pomroy, 1993), giving a range of annual primary production from 79 g C m^{-2} yr^{-1} in the turbid and strongly stirred coastal waters of eastern England by way of 100 - 119 g C m^{-2} yr^{-1} in the offshore seasonally stratified waters to 261 g C m^{-2} yr^{-1} in the coastal waters west of Jutland. The mean for the areas south of 55°30'N was just under 150 g C m^{-2} yr^{-1}.

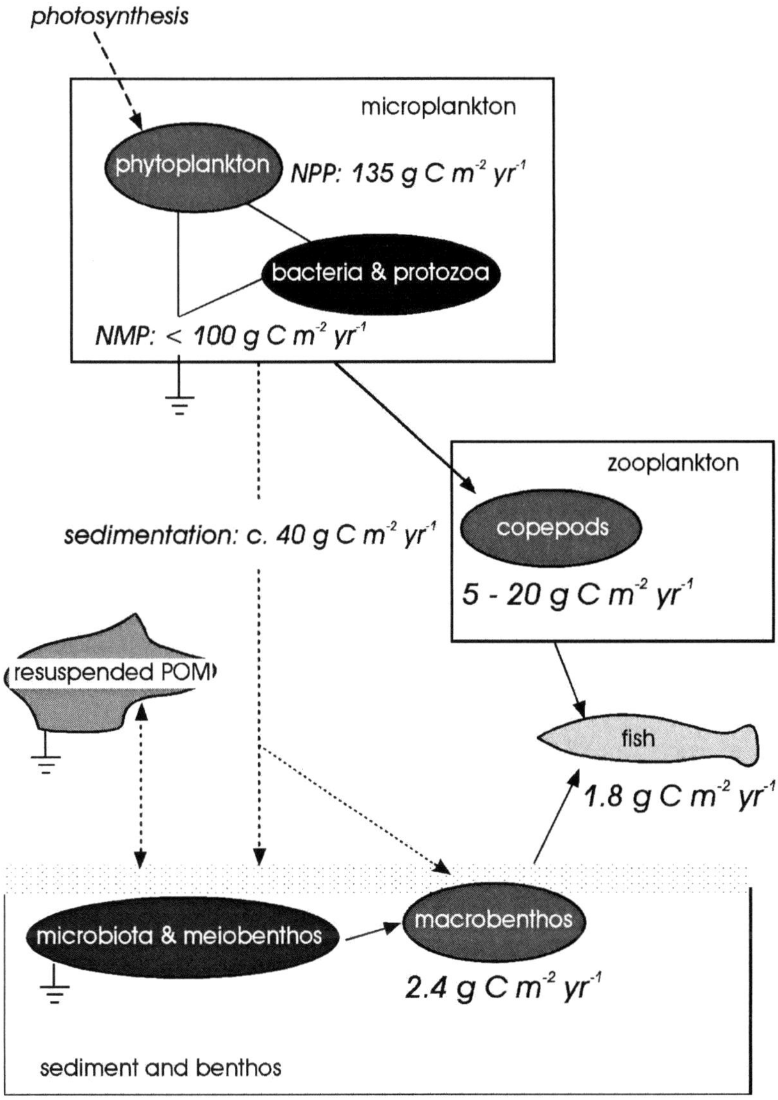

Figure 26.6 Simplified diagram of the production and fate of organic matter in the North Sea. The earthing symbol shows the main respiratory sinks of organic carbon. NPP is 'Net Phytoplankton (or, Primary) Production'; NMP is 'Net Microplankton Production', which takes account of losses by pelagic microbial consumers. Standard methods measure more than NPP, or between NPP and NMP, depending on duration of incubation. At shallow and tidally energetic sites, resuspended POM may be a large part of total organic matter in the water column. Labile POM is probably not transported over large distances, but its export over the shelf break (and into the Norwegian Trench) may be an important sink for adjacent waters.

Estimates of production at higher trophic levels are at least equally uncertain because of local, regional, seasonal and inter annual variability. North Sea Task Force (1993) reported that "the average annual production of copepods in the North Sea is 5–20 g C m^{-2} yr^{-1}, macrobenthos production is about 2.4 g C m^{-2} yr^{-1}, and fish production is about 1.8 g C m^{-2} yr^{-1}." Fig. 26.6 completes the food web (albeit in very simplified form: cf. Hardy, 1924). Phytoplankton are shown here as part of a microplankton compartment, which also includes the heterotrophic bacteria and pelagic protozoa that are the first beneficiaries of primary production. Net microplankton primary production (NMP) is lower than net phytoplankton primary production (NPP) because of the extra losses of photosynthetically-fixed organic carbon in the metabolism of the bacteria and protozoa (see Tett, 1990; Lee et al., 2002). A substantial fraction of NMP reaches the seabed, either in sinking microplankters or in the faeces of animals, and its mineralisation is responsible for sediment oxygen demand (SOD) of 2–20 mmol O_2 m^{-2} d^{-1}. The main mineralisers are the microbiota and meiofauna of the sediment; larger benthic animals, feeding on the smaller organisms or on directly sedimented POM, harness only a small part of the input. It is becoming increasingly apparent that much of the sedimented POM is mineralised not in the consolidated sediment, but in a loosely consolidated fluff layer, which may at times be largely resuspended into the water column (e.g. Jago et al., 2002, and references therein).

Permanent deposition of suspended particulate matter is in the North Sea restricted to some shallow areas in the southeastern part and, above all to the eastern part of the Norwegian Trench (the Skagerrak) and to the northern part of the Kattegat. In most parts of the North Sea the sedimented organic material becomes decomposed instead of buried, whereas the non-organic material is transported to the deposition areas. This implies that the organic content of the accumulated sediments is fairly low in most parts of the North Sea.

4.2 The Baltic Sea

The Baltic Sea is today about as productive as the North Sea, because nutrient inputs have increased by 4 and 8 times for nitrogen and phosphorus, compared to a century ago (Larsson et al., 1985). The overall primary production is now about 150 g C m^{-2} yr^{-1}, with a pronounced south-north gradient from 200 g C m^{-2} yr^{-1} in the Kattegat/Belt Sea and southern Baltic proper to 50 g C m^{-2} yr^{-1} in the Bothnian Sea and less than 20 g C m^{-2} yr^{-1} in the Bothnian Bay (Wasmund et al., 2001). The increase in productivity has been most pronounced in coastal areas, close to river mouths and large point sources of nutrients. Increased overall levels in the off shore regions Kattegat, Belt Sea, Baltic proper and Bothnian Sea and the Gulfs of Finland and Riga are also evident. Production estimates vary between a factor 1.5–2, compared to levels found or calculated for the 1950's or earlier (see Stigebrandt, 1991; Elmgren, 1989; Wasmund et al., 2001). Agricultural land has been estimated to account for about half of the total N load and a fifth of the P load to the Baltic. After the independence in 1990, the fertilizer use in Estonia, Latvia and Lithuania dropped drastically to levels of the 1950s and the yield of most agricultural products decreased with almost 50%. In spite of these drastic changes, there have been almost no changes in nutrient runoff to the Baltic, at least until 1998. Löfgren et al. (1999) attribute this to large nutrient pools, still remaining in the previously heavily fertilized soils that will continue to leak nutrients for many decades, before substantial decreases in runoff will be seen. Although the riverine input is the major source for nutrients, the atmospheric deposition of nitrogen directly over the sea could represent about a third of the total input. Only a minor fraction of the phosphorus load is coming through atmospheric deposition. Recent calculations of nitrogen deposition to the Baltic show a downward trend in deposition for the last decade.

Nutrient budgets provide estimates of the relative importance of external sources, advection to adjacent areas and of internal processes for the maintenance of nutrient concentrations in a sea area as shown in Fig. 26.7 for the N and P budget of the major Baltic Sea Basins.

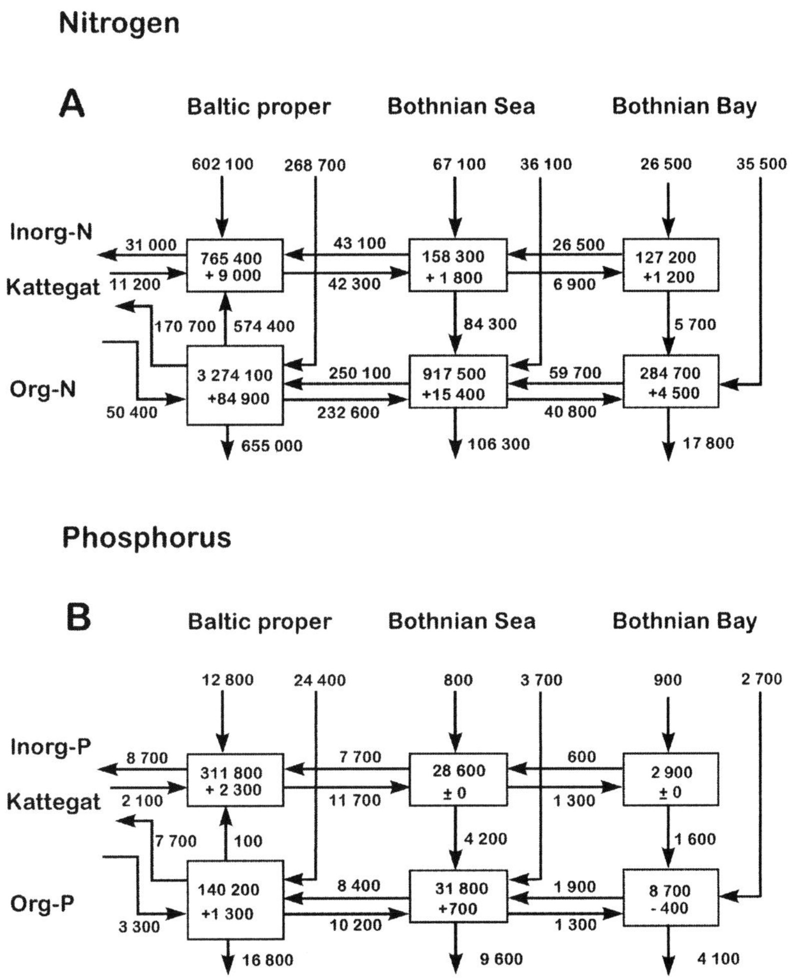

Figure 26.7 Nitrogen and phosphorus budgets for 1975–91 of the major basins of the Baltic Sea. Flows and changes of pools in the water column are expressed in tons yr^{-1}. (Modified from Wulff et al., 2001).

In contrast to the North Sea, there is a net export, from the Baltic proper to Kattegat of about 14% and 20% of the load of N and P, to the basins inside the Danish straits. The exceeding inputs are either causing an increase in concentrations or a loss via denitrification (N) or an accumulation in bottom sediments (N and P). The residence times vary between 2 and 6 years for organic N and between

1 and 4 years for DIN and vary between 1.6 to 3.2 years and between 1.9 and 3.2 years for the DIP and organic P fractions. The residence times for total N and P for the entire Baltic (inside the Danish Straits) are 5 and 11 years, respectively. A closer inspection of these budgets also shows differences that can be related to differences in the relative importance of various biogeochemical processes. The northernmost basin (Bothnian Bay) retains a much larger fraction of the external P inputs than the southernmost basin (Baltic proper). This has been attributed to the efficient P sink in the well oxidized iron rich sediment of the oligotrophic Bothnian Bay, while the partly reduced sediments of the more eutrophic Baltic proper are less efficient. On the other hand, the N losses occur primarily in the Baltic proper while a much larger fraction of the inputs are exported from the Bothnian Bay. N losses through denitrification plays a minor role in the northernmost oligotrophic basin where most organic matter is decomposed aerobically while the opposite is true for the Baltic proper (Wulff et al., 2001).

These budgets thus yield a system-level principal view on how nutrient concentrations and ratios will change in relation to inputs. This is illustrated in Fig. 26.8, and consistent with our understanding of key processes (Wulff et al., 2001). With increasing inputs, primary production and sedimentation will increase, as are illustrated in the three panels, from left to right in the figure. Low sedimentation means that a limited amount of organic matter will be decomposed in the sediments. This means that the sediment surface will remain oxidized and function as an efficient P-sink. Most of the organic matter will be decomposed aerobically and only a small amount will be buried deeper into the anoxic part of the sediment. Thus, denitrification, i.e. anaerobic decomposition of organic matter using nitrate, is probable of little importance here. More N, compared to P will then be recycled and the primary production will be P-limited. With increasing production and sedimentation of organic matter, oxygen consumption in the sediments will increase. If the sediment surface is still oxic, P-adsorption will occur but to a lesser extent and more is recycled into the water column. On the other hand, a larger fraction of organic matter is buried and decomposed anaerobically. Denitrification is probably highly efficient in such an environment where high concentration of nitrate from the surface layer, can diffuse over a short distance and supply the decomposing bacteria with electron acceptors. Thus, it is likely that the efficiency of the P-sink will decrease and the N-sink will increase with increasing organic load to the sediment. In such a system, excess P, in comparison to N, will be recycled into the water and primary production is likely to be N-limited. In Fig. 26.8, an additional N source; N_2-fixation by cyanobacteria is favored under such conditions, i.e. with high P but low N concentrations in the water.

Finally, the load of organic matter has increased to such a degree that the increased decomposition has caused the redoxcline to move into the water column. The efficiency of both the P and N sinks of the sediments will decrease drastically. In this case, both nutrients will be recycled into the water column, supplying primary production and accelerating eutrophication.

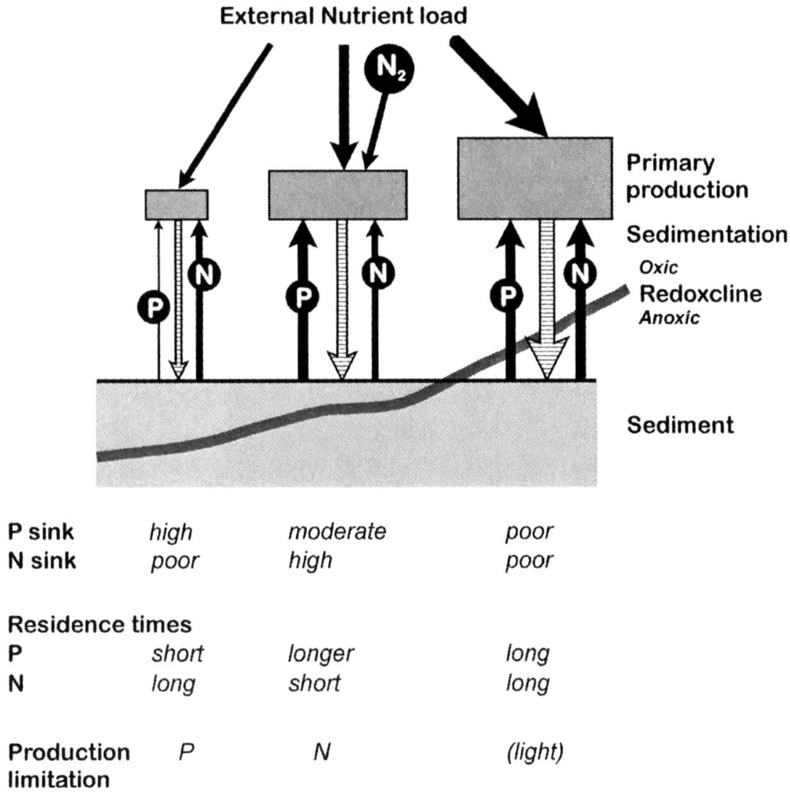

Figure 26.8 A schematic illustration of how the cycles of nitrogen and phosphorus in a shallow marine ecosystem will change in response to increase in external loads (left to right). (Modified from Wulff et al., 2001).

The internal physical transports are consistent with the budgets results. The net upwelling of deep water is only a few meters per year. The recirculation of nutrients from the sub-surface layers during winter convection is ineffective due to the haline stratification, which obstructs deep reaching convection. The nutrient levels in the deep water of the Baltic proper are considerably higher than in the inflowing deep water due to remineralization and slow renewal the basin water. Evidence are now accumulating that the increase in nutrient levels, as a direct effect due to anthropogenic inflows, occurred primarily between 1950–1970. The variations in concentration thereafter seem to be regulated to a larger extent by internal biogeochemical processes (Rahm et al., 1996; Conley et al., 2002). For instance, the release and uptake of phosphate from anoxic and oxidized sediments have, during the 1990's been considerably larger (~3 times) than the external load from land.

The Baltic Sea thus presents large gradients in nutrient dynamics and in the relative importance of different processes. Before the onset of production in spring the surface N:P ratio is 40–60 in the northernmost part of the Gulf of Bothnia, whereas it is typically 2–4 in offshore basins of the Baltic proper. There variations are to a large extent attributed to internal processes rather than variation in the ratio of the external load (Wulff et al., 1990). Compared to the Redfield 16/1 N/P ratio for plankton uptake, the ratios indicate phosphorus limitation in the Bothnian Bay and N limitation in the Baltic proper. The disappearance of DIN after the phytoplankton spring bloom while there are still DIP in the surface water of the Baltic proper, creates an competitive advantage for cyanobacteria over other phytoplankton groups by their ability to fix atmospheric nitrogen. Cyanobacterial bloom are indeed a conspicuous feature of the late summer in the Baltic proper, creating foul smelling accumulation when drifting ashore and sometimes even toxic (Finni et al., 2001). Cyanobacterial nitrogen fixation also represents a substantial internal nitrogen source primarily for the Baltic proper, about 400 kt N yr^{-1}, corresponding to about half that from anthropogenic sources in the drainage basin (Stålnacke et al., 1999; Larsson et al., 2001).

The magnitude of cyanobacterial blooms in the Baltic Sea during the last few decades has been attributed to enhanced eutrophication increased due to anthropogenic forcing, Kahru et al. (1994). However, the intensity of the accumulation events varies inter-annually and climate variations on decadal time scales are probably important. For example, cyanobacterial blooms in the Baltic reached a very low level in mid 1980's but increased again in the 1990's. Particular strong bloom events occurred in recent years. The probable cause is still under debate but the availability of phosphate at low nitrogen concentration is probably a major factor. Blooms were reported in the 19th century and, as sedimentary records showed, occurred even earlier, before human activities have begun to load the Baltic Sea with nutrients. Cyanobacterial blooms may therefore represent a natural phenomenon that is modulated by environmental changes and anthropogenic activities, Bianchi et al. (2000); Westman et al. (2003).

It is interesting to note that, whereas the primary production is of about the same magnitude in the Baltic and North Seas, the large-scale cycling of nutrients differ completely. The horizontal advection of nutrients is of first order importance for the production in the North Sea whereas the cycling between the deep basin water and the surface layer accounts for the largest fluxes in the Baltic Sea.

Areas with frequent low oxygen concentration are found in the south-eastern corner of the North Sea, in some fjords along the Norwegian and Swedish coasts in the Skagerrak, in the shallow parts of the southern Kattegat, in some Danish fjords and bays, in fjords and archipelagos along the rim of Baltic proper and, above all, in the deep basins of the Baltic proper. Even if most of the exemplified areas have shown periods of anoxia in the past, the anoxic regions have become more widespread and the anoxic periods more frequent due to anthropogenic input of nutrients (Æertebjerg et al., 2003). Anoxic events have profound effects on the cycling of N and P as discussed earlier and has eliminated all macroscopic fauna in the deeper part of the Baltic proper. The renewals with oxygen rich deep water are usually of such short duration that the establishment of a macrobenthic community is not possible. On the other hand, eutrophication has caused an increase in benthic biomass on shallower bottoms, Cederwall and Elmgren (1980), so that the total biomasses are considerably higher than before eutrophication influenced this ecosystem.

Different regions of the North Sea and the Baltic Sea feature different ecosystem dynamics mainly as result of differences in physical environment and nutrient supply. We will next review the bio-geo-chemical interaction processes in these different subregions.

5. Dominating interactive processes in different ecosystem regions

Except at their shallow margins, the ecosystems of the North and Baltic Seas are primarily fuelled by the photosynthesis of pelagic micro-algae and cyanobacteria. In addition to a supply of nutrients, this primary production also needs enough light for the rate of manufacture of organic compounds to exceed the local loss rate. During mid-winter, however, the sea surface at these latitudes receives only a few percent of midsummer daylight, and this light is spread through a deep or a turbid water column. At this time, therefore, decay and mineralization processes dominate the water column, phytoplankton production and biomass are low, and nutrient levels increase to annual maxima in February. It is this season, when nutrient elements are predominantly in dissolved inorganic forms, that is mapped in Fig. 26.9.

The primary production accelerates in spring when the compensation depth exceeds the depth of the surface mixed layer (SML). The compensation depth is determined by the intensity of the incoming light and the amount of particles in the water column. The SML depends on the strength on the wind, on the salinity stratification, if any, and on the thermal stratification induced by the surface heating. In some shallow areas the water column is kept well mixed by bottom stress generated turbulence due to strong tidal currents. Different physical environments are encountered within the borders of the Baltic and North Seas, and the primary production responds in different ways to the seasonal variation of the forcing dependent on the local conditions. Four type regions will be discussed in what follows.

In northern and central part of the North Sea the salinity stratification is of no importance, implying winter convection to the bottom and a layering of the water column during the summer season (see Section 5.1). The spring bloom starts when the combination of the increased heating and reduced wind mixing forms a shallow enough SML. Similar conditions are encountered in the large basins of the Baltic Sea (see Section 5.2). An important difference is, however, the halocline at a depth of some 40–70 m in the Baltic, which sets the limit for the winter convection in those parts where the bottom is found at greater depth. Seasonal variations in the vertical fluxes of nutrients have large implications on the primary production in this kind of systems. In large parts of the southern and western coastal region of the North Sea tidal currents are strong enough to keep the water column vertically mixed despite the incoming heat in summer. The dynamics of this kind of regions will be discussed in Section 5.3. The last type of region to be discussed is the region of the Baltic Sea outflow in the Kattegat, and the continuation as a coastal current along the Swedish and Norwegian coasts in the Skagerrak (see Section 5.4). The stratification is here basically determined by salinity. Seasonal heating and cooling only strengthens or weakens the stratification. Typically the salinity stratification is very shallow and the acceleration of the primary production is trigged by incoming light. In addition there are regions like fjords, archipelagos, straits, estuaries and regions of freshwater influence outside large rivers. These comparatively small-scale regions will not be discussed in the present review. They are, however, of importance to the large-scale dynamics of the Baltic and North Seas since much of the input of nutrients and other material become filtered when it passes through these areas before entering the open sea. Some of the large archipelago areas in the Baltic Sea also act as effective sinks for nutrients from the open sea due to the presence of anoxic basins and sediments.

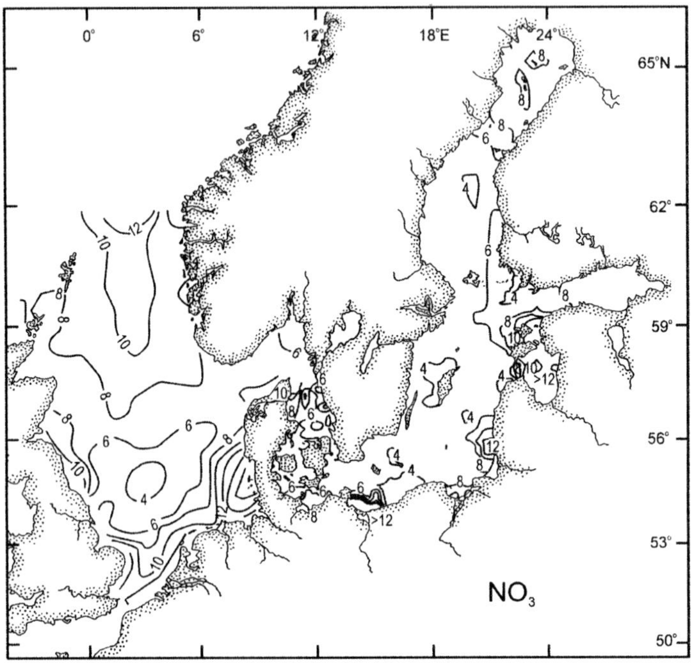
0°
6°
12°
18°E
24°
65°N
62°
59°
56°
53°
50°
NO3

(a)

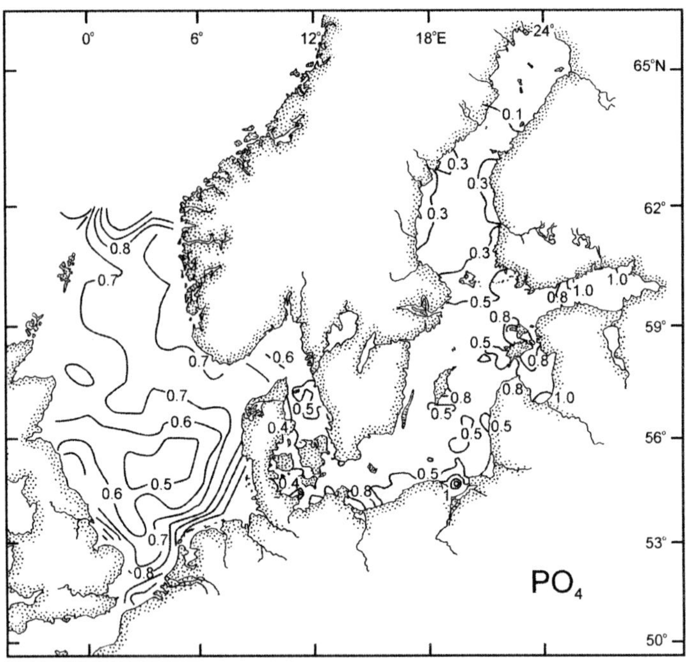
0°
6°
12°
18°E
24°
65°N
62°
59°
56°
53°
50°
PO4

(b)

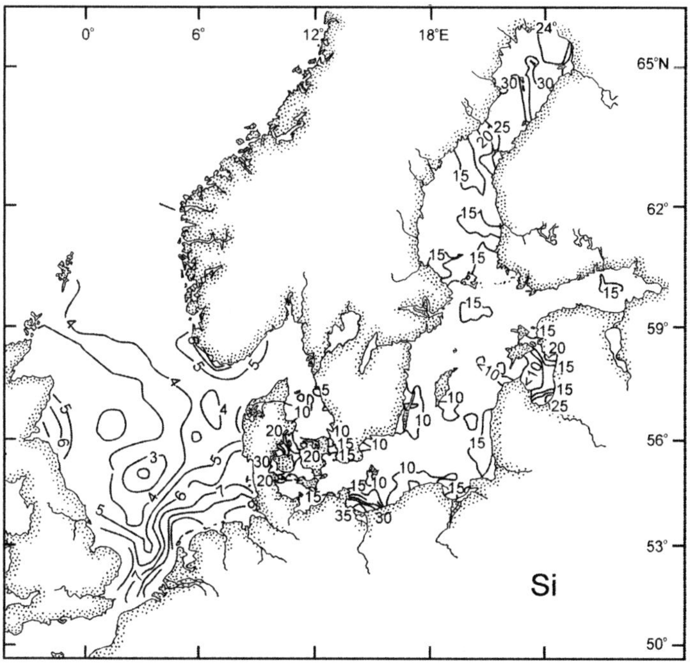

(c)

Figure 26.9 Winter distribution of nitrate, phosphate and silicate in surface (μM) waters of the Baltic and North Seas. The Baltic Sea parts of the figures are based on data from the extensive monitoring programs of the Baltic (for details see *http://data.ecology.su.se/Baltic/*), and represent January-February averages for 1990–2000. The North Sea parts are redrawn from OSPAR Commission (2000), source: Radach and Gekeler, (1997), and represent February averages for 1993–1996.

Looking at the region from another point of view we can recognise three main spatial gradients that control the ecology of the North and Baltic Seas. The first is the gradient from ocean-influenced conditions in the northern North Sea, over freshwater-influenced conditions in the southern and eastern parts of the North Sea and into the freshwater-dominated conditions in the Baltic. The second, within the North Sea, is a northwards increase in depth, which determines the kind of dominant zooplankter, in turn influencing abundance and seasonality of phytoplankton. The existence of large-scale fjordic features (the Norwegian trench, the Baltic Sea itself) complicates this gradient, providing the greatest depths and traps for sinking material. The third gradient is the tidal regime, with highest rates of tidal energy dissipation in the southern and western parts of the North Sea and lowest rates in the east and especially in the Skagerrak-Kattegat channel and Baltic Sea.

Superimposed on these gradients are smaller-scale features, especially tidal mixing fronts, river plumes, haline fronts, estuarine-type circulations in river outlets, and local upwellings. In the remaining parts of this section, we describe some ecosystem characteristics in different parts of the North and Baltic Seas, based on studies at particular (and, we hope, typical) sites.

5.1 The seasonally stratified northern and central North Sea

A regime of seasonal stratification occupies much of the central and northern North Sea. Where tidal stirring is low, local heating brings about thermal stratification during April, resulting in a Summer SML that is 40 – 50 m deep in the north and 20 - 30 m deep in the central North Sea. Heat loss causes convection during autumn, which deepens this layer by eroding the thermocline and entraining deep water, until it merges with the bottom mixed layer (BML) at some time between October and December. The regime in the north is influenced by advection of heat, nutrients and organisms from the Atlantic Ocean. The North Sea has also an interior (Huthnance, 1995), in which water resides long enough for biogeochemistry and ecosystem properties to be strongly influenced by local factors. This is apparent in the minima in the maps of winter nutrient distribution in the North Sea (Fig. 26.9), although with slow replacement by water of Atlantic origin, which has lost much of its oceanic characteristics. Fronts, and the topographically-steered circulation, restrict the diffusion of pollutants and enhanced nutrients into this region from coastal waters.

The seasonal cycle of phytoplankton abundance may be called 'classical' because it has long featured in textbooks. It involves a spring diatom bloom which draws on winter nutrients, followed by a summer period of lower biomass and flagellate dominance in the nutrient-depleted SML and often an autumn bloom triggered by nutrients entrained into the deepening SML from the thermocline and the bottom mixed layer (BML). The spring bloom occurs during April in much of the North Sea. In the salinity stratified parts of the eastern North Sea it often starts earlier in February-March, and sometimes even in January. The spring bloom ceases when limiting winter nutrients have been converted into biomass and when grazing by planktonic animals, or sinking, begins to remove the algal biomass. After the spring bloom, continued production depends mainly on the local recycling of nutrients within the SML and on an input of new nutrients from vertical mixing and upwelling. The key determinants are winter nutrients, which fuel the spring bloom and then are largely exported from the SML into deeper water (through the sinking of phytoplankton, detritus and the faeces and vertical migration of zooplankton). Since summer production much depends on recycled nutrients, this part does not count in the total productivity budget (of so-called new production).

The seasonal cycle of pelagic production in these waters will be discussed in relation to two sites (Table 26.2). One site, at about 59°N 01°E and with a water depth of 100 – 110 metres, was studied in detail during spring 1976 (the FLEX project) and autumn 1998 (the PROVESS project); references are given by Lee et al. (2002). A second (CS) site at about 56°N 1°E, in a water depth of about 70 m, was studied intensively during the 'North Sea Programme' (NSP) in 1988–89; details and references are given in Charnock et al. (eds., 1994) and by Tett and Walne (1995).

TABLE 26.2:
Two sites in the seasonal thermally stratified North Sea

	northern	central	
Project	FLEX, PROVESS N	CS of NSP	
Latitude (N)	59°	56°	
Longitude (E)	1°	1°	
site depth	110	83 (region: 69)	M
winter NO	10	5	μM
winter PO^{3-}_4	0.7	0.5	μM
winter SiO_2	5	3	μM
N:Si	2.0	1.7	
N:P	14	10	
chl max during spring bloom	11	5	mg m^{-3}
chl max/winter NO_3^-	1.1	1.0	mg chl/mmol N
observed NMP (region)	100 to 125	100	g C m^{-2} yr^{-1}
observed NMP at site		115*	g C m^{-2} yr^{-1}
characteristic copepod	*Calanus*	small copepods	

Data for one or more years from 1976 – 1998 (there is no evidence of systematic changes in this region); for sources see: north – Lee et al. (2002), central –Tett and Walne (1995).

According to the classical paradigm, the winter condition of low phytoplankton abundance is a result of the thickness of the SML exceeding the 'critical depth' at which phytoplankton photosynthesis is in balance with immediate losses (Sverdrup, 1953; Tett, 1990). However, the Sverdrup condition can be satisfied without stratification when the depth of mixing is limited by the sea bed, and the water is clear. This seems to be the case in the North Sea: at station CS, chlorophyll concentration remained at 0.5 mg m^{-3}, or higher, throughout the winter; during the FLEX study, chlorophyll was observed to increase, and nitrate to decrease, prior to thermal stratification.

Nevertheless, the onset of stratification spurs the period of rapid phytoplankton growth, which is recognized as the spring bloom. The bloom was short-lived at CS in 1989 and indeed only observed as a result of a deployment of a recording fluorometer; the peak chlorophyll here was only about 5 mg m^{-3}, corresponding to maximum winter nitrate of 5 μM. Chlorophyll reached 12 mg m^{-3} during the FLEX study in 1976; the result of higher levels of winter nitrate (10 μM), and the bloom lasted for several weeks. In both cases the spring phytoplankton was dominated by chain-forming centric diatoms, yet careful examination of floristic data (Mills et al., 1994) suggests that the diatom peak arose from a more mixed phytoplankton containing phytoflagellates and small cyanobacteria, and returned to this state after the bloom was ended. It may be that populations of the smaller organisms were held in check by protozoan (especially heterotrophic dinoflagellate) grazing, allowing diatoms to take sole advantage of increasing illumination until the diatom bloom was ended by a combination of nutrient limitation, sinking, and increasing mesozooplankton grazing. Limitation by exhaustion of dissolved silica is also plausible in the case of the northern North Sea, where winter ratios of N:Si may exceed 2:1. As their abundance increases, aggregation of diatom chains may increase the sinking flux, providing a large input of fresh organic matter to the seabed.

Summer chlorophyll concentrations in the SML can be lower than those observed during winter. Although there is relatively little information about floristic composition, it seems likely that small phytoflagellates are the characteristic autotrophs. These include cryptomonads, green flagellates, silicoflagellates and smaller autotrophic dinoflagellates such as *Prorocentrum;* coccolithophorid blooms have been reported, but *Chrysochromulina*-like prymnesiophytes may be less common here (see Novarino et al., 1997). At the northern site in 1998 small flagellates remained the main component of phytoplankton during the early autumn period of increasing chlorophyll. At station CS, however, they were replaced by diatoms. During summer the thermocline at station CS sometimes displayed a midwater chlorophyll maximum, in which large autotrophic dinoflagellates (especially *Ceratium* spp.) were important. They may be using vertical migration to exploit nutrient gradients in the thermocline. Weak vertical mixing in this layer, and occasional entrainment of lower thermocline water as a result of thermocline deepening after storms, serves to fuel some new production during summer (Sharples and Tett, 1994). However, it seems likely that most production at this time is recycled through a microbial loop in which ciliates and small zooflagellates graze bacteria and small flagellates.

Sediment trap measurements during FLEX suggest that sedimentation of spring bloom diatoms provides the main organic input to the seabed. However, much of this material forms a loose 'fluff' layer atop the consolidated sediments, and is periodically resuspended into the BML, where a large part of its mineralization takes place. The combination of water-column and sea-bed mineralization removes oxygen from the BML which is trapped beneath the thermocline during the months of summer and early autumn. By late September, 1998, about 40 mmol O_2 m^{-3} had been removed from this layer at the northern site, implying the mineralization of about 20 g C m^{-2} since the onset of stratification. Annual mineralization in the sediment at CS was estimated as about 50 g C m^{-2} during the NSP. Wide-area surveys give the annual mean macrofaunal biomass as 3.5 g organic weight m^{-2} (about 1.4 g C m^{-2}) north of the 100 m depth contour in the North Sea, with deposit feeders being the main component (Heip et al., 1992; Künitzer et al., 1992). The organic content of the upper sediments found at CS and the north site was about 1% (of dry weight), or several kilograms of organic carbon per square metre in each 1 cm layer of sediment. Comparison with the annual mineralization suggests that almost all of this carbon is highly refractory.

Despite lower levels of winter nutrients, observed net primary production at CS appears to be about the same as that at the northern site, between 100 and 125 g C m^{-2} yr^{-1}. The data for mineralization, and modelling studies, suggests that at least 20%, and up to 50% of NMP reaches the BML and the seabed in the stratified North Sea. Some of the remainder is mineralised in the SML and thermocline, but a part that is likely to be 50% or more, passes to mesozooplankton. These animals are, typically, copepods, which may however be feeding as much on protozoans as on phytoplankters. In the northern North Sea the dominant is the large copepod *Calanus finmarchicus,* together with euphausiids. *Calanus* appear at CS in summer, but form only part of the copepod biomass here; the dominants are the smaller forms of the genera *Acartia, Centropages, Pseudocalanus* and *Temora.* The small copepods probably overwinter locally, only a small proportion surviving until spring, and although their intergenerational time is less than that of *Calanus,* grazing pressure develops slowly during the spring and summer. Protozan grazing of flagellates is likely to be supplemented by salps of oceanic origin during late summer in the northern North Sea. In the central North Sea this grazing niche is filled by (much smaller) appendicularians.

5.2. Southern and western coastal region of the North Sea

This sub-area takes in the southern and western boundary waters, which are those inside a line approximately following the 40 m depth contour that runs parallel to the Scottish and Northumbrian coasts, and then across the southern North Sea from Yorkshire in England to Jutland in Denmark. The contour corresponds roughly to the transition from tidally stirred to stratified waters predicted by the "h u^{-3} model" (h is the depth and u is the amplitude of the tidal current, see discussion in Rodhe (1998) and marks the approximate position where thermal fronts occur. However, as shown by Fig. 24.6 in Rodhe (1998), the fronts of the North Sea are only weakly linked to topography and lie in a broad belt. Inside the frontal belt lie waters that are typically shallow, well stirred, and turbid because of suspended sediments, coastal erosion, and river discharges, but which can also include 'regions of freshwater influence', ROFIs, with intermitted haline stratification, and some deeper yet well-mixed holes. These coastal waters are subject to dilution with nutrient-rich freshwaters, the greatest enrichment being apparent in the inner German Bight.

Paradoxically, it is probably the southern parts of the North Sea, which are more effectively light-limited despite of the shallow depth, because tidal and wave stirring here increases turbidity (Tett and Walne, 1995).

The abundance and productivity of phytoplankton in this sub-area can fluctuate irregularly between low and high values, depending on water transparency and stratification; in some parts and years, phytoplankton growth commences early, corresponding to a spring bloom; in other parts or years maximum biomass is not reached until summer, if at all. Although diatoms and small phytoflagellates are common throughout the productive season, this is also a region in which blooms of the colonial prymnesiophyte flagellate *Phaeocystis pouchetii* are typical of late spring, and red tides of dinoflagellates (including the large, heretotrophic, *Noctiluca scintillans*) can occur in summer. There is little doubt that these phenomena have been stimulated by human influences; nevertheless, both are likely to be natural features of this region. *Phaeocystis* blooms have been known since the 19th century and red tides are the result of dinoflagellate accumulation at frontal convergences. Both these phenomena have been classified as 'Harmful Algal Blooms' (HABs), but so have occurrences of the toxic dinoflagellates of the genera Alex-*andrium* and *Dinophysis* which do not noticeably discolour the sea but which do result in filter-feeding shellfish accumulating sufficient toxin to pose a threat of paralytic shellfish poisoning (PSP, caused especially by *A. tamarense*) or diarheëtic shellfish poisoning (DSP, caused by *Dinophysis* spp.) when humans or seabirds eat mussels or oysters. Epidemics of PSP have been known for several centuries along the northwestern shore of the North Sea, occurring sporadically and sometimes with lethal consequences until these were prevented by regular monitoring of shellfish and closure of the fishery. It is possible that the over-wintering cysts of *Alexandrium* accumulate in sediments beneath coastal fronts, or perhaps in estuaries such as that of the river Forth in Scotland, which thus form seedbeds for HABs

(Joint et al. 1997). In contrast, *Dinophysis* is more typical of continental coastal waters (Novarino et al., 1997).

Given the great variability of plankton biomass in this region, it is not possible to show a characteristic seasonal cycle, although these have been compiled for particular locations. Observed climatologies (Table 26.3) for ICES boxes (a division of the region used by the International Council for Exploration of the Sea) can be confusing as they combine locations with different spatial characteristics as well as great inter-annual variability. Nevertheless, they are useful for showing the greater phytoplankton biomass and primary productivity in this region, together with spatial differences within it—especially those that follow from the east-west gradient of increasing tidal stirring.

TABLE 26.3.
(climatological) productivity in the North Sea inshore of the main tidal mixing front (thermally stratified Central North Sea included for comparison).

	Central North Sea	English coastal and western Southern Bight	Eastern Southern Bight: Rhine ouflow region	German Bight and west Jutland coastal	
ICES region	7' or 7a	3" or 3b	4	5	
mean water depth	67	40	28	23	m
max chl, spring[1]	2.5	5	12	10	mg chl m^{-3}
mean chl, summer[2]	0.8	1.5	2.5	3.5	mg chl m^{-3}
observed N(PorM)P [3]	100	79	199	261	g C m^{-2} yr^{-1}
observed N(PorM)P [4]	82	-	254	-	g C m^{-2} yr^{-1}

[1] greatest 83% quantile of monthly values (March-May) given by Moll (1998) [2] average of monthly medians (June-August) given by Moll (1998); [3] Joint and Pomroy (1993), [4] Klein and van Burren (1992).

Although anthropogenically increased nutrients play some part in these generally increased levels of phytoplankton abundance and production another part is explained by hydrographic conditions. We will examine this argument briefly using results from the PROVESS south site in Netherlands waters in 1999 and from sites sampled by the NSP in 1988–89 (Table 26.4).

Near the East Anglian coast of England, at NSP site AB, the water is well mixed due to strong tidal streaming, and is very turbid because of SPM, including that derived from coastal erosion. The spring bloom, dominated by Phaeocystis, did not occur until May in 1989 (Mills et al., 1994) and may indeed have been advected from shallower water. Chlorophyll concentration exceeded 4 mg m^{-3} during May and June sampling visits, but was less than 2 mg m^{-3} during the rest of the 1988–89 study period, without obvious seasonality.

The PROVESS south site (at roughly 52°30'N 4°E) in 20 m water off the Netherlands coast, was studied intensively during April 1999 (Howarth et al., 2002, and references therein). u^{-3}/h (see Section 5.2) was about 60 cm^2 s^{-3}, yet there was sometimes intermittent haline stratification as the Rhine plume extended further offshore. The phytoplankton was mainly diatoms and *Phaeocystis* colonies. Chlorophyll fluctuated rapidly during this period, between 10 and 30 mg m^{-3} near the sea bed and between 2 and 40 mg m^{-3} near the sea surface. The high mean biomass would seem to be a result of the nutrient enrichment of these waters, and perhaps also the comparatively low abundance of mesozooplankton and weak impact on large cells and algal colonies. Calculations of the magnitude of various processes (Wild-Allen et al., 2002) suggests that most of the biomass fluctuation was the result of physical processes—mixing or advection of strong lateral gradients, and resuspension, vertical mixing and rapid sinking, the balance depending on the strength of stratification—rather than the growth of phytoplankton.

TABLE 26.4.
Some data for selected sites from the NSP (CS data included for comparison)

	CS	BL	BO	AB	
Latitude (N)	55°30'	54°03'	53°40'	52°42'	
Longitude (E)	0°55'	5°40'	6°10'	2°25'	
	stratified	frontal	mixed	mixed	
Depth	81	41	25		m
Intensity of tidal stirring (u^3/h)[1]	2.7	18	34	137	$cm^2\ s^{-3}$
background diffuse attenuation[2]	0.08 to 0.11	0.32 to 0.23	0.38 to 0.29	0.59 to 0.36	m^{-1}
near-surface NO_3^-, winter[3]	5	10	31	12	µM
near-surface NO_3^-, summer[3]	0.2	0.7	2.5	0.4	µM
max. near-surface chlorophyll	5 (spring)	6 (summer)	50 (summer)		$mg\ m^{-3}$
annual N(MorP)P[4]	115	168	139		$g\ C\ m^{-2}yr^{-1}$

Data from Tett et al. (1993) and Tett and Walne (1995). [1]ratio of m_2 tidal amplitude cubed to water column depth; [2]minimum value of diffuse attenuation estimated from subsurface PAR measurements during winter and summer cruises; [3]winter is Jan-Feb; summer is Jul-Aug; [4] from Joint and Pomroy (1993).

Thus, near-shore waters tend to be highly variable, and do not correspond well to the mixed region of what might be called the 'classical' paradigm for the distribution of phytoplankton at tidal mixing fronts. Equally, whereas frontal convergences in the frontal band of the North Sea are sometimes associated with high biomass in summer, the band is too broad to exhibit the strong biomass enhancement that is characteristic of classical fronts. In the case of the North Sea, a pattern emerges only after statistical analysis (Tett and Walne, 1995). Nevertheless, the frontal band does exhibit the greatest annual primary production and the greatest overall abundance of mesozooplankton.

5.3. The Skagerrak-Kattegat and the Norwegian coastal current

The region of the Baltic outflow, i.e. the Danish straits, the Kattegat and the region of the low-saline current along the Swedish and Norwegian coasts, is characterized by a shallow salinity stratification. This stabilizes the water column throughout the year, allowing the illumination conditions for phytoplankton growth to occur early, often in late winter while the water is still cold. In the southern Kattegat, for example, the spring bloom occurs in March-April (Richardson and Christoffersen, 1991). As an implication the low-saline surface water in the coastal current can be completely depleted of nutrients long before the onset of the spring bloom in the open-sea areas discussed in Section 5.1, and under pristine conditions summer concentrations of chlorophyll would be expected to be low.

However, additional nutrients reach the euphotic zone in summer both by way of natural processes and from anthropogenic sources. Consequentially, summer chlorophyll concentrations tend to be moderately high, averaging 1- 2 mg m^{-3}, but with much variability in some parts.

An important feature of this region is that the stratification stays shallow during the summer, typically some 10–15 m, whereby the thermal and the haline stratification coincide. This is in contrast to what is found in the seasonally stratified parts of the North Sea (Section 5.1), where the surface mixing is limited by a thermocline at twice (or more) that depth. Sunlight can thus penetrate into the underlying water and give rise to photosynthesis below the mixed surface layer. Especially in the Skagerrak, a subsurface chlorophyll maximum is a prominent feature. One source of nutrients for this maximum is the intense upwelling of nutrient-rich water of Atlantic origin, which is indicated by a doming structure of the stratification. Another source is nutrient-enriched water from the southern parts of the North Sea, which interleaves with the other water masses. The Kattegat-Skagerrak front is also associated with increased phytoplankton and primary production (Richardson, 1985), perhaps because of local upwelling. The coastal waters of the Kattegat are about twice as productive as the offshore waters (Carstensen et al., 2003), probably because of stronger mixing here as well as terrestrial inputs.

In modern times, the nutrient content of the surface waters is much increased by direct anthropogenic nutrients and in river discharges, and there seems little doubt that these have increased the annual totals of primary production from less than 100 g C m^{-2} yr^{-1} during the 1950s to 135–165 g C m^{-2} yr^{-1} during 1981–2000 in the Kattegat (Rydberg et al., 2003). The spatial variation is large, with much higher values in some coastal regions and also in the Belt Sea (195 g C m^{-2} yr^{-1}). Lowest production is found in open Kattegat waters (90 g C m^{-2} yr^{-1}). For data after 1980 they also found a co-variation between the annual load of nutrients, especially total nitrogen, and the annual mean primary production. Riverine and Baltic contributions are less during summer, and direct atmospheric deposition of nitrogen (Asman et al., 1995) may be a particular important factor at this time of year.

Winter nutrient concentrations in the Skagerrak and Kattegat vary greatly from year to year, as a result, especially, of fluctuations in the balance between North Sea and Baltic waters. To this interannual variability is added higher frequency variability due to local and expecially atmospheric inputs. Given the existence of frontal convergences and midwater phytoplankton maxima, well suited to dinoflagellates and flagellates, it is perhaps unsurprising that the Skagerrak-Kattegat region is subject to irregular blooms of algae which are sometimes harmful and which may be carried in, or may seed, the waters of the Norwegian Coastal Current. A well-known example is the1988 bloom of the toxic flagellate *Chrysochromulina polylepis* (Nielsen and Richardson, 1990) and the HABs of Chattonella in 1998, 2000, 2001 and 2002.

5.4 The Baltic Sea

The Baltic proper and parts of the Gulf of Bothnia have relatively deep permanent salinity stratifications. The surface layer becomes mixed during winter and thermally stratified with a thermocline at 15–25 m during summer. The physical limitations for the primary production are thus similar to those in the central North Sea.

For a detailed description of the pelagic community in the Baltic See, see i.e. Granéli et al (1990); Kivi et al. (1993); Hagström et al. (2001). Here we will just give a brief outline with emphasis on what differs in the Baltic from the North Sea.

The diatom/dinoflagellate spring bloom, commencing in March-April in Baltic proper and as late as in May in the Bothnian Sea, depletes the pool of DIN in the entire water column down to the halocline. However appreciable amount of DIP remains in the Baltic proper and Kattegat, demonstrating the N limitation of the bloom. In the Bothnian Sea, both N and P are depleted and in the Bothnian Bay, no spring bloom occurs since there is no winter accumulation of the limiting P. Here, enhanced primary production does not occur until late summer when heterotrophic activity has enhanced nutrient release from the large input of organic matter that occurs in this region. A conspicuous difference between the Baltic region and the North Sea is the absence of zooplankton feeding of the spring bloom. In the Baltic, the zooplankton populations are to a large extent recruited each spring from resting eggs, hibernating in bottom sediments. The spring bloom is thus not well utilized by pelagic herbivores and to a large extent deposited on the bottoms, representing the major food input to the benthic fauna.

The pelagic communities during summer are characterized by a dominance of small organisms, less sedimentation, increased heterotrophic activity and nutrient recycling. Production of heterotrophic bacteria feeds the food chain of flagellates, ciliates to large zooplankton. The net amount of biomass that is transferred via the 'microbial loop' is actually small since the carbon flow is largely a loop of respiratory losses (Hagström et al., 1988). The dependency on available mineral nutrients or substrate for regeneration of nutrient is valid not only for phytoplankton but also for heterotrophic bacteria as well (Hagström et al., 2001). During summer, lack of mineral nutrients and high predation of bacteria causes an accumulation of organic carbon (DOC), particularly in the northern more oligotrophic basins. The very low concentrations of DIN but still appreciable amount of DIP that remains after the spring bloom, particularly in the Gulfs of Riga and Finland and the Baltic proper, give nitrogen fixing cyanobacteria a competitive advantage over other autotrophs. The reasons why cyanobacterial bloom do not extend into the Kattegat, also with a low N:P ratio, is less clear but higher sulphate concentrations at higher salinities, does inhibit the molybdenum assimilation that is essential for cyanobacteria (Marino et al., 2003).

During autumn, mixing brings up nutrient to the surface and conditions thus favour an autumn bloom. This is a regular phenomenon in Kattegat and southern Baltic but rarer further north where low light conditions at these high latitudes may hinder a bloom if the autumn mixing is late.

5.5. *The filtering by mussels in Öresund as an example of a small-scale interaction process with large-scale implications on the ecosystem.*

Benthic suspension-feeders like mussels (*Mytilus edulis*) can have a dominant influence on the flux of nutrients, and thus on the ecosystem as a whole. An interesting example is found in Öresund, where vast areas of the 8-m deep sill are almost completely covered by mussels. In an investigation by Haamer and Rodhe (2000) it was shown that the mussels at the sill where capable of filtering all water passing through this connection between the Baltic and the North Seas (on average about 20, 000 m^3s^{-1}). The mussels contributed to give a large roughness to the seabed. The vigorous turbulence in the flow over the rough sea floor then ensured that all water passing over this sill was forced in contact with the mussel bed, where it became filtered by the mussels. Effects of this filtering are illustrated in Fig. 26.10.

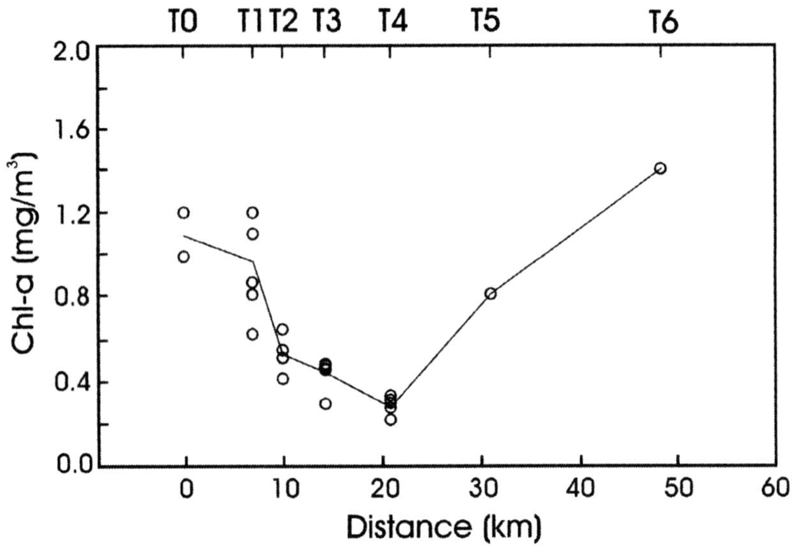

Figure 26.10 Observations of Chl-a concentration along a transect following the water passing over the mussel bed at the sill in Öresund (stations T1 to T4) and also as surface water in the deep part of Öresund north of the sill (stations T4 to T6). (Modified from Haamer and Rodhe, 2000.)

Following the water over the mussel bed (from T1 to T4 in Fig. 26.10) the observations indicate the withdrawal of phytoplankton from the water. The passing over the bed lasted for about 10 h. The rise in Chl-a content after the passage of the sill (stations T4 to T6) indicates the recover of the planctonic biomass over the deep parts of Öresund. The travel time for the water was here a day or two. What is not shown in the figure is that there is a qualitative change in the biomass towards much smaller species after the passage of the sill (Norén et al., 1999). A very marked effect of the filtering by mussels is that the water in the area is extremely clear. Therefore, one can find flourishing fields of eelgrass (*Zostera marina*) down to 7 m deep and *Laminaria sacclarina* down to 14 m in Öresund. The input of ammonium from mussel banks due to the metabolism of the mussels is considerable. Haamer and Rodhe (2000) estimated the total release of NH4-N to be 0.7 t h^{-1} during a situation with northward flow from the sill into Öresund. This indicates an input which has a magnitude comparable to the nitrogen demand for the total production in Öresund, 1.3 t h^{-1}, calculated using an estimate of the yearly total production, 91 g C m^2 y^{-1}, by Mattsson (1996). The net export of nitrogen from the Baltic has been estimated to about 2.3 t h^{-1} (20000 t y^{-1}, see Section 4 of the present study). These rough estimates indicate that the filtering by mussels in the Danish straits constitutes a first order physical-bio-chemical interaction process. Already

Kautsky and Wallentinus (1980) pointed upon the importance of the filtering by mussels in the Baltic Sea for the nutrient dynamics.

6. Human exploitation and problems

Human activities have put pressure on the environment of the Baltic and North Seas. Problems related to the intense fishing industry are probably the most serious ones. These include e.g. fishing over sound biological limits for some important target species and seabed disturbances. Input of nutrients has changed conditions of the primary production implying quantitative and qualitative changes of the biota. Both these classes of problems are probably to a large extent reversible and will hopefully in the not to distant future be solved, though they are today severe. More serious are probably problems related to input of different hazardous trace elements, which can stay in the environment for long times. An assessment of human negative effects of different human activities is found in Table 26.5 where human pressures are classified according to their seriousness for the North Sea environment. Most of these pressures are also put on the Baltic Sea environment, with almost the same classification. An important difference between the two seas is however related to the difference in time scale for water exchange. The Baltic Sea will feel the effects of anthropogenic input of e.g. nutrient a long time after a reduction. Here plays also the important buffering effect by the sediments in the deep basins an important role. The North Sea, on the other hand, will respond "immediately". The reason being the short time scale for the water exchange and also the very limited regions from which sedimented material will return to the surface layer on time scales larger the a year. Most of the accumulation of material occurs in the deep parts of the Norwegian Trench and that material will probably not re-enter into the circulation of water masses in the North Sea. Many changes have been observed in North Sea pelagic and benthic communities during the 20th century, especially during the final decades. In most cases it has not proven easy to identify the causes of these changes, causes that include natural climatic variation as well as increased anthropogenic inputs and disturbances due to fishing (Tett and Mills, 1991; Kaiser, 1998; OSPAR Commission, 2000). To illustrate these points we will briefly consider two of the most heavily human-impacted parts of the North Sea.

One of these regions is the inner German Bight of the North Sea. As a result of anthropogenic enrichment of water from the Elbe and other rivers discharging into the Bight (Radach, 1992; Hickel et al., 1993), winter N:Si ratios near the island of Helgoland, which were in the range 1–2 in the late 1960s, increased to 4–8 in the mid-1990s (Radach et al., 1990). The ratio change was the result of increasing DAIN and decreasing silica. Changes in N:P have been more complex, with winter values in the range 16–32 in the early 1960s falling into the range 4–16 in the 1970s and increasing again to 16–32 in the mid-1980s.

According to Hickel et al. (1993), starting in "the early sixties, phosphate concentrations rose for about a decade, levelling off to about twice the former concentrations for another decade, and then decreasing (since 1982) as a result of phosphate-reducing measures. Nitrate concentrations, however, have only increased since 1980/81, following Elbe river flood events. In 1987, three times the former concentrations were reached."

TABLE 26.5.
Priority classes of human pressure on the environment in the North Sea, as shown in OSPAR Commission (2000).

Class*	Human pressure	Category
A	Removal of target species by fisheries	fisheries
	Inputs of trace organic contaminants (other than oil and PAHs) from land	trace organic contaminants
	Seabed disturbances by fisheries	fisheries
	Inputs of nutrients from land	nutrients
	Effects of discards and mortality of non-target species by fisheries	fisheries
	Input of TBT and other antifouling substances by shipping	trace organic contaminants
B	Input of oil and PAHs by offshore oil and gas industry	oil and PAHs
	Input of oil and PAHs by shipping	oil and PAHs
	Input of other hazardous substances (other than oil and PAHs) by offshore oil and gas industry	other hazardous substances
	Inputs of heavy metals from land	heavy metals
	Inputs of oil and PAHs from land	oil and PAHs
	Introduction of non-indigenous species by shipping	biological impacts
	Input of other hazardous substances (other than oil, PAHs and antifouling) by shipping	other hazardous substances
	Introduction of cultured specimen, non-indigenous species and diseases by mariculture	biological impacts
	Inputs of microbiological pollution and organic material from land	biological impacts
C	Input of litter specific to fisheries	litter and disturbance I
	Physical disturbance (e.g. seabed, visual, noise, pipelines) by offshore oil and gas industry	litter and disturbance I
	Input of litter by shipping	litter and disturbance I
	Dispersion of substances by dredging and dumping of dredged material	dredging and dumping
	Dumping of (chemical) ammunition by military activities	dredging and dumping
	Constructions in the coastal zone (incl. artificial reefs) by engineering operations	engineering operations
	Input of chemicals (incl. antibiotics) by mariculture	mariculture
	Mineral extraction (e.g. sand, gravel, maërl) by engineering operations	engineering operations
	Input of nutrients and organic material by mariculture	mariculture
	Physical disturbance by dredging and dumping of dredged material	dredging and dumping
	Inputs of radionuclides from land	radionuclides
D	Physical disturbance (e.g. noise, visual) by shipping	litter and disturbance II
	Input of litter by recreation	litter and disturbance II
	Physical disturbance (e.g. seabed, noise, visual) by military activities	litter and disturbance II
	Physical disturbance (e.g. noise, visual) by recreation	litter and disturbance II
	Power cables (electromagnetic disturbances) by engineering operations	litter and disturbance II
	Dumping of inert material (e.g. wrecks, bottles)	litter and disturbance II

*Human pressures are ranked according to their relative impact on the Region II ecosystem, including sustainable use. While the Division in the four classes A-D was established firmly, ranking within classes was not considered to be significant. Class A = highest impact; Class B = upper intermediate impact; Class C = lower intermediate impact; Class D = lowest impact.

The nutrient changes have been associated both with an increase in the total stock of phytoplankton (Gillbricht, 1988; Radach et al., 1990; Hickel et al., 1993) and (Gillbricht, 1988) with changes in the seasonal succession.

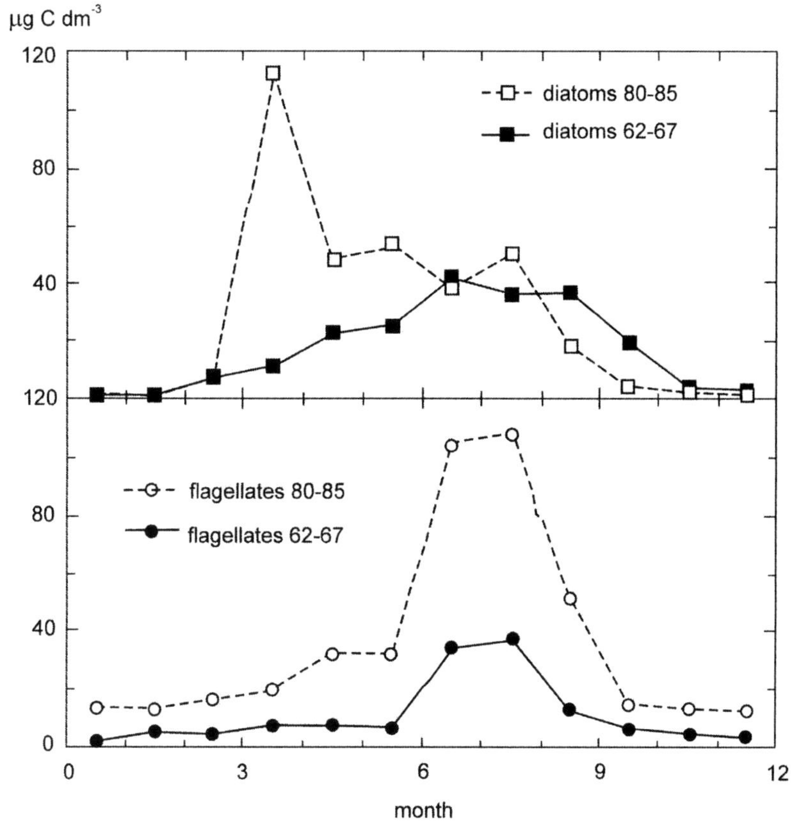

Figure 26.11 Seasonal cycles of phytoplankton components at Helgoland. Redrawn from Gillbricht (1988). 'flagellates' includes dinoflagellates, and perhaps also some zooflagellates.

In the 1960s, diatom abundance peaked in July, followed rapidly by a peak in flagellates reaching almost the same biomass (Fig. 26.11). In the early 1980s, diatoms peaked in April at nearly three times the old biomass, but the flagellate peak remained in August, also reaching nearly 3 times its previous level. The flagellates included dinoflagellates, especially species of *Ceratium* and "small naked forms" (Radach et al., 1990). The change has persisted at least until 1994 (Hickel, 1998; Barkmann et al., 2002)

However, explaining the changes between the 60's and the 80's as a sole result of nutrient enrichment is open to several objections. Hickel et al. (1993) points out that, in the control of phytoplankton growth, "hydrographical factors possibly dominate. Additional nutrient input by Elbe river floods did not always result in elevated phytoplankton stocks near Helgoland, while extended periods of vertical density stratification of the German Bight water caused large plankton blooms."

The phytoplankton at Helgoland may have been participating (Hickel, 1998) in long-term changes affecting the North Sea as a whole, and resulting from climate change as much as nutrient enrichment (Tett and Mills, 1991; Reid et al., 2001).

The second region is that of the Skagerrak and Kattegat. Rosenberg et al. (1996) summarise changes as (in some parts): "increased nutrient concentrations, increased occurrence of fast-growing filamentous algae in coastal areas affecting nursery and feeding conditions for fish, declining bottom water oxygen concentrations with negative effects on benthic fauna, and sediment toxicity to invertebrates also causing physiological responses in fish." These are ascribed to eutrophication and anthropogenic toxins. Evidence for eutrophication includes that reported by Richardson and Heilmann (1995) and mentioned in section 5.3. A number of workers have documented changes in the macrobenthic fauna, the initial and influential study being that of Pearson et al.(1985).

The macrobenthos of the Kattegat had been sampled by C.G.J. Petersen in 1911–1912. Petersen's stations were resampled in 1984 by Pearson et al. using equipment and techniques similar to those used in the original survey. Pearson et al. found that, compared with the situation in 1911–12, biomass had decreased at many stations, mostly due to decreases in the populations of the large echinoid *Echinocardium.* Deposit feeders had decreased, and suspension-feeders and carnivores had increased in biomass and dominance. The amphiuroid *Amphiura filiformis* had increased at most stations. The possible causes of these changes include changed fish predation pressure, direct effects of trawling, climate change, and eutrophication, with the latter seeming the most likely cause.

Other work includes that of Göranson (2002) who in 1990 revisited Petersen's stations in the Oresund, between the Kattegat and the Baltic, observing changes similar to those seen by Pearson et al. and concluding that "Some of these changes seem to have taken place during the last decades, parallel with increased nutrient load to the coastal waters, with subsequent high sedimentation of organic material and oxygen deficiency. High concentrations of different xenobiotics, e.g. organic tin compounds, might explain why some species of gastropods are very rare today. Trawling or changed hydrographical conditions are not likely to have caused the observed great structural changes."

7. Concluding remarks

7.1. *Old and new technologies*

Several new technologies can aid study of physical-chemical-biological interactions in regions such as the North and Baltic Seas. One of these, ecosystem modelling, has been drawn on substantially in this chapter. Another, the continuous *in situ* recording of chemical and biological variables, has made a smaller contribution. The third, remote sensing, has so far been only of limited use. In this section we will briefly comment on the recent and likely future worth of these technologies for studies of our waters. We will conclude by referring to an older technology - that of the Continuous Plankton Recorder CPR) - which, despite limitations, still can shed much light on long-term and wide-area changes in the North Sea.

Franz et al. (1991) summarized earlier physical-biological models, and recent 3D models of the North Sea have been reviewed by Moll and Radach (2003). They include ECOHAM1 and NORWECOM (Moll, 1998; Skogen and Moll, 2000). Models, such as COHERENS (Luyten et al., 1999), capable of high and accurate resolution of horizontal variation are essential for the parts of the North Sea influenced by river plumes. Over much of the North Sea, however, vertical structure and seasonality are the key elements. These can be well emulated by depth-resolving 1-D models, such as PROWQM (Lee et al., 2002), which combines benthic resuspension and mineralisation as well as water column microbial processes. In some cases, simple 2-layer models have been used succesfully (Tett and Walne, 1995). ERSEM (Baretta et al., 1995; Baretta-Bekker and Baretta, 1997) is the most complete ecosystem model so far implemented for the North Sea. ERSEM has been used to examine regional and long-term variability in phytoplankton and primary production using a 2-layer compartmentalisation (Pätsch and Radach, 1997). Recently (Proctor et al., 2003), it been applied with improved horizontal and vertical resolution to estimate nutrient fluxes over the entire north-west European continental shelf. When perfected, such modelling should greatly improve the reliability of North Sea nutrient budgets, inaccuracies in which were briefly discussed in section 4.1.

Results from some of these studies were drawn on in Section 5. The models not only interpolate between sometimes sparse observations but also give unambiguous values of primary production—either of NMP (net microplankton production), calculated from the product of microplankton growth rate and biomass in COHERENS and PROWQM) or NPP (net phytoplankton production), from phytoplankton growth rate and biomass, in the other models. Microplankton comprises phytoplankton and associated pelagic microheterotrophs, and NMP may give a better estimate of organic production available to higher trophic levels.

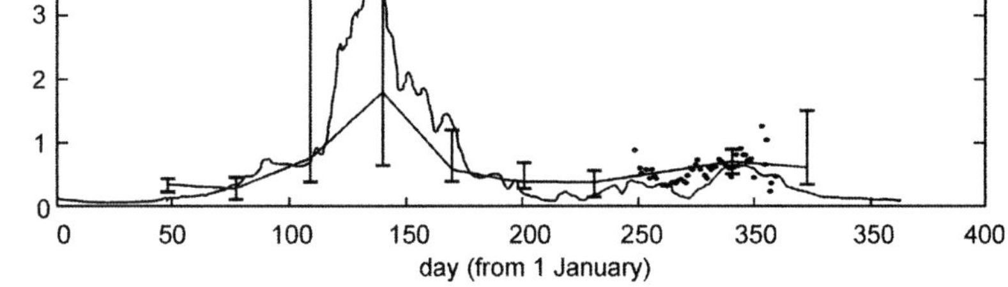

Figure 26.12 Phytoplankton seasonal cycle in the northern North Sea. The diagram shows mean chlorophyll concentration, 0–30 m, in ERSEM box 1 - straight line segments connecting monthly means, with vertical bars showing 17–83% range, taken from the NOWESP climatology in Moll (1998). Also shown (as points) are observations at the PROVESS North Station during 1998, most made by a moored recording fluorometer (Lee et al., 2002). The continuous line is a simulation of 1998 with PROWQM, in which physical forcing has been improved from that of Lee et al. (2002).

Fig. 26.12 presents chlorophyll concentrations from the ship-derived NOWESP climatology of Moll (1998) along with a simulation by PROWQM and moored fluorometer data for the PROVESS site in the Northern North Sea (Lee et al., 20002) discussed in section 5. Deployment of moored fluorometers commenced during 1988–89 in the southern and central North Sea (Mills et al., 1994). Subsequent offshore deployments, and those of more recently developed moored nutrient analysers, have often been frustrated by instrumental problems as well as the ship costs of regular visits to the moorings. However, the next few years should see better descriptions of offshore cycles as biological and chemical data from moorings are analysed and new deployments made of more reliable instruments.

It might be thought that the costs and difficulties of offshore biological observations in the North Sea could be evaded by satellite remote sensing, by analogy with the use of AVHRR sea surface temperature images to show the position of tidal mixing fronts (see Rodhe, 1998). In their satellite colour atlas of the North Sea, Holligan et al. (1989) used CZCS (Coastal Zone Colour Scanner) images to show the existence of features such as fronts, phytoplankton blooms, and regions of high SPM (suspended particulate matter), without, however, being able adequately to quantify their content of chlorophyll or SPM. However, the North Sea is a difficult place for remote sensing of sea colour. High incidence of cloud cover and low sun angle make it difficult to obtain data. The correction of remotely measured radiances for atmospheric attenuation is complicated because the North Sea lies under a region of competing air masses, which are themselves influenced by anthropogenic processes. Because of its freshwater content and strong tidal stirring, much of the southern North Sea exemplifies the most complicated optical water type in which dissolved yellow substances (Warnock et al. 1999), SPM, and phytoplankton chlorophyll all contribute to the optical signal (Aarup et al., 1990; Doerffer and Fischer, 1994; Jorgensen, 1999; Wild-Allen et al., 2002). The results of applying global algorithms should thus be viewed with suspicion, and, in our opinion, no reliably quantitative, remotely sensed, data are yet available for most North Sea pelagic biological conditions. New satellite-born sensors such as SeaWiFs and MERIS, as well as the availability of more sea-truth and better (and regionally-specific) algorithms, may improve this situation. More progress has been made in the case of the Baltic (e.g. Kratzer et al., 2003), where SPM concentrations are lower.

The CPR survey (Brander et al., 2003; Warner and Hays, 1994) uses the 1920's technology of the Hardy Plankton Recorder, towed behind ships of opportunity, and filtering water through moving, finely woven, cloth. The CPR takes and preserves smaller zooplankters and larger phytoplankters quite well, and its gauze absorbs algal pigments to give a quantifiable 'green colour'. Although it is often difficult to relate CPR results to observations from ships, moorings, or satellites, much effort has gone to make the CPR data internally consistent and thus useful for monitoring trends over long periods and wide areas. These areas include the temperate and subpolar North Atlantic as well as the North Sea, but do not extend into the Baltic Sea.

7.2. *Trends*

The "ocean climate" of the Baltic and North Seas has varied during the last century. This includes variations in salinity, temperature and ice cover. However, the structure and importance of the variations differ considerably between the two seas. The Baltic Sea has a local heat budget and the temperature has no year-to-year memory. Thus, heat content and sea ice extent respond to the year-to-year variations of the climate forcing and these are to a large extent related to the NAO. From a sea-ice point of view the Baltic Sea is a part of the northern hemisphere "marginal ice zone", and the ice extent is thus very sensitive to changes in the climate (c.f. Section 2.2).

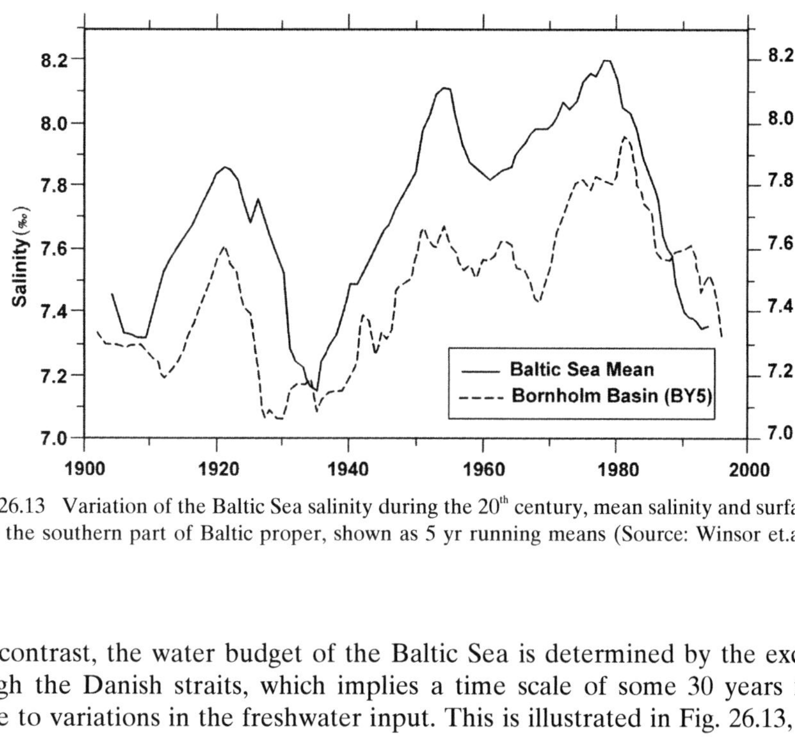

Figure 26.13 Variation of the Baltic Sea salinity during the 20th century, mean salinity and surface layer salinity the southern part of Baltic proper, shown as 5 yr running means (Source: Winsor et.al., 2001, 2003).

In contrast, the water budget of the Baltic Sea is determined by the exchange through the Danish straits, which implies a time scale of some 30 years for response to variations in the freshwater input. This is illustrated in Fig. 26.13, where the variations of the mean salinity are shown together with the variations of the surface layer salinity in the Bornholm basin. Note the similarity between the mean salinity and the actual salinity in a specific region. This means that the temporal variations shown in Fig. 26.13 can be more or less accurately transformed to an in-out displacement of the horizontal salinity gradient in the Baltic Sea shown in Fig. 26.2. Transformed in this way, the 1-psu amplitude shown in Fig. 26.13 corresponds to a horizontal displacement of the salinity pattern of about half the length of the Baltic proper. The biological implications of these variations are not well known, but because many marine organisms in the Baltic Sea are stressed by the very low salinity, it is likely that they are sensitive to such changes in salinity.

The waters in the North Sea, having residence times of a year or so, respond much faster to changes in the atmospheric forcing. The most important variations, both in temperature, salinity and other quantities depending on water exchange and vertical mixing are related to the NAO. This has impact on e.g. the inflow of nutrients with the Atlantic inflow to the north, the intensity of the horizontal circulation and the spreading of different water masses within the North Sea. Some of the ecosystem consequences of these changes have been demonstrated by the CPR. Reid et al. (2001) reported that, "after 1987, Phytoplankton Colour... measured on samples taken by the... CPR in the North Sea increased substantially, both in level and seasonal extent, compared to earlier years since 1946. Many species of phytoplankton and zooplankton showed marked changes in abundance at about the same time.". Hydrodynamic model simulations forced by observed wind suggest a substantial increase in oceanic inflow during December to March, from 1988 onwards. "From 1988 onwards, the NAO index, the pressure difference between Iceland and the Azores, increased to the highest positive level observed in this century. Positive NAO anomalies are associated with stronger and more southerly tracks of the westerly winds and higher temperatures in western Europe."

Beaugrand et al. (2002) used CPR data to show a northwards shift of more than 10° in warm-water assemblages of calanoid copepods in the eastern North Atlantic Ocean and European shelf seas including the North Sea, and apparently related both to the NAO and the global warming trend of increasing northern hemisphere temperatures. Beaugrand et al. (2003) considered the decline in cod biomass and recruitment in the North Sea since the late 1960s, and showed that, in addition to the effects of overfishing, climatically-linked fluctuations in plankton have resulted in long-term changes in cod recruitment in the North Sea. They concluded that rising temperature since the mid-1980s have modified the plankton ecosystem in a way that reduces the survival of young cod.

Although CPR green colour is hard to relate to direct measurements of phytoplankton chlorophyll, and although Radach and Gekeler (1996) and Pätsch and Radach (1997) found no trends in chlorophyll and nutrients measured during research cruises in the northern North Sea, the evidence for decadal-period changes in the zooplankton of the North Sea seems convincing. Even if those of the last few decades are mostly due to the NAO, they are, at least, strongly suggestive of the scale of changes that might be expected to result from global warming.

As discussed in section 6, the North Sea ecosystems are additionally under pressure from human activities, in particular from the two-pronged impact of fisheries (on the upper levels of the pelagic food web and on the stability of benthic communities), from nutrients in the coastal regions influenced by freshwater, and in a less well known way, from pollution by synthetic chemicals (Mills and Tett, 1991; OSPAR Commission, 2000). Nevertheless the North Sea's ecosystems appear to be robust. Although they have clearly been perturbed from their state at the beginning of the 20th century, and although we should expect substantial changes in species composition at all trophic levels during the 21st century, the ecosystems do not appear to be seriously degraded in their functioning, and should remain that way so long as present and proposed measures to reduce anthropogenic inputs of nutrient and synthetic pollutants are enforced. Equally important will be to reduce fisheries impact on target species and the seabed, ideally through setting aside substantial parts of the North Sea where fishing is not allowed.

The Baltic Sea ecosystem is still in a highly perturbed state. There have been substantial improvements in the heath and populations of top predators like seals and sea eagles since the bans of some organochlorines such as DDTs in the late 1980's. However, loads of heavy metals are still high and the levels of dioxin in some fish species are still higher than the EU limits for food consumption. The intense shipping in the Baltic Sea accounts for approximately 15% of all maritime traffic around the world (HELCOM 2003). Many of the oil tankers operating in this area are still only single-hulled and the risk of a major oil spill is imminent, considering the difficult climate and navigational problems in this region. Minor 'accidental' oil spills along the major sea routes occur almost daily and are seriously damaging the large overwintering sea-bird populations, feeding in the same regions.

Total fish catches are today 10 times higher than a century ago, due to more efficient fishing methods, increased production caused by eutrophication and less competition from seals. But fishing has caused massive alteration in the food web, where the overexploitation of cod has resulted in increasing population of less valuable clupeid prey species, particularly sprat (Harvey et al (2003). The Baltic cod population is close to extinction. The situation is worsened by climate and eutrophication effects, which have led to the anoxic conditions in the deep basins, which are the only regions where this marine species can reproduce.

There has been a substantial decrease in nutrient emissions from the Baltic Sea countries, particularly from phosphorus point sources in industries and municipalities. The reduction has been less clear for nitrogen, although substantial reduction in fertilizer-use and livestock holdings, nor has any clear reduction occurred in atmospheric depositions. The nutrient concentration in the Baltic is however still high, and the deep basins of the Baltic proper remain in an anoxic state, but the increasing trend have levelled off.

As we write these closing words, newspapers in the UK are giving prominence to the discovery of high levels of dioxins in salmon farmed in North-West Scotland (Hites et al., 2004). It seems possible that this is the result of the salmons' diet of meal made from fish such as sand-eels caught in the shallower parts of the North Sea and themselves carrying a high burden of toxin from dioxins absorbed onto sediments. Although data on such sediment contamination is scarce (OSPAR Commission, 2000), it does suggest high concentrations in sediments in regions exemplified by industrially polluted fjords in S.E. Norway and the estuaries of the Rhine and Scheldt. "Even off the south-eastern open sea coast of Norway, some 20 km from the waste outfall, the local background was exceeded 5 – 100 times. . . . The presence of dioxins in the tissue of several seafood species at some Norwegian locations reflects the high values in sediments."

This pollution is unlikely to be limited to the regions named. While it might be expected to decrease as anthropogenic production of dioxins is restricted, dioxin transfer by way of salmon to humans is an indirect result of both climate change and over-fishing of cod, which has resulted in the switching of North Sea fisheries effort towards the industrial fishery. This illustrates the complex, and sometimes far-reaching, interactions that must be taken into account when humans make use of coastal ecosystems for conflicting purposes.

For the enclosed Baltic Sea, dioxin levels in fish are very high, still on the levels found in the 1980's. It is not allowed to use fish as animal food, according to EU-rules, and the dioxin levels in fish are too high for human consumption as well.

However, Sweden and Finland have succeeded in getting an exception from the EU-rules with the argument that the consumers in these countries rarely eat fish!

Acknowledgements

Part of the work by Johan Rodhe has been funded by MISTRA and SMHI through the Swedish regional climate-modelling program (SWECLIM). Some of Paul Tett's work was funded by the UK NERC (GST/02/268) as part of the 1988–89 North Sea Project and by the European PROVESS (MAS3-CT97–0159) and OAERRE (EVK3–1999–00015) projects. Part of the work by Fred Wulff has been funded by MISTRA through the MARE program and by the Swedish Environment Protection Agency. We are grateful to Agneta Malm for help with many of the figures.

Bibliography

Aarup, T., S. B. Groom and P. M. Holligan. (1990). The processing and interpretation of North Sea CZCS imagery. Netherlands Journal of Sea Research, 25, 3–9.

Æertebjerg, G., J. H. Andersen and O. S. Hansen (Eds.) (2003) Nutrients and Eutrophication in Danish Marine Waters. A challenge for science and management. Danish Environmental. Danish Environmental Protection Agency and National Environmental Research Institute, 126 pp.

Asman, W. A. H., O. Hertel, R. Bercowicz, J. Christensen, E. H. Runge, L. L. Sorensen, K. Granby, H. Nielsen, B. Jensen, S. E. Gryning, A. M. Sempreviva, S. Larsen, P. Hummelshoj, N. O. Jensen, P. Allerup, J. Jorgensen, H. Madsen, S. Overgaard and F. Vejen. (1995). Atmospheric nitrogen input to the Kattegat. Ophelia, 42, 5–28.

Baretta-Bekker, J. G. and J. W. Baretta. Eds. (1997). European Regional Seas Ecosystem model II. Journal of Sea Research 38, 169–436.

Baretta, J. W., W. Ebenhöh, and P. Ruardij. (1995). The European Regional Seas Ecosystem Model, a complex marine ecosystem model. Netherlands Journal of Sea Research, 33, 233–246.

Beaugrand, G., P. C. Reid, F. Ibañez, J. A. Lindley and M. Edwards. (2002). Reorganization of North Atlantic Marine Copepod Biodiversity and Climate. Science, 296, 1692–1694.

Bianchi, T. S., E. Engelhaupt, P. Westman, T. Andren, C. Rolff and R. Elmgren. (2000). Cyanobacterial blooms in the Baltic Sea: Natural or human-induced. Limnology and Oceanography 45 (3), 716–726.

Billen, G., C. Lancelot and M. Meybeck. (1991). N, P and Si retention along the aquatic continuum from land to ocean. Ocean Margin Processes in Global Change, ed. Mantoura, R. F. C., J-M. Martin and R. Wollast. John Wiley and Sons Ltd., 19–44.

Blomqvist, S. and U. Larsson. (1994). Detrial bedrock elementsas tracers of settling resuspended particulate matte in a coastelareaof the Baltic sea. Limnol. Oceanogr. 39(4), 880–896.

Brander, K. M., R. R. Dickson, and M. Edwards. (2003). Use of Continuous Plankton Recorder information in support of marine management: applications in fisheries, environmental protection, and in the study of ecosystem response to environmental change. Progress in Oceanography, in press.

Brockmann, U., B. Heyden, M. Schütt, A. Starke, D. Topcu, K. Hesse, N. Ladwig and H. Lenhart (2002). Assessment criteria for Eutrophication Areas - Emphasis German Bight -. University of Hamburg, report no. UBA-FB-000338, 109+140pp.

Carstensen, J., D. Conley and B. Muller-Karulis. (2003). Spatial and temporal resolution of carbon fluxes in a shallow coastal ecosystem, the Kattegat. Marine Ecology - Progress Series, 252, 35–50.

Cederwall, H. and R. Elmgren (1980). Biomass increase of benthic macrofauna demonstrates eutrophication of the Baltic Sea. Ophelia (Suppl. 1): 287–304.

Charnock, H., K.R. Dyer, J.M. Huthnance, P.S. Liss, J.H. Simpson, and P.B. Tett, ed. (1994). Understanding the North Sea System, Chapman & Hall, London, 176 pp.

Conley, D. J., C. Humborg, L. Rahm, O. P. Savchuk and F. Wulff. (2002). Hypoxia in the Baltic Sea and basin-scale changes in phosphorus biogeochemistry. Environmental Science and Technology, 36(24), 5315–5320.

Doerffer, R. and J. Fischer. (1994). Concentrations of chlorophyll, suspended matter, and gelbstoff in Case-II waters derived from satellite Coastal Zone Color Scanner data with inverse modeling methods. Journal of Geophysical Research-Oceans, 99, 7457–7466.

Eisma, D. and J. Kalf. (1987). Dispersal, concentration and deposition of suspended matter in the North Sea. Journal of the Geological Society, 144, 161–178.

Elmgren, R. (1989). Man's impact on the ecosystem of the Baltic Sea: Energy flows today and at the turn of the century. Ambio 18(6), 326–332.

Finni, T., K. Kononen and R. Olsonen. 2001. The history of cyanobacterial blooms in the Baltic Sea. Ambio 30(4–5), 172–178.

Flodén, T. and B. Winterhalter. 1981. Pre-Quaternary geology of the Baltic Sea. In: The Baltic Sea. Voipio, A. (Ed.), 1–54.

Fransz, H. G., J. P. Mommaerts and G. Radach. (1991). Ecological modelling of the North Sea. Netherlands Journal of Sea Research, 28, 67–140.

Gillbricht, M. (1988). Phytoplankton and nutrients in the Helgoland region. Helgolander Meeresuntersuchungen, 42, 435–467.

Granéli, E., K. Wallström and U. Larsson. (1990). Nutrient limitation of primary production in the Baltic Sea area. Ambio 19(3), 142–151.

Göransson, P. (2002). Petersen's benthic macrofauna stations revisited in the Öresund area (southern Sweden) and species composition in the 1990s - signs of decreased biological variation. Sarsia, 18, 263–280.

Haamer, J and J. Rodhe. (2000). Mussel *Mytilus edulis* filtering of the Baltic-Sea outflow through the Öresund - an example of a natural large-scale ecosystem restoration. Journal of Shellfish Res. 19 (1), 413–421.

Hagström, Å., F. Azam, et al. (1988). Microbial loop in an oligotrophic pelagic marine ecosystem: possible roles of cyanobacteria and nanoflagellates in the organic fluxes. Mar. Ecol. Prog. Ser. 49., 171–178.

Hagström, Å .. , F. Azam, J. Kuparinen and U.-L. Zweifel. (2001). Pelagic plankton growth and resource limitation in the Baltic Sea. *In* A Systems Analysis of the Baltic Sea. F. Wulff, L. Rahm and P. Larsson (Eds). Berlin, Heidelberg, Springer-Verlag. 460 pp.

Harvey, C. P., S. P. Cox, T. E. Essington, S. Hansson and J. F. Kitchell. (2003). An ecosystem model of food web and fisheries interactions in the Baltic Sea. ICES Journal of Marine Science, 60, 939–950.

Heip, C., D. Basford, J. A. Craeymeersch, J. M. Dewarumez, J. Dörjes, P. A. J. de_Wilde, G. C. A. Duineveld, A. Eleftheriou, P. Herman, U. Niermann, P. Kingston, A. Künitzer, E. Rachor, H. Rumhor, K. Soetaert and T. Soltwedel. (1992). Trends in biomass, density and diversity of North Sea macrofauna. ICES Journal of Marine Science, 49, 13–22.

HELCOM (2002). Environment of the Baltic Sea Area 1994–1998. Balt. Sea Environ. Proc. No. 82B,. 216 pp.

HELCOM (2003). The Baltic Marine Environment 1999–2002. Balt. Sea Environ. Proc. No. 87, 47 pp.

Hickel, W. (1998). Temporal variability of micro- and nanoplankton in the German Bight in relation to hydrographic structure and nutrient changes. ICES Journal of Marine Science, 55, 600–609.

Hickel, W., P. Mangelsdorf and J. Berg. (1993). The human impact in the German Bight - eutrophication during 3 decades (1962–1991). Helgolander Meeresuntersuchungen, 47, 243–263.

Hites, R. A., J. A. Foran, D. O. Carpenter, M. C. Hamilton, B. A . Knuth,.and S. J Schwager,. (2004). Global Assessment of Organic Contaminants in Farmed Salmon. Science, 303, 226–229.

Holligan, P. M., T. Aarup and S. B. Groom (1989). The North Sea: Satellite colour atlas. Continental Shelf Research, 9, 667–765.

Howarth, M. J., J. H. Simpson, J. Sündermann and H. van Haren. (2002). Processes of Vertical Exchange in Shelf Seas (PROVESS). Journal of Sea Research, 47, 199–208.

Huthnance, J. M. (1995). Circulation, exchange and water masses at the ocean margin: the role of physical processes at the shelf edge. Progress in Oceanography, 35, 353–431.

Jago, C. F., S. E. Jones, R. J. Latter, R. R. McCandliss, M. R. Hearn and M. J. Howarth. (2002). Resuspension of benthic fluff by tidal currents in deep stratified waters, northern North Sea. Journal of Sea Research, 48, 259–269.

Joint, I. and A. Pomroy. (1993). Phytoplankton biomass and production in the southern North Sea. Marine Ecology - Progress Series, 99, 169–182.

Joint, I., J. Lewis, J. Aiken, R. Proctor, G. Moore, W. Higman and M. Donald. (1997). Interannual variability of PSP outbreaks on the north east UK coast. Journal of Plankton Research, 19, 937–956.

Jonsson P., R. Carman and F. Wulff. (1990). Laminated sediments in the Baltic – a tool for evaluating nutrient mass balances. Ambio 19, 152–158.

Jonsson, P. and R. Carman. (1984). Changes in deposition of organic matter and nutrients in the Baltic Sea during the twentieth century. Mar. Pollut. Bull. 28, 417–426.

Jorgensen, P. V. (1999). Standard CZCS Case 1 algorithms in Danish coastal waters. International Journal of Remote Sensing, 20, 1289–1301.

Kahru, M., U.Horstmann, and O. Rud (1994). Satellite detection of increased cyanobacteria in the Baltic Sea: Natural fluctuation or ecosystem change? Ambio 23, 469–472.

Kaiser, M. J. (1998). Significance of bottom-fishing disturbance. Conservation Biology, 12, 1230–1235.

Kautsky, N. and I. Wallentinus. (19809. Nutrient release from a Baltic Mytilus-red alga community and its role in bentic and pelagic productivity. Ophelia, Suppl. 1, 17–30.

Kivi, K., S. Kaitala, H. Kuosa, J. Kuparinen, E. Leskinen, R. Lignell, B. Marcussen and T. Tamminen (1993). Nutrient limitation and grazing control of the Baltic plankton community during annual succession. Limnol. Oceanogr. 38(5), 893–905.

Klein, A. W. O. and J. T. van Buuren. (1992). Eutrophication of the North Sea in the Dutch coastal zone 1976–1990. Ministry of Transport, Public Works and Water Management, Tidal Waters Division, The Hague, Report, 70pp.

Kratzer, S., B. Håkansson and C. Sahlin (2003). Assessing Secchi and Photic Zone Depth in the Baltic Sea from Satellite Data. Ambio, 32, 577–585.

Künitzer, A., D. Basford, J. A. Craeymeersch, J. M. Dewarumez, J. Dörjes, G. C. A. Duineveld, A. Eleftheriou, C. Heip, P. Herman, P. Kingston, U. Niermann, E. Rachor, H. Rumhor. and P. A. J. de Wilde. (1992). The benthic infauna of the North Sea: species distribution and assemblages. ICES Journal of Marine Science, 49, 127–143.

Larsson, U., R. Elmgren and F. Wulff. (1985). Eutrophication and the Baltic Sea - Causes and Consequences. Ambio 14(1), 9–14.

Larsson, U., S. Hajdu, J. Walve and R. Elmgren. (2001). Baltic Sea nitrogen fixation estimated from the summer increase in upper mixed layer total nitrogen. Limnology and Oceanography. 46(4): 811–820.

Lee, J.-Y., P. Tett, K. Jones, S. Jones, P. Luyten, C. Smith and K. Wild-Allen. (2002). The PROWQM physical-biological model with benthic-pelagic coupling applied to the northern North Sea. Journal of Sea Research, 48, 287–331.

Löfgren, S., A. Gustafson, S. Steineck and P. Stålnacke (1999). Agricultural development and nutrient flows in the Baltic States and Sweden after 1990. Ambio 28, 320–327.

Luyten, P. J., J. E. Jones, , R. Proctor, A. Tabor, P. Tett and K. Wild-Allen. (1999). COHERENS - a coupled hydrodynamical-ecological model for regional and shelf seas: user documentation. Management Unit of the Mathematical Models of the North Sea, Brussels.

Marino R., R.W. Howarth, F. Chan, J.J. Cole and G.E. Likens. (2003). Sulfate inhibition of molybdenum-dependent nitrogen fixation by planktonic cyanobacteria under seawater conditions: a non-reversible effect. Hydrobiologia 500(1–3): 277–293.

Mattsson, J. (1996). Oceanographic studies of transport and oxygen conditions in the Öresund. Ph.D. Thesis. Department of Oceanography, Göteborg University, Sweden.

Mills, D. K., P. Tett and G. Novarino. (1994). The Spring Bloom in the south-western North Sea in 1989. Netherlands Journal of Sea Research, 33, 65–80.

Moll, A. (1998). Regional distribution of primary production in the North Sea simulated by a three-dimensional model. Journal of Marine Systems, 16, 151–170.

Moll, A. and G. Radach. (2003). Review of three-dimensional ecological modelling related to the North Sea shelf system. Part 1: models and their results. Progress in Oceanography, 57, 175–217.

Nielsen, A. and K. Richardson. (1990). *Chrysochromulina polylepis* bloom in Danish, Swedish and Norwegian waters, May-June 1988 - an analysis of extent, effects and causes. Eutrophication and algal blooms in North Sea coastal zones, the Baltic and adjacent areas: prediction and assessment of preventive actions, ed. Lancelot, C., Billen, G. and Barth, H., Commission of the European Communities, Brussels, Water Pollution Research Report, 12, 11–25.

Norén, F. J. Haamer and O. Lindahl. (1999). Changes in the plankton community passing a Mytilus edulis mussel bed. Mar. Ecol.-Prog. Ser. 191, 187–194.

North Sea Task Force. 1993. North Sea quality status report 1993, Oslo and Paris Commissions, London. Olsen and Olsen, Fredensborg, Denmark, 132+vii pp.

Novarino, G., D. K. Mills and F. Hannah. (1997). Pelagic flagellate populations in the southern North Sea, 1988–89. I. Qualitative observations. Journal of Plankton Research, 19, 1081–1109.

OSPAR Commission, 2000. Quality Status Report 2000, Region II – Greater North Sea. OSPAR Commission, London, 136 + xiii pp.

Pätsch, J. and G. Radach. (1997). Long-term simulation of the eutrophication of the North Sea: temporal development of nutrients, chlorophyll and primary production in comparison to observations. Journal of Sea Research, 38, 275–310.

Pearson, T. H., A. B. Josefson and R. Rosenberg. (1985). Petersen's benthic stations revisited. I. Is the Kattegat becoming eutrophic? Journal of Experimental Marine Biology and Ecology, 92, 157–206.

Proctor, R., Holt, J. T., Allen, I. and Blackford, J. (2003). Nutrient fluex and budgets for the North West European Shelf from a three-dimensional model. The Science of the Total Environment, 314–316, 769–785.

Radach, G. (1992). Ecosystem functioning in the German Bight under continental nutrient inputs by rivers. Estuaries, 15, 477–496.

Radach, G. and J. Gekeler. (1996). Annual cycles of horizontal distributions of temperature and salinity, and of concentrations of nutrients, suspended particulate matter and chlorophyll on the Northwest European shelf. Deutsche Hydrographische Zeitschrift, 48, 261–297.

Radach, G., J. Berg. and E. Hagmeier. (1990). Long-term changes of the annual cycles of meteorological, hydrographic, nutrient and phytoplankton time series at helgoland and at LV ELBE 1 in the German Bight. Continental Shelf Research, 10, 305–328.

Radach, G. and J. Genkeler. 1997. Gridding of the NOWESP data sets, Nr.27, Berichte aus dem Zentrum für Meeres- und Klimaforschung. Reihe B: Ozeanografie, Institut für Meereskunde, Hamburg, 375 pp.

Rahm, L., D. Conley, P. Sanden, F. Wulff and P. Stålnacke. 1996. Time series analysis of nutrient inputs to the Baltic sea and changing DSi:DIN ratios. Mar. Ecol.-Progr. Ser. 130(1–3), 221–228.

Reid, P. C., M. D. Borges and E. Svendsen. (2001). A regime shift in the North Sea circa 1988 linked to changes in the North Sea horse mackerel fishery. Fisheries Research, 50, 163–171.

Richardson, K. (1985). Plankton distribution and activity in the North Sea/Skagerrak-Kattegat frontal area in April 1984. Marine Ecology Progress Series, 26, 233–244.

Richardson, K. and A. Christoffersen. (1991). Seasonal distribution and production of phytoplankton in the southern Kattegat. Marine Ecology Progress Series, 78, 217–227.

Richardson, K. and J. P. Heilman. (1995). Primary production in the Kattegat: past and present. Ophelia, 41, 317–328.

Rodhe, J. (1998). The Baltic and North Seas: a process-oriented review of the physical oceanography. In: Robinson, A.R. and K. Brink (eds.) The Sea. John Wiley and Sons, New York, 699–732.

Rosenberg, R., I. Cato, L. Forlin, K. Grip and J. Rodhe. (1996). Marine environment quality assessment of the Skagerrak-Kattegat. Journal of Sea Research, 35, 1–8.

Rydberg, L., G. Ærtebjerg and L. Edler. (2003). Temporal development of primary production in the Baltic Entrance region: trends and variability. Submitted to J. Sea Res.

Seitzinger, S. P. and A. E. Giblin. (1996). Estimating denitrification in North Atlantic continental shelf sediments. Biogeochemistry, 35, 235–260.

Sharples, J. and P. Tett. (1994). Modelling the effect of physical variability on the midwater chlorophyll maximum. Journal of Marine Research, 52, 219–238.

Skogen, M. D. and Moll, A. (2000). Interannual variability of the North Sea primary production: comparison from two model studies. Continental Shelf Research, 20, 129–151.

Stigebrandt, A. (1991). Computations of oxygen fluxes through the sea surface and the net production of organic matter with application to the Baltic and adjacent seas. Limnol. Oceanogr. 36(3), 444–454.

Stålnacke, P., A. Grimvall, K. Sundblad and A. Tonderslki. 1999. Estimation of riverine loads of nitrogen and phosphorus to the Baltic Sea, 1970–1993. Environ. Monit. Assess. 58(2), 173–200.

Sundermann J, S. Beddig, J. Huthnance and C.N.K. Mooers. (2001). Impact of climate change on the coastal zone: discussion and conclusions Climate Res. 18 (1–2), 1–3 2001)

Sverdrup, H. U. (1953). On conditions for the vernal blooming of phytoplankton. Journal du Conseil, 18, 237–295.

Tett, P. (1990). The Photic Zone. Light and Life in the Sea, ed. Herring, P. J., Campbell, A. K., Whitfield, M. and Maddock, L., Cambridge University Press, Cambridge, U.K., 59–87.

Tett, P. and D. Mills. (1991). The plankton of the North Sea - pelagic ecosystems under stress? Ocean and Shoreline Management, 16, 233–257.

Tett, P., I. Joint, D. Purdie, M., Baars, S. Oosterhuis, G. Daneri, F. Hannah, D. K. Mills, D. Plummer, A. Pomroy, A. W. Walne and H. J. Witte. (1993). Biological consequences of tidal stirring gradients in the North Sea. Philosophical Transactions of the Royal Society of London, A340, 493–508.

Tett, P. and A. Walne. (1995). Observations and simulations of hydrography, nutrients and plankton in the southern North Sea. Ophelia, 42, 371–416.

Tett, P., D. Hydes and R. Sanders. (2003). Influence of nutrient biogeochemistry on the ecology of North-West European shelf seas. Biogeochemistry of Marine Systems, ed. Schimmield, G. and K. Black, Sheffield Academic Press Ltd, Sheffield, 293–363.

Warner, A. J. and Hays, G. C. (1994). Sampling by the Continuous Plankton Recorder. Progress in Oceanography, 34, 237–256.

Warnock, R. E., Gieskes, W. W. C. and van Laar, S. (1999). Regional and seasonal differences in light absorption by yellow substance in the Southern Bight of the North Sea. Journal of Sea Research, 42, 169–178.

Wasmund, N., A. Andrushaitis, E. Lysiak-Pastuszak, B. Muller-Karulis, G. Nausch, T. Neumann, H. Ojaveer, I. Olenina, L. Postel and Z. Witek. 2001. Trophic status of the south-eastern Baltic Sea: A comparison of coastal and open areas. Estuarine Coastal and Shelf Sc. 53(6), 849–864

Westman, P., J. Borgendahl, T. S. Bianchi and N. H. Chen (2003). Probable causes for cyanobacterial expansion in the Baltic Sea: Role of anoxia and phosphorus retention. Estuaries 26(3), 680–689.

Wild-Allen, K., A. Lane and P. Tett. (2002). Plankton, sediment and optical observations in Netherlands coastal water in spring. Journal of Sea Research, 47, 303–315.

Winsor, P., J. Rodhe and A. Omstedt. (2001). Baltic Sea ocean climate: an analysis of 100 yr of hydrographic data with focus on the frewshwater budget. Clim. Res. Vol. 18, 5–15.

Winsor, P., J. Rodhe and A. Omstedt. (2003). Erratum. Baltic Sea ocean climate: an analysis of 100 yr of hydrographic data with focus on the frewshwater budget. Clim. Res. 25 (2), 183.

Wulff, F., L. Rahm, I. Hallin and J. Sandberg. (2001). A nutrient budget model of the Baltic Sea. In: A Systems Analysis of the Baltic Sea. F. Wulff, L. Rahm and P. Larsson (Eds.). Berlin, Springer-Verlag. 148, 353–372.

Wulff, F., L. Rahm and A. Stigebrandt. (1990). Nutrient dynamics of the Baltic Sea. Ambio 19, 126–133.

Chapter 27. ICELAND, FAROE AND NORWEGIAN COASTS (21,E)

EILIF GAARD

Faroese Fisheries Laboratory, P.O. Box 3051, FO-110 Tórshavn, Faroe Islands,

ASTTHOR GISLASON

Marine Research Institute, Box 1390, IS-121, Reykjavik, Iceland

WEBJØRN MELLE

Institute of Marine Research, P.O. Box 1870, Nordnes, NO-5817 Bergen, Norway

Contents

1. Introduction

The Icelandic, Faroese and Norwegian shelves (Fig. 27–001) are influenced by a persistent inflow of relatively warm Atlantic Water of a southern origin. Cooling and sinking of surface water in the Arctic Ocean and the Nordic seas leads to formation of cold Norwegian Sea Deep Water at depths between 500 and 1000 m and a deep overflow of this water across the Greenland-Scotland Ridge, into the North Atlantic. To balance this overflow and various surface outflows, there is a persistent inflow of warm water in the upper layers, from the North Atlantic into the Nordic Seas (Hansen *et al.*, 1998; Hansen and Østerhus, 2000; Østerhus *et al.*, 2001). This inflow has a decisive influence on the oceanic climate, and the biological productivity on the shelves (Skjoldal *et al.*, 1992; Astthorsson and Vilhjalmsson, 2002). It is generally also considered to be the main reason for the relatively mild climate of large parts of the Arctic Mediterranean as well as much of northern Europe (Rahmstorf and Ganopolski, 1999). There is, however, evidence, that the

The Sea, Volume 14, edited by Allan R. Robinson and Kenneth H. Brink
ISBN 0–674–01527–4 ©2005 by the President and Fellows of Harvard College

exchange of water between the North Atlantic and the Arctic Mediterranean has been decreasing during the last 50 years (Hansen *et al.*, 2001; Østerhus *et al.*, 2001).

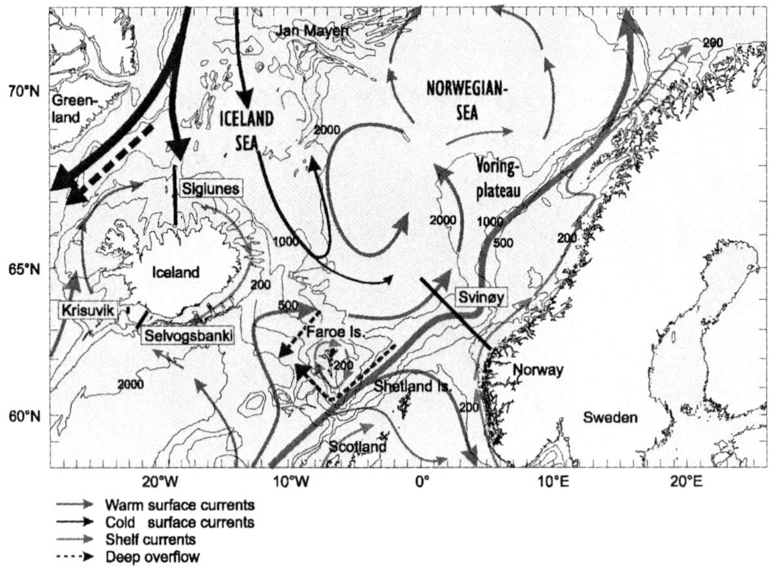

Figure 27.1 Bottom topography and main features of the shelf and oceanic currents in the Iceland-Faroes-Norway region.

Phytoplankton production is seasonal and lasts from March-May to around September (Heath *et al.,* 2000a). Generally primary production starts earlier in spring on the shelves than offshore. Abundance of phytoplankton highly affects the reproduction and growth of zooplankton on the shelves, and hence, the zooplankton abundance and diversity are also seasonal. During winter the abundance of zooplankton is very low on the shelves and in the upper offshore layer above 500 m, but during spring and summer the numbers of species and individuals increase very much (Gaard, 1999; Gislason and Astthorsson, 1995, 1998a; Østvedt, 1955). The species diversity of the zooplankton is significantly higher on the shelves than offshore and contains a mixture of neritic holo- and meroplankton and oceanic plankton (Gaard, 1999; Gislason and Astthorsson, 1995).

The shelves are influenced much by a common large resource of the copepod *Calanus finmarchicus.* During winter this copepod resides in diapause as stages C4 and C5 in the deeper layers of the North Atlantic and the Nordic Seas (*e.g.* Östvedt, 1955; Hirche, 1996; Ingvardsdóttir *et al.,* 1999, Gislason and Astthorsson, 2000; Heath *et al.,* 2000a). In late winter and spring it ascends and produces a new generation (Niehoff *et al.,* 1999; Gaard and Hansen, 2000; Gislason and Astthorsson, 2000; Niehoff and Hirche, 2000) and is also advected onto the shelves where it is a dominant secondary producer and an important food item for larval and juvenile fish and for large quantities of pelagic fish. The spring reproduction of *C. finmarchicus* is closely related to the phytoplankton spring bloom (Niehoff and Hirche, 2000; Gislason and Astthorsson, 1996; Gaard, 2000). Also euphausiids are abundant in the Nordic Seas and the adjacent shelves, and, similarly to *C. finmarchicus,* the growth and reproduction of the euphausiids is closely coupled to the growth of phytoplankton (*e.g* Einarsson, 1945: Mauchline, 1980; Dalpadado and Skjoldal, 1996; Astthorsson and Gislason, 1997a; Dalpadado *et al.,* 1998a).

At higher trophic levels, pelagic fish have spawning-feeding migrations between the shelves and the Nordic Seas, for example Atlantic salmon (Hansen and Jacobsen, 2000; Jacobsen and Hansen, 2000a; 2000b; 2001), Norwegian spring spawning herring (*e.g.* Misund *et al.,* 1998; Dalpadado *et al.,* 1998b), capelin (Vilhjálmsson, 1994); blue whiting (Bailey, 1982; Bjelland and Monstad, 1997) and mackerel (Reid *et al.,* 1997; Belikov *et al.,* 1998). These migratory fishes transport organic production between the Nordic Seas and the adjacent shelves, due to their migrations between spawning areas, nursery areas and feeding areas.

2. Icelandic shelf

2.1. Topography, currents and water masses

Iceland is located at the junction of two submarine ridge systems, the Mid-Atlantic Ridge and the Greenland-Scotland Ridge (Fig. 27–001). The ridges have important influences on the ocean circulation and the distribution of water masses around Iceland, and therefore also on the distribution of organic production in Icelandic waters (Stefánsson, 1962; Stefánsson and Ólafsson, 1991; Valdimarsson and Malmberg, 1999). The Greenland-Scotland Ridge is particularly important in this respect, since it acts as a barrier that constrains flow between the relatively warm waters of the North Atlantic and the cold arctic deep waters of the Iceland and the Norwegian Seas.

On the shelf south and west of Iceland the water is relatively warm Atlantic Water which flows towards the south and west coasts with the North Atlantic Current, whereas on the shelf north and east of the country the water masses are formed by mixing and local modifications of warm Atlantic water originating from off the south and west coasts and cold Polar and Arctic Water carried by respectively the East Greenland and East Icelandic currents from the north (Fig. 27–001). Closest to the coast there is less saline Coastal Water that is Atlantic or Atlantic/Arctic Water, diluted by fresh water from the land (Stefánsson, 1962).

2.2. Primary production

In Icelandic shelf areas, diatoms of the genera *Thalassiosira* and *Chaetoceros* tend to dominate the spring bloom (Thórdardóttir and Gudmundsson, 1998). North of Iceland the prymnesiophyte *Phaeocystis pouchetii* may also be abundant during spring (Gunnarsson *et al.,* 1998). After the spring bloom, dinoflagellates (mainly the genera *Ceratium* and *Protoperidinium*) may increase in abundance, although diatoms continue to be relatively abundant (H. Gudfinnsson and K. Gudmundsson, unpublished data). During the autumn, there may be another bloom of diatoms of some of the same genera as in spring, but dinoflagellates can still be abundant (Thórdardóttir and Gudmundsson, 1998). The relative abundance of dinoflagellates is thus generally higher during the autumn bloom compared to the spring bloom.

Long term monitoring of primary production in Icelandic waters shows that the annual primary production is on average higher over the shelves (~220 g C m^{-2} yr^{-1}) than farther out (~150 g C m^{-2} yr^{-1}) (Fig. 27–002; Thórdardóttir, 1994). The most likely reason is that the shelf areas are more nutrient rich than the oceanic ones as they are bathed by nutrient rich deep water that is brought to the surface by upwelling currents (Thórdardóttir, 1994). Over the shelf, the annual primary production is on average higher in the southern and western regions most influenced by Atlantic Water than in the northern and eastern ones where the water is a mixture of Atlantic and Arctic Waters (Thórdardóttir, 1994). This difference is related to the fact that in the southern and western regions the water tends to be less strongly stratified than in the northern and eastern ones where the admixture of Polar or Arctic Water may create a relatively strong stratification. Therefore the renewal of nutrients into the euphotic layer is generally more efficient and the plant production higher in the southern and western regions compared to the northern and eastern ones. Further, the concentration of nutrients is generally higher in the Atlantic Water than in the Atlantic-Arctic mixture (Stefánsson and Ólafsson, 1991). Over the southern and western shelf, the production tends to decrease outwards from land, whereas over the northern shelf the primary productivity decreases from west to east, *i.e.* in the direction of the diminishing admixture of Atlantic Water. Another feature that is very evident is the relatively high primary productivity in the frontal areas northwest and southeast of Iceland, where the mixing of warm and cold water masses leads to a relatively high nutrient content of the surface waters, which in turn leads to increased phytoplankton production.

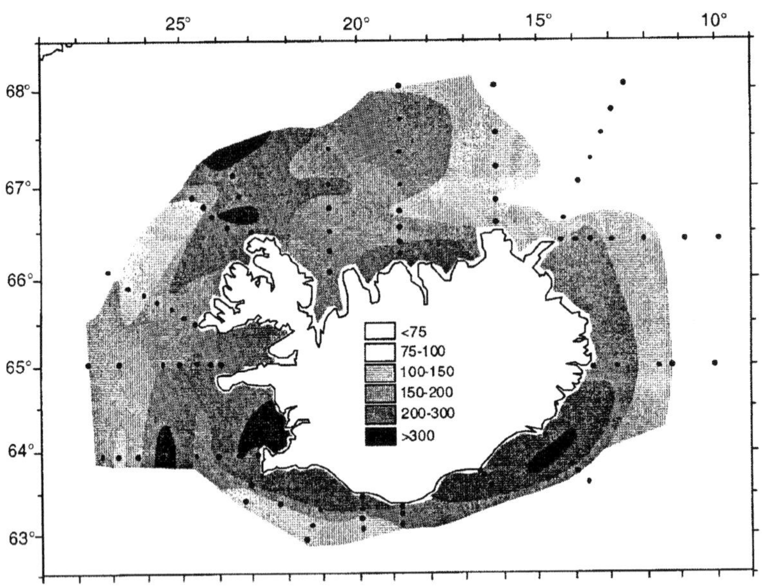

Figure 27.2 Average annual production in Icelandic waters (g C m^{-2} $year^{-1}$). The values are based on annual measurements carried out during the years 1958–1982 (Redrawn from Thórdardóttir, 1994).

In Icelandic waters, the growth of phytoplankton generally starts in late March-early April and it usually culminates in May (Gudmundsson, 1998). However, the timing of bloom initiation varies both interannually and between regions, mainly depending on the development of surface water stability in spring. As a result of freshwater efflux, the waters nearest to the coast south and west of Iceland become stratified earlier than the ones farther from the coast where warming of surface waters creates stratification. Therefore, the phytoplankton spring bloom on the shelf south of Iceland tends to start earlier near the shore (mid to late March) than farther from land (early April) (Thórdardóttir, 1994; Gudmundsson, 1998), while peak values are generally reached in May in both environments (Gudmundsson, 1998). On the other hand, on the northern shelf, an onshore-offshore gradient in the timing of surface water stratification is generally not as evident as off the south and west coasts. The reason is that near the shore the effects of freshwater runoff are relatively small. Therefore the primary production may begin at similar times for different distances from the coast (early to late April) (Stefánsson and Ólafsson, 1991; Gudmundsson, 1998). In years with relatively great influx of Arctic Water onto the northern shelf, the spring bloom may even start earlier offshore than closer to land as pycnocline formation in the more oceanic areas may be related to ice melting.

2.3. Zooplankton

Over the Icelandic continental shelf, copepods dominate the mesozooplankton in terms of numbers, generally comprising >80% of net-caught zooplankton (Fig. 27–003) (Astthorsson *et al.,* 1983; Gislason and Astthorsson, 1995; 1998a; Astthorsson and Gislason, 1992; 1999). In terms of biomass *Calanus finmarchicus* is generally the most abundant species both in the Atlantic waters south and west of Iceland and in the mixed Arctic-Atlantic waters north of the island. However, its biomass is generally higher in the warm waters south and west of Iceland than in the colder waters off the north and east coasts (Fig. 27–003) (Gislason, 2005). This indicates that the warmer waters off the south and west coasts provides a more favourable environment for the growth and development of *C. finmarchicus* than the colder waters off the north and east coasts.

In addition to *C. finmarchicus,* the copepods *Pseudocalanus* spp., *Acartia longiremis,* and *Oithona* spp. occur regularly over the shelf around Iceland. The copepods *Temora longicornis* and *Centropages hamatus,* and the cladocerans *Evadne nordmanni* and *Podon leuckarti* are more abundant off the south and west coasts than off the north and east coasts, whereas the reverse is true for the copepods *Metridia longa, C. hyperboreus* and *C. glacialis,* and the chaetognath *Sagitta elegans* (Gislason and Astthorsson, 1995; 1998a; Hallgrímsson, 1954; Jespersen 1940). Among the meroplankton, cirripede larvae are generally the most numerous group, being most abundant in May-June (Astthorsson *et al.,* 1983; Gislason and Astthorsson, 1995).

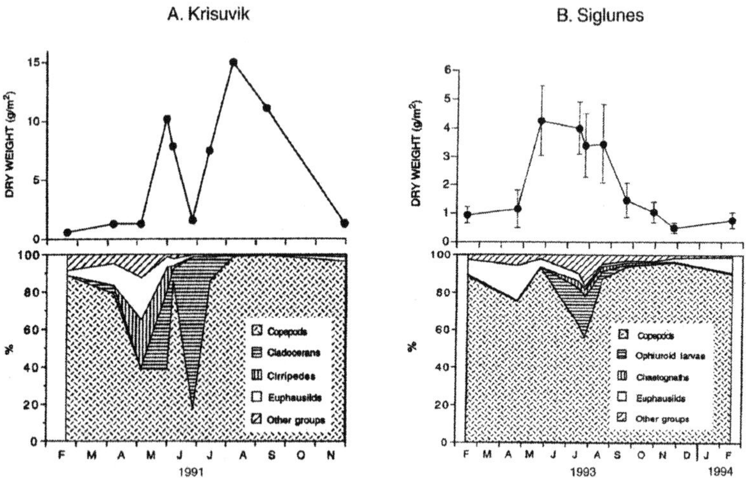

Figure 27.3 Seasonal variations in dry weight of total zooplankton (upper panel) and relative frequency by numbers of most numerous zooplankton taxa (lower panel) southwest (A) and north (B) of Iceland. The values in (A) are means from two stations off Krisuvik (See Fig. 27–001). The values in (B) are averages from 8 stations on the Siglunes-transect and vertical lines denote standard error. Mesh size of the nets 335 μm. Note different scales for dry weight. (From Gislason and Astthorsson, 1995, 1998a).

On the shelf south and west of Iceland, seasonal variations in the abundance of zooplankton are characterised by low winter values and two peaks in both total numbers and total biomass during the summer, the first in May-June and the second in July-August (Fig. 27–003) (Hallgrímsson, 1954; Gislason and Astthorsson, 1995; Gislason *et al.*, 2000). In contrast, in the colder waters northwest, north and east of Iceland, both in the fjords and in the more exposed shelf areas, there is only one maximum during the summer (Fig. 27–003) (Astthorsson and Gislason, 1992; Gislason and Astthorsson, 1998a; Kaasa and Gudmundsson, 1994). These seasonal changes partly reflect the different life histories of *C. finmarchicus* in these environments, with two main recruitment events of the *C. finmarchicus* stock south of Iceland and one in the colder waters off the northwest, north and east coasts (Gislason and Astthorsson 1996, 1998a; Gislason *et al.*, 2000). The reason for the observed differences in seasonal dynamics of *C. finmarchicus* in the different environments around Iceland is most probably related to the different temperature regimes in these environments, with temperatures in the surface layers during spring and summer being much higher south of the country (~6–12°C) than in the north (~1–6°C). However, other factors, such as advection, feeding conditions and predatory impact may also play a part.

From the offshore overwintering areas, *C finmarchicus* invades the southern and western shelves, probably mainly from the south, *i.e.* from the Iceland Basin. However, import from the Irminger Sea may also be important, especially onto the western shelf. The most likely springtime sources of *C. finmarchicus* to the shelf waters north and northeast of Iceland are the deep oceanic waters of the Iceland Sea. Later during the summer, the northern shelf areas may also be invaded by

Calanus advected by the coastal current and the Irminger current from the southern and western shelves.

The spring spawning of *C. finmarchicus* is closely coupled to the development of the phytoplankton. Thus, due to the similar timing of the spring bloom in the Atlantic water south of Iceland and in the Subarctic waters north of the island, and in spite of the much higher spring surface water temperatures in the former region compared to the latter, the spring spawning of *C. finmarchicus* begins at similar time, *i.e.* April-May in both areas (Hallgrímsson, 1954; Gisalson, 2005 Gislason and Astthorsson, 1995; 1996; 1998a; Gislason *et al.*, 1994; 2000). Off the south and west coasts, the spawning of *C. finmarchicus* usually begins near the coast and then progresses to the deeper waters, thus following the development of the spring bloom which, as noted previously, tends to be delayed with increasing distance from the coast. Off the north coast, where the seasonal development of the phytoplankton may be similar for different distances from the coast, an onshore-offshore variability in the spawning of *C. finmarchicus* is not evident. In the fjords off the northwest and north coasts, the spring spawning of *C. finmarchicus* appears to occur at a similar time (March-April) as the phytoplankton spring bloom (Astthorsson and Gislason, 1992; Kaasa and Gudmudsson, 1994).

Two euphausiid species, *Thysanoessa raschi* and *T. inermis* are common in the fjords and over the banks (Einarsson, 1945; Astthorsson, 1990; Gislason and Astthorsson, 1995; Astthorsson and Gislason, 1997a). *T. raschi* is mostly confined to fjords and bays while *T. inermis* mainly inhabits the coastal banks. While the spawning of the euphausiids is linked to the spring bloom they have different life histories in the different environments around Iceland. Thus, *T. inermis* matures and spawns as one year old in the warm waters over the shelf off the south and west coasts, whereas they do not become mature until two years old in the colder waters off the northwest and north coasts (Einarsson, 1945; Astthorsson, 1990; Astthorsson and Gislason, 1997a). *T. raschi* has been reported to spawn both as one year and two years old in Isafjord deep on the northwest coast (Astthorsson, 1990). In addition to these species, the euphausiids *M. norvegica* and *T. longicaudata* are regularly found on the banks. However, their main areas of distribution are located over the shelf edge or in the oceanic areas beyond the shelf. Euphausiids may sometimes make up a significant fraction of the total zooplankton biomass over the shelf and they may play an important trophodynamic role in Icelandic shelf waters.

Long-term monitoring of zooplankton biomass around Iceland has been carried out since 1960 (Astthorsson and Gislason, 1995; 1998; Astthorsson *et al.,* 1983; Gislason and Astthorsson, 1998b; Beare *et al.*, 2000). The investigations are carried out along transects that extend perpendicularly from the coastline and beyond the shelf, and they therefore integrate conditions both over the banks and beyond them. However, as most of the stations lie over the shelves, it is suggested that the observed variations may to a large extent reflect changes in biomass and productivity over the banks. The results from these investigations have shown that, with the exception of the mid 70's, the biomass has fluctuated more or less synchronously all around the country with maxima at intervals of 7–10 years (Fig. 27–004) (Astthorsson and Gislason, 1995; Beare *et al.,* 2000).

On the shelf north of Iceland the variations are mainly related to the inflow of Atlantic Water (Astthorsson and Gislason, 1995; 1998), which in turn is influenced by the frequency of northerly and southerly winds north of the island (Stefánsson, 1962; Ólafsson, 1999). Thus, in warm years, when the flow of Atlantic Water onto the northern shelf is high, the zooplankton biomass is almost two times higher than in cold years, when this inflow is not as evident (Astthorsson and Gislason, 1998). As discussed by Astthorsson and Gislason (1998) and Astthorsson and Vilhjalmsson (2002) the reason for this may be the better feeding conditions of the zooplankton due to increased primary production in warm years, advection of zooplankton with the Atlantic Water from the south, and faster temperature dependent growth of the zooplankton in warm years.

The links between climatic variability and zooplankton productivity (as expressed in zooplankton biomass during spring) are less evident on the southern shelf than on the northern one. The growth of phytoplankton and key zooplankters are closely related, and therefore it is likely that the variability is related to the timing and magnitude of the primary productivity, which off the south coast are heavily influenced by the freshwater efflux from rivers and by wind force and direction.

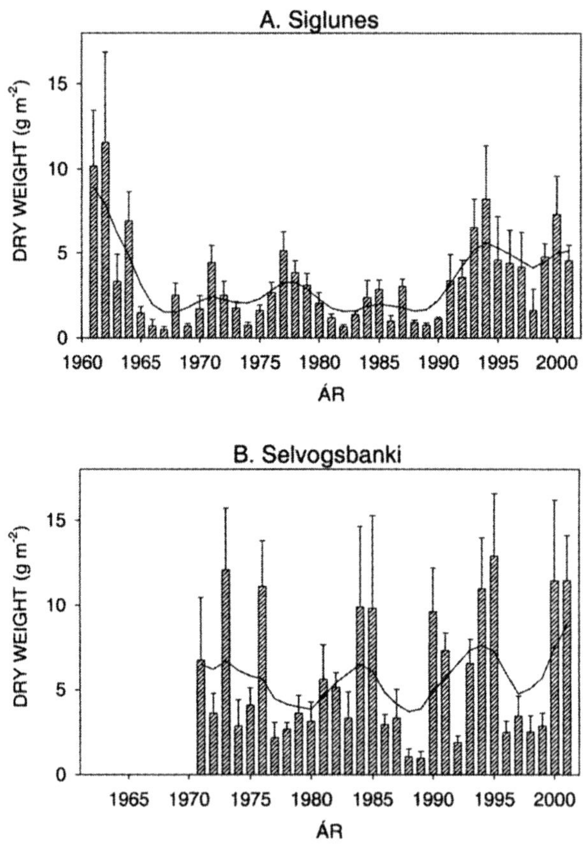

Figure 27.4 Variations in zooplankton biomass (g dry weight m^{-2}, 0–50 m) in spring at Siglunes (A) and Selvogsbanki (B) transects (See Fig. 27–001). The columns show means for all stations at the respective transects and the vertical bars denote standard error. The curved line shows 7 years running mean. (From Anonymous, 2001, and additional data).

2.4 *Biological resources*

The total catch of fish and invertebrates from Icelandic waters has risen from about 200 000 tonnes near the beginning of the 20th century to about 1 500 000 tonnes near its end (Jakobsson and Stefansson, 1998). Demersal fish are economically most important, and during the last five decades (1950–2000) the average catches have fluctuated from about 450 000 tonnes to around 850 000 tonnes. The annual harvest of pelagic fish in the same period has been between ~100 000 and ~1 600 000 tonnes. The catches of invertebrates increased significantly in the early 1980's and since then have ranged between ~20 000 and ~80 000 tonnes (Anonymous, 2002).

Most of the commercial fish stocks have their main spawning grounds in the warm Atlantic water off the south and west coasts, while the nursery areas are located in the colder waters off the north coast.

The economically most important demersal fish species inhabitating Icelandic waters are cod, haddock, saithe, redfish and Greenland halibut. Among these, the Icelandic cod is the most important, with annual yields that during the last five decades (1950–2000) have varied between ~170 000 and ~540 000 tonnes (Anonymous, 2002). The main spawning grounds of the cod are situated on the southern and western shelves, but there are smaller spawning areas in inshore regions all around the country. Larvae hatched on the shelf areas south and west of the country are transported by the clockwise coastal current and the Irminger current to the nursery areas that are located off the northwest, north and northeast coasts. During the drifting phase of the life cycle in spring and summer the larvae feed mainly on early stages of copepods (Jonsson and Fridgeirsson, 1986; Thorisson, 1989). On the nursery areas, the 0-group cod feeds pelagically until early autumn, when the cod gradually settle near the bottom. The cod spends the first 4–8 years of its life on the nursery areas, but when sexually mature it migrates back to the spawning grounds to spawn.

The herring and the capelin are the most important pelagic fish in Icelandic waters. Historically, the fishery for herring in Icelandic waters has been based on three different stocks, two of which spawn off the Icelandic south and west coasts (Icelandic spring and summer spawners) and one that spawns off the mid coast of Norway (Norwegian spring spawners). The catches of herring reached a peak of ~600 000 tonnes during the mid 1960's, when the catches mainly comprised the Norwegian spring spawners. After the collapse of this stock in the late 1960's the herring fishery on the Icelandic shelf has been mainly directed at the Icelandic summer spawners. During the last decade (1990–2000) the average herring catch from Icelandic waters has been around ~100 000 tonnes (Anonymous 2002).

After the collapse of the herring in the late 1960's, the capelin replaced the herring as the largest pelagic fish resource in Icelandic waters. Since the fishery started in the late 1960's the average stock size has been ~1 900 000 tonnes (Vilhjálmsson, 1994). However, the stock size has fluctuated considerably, with lows in the early 1980's and the early 1990's. These fluctuations appear to be related to variations in zooplankton productivity, rather than overfishing (Astthorsson and Gislason 1998; Jakobsson and Stefansson, 1998). Reflecting the variable stock size, the catches have been variable, but during the last decade the annual harvest from this stock has been around ~900 000 - 1 000 000 tonnes (Anonymous,

2002). As with cod, the main spawning grounds of the capelin are located in relatively shallow areas near the south and west coasts (Vilhjálmsson, 1994). From the spawning areas larvae and juveniles drift with currents in a clockwise direction to the bank areas west north and east of Iceland. During the winter the juveniles are mainly found over the banks, but in spring and summer they migrate into the Iceland Sea where they prey heavily on copepods and euphausiids (Astthorsson and Gislason, 1997b). During autumn the capelin concentrate near the continental shelf slope off the north coast, and in late winter the maturing capelin start the return migration towards the south coast for spawning. In this way a huge zooplankton biomass in the Iceland Sea is converted into fish biomass and transported onto the shelves.

Three invertebrate species are commercially exploited in Icelandic waters, i.e. the northern shrimp, the Nephrops and the Icelandic scallop. Of these species the northern shrimp is the most important, with a fishery that has increased rapidly during the last 20 years, from around 10 thousand tonnes in the mid 1980's to about 70 thousand tonnes in the late 1990's (Anonymous, 2002). The main shrimp fishing grounds are situated off the north and northwest coasts of Iceland.

2.4 *Trophic interactions*

The capelin plays a central role in the pelagic food web of Icelandic waters as it is the major plankton feeding species in the ecosystem, while at the same time being an important food source for a large variety of fishes, such as cod, haddock, saithe and even flatfishes such as Greenland halibut and long rough dab. The capelin is particularly important as food for cod, as it constitutes on an annual basis around half of the food of 4 years and older cod (Pálsson, 1983). The importance of the capelin as prey of cod is further demonstrated by the fact, that the collapse of the capelin stock in the beginning of the 1980's led to a 20–30% reduction in the mean weight at age of cod, indicating that the cod was not able to compensate for the lack of capelin by eating other food items (Magnússon and Pálsson, 1991; Vilhjalmsson, 1997).

It is also known that cod feed heavily on deep-water shrimp, thus influencing the abundance of the shrimp stock (Magnússon and Pálsson, 1991). However, in this case the predator- prey relationship appears to be of the top down kind, i.e. the cod influences the growth and abundance of the shrimp, whereas the growth of cod appears to be relatively unaffected by the size of the shrimp stock (Stefansson et al., 1998).

To summarize, these three stocks interact in such a way, that the growth of cod is positively related to that of the capelin, whereas the recruitment of the deep-water shrimp is negatively affected by the size of the cod stock. These interactions are presently taken into consideration in management strategies in order to get an optimal harvest of these stocks. As described above, the capelin is a key species in the system, and its growth is dependent on that of the zooplankton. There is some evidence, that the growth of zooplankton is in turn related to hydrographic conditions (Astthorsson and Gislason, 1998; Astthorsson and Vilhjalmsson, 2002), and therefore it is important also to consider hydrographic conditions and plankton productivity in the harvesting strategy for these stocks.

It may be added that sea birds, seals and whales are also important predators in the Icelandic marine ecosystem. According to calculations made by Sigurjonsson and Vikingsson (1998) the total food consumption of whales in Icelandic and adjacent waters is around 6.3 million tonnes, i.e. more than three times the total landings of the Icelandic fishing fleet. About 2 million tonnes of the whale consumption is comprised of fish, most of which may be capelin. Thus it is conceivable that consumption by whales may have great impact on the yield of fish resources.

3. *Faroe shelf*

3.1. *Topography, currents and water masses*

Similar to the Icelandic shelf, the Faroe shelf is located on the Scotland-Greenland Ridge, and hence has the Nordic Seas to the north and east and Atlantic to the west and south. (Fig. 27–001). In the upper layers (above ~500 m depth) the shelf and slope is surrounded by warm Atlantic water, which basically flows in northeasterly direction (Hansen *et al.*, 1998; Turrell *et al.*, 1999; Hansen and Østerhus, 2000). Below about 500 m depth, the Faroe shelf is almost entirely surrounded by cold Norwegian Sea Deep Water (NSDW) which flows southwest through the Faroe-Shetland (Turrell *et al.*, 1999; Hansen and Østerhus, 2000) and continues northwards through the Faroe Bank Channel to the west of the Faroes (Fig. 27–001) (Johnson and Sanford, 1992; Hansen and Kristiansen, 1999; Hansen *et al*, 2001).

On the Faroe shelf, extremely strong tidal currents (Hansen, 1992) lead to intense mixing, resulting in homogeneous water mass in the shallow shelf areas. The well-mixed shelf water is separated from the offshore water by a persistent tidal front, which surrounds the shelf at about the 100–130 m bottom depth contour (Gaard *et al.*, 1998). In addition, residual currents have a persistent clockwise circulation around the islands (Gaard and Hansen, 2000). The flushing time of the shelf water is highly variable, and on average it is estimated to be about 2.5–3 months on the shallow parts of the shelf (Gaard and Hansen, 2000; Gaard, 2003). The area inside the tidal front is 8000–10000 km^2.

These hydrographic facts lead to maintenance of an ecosystem on the Faroe shelf that is quite different from the offshore environment with distinct planktonic communities, benthic fauna and several fish stocks. Furthermore, a large number of sea birds breed on the Faroe Islands and depend on the biological production on the shelf for food (Gaard *et al*, 2002). Hence the Faroe shelf water can be considered as a separate neritic ecosystem. It is, however, highly influenced by variable advection of offshore water onto the shelf.

3.2. *Phytoplankton*

The strong turbulence the Faroe shelf water supports a typical diatom-dominated plankton community. Similar to the Icelandic and Norwegian shelves diatoms (mainly of the genera *Thalassiosira, Chaetoceros* and *Rhizosolenia*) are commonly dominant during spring bloom. In summers with relatively high nutrient concentrations diatoms usually remain dominant. However in summers with low nutrient concentrations, they may be replaced by smaller flagellates, *e.g.* the prymnesiophyte, *Phaeocystis pouchetii* and coccolothophorids (Gaard, 1996; Gaard *et al.*, 1998).

Due to the retention of water on the Faroe shelf the amounts of nutrients are limited and primary production therefore strongly affects the decrease in nutrient concentrations in the ecosystem during summer. The nutrient concentrations may decrease down to very low levels in the shelf water. This decrease in nutrients is, however, variable between years (Fig. 27–005). The variable decrease in nutrient concentrations seems mainly to be a result of variable new primary production, more than the variable influx (Gaard *et al.,* 1998). Based on observed decrease in nitrate plus a calculated mean net influx of nutrients into the shelf water (Gaard and Hansen, 2000; Gaard, 2003), an approximate index of a potential new primary production (Dugdale and Goering, 1967) has been calculated (Fig. 27–006) (Gaard *et al.,* 2002; Gaard, 2003). This index shows a fluctuation of the potential new primary production by more than a factor of five from 1990 to 2001. The timing of the spring bloom and the annual phytoplankton peak also varies significantly between years, and generally the years with early onset of the primary production during spring are the same years that show high primary production (low nutrients) and *vice versa* (Fig. 27–005) (Gaard, 2003).

Due to the limited amounts of nutrients in the ecosystem, there is a maximum limit for new primary production. In productive years (low nutrient concentrations), the new primary production seems to reach that limit (Fig. 27–005).

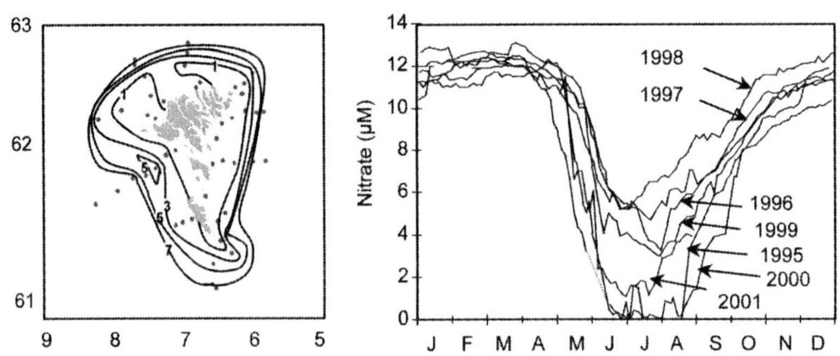

Figure 27.5 Nitrate (µM) at 40 m depth around the Faroes on 23 June-1 July 2000 (left) and nitrate concentrations at a fixed, shallow station on the central Faroe shelf, 1995–2001. (From Gaard, 2003, and additional data).

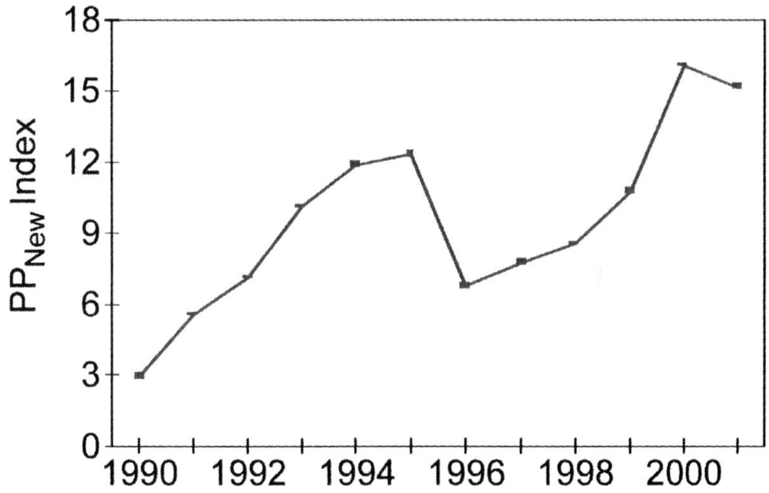

Figure 27.6 Calculated index of potential new primary production on the Faroe shelf, between spring and 26 June 1990–2001.

3.3. Zooplankton

In most years, the species composition on the shelf is quite different from that in the surrounding oceanic environment. While the oceanic environment is dominated by mainly the copepod *Calanus finmarchicus,* the zooplankton on the shelf contains a mixture of oceanic and neritic zooplankton species. Of copepods, especially *Acartia longiremis* and *Temora longicornis* may be abundant during summer (Fig. 27–007). In addition, other neritic copepods, cladocerans, chaetognaths and appendicularians are commonly found. Meroplanktonic larvae from the benthic fauna also occur, with cirripedia larvae usually by far the most abundant in spring. Furthermore, fish eggs and larvae belonging to the local stocks may be quite abundant on the shelf during spring and early summer. Oceanic zooplankton are also advected onto the shelf. The relative abundance of oceanic species is highly variable from year to year (Gaard, 1999; Gaard and Hansen, 2000).

There is considerable seasonal variation in zooplankton species composition, abundance and development on the Faroe shelf, and to a large degree, this is related to the seasonal production cycle of phytoplankton (Gaard, 1999). During winter, very few zooplankton individuals are found on the shelf (Fig. 27–007). In early spring the composition and abundance of zooplankton on the Faroe shelf changes considerably. *C. finmarchicus* is recruited onto the shelf from the overwintering areas offshelf in the Norwegian Sea, and in the channels east and west of the Faroe Islands (Heath and Jónasdóttir, 1999; Heath *et al.*, 2000b; Gaard and Hansen, 2000), and may start spawning prior to the spring bloom (Gaard, 2000). Also meroplanktonic larvae (mainly barnacle larvae) may be abundant during spring.

As the phytoplankton biomass increases during spring, the neritic copepod species increase in number. Also jellies are common during summer and especially *Aurelia aurita* may be quite abundant. By September-October, when the phytoplankton biomass decreases to low winter levels, the copepods decrease in abundance.

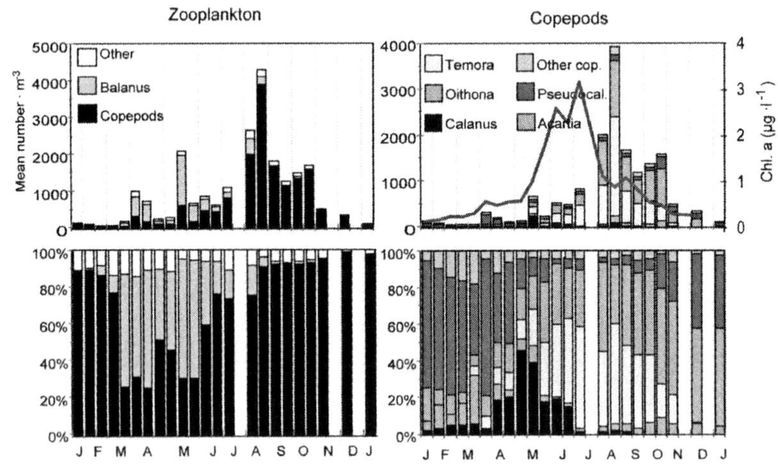

Figure 27.7 Mean absolute (upper) and relative (lower) abundance of the dominant zooplankton (left) and copepods (right) and chlorophyll *a* concentrations on the Faroe shelf during 1997. Copepod nauplii are not included.

There is high interannual variability in the zooplankton species composition, abundance and biomass on the Faroe shelf. This variability seems to be affected by variable food availability (phytoplankton) and also by variable advection of *C. finmarchicus* from offshore. For instance, during the late 1980's and the beginning of the 1990's, the ecosystem was dominated by *C. finmarchicus,* while neritic zooplankton was of minor importance. Differences between the shelf area and the surrounding offshore environment were generally small. However, during the early 1990s, the species composition on the Faroe shelf changed, and the area gradually became more neritic. *C. finmarchicus,* which still was the dominant copepod outside the tidal front, gradually became less abundant inside the tidal front while neritic copepod species, mainly *A. longiremis* and *T. longicornis,* increased in numbers (Gaard, 1999). The most likely reasons for this variability are a combination of variable advection of *C. finmarchicus* from the offshore environment, variable production of zooplankton on the shelf and variable predation on the zooplankton in the ecosystem.

C. finmarchicus dominates the zooplankton biomass during spring and summer and therefore generally reflects the total zooplankton biomass. During the 1990s the zooplankton biomass on the shelf in summer fluctuated by a factor of 10, while it remained relatively constant in the oceanic environment outside the tidal front (Gaard, 2003).

The macrozooplankton around the Faroe shelf is mainly represented by the euphausiids *Meganyctiphanes norvegia* and *Thysanoessa longicaudata.* However, they are much more abundant in the offshore area outside the tidal front than on the shelf.

3.5 Fish

The total annual catch of fish and invertebrates in Faroese waters (mean 1997–2002) is around 500,000 tonnes. The majority is pelagic fish (mainly blue whiting) caught during feeding migrations in the offshore areas.

The Faroe shelf is a habitat for several local demersal fish stocks. The most abundant species are cod, haddock, saithe, sandeel, and Norway pout. The annual catches of demersal fish on the Faroe shelf and slope is around 100,000–130,000 tonnes. Cod, haddock and saithe are the most important commercial species, quantitatively as well as economically.

The main spawning season for the fish on the shelf is spring, between February and May. The eggs and larvae are advected clockwise and then dispersed around the shelf area with the currents (Jákupsstovu and Reinert, 1994) where they feed on zooplankton during spring and summer (Gaard and Steingrund, 2001; Gaard and Reinert, 2002).

When the larvae are about 3–4 months old, at lengths of about 4 cm, the pelagic phase of most juveniles is over. Saithe and cod migrate into the fjords and sounds, while haddock and Norway pout make the transition to a predominant demersal habit on the plateau and the banks at depths of about 90–200 m (Joensen and Tåning, 1970). For saithe this occurs in May and for the other three species in July. From an age of about 2 years the cod mostly inhabitat the shelf. The saithe move to the slope at age of about 3 years.

Statistics for the Faroe Plateau cod and haddock show, that despite a marked increase in fishing effort during the 20th century, the landings have not increased correspondingly. The long-term landings of the cod usually have fluctuated between 20 000 and 40 000 tonnes and of haddock between 15 000 and 25 000 tonnes. The catches of these two fish stocks therefore have for a long time reached the limit for long-term production within the ecosystem. Consequently, it is likely that the catches reflect interannual variability in production of these fish stocks.

Although the long-term fish landings of the Faroe Plateau cod and haddock have been relatively stable, there have nevertheless been periods with some quite large fluctuations. With the exception of the World War II period (1939–1945), the most pronounced variability was a quite dramatic decrease in catches of both species during the beginning of the 1990s when the landings reached the lowest values ever recorded (~6000 and ~4000 tonnes respectively in 1993). However, after the decline, the catches of both species increased rapidly to well above the long-term averages for both species.

The variability in catches of these two species is largely a result of a combination of variable recruitment and growth rates (Gaard *et al.,* 2002). Long-term relations between cod and haddock recruitment and growth (expressed as weight at age) have furthermore demonstrated that periods with high weight-at-age occur simultaneously with periods with good recruitment of 2- year old fish (Fig. 27–008). This indicates that interannual variability in food production for fish exceeds variability in predation pressure. The ecosystem therefore seems to go through periods with low recruitment and production levels and other periods with higher recruitment and production in several trophic levels.

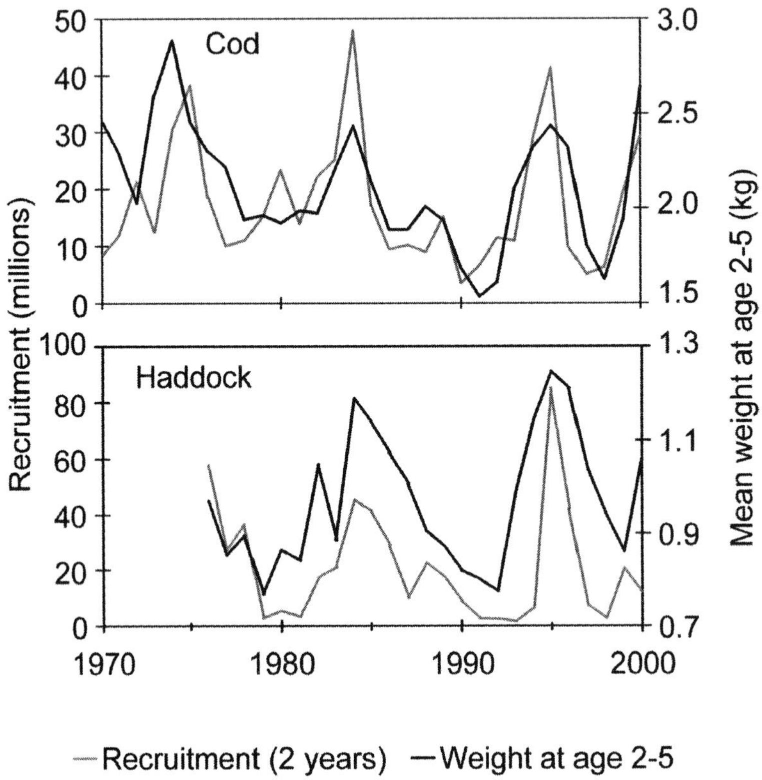

Figure 27.8 Relationship between recruitment of 2 years old cod and haddock and the mean weight of 2–5 years old cod during 1970–2001 and haddock during 1976–2001 on the Faroe shelf. (From Gaard *et al.*, 2002, and additional data).

Variable recruitment of the Faroe Plateau cod and haddock may be due to variable spawning stocks (Jákupsstovu and Reinert, 1994) survival during their pelagic larval and juvenile phases (Gaard and Steingrund, 2001; Gaard and Reinert, 2002) or survival during the first one-two years of the demersal stage (Steingrund and Gaard, 2003). Variable growth rates seem to be correlated with variable feeding conditions (Gaard *et al.*, 2002). On the other hand, temperature does not correlate with either recruitment or with growth of cod and haddock on the Faroe shelf (Planque and Frédou, 1999; Gaard *et al.*, 2002). Both species show diversity in prey items, and predate on benthic fauna as well as fish, with fish being a somewhat more prevalent prey item for cod than for haddock (Rae, 1967; Du Buit, 1982). Of the fish, sandeel is an important food species in the shallow areas. When sandeel is abundant, it is their preferred food item, and variability in their growth rates, and apparently also recruitment, largely derives from variable abundance of sandeel (Gaard *et al.*, 2002).

3.6. *Trophic interactions*

A clear relationship has been observed between the trophic levels in the Faroe shelf ecosystem, from primary production to the higher trophic levels (fish and seabirds) (Gaard *et al.*, 2002). At the lower trophic levels, there is a clear inverse relationship between nitrate loss (potential new primary production) during spring and summer and zooplankton biomass (variable abundance of *Calanus finmarchicus*). The most productive years for higher trophic levels are those with high primary production (and low zooplankton biomass). The variability in zooplankton biomass is due to variable abundance of *C. finmarchicus*, which most likely is to a combination of variable advection of *C. finmarchicus* from the offshore environment and variable predation on the zooplankton in the ecosystem.

Recruitment of 2-year old cod and haddock and their growth also largely cofluctuate (Fig. 27–008), and variable production of food seems to be a main determinant for the observed variation. There is a good interannual relationship between primary production and higher trophic levels, including fish and seabirds. In particular, the cod responds clearly to variable primary production, and shows very high correlation between primary production and total cod production in the ecosystem (Steingrund *et al.*, 2003; Steingrund and Gaard, 2003). Although benthic production is important food source the demersal fish, sandeel seems to be even more important. It is a main link between the zooplankton production and higher trophic levels, and when it is abundant, it is a main food sources for fish and seabirds in the ecosystem (Gaard *et al.*, 2002).

Fish that have their main habitat on the shelf slope (*e.g.*) mature saithe and the large individuals of cod and haddock predate largely on krill, blue whiting and norway pout.

4. Norwegian Shelf

4.1. *Topography, currents and water masses*

The Norwegian coast borders three economically important seas: The North Sea, the Norwegian Sea and the Barents Sea. The topography and width of the Norwegian continental shelf varies along the coast (Fig. 27–001). The shelf off southern and southwestern Norway is narrow because the Norwegian Deep (>400 m deep), starting at about 62°N, cuts a deep trench from the Norwegian Sea into the North Sea, close to the Norwegian coast. The shelf from 62–63° N and from 68–70° N is shallow (~200 m) and rather narrow, while the shelf off mid-Norway is deep (~500 m) and wide. The Barents Sea, north of northern Norway, is a sub-arctic, shelf sea, mostly less than 300 m deep. A trench, more than 400 m deep in the west, cuts into the Barents Sea from the Norwegian Sea. Bottom depths on the shelf south of this trench are generally 200–300 m deep. An important feature of the Norwegian shelf is the coastal banks, usually less than 200 m deep. Important as well, is the sheltered area found behind the numerous islands along the coast, and the deep and long Norwegian fjords.

The North-Atlantic current, an extension of the Gulf Stream, enters the Norwegian Sea between Shetland and Iceland. Further north on its way through the eastern parts of the Norwegian Sea, it is called the Norwegian Atlantic Current

(NAC). This current brings warm and saline water to the region and has major impact on the marine and terrestrial climate of Norway. The NAC flows north-eastwards in the Norwegian Sea until it hits the Norwegian shelf slope at about 62° N (Fig. 27–001). Then the main branch of the NAC flows northwards along the shelf break, while to the south a smaller branch enters the North Sea through the Norwegian Deep. In the centre of the NAC, at about 50 m depth, temperatures of more than 8°C, salinities above 35.2 and current speeds of more than 30 cm s^{-1} prevail (Mork and Blindheim, 2000; Orvik et al., 2001). The NAC branches off into the Barents Sea north of the coastal bank, Tromsøflaket, and then leaves into the Arctic Ocean through the Fram Strait.

The Norwegian Coastal Current (NCC) is a surface current of light, low saline water (salinity 32–35). The NCC originates in the Baltic Sea and the southern North Sea. During its flow along the Norwegian coast, from the Swedish to the Russian border, the NCC also receives input of fresh water from Norwegian run-off. Due to mixing with the deeper and more saline Atlantic water, the salinity of the coastal water increases towards north.

The front between Atlantic and Coastal water masses is generally situated close to the shelf break. Atlantic water, which is characterised by a salinity above 35, is heavier than Coastal water and dives beneath the Coastal water at the front. Over deeper parts of the Norwegian shelf and in deep trenches cutting through the shelf between the coastal banks, Atlantic water enters the shelf as deep water. At the surface a wedge of light Coastal water spreads out from the coast. The extension of the surface wedge depends strongly on the wind direction (Sætre et al., 1988). During periods of northerly winds the Coastal water masses are driven away from the coast, spreading into the open ocean as a thin layer of surface water. The transition depth between light surface water and the dense Atlantic water beneath forms a strong permanent pycnocline. Thus, during periods with northerly winds, the depth of the wind mixed layer is much reduced which strongly influences the timing and dynamics of primary production. As the wind forces Coastal water away from the coast Atlantic water will replace the Coastal water masses from beneath, entering the shelf and fjords.

Temperature in the coastal water masses is at a minimum in March-April and reaches a maximum in August (Fig. 27–009). The start of the warming occurs at approximately the same time in the south and in the north (Aure and Strand, 2001). Salinity is highest during winter and reaches a minimum during the summer months, June-August, due to increased freshwater runoff (Fig. 27–009). Long-term time series of temperature and salinity from fixed coastal stations have revealed decreasing winter salinity and increasing summer temperatures over the last 3–4 decades (Aure and Strand, 2001). These changes coincide with an increase in the winter North Atlantic Oscillation index (NAO), reflecting more persistent and stronger south-westerly winds in the northeast Atlantic (Hurrell, 1995).

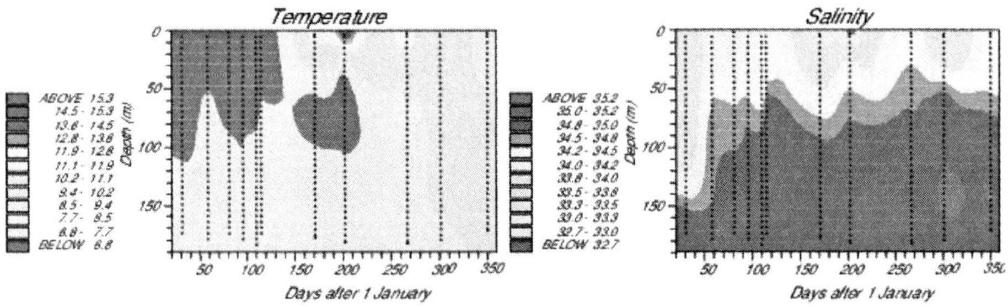

Figure 27.9 Temperature and salinity on a shelf station (180 m bottom depth) off Møre in 1997 (see Fig. 27.1).

4.2 *Primary production*

Diatoms of the genera *Chaetoceros, Thalassiosira* and *Skeletonema* together with the flagellate *Phaeocystis pouchetii* usually dominate the phytoplankton spring bloom in fjords and coastal waters of western and northern Norway (Heimdal, 1974; Braarud and Nygaard, 1980; Rey, 1993; Hegseth *et al.*, 1995). During summer and in the autumn dinoflagellates of the genera *Protoperidinium, Dinophysis* and *Ceratium* become more abundant. When there is an autumn bloom, diatoms usually dominate the phytoplankton community again. The most common diatoms during autumn blooms are of the genera *Chaetoceros, Skeletonema, Pseudonitzschia* and *Proboscia.*

On the shelf between 62° and 72°N the annual primary production ranges between 90 and 120 g C m^{-2} y^{-1}, lowest in the south and north and highest at middle latitudes. Average daily primary production from early March to early May ranges between < 0.2 g C m^{-2} d^{-1} in March to > 3.5 g C m^{-2} d^{-1} in May (Rey, 1981).

Nutrient concentrations, in coastal surface waters between 62° and 64°N peak in April (Fig. 27–010). The maximum nitrate concentrations in coastal water are lower than in Atlantic water, which is mainly because the water column over the shelf is permanently stratified and not prone to winter convection of nutrient rich water from greater depths. This means that the amount of nitrate available for new production (e.g. Rey *et al.*, 1987) is about 8 μM in coastal water as compared to about 13 μM in Atlantic water (Fig. 27–010).

The timing and dynamics of the phytoplankton bloom are very different in Atlantic and coastal water (Rey *et al.*, 1987; Skjoldal *et al.*, 1987; Melle and Skjoldal, 1998). The development of a bloom is dependent of a stratified water column (Sverdrup, 1953). In Atlantic water stabilisation of the water column is due to the formation of a thermocline by atmospheric heating of the surface layer. On the continental shelf, where coastal water sits on top of Atlantic water, the water column is permanently stratified. Under such conditions the initiation of the bloom is triggered by increased solar irradiance towards the end of the winter. The shallower the pycnocline, the earlier the bloom starts, because it takes less irradiance to promote net growth in the algae stocks. The shallowest pycnocline is found at the front towards Atlantic water, situated above the shelf edge. Therefore the bloom tends to start at the front. During periods of northerly winds, when the surface layer of coastal water is forced westward, the pycnocline is especially shallow and the bloom correspondingly early. Over the shallow coastal banks the sea bottom will have the same effect as the pycnocline on the timing of the bloom. Also the freshwater layer in the fjords promotes an early bloom compared to Atlantic water off-shelf.

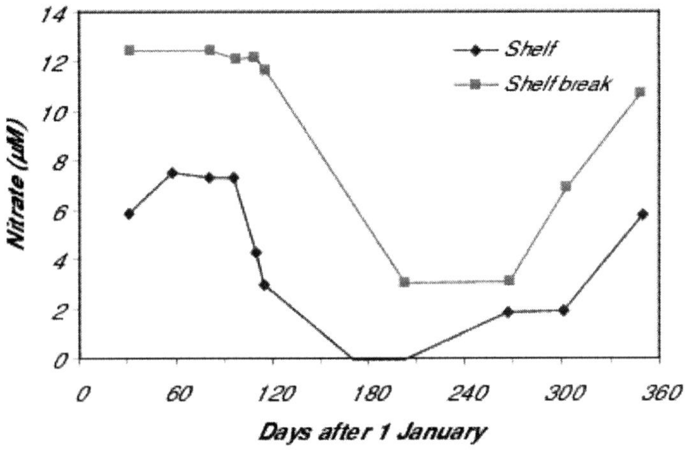

Figure 27.10 Nitrate concentrations (10 m) on a shelf and a shelf break station off Møre in 1997 (see Fig. 27–001).

The phytoplankton spring bloom usually starts in early April on the shelf at Møre (ca. 62–64°N), in late April off the shelf break at Møre (Fig. 27–011), and in the middle of May at Weather Station Mike in the Norwegian Sea (Fosså, 2002). In Atlantic water in the Barents Sea the bloom starts in the middle of May and on the continental shelf off northern Norway in early April (Rey, 1993; Melle and Skjoldal, 1998). The bloom in the northern Norwegian fjords usually starts in early April (Hegseth *et al.*, 1995). Thus, latitudinal differences in the timing of the bloom are small compared to the differences between shelf and off-shelf waters.

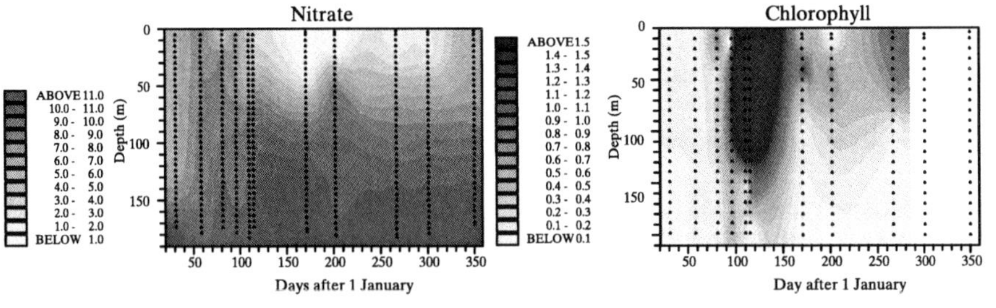

Figure 27.11 Nitrate (µM) and chlorophyll (µg l^{-1}) on a shelf station (180 m bottom depth) off Møre in 1997 (see Fig. 27–001).

4.4 *Zooplankton*

During spring and summer more than 80% of the biomass of zooplankton in coastal water is *Calanus helgolandicus* and *C. finmarchicus. C. helgolandicus* is most abundant in the south, in southern Norwegian fjords and especially during late summer and in the autumn. *C. finmarchicus* is more abundant farther north (Wiborg, 1954; Melle *et al.,* 1993). By numbers, however, the large *C. finmarchicus* only dominates the zooplankton community for brief periods during spring and early summer (Fig. 27–012). During late summer and autumn, when *C. finmarchicus* in off shelf waters descends for overwintering at depths of 500–2000 m (Østvedt, 1955; Heath *et al.,* 2000b), the coastal stock is flushed off the shelf or wiped out by predation (Slagstad and Tande, 1996). The smaller copepods such as *Oithona* spp. and *Microcalanus* spp. and the omnivore *Metridia* spp. are the major contributors to zooplankton biomass during winter and late summer when *C. finmarchicus* has left the surface waters for overwintering (Fig. 27–012). During spring *C. finmarchicus* is recruited to the shelf waters from the overwintering areas off shelf and, probably, from fjord stocks (Ruud, 1929). It is not clear whether recruitment to the shelf occurs when *C. finmarchicus* is in adult or larval stages (Ruud, 1929; Slagstad and Tande, 1996). During spring Echinoderm larvae may dominate the shelf zooplankton community. The most abundant gelatinous zooplankton species on the shelf is *Aglantha digitale.* Occasionally salps and southerly species of jellies, such as *Apolemia uvaria* (Fosså, 2002), may invade the coastal waters. During spring fish eggs and larvae may also be abundant members of the zooplankton community, especially on the spawning grounds of the large fish stocks (Fossum and Øiestad, 1991).

The contribution of euphausiids to the zooplankton biomass is often underestimated due to the lack of proper sampling gears. The most abundant euphausiids on the shelf are *Meganyctiphanes norvegica* and *Thysanoessa inermis* (Melle *et al.,* 1993). On the shelf, *M. norvegica* has a two-year life cycle, while *T. inermis* probably has a one year life cycle (Melle *et al.,* 1993; Einarson, 1945). They are both spawning on the shelf during spring (Melle *et al.,* 1993). *M. norvegica* and *T. inermis* are known to feed both on algae and copepods (Båmstedt and Karlson, 1998; Onsrud and Kaartvedt, 1998; Kaartvedt *et al.,* 2002).

The female *C. finmarchicus* shed their eggs freely into the water masses (Melle and Skjoldal, 1989). The eggs sink about 30 m d^{-1}, and at the temperatures prevailing on the shelf during spring most of them may reach the bottom before hatching (Melle and Skjoldal, 1989; Knutsen *et al.,* 2001). It is not clear how this may influence the recruitment success of *C. finmarchicus* on the shelf compared to its natural habitat in the deep waters of the Norwegian Sea (Melle, 1998). In coastal waters and in the Barents Sea shelf waters there is a close relationship between the spawning of the first generation of eggs of *C. finmarchicus* and the phytoplankton spring bloom (Diel and Tande, 1992; Melle and Skjoldal, 1998). The initiation of egg production in *C. finmarchicus* is earlier on the shelf than in off-shelf waters (Melle and Skjoldal, 1998), a feature explained by water mass stabilisation and phytoplankton bloom dynamics.

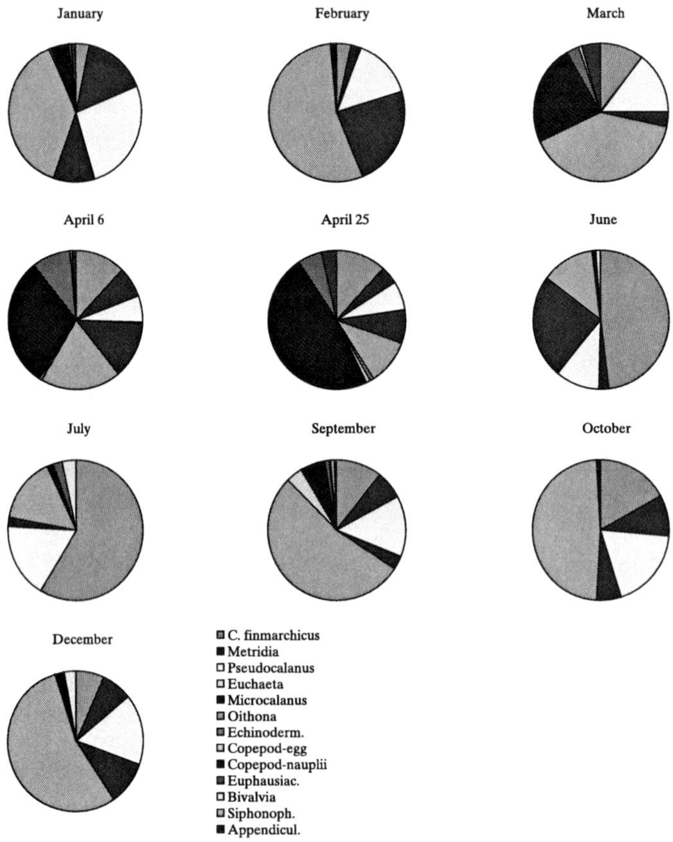

Figure 27.12 Relative abundance of zooplankton species at a shelf station (180 m bottom depth) off Møre in 1997 (see Fig. 27–001).

C. finmarchicus on the shelf produces three generations or more in the south and possibly only one in the north (Wiborg, 1954). The first generation on the shelf off Møre recruits into the first copepodite stage in the beginning of May (Fig. 27–013). Two generations are produced on the shelf off Møre (Fig. 27–013).

The zooplankton biomass on the shelf in May varies from year to year and from 1998, onwards, a declining trend has been observed (Fig. 27–014). While variations of zooplankton biomass in the off shelf waters are correlated with the North Atlantic Oscillation winter index (Melle *et al.* 2003), no correlation between climate indices and zooplankton biomass has been found on the shelf.

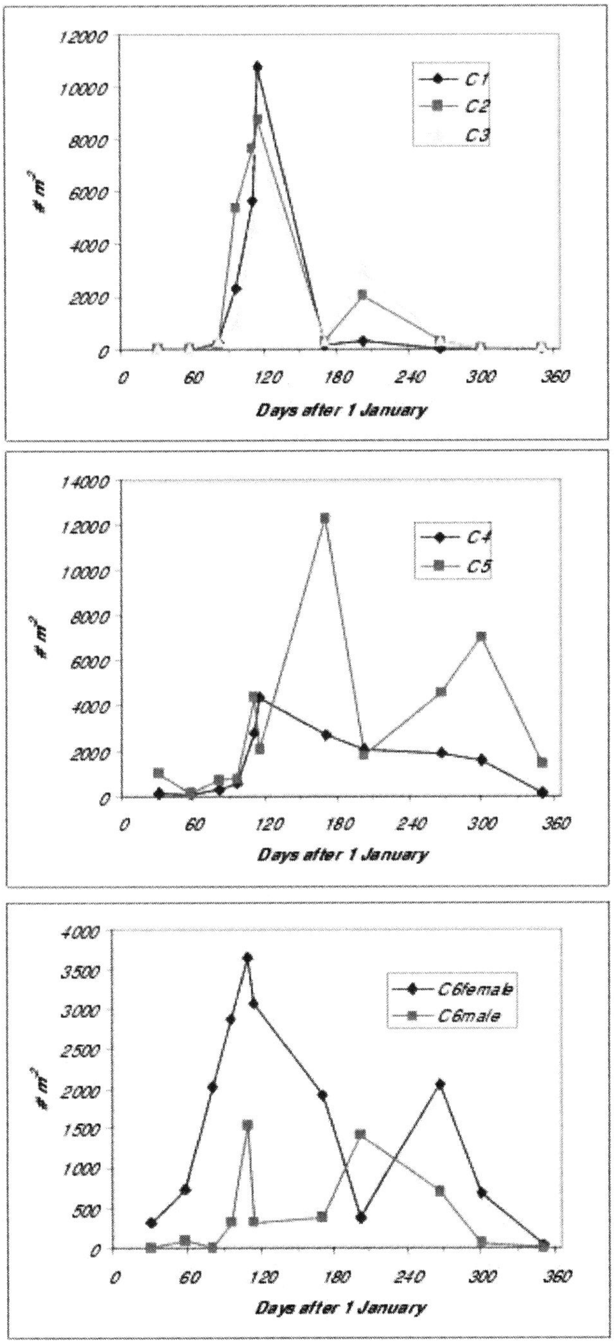

Figure 27.13 Abundance of *Calanus finmarchicus* on a shelf station (180 m bottom depth) off Møre in 1997 (see Fig. 27–001). Copepodite stages 1–6 (C1-C6), male and females.

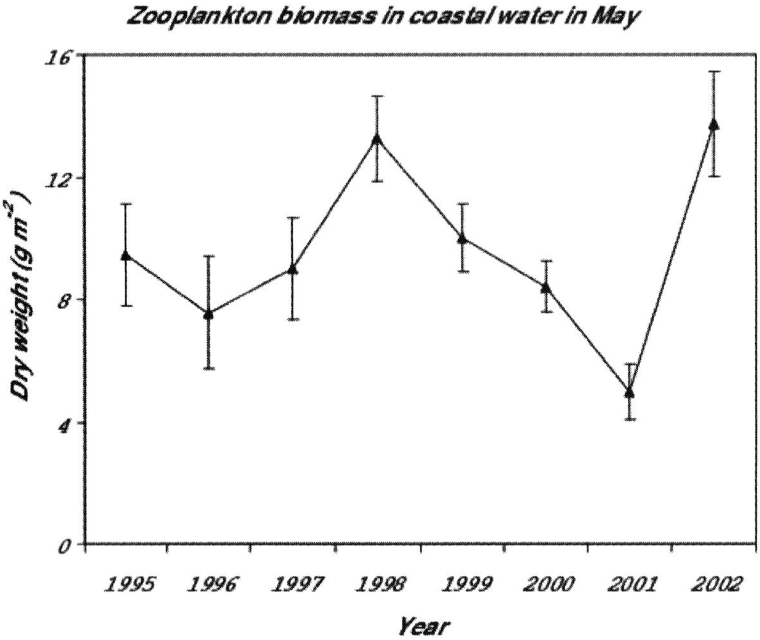

Figure 27.14 Average zooplankton biomass in May in Norwegian coastal water. Error bars: 95% confidence limits. (After Melle *et al.,* 2003).

4.4 *Fish*

In 2000 the annual catch of fish and shellfish in the Norwegian and Barents Seas, including the Norwegian shelf, was about 4 mill tonnes pelagic fishes, 1 mill tonnes demersal fishes and 30 000 tonnes crustaceans and scallops (Iversen, 2002). The most important species in the catches are cod (0.4 mill tonnes), herring (0.8 mill tonnes), blue whiting (1.4 mill tonnes), capelin (0.5 mill tonnes), and shrimps (83 000 tonnes) (Iversen, 2002).

After a period of low stock size in the late 1970's and 1980's the Northeast Arctic cod stock recovered during the 1990's (Iversen, 2002). Recently another fall in stock size has occurred in spite of improved assessment tools and experience. At present the stock has increased somewhat from a minimum below 1.5 mill tonnes. It is not clear to what extent the variations in stock size are due to fishery or environmental variability (Hamre and Hatlebakk, 1998; Ottersen and Loeng, 2000).

The main nursery areas of cod are located in the Barents Sea. They spawn along the coast from 60–70°N, and the most important spawning area is the Lofoten area. In the Lofoten area the median spawning date for cod is 1 April, and they spawn in the transitional layer between coastal and Atlantic water at more than 100 m depth, and always in water close to 4°C. The eggs are positively buoyant and are soon found in the wind mixed layer (Sundby, 1991). After hatching the larvae start feeding on nauplii of *C. finmarchicus* (Ellertsen *et al.,* 1989). During larval and post-larval development, the recruits drift northwards into the Barents Sea, feeding on progressively older developmental stages of *C. finmarchicus* (Helle, 1994).

Differences in year class strength in cod are large (Fossum and Øiestad, 1991), and the factors determining year class strength are largely unknown. However, from numerous investigations during the 20th century, we have learned that recruitment in cod is good when larvae are advected into the Barents Sea, the timing of the first feeding event of the larvae matches the production of *C. finmarchicus* nauplii, growth during the post-larval stage is good and the abundance of predators, including older year classes of cod, in the nursery area is low (Fossum and Øiestad, 1991; Sundby *et al.,* 1989, Hamre and Hatlebakk, 1998, Ottersen and Loeng, 2000).

After the collapse in the stock of Norwegian spring spawning herring during the late 1960's, more than twenty years went past until the stock recovered during the late 1980's, and the adult stock resumed its feeding migration into the central and western Norwegian Sea (Dragesund *et al.* 1997). At present the spawning stock of herring, some years more than 10 mill. tonnes, overwinters in a single fjord in the Vestfjorden area, northern Norway, from October to February (Slotte, 1998). The herring spawn benthic eggs in March and April on several spawning grounds situated on the coastal banks between 58–68°N, the most important ones being off Møre. The main hatching period is late March and early April. From April to July the herring larvae are found drifting with the Norwegian coastal current and the shelf break arm of the Norwegian Atlantic current into the nursery areas in the Barents Sea. The main food of herring larvae is eggs and nauplii of *Calanus finmarchicus.* After 3–5 years the adolescent herring leave the Barents Sea and join the adult part of the stock during its summer feeding migration to the Norwegian Sea. Recruitment success seems to vary with the same environmental factors as mentioned for the cod stock (Slotte, 1998; Ottersen and Loeng, 2000, Toresen 2002). Recruitment success in herring is extremely variable, which means that fishery will be strongly dependent on the rich year classes typically occurring every tenth year or so. In some years, such as 1983, exceptionally strong year classes of both cod and herring are produced.

5. *Conclusions*

Many similarities, but certainly also differences characterise the three shelf systems. Nowhere in the world is the climate as warm at such high latitudes. The reason is a continuous inflow of warm Atlantic water into the Nordic Seas. This inflow affects the biological processes, diversity and production substantially.

The shelf systems are characterized by large seasonal differences in the productivity of phyto- and zooplankton. However, due to the different hydrographic features, the timing of the spring bloom is different on the three shelves. Due to coastal water overlaying Atlantic water, the water column on the Norwegian shelf to a large extent is permanently stratified, mainly at the shelf break front. Over the coastal banks bottom depth is less than 200 m, and the spring bloom tends to start at the shelf break front and over the banks. The spring bloom on the Norwegian shelf usually starts in early April.

On the Icelandic shelf freshwater runoff from land similarly is important for the formation of a pycnocline, particularly off the south and west coasts. On the northern and eastern shelves, the timing and magnitude of the spring bloom may depend more on the variability in the influx of Atlantic and Polar Water to the northern and eastern regions rather than freshwater runoff from land.

On the Faroe shelf, however, the water column is usually mixed due to very strong tidal currents and practically no stratification occurs. The depth of the mixed layer is therefore equal to the bottom depth (70–80 m in average). The timing of the spring bloom is highly variable between years and occurs in most years in early-mid May.

Outside the shelves the spring bloom depends on development of summer thermocline. This usually occurs around mid May in most areas. However, in areas seasonally covered by sea ice, water column stabilisation, and therefore spring bloom development, may depend on ice melting and halocline formation. Generally, this occurs earlier than thermocline formation due to warming of surface water.

The timing of the secondary production depends highly on the primary production and usually zooplankton production starts significantly earlier on the shelves than in the Atlantic water offshore. The species composition on the shelves is quite similar during spring and summer, with a mixture of neritic and oceanic copepods and meroplanktonic larvae. During spring and early summer the most important single species is the oceanic copepod *Calanus finmarchicus* that is advected onto the shelves from offshore. Although pre-bloom reproduction certainly occurs in this species, its fecundity largely depends on food availability. On all three shelves the spring spawning starts earlier on the shelf than farther out, and this is most likely due to the earlier onset of primary production on the shelves. *C. finmarchicus* is most abundant during spring and early summer. In summer and autumn the part of the stock that is not demised by predation is flushed out of the shelves, where the animals descend into deep cold water for diapause. After this, the zooplankton community on the shelves may be characterized by smaller copepods such as *Oithona* spp., *Microcalanus* spp. and *Pseudocalanus* spp.

Sedimentation of phytoplankton onto the seabed is seasonal, depending on the new primary production and grazing. The new primary production occurs mainly during the spring bloom and this time of the year has the highest potential for input of organic material to the sediments. Furthermore, the dominant algae of the spring blooms are large fast-sinking diatoms. High zooplankton grazing and strong pycnoclines may, however, postpone or prevent increased vertical losses (Wassmann, 1998). Large overwintering stocks of copepods may control the accumulation of phytoplankton biomass during the spring bloom (Bathmann *et al.*, 1990). Also the timing of copepod seasonal cycles and the onset of the bloom are important for the fate of the vernal bloom in the Nordic seas and the adjacent shelves i.e. sedimentation or retention in the upper layer.

All three shelves show long-term variations in production and biomass of plankton and nekton. These changes do not necessarily co-fluctuate, but often they do. Cod, which is a main demersal fish species on all three shelves, show fluctuation in reproduction and growth that may largely be influenced by local environmental factors (*e.g.* advection of larvae temperature, and food availability). Cod stocks on all three shelves have their main spawning grounds in areas that are assumed to be main areas for invasion of *C. finmarchicus* in spring, and in all three regions *C. finmarchicus* eggs and nauplii are shown to be the principal food items for the first feeding cod larvae (Thorisson, 1989; Fossum and Øyestad, 1991; Gaard and Steingrund, 2001). Recruitment and growth in all three regions may be affected by hydrographic factors. High inflow of warm Atlantic water north of Iceland and into the Barents Sea seems to affect cod growth and recruitment positively. For instance, years with high inflow of warm water north of Iceland seems to result in high primary production, high secondary production, and high cod growth rates through high production of capelin, which is its principal food item (Astthorsson and Vilhjalmsson, 2002). Similarly high inflow of warm water to the Barents Sea seems to coincide with good conditions for survival and transport of larval fish and zooplankton (Rey *et al.*, 1987; Skjoldal et al., 1992). On the Faroe shelf a clear relationship between primary production and cod recruitment and growth has been identified (Gaard *et al.*, 2002; Steingrund and Gaard, 2003). In that system sandeel is the principal food item for cod and haddock. Production in the system seems to be vulnerable to variable exchange rates of the shelf water.

The shelf systems, described in this chapter, are vulnerable to short term hydrographic and ocean climate variability. Therefore we cannot discount that extensive ecological alterations would follow a future global change in climate. There may be evidence of decreasing thermohaline ventilation in the Nordic Seas following recent years surface temperature increase (Hansen et al., 2001). If that should happen in the future, the ocean climate on the Iceland, Faroe and Norwegian shelves may change considerably, since ventilation of deep-water formation is regarded as a driving force for the Atlantic inflow to the region.

Bibliography

Anonymous, 2001. Þættir úr vistfræði sjávar 2000. Environmental conditions in Icelandic waters 2000. Hafrannsóknastofnun fjölrit, 83, 41 pp.

Anonymous, 2002. Nytjastofnar sjávar 2001/2002. Aflahorfur fiskveiiári 2002/2003. (State of marine stocks in Icelandic waters 2001/2003. Prospects for the quota year 2002/2003. In Icelandic. English summary). Hafrannsóknastofnunin, Fjölrit, 88, 196 pp.

Astthorsson, O. S. 1990. Ecology of the euphausiids *Thysanoessa raschi, T. inermis* and *Meganyctiphanes norvegica* in Ísafjord-deep, northwest-Iceland. Mar. Biol., 107: 147–157.

Astthorsson, O. S. and Gislason, A. 1992. Investigations on the ecology of the zooplankton community in Isafjord-deep, northwest Iceland. Sarsia, 77: 225–236.

Astthorsson, O. S. and Gislason, A. 1995. Long-term changes in zooplankton biomass in Icelandic waters in spring. ICES J .. Mar. Sci., 52: 657–688.

Astthorsson, O. S. and Gislason, A. 1997a. Biology of euphausiids in the subarctic waters north of Iceland. Mar. Biol., 129: 319–330.

Astthorsson, O. S. and Gislason, A. 1997b. On the food of capelin in the subarctic waters north of Iceland. Sarsia, 82: 81–86.

Astthorsson, O. S. and Gislason, A. 1998. Environmental conditions, zooplankton and capelin in the waters north of Iceland. ICES J. Mar. Sci., 55: 808–810.

Astthorsson, O. S. and Gislason, A. 1999. Inter-annual variation in abundance and development of *Calanus finmarchicus* in Faxaflói, West Iceland. Rit Fiskideild., 16: 131–140.

Astthorsson, O. S. and Vilhjalmsson, H. 2002. Iceland shelf LME: Decadal assessment and resource sustainability. *In:* L. Sherman and H-R. Skjoldal (eds). Large Marine Ecosystems of the North Atlantic changing states and sustainability: 219–243.

Astthorsson, O. S., Hallgrímsson, I. and Jónsson, G. S. 1983. Variations in zooplankton densities in Icelandic waters in spring during the years 1961–1982. Rit Fiskideild., 7: 73–113.

Aure, J. and Strand, Ø. 2001. Havforskingsinstituttets termografstasjoner. Hydrografiske normaler og langtidsvariasjoner i norske kystfarvann mellom 1936 og 2000. (Hydrographic normals and long - term variations at fixed surface layer stations along the Norwegian coast from 1936 to 2000) Fisken Havet, 13 - 2001. Institute of Marine Research.

Bailey, R. S. 1982. The population biology of blue whiting in the North Atlantic. Adv. Mar. Biol., 19: 257–355.

Bathmann, U. V., Peinert, R., Noji, T. T., and Bodungen, B. von 1990. Pelagic origin and fate of sedimenting particles in the Norwegian Sea. Progress in Oceanography, 24: 117–125.

Båmstedt, U. and Karlson, K. 1998. Euphausiid predation on copepods in coastal waters of the Northeast Atlantic. Mar. Ecol. Prog. Ser., 172:149–168.

Beare, D. J., Gislason, A., Astthorsson, O. S., and McKenzie, E. 2000. Assessing long-term changes in early summer zooplankton communities around Iceland. ICES J. Mar. Sci., 57: 1545–1561.

Belikov, V., Jákupsstovu, S. H., Shamrai, E. and Thomsen, B. 1998. Migration of mackerel during summer in the Norwegian Sea. ICES C.M. 1998/AA:8.

Bjelland, O. and Monstad, T. 1997. Blue whiting in the Norwegian Sea, spring and summer 1995 and 1996. ICES C.M. 1997/CC: 15.

Braarud, T. and Nygaard, I. 1980. Phytoplankton observations in offshore Norwegian coastal waters between 62°N and 49°N: 2. Diatom societies from Møre to Vesterålen, March-April. Sarsia, 65: 93–114.

Dalpadado, P. and Skjoldal, H. R. 1996. Abundance, maturity and growth of the krill species *Thysanoessa inermis* and *T. longicaudata* in the Barents Sea. Mar. Ecol. Prog. Ser., 144: 175–183.

Dalpadado, P., Ellertsen, B., Melle, W. and Domnanes, A. 1998b. Food and feeding conditions and prey selectivity af herring (*Clupea harengus*) through its feeding migrations from coastal areas of Norway to the Atlantic water masses of the Nordic Seas. ICES C.M. 1998/R:2.

Dalpadado, P., Ellertsen, B., Melle, W. and Skjoldal, H. R. 1998a. Summer distribution patterns and biomass of macrozooplankton and micronekton in the Nordic Seas. Sarsia, 83: 103–116.

Diel, S. and Tande, K. 1992. Does the spawning of *Calanus finmarchicus* in high latitudes follow a reproducible pattern? Mar. Biol., 113:21–32.

Dragesund, O., Johannessen, A. and Ulltang, O. 1997. Variation in migration and abundance of Norwegian spring spawning herring (*Clupea harengus* L.). Sarsia, 82: 97–105.

Du Buit, 1982. Essai sur la predation de la morue (*Gadus morhua* L.), L'eglefin (*Melanogrammus aeglefinus* (L.) et du lieu noir (*Pollachius virens* (L.)) aux Faeroe. Cybium, 6: 3–13.

Dugdale R. C. and Goering, J.J. 1967 Uptake of new and regenerated forms of nitrogen in primary productivity. Limnol. Oceanogr. 12:196–206.

Einarsson, H. 1945. Euphausiacea I. Northern Atlantic species. Dana Report, 27: 1–191.

Ellertsen, B., Fossum, P., Solemdal, P. and Sundby, S. 1989. Relation between temperature and survival of eggs and first-feeding larvae of northeast Arctic cod (*Gadus morhua* L.). Rapp. P.-V. Reun. Ciem., 191: 209–219.

Fosså, J. H. 2002. Havets miljø 2002. Fisken Havet, særnr. 2–2002. Institute of Marine Research.

Fossum, P. and Øiestad, V. 1991. De tidlige livsstadiene hos fisk I mote med trusselen fra petroleumsvirksomheten. Sluttrapport fra Havforskningsinstituttets egg og larveprogram – HELP (1985–1991). Institute of Marine Research, Bergen.

Gaard, E. 1996. Phytoplankton community structure on the Faroe shelf. Fróðskaparrit, 44: 95–106.

Gaard, E. 1999. Zooplankton community structure in relation to its biological and physical environment on the Faroe Shelf, 1989–1997. J. Plankton Res., 21: 1133–1152.

Gaard, E. 2000. Seasonal abundance and development of the copepod *Calanus finmarchicus* in relation to phytoplankton and hydrography on the Faroe shelf. ICES J. Mar. Science, 57: 1605–1611.

Gaard, E. 2003. Plankton variability on the Faroe shelf during the 1990s. ICES J. Mar. Sci., 219:182–189.

Gaard, E. and Hansen, B. 2000. Variations in the advection of *Calanus finmarchicus* onto the Faroese shelf. ICES J. Mar. Sci., 57: 1612–1618.

Gaard, E. and Reinert, J. 2002. Pelagic cod and haddock juveniles on the Faroe Plateau: Distribution, diets and feeding habitats. Sarsia, 87: 193–206.

Gaard, E. and Steingrund, P. 2001. Cod spawning and larval advection and feeding on the Faroe shelf. Fróðskaparrit, 48: 87–103.

Gaard, E., Hansen, B. and Heinesen, S. H. 1998. Phytoplankton variability on the Faroe shelf. ICES J. Mar. Sci., 55: 688–696.

Gaard. E., Hansen, B., Olsen, B and Reinert, J. 2002. Ecological features and recent trends in physical environment, plankton, fish stocks and sea birds in the Faroe plateau ecosystem. *In:* K. Sherman and H-R Skjoldal (eds). Large Marine Ecosystems of the North Atlantic changing states and sustainability: 245–265.

Gislason, A. 2005. Seasonal and spatial variability in egg production and biomass of *Calanus finmatchicus* around Iceland. Mar. Ecol. Prog. Ser., 286: 177–192.

Gislason, A., and Astthorsson, O. S. 1995. Seasonal cycle of zooplankton southwest of Iceland. J. Plankton Res., 17: 1959–1976.

Gislason, A., and Astthorsson, O. S. 1996. Seasonal development of *Calanus finmarchicus* along an inshore-offshore gradient southwest of Iceland. Ophelia, 44: 71–84.

Gislason, A., and Astthorsson, O. S. 1998a. Seasonal variations in biomass, abundance and composition of zooplankton in the subarctic waters north of Iceland. Polar Biol., 20: 85–94.

Gislason, A. and Astthorsson, O.S. 1998b. Variability in the population structure of *Calanus finmarchcus* in Icelandic Waters in spring. ICES J. Mar. Sci., 55: 811–813.

Gislason, A. and Astthorsson, O. S. 2000. Winter distribution, ontogenetic migration, and rates of egg production of *Calanus finmarchicus* southwest of Iceland. ICES J. Mar. Sci., 57: 1727–1739.

Gislason, A., and Astthorsson, O. S., and Gudfinnsson, H. 1994. Phytoplankton, *Calanus finmarchicus,* and fish eggs southwest of Iceland 1990–1992. ICES Marine Science Symposia 198: 423–430.

Gislason, A, Astthorsson, O. S., Petursdottir, H., Gudfinnsson, H., and Bödvarsdottir, A. R. 2000. Life cycle of *Calanus finmarchicus* south of Iceland in relation to hydrography and chlorophyll *a.* ICES J. Mar. Sci., 57: 1619–1627.

Gudmundsson, K. 1998. Long-term variation in phytoplankton productivity during spring in Icelandic waters. ICES J. Mar. Sci., 55: 635–643.

Gunnarsson, K., Jónsson, G., and Pálsson, Ó. K. 1998. Sjávarnytjar við Ísland. (Marine resources around Iceland. In Icelandic). Reykjavík, Mál og menning, 280 pp.

Hallgrímsson, I. 1954. Noen bemerkninger om Faxaflóis hydrografi og zooplanktonbestand in 1948. (Aspects of the hydrography and zooplankton of Faxafloi Bay during 1948. In Norwegian). M. Sc. dissertation, University of Oslo, Norway, 87 pp.

Hamre, J. and Hatlebakk, E. 1998. System model (Systmod) for the Norwegian Sea and Barents Sea. In: Rødseth, T. (ed.). Models for multispecies management. Physica-Verlag. Heidelberg.

Hansen B. and Kristiansen, R. 1999. Variations of the Faroe Bank Channel overflow. Rit Fiskideild., 16: 13–22.

Hansen, B. 1992. Residual and tidal currents on the Faroe Plateau. ICES CM/C:12, 18 pp.

Hansen, B. and Østerhus, S. 2000. North Atlantic-North Seas exchanges. Progr. Oceanogr., 45: 109–208.

Hansen, B., Stefánsson, U. and Svendsen, E. 1998. Iceland, Faroe and Norwegian Coasts. Coasal segment (21,E). *In:*. Allan R. Robinson and Kenneth H. Brink (eds). The Sea, Volume 11: 733–758.

Hansen, B., Turrell, B. and Østerhus, S. 2001. Decreasing overflow from the Nordic seas into the Atlantic Ocean through the Faroe Bank Channel since 1950. Nature, 411: 927–930.

Hansen, L.P., Jacobsen, J.A. 2000. Distribution and migration of Atlantic salmon, *Salmo salar* L., in the sea. In: Mills, D. (ed.). The ocean life of Atlantic salmon, Fishing News Books, Blackwell Science: 75–87.

Heath, M. R. and Jónasdóttir, S. H. 1999. Distribution and abundance of overwintering *Calanus finmarchicus* in the Faroe-Shetland Channel. Fish. Oceanogr., 8(Suppl. 1): 40–60.

Heath, M.R., Astthorsson, O.S., Dunn, J. Ellertsen, B., Gaard, E., Gislason, A., Gurney, W.S.C., Hind, A.T., Irigoien, X., Melle, W., Neihoff, B., Olsen, K., Skreslet, S., and Tande, K.S. 2000a. Comparative analysis of *Calanus finmarchicus* demography at locations around the Northeast Atlantic. ICES J. Mar. Sci., 57: 1562–1580.

Heath, M. R., Fraser, J. G., Gislason, A., Hay, S. J., Jónasdóttir, S. H. and Richardson, K. 2000b. Winter distribution of *Calanus finmarchicus* in the Northeast Atlantic. ICES J. Mar. Sci., 57: 1628–1635.

Hegseth, E.N., Svendsen, H. and Hellum, C.Q. 1995. Phytoplankton in fjords and coastal waters of northern Norway: Environmental conditions and dynamics of the spring bloom. In: Skjoldal, H.R., Hopkins, C., Erikstad, K.E. and Leinaas, H.P. (eds). Ecology of Fjords and Coastal Waters: Proceedings of the Mare Nor Symposium on the Ecology of Fjords and Coastal Waters, Tromsoe, Norway, 5–9 December, 1994, Elsevier Science B.V., P.O. Box 211 Amsterdam 1000 AE Netherlands, 1995, pp. 45–72.

Heimdal, B.R. 1974. Composition and abundance of phytoplankton in the Ullsfjorden area, North Norway. Astarte. 7: 17–42.

Helle, K. 1994. Distribution of early juvenile Arcto-Norwegian cod (*Gadus morhua* L.) in relation to food abundance and watermass properties. ICES Mar. Sci. Symp. 198: 440–448.

Hirche, H.-J. 1996. Diapause in the marine copepod, *Calanus finmarchicus* – a review. Ophelia, 44: 129–143.

Hurrell, J.W. 1995. Decadal trends in the North Atlantic Oscillation, regional temperatures, and trecipitation. Science, 269: 676–679.

Ingvarsdóttir, A., Houlihan, F., Heath, M. R. and Hay, S. J. 1999. Seasonal changes in respiration rates of copepodite stage V *Calanus finmarchicus* (Gunnerus). Fish. Oceanogr., 8 (Suppl. 1): 73–83.

Iversen, S.A. 2002. Havets ressurser. Fisken Havet, særnr. 1–2002. Institute of Marine Research, Bergen, Norway.

Jacobsen, J.A. and Hansen, L.P. 2000a. Distribution and migration of Atlantic salmon, *Salmo salar* L., in the sea. *In:* Mills, D. (ed.). The ocean life of Atlantic salmon: Environmental and biological factors influencing survival. Fishing News Books, Blackwell Science: 75–82.

Jacobsen, J.A. and Hansen, L.P. 2000b. Feeding habits of Atlantic salmon at different life stages at sea. *In:* D. Mills (ed.). The ocean life of Atlantic salmon: Environmental and biological factors influencing survival. Fishing News Books, Blackwell Science:170–192.

Jacobsen, J.A. and Hansen, L.P. 2001. Feeding habits of wild and escaped farmed of Atlantic salmon, *Salmo salar* L., in the Northeast Atlantic. ICES J. Mar. Sci. 58: 916–933.

Jakobsson, J. and Stefansson, G. 1998. Rational harvesting of the cod-capelin-shrimp complex in the Icelandic marine ecosystem. Fisheries Research 37:7–21.

Jákupsstovu, S. H. and Reinert, J. 1994. Fluctuations in the Faroe Plateau cod. ICES Mar. Sci. Symp., 198: 194–211.

Jespersen, P. 1940. Investigations on the quantity and distribution of zooplankton in Icelandic waters. Medr Kommn Havunders. (Ser. Plankton), 3(5): 1–77.

Joensen J. S. and Tåning, V. 1970. Marine and Freshwater Fishes. Vald Petersen Bogtrykkeri, Copenhagen, 241 pp.

Johnson, G. C. and Sanford, T. B. 1992. Secondary circulation in the Faroe Bank Channel Outflow. J. Phys. Oceanogr., 22: 927–933.

Jónsson, E., and Fridgeirsson, E. 1986. Observations on the distribution and gut contents of fish larvae and environmental parameters south-west of Iceland. ICES CM 1986/L: 36, 22 pp.

Kaartvedt, S., Larsen, T., Hjelmseth, K. And Onsrud, M.S.R. 2002. Is the omnivorous krill *Meganyctiphanes norvegica* primarily a selectively feeding carnivore? Mar. Ecol. Prog. Ser. 228: 193–204.

Kaasa, Ø., and Gudmundsson, K. 1994. Seasonal variations in the plankton community in Eyjafjörður, North Iceland. ICES CM 1994/L: 24: 15 pp.

Knutsen, T., Melle, W. and Calise, L. 2001. Determining the mass density of marine copepods and their eggs with a critical focus on some of the previously used methods. J. Plankton Res., 23: 859–873.

Magnusson, K.G. and Palsson, O.K., 1991. Predation-prey interactions of cod and capelin in Icelandic water. ICES marine Science Symposium 193:153–170.

Mauchline, J. 1980. The biology of euphausiids. Adv. Mar. Biol., 18: 373–681.

Melle, W. 1998. Reproduction, life cycles and distributions of *Calanus finmarchicus, C. glacialis* and *C. hyperboreus* in relation to environmental conditions in the Barents Sea. Dr. scient. thesis. Department of Fisheries and Marine Biology. University of Bergen, Norway.

Melle, W. and Skjoldal, H.R. 1989. Zooplankton reproduction in the Barents Sea: Vertical distribution of eggs and nauplii of *Calanus finmarchicus* in relation to spring phytoplankton development. In: J.S. Ryland and P. Tyler (eds.) Reproduction, Genetics and Distributions of Marine Organisms: 137–145. Olsen and Olsen, Copenhagen.

Melle, W. and Skjoldal H.R. 1998. Reproduction and development of *Calanus finmarchicus, C. glacialis* and *C. hyperboreus* in the Barents Sea. Mar. Ecol. Prog. Ser., 169: 211–228.

Melle, W., Knutsen, T., Ellertsen, B., Kaartvedt, S. and Noji, T. 1993. Økosystemet i østlige Norskehavet; sokkel og dyphav. Havforskningsinstituttet. Rapport fra Senter for Marint Miljø 4. ISSN 0804–2128.

Melle, W., Mork, K.A., Holst, J.C. and Rey, F. 2003. The Norwegian Sea; Hydrography, plankton and herring feeding. Status report April 2002. Working document to ICES Northern Pelagic and Blue Whiting Working Group, Copenhagen, Denmark, 29 April – 8 May 2003.

Misund, O. A., Vilhjálmsson, H., Jákupsstovu, S. H., Røttingen, I., Belikov, S., Astthorsson, O., Blindheim, J., Jonsson, J., Krysov, A., Malmberg, S. A., and Sveinbjørnsson, S. 1998. Distribution, migration and abundance of Norwegian Spring spawning herring in relation to the temperature and zooplankton biomass in the Norwegian Sea as recorded by coordinated surveys in Spring and Summer 1996. Sarsia, 83: 117–127.

Mork, K.A. and Blindheim, J. 2000. Variations in the Atlantic inflow to the Nordic Seas, 1955–1996. Deep-Sea Res. I, 47: 1035–1057.

Niehoff, B. and Hirche, H.-J. 2000. The reproduction of *Calanus finmarchicus* in the Norwegian Sea in spring. Sarsia, 85: 15–22.

Niehoff, B., Klenke. U., Hirche, H.-J., Irigoien, X., Head, R. and Harris, R. 1999. A high frequency time series at Weathership M, Norwegian Sea, during the 1997 spring bloom: The reproductive biology of *Calanus finmarchicus*. Mar. Ecol. Prog. Ser., 176: 81–91.

Ólafsson, J. 1999. Connections between oceanic conditions off N-Iceland, Lake Mývatn temperature, regional wind direction variability and the North Atlantic Oscillation. Rit Fiskideild, 16: 41–57.

Onsrud, M.S.R and Kaartvedt, S. 1998. Diel vertical migration of the krill *Meganyctiphanes norvegica* in relation to physical environment, food and predators Mar. Ecol. Prog. Ser., 171: 209–219.

Orvik, K.A., Skagseth, Ø. and Mork, M. 2001. Atlantic inflow to the Nordic Seas: current structure and volume fluxes from moored current meters, VM-ADCP and SeaSoar-CTD observations, 1995–1999. Deep-Sea Res. I, 48: 937–957.

Østerhus, S., Turrell, W. R. Hansen, B., Lundberg, P. and Buch, E. 2001. Observed transport estimates between the North Atlantic and Arctic Mediterranean in the Iceland-Scotland relation. Polar Res., 20: 169–175.

Østvedt, O. J. 1955. Zooplankton investigations from Weathership M in the Norwegian Sea, 1948–1949. Hvalrd. Skr., 40: 1–93.

Ottersen G. and Loeng H. 2000. Covariability in early growth and year-class strength of Barents Sea cod, haddock and herring: The environmental link. ICES J. Mar. Sci. 57: 339–348.

Palsson, O.K., 1983. The feeding habits of demersal fish species in Icelandic waters. Rit Fiskideildar, 7:1–60.

Planque, B. and Frédou, T. 1999. Temperature and the recruitment of Atlantic cod (*Gadus morhua*). Can. J. Fish. Aquat. Sci., 56: 2069–2077.

Rae, B. B. 1967. The food of cod on Faroese grounds. Mar. Res., 6: 1–23.

Rahmstorf, S. and Ganopolski, A. 1999. Long-term global warming scenarios computed with and efficient coupled climate model. Clim. Change, 43: 353–367.

Reid, D. G., Turrell, W. R., Walsh, M., and Corten, A., 1997. Cross-shelf processes north of Scotland in relation to southerly migration of western mackerel. ICES J. Mar. Sci., 54: 168–178.

Rey, F. 1981. Primary production estimates in the Norwegian Coastal Current between 62°N and 72°N. In: Sætre, R. and Mork, M. (eds.). The Norwegian Coastal Current. Proceedings from the Norwegian Coastal Current Symposium. Geilo, 9–12 September 1980. Volume II. University of Bergen 1981. Reklametrykk A.s. Bergen.

Rey, F. 1993. Phytoplankton and its primary production in the northern Barents Sea. Fisken Havet 10, 39 pp.

Rey, F., Skjoldal, H.R. and Slagstad, D. 1987. Primary production in relation to climatic changes in the Barents Sea. In: Loeng, H. (ed). The effect of oceanographic conditions on distribution and population dynamics of commercial fish stocks in the Barents Sea. Institute of Marine Research. Pp. 29–46.

Ruud, J. T. 1929. On the biology of the copepods off Møre 1925–1927. Rapp P-v Cons Perm int Explor Mer 56:1–84

Sætre, R., Aure, J. and Ljøen, R. 1988. Wind effects on the lateral extension of the Norwegian coastal water. Cont. Shelf Res., 8: 239–253.

Sigurjonsson, J., Vikingsson, G.A., 1998. Seasonal abundance of and estimated food consumption by cetaceans in Icelandic and adjacent waters. Journal of Northwest Atlantic Fishery Science. 22:271–287.

Skjoldal, H. R., Gjøsæter, H.and Loeng, H., 1992. The Barents Sea ecosystem in the 1980s: Ocean climate, plankton, and capelin growth. ICES Mar. Sci. Symp., 195: 278–290.

Skjoldal, H.R., Hassel, A., Rey, F. and Loeng, H. 1987. Spring phytoplankton development and zooplankton reproduction in the central Barents Sea in the period 1979–1984. In: Loeng, H. (ed). The effect of oceanographic conditions on distribution and population dynamics of commercial fish stocks in the Barents Sea. Institute of Marine Research. Pp 59–89.

Slagstad D. and Tande K. S. 1996. The importance of seasonal vertical migration across the shelf transport of *Calanus finmarchicus*. Ophelia 44: 189–205.

Slotte, A. 1998. Spawning migration of Norwegian spring spawning herring (*Clupea harengus* L.) in relation to population structure. Where and when do they spawn? Dr. Scient. Thesis. Department of fisheries and marine biology. University of Bergen. Norway.

Stefánsson, U. 1962. North Icelandic waters. Rit Fiskideild., 3: 1–269.

Stefánsson, U., and Ólafsson, J. 1991. Nutrients and fertility of Icelandic waters. Rit Fiskideild., 7: 1–56.

Stefansson, G., Skuladottir,U., Steinarsson.B., 1998. Aspects of the ecology of a Boreal system. ICES Journal of Marine Science. 55:859–862

Steingrund P., Ofstad, L. H., and Olsen, D. H. 2003. Effect of recruitment, individual weights, fishing effort, and fluctuating longline catchability on the catch of Faroe Plateau cod (*Gadus morhua* L.) in the period 1989–1999. ICES J. marine Science Symposia, 219: 418–420.

Steingrund, P. and Gaard., E. 2003. Relationship between phytoplankton production and cod production on the Faroe shelf. ICES CM 2003/P:40.

Sundby, S. 1991. Factors affecting the vertical distribution of eggs. ICES Mar. Sci. symp., 192: 33–38.

Sundby, S., Bjørke, H., Soldal, A.V. and Olsen, S. 1989. Mortality rates during the early life stages and year class strength of the Arcto-Norwegian cod (*Gadus morhua* L.). In: Sundby, S. (ed.). Year class variations as determined from pre-recruit investigations. Part 1, 2. Proceedings from the second workshop under the cooperative programme of fisheries research between the institutions in Seattle, Nanaimo and Bergen, held in Bergen 28.-30. September 1988., Inst. Mar. Res., Bergen (Norway), 1989, pp. 477–500.

Sverdrup, H.U. 1953. Conditions for the vernal blooming of phytoplankton. J. Cons. Perm. Int. Explor. Mer. 18: 287–295.

Thórdardóttir, Th. 1994. Plöntusvif og frumframleiðni í sjónum við Ísland. (Phytoplankton and primary production in Icelandic waters. In Icelandic) *In:* Stefánsson, U. (ed.). Íslendingar hafið og auðlindir þess. (Icelanders, the sea and its resources). Vísindafélag Íslendinga (Societas scientarium Islandica), Reykjavík: 65–88.

Thórdardottir, Th., and Gudmundsson, K. 1998. Plöntusvif. (Phytoplankton. In Icelandic). Námsgangastofnun og Hafrannsóknastofnunin, Reykjavík: 1–12.

Thorisson, K. 1989. The food of larvae and pelagic juveniles of cod (*Gadus morhua* L.) in the coastal waters west of Iceland. Rapports et Procés-Verbaux des Réunions du Conseil International pour l'Exploration de la Mer, 191: 264–272.

Toresen, R. 2002. Environmental influence on herring (*Clupea harengus* L.) abundance. An environmental approach for the understanding of herring stock fluctuations and its application in management. Dr. scient. thesis. Department of fisheries and marine biology. University of Bergen, Norway.

Turrell, W. R., Slessner, G., Adams, R. D., Payne, R. and Gillibrand, P. A. 1999. Decadal variability in the composition of Faroe Shetland Channel Bottom Water. Deep-Sea Res. I, 46: 1–25.

Valdimarsson, H., and Malmberg, S.-A. 1999. Near-surface circulation in Icelandic waters derived from satellite tracked drifters. Rit Fiskideild., 16,: 23–39.

Vilhjálmsson, H. 1994. The Icelandic capelin stock. Capelin, *Mallotus villosus* (Muller) in the Iceland – Greenland – Jan Mayen area. Rit Fiskideildar, 13:1–281.

Vilhjálmsson, H. 1997. Interactions between capelin *(Mallotus villosus)* and other species and the significance of such interactions for the management and harvesting of marine ecosystems in the northern North Atlantic. Rit Fiskideildar, 15:32–63.

Wassmann, P. 1998. Pelagic retention versus export food chains: Processes controlling sinking loss from marine pelagic systems. Hydrobiologica, 363: 29–57.

Wiborg, K. P. H. 1954. Investigations on zooplankton in coastal and offshore waters off western Northwestern Norway. Rep. Norw. Fish Invest. 11(1).

Chapter 28. LAPTEV AND EAST SIBERIAN SEAS (23,P)

SERGEY PIVOVAROV

Arctic and Antarctic Research Institute (AARI)

JENS A. HÖLEMANN

Alfred Wegener Institut für Polar und Meeresforschung (AWI)

HEIDEMARIE KASSENS

Forschungzentrum für marine Geowissenschaften (GEOMAR)

DIETER PIEPENBURG

Akademie der Wissenschaften und der Literatur Mainz,
c/o Institut für Polarökologie der Universität Kiel

MICHAEL K. SCHMID

Alfred-Wegener-Intitute für Polar-und Meeresforschung (AWI),
c/o Institut für Polarökologie der Universität Kiel

Contents

1. Introduction

Key words usually used to characterize the environment of the Laptev and East Siberian seas are: shallow, ice-covered, and extremely cold. However, the Arctic bitter cold and heavy ice could not stop the first polar explorers from discovering and describing for us that wonderful and rich world. Russian Cossacks, followed by

The Sea, Volume 14, edited by Allan R. Robinson and Kenneth H. Brink
ISBN 0–674–01527–4

navy officers and hydrographers, explored the Arctic seas, no doubt with the main aim to discover new lands, new transport routes, to annex new territories to the Russian Empire, and they described the environment of the seas, coasts, and islands. They left their names on the maps and charts and on the many graves, which were scattered along the coast. Besides the Russians, the eastern Arctic seas were explored by the Sweden (Nordenskøeld, *Vega,* 1878–1879), American (De Long, *Jannette,* 1879–1881), and Norwegian (Nansen, *Fram,* 1893; Amundsen, *Maud,* 1922–1924) polar expeditions.

The Soviet Union organized many interdisciplinary polar expeditions after the Great October Revolution. The main purposes were to ensure a security of navigation through the Northern Sea Route and to explore mineral resources. Many stationary observatories and scientific stations were founded on the coasts of the Arctic seas. The main efforts were focused on ice studies in conjunction with hydrographic conditions. Physical, chemical and biological properties of ice and water column were studied intensively. The observations obtained became a foundation for future studies. Many famous Russian (soviet) professors published a great number of papers and books concerning hydrography, climate, ice conditions, bottom sediments and biota in the Arctic seas and in the Eastern sector in particular at that time (see the reviews in the *Problems of the Arctic and Antarctic,* Vol. 70, 1995).

The studies of the Laptev and East Siberian seas were continued after the Second World War. The Golden Era of Russian Arctic research was from 1950 till 1990. Two or three expeditions worked in each sea every summer and winter. These were, so called, *Ice Patrols* at sea in summer, the airborne expeditions *Sever* in winter, and ice camps all year round. Unfortunately, it is not possible to describe these expeditions as interdisciplinary because oceanographers, biologists and geologists explored the seas separately. The Arctic and Antarctic Research Institute (AARI) studied sea ice, hydrometeorological conditions, and some nutrients as indicators of water mass origin, and to monitored pollutants distributions. Zoological and Biological Institutes (ZIN and BIN) and Shirshov Institute of Oceanography of the Russian Academy of Science studied biological processes. The All-Union Research Institute for Geology and Mineral Resources of the World Ocean (VNIIO) studied bottom sediments and geological processes. But, unfortunately inter-institutional cooperation was not a priority. There were many examples, when biologists or geologists tried to explain their own interesting findings by complexity of oceanography and regretted that appropriate observations had not been collected. Nevertheless, the Russian scientific community has a long tradition in working on the Eurasian shelf seas because of oil, gas and mineral resources found there and the economic advantages of the Northern Sea Route. Relevant institutions have accumulated a tremendous amount of experience in Arctic research. Much data and numerous papers about the Siberian shelf seas have been published, but they have appeared primarily in Russian reports or journals, which have not easily been available for the Western world. These areas were closed to international exploration for a number of decades.

Actually, real interdisciplinary studies at the Eastern shelf began ten years ago, thanks to international cooperation. Very interesting new results were obtained in the international expeditions SPASIBA (1989–1991). Multidisciplinary team studied river influx into the Laptev Sea and then published their results in a special volume of the *Marine Chemistry*. The Russian-German ESARE (1992) expedition operated in the eastern part of the Laptev Sea in winter (Dethleff et al., 1993). The great progress in interdisciplinary studies in the Laptev Sea was achieved owing to the bilateral research project *Laptev Sea System* (1993–2003) that was aimed to the interpretation of paleoclimatic records in conjunction with the study of modern processes for better understanding and prediction of the Arctic System (Thiede et al., 1999). The project comprised marine and terrestrial observations as well as a suite of modeling experiments and theoretical considerations. Ten expeditions to the inner Laptev Sea in the frame of the multi-disciplinary research program were carried out. These expeditions were accompanied by both RV *Polarstern* cruises and the land-based Lena-Yana expeditions (Fütterer, 1994; Rachor, 1997; Rachold, 1999). Some of them were unique. For example, the expedition TRANSDRIFT III (1995) aboard the icebreaker *Kapitan Dranitsyn* focused on processes which occur during the extreme change from the ice-free conditions in late summer to the onset of freeze-up during autumn (Kassens et al., 1997), and the TRANSDRIFT IV (1996) expedition studied the Lena River break-up and its influence on the environmental system of the Laptev Sea (Kassens et al., 1998). For the first time, a major comprehensive research program combined the efforts of many Russian and German institutions and addressed both oceanic and terrestrial processes, and their consequences for marine and terrestrial biota, landscape evolution as well as land-ocean interactions. Extensive studies of the atmosphere, sea ice, water column, and sea-floor on the Laptev Sea Shelf, as well as of the vegetation, soil development, carbon cycle, permafrost behavior and lake hydrology were performed during the recent years (Thiede et al., 1999).

2. Geographical location, bottom topography, climate, ice, river run-off, key physical processes

2.1. Geographical location and bottom topography

The Laptev and East Siberian seas are marginal open seas of the Arctic Ocean and occupy the central and eastern parts of the vast Siberian (Eurasian) shelf, which is the broadest one on Earth (Fig. 28.1). Northern boundaries of the seas are conventional. That is why the Laptev Sea includes the continental slope and a part of the deep Arctic Basin, while 72% (Timokhov, 1994) of the sea is located on the extremely shallow shelf. The total area of the sea is 662,000 km^2 (Atlas of the Arctic, 1985). The East Siberian Sea shelf makes up 96% of the sea area (913,000 km^2). Both of the seas are located within the Polar Circle. The coastal line is indented with many gulfs. The largest of them are Khatanga, Anabar, Olenek, Buor-Khaya, Yana gulfs in the Laptev Sea, and Khroma, Indigirka, Kolyma and Chauna gulfs in the East Siberian Sea.

The Laptev Sea shelf is very shallow. Depths less than 50 m dominate over the southern part of the shelf. Submarine valleys are the main and very important relief features of the sea because they control water mass redistribution and sedimentation rate. They comprise a series of depressions 25–45m in depth. A vast shoal (bank) separates the Lena and Yana underwater valleys in the eastern part of the Laptev Sea. This shoal comprises the remains of eroded and recently submerged islands. The islands were mapped and named only a little more than one hundred years ago (Zigarev and Sovershaev, 1984). This is an example of the intensive erosion of the shore, which is an object of interdisciplinary studies nowadays (Are, 1999; Are et al., 2002).

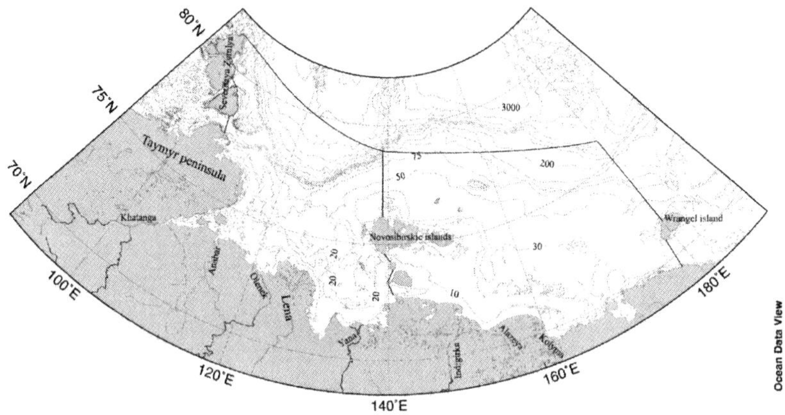

Figure 28.1 Geographical boundaries and bottom topography of the Laptev and East Siberian seas. Thin solid lines with numbers indicate isobaths according to IBCAO (International Bathymetric Chart of the Arctic Ocean) bathymetry data (Jakobsson et al., 2000). Thicker solid lines show conventional boundaries of the seas according to the Atlas of the Arctic (1985). Coastlines and bathymetry were taken from the Ocean Data View software (Schlitzer, 2002).

The East Siberian Sea floor is a smooth plain dipping gently to the north and east. The shelf break occurs at the water depth of 50–60 m as far as 350–400 km from the coast. The submarine valley of the Indigirka River stretches northward in the western part of the sea. The relatively narrow and deep underwater valley of the Kolyma River extends along the Siberian coast and turns northward as it widens near Wrangel Island.

2.2. *Climate*

The severe arctic climate of the seas is characterized by a long and extremely cold winter (mean January air temperature is up to -32 °C) and a short and rather cold summer (mean August air temperatures are from 0 to 5 °C). Cyclones visit this region rarely in winter, and they are not active and deep. An average wind speed over the seas is about 5 m s^{-1}. The region is calm even in summer when atmospheric circulation is weak with storms occurring for two or three days per month. Cloudiness is slight in winter, but it usually covers the entire sky in summer. The precipitation rate is low, 150–300 mm per year with most of the precipitation falling in summer. The relative humidity is rather high, 95–98% so fogs are quite frequent, especially in the ice-covered regions (Timokhov, 1994; Pavlov, 1998).

2.3. *Ice conditions*

Sea ice influences the climate, oceanography, biology as well as the geology of the area to a high degree. Ice significantly reduces the heat flux between ocean and atmosphere; through its high albedo it has a strong influence on the radiation budget. It controls the thermohaline circulation driven by ice formation and melting. Sea ice is a key geological agent in the Arctic, it transports large amounts of the sediments incorporated into the ice in the shelf area (Pfirman et al., 1990; Lindemann et al., 1999). Sea ice cover is one of the most sensitive expressions of the modern climate regime with its extreme seasonal fluctuations being a good indicator of potential instability (Thiede, 1996).

New ice formation depends on hydrometeorological and ice conditions present at the beginning of the fall. The dates of the onset of freezing vary dramatically; ice can be formed even in summer among the remaining ice patches. As a rule, new ice appears at the edge and among the existing ice floes, in shallow near shore areas and estuaries. It means that freeze-up starts both from the north and south of the seas at the same time. Autumn storms play a key role in ice formation. Strong winds induce a vertical mixing and a rapid cooling of the surface layer. Ice formation is more intensive and occupies larger areas after the storms. Air temperature and hydrographic conditions, which are changing in space and time, control the ice growth. At the end of winter, ice thickness reaches 2 m (Zakharov, 1996). The shallow parts of the seas (depth less than ca. 20 m) are covered by fast ice. The width of the fast ice depends on bottom topography and is controlled by offshore winds (Zakharov, 1966; 1996). A perennial polynya, hundreds of kilometers wide, borders the very smooth fast ice (Fig. 28.2). *In this body of open water, rapidly forming ice is continuously advected offshore, making the Laptev Sea the single the major ice factory for the Arctic Ocean and Transpolar Drift* (Reimnitz, 1994). Ice outflows from the Laptev and East Siberian seas are 540 and 150 km^3 respectively (Timokhov, 1994). *With summer warming, this winter factory turns into an area of heat gain, aiding the retreat of the ice edge to a much higher latitude and greater distance (500 km) from the mainland than in the Beaufort Sea* (Reimnitz et al., 1993; Reimnitz, 1994). The fast ice massifs thaw at the same places where they have been formed. A huge amount of ice, comparable with the riverine freshwater influx, melts in the seas in summer (Zakharov, 1996; Rigor and Colony, 1997).

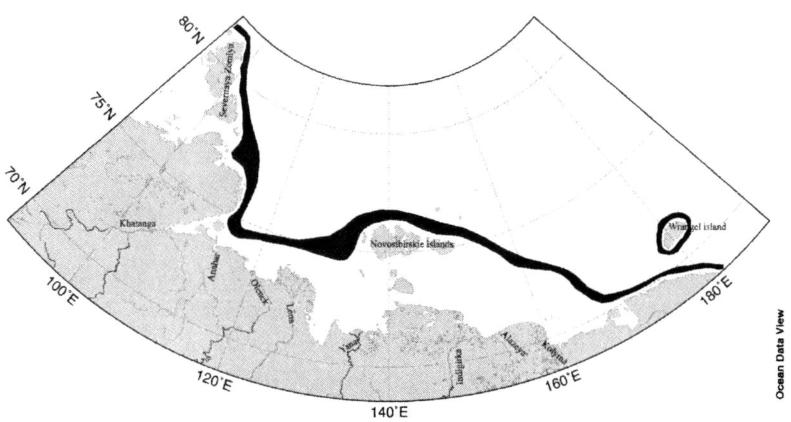

Figure 28.2 Location of the Great Siberian polynya in the Laptev and East Siberian seas according to Zakharov (1996).

2.4. *River run-off*

Pulsing river run-off impacts the seas. This freshwater input together with melt water from the sea ice is of critical importance in influencing the circulation pattern, ecosystem, sediment dynamic, and geochemistry of the seas. The most important contributors are the Siberian rivers Lena (520 km^3), Khatanga (105 km^3), Anabar (25.2 km^3), Olenek (40 km^3), Yana (37.5 km^3), Indigirka (56.7 km^3), Alazeya (10.2 km^3), and Kolyma (102 km^3) (Atlas of the Arctic, 1985; Ivanov and Piskun, 1999). Total river run-off is estimated to be about 1000 km^3 per year. These rivers transport millions of tons of dissolved and particulate material (i.e., chemical elements, nutrients, siliciclastic and organic matter, etc.) onto the shelf where it is accumulated or further transported by different mechanisms (sea ice, icebergs, turbidity currents, etc.) (Stein et al., 2001). The river influx changes dramatically in seasonal circles. Small, medium, and even some branches of the large rivers are completely frozen down to the bottom in winter. From 70 up to 95% of the river influx occurs during a short Arctic summer (Ivanov and Piskun, 1995).

2.5. *Water circulation*

The surface circulation pattern is extremely variable in the shallow Laptev and East Siberian seas both in summer and in winter, and it is intimately connected with atmospheric processes. In spite of large mesoscale variability in surface currents caused by changes in the wind field, it is possible to identify permanent currents in the seas. In general, a very slow counterclockwise (cyclonic) water circulation is observed at the surface of the seas in summer (Pavlov, 1998). There are more or less permanent coastal currents directed eastward along the shore. Very little is known about the currents in the Intermediate and Bottom Structural Zones.

2.6. *Waves*

Wave heights are usually small along the navigation route in the Laptev and East Siberian seas. Ice cover, shallowness, and strong density stratification inhibit wave development. Steep short waves with a height of less than 1.5 m are most common. Heavy storms are unusual events for this region, however, strong eastern winds can generate waves with a height of more than 3 m in late September when the ice edge retreats far to the north and the southern parts of the seas are free of ice (Timokhov, 1994; Pavlov, 1998).

2.7. *Tides*

Semidiurnal tides prevail in the Laptev and East Siberian seas. The tides are induced mainly by waves coming from the Arctic Basin. The tidal waves strongly deform and gradually attenuate in the shallow shelf due to a large difference of depth at the continental slope. The tidal range decreases from 1.5 m in the north (near the New Siberian Islands and Wrangel Island) to 0.1m near the coast of mainland. So, as a rule, the tidal range is very small at the shore. Tidal current velocities are usually less than 0.2 m s^{-1}. However, there are some exceptions. According to Earth's rotation, the tidal waves deviate to the right from the direction of propagation that induces an increasing tidal range along the western coasts of these seas. Tidal range increases up to 2.5 m near the entrance of the Khatanga Gulf (the western part of the Laptev Sea). Enhanced tidal range is observed in the western part of the East Siberian Sea near the Sannikov and Dmitry Laptev Straits (Timokhov, 1994).

2.8. *Surges*

Sea level fluctuations caused by surges are very common along the coast of the seas. Surges increase sea level up to 2 m and more. The wind surge effect is inversely proportional to depth. The largest surges are observed in shallow regions of the eastern Laptev Sea and in the Kolyma estuary. Their ranges are five to twenty times greater than tidal fluctuations of sea level. Surges decrease to the north along the coasts of islands and the mainland. The sea level fluctuations caused by surges are a little more than 1m at the coasts of the Taymyr Peninsula and the New Siberian Islands but they exceed the tidal range there nevertheless (Timokhov, 1994; Pavlov, 1998).

3. Water column structure

A clear understanding of a water body structure is necessary for modeling and prediction of entire sea ecosystem. It was mentioned earlier that there are thermohaline, hydrochemical, biological, and other structures. It is now obvious that a water column has only one structure, and it should be studied in a interdisciplinary manner, combining physics, chemistry, geochemistry and biology.

As a result of changeable hydrometeorological, ice, and biological conditions, a multilayered and mosaic water column structure is formed in the Laptev and East Siberian seas. Water mass (a volume of water having *common formation history* (Tomczak, 1999) and *possesses for a long time almost constant and continuous distributions of physical, chemical, and biological characteristics, constituting a united complex* (Dobrovolsky, 1961)) is a basic element of the water column. Water masses with different properties are formed in the same regions of the seas during different seasons. Small volumes and a short life span characterize the water masses in the shallow Arctic seas. A great number of water masses could be separated into several types according to their position in the structural zones, and the place, and time of their formation.

The water column in the seas can be subdivided vertically into three structural zones, namely: Surface, Intermediate and Bottom. The structural zones consist of water masses appurtenant to corresponding types (surface, intermediate and bottom). Transport and transformation of substances, including nutrients and contaminants, in these structural zones occur in different ways.

3.1. *Surface Structural Zone (SSZ)*

The SSZ is the most active area where energy and material is transformed by interactions between sea, ice, and atmosphere, and is a region where biological processes are very important. It is the most dynamic as well because of its own characteristics change during the annual cycle and the movements of water masses caused by winds, tides, gravity, and the Earth's rotation. The thickness of the SSZ in the well-stratified seas is usually 5–10 m and changes seasonally.

Two different regions in the Laptev Sea can be distinguished according to temperature and salinity distributions, and chemical and biological properties of the surface water masses. River plumes occupy the southern and eastern regions of the sea while the Arctic Basin influences the northwestern part of the sea. It is considered that the silicate distribution at the surface (Fig. 28.3a) is a good indicator of the river plume in summer (Rusanov et al., 1979). The 10 μmol l^{-1} silicate isoline is accepted to be a boundary of the river plume that is also characterized by low salinity (less than 25), low oxygen saturation (95 – 100%), and high concentrations of suspended matter (0.5–0.8 mg l^{-1} at the outer shelf and up to 70 mg l^{-1} near the Lena River Delta). They can be distinguished by many other physical, chemical and biological properties (e.g. water color and smell). Bottom sediment composition and distribution of benthic species reflect an average location of the river plume.

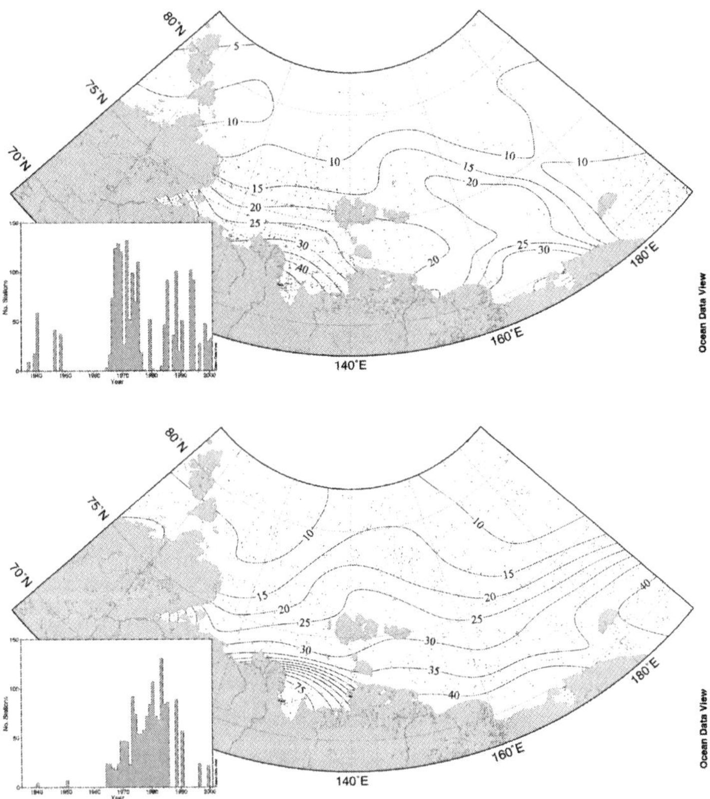

Figure 28.3 Averaged silicate (μmol l^{-1}) distributions at the surface of the Laptev and East Siberian seas in summer (August-September) (a), and in winter (March-May) (b) for the years 1922–2000. Thin solid lines with numbers indicate isolines of silicate concentrations. High silicate concentrations at the surface of the seas in summer indicate a riverine influence. Black points show locations of the oceanographic stations data of which were used for averaging and gridding. Data source is the US-Russian Hydrochemical Atlas of the Arctic Ocean (Colony et al., 2002). Windows at the lower left corners of the maps show temporal distributions of the stations. Ocean Data View software (Schlitzer, 2002) was used for gridding and map construction.

According to the silicate distribution (Fig. 28.3a), the East Siberian Sea is influenced by river run-off to a high degree too. The Kolyma River plume is visible on the map, however huge amounts (ca. 250 km^3) of fresh water are imported from the Laptev Sea.

Nitrate and phosphate are almost consumed in the SSZ in summer. Nitrate concentrations are near zero in the both parts of the Laptev Sea though phosphate concentrations range between 0 and 0.2 $\mu mol\ l^{-1}$. Increased nitrate (up to 1 $\mu mol\ l^{-1}$) and phosphate (up to 0.5 $\mu mol\ l^{-1}$) concentrations can be observed near river mouths in summer.

Ice margins are special regions where surface water masses are characterized by extremely high oxygen saturation (more than 120%), and pH, and alkalinity anomalies caused by continuous phytoplankton bloom. These regions are of great ecological significance because of high primary production and special conditions for marine inhabitants. However it should be noted that the geographical location of an ice margin changes dramatically in summer.

The SSZ is enriched by nutrients in winter as a result of mixing with intermediate waters. An average silicate concentration at the sea surface in winter is higher than in summer (Fig. 28.3). Water-cooling and the brine injection accompanying ice formation cause deep convection, which increases the thickness of the surface mixed layer. The convection is able to destroy the halocline in some shallow regions and penetrate down to the bottom. In this case bottom sediments and nutrient rich pore waters are involved in the effects of mixing. Winter water masses formed under fast ice, in polynya, and under drift ice are characterized by different physical, chemical and biological properties. A polynya is a natural ventilator for the seas, and their water masses are very well saturated by oxygen since ice does not hinder gas exchange between sea and atmosphere.

3.2. *Intermediate Structural Zone (ISZ)*

The ISZ in the shallow Arctic seas consists of just the halocline, a layer up to 20 m of thickness with strong salinity gradients. It has a multi-layered structure itself and consists of water masses of different origin. Data of the summer oceanographic surveys show that it is possible to recognize in the halocline the intermediate water masses, which have been formed in winter, in spring, and in summer according their temperature and chemical properties. Extremely low temperature (less than -1.5 °C) is a distinguishing characteristic of the intermediate winter water masses. The intermediate spring water masses are characterized by low temperature (sometimes less than -1.0 °C), very low nutrient concentrations, and high oxygen saturation (more than 100%). An intermediate oxygen maximum indicates the location of the spring water masses, which are formed at the surface of the sea before ice melting. The intermediate summer water masses are characterized by relatively high positive temperatures (up to 5 °C), low nutrient concentrations, and sufficient oxygen saturation (up to 95%). They are formed at the surface and penetrate to the intermediate depth at the frontal zone that separates the river plume from surrounding water masses. An intermediate temperature maximum is a remarkable feature of the river plume region in the late summer and autumn when the surface mixed layer is cooled.

3.3. Bottom structural zone (BSZ)

Many different types of water masses can be distinguished in the BSZ according to their physical, chemical and biological properties. For example, stagnant bottom water masses are formed in the river plume areas of the Arctic seas under fast ice in winter. Low temperature (-1.2 – -1.5 °C), extremely low oxygen saturation (30–50%), and high nutrient concentrations characterize them. The intensive influx of riverine organic matter and limited ventilation are the reasons for the oxygen deficit. In winter, the fast ice prevents gas exchange with the atmosphere; in summer, the halocline insulates the bottom water masses from the surface waters. In the BSZ of the Buor-Khaya Bay (the Laptev Sea), even hydrogen sulfide has been found in certain years and these conditions negatively affect the benthic community (Sidorov and Gukov, 1992). Generally, the water masses accumulate in small pits, underwater valleys, and other seabed depressions where advective ventilation is suppressed. They are not really stagnant water masses. When the water masses overfill the seabed depressions, they spread according to their density and can penetrate into different structural zones. Oceanographic transects across underwater valleys in the Laptev seas showed that, as a rule, the water masses spread toward the shelf edge along the eastern slopes of the valleys as a result of the Earth's rotation (Fig. 28.4). Two different water masses can be found at the same depth in the Bottom Structural Zone in the underwater valley. They spread in opposite directions. The data obtained during the Russian-German multidisciplinary expedition TRANSDRIFT-V in the Laptev Sea in the summer 1998 (Kassens et al., 1998) allowed us to trace the stagnant bottom water masses from the region of their formation to the continental slope (Fig. 28.5). Strikingly, there is a possibility of upwelling of these nutrient rich water masses near the shelf break. According to their density, the bottom water masses first rise to the Intermediate Structural Zone and then to the Surface Structural Zone where physical, chemical, and biological properties of the water masses change rapidly. The remains of these nutrient rich bottom water masses, after such transformation, serve as an additional source of nutrients for phytoplankton in the ice marginal zone where nutrients are completely consumed in summer.

In the East-Siberian Sea, oxygen-depleted bottom water masses also settle in depressions on the sea floor. However, the mechanism of their formation has not been adequately studied yet. It is possible that the organic matter that consumes the bottom-layer dissolved oxygen enters the sea from the Pacific Ocean through the Chuckhi Sea or it is produced in large quantities at the shelf and at the edge of the drift ice. Bottom water masses of this type are denser than the stagnant bottom water masses in the Laptev Sea. They spread from the regions of their formation and penetrate into the Intermediate Structural Zone of the Canadian and Makarov Basins as a part of a very complex assembly of the Pacific waters. Physical, chemical and biological properties of the Pacific water masses are preserved in the ISZ for a long time. That allows tracing them from the Chukchi Sea up to the Fram Strait (Rusanov et al., 1979, Jones et al., 1991, Jones et al., 1998).

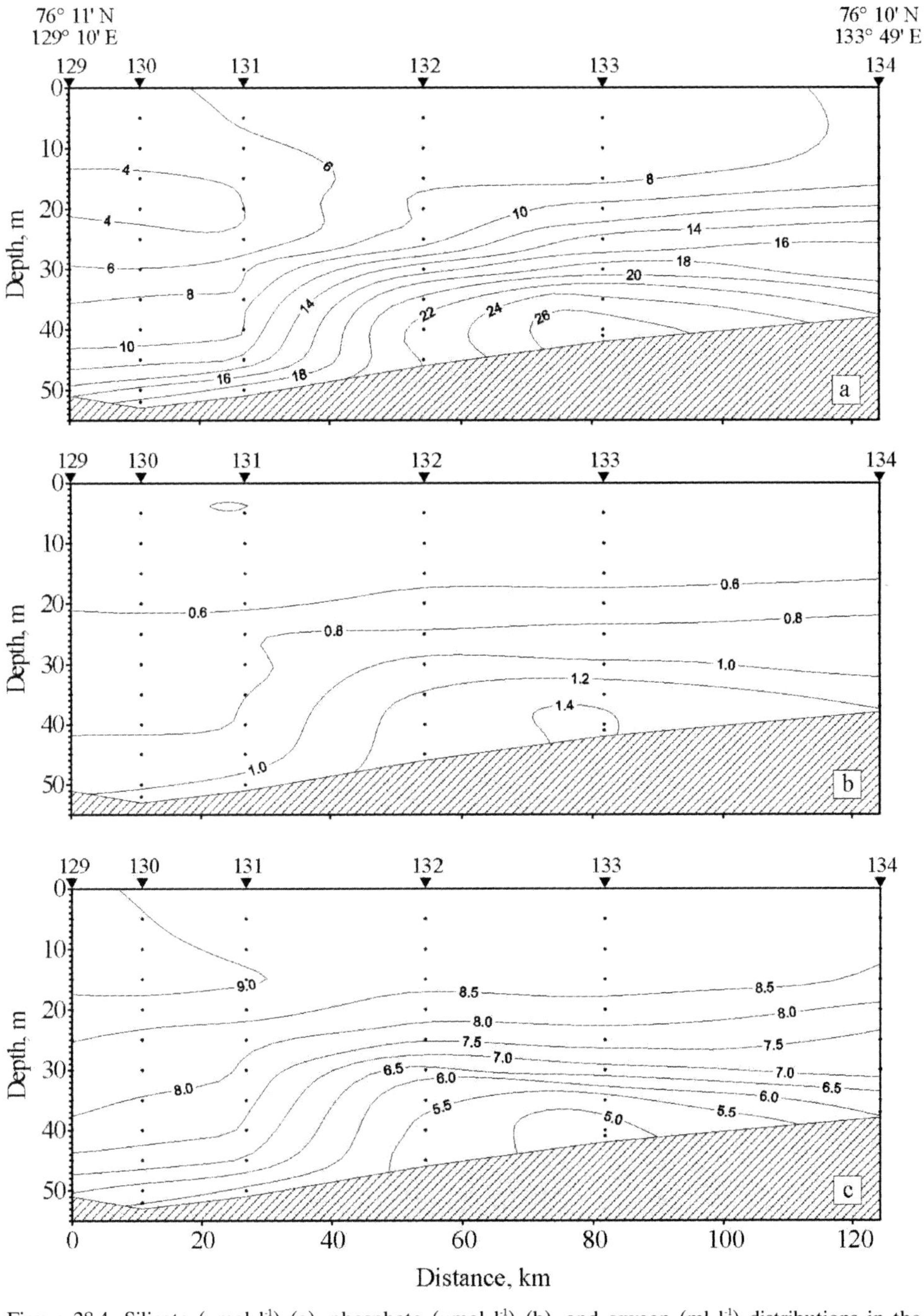

Figure 28.4 Silicate (µmol l^{-1}) (a), phosphate (µmol l^{-1}) (b), and oxygen (ml l^{-1}) distributions in the transect across the Lena underwater valley in the Laptev Sea. RV Polarstern, RK14/1b, 10–11 September 1998. High nutrient concentrations and low oxygen concentrations near the bottom indicate a location of the stagnant bottom water masses.

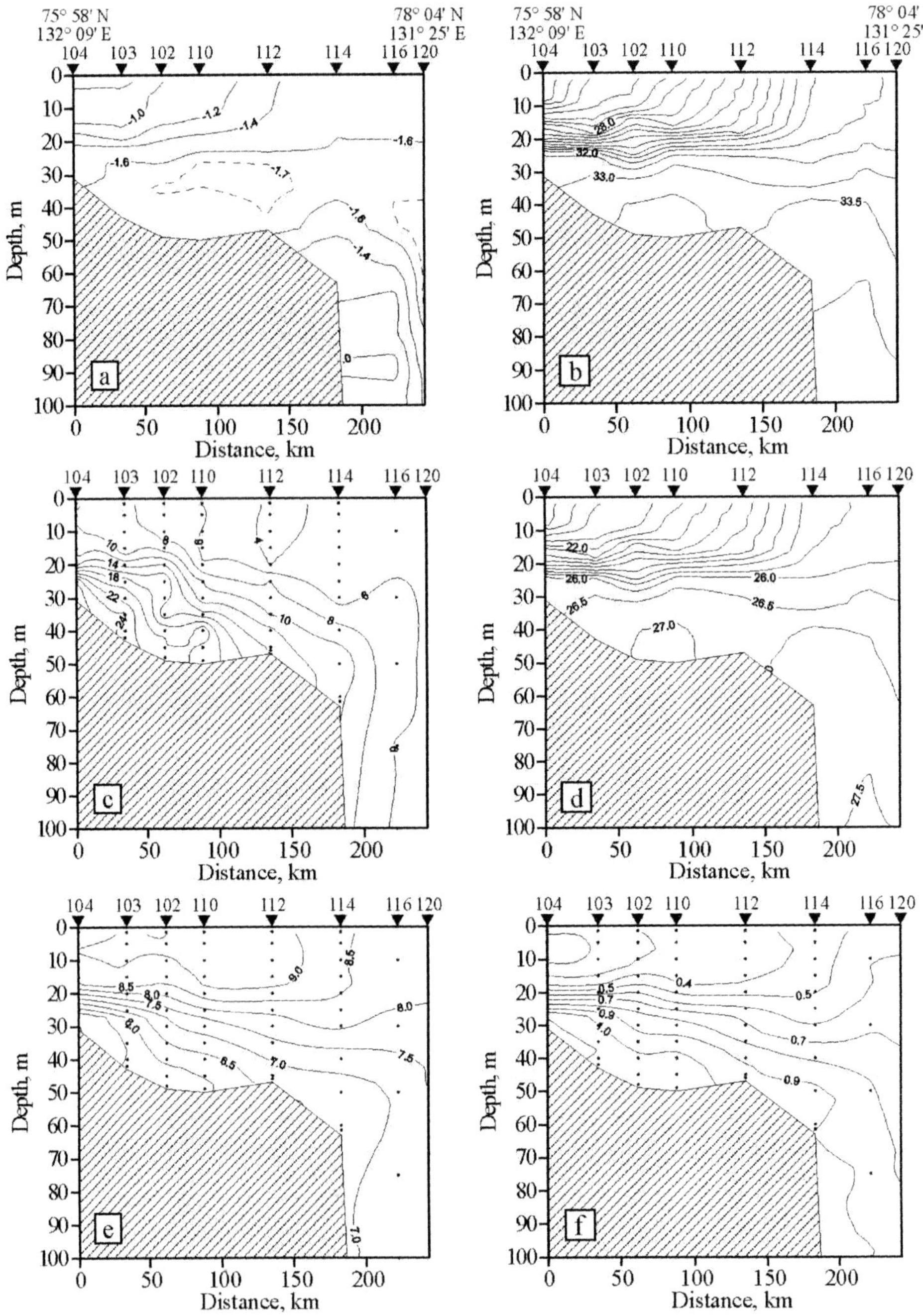

Figure 28.5 Potential temperature (°C) (a), salinity (b), silicate (μmol l^{-1}) (c), density (SIGMAT) (d), oxygen (ml l^{-1}) (e), and phosphate (μmol l^{-1}) (f) distributions in the transect across the continental slope in the Laptev Sea. RV Polarstern, ARK14/1b, 05–08 September 1998.

4. Sources and present-day transport pathways of dissolved and particulate matter on the Laptev Sea and East Siberian Sea shelves

During the last few decades a number of interdisciplinary research projects (SPASIBA, Laptev Sea System, Laptev Sea System - 2000; Ecology of the Marginal Seas of the Eurasian Arctic) have focused on the study of modern environmental processes and the Late Quaternary paleoenvironmental history of the Laptev Sea. Overviews of the results have been published in a number of articles and books (Martin et al., 1993, Kassens et al., 1995; Kassens et al., 1999). In contrast only little information has been published for the East Siberian Sea. Thus, this chapter focuses mainly on the scientific results that were obtained in the Laptev Sea.

Recently published articles on modern environmental processes in the Laptev Sea were mainly addressed to two major issues: (1) the transport processes and pathways of dissolved and particulate substances across the shelf, and (2) the importance of shelf processes for the modification of chemical constituents, especially organic carbon. The main aim of this chapter is to summarize the new results of the scientific research connected to these issues and to give an overview of the transport processes that are unique for shallow Siberian shelf seas, which are ice-covered for nearly ten months a year.

4.1. Input, transport and deposition of sediments, organic carbon and trace elements

Paleoenvironmental reconstructions revealed that the modern depositional environment of the Laptev Sea was probably established no earlier than 5000 yr B.P., after the Holocene sea level highstand was reached (Bauch et al., 2001a). On the inner shelf of the Laptev Sea total sediment input has remained rather constant since about 4000 yr B.P. (Bauch et al., 2001b). The inundated areas now form large shoals that are covered by relict and palimpsest sandy sediments (Gukov, 1999; Klenova, 1962). Recent sediment deposition on the inner and central shelf has been mainly connected to depressions in shelf topography. But even in the depocenter near the major outlet of the Lena River, the average sedimentation rate of the last 5 ka was not higher than 30 cm/ky (Bauch et al., 2001b).

Modern sedimentary environments of the Laptev and East Siberian seas are controlled by sediment input from rivers (reviewed in Gordeev, 2000), and coastal and seafloor erosion (Are, 1999). In particular, the riverine input to the Laptev and East Siberian seas is characterized by a strong seasonal variability with a maximum discharge of freshwater and sediments during June (Ivanov and Piskun, 1999; Pivovarov et al., 1999). Gordeev (2000) estimated a total riverine input of $25.1 \cdot 10^{6}$ t a^{-1} of suspended matter into the Laptev Sea and 33.610^{6} t a^{-1} into the East Siberian Sea. Sediment budget calculations based on coastal retreat rates have shown that sediment input due to shore erosion can exceed 5010^{6} t a^{-1} (Rachold et al., 2000).

Environmental forcing factors, i.e. atmospheric circulation, sea ice cover and river runoff, mainly affect the shallow water environment of the Laptev Sea. Especially the predominance of cyclonic or anticyclonic atmospheric circulation over the Arctic influences the current system and the distribution of river runoff on the shelf (Proshutinsky and Johnson, 2001). As a consequence the transport of sediments and the sedimentation processes are also strongly affected by different regimes of atmospheric circulation and ice cover. New data show that this effect starts as soon as the Laptev Sea flow polynya opens up during winter (Dmitrenko et al., 2001). New long-term measurements with bottom-moored instruments provide strong evidence for the assumption that modern shelf sediment transport is mainly wind-forced and connected to the N-S running submarine valleys on the shelf of the eastern Laptev Sea (Wegner et al. 2003; Wegner et al., accepted). In these submarine valleys suspended sediments are transported in a distinct bottom nepheloid layer, a layer of increased suspended matter concentration (SPM) with up to 12 m thickness, which is strongly influenced by the prevailing atmospheric circulation and the ice cover (Burenkov et al., 1997; Dmitrenko et al., 2001; Wegner et al., 2003). Calculations of the net horizontal sediment flux during the ice-free period revealed that the main transport within the bottom nepheloid layer in the submarine valleys is directed towards the inner shelf (Wegner et al., 2003) (Fig. 28.6). With respect to the sediment export from the eastern Laptev Sea shelf into the deep Arctic Ocean, Wegner et al. (accepted) assume that during the ice-free period most of the material derived from riverine input is trapped within a quasi-estuarine circulation system. This pattern of sediment transport might also explain the low average sedimentation rates that were observed on the outer shelf and the slope of the Laptev Sea (<10 cm/ky).

If the net sediment transport in the bottom nepheloid layer is directed towards the central and inner shelf and if we further take into account that large areas of the Laptev Sea are covered by relict sediments and lag deposits (Viscosi-Shirley et al., 2003b) with no present-day sediment deposition, we can draw the conclusion that the sediment input into the Laptev Sea is not balanced by sediment deposition on the shelf and long-range export through the water column to the Arctic basins. This discrepancy in sediment budgets may be explained by an eastward sediment transport to the East Siberian Sea that follows the Siberian Coastal Current (Viscosi-Shirley et al., 2003b) and a long-range transport of sediments by sea ice (Wegner et al., accepted).

Evidence has accumulated over the past two decades demonstrating that the entrainment of sediments into sea ice is a common phenomenon on the shallow Siberian shelves (Reimnitz, 1993; Eicken et al., 1997). It was shown that even under calm weather conditions, the freeze-up during October seems to be an important time period for sediment transport by sea ice from the shallow shelf areas of the Laptev Sea towards the Arctic Basin (Lindemann et al., 1999). Combining field measurements, remote sensing and numerical modelling, Eicken et al. (2000) could identify the shallow shelf near the New Siberian Islands as a key site for ice entrainment and a basin-wide dispersal of sediments by sea ice. They documented a total ice-bound sediment export of 18.510^6 t for one entrainment event in 1994/95. Another possible mechanism for the formation of sediment-laden sea ice is the resuspension of fine-grained bottom sediments in the polynya area and the subsequent entrainment of these sediments into the newly forming ice (Pfirman et al., 1990; Nürnberg et al., 1994; Rigor and Colony, 1997, Eidsvik, 2000). Recent studies have shown that the Laptev Sea is one of the major source areas for sea ice in the transpolar drift system (Rigor and Colony, 1997; Alexandrov, et al., 2000) and a center of sediment entrainment by ice (Pfirman et al., 1990; Darby, 2003). However, the general idea that suspension freezing during sea ice formation in the winter polynya is the dominant sediment entrainment process in the Laptev Sea is in conflict with the field observations of Dmitrenko et al. (2001) who have shown that even during winter the strong density stratification of the water column especially in the eastern Laptev Sea prevents convection to penetrate down to the seafloor. Thus resuspension of fine-grained bottom sediments accompanied by suspension freezing beneath the polynya is unlikely to occur in the eastern Laptev Sea. This supports the hypothesis that the fall freeze-up (October) might also be an important and yet underestimated period for the formation and export of sediment-laden sea ice.

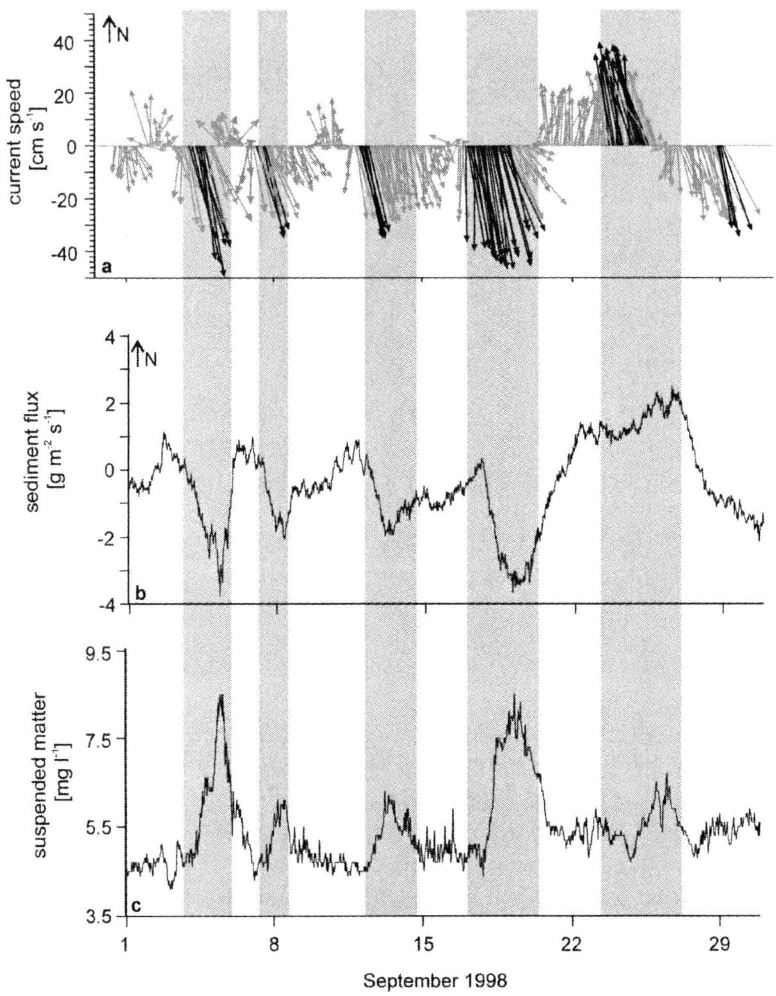

Figure 28.6 Bottom current speed (a), with the black vectors indicating current speeds exceeding the threshold velocity for incipient grain motion, sediment flux (b), and suspended matter concentration (c) at 4 m above bottom at a bottom-mooring station (75°09' N, 130°50' E) in the Eastern Lena Valley on the Laptev Sea shelf during September 1998. The grey bars indicate periods of wind-induced bottom currents.

The role of sea ice as a transport vehicle for the basin wide dispersal of sediments is also one of the unknown key characteristics of the Laptev Sea and East Siberian Sea budget of organic carbon. The average riverine input of total organic carbon (TOC) is about 5.1610^{6} t a^{-1} in the Laptev Sea and 1.7810^{6} t a^{-1} in the East Siberian Sea (Gordeev, 2000). The contribution of coastal erosion to the input of organic carbon is still unknown. Biomarker studies (Peulvé et al., 1996; Fahl and Stein, 1999, Bauch et al., 1999) and maceral analysis (Boucsein and Stein, 2000) of sediments and suspended matter from the Laptev Sea have shown that the particulate organic carbon is predominantly of terrigenous origin (Kattner et al., 1999). Although recent studies have accumulated new information about various sources and pathways of organic carbon in arctic shelf seas, we are still far away from establishing detailed budgets of the carbon input, cycling and deposition of allochthonous organic carbon.

The incorporation of sediments into newly formed ice is not only important for the transport of sediments and organic carbon. Arctic sea ice also plays a crucial role for the large-scale transport and cycling of trace elements (Rigor and Colony, 1997) and radionuclides (Meese et al., 1997). Within the framework of an interdisciplinary field study of freeze-up processes in the Laptev Sea, Hölemann et al. (1999b) observed that the concentrations of dissolved Mn, Fe, Zn, Cd, and Pb in newly formed sediment-laden ice were up to 40 times higher than the dissolved concentrations that were measured in sea water and freshwater in the region of ice formation. The elevated concentrations of dissolved trace metals in the newly formed ice were probably caused by a remobilization of trace elements from the ice-rafted sediment particles. This mechanism can play an important role for the dispersal of trace elements through the arctic environment. As an example, Winter et al. (1997) have shown that the primary source of rare earth elements and Pb for the dissolved reservoir in arctic seawater is not river water and that ice rafted debris, by dissolution or exchange processes, is an important source of trace elements for seawater in ice-covered oceans. This is supported by Measures (1999) who observed that Arctic Ocean surface waters with the highest reactive Al and reactive Fe values appear to coincide in many cases with the presence of high concentrations of ice-rafted sediments. The author presumed that the seasonal melting of ice containing rafted sediments adds particulate and dissolved trace metals to arctic surface waters.

However, the present-day concentrations of dissolved and particulate trace elements in the waters of the Laptev Sea are comparable to or even lower than in most major rivers of the world (Martin et al., 1993; Hölemann et al., 1995; Guieu et al., 1996; Gordeev, 2000; Hölemann et al., accepted). Also, the geochemical analysis of surficial and ice-bound sediments from the Laptev Sea showed no indication of an anthropogenic perturbation of the trace metal inventory (Nolting et al., 1996; Hölemann et al., 1999b). The fact that the Laptev Sea is still a pristine environment - if compared to low-latitude shelf seas - has also been confirmed by studies that show no signs of anthropogenic oil pollution in sediments (Zegouagh et al, 1998), only low levels of polychlorinated biphenyls and chlorinated pesticides (Utschakovski, 1998), and also low levels of dissolved and particulate anthropogenic radionuclides (Pavlov et al., 1999).

4.2. Geochemical and geological characteristics of surficial sediments as indicators for modern transport processes

Geochemical and mineralogical characteristics of surficial sediments indicate that the present-day sedimentary regime on the Laptev Sea shelf is dominated by the fluvial input of the rivers draining the Siberian platform (Hölemann et al, 1999b; Wahsner et al., 1999; Viscosi-Shirley et al., 2003a; Viscosy-Shirley et al., 2003b). The sediment input by the rivers Lena in the eastern and Khatanga in the western Laptev Sea especially control the mineralogical and geochemical compositions of surficial sediments. The main geochemical difference between both rivers is caused by the differing geology of the hinterland. The Khatanga River, like the Yenisei River, drains a plateau with basalts. Basalts have a different mineralogical and geochemical composition if compared to average crustal abundances (Rachold, 1999; Eisenhauer et al., 1999). The Lena River drains an area with different sedimentary, metamorphic, and magmatic units. Thus the geochemical composition of sediments from the Lena River resembles a more or less average crustal signal.

Interdisciplinary studies using the difference between both river systems as an indicator for transport pathways pointed out that the suspended matter discharge from the Lena River is mainly deposited on the southeastern Laptev Sea shelf (Silverberg, 1972; Wahsner et al., 1999; Lisitzyn, 1996; Hölemann et al., 1999a; Peregovich et al., 1999; Fahl et al., 2001; Viscosi-Shirley et al., 2003a). This conclusion is also supported by the relative abundance pattern of freshwater diatoms (Cremer, 1999) and chlorococcalean algae (Kunz-Pirrung, 2001) in surficial sediments, which also point to the conclusion that the deposition of riverine material is highest in the southeastern Laptev Sea near the delta of the Lena River. On the other hand, as a result of the strong density stratification of the water column of the Laptev Sea the riverine suspended matter is dispersed above the pycnocline over large areas of the eastern Laptev Sea shelf, especially during the high discharge period in June (Létoille et al., 1993; Burenkov et al., 1997; Pivovarov et al., 1999; Wegner et al., accepted).

4.3. The importance of shelf processes for the modification of dissolved organic matter and trace metals

River runoff widely determines the signatures of dissolved trace elements and dissolved organic matter (DOM) in the Laptev Sea. But only few studies have focused on the biogeochemical and sedimentological processes in the mixing zone of the Lena River waters and the Laptev Sea waters (Guieu et al., 1996; Garnier et al., 1996; Cauwet and Sidorov, 1996; Kattner et al., 1999; Levasseur et al., 2000; Hölemann et al., accepted). Cauwet and Sidorov (1996), Kattner et al. (1999) and Dittmar and Kattner (2003) pointed out that the distribution of dissolved organic carbon (DOC) in the Laptev Sea is largely controlled by conservative mixing between freshwater and marine waters. A conservative mixing was also observed for dissolved Zn (Guieu et al., 1996). All other elements showed a decrease (Fe) or a strong increase (Cu, Ni, Cd) in dissolved concentration within the mixing zone. Guieu et al. (1996) stated that this behaviour is mainly controlled by particle-dissolved phase interaction. Also, the behaviour of dissolved osmium, which rapidly decreases in the mixing zone of the Lena River, was explained by the fact that the osmium becomes adsorbed to suspended particles (Levasseur et al., 2000).

A general drawback of the studies on the mixing process in arctic seas is that the river discharge is highly variable during the year and that most of the observations were carried out during the ice-free period in the late summer, nearly two months after the high-discharge period in June. Thus, more detailed studies on the mixing behaviour of various chemical constituents should be carried out during the high-discharge period. One of the first studies on discharge and transport of trace elements during the high-discharge period was carried out by Hölemann et al. (accepted). The authors showed that in the Lena River during the spring high flow (freshet) dissolved concentrations of metals were significantly higher than those reported during the rest of the year (Fig. 28.7). The dissolved metals were transported within the Laptev Sea in an under-ice fresh water layer spreading seaward and resulting in a long-range transport without mixing processes.

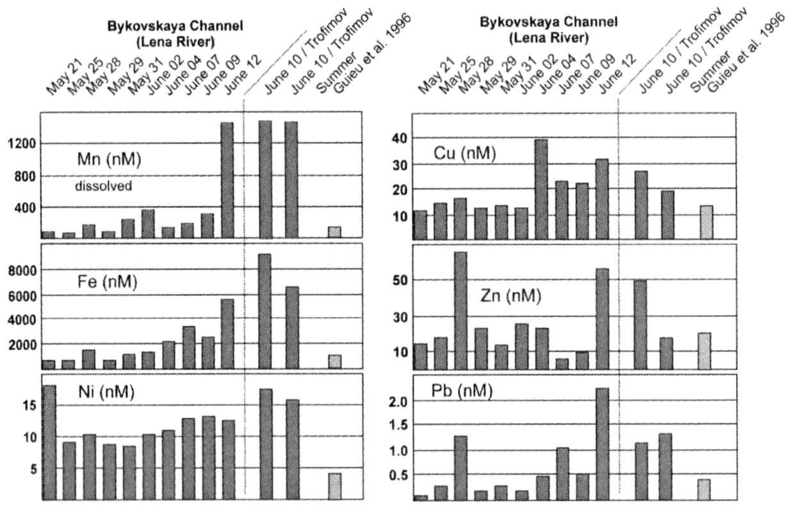

Figure 28.7 Dissolved concentrations of Mn, Fe, Ni, Cu, Zn, and Pb in the Lena River water (Bykovskaya and Trofimovskaya channels in the Lena Delta) before (21 and 25 May) and during the initial phase of the spring freshet (after May 25). The light grey bars indicate summer concentrations published by Guieu et al. (1996).

5. Recent studies of the Laptev and East-Siberian seas ecosystems

5.1. Distribution and development of the microalgal community in the Laptev Sea kryo-pelagic system during autumnal freeze-up

During the autumnal freeze-up period, sea ice formation in the shallow Laptev Sea results in the development of a coastal belt of fast ice and a zone of pack ice, separated from each other by a polynya. Several mechanisms such as wave pumping or scavenging by ice crystals are considered to be responsible for the incorporation of suspended matter and organisms into the newly forming ice sheet.

During the TRANSDRIFT III expedition in October 1995 (Kassens et al., 1997; Juterzenka and Knickmeier, 1999; Tuschling, 2000), samples were taken from the water column as well as various types of newly forming sea ice and analyzed for Chlorophyll *a* as an indicator of algal biomass as well as abundance and species composition of microalgae. *In situ* fluorescence measurements were carried out along two transects from the coast east of the Lena Delta towards the central Laptev Sea east of Kotelnyy. At the beginning of the sampling period, the Laptev Sea was ice-free and subsequently freezing up, to be almost entirely ice-covered at the end of October. Ice samples analyzed for Chlorophyll *a* included grease ice, nilas and young ice (≤ 25 cm). Biomass, as expressed by integrated pigment concentrations, varied considerably in the water column as well as in new and young ice samples, although the Chl *a* content increased as soon as ice floes were formed. At some stations, *in situ* fluorescence showed near-bottom maxima, which may be related to resuspension processes within the bottom nepheloid layer.

Phytoplankton community studies revealed distinct differences between the three investigated seasons. Phytoplankton blooms could already be observed at the northern stations during spring, whereas Chl *a* concentrations at the southern stations were low. Most of the phytoplankton species were euryhaline forms with an arctic-boreal distribution. In spring and summer diatoms prevailed in the area. The percentage of dinoflagelates grew higher during the seasons and peaked in autumn. A distinct correlation between the biomass of the smaller phytoplankton species, Chl *a* contents and primary production rate could be proven by means of a factor analyses. The influence of the river water is evident for species composition and distribution as well as biomass and production rates.

5.2. Zooplankton communities in the Laptev Sea

The distribution and population structure of zooplankton (abundance, biomass, ontogenetic stages) in the Laptev Sea were studied during the Russian-German TRANSDRIFT I expedition in August/September 1993 and during TRANSDRIFT III in October 1995 (Kassens and Karpiy, 1994; Kassens et al., 1997). In late summer 1993 and autumn 1995 a total of 42 zooplankton species/taxa were identified. Copepods contributed 17 species (calanoids, cyclopoids, poecilostomatoids, harpacticoids). Other crustacean taxa were amphipods, decapods, ostracods, mysids and cumaceans. Additional taxa included hydrozoans, ctenophores, appendicularians as well as juvenile stages of polychaetes and echinoderms (Kosobokova et al., 1998). Both expeditions showed maximum abundances in the northeast of the Lena Delta. This zooplankton was dominated by *Drepanopus bungei, Acartia* sp. and *Limnocalanus grimaldii. D. bungei* was also the most important biomass species in this area. Other areas of the investigated region exhibited a high local variability in abundance and biomass. Multivariate analyses separated three main clusters for the zooplankton in late summer 1993 (provinces *Southeast, Northeast* and *Northwest*), while two major clusters were distinguished for autumn 1995 (provinces *East* and *West*). In conclusion, the results demonstrate the strong impact of abiotic factors on the zooplankton communities in this unique ecosystem.

5.3. Distribution and ecology of selected Zooplankton species

The distribution, abundance and composition of smaller zooplankton species has been investigated and discussed within the context of main abiotic factors. Samples have been taken in August/September 1999. Overall abundances have been high in the eastern Laptev Sea but exhibited variations of 815 individuals m^{-3} and 9,177 m^{-3} near the Lena River mouth. Biomasses were highest in the south western part of the Laptev Sea (25.3 mg dry mass m^{-3}) and lowest in the east (1.6 mg dry mass m^{-3}). This discrepancy in high abundances but low biomasses in the east is due to smaller neritic species as *Drepanopus bungei, Pseudocalanus major* and *Acartia longiremi,* whereas in the west large copepods of the genus *Calanus* prevail. It is noticeable that regionally the Laptev Sea can be highly productive. The combination and analyses of abundance, biomass and distribution of the zooplankton with hydrographical parameters such as salinity and temperature in addition to Chlorophyll *a* revealed a governing influence of the fluvial Lena waters on the distribution of the zooplankton.

Biochemical analyses helped to investigate the feeding ecology of selected zooplankton species. It is known that many organisms develop adaptations to the relatively long period of short food supply. Copepods are capable of building and storing high concentrations of wax esters or other lipids during summer when the nutrient supply is high. Copepods revealed high concentrations of wax esters, other zooplankton like amphipods, mysids or chaetognaths stored energy as triaclglycerines. Total lipid contents as well as the high concentration of wax esters pointed to a considerable accumulation of lipids already in August/September. Marker lipids revealed a diatom-dominated diet of the investigated zooplankton species. While the Calanus species seem to be mostly herbivorous, other copepods seem to have an opportunistic lifestyle.

5.4. *Benthos*

As a result of six joint expeditions to the Laptev Sea in 1993, 1994 and 1995, 150 invertebrate species were added to the known list of species from this high-Arctic region, comprising now 1235 species in total (Petryashov et al., 1999; Piepenburg and Schmid, 1997). Most of these species (987) are macrozoobenthic. The geographic distribution of macrobenthic species numbers and biomass exhibited a pronounced increase to the north up to 77 °N. In terms of biogeographic composition, Boreal-Arctic species accounted for most of the species, primarily widespread Boreal-Arctic species in the southern and central Laptev Sea at depths from 10 to 75 m and primarily Atlantic Boreal-Arctic species in the northern Laptev Sea at depths of 90 to 1100 m. Only in depths of more than 2000 m, as well as in estuarine and brackish coastal waters, Arctic forms had significant species shares. On sandy and silty bottoms, nine major communities and two trophic zones were distinguished. The distribution of species numbers, biomass and faunal composition is discussed in relation to water depth, seabed sedimentology and near-bottom water salinities. The latter are supposed to be the major environmental factors regulating the distribution and structure of macrobenthic communities. An ecological-biogeographical zonation scheme is proposed, combining the current information about the distribution of macrobenthic communities and brackish water masses (Sirenko, 1999).

The depth zonation of the communities is presumably governed by depth correlated environmental factors. The faunistic composition in the shallower part of the Laptev Sea is determined mainly by physical disturbances such as freshwater and sediment input or by direct ice scouring from anchor ice or small icebergs. Low species numbers and diversity indicate the existence of physical disturbances. Communities of the intermediate and deeper parts seem to be influenced mainly by sediment structure and hydrographical conditions. The more estuarine conditions in the south with low salinity and high sediment loaded waters supported mostly detritovorous bivalves. It is assumed that high sedimentation rates affect the existence of suspension-feeding organisms. Highest diversities were found in areas where two water masses meet.

The epibenthic megafauna of the high-Arctic Laptev Sea shelf was investigated in August/September 1993 and October 1995 (Kassens et al., 1997; Piepenburg and Schmid, 1997). At 13 stations in water depths between 14 and 45 m, a series of 5 to 29 photographs, each depicting about 1 m^2 of the seabed, was taken to assess epifaunal distribution patterns and abundances. Furthermore, the population biomass of dominant brittle stars was estimated by combining abundance values with size-mass relationships and size frequencies established by measuring specimens on scaled photographs. A total of 13 epibenthic species were identified. Species numbers per station were low, ranging between one and six. Total epibenthic abundances, averaging 173.7 ind m^{-2}, ranged considerably between 0.1 and 579.5 ind m^{-2}. Except for some stations on shallow shelf banks < 20 m that were characterized by bottom-water salinities < 30 due to fluvial dilution, the brittle star *Ophiocten sericeum* dominated the megabenthic shelf assemblages. At the flanks of sunken Pleistocene river valleys in depths > 30 m, they reached maximum density and biomass values of 566 ind m^{-2} and 1.5 g ash-free dry mass m^{-2}, respectively. At some sites, the brittle star *Ophiura sarsi* occurred in abundances of up to 35 ind m^{-2} and attained a biomass of 3.8 g AFDM m^{-2}. Of local importance were the sea cucumber *Myriotrochus rinckii* (up to 70 ind m^{-2}) and the bivalve *Arctinula greenlandica* (up to 33 ind m^{-2}). All other species were recorded with distinctly lower densities (≤ 1 ind m^{-2}). Gross estimates of population respiration and production of dominant brittle stars suggest that their organic carbon demand may amount to a pooled average of about 4 mg C m^{-2} d^{-1} in the Laptev Sea, locally even to a maximum of > 10 mg C m^{-2} d^{-1}. This finding indicates that a substantial portion of the energy flow in this high-Arctic shelf ecosystem may be channelled through dense brittle star assemblages (Piepenburg, 2000).

Due to its remoteness and 8-month ice coverage, knowledge has been limited on ecology, turnover processes, coupling of marine communities in the Laptev Sea. Taxonomic and zoogeographical questions have been addressed and (several hundred benthic and zooplankton species could be identified) abundance and biomass composition wherever possible gave valuable insights about community structure and distribution patterns. Community structure analysis revealed three distinct faunal provinces in the western, northwestern and the Lena Delta region. Those provinces showed in the phytoplankton and zooplankton communities as well as in the benthos giving a first clue about a tight benthic-pelagic coupling (see also above chapters). In a second step, process-oriented studies have been carried out to investigate biological carbon sources as phytoplankton production, its transformation in the pelagial, and degradation in the benthos. High chlorophyll *a* contents in the sediment hint to rapid sinking of phytoplankton blooms to the benthos making high quality food available for benthic organisms as well. Already early in the year phytoplankton production is possible in the polynya region which is dividing the fast and sea ice from east to west app. 200 km north of the Lena Delta. Hence this year round ice-free water is very important for all biological processes.

Bibliography

Alexandrov, V.Y., T. Martin, J. Kolatschek, H. Eicken and M. Kreyscher, 2000. Sea ice circulation in the Laptev Sea and ice export to the Arctic Ocean: Results from satellite remote sensing and numerical modelling. *J. Geophys. Res.*, **105(C5),** 17143–17159.

Are, F., 1999. The role of coastal retreat for sedimentation in the Laptev Sea. In *Land-Ocean Systems in the Siberian Arctic: Dynamics and History,* Kassens, H., H.A. Bauch, I.A. Dmitrenko, H. Eicken, H.-W. Hubberten, M. Melles, J. Thiede and L.A. Timokhov, eds. Springer Verlag, Berlin, pp. 287–295.

Are, F., M.N. Grigoriev, H.-W. Hubberten, V. Rachold, S. Razumov and W. Schneider, 2002. Comparative shoreface evolution along the Laptev Sea coast. *Berichte zur Polarforschung,* **70,** 135–150.

Atlas of the Arctic, 1985. Izdatelstvo GUNIO, Leningrad, Russia, 204 p.

Bauch, H.A., H. Kassens, H. Erlenkeuser, P.M. Grootes and J. Thiede, 1999. Depositional environment of the Laptev Sea (Arctic Siberia) during the Holocene. *Boreas,* **28** (1), 194–204.

Bauch, H.A., T. Müller-Lupp, R.F. Spielhagen, E. Taldenkova, H. Kassens, P.M. Grootes, J. Thiede, J. Heinemeier and V.V. Petryashov, 2001a. Chronology of the Holocene transgression at the northern Siberian margin. *Global and Planetary Change,* **31** (1–4), 125–139.

Bauch, H.A., H. Kassens, M. Kunz-Pirrung, O.D. Naidina and J. Thiede. 2001b. Composition and flux of Holocene sediments on the eastern Laptev Sea shelf, Arctic Siberia. *Quaternary Research,* **55** (3), 344–351.

Boucsein, B. and R. Stein, 2000. Particulate organic matter in surface sediments of the Laptev Sea (Arctic Ocean): application of maceral analysis as organic-carbon-source indicator. *Marine Geology,* **162,** 573-586.

Burenkov, V.I., V.M. Kuptzov, V.V. Sivkov and V.P. Shevchenko, 1997. Spatial distribution and size composition of suspended matter in the Laptev Sea in August-September 1991. *Oceanology,* **37** (6), 831–837.

Cauwet, G. and I. Sidorov, 1996. The biogeochemistry of Lena River: organic carbon and nutrients distribution. *Marine Chemistry,* **53,** 211–227.

Colony, R., E. Nikiforov, S. Pivovarov, O. Pokrovsky, S. Priamikov, L. Timokhov, 2002. Data Base and Atlas of Hydrochemistry for the Arctic Ocean. Ocean Sciences Meeting, 11–15 February 2002 Honolulu, Hawaii, USA.

Cremer, H., 1999. Distribution patterns of diatom surface sediment assemblages in the Laptev Sea (Arctic Ocean). *Marine Micropaleontology,* **38,** 39–67.

Darby, D.A., 2003. Sources of sediment found in sea ice from the western Arctic Ocean, new insights into processes of entrainment and drift patterns. *Journal of Geophysical Research,* **108(C8),** 3257, doi:10.1029/2002JC001350.

Dethleff, D., D. Nurnberg, E. Reimnitz, M. Saarso and Y.P. Savchenko, 1993. East Siberian Arctic Region Expedition '92: The Laptev Sea - its significance for Arctic sea-ice formation and transpolar sediment flux. *Berichte zur Polarforschung,* **120,** 1–44.

Dittmar, T. and G. Kattner, 2003. The biogeochemistry of the river and shelf ecosystem of the Arctic Ocean: a review. *Marine Chemistry,* **83 (3–4),** 103–120.

Dmitrenko, I., J.A. Hölemann, S. Kirillov, C. Wegner, V.A. Gribanov, S.L. Berezovskaya and H. Kassens, 2001. Thermal regime of the Laptev Sea bottom layer and affecting processes. *Earth Cryosphere,* **3,** 40–55. (in Russian with English abstract)

Dobrovolsky, A.D., 1961. The Definition of Water Masses. *Oceanology,* **1**(1), 12–24. (in Russian)

Eicken, H., E. Reimnitz, V. Alexandrov, T. Martin, H. Kassens and T. Viehoff, 1997. Sea ice processes in the Laptev Sea and their importance for sediment export. *Continental Shelf Research,* **17,** 205–233.

Eicken, H., J. Kolatschek, J. Freitag, F. Lindemann, H. Kassens and I. Dmitrenko, 2000. A key source area and constraints on entrainment for basin-scale sediment transport by Arctic sea ice. *Geophys. Res. Lett.,* **27** (13), 1919–1922.

Eidsvik, K.J., 2000. Coagulation of suspended sediments and ice crystals below leads. *Cold Regions Science and Technology,* **31,** 119–131.

Eisenhauer, A., H. Meyer, V. Rachold, T. Tütken, B. Wiegand, B.T. Hansen, R.F. Spielhagen, F. Lindemann and H. Kassens, 1999. Grain size separation and sediment mixing in Arctic Ocean sediments: evidence from the strontium isotope systematic. *Chemical Geology,* **158,** 173–188.

Fahl, K. and R. Stein, 1999. Biomarkers as organic-carbon-source and environmental indicators in the Late Quaternary Arctic Ocean: problems and perspectives. *Marine Chemistry,* **63,** 293–309.

Fahl, K., H. Cremer, H. Erlenkeuser, H. Hanssen, J. Hölemann, H. Kassens, K. Knickmeier, K. Kosobokova, M. Kunz-Pirrung, F. Lindemann, E. Markhaseva, S. Lischka, V. Petryashov, D. Piepenburg, M. Schmid, M. Spindler, R. Stein and K. Tuschling, 2001. Sources and pathways of organic carbon in the modern Laptev Sea (Arctic Ocean): implications from biological, geochemical and geological data. *Berichte zur Polarforschung,* **69,** 193–205.

Fütterer, D.K., 1994. Die Expedition ARCTIC '93. Der Fahrtabschnit ARK-IX/4 mit FS Polarstern 1993. *Berichte zur Polarforschung,* **149,** 1- 244.

Garnier, J.-M., J.-M. Martin, J.-M. Mouchel and K. Sioud, 1996. Partitioning of trace metals between the dissolved and particulate phases and particulate surface reactivity in the Lena River estuary and the Laptev Sea (Russia). *Marine Chemistry,* **53,** 269–283.

Gordeev, V.V., 2000. River input of water, sediment, major ions, nutrients and trace metals from Russian territory to the Arctic Ocean. In *The freshwater budget of the Arctic Ocean,* Lewis, E.L., E.P. Jones, P. Lemke, T.D. Prowse and P. Wadhams, eds. Kluwer Academic Publisher, Nato Science Series, **70,** pp. 297–322.

Guieu, C., W.W. Huang, J.-M. Martin and Y.Y. Yong, 1996. Outflow of trace metals into the Laptev Sea by the Lena River. *Marine Chemistry,* **53,** 255–267.

Gukov, A., 1999. *Ecosystem of the Siberian Polynya.* Scientific World, Moscow, Russia. (in Russian)

Hölemann, J.A., M. Schirmacher and A. Prange, 1995. Transport and distribution of trace elements in the Laptev Sea: First results of the Transdrift expeditions. *Reports on Polar Research,* **176,** 297–302.

Hölemann, J.A., M. Schirmacher, H. Kassens and A. Prange, 1999a. Geochemistry of surficial and ice-rafted sediments from the Laptev Sea (Siberia). *Estuarine, Coastal and Shelf Science,* **49,** 45–59.

Hölemann, J.A., M. Schirmacher and A. Prange, 1999b. Dissolved and particulate trace elements in newly formed ice from the Laptev Sea (Transdrift III, October 1995). In *Land-ocean systems in the Siberian Arctic: Dynamics and history,* Kassens, H., H.A. Bauch, I.A. Dmitrenko, H. Eicken, H.-W. Hubberten, M. Melles, J. Thiede and L.A. Timokhov, eds. Springer Verlag, Berlin, pp. 101–112.

Hölemann, J.A., M. Schirmacher and A. Prange (accepted). Outflow of particulate and dissolved trace elements to the Laptev Sea during the spring high flow of the Lena River (Arctic Siberia). *Estuarine, Coastal and Shelf Science.*

Ivanov, V.V. and A.A. Piskun, 1999. Distribution of river water and suspended sediment loads in the deltas of rivers in the basins of the Laptev and East-Siberian seas. In *Land-ocean systems in the Siberian Arctic: Dynamics and history,* Kassens, H., H.A. Bauch, I.A. Dmitrenko, H. Eicken, H.-W. Hubberten, M. Melles, J. Thiede and L.A. Timokhov, eds. Springer Verlag, Berlin, pp. 239–250.

Jakobsson, M. et al., 2000. International Bathymetric Chart of the Arctic Ocean (IBCAO), 2000. http://www.ngdc.noaa.gov/mgg/bathymetry/arctic/arctic. html.

Jones, E. P., L. G. Anderson and D. W. R. Wallace, 1991. Tracers of near surface, halocline and deep waters in the Arctic Ocean: implications for circulation. *Journal of Marine Systems,* **2,** 241–255.

Jones, E. P., L. G. Anderson and J. H. Swift, 1998. Distribution of Atlantic and Pacific waters in the upper Arctic Ocean: implications for circulation. *Geophys. Res. Lett.,* **25,** 765–768.

Juterzenka, K. V. and K. Knickmeier, 1999. Chlorophyll a distribution in water column and sea ice during the Laptev Sea Freeze-up Study in autumn 1955. In *Land-ocean systems in the Siberian Arctic: Dynamics and history,* Kassens, H., H.A. Bauch, I.A. Dmitrenko, H. Eicken, H.-W. Hubberten, M. Melles, J. Thiede and L.A. Timokhov, eds. Springer Verlag, Berlin, pp. 153–160.

Kassens, H. and V.Y. Karpiy, 1994. Russian-German cooperation: The TRANSDRIFT I expedition to the Laptev Sea. *Berichte zur Polarforschung,* **151,** 168 p.

Kassens, H., D. Piepenburg, J. Thiede, L.A. Timokhov, H.-W. Hubberten and S. Priamikov, 1995. German Russian Cooperation: Laptev Sea System. *Berichte zur Polarforschung,* **176,** 387 p.

Kassens, H., I. Dmitrenko, L. Timokhov and J. Thiede, 1997. The TRANSDRIFT III Expedition: freeze-up studies in the Laptev Sea. *Berichte zur Polarforschung,* **248,** 1–192.

Kassens, H., I. Dmitrenko, V. Rachold, J. Thiede and L. Timokhov, 1998. Russian and German scientists explore the Arctic's Laptev Sea and its climate system. *EOS,* American Geophysical Union, **79** (27), pp. 317, 322–323.

Kassens, H., H.A. Bauch, I.A. Dmitrenko, H. Eicken, H.-W. Hubberten, M. Melles, J. Thiede and L.A. Timokhov, eds., 1999. *Land-Ocean Systems in the Siberian Arctic: Dynamics and History.* Springer Verlag, Berlin.

Kattner, G., J.M. Lobbes, H.P. Fitznar, R. Engbrodt, E.-M. Nöthig and R.J. Lara, 1999. Tracing dissolved organic substances and nutrients from the Lena River through Laptev Sea (Arctic). *Marine Chemistry,* **65,** 25–39.

Klenova, M.V., 1962. *Sediments of the Siberian Arctic basin according to material of the drift of the Sedov.* Izdatelstvo AN SSSR, Moscow, Russia. (in Russian)

Kosobokova, K. N., H. Hanssen, H.-J. Hirche and K. Knickmeier, 1998. Composition and distribution of zooplankton in the Laptev Sea and adjacent Nansen basin during summer, 1993. *Polar Biol.,* **19,** 63–76.

Kunz-Pirrung, M., 2001. Dinoflagellate cyst assemblages in surface sediments of the Laptev Sea region (Arctic Ocean) and their relationship to hydrographic conditions. *Journal of Quaternary Science,* **16** (7), 637-649.

Levasseur, S., V. Rachold, J.-L. Birck and C.J. Allègre, 2000. Osmium behavior in estuaries: the Lena River example. *Earth and Planetary Science Letters,* **177,** 227–235.

Létoille, R., J.M. Martin, A.J. Thomas, V.V. Gordeev, S. Gusarova and I.S. Sidorov, 1993. 18-O abundance and dissolved silicate in the Lena delta and Laptev Sea (Russia). *Marine Chemistry,* **43,** 47–64.

Lindemann, F., J.A. Hölemann, A. Korablev and A. Zachek, 1999. Particle entrainment in newly forming sea ice - Freeze-up studies in October 1995. In *Land-ocean systems in the Siberian Arctic: Dynamics and history,* Kassens, H., H.A. Bauch, I.A. Dmitrenko, H. Eicken, H.-W. Hubberten, M. Melles, J. Thiede and L.A. Timokhov, eds. Springer Verlag, Berlin, pp. 113–124.

Lisitzyn, A.P., 1996. *Oceanic sedimentation: lithology and geochemistry.* American geophysical Union, Washington, D.C.

Martin, J.M., D.M. Guan, F. Elbaz-Poulichet, A.J. Thomas and V.V. Gordeev, 1993. Preliminary assessment of the distributions of some trace elements (As, Cd, Cu, Fe, Ni, Pb, and Zn) in a pristine aquatic environment: the Lena River estuary (Russia). *Marine Chemistry,* **43,** 185–199.

Measures, C.I., 1999. The role of entrained sediments in sea ice in the distribution of aluminium and iron in the surface waters of the Arctic Ocean. *Marine Chemistry,* **68,** 59–70.

Meese, D.A., E. Reimnitz, W.B. Tucker III, A.J. Gow, J. Bischof and D.A. Darby, 1997. Evidence for radionuclide transport by sea ice. *The Science of the Total Environment,* **202,** 267–278.

Nolting, R.F., M. van Dalen and W. Helder, 1996. Distribution of trace and major elements in sediment and pore waters of the Lena Delta and Laptev Sea. *Marine Chemistry,* **53,** 285–299.

Nürnberg D., I. Wollenburg, D. Dethleff, H. Eicken, H. Kassens, T. Letzig, E. Reimnitz and J. Thiede, 1994. Sediments in Arctic sea ice: Implications for entrainment transport and release. *Marine Geology,* **119,** 185–214.

Pavlov, V.K., 1998. Features of the structure and variability of the oceanographic processes in the shelf zone of the Laptev and East-Siberian seas. In *The Sea. Volume 11. The global coastal ocean. Regional Studies and Syntheses.* Robinson, A.R. and Brink K.H. eds. John Wiley & Sons, Inc., USA, pp. 759–787.

Pavlov, V.K., V.V. Stanovoy and A.I. Nikitin, 1999. Possible causes of radioactive contamination in the Laptev Sea. In *Land-ocean systems in the Siberian Arctic: Dynamics and history,* Kassens, H., H.A. Bauch, I.A. Dmitrenko, H. Eicken, H.-W. Hubberten, M. Melles, J. Thiede and L.A. Timokhov, eds. Springer Verlag, Berlin, pp. 65–72.

Peregovich, B., E. Hoops and V. Rachold, 1999. Sediment transport to the Laptev Sea (Siberian Arctic) during the Holocene - evidence from the heavy mineral composition of fluvial and marine sediments. *Boreas,* **28** (1), 205–214.

Petryashov, V.V., B.I. Sirenko, A.A. Golikov, A.V. Novozhilov, E. Rachor, D. Piepenburg and M. K. Schmid, 1999. Macrobenthos distribution in the Laptev Sea in relation to hydrology. In *Land-ocean systems in the Siberian Arctic: Dynamics and history,* Kassens, H., H.A. Bauch, I.A. Dmitrenko, H. Eicken, H.-W. Hubberten, M. Melles, J. Thiede and L.A. Timokhov, eds. Springer Verlag, Berlin, pp. 169–180.

Peulvé, S., M.-A. Sicre, A. Saliot, J.W. De Leeuw and M. Baas, 1996. Molecular characterization of suspended and sedimentary organic matter in an Arctic delta. *Limnology and Oceanography,* **41,** 488–497.

Pfirman, S., M.A. Lange, I. Wollenburg and P. Schlosser, 1990. Sea ice characteristics and the role of sediment inclusions in deep-sea deposition: Arctic – Antarctic comparison. In *Geological History of the Polar Oceans: Arctic versus Antarctic,* Bleil, U. and J. Thiede, eds. NATO ASI Ser., **C308,** pp. 187–211.

Piepenburg, D., 2000. Arctic brittle stars (Echinodermata: Ophiuroidea). *Oceanogr. Mar. Biol. Annual Rev.,* **38,** 189–256.

Piepenburg, D., and M.K. Schmid, 1997. A photographic survey of the epibenthic megafauna of the Arctic Laptev Sea shelf: distribution, abundance, and estimates of biomass and organic carbon demand. *Mar. Ecol. Prog. Ser.,* **147,** 63–75.

Pivovarov, S.V., J.A. Hölemann, H. Kassens, M. Antonow and I. Dmitrenko, 1999. Dissolved oxygen, silicon, phosphorous and suspended matter concentrations during the spring breakup of the Lena River. In *Land-ocean systems in the Siberian Arctic: Dynamics and history,* Kassens, H., H.A. Bauch, I.A. Dmitrenko, H. Eicken, H.-W. Hubberten, M. Melles, J. Thiede and L.A. Timokhov, eds. Springer Verlag, Berlin, pp. 251–264.

Proshutinsky, A.Y. and M. Johnson, 2001. Two regimes of the arctic's circulation from ocean models with ice and contaminants. *Marine Pollution Bulletin,* **43** (1–6), 61–70.

Rachold, V., 1999. Major, trace and rare earth element geochemistry of suspended particulate material of East Siberian rivers draining to the Arctic Ocean. In *Land-ocean systems in the Siberian Arctic: Dynamics and history,* Kassens, H., H.A. Bauch, I.A. Dmitrenko, H. Eicken, H.-W. Hubberten, M. Melles, J. Thiede and L.A. Timokhov, eds. Springer Verlag, Berlin, pp. 199–222.

Rachold, V., M.N. Grigoriev, F.E. Are, S. Solomon, E. Reimnitz, H. Kassens and M. Antonow, 2000. Coastal erosion vs. riverine sediment discharge in the arctic shelf seas. *International Journal of Earth Sciences (Geolische Rundschau), 89,* 450–460.

Rachor, E. (ed.), 1997. Scientific cruise report of the Arctic Expedition ARK-XI/1 of RV Polarstern in 1995. *Berichte zur Polarforschung,* **226,** 1–157.

Reimnitz, R., 1994. The Laptev Sea shelf ice regime from a western perspective. *Berichte zur Polarforschung,* **144,** 45–47.

Reimnitz, E., M. McCormick, K. McDougall and E. Brouwers, 1993. Sediment export by ice rafting from a coastal polynya, Arctic Alaska, U.S.A. *Arctic and Alpine Research,* **25,** 83–98.

Rigor, I. and R. Colony, 1997. Sea-ice production and transfer of pollutants in the Laptev Sea, 1979–1993. *The Science of the Total Environment,* **202,** 89–110.

Rusanov, V. P., N. I. Yakovlev and A. G. Buinevich, 1979. The hydrochemical regime of the Arctic Ocean. *Proceedings of AARI,* **355,** 1–144. (in Russian)

Sidorov, I.S. and A.Y. Gukov, 1992. The influence of the oxygen regime on conditions for zoobenthos in coastal regions of the Laptev Sea. *Oceanology,* **32**(5), 902–904. (in Russian)

Silverberg, N., 1972. *Sedimentology of the surface sediments of the East Siberian and Laptev Sea.* University of Washington, USA, 185 p.

Sirenko, B.I., 1999. Arctic marine fauna (from the expeditions of the Zoological Institute of the Russian Academy of Sciences). *Marine Biology,* **24**(6), 341–350. (in Russian)

Schlitzer, R., 2002. Ocean Data View. http://www.awi-bremerhaven.de/geo/odv

Stein, R., B. Boucsein, K. Fahl, T. Garcia de Oteyza, J. Knies and F. Niessen, 2001. Accumulation of particulate organic carbon at the Eurasian continental margin during late quaternary times: controlling mechanisms and paleoenvironmental significance. *Global and Planetary Change,* **31,** 87–104.

Thiede, J., 1996. Arctic paleoceanography – quo vadis? *Berichte zur Polarforschung,* **212,** 19–35.

Thiede, J., L. Timokhov, H. A. Bauch, D. Bolshiyanov, I. Dmitrenko, H. Eicken, K. Fahl, A. Gukov, J. Hölemann, H. W. Hubberten, K. v. Juterzenka, H. Kassens, M. Melles, V. Petryashov, S. Pivovarov, S. Priamikov, V. Rachold, M. Schmid, C. Siegert, M. Spindler, R. Stein and Scientific Party, 1999. Dynamics and history of the Laptev Sea and its continental hinterland: A summary. In *Land-ocean systems in the Siberian Arctic: Dynamics and history,* Kassens, H., H.A. Bauch, I.A. Dmitrenko, H. Eicken, H.-W. Hubberten, M. Melles, J. Thiede and L.A. Timokhov, eds. Springer Verlag, Berlin, pp. 695–711.

Timokhov, L.A., 1994. Regional characteristics of the Laptev and East Siberian seas: climate, topography, ice phases, thermohaline regime, circulation. *Berichte zur Polarforschung,* **144,** 15–31.

Tomczak, M., 1999. Some historical, theoretical and applied aspects of quantitative water mass analysis. *Journal of Marine Research,* 57, 275–303.

Tuschling, K., 2000. Phytoplankton ecology in the arctic Laptev Sea. *Berichte zur Polarforschung,* 347, 1–144.

Utschakowski, S., 1998. Anthropogenic organic trace compounds in the Arctic Ocean. *Berichte zur Polarforschung,* **292,** 141 p. (in German with English abstract)

Viscosi-Shirley, C., K. Mammone, N. Pisias and J. Dymond, 2003a. Clay mineralogy and multy-element chemistry of surface sediments on the Siberian-Arctic shelf: implications for sediment provenance and grain size sorting. *Continental Shelf Research,* **23,** 1175–1200.

Viscosi-Shirley, C., N. Pisias and K. Mammone, 2003b. Sediment source strength, transport pathways and accumulation patterns on the Siberian-Arctic's Chukchi and Laptev shelves. *Continental Shelf Research,* **23,** 1201–1225.

Wahsner, M., C. Müller, R. Stein, G.V. Ivanov, M.A. Levitan, E. Shelekova and G. Tarasov, 1999. Clay-mineral distribution in surface sediments of the Eurasian Arctic Ocean and continental margin as indicator for source areas and transport pathways - a synthesis. *Boreas,* **28** (1), 215–233.

Wegner, C., J.A. Hölemann, S. Kirillov, K. Tuschling, E. Abramova and H. Kassens, 2003. Suspended particulate matter on the Laptev Sea shelf (Siberian Arctic) during ice-free conditions. *Estuarine, Coastal and Shelf Science,* **57,** 55–65.

Wegner, C., J.A. Hölemann, I. Dmitrenko, S. Kirillov and H. Kassens (accepted). Seasonal variations in Arctic sediment dynamics - evidence from one-year records in the Laptev Sea (Siberian Arctic). Global and Planetary Change.

Winter B.L., C.M. Johnson and D.L. Clark, 1997. Strontium, neodymium, and lead isotope variations of authigenic and silicate sediment components from the Late Cenozoic Arctic Ocean: Implications for sediment provenance and the source of trace metals in seawater. *Geochimica et Cosmochimica Acta,* **61,** 4181–4200.

Zakharov, V.F., 1966. The role of flaw leads off the edge of the fast ice in the hydrological and ice regime of the Laptev Sea. *Oceanology,* **6,** 815–821.

Zakharov, V. F., 1996. *Sea ice in the climatic system.* Hydrometeoizdat, St. Petersburg, Russia. (in Russian)

Zegouagh, Y., S. Derenn, C. Largeau, G. Bardoux and A. Mariotti, 1998. Organic matter sources and early diagenetic alterations in Arctic surface sediments (Lena River delta and Laptev Sea, Eastern Siberia), II. Molecular and isotopic studies of hydrocarbons. *Organic Geochemistry,* **28** (9/10), 571–583.

Zigarev, L.A. and V.A. Sovershaev, 1984. Termoabrazionnoe razrushenie arcticheskih ostrovov (Thermoabrasional destruction of Arctic islands). In *Beregovye processy v kriolitozone.* Nauka, Novosibirsk, Russia, pp. 31–38 (in Russian).

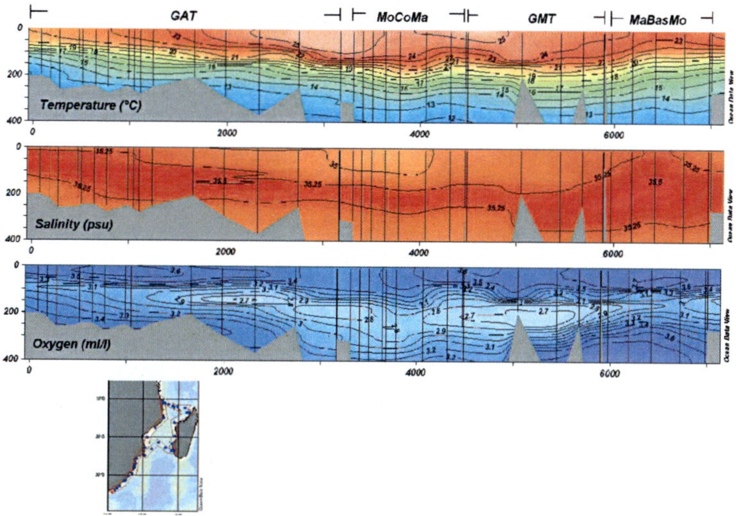

Figure 20.9 Hydrographic characteristics of shelf waters along the east coast of South Africa and the continental shelves in the Mozambique Channel. (With special permission of M. Roberts.) GAT is the section from Port Elizabeth in the south along the east coast of South Africa and of Mozambique; MoCoMa represents the zonal section from the coast of Mozambique to the west coast of Madagascar via the Comores; GMT a poleward section along the west coast of Madagascar and MaBasMo a zigzag section from the southern point of Madagascar to the island of Bassas da India and on to Maputo on the east coast of Mozambique.

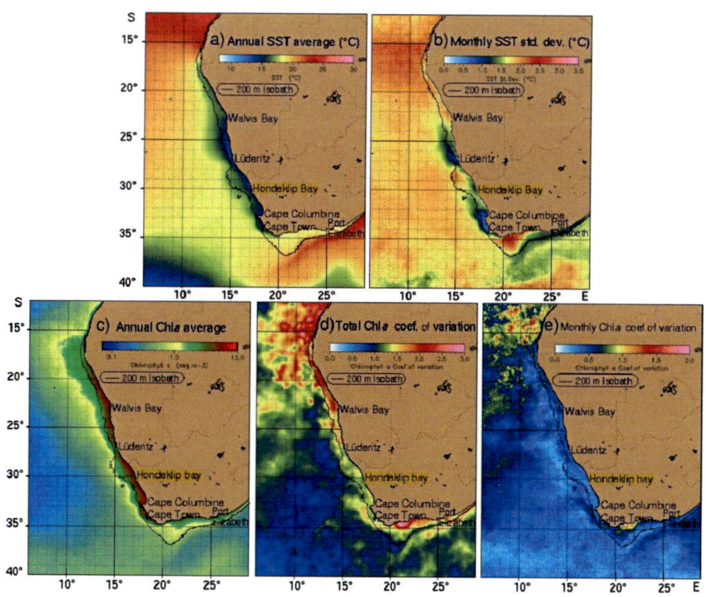

Figure 21.2 a) Annual average SST and b) mean monthly SD for SST computed from monthly composites for the period 1982 – 1999; c) annual average Chlorophyll and d) total CV of chlorophyll computed from 5-day weighted composites from Sept. 1997 – April 2002; e) mean monthly CV of chlorophyll computed from monthly composites (From Demarcq *et al.*, 2003).

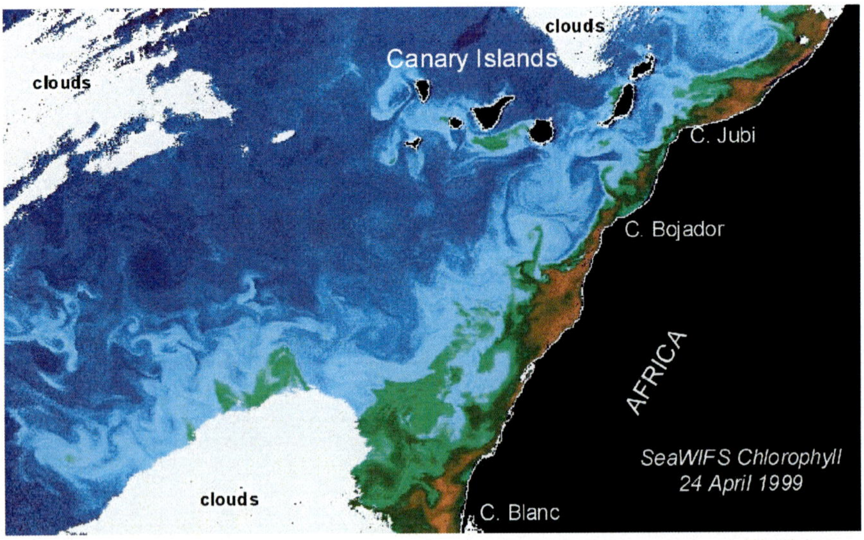

Figure 23.5 SeaWIFS chlorophyll image from the Canary Islands region (24 April, 1999), showing the presence of numerous cyclonic and anticyclonic eddies downstream the islands. The eddy field extends southward down to the latitude of Cape Blanc, where the Canary Current detaches from the coast and flows westward. As a reference, green colour represents chlorophyll values >1 mg m^{-3} and red colour corresponds to values >3 mg m^{-3}

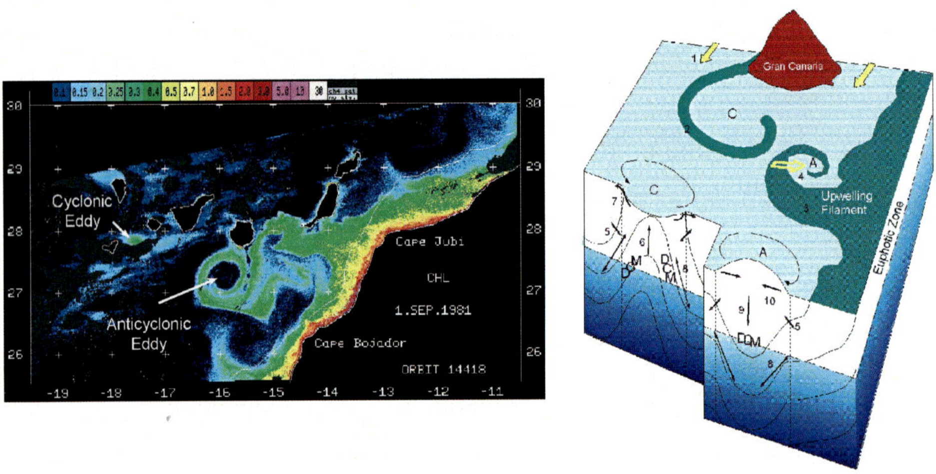

Figure 23.8 (a) CZCS chlorophyll image from the Canary Islands region (1 September 1981), showing the entrainment of high-chlorophyll water from an upwelling filament by an anticyclonic island eddy. (b) Schematic diagram showing vertical and horizontal fluxes associated with island-generated cyclonic (C) and anticyclonic (A) eddies downstream Gran Canaria Island. (1) Mean surface flow and wind direction; (2) advection of high-chlorophyll coastal water during cyclonic eddy formation; (3) entrainment of high-chlorophyll water upwelled from the African coast by anticyclonic eddy; (4) lateral exchange of chlorophyll between the periphery and the eddy centre; (5) diapycnal mixing; (6) upwelling or uplifting of the thermocline; (7) outward radial advection; (8) isopycnal mixing; (9) downwelling or deepening of the thermocline; (10) inward radial advection. DCM = Deep Chlorophyll Maximum. (Panel b, adapted from Arístegui et al., 1997)

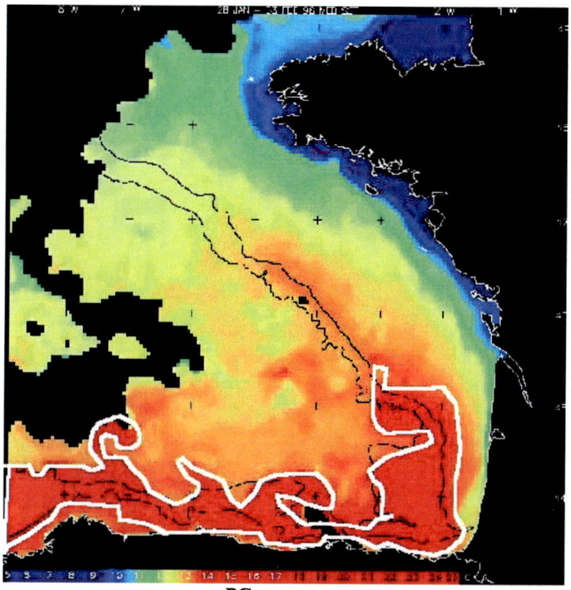

PC (A)

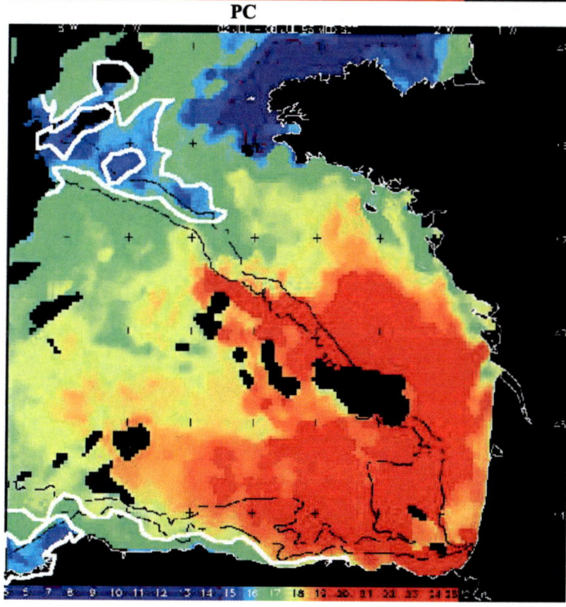

UPW (B)

Figure 24.8 Mean features of Bay of Biscay SST in winter (A) and summer (B). Strong Poleward Current (warm feature marked with white line) and cold water over the inner slope on the Armorican shelf (North France) in winter and strong warming in the inner Bay of Biscay and cold signal due to upwelling in the western Cantabrian Shelf and in the north part of the Bay of Biscay Ushant front in summer are presented. Mesoscale activity is noted in both periods. Satellite images were received by the NERC Dundee Satellite Receiving Station and processed by Peter Miller at the Plymouth Marine Laboratory Remote Sensing Group (www.npm.ac.uk/rsdas/).

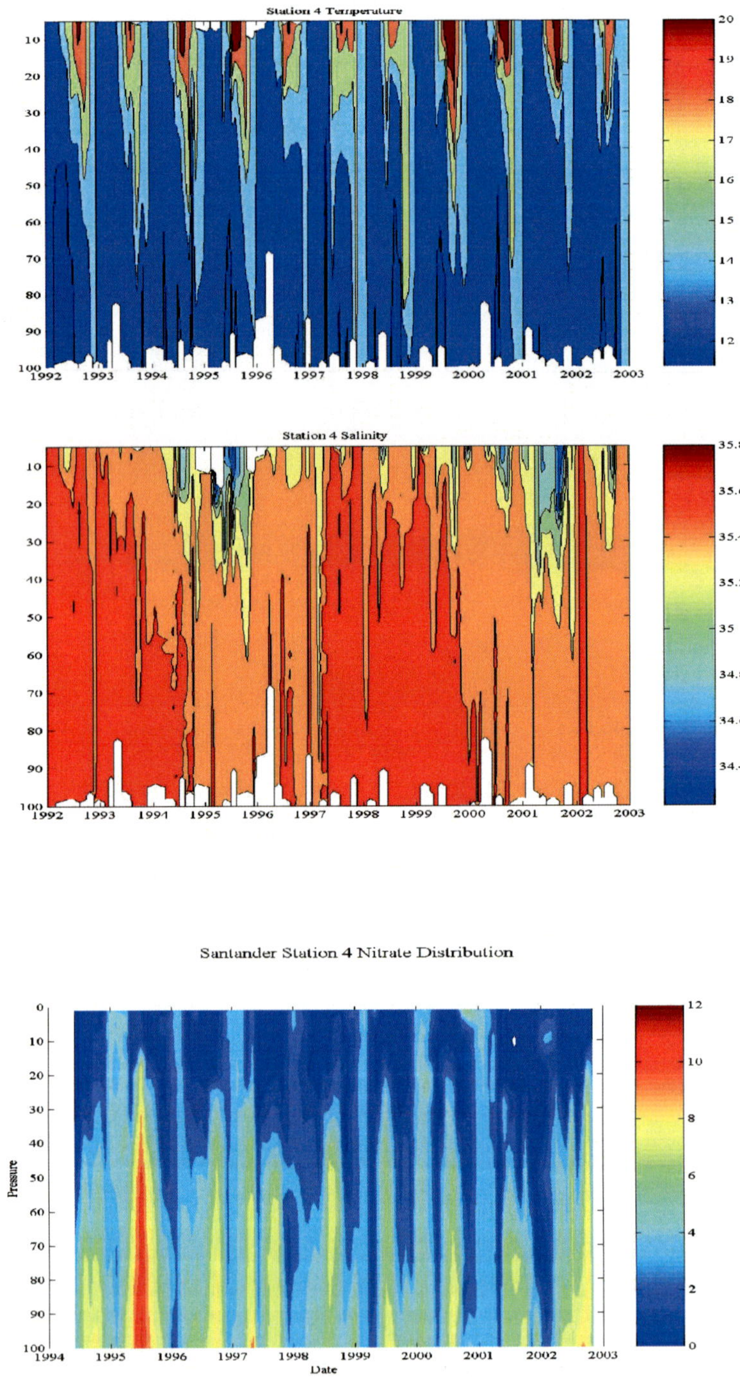

Figure 24.10 Seasonal and interannual variations in temperature from 1991 to 2003 (A), salinity from 1991 to 2003 (B) and nitrates from 1993 to 2003 (C) over the shelf off Santander (43° 34.5'N, 3° 47'W, southern Bay of Biscay).

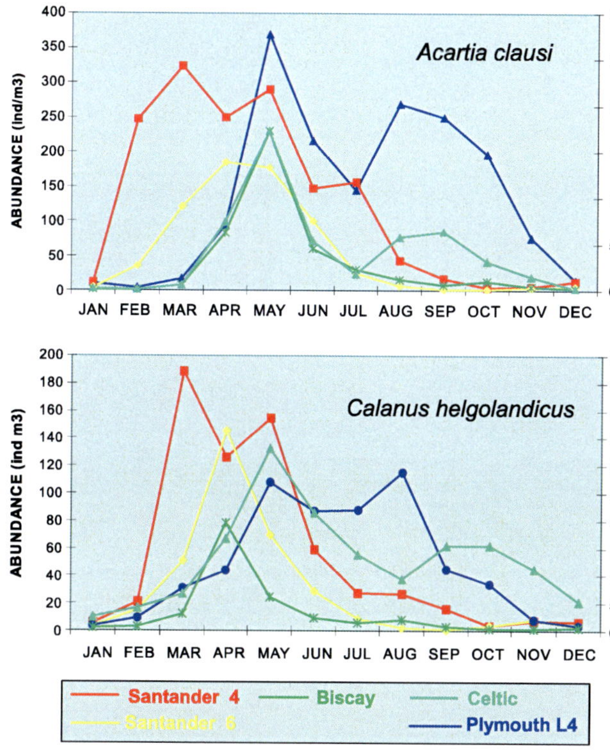

Figure 24.19 Mean monthly abundance (1992–1999) of *Acartia spp.* (upper panel) and *Calanus helgolandicus* (lower panel) over the shelf off Santander (dark grey), an oceanic station off Santander (white), open ocean in the Bay of Biscay (light grey , from CPR), Celtic Sea (light grey , from CPR) and over the shelf off Plymouth (black). The left axis represents abundance off Santander and Plymouth, the right axis abundance from CPR results in the Bay of Biscay and Celtic Sea (Valdés et al., 2001).

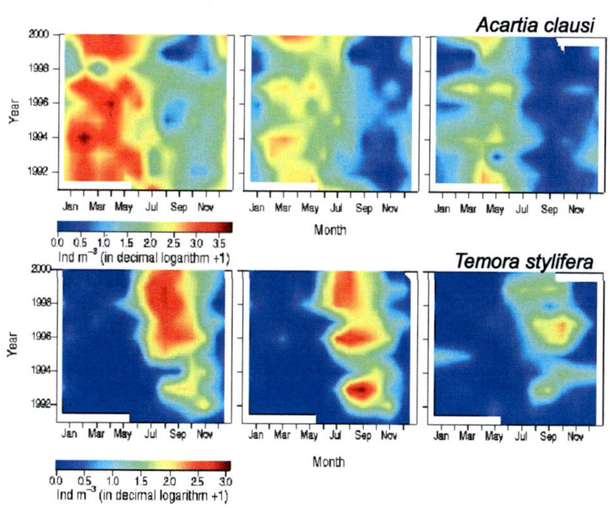

Figure 24.20 Year-to-year variability of *A. clausi* (upper panel) and *Temora stylifera* (lower panel) at the coastal, shelf and oceanic stations off Santander during the time series 1992–2000.

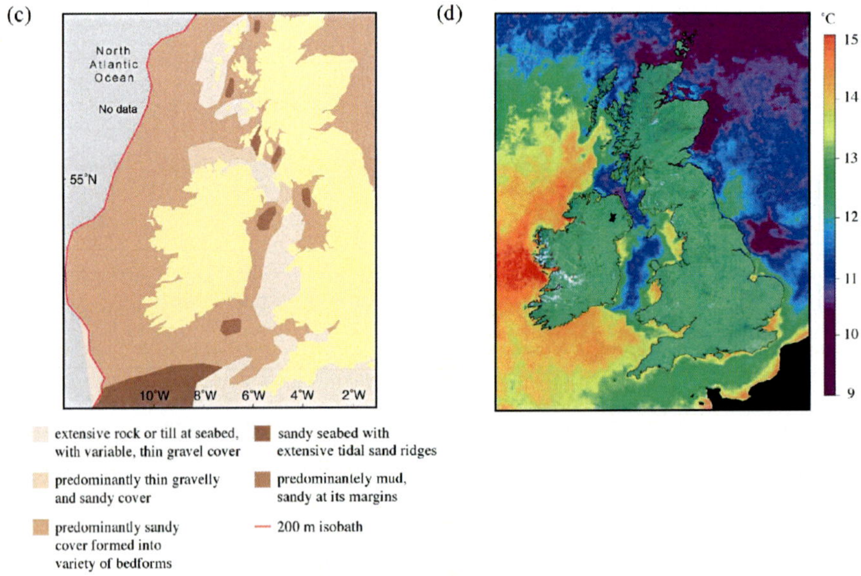

Figure 25.1 (c) Bottom sediment type. (d) A composite sea surface temperature image for June 1997. Sediment distribution in (c) after OSPAR Commission 2000. Image (d) courtesy of Peter Miller, Remote Sensing Group, Plymouth Marine Laboratory, UK.

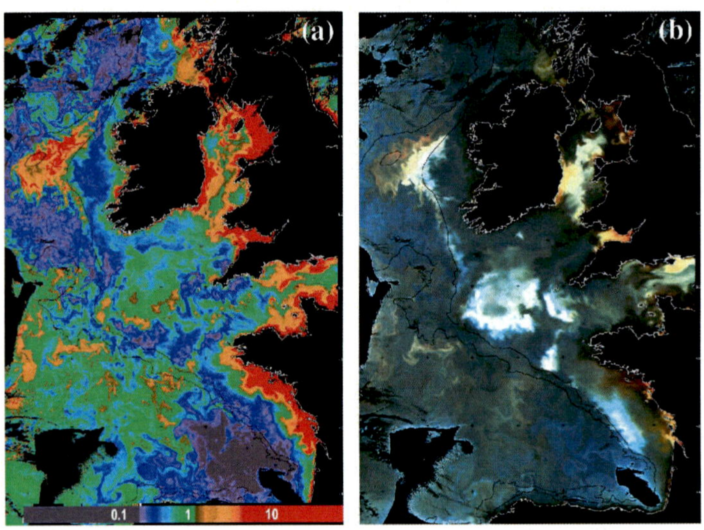

Figure 25.3 SeaWiFS images for the Celtic Seas, 18 May, 1998, supplied by S. Groom (Plymouth Marine Laboratory).

a) Chlorophyll image showing the distribution of surface phytoplankton in late spring about one month after the onset of stratification in the Celtic Sea.

b) Colour composite image showing strong reflectance due to extensive blooms of the coccolithophore *E. huxleyi* in the western Celtic Sea and to the west of Ireland, and due to suspended sediments in the regions of strong tidal mixing in the Irish Sea and the English Channel. Under both conditions algorithms for the determination of chlorophyll tend to break down so that the estimates of phytoplankton abundance in panel (a) for the high reflectance areas are likely to be in error.

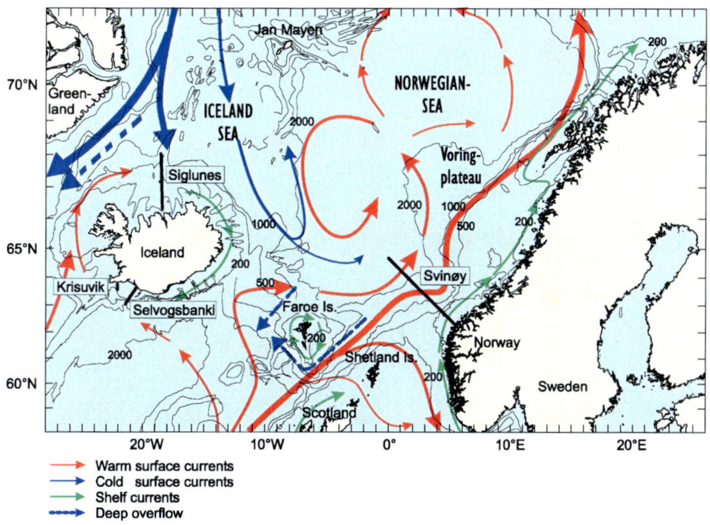

Figure 27.1 Bottom topography and main features of the shelf and oceanic currents in the Iceland-Faroes-Norway region.

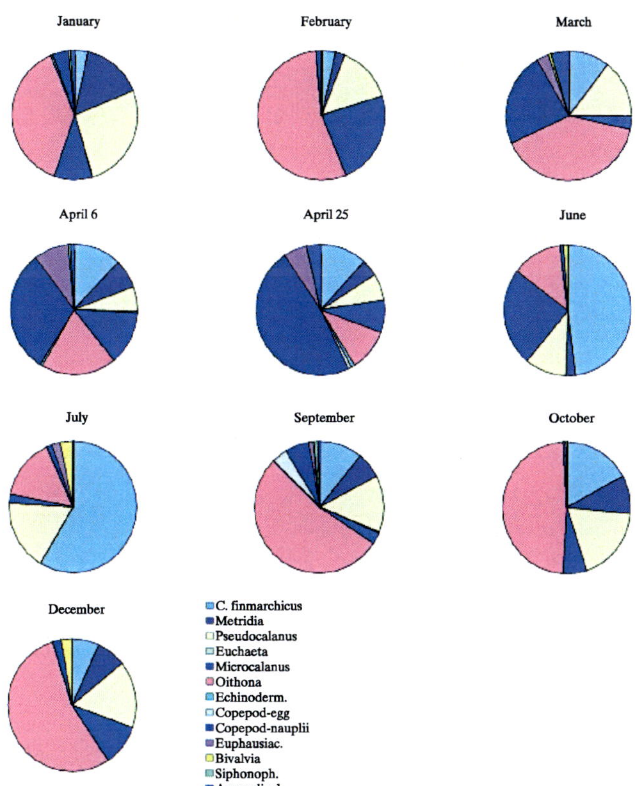

Figure 27.12 Relative abundance of zooplankton species at a shelf station (180 m bottom depth) off Møre in 1997 (see Fig. 27–001).

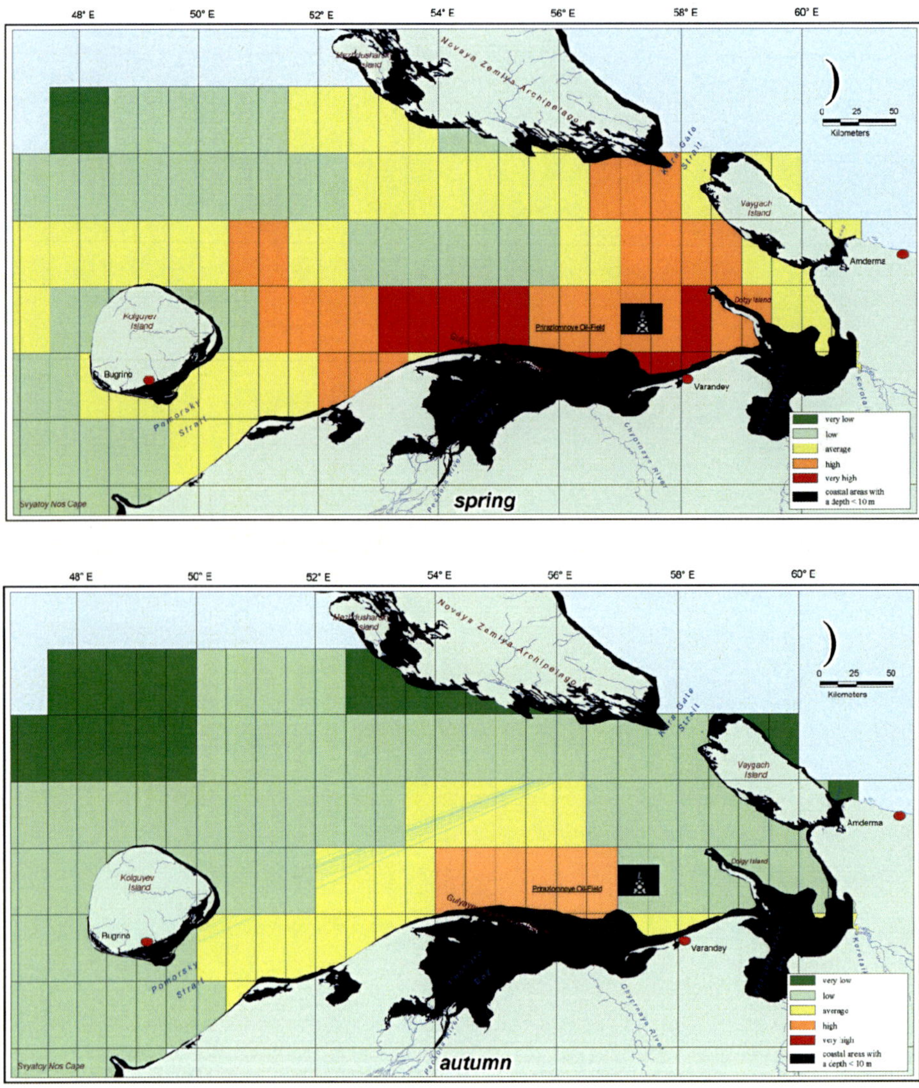

Figure 29.20 Potential sensitivity of the Pechora Sea to accidental oil spills by the integral biological characteristics in spring (late April–June: development of plankton communities, spawning of herring, spring migration of fish, spring migration and pre- and nesting concentrations of birds, pupping and molting of seals) and autumn (September-October: autumn migration of fishes, birds and mammals to the wintering grounds)

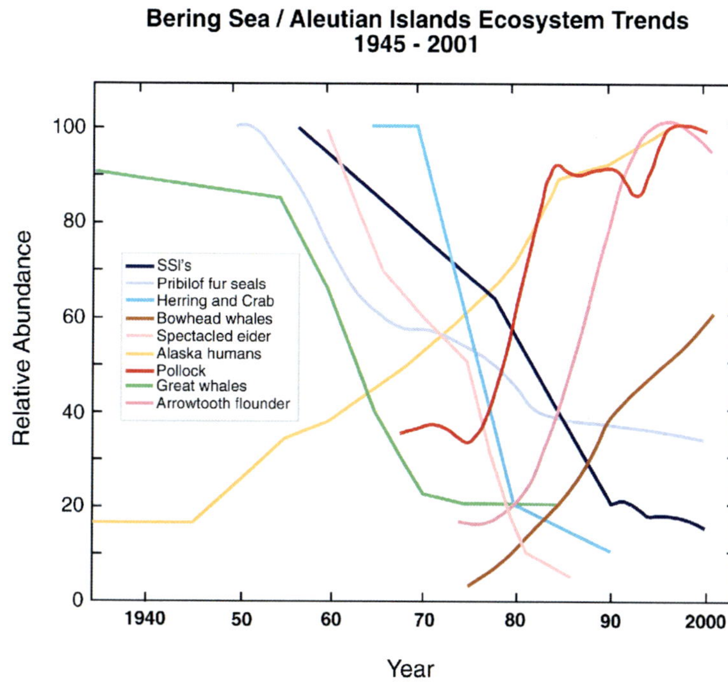

Figure 30.5 Schematic of ecosystem changes in the eastern Bering Sea showing trends of the relative abundance of select marine mammals, fish, and shellfish in the eastern Bering Sea. (after NRC 2003)

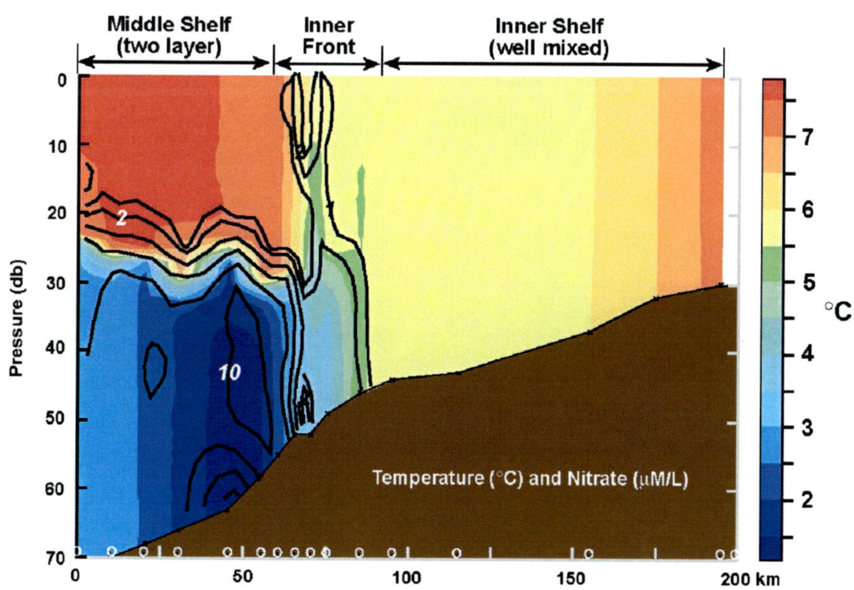

Figure 30.7 Contours of temperature (color; °C) and nitrate (black lines; mmoles l^{-1}). The locations of the CTD stations are marked along the bottom axis. Nutrients were measured on each cast. (From Kachel *et al.,* in 2002).

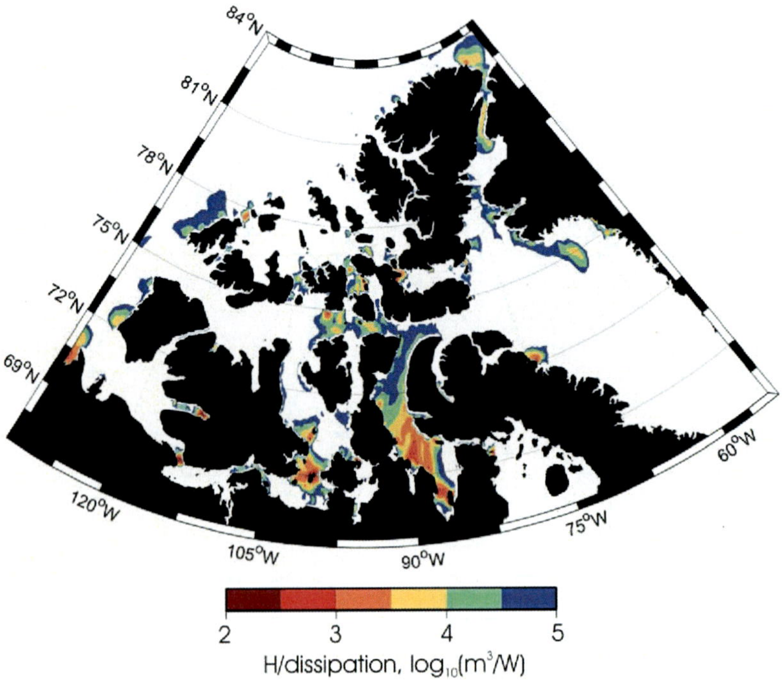

Figure 31.7 Map showing the tidal mixing parameter of the H/U^3 (adapted from Greenberg, pers. comm.)

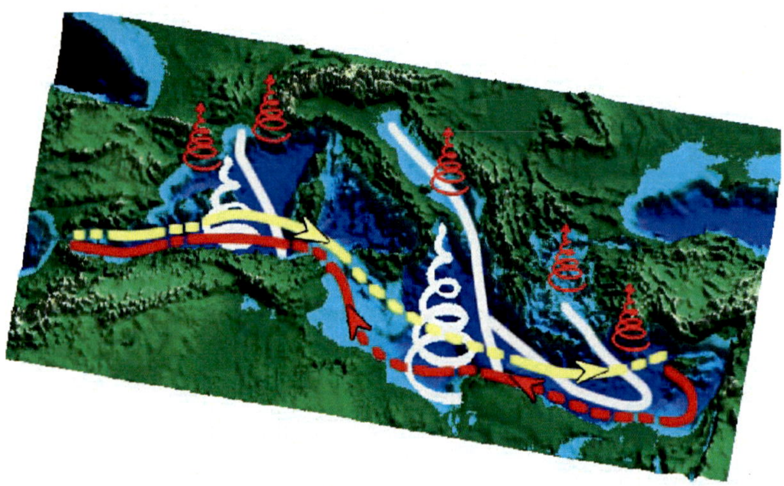

Figure 32.13 The three conveyor belts of the Mediterranean Sea. The first, dashed and in the zonal direction, is the surface-intermediate water masses circulation that is forced by Gibraltar and the Levantine Intermediate Water (LIW) formation processes occurring in the Northern Levantine basin. The second is the meridional vertical circulation in the Eastern Mediterranean, that is forced by deep water formation in the Southern Adriatic and Aegean Seas. The third is the meridional circulation in the western Mediterranean, forced by open ocean deep water convection in the Gulf of Lions. The spirals indicate the preferential site for strong heat losses from the ocean to the atmosphere during winter time.

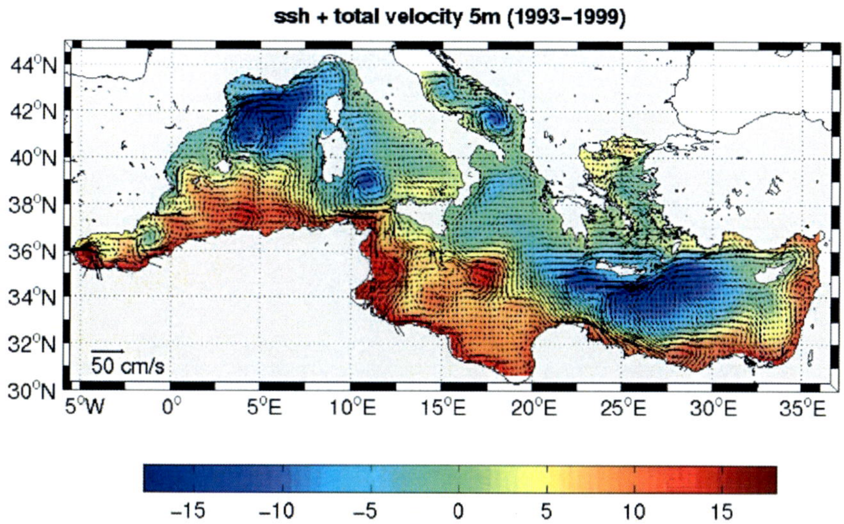

Figure 32.14 The mean sea surface height (ssh) from a 1993–1999 model simulation together with the surface velocity field. The grey bar is in cm and the basin average ssh is set equal to zero.

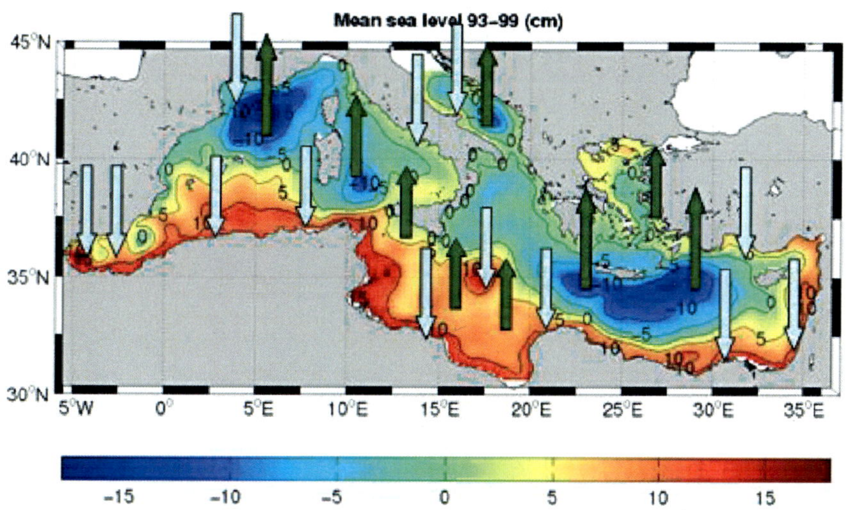

Figure 32.15 The time mean sea surface height of Fig. 32. 14 shown here only with amplitude contours. Arrows indicate the large scale vertical motion direction. The arrows start and point to the specific areas where the motion has the largest amplitude.

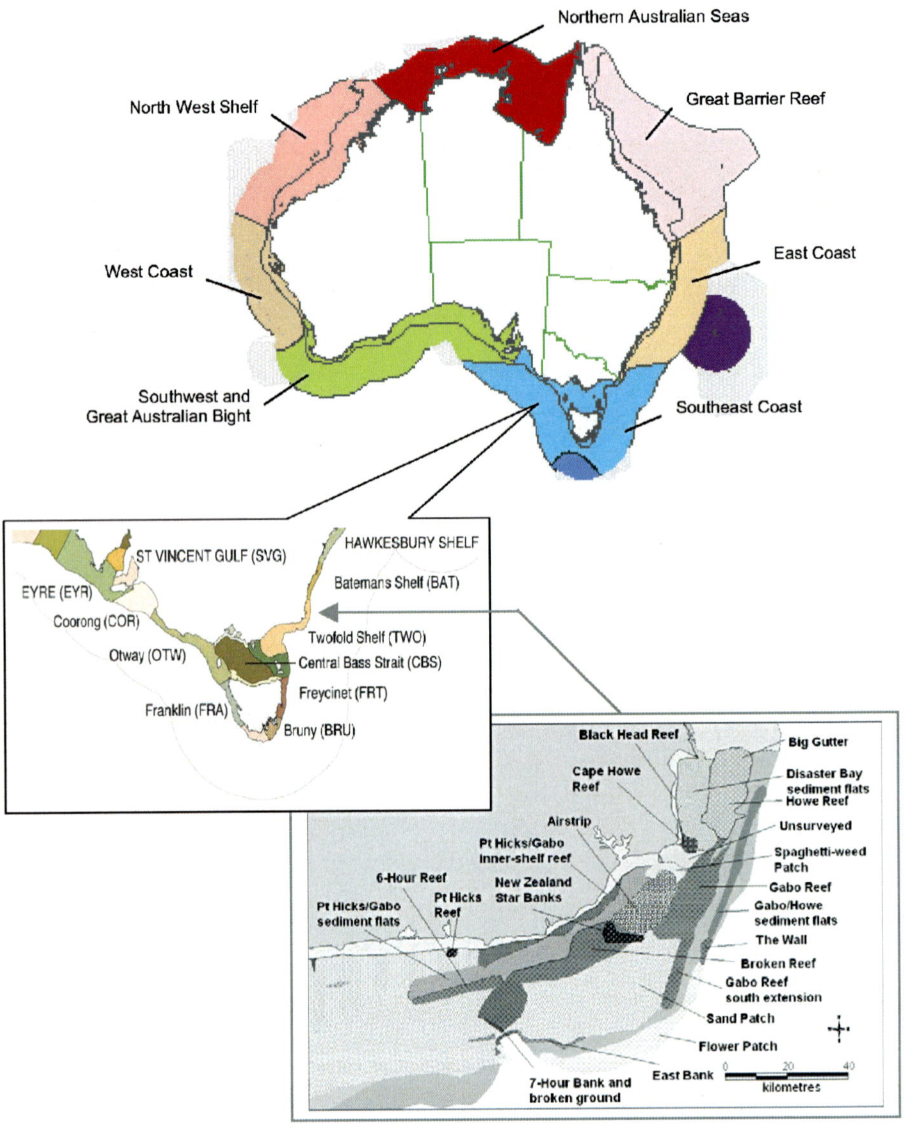

Figure 35.9 A Large Marine Ecosystem (LME) regionalisation for the seas surrounding the Australian continent (upper, The Marine Science and Technology Plan Working Group 1999); the southeast sector of a meso-scale regionalisation referred to as the Interim Marine and Coastal Regionalisation of Australia (IMCRA) for the continental shelf (center left, Interim Marine and Coastal Regionalisation for Australia Technical Group 1998); and a still smaller scale mapping of seabed habitats over part of the south-eastern continental shelf (lower right, adapted from Bax and Williams 2001).

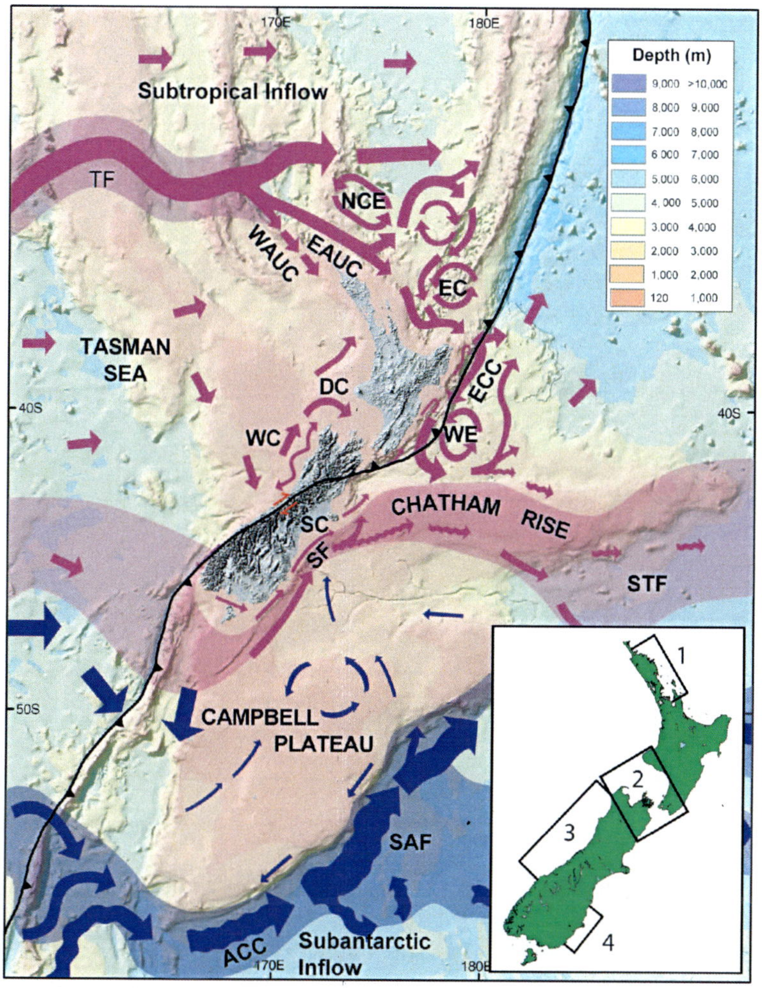

Figure 36.1 New Zealand region bathymetry, major currents, and the location of the major plate boundary (after Carter et al. 1998). ACC = Antarctic Circumpolar Current, DC = D'Urville Current, EAUC = East Auckland Current, EC = East Cape Eddy, ECC = East Cape Current, NCE = North Cape Eddy, SAF = Subantarctic Front, SF = Southland Front, STF = Subtropical Front, TF = Tasman Front, WE = Wairarapa Eddy, WAUC = West Auckland Current. Inset showing the four case studies mentioned in section 2. 1 = Northeast shelf, 2 = Greater Cook Strait, 3 = Western South Island shelf, 4 = Otago shelf.

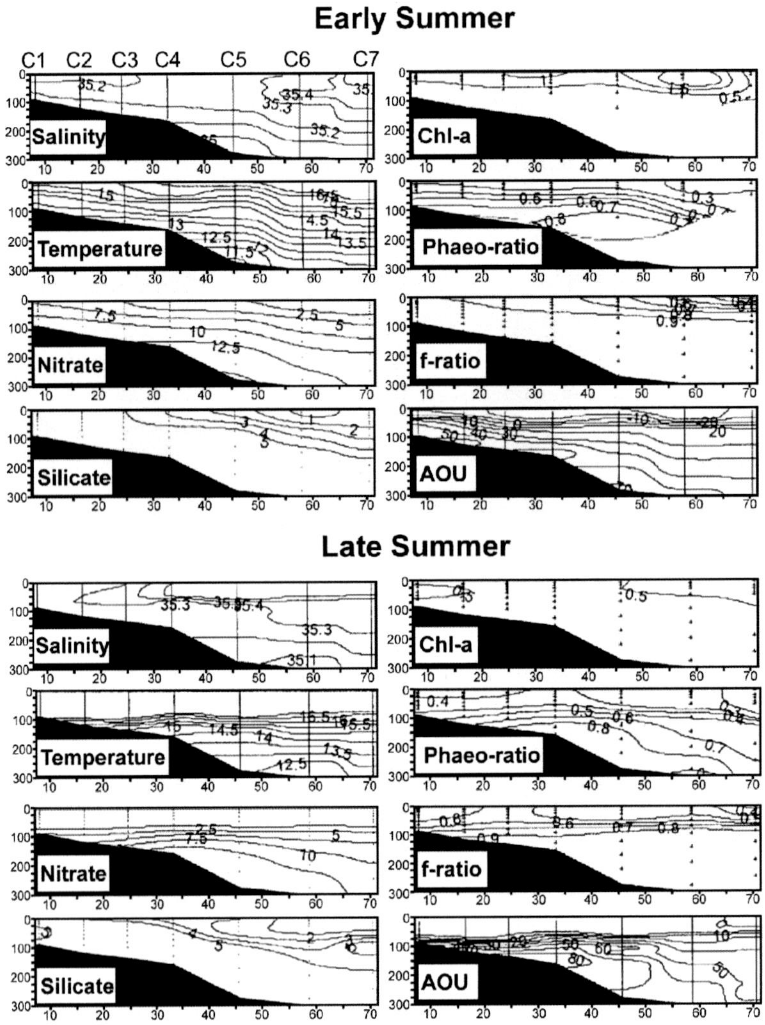

Figure 36.6 Cross sections of salinity, temperature, nitrate, silicate, chlorophyll a, phaeophytin ratio, f-ratio, Apparent Oxygen Utilisation in early and late summer (from Zeldis, 2004). Location of transect indicated on Fig. 36.5.

Figure 36.19 Aerial photo of *Munida gregaria* aggregation at a headland front south east of Taiaroa Head, taken from 1100 m altitude. (from Zeldis and Jillett, 1982).

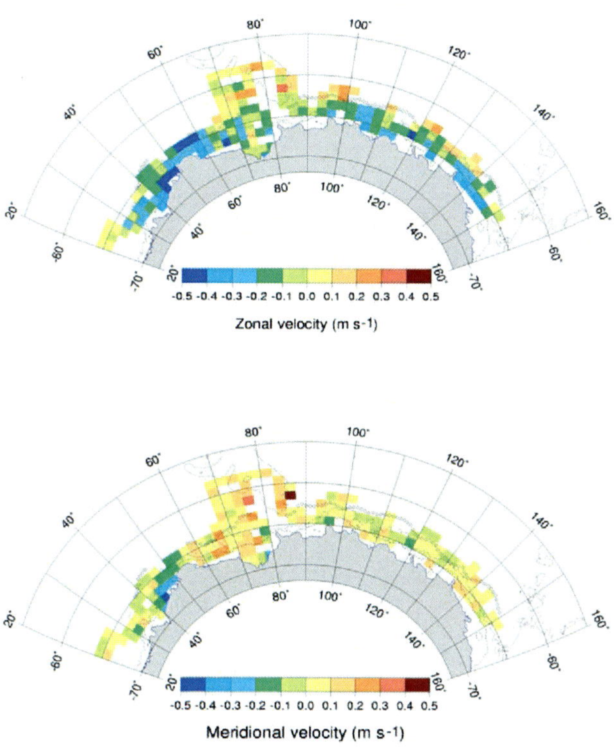

Figure 37.4 Sea ice drift velocities off East Antarctica. Figure 37. From Heil and Allison (1999).

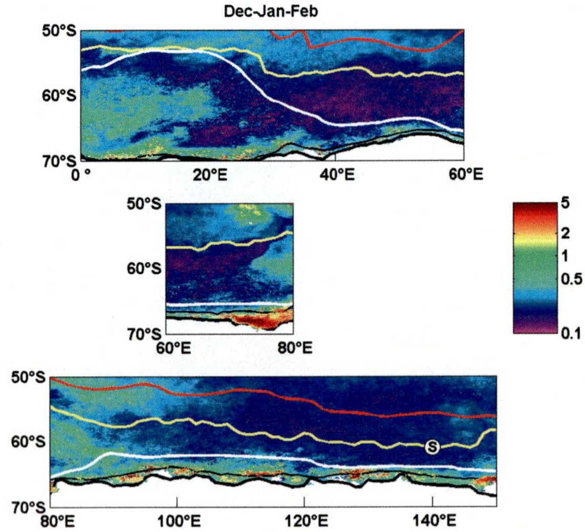

Figure 37.7 Mean summer chlorophyll (Dec-Jan-Feb, SeaWiFS data spanning Dec-1997 to Feb-2003) for the region of interest, split into three subsections: The SE Atlantic/SW Indian sector (top), Prydz Bay (middle) and the region south of Australia (bottom). Also plotted are the 1000m isobath (solid white line), the southern boundary of the ACC (SBACC; dotted line), the maximum extent of winter sea ice (dashed line) and the polar front (solid line). The location of the SOIREE experiment is indicated by `S'.

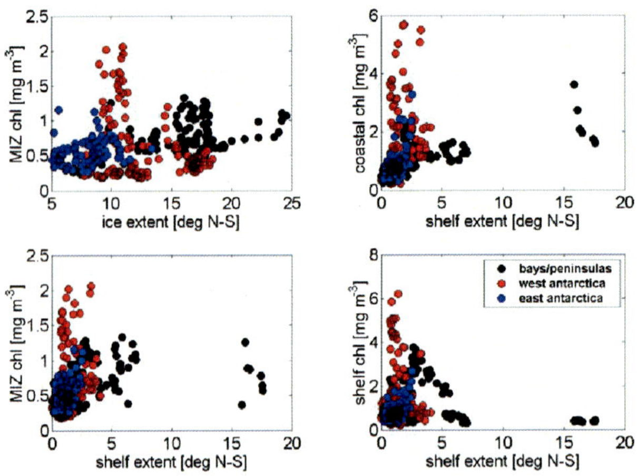

Figure 37.8 Relationship between the area bounded by a given physical feature and the mean chlorophyll within that area. Mean chlorophyll data depicted are the Dec-Jan-Feb average for December 1997 through February 2003 (as for Figure 7). Each data point represents a 1° bin (i.e. n = 360). The data are presented by area, with sub-regions defined at (1) Bays and Peninsulas (Prydz Bay, Ross Sea, Weddell Sea and the Antarctic peninsula, plotted as open circles), (2) East Antarctica (80°E to 170°E, plotted as + symbols) and (3) the remainder of the coast (plotted as open triangles). The four panels depict: (a) Relationship between mean chlorophyll within the MIZ and the north-south extent of the MIZ. (b) Relationship between mean chlorophyll south of the SBACC and the extent of the continental shelf. (c) Relationship between mean chlorophyll within the MIZ and the extent of the continental shelf, and (d) Relationship between mean chlorophyll at the shelf break and the extent of the continental shelf.

Chapter 29. ECOSYSTEM OF THE BARENTS AND KARA SEAS, COASTAL SEGMENT (22,P)

MIKHAIL YU. KULAKOV AND VLADIMIR B. POGREBOV

Arctic and Antarctic Research Institute

SERGEY F. TIMOFEYEV

Murmansk Marine Biological Institute of the Kola Science Centre of Russian Academy of Sciences

NATALIA V. CHERNOVA

Zoological Institute of Russian Academy of Sciences

OLGA A. KIYKO

All-Russia Research Institute for Geology and Mineral Resources of the World Ocean

Contents

1. Introduction

The interest in the ecosystems of the Barents and Kara seas has increased recently in connection with growing anthropogenic impact. There are many investigated oil

The Sea, Volume 14, edited by Allan R. Robinson and Kenneth H. Brink
ISBN 0–674–01527–4

and gas fields on the shelf of the Barents and Kara seas. However, the hard ice regime and severe natural and climatic conditions of the region to be developed put forward a number of complicated engineering problems, including the protection of biota. To solve them there is a need for reliable and complete information on the environment.

The structure and function of coastal ocean ecosystems differ greatly among continental shelves, being driven largely by differences in net primary production that are ultimately determined by the interplay of many factors such as boundary currents, shelf geometry, river runoff, upwelling and water and sediment chemistry that are unique to each shelf margin (Alongi, 2003). The shelf regions of the Barents and Kara seas are a link between the North Atlantic and the Arctic Ocean. The hydrography is characterized by frontal structures, transformation and mixing processes, resulting from the penetration of warm and saline North Atlantic Water from western boundary and abundant river runoff from east.

In Russian oceanographic and biological sciences the Barents and the Kara Seas are traditionally described separately due to significant differences of their oceanographic regime and ecosystems types. The Barents Sea is influenced by warm and saline Atlantic waters to a significantly greater extent as compared with the Kara Sea. Wide development of the polar front phenomenon and vertical water circulation are a basis of high biological productivity in the sea and high richness of its pelagic and bottom life. Biota of the Barents Sea in general is characterized by mixture of boreal and arctic elements. Quantitative variability of flora and fauna is highest in the southwestern part of the sea, and drops in significantly in the northward and eastward directions. Large quantities of heat-loving species penetrate here with warm currents from the Atlantic. Macrophytes in the coastal zone are diverse. Fishery is based mainly on cods, herrings, plaices, and capelin.

The Kara Sea is almost closed from Atlantic influence, and the biota of the sea includes much more arctic elements. Phytoplankton vegetation is almost absent about 10 months per a year, and biological productivity is low. Biodiversity in the Kara Sea is approximately twice as low as compared to the Barents Sea. Littoral and sub-littoral zones (due to rough ice conditions) are especially poor of life. Macrophyte vegetation in the coastal zone is actually absent. Large estuarine and freshened water areas are inhabited by specific complex of euryhaline organisms and are very typical for the sea. Fishery exists predominantly in the estuarine areas and is based mainly on whitefishes.

2. Oceanography

The Barents and Kara seas are semi-enclosed marginal seas of the Arctic Ocean that lie between the northern coast of Europe and four archipelagos: Svalbard, Franz Josefs Land, Novaya Zemlya and Severnaya Zemly (Fig. 29.1). According to (Gorshkov, 1980), the average depth of the Barents Sea comprises 199 m, its area is equal to 1.417 million km^2 and the greatest depth in the Medvezhinsky Trough is slightly more than 500 m. The average depth of the Kara Sea is 111 m and its area comprises 883 thousand km^2 with a maximum depth (620 m) in the northern part of St. Anna Trough.

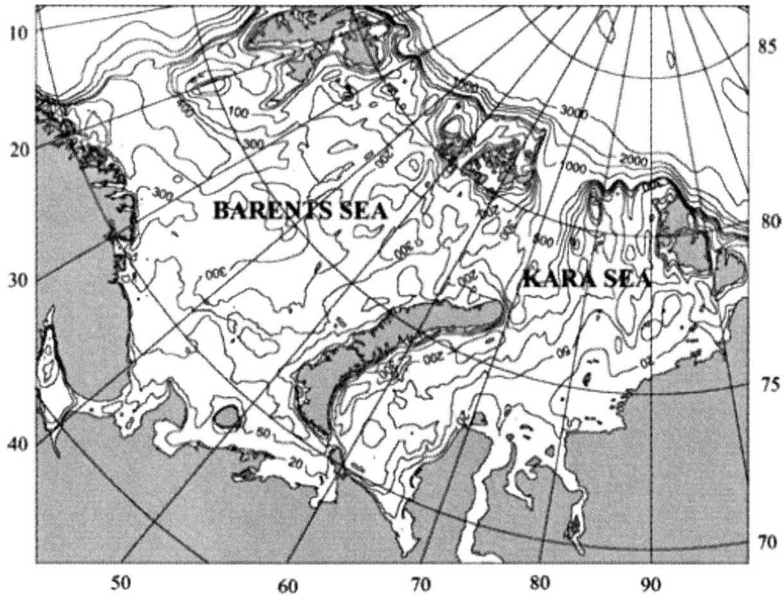

Figure 29.1 Coastal geometry and bottom topography of the Barents and Kara seas

The seabed relief of the Barents Sea, which is significantly deeper than the Kara Sea, is characterized by strong dissection. Many relief features are attributed to the young age of the sea basin that has attained its modern style not more than 10 thousand years ago resulting from the ocean transgression. Large structural components of seabed relief are of tectonic origin with the relict relief forms and the ancient river network being superimposed on them.

A deep depression of the St. Anna trough separates the Barents Sea shelf from the Kara shelf. The relief of the northern Kara Sea is very irregular due to bedrock outcrops. The sea is abundant in islands. Here, in addition to St. Anna Trough, Voronin Trough and the accumulative elevations including the Central Kara Rise are distinguished.

The main features of the oceanographic regime of the Barents and Kara seas are primarily governed by its high latitude. These seas are situated completely above the Arctic Circle. Hence, during winter, solar radiation flux is absent, whereas, in summer, it is comparatively small. The polar night duration increases from 50–60 days at the southern part of the Barents Sea to 80–100 days in the northern part of the seas. The average annual air temperatures at the southwestern part of region is about 2° C. Further north, it decreases to –6 to -10° C at latitude 77° in the Barents Sea and to –11 to -14° C in the northern Kara Sea (Gorshkov, 1980).

The Barents and Kara Seas due to atmospheric circulation features are characterized by the monsoon distribution of the prevailing wind directions–south quarter in winter and north quarter in summer. The average annual speed over the seas ranges from 6 to 7 m/s and only in the vicinity of Cape Zhelaniya, it becomes higher (7.8 m/s). The maximum wind speed over a year is practically, at all coastal stations, equal to or greater than 40 m/s being mainly observed in the autumn-winter period. In summer, the maximums decrease to 20 m/s in July and only near Cape Zhelaniya, they can comprise 40 m/s as in winter (Hydrometeorological conditions ..., 1985, 1986).

The Barents Sea is primarily a throughflow region for the considerable transport of water masses (about 2 Sv) from the Greenland-Norwegian Sea towards the Arctic Ocean (Blindheim, 1989; Ingvaldsen et al., 2002). An eastward flowing branch of the Norwegian Atlantic Current reaches the Barents Sea via the Bear Island Trough. The interference of Atlantic and Arctic water masses in the central Barents Sea leads to a pronounced Polar Front south and east of Bear Island (Fig. 29.2). In the eastern Barents Sea the water has become transformed from warm saline water to cold, less saline intermediate and bottom water. This transformation happens through mixture of cooled Atlantic Water with cold brine-enriched shelf water generated west of Novaya Zemlya, and possibly also at the Central Bank. The moderately cold, low salinity mixture continues to the Kara Sea without further change. According Schauer et al. (2003), in 1991/1992 the water transport through the northeastern Barents Sea was between 0.6 Sv in summer and 2.6 Sv in winter towards the Kara Sea and between 0.0 and 0.3 Sv towards the Barents Sea with 11-month averages of 1.5 Sv and 0.1 Sv, respectively.

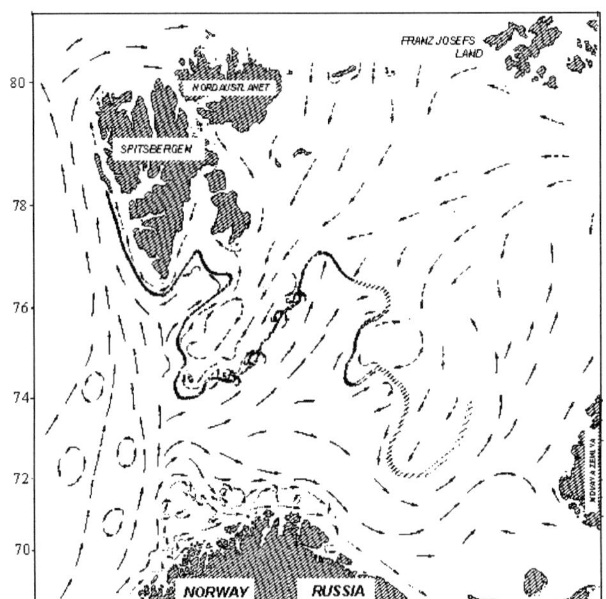

Figure 29.2 Surface currents and Polar Front position in the Barents Sea (Loeng, 1988)

River runoff into the Barents Sea is rather small. The Pechora River is the largest river of the Barents Sea basin. Its mean annual runoff is up to 134 km^3 (Joint US-Russian Atlas .. , 1998). The rivers of Scandinavia and the Kola Peninsula together give about 10% of the runoff. In the consideration of the continental runoff to the Barents Sea one should take into account its water exchange with the White Sea, into which quite large rivers Severnaya Dvina, Mezen' and Onega fall. This aggregated annual runoff is up to 135 km^3. The river runoff significantly affects the hydrological conditions of the Barents Sea only in the south-eastern part of the sea.

According Hanzlick and Aagaard (1980), the importance of the Kara Sea to the Arctic Ocean derives from two features. First, it receives more than one third of the total freshwater discharged into the Polar Basin, and the conditioning and cycling of this freshwater has consequences for the salinity stratification of the Arctic Ocean (Aagard and Coachman, 1975). Second, the Kara Sea appears to be the site of significant transformation of Atlantic Water, with a large attendant vertical heat flux (Timofeyev, 1962).

The Kara Sea receives about 55% (1350 km^3/year) of the total river runoff discharged to the entire Siberian Arctic (Soviet Arctic, 1970, Ivanov et al., 1984, Pavlov and Pfirman, 1995). The annual discharge of the Ob' River is 530 km^3, the Yenisey River, 605 km^3, the other rivers add up to 190 km^3. The continental discharge could fill the entire volume of the Kara Sea within only 73 years. The total watershed area of the Kara Sea comprises 5.57 million km^2 and Ob' and Yenisey carry about 221 mill. tons of suspended matter and almost 8 mill. tons of organic mater (particulate and dissolved) per year into the Arctic Ocean (Gordeev et al., 1996).

The main feature in the sea surface temperature distributions is a general decrease in temperature from west to east (Fig. 29.3). The western region of the Barents Sea, influenced by the warm Atlantic waters, is characterized by positive temperature values in the surface layer and near the bottom. The highest temperature in the coastal zone of the western Barents Sea is up to 16° C (Hydrometeorological conditions ..., 1985). In winter, sea surface temperature is generally less than 3° C. At the same time, the northeastern part of the Kara Sea, directly influenced by the Arctic Basin, has a significantly smaller range of temperature oscillations. The sea surface temperature in this region is lower than 0° C during the whole year. Because relatively warm water of Atlantic origin penetrates into the Kara Sea through the St. Anna and Voronin troughs, water temperature increases at about 50 to 70 m water depth, reaching maximum values of 1.0 to 1.5° C (Hanzlick and Aagaard, 1980, Pavlov and Pfirman, 1995)

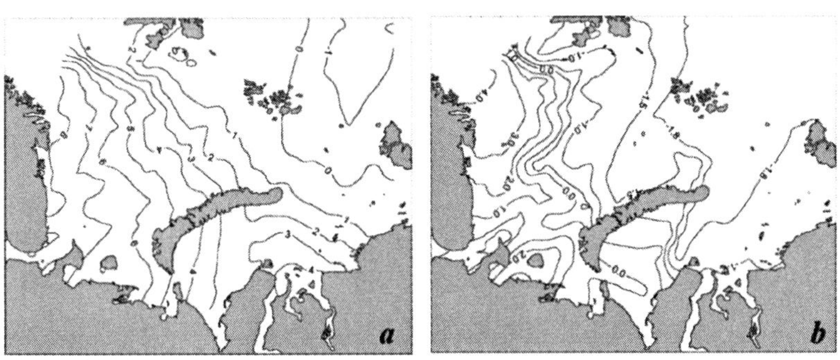

Figure 29.3 Mean multiyear distribution of the surface water temperature for summer (a) and winter (b).

The most prominent features of salinity distribution in the Barents Sea are its high values during the whole year (Fig. 29.4). It is most clearly defined in the Atlantic water mass (up to 35.0 ‰). Salinity values in the coastal zone are somewhat lower, and it is caused by peculiarities of freshwater balance and income of freshened waters from the White Sea.

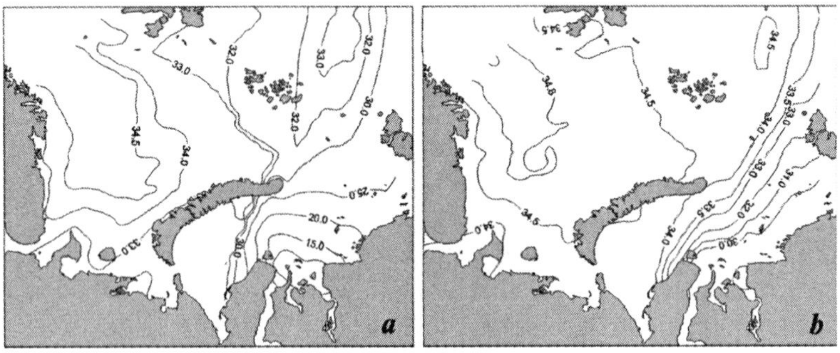

Figure 29.4 Mean multiyear distribution of the surface water salinity for summer (a) and winter (b).

In the Kara Sea, the salinity distribution exhibits pronounced seasonality due to fluctuations in river runoff as well as ice formation and melting. In summer a strong frontal zone develops where river runoff meets the more saline shelf water. Horizontal salinity gradients can reach value 1.5 ppt per 1 km, and vertical 5 ppt per 1 m in the open sea and 8 ppt per 10 cm in the Ob' and Yenisey estuaries (Kulakov and Stanovoy, 2002). The lowest salinities (several parts per thousand) are observed in the vicinity of the Ob' and Yenisey estuaries, while salinity in the western part of the sea are greater then 32 ppt. In winter, river runoff decreases. At the same time, ice formation and consequent brine-release cause salinity to increase. In the southwest, except for regions directly adjacent to the river mouths, salinity approaches 25 to 30 ppt. To the north and near Novaya Zemlya, the salinity is generally about 34 ppt.

Due to income of warm Atlantic waters brought by the North-Atlantic current, the Barents Sea never (even in the most severe winters) is covered completely with ice (Fig. 29.5). This is its basic difference from other seas of the Arctic shelf. One of the main features of the sea is the significant interannual and seasonal variability of its ice cover extent. The greatest ice cover extent is seen usually in the middle of April; the least at the end of August, and first half of September. In the extremity of the most severe winters more than 90% of the sea surface is covered with ice, and in especially warm winters the greatest ice cover extent even in April does not exceed 55–60% (Hydrometeorological conditions ..., 1985).

In August–September anomalously warm years the sea completely frees from ice, and in anomalously cool years the ice cover these months is stored on 40–50% of its area, placing mainly in northern regions. The location of the ice edge during summer can vary by hundreds of kilometers from year to year, and there is also variability on longer time scales, up to century scale, correlated with North Atlantic Oscillation (Vinje, 2001). These variations reflect the inter-annual dynamics of inflowing Atlantic Water and atmospheric forcing.

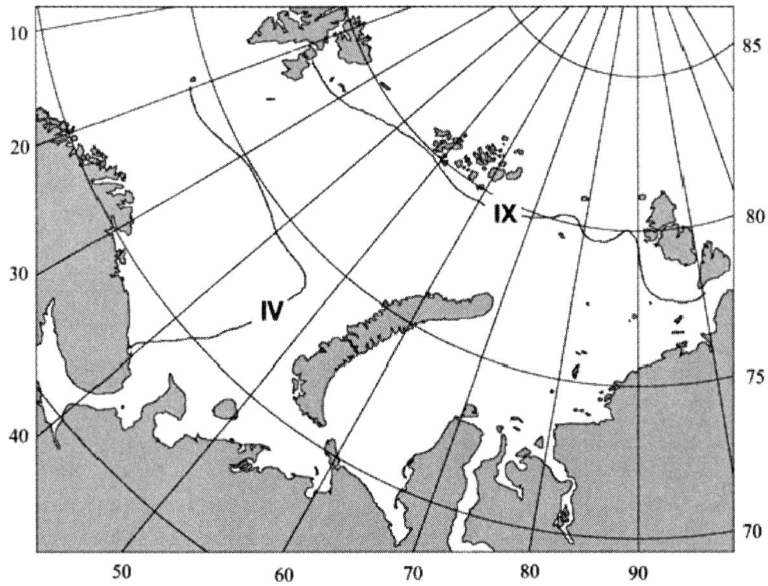

Figure 29.5 Mean multiyear position of the ice edge for April (IV) and September (IX).

In the Barents Sea, ice formed within the limits of the sea usually predominates. But some years in a northwest part of the sea in the winter the old ice from Polar Basin through a channel between Svalbard and Franz Jozef Land enter. Quite often in a north-east part of the sea the thick ice from northern part the Kara Sea are brought. This particular ice enters in the winter from the White Sea, and also from a southwest part of the Kara Sea through Kara Gate.

Ice formation begins in the Kara Sea in September in the north and in October in the south. From October to May almost the entire sea is covered with ice of different types and ages (Soviet Arctic, 1970, Pavlov and Pfirman, 1995). Fast ice occupies the coastal zone, and its development is patchy. Typically in July, the fast-ice band breaks up and disintegrates into separate floes that sometimes persist throughout the year, in the form of Severnaya Zemlya massif. Seaward of the fast ice zone in winter is typically a region that is either ice-free or has young ice. Floe thickness of first year ice reaches maximum in May of 1.5 to 2 m.

The ice distribution in the spring and summer depends on the winds and resultant surface ocean currents. According to Zakharov (1976) the Kara Sea discharges ice to the central Arctic about 180,000 km^2/year (approx. equal to 170 km^3/year). Most of ice export from the Kara Sea to the Arctic Basin occurs in winter. According Zubakin (1987), in winter approximately 140–198 km^3/year of ice exported from the Kara Sea to the Barents Sea through the strait between FJL and Novaya Zemlya. The estimated net flux from the Barents to the Kara Sea through Karskiye Vorota Strait is about 16.8 km^3/year.

3. Sediments

Grain-size analysis of seabed sediments of the Western Arctic Shelf shows that the occurrence of monogranular or pure deposits (containing more than 75% sand, silt or clay) does not exceed 1 %, either by space or mass (Gurevich, 1995). As regards total mass, the figures for bigranular or transitionary deposits, especially those having relatively silty composition, are: sandy silt - >10%, silty sand - > 12%, silty clay - > 17%, clayey silt–nearly 47 5% (see also Figs. 29.6 and 29.7). Triangular or mixed sand-silt-clay mixtite, practically absent from the White Sea, is widespread in northern parts of the Barents Sea and the Kara Sea. Polygranular bouldery-pebbly and gravelly polymixtites are only present in appreciable quantities in the Barents Sea.

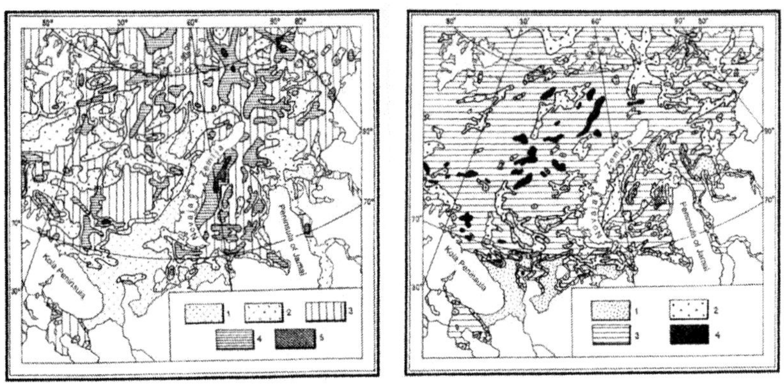

Figure 29.6 The content (%) of the sand (left, 1 - < 10; 2 – 10–25; 3 – 25–50; 4 – 50–75; 5 - > 75.) and silty (right, 1 - < 25; 2 –25–50; 3 – 50–75; 4 - > 75) fraction in the modern deposits on the Western Arctic Shelf.

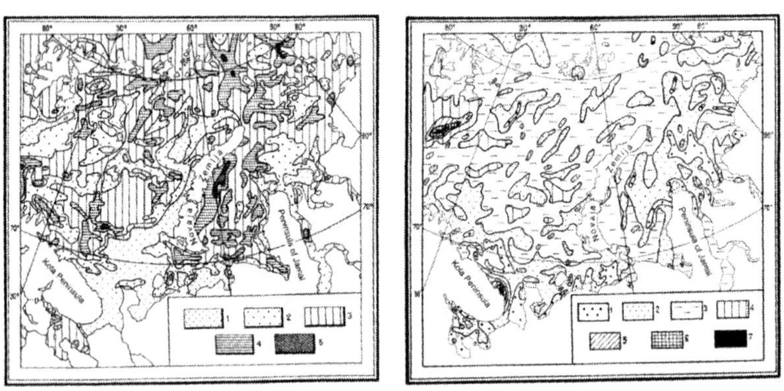

Figure 29.7 The content (%) of the clay (left, 1 - < 10; 2 – 10–25; 3 – 25–50; 4 – 50–75; 5 - > 75) and biogenic components (right, 1 - < 1; 2 – 1–5; 3 – 5–10; 4 – 10–30; 5 – 30–50; 6 – 50–70; 7 - > 70) fractions in the modern deposits on the Western Arctic Shelf.

The biogenic components of modern sediments on the Western Arctic Shelf are carbonate and siliceous remains and organic material originating from animals and plants (Gurevich, 1995). The total content of biogenic components (Fig. 29.7) is usually only exceeded by the terrigenic components. A significant proportion of the organic material of deposits in near-shore areas is a humus compound of vegetal origin. There is a very high content of sapropelic components in organic material near polar fronts. The background content of chloroformic bituminoids is less than 0.007%, but this is enriched to 0.1–03% in fine-grained sediments of frontal zones. The biogenic carbonate content in the modern sediments is generally low, and carbonate sedimentation is uncommon in polar regions in general. Across 65% of the Western Arctic Shelf it measures no more than 1%, and in 32% of the area it ranges 1 and 5% (Gurevich, 1995). The biogenic siliceous components of the Western Arctic Shelf sediments are mainly represented by diatom shells and sponge spicules. White, glass-like, "thick felt" derived from spicules of siliceous sponges was noted several times in south-western areas of the Barents Sea. Increasing contents of amorphous silica reach 2–3% or more of the total mass of modern deposits in these areas.

4. Phytoplankton

4.1. The Barents Sea

Composition. There are 308 phytoplankton species in the Barents Sea, including Bacillariophyta - 167, Dinophyta - 119, Chlorophyta - 10, Chrysophyta - 8, Xanthophyta - 3, Euglenophyta - 1 (Makarevich and Larionov, 1992).

Biogeographical structure. According to taxonomic composition and quantitative indices of phytoplankton development, the Barents Sea is divided into three provinces: Boreal Polar Province–the northern part of the sea (the Arctic waters), Atlantic Subarctic Province–the southern part affected by the Atlantic water bodies, and Novaya Zemlya Province–south-eastern shallow part of the sea (the Pechora Sea; Longhurst, 1998; Druzhkov and Makarevich, 1999). Arcto-boreal and cosmopolitan species make up 70–80 % of the phytoplankton species in the all provinces; the warmest years are marked with entering of boreal and even tropical-boreal species.

Seasonal dynamics. Seasonal development of the Barents Sea phytoplankton is thoroughly studied for the Atlantic Subarctic Province, only, and is characterized by the all principal features which are typical for boreal marine ecosystems (Fig. 29.8). There are the following features of the seasonal succession of the Barents Sea phytoplankton: (i) very low number in winter; (ii) predomination of Bacillariophyta representatives in spring; (iii) unobligatory (facultative) autumn peak. The start of the phytoplankton development depends on the rate of insolation and water column stabilization (Druzhkov *et al.*, 1997). At first phytoplankton develops in the region of the ice edges (Polar front) and in the coastal zone, and only then the development is recorded in the open sea (Slagstad and Støle-Hansen, 1991). In the Novaya Zemlya Province the spring phytoplankton development also begins under ice at water temperature below 0° C in April, and is associated with polynyas (Pautova, Vinogradov, 2001).

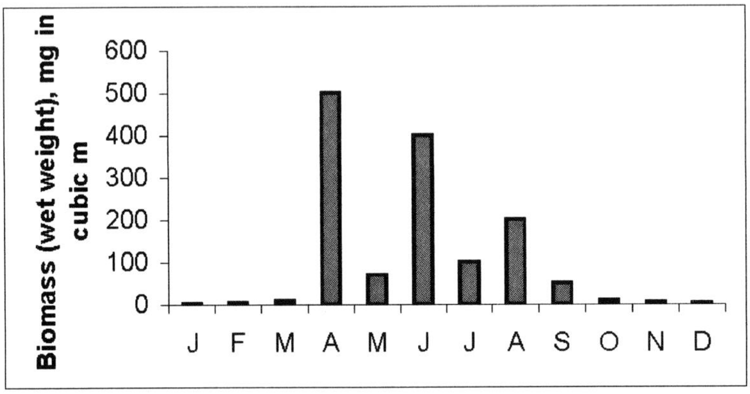

Figure 29.8 Seasonal dynamic of the phytoplankton in the southern part of the Barents Sea (Atlantic Subarctic Province; by Druzhkov *et al.*, 1997).

Long-term dynamics. Long-term phytoplankton dynamics depends on climate dynamics. In warm years the delay of phytoplankton development is recorded in the Atlantic waters (Atlantic Subarctic Province); the colder years are characterized with more early development; the shift in maximal phytoplankton development can make 2–3 weeks (Rey *et al.*, 1987). Abundance and biodiversity of the phytoplankton in warm years can be twice more than in cold years (Matishov *et al.*, 2000).

Sedimentation. In spring phytoplankton makes up about a half of all sediments from the euphotic zone. At this time the rate of sedimentation (due to phytoplankton, only) is similar in the Atlantic and Arctic waters, and makes up about 200 mgC m^{-2} per day (Olli *et al.*, 2002).

4.2. The Kara Sea

Composition. There are 264 phytoplankton species in the Kara Sea, including Bacillariophyta – 148, Dinophyta – 89, Chrysophyta – 9, Cyanophyta – 9, Chlorophyta – 7, Xanthophyta – 1 and Haptophyta – 1 (Makarevich and Koltsova, 1989; Druzhkov and Makarevich, 1999; Druzhkov *et al.*, 2001).

Biogeographical structure. In respect to taxonomic composition and quantitative indices of phytoplankton development the Kara Sea is divided into 5 regions or provinces (Usachev, 1968): (i) high latitude Arctic – north-eastern part of the sea; (ii) central – affected by the Barents Sea; (iii) Novaya Zemlya Province – practically the whole south-west region of the sea (Druzhkov, Makarevich, 1999); (iv) the region affected by the Ob-Yenisey river system (estuary Arctic interzonal province; Skarlato and Golikov, 1985); (v) the sea region adjacent to the western coast of the Severnaya Zemlya archipelago.

Abundance and production. In spring-summer phytoplankton biomass varies in superficial water layer from 6.8 g m^{-3} in the north (high latitude Arctic and central provinces) (Usachev, 1968; Druzhkov *et al.*, 2001) up to 1.2–1.4 g m^{-3} (Ilyash, Koltsova, 1981; Matishov *et al.*, 2001a) and even 5.5 g m^{-3} (Usachev, 1968) in the zone of the rivers Ob and Yenisey confluence. Phytoplankton biomass in the Novaya Zemlya Province can reach 24 g m^{-2} (in the layer 0–100 m; Druzhkov *et al.*, 2001). The phytoplankton production (primary production) differs in different regions (provinces; Vedernikov *et al.*, 1994): 39–359 (104 at an average) mgC m^{-2} per day in the south-western region of the sea (the Novaya Zemlya province), 29–147 (64, on average) mgC m^{-2} per day in the central shallow region, 25–63 (47, on average) mgC m^{-2} per day in the Ob estuary and 107–312 (224, on average) mgC m^{-2} per day in the Yenisey Bay (estuary Arctic interzonal province). The annual phytoplankton production for the whole sea is 133.5–160.2 mgC m^{-2} (Galkina *et al.*, 1994), or 14×10^6 t C (Romankevich and Vetrov, 2001).

Seasonal dynamics. The Kara Sea is covered with ice for many months in year. Hence phytoplankton seasonal dynamics still remains unstudied. Earlier phytoplankton development was considered to start in June (Usachev, 1968), but the data obtained in February-May, 1996–1997 (during expeditions on board the nuclear ice-breakers) demonstrated that vegetation season in the Kara Sea starts in April (phytoplankton biomass makes up 218 mg m^{-3}; Makarevich, 1998).

Long-term dynamics. We have no enough data to make conclusions about long-term phytoplankton dynamics in the Kara Sea. It should be noted, however, that during "warmening" of the Arctic (in 1934) the focus of maximal phytoplankton development in August-September was situated in the more northern region (about 80° N, 75° E), than in 1945 (76° N, 80° E), 1981 (75–76° N, 65° E) and 1991 (71° N, 60° E; Usachev, 1968; Matishov *et al.*, 2000).

5 Zooplankton

5.1. The Barents Sea

Composition. About 200 zooplankton taxa are recorded in the Barents Sea. Biodiversity of the zooplankton varies in different regions of the sea (Fig. 29.9): 180 taxa in Atlantic waters, 64 in Arctic waters, 100 – in the south-eastern shallow part of the sea (Pechora Sea; Timofeev, 2000; Troshkov and Gnetneva, 2000).

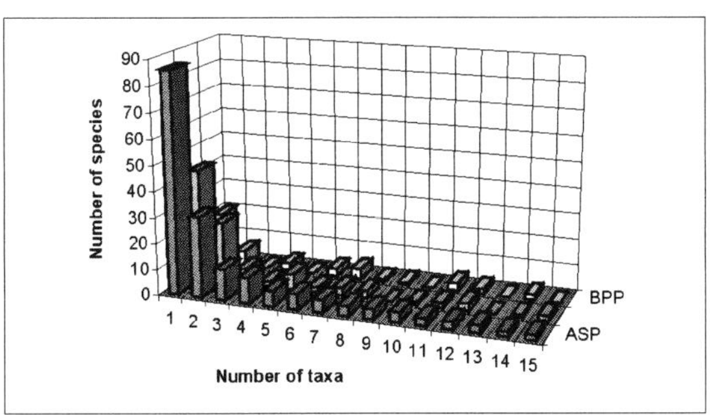

Figure 29.9 Number of zooplankton species in different parts of the Barents Sea. BPR – Boreal Polar Province, ASP – Atlantic Subarctic Province, NZP – Novaya Zemlya Province. 1 – Copepoda, 2 – Coelenterata, 3 – Decapoda (larvae), 4 – Rotatoria, 5 – Cirripedia (larvae), 6 – Euphausiacea, 7 – Hyperiidae, 8 – Ctenophora, 9 – Polychaeta, 10 – Mysidacea, 11 – Gastropoda, 12 – Appendicularia, 13 – Ostracoda, 14 – Chaetognatha, 15 – Cladocera.

Biogeographical structure. With respect to biogeographical composition of the zooplankton and level of the species predomination over the Barents Sea, three regions may be identified; their boundaries coincide with those of the biogeographical provinces for phytoplankton. Boreal oceanic species *Calanus finmarchicus* predominates in the Atlantic Subarctic Province; *Calanus glacialis* inhabiting Arctic shelf predominates in the Boreal Polar Province, and widespread coastal species *Pseudocalanus minutus, Oithona similis* and larvae of benthic invertebrates (especially of Bivalvia) predominate in the Pechora Sea. Increase in input of Atlantic waters (for example, during "warmening" of the Arctic in 1930s) may be marked with occurrence of some subtropical zooplankton species (for example, euphausiids *Stylocheiron maximum*).

Abundance. The level of zooplankton biomass in the Barents Sea is determined with the two factors: temperature regime and press from planktonophagous fishes). In the western region of the sea (Atlantic Subarctic Province) during the spring-summer period the average zooplankton biomass in the water layer 0–50 m makes up about 100–200 mg m^{-3} during cold years, and 200–400 mg m^{-3} in warm years. In the northern region of the Barents Sea (Boreal Polar Province) in summer zooplankton biomass in the same water layer doesn't exceed 50–100 mg m^{-3} (Nesterova, 1990; Tereschenko *et al.,* 1994). In the south-eastern region of the Barents Sea (Pechora Sea) zooplankton biomass varies from 100 to 300 mg m^{-3} during ice-free period (June-August) (Troshkov, Gnetneva, 2000).

Communities and trophic structure. In the Atlantic waters (Atlantic Subarctic Province) 75–90% of spring-summer zooplankton biomass consist of herbivorous *Calanus finmarchicus,* in the Arctic waters - of herbivorous *Calanus glacialis.* In the coastal waters and in the Pechora Sea the great role is played with herbivorous planktonic larvae of benthic invertebrates (which constitute up to 50% of zooplankton biomass in some regions). There are following carnivorous among zooplankton species: chaetognaths (Chaetognatha), hydromedusae (Coelenterata) and pelagic amphipods (Amphipoda, Hyperiidea; Timofeev, 2000).

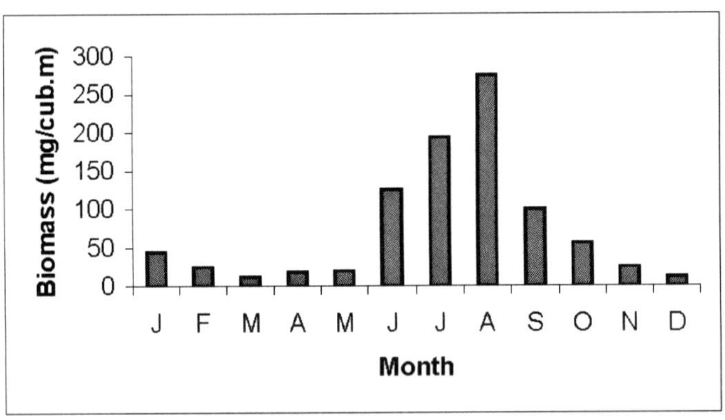

Figure 29.10 Seasonal dynamic of zooplankton biomass in the southern part of the Barents Sea (Atlantic Subarctic Province; by Zelikman, 1977).

Seasonal dynamics. The Barents Sea is a reservoir with pronounced seasonal cycles in production of organic matter in pelagic zone. The seasonal cycles in the zooplankton biomass are mainly caused by oceanographical regime of the sea and trophic factors (occurrence of food–phytoplankton, and press by consumers; Manteufel, 1941; Zelikman, 1977). Seasonal dynamics of zooplankton is only studied for the regions affected by Atlantic water bodies (ice-free in winter; Fig. 29.10). Decrease in biomass, which begins in summer, is connected with press from consumers of the primary zooplankton species *Calanus finmarchicus,* by ctenophores, jellyfishes, chaetognaths and planktonophagous fishes, and with transfer of the crustaceans into eastern and northern regions of the sea.

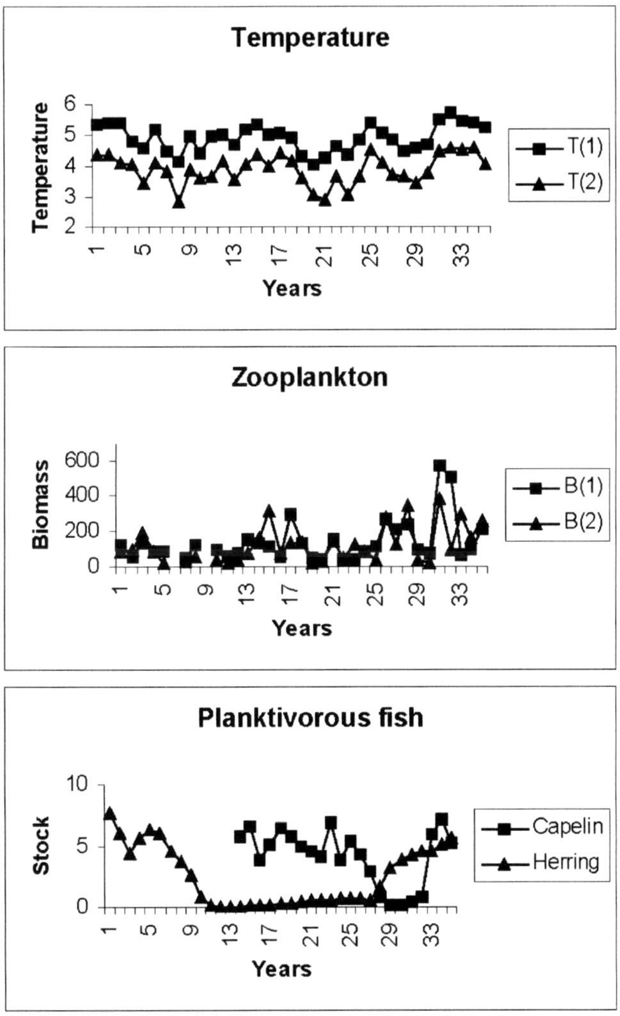

Figure 29.11 Long-term dynamic of zooplankton biomass in the south-western part of the Barents Sea (Atlantic Subarctic Province; by Timofeev, 1997, 2001). 1 – cold year, 2 – "normal" year, 3 – warm year, 4 – collapse of the herring stock, 5 – collapse of the capelin stock. III – transect from the Nord Cape to open sea, VI – Kola transect (along 33°30′ E).

Long-term dynamics of zooplankton biomass in the western part of the sea (Atlantic Subarctic Province) depends on the two processes (Timofeev, 1997, 2001): (i) intensity in inflow of Atlantic water (it determines the amount of crustaceans *Calanus finmarchicus* transported with these water bodies); and (ii) press from planktonophagous fishes (Fig. 29.11).

5.2. The Kara Sea

Composition. Eighty zooplankton taxa (with exception of Protozoa and planktonic larvae of benthic animals) have been recorded in the Kara Sea. Taxonomic diversity in different regions of the Kara Sea is very similar: south-western region – 43, south-eastern region – 44 and northern region – 40 species (Timofeev, 1989; Halsband and Hirche, 1999; Fetzer *et al.,* 2002). In the south-eastern, shallow and freshened region of the Kara Sea the central role is played by freshwater Cladocera and Copepoda, and larvae of benthic invertebrates.

Biogeographical structure. In respect to biogeographical composition of zooplankton and species predomination the Kara Sea may be divided into three regions: (i) south-western, where *Calanus glacialis* inhabiting Arctic shelf predominates; (ii) south-eastern, where the central role is played by the Arctic endemic, neritic and brackish-water species *Drepanopus bungei;* (iii) northern, predominated by representatives of genus *Calanus* (*C. glacialis, C. finmarchicus* – boreal North-Atlantic species, *C. hyperboreus* - Arctic oceanic species).

Abundance. During ice-free season distribution of zooplankton abundance and biomass is determined by oceanographic conditions. Zooplankton biomass in the south-western region of the Kara Sea varies from 50 to 300 mg m^{-3} (Ponomareva, 1957; Fomin and Petrov, 1985; Vinogradov, M.E. *et al.,* 1994 a), in the south-eastern region (Ob-Yenisey shallow) – 100–1000 mg m^{-3} (Ponomareva, 1957; Vinogradov *et al.,* 1994 b). There are no data on zooplankton biomass for the northern region of the Kara Sea.

Communities and trophic structure. In the south-western and northern regions of the Kara Sea zooplankton is predominated by genus *Calanus,* typical herbivorous species which fattening seasons fall on spring-summer phytoplankton development. In these regions carnivorous are mainly represented with Chaetognatha and Coelenterata (Timofeev, 1989). In the south-eastern part of the Kara Sea omnivorous species predominate; their diet consists of phytoplankton and suspended organic matter transported with flow of rivers Ob and Yenisey.

Seasonal dynamics. Zooplankton seasonal dynamics has only been studied in the southern region of the Kara Sea (the Dikson harbor, 1955–1956; Chislenko, 1972). From November till May-June zooplankton biomass is very little and doesn't exceed 30 mg m^{-3}. June is marked with drastic increase, annual maximum is recorded in August-September (up to 375 mg m^{-3}), whereas October is characterized with dramatic decrease in biomass (Fig. 29.12). In the south-western region of the Kara Sea zooplankton biomass in winter doesn't exceed 30–40 mg m^{-3} (Vinogradov *et al.,* 2001).

Long-term dynamics. We have no enough data on the zooplankton long-term dynamics in the Kara Sea. It should only be noted that in 1936 (during "warmening" of the Arctic) occurrence of indicator species of Arctic water bodies was significantly less than in 1981 (Matishov *et al.,* 2000).

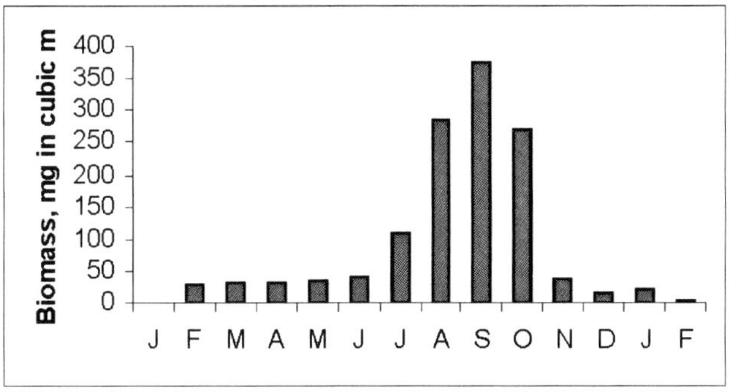

Figure 29.12 Seasonal dynamic of zooplankton biomass in the Dikson harbor, Kara Sea (by Chislenko, 1972).

6. Bottom algae and invertebrates

6.1. Biological diversity, biomass and structure

The number of known species of benthic invertebrates decreases from west to east: Barents Sea–2499 species; Kara Sea–1580 species (Northern Sea Route, 1998). The highest benthic biodiversity in the Barents Sea is observed on shoal and hard bottom near the coast of Kola Peninsula and archipelagos, the lowest is in its southwest deepwater part. In the Kara Sea the highest species diversity is registered on hard bottoms and small depths along the Novaya Zemlya coast, in Karskiye Vorota and Yugorskiy Shar Straits. The lowest biodiversity is typical for regions influenced by the Ob and Yenisey River discharge. It is also rather low in northern (deepwater) regions of the Novaya Zemlya Trough.

Distribution of bottom organism biomass (Fig. 29.13) is similar to that of biodiversity. In absolute expression, macrobenthos biomass varies from 0.1 g m^{-2} up to 12 kg m^{-2} in the Barents Sea and from 1.5 up to 400 and more g m^{-2} - in the Kara Sea (Kiyko and Pogrebov, 1997 a). The highest values of biomass as those of species diversity correspond to shallow depths along the coast. Thus, near Svalbard and in the southeast part of Spitsbergen Bank the biomass ordinary exceed 1.5–2 kg m^{-2}, with main contribution made by sponges and bivalves. Average biomass up to 300 g m^{-2} on the Central Bank in the Barents Sea is determined by high density of sea urchins of the *Strongylocentrotus sp.* and, on some sites, by sea cucumbers of *Trochostoma sp.* In the southern part of the Barents Sea, on the Murman, Geese, Kanin Banks, Kolguyev Island and Pechora Sea shoals the biomass averages 100–300 g m^{-2}. Bivalves and barnacles form its base here. The maximum biomasses for the Murman slope, exceeding 1.5 g m^{-2}, are obtained near the offshore Seven Islands owing to aggregations of mollusks of *Chlamys islandica* and *Modiolus modiolus.* From the Barents Sea side of Novaya Zemlya, benthic biomass on some sites also exceeds 1 kg m^{-2}. Bivalves, hydroids and barnacles play the main role here. The highest biomass is registered in the Karskiye Vorota and Yugorskiy Shar Straits where, due to mussels and red algae, the total benthic biomass reaches 10–12 g m^{-2}. The Kara Sea side of Novaya Zemlya is characterized by biomass 200–400 g m^{-2}. The main roles here are played by sea urchins of *Strongylocentrotus sp.* and mollusks of *Astartidae spp.* Biomasses up to 300 g m^{-2} are observed in the Kara Sea as well in the region of the Baydaratskaya Inlet. Mainly they are formed by mollusks of *Serripes groenlandicus, Ciliatocardium ciliatum* and *Astartidae spp.*

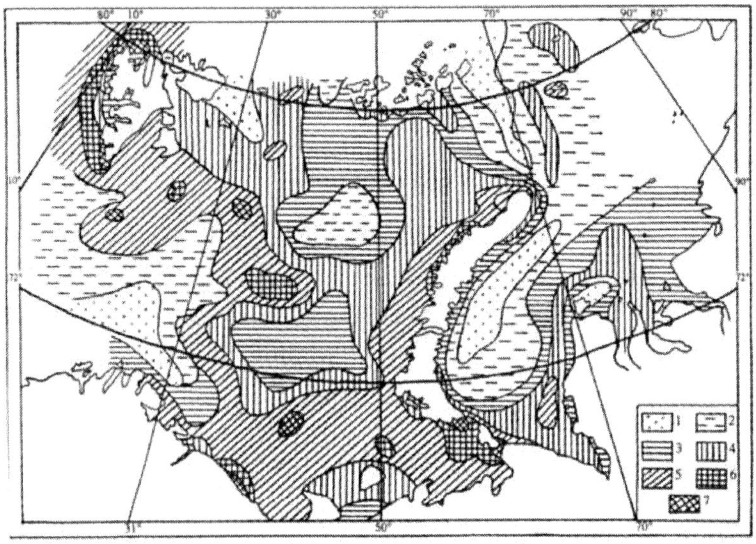

Figure 29.13 Distribution of total benthic biomass (g m^{-2}) on the Russian West Arctic shelf according to the result of 1991–1994 cruises. 1 – less than 10; 2 – 10–25; 3 – 25–50; 4 – 50–100; 5 – 100–300; 6 – 300–500; 7 – more than 500.

Regions with especially low biomass (no more than 25 g m^{-2}) enter the Barents Sea from the west, occupying the Bear Island Trough and the north-western part of the Norwegian Trough. Regions with reduced biomass are located also in the central deepwater part of the Barents Sea: Central and Northeastern Deeps and, partly, on the Central Plateau. In the Kara Sea the minimum biomasses (no more than 10 g m^{-2}) are registered in the deepest parts of the Novaya Zemlya Trough and the Saint Anna Trough. The maps of biomass, to a great extent, may be interpreted by bathymetry and bottom sediment maps: regions of increased biomass correspond to the bottom relief rise and mainly coincide with regions of hard bottom and strong current. The opposite environment would characterize the biomass minimum.

A decisive influence on the results of division of the studied area into trophic zones (Figs. 29.14 and 29.15) is rendered by morphological, lythological and hydrodynamic peculiarities of areas. Areas with abundance of sessile and motile filter-feeders (SFF and MFF) usually coincide with the largest relief bulgings while those with prevailing surface and subsurface deposit-feeders or burrowing (SDF and BDF) are typical for the largest flexures and depressions. The greatest part of the Barents Sea central regions is occupied by BDF zone (Fig. 29.14). Regions of SFF and MFF prevailing are found off Svalbard, Franz Josef Land, Novaya Zemlya and Murman, as well as on the Spitsbergen and Kanin Banks and the Kolguev

Island and Pechora Sea shoals. The largest SDF zone in area is registered on the northeast of the Barents Sea. In the Kara Sea the largest area is occupied by SDF zone (Fig. 29.15). Zone of BDF prevailing is situated mainly in deepwater regions of the Novaya Zemlya Trough. Areas with SFF prevailing enveloped by a narrow band the Novaya Zemlya coast, occupies the Karskiye Vorota and Yugorskiy Shar Straits. MFF zone is observed along the Novaya Zemlya coast, occupying the Baydaratskaya Inlet, spreading along the west and north coasts of Yamal and wedging north-ward.

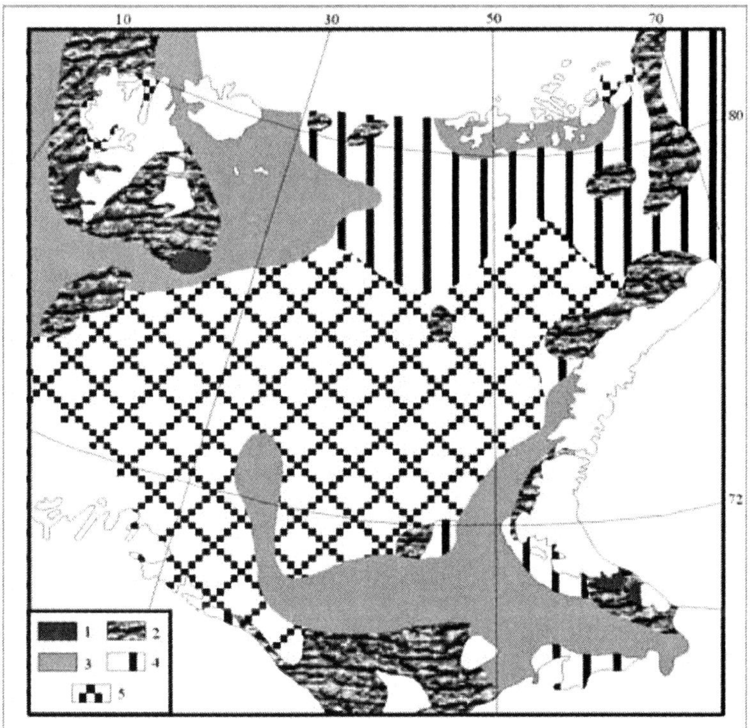

Figure 29.14 Trophic zones, singled out on the area of the Barents Sea shelf according to the result of 1991–1994 cruises. 1 – autotrophs; 2 – sessile filter-feeders; 3 – motile filter-feeders; 4 – surface deposit-feeders; 5 – subsurface deposit-feeders.

Autotrophs (AT) are prevailing at small depths in the straits, off the mainland and island. It may be assumed that the areas where AT are prevailing are more numerous but they are located on shoals that are rather difficult for studying aboard a large vessel. Sites with carnivores and herbivores prevailing are dispersed in patches all over the investigated area. This group is not represented at our scheme since it is considered to be a migration component of fauna and depends in its distribution upon distribution of other benthic groups (Kuznetsov, 1970).

On the whole, trophic zonality of the Kara Sea is less patchy in comparison with that of the Barents Sea, which may be due to the more simple morphology of the

Kara Sea bottom. Most distinctly trophic zonality is observed in western part of the Kara Sea where regular interchange of AT-SFF-MFF-SDF-BDF belts may be seen clearly along the depth gradient.

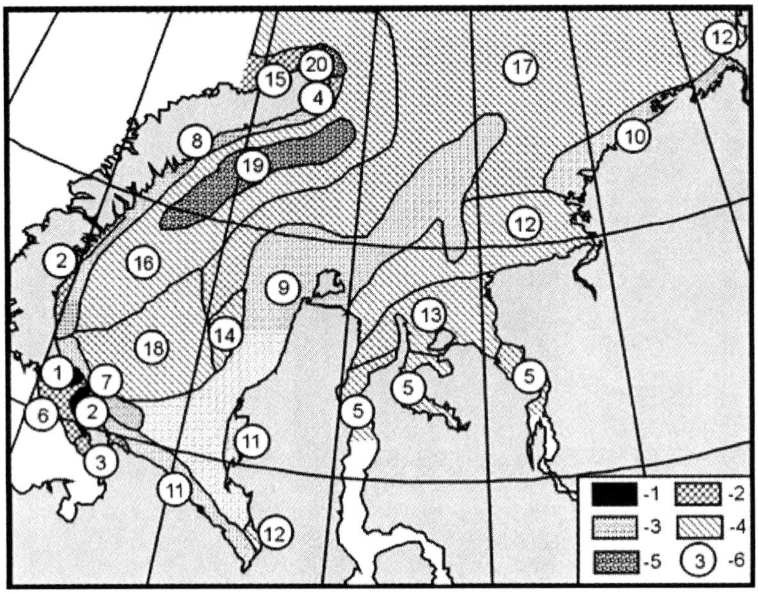

Figure 29.15 Trophic zones and bottom communities, singled out on the Kara Sea shelf. 1 – autotrophs; 2 – sessile filter-feeders; 3 – motile filter-feeders; 4 – surface deposit-feeders; 5 – subsurface deposit-feeders; 6 – bottom communities: 1 – Rhodophyta var.; 2 – Lithothamnion sp.; 3 – Spongia var. + *Mytilus edulis;* 4 – Spongia var. + *Strongylocentrotus sp.;* 5 – *Saduria sibirica;* 6 – *Balanus balanus* + *Balanus crenatus;* 7 – *Arca glacialis;* 8 – *Astarte crenata;* 9 – *Tridonta borealis;* 10 – *Tridonta borealis* + *Ciliatocardium ciliatum;* 11 – *Serripes groenlandicus* + *Ciliatocardium ciliatum;* 12 – *Portlandia arctica;* 13 – *Portlandia aestuariorum;* 14 – *Macoma calcarea;* 15 – *Mya truncata* + *Nuculana pernula;* 16 – *Ophiopleura borealis;* 17 – *Ophiocten sericeum;* 18 – *Ophiopleura borealis* + *Ophiocten sericeum;* 19 – *Elpidia glacialis* + *Ophiopleura borealis;* 20 – *Trochostoma sp.*

By analyzing the community structure of the Barents and Kara Seas it is shown that dominating species biodiversity here is comparatively high while the area of communities, formed by them, as well as their share in total benthic biomass, varies in a wide range. The most diverse forms among them are bivalves, echinoderms and polychaetes. A large area in the central part of the Barents Sea is occupied by communities of worms (polychaetes and sipunculides *Golfingia sp.*) and sea cucumber *Trochostoma sp.* (Fig. 29.16). In the southern part of the Barents Sea, on Kanin Bank, the Kolguyev Island and Pechora Sea shoals the community of bivalves is a typical one. Such clam species as *Tridonta borealis, Serripes groenlandicus, Ciliatocardium ciliatum* and *Macoma calcarea* prevail here. Communities of *Strongylocentrotus sp., Chlamis islandica, Balanus sp., Lithothamnion sp.* are found usually near the coasts of Svalbard, Novaya Zemlya, Kola Peninsula, Vaygach Island and on the shoals of Geese and Central Banks. The largest areas in the Kara Sea are occupied by the brittlestar community: *Ophiopleura borealis, Ophiocten sericeum,* and bivalves: *Tridonta borealis, Portlandia arctica,* and *P. aestuariorum* (Fig. 29.15).

Figure 29.16 Bottom communities, singled out on the Barents Sea shelf according to the results of cluster analysis. 1 –*Ophiopleura borealis* + *Hormosina globulifera;* 2 –Polychaeta + Sipunculoidea (*Golfingia sp.*); 3 –*Trochostoma sp.;* 4 – *Elliptica elliptica* + *Astarte crenata;* 5 – *Brisaster fragilis;* 6 – soft bottom community adjacent to Svalbard; 7 – community of Saint Anna Trough slopes; 8 – *Strongylocentrotus sp.* + *Ophiopholis aculeata;* 9 – shoal community of sessile filter-feeders adjacent to Svalbard; 10 – shoal community of sessile filter-feeders on *Lithothamnion sp.;* 11 – shoal community adjacent to western coast of Novaya Zemlya and Vise Island; 12 – *Tridonta borealis;* 13 – *Ciliatocardium ciliatum; Macoma calcarea* + *Serripes groenlandicus;* 14 –Bivalvia; 15 – *Macoma fusca.*

6.2. Long-term benthic population changes

A general review of obtained data had not shown any striking abnormalities in benthos status and was mainly indicative of normal natural features over the Barents and Kara Seas area. At almost all studied sites of the Kara Sea, composition, abundance and bottom population structure were just the same as described earlier. In any case, data for their statistical analysis appeared to be unavailable.

For the Barents Sea, the results of our survey and of ones carried out earlier, some differences, which were statistically significant, were observed. The results of the statistical assessment of long-term changes in the bottom communities of the Barents Sea fulfilled on the basis of three-way ANOVA show that the total benthic biomass value has no significant differences compared to that registered in 1920–1930ies (Brotskaya and Zenkevich, 1939; Kiyko and Pogrebov, 1997a). Negligible distinction, observed at some sites, most likely should be regarded as being engendered by natural fluctuations of benthic species population density (if they are also not caused by the errors of random sampling, what is rather probable). The total biomass values, obtained during the surveys of the end of 1960's (Antipova, 1975), significantly differ both from the values of 1920–1930's and from our values of 1991–1994's (for 1968–1970's notes significant reduction in benthic biomass all over the Barents Sea; Antipova, 1975; Kiyko and Pogrebov, 1997a). Biomass lowering on the south has reached 50–70% of the previous values and in the richest southeastern part of the Barents Sea–40–60%. Decrease in abundance has affected primarily arctic-boreal species, which from the base of the Barents Sea bottom dwellers (Zenkevich, 1963).

According to Yu.I. Galkin (1987) an opinion to which we also subscribe, the decrease in biomass, registered in collections of 1968–1970's, is the result of a prolonged rise of temperature which was observed in the Barents Sea in 1940–1960's and had told on arctic-boreal species especially negatively.

Further studies fulfilled in the southeastern segment of the Barents Sea in 2000–2002 had revealed decrease in the benthic biomass as compared to the 1991–1994 data. The average biomass decrease was two-fold and in some areas 3–5-fold (Kiyko *et al.,* 2002). At the same time, benthos composition and structure of both survey periods were quite similar. Statistical analysis of data has shown that significant lowering of benthic organism size took place. That caused decrease in total biomass while the community structure remained the same. The revealed long-term dynamic could be caused by climatic changes, natural fluctuations of the population abundance, geochemical processes or man-induced impacts. However, definite knowledge of real causes of these changes may be obtained only after additional studies.

Summarizing all the result obtained during the study of north-western shelf of Russian arctic in 1991–1994, we can regard contemporary ecological state of the Barents and Kara Sea benthos (except some coastal regions) as being close to the average long-term norm.

Considerable man-induced disturbances of macrobenthic structure (in comparison with biological norm for studied abiotic environments) on trans-regional level were discovered only in rare cases in the Kola Bay and in the vicinity of the Novaya Zemlya coasts. First of all, they manifested themselves in a decrease in biodiversity and biomass of bottom dwellers. However, confirmation of cause-and-effect relationships between disturbances of the benthic structure and level of human impact in particular regions needs more thorough research.

6.3. Pollutant concentrations in bottom organisms

The comparative estimate of persistent organic pollutant (POP), trace metal (TM) and radionuclide (RN) concentrations in benthic invertebrates and algae of various region of the Barents and Kara Seas shelf did not reveal any meaningful anomalies (Kiyko and Pogrebov, 1997b). Their variation was only dependent on taxonomic classification and was practically unaffected by the location of sampling site in the water basin.

POP concentrations (ng g^{-1} wet weight) in benthic organisms of the Barents Sea are as follows: alpha-HCH – 0–32.4; gamma-HCH – 0–39.0; DDT – 0–23.3; PCB – 0–11.4. TM concentrations (mg g^{-1} dry weight) in benthic organisms of the Barents Sea varies within the following limits: Zn – 10–173; Fe – 11–663; Sn – 0.2–2.8; Mn – 0.4–44.3; Ni – 0.4–2.9; Cu – 0.3–2.1; Cd – 0.1–2.2; Pb – 0.2–2.0; Co – 0.1–1.1 (Kiyko and Pogrebov, 1997b). No present of gamma-emitting man-made RN has been registered in any one sample at any one station in 1992 (Kiyko and Pogrebov, 1997b). The results of sample radiometric analysis are indicative of the background content of gamma-emitting RN (on the level of 10 Bq kg^{-1} of dry weight), and correlation between the registered rate of count of beta-particles and K-40 content in the preparations indicates that it is particularly this RN which is responsible for basic contribution into the total beta-activity. Gamma-active K-40 (Bq kg^{-1} dry weight) and beta-activity in benthic specimens of the Barents Sea are as follows: K-40 – 1–114; beta particles counting fate (l/S) <0.2–1.43 (Kiyko and Pogrebov, 1997b). Measurements of radionuclide concentrations in biological specimen from Chernaya Inlet (where nuclear tests took place) testified to absence of their significant bioaccumulation by benthic and invertebrates (Pogrebov *et al.*, 1997). On the whole, the results of radionuclide concentration measurements were in the limits of their variability known from literature (Matishov *et al.*, 1994).

The comparison between the pollutant concentration in the benthic organisms of the Barents Sea and the literature data on other areas may suggest that in our studies the POP and TM concentrations are basically much lower and are only comparable to similar concentrations in areas referred to as background. On the other hand, according some published data (Moore and Ramamurti, 1987), the levels of zinc and cadmium in the bottom dweller tissues are close to these from the impacted areas.

7. Fishes

7.1. The Barents Sea

Ichthyofauna of the Barents Sea counts not less than 150 species of fishes and fish-like vertebrates (Andriashev, 1954), which belong to 53 families. The most specious are families Gadidae (18 species), Zoarcidae (13), Cottidae (12), Pleuronectidae (9), and Salmonidae (7). Approximately a third of total number of species are rare or threatened species. About 90–95 species occur permanently, and 30–35 of them is used commercially (Rass, 1993).

Vast estuarine areas are not typical for the Barents Sea, with exception of the south-eastern part, which is freshened by the Pechora River outflow. Thus, a brackish-water complex of fish species is represented in the sea insignificantly. Only a few freshwater species coming into the brackish waters near mouths of rivers occur (burbot, pike, ide, whitefish).

Some anadromous fish species also present, life cycle of which is related partly with fresh waters. They reproduce in fresh waters but feed in the sea. They are valuable salmonids (Atlantic salmon, sea trout, Arctic char, introduced pink salmon) and also whitefish and sea lamprey. Anadromous fishes are fished mainly in rivers during their spawning migrations. At present the most important is salmon fishery. Sea trout, Arctic char, and whitefish are listed practically in all rivers where fishery control exists, but they are collected as by-catch only. Knowledge about sea period of life of sea trout, Arctic char, and whitefish in the Barents Sea is poor. Sea trout and whitefish make short migrations into the sea, whereas the Arctic char make much longer sea migrations, almost similar with the Atlantic salmon.

Majority of fishes of the Barents Sea are marine species, which spend the whole life in salt water and reproduce here. Types of their life cycles are connected with their relations to temperature conditions. Three types of species can be distinguished. Many warm-loving boreal species come into the Barents Sea sporadically, in "warm" periods and when inflow of warm Atlantic waters enlarges (blue whiting, whiting, Atlantic argentine and others). Fishes of this group do not make large schools in the Barents Sea and are not commercial here. Group of boreal-arctic species reproduce mainly in more southern areas (coast of northern Norway, Lofoten Islands), and use highly productive areas of the Barents Sea as feeding grounds (Atlantic cod, haddock, pollack, Atlantic herring, capelin, red-fishes). They reach northern and easternmost areas of the sea during feeding migrations. The size of their feeding grounds and prolongation of migrations depend greatly from temperature conditions. For example, in "warm" and "normal" years (1993–1996) Atlantic cod make its main summer-autumn feeding migration into southern and eastern parts of the Sea but in "cold" years (1997–1998) they migrate into the area of the Bear Island and Spitsbergen. Species of Arctic fauna (Arctic cod, *Eleginus navaga,* some *Lycodes*) are related in their distribution with arctic cold waters. In periods of cold snap the area of their distribution in the Barents Sea greatly enlarges.

In relation to habitat preference the following groups can be distinguished: pelagic, which live in mid-water (capelin, Atlantic herring, Arctic herring *Clupea pallasii suworovi*) and benthic or bentho-pelagic, which live on bottom or near bottom (Atlantic cod, pollack, haddock, wolf-fishes, plaices, halibut, red-fishes). Polar cod living among arctic ice is called kryopelagic species. Pelagic capelin, Polar cod are abundant fishes and play a large role in the ecosystem, as many carnivorous fishes, sea birds and marine mammals prey on them.

Strong fishery exists mainly in the southern, western and central parts of the sea. About 40 fish species regularly occur in trawl catches (Karamushko et al., 2001), among them 7 species of Pleuronectidae, 6 species of Gadidae, 4 species of Rajidae. Pelagic species comprise about 19.5%, benthic and bentho-pelagic comprise 80.5%. Fishery is based on Atlantic cod *Gadus morhua,* capelin *Mallotus villosus,* haddock *Melanogrammus aeglefinus,* Polar cod *Boreogadus saida* and plaice *Pleuronectes platessa.* Pollack *Pollachius virens,* Atlantic herring *Clupea harengus,* red-fishes (*Sebastes marinus* and *S. mentella*), three species of wolf-fishes (*Anarhichas denticulatus, A. lupus* and *A. minor*), European plaice *Hippoglossoides platessoides limandoides* are also important. At present, stocks of all commercially used fishes are subjected to over-fishing. Fishery statistic of the International Council for the Exploration of the Sea (ICES) is carried on for the Barents Region, which includes besides the Barents Sea itself also the northeastern part of the Norwegian Sea and West Spitsbergen. In the Barents Region the fishery was most intensive in 1960–1970's. The total catches of fish increased from 2.3–2.8 to 4.3–4.6 million tons per year, what comprised about 5% of the total world catches. Then fish catches decreased significantly. In 1997–1998 about 2.3–2.7 million tons of fishes per year was caught in the Barents Region (Borisov et al., 2001). Accounting stock of all biological resources (invertebrates inclusive) in the Barents Region is estimated at approximate level of 28 million tons, with variations from 33–35 million tons (in 1950 and 1954) to 5.6–8 million tons (1986 and 1989; Borisov et al., 2001). Content of pollutants and radionuclides in fishes caught in the Barents Sea at present is significantly lower than the levels of maximum allowable concentration (Ilyin, 2001; Matishov et al., 2001b).

7.2. The Kara Sea

Fish fauna of the Kara Sea (including bays, inlets and mouth areas of numerous rivers) counts about 83 species of fishes and fish-like vertebrates (Esipov, 1952; Andriashev, 1954; Andriashev and Chernova, 1994), which belong to 28 families. The most specious are families Zoarcidae (14 species), Cottidae (12), Coregonidae (9). Among them 54 species (or 62.1%) are marine, 14 (16.1%) are semi-anadromous, 1 euryhaline, 18 (20.7%) are freshwater species which occur in deltas of Kara, Ob, Yenisey, Pyasina rivers and brackish water areas nears other streams entering the sea.

Distinguishing feature of the Kara Sea fish fauna as compared with the Barents Sea is quite large number of freshwater fishes, which occur offshore. This is due to much larger influence of freshened waters from outflow of numerous rivers. Characteristic is also a small number of "guest" warm-loving species coming from the Barents Sea (Atlantic cod, Atlantic herring, haddock, lump-sucker *Cyclopterus lumpus*).

Biology of anadromous and semi-anadromous fishes during marine period of their life is purely studied in the area. It is known that Siberian sturgeon *Acipenser baeri stenorhynchus,* whitefishes *Coregonus nasus, C. sardinella, C. muksun* and *Stenodus leucichthys nelma* feed inside bays and inlets. Only *C. autumnalis* spread more wide in the coastal zone of the sea.

Marine fishes belong to 15 families: Squalidae (1 species), Rajidae (1), Clupeidae (1), Osmeridae (1), Myctophidae (1), Cottidae (11), Cottunculidae (1), Agonidae (2), Cyclopteridae (3), Liparidae (6), Lumpenidae (3), Zoarcidae (14), Ammodytidae (1), Gadidae (5), Pleuronectidae (3), total number is 54 species. The most specious are families Zoarcidae and Cottidae, which count 25 species (or 46.3%).

Arctic species predominate, which permanently live and reproduce in water with negative temperatures (to –1.96°C; *Gymnelus esipovi, G. andersoni* and other). Some of these species are distributed circumpolarly in the Arctic (*Lycodes polaris, Triglopsis quadricornis polaris, Aspidophoroides olriki, Liopsetta glacialis, Boreogadus saida, Leptagonus decagonus*). Endemic species in the sea are absent.

Benthic and near-bottom species predominate. The Polar cod is cryopelagic abundant species which play important role in food webs. In relation to vertical distribution three categories of fishes can be distinguished. Fishes of coastal shallow waters occur at a depth less than 50 meters (*Myoxocephalus scorpius, Triglopsis quadricornis polaris, Artediellus scaber, Liparis tunicatus*). Next group of species occur at a depth of a few meters to 400–500 meters (*Icelus bicornis, Liparis fabricii, Lycodes rossi, L. pallidus, Gymnelus andersoni, Boreogadus saida*). Deepwater species occur at a depth from 100–250 to 700 m (*Triglops pingelii, Careproctus reinhardti, Cottunculus sadko, Leptagonus decagonus, Lycenchelys sarsi, Lycodes seminudus*). Fishes of the two latter groups predominate. The shallow water coastal zone is poorly inhabited by fishes due to heavy ice present here permanently during almost the whole year. So conditions for fishes are adverse. Reach fish population in coastal zone occur only in summer time in warm parts of bays near mouths of rivers (Ob, Yenisey Rivers and other).

In relation to salinity three categories of fishes can be distinguished. Highly-stenohaline species live in water under salinity from 33.3–34.5 to about 35‰ (*Triglops pingelii, Cottunculus sadko, Leptagonus decagonus, Lycodes seminudus, L. rossi, L. pallidus, Gymnelus esipovi, G. andersoni*). Moderately-stenohaline species occur under salinity from 29 to 35‰ (*Myoxocephalus scorpius, Triglopsis quadricornis polaris, Artediellus scaber, Gymnelus knipowitschi*), e.g. all coastal forms belong to this group. Euryhaline species occur under salinity from 29 to 35‰ (*Gymnocanthus tricuspis, Icelus bicornis, I. spatula, Lycodes agnostus, Boreogadus saida*).

Quantitative studies of the Kara Sea fish fauna are absent. The following species are numerous in trawl catches made in the offshore shelf areas: *Icelus bicornis, Boreogadus saida, Gymnelus andersoni, Gymnelus knipowitschi, Lycodes pallidus, Artediellus scaber, Triglops pingelii, Gymnocanthus tricuspis.* In the coastal zone *Triglopsis quadricornis polaris* is common. Food of *Coregonus autumnalis* in the Ob Bay consists mainly from this species. The most important item in the food webs in the open sea is Polar cod.

In general, the level of knowledge on the Kara Sea fish is many times less than that of the neighboring Barents Sea. It depends on the lesser fishery importance of the region and its more unfavorable ice conditions. Ichthyological studies have been conducted here mainly incidentally together with hydrobiological or other researches. The fishes of estuarine areas are better known. Offshore fauna is studied better than in the coastal zone, where trawling is not possible. Eastern coast of Novaya Zemlya ichthyologically is almost unstudied. Reproduction of fishes in the Kara Sea and other aspects of their biology are fragmentary, eggs and larvae for 18 species are only known.

Commercially valuable in the sea are semi-anadromous and anadromous species fished in estuarine waters. They are Siberian sturgeon, and several forms of whitefishes.

8. Food webs

8.1. The Barents Sea

The Barents Sea ecosystem is characterized by branchy interactions between pelagic, bottom and terrestrial communities. Main processes are realized in the water column (Zenkevich, 1963; Marti and Martinsen, 1969). Major part of organic compounds is produced here and is consumed by bottom communities, fishes and birds. Due to that, food webs of the water column are studied better as compared to the seabed communities.

Food webs of the Barents Sea pelagic system are comparatively short and simple. In general, trophic structure of the pelagic community in the Atlantic Subarctic Province may be described as follows: phytoplankton – zooplankton (*Calanus* and krill) – capelin (herring) – cod (Fig. 29.17). Attempts to evaluate the energy flow in the pelagic food webs quantitatively were made more than once (Marti and Martinsen, 1969; Timofeev, 1990, 1996; Nöthig et al., 1994; Sakshaug et al., 1994). However, due to considerable long-term variability of the plankton and fish production, intensity of fishery and other factors, this evaluation should be considered to be only of historic value (as old photos). For example, up to 1980's, cod was feeding mainly on capelin but in 1980's (due to collapse of the capelin stock caused by over-fishing) it was forced to feed on krill and its own fry (cannibalism; Orlova and Matishov, 1993). In its turn, decrease in the capelin resources caused increase in the large plankton species abundance (e.g. *Temisto sp.;* Dalpadado et al., 2002), which feed on *Calanus*. As a result, redistribution of the energy flows occurred and trophic structure changed leading to the strengthening of the levels nearest to the phytoplankton.

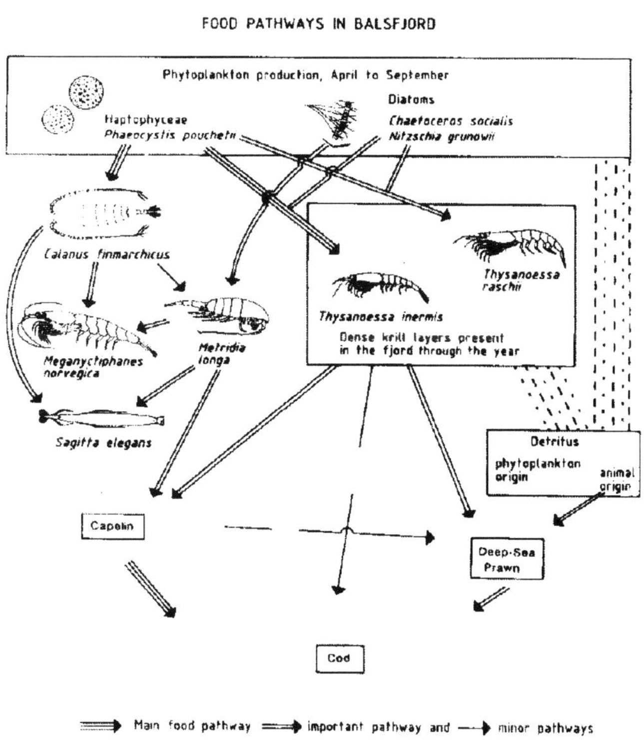

Figure 29.17 The food pathways in Balsfjord, Northern Norway (by Falk-Petersen et al., 1990)

Capelin *Mallotus villosus,* Polar cod *Boreogadus saida,* Atlantic herring *Clupea harengus* and also young of many other Barents-sea fishes are zooplankton feeders. Predominanting benthic and near-bottom fishes feed mainly on benthic and near-bottom organisms. Wolf-fishes (*Anarhichas denticulatus, A. lupus* and *A. minor*) prey predominantly on echinoderms, mollusks, crustaceans and fishes (Orlova et al., 1989). European plaice *Hippoglossoides platessoides limandoides* and other pleuronectids feed usually on polychets, echinoderms and mollusks (Berestovsky, 1989). Haddock *Melanogrammus aeglefinus* used more often krill and shrimps, in lesser extent polychets, echinoderms and bivalves, also capelin and other fishes (Kovtsova et al., 1989). Cod *Gadus morhua* prefer to eat various fishes (herring, capelin, young cod and haddock, polar cod, young places, *Lumpenus, Lycodes,* Cottidae), and big crustaceans, but use totally about 200 feeding objects (Treska... , 1996). Red-fishes prey mainly on herring, capelin, young cod-fishes, also as on crustaceans Euphausiidae, Hyperiidae, *Pandalus sp.* (Barsukov et al., 1986).

Many of sea birds species (*Fratercula arctica, Rissa tridactyla, Cepphus grille, Uria lomvia* and others) feed on school pelagic fishes (capelin, herring, polar cod), also they prey on many near-bottom fish species living in sublittoral zone and shallow shelf water (young cod-fishes, wolf-fishes, *Lumpenus sp.,* Cottidae sp., *Liparis sp., Lycodes sp.* etc.). Some other species of sea birds prey mainly on benthic invertebrates (Krasnov et al., 1995).

Bearded seal *Erignathus barbatus* feeds usually on big marine invertebrates. Ringed seal *Phoca hispida* consumes mainly fish and shrimps. Harp seal *Histrophoca groenlandica,* Common seal *Phoca vitulina,* Grey seal *Halichoerus grypus,* Hooded seal *Cystophora cristata* also used fishes and big invertebrates. Walrus *Odobenus rosmarus* which still inhabit the Frantz Joseph Land, Spitsbergen and Novaya Zemlya areas, prefers big forms of the bivalve mollusks (Atlas..., 1980). The ringed seals and young walruses are important as the main food for polar bears *Ursus maritimus.*

Whales of the Subordo Mustacoceti (baleen whales) feeds mainly on zooplankton. From this group, the Blue Whale *Balaenoptera musculus,* Fin Whale *Balaenoptera physalus,* Seiwal *Balaenoptera borealis,* Minke Whale *Balaenoptera acutorostrata,* Humpbacked whale *Megaptera novaeangliae* and Bowhead whale *Balaena mysticetus* occur sporadically in the Barents Sea (Atlas..., 1980).

Bottle-nosed dolphin *Tursiops truncates* feed mainly on benthic invertebrates and fishes. Atlantic White-sided Dolphin *Lagenorhynchus acutus* preys usually on school pelagic fishes (herring, salmon) and squids. White-beaked dolphin *Lagenorhynchus albirostris* catch usually cod, capelin, navaga, herring, squids. Carnivorous Killer whale *Orcinus orca* kill seals, walruses, whales, but feeds also on fishes (cod, capelin, salmons, skates, halibuts) and squids. Long-finned Pilot Whale *Globicephala melaena,* Harbour Porpoise *Phocoena phocoena,* Giant sperm whale *Physeter catodon,* Beluga Whale *Delphinaptesus leucas,* Narval Whale *Monodon monoceros* feed usually on fish (herring, navaga, Atlantic cod, polar cod, places, skates and others) and squids (Atlas..., 1980).

8.2. The Kara Sea

Food webs in the Kara Sea Region are well studied mainly for the Ob and Yenisey estuaries. Feeding preferences of the most abundant species leaving here are as follows (Pogrebov et al., 2001; see also Fig. 29.18). The Siberian sturgeon is a benthos-eating fish. Its main food is tendipedid larvae and larvae of Mayflies and buffalo gnats. It also feeds on amphipods, oligochaetes, and isopods. Isopods *Saduria entomon* comprise up to 90% of its total food in the Yenisey Bay.

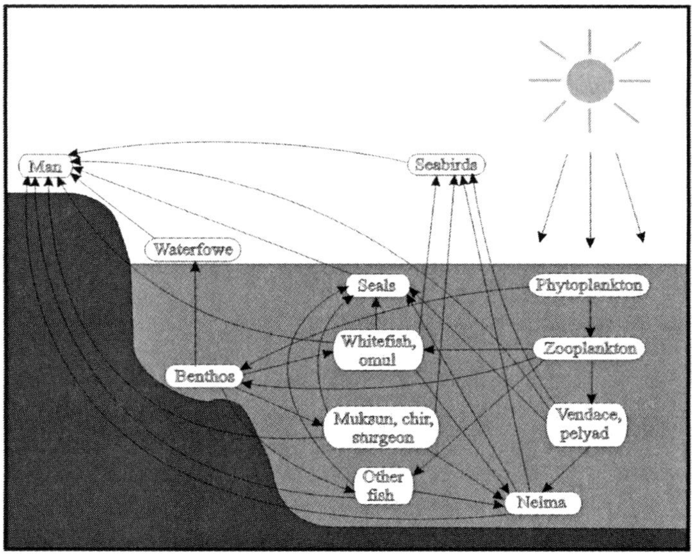

Figure 29.18 Main food webs in the Yenisey Estuary (by Pogrebov et al., 2001)

The vendace and pelyad preys mainly on zooplankton. Mysidaceans, cladocerans, and copepods are prevailing in the vendace food in the bays, whereas in delta tendipedid larvae and amphipods are leading. Crustaceans, mainly cladocerans, and, to a lesser degree, insect larvae, are predominating in the pelyad food during summer, though benthic invertebrates may be found in stomachs of this species in winter time and fish eggs in autumn. The whitefish and omul preys both on zooplankton and benthos. Young whitefish is feeding mainly on cladocerans and copepods, while adults, besides above-mentioned tendipedid larvae and amphipods, –on small mollusks and adult insects. Amphipods and mysidaceans comprise the main food of adult omul. Copepods and cladocerans are important only for the young of this species. The muksun and chir are mainly benthos-eating fishes. During summer, muksun feeds on amphipods and isopods *Saduria entomon,* and, to a lesser degree, on diatom algae. Its fry also uses cladocerans, copepods, tendipedid larvae, and amphipods in the course of migration in the deltas. The chir food comprises tendipedid larvae, small mollusks, and near-bottom crustaceans. Its fry feeds on plankton in the first months of life but by the autumn time bottom dwellers begin to prevail in its ration. The nelma is typical predator, though its young are preying on zooplankton, benthos, and adult insects during the first three yours of life. From an age of 4 to 5 years, main food of nelma comprises the fry of whitefish, muksun, burbot and own fry, as well as adult tugun, vendace, grayling, and nine-spined stickleback. From the other fish inhabiting estuaries, the smelt feeds on zooplankton (cladocerans and copepods) at the early stages of development, and on amphipods and especially mysidaceans, being adult. Role of fry in feeding is also significant. The burbot and pike are mainly predators. The majority of other fish species feeds on both plankton and benthos. In general, the proportion of benthos-eating fishes estimated for the study area is 43% of the total fish fauna, of plankton-eating fishes 18%, whereas predators compose 20% of the fish fauna.

In offshore areas the Polar cod and *Liparis cf. fabricii* are the main cryopelagic abundant species, which play important role in food webs as zooplankton feeders. Cottid, zoarcid and agonid fishes consume mainly benthic and near-bottom organisms.

The role of fish is significant in seabird diets. Such true marine birds as guillemots and seagulls are among the most significant fish consumers, whereas birds of other orders (*e.g.*, waterfowl and water birds) prey mainly on benthic algae and invertebrates.

In addition to fish, some marine mammal species also inhabit the bays. The white whale is the most abundant among the cetaceans, whereas the widely distributed ringed seal is the most common species among the pinnipeds. Polar bears may be observed in the study area on rare occasions. The white whales are usually seen in the near-shore waters during the ice-free period. In some areas these whales undertake extensive seasonal migrations. The white whale feeds on squid, benthic crustaceans and fish, in particular Arctic cod. They are important for ecology of the area as a primary marine predator. The circumpolar ringed seal is inhabitant of the permanent pack ice but congregate on landfast ice for breeding. Fish, pelagic amphipods, shrimps and other crustaceans make up the bulk of the diet. The ringed seal is important as a predator and as the main food for polar bears. The polar bear visits the bays in winter. The polar bear lives mainly on ringed seals and partly on bearded seals. They also feed on other seals, walruses, white whales, carcasses and whatever they find of birds, eggs, etc. The polar bear is at the top of the Arctic marine food chain. The size of the population has an effect on the size of the population of ringed and bearded seals.

9. Radioactive contamination of seabed sediments and biota's reply

Considerable concern, not only of specialists, but of all the world community in the recent past was connected with the environmental state of the nuclear test sites and sites of radioactive waste (RAW) disposal off Novaya Zemlya. Novaya Zemlya Test Site experienced 132 nuclear explosions (Adushkin and Krasilov, 1993), and RAW of the total radioactivity 2400 kCi was damped here on shelf eastwards archipelago (Facts, 1993; see also Fig. 29.19).

9.1. The Barents Sea

Underwater nuclear explosions in Chernaya Inlet (see Fig. 29.19) were conducted in 1955, 1957 and 1961 (the first of them at a depth of 50 m). Measurements taken in 1992 had shown that Cs-137 concentration in the inlet bottom sediments was 80 Bq kg^{-1} while the average value all over the Barents Sea was less than 5 Bq kg^{-1} (Smith et al., 1995). Radioactivity of Pu-239, 240 made up 2500–11000 Bq kg^{-1} against background level 0–3 Bq kg^{-1}. Co-60 (140 Bq kg^{-1}) was also registered.

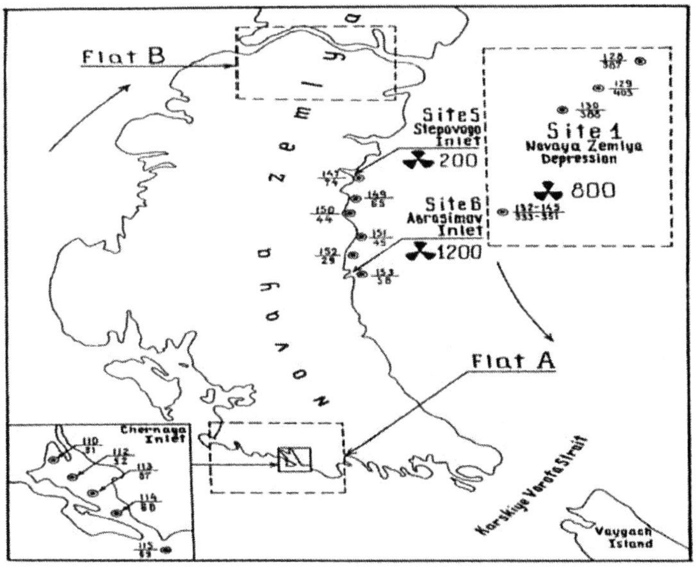

Figure. 29.19. Scheme of location of the Novaya Zemlya Nuclear Test Site Flats (Adushkin and Krasilov, 1993), radioactive waste dump sites and benthic stations, sampled off Novaya Zemlya during the ecological cruise of 1993. Numbers under the names of dump sites correspond to the expert estimation of maximum possible radioactivity of all forms of solid radioactive waste to the moment of dumping, kCi. In stations, designation numbers above the line correspond to the station ordinal numbers, below the line–to the depth, in metres Arrows corresponds to the directions of prevailing currents.

According to the results of sampling provided in Chernaya Inlet in recent decade (Pogrebov et al., 1997), the macrobenthos biodiversity at investigated sites was rather high and comprised 34 species per single quantitative sample while the biomass averaged 245.8 g m^{-2}. By species diversity the first places belonged to Polychaeta and Bivalvia. Along the depth gradient, the total benthic biomass decreased from 384 g m^{-2} at 31 m to 106 g m^{-2} at 58–69 m. By prevailing trophic group the community belonged to subsurface deposit-feeder zone (60% of total biomass). In general, no deviation from statistically averaged state of bottom dwellers in Chernaya Inlet as compared to the adjacent areas of the Barents Sea could be noted.

In meiofauna of the studied area representatives of all the main taxa of this group were found. Meiobenthos abundance varied within 108–5426 thousand individuals m^{-2}, biomass–within 2–112 g m^{-2}. Foraminifera dominated. Minimum values of meiobenthos abundance were observed on stations, located in the center of the inlet, on silty and clayey seabed sediments. Averaged characteristics of the nearest offshore differed from those of the inlet: (i) meiobenthos density here was three times lower and biomass–five times lower than in Chernaya Inlet; (ii) dominating species and community structure were different; (iii) community variability turned out to be smaller. However, we are far from taking responsibility to correlate the revealed differences with the consequences of nuclear tests, particularly as no abnormalities in the meiobenthic organism morphology could be found.

Microbenthic fauna of Chernaya Inlet was not found to be poor. However it was difficult to overlook pauperisation of the inlet Ciliophora, both by the species number and frequency of occurrence. Thus, on the same number of stations, at similar depth and seabed sediments in the adjacent areas of the Barents Sea, no less than 10 species of these ciliate with 100% occurrence were recorded. At the same time in Chernaya Inlet only 4 species were recorded with occurrence rate of 40%. In the inlet itself infusorians were discovered only on the shallowest station located at its top. Morphology of Ciliophora individuals, belonging to gen. *Euplotes,* which were met here (5 specimens), was strongly changed: form of their body and basal granules had significant divergence from norm. As a result, movement of animals in water also differed from normal. According to the results of laboratory measurements, content of Cs-137 in seabed sediments of Chernaya Inlet on shallow-water stations in the inlet and deep-water stations in the vicinity of the inlet was similar to the background level (6 Bq kg^{-1} of air-dry sediment). On the deep-water stations of the inlet content of Cs-137 in sediments increased more than 50 times (up to 328 Bq kg^{-1}). It is another matter that Cs-137 concentration even as much as 328 Bq kg^{-1} cannot be considered high enough for the explanation of revealed phenomenon. However, if one would: (i) consider radio-caesium as a marker of all anthropogenic radionuclides; (ii) high concentration of Pu-239, 240, measured here earlier (it exceeds the background level by three-four orders of magnitude); (iii) take into account its high toxicity (10 thousand times as high as that of arsenic), then the possible cause of the infusorians state in Chernaya Inlet may be linked with the increase of man-made radionuclide concentrations in the inlet sediments with high probability. So, interpretation of their absence on deep-water stations, and especially of their morphological changes on shallow-water stations, by specific hydrological and hydrochemical regime of the inlet may hardly be regarded as logical. It is interesting to note that in the flagellate fauna of Chernaya Inlet no disturbances were found.

9.2. *The Kara Sea*

The largest in total activity RAW disposals in the Kara Sea were conducted in the Novaya Zemlya Trough, Abrosimov and Stepovogo Fjords (see Fig. 29.19). Named areas represent sites of RAW disposal with radioactivity more than 90% of total for the whole Kara Sea. According to the computations of the Livermore Laboratory scientists (Mount and Sheaffer, 1994), radionuclide inventories of dumped objects in the Novaya Zemlya Trough, Abrosimov and Stepovogo Fjords for the time of disposal are estimated at 213–811, 663–2300 and 187–191 kCi respectively.

According to the results of sampling provided in the Novaya Zemlya Trough (Pogrebov *et al.,* 1997), Polychaeta here were predominant by species diversity and abundance. By dominating trophic group community belonged to the subsurface deposit-feeder zone (51% of total biomass). Comparison of our data with those of previous (Filatova and Zenkevich, 1957; Antipova and Semenov, 1989) reveled similarity between the results of the 1927–1945 and 1993 surveys, and their dissimilarity with the results with the results of 1975 survey. Thus, average biomasses of the 1927–1945 and 1993 surveys were equal to 13 and 16 g m^{-2} respectively, while biomass of the 1975 survey averaged only 2 g m^{-2}. Dissimilarity of dominating species and benthic structure could also be considered as "high". We explain differences as sampling errors (small number of replicates in all studies) rather than fluctuation of benthic species populations. We have no evidence that the observed differences were caused by changes of radiological situation.

Taxonomic composition of meiofauna was typical for the region. Abundance of meiobenthic organisms varied from 15 to 948 thousand individuals m^{-2}, biomass – from 0.2 to 9.4 g m^{-2}. At two stations Nematoda were prevailing by abundance; in other cases Foraminifera were dominating by abundance. The same Foraminifera were prevailing on such stations by biomass too. Facts testifying to possible impact of RAW disposal on the meiofauna of the studied area were not revealed.

Microbenthic communities of the Novaya Zemlya Trough were the richest from all studied. No abnormalities in the organism state were revealed.

Along the Novaya Zemlya eastern coast, first place with respect to the species number belonged to Polychaeta and Bivalvia, with respect to biomass – to Rhodophyta and Bivalvia. On stations with a large share of hard substratum: (i) flora and fauna composition was more diverse; (ii) total benthic biomass was higher (up to 399 g m^{-2}); (iii) dominating of *Lithotamnion* sp. was more pronounced; (iv) abundance of other red algae increased; (v) in trophic structure autotrophs were leading. On silt and silty sand: (i) species diversity was lower; (ii) biomass decreased (up to 14 g m^{-2}); (iii) *Astarte crenata* were dominating; (iv) role of other bivalves increased; (v) in trophic structure motile filter-feeders and surface deposit-feeders were prevailing.

Meiobenthic composition of sites along the Novaya Zemlya eastern coast was similar to that described for the two previous sites. Organism abundance varied from 184 to 3095 thousand individuals m^{-2} and biomass varied from 2.3 to 62.3 g m^{-2}. On the majority of stations Foraminifera prevailed. On two stations nematodes were highest in abundance. On one of the stations turbellarians were highest in biomass. Along the depth gradient, quantity of meiobenthos decreased: from 40 to 70 m by 0.6 million individuals m^{-2} and by 13 g m^{-2} on each 10 m of depth. Abnormal individuals in meiobenthic samples from the Novaya Zemlya shoal were not recorded.

Analysis of Protozoa distribution along the Novaya Zemlya eastern coast revealed no ecological peculiarities. Composition and number of species per sample of both infusorians and flagellates varied insignificantly. Appreciable correlation of these indices with depth or seabed sediment composition was not observed. No morphological abnormalities of individuals or their behavioral changes were noted.

10. Issues perspective for further studies

Taking into account the development of oil and gas extraction and transportation in the Barents and the Kara Sea coastal and offshore regions in the nearest future, one should consider the following issues to be the most perspective for further biological studies (Kiyko and Pogrebov, 1999): (i) study of biological effects caused by low environment contamination (in nature); (ii) elaboration of a concept of the marine ecosystem adaptation syndrome caused by natural and man-made impacts; (iii) prognostic modeling of biota changes caused by the changes in environment;

(iv) study of biological effects caused by oil pollution and subsequent rehabilitation succession in typical cases.

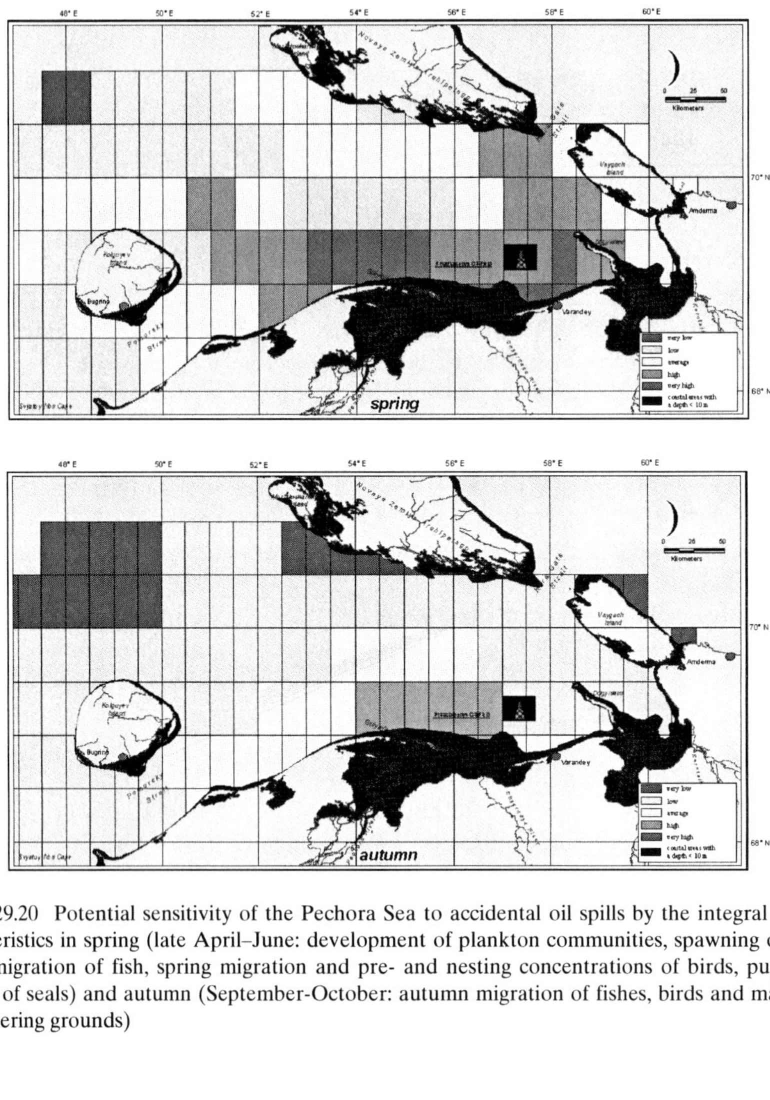

Figure 29.20 Potential sensitivity of the Pechora Sea to accidental oil spills by the integral biological characteristics in spring (late April–June: development of plankton communities, spawning of herring, spring migration of fish, spring migration and pre- and nesting concentrations of birds, pupping and molting of seals) and autumn (September-October: autumn migration of fishes, birds and mammals to the wintering grounds)

Practically significant are the studies connected with the integral assessment of sensitivity of marine and coastal biota to the offshore oil-field development operations (Pogrebov and Puzachenko, 2001, 2003). This GIS-based technology of mapping uses ranked input data on seasonal distribution of main marine and coastal ecosystem components (plankton, benthos, fishes, birds and marine mammals) and as an output one may get integral seasonal maps showing comparative spatial and temporal sensitivity of the study area. A preference is usually given to quantitative data, and high attention is paid to the seasonal dynamics. Data on the distribution of organisms presented in the form of by-season maps for each species or a group of organisms are used as a basis. The maps are digitized and presented as individual GIS "layers". As an individual "layer", a regular grid is constructed whose cell size is determined based on the minimum contour size on the maps. Data on the object presence or the indicators characterizing its abundance are written at the table margins. Data on the object distribution are multiplied by the sensitivity coefficients and summed up for all objects. For compiling the end maps the sensitivity values are ranked according to a 5-point scale (from green to red, the latter characterizing the highest sensitivity). Ranking is made simultaneously for all seasons so that the assessment scale took into account the indicator variability over the year. The results of such an approach usage for the southeastern Barents Sea are presented in Fig. 29.20 (the analysis includes evidence on the distribution of about 30 species and groups of organisms for the water area considered).

As a result, it is shown that the so-called Pechora Sea is the most sensitive area of the region to the accidental oil spills in all seasons. The least sensitive period is spring with winter being the least sensitive. The potential sensitivity zone occupies the maximum area in the winter-spring period and the minimum area in spring. The most sensitive locations are in the vicinity of Gulyayevskiye Koshki, Dolgy, Bolshoy and Maly Zelentsy and Matveyev Islands and in the western part of the Pechora Sea coast. On the Kola Peninsula, the sensitive water areas are in the vicinity of bird colonies. The especially valuable objects also include Kildin Island, Motovskoy Bay, archipelagos of Seven Islands and the Ainovy Islands.

Acknowledgements

The study is carried out within the framework of routine AARI programs and partly sponsored by the grant of European Commission (project ESTABLISH, contract No. ICA2-CT-2000–10008).

Bibliography

Aagard, K. and L. K. Coachman, 1975. Toward an ice-free Arctic Ocean. Eos Trans. AGU, 56(7), 484 – 486.

Adushkin, V. and G. Krasilov, 1993. Novaya Zemlya Test Site and the problem of the radioactive pollution of the Polar Ocean. In *Radioactivity and Environmental Security in the Oceans:* Proceedings. *Woods Hole,* 27–36.

Alongi, D.M., 2003. Ecosystem types and processes. *The Sea,* Vol. 13. (submitted).

Andriashev, A.P., 1954. Fishes of the northern seas of the USSR. Izd-vo AN SSSR, *Moscow-Leningrad,* 1–567. (In Russian).

Andriashev, A.P. and N.V. Chernova, 1994. Annotated list of the arctic and adjacent areas fishes // *Problems of Ichthyology,* **30,** 435–456. (In Russian).

Antipova, T.V., 1975. Distribution of the Barents Sea benthos biomass. *Transactions of the Polar Institute for Fisheries and Oceanography,* **35,** 121–124. (In Russian).

Antipova, T.V. and V.N. Semenov, 1989. Composition, distribution and benthic communities of the southwestern areas of the Kara Sea. In *Ecology and bioresources of the Kara Sea.* Kola Science Center of USSR Academy of Sciences, *Apatity,* 145–178. (In Russian).

Atlas of sea mammals of the USSR. 1980. *Moskva:* Pischevaya promishlennost', 183 pp. (In Russian).

Barsukov, V. V., L.M. Shestova and N. V. Mukhina. 1996. Red-fishes of the genus Sebastes. In *Ichthyofauna of the Barents Sea and its environmental conditions.* Kola Science Center of USSR Academy of Sciences, *Apatity,* 48–54. (In Russian).

Berestovsky, E.G., 1989 Feeding and growth of the *young European plaice Hippoglossoides platessoides limandoides* in some areas of the Barents Sea. Trophic relationships of the benthic organisms and near-bottom fishes of the Barents Sea. Kola Science Center RAS, *Apatity,* 14–26. (In Russian).

Blindheim, J., 1989. Cascading of Barents Sea water into the Norwegian Sea. Rapp. P.-V. Reun. Cons. Int. Explor. Mer. 188, 49–58.

Borisov V.M., V.P. Ponomarenko, V.N. Semenov. 2001. Biological resources of the Barents Sea and fishery at the second half of the XX century. In *Ecology of commercially important fishes of the Barents Sea.* Matishov G. G., V. V. Denisov, A .D. Chinarina (Eds.). Kola Science Center RAS, *Apatity,* 139–195. (In Russian).

Brotskaya, V.A. and L.A. Zenkevich, 1939. Quantitative registration of the Barents Sea bottom fauna. *Transactions of the All-Union Institute for Fisheries and Oceanography,* **4,** 5–126. (In Russian).

Chislenko, L.L. 1972. Zooplankton of the Dikson Bay (Kara Sea). In *Geographical and seasonal variability of marine planktonic organisms* (Zh.A. Zvereva, Ed.), Nauka, *Leningrad,* 228–238. (In Russian).

Dalpadado P., B. Bogstad, H. Gjøsaeter, S.Mehl and H.R. Skjoldal, 2002. Zooplankton-fish interactions in the Barents Sea. In *Large Marine Ecosystems of the North Atlantic* (Sherman K., Skjoldal H.R., Eds.), Elsevier, *Amsterdam e.a.,* 269–291.

Druzhkov, N.V., L.L. Kuznetsov, O.N. Baitaz and E.I. Druzhkova, 1997. The seasonal cyclic processes in the coastal pelagic ecosystems of North Europe (on the example of Central Murman, Barents Sea). In *Plankton of the Sea of the Western Arctic* (G.G. Matishov, Ed.), Kola Science Center RAS, *Apatity,* 145–178. (In Russian).

Druzhkov, N.V. and P.R. Makarevich, 1999. Comparison of the phytoplankton assemblages of the south-eastern Barents Sea and south-western Kara Sea: phytogeographical status of the regions. *Botanica Marina,* **42,** 103–115

Druzhkov, N.V., P.R. Makarevich and E.I. Druzhkova, 2001. Phytoplankton in the south-western Kara Sea: composition and distribution. *Polar Research,* **20** (1), 95–108.

Esipov, V.K., 1952. Fishes of the Kara Sea. Izd-vo Akad. Nauk, *Moscow-Leningrad,* 1–145. (In Russian).

Facts and problems related to radioactive waste disposal in seas adjacent to the territory of the Russian Federation (Materials of the Governmental Commission), 1993. *Moscow,* 1–102. (In Russian).

Falk-Petersen S., C.C.E. Hopkins and J.R. Sargent, 1990. Trophic relationships in the pelagic, Arctic food web. In Trophic relationships in the marine environment (Barnes M. and R.N. Gibson, Eds.), Aberdeen University Press, *Aberdeen,* 315–333.

Fetzer, I., H.-J. Hirche and E.G. Kolosova, 2002. The influence of freshwater discharge on the distribution of zooplankton in the southern Kara Sea. *Polar Biology,* **25,** 404–415.

Filatova, Z.A. and L.A. Zenkevich, 1957. Quantitative distribution of the Kara Sea bottom fauna. *Transactions of All-Union Hydrobiological Society,* **8,** 3–67. (In Russian).

Fomin, O.K. and V.S. Petrov, 1985. The role of environmental factors in the distribution of plankton biomass in the Kara Sea. In *Nature and industry of the North.* Vol. 13. (G.S. Biskae *et al.,* Eds.), Knizhnoe izd., *Murmansk,* 35–45. (In Russian).

Galkin, Yu.I., 1987. Climate fluctuations and long-term changes of the Barents Sea benthos biomass. In *The Arctic and Antarctic Biological Resources.* Nauka, *Moscow,* 90–122. (In Russian).

Galkina, V.N., A.D. Rura and S.Y. Gagayev, 1994. Phytoplankton and its production in the Chaun Bay of the East Siberian Sea. In *Ecosystems and the flora and fauna of the Chaun Bay of the East Siberian Sea.* Part 1. (A.N. Golikov, Ed.), Zoological Institute RAS, *Saint-Petersburg,* 112–120. (In Russian).

Gordeev, V. V., J. M. Martin, I. S. Sidorov, and M. V. Sidorova, 1996. A ressessment of the Eurasian river input of water, sediment, and nutrients to the Arctic Ocean. American Journal of Science, Vol. 296, 664 - 691.

Gorshkov S. G. (editor), 1980. World Ocean Atlas, Vol. 3, Arctic Ocean, Pergamon Press, Oxford, 184.

Gurevich, V.I., 1995. Recent sedimentogenesis and environment on the Arctic shelf of Western Eurasia. Norsk Polarinstitutt. Meddeleser Nr. 131. Oslo, 1–92.

Halsband, C. and H.-J. Hirche, 1999. The role of mesozooplankton for the transformation of organic matter. *Berichte zur Polarforschung,* **300,** 45–50.

Hanzlick, D., and K. Aagaard, 1980. Freshwater and Atlantic water in the Kara Sea. J. Geoph. Res. Vol. 85, No. C9, 4937 – 4942.

Hydrometeorological conditions of the shelf zone of the USSR's seas, 1985. Vol. 6. Barents Sea, Leningrad, Hydrometeoizdat, 263. (In Russian).

Hydrometeorological conditions of the shelf zone of the USSR's seas, 1985. Vol. 7. Kara Sea, Leningrad, Hydrometeoizdat, 278. (In Russian).

Ilyash, L.V. and T.I. Koltsova, 1981. The phytoplankton of the Yenisey Bay. *Gydrobiologicheskii Zhurnal,* **17** (3), 3–8. (In Russian).

Ingvaldsen, R., H. Loeng, and L. Asplin, 2002. Variability in the Atlantic inflow to the Barents Sea based on one-year time series from moored current meters. Continental Shelf Research, 12 (9), 1043–1058.

Ilyin G.V. 2001. Accumulation of pollutants in tissues and organs of the Barents Sea fishes. In *Ecology of commercially important fishes of the Barents Sea.* Matishov G. G., V. V. Denisov, A .D. Chinarina (Eds.). Kola Science Center RAS, *Apatity,* 196–216. (In Russian).

Ivanov, V. V., V. P. Rusanov, O. I. Gordin and I. V. Osipova, 1984. Year to year variability of the river water spreading in the Kara Sea. *Trudy AARI,* Vol. 368, 74 - 81. (In Russian).

Joint American-Russian Atlas of the Arctic Ocean, Summer, 1998 (on CD-ROM), National Snow and Ice Data Centre, Environmental Working Group, Boulder, Colorado.

Karamushko O.V., E.G. Berestovskiy, L.I. Karamushko, O.Yu. Yunacheva. 2001. Some aspects of biology of the commercial fishes in 1993–1998. In *Ecology of commercially important fishes of the Barents Sea.* Matishov G. G., V. V. Denisov, A .D. Chinarina (Eds.). Kola Science Center RAS, *Apatity,* 13–138. (In Russian).

Kiyko, O.A. and V.B. Pogrebov, 1997 a. Long-term benthic population changes (1920–1930s - present) in the Barents and Kara Seas. *Marine Pollution Bulletin,* **35,** 322–332.

Kiyko, O.A. and V.B. Pogrebov, 1997 b. Persistent organic pollutant, trace metal and radionuclide concentrations in bottom organisms of the Barents Sea and adjacent areas. *Marine Pollution Bulletin,* **35,** 340–344.

Kiyko, O.A. and Pogrebov, V.B. 1999. Tendencies and perspectives of the ecological studies in the Barents Sea. In *Barents Sea Impact Study. Proceedings of the First International BASIS Conference.* Munster, Germany, 404–405.

Kiyko, O.A., V.B. Pogrebov and B.G. Vanshtein, 2002. Stability and variability of structural parameters of benthic communities under climatic changes and anthropogenic impact. In *Fifth Workshop on Land Ocean Interactions in the Russian Arctic (LOIRA), Moscow,* 52–53.

Kovtsova, M.V., Antonov S.G. and Orlova E.L. 1989. Feeding of the haddock *Melanogrammus aeglefinus* (L.) in the Barents Sea and its fat-content dynamic. Trophic relationships of the benthic organisms and near-bottom fishes of the Barents Sea. Kola Science Center RAS, *Apatity,* 27–36. (In Russian).

Krasnov, Yu.V., G. G. Matishov, K. V. Galaktionov, T. N. Savinova. 1995. Colonial sea birds of the Murman coast. Sankt-Petersburg: Nauka, 224 pp. (In Russian).

Kulakov M, V. Stanovoy, 2002. Frontal zones in the Kara Sea: observation and modeling. / *Proc. Of The 11th International Biennial Conference on Physics of Estuaries and Coastal Seas,* Hamburg, Germany, September 17–20, 393 – 397.

Kuznetsov, A.P. 1970. Features of the bottom invertebrate trophic group distribution in the Barents Sea. *Transactions of the Institute for Oceanology,* **88,** 5–80. (In Russian).

Loeng, H., 1988. The influence of climate on biological conditions in the Barents Sea. NAFO Scient. Coun. Meet. Doc. 88/83: 1–19.

Longhurst, A., 1998. Ecological geography of the sea. Academic Press, San Diego *et al.*, 1–398.

Makarevich, P.R., 1998. The vernal state of the microphytoplankton community in the ice-covered areas of the south-eastern Barents and the south-western Kara Seas. In *Biology and oceanography of the Kara and Barents Seas (along the Northern Marine Route).* Kola Science Center AS USSR, *Apatity,* 138–149. (In Russian).

Makarevich, P.R. and T.I. Koltsova, 1989. History of investigations and modern status of the phytoplankton. In *Ecology and biological resources of the Kara Sea* (G.G. Matishov *et al.,* Eds.), Kola Science Center AS USSR, *Apatity,* 38–45. (In Russian).

Makarevich, P.R. and V.V. Larionov, 1992. Taxonomic composition of phytoplankton and history of the phytoplankton studies in the Barents Sea. *In Phytoplankton of the Barents Sea* (L.L. Kuznetsov, Ed.), Kola Science Center RAS, *Apatity,* 17–51.

Manteufel, B.P., 1941. Plankton and herring in the Barents Sea. *Trudy PINRO,* **7,** 125–218. (In Russian).

Marti Yu.Yu. and G.V. Martinsen, 1969. Problems of the forming and utilization of biological production in the Atlantic Ocean. Pischevaya Promyshlennost, *Moscow,* 1–267. (In Russian).

Matishov, G.G., N.V. Druzhkov, P.R. Makarevich, and V.V. Larionov, 2001a. The effects of the freshwater phytoplankton on the biological productivity in the south part of the Kara Sea (Ob and Yenisey River area). *Doklady Akademii Nauk,* **378** (3), 424–426. (In Russian).

Matishov, G.G., P.R. Makarevich, S.F. Timofeev, *et al.,* 2000. Biological atlas of the Arctic Seas 2000: Plankton of the Barents and Kara Seas. Kola Scientific Center RAS, National Oceanographic and Atmospheric Administration, *Murmansk-Silver Spring.* 1–348 + CD: NOAA Atlas NESDIS 39.

Matishov, G.G., D.G. Matishov, A.A. Namyatov. 2001b. Radionuclides in tissues of the Barents Sea fishes. In *Ecology of commercially important fishes of the Barents Sea.* Matishov G.G., V.V. Denisov, A.D. Chinarina (Eds.). Kola Science Center RAS, *Apatity,* 217–228. (In Russian).

Matishov, G., D. Matishov, E. Shchipa, and K. Rissanen, 1994. Radionuclides in the ecosystem of the Barents and Kara Sea Region. Kola Science Center RAS, *Apatity,* 1–238.

Moore, J.V. and S. Ramamurti, 1987. Heavy metals in near-bottom water. *Moscow,* 1–285. (In Russian).

Mount, M.E. and M.K. Sheaffer, 1994. Estimated inventory of radionuclides in FSU naval reactors dumped in the Kara Sea. *Arctic Research of US,* **8,** 160–178.

Nesterova, V.N. 1990. Plankton biomass on the drift of cod larvae. PINRO, *Murmansk,* 1–64. (In Russian).

Northern Sea Route (1998) Dynamic Environmental Atlas. Brude O.W *et al.,* Eds.), Norway: Norsk Polarinstitute, 1–58.

Nöthig E.-M., D. Piepenburg and E. Rachor, 1994. Carbon flux and pelago-benthic coupling in the northern Barents Sea in summer 1991. In Polar research – still a challenge: European week for scientific culture (Rachor E., Ed.), Alfred-Wegener-Institut für Polar und Meeresforschung, Bremerhaven, 28–29.

Olli K., R.C. Wexels, P. Wassmann, T. Ratkova, E. Arashkevich and A. Pasternak, 2002. Seasonal variation in vertical flux of biogenic matter in the marginal ice zone and the central Barents Sea. *J. Mar. Syst.,* **38** (1–2), 189–204.

Orlova, E.L., Berestovsky E.G., Karamushko L.I. and Savelova E.A. 1989. Age-related changes in feeding of the Barents-sea wolf-fishes *Anarhichas lupus, A. minor, A. denticulatus.* Trophic relationships of the benthic organisms and near-bottom fishes of the Barents Sea. Kola Science Center RAS, *Apatity,* 4–13. (In Russian).

Orlova E.L. and G.G. Matishov, 1993. Structural and functional roles of the cod in the Barents Sea ecosystem. Kola Science Center RAS, *Apatity,* 1–164. (In Russian).

Pautova, L.A. and G.M. Vinogradov, 2001. South eastern Barents Sea phytoplankton in April 2000. *Oceanology,* **41,** 224–230. (In Russian).

Pavlov, V. K., and S. L. Pfirman, 1995. Hydrographic structure and variability of the Kara Sea: Implications for pollutant distribution. *Deep-Sea Res.* II, Vol. 42, No. 6, 1369 - 1390.

Pogrebov, V.B., S.I. Fokin, V.V. Galtsova and G.I. Ivanov, 1997. Benthic communities as influenced by nuclear testing and radioactive waste disposal off Novaya Zemlya in the Russian Arctic. *Marine Pollution Bulletin,* **35,** 333–339.

Pogrebov, V.B., T.V. Panteleimonov and O.A. Kiyko, 2001. Composition, structure and ecological state of the Yenisey Estuary aquatic biota and its utilization by man. Deliverable Report for the EC Copernicus Project "ESTABLISH". Contract No. ICA2-CT-2000–1008, NRPA, *Østerås,* 1–89.

Pogrebov, V.B. and A.Yu. Puzachenko, 2001. Integral assessment of potential biota sensitivity to offshore oil-field development operations // *Coastline.* No. 1. P. 10–11.

Pogrebov, V.B. and A.Yu. Puzachenko, 2003. Ecological sensitivity of the Barents, White, Baltic, Black and Caspian Seas to oil extraction and transportation: comparative analysis // Proceedings of International Conference "Russian Arctic Offshore - 2003". Saint Petersburg, 389–393.

Ponomareva, L.A. 1957. Zooplankton of the western part of the Kara Sea and the Baidara Bay. *Transactions of the P.P.Shirshov Institute of Oceanology,* **20,** 228–245. (In Russian).

Rass, T.S., 1993. Fishery resources of the Barents Sea and the cause of their changes. *Oceanology,* **33** (4), 551–557. (In Russian).

Rey, F., H.R. Skjoldal and D. Slagstad, 1987. Primary production in relation to climate changes in the Barents Sea. In *The effect of oceanographic conditions on distribution and population dynamics of commercial fish stocks in the Barents Sea* (H. Loeng, Ed.), Institute of Marine Research, *Bergen,* 29–46.

Romankevich, E.A. and A.A. Vetrov, 2001. Cycle of carbon in the Russian Arctic Seas. Nauka, *Moscow,* 1–302. (In Russian).

Sakshaug, E., A. Bjørge, B. Gulliksen, H. Loeng and F. Mehlum, 1994. Structure, biomass distribution, and energetics of the pelagic ecosystem in the Barents Sea: A synopsis. *Polar Biology* **14,** 405–411.

Schauer, U., H. Loeng, B. Rudels, V. K. Ozmidov, and W. Dieck, 2003. Atlantic Water flow through the Barents and Kara Seas. Deep-Sea Research (in press).

Skarlato, O.A. and A.N. Golikov, 1985. Arctic Ocean: Biology. In *Arctic and South Oceans* (A.F. Treshnikov and S.S. Salnikov, Eds.), Nauka, *Leningrad,* 102–119. (In Russian).

Slagstad, D. and K. Støle-Hansen, 1991. Dynamics of plankton growth in the Barents Sea: model studies. *Polar Research,* **10,** 173–186.

Smith, J.N., K.M. Ellis, S. Forman, L. Polyak, G. Ivanov, D. Matishov, L. Kilius and S.B. Moran, 1995. Radionuclide sources and transport pathways in the Arctic Ocean. In *Arctic Nuclear Waste Assessment Program Workshop:* Abstracts. *Woods Hole,* no pages.

Soviet Arctic, 1970, Seas and islands of the Arctic Ocean, Moscow, Nauka, 526.

Tereshchenko, V.V., G.I. Nesvetova, V.N. Nesterova and O.V. Titov, 1994. Characteristics of oceanographic conditions and plankton in the central zone of the Barents Sea. PINRO, *Murmansk,* 1–74. (In Russian).

Timofeev, S.F. 1989. Pelagy architectonic in the Kara Sea. In *Ecology and biological resources of the Kara Sea* (G.G. Matishov *et al.,* Eds.), Kola Science Center AS USSR, *Apatity,* 86–93. (In Russian).

Timofeev S.F., 1990. Modern state and future investigations of the plankton communities of the Barents Sea. In Ecology and biological productivity of the Barents Sea (Matishov G.G. *et al.,* Eds.), Nauka, *Moscow,* 47–54. (In Russian).

Timofeev S.F., 1996. Structural and functional analysis of the plankton communities in the southern part of the Barents Sea. In *Pelagic ecosystems of the Western Arctic Seas* (Matishov G.G., Ed.), Kola Science Center RAS, *Apatity,* 95–100. (In Russian).

Timofeev, S.F. 1997. Zooplankton of the Barents Sea. In *Plankton of the West Arctic seas* (G.G. Matishov, Ed.), Kola Scientific Center RAS, *Apatity,* 266–295. (In Russian).

Timofeev, S.F. 2000. The ecology of the marine zooplankton. Murmansk State Pedagogical Institute, *Murmansk*, 1–216. (In Russian).

Timofeev, S.F. 2001. Influences of climatic factors on the variability of zooplankton biomass in different parts of the Barents Sea. In *Long-term changes of the Arctic marine ecosystems* (G.G. Matishov *et al.*, Eds.), Kola Scientific Center RAS, *Apatity*, 33–49. (In Russian).

Timofeyev, V. T., 1962. Vliyanniya glubinnikh Atlanticheskikh vod na obrazovanniya i tayanniya lida v moryakh Karskom i Laptevikh, *Okeanologiya*, 2(2), 221 – 225. (In Russian).

Treska Batentseva morya [Cod of the Barents Sea]. 1996. Boytsov, V.D., N.I. Lebed', V.P. Ponomarenko, I.Ya. Ponomarenko, V.V. Tereschenko, V.L. Tretjak, M.S. Shevelev and N.A. Yaragina. Murmansk: PINRO. 285 pp. (In Russian).

Troshkov, V.A. and L.V. Gnetneva, 2000. Zooplankton of the southeastern Barents Sea. In *Biological resources of the Russian Arctic littoral* (V.P. Ponomarenko and V.M. Zelenkov, Eds.), VNIRO Publ., *Moscow*, 143–150. (In Russian).

Usachev, P.I. 1968. The phytoplankton of the Kara Sea. In *Plankton of the Pacific Ocean* (G.I. Semina, Ed.), Nauka, *Moscow*, 6–28. (In Russian).

Vedernikov, V.I., A.B. Demidov and A.I. Sudbin, 1994. Primary production and chlorophyll in the Kara Sea in September 1993. *Oceanology*, **34** (5), 693–703. (In Russian).

Vinje, T., 2001: Anomalies and trends of sea-ice extent and atmospheric circulation in the Nordic Seas during the period 1864–1998. *J. Climate 14*, 255–267.

Vinogradov, G.M., N.V. Druzhkov, E.F. Marasayeva, and V.V. Larionov, 2001. Mesozooplankton under ice in the Pechora and Kara Seas during the winter-spring period of 2000. *Oceanology*, **41** (5), 728–735. (In Russian).

Vinogradov, M.E., G.M. Vinogradov, G.G. Nikolaeva and V.S. Khoroshilov, (1994 a). The mesoplankton of the western Kara Sea and the Baidara Bay. *Oceanology*, **34** (5), 709–715. (In Russian).

Vinogradov, M.E., E.A. Shushkina, L.P. Lebedeva and V.I. Gagarin, 1994 b. Mesoplankton in the eastern part of the Kara Sea and Ob and Yenisei River estuaries. *Oceanology*, **34** (5), 716–723. (In Russian).

Zakharov, V. F., 1976. Cooling of the Arctic and the ice cover of the Arctic seas. *Trudy AARI*, Vol. 337, 96. (In Russian).

Zelikman, E.A. 1977. Plankton communities of the Arctic pelagic zone. In *Oceanology. Ocean biology. Vol. 2. Biological productivity of the ocean* (M.E. Vinogradov, Ed.), Nauka, *Moscow*, 43–55. (In Russian).

Chapter 30. PHYSICAL FORCING OF ECOSYSTEM DYNAMICS ON THE BERING SEA SHELF (24,P)

P. J. STABENO

Pacific Marine Environmental Laboratory

G. L. HUNT, JR.

University of California

J. M. NAPP

Alaska Fisheries Science Center

J. D. SCHUMACHER

Two Crow Consulting

Contents

1. Introduction

The Bering Sea (Fig. 30.1) is a northern extension of the North Pacific Ocean, and is the world's third-largest semi-enclosed sea. A continental shelf underlies about one-half of the Bering Sea, consisting of a broad (>500 km) eastern portion between the Alaska Peninsula and Cape Navarin, and a narrower (<100 km) western portion between Cape Navarin and Kamchatka Strait. Mean northward transport (~0.8 x 10^6 m^3 s^{-1}) through Bering Strait is driven by an ~0.5 m sea level difference between the Bering Sea and Arctic Ocean (Coachman, 1993). This transport pro-

The Sea, Volume 14, edited by Allan R. Robinson and Kenneth H. Brink
ISBN 0–674–01527–4

vides the only connection and exchange of water between the Pacific and Atlantic Oceans in the northern hemisphere. This northward flow requires transport onto the shelf of nutrient-rich slope waters from the Bering Slope Current, the eastern boundary current of the cyclonic gyre that dominates circulation in the Bering Sea gyre (Stabeno *et al.,* 1999). Over the northern shelf, nutrient-rich water up-wells due to shoaling topography, thereby stimulating primary production (Nihoul *et al.,* 1993). Ice forms predominantly in the polynyas over the northern shelf. The cold, saline water formed from brine rejection sinks and flows northward through Bering Strait. Globally, this water plays a role in maintaining the Arctic Ocean halocline and hence ventilation of the deep waters (Aagaard *et al.,* 1985).

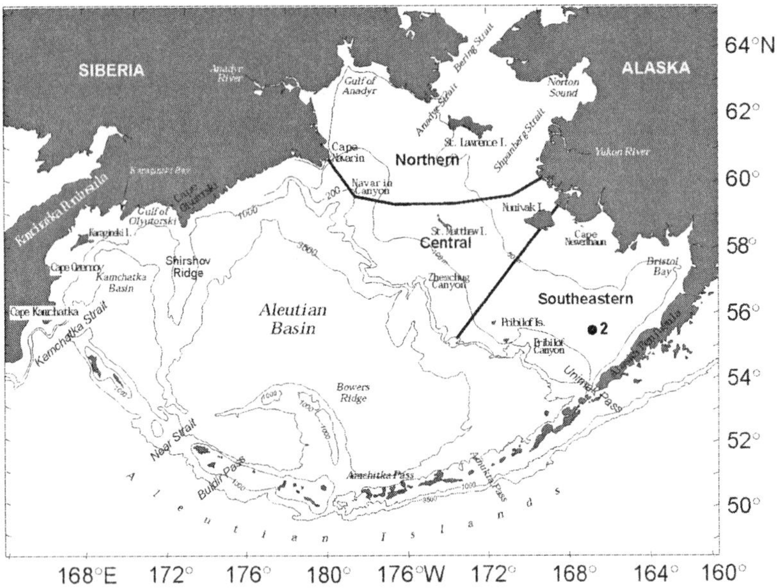

Figure 30.1 Geography and place names of the Bering Sea. The location of the M2 is indicated. Depth contours are in meters. (From Schumacher and Stabeno, 1998 or Stabeno *et al.,* 1999.)

The eastern Bering Sea is home to a rich variety of biota, including the world's most extensive eelgrass beds; at least 450 species of fish, crustaceans, and mollusks; 50 species of seabirds; and 25 species of marine mammals. It has been classified as its own bio-geographic province (Longhurst, 1998) and as a Class II, moderately high (150–300 $gC/m^2/yr$) productivity, ecosystem based on satellite remote sensing estimates (www.edc.uri.edu/lme/text/east-bering-sea.htm), although primary production estimates for sub-regions are much higher (e.g., Springer and McRoy, 1993). The abundant fish and wildlife of the Bering Sea have supported the lives and livelihoods of Asians and North Americans since prehistoric times. For the 65,000 Alaskan natives living on the shores of the eastern Bering Sea (NRC 1996), subsistence is not just a means of providing food; it is the wellspring of spiritual and cultural traditions. Presently (2002), the U.S. Bering Sea fishery provides about 40% of the U.S. and about 5% of the world harvest of fish and shellfish with walleye pollock (*Theragra chalcogramma*) comprising much of the fish landings; the tanner crab (*Chionoecetes bairdi*) fishery currently is the largest crustacean fishery (by weight) in the U.S. Bristol Bay supports the world's largest sockeye salmon (*Onchorhynchus nerka*) fishery. In addition to supporting a large portion of the nation's fishery production, the Bering Sea supports ~80% of the U.S. nesting seabird population, comprising approximately 36 million birds. Furthermore, many endemic species such as red-legged kittiwakes (*Rissa brevirostris*) and whiskered auklets (*Aethia pygmaea*) are found in the Bering Sea and further highlight the significance of this region. Prior to the 1960s, there were much larger populations of northern fur seals (*Callorhinus ursinus*), Steller sea lions (*Eumetopias jubatas*), whales and birds. The U.S. government has listed the Steller sea lion and spectacled eider (*Somateria* fischeri) as endangered species, sea otters (*Enhydra lutris*) as threatened and there are declines of some salmon populations that have adversely affected commercial and subsistence harvest. This region's wetlands, coastlines and islands provide globally significant habitats for many additional wildlife species, and its natural history holds answers to important questions about world history including the introduction of hominids to North America.

The current state of this ecosystem is a product of its history that includes substantial human and natural impacts. Western knowledge of the Bering Sea's rich ecosystem began with Vitus Bering's voyage in 1741 (Ford, 1966). Over the next two centuries, commercial exploitation brought the Steller sea cow to extinction, and the sea otter, fur seal, walrus and bowhead whale nearly to extinction (Fay, 1981). In the early 1950s, the "commercial fishing period" began and that, together with the removal of marine mammals (until the early 1970s), had a marked impact on the composition of the ecosystem (NRC, 1996). These factors form the basis of the Trophic Cascade Hypothesis (Merrick, 1997; NRC, 1996). As the large populations of mammals (particularly whales) and fish were removed, there was an increased amount of food (zooplankton and small fish) for other vertebrate predators. There is little doubt that the over-exploitation and reductions of these populations, together with a regime shift in the physical environment, led to a switch in the late 1960s and early 1970s to a system dominated by pollock, which has persisted since then (NRC, 1996).

Much of the early oceanographic research was conducted to address questions relating to the international fishery (Takenouti and Ohtani, 1974; Arsenev, 1967; Hood and Kelley, 1974; and Favorite *et al.,* 1976). In the early 1950s, Hokkaido University began annual training cruises to the Bering Sea and established what has become one of the longest physical and biological time series available in the region (e.g., Sugimoto and Tadokoro, 1997). Between the mid-1970s and late 1980s, the focus changed to resource assessment on the eastern shelf as part of the Outer Continental Shelf Environmental Assessment Program (OCSEAP; Hood and Calder, 1981). Annual groundfish and crab assessment continues to this day by NOAA/National Marine Fisheries Service. OCSEAP was followed by ecosystem research over the southeastern (PROBES; Processes and Resources of the Bering Sea: Hood, 1999) and northern shelf (ISHTAR; Inner Shelf Transfer and Recycling in the Bering and Chukchi Seas: McRoy, 1999), and examination of ice related phenomenon during the Bering Sea Marginal Ice Zone Experiment (MIZEX: Muench, 1983). During the 1990s, research focused again on fisheries and the influence of the physical factors on the ecosystem. Major contributions included those from the Fisheries Oceanography Coordinated Investigations program (FOCI: Schumacher and Kendall, 1995); NOAA's Coastal Ocean Programs, Bering Sea FOCI (BSFOCI: Bailey *et al.,* 1999, Brodeur *et al.,* 1999b; Napp *et al.,* 2000) and the Southeast Bering Sea Carrying Capacity program (SEBSCC: Stabeno and Hunt, 2002; Macklin *et al.,* 2002); the National Science Foundation (Polar Programs) "Study of prolonged production, trophic transfer, and processes at the Bering Sea inner front" (hereafter called Inner Front Study: Stabeno and Hunt, 2002) and a program to conduct long-term ecological research on marine ecosystems in the Arctic and Pacific Oceans (BERPAC: Tsyban, 1999). Alexander (1999) provides a review of many of the interdisciplinary studies of the Bering Sea.

Features of the Bering Sea (including solar radiation, atmospheric phenomenon, sea ice cover, and water column structure and temperature, and biological production) vary over a wide range of time scales. These include strong seasonal and annual signals and smaller, but potentially important longer-period fluctuations. Over the vast region of the eastern shelf, most of the physical phenomena also vary greatly with both latitude and longitude. The variability in physical features, together with the flux of nutrient-rich slope waters onto the eastern shelf, shapes one of the world's most productive ecosystems (Walsh *et al.,* 1989). Observations suggest that a highly productive region also exists over the edge of the continental shelf (Springer *et al.,* 1996). This "green belt" extends around the entire perimeter of the continental shelf and estimates of both primary and secondary production in this belt are more than 60% greater than production over the adjacent shelf waters.

In this chapter, we first present a conceptual model of the pathways by which changes in atmospheric phenomenon can influence sea ice, oceanic features and biota. While some of the pathways may be specific to the Bering Sea, much is directly applicable to other subarctic seas. We then present the salient features of the physical environment summarized from Schumacher and Stabeno (1998), updated to include more recent findings (e.g., Kachel *et al.,* 2002; Overland *et al.,* 1999a; Stabeno *et al.,* 2002). Our focus here is on interdisciplinary ocean science, with a particular focus on several phenomena whose influence on biota has been established: water temperature, transport and turbulence. Following Schumacher and Stabeno (1998), we partition the continental shelf of the Bering Sea into eastern (primarily U.S. waters) and western regions, with the former further differentiated into southeastern, central and northern sub-regions. The following sections describe processes of the eastern Bering Sea shelf and those over the northern and western shelf. We conclude with a discussion of suggestions for future research.

2. Pathways from Physics to Biology

The complex pathways that weave together the biotic and abiotic components of the Bering Sea ecosystem are not all known, nor are all the interactions among the various components well understood. Yet, identifying and understanding mechanisms that transfer climate change in the atmosphere and ocean to biota is essential if we are to comprehend ecosystem dynamics (Francis *et al.,* 1998). Fluctuations in the physical environment can impact the ecosystem through both changes in the nutrient-phytoplankton-zooplankton sequence (i.e., bottom-up control), and/or by altering habitat or some other condition that results in changes in abundance and/or spatial composition of higher trophic level animals (i.e., top-down control). Temperature, turbulence and transport are mechanisms through which the physics influences the biology. These, in turn, respond to wind and, for high latitude seas, sea ice. All of these phenomena are included in a conceptual model (Fig. 30.2) that provides a schematic of the pathways in which changes in the physical environment impact biota. While the sequence is generally thought of as unidirectional starting with changes in the atmospheric climate driving those in the ocean and on to biota, changes in oceanic environment (e.g. heat fluxes) can alter atmospheric features (a feedback loop). In this conceptual model, the direct influence of the physical environment is a function of trophic level; the arrows between levels represent both biological interactions and the communication of physical influences.

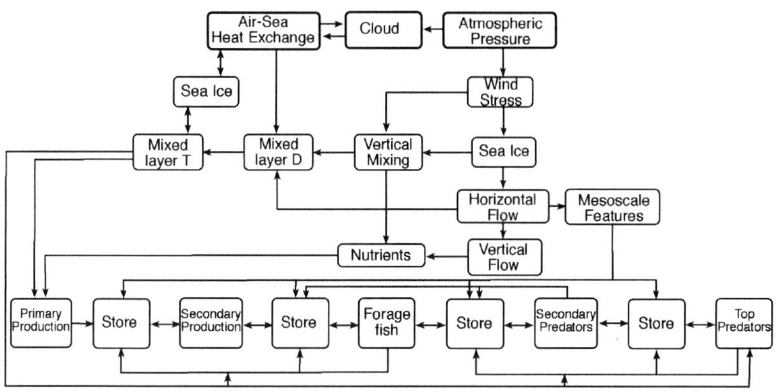

Figure 30.2 Pathways whereby changes in physical environment influence the biological processes (after Schumacher *et al.,* 2003). Sea ice appears twice because it represents pathways for different processes.

The pathways in the conceptual model also include biological interactions (e.g. density dependence) that can be a dominant factor in population dynamics; for example, cannibalism has a marked influence on recruitment processes of pollock in the eastern Bering Sea (Livingston and Methot, 1998). In a recent examination of variations in the strength of pollock year classes (Wespestad *et al.,* 2000), physical and biological mechanisms set survival. For instance, the availability of young pollock as prey for adult fish may be related to transport of larval stage animals. A switch model of larval pollock dynamics (e.g., Napp *et al.,* 2000) also highlights that physical and biological factors interact to influence survival.

Understanding the influence of changes in the physical environment on the ecosystem is confounded by many factors. Physical conditions favorable for one life history stage may be detrimental for another. In addition, many marine populations are likely to respond to change in a nonlinear fashion (e.g., Bailey *et al.,* 2002; Cury *et al.,* 1995; Stenseth *et al.,* 2002). For example, it is hypothesized that during cold years in the Bering Sea, the spatial domain of age-1 pollock over the shelf is restricted to warmer waters (a behavioral response to temperature) and that increased local densities of these fish result in higher predation mortality by older fish (Ohtani and Azumaya, 1995). Cold winters with extensive ice cover can also provide conditions that support rapid increases of primary production associated with the presence of sea ice. Due to the cold-water temperatures and related low physiological rates, the limited zooplankton biomass is unable to consume a significant portion of the primary production. Thus, most of the carbon sinks to the sea floor feeding the benthic food web (Walsh and McRoy, 1986). In addition, it has been demonstrated that nitrate uptake on the southeastern Bering Sea shelf has a nonlinear relationship to wind-induced mixing, and timing of storms relative to the phase of the production system (i.e., respiration or nutrient limited period) is critical (Sambrotto *et al.*, 1986).

3. Physical Features

3.1 Direct Solar Forcing

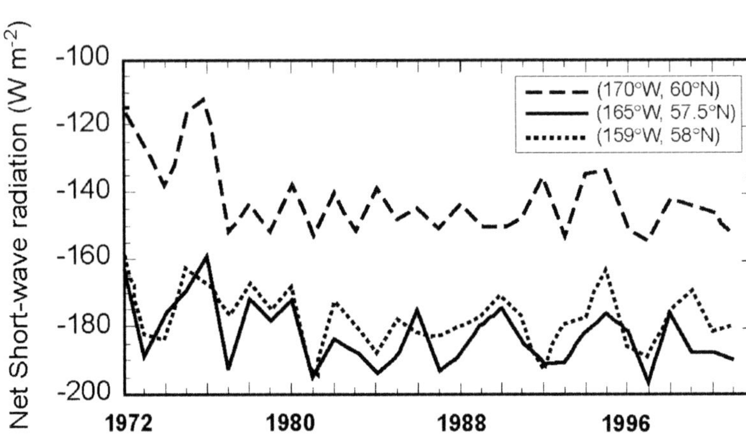

Figure 30.3 Average net short wave radiation (NSWR) from April 1–June 30 (Wm^{-2}). Note negative indicates a downward flux.

While many environmental features have been well documented (e.g., Schumacher and Stabeno, 1998; Stabeno *et al.,* 2001; Hunt *et al.,* 2002), temporal and spatial variations in the amount of solar radiation and its potential impact on ecosystem dynamics has not been well addressed, yet such variations are evident in time series of net short wave radiation (Fig. 30.3). The effect of latitude is apparent in these time series, with the northernmost station receiving approximately 19% less energy. The other two net short-wave radiation time series, which are at similar latitudes, appear coherent. The northern and the two southern time series show similar patterns of variability in 1972–1983 and 1995–2001. During the intervening period the time series appear to be out of phase by ~180 degrees with the magnitude of variability of the northern series much less than that seen in the southern time series. While the two series at approximately the same latitude are often similar in magnitude, there are some years (e.g., 1995) when they differ by more than 10 Wm^{-2}. Some of the longer-term variation may be due to the sun spot cycle that results in a solar flux variation of ~0.1% or ~2 Wm^{-2} at sea level. It is likely, however, that regional differences in cloud cover account for the majority of this signal. It is not presently known how these differences influence ecosystem dynamics, although the latitudinal differences relate to a shorter more intense period of production over the northern shelf than exists over the southeastern shelf. This likely contributes to the observed differences in the dominant pathway of carbon cycling (pelagic versus benthic): the northern shelf is predominately a benthic system whereas the southeastern shelf may be either pelagic or benthic depending on timing of spring blooms (Walsh and McRoy, 1986; McRoy, 1993).

3.2 Atmospheric Features

Recent research (e.g., Minobe, 1999, 2002; Hare and Mantua, 2000) has highlighted atmospheric variations that exist on multi-decadal time scales (typically, 10 to 70 years), which cause significant alterations throughout the ecosystem. These changes, known as regime shifts, are seen in the biological environment primarily as changes in community composition and/or biomass of a given species. The mechanisms that initiate a climate regime shift, however, are presently unknown. In a review of how climate variability impacts biota in the eastern Bering Sea, Schumacher and Alexander (1999) identify the following four candidates for forcing interdecadal signals in that region. First, changes in solar activity are correlated with temperature and pressure in the atmosphere over the North Pole (Labitzke and van Loon, 1988), and may provide a forcing mechanism for decadal oscillations in the coupled air–ice–sea system in the northern hemisphere (Ikeda, 1990). The lunar nodal cycle of the moon (period of 18.6 yrs) affects mixed layer depths and has been linked to oceanic conditions in the North Pacific Ocean (Royer, 1998; Royer *et al.,* 2001). Third, atmospheric interactions exist between the Southern Oscillation and the Aleutian Low as proposed by Wooster and Hollowed (1995). Finally, Latif and Barnett (1994) suggest that an unstable air-sea interaction exists between circulation in the North Pacific subtropical gyre and the Aleutian Low pressure system. Since 1999, other candidates have been proposed, including changes in mid-latitude wind stress (Parrish *et al.,* 2000), subtropical oceanic Rossby waves (Jin *et al.,* 2001), and subsurface changes in the ocean south of the equator as part of an oscillation intrinsic to the equatorial and southern tropical Pacific Ocean (Giese *et al.,* 2002).

The primary atmospheric pressure features influencing the Bering Sea include weather patterns in the tropical South Pacific (El Nino-Southern Oscillation: ENSO), North Pacific (Pacific-North America: PNA) and Arctic (Arctic Oscillation: AO). The mode of connectivity between these hemispheric-scale features and the regional weather appears to be mainly a perturbation in the magnitude, pathway, and frequency of storm passage along the Aleutian Island chain (Stabeno *et al.,* 1999). The frequent migration of storms results in a statistical feature known as the Aleutian Low. During summer, with its long periods of daylight and high solar radiation, the Aleutian Low is typically weak and the weather benign. During winter, a marked change occurs in atmospheric pressure fields. High sea level pressure (Siberian High) dominates Asia, while the Aleutian Low deepens and dominates weather over the North Pacific and Bering Sea. The juxtaposition of these features results in strong, frigid winds from the northeast. The frequency and intensity of storms in the southern Bering Sea decreases temporally from winter to summer, but the frequency also decreases with increasing latitude (Overland, 1981; Overland and Pease, 1982).

3.3 Hydrographic Features

3.3.1 Southeastern Bering Sea

The physical features of the oceanic and shelf region of the eastern Bering Sea were reviewed by Schumacher and Stabeno (1998) and Stabeno *et al.* (1999). Over the southeastern shelf during summer, three distinct cross shelf domains exist which are characterized by water column structure, currents and biota (Cooney and Coyle, 1982; Coachman, 1986; Schumacher and Stabeno, 1998). These are the coastal (depth <50 m and characterized by weak stratification), middle shelf (50–100 m deep, characterized by a wind-mixed surface layer abutting a tidally-mixed bottom layer), and outer shelf (100–180 m deep, mixed upper and lower layers separated by a layer with slowly increasing density). Over much of the southeastern shelf during summer, temperature is an excellent indicator of vertical water-column structure. The primary source of freshwater is the melting of sea ice; the sea ice is advected southward during the winter and early spring. During early spring, the combination of tidal and wind mixing tend to weaken the halocline. For waters < 30 m deep, tidal mixing energy tends to stir the entire water column. In the deeper waters (30 < z <90 m), there is some vertical structure in salinity, but it is typically < 0.3 psu (Overland et al, 1999b). It is only in late spring and summer when the storms weaken and solar heating begins, that pronounced vertical stratification occurs over the middle domain. During winter, the domains are poorly defined since storms mix the water column to >90m.

These summer (April-October) domains are separated by a system of transitional zones or fronts (Coachman, 1986; Iverson *et al.,* 1979; Schumacher and Stabeno, 1998). The shelf-break front separates the outer shelf from slope waters; the broad middle shelf transition zone lies between outer and middle shelf waters; and an inner (structure) front separates the well-mixed coastal waters and the two-layered middle shelf domain. Knowledge of the characteristics of the inner front was recently refined: the inner front is wider than previously thought and its location varies by tens of kilometers (Kachel *et al.,* 2002), rather than being relatively fixed to the 50-m isobath as earlier hypothesized. The balance of wind and tidal energy plays a major role in shaping the vertical structure of both the coastal and middle shelf domains (Schumacher and Stabeno, 1998; Coachman, 1986). The domains provide unique habitats for biota. For example, the meso-zooplankton community in the two shallower domains is comprised primarily of the small to medium-sized copepods, whereas in the outer shelf domain and oceanic region, large copepods dominate (Cooney and Coyle, 1982; Smith and Vidal, 1986).

3.3.2 Northeastern Bering Sea

We consider the northern shelf to be that portion of the eastern shelf north of ~62°N where changes in topography, tidal energy, and river discharge (primarily from the Yukon River) modify the boundaries between domains. The width of both the coastal and middle shelf domains increase. Nearly the entire shelf east of the Anadyr Strait has depths <50 m, and much of the Gulf of Anadyr lies in the middle domain. North of Nunivak Island, the inner front moves to the vicinity of the 30-m isobath as tidal mixing energy decreases (Schumacher and Stabeno, 1998). In the vicinity of the Yukon River delta, however, the substantial freshwater discharge can result in stratification in waters <30 m. The water column (generally <20 m) in Norton Sound typically exhibits a two-layered structure during summer as opposed to the coastal domain in the southeast, which is at most weakly stratified. During winter, strong heat and salt fluxes result in a vertically mixed water column (Muench *et al.*, 1981).

Across the shelf south of St. Lawrence Island, three water masses exist: Alaskan Coastal, Bering Shelf, and Anadyr (Coachman *et al.*, 1975). The accompanying regional salinity field is characterized by a zonal gradient with salinity increasing from east to west (Coachman *et al.*, 1975; Schumacher *et al.*, 1983). In water where the temperature is <0°C, the salinity between ~30 to 60 m increases by 2.5 psu from east to west across the shelf. During ice-free conditions, salinity from St. Lawrence Island to the Gulf of Anadyr seldom exceeds 32.8 psu; however, during ice formation values >34.2 psu occur. In the Gulf of Anadyr, salinity during summer ranges between 33.0 and 33.5 psu. The saline waters which flow northwestward across the mouth of the Gulf of Anadyr carry relatively warm temperatures and nutrients which suggests the presence of outer shelf water. North of St. Lawrence Island, all three water masses are present and can be identified as they flow northward through Bering Strait (Coachman *et al.*, 1975)

3.3.3 Western Bering Sea

A system of three hydrological zones exists over the western shelf that are somewhat analogous to those found on the eastern shelf (Verkhunov, 1994). These hydrological zones (coastal, transitional and oceanic) are easily distinguished in temperature/salinity diagrams (Khen, 1999). The coastal zone has low (<31.0 psu) salinity surface water and a strong pycnocline. Due to a less pronounced halocline, the transitional zone exhibits a more weakly two-layered structure than the coastal zone. In both zones, a strong seasonal thermocline develops over summer. The oceanic zone is identified by a three-layered vertical structure with relatively warm (~1.0°C) bottom temperatures that indicate the presence of slope waters.

While there is some similarity to the hydrographic structure of the eastern shelf, fundamental differences exist. The locations of the zones over the western shelf are not stationary, although they can at times be associated with depth contours as occurs on the eastern shelf (Verkhunov, 1994). The western shelf is relatively narrow and divided by peninsulas into three separate and somewhat isolated gulfs. The immense width of the eastern shelf eliminates the direct influence of gyre circulation on all but the outer domain. Such is not the case on the western side, where the Kamchatka Current, the western boundary current of the Bering Sea gyre, has a profound impact on the location of hydrographic zones. When this current moves shoreward and flows over the continental slope, all the zones are compressed and their frontal divisions are shifted shoreward (Khen, 1999).

3.4 Circulation

A schematic of the general circulation is shown in Fig. 30.4. The portion of the Alaskan Stream that flows through Aleutian passes, especially Amchitka and

Amukta Passes, forms the eastward flowing Aleutian North Slope Current (ANSC: Reed and Stabeno, 1999; Stabeno *et al.,* 1999). The ANSC provides the main source of the Bering Slope Current (BSC), which is often characterized as variable flow replete with eddies and meanders, although at times it appears as a more organized northwestward flowing current (Stabeno *et al.,* 1999). The BSC separates from the slope, resulting in a broad weak westward flow across the basin. Along the west coast the current intensifies, forming the southward flowing Kamchatka Current. Exchange between the slope and eastern shelf likely varies depending upon which mode of the BSC is dominant. Transport in the ANSC/BSC system is highly variable ranging from ~2 x 10^6 m^3 s^{-1} to ~9 x 10^6 m^3 s^{-1} (Stabeno et al., 1999). The importance of these slope currents to ecosystem dynamics of the eastern shelf is threefold. First, they transport material northward along the slope that can be advected onto the shelf through various onshelf mechanisms (Schumacher and Stabeno, 1994; Stabeno *et al.,* 2001). Second, eddies, which are common seaward of the shelf break (Schumacher and Reed, 1992), apparently provide a temporary habitat that favors survival of larvae (Schumacher and Stabeno, 1994) and are regions of enhanced primary production (Mizobata *et al.,* 2002). Finally, the vertical temperature structure of the slope current with its warm subsurface maximum is potentially important to fish stocks (Reed, 1995).

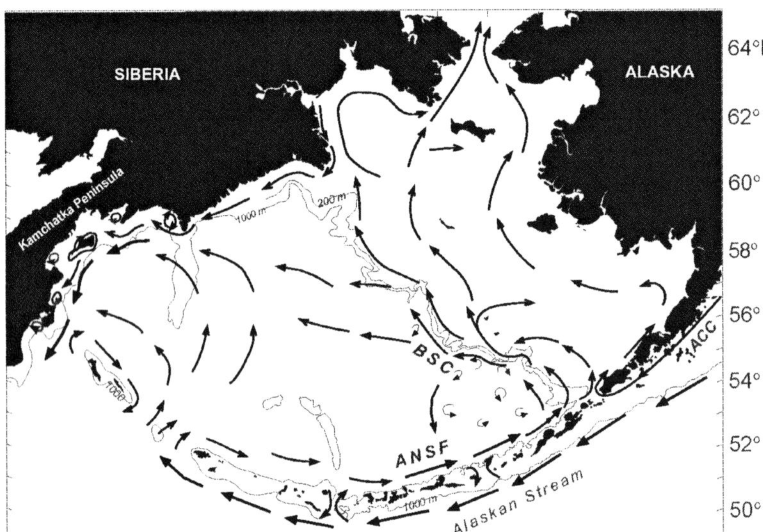

Figure 30.4 Schematic of mean circulation in the upper 40 m over the basin and shelf (from Stabeno *et al.,* 1999). The arrows with large heads represent currents with mean speeds >50 cm s^{-1}. The Alaskan Stream, Aleutian North Slope Current (ANSC), Bering Slope Current (BSC) and Kamchatka Current are indicated. Depth contours indicate 1000 m isobath and in the Bering Sea the 200m isobath.

Low frequency flow on the shelves is much weaker than in the basin. Over the eastern shelf, the net flow from late spring through mid autumn is northward. Enhanced, more organized flow occurs along the 100 m isobath and the 50m isobath. Both of these coincide with frontal or transition zones on the shelf. The flow along the 50m isobath, which we call the Bering Coastal Current (BCC), is weak (1– 5 cm s^{-1}). The BCC consists of waters from the Gulf of Alaska's Alaska Coastal Current (ACC) that flows through Unimak Pass with a transport of ~ 0.3 x 10^6 m^3 s^{-1} (Schumacher and Stabeno, 1998; Stabeno after al, 2002). Some of this transport flows along the Alaska Peninsula and the west coast of Alaska where freshwater is added from widely distributed riverine sources. Unlike other buoyancy driven flows that are strengthened and compressed to the coastline by alongshore wind stress (e.g., the ACC; Stabeno et al., 1995), the BCC is strongest in the vicinity of the 50 m isobath, which can be more than 100 km offshore (Schumacher and Stabeno, 1998; Schumacher and Kinder, 1983). After flowing around the entire perimeter of the eastern shelf, the BCC exits the Bering Sea through Bering Strait.

While some of the transport through Unimak Pass forms the BCC, the remainder flows northwestward in the vicinity of the 100m isobath to the Pribilof Islands. The strongest mean current over the central and southeastern shelf is found south of St. George Island, where the outer shelf narrows and the bottom slope increases and thus narrowing the flow along the 100m isobath. Daily average currents can exceed 40 cm s^{-1}. The mean flow over the rest of the central and southeastern shelf is weak (generally <5 cm s^{-1}). The currents over the northern shelf are stronger, especially the Anadyr Current which is the primary source of water flowing through Bering Strait. During winter, the flow becomes less organized over the shelf as the frontal structures break down and strong winter storms mix the water to the bottom.

Tidal currents are a major circulation feature of the southeastern shelf, but over the northern and western shelves their importance is greatly diminished (Schumacher and Stabeno, 1998; Stabeno et al, 1999; Kowalik and Stabeno, 1999). Over the southeastern shelf, tides are the dominant source of kinetic energy and they typically mix the coastal waters (z < ~ 50 m) and the lower 30 – 40 m of the deeper portions of the shelf. In addition, their interaction with bathymetric features can result in residual flow, particularly in canyons (Schumacher and Reed, 1992; Kowalik and Stabeno, 1999) and around islands such as the Pribilof Islands (Kowalik and Stabeno, 1999 Stabeno after al. 1999).

3.5 Sea Ice

The eastern Bering shelf is a marginal ice zone, and sea ice markedly impacts the ecosystem. Ice-free conditions typically exist from June through October. Formation of sea ice generally begins in the polynyas of the northern shelf in November, with maximum ice extent occurring as early as January or as late as April, but more typically in March. The mechanism of ice formation for the eastern Bering Sea has been described by analogy to a "conveyor belt" (Overland and Pease, 1982). Ice is produced along leeward (south-facing) coasts of the northern shelf, and is driven southward by wind to the vicinity of its thermodynamic limit where it melts. This limit advances southward as cold northerly winds and ice-melt cool the shelf waters. The amount of production and advection of ice depends upon which storm track dominates, with greatest ice production occurring in years when the Aleutian Low is well developed and displaced eastward so that storms migrate along a primary storm track south of the Alaska Peninsula. Satellite observations of ice cover show ~40% variation about the mean (Niebauer, 1998). Using the same sea ice data set, but resolved along a longitude, Wyllie-Echeverria (1995) showed that other characteristics also exhibit large interannual variability. These characteristics include: duration of ice at its southern extent (3–15 weeks), time of retreat from the southernmost extent (between February and June), and the number of weeks that ice remained over the middle shelf (3–28 weeks, with a mean of 20).

Changes in characteristics of sea ice over the southeastern shelf (including the time of arrival, departure and persistence), indicate that the most extensive ice years coincided with the negative phase of the Pacific Decadal Oscillation (the first mode of decadal variability in sea surface temperature over the North Pacific Ocean), although considerable interannual variability occurs (Stabeno *et al.*, 1998 and 2001). These authors characterized temporal variability in spatial patterns by dividing the time series of ice observations into three subsets according to generally agreed upon periods: 1972–76 (cold), 1977–88 (warm), and 1989–98 (weaker cold). A marked difference exists in persistence and spatial distribution of ice between the first and the latter two periods. During the cold period, ice covered the shelf out to and over the slope, and remained around St. Paul Island for more than a month. During the later years, ice did not extend as far seaward and its residence time was typically 2–4 weeks less than during the cold period. The differences between the two latter regimes are more subtle, but still evident. Between the mid-shelf and slope northwest of the Pribilof Islands, ice remained for 2–4 weeks longer during 1989–1998 than during 1977–1988. North and west of St. Lawrence Island and along the coast of Alaska north of Kuskokwim Bay, there were 1–2 less weeks of ice cover in 1989–1998 than in 1977–1988. Changes in the Arctic Oscillation and the attendant changes in wind patterns are likely the causal mechanism for these recent alterations in sea ice (Stabeno and Overland, 2001). In addition, marked differences existed in the ice distribution and extent along the Alaskan Peninsula. During the cold period, ice extended seaward nearly to Unimak Pass, whereas in the other periods contours of ice persistence were constrained or limited to inner Bristol Bay. This pattern is likely related to variations of inflow and/or temperature of shelf waters from the Gulf of Alaska, which flows through Unimak Pass onto the Bering Sea shelf (Schumacher and Stabeno, 1998).

Cold northerly winds, which advect the ice southward, and melting play a critical role in fluxes of heat and salt, and in generation of both baroclinic flow and the cold lower layer (cold pool) which persists all summer over the middle shelf domain (Schumacher and Stabeno, 1998; Wyllie-Echeverria and Ohtani, 1999). The positive buoyancy from melting ice initiates both baroclinic transport along the marginal ice zone and stratification. Cooling and mixing associated with ice-advance help to condition the entire water column over the middle shelf domain (Stabeno *et al.,* 1998). With subsequent seasonal heating of the upper layer, the lower layer becomes insulated and temperatures often remain below 2°C (Reed, 1995). The annual area of this cold pool varies by ~2.0×10^5 km^2 between maximum and minimum extent.

4. Processes in the Southeastern Bering Sea

Natural fluctuations in climate dramatically influence biota in the Bering Sea (e.g., Napp and Hunt, 2001). Trends observed in 11 species (fish, marine birds, and marine mammals) found in the eastern Bering Sea/Aleutian Islands ecosystem show marked changes in relative abundance (Fig. 30.5). Using 100 physical and biological time series (29 of these from eastern Bering Sea marine biota), Hare and Mantua (2000) showed a significant change occurred in 1976/77 and to a lesser degree in 1989. While the mechanisms that link climate to biota were not addressed, it appears that a shift in the decadal patterns of climate (indicated by changes in the Pacific Decadal Oscillation and the Arctic Oscillation) contributed to changes in biota. More recently the change in the Pacific Decadal Oscillation (PDO) in the late 1990s has been interpreted as a regime shift (McFarlane *et al.*, 2000; Macklin *et al.*, 2002; Peterson and Schwing, 2003) or more recently as a shift in the dominate mode of variability in sea surface temperature (Bond et al., 2004). It is not necessary for the climate shift to be immediately manifest as biological change; shifts in the demersal fish communities are characterized by lags from the climate/ocean regime shift (Conners *et al.*, 2002) and buildup of demersal fish stocks can exert top down control on populations of commercially important fish well after the climate shift (Bailey, 2000).

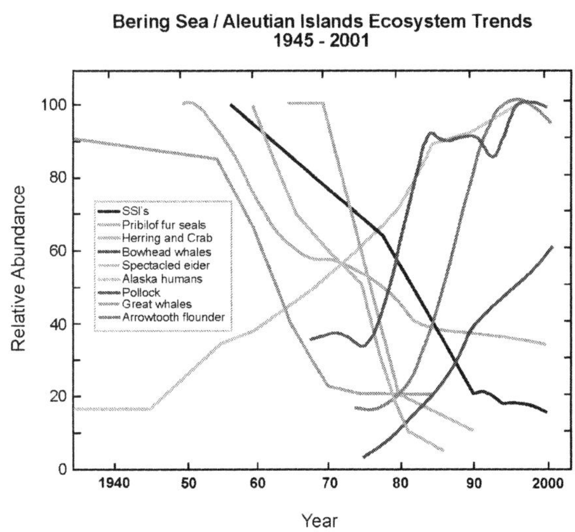

Figure 30.5 Schematic of ecosystem changes in the eastern Bering Sea showing trends of the relative abundance of select marine mammals, fish, and shellfish in the eastern Bering Sea. (after NRC 2003)

In the mid to late 1990s, the Eastern Bering Sea exhibited a host of noteworthy changes in physical and biological conditions (Kruse, 1998; Vance *et al.*, 1998; Tynan, 1998; Baduini *et al.*, 2001; Napp and Hunt, 2001; Hunt *et al.*, 1999; Stabeno *et al.*, 2001; Stabeno and Overland, 2001; Macklin *et al.*, 2002; Napp *et al.*, 2002; Olson and Strom, 2002): the first recorded major Bering Sea coccolithophorid blooms (1997, followed by blooms in 1998 through 2001), a large die-off of shearwaters (1997), salmon returns far below predictions (1997 and 1998), unusually warm summer sea surface temperatures (1997), unusually high ocean heat content (1998), and a decrease in the onshore transport of slope water (1997). While in the eastern Bering Sea, the abundance of pollock decreased (but continued to dominate biomass), a substantial increase occurred in gelatinous zooplankton (Brodeur *et al.,* 1999a, 2002), arrowtooth flounder and other flatfish (Wilderbuer *et al.,* 2002). Further, populations of Steller sea lions have continued their 30-year decline throughout the eastern Bering Sea. Below we address processes where the ocean temperature is directly related to sea ice processes and distribution, and to either wind or tidally generated turbulence and transport.

4.1. Processes Related to Sea Ice

The pathways model (Fig. 30.2) considers sea ice as an important parameter, particularly fluctuations in extent, time of advance and subsequent melt-back, which are among the most striking physical phenomena (Schumacher and Stabeno, 1998) that also have profound impacts on biota (e.g., Hunt *et al.,* 2002; Schumacher *et al.,* 2003). Sea ice also provides habitat (e.g., haul outs) for some marine mammals, and once adequate solar radiation is available (early March), an under ice phytoplankton bloom occurs (Stabeno *et al.,* 2001). The recent Oscillating Control Hypothesis (OCH; Hunt *et al.,* 2002) relates climate change to energy flow and the timing of the plankton blooms, i.e., the timing of sea ice presence and melt-back is a critical aspect.

The interrelationships among the presence of sea ice, water column temperature, stability, and phytoplankton blooms (Fig. 30.6) have been observed from a long-term monitoring site on the middle shelf of the southeastern Bering Sea (Stabeno *et al.,* 2001, Hunt *et al.,* 2002). When ice is present in or after mid March, a strong peak occurs in chlorophyll fluorescence under the ice (e.g., 1995, 1997). The level of incoming short wave radiation prior to this time is likely insufficient to initiate a bloom of most phytoplankton species. The ice itself is relatively "thin" first year ice (<1 m) and has very little snow cover to attenuate or reflect incident radiation. When there is no ice present at this monitoring site (e.g., 2001), or the ice retreats before mid March (e.g., 1998, 2000), the bloom occurs in May or June. In 1999, sea ice was present at the site in late March, and returned in early May. As a result, there was an initial bloom in late March and another weaker and prolonged period of elevated fluorescence in late May and June.

The spring bloom of phytoplankton associated with the sea ice accounts for 10–65% of the total annual primary production (Niebauer *et al.,* 1995). A large proportion of the primary production from the sea-ice associated bloom eventually falls, unused, to the sea floor. This occurs, because cold temperatures likely impact the ability of micro- and meso-zooplankton grazers to effectively utilize the increase in production. Estimated total zooplankton production in 1999, a cold year, was 8–52% that of the two previous years when winter temperatures were average to above average (Coyle and Pinchuk, 2002a). While the total zooplankton production is dominated by small species that respond positively to temperature, at least one key species over the middle domain (*Calayus marshallae*) has higher standing stocks in colder years (Baier and Napp, 2003. In years when sea ice either does not exist over the southeastern shelf or retreats prior to the time when adequate light

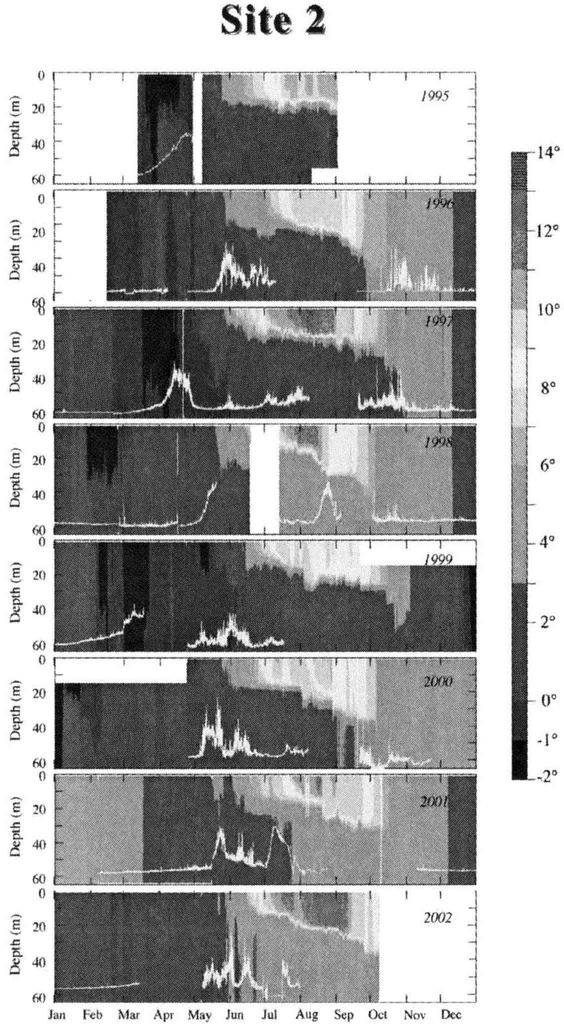

Figure 30.6 Time series of temperature contours at the long-term monitoring site in the middle domain (Site 2). Areas of black indicate cold water resulting from the presence of melting sea ice. The yellow line near the bottom of each panel indicates fluorescence at 11–13 m depth. For each year, the fluorometer tracts have been scaled to the highest value in that year. Gaps in this record occur due to fouling of the instrument or loss of mooring. (After Stabeno and Hunt, 2002.)

exists for a bloom (mid March), the spring bloom is delayed until later in the season when thermal stratification of the water column occurs (Stabeno *et al.,* 2001). Under this scenario, phytoplankton biomass accumulates in the water column, but a much larger fraction of the primary production is utilized by zooplankton. The frequency of alternation between cold and warm regimes and the lengths of the individual stanzas may determine whether higher trophic levels are predominantly controlled by bottom-up or top-down mechanisms (Hunt *et al.,* 2002; Hunt and Stabeno, 2002). In addition to how ice effects bottom up control of energy flow in the pelagic ecosystem, ice and its attendant cold pool directly influence the spatial distributions of higher trophic level biota (Ohtani and Azumaya, 1995; Wyllie-Echeverria and Ohtani, 1999; Brodeur *et al.,* 1999b). As previously mentioned, avoidance of the cold pool often has the affect increasing predation and cannibalism on larval and juvenile fishes by increasing the spatial overlap (or separation) of predator and prey.

4.2 Processes Related to Turbulence

In the Bering Sea, wind stress is another important physical force that strongly influences biological production as well as the flux of heat and nutrients (Fig. 30.2). Wind mixing largely determines the timing of the spring phytoplankton bloom when ice is not present, and helps to set annual new production through controlling the resupply of nitrate into the euphotic zone during summer (Sambrotto, *et al.,* 1986). Over the southeast shelf, the effect of wind mixing (through changes in water column stability) on nitrate uptake (primary production) can either decrease uptake during the respiration-limited phase or increase nitrate uptake once nutrients are limiting through resupply from deeper waters (Sambrotto *et al.,* 1986). Not included in the pathway model are tides that strongly influence the southeastern Bering Sea. These two sources of turbulence are important to the region's biology and sediment processes.

The Inner Front Study, which began in 1997, focused on the importance of physical processes of the inner front to prolonged primary production. Results from Kachel *et al.* (2002) not only enhanced our knowledge of the inner front's physical characteristics, but they also elucidated some of its biological importance. During summer, nutrients could be pumped into the euphotic zone at the front , thereby stimulating production. The effectiveness of this process depends on two factors. First, a deep (sub-pycnocline) reservoir of nutrients must exist, and this is usually found within the middle shelf's bottom layer, the cold/cool pool. Typically, high concentrations of nutrients exist in the bottom layer over the middle shelf. An exception was 1997, when two factors (late spring storm and a shallow summer mixed layer) conspired to deplete nutrients from the cold pool. The strong, late spring storm vertically mixed the water column to a depth >50m. Since nutrients in the euphotic zone had been depleted by an earlier, ice-associated, phytoplankton bloom, this caused the nutrient concentration in the cold pool to be reduced by almost half. Next, the very shallow mixed layer during the summer, allowed a phytoplankton bloom to occur beneath the surface mixed layer, which slowly consumed the nutrients in the cold pool (Stabeno *et al.,* 2001; Stockwell *et al.,* 2001). In addition to nutrients being available, sufficient mixing energy must occur to erode the extant vertical stratification. This can happen either by intensification of wind mixing and/or tidal mixing (throughout the fortnightly cycle); both of these processes move the inner front seaward. When this occurs, nutrients are made available to the frontal region, and are mixed upward where they can be utilized by phytoplankton. An example of this process can be seen in temperature and nitrate data collected across the inner front (Fig. 30.7). High concentrations occur within the cold/cool pool while those in the upper layer of the middle shelf and in the coastal domain are low. At the inner front, vertical finger-like structures with elevated concentrations of nitrate coincide with the 5.5°C isotherm. This structure was observed several days after an intense (wind speeds >14 m s^{-1}) storm passed through the region and tidal currents were near their fortnightly maxima (Kachel *et al.*, 2002).

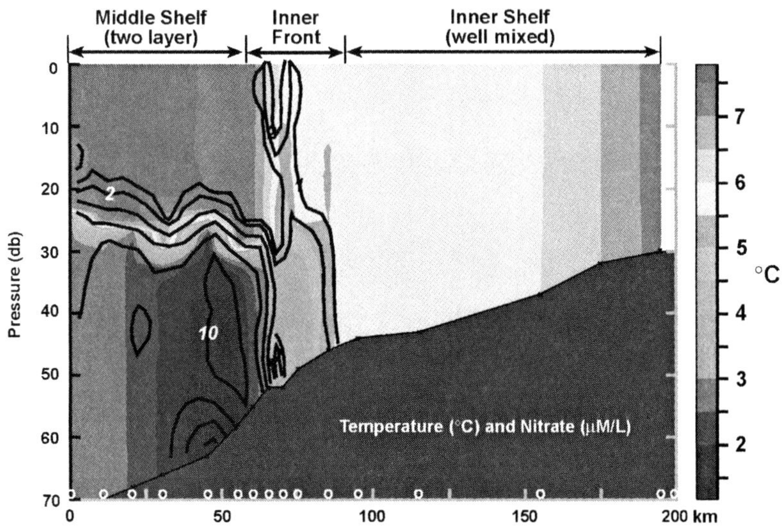

Figure 30.7 Contours of temperature (color; °C) and nitrate (black lines; mmoles l^{-1}). The locations of the CTD stations are marked along the bottom axis. Nutrients were measured on each cast. (From Kachel *et al.*, in 2002).

Turbulence also can directly affect the efficiency of larval feeding. Megrey and Hinckley (2001) used a process-oriented individual based model (IBM) of larval pollock that incorporated a turbulence-contact rate-feeding success mechanism, thus relating wind-generated turbulence to feeding. Output from the model agreed with hydrodynamic theory, with a well-defined peak in consumption at intermediate wind speeds (MacKenzie *et al.*, 1994). The functional form of wind speed versus maximum consumption was determined by a quadratic fit to model results. Optimum feeding (540 μg dry weight per day per individual) occurred at a wind speed of ~7 m s^{-1}. At wind speeds >9.5 m s^{-1} increased turbulence negatively affected feeding, and at wind speeds <4.8 m s^{-1} feeding was less than optimal because turbulence was not sufficient to enhance contact rate. Fish and plankton can regulate their exposure to turbulence by adjusting their vertical position in the water column (Incze *et al.*, 2001; Olla and Davis, 1990). In the Gulf of Alaska, survival of larval pollock cohorts has been tied to coincidence of the critical first feeding period and calm weather (Bailey and Macklin, 1994).

4.3 Processes Related to Transport

Transport, which is indicated in the pathway model as changes in horizontal flow, is critical to many aspects of ecosystem dynamics, including advection of nutrients and plankton. Several studies suggest a connection between changes in climate and changes in recruitment of fish and shellfish species (e.g., Incze *et al.*, 1987, Rosenkranz, et al., 2001, Wilderbuer *et al.,* 2002,). One mechanism often cited is a change in the direction of transport of the planktonic stages (towards or away from favorable habitat or predators). Wespestad *et al.* (2000) related changes in climate, and their regional impact on wind fields and transport of larval pollock in an attempt to identify sources of recruitment variability.

Recent studies have shown this shelf has considerable north-south variability. The southeastern shelf contains three distinct regions or latitudinal zones along the 70-m isobath: a strong two-layered system with cold/cool pool that lies south of about 57°N, an intermediate zone consisting of more well-mixed water; and a two-layered system with the northern cold pool that extends northward from about 58°N (Stabeno *et al.,* 2002). In the bottom layer over the middle shelf, distributions of salinity and nutrients (nitrate) appear correlated. The greatest salinities and nitrate concentrations occurred in the northernmost zone and may represent the impact of across-shelf transport due to the mean flow in this region (Reed, 1998). Slope/outer shelf domain plankton taxa have been observed near the inner front as further evidence of this cross-shelf transport (Coyle and Pinchuk, 2002b).

Transport has important ramifications for dissolved and planktonic material, including larval fish (Wespestad *et al.,* 2000; Wilderbuer *et al.,* 2002) and crabs (Rosenkranz *et al.,* 2001). Wespestad *et al.* (2000) used a simple wind drift model to generate trajectories of pollock eggs and larvae over the southeastern shelf. After a time period related to larval development, the young pollock were subjected to cannibalism by age-2 and older adults. While this pattern was consistent with some year-classes of high recruitment, it did not fit the recruitment every year. One caveat is that the wind drift model only applies to the upper few meters of the water column, and both pollock eggs (Kendall, 2001) and larvae exist deeper in the water column (Napp *et al.,* 2000). In addition, using a single initial point for the eggs/larvae is an oversimplification of the actual spawning regions for pollock (Hinckley, 1987).

To address these disparities, a different transport model was used to simulate transport. The North-eastern Pacific Regional Ocean Model System (NEPROMS) was selected to simulate drifter trajectories that more closely simulate pollock egg and larval transport (D. Ridgi, personnel communications). The simulated pollock eggs were initialized in a grid, which contained the initial point used by Wespestad *et al.* (2000), but the drifter initial positions are denser near the surface, replicating egg distribution data collected in the Bering Sea (Kendall, 2001). A prominent feature in this region (north and east of Unimak Pass) is the confluence of flow of the Alaska Coastal Current into the Bering Sea through Unimak Pass and the upslope flow of the ANSC in Bering Canyon (Stabeno *et al.,* 2002). The initial drifter positions were a seven by seven grid with horizontal separations of about 10 km at the confluence of two flows. Vertically, there were 15 drifters initialized at each grid point to a maximum depth just over 40 m. Drifters were released on April 1 of each year and tracked for 90 days.

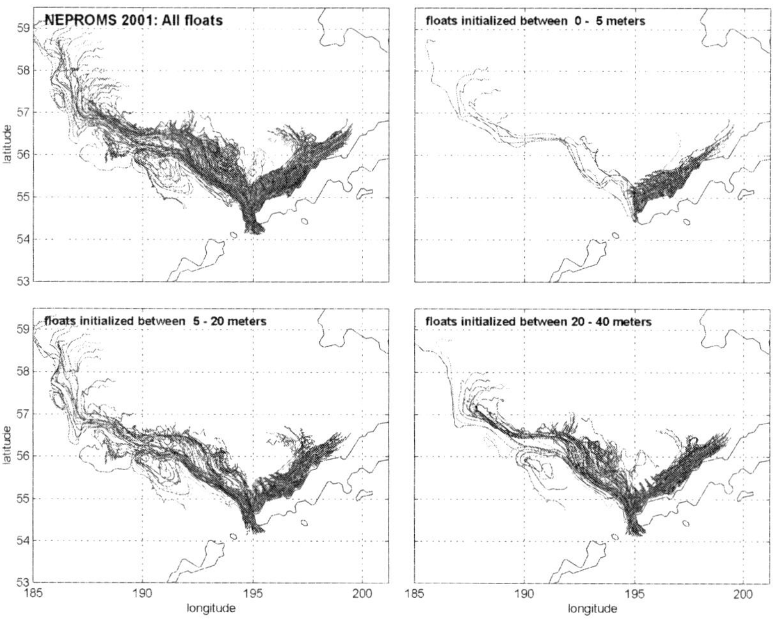

Figure 30.8 Trajectories simulated for a 90-day period starting on 1 April 2001. The upper left panel shows all drifters, while the other panels show drifters divided as a function of initial release depth. Note that with increasing depth there is a greater tendency for drifters to flow northwestward and even become involved in the eddy-like circulation associated with the Bering Slope Current. This suggests a mechanism for planktonic transport between oceanic and shelf waters.

Simulations were conducted for the period 1997–2001. In all years there was a tendency for drifters to move either toward the northeast along the Alaska Peninsula, or toward the northwest along the 100 m isobath. This result is in agreement with a schematic of mean circulation (Fig. 30.4) and satellite tracked drifter trajectories (Schumacher and Stabeno, 1998). Recently, abundances of age-0 pollock in and around the inner front at Cape Newenham were estimated to be the same order of magnitude as those around the Pribilof Islands (Coyle and Pinchuk, 2002b). Model trajectories of drifters from 1997 show that a subset of the drifters that began by following the 100 m isobath and then veered to the northeast. In 2000, trajectories revealed a strong turning to the northwest of trajectories that had been moving along the Alaska Peninsula. Fig. 8 shows trajectories for 2001. The simulations also suggest the importance of interannual variations in transport. They also suggest how dependent the end points of the "larval drift" are to small variations in both horizontal (order 10 km) and vertical (order 5–10 m) initial positions. Numerous trajectories generated by a model, which includes only wind drift (OSCURS: Wespestad *et al.,* 2002) were toward the northeast along the Alaskan Peninsula, as were the majority of the NEPROMS simulations in the upper five meters of the water column (Fig. 30.8). With deeper initial start points, however, comes a stronger divergence of the trajectories. In the 5–20 m and 20–40 m release bins there were many drifters that followed the 100 m isobath to the northwest, with some even moving through Unimak Pass into the North Pacific

Ocean before turning back. Further examination is required to determine the environmental parameters that resulted in the interannual differences in trajectories, and therefore to help understand how transport affects year class strength of pollock and other plankton.

5. Processes in the Northern Shelf of the Eastern Bering Sea

While many of the observations of recent changes come from the southeastern shelf, similar ecosystem-level change has been observed in the northern Bering Sea. There are indications of changes in benthic biomass south of St. Lawrence Island, and in the Chirikov Basin between Lawrence Island and Bering Strait (Grebmeier and Cooper, 2002). Studies begun in the mid-1980s have shown declines in the biomass (Grebmeier, 1993; Sirenko and Koltun, 1992; Grebmeier and Dunton, 2000), and mean sizes of the dominant bivalves in the area (Grebmeier and Cooper, 2002). Sediment respiration rates, which indicate carbon loading to the sea floor, have also declined since the late 1980s. Seasonal patterns of sediment chlorophyll concentrations show that deposition of carbon in this area is related to the ice-edge spring bloom (Cooper *et al.,* 2002), thus any change in the timing of ice retreat during the late winter/early spring will likely have a major impact on this ecosystem. Although commercial fishing may have played a role in changes to the southern Bering Sea ecosystem, there is little commercial fishing on the northern shelf where changes in benthic faunal populations and declines in dominant fauna have occurred, resulting in a cascade effect on higher trophic levels (Grebmeier and Dunton, 2000; Grebmeier and Cooper, 2002).

Ecosystem processes related to sea ice are dominant features of the northern shelf. In years with extreme occurrences of sea ice (e.g., extremely high in 1975/76 and extremely light in 1978/79), sea ice can be present at 62°N over the northern shelf from mid-November until mid-June, or only from late January to late April. While the presence of sea ice provides a substrate for marine mammals, it also limits at-sea observations. In other regions, under-ice algae provide a wintertime food source for zooplankton during periods of reduced water column production (Toureaugu and Runge, 1991). Most of our knowledge of this high latitude ecosystem comes from summertime observations. Primary production here is not consumed by pelagic secondary consumers (Coyle and Cooney, 1988, Springer and McRoy, 1993), but rather by a rich macro-benthic community. Large populations of benthic, rather than pelagic-feeding marine mammals and birds serve as apex predators in this food chain (Grebmeier and Cooper, 1995).

The primary physical process that results in high primary production over the northern shelf is the onshelf transport of nutrient-rich waters, which are then advected across the shelf. This current, known as the Anadyr Current (Shuert and Walsh, 1993), flows through Anadyr Strait and then northward through Bering Strait (Schumacher and Stabeno, 1998). As the bathymetry shoals (<40 m) in the northern Gulf of Anadyr, nutrients enter the euphotic zone and a production-deposition center is formed (Coachman, 1993). Another center is located over the Chirikov basin north of St. Lawrence Island. As the Anadyr Current flows between St. Lawrence Island and Siberia, the bathymetry deepens (50–60 m) and bottom-generated turbulence results in strong vertical mixing. This tends to mix phytoplankton below the critical depth and interrupts the bloom in the strait

(Springer and McRoy, 1993). As the current speed is reduced over Chirikov Basin, a region of extremely high primary production occurs (12–16 $gCm^{-2} day^{-1}$; Springer and McRoy, 1993). The strength of the Anadyr Current is related to the magnitude of flow through Bering Strait. This net northward flow is driven by the difference between sea level in the North Pacific and Arctic Oceans (0.4–0.5 m). On short time scales, this northward flow is modified by wind generated coastal changes in sea level (Aagaard *et al.*, 1985; Overland and Roach, 1987). During weak winds of summer, northward transport is greatest (Coachman, 1993) and provides the strong flux of nutrients for primary production. The Anadyr Current also transports a high biomass of large oceanic copepods from the slope regions of the basin onto the northern shelf (Springer *et al.*, 1989). In the Chirikov Basin, these support high numbers of planktivorous auklets (Springer and Roseneau, 1985; Springer *et al.*, 1987) that forage in stratified Bering shelf water (Hunt *et al.*, 1990) or in frontal regions on aggregations of these copepods (Hunt and Harrison, 1990). This combination of biophysical coupling supports a pelagic seabird community far from the origin of the zooplankton on which they are dependent.

Flow along the Alaskan coast of the northern Bering Sea is a combination of river runoff (mainly the Yukon River) and a continuation of the BCC of the southeastern Bering Sea (Schumacher and Stabeno, 1998). Production here has been identified as typical of shallow shelves elsewhere: once nutrients are exhausted during an initial bloom, production is low and only ~10% of that for Chirikov Basin and the Gulf of Anadyr (Springer and McRoy, 1993). The east–west gradient in water properties, nutrients, primary production and fauna (transported in the Anadyr Current) result in two different community structures, a rich benthic system to the west and a less productive eastern system (McRoy, 1993). Benthic macro faunal biomass ranges from ~30–60 gCm^{-2} for the western portion and <10 gCm^{-2} in the Alaskan coastal waters.

Over the northern shelf, regions exist that are usually ice-free throughout wintertime, and these regions are known as polynyas. These features often are found on down-wind facing coasts (e.g., the south coast of St. Lawrence Island). In the polynyas ice formation is caused by frigid winds blowing from the northeast. This and the resulting brine rejection are important features of the physical environment. As ice is formed, brine is released and sinks to the bottom. The current field generated by this process results in both offshore flow and sometimes a reversal of the mean eastward flow of the branch of the Anadyr Current that flows south of St. Lawrence Island (Schumacher *et al.*, 1983). This wintertime set of physical processes apparently affects sediment patterns, e.g., surface C/N ratios, total organic carbon, sediment oxygen uptake, and benthic biomass (Grebmeier and Cooper, 1995). As a result of the enhanced production, the region southwest of St. Lawrence Island has long been a favorite feeding ground for gray whales.

6. Processes of the Western Shelf of the Bering Sea

Recent changes have been observed in the ecosystem of the western Bering Sea (Radchenko *et al.*, 2001). Although the Bering Sea responds to climate shifts, a significant portion of the variability occurs on year-to-year time scales, which can make the identification of regime shifts difficult. This year-to-year variability is clearly evident in recent years. For instance, 1999 was an unusually cold year, characterized by negative anomalies in atmosphere and sea surface temperatures, and extensive and prolonged sea ice cover. In contrast, positive sea surface temperature anomalies were predominant throughout the northwestern shelf of the Bering Sea in 1997 (spring and summer) and 1998 (summer). In response to the physical conditions in 1999, some biological processes were less active and/or delayed. During 1999, the return of Pacific salmon to streams in northeastern Kamchatka was approximately two weeks later than typical.

Over the continental shelf of the western Bering Sea, the importance of transport to distribution of gonatid squid (*Berryteuthis magister*) has recently been investigated (Arkhipkin *et al.*, 1998). This squid is one of the most abundant in the North Pacific Ocean/Bering Sea and is exploited commercially by both Russian and Japanese bottom trawl fisheries. Distinct patterns of size composition, age structure and growth occur over the western shelf, and given the life history of this squid, time-dependent transport of planktonic stages provides the most likely explanation (Arkhipkin *et al.*, 1998). While much of the counter-clockwise current over the Bering Sea basin is related to inflow of the Alaskan Stream through passes in the Aleutian Island Chain (Overland *et al.*, 1994, Stabeno *et al.*, 1999, Stabeno *et al.*, 2005), seasonal changes in wind stress can alter transport in the gyre by ~50% (Bond *et al.*, 1994). As a result of this strong seasonal signal, regional shelf edge current patterns in the western Bering Sea apparently have two modes: a strong along slope mode and an eddy mode that exists as the current relaxes (Verkhunov and Tkachenko, 1992). As noted in Section 3.4, this is also the case in for the eastern BSC. Associated with these modes are two primary transport routes for the juvenile squid: winter and spring hatched juveniles are first transported into the eastern region of the shelf and then westward along the shelf; spring/summer hatched juveniles are transported directly across the basin into the western region of the shelf (Verkhunov and Tkachenko, 1992).

In an analysis of long-term fluctuations in several species of pelagic and benthic fishes of the western Bering Sea, Naumenko (1996) notes the correspondence between changes in fish biomass and conditions in the ecosystem. Using catch data, he discerned that the fish community could be divided into four periods: 1958–1964 when herring dominated, 1965–1974 a transitional period when no single species dominated and stocks were highly variable, 1975–1987, when pollock dominated in both biomass and abundance, and 1987–1993 (last data set presented) when pollock declined and groundfish populations increased. While some of the fluctuations could be accounted for by fishing mortality, it appeared that environmental conditions have played an important role in the fluctuations. The first connection Naumenko (1996) made was between bottom water temperatures in fall and surface temperatures in spring and summer, with the four fish community periods. The first period was moderately warm, the second abnormally cold, the third abnormally warm. The fourth period appeared to be one of transition. He notes that changes in the biomass of zooplankton closely correlated with the temperature changes, with zooplankton biomass being greatest during the warm period and lowest in the cold period. Note the correspondence of these observations to the OCH (discussed in section 7): cold regime implies low zooplankton biomass and warm regime high biomass. Changes in recruitment of pollock in the western Bering Sea have also been attributed to changes in the climate/oceanic regime

(Balykin, 1996), although the processes that might link the climate/oceanic system to fish survival were not discussed.

7. Models of Ecosystem Dynamics

Efforts to model the eastern Bering Sea ecosystem were reviewed by Francis *et al.* (1999). Models of ecosystem dynamics of the southeastern Bering Sea shelf span the full spectrum from conceptual models (e.g., the OCH, Hunt *et al.,* 2002; the Trophic Cascade Hypothesis, Merrick, 1995; NRC 1996), to trophic or food web models (e.g., Laevastu and Larkins, 1981; Trites *et al.,* 1999), to models that explicitly include links between specific ecological processes, such as physical processes and primary/secondary production (Hood, 1999; Walsh and McRoy, 1986). After providing some essential background information on the ecosystem, we focus on the OCH conceptual model. This model encompasses most of the previous results and ideas regarding how the ecosystem of the southeastern Bering Sea functions, and it extends the conceptual linkages of processes, which were initially developed in the PROBES study (Walsh and McRoy, 1986).

Currently, pollock is the most abundant fish species harvested in the Bering Sea, accounting for >65% of the total groundfish biomass; during the 1980s their total biomass exceeded 20 million metric tons (Napp *et al.,* 2000). The biomass trends in the eastern Bering Sea (1979 to 1998: Schumacher *et al.,* 2003) show that while the total biomass of pollock in the 1990s was less than in the 1980s, they still dominated biomass in any of the trophic guilds which include marine birds, mammals, other fishes and crabs. Walleye pollock is a nodal species in the food web (NRC Report, 1996) with juveniles being the dominant prey of fishes (including adult pollock), seabirds, and marine mammals (Springer and Byrd, 1989; Livingston, 1993). The abundance of pollock is determined by episodic occurrence of strong year-classes. Such a strong year-class occurred in 1978, the remaining year-classes fluctuating about the long-term mean abundance (Fig. 30.9).

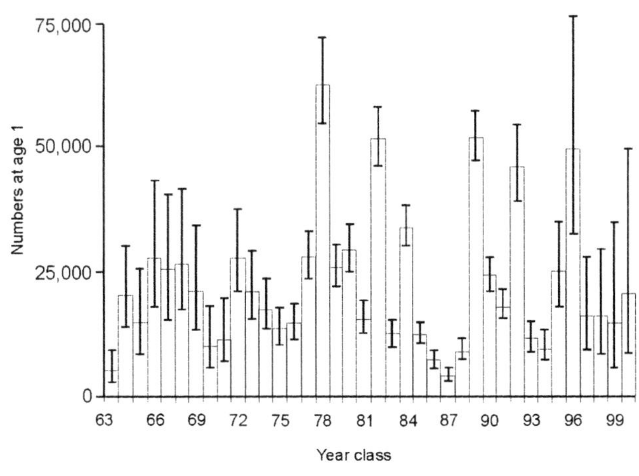

Figure 30.9 Time series of year class abundance of age-1 of pollock. Error bars represent ~95% confidence interval. (From Ianelli *et al.,* 2001.)

The observational basis for the OCH includes many of the elements of the pathway model, including sea ice, wind generated turbulence, water column temperature and chlorophyll fluorescence (Fig. 30.10). The observations of biota included zooplankton, jellyfish, and four species of fish, marine mammals and sea birds. After an analysis of these data, Hunt *et al.* (2002) formulated the OCH. The OCH recognizes that late retreat of sea ice with the attendant cold temperatures in the water column and early, short phytoplankton blooms are hallmarks of cold regimes. The associated impacts on biota include reduced survival of fish eggs (Blood, 2002) and diminished production of zooplankton prey for larval fish. Under this set of conditions, the hypothesis predicts that recruitment of pollock will be nominal or weak, and strong year-classes are not expected. Bottom-up processes dictate the flow of energy through the ecosystem during a cold regime (Fig. 30.11). Cold water-column temperatures can directly impact distributions of some forage fish species. The OCH allows that pinnipeds and piscivorous seabirds may thrive, even under cold conditions, if the population centers of forage fish change and become more available as prey. During years when sea ice is either not present or retreats before there is sufficient net short wave radiation to initiate a bloom, the spring bloom occurs later than during the cold regime, and water column temperatures are warmer. Under this set of conditions, the spring bloom is expected to be prolonged and zooplankton production is expected to be strong, resulting in readily available prey for larval and juvenile fish. The potential then would be high for strong year-classes of pollock and other piscivorous fish (Fig. 30.11).

There are years when the observed regime conditions (e.g., 1976 – cold with low recruitment; 1996 – warm with high recruitment) and estimated pollock production fit the outcomes predicted by the OCH; however, there are also years when the fit is poor (e.g., 1987 – warm with low recruitment, 1992 – cold with high recruitment). The history of the predator field (i.e., top-down process) on young pollock, therefore, must also be taken into account. The marked predatory impact of adult pollock on age-1 and younger fish, together with other fish (e.g., arrowtooth flounder and Pacific cod) is well documented (Livingston *et al.,* 1999; Livingston and Methot, 1998). The OCH weaves in such biological mechanisms as follows. When there is a sequence of warm regime years, recruitment is above average and the populations of adult predatory fish will eventually increase to a point where the control of future year-class strength is mainly a top-down process. This switch to top-down dominance may have already occurred for the pollock population in the Gulf of Alaska (Bailey, 2000). There, as the arrowtooth flounder population markedly increased, the pollock population underwent a severe decline and the "critical period" for determining year class strength switched from the egg and larval stages to the juvenile period. The hypothesis predicts that as predation becomes greater, the abundance of young pollock and forage fish decline and zooplankton become available for other populations (e.g., jellyfish, salmon, baleen whales). In addition, the reduction in abundance of forage fish could cause declines in populations and/or productivity of pinnipeds and piscivorous seabirds (Hunt and Stabeno, 2002).

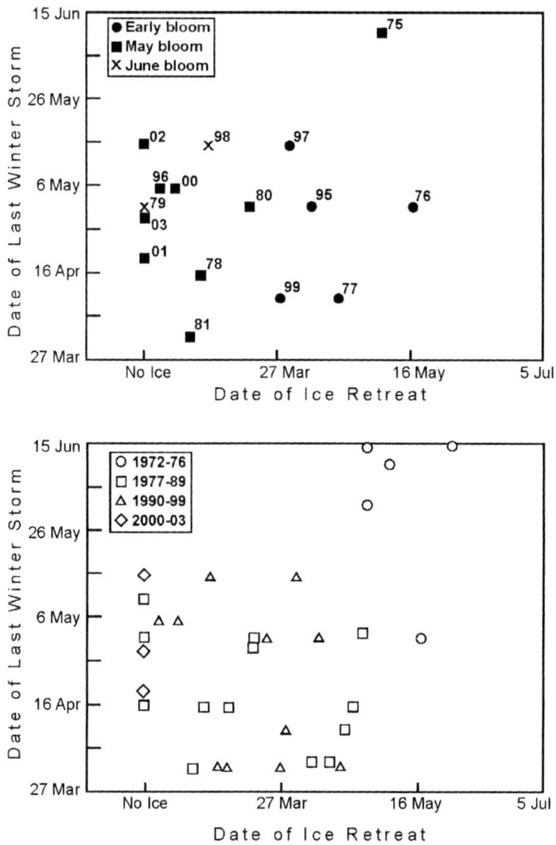

Figure 30.10 A synthesis of physical phenomena that help dictate the timing of the spring phytoplankton bloom of the southeastern Bering Sea shelf. The top panel includes the timing of the last retreat of sea ice, the timing of the shift between winter and summers wind conditions, and their effect on the timing of the spring bloom over the southern portion of the middle domain. Note that ice-related or early blooms occur in years when the ice retreat comes in mid March or later. In 1975 there was an ice-related bloom in May. The date of the last winter storm was defined as when the wind speed cubed fell below 2500 $m^3 s^{-3}$ for the summer. The winds were measured at St. Paul Island. (From Hunt *et al.*, 2002) The lower panel shows the relationship between ice retreat and date of last winter storm with the years divided into regimes. If the relationship between timing of ice retreat and spring phytoplankton bloom hold, then the timing of the spring bloom can be inferred for each regime.

While decadal time scale or regime shifts have been identified for the Bering Sea (e.g., Stabeno *et al.*, 2001; Minobe, 2002; Hare and Mantua, 2000), much of the variance of physical variables lies in the interannual frequency band. The OCH addresses the impact of annual changes in the following manner. When a cold year occurs in a warm regime, the inertia of the large biomass of predatory adult fish maintains a top-down energy flow. If sufficiently large year-classes of piscivorous predators (e.g., pollock) result from one or two warm years in a cold regime, the control could switch from bottom-up to top-down. Variations in forage fish and pollock year-class strength can also occur within a regime as a result of the redistribution of their predators.

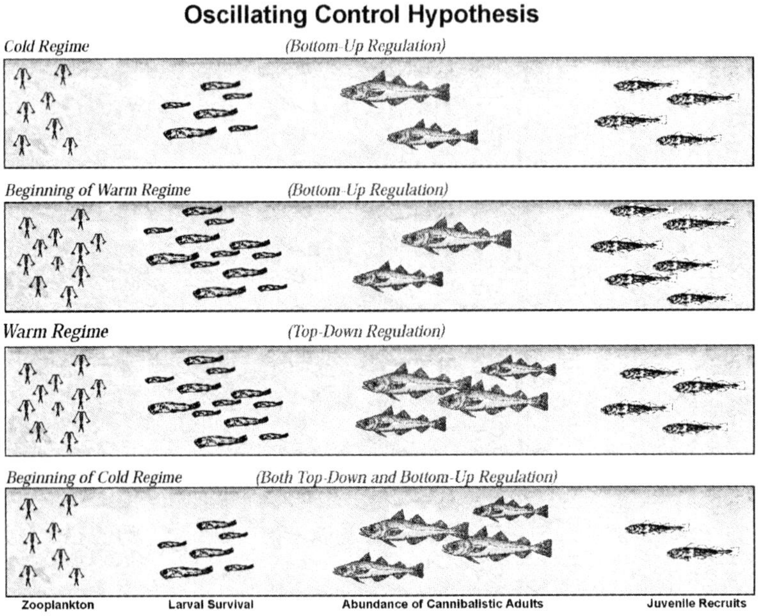

Figure 30.11 Schematic of the OCH. The number of pollock recruits affects not only pollock population dynamics, but also the availability of age-1 pollock to predators such as marine birds and mammals. Top Panel: In a cold regime, copepod production is limited by water temperature and larval/juvenile pollock survival is prey limited. Upper Middle Panel: At the beginning of a warm regime, copepods flourish and provide ample food resources to support strong survival of larval and juvenile pollock; juvenile pollock survival is high because there are few large piscivorous fish to consume them. Lower Middle Panel: In a warm regime, the production of copepod food for larval and juvenile pollock remains strong, but survival of juveniles and recruitment is limited by cannibalism from the growing biomass of adult pollock and other predators such as Pacific cod and arrowtooth flounder. Bottom Panel: At the beginning of a cold regime, copepod production is limited and larval and juvenile pollock are food limited. In addition, cannibalism by adult pollock and other predators further limits recruitment. Thus, there are asymmetries in how pollock recruitment will respond to changes from a cold to a warm regime versus from a warm to a cold regime.

8. Future Directions

A substantial amount of information has been compiled which identifies detailed needs for future interdisciplinary research in coastal waters of the eastern Bering Sea (e.g., NRC, 1996; Draft Bering Sea Ecosystem Research Plan, 1998; Springer, 1999). Further, Hunt *et al.* (2002) present seven testable relationships among physical and biological processes that would lead toward substantiation or refutation of the OCH. Of these, five require comprehensive monitoring of biophysical features of the shelf.

At present, there are several monitoring efforts underway in the Bering Sea. Among the most notable are the annual trawl and acoustic surveys conducted by the National Marine Fisheries Service. The ensuing time series of water temperature, and biomass/abundance of many fish and other marine biota form the basis of numerous publications regarding the Bering Sea ecosystem Ianelli *et al.,* 2001. Bird (Dragoo *et al.,* 2000) and marine mammal counts (National Marine Mammal

Laboratory, Seattle, WA) are being conducted at many sites in the region and supplement the survey observations. Sea ice is well monitored by satellite observations; water column currents and properties, chemistry, and ichthyo-zooplankton populations, however, are greatly under-sampled. When we consider that the eastern shelf is approximately the same size as the states of California, Oregon, and Washington combined and that at present there is only one continuously occupied site (1995–present, Site 2 in the middle shelf of the southeastern Bering Sea) where biophysical moored instruments are maintained, it is amazing we have learned as much as we have. Recent re-analyses of long biological time series exemplify the great degree of inherent variability on the Bering Sea shelf that must be taken into account when attempting to design new monitoring studies (Napp *et al.,* 2002). We note that the time-series from Site 2 (Fig. 30.6) were a keystone in formulating the OCH and have formed the basis for more than a dozen publications and supportive data for many others. A more comprehensive, permanent monitoring program is needed for Alaskan coastal waters.

The realization that there is value to the traditional knowledge and wisdom (TKW) of the local Native Peoples has come slowly, but is now gaining recognition (NRC, 1996). Native people rely on their ability to see the patterns in all elements of their environment: the weather, sea ice, fish, marine mammals and birds for their existence. This knowledge is now finding its way into scientific literature (e.g., Pungowiyi, 2000; Noongwook, 2000; Huntington, 2000). While TKW is not quantitative, it has the potential to make contributions to both retrospective and current studies. Its use as an integrative tool will help us to better understand the ecosystem of the Bering Sea. Further, it is obvious that the Native Peoples could provide a valuable component for monitoring the vast Bering Sea (e.g., Harwood *et al.,* 2002).

On the national level, the National Ocean Research Leadership Council (NORLC), a Cabinet-level group of 14 Federal agencies, has designated the establishment of the U.S. Integrated Ocean Observing System (IOOS) as its highest priority. The process was accelerated in 2000 with the creation of Ocean.US, a federal interagency office established by the National Oceanographic Partnership Program (NOPP). To date, nine agencies have agreed to participate in the Ocean.US endeavor. Establishment of an IOOS is driven by both national and regional (multi-state) priorities for data and informational products. Among the objectives of this proposed program are: to improve predictions of climate change and its effects on coastal communities, to protect more effectively and restore healthy coastal marine ecosystems, and to enable the sustained use of marine resources.

In addition to observations, expansion of the modeling effort is necessary to better understand dynamics of the Bering Sea ecosystem. Modeling can be portioned into two aspects. Hydrodynamic models can treat biological components (e.g., pollock eggs and larvae) as "floats" so that the model becomes an individual based model of ecosystem dynamics containing many of the elements of the conceptual pathway model (e.g., Walsh and McRoy, 1986). Another approach is the application of the ECOSIM model (e.g., Walters *et al.,* 1999; Aydin, 2002), which is a population-scale quantitative food web model that adheres to a mass-balance (not necessarily equilibrium), where the linkage to physics can enter though bottom-up forcing on primary production. Both approaches have value and both suffer because of the complex nature of using them as simple tools to examine biophysical processes and mechanisms. We believe that models and/or their informational products need to become more "user friendly" and generally available for all investigators.

How can the NEPROMS become a user-friendlier tool? The answer to this question could take the form of developing informational products from model simulations. Computed and averaged (perhaps on a daily time step) water property and velocity field distributions could be stored. Coupling these averaged fields with an interactive interface, which would allow a naive user to choose particular simulations and then access them in selected time/space domains, would permit far more research into questions of how the ecosystem functions than is possible today.

The Bering Sea supports a rich and productive ecosystem, with strong physical forcing and a well-defined climate signal. Economically, it is extremely valuable, providing nearly half of the US catch of fish and shellfish. It provides sustenance and the basis for important cultural traditions of native communities throughout the region. Although its economic importance has been acknowledged for some time, long-term biological and physical observations do not adequately cover the vast area of the Bering Sea. New technologies of satellite observation, moored autonomous samplers, and autonomous vehicles may help to solve the problems of under sampling. Multi-national research endeavors in this region would also help us to obtain sufficient sample densities to describe the mechanisms that structure this ecosystem. The Bering Sea ecosystem is ripe for description and improving our understanding of the biophysical mechanisms that lead to its high productivity.

Acknowledgments

Much of the material presented in the Chapter is taken from three National Science Foundation research programs (PROBES, ISHTAR and INNER FRONT), two National Oceanic and Atmospheric Administration (NOAA) Coastal Ocean Programs studies (Bering Sea Fisheries Oceanography Coordinated Investigations and Southeast Bering Sea Carrying Capacity), and from NOAA's ongoing Fisheries Oceanography Coordinated Investigations program conducted jointly by the Pacific Marine Environmental Laboratory and the Alaska Fisheries Science Center, both in Seattle, WA. J. Schumacher acknowledges funding from the PMEL/FOCI program for his time in writing this Chapter, and G. Hunt received support from NSF Grant OPP-9617287. NOAA's Coastal Ocean Program sponsored some of this research and this chapter is contribution FOCI-B451 to Fisheries-Oceanography Coordinated Investigations, and is contribution #2529 from PMEL.

Bibliography

Aagaard, K., L.K. Coachman, and E. Carmack, 1985. On the halocline of the Arctic Ocean. *J. Geophys Res.*, **90**, 4833–4846.

Alexander, V., 1999. Interdisciplinary studies of the Bering Sea. In: *Dynamics of the Bering Sea: A Summary of Physical, Chemical, and Biological Characteristics, and a Synopsis of Research on the*

Bering Sea, T.R. Loughlin and K. Ohtani (eds.), North Pacific Marine Science Organization (PICES), University of Alaska Sea Grant, AK-SG-99–03, 683–686.

Arkhipkin, A.I., V.A. Bizikov and A.V. Verkhunov, 1998. Distribution and growth in juveniles of the squid *Berryteuthis magister* (Cephalopoda, Gonatidae) in the western Bering Sea. *Sarsia* **83,** 45–54.

Arsenev, V.S., 1967. Currents and water masses of the Bering Sea. *Nuka,* Moscow, 135 pp.

Aydin, K. 2002. The use of ecosystem models to investigate multispeices management strategies for capture fisheries, pp. 33–38. In *Fish. Cent. Res. Rep.,* T. Pitcher and K. Cochrane (eds.).

Baduini, C.L., K.D. Hyrenbach, K.O. Coyle, A. Pinchuk, V. Mendenhall, and G.L. Hunt, Jr., 2001. Mass mortality of short-tailed shearwaters in the southeastern Bering Sea during summer 1997. *Fish. Oceanogr.* **10,** 117–130.

Baier, C. T. and J. M. Napp, 2003. Climate-induced variability in *Calanus marshallae* populations. *J. Plankton Res.,* **25,** 771–782.

Bailey, K.M., 2000. Shifting control of recruitment of walleye pollock *Theragra chalcogramma* after a major climate and ecosystem shift. *Mar. Ecol. Prog. Ser.,* **198,** 215–224.

Bailey, K.M. and S.A. Macklin, 1994. Analysis of patterns in larval walleye pollock *Theragra chalcogramma* survival and wind mixing events in Shelikof Strait, Gulf of Alaska. *Mar. Ecol. Prog. Ser.,* **113,** 1–12.

Bailey, K.M., T.J. Quinn, W.S. Grant, and P. Bentzen, 1999. Population structure and dynamics of walleye pollock, *Theragra chalcogramma. Adv. Mar. Biol.,* **37,** 179–255.

Bailey, K.M., L. Ciannelli, N.A. Bond, A. Belgrano, and N. Chr. Stenstreth 2005. Recruitment of walleye pollock in a complex physical and biological ecosystem, *Prog. Oceanogr.,* in revision.

Balykin, 1996. Dynamics and abundance of Western Bering Sea Walleye pollock. In: *Ecology of the Bering Sea: A review of Russian literature,* Mathiesen, O.A. and K.O. Coyle (eds.), University of Alaska Sea Grant Program Report No. 96–01, Fairbanks , AK, 177–182.

Blood, D. M., 2002. Low-temperature incubation of walleye pollock (*Theragra chalcogramma*) eggs from the southeast Bering Sea shelf and Shelikof Strait, Gulf of Alaska. *Deep Sea Res. II,* **49:** 60695–6108.

Bond, N.A, J.E. Overland and P. Turet, 1994. Spatial and temporal characteristics of the wind forcing of the Bering Sea. *J. Climate,* **7 ,** 1119–1130.

Bond, N.A., J.E. Overland, M. Spillane, and P.J. Stabeno, 2004. Recent shifts in the state of the North Pacific. *Geophys. Res. Lett., 30,* 2183, doi: 10.1029/2003GL018597.

Brodeur, R.D., C.E. Mills, J.E. Overland, G.E. Walters, and J.D. Schumacher, 1999a. Evidence for a substantial increase in gelatinous zooplankton in the Bering Sea, with possible links to climate change. *Fish. Oceanogr.,* **8,** 296–306.

Brodeur, R.D , M.T. Wilson, G.E. Walters, and I.V. Melnikov, 1999b. Forage fishes in the Bering Sea: distribution, species associations and biomass trends. In: *Dynamics of the Bering Sea: A Summary of Physical, Chemical, and Biological Characteristics, and a Synopsis of Research on the Bering Sea,* T.R. Loughlin and K. Ohtani (eds.), North Pacific Marine Science Organization (PICES), Univ. of Alaska Sea Grant, AK-SG-99–03, 509–536.

Brodeur, R.D., H. Sugisaki, and G.L. Hunt, Jr., 2002. Increases in jellyfish biomass in the Bering Sea: Implications for the ecosystem. *Mar. Ecol. Prog. Ser.,* **233,** 89–103.

Coachman, L.K., 1986. Circulation, water masses, and fluxes on the southeastern Bring Sea shelf. *Cont. Shelf Res.,* **5,** 23–108.

Coachman, L.K., 1993. On the flow field in the Chirikov basin. *Cont. Shelf Res.,* **13,** 481–508.

Coachman L.K., K. Aagaard, and R.B. Tripp, 1975. Bering Strait: The regional physical oceanography. Univ. Wash. Press, Seattle, 172 pp.

Conners, M.E., A.B. Hollowed, and E. Brown, 2002. Retrospective analysis of Bering Sea bottom trawl surveys: regime shift and ecosystem reorganization. *Prog. Oceanogr.,* **55,** 209–222.

Cooney, R.T., and K.O. Coyle, 1982. Trophic implications of cross-shelf copepod distributions in the southeastern Bering Sea. *Mar. Biol.,* **70,** 187–196.

Cooper, L.W., J.M. Grebmeier, I.L. Larsen, V.G. Egorov, C. Theodorakis, H.P. Kelly, and J.R. Lovvorn, 2002. Seasonal variation in sedimentation of organic materials in the St. Lawrence Island polynya region, Bering Sea. *Mar. Ecol. Prog. Ser.,* **226,** 13–26.

Coyle, K.O., and R.T. Cooney, 1988. Estimating carbon fluxes to pelagic grazers in the ice-edge zone of the eastern Bering Sea. *Mar. Biol.,* **98,** 299–306.

Coyle, K.O., and A.I. Pinchuk, 2002a. Climate-related differences in zooplankton density and growth on the inner shelf of the southeastern Bering Sea. *Prog. Oceanogr.,* **55,** 177–194.

Coyle, K.O., and A.I. Pinchuk, 2002b. The abundance and distribution of euphausiids and zero-age pollock on the inner shelf of the southeast Bering Sea near the Inner Front in 1997–1999. *Deep Sea Res. II,* **49,** 6009–6030.

Cury, P., C. Roy, R. Mendelssohn, A. Bakun, D.M. Husby, and R.H. Parrish, 1995. Moderate is better: exploring nonlinear climate effects on the Californian northern anchovy (*Engraulis modax*). *Canadian Special Publication of Fisheries and Aquatic Sciences,* **108,** 417–424.

Draft Bering Sea Ecosystem Research Plan, 1998. Copies available from Pat Livingston, AFSC/NMFS, Bldg. #4, 7600 Sand Point Way NE, Seattle WA, 98115–0070, 58 pp.

Dragoo, D.E., G.V. Byrd, and D.B. Irons, 2000. Breeding status and population trends of seabirds in Alaska in 1999. *U.S. Fish and Wildl. Serv. Report AMNWR* **2000/02,** 61 pp.

Favorite, F., A.J. Dodimead, and K. Nasu, 1976. Oceanography of the subarctic Pacific region, 1960–71. *Int. N. Pac. Fish. Comm. Bull.* **#33,** 187 pp.

Fay, F.H., 1981. Marine mammals of the Eastern Bering sea shelf: An overview. In: *The Eastern Bering Sea Shelf: Oceanography and Resources, Vol. II,* D.W. Hood and J.A. Calder (eds.), 807–811.

Ford, C., 1966. Where the sea breaks its back: The epic story of early naturalist George Steller and the Russian exploration of Alaska. Alaska Northwest Books, Portland, OR. 206 pp.

Francis, R.C., S.R. Hare, A.B. Hollowed, and W.S. Wooster, 1998. Effects of interdecadal climate variability on the oceanic ecosystems of the NE Pacific. *Fish. Oceanogr.,* **7,** 1–21.

Francis, R.C., K. Aydin, R.L. Merrick, S. Bollens, 1999. Modeling and management of the Bering Sea Ecosystem, pp. 409–433 In *Dynamics of the Bering Sea,* Loughlin and Ohtani (Eds.).

Giese, B.S., S.C. Urizar, and N. Fuckar, 2002. Southern Hemisphere origins of the 1976 regime shift. *Geophys. Res. Lett.,* **29**(2), 10.1029/2001GL013268.

Grebmeier, J.M., 1993. Studies of pelagic-benthic coupling extended into the Soviet continental shelf in the northern Bering and Chukchi Seas. *Cont. Shelf Res.,* **13,** 653–668.

Grebmeier, J.M, and L.W. Cooper, 1995. Influence of the St. Lawrence Island polynya upon the Bering Sea benthos. *J. Geophys. Res.,* **100,** 4439–4460.

Grebmeier, J.M., and L.W. Cooper, 2002. Benthic processes in the Bering Strait region of the Arctic: temporal/spatial variability and global change. *Eos Trans. Amer. Geophys. Union,* **83**(4), Ocean Sciences Meeting Supplement, Abstract OS32M-01.

Grebmeier, J.M., and K.H. Dunton, 2000. Benthic processes in the Northern Bering/Chukchi Seas: status and global change. In: *Impacts of Changes in Sea Ice and Other Environmental Parameters in the Arctic,* Huntington, H.P. (ed.), Marine Mammal Commission, Bethesda, Maryland, 61–71.

Hare, S.R., and N.J. Mantua, 2000. Empirical evidence for North Pacific regime shifts in 1977 and 1989. *Prog. Oceanogr.,* **47,** 103–146.

Harwood, L.A., P. Norton, B. Day, and P.A. Hall, 2002. The harvest of Beluga Whales in Canada's western Arctic: Hunter-based monitoring of size and composition of the catch. *Arctic,* **55,** 10–20.

Hinckley, S., 1987. The reproductive biology of walleye pollock, *Theragra chalcogramma,* in the Bering Sea, with reference to spawning stock structure. *Fish. Biol.,* **85,** 481–498.

Hood, D.W., 1999. PROBES: Processes and Resources of the eastern Bering Sea shelf. In: *Dynamics of the Bering Sea: A Summary of Physical, Chemical, and Biological Characteristics, and a Synopsis of*

Research on the Bering Sea, T.R. Loughlin and K. Ohtani (eds.), North Pacific Marine Science Organization (PICES), University of Alaska Sea Grant, AK-SG-99–03, 697–712.

Hood D.W., and J.A. Calder, 1981. The eastern Bering Sea Shelf: oceanography and resources. University Wash. Press, Seattle. 520 pp.

Hood, D.W., and E.J. Kelley, 1974. Oceanography of the Bering Sea. Occasional Publication 2, Inst. of Marine Sci., Univ. of Alaska, Fairbanks, 623 pp.

Hunt, Jr., G.L., and N.M. Harrison, 1990. Foraging habitat and prey taken by Least Auklets at King Island, Alaska. *Mar. Ecol. Prog. Ser.,* **65,** 141–150.

Hunt, Jr., G.L., N.M. Harrison and T. Cooney, 1990. Foraging of Least Auklets: The influence of hydrographic structure and prey abundance. *Studies in Avian Biol.* **14:** 7–22.

Hunt, Jr., G.L., and P.J. Stabeno, 2002. Climate change and the control of energy flow in the southeastern Bering Sea. *Prog. Oceanogr.,* **55,** 5–22.

Hunt, Jr., G.L., C.L. Baduini, R.D. Brodeur, K.O. Coyle, N.B. Kachel, J.M. Napp, S.A. Salo, J.D. Schumacher, P.J. Stabeno, D.A. Stockwell, T.E. Whitledge, and S.I. Zeeman, 1999. The Bering Sea in 1998: The second consecutive year of extreme weather-forced anomalies. *Eos Trans. AGU,* **80*(47),*** 561, 565–566.

Hunt, Jr., G.L., P. Stabeno, G. Walters, E. Sinclair, R. Brodeur, J.M. Napp, and N. Bond, 2002. Climate change and control of the southeastern Bering Sea pelagic ecosystem. *Deep-Sea Res. II, Topical Studies in Oceanography,* **49,** 5821–5853.

Huntington, H.P., 2000. Traditional knowledge of the ecology of Belugas, *Delphinapterus leucas,* in Cook Inlet, Alaska. *Mar. Fish. Rev.,* **62,** 134–140.

Ianelli, J.N., T. Buckley, T. Honkalehto, N. Williamson, and G. Walters, 2001. Bering Sea-Aleutian Islands Walleye Pollock Assessment for 2002. In: *Stock assessment and fishery evaluation report for the groundfish resources of the Bering Sea/Aleutian Islands regions,* North Pac. Fish. Mgmt. Council, Anchorage, AK, section 1, 1–105.

Ikeda, M., 1990. Decadal oscillations of the air-ice-sea system in the northern hemisphere. *Atmosphere-Ocean,* **28,** 106–139.

Incze, L.S., D.A. Armstrong, and S.L. Smith, 1987. Abundance of larval Tanner crabs (*Chionoectes* spp.) in relation to adult females and regional oceanography of the southeastern Bering Sea. *Can. J. Fish. Aquat. Sci.,* **44,** 1143–1156.

Incze, L.S., D. Hebert, N. Wolff, N. Oakey, and D. Dye, 2001. Changes in copepod distributions associated with increased turbulence from wind stress. *Mar. Ecol. Prog. Ser.,* **213,** 229–240.

Iverson, R.L., L.K. Coachman, R.T. Cooney, T.S. English, J.J. Goering, G.L. Hunt, Jr., M.C. McCauley, C.P. McRoy, W.S. Reesburgh, and T.E. Whitledge, 1979. Ecological significance of fronts in the southeastern Bering Sea. In: *Ecological Processes in Coastal and Marine Systems,* R.J. Livingston (ed.), Plenum Press, New York, 437–466.

Jin, Fei-Fei, M. Kimoto, and X. Wang, 2001. A model of decadal ocean-atmosphere interaction in the North Pacific basin. *Geophys. Res. Lett.,* **28,** 1531–1534.

Kachel, N.B., G. Hunt, S.A. Salo, J.D. Schumacher, P.J. Stabeno, and T.E. Whitledge, 2002. Characteristics of the Inner Front of the Southeastern Bering Sea. *Deep Sea Res. II, Topical Studies in Oceanography,* **49,** 5889–5909.

Kendall, Jr., A.W., 2001. Specific gravity and vertical distribution of walleye pollock (*Theragra chalcogramma*) eggs. AFSC Processed Report 2001–01.

Khen, G.V., 1999. Hydrography of the western Bering Sea shelf water. In: *Dynamics of the Bering Sea: A Summary of Physical, Chemical, and Biological Characteristics, and a Synopsis of Research on the Bering Sea,* T.R. Loughlin and K. Ohtani (eds.), North Pacific Marine Science Organization (PICES), Univ. of Alaska Sea Grant, AK-SG-99–03, 161–176.

Kowalik, Z., and P. Stabeno (1999): Trapped motion around the Pribilof Islands in the Bering Sea. *J. Geophys. Res.,* **104** (C11), 25,667–25,684.

Kruse, G.H., 1998. Salmon run failures in 1997–1998: A link to anomalous ocean conditions? *Alaska Fish. Res. Bull.*, **5,** 55–63.

Labitzke, M., and H. van Loon, 1988. Associations between the 11-year solar cycle, the QBO and the atmosphere. I, The troposphere and stratosphere of the northern hemisphere in winter. *J. Atmos. Terr. Phys.*, **50,** 197–206.

Laevastu, T., and H.A. Larkins, 1981. Marine fisheries ecosystems: Its quantitative evaluation and management. *Fishing News Books,* Surrey, England, 162 pp.

Latif, M., and T. Barnett, 1994. Causes of decadal climate variability over the over the North Pacific and North America. *Science,* **266,** 634–637.

Livingston, P.A., 1993. The importance of predation by groundfish, marine mammals, and birds on walleye pollock *Theragra chalcogramma* and Pacific herring *Clupea pallasi* in the eastern Bering Sea. *Mar. Ecol. Prog. Ser.,* **102,** 205–215.

Livingston, P.A., and R.D. Methot, 1998. Incorporation of predation into a population assessment model of eastern Bering Sea walleye pollock. In: *Fishery Stock Assessment Models,* Alaska Sea Grant College Program, AK-SG-98–01, 16 pp.

Livingston, P.A., L.L. Low, and R.J. Marasco, 1999. Eastern Bering Sea ecosystem trends. In: *Large Marine Ecosystems of the Pacific Rim: Assessment, Sustainability, and Management,* Q. Tang and K. Sherman (eds.), Blackwell Science, Boston, 140–162.

Longhurst, A.R., 1998. *Ecological Geography of the Sea.* Academic Press San Diego, 398 pp.

MacKenzie, B.R., T.J. Miller, S. Cyr, S., and W.C. Leggett, 1994. Evidence for a dome-shaped relationship between turbulence and larval fish ingestion rates. *Limnol. Oceanogr.,* **39,** 1790–1799.

McRoy, C.P., 1993. ISHTAR, the project: an overview of Inner Shelf Transfer And Recycling in the Bering and Chukchi seas. *Cont. Shelf Res.,* **13,** 473–480.

McRoy, C.P, 1999. Water over the Bridge: A summing up of the contributions of the ISHTAR Project in the Northern Bering and Chukchi Seas. In*: Dynamics of the Bering Sea: A Summary of Physical, Chemical, and Biological Characteristics, and a Synopsis of Research on the Bering Sea,* T.R. Loughlin and K. Ohtani (eds.), North Pacific Marine Science Organization (PICES), Univ. of Alaska Sea Grant, AK-SG-99–03, 687–697.

Macklin, S.A., V.I. Radchenko, S. Saitoh, and P.J. Stabeno, 2002. Variability in the Bering Sea ecosystem. *Prog. Oceanogr.,* **55,** 1–4.

McFarlane, G.A., J.R. King, and R.J. Beamish, 2000. Have there been recent changes to climate? Ask the fish. *Prog. Oceanogr.,* **47,** 147–169.

Megrey, B.A., and S. Hinckley, 2001. Effect of turbulence on feeding of larval fishes: A sensitivity analysis using an individual-based model. *ICES J. Mar. Sci.,* **58,** 1015–1029.

Merrick, R., 1995. The relationship of the foraging ecology of Stellar sea lions (*Eumetopias jubatus*) to their population decline in Alaska. Ph.D. Thesis, Univ. Washington, Seattle, Washington, USA, unpublished.

Merrick, R.L., 1997. Current and historical roles of apex predators in the Bering Sea ecosystem. *J. Northwest Atlantic Fish. Sci.,* **22,** 343–355.

Minobe, S., 1999. Resonance in bidecadal and pentadecadal climate oscillations over the North Pacific: Role in climate regime shifts. *Geophys. Res. Lett.,* **26**, 853–858.

Minobe, S., 2002. Interannual to interdecadal changes in the Bering Sea and concurrent 1998/99 changes over the North Pacific. *Progr. Oceanogr.* , **55,** 45–64.

Mizobata, K., S.I. Saitoh, A. Shiomoto, T. Miyamura, N. Shiga, K. Imai, M. Toratani, Y. Kajiwara, and K. Sasaoka, 2002. Bering Sea cyclonic and anticyclonic eddies observed during summer 2000 and 2001. *Prog. Oceanogr.,* **55,** 65–75.

Muench, R.D. (ed.), 1983. Marginal Ice Zones, *J. Geophys. Res.,* **88,** 2713–2966.

Muench, R.D., R.B. Tripp, and J. D. Cline, 1981. Circulation and hydrography of Norton Sound. In: *The Eastern Bering Sea Shelf: Oceanography and Resources,* Vol.1, D.W.Hood and J.A. Calder (eds.), U.S. Government Printing Office, Washington D.C. (distributed by the University of Washington Press, Seattle), 77–93.

Napp, J.M., and G.L. Hunt, Jr., 2001. Anomalous conditions in the southeastern Bering Sea, 1997: linkages among climate, weather, ocean, and biology. *Fish. Oceanogr,* **10**, 61–68.

Napp, J.M., A.W. Kendall, Jr., and J.D. Schumacher, 2000. A synthesis of biological and physical processes affecting the feeding of larval walleye pollock (*Theragra chalcogramma*) in the Eastern Bering Sea. *Fish. Oceanogr,* **9,** 147–162.

Napp, J.M., C.T. Baier, K.O. Baier, R.D. Brodeur, N. Shiga, and K. Mier, 2002. Interannual and decadal variability in zooplankton communities of the southeast Bering Sea shelf. *Deep-Sea Res. II,* **49,** 5991–6008.

Naumenko, N.I., 1996. Long-term fluctuations in the ichthyofauna of the Western Bering Sea. In: *Ecology of the Bering Sea: A review of Russian literature,* Mathiesen, O.A. and K.O. Coyle (eds.), University of Alaska Sea Grant Program Report No. 96–01, Fairbanks , AK, 143–158.

Niebauer, H.J., 1998. Variability in the Bering Sea ice cover as affected by a regime shift in the North Pacific in the period 1947–1996. *J. Geophys. Res.,* **103,** 27,717–27,737.

Niebauer, H.J., V. Alexander, and S.M. Henrichs, 1995. A time-series study of the spring bloom at the Bering Sea ice edge I: Physical processes, chlorophyll and nutrient chemistry. *Cont. Shelf Res.,* **15,** 1859–1878.

Nihoul, J.C.J., P. Adam, P. Brasseur, E. Deleersnijder, S. Djenidi, and J. Haus, 1993. Three-dimensional general circulation model of the Bering-Chukchi shelf. *Cont. Shelf Res.,* **13,** 509–542.

Noongwook, G., 2000. Native observations of local climate around St. Lawrence Island. In: *Impacts of Changes in Sea Ice and Other Environmental Parameters in the Arctic,* Huntington, H.P. (ed.), Report of the Marine Mammal Commission Workshop, Marine Mammal Commission, Bethesda, Maryland, 21–24.

NRC Report, 1996. *The Bering Sea ecosystem.* National Academy Press, 307 pp.

NRC (2003): The decline of the Steller sea lion in Alaskan waters: Untangling Food Webs and Fishing Nets. National Academy Press, Washington, D.C.204 pp.

Ohtani, K., and T. Azumaya, 1995. Influence of interannual changes in ocean conditions on the abundance of walleye pollock (*Theragra chalcogramma*) in the eastern Bering Sea. In: *Climate Change and northern fish populations,* R.J. Beamish (ed.), *Can. Spec. Pub. Fish. Aquat. Sci.,* **121,** 87–95.

Olla, B.L., and M.W. Davis, 1990. Effects of physical factors on the vertical distribution of larval walleye pollock *Theragra chalcogramma* under controlled conditions. *Mar. Ecol. Prog. Ser.,* **63,** 105–112.

Olson, M.B., and S.L. Strom, 2002. Phytoplankton growth, microzooplankton herbivory and community structure in the southeast Bering Sea: insight into the formation and temporal persistence of an *Emiliania huxleyi* bloom. *Deep-Sea Res. II, Topical Studies in Oceanography,* **49,** 5969–5990.

Overland, J.E., 1981. Marine climatology of the Bering Sea. In: *The Eastern Bering Sea Shelf: Oceanography and Resources,* Vol.1, D.W. Hood and J.A. Calder (eds.), U.S. Government Printing Office, Washington D.C., (distributed by the University of Washington Press, Seattle), 15–30.

Overland, J.E., and C.H. Pease, 1982. Cyclone climatology of the Bering Sea and its relation to sea ice extent, *Mon. Weather Rev.,* **110,** 5–13.

Overland, J.E., and A.T. Roach, 1987. Northward flow in the Bering and Chukchi Seas. *J. Geophys. Res.,* **92**(C7), 7097–7105.

Overland, J.E., M.C. Spillane, H.E. Hurlburt, and A.J. Wallcraft, 1994. A numerical study of the circulation of the Bering Sea basin and exchange with the North Pacific Ocean. *J. Phys. Oceanogr.,* **24,** 736–758.

Overland, J.E., J.M. Adams, and N.A. Bond, 1999a. Decadal variability of the Aleutian Low and its relation to high-latitude circulation. *J. Climate,* **12,** 1542–1548.

Overland, J.E, S.A. Salo, L.H. Kantha and C.A. Clayson, 1999b. Thermal stratification and mixing on the Bering Sea shelf. In: *Dynamics of The Bering Sea,* Loughlin, T. R., and K. Ohtani (eds.), Alaska Sea Grant College Program, Fairbanks, AK, 129–146.

Parrish, R.H., F.B. Schwing, and R. Mendelssohn, 2000. Mid-latitude wind stress: the energy source for climate shifts in the North Pacific Ocean. *Fish. Oceanogr,* **9,** 224–238.

Peterson, W. T. and F. B. Schwing, 2003. A new climate regime in northeast Pacific ecosystems. *Geophys. Res. Letters,* **30:** OCE 6–1 to 6–4.

Pungowiyi, C., 2000. Native observations of change in the marine environment of the Bering Sea. In: *Impacts of Changes in Sea Ice and Other Environmental Parameters in the Arctic,* Huntington, H.P. (ed.), Report of the Marine Mammal Commission Workshop, Marine Mammal Commission, Bethesda, Maryland, 18–20.

Radchenko, V.I., G.V. Khen, and A.M. Slabinsky, 2001. Cooling in the western Bering Sea in 1999: quick propagation of La Niña signal or compensatory processes effect? *Prog. Oceanogr.,* **49,** 407–422.

Reed, R.K., 1995. On the variable subsurface environment of fish stocks in the Bering Sea. *Fish. Oceanogr.,* **4,** 317–323.

Reed, R.K., 1998. Confirmation of a convoluted flow over the Eastern Bering Sea Shelf. *Cont. Shelf Res.,* **18,** 99–103.

Reed, R.K., and P.J. Stabeno, 1999. The Aleutian North Slope Current. In: Loughlin, T.R. and K. Ohtani, eds. *Dynamics of The Bering Sea.* University of Alaska Sea Grant, AK-SG-99–03, 177–192.

Rosenkranz, G.E., A.V. Tyler, and G.H. Kruse. 2001. Effects of water temperature and wind on recruitment of Tanner crabs in Bristol Bay, Alaska. *Fisheries Oceanography* **10,** 1–12.

Royer, T. C. 1998. Coastal Processes in the northern North Pacific. In: *The Sea: Vol. 11—The Global Coastal Ocean: Regional Studies and Synthesis*, John Wiley & Sons, Inc., New York, NY, 395–414.

Royer, T.C., C.E. Gorsch, and L.A. Mysak, 2001. Interdecadal variability of Northeast Pacific coastal freshwater and its implications on biological productivity. *Prog. Oceanogr.,* **49,** 95–111.

Sambrotto R.N., H.J. Niebauer, J.J. Goering, and R.L. Iverson, 1986. Relationships among vertical mixing nitrate uptake and phytoplankton growth during the spring bloom in the southeast Bering Sea middle shelf. *Cont. Shelf Res.,* **5,** 161–198.

Schumacher, J.D., and V. Alexander, 1999. Variability and role of the physical environment in the Bering Sea ecosystem. In: *Dynamics of the Bering Sea: A Summary of Physical, Chemical, and Biological Characteristics, and a Synopsis of Research on the Bering Sea,* T.R. Loughlin and K. Ohtani (eds.), North Pacific Marine Science Organization (PICES), Univ. of Alaska Sea Grant, AK-SG-99–03, 147–160.

Schumacher, J.D., and T.H. Kinder, 1983, Low frequency currents over the Bering Sea Shelf. *J.Phys.Oceanogr,* **13,** 607–623.

Schumacher, J.D., and A.W. Kendall, Jr., 1995. Fisheries Oceanography: Walleye Pollock in Alaskan Waters. *Reviews Geophys., Suppl.,* **33,** 1153–1163.

Schumacher, J.D., and R.K. Reed (1992): Characteristics of currents near the continental slope of the eastern Bering Sea. *J. Geophys. Res., 97,* 9423–9433.

Schumacher, J. D. and P. J. Stabeno, 1994. Ubiquitous eddies of the Eastern Bering Sea and their coincidence with concentrations of larval pollock. *Fish. Oceanogr.,* **3,** 182–190.

Schumacher, J.D., and P. J. Stabeno, 1998. Continental shelf of the Bering Sea. In: *The Sea: Vol. 11—The Global Coastal Ocean: Regional Studies and Synthesis,* John Wiley & Sons, Inc., New York, NY, 789–822.

Schumacher, J.D., K. Aagaard, C.H. Pease, and R.B. Tripp, 1983. Effects of a shelf polynya on flow and water properties in the northern Bering Sea. *J. Geophys. Res.,* **88,** 2723–2732.

Schumacher, J.D., N.A. Bond, R.D. Brodeur, P.A. Livingston, J.M. Napp, and P.J. Stabeno, 2003. Climate Change in the Southeastern Bering Sea and Some Consequences for Biota. In: *Large Marine Ecosystems of the World: Trends in Exploitation, Protection, and Research.* G. Hempel, K. Sherman (eds.), Amsterdam: Elsevier Science, 17–40.

Shuert, P.G., and J.J. Walsh, 1993. A coupled physical-biological model of the Bering-Chukchi seas. *Cont. Shelf Res.,* **13,** 543–573.

Sirenko, B.I., and V.M. Koltun, 1992. Characteristics of the benthic biocenoses of the Chukchi and Bering Seas. In: *Results of the third joint U.S.-U.S.S.R. Bering and Chukchi Sea Expedition (BERPAC)*, P.A. Nagel (ed.), U.S. Fish and Wildlife Service, Washington, DC, 251–261.

Smith, S.L., and J. Vidal, 1986. Variations in the distribution, abundance, and development of copepods in the southeastern Bering Sea in 1980 and 1981. *Cont. Shelf Res.*, **13,** 215–240.

Springer, A.M., 1999. Summary, conclusions and recommendations. In: *Dynamics of the Bering Sea: A Summary of Physical, Chemical, and Biological Characteristics, and a Synopsis of Research on the Bering Sea,* T.R. Loughlin and K. Ohtani (eds.), North Pacific Marine Science Organization (PICES), Univ. of Alaska Sea Grant, AK-SG-99-03, 777–800.

Springer, A.M., and G.V. Byrd, 1989. Seabird dependence on walleye pollock in the Eastern Bering Sea. In: *Proceedings of the International Symposium on the Biology and Management of Walleye Pollock,* Alaska Sea Grant, AK-SG-89-01, 667–677.

Springer, A.M., and C.P. McRoy, 1993. The paradox of pelagic food webs in the northern Bering Sea—III. Patterns of primary production. *Cont. Shelf Res.* **13,** 575–600.

Springer, A.M., E.C. Murphy, D.G. Rosenau, C.P. McRoy, and B.A. Cooper, 1987. The paradox of pelagic food webs in the northern Bering Sea—I. Seabird food habits. *Cont. Shelf Res.,* **7,** 895–911.

Springer, A.M., C.P. McRoy, and K.R. Turco, 1989. The paradox of pelagic food webs in the northern Bering Sea—II. Zooplankton communities. *Cont. Shelf Res.,* **9,** 359–386.

Springer, A.M., C.P. McRoy, and M.V. Flint, 1996. The Bering Sea Green Belt: shelf-edge processes and ecosystem production. *Fish. Oceanogr.* **5,** 205–223.

Springer, A.M. and D.G. Roseneau, 1985. Copepod-based food webs: auklets and oceanography in the Bering Sea. *Marine Ecology Progress Series* **21,** 229–237.

Stabeno, P.J., and G.L. Hunt, 2002. Overview of the Inner Front and Southeast Bering Sea Carrying Capacity Programs. *Deep Sea Res. II: Topical Studies in Oceanogr.* **49,** 6157–6168.

Stabeno, P.J., and J.E. Overland, 2001. Bering Sea shifts toward an earlier spring transition. *Eos, Trans. AGU,* **29,** 317, 321.

Stabeno, P.J., R.K. Reed, and J.D. Schumacher, 1995: The Alaska Coastal Current: Continuity of transport and forcing. *J. Geophys. Res.,* **100,** 24772485.

Stabeno, P.J., J.D. Schumacher, R.F. Davis, and J.M. Napp, 1998. Under-ice observations of water column temperature, salinity and spring phytoplankton dynamics: Eastern Bering Sea shelf, 1995. *J. Mar. Res.,* **56,** 239–255.

Stabeno, P.J., J.D. Schumacher, and K. Ohtani, 1999. The physical oceanography of the Bering Sea. In: *Dynamics of the Bering Sea: A Summary of Physical, Chemical, and Biological Characteristics, and a Synopsis of Research on the Bering Sea,* T.R. Loughlin and K. Ohtani (eds.), North Pacific Marine Science Organization (PICES), Univ. of Alaska Sea Grant, AK-SG-99-03, 1–28.

Stabeno, P.J., N.A. Bond, N.B. Kachel, S.A. Salo, and J.D. Schumacher, 2001. On the temporal variability of the physical environment over the southeastern Bering Sea. *Fish. Oceanogr,* **10,** 81–98.

Stabeno, P.J., N.B. Kachel, M. Sullivan, and T.E. Whitledge, 2002. Variability along the 70-m isobath of the southeast Bering Sea. *Deep Sea Res., II: Topical Studies in Oceanography,* **49,** 5931–5943.

Stabeno, P.J., D.K. Kachel, and M.A. Sullivan, 2004. Observations from moorings in the Aleutian passes: Temperature, salinity, and transport. *Fish. Oceanogr.,* in press.

Stenseth, N.C., A. Mysterud, G. Ottersen, J.W. Hurrell, K.-S. Chan, and M. Lima, 2002. Ecological effects of climate fluctuations. *Science,* **297,** 1292–1296.

Stockwell, D.A., T.E. Whitledge, S.I. Zeeman, K.O. Coyle, J.M. Napp, R.D. Broduer, A.I. Pinchuk, and G.L. Hunt, Jr., 2001. Anomalous conditions in the southeastern Bering Sea, 1997: Nutrients, phytoplankton, and zooplankton. *Fish. Oceanogr.,* **10,** 99–116.

Sugimoto, T., and Tadokoro, K., 1997. Interannual–interdecadal variations in zooplankton biomass, chlorophyll concentration and physical environment in the subarctic Pacific and Bering Sea. *Fish. Oceanogr.,* **6,** 74–93.

Takenouti, Y., and K. Ohtani, 1974. Currents and water masses in the Bering Sea: A review of Japanese work. In: *Oceanography of the Bering Sea,* D.W. Hood and E.J. Kelley (eds.), Occasional Publication Number 2, Institute of Marine Science, University of Alaska, Fairbanks, 39–57.

Tourangeau, S., J.A. Runge, 1991. Reproduction of *Calanus glacialis* under ice in spring in southeastern Hudson Bay, Canada. *Mar. Biol.,* **108,** 227–233.

Trites, A.W., P.A. Livingston, M.C. Vasconcellos, S. Mackinson, A.M. Springer, and D. Pauly, 1999. Ecosystem change and the decline of marine mammals in the Eastern Bering Sea: testing the ecosystem shift and commercial whaling hypothesis. *Fisheries Centre Reports,* **7,** 98 pp.

Tsyban, A.V., 1999. The BERPAC Project: development and overview of the ecological investigation in the Bering and Chukchi Seas. In: *Dynamics of the Bering Sea: A Summary of Physical, Chemical, and Biological Characteristics, and a Synopsis of Research on the Bering Sea,* T.R. Loughlin and K. Ohtani (eds.), North Pacific Marine Science Organization (PICES), University of Alaska Sea Grant, AK-SG-99-03, 697–712.

Tynan, C.T., 1998. Coherence between whale distributions, chlorophyll concentration, and oceanographic conditions on the southeast Bering Sea shelf during a coccolithophore bloom, July–August, 1997. *Eos, Trans. AGU,* **79,** 127.

Vance, T.C., J.D. Schumacher, P.J. Stabeno, C.T. Baier, T. Wyllie-Echeverria, C.T. Tynan, R.D. Brodeur, J.M. Napp, K.O. Coyle, M.B. Decker, G.L. Hunt, Jr., D. Stockwell, T.E. Whitledge, M. Jump, and S. Zeeman, 1998. Aquamarine waters recorded for first time in eastern Bering Sea. *Eos Trans. AGU,* **79,** 121, 126.

Verkhunov, A.V., 1994. Thermohaline characteristics of the shelf front in the western part of the Bering Sea, *Oceanology,* **34,** 356–369.

Verkhunov, A.V., and Y.Y. Tkachenko, 1992. Recent observations of variability in the Western Bering Sea Current system. *J. Geophys. Res.,* **97,** 14,369–14,376.

Walsh, J.J., and C.P. McRoy, 1986. Ecosystem analysis in the southeastern Bering Sea. *Cont. Shelf Res.,* **5,** 259–288.

Walsh, J.J., C.P. McRoy, L.K. Coachman, J.J. Georing, J.J. Nihoul, T.E. Whitledge, T.H. Blackburn, P.L. Parker, C.D. Wirick, P.G. Shuert, J.M. Grebmeier, A.M. Springer, R.D. Tripp, D.A. Hansell, S. Djenidi, E. Deleersnijder, K. Henriksen, B.A., Lund, P. Andersen, F.E. Muller-Karger, and K. Dean, 1989. Carbon and nitrogen recycling with the Bering/Chukchi Seas: source regions for organic matter effecting AOU demands of the Arctic Ocean. *Prog. Oceanogr.,* **22,** 277–359.

Walters, C.J, V. Christensen, D. Pauly, and J. Kitchell, 1999. ECOSIM II: Representation of trophic interaction effects in a population dynamics model. *Ecosystems,* **3,** 70–83.

Wespestad, V.G., L.W. Fritz, W.J. Ingraham, Jr., and B.A. Megrey, 2000. On relationships between cannibalism, climate variability, physical transport and recruitment success of Bering Sea walleye pollock, *Theragra chalcogramma. ICES J. Mar. Sci.,* **57,** 272–278.

Wilderbuer, T.K., A.B. Hollowed, W.J. Ingraham, Jr., P.D. Spencer, M.E. Conners, N.A. Bond, and G.E. Walters, 2002. Flatfish recruitment response to decadal climate variability and ocean conditions in the eastern Bering Sea. *Prog. Oceanogr.,* **55,** 235–247.

Wooster, W.S., and A.B. Hollowed, 1995. Decadal-scale variations in the eastern subarctic Pacific. I. Winter ocean conditions. *Can. Spec. Publ. Fish. Aquat. Sci.,* **121,** 81–85.

Wyllie-Echeverria, T., 1995. Sea-ice conditions and the distribution of walleye pollock and the Bering and Chukchi Sea shelf. In: *Climate Change and Northern Fish Populations,* R.J. Beamish (ed.), *Can. Spec. Publ. Fish. Aquat. Sci.,* **121,** 87–95.

Wyllie-Echeverria, T. and K. Ohtani, 1999. Seasonal sea ice variability and the Bering Sea ecosystem. In: *Dynamics of the Bering Sea: A Summary of Physical, Chemical, and Biological Characteristics, and a Synopsis of Research on the Bering Sea,* T.R. Loughlin and K. Ohtani (eds.), North Pacific Marine Science Organization (PICES), Univ. of Alaska Sea Grant, AK-SG-99-03, 435–452.

Zheng, J., G.H. Kruse, and D.A. Ackley, 2001. Spatial distribution and recruitment patterns of Snow crabs in the eastern Bering Sea. In: *Spatial process and management of marine populations,* Alaska Sea Grant College Program, AK-SG-01-02, 233–255.

Chapter 31. Oceanography of the Northwest Passage (26,P)

F.A. McLaughlin and E.C. Carmack

Institute of Ocean Sciences, 9860 W. Saanich Road, Sidney, B.C., V8L 4B2 Canada

R.G. Ingram and W. J. Williams

Department of Earth and Ocean Sciences, University of British Columbia, 6339 Stores Road, Vancouver, B.C., V6T 1Z4 Canada

C. Michel

Freshwater Institute, 501 University Crescent, Winnipeg, Manitoba, Canada

Contents

1. Introduction

One of the greatest quests in exploration was the European search for the Northwest Passage (NWP), an oceanic short-cut from the Atlantic to the Pacific via the Canadian Arctic Archipelago (CAA; Figure 1). This same passage is now scientifically recognized to be a key pathway of oceanic waters that move in the opposite direction, from the Pacific and Arctic oceans into the North Atlantic via Baffin Bay and the Labrador Sea. It is our intent to review the general features of the CAA with particular focus on the NWP. We summarize basic physical, chemical and biological features of the region, and examine the dynamical constraints governing the movement of water masses through the system. Data from the *direct* (deepwater) NWP is presented: that is, the West-East passage that connects M'Clure Strait with Lancaster Sound via Barrow Strait. This differs from the more southerly *practical* NWP via Coronation Gulf, which allows easier ice navigation but also presents dangers due to shallow waters.

The Sea, Volume 14, edited by Allan R. Robinson and Kenneth H. Brink
ISBN 0-674-01527-4

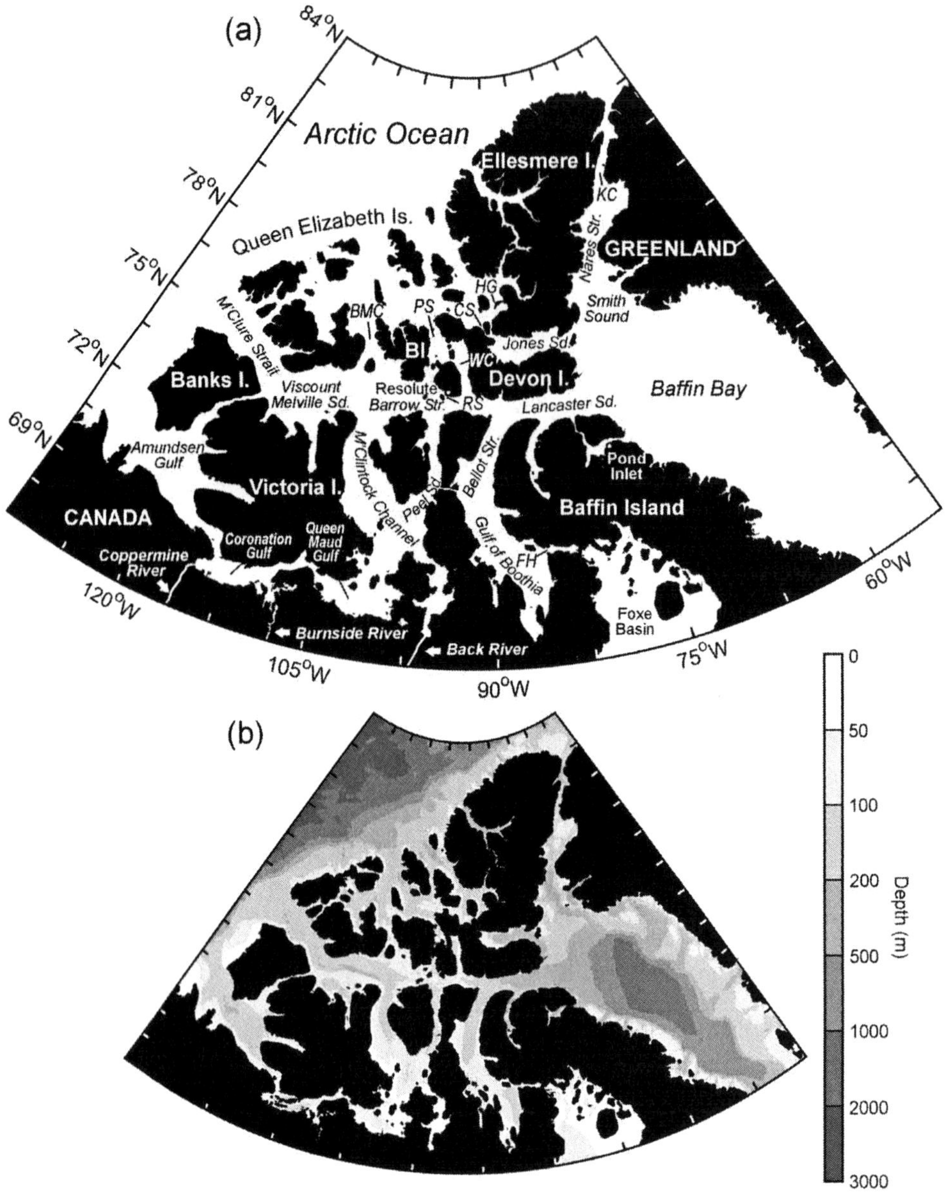

Figure 31.1: Maps of the Canadian Arctic Archipelago (CAA) showing (a) place names and (b) bathymetry. Hell Gate, Cardigan Strait, Penny Strait, Byam Martin Channel, Resolute Strait, Kennedy Channel and Fury and Hecla Strait are indicated by HG, CS, PS, BMC, RS, KC and FH, respectively. Bathurst Island is indicated by BI. Resolute and Pond Inlet locations are identified by white circles.

The quest for the NWP began with the first voyage of Martin Frobisher in 1576, less than 90 years after the voyage of Columbus, and reached its apex with the tragic voyage of Sir John Franklin in 1845–48 (Berton, 1988; Delgado, 1999). The ensuing search for Franklin resulted in the charting of the NWP by the mid-1850s, a feat driven mainly by the dogged persistence of his wife Lady Jane Franklin. The eleven-year search for Franklin was unsuccessful however and ultimately the Admiralty Prize for discovery of the direct NWP was awarded to Robert M'Clure aboard the *Investigator,* despite the fact that his ship was lost *en route* and he had to walk the final distance. The practical NWP was established by Amundsen's 1903–06 voyage aboard the *Goja* and is today the major supply- route for the small northern communities of the CAA. Despite M'Clure's accomplishment, the extreme environmental conditions encountered by the early explorers left doubt as to the economic significance of the NWP. Indeed, the severity of the voyage led Johann Miert, who served on M'Clure's expedition as ship's interpreter, to state that the 'now-discovered' NWP was:

> "...without significance and useless for navigation as long as the climate in these parts is so severe and the sea so covered with ice..."

Will climate evolution prompt a re-evaluation of Johann Miert's words? Will global warming alter the accessibility of the NWP? The palaeo-record clearly shows that climate variability has had a large impact on human settlement and the use of resources in the CAA over the past 3,000 - 4,000 years (cf. Schledermann, 1990; McGhee, 1996; Dyke et al., 1996; Dyke and Savelle, 2001). Archaeological artefacts demonstrated that pre-historic settlement was largely determined by the extent and duration of sea-ice cover, enabling some populations (e.g. the Dorset seal hunters) and limiting others (e.g. the Thule bowhead whale hunters). In addition, the presence of recurrent open-water areas (polynyas) allowed hunters access to marine mammals. Fluctuations in climate greatly conditioned areas and dates of settlement, thus demonstrating the vital links among humans, the marine ecosystem and the climate, as well as providing a valuable historical socio-economic lesson.

2. Regional Description

2.1 Hypsometry and Channel Characteristics

The CAA is a large, island-studded continental shelf and vast portions of this area remain oceanographically unsurveyed. The total surface area, including the Queen Elizabeth Shelf, Foxe Basin and Hudson Strait, is ~ 2.9 x 10^6 km^2; the area of the marine component is ~ 1.5 x $10^6 km^2$ and its volume is 0.38 x $10^6 km^3$. A summary of the area and volume of the marine CAA in comparison to the world ocean and Arctic Ocean is shown by the hypsometric curves in Figure 2 (see also Jacobsson, 2002). Here the significance of the large shelf-to-basin ratio for both the entire Arctic Ocean and the sector defined as the CAA is evident. The areal distribution of depth is especially evident in the log-depth representation which shows that about 70% of the CAA is shallower than 500 m.

The direct NWP is one of three main routes connecting the Arctic Ocean to the Labrador Sea and North Atlantic. On the west, the sill depth between the Arctic Ocean and M'Clure Strait is ~375m then water depths gradually increase and reach depths ~550 m in the sub-basin of Viscount Melville Sound. The limiting sill of the NWP is located further east in Barrow Strait where the depth is ~125 m. Continuing eastward water depths again increase gradually to ~ 500m in Lancaster Sound, then increase rapidly to over 2000 m in the center of Baffin Bay. The other two routes connecting the Arctic Ocean to the Labrador Sea are Cardigan Strait/Hell Gate and Nares Strait and their limiting sill depths are 85m and 220m, respectively. Melling (2000) demonstrates the relevance of these key routes in a comparison with Bering Strait, showing that a flow of only 4 cm s^{-1} through the CAA would deliver the same amount of water that enters through Bering Strait (about 1 Sverdrup with an average current speed of 32 cm s^{-1}; 1 Sv = $10^6 m^3 s^{-1}$).

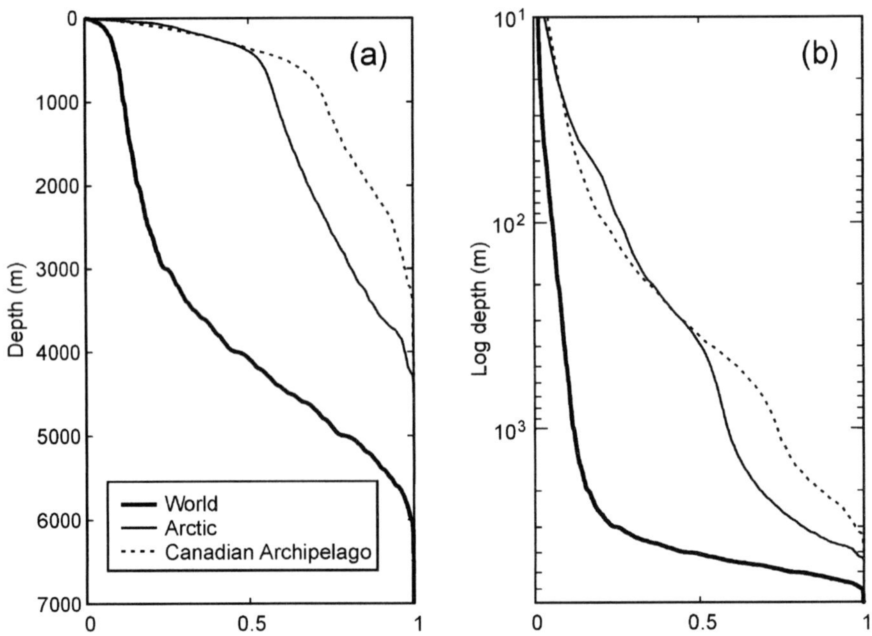

Figure 31.2 Hypsometric curves showing (a) normalized area versus depth curves for the world ocean, the Arctic Ocean and the CAA, and (b) the same data plotted as log depth.

2.2 *Regional Hydrology*

Freshwater input to the channels of the CAA includes advection from the relatively fresh upper layers of the Arctic Ocean, ice melt, local river discharge and net precipitation. These sources control stratification within the CAA and are an important component of the Arctic Ocean's freshwater budget (cf. Aagaard and Carmack, 1989). The two main drainage areas that supply local river discharge are the Arctic mainland coast east of the Mackenzie Basin to Fury and Hecla Strait (~ 0.6 km^2) and the archipelago islands west of Fury and Hecla Strait (~ 0.9 km^2). The major mainland rivers that directly affect the CAA include the Coppermine, Bathurst and Back. However, only a small fraction of these watersheds are gauged. Vuglinsky (1997) estimated the ungauged discharge from the North American arctic islands to be 211 $km^3 yr^{-1}$, identifying the potential importance of ungauged rivers. Walker (1977) also estimated the total fresh water discharge to the CAA (excluding Foxe Basin and Hudson Strait) at 219 $km^3 yr^{-1}$. Immediately west of the

CAA, the Mackenzie River drains a watershed of 1.7 x 10^6 km^2 and discharges approximately 340 $km^3 yr^{-1}$ into the Beaufort Sea. The volume of fresh water subsequently delivered into the CAA from this source is major but unquantified.

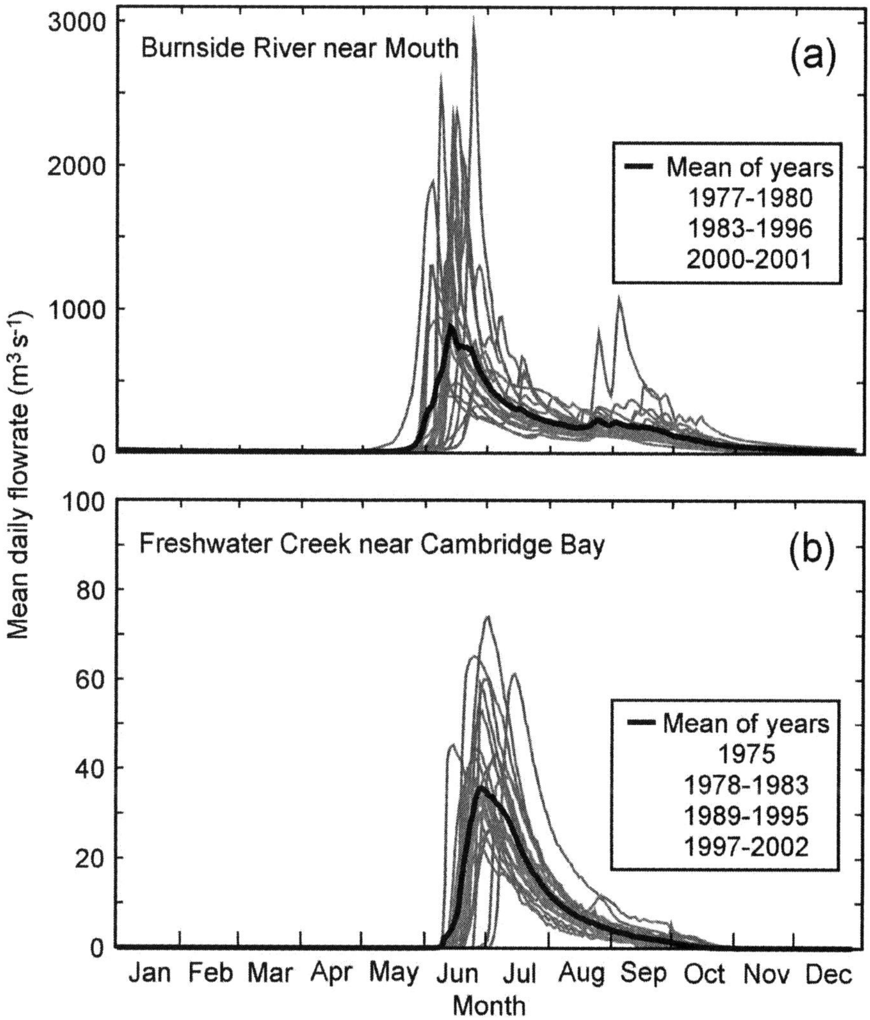

Figure 31.3 Annual river discharge ($m^3 s^{-1}$) for (a) the Burnside River, Arctic mainland and (b) Fresh Water Creek, Victoria Island.

Arctic rivers typically exhibit large seasonal and annual variation in discharge. Rivers within the CAA draw from watersheds ranging in latitude from 55 – 82 °N and the variety of landscape (vegetation, elevation and terrain) and climate (temperature and precipitation) within this latitude range places additional variability on the hydrological retention characteristics of these diverse drainage systems (cf. Bjornsson et al., 1995). Rivers with major headwater lakes will exhibit different annual discharge patterns than those without. The Mackenzie, for example, draws water from the Athabaska, Great Slave and Great Bear lakes and moderate wintertime flows are maintained. Other large mainland rivers (Coppermine, Burnside and Back) drain tundra comprised of numerous, small lakes and thus wintertime discharge is diminished. Nivel, or meltwater, rivers of the high arctic islands are frozen solid for most of the year and exhibit extreme seasonal runoff events (see Woo et al., 2000). Nivel rivers in the high Arctic only flow for a brief time in summer, whereas the Mackenzie River exhibits about a five-fold difference in flow between summer and winter. A comparison of discharge from the Burnside River (Arctic mainland) and Fresh Water Creek (Victoria Island), shown in Figure 3, illustrates the difference between a tundra-draining river and a nivel river, respectively. The freshets of both rivers occur in early summer due to the melting of snow, however discharge from the Burnside commences approximately two weeks earlier and is over ten-fold larger. The Burnside discharge also has a second peak in the autumn in some years due to storms prior to freeze-up. Both rivers show far greater interannual variability than the Mackenzie. A comprehensive overview of hydrological information in the CAA is presented in Prowse and Flegg (2000) and the water balance of Hudson Bay is reviewed by Prinsenberg (1984).

2.2 Meteorology and Sea Ice

The CAA experiences great variability in solar radiation over the annual cycle: for example at Resolute, north of Barrow Strait, there is 24-hour darkness between November 6 and February 5 and 24-hour daylight between April 29 and August 13. Although radiation values are relatively high in summer, the high reflectance of snow and ice reduce absorption levels. Air temperatures are very cold during winter with mean temperatures below -30 °C and cool in summer with mean temperatures near 4 °C. Winds are usually about 5 cm s^{-1} and from the northwest. The meteorology in the western part of the CAA is similar to that found in the Beaufort Sea and varies with the large-scale atmospheric fluctuations of the Arctic Oscillation (Barber and Hanesiak, 2004). In winter this region is influenced by a semi-permanent high pressure system located over the Beaufort Sea, and in summer, storms move in from the west, southwest, and north (Agnew and Silas, 1995). In the central and eastern CAA, wind forcing and weather conditions are linked to local topographic features and conditions in Baffin Bay (Ingram and Prinsenberg, 1998; Ingram et al, 2002). Although the Baffin Bay region is influenced by the high frequency of storms that track across North America (Zishka and Smith, 1980), the central CAA is protected by the mountainous terrain of Ellesmere, Devon, and Baffin islands (Agnew and Silas, 1995).

Sea ice is generally present year-round in the CAA and undergoes large seasonal changes in distribution. The onset of freezing occurs from late September to late October along the NWP and break-up begins in mid-June in the east and late August-early September in the west (Canadian Ice Service, 2002). Sea ice statistics for the period 1971–2000 have been compiled by the Canadian Ice Service (2002) and show that the minimum sea ice cover for both the practical and direct NWP occurs in early September. A useful statistical property is the "frequency of presence of sea ice" which shows the likelihood that ice - of concentration equal to or greater than 1/10 - would occur at a particular location. This statistical property, presented in Figure 4a for 10 September, shows that ice is always encountered in the Arctic Ocean north of the CAA. In the western region of the NWP there is a 50–100% chance of encountering ice and in the eastern part there is a less than 50% chance of encountering ice. Coronation gulf and parts of Baltin Bay are always ice free.

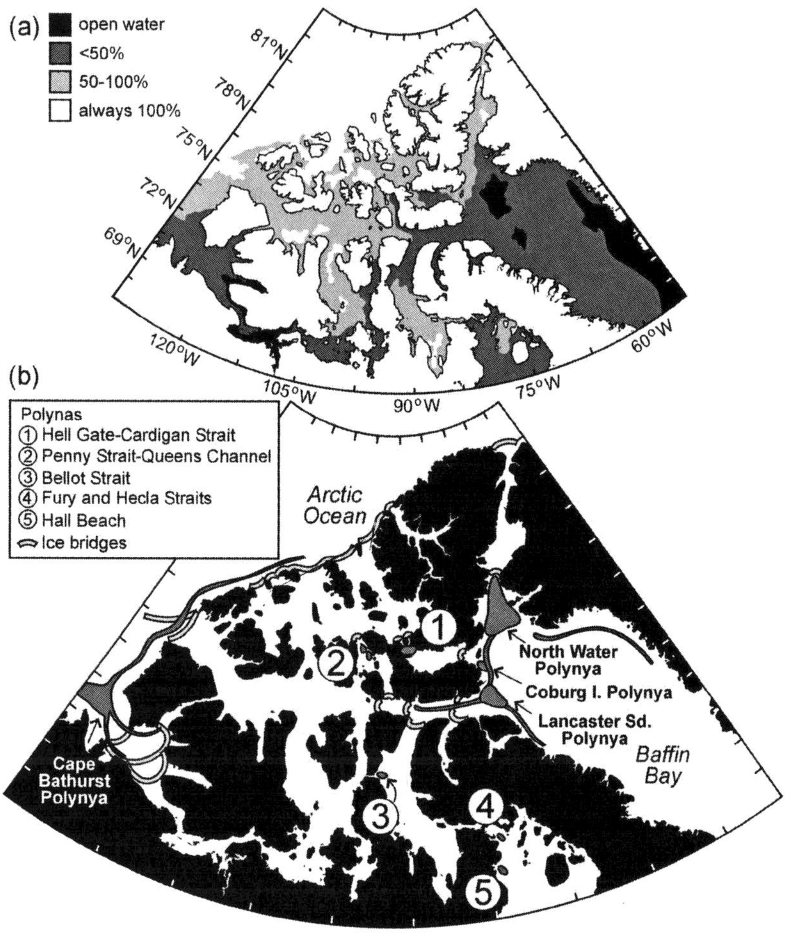

Figure 31.4 Ice climatology and structures in the CAA: (a) Frequency of presence of sea ice (%) that ice is present on 10 September for the period 1971–2000, adapted from Canadian Ice Service (2002); and (b) locations of recurrent ice bridges and polynyas, adapted from Topham et al.(1983) and Melling (2000).

As a result the flux of freshwater through the CAA in the form of ice has a strong seasonal signal. Typical values from mobile pack ice flow are 5–10 cm s^{-1} in the west and north and 10–15 cm s^{-1} in the east and central channels of the CAA (cf. Melling, 2000). Heavily deformed floes of pack ice from the Arctic Ocean west of M'Clure Strait and north of the Queen Elizabeth Islands also enter the NWP and supplement freshwater flux. A comprehensive analysis and discussion of sea ice motion and physical forcing in the northern CAA is given in Melling (2002).

The thickness of first year sea ice in the CAA varies from maximum values (away from ridging areas) of 2.5 m in the northern to 2.0 m in the southern sections (Canadian Ice Service, 2002). Multi-year ice can reach a thickness of 3–5 m. Although there is less interannual variability in ice cover extent here than in most other Arctic areas (Comiso, 2003), there is more open water during late spring-early fall in the eastern regions of the CAA than in the west (Agnew and Howell, 2003). Early break up in Lancaster Sound is caused by a combination of the dominant winds and currents, which advect ice eastward into Baffin Bay. Sea ice formation in early fall may also be delayed by the advection of warmer and saltier water from Baffin Bay into the entrance of Lancaster Sound (Prinsenberg and Hamilton, in press).

Between freeze-up in January and break-up in late July the ice is generally immobilized by attachment to the land (landfast) and by establishment of stable arches across passages (Figure 4b). Another important characteristic of winter ice cover is the occurrence of polynyas, which, due to physical forcing, are recurrent areas of open-water, or of much reduced sea ice cover, that would otherwise be fully covered by sea ice (Dunbar, 1969). Polynyas are generally classified by their mechanism of formation: either as sensible heat polynyas, formed by vertical mixing of warm deep water; or latent heat polynyas, formed by wind-driven removal of ice in the lee of islands and ice bridges (cf. Smith et al, 1990 and Melling, 2000). The presence of a polynya reflects intense physical forcing (e.g. mixing, winds) which, in turn, strongly affects biological processes by influencing the upwelling of nutrients, underwater light climate, stratification and the timing of seasonal productivity events (cf. Stirling, 1980; Tremblay et al., 2002a). The largest polynyas in the CAA are located in Smith Sound (North Water), Lancaster Sound and Amundsen Gulf. Others, influenced by strong tidal flows and vertical mixing, are found in various straits of the CAA: Hell Gate, Cardigan, Penny, Bellot and Fury and Hecla (Figure 4b). A case study of the physical forcing of an arctic polynya is given by Topham et al. (1983) and an introduction to biological processes in the North Water polynya is given by Deming et al. (2002).

3. Physical and Chemical Oceanography

3.1 Mean-Flow Characteristics

Early oceanographic descriptions of the CAA were produced by Bailey (1957), Collin (1962) and Walker (1977). Flow through the CAA, from the Pacific to Arctic to Atlantic oceans, is due to the higher (steric) sea-level of the Pacific. This sea level difference occurs because Pacific waters are fresher, lighter and, assuming a level of no-motion among the three ocean basins, the Arctic is thought to be 0.15 m higher than the Atlantic (Stigebrandt, 1984). The steric drop across the CAA from the Arctic to Baffin Bay (relative to 250 db) was calculated by Muench (1971) to be 0.3 m.

The flow of water through the CAA from the Arctic to the Atlantic Ocean is neither well known nor dynamically well understood (Melling, 2000). Nevertheless, the general near-surface circulation of the CAA as we know it is shown in Figure 5. Arctic surface waters flow westward and south-westward via channels in the CAA into Baffin Bay and, via Foxe Basin and Hudson Strait, into the Labrador Sea. There are a small number of narrow, shallow straits through which all flow must pass. These are Kennedy Channel, Barrow Strait, Wellington Channel, Cardigan Strait/Hell Gate and Bellot Strait. Barrow Strait, for example, has a minimum cross-sectional area of 7.9 x 10^6 m^2 (about 2.5 times the area of Bering Strait) and a mean depth of 152 m. The dynamics of flow in these straits are complex, involving rotational hydraulics, flows and counter flows, and tidal mixing. These processes are discussed below.

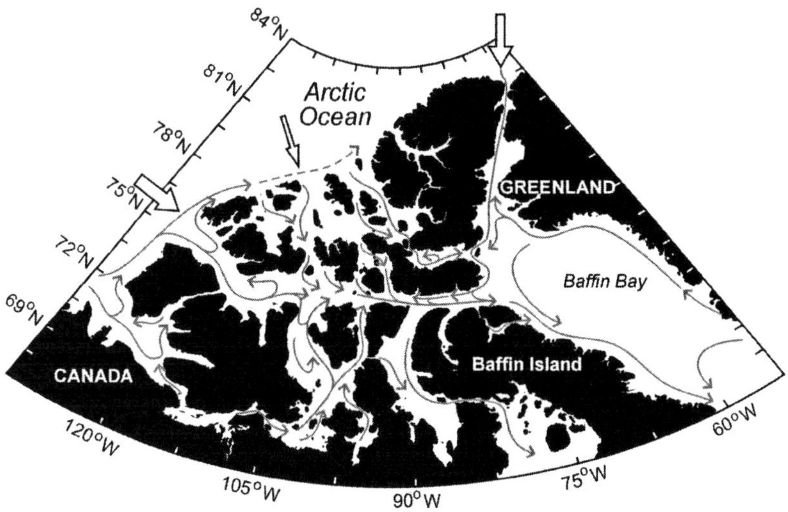

Figure 31.5 Schematic of mean near-surface circulation in the CAA, adapted from Ingram and Prinsenberg (1998).

Estimating flow through the CAA is difficult since few data are available and direct measurement is challenging. Total transport estimates of the flow through Lancaster, Jones and Smith Sounds were made by Collin (1962) who, using the dynamic method, found values ranging from 0.7 to 1.7 Sv. Applying the same technique, Muench (1971) estimated a mean southward flow in northern Baffin Bay of 2.1 Sv, although this number included the return flow of the West Greenland Current. Rudels (1986) used a two-layer approximation to compute a total transport of 1 Sv through the CAA. Fissel et al. (1988) synthesized the results of the few direct current measurements in major channels of the CAA and found a net transport of 1.7 Sv. Prinsenberg and Bennett (1989a) collected current meter and hydrographic data at the northern entrance of Peel Sound and obtained a transport contribution of 0.17 Sv.

To address the needs of climate prediction and modelling, an estimate of the freshwater transport through the CAA is needed. Here it is important to distinguish between total and freshwater transport because the latter is of particular climate significance (Carmack, 2000). Freshwater is delivered to the Arctic by atmospheric transport and by river and ocean inflows. As mentioned above, river inflow was estimated by Vuglinsky to be 4270 km^3 yr^{-1} and, of this, 711 km^3 yr^{-1} was from arctic islands. Based on a mean throughflow salinity of S=34.2 and a reference salinity of S=34.8, Aagaard and Carmack (1989) estimated that the major oceanic freshwater influx to the Arctic was 1670 km^3 yr^{-1} through Bering Strait. They estimated outflow through Fram Strait to be 2790 km^3 yr^{-1} as ice and 820 km^3 yr^{-1} as water and outflow through the CAA to be 155 km^3 yr^{-1} as ice and 920 km^3 yr^{-1} as water. The importance of freshwater flow through Bering Strait and the CAA on thermohaline circulation in the North Atlantic has been recently demonstrated in global climate models by Goosse et al. (1997) and Wadley and Bigg (2002).

3.2 Flow dynamics

3.2.1 Buoyancy-driven Flow Characteristics

The Arctic Ocean and its marginal seas are essentially a large estuarine system wherein the inputs and disposition of freshwater components provide thermohaline (buoyancy) forcing (cf. Stigebrandt, 1984; Carmack, 1986). A key parameter governing buoyancy-driven flow is the internal Rossby radius of deformation, $R_i = g'H^{½}/f$, where $g' = g\Delta\rho/\rho$ is reduced gravity, g is gravity, ρ is density, H is layer depth and f is the Coriolis parameter. In the Arctic, owing to its high latitude and a relatively weakly stratified water column, the R_i of a buoyant upper layer is quite small, about 5–10 km. Most of the straits in the CAA are much wider than this, allowing the formation of complex flow structures across channels. For example, distinct and oppositely-directed buoyancy currents on either side of the strait are possible. Accordingly, the CAA cannot be regarded as a network of channels that simply convey water from the Arctic Ocean.

It is likely that most rivers flowing into the CAA generate surface-trapped plumes (e.g. Ingram and Larouche, 1987). Each plume can either form a buoyancy-boundary current against the coast directly, or drive a cyclostrophically-balanced 'pool' that collects near the river mouth and subsequently leaks to form a reduced coastal current (Fong and Geyer, 2001; Yankovsky and Chapman, 1997). The buoyant water pool is susceptible to advection by mean flow and wind forcing. For example, Yanovsky (2000) shows that a mean flow opposing the coastal current can cause the buoyant pool to be 'shed', as an eddy, and advected downstream, while another buoyant pool subsequently forms at the river mouth. In another example Fong and Geyer (2001) shows that an along-shore current in the direction of Kelvin wave propagation tends to cause the buoyant pool to be advected downstream and become elongated against the coast, essentially forming a localised coastal current. Such a current is enhanced by Ekman transport under downwelling-favourable wind conditions, but under upwelling-favourable conditions the buoyant pool is pushed offshore instead. Wind forcing within the islands of the Archipelago may cause buoyant plumes to come into contact with the opposite side of the strait and thereby reverse the direction of flow (Arfeuille, 2001). Varied configurations of such plume-spreading are evident in satellite images of the Mackenzie River plume.

Much of the year the Archipelago is covered with sea ice and this has two distinct effects on plume dynamics. When ice concentration is low enough for the ice to be mobile, the enhanced drag between water and ice can effectively increase wind-stress and Ekman transport. When ice is immobile however, it acts as a rigid lid and the surface boundary layer beneath the ice will tend to widen the plume and dissipate the flow.

Buoyant plume water flowing across a strait due to Ekman transport is likely to augment the coastal current on that side of the strait in a similar fashion to the 'leakage' of the buoyant pool of a surface-trapped plume. Buoyant water can also cross a strait by forming eddies. Two mechanisms are known to exist for this. Firstly if the plume is sufficiently wide so that the Froude number $F_d = U / (g'H)^{1/2}$ (where U is speed, H is depth and $g' = g\Delta\rho/\rho$ is reduced gravity and ρ is density) is small (<1), then the surface- trapped coastal current is likely to form eddies through a baroclinic instability mechanism (Griffiths and Linden, 1981; Williams, 2003). These eddies cause an offshore flux of the buoyant water and a broadening of the plume. Secondly, topographically induced eddies are likely to occur in a coastal current at sharp bends in the coastline such as at the entrance to straits, where the radius of curvature of the coast is less than the Rossby radius of deformation (LeBlond, 1980; Klinger, 1994a). Here eddies can grow (Klinger, 1994b; Cenedese & Whitehead, 2000) in a similar fashion to the pool of buoyant inflow water for a surface-trapped plume. These eddies may detach from the coastal current (Cenedese & Whitehead, 2000) or may grow large enough to bridge the gap across the strait, allowing the coastal current to continue both along the straight and across it to the next island. In some cases, such as Bellot Strait, the gap between islands may be significantly less than the Rossby radius of deformation and this eddy formation process may not be needed for the coastal current to cross the gap. Flow through the Bellot Strait is still expected, however, because the sea-surface height is elevated near the coast, due to the presence of boundary currents.

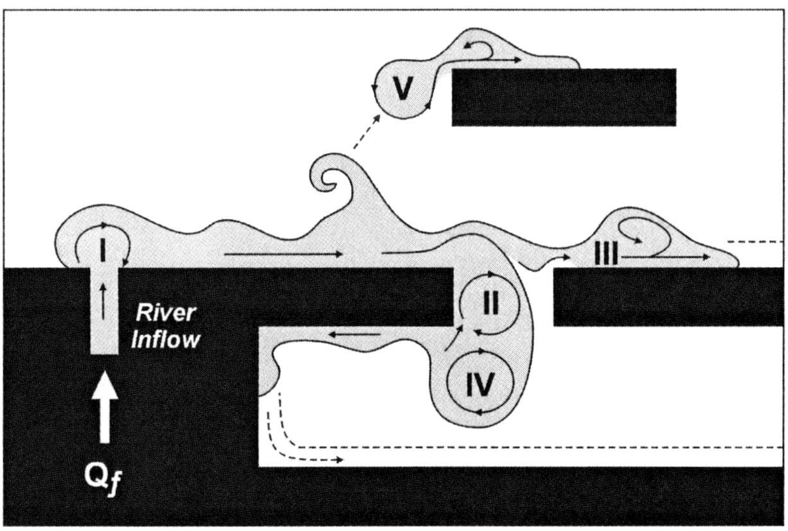

Figure 31.6 Cartoon of mechanisms affecting buoyancy-driven flow among the islands and gaps in the CAA. See text for explanation and discussion.

A cartoon illustrating these mechanisms affecting the dynamics and transport of buoyancy-driven flow in regions of complex topography is shown in Figure 6. The initial buoyancy input by river inflow (Q_f) forms a cyclostrophically-balanced pool (I), which subsequently leaks to form a buoyancy-boundary current (BBC) along the right-hand shoreline. Instabilities along this flow may form eddies which subsequently detach and migrate offshore and, in the presence of islands, reattach to form a secondary BBC (V). When the primary BBC encounters a gap or a sharp corner an eddy is formed within the gap (II). If the eddy is becomes large enough to bridge the gap it too may re-attach and allow the BBC to continue along the right-hand shore (III). Depending on the topography south of the gap this eddy (II) may prompt development of a paired eddy (IV) to the south, which subsequently flows along its right-hand shoreline.

3.2.1 Channel Dynamics

It has long been recognised that flow through the irregular topography of the CAA is complex (LeBlond, 1980). Within the NWP, Barrow Strait is a constriction through which all flow must pass, being the shallowest (125 m deep) and narrowest (~52 km wide) part. A simple barotropic, geostrophically-determined model of flow through a strait, driven by a sea-level difference between the ends of the strait, is given by Toulany and Garrett (1984). Because sea level is higher in the Arctic than the Atlantic, this model would prescribe an eastward flow through Barrow Strait such that the sea level difference across the Strait would be equal to the sea level difference between the ends of the Strait. The water column in Barrow Strait is stratified however (Prinsenberg and Bennett, 1987; also see Figure 10) and therefore baroclinic dynamics also contribute to flow in the Strait. Rudels (1986), in comparing stratification in the Canada Basin with that in Baffin Bay (above the sill depth of Barrow Strait), found evidence of flow reversal above sill depth. That is, the along-channel pressure gradient in the upper water column is towards Baffin Bay, in keeping with the sea-level difference, whereas the pressure gradient near the sill depth of Barrow Strait is towards the Canada Basin. A geostrophic model allowing maximum flow in the strait then leads to a flow of fresher, surface water towards Baffin Bay and flow at depth towards the Arctic. Flows of this type within a channel resemble rotating "lock exchange flow" (Whitehead, 1998; Hunkins and Whitehead, 1992) where separate reservoirs of different density are joined by a channel and equal and opposite flows from each reservoir are exchanged through the channel. Hydraulic control then gives the volume flux of the exchange as proportional to $g'H^2/f$, for a dynamically wide channel, and should also prescribe the mixing within the channel. The condition of equal exchange is not strictly necessary within Barrow Strait however because any volume flux imbalance can be corrected via exchange elsewhere in the Arctic system. If such hydraulic control occurs in Barrow Strait, the exchange flow should increase during summer, when g' is large due to freshwater influx, and decrease during winter when g' is smaller and sea-ice provides additional friction to surface flow (cf. Melling, 2000). Tidal flows in Barrow Strait will further complicate matters and may lead to a complex, time-dependent hydraulic exchange flow, such as described by Helfrich (1995), where exchange is enhanced by the tides and complete information of the entire strait geometry is needed to determine the flow.

Frictional effects can also be an important factor governing flow through the NWP. For frictionally-controlled barotropic flow through a strait, the requirement that sea-level difference across the strait due to geostrophy be equal to the sea-level difference along the strait necessary to maintain flow against friction leads to development of a parameter fWH/LC_dU, where f is the Coriolis parameter, W is the width of the channel, L the length of the channel, H the depth, C_d a coefficient of drag and U the flow speed (H. Johnson and C. Garrett, pers comm.). For long narrow channels in the CAA, frictional decay of barotropic flow may be significant, especially during winter when sea ice effectively doubles friction. Also, at choke points in the CAA such as Barrow Strait, tidal flows are amplified and cause increased mixing. In this case, the flow speed U would be augmented by a larger tidal velocity and thus increase the effects of friction. Frictional effects on exchange flow were also considered by Johnson and Ohlsen (1994) who found that the addition of frictional boundary layers to the top and bottom walls of the channel and the fluid interface cause a significant flow reduction that is curiously independent of channel length.

Increased mixing within straits by shear flow and tides also lessens stratification and g', thereby reducing buoyancy forcing and affecting geostrophic flow and any hydraulic controls. In an extreme case, sill regions can be fully mixed if the tidal mixing parameter H/U^3 is sufficiently small. The removal of the baroclinic structure within the strait would then block along-strait baroclinic geostrophic flow.

3.3 Tides and Tidal Mixing

Tidal energy enters the CAA predominantly from the Atlantic Ocean and is mainly semi-diurnal. As a result waters transiting the NWP are significantly modified by tidally-driven mixing and, in the vicinity of Barrow Strait, tidal currents are especially strong, reaching 50–150 cm s^{-1} (Prinsenberg and Bennett, 1989b). Numerical models have been used by Kowalik and Proshutinsky (1993), Lyard (1997) and Greenberg (pers. comm.) to estimate tidal current amplitudes in the CAA. Maximum tidal currents due to the M2, S2, K1, and O1 constituents in the NWP were typically 0.2- 0.4 m s^{-1}, except south of Bathurst Island, where maximum values reached 1–1.5 m s^{-1}. Observational descriptions for northern Baffin Bay are given by Fissel (1982).

A useful parameter for evaluating the effectiveness of tidal mixing is H/U^3 where H is depth and U is tidal velocity (Simpson and Hunter, 1974) and low values indicate strong mixing. The most intense areas of mixing in the CAA are the Gulf of Boothia, Queen Maud Gulf, Dolphin and Union Strait, Barrow Strait and Hell Gate/Cardigan Strait (Figure 7). As a result the water column in Barrow Strait is more homogeneous than in the reservoirs to the east or west and thus the surface density has a local maximum (see below). Tidal mixing also enhances the vertical flux of nutrients and thus increases productivity of ice algae (cf. Cota and Horne, 1989).

The occurrence of critical latitudes in the CAA further contributes to the efficiency of tidal mixing (Furevik and Foldvik, 1996; Kulikov et al., 2004). At these critical latitudes (71 °N, 74.5 °N and 85.7 °N for the N_2, M_2 and S_2 tides) semi-diurnal tidal constituents resonate with inertial waves, causing a thickening of the bottom Ekman layer of the clockwise rotary tidal component and local trapping of baroclinic tides generated at topographic discontinuities. It is especially noteworthy that the direct NWP is located along the M2 critical latitude where semi-diurnal tides are strong and therefore the effect of boundary friction on mixing

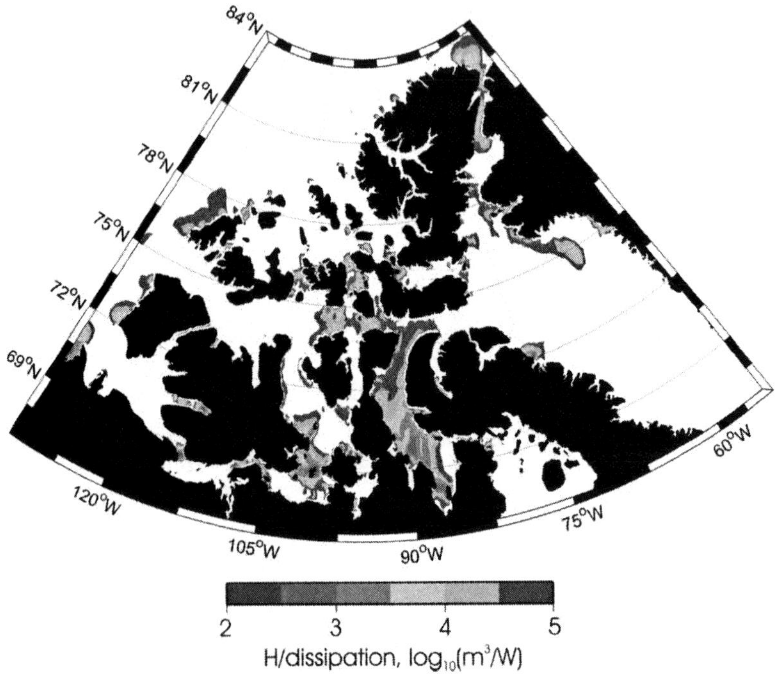

Figure 31.7 Map showing the tidal mixing parameter of the H/U^3 (adapted from Greenberg, pers. comm.)

may be enhanced (cf. Stronach et al., 1987).

Prinsenberg (1988) showed that significant changes in tidal amplitude and phase for both diurnal and semi-diurnal components, occurred in Hudson Bay in the presence of sea ice. Although a similar response is expected in the NWP, there are few tidal observations to confirm whether this occurs.

3.4 Water mass transformation

The Arctic Ocean receives water from the Pacific and Atlantic oceans via Bering and Fram straits, respectively. The CAA, in turn, receives water from the Canada Basin through Amundsen Gulf, M'Clure Strait and straits along the Queen Elisabeth Islands and also from the Makarov Basin and the Lincoln Sea via Nares Strait. In the Canada Basin waters of Pacific-origin, being relatively fresh, occupy the upper ~ 200 m of the water column and overlie Atlantic-origin waters. This upper 200 m contains three water mass components: a seasonal mixed layer, Pacific-origin summer water and Pacific-origin winter water. Atlantic-origin waters are found between 200–1600 m and consist of lower halocline, Fram Strait Branch and Barents Sea Branch components. Deep waters occupy the remainder of the Canada Basin water column.

The direct NWP receives inflowing arctic waters from the Canada Basin via M'Clure Strait on the west and Byam Martin Channel and Penny Strait to the north. Inflow from the Canada Basin is restricted by a shallow sill (~125 m) at Barrow Strait that, oceanographically, separates the NWP into its western and eastern parts. Canada Basin waters (depth ~ 3500 m) are separated from Viscount Melville Sound (depth ~ 600 m) by a deep sill (~ 375) and m'Clure Strait (depth ~ 475m). Canada Basin waters (depth ~ 3500 m) are separated from M'Clure Strait waters (depth ~ 475 m) by a deep sill (~375 m) and further east is a sub-basin in Viscount Melville Sound (depth ~ 600 m). Both Pacific and Atlantic-origin waters are found in M'Clure Strait and Viscount Melville Sound. The direct throughflow of Atlantic-origin water, however, is strongly constrained by the shallow sill depth at Barrow Strait and by the mixing processes discussed above. Nutrient concentrations, being higher in Pacific-origin waters, are used to identify water masses in the NWP. Codispoti and Owens (1975) used silicate and phosphate measurements to estimate that ~60% of the waters in Lancaster Sound were from Bering Strait. Similarly, Jones et al. (2003) used the nitrate to phosphate relationship to determine that almost all of the water flowing through Lancaster Sound was of Pacific-origin.

During transit through the NWP, Pacific- and Atlantic-origin water masses undergo mixing and geochemical modification, as seen in Figures 8 and 9 where data from the Canada Basin is compared to data from five locations across the NWP, from west to east: M'Clure Strait, Viscount Melville Sound, Barrow Strait, Lancaster Sound and Baffin Bay. In the Canada Basin, the seasonal mixed layer in summer contains river water and sea-ice melt and is ~ 40 m thick (Figure 8). Temperatures are warm, salinities are low ($S>24$), nitrate concentrations are zero (limiting nutrient) and oxygen concentrations are high (Figure 9). Below lie Pacific-origin waters that have entered via Bering Strait in summer and winter (Coachman and Barnes, 1961). Summer waters are found from ~ 40–75 m and are characterised by relatively warm temperatures at $31<S<32$, increasing nutrient and decreasing oxygen concentrations. Winter waters, found from 125–180 m, identified by a temperature minimum near $S=33.1$ and a maximum in nutrient concentrations. The transition to Atlantic-origin waters begins near ~200 m. The core of lower halocline water is located near ~225 m at $34.0<S<34.4$ and is characterised by increasing temperature, decreasing nutrients and an oxygen minimum. The core of Fram Strait Branch is identified by a deep (~ 450 m) temperature maximum (0.4 °C) and deep nutrient minimum near $S=34.8$. The underlying Barents Sea Branch waters contain higher concentrations of oxygen and nutrients and its core is found near $S=34.88$ and 900 m (not shown).

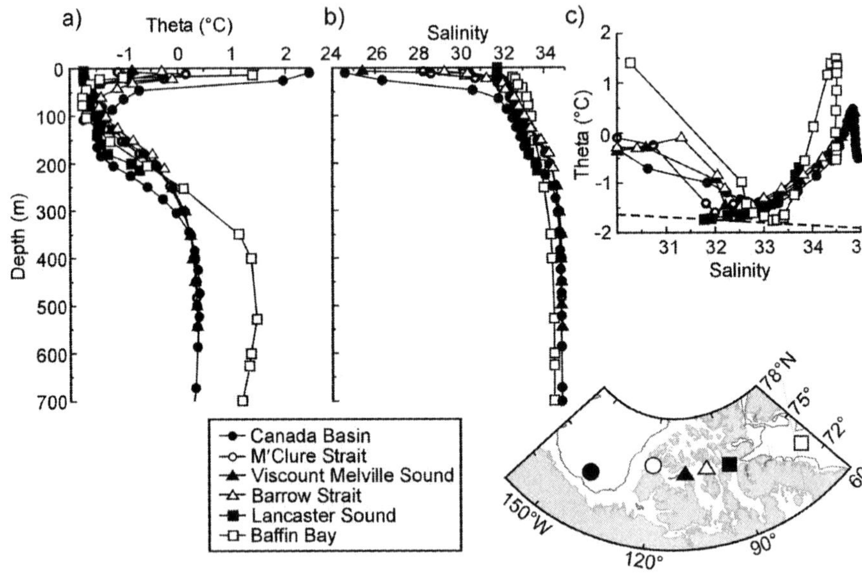

Figure 31.8 Vertical profiles of (a) temperature (T), (b) salinity (S), and (c) T/S correlation plots at six sites from the Canada Basin to Baffin Bay. See color plate for detail.

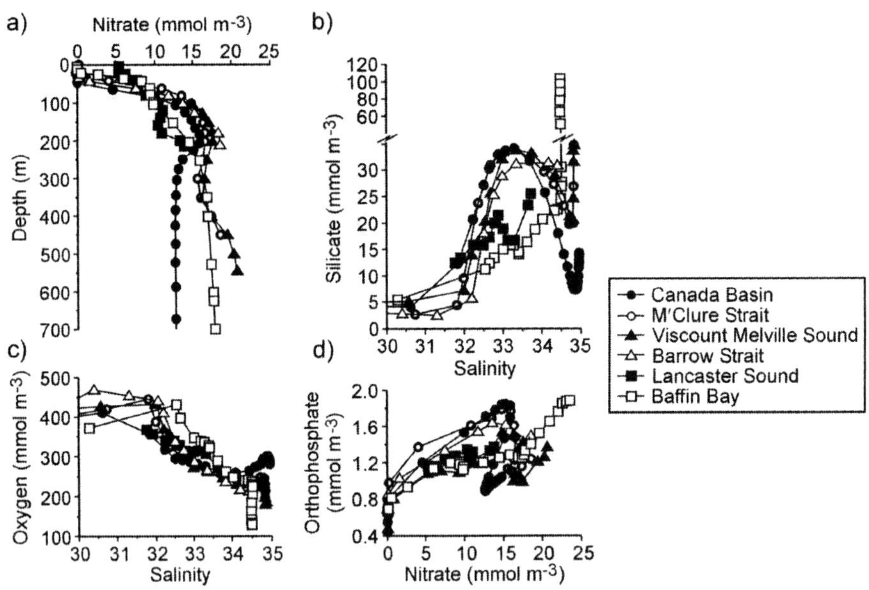

Figure 31.9 Vertical profile of (a) Nitrate, and property-salinity plots of (b) Silicate (c) Oxygen and (d) Nitrate/Phosphate correlation diagrams at six sites from the Canada Basin to Baffin Bay. See clolor plate.

In the NWP the base of the seasonal mixed layer becomes progressively shallower, colder and more saline moving east from the Canada Basin to Barrow Strait. The depth of nitrate depletion decreases from ~50 m in the Canada Basin to 30 m in M'Clure and Barrow straits. In Lancaster Sound however, surface layer nitrate concentrations were above 5 mmol/m^3 due to enhanced mixing as descibed above. Pacific-origin winter waters are identified by a temperature minimum in the western NWP but further eastward this feature disappears and the nutrient maximum is used instead as a tracer of Pacific-origin water. From M'Clure to Barrow Strait the nutrient maximum is ~50 m shallower than outside in the Canada Basin and occurs at progressively higher salinities as a result of turbulent mixing (33.2<S<33.3 near 125 m; Figure 9a and 9b). Nutrients identify three layers. In the upper layer, which shallows from west to east, nitrate is zero and silicate and phosphate remain above zero. In the mid-depth nutrient maximum layer near S=33.1, nitrate concentrations are similar from the Canada Basin to Barrow Strait, then are much lower in Lancaster Sound and Baffin Bay. Silicate and phosphate concentrations decrease progressively from the Canada Basin to Baffin Bay. In the near bottom layers nutrients increase due to regeneration of particulate matter settling from the euphotic zone. Thus the general pattern for pacific-origin waters transiting the NWP is the decrease in concentration of the mid-depth nutrient maximum layer and the concentration increase in the near-bottom layer. Waters in Lancaster Sound are noticeably different (i.e. presence of a local nutrient and temperature maximum at S=32.8), due to mixing caused by the physical processes described above.

Atlantic-origin waters are found in the NWP where water depths are greater than ~135 m. They are identified by temperatures increasing with depth (Figure 8a), by a change in slope in salinity profiles (Figure 8b) and by a decrease in nutrient concentration (Figure 9b). In the lower halocline at S~34, temperatures are warmer than in the Canada Basin and are progressively warmer as one moves eastward across the NWP. Temperature increases for two reasons: heat transport from below and the absence of lateral intrusions of shelf-water at the freezing temperature (Melling et al., 1984). The lower halocline's oxygen minimum at 34.0<S<34.4 is much lower in concentration than in the Canada Basin and is lowest in Barrow Strait (Figure 9c). Nutrient concentrations at these salinities are higher than in the Canada Basin which, taken together with the decrease in oxygen, indicate that nutrients are regenerated locally.

Fram Strait Branch waters in M'Clure Strait and Viscount Melville Sound are slightly colder than in the Canada Basin. Below ~350 m and at S>34.8, oxygen concentrations decrease and nutrient concentrations increase with increasing depth, indicative of nutrient regeneration. These data, together with CFC data showing the absence of CFC-113 in Viscount Melville Sound deep waters, suggest that M'Clure Strait and Viscount Melville Sound sub-basins are not renewed regularly by inflow from the Canada Basin (McLaughlin and Carmack, 1998; in prep.).

The nitrate to phosphate relationship, shown in Figure 9d, illustrates the modification of water masses while flowing through the NWP. In the Canada Basin there is a distinctive "Z-like" pattern, where the upper line denotes the range in Pacific-origin water concentrations and the lower line, much shorter, the range of Atlantic-origin water concentrations. In M'Clure Strait the pattern is similar but shows the effects of mixing between Pacific and Atlantic-origin water by the change in the slope of the connecting line and of regeneration by the increase in concentration of both nutrients in deeper waters. At Barrow Strait the Z-like pattern has been strongly altered by tidal mixing, and further east in Lancaster Sound and Baffin Bay the separation between the two parallel lines has decreased significantly.

Waters in the western NWP are clearly influenced by inflow of Pacific and Atlantic-origin waters from the Canada Basin. In the eastern NWP, however, Lancaster Sound waters appear to be more influenced by waters flowing in from Baffin Bay in the east. Jones and Coote (1980) observed that eastern NWP waters are a mixture of eastward flow from the Canada Basin, southward flow from the Arctic Ocean via Nares Strait and Jones Sound and northward flow of Atlantic-origin water via the western coast of Greenland. They associated higher nutrient concentrations on the southern side of Barrow Strait with eastward flowing Canadian Basin waters and, as Figures 9a and 9b show, Pacific-origin waters are found in the upper 100m. Flow into Lancaster Sound from the north was also demonstrated by direct current meter measurements (Lemon and Fissel, 1982). Temperature and salinity data from Lancaster Sound (Figure 8) show that surface waters are colder and more saline and the slope of the T/S plot at $S>33$ is steeper than at locations in the western NWP, all suggesting that waters in the center of Lancaster Sound have travelled via a different pathway. Nutrients are regenerated in Lancaster Sound below ~175 m and $S>33.2$ in Atlantic-origin waters as indicated by the increase in nutrient and decrease in oxygen concentrations (Figures 9a, 9c).

Both Pacific- and Atlantic-origin water mass properties are evident in northern Baffin Bay. The small nutrient maximum at S~33.2 indicates the presence of Pacific-origin water in the upper 50–60 m and therefore outflow from the Arctic Ocean. The temperature maximum, indicative of Atlantic-origin water, is much higher (i.e. warmer) and found at a much lower salinity than the equivalent maximum observed in the Canada Basin, which indicates the presence of Atlantic-origin waters that have followed a different pathway, i.e. via Davis strait (Bacle et al., 2002). Nutrient regeneration is evident in waters below 500 m in the Baffin Bay sub-basin, as was observed by Jones et al. (1984). Silicate is especially enriched, which Michel et al. (2002) and Tremblay et al. (2002b) attributed to the large export of biogenic silica from sinking diatoms and their dissolution at depth.

As discussed above, the transport of freshwater from the Arctic Ocean to the North Atlantic is of particular importance in the balance of the global freshwater budget. In east-west channels freshwater flow is predominant along the southern shoreline. For example, a salinity section across Barrow Strait (Figure 10) illustrates the presence of a buoyancy-boundary current located in the upper 30 m on the southern side of the channel. In north-south channels, the freshwater flow is predominantly located on the western side. For example, Tan and Strain (1980) used the ratio of $^{18}O/^{16}O$ to distinguish sea ice melt from its "lighter" meteoric (precipitation, land run-off and glacial melt) counterpart and showed its presence on the western side of Davis Strait. They also determined that sea ice melt penetration varied from 50 m in Baffin Bay to 140 m in Lancaster Sound. Bedard et al. (1981) also showed that waters in the CAA were diluted with sea ice melt.

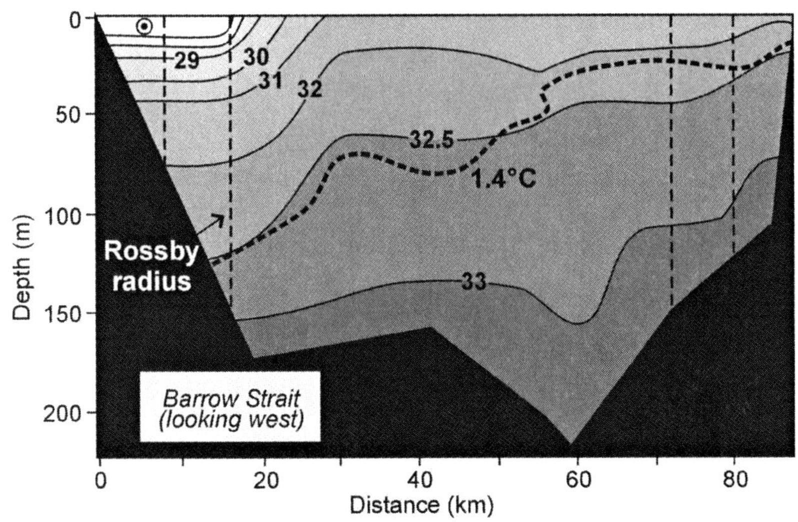

Figure 31.10 Section across Barrow Strait showing typical flow within a broad channel of the CAA. The vertical dashed lines illustrate the size of the Rossby radius as measured from the coast. The circle with a dot in the centre represents the narrow, shallow bouyancy boundary current, and the direction of flow is toward the reader.

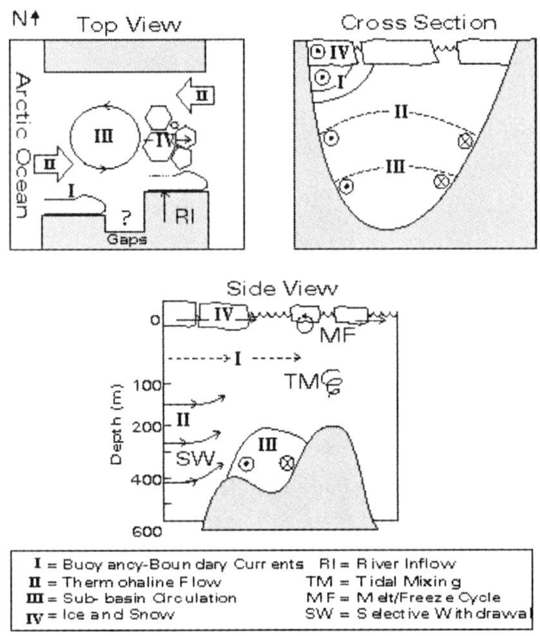

Figure 31.11 Cartoon showing idealised modes of flow within a broad channel of the CAA. Circles with a dot in the center denote flow toward the reader; circles with an x in the center denote flow away from the reader, (into the page).

In summary, flow through the NWP cannot be regarded as simple channel flow and a conceptual model of flow regimes is shown in Figure 11. Three distinct flow regimes are identified: a shallow, buoyancy-boundary current located along the southern (right-hand) shore; a thermohaline flow further offshore; and if the channel is sufficiently wide and deep, sub-basin circulation of waters below sill depth. Of these, the buoyancy-boundary current is most rapid. Sub-basin waters are generally trapped within cyclonic gyres. CFC profiles indicate that renewal of deeper waters is on the order of many decades and this permits substantial nutrient regeneration to occur (McLaughlin and Carmack, in preparation). Another point, not shown, is that the major fraction of freshwater transport occurs in the upper tens of meters where waters are transported more rapidly than at depth.

4. Biological Perspectives

Low temperatures, extreme seasonal variation in light regime, and the presence of a permanent or seasonal ice cover are obvious constraining factors for biological production in the Arctic Archipelago. The presence of polynyas, shore lead systems and ice edges also play key roles, influencing the abundance and distribution of marine mammals and seabirds. Because of these factors, production in the CAA is considered to be highly pulsed and highly variable over space and time. Although various data sets exist in technical or data reports (the so-called grey literature), little biological data have been published for many areas of the NWP. Most of the published studies, reported here, were carried out in the eastern part of the NWP, including Resolute Passage, Lancaster Sound and Barrow Strait.

During late spring and early summer, large numbers of seabirds and marine mammals congregate at the edges of landfast ice in the eastern NWP (e.g. Bradstreet, 1979 and 1980; Finlay et al., 1980; McLaren, 1982). Arctic cod plays a central role in the diet of the large colonies of thick-billed murres (> 650,000 adults and juveniles), northern fulmars (>450,000 adults and juveniles), black-legged kittiwakes (300,000 adults and juveniles) and black guillemots (>50,000 adults and juveniles) that come to breed in Lancaster Sound in summer (Welch et al., 1992). The amphipod *Parathemisto* spp. also constitutes an important part of the thick-billed murre's diet and can represent, by weight, the equivalent of arctic cod (Bradstreet and Cross, 1982). As for marine mammals, narwhal, beluga, walrus, killer whale, polar bear, and three species of seals (beaded, harp and ringed) are found in abundance in the Lancaster Sound region. Again, arctic cod constitute a key component of the diet of many of these mammals, thereby playing a key role in the flow of energy through the marine food web (Fig. 12, adapted from Welch et al., 1992). Overall, an estimate of 148,000 tonnes of arctic cod would be consumed annually by seabirds and marine mammals in the Lancaster Sound region. According to Welch et al.'s (1992) estimate, 16% would be channelled to the four species of seabirds mentioned above, 47% to ringed seals, and the rest consumed by narwhal, beluga and harp seals (see their Table 8).

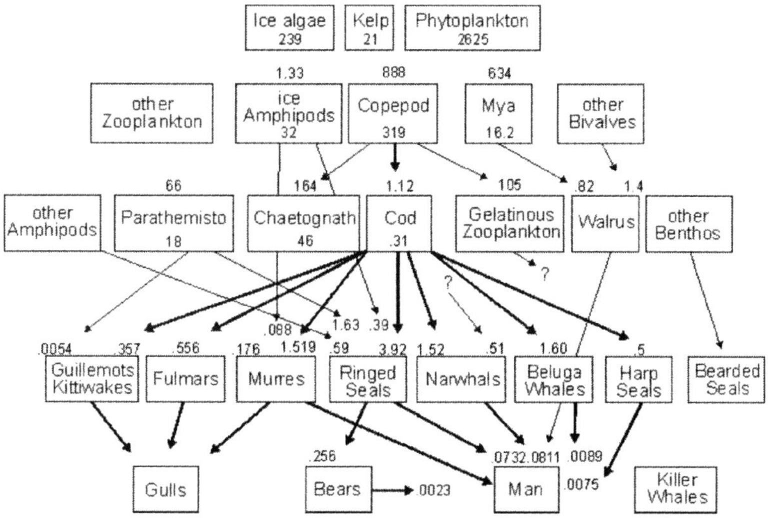

Figure 31.12 Schematic of energy flow (in $kJm^{-2}yr^{-1}$) through the marine food within Lancaster Sound. Numbers at the top of boxes are ingestion; numbers within boxes are growth. Heavy arrows highlight energy flow mediated by arctic cod (adapted from Welch et al., 1992).

In order to sustain the large populations of marine mammals and seabirds observed in the area, Lancaster Sound waters, appear to be able to support a high level of primary production. In a study comparing benthic communities in Lancaster Sound, Pond Inlet and Baffin Bay areas, Thomson (1982) reported that the mean biomass of benthic animals at 50–100 m in Lancaster Sound is among the highest recorded for the Arctic. He attributed this high biomass to the high productivity of the region and high current speed. Primary production in the Lancaster Sound region is estimated (within a factor of two), to be about 60 g C m^{-2} fixed annually of which approximately 90% is contributed by phytoplankton, 10% by ice algae, and a small amount by kelp (Welch et al., 1992).

A phytoplankton bloom typically develops in July-August after ice melt causes stratification of the surface layer, or at marginal ice zones. Chlorophyll *a* (chl *a*) concentrations near Resolute Passage reached 5 mg m^{-3} in July following ice melt (Conover et al., 1999). In the same area, Fortier et al. (2002) reported the unusual occurrence of an early under-ice phytoplankton bloom during the month of June, due to rapid and early snow melt and release of ice algae. On average, phytoplankton chl *a* standing crop appears to be moderate and comparable to other Arctic regions (69 mg m^{-2} in the top 35 m), although a strong subsurface chl *a* maxima (up to 18 mg m^{-3}) has been observed in association with the pycnocline due to local ice melt (Borstad and Gower, 1984). The distribution of chl *a* shows large spatial heterogeneity and reveals the presence of "biological hot spots" where high localized primary production is associated with physical mixing processes that provide abundant nutrients for phytoplankton growth. The fact that major seabird colonies in the Lancaster Sound area are located near semi-permanent ice edges or energetic eddy zones where strong oceanographic fronts are common (McLaren, 1982) demonstrates the importance of these "biological hot spots" for the feeding of marine fauna.

In a trophic diet study conducted at Pond Inlet, Bradstreet and Cross (1982) concluded that copepods and amphipods dominate the diet of arctic cod, regardless of the age class of the fish or whether they were collected inshore or offshore. They also reported that ice algae species dominate the diet of all amphipod and copepod species studied, with ice algae genera *Nitzschia* and *Navicula* being most abundant. These results reflect the importance of ice algae to herbivorous grazers who, in turn by being consumed, subsequently channel energy to arctic cod, marine birds and mammals. The role and contribution of ice algae is more than just its relative biomass because it also extends the grazing season of herbivores and concentrates food resources on or near a two-dimensional environment (Bradstreet and Cross, 1982). Indeed, under-ice copepods have been shown to feed upon ice algae during the algal growth season (Conover et al., 1986; Head et al., 1988). A carbon budget of ice algae production and export for a 60-d period in spring-summer in Resolute Passage showed that ca. 60% of the ice algae production was transferred to water column zooplankton grazers, mainly *Pseudocalanus* spp, which fed on ice algae in or released from the ice (Michel et al., 1996). They suggested that the timing of processes that drives the release of ice algae from the under-ice surface, rather than the actual period of ice algal production, may be critical for the feeding success of pelagic grazers at a time when the phytoplankton crop is at a minimum. In a later study, Fortier et al. (2002) showed high interannual variability in the timing, magnitude and nature of under-ice vertical fluxes of organic matter and related this variability to meterological forcing events that controlled snow thickness and to differences in the zooplankton community.

First-year sea ice covers most of the NWP and can support a high production of ice algae, with values up to 23 g C m^{-2} y^{-1} in Resolute Passage (Smith et al., 1988). Here a complex microbial under-ice community was observed, dominated by pennate diatoms (e.g. Michel et al., 1996), but also comprised of active protozoa (Laurion et al., 1995), nematodes (Riemann and Sime-Ngando, 1997), and high DOC concentrations (Smith et al., 1997). Light is often considered as the most important factor in driving primary production in the ice (e.g. Smith et al., 1988), however nutrient availability can also play a key role for ice algae production. During the spring-summer transition period, the period of ice algal growth, the optical properties of sea ice exhibit considerable seasonal, spatial and spectral variability. Albedo, the ratio of reflected to incident shortwave radiation (Maykut, 1985), varies greatly, from 0.6–0.8 in snow-covered ice to 0.2–0.4 in melt ponds, and this causes large temporal and horizontal variability in the amount of radiation that reaches the ice algae (Perovich et al., 1998). And, because snow cover strongly attenuates the photosynthetically active radiation (PAR), its distribution largely influences ice algal biomass (Welch and Bergmann, 1989). Smith et al. (1988) showed that the biomass maximum attained by ice algae under natural snow cover in Resolute Passage was equivalent to the maximum imposed by self-shading, with values ranging from 77 to 225 mg chl a m^{-2}. The presence of ice algae and other particulate or dissolved material in the ice also influences the amount and spectral composition of PAR (Legendre and Gosselin, 1991) and attenuation of UV radiation through the ice (Belzile et al., 2000). As a result there is very low irradiance beneath the ice and under-ice phytoplankton biomass are typically < 0.5 mg chl *a* m^{-3} (Conover et al., 1999).

Nutrient replenishment at the ice-water interface has been shown to occur during neap tides in Resolute, fuelling ice algae production that would otherwise be nutrient-limited (Cota et al., 1987). Current measurements in the same area suggest that internal waves could also support nutrient replenishment (Marsden et al., 1994). High phosphate and silicate concentrations in the Arctic Ocean surface layer waters exiting through Lancaster Sound and Jones Sounds (Jones and Coote, 1980), together with physical mixing processes bringing nutrients back to the ice interface most likely contribute to explain the high ice algae biomass observed in this region (max. bottom ice chl *a* of ca. 160 mg m^{-2}, Michel et al., 1996 and 250 mg m^{-2}, Smith et al., 1988).

In summary, for the eastern part of the NWP, Welch et al., (1992) suggest that four copepod species, *Pseudocalanus acuspes, Calanus hyperboreus, Calanus glacialis,* and *Metridia longa,* dominate the energy flow through herbivorous grazers and that arctic cod constitute a major trophic link within the marine ecosystem. Ice algae are believed to play a more important role than simply their primary production levels would suggest. Spatial heterogeneity in phytoplankton distributions results in "biological hot spots" where high primary production support abundant populations of marine birds and mammals. Inter-annual fluctuations in the abundance of certain species, e.g. *Pseudocalanus* spp. whose abundance can vary tenfold, are believed to have little influence on the channelling of energy through the food chain (Welch et al., 1992). Inter-annual fluctuations in other species however can strongly impact their predators. For example, polar bears depend almost exclusively on ringed seals for their diet (Stirling and Oritsland, 1995). Ringed seal abundance shows strong inter-annual variability (Stirling et al., 1982; Harwood and Stirling, 1992) because their abundance is linked to ice conditions (Kingsley et al., 1985), which consequently has a significant impact on the distribution, reproductive success, and survival of polar bears. Stirling (2002) describes the relationship between polar bears, seals and sea ice conditions in Amundsen Gulf and showed that heavy ice conditions in the mid-1970s and mid-1980s caused significant declines in productivity of ringed seals, which led to decreased populations of polar bears.

There are fewer observations for the western NWP and other regions of CAA. Beluga whales belonging to the western Beaufort Sea stock were observed in M'Clure Strait, Viscount Melville Sound, and Amundsen Gulf (Richard et al., 2001; Harwood and Smith, 2002). In summer hundreds, perhaps thousands, of male belugas from this stock travel to Viscount Melville Sound where they likely feed on arctic cod and benthic flat fish. Although this does not occur yearly it indicates Viscount Melville Sound may be another "hot spot". McLaughlin and Carmack (1998), using geochemical correlations, estimated new production in Viscount Melville Sound, a region mostly ice-covered, was ~ 40 gC m^{-2} yr^{-1}.

Although the CAA is believed to be a pristine and isolated environment, the atmospheric and oceanic transport of contaminants from southern industrial regions northward into the Arctic and subsequent entry into the marine food web is well documented (e.g. Muir et al., 1999; Fisk et al., 2003). Some contaminants, such as organochlorines and mercury, biomagnify in the marine food web, showing that more than just carbon and nutrients are transferred to upper trophic levels. Following the ban of organochlorines (OCs) in the US, Canada and Western Europe OC concentrations declined rapidly in the 1980s but in the 1990s the rate of decrease slowed or stopped in many species. The slowest decline is in beluga and polar bears and the most rapid is in ringed seals and seabirds. There are large regional differences in specific OC distributions. For example, hexachlorobenzene and hexachlorocyclohexane (HCH) concentrations in zooplankton were higher in Amundsen Gulf than in Barrow Strait, reflecting the effects of long-range atmospheric transport and geographic proximity to Asian sources, but toxaphene concentrations were higher in the east (Hoekstra et al., 2002). Total PCBs are the dominant OC contaminant in Arctic cod (Buckman et al., 2002) and also seabirds (Buckman et al., 2002). PCB concentrations in seals are lower in the west, and HCH concentrations are higher, reflecting trends at lower trophic levels. Generally, concentrations of PCBs, DDT, chlordane-related compounds and toxaphene in marine biota are slightly higher in the eastern Canadian Arctic, consistent with circumpolar trends. Only HCH concentrations are higher in the west. With the exception of HCH, OC levels in ringed seal, polar bears and seabirds are lower in the Canadian Arctic than in the European and Russian Arctic.

Mercury concentrations in Arctic biota have increased 10–17-fold from preindustrial time to the present but temporal trends are regional and species specific. For example mercury levels in Beaufort Sea beluga have increased 4-fold in the past ten years but only 2.5-fold in western Hudson Bay. Levels of mercury have slowly but steadily increased over the last thirty years in Lancaster Sound seabird eggs. Mercury concentrations in ringed seal livers from Pond Inlet have increased three-fold over the past thirty years but no significant increase was observed in Amundsen Gulf ringed seal livers. Cadmium concentrations also vary regionally and are higher in the east.

Tynan and deMaster (1997) discuss the effect of decreasing sea ice extent in the Arctic Ocean and its potential impact on marine mammals. Sea ice variations may alter the seasonal distributions, geographic ranges, patterns of migration, nutritional status, reproductive success, and ultimately the abundance and stock structure of some species. Seals may be more affected than others because of their dependence on the presence of sea ice. Likewise, species (e.g. arctic cod) that depend on the magnitude of productivity at ice-edges may be negatively affected and as the ice edge retreats northward beyond the shelf to locations over the slope and basin the benthos will be affected (see Loeng et al., in press). Macdonald et al (2003) indicate the importance of climate change on the distribution and abundance of different species and on contaminant burdens.

5. Summary and Outlook

The CAA has long been considered an oceanographic backwater by both observations and modellers, largely because of logistical difficulties, dynamical complexity, economic insignificance and a general lack of appreciation for its connectivity to the global system. It is now recognized, however, that the CAA plays a critical role in global climate, mainly through the export of ice and low salinity waters to the convective centers of the North Atlantic (see Weaver, et. al., 1999, for review). Abrupt changes in global climate during the last glacial epoch, such as Heinrich events, have been explained by releases of massive volumes of glacial ice from the CAA (see Bard, 2002 for synthesis). On decadal time scales, fluctuations in the export of fresh water from the Arctic Ocean may alter conditions downstream in the Labrador Sea and North Atlantic (Belkin et al., 1998; Proshutinsky et al., 2002) and as far south as the shelf waters of the Middle Atlantic Bight (Chapman and Beardsley, 1989). Pioneering work by Stigebrant (1981) and Rudels (1987) provide a framework for evaluating such fluxes. Recent models have demonstrated the climatic importance of freshwater transport from the Pacific to the Atlantic via Bering Strait (Shaffer and Bendtsen, 1994) and via the CAA (Goosse et al., 1997; Wadley and Bigg, 2002).

Processes within the CAA may impact global climate, but, global climate change may also alter conditions with the CAA: both aspects of this "two-way street" must be considered. The consequences of global climate evolution on ice and oceanographic conditions in the CAA under scenarios of greenhouse gas (GHG) warming are especially topical. Such changes will clearly impact transportation, resource development and human settlement. Most coupled air/ice/sea GCMs predict a warming of the CAA and a significant reduction in ice cover through the 21st century (cf. Arctic Climate Impact Assessment, 2004). However, much uncertainty exists. As noted by Melling (2002), a thinning and reduction in landfast ice in the CAA may simply allow a greater advection of multi-year floes from the Arctic Ocean into the CAA which would act as an impediment to marine transportation. Increased precipitation, also predicted under GHG warming scenarios, may lead to thicker snow-on-ice, and this insulating effect may impact ice thickness more than changes in temperature (Brown and Cote, 1992). Due to increased net precipitation in the high-latitudes, some models predict a counter-intuitive slowing-down of the thermohaline circulation, and a subsequent cooling of northern Europe (Ganopolski et al., 1998). Such uncertainty in both the sign and magnitude of change can only be reduced by better understanding and parameterization of the processes and feedbacks that govern the high-latitude disposition and export of freshwater components (Hopkins, 2001).

The ecological impacts of anticipated changes in the CAA deserve special attention. Carmack and McLaughlin (2001) argue that climate change in regions currently covered by ice will affect the structure of marine ecosystems through both bottom-up and top-down reactions. Bottom-up changes in trophic structure would be the result of changes in ice cover extent that affect underwater light, mixing and nutrient fluxes into the euphotic zone. For example, Brodeur et al. (1999) documented an immense increase in gelatinous zooplankton in the Bering Sea concomitant with extreme weather anomalies. Top-down changes would be the result of changes in the distribution and abundance of ice-dependant predators at various levels within the food- web and downward cascading of grazing effects. For example, Tynan and deMaster (1997) discuss how cod, seals, whales, walrus and bears may all be impacted by changes in ice cover.

One might hope that the passion of the 19th century explorers to map the NWP will be re-kindled in the curiosity of 21st century scientists to understand and model this complex and important system.

Acknowledgments

This work was supported by the Department of Fisheries and Oceans Canada, Government of Canada Climate Change Action Fund and Natural Sciences and Engineering Research Council of Canada. We thank Trish Kimber and Sarah

Zimmerman for their help in preparing the figures, Angus Pippy for the river flow data and Frederic Dupont and David Greenberg for providing Figure 7. We also extend our appreciation to Humfrey Melling for his many helpful suggestions during the preparation of this chapter.

Bibliography:

Aagaard, K. and E. C. Carmack, 1989. On the role of sea-ice and other freshwater in the Arctic Circulation. *J. Geophys. Res.*, **94,** 14,485–14,498.

Arctic Climate Impact Assessment: An Assessment of Consequences of Climate Variability and Change and the Effects of Increased UV in the Arctic Region, 2004. Arctic Council, University of Alaska, Fairbanks, in press.

Agnew, T. and A. Silas, 1995. Spring seasonal climate variability in the central Canadian Arctic Islands. *Annals of Glaciology,* **21,** 330–336.

Agnew, T. and S. Howell, 2003. The use of operational ice charts for evaluating passive microwave ice concentration data. *Atmosphere-Ocean,* **41,** 317–331.

Arfeuille, G., 2001. On the freshwater transport through the southwest Canadian Arctic Archipelago due to buoyancy and wind forcing. M. Sc. Thesis, University of Victoria, Victoria, British Columbia, Canada.

Bacle, J., E.C. Carmack and R.G. Ingram, 2002. Water column structure and circulation under the North Water during spring transition: April-July 1998. *Deep-Sea Res. II,* **49,** 4893–4908.

Bailey, W.G., 1957. Oceanographic features of the Canadian archipelago. *J. Fish. Res. Bd. Can.,* **14,** 731–769.

Barber, D. and J. Hanesiak, 2004. Meteorological forcing of sea ice concentrations in the southern Beaufort Sea over the period 1979 to 2000. *J. Geophys. Res.,* **109,** C06014, doi:10.1029/2003JC002027.

Bard, E., 2002. Climate shock: abrupt change over millennial time scales. *Physics Today,* **55,** 32–39.

Bedard, P., C. Hillaire and P. Pagé, 1981. ^{18}O modelling of freshwater inputs in Baffin Bay and Canadian Arctic coastal waters. *Nature,* **293,** 287–289.

Belkin, I.M., S. Levitus, J. Antonov, and S. Malmberg, 1998. "Great Salinity Anomalies" in the North Atlantic. *Progr Oceanogr.,* **41,** 1–68.

Bjornsson, H, L.A. Mysak and R. D. Brown, 1995. On the interannual variability of precipitation and runoff in the Mackenzie drainage basin. *Clim. Dyn.,* **12,** 67–76.

Belzile, C., S. C. Johannessen, M. Gosselin, S. Demers and W. L. Miller, 2000. Ultraviolet attenuation by dissolved and particulate constituents of first-year sea ice during late spring in an Arctic polynyas. *Limnol. Oceanogr.,* **45,** 1265–1273.

Berton, P., 1988. *The Arctic Grail: The quest for the North West Passage and the North Pole, 1818–1909.* McClelland and Stewart, Toronto, 672 pp.

Borstad, G.A. and J.F.R. Gower, 1984. Phytoplankton chlorophyll distribution in the eastern Canadian Arctic. *Arctic,* **37,** 224–233.

Bradstreet, M.S.W.,1979. Thick-billed murres and black guillemots in the Barrow Strait area, N.W.T., during spring: distribution and habitat use. *Can. J. Zoo.,* **57,** 1789–1802

Bradstreet, M.S.W., 1980. Thick-billed murres and black guillemots in the Barrow Strait area, N.W.T., during spring: diets and food availability along ice edges. *Can. J. Zoo.,* **58,** 2120–2140

Bradstreet, M.S.W. and W. Cross, 1982. Trophic relationships at High Arctic ice edges. *Arctic,* **35,** 1–12.

Brown, R.D. and P.Cote, 1992. Interannual variability of land-fast ice thickness in the Canadian high Arctic, 1950–1989. *Arctic,* **45,** 272–284.

Brodeur, R.D., C.E. Mills, J.E. Overland, G.E. Walters and J.D. Schumacher, 1999. Evidence for a substantial increase in gelatinous zooplankton in the Bering Sea, with possible links to climate change. *Fish. Oceanogr.*, **8,** 296–306.

Buckman, A.H., R.J. Norstrom, K.A. Hobson, N.J. Karnovsky, J. Duffe and A.T.Fisk, 2002. Variables influencing persistent organochlorine pollutants in seven species of Arctic seabirds from northern Baffin Bay. *Arch. Environ. Toxicol. Chem.*, submitted.

Canadian Ice Service, 2002. *Sea Ice Climatic Atlas, Northern Canadian Waters 1971–2000,* Minister of Public Works and Government Services of Canada, Ottawa, 48 p. and 200 figs.

Carmack, E. C., 1986. Circulation and mixing in ice-covered waters. In *The Geophysics of Sea Ice,* N. Untersteiner, ed., Plenum, New York, pp. 641–712.

Carmack, E.C .. , 2000. The Arctic Ocean's freshwater budget: sources, storage and export. In *The Freshwater Budget of the Arctic Ocean,* E.L. Lewis, E.P. Jones, P. Lemke, T.D. Prowse and P. Wadhams, eds., Kluwer Academic Publishers, Dordrecht., pp. 91–126.

Carmack, E.C. and F.A. McLaughlin, 2001. Arctic Ocean change and consequences to biodiversity: a perspective on linkage and scale. *Mem. Nat. Inst. Polar Res.,* **54,** 365–375.

Cenedese, C. and J.A. Whitehead, 2000. Eddy Shedding from a Boundary Current around a Cape over a Sloping Bottom. *J. Phys. Oceanogr.,* **30,** 1514–1531.

Chapman, D. C. and R. C. Beardsley, 1989. On the Origin of Shelf Water in the Middle Atlantic Bight. *J. Phys. Oceanogr., **19,*** 384–391.

Coachman, L.K. and C.A. Barnes, 1961. The contribution of Bering Sea water to the Arctic Ocean. *Arctic,* **14,** 147–161.

Codispoti, L.A. and T.G. Owens, 1975. Nutrient transport through Lancaster Sound in relation to the Arctic Oceans reactive silicate budget and the outflow of Bering Strait waters. *Limnol. Oceanogr.,* **20,** 115–119.

Collin, A.E., 1962. Oceanographic observations in the Canadian Arctic and the adjacent Arctic Ocean. *Arctic,* **15,** 194–201.

Conover, R.J., A.W. Hermann, S.J. Prinsenberg and L.R. Harris, 1986. Distribution of and feeding by the copepod *Pseudocalanus* under fast ice during the arctic spring. *Science,* **232,** 1245–1247.

Conover, R.J., N. Mumm, P. Bruecker, S. MacKenzie, 1999. Sources for urea in Arctic seas: seasonal fast ice? *Mar. Ecol. Prog. Ser.,* **179,** 55–69.

Comiso, J., 2003. Large scale characteristics and variability of the global sea ice cover. In *Sea Ice,* D.N. Thomas and G.S. Dieckmann, eds.Blackwell, London, pp. 112–142.

Cota, G.F. and E.P.W. Horne, 1989. Physical control of arctic ice algal production, *Mar. Ecol. Prog. Ser.,* **52,** 111–121.

Cota, G.F., S.J. Prinsenberg, E.B. Bennet, J.W. Loder, M.R. Lewis, J.L. Anning, N.H.F. Watson, and L.R. Harris, 1987. Nutrient fluxes during extended blooms of arctic ice algae. *J. Geophys. Res.,* **92,** 1951–1962.

Delgado, J. P., 1999. *Across the top of the world: the quest for the Northwest Passage.* Douglas and McIntyre,Vancouver and Toronto, 228 pp.

Deming, J.W., L. Fortier and M Fukuchi, 2002. The international North Water polynya study (NOW): a brief overview. *Deep-Sea Res.* II, **49,** 4887–4892.

Dyke, A.S., J. Hooper and J.M. Savelle,1996. A history of sea ice in the Canadian Arctic Archipelago based on post-glacial remains of the bowhead whale (*Balaena mysticetus*). *Arctic,* **49,** 235–255.

Dyke, A.S. and J.M. Savelle, 2001. Holocene history of the Bering Sea bowhead whale (*Balaena mysticetus)* in its Beaufort Sea summer grounds off southwestern Victoria Island, western Canadian Arctic. *Quant. Res.,* **55,** 371–379.

Dunbar, M.,1969. The geographical position of the North Water. *Arctic,* **22,** 438–441.

Finlay, K.J., Davis, R.A. and H. Silverman, 1980. Aspects of the narwhal hunt in the eastern Canadian High Arctic. *Report of the International Whaling Commission,* **30,** 459–464.

Fisk, A.T., K. Hobbs and D.C.G. Muir, eds, 2003. *Canadian Arctic Contaminants Assessment Report II: Contaminant levels, trends and effects in the biological environment.* Indian and Northern Affairs Canada, Ottawa, 175 pp.

Fissel, D.B., 1982. Tidal currents and inertial oscillations in northwestern Baffin Bay, *Arctic,* **35,** 201–210.

Fissel, D. B., D. D. Lemon, H. Melling, and R. A. Lake, 1988. Non-tidal flows in the Northwest Passage, *Can. Tech. Rep. Hydrogr. Ocean Sci.,* **98,** Inst. of Ocean Sci., Sidney, B. C., Canada, 143 pp.

Furevik, T. and A. Foldvik, 1996. Stability at critical M2 latitude in the Barents Sea. *J. Geophys. Res.,* **101,** 8823–8838.

Fong, D. A. and W. R. Geyer, 2001. Response of a river plume during an upwelling favorable wind event. *J. Geophys. Res., **106,*** 1067–1084.

Fortier, M., L. Fortier, C. Michel and L. Legendre, 2002. Climatic and biological forcing of the vertical flux of biogenic particles under seasonal Arctic sea ice. *Mar. Prog. Ecol. Ser.,* **225,** 1–16.

Ganopolski, A., S. Rahmstorf, V. Petoukhov and M. Claussen, 1998. Simulation of modern and glacial climates with a coupled global model of intermediate complexity. *Nature,* **391,** 351–356.

Goosse, H., T. Fichefet and J.-M. Campin, 1997. The effects of the water flow through the Canadian Archipelago in a global ice-ocean model. *Geophys. Res. Lett.,* **24,** 1507–1510.

Griffiths, R.W. and Linden, P.F., 1981. The stability of buoyancy driven currents. *Dynam. Atmos. Oceans,* **5,** 281–306.

Harwood, L.A. and I. Stirling, 1992. Distribution of ringed seals in the southeastern Beaufort Sea during late summer. *Can. J. Zool.,* **70,** 891–900.

Harwood, L. and T. Smith, 2002. Whales of the Inuvialuit settlement region in Canada's western Arctic: An overview and outlook. *Arctic,* **55** (suppl. 1), 77–93.

Head, E.J.H., A. Bedo, and L.R. Harris, 1988. Grazing, defecation and excretion rates of copepods from some inter-island channels of the Canadian Arctic archipelago. *Mar. Biol.,* **99,** 333–340.

Helfrich, K. R., 1995. Time-Dependent Two-Layer Hydraulic Exchange Flows. *J.Phys. Oceanogr.,* **25,** 359–373.

Hoekstra , P.F., T.M. O'Hara, C. Teixeria, S. Backus, A.T. Fisk and D.C.G. Muir, 2002. Spatial trends and bioaccumulation of organochlorine pollutants in marine zooplankton from the Alaskan and western Canadian Arctic. *Environ. Toxicol. Chem.,* **21,** 575–583.

Hopkins, T.S., 2001. Thermohaline feedback loops and the Natural Capital. *Sci. Mar.,* **65,** 231–236.

Hunkins K. and J.A. Whitehead, 1992. Laboratory simulation of exchange flow through Fram Strait. *J. Geophys. Res.,* **97,** 11,299–11,321.

Ingram, R.G. and P. Larouche, 1987. Variability of an under-ice river plume in Hudson Bay. *J. Geophys. Res.,* 92, 9541–9547.

Ingram, R.G. and S. Prinsenberg, 1998. Coastal oceanography of Hudson Bay and surrounding eastern Canadian Arctic waters. In *The Sea,* Vol. 11: The Global Coastal Ocean, Regional studies and syntheses, A.R. Robinson and K.H. Brink, eds. John Wiley and Sons, Inc., New York, pp 835–862.

Ingram, R.G., J. Bacle, D.G. Barber, Y. Gratton and H. Melling, 2002. An overview of physical processes in the North Water. *Deep Sea Res. II,* **49,** 4893–4906.

Jacobsson, M., 2002. Hypsometry of the Arctic Ocean and its constituent seas. *Geochem. Geophys. Geosys.,* 3, 10.1029/2001GC000302.

Johnson, G.C and D.R. Ohlson, 1994. Frictionally modified rotating hydraulic channel exchange and ocean outflows. *J. Phys. Oceanogr.,* **24,** 66–78.

Jones, E.P. and A.R. Coote, 1980. Nutrient distributions in the Canadian Archipelago: Indicators of summer water mass and flow characteristics. *Can. J. Fish. Aquat. Sci,* **37,** 589–599.

Jones, E.P., D. Dryssen and A.R. Coote, 1984. Nutrient regeneration in deep Baffin Bay with consequences for measurements of the conservative tracer NO and fossil fuel CO_2 in the oceans. *Can. J. Fish Aquat. Sci.,* **41,** 30–35.

Jones, E.P., J.H. Swift, L.G. Anderson, M. Lipizer, G. Civitarese, K.K. Falkner, G. Kattner and F. McLaughlin, 2003. Tracing Pacific water in the North Atlantic Ocean, *J. Geophys. Res.,* **108,** 13:1–13:10.

Kingsley, M., I. Stirling and W. Calvert, 1985. The distribution and abundance of seals in the Canadian high Arctic. *Can. J. Fish. Aquat. Sci.,* **42,** 1189–1210.

Klinger, B.A., 1994a. Inviscid current separation from rounded capes. *J. Phys.Oceanogr.,* **24,** 1805–1811.

Klinger, B.A., 1994b. Baroclinic eddy generation at a sharp corner in a rotating system. *J. Geophys. Res.,* **99,** 12515–12531.

Kowalik, Z. and A.Y. Proshutinsky, 1993. Diurnal tides in the Arctic Ocean. *J. Geophys. Res,* **98,** 16449–16468.

Kulikov, E.A., A.B. Rabinovich and E.C. Carmack, 2004. Barotropic and baroclinic tidal currents on the Mackenzie shelf break in the southeastern Beaufort Sea. *J. Geophys. Res.,* in press.

Laurion, I., S. Demers and A.F. Vézina, 1995. The microbial food web associated with the ice algal assemblage: biomass and bacterivory of nanoflagellate protozoans in Resolute Passage (High Canadian Arctic). *Mar. Ecol.Prog. Ser.,* **120,** 77–87.

LeBlond, P.H., 1980. On the surface circulation in some channels of the Canadian Arctic Archipelago. *Arctic,* **33,** 189–197.

Lemon, D.D. and D.B. Fissel., 1982. Seasonal variations in currents and water properties in northwestern Baffin Bay, 1978 and 1979. *Arctic,* **35,** 211–218.

Legendre, L. and M. Gosselin, 1991. In situ spectroradiometric estimation of microalgal biomass in first-year sea ice. *Polar Biol.,* **11,** 113–115.

Loeng, H., K. Brander, E. Carmack, S. Denisenko, K. Drinkwater, B. Hansen, K. Kovacs, P. Livingston, F. McLaughlin and E. Sakshaug, 2004. Marine Systems. In *Arctic Climate Impact Assessment: An Assessment of Consequences of Climate Variability and Change and the Effects of Increased UV in the Arctic Region.* Arctic Council, University of Alaska, Fairbanks, in press.

Lyard, F.H., 1997. The tides of the Arctic Ocean from a finite element model. *J. Geophys. Res.,* **102,** 15611–15638.

Macdonald, R., T. Harner and J. Fyfe, 2003. The interaction of climate change with contaminant pathways to and within the Canadian Arctic. *Canadian Arctic Contaminants Assessment Report II,* Physical Environment, T. Bidleman, R.W. Macdonald and J. Snow, eds. Indian and Northern Affairs Canada, Ottawa, p. 225–279.

Marsden, R.F., R.G. Ingram and L. Legendre, 1994. Currents under land-fast ice in the Canadian Archipelago Part 2: Vertical mixing. *J. Mar. Res.,* **52:** 1037–1049.

Maykut, G.A., 1985. The ice environment, In: Horner, R.A. (ed.) *Sea ice biota.* CRC Press, Boca Raton, Florida. pp. 21–82.

McGhee, R., 1966. *Ancient peoples of the Arctic.* UBC Press, in association with the Canadian Museum of Civilization, Vancouver.

McLaren, P.L. 1982. Spring migration and habitat use by seabirds in eastern Lancaster Sound and western Baffin Bay. *Arctic,* **35,** 88–111.

McLaughlin, F.A. and E.C. Carmack, 1998. Sub-basin circulation in the Canadian Arctic Archipelago. *EOS,* Trans., **79,** F399.

McLaughlin, F. A., and E. C. Carmack, in prep. Flow regimes in the Canadian Arctic Archipelago: Evidence from geochemical tracers.

Melling, H., 2000. Exchanges of freshwater through the shallow straits of the North American Arctic. In *The Freshwater Budget of the Arctic Ocean,* E.L. Lewis, E.P. Jones, P. Lemke, T.D. Prowse and P. Wadhams, eds. Kluwer Academic Publishers, Dordrecht., 479–502.

Melling, H., 2002. Sea ice of the northern Canadian Arctic Archipelago. *J. Geophys Res.,* **107,** 3181, 10.1029/2001JC001102.

Melling, H., R.A. Lake, D.R. Topham and D.B. Fissel, 1984. Oceanic thermal structure in the western Canadian Arctic. *Cont. Shelf Res.,* **3,** 233–258.

Michel C., M. Gosselin and C. Nozais, 2002. Preferential sinking export of biogenic silica during the spring-summer in the North Water Polynya (northern Baffin Bay): temperature or biological control? *J. Geophys. Res.,* **107,** 10.1029/2000JC000408

Michel, C., Legendre, L., R. G. Ingram, M. Gosselin, and M. Levasseur, 1996. Carbon budget of sea-ice algae in spring: Evidence of a significant transfer to zooplankton grazers. *J. Geophys. Res.,* **101,** 18,345–18,360.

Muench. R.D., 1971. The physical oceanography of the northern Baffin Bay Region. *North Water Project Scientific Report No.* **1,** Arctic Institute of North America, University of Calgary, Calgary, Alberta, Canada., 150 pp.

Muir, D., B. Braune, B. DeMarch, R. Norstrom, R. Wagemann, L. Lockart, B. Hargrave, D. Bright, R. Addison, J. Payne and K. Reimer, 1999. Spatial and temporal trends and effects of contaminants in the Canadian Arctic marine ecosystem: a review. *Sci. Tot. Envir.,* **230,** 83–144.

Perovich, D.K., C.S. Roesler and W.S. Pegau, 1998. Variability in Arctic sea ice optical properties. *J. Geophys. Res.,* **103,** 1193–1208.

Prinsenberg, S.J., 1984. Freshwater contents and heat budgets of James Bay and Hudson Bay. *Cont. Shelf. Res,* **3,** 191–200.

Prinsenberg, S.J., 1988. Damping and phase advance of the tide in western Hudson Bay by the annual ice cover. *J. Phys Oceanogr.,* **18,** 1744–1751.

Prinsenberg, S.J. and E.B. Bennett, 1987. Mixing and transports in Barrow Strait, the central part of the Northwest Passage. *Cont. Shelf Res.,* **7,** 913–935.

Prinsenberg, S.J. and E.B. Bennett, 1989a. Transport between Peel Sound and Barrow Strait in the Canadian Arctic. *Cont. Shelf Res.,* **9,** 427–444.

Prinsenberg, S.J. and E.B. Bennett, 1989b. Vertical variations of tidal currents in shallow land fast ice-covered regions. *J. Phys. Oceanogr.,* **19,** 1268–1278.

Prinsenberg, S. and J. Hamilton, (in press). Monitoring the volume, freshwater and heat fluxes passing through Lancaster Sound in the Canadian Arctic Archipelago. *Atmophere-Ocean.*

Proshutinsky, A., R.H. Bourke and F.A. McLaughlin, 2002. The role of the Beaufort Gyre in Arctic climate variability: seasonal to decadal time scales. *Geophys. Res. Letts.,* **29,** doi:10.1029/2002GL015847.

Prowse, T.D. and P.O. Flegg, 2000. The magnitude of river flow to the Arctic Ocean: dependence on contributing area. *Hydrolog. Proc.,* **14,** 3185–3188.

Richard, P.R., A.R. Martin and J.R. Orr, 2001. Summer and autumn movements of belugas of the eastern Beaufort Sea stock. *Arctic,* **54,** 223–236.

Riemann, F. and T. Sime-Ngando, 1997. Note on sea-ice nematodes (Monhysteroida) from Resolute Passage, Canadian High Arctic. *Polar Biol.,* **18,** 70–75

Rudels, B., 1986. The outflow of polar water through the Arctic Archipelago and the oceanographic conditions in Baffin Bay. *Polar Res.,* **4,** 161–180.

Schledermann, P., 1990. *Crossroads to Greenland,* Arctic Institute of North America, Calgary, 364 p.

Shaffer, G. and J. Bendtsen, 1994. Role of the Bering Strait in controlling North Atlantic ocean circulation and climate. *Nature,* **367,** 354–357.

Simpson, J.H. and J.R. Hunter, 1974. Fronts in the Irish Sea. *Nature,* **250,** 404–406.

Smith, R.E.H., M. Gosselin, S. Kudoh, B. Robineau, and S. Taguchi, 1997. DOC and its relationship to algae in bottom ice communities. *J. Mar. Syst.,* **11,** 71–80.

Smith, R.E.H., J. Anning, P. Clement, and G. Cota, 1988. Abundance and production of ice algae in Resolute Passage, Canadian Arctic. *Mar. Ecol. Prog. Ser.*, **48,** 251–263.

Smith, S.D., R.D. Muench and C.H. Pease, 1990. Polynyas and leads: an overview of physical processes and environment. *J.Geophys. Res.*, **95,** 9461, 9479.

Stigebrandt, A., 1981. A model for the thickness and salinity of the upper layer in the Arctic Ocean and the relationship between the ice thickness and some external parameters. *J. Phys. Oceanogr.*, **11,** 1407–1422.

Stigebrandt, A., 1984. The North Pacific: a global-scale estuary. *J. Phys. Oceanogr.*, ***14,*** 464–470.

Stirling, I, 1980. The biological importance of polynyas in the Canadian Arctic. *Arctic,* **33,** 303–315.

Stirling, I., M. Kingsley and W. Calvert, 1982. The distribution and abundance of seals in the eastern Beaufort Sea, 1974–1979. *Can. Wildlife Service Occasional Paper* **47,** 25p.

Stirling, I. and N.A Oritsland, 1995. Relationships between estimates of ringed seal (*Phoca hispida*) and polar bear (*Ursus maritimus*) populations in the Canadian Arctic. *Can. J. Fish. Aquat. Sci.*, **52,** 2594–2612.

Stirling, I., 2002. Polar bears and seals in the eastern Beaufort Sea and Amundsen Gulf: A synthesis of population trends and ecological relationships over three decades. *Arctic* **55,** (suppl. 1), 59–76.

Stronach, J.A., J.A. Helbig, S.S. Salvador, H. Melling and R.A. Lake, 1987. Tidal elevations and tidal currents in the Northwest Passage. *Can. Tech. Rep. Hydrog. Ocean Sci.*, **97,** 329 pp.

Tan, F.C. and P.M. Strain, 1980. The distribution of sea ice meltwater in the eastern Canadian Arctic. *J. Geophys. Res.*, ***85,*** 1925–1932.

Thomson, D.H., 1982. Marine benthos in the Eastern Canadian High Arctic: miltivariate analyses of standing crop and community structure. *Arctic,* **35,** 61–74.

Topham, D.R., R.G. Perkin, S.D. Smith, R.J. Anderson and G. Den Hartog, 1983. An investigation of a polynya in the Canadian archipelago, 1. Introduction and Oceanography. *J. Geophy.Res.*, ***88,*** 2888–2899.

Toulany, B. and C. Garrett, 1984. Geostrophic control of fluctuating barotropic flow through straits. *J. Phys. Oceanogr.*, ***14****(4),* 649–655.

Tremblay, J-E, Y. Gratton, J. Fauchot, and N.M. Price, 2002a. Climatic and oceanic forcing of new, net, and diatom production in the North Water. *Deep Sea Res. II,* **49,** 4927–4946.

Tremblay J-E., Y. Gratton, E.C. Carmack, C.D. Payne and N.M. Price, 2002b. Impact of the large-scale Arctic circulation and the North Water Polynya on nutrient inventories in Baffin Bay. *J. Geophys. Res.*, **107,** 10.1029/2000JC000595.

Tynan, C.T. and D.P. deMaster, 1997. Observations and predictions of Arctic climate change: potential effects on marine mammals. *Arctic,* **50,** 308–322.

Vuglinsky, V.S., 1997. River inflow to the Arctic Ocean: Conditions of formation, time variability and forecasts. In *Polar Processes and Climate,* ACSYS, Orcas Island, Washington, pp. 275–276.

Wadley, M.R. and G.R. Bigg, 2002. Impact of flow through the Canadian Archipelago and Bering Strait on the North Atlantic and Arctic circulation: an ocean modelling study. *Q. J. R. Meterol. Soc.*, **128,** 2187–2203.

Walker, E.R., 1977. Aspects of the oceanography in the Archipelago, Institute of Ocean Sciences, Sidney, B.C., unpublished data report, 186 p.

Weaver, A.J., C.M. Bitz, A.F. Fanning and M.M. Holland, 1999. Thermohaline circulation: High latitude phenomena and the difference between the Pacific and Atlantic. *Annu. Rev. Earth Planet. Sci.*, **27,** 231–285.

Welch, H.E. and M.A.Bergmann, 1989. Seasonal development of ice algae and its prediction from environmental factors near Resolute, NWT, Canada. *J. Fish. Aqua. Sci.*, **46,** 1793–1804.

Welch, H.E., M.A. Bergmann, T.D. Siferd, K.A. Martin, M.F. Curtis, R.E. Crawford, R.J. Conover, and H. Hop, 1992. Energy flow through the marine ecosystem of the Lancaster Sound region. Arctic Canada, *Arctic,* **45,** 343–357

Whitehead, J.A.,1998. Topographic control of oceanic flows in deep passages and straits. *Reviews Geophys.,* **36**(3), 423–440.

Williams, W.J., 2003. Idealized modeling of seasonal variation in the Alaska Coastal Current. Ph.D. thesis, University of Alaska, Fairbanks, Alaska.

Woo, M.K., Marsh, P. and Pomeroy, J.W., 2000. Snow, frozen soils and permafrost hydrology. *Hydrolog. Proc.,* **14,** 1591–1611.

Yankovsky A.E. and D.C. Chapman, 1997. A simple theory for the fate of Buoyant Coastal Discharges. *J. Phys.Oceanogr.,* **27,** 1386–1401.

Yankovsky A.E., 2000. The cyclonic turning and propagation of buoyant coastal discharge along the shelf. *J. Mar.Res.,* **58,** 585–607.

Zishka, K.M. and P.J. Smith, 1980. The climatology of cyclones and anticyclones over North America and surrounding ocean environs for January and July, 1950–70. *Monthly Weather Review,* **108,** 387–401.

Chapter 32. THE PHYSICAL, SEDIMENTARY AND ECOLOGICAL STRUCTURE AND VARIABILITY OF SHELF AREAS IN THE MEDITERRANEAN SEA (27,S)

N. PINARDI AND M. ZAVATARELLI

University of Bologna, Corso di Scienze Ambientali, Ravenna, Italy

E. ARNERI

ISMAR-CNR, Ancona, Italy

A. CRISE

Istituto nazionale di oceanografia e geofisica sperimentale, Trieste, Italy

M. RAVAIOLI

ISMAR-CNR, Bologna, Italy

Contents

1. Introduction

The article reviews the most recent findings on the interdisciplinary characteristics of the Mediterranean Sea shelf areas, from the Northern to the southern shelf regions. First of all a classification of the shelves is carried out in terms of shelf extension, sedimentary and physical flow field properties. Secondly, the ecosystem functioning in the pelagic compartment is described as a function of the hydrodynamic regimes, as well as in relationship to land derived river inputs, if present.

The Sea, Volume 14, edited by Allan R. Robinson and Kenneth H. Brink
ISBN 0–674–01527–4 ©2005 by the President and Fellows of Harvard College

The shelf ecosystem is analysed up to the level of fish stocks and recruitment, making connections with lower trophic level dynamics and physical conditions.

The aim is to start a phenomenological and methodological synthesis toward the understanding of the coupled physical, sedimentary and ecological processes in the Mediterranean shelf areas. In order to do so, the physical, sedimentary and ecological subsystems are reviewed for several shelf system study cases. The main aim is to discuss in these subsystems the coupling between physical-chemical-biological and sedimentary processes at the level of ecosystem functionality in the Mediterranean Sea coastal areas. The main time scale of interest is the seasonal that captures the largest changes in amplitude for all the system state variables.

Six different shelf regimes are chosen to describe the characteristics of the processes present in the Mediterranean shelves. We selected two northern shelves, the Gulf of Lions and the Northern Adriatic Sea and three in the south, the Algerian, Egyptian and Israeli shelves. In addition, we describe the Sicily Strait that represents an extended shelf area without any relevant river input and with distinct characteristics from the other sites since it is a large channel area that allows the exchange of water masses between the Eastern and Western Mediterranean sub-basins.

2. The shelf scales of the Mediterranean

The shelf areas of the world ocean as well as the Mediterranean are characterized first of all by their morphological extension, the connection with the land derived inputs and the open ocean currents. The land-derived inputs are determined by the flow rate or river runoff that is strongly forced by precipitation. Thus the shelf extension and the precipitation regimes are the background environmental factors that will allow a first classification of the Mediterranean shelf areas. In the following we will describe the shelf extension and the hydrological cycle of the Mediterranean Sea with the aim of classifying the main characteristics of the Mediterranean shelf areas on the basis of these two basic environmental field variables.

2.1 Shelves of the basin

The Mediterranean shelf areas are represented by a bathymetric chart in Fig. 1a. In this picture the 200 meters isobath has been chosen to define the seaward boundary of the continental shelf. The geographical names are shown in Fig.1b for future reference. Analysing Fig. 1a the impression is that the Mediterranean shelf areas are mainly narrow bands around deep basins. However, Manzella (1992) reports a shelf relative extension of 20% in Mediterranean against 7,6% for the World Ocean which is confirmed by our new computations. Thus shelf dynamics is relatively more important in the Mediterranean than in the rest of the world ocean.

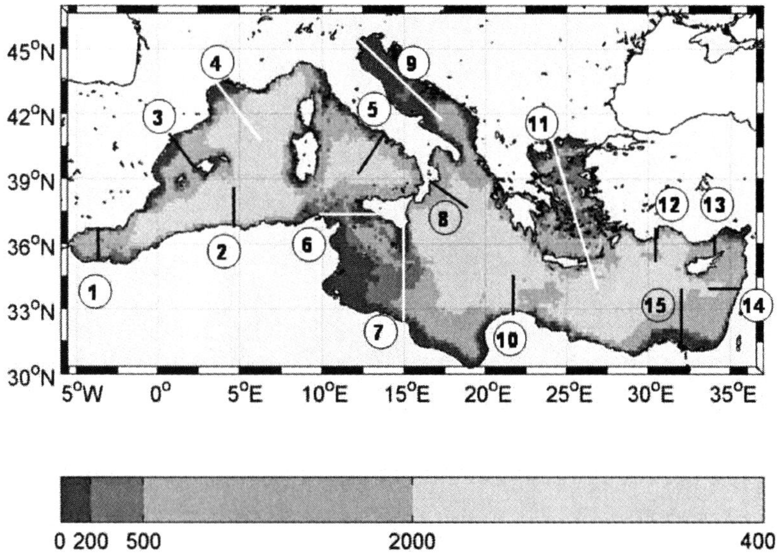

Figure 32.1a Shelf areas are indicated by different grey colors indicated in the palette. The numbers indicate 15 transects that are used to compute the topographic slopes in Table 1.

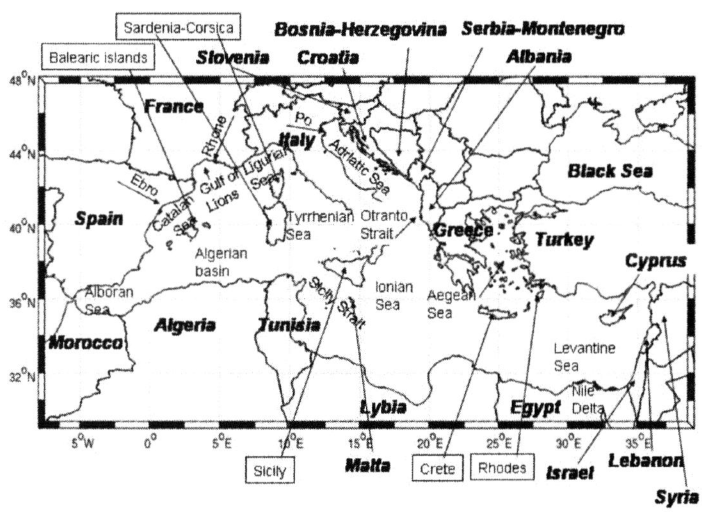

Figure 32.1b The coastlines and political geography of the Mediterranean Sea together with the major Sea names.

Taking as a reference Fig. 1a, it is evident that the major differences in shelf extension are between the northern and southern shores. The narrow and steep shelves of the Moroccan, Algerian and Libyan coasts contrast with the relatively extended shelves of the northern regions. There are exceptions to this north-south rule: for the northern coasts, narrow shelves are found along part of the Turkish coasts, in the Aegean Sea, in the Ligurian and the northern Alboran Sea. In the southern coasts, extended shelves are found in the Tunisian shelf and near the Nile Delta. However, the Tunisian shelves are part of the Sicily Strait, which is an exception to many rules, as we will discuss later. Other very narrow shelves are along the eastern sides of the Mediterranean sub-basins, such as the Israeli and Syrian coasts, as well as the eastern part of the Aegean, the Adriatic Sea and the Tyrrhenian Sea.

The island's shelves are exceptions to a north-south rule. Most of them, are connected to the mainland by the 500 meters isobath. The islands shelves are different mainly between the western and eastern Mediterranean sub-basins. Eastern islands are dominated by narrow shelves (Crete and Cyprus) while more extended shelves are found in the western Mediterranean islands (Mallorca, Corsica and Sardinia). Moreover for the island of Sardinia, shelves are more extended on the western than the eastern side.

We continue our classification of the shelf areas by looking at the different topographic gradients of the shelves. We calculated such gradients along fifteen transects indicated in Fig. 1a and results are presented in Table 1.

TABLE 32.1

The topographic gradients along the 15 transects depicted in Fig. 1a in (km/km) and the approximate distance from the coast of the sea-ward limit of the continental shelf, chosen to be the 200 meters contour isobath. Some transects connect two continental boundaries and then two different shelf extensions and gradients are calculated and the transect number is repeated.

Transect name	Shelf gradient (km/km)	Shelf extension from coasts in km	Comments
1-Morocco	1.7E-01	20-30	
1-Spain	4.2E-02	10-20	
2-Algeria	5.2E-01	10-20	
3-Spain	6.2E-02	50-60	
3-Mallorca	5.4E-02	10-20	
4-France	7.4E-02	70-80	
5-Italy	1.0E-01	10-20	
6-Tunisia	4.7E-02	110-120	This shelf area contains a 'deep pit'
6-Italy	4.9E-02	30-40	
7-Libya	1.1E-01	10-20	
7-Italy	1.5E-02	140-150	
8-Italy	1.7E-01	10-20	
9-Italy	7.0E-03	330-340	This shelf area is considered without the middle Adriatic
9-Italy	1.1E-02	520-530	This shelf area contains the pits of the middle Adriatic
10-Libya/Egypt	3.2E-01	10-20	
11-Greece	8.0E-02	40-50	
12-Turkey	1.8E-01	10-20	
13-Turkey	1.8E-01	10-20	
13-Cyprus	not determined	< 10	
14-Israel	not determined	< 10	
15-Egypt	4.7E-02	70-80	

The gradient values are calculated with a grid spacing of 10 km along the transects and then an average value is obtained for each shelf transect. Most regions of Table 1 have shelves extending only for 10–30 km from land (12 shelves out of 20 contained in the 15 transects) with gradients exceeding 0.01 (km/km). The wide shelf areas extend from 30 to 150 km offshore and they are all directly connected to major river runoffs (Spanish-Catalan coast, the Gulf of Lions region, the Adriatic, the Northern Aegean Sea, the Nile coastal area) except for the Sicily Strait. The Adriatic Sea is an exception since its shelf can be considered to extend for about 500 km along the longitudinal axis of the basin with an interruption due to the presence of the Middle Adriatic depressions (about 250 meters deep and located in Fig. 2 at about 320 km from the northern end of the Adriatic). The shelf break practically occurs between the Middle and the Southern Adriatic, as we will see later.

In Fig. 2 we show some of the transects of Fig. 1a that are chosen as study cases for the present investigation. The narrow shelves of the Israeli and Algerian coasts are similar in shape to each other (for this reason, we have shown only transect 2, the Algerian transect), while the extended shelves are different and supposedly they are affected by very different physical, sedimentological and ecosystem processes.

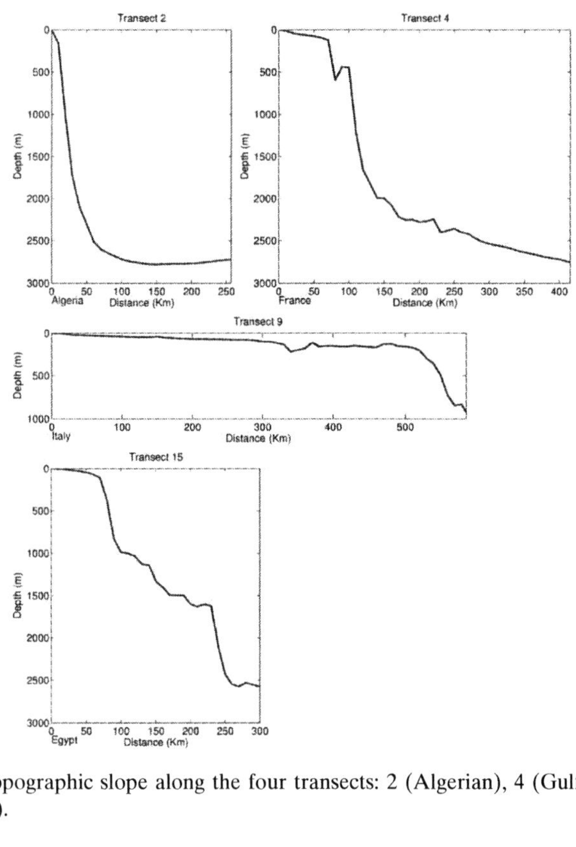

Figure 32.2 The topographic slope along the four transects: 2 (Algerian), 4 (Gulf of Lions), 9 (Adriatic) and 15 (Egypt).

The concept of narrow shelves has been recently reviewed by Sanchez-Arcilla and Simpson (2002). They postulate that ‘narrow shelves may be regarded as regions constrained by two “boundary layers” (one in the near-shore and the other on the slope) that merge and control processes across the continental shelf’. They also add that ‘in narrow shelves, therefore, the near-shore and slope processes effectively couple these regions to the shelf dynamics, while in the wide shelf areas the three areas will behave in a more uncoupled manner’. Sanchez-Arcilla and Simmons (2002) note that the nonlinear coupling between near-shore and slope processes can be large when the local Rossby number becomes larger than a threshold value, let’s say:

$$Ro \quad \frac{U}{f_o L} \quad 0.1$$

where L is the shelf width, f_o is the Coriolis parameter and U is the along slope current amplitude.

Supposing that everywhere in the Mediterranean the along slope currents are of the order of $20 \quad 30 \ \frac{cm}{s}$, $f_o \quad 1.10 \ ^4 s \ ^1$, the critical shelf width for strong nonlinear coupling between near-shore and slope processes is:

$$L \quad 10\frac{U}{f_o} \quad 20 \quad 30 \quad km \tag{1}$$

Thus all the narrow shelves described in Table 1 with extensions less than 20–30 km are narrow from both a morphological and a dynamical point of view. Sanchez-Arcilla and Simpson (2002) classified the Ebro river shelf area as a mixture of narrow and extended shelves. Our transect 3, shown in Fig. 1a and Table 1, falls between the two chosen by Sanchez-Arcilla and Simpson (2002) and it is defined as well as an ‘extended shelf area’.

To conclude this section we show in Fig. 3 the Sicily Strait transects 6 and 7 since they are very peculiar. The westernmost transect contains two deep troughs that allow the intermediate waters of the Eastern Mediterranean to propagate westward. In addition, on the Tunisian side of the Strait the shelf break is interrupted by a distinct trough that is occupied by the current flowing eastward, carrying waters of Atlantic origin. Thus the continental shelf areas of Tunisia and Sicily in the transect 6 are well separated and different in character event tough they are both extended. The easternmost transect of Fig. 3 shows the steep continental shelf break and narrow shelf on the Libyan side, which partially confirms the rule that southern shelves are steeper than northern ones also in the Sicily Strait. The Italian side of the same transect has an extended shelf area partially connected to the Malta island shelf. In conclusion, the Sicily Strait offers a complex shelf area scenario, with the continental shelf interrupted by deep troughs and a unique, very extended shelf area on the Tunisian side of the Strait.

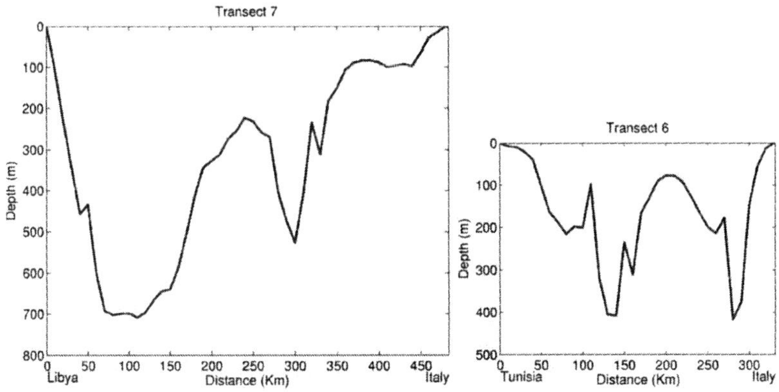

Figure 32.3 The topography along transects 7 (left panel) and 6 (right panel) delimiting the Sicily Strait.

2.2 Hydrological cycle of the basin

As noted before, most of the extended shelf areas have relevant river runoff inputs. Thus, to initially characterize the shelf areas, we need to concentrate on the hydrological cycle of the Mediterranean basin. The difference between the northern and southern shelf characteristics will be expressed here in terms of precipitation regimes over the Mediterranean wide area. The precipitation regime over land in turn determines the river runoff and by consequence the shelf sedimentary structure and composition. The area in which precipitation is analyzed encompasses the river catchments excluding the Nile catchment that extends further south of 25 N, in the tropical African wind belt.

The hydrological cycle in the Mediterranean area has been studied extensively in the eighties (Peixoto et al., 1982) and a very recent study has revisited the hydrological balance of the Mediterranean from recent meteorological data sets (Mariotti et al, 2002, hereafter called MSZL).

MSZL revisited the various components of the hydrological cycle focusing on the ocean-atmosphere branch. In Fig. 4 we show the large latitudinal gradient in precipitation from the NCEP 50 years re-analysis data set (Kalnay et al., 1996). The precipitation latitudinal gradient is more evident during the summer months while during winter the precipitation maxima are centred on the sea, always embedded in a positive northward gradient.

MSZL computed a new evaporation (E) minus precipitation (P) budget from the NCEP re-analyses, giving a range of $E \quad P \quad 603 \quad 699 \frac{mm}{yr}$ for the entire Mediterranean Sea. This is somewhat lower than published before from the data in the sixties and seventies, due to decadal variability connected to NAO. MSZL discuss the precipitation-NAO teleconnection and find significant correlations between the precipitation regime over the Mediterranean and the NAO index, as we will discuss later.

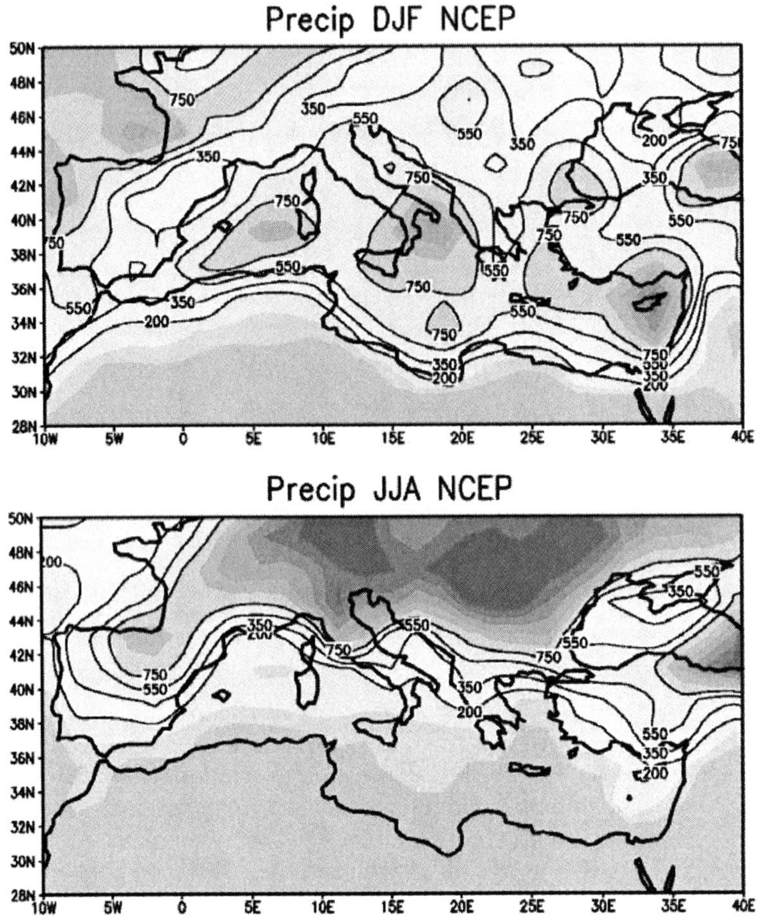

Figure 32.4 The climatological precipitation over the Mediterranean area computed by MSZL from the NCEP re-analysis data set. Upper panel: the December-January-February-DJF mean, lower panel: the June-July-August-JJA mean (units mm yr^{-1}).

The Runoff (R) time mean budget for the Mediterranean is connected to the precipitation regime and it can be deduced considering the following steady state balance:

$$R \quad E \quad P \quad G \quad B \tag{2}$$

where G is the net Gibraltar inflow and B is the Black Sea inflow and R the total runoff in the basin. Considering $B \quad 75\frac{mm}{yr}$ (corresponding to a net inflow of $6.\ 10^3\frac{m^3}{s}$ assuming an area of $2.510^{12}m^2$ as in MSZL) and $G \quad 500\frac{mm}{yr}$ (or $G \quad 0.04 \ Sv$) we obtain the range

$$R \quad 28 \quad 124 \frac{mm}{yr} \tag{3a}$$

or

$$R \quad 2200 \quad 9800 \frac{m^3}{s} \tag{3b}$$

for the two extreme values of E-P range estimated by MSZL.

MSZL does not include a detailed breakdown of the land hydrological budget in terms of river runoff but they take a basin mean runoff value of 100 mm/yr or equivalently 8. $10^3 \frac{m^3}{s}$ which is well within our estimate (3b). This contribution is about 10% of the value of precipitation and evaporation but it is important during spring, becoming relatively comparable to the E-P budget. It is interesting to note that the traditional value for the runoff given by Anati and Gat (1989) is quite larger, of the order of 14. $10^3 \frac{m^3}{s}$ but the source of these estimates in unknown. Recently, Struglia et al. (2004) revisited the river discharges into the Mediterranean Sea, utilizing different historical data sets. Keeping in mind all the possible shortcomings of these data sets, their total annual discharge estimate is 8.1 $10^3 \frac{m^3}{s}$ which is within the limits of our indirect calculation. Boukthir and Barnier (2000) give a value of 11. $10^3 \frac{m^3}{s}$ using a different data set. Thus we can say that the uncertainty in the river discharge coming from the best existing data sets is of several thousands of $\frac{m^3}{s}$.

The distribution of runoff for the climatology (January and May) is given in Fig. 5. All the northern rivers have a late winter-spring and autumn maximum discharge regime connected to the ice-snow melting cycle and the precipitation maxima during autumn (not shown). The pattern of runoff magnifies the north-south gradient already seen in the precipitation pattern but the quality of data for the runoff is more questionable. In our data set, all the African and the Middle East rivers are absent. Struglia et al. (2004) show instead that the Northern African rivers plus the Middle East rivers can account for 2.4 $10^3 \frac{m^3}{s}$ about 25% of the total runoff. They however include a Nile runoff (1.2 $10^3 \frac{m^3}{s}$) that is much larger than the one we show in Table 2 based upon more recent data. They also document a much smaller influence of Adriatic rivers except the Po, because they do not consider the Croatian rivers and Albanian rivers that could account for a large part of the fresh water discharge in the Adriatic Sea (Raicich, 1996).

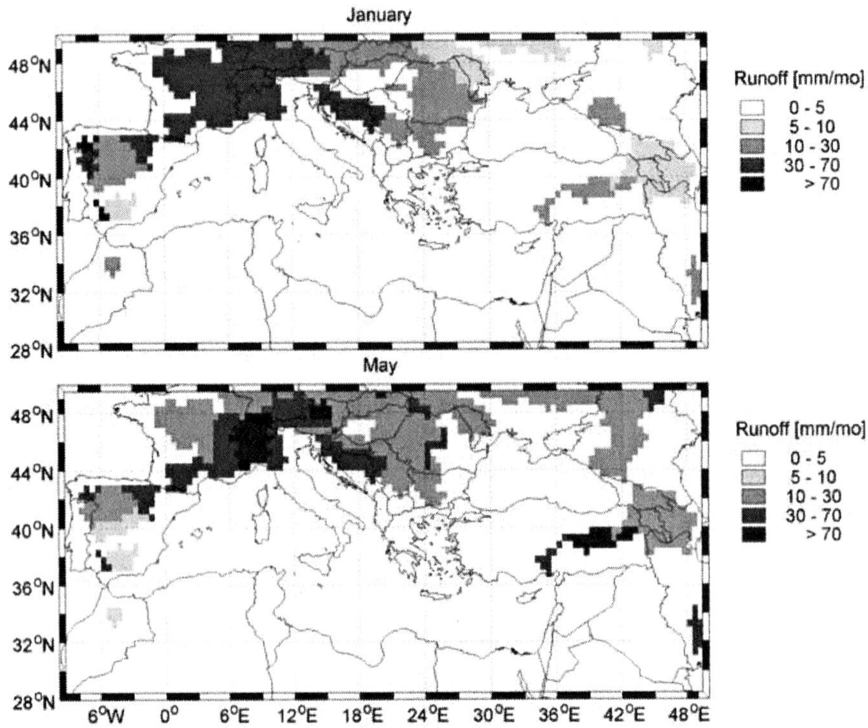

Figure 32.5 January and May mean observed runoff values from *http://www.grdc.sr.unh.edu/*(units mm mo^{-1}).

The annual mean runoffs by the major rivers and areas are reported in Table 2 for future reference. We use the largest estimate of the Adriatic river runoff by Raicich (1996) but our conclusions will not be changed by using a smaller estimate. The overall runoff budget (3b) can be made up for its largest part by summing solely the major rivers localized along the northern coasts (the Ebro, the Rhone, the Po, the other Adriatic rivers, the northern Aegean rivers). This result is also confirmed by Struglia et al. (2004): they show that European continental discharges can account for over 75% of the annual runoff budget.

TABLE 32.2
The major Mediterranean rivers and areas runoff values
(major is considered to be a runoff greater than $100\frac{m^3}{s}$).

River	Mean annual runoff (m3/sec)	Reference
Ebro	150	Cruzado et al., 2002
Rhone	1690	http://www.grdc.sr.unh.edu
Po	1585	Raicich (1994)
All other Adriatic rivers	4091	Raicich(1994)

TABLE 32.2
The major Mediterranean rivers and areas runoff values
(major is considered to be a runoff greater than $100\frac{m^3}{s}$).

River	Mean annual runoff (m3/sec)	Reference
Northern Aegean rivers	515	Kourafalou et al. (2003)
Nile	110	Hamza(2003)
All other Nile Delta runoff	430	Hamza(2003)
Total	8571	

A particular attention has to be put in the Nile runoff that has been totally regulated by the Aswan dam and it is now lower than other river runoffs. The sedimentary remnants of the present interglacial period Nile runoff form the present day extended Egyptian continental shelf. The same sedimentary remnants could be also present in the Tunisian shelf where there is evidence that, between 9 and 7 thousand years ago, there was higher precipitation due to stronger teleconnections with the Arctic Oscillation Index (Arz et al., 2003), which supposedly gave rise to high runoff along the northern African coastal areas.

Struglia et al. (2004) document the seasonal changes in the annual mean runoff that amount to $5\ 10^3\frac{m^3}{s}$ with minimum values reached during July-August-September for all the rivers except the Nile, which is now totally regulated. This large seasonal variability is a major forcing of the Mediterranean coastal areas, which has profound influences on the shelf dynamics.

If we consider the NAO-precipitation correlation and variability, MSZL estimate that the E-P budget could vary on a decadal frequency by $100\frac{mm}{yr}$ which is of the same order of magnitude of the mean runoff value deduced before from the steady state budget calculation (3a). The NAO index (Hurrell, 1995) exhibits a large negative correlation with precipitation over the Mediterranean region on interannual and decadal time scales (MSZL compute a correlation coefficient of -0.84). This means that the decadal/interannual variability of the sedimentary and ecosystem dynamics in the shelf areas of the Mediterranean, dominated by runoff regimes, can be as large as the seasonal variability. On interannual time scales, Josey (2003) computes that changes in the E-P budget can be one order of magnitude less than the decadal changes. These E-P changes are mainly due to precipitation variability, which in turn will affect the river runoff.

Another remarkable result about precipitation variability is that a systematic increase in winter droughts, all over Italy has been documented by Brunetti et al. (2000). The proportion of dry days has increased by almost 50% with respect to the previous 30 years average (see Fig. 5 of Brunetti et al., 2000). The change has been abrupt from the beginning of the 80's until the end of 2000. This means that the river regimes have also changed, especially for large river catchment systems such the Ebro, Rhone and Po.

Struglia et al. (2004) discuss directly the runoff variability correlated with the NAO index. It is found that several rivers are significantly anti-correlated with NAO winter index and they loose the anti-correlation going toward spring and summer. They also argue that this anti-correlation is strictly related to the anti-correlation of precipitation with NAO. Struglia et al. (2004) compute then the changes in discharge due to the constant increase of the NAO index from the middle of the '70s to the '90s computing a decrease in total discharge in the basin of about 1.7 $10^3 \frac{m^3}{s}$ which is 17% percent of the winter mean budget and that is concentrated on the Rhone, Ebro and Po rivers.

In conclusion, the atmospheric precipitation distribution determines the runoff budget of the basin that is determined for 75% or more of the total amount by the northern river discharges and the remaining by the runoff on the southern shores. Together with the shelf extension, this characterizes the overall structure of the Mediterranean shelf areas that mainly consists of narrow shelf areas without river runoff in the southern shores and extended shelf areas with runoff in the northern regions.

The river runoff spatial distribution could be also partially responsible for the surface chlorophyll latitudinal gradient: the northern shores are fertilized by the nutrient inputs (see section 6) and local chlorophyll maxima are established in every near-shore coastal area under river runoff influence (ROFI- areas or Region of Fresh Water Influence). As it can be seen in Fig. 6, the northern shores show higher chlorophyll concentrations with respect to the southern shores and local maxima appear near the northern river outflows. The other two local maxima, on the Tunisian and Egyptian shelves, are mainly connected to non-photosynthetic suspended matter.

Crispi et al. (2001) studied the longitudinal/latitudinal gradient in chlorophyll with a box model that demonstrated the importance of river loads in the maintenance of the chlorophyll gradient of Fig. 6. In particular, they point out that the western basin is generally more productive because it has larger nutrient inputs due to the presence of both the Ebro and Rhone river runoffs and the Gibraltar inflow. We argue here that this demonstrates an extremely active role played by the shelf regions of the basin on the biochemical fluxes of the basin, driving the large scale north-south gradients at the same level of importance of the Gibraltar inflow.

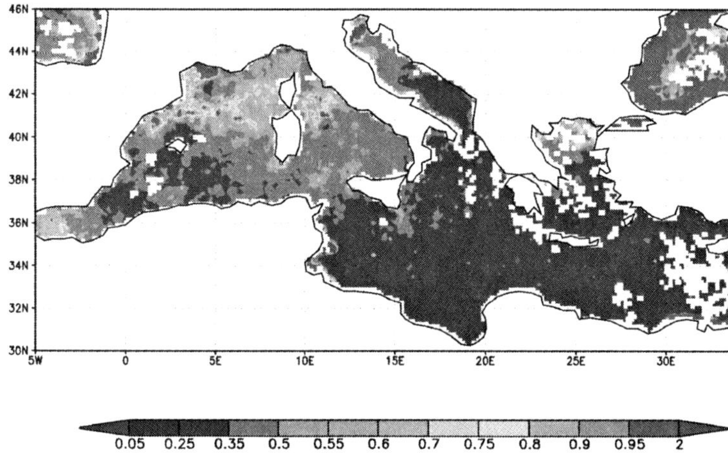

Figure 32.6 The SeaWiFs surface chlorophyll analysed for the winter of year 2000 (white areas are cloud covered).

3. Sedimentary structures in the shelf areas

Sediment processes on the shelf are heavily influenced by topographic structures such as slopes and canyons and by river runoff. Seasonal variations in the atmospheric forcing influence the marine currents on the shelf that are responsible for transferring the suspended particulate to the sediments. The sedimentation rate and the sediment composition depend on all these environmental factors. In this section, we will take a detailed look at sedimentary structures in the shelf areas identified here as case studies. In the western Mediterranean, we will consider the Algerian and Gulf of Lions continental shelves; in the central Mediterranean, the Adriatic and Sicilian shelves and in the eastern Mediterranean, the Nile Delta zone and the Israeli coasts.

3.1 Gulf of Lions

The continental margin of the North West Mediterranean is characterised by a relatively extended continental shelf (see section 2) with a well-defined edge at a depth of 100–200 m. The shelf is eroded by numerous undersea canyons. The principal sources of sediments are the main rivers, the Rhône and Ebro, and the atmosphere. The biogenic productivity of surface waters also plays a major role.

Holocene deposits in the Gulf of Lions can be divided into three separate zones according to their thickness. The first zone extends from the Rhône delta to the Têt (western side of the Gulf) and is characterised by large sediment accumulations with a thickness between 30 and 50 m (Got and Aloisi, 1990). The second zone covers a large part of the middle shelf and the thickness drops from 30 m to 5 m. The third zone occupies the outer shelf at a depth between 90 and 200 m and is characterised by thin Holocene deposits and outcrops of relict sand representing the coastline during the last phase of eustatic lowering, remoulded during the first phase of rising sea level (Figure 7a, Durrieu de Madron et al., 2000).

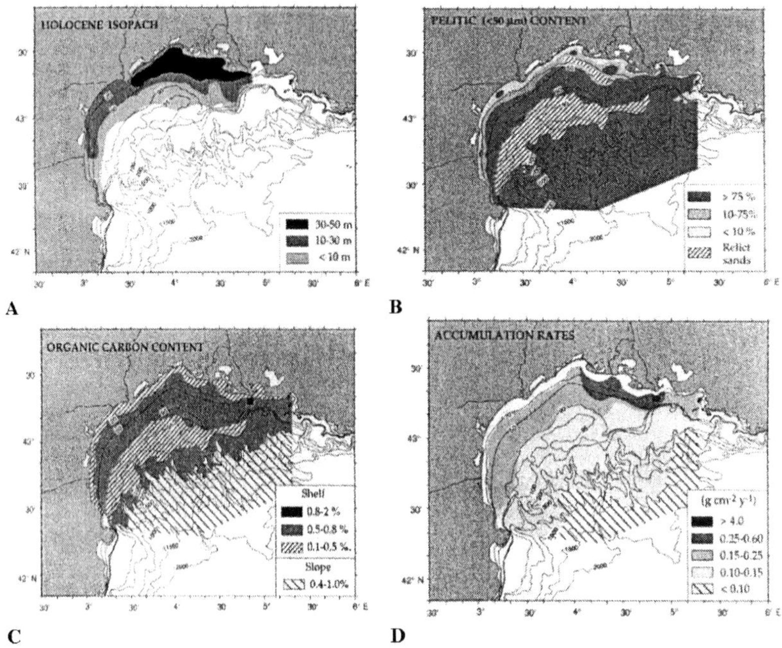

Figure 32.7 a Thickness of the Holocene sediments in the Gulf of Lions (Got and Aloisi, 1990). b Pelitic content in Holocene sediments in the Gulf of Lions (Got and Aloisi, 1990; Gensous et al., 1993). c Organic carbon content in Holocene sediments in the Gulf of Lions (Durrieu de Madron et al., 2000). d Accumulation rates in the Holocene sediments of the Gulf of Lions (Buscail and Germain, 1997; Durrieu de Madron et al., 2000, Zuo et al., 1991 and 1997).

3.1.1 Grain size

The grain size of the inner shelf sands decreases seawards, with the muds of the middle shelf occurring at a depth of about 20–30 m. There is, in fact, a correlation between grain size and the attenuated effect of wave motion seawards (Jago and Barusseau, 1981). The only exceptions are the prodelta deposits accumulating at river mouths and consisting predominantly of silty mud. These sediment accumulations are due to the fact that precipitation is large in the autumn and spring leading to a significant flow of terrigenous sediments into coastal marine environments. In this coastal environment, there is a high seasonal variation of geochemical parameters and discontinuities in the typical prodelta sediment sequence are due to physical-chemical rather than sedimentological variations. In the case of the river Têt, the surface layers of the prodelta deposits are silty (60 μm median) and poor in carbonates (5%) and organic carbon (0.5%). The sediment water content in the surface layer is high (45–50%), compared to about 30% in underlying layers. The grain size of the silt is graduated as a result of hydrodynamic effects, but a relationship between the intense bioturbation produced by a crustacean (Thalassinidean) and an increase in the coarser particle fraction has also been noted. In the spring, when this crustacean is excavating tunnels, the finer particles are eliminated in the overlying water, while the coarser particles accumulate in the tunnels (Buscail et al., 1995). Deposits on the outer shelf and slope are predominantly pelitic (Figure 7b).

3.1.2 Organic carbon

The organic carbon content varies according to the type of sediments. It is generally low in the sandy deposits of the inner shelf (< 0.3%) and the relict sands (0.25–0.5%) (Buscail et al., 1995; Durrieu de Madron et al., 2000). In the muddy littoral deposits (between 40 and 100 meters depth) values remain low (0.4–0.6%) and drop to 0.3% in deep waters. The highest values are found in the muddy prodelta deposits with values between 1–2.2% (Figure 7c and Table 3).

TABLE 32.3
Figures indicating the organic carbon content in the delta zone of the principal rivers in the Gulf of Lions according to various authors.

River	Organic carbon (%)	Notes	Reference
Aude	1.1–1.4		Buscail et al., 1995
Rhône	1.1–1.6		Buscail et al., 1995
	1–2		Durrieu de Madron et al., 2000
	2.17		Zuo et al., 1997
Têt	0.5	in autumn the deposits are enriched with organic carbon (2.5%), pelite (92%) and fine bioclastic carbonates (20%)	Buscail et al., 1995

The same variations occur for other trace elements, confirming that more than 90% of particulate matter coming from the rivers is deposited in the prodelta, shelf and slope zones, with only a small quantity transported into the basin through the canyons.

These hydrodynamically active prodelta environments demonstrate that carbon degradation predominates over burial. Land and sea sources are important, but 80% of the organic carbon flow is degraded at the interface on the continental shelf (Buscail et al., 1990).

3.1.3 Sedimentation and accumulation rates

It is difficult to determine the sedimentation rate because there are low gradients in the Pb activity profile of core samples taken at river mouths due to the rapid sediment deposition and mixing (Zuo et al., 1997). The sediments consist of silt, silty shale and shale and a linear dependence can be noted (with a correlation coefficient of $R=0.87$) between the sedimentation rate and depth. The sedimentation rate in the prodelta zone is very high and near the mouth of the river Rhône may reach a few centimetres per year (5–9 cm/year). The same trend can be observed in the area of the Ebro and Têt rivers, although the sedimentation and mixing rates are lower as a result of the lower sediment transport (sedimentation rate in the Têt prodelta is estimated to be 0.1 cm/year, Courp and Monaco, 1990).

In the shelf zone, the sedimentation rate is between 10–60 cm/100years (Zuo et al., 1991). The sedimentation rate enables the sediment accumulation to be estimated. For an area of 15340 km^2, the sediment accumulation is about 10x106 ton/year. This estimate does not include prodelta areas. In prodelta areas, considering an area of 150 km^2, a sedimentation rate of 5–9 cm/year (see above) and a mean density of 1.60 g cm^3, the accumulation rate amounts to 2–3x106 ton/year (Zuo et al., 1997).

According to Durrieu de Madron et al. (2000) the mean accumulation rate at the Rhône mouth amounts to 40 $\frac{g}{cm\ \ yr}$. This value decreases rapidly seawards. In the distal part of the Rhône prodelta (20 km from the coast), the mean value is 0.4 $\frac{g}{cm\ \ yr}$. The values are even lower for the rest of the shelf with a mean of 0.15 $\frac{g}{cm\ \ yr}$. Figure 7d and Table 4 gives the accumulation rates from a number of authors. The discrepancies between the values in Table 4 are due to the fact that Zuo et al. consider the last 100 years, while Got and Aloisi consider the whole Holocene period (10,000 years), but the general sediment transport and deposition pattern is confirmed by both authors.

TABLE 32.4
Sedimentation rate and flows in the North West Mediterranean according to various authors.

Area	Sed. rate (cm/100years)	Flow (g/cm^2) /100years	Accumulation (10^6ton)	Reference
Gulf of Lions (15340 km2)	7–41	5–31	10+-4	Zuo et al., 1991 and 1997
NW Mediterranean (280000 km2):				Zuo et al., 1997
Shelf (10%)	6 (2–12)	4	11	
Slope and basin	1–2	0.9	23	
Tot.			34±15	
NW Mediterranean (280000 km2):				Got and Aloisi, 1990
Shelf (10%)				
Holocene deposits:				
(2.6x10–11 m3)			44	
Slope and basin			20	
Tot.			64	

It is interesting to note that the Gulf of Lions, occupying just 5% of the total surface of the North West Mediterranean, represents about 30% of total sedimentation.

3.2 The Algerian continental shelf

The Algerian continental shelf is narrow (see section 2). In the bays and gulfs, the shelf extends for 10–20 km, such as in the Gulf of Oran, the Gulf of Arzew and the Gulf of Bône, where the edge of the shelf is located at a depth of 120 - 150 m.

Climatic and geodynamic variations during the Pliocene and Quaternary led to the temporal and spatial superimposition of two types of marine sediments; 1) autochthonous biogenic carbonate sediments deposited mainly during wet periods with abundant land plant cover and scarce rill erosion of continental zones; 2) allochthonous terrigenous sediments deposited during semi-arid periods such as the current African climate as a result of rill erosion of continental zones with scarce plant cover (Pauc, 1991). The carbonate sediments consist of silt and biogenic calcareous sands, composed mainly of bioclastic fragments, calcareous algae, bryozoans, pelecypods and gastropods (Caulet, 1972).

In the Gulf of Arzew, biogenic carbonates are distributed in the western part from Cape Carbon to the sea off Mostaganem. Towards the west, this facies is largely buried under more recent sediments (Figure 8a). Current terrigenous sediments come mainly from the River Chèliff, which drains silty and shaly debris. The finest fraction of these sediments (<2μm) is distributed towards the west under the influence of currents; the silty fraction (>10μm), is, on the other hand, distributed along the coastal belt at a depth of 40 m. The total organic carbon (TOC) content amounts to between 0.2 and 1.2% with the highest values occurring in the pelitic fraction (Figure 8b).

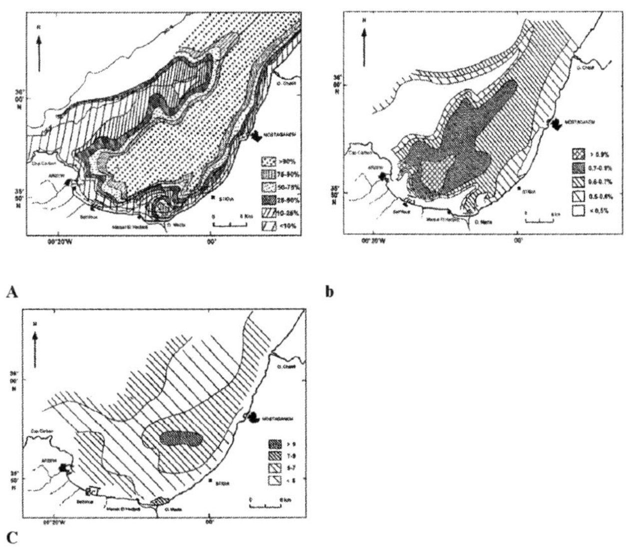

Figure 32.8 a Pelite distribution in surface sediments in the Gulf of Arzew (South West Mediterranean) (Buscail et al., 1999). b Total organic carbon distribution in surface sediments in the Gulf of Arzew (South West Mediterranean) (Buscail et al., 1999). c Carbon/nitrogen (C/N) ratio distribution in surface sediments in the Gulf of Arzew (South West Mediterranean) (Buscail et al., 1999). Reprinted with permission.

The nitrogen content varies between 0.05 and 0.35% with the highest values occurring near the port of Arzew, and the lowest in the coarser sediments near Cape Carbon. The C/N ratio varies from 5 to 7 (figure 8c). In the fine fraction (<40μm) the ratio is slightly higher (Buscail et al., 1999). This ratio is indicative of the origin and the degree of evolution of organic material present in the sediment. Values of C/N <6, found in these areas, indicate the presence of organic material of essentially marine origin (Monoley et al., 1991; Müller and Suess, 1979), although interpretation of the C/N ratio is somewhat delicate as nitrogen is more sensitive than carbon to degradation and a high ratio may thus also indicate the presence of degraded or already highly mineralised organic debris.

The same distribution, with the same facies, also occurs in the Bay of Algeri (Maouche, 1987). The TOC content (0.3–1.5), nitrogen values and C/N ratio are similar to those found in the Gulf of Arzew.

3.3 Adriatic Sea

The Adriatic itself is an epicontinental sea, in other words, a semi-closed basin within a continent. The eastern Adriatic coast is generally high and rocky, while the western coast is low and largely sandy. The northern part of the basin, by convention bounded to the south by the transect approximately at 43.5° N (see Fig. 9), is shallow, with a mean depth of 35 m and a slight slope (0.35 m km-1) towards the south east (see section 2). The middle Adriatic has moderately deep waters (on average 140 m) with two depressions dropping down to 260 m. South of the depressions is the morphological elevation Pelaguza sill (see Fig. 9), oriented in a northeast – southwest direction and formed during the Quaternary. This represents really the shelf break for the Adriatic Sea. The south Adriatic extends south of the latitude of 42° N to the threshold of the Strait of Otranto and is characterised by a large depression more than 1,200 m deep (Fig. 3.3). Exchange of water with the Mediterranean Sea occurs through the Straits of Otranto with a threshold 800 m.

Sediment dynamics forces a distribution of the coarser sediments close to the coasts and the formation of a band of fine sediment accumulation approximately parallel to the coast further offshore (Wang and Pinardi, 2002). Outward from the coasts, there are outcrops of relict sands from the Versilia transgression (Flandrian) covering a large part of the shelf with an irregular morphology inherited from the sub-aerial environment of the last ice age. A large number of rivers flow into the Adriatic, transporting suspended material. Of particular importance are the river Po in the northern basin and the group of Albanian rivers in the southern basin.

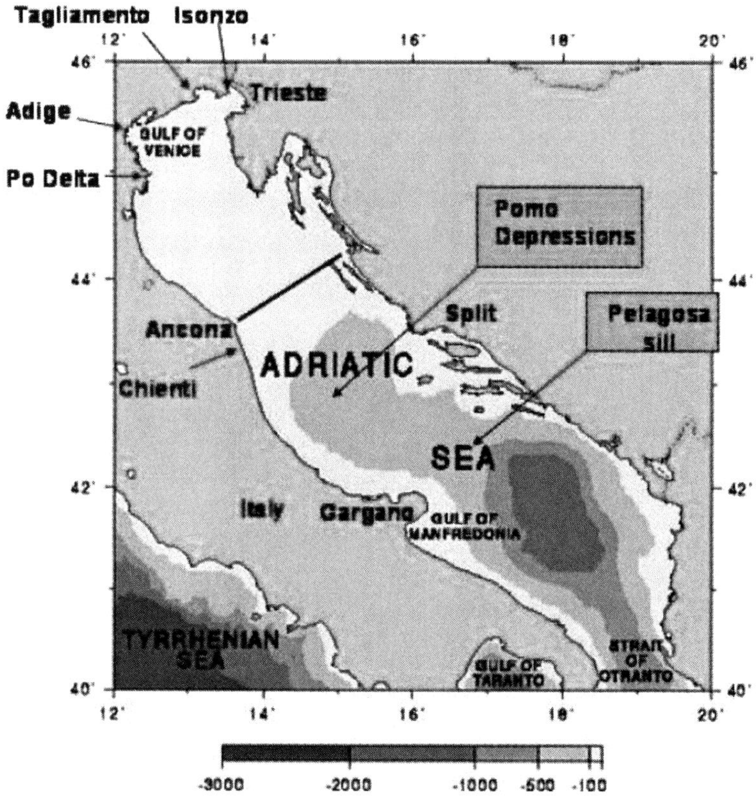

Figure 32.9 The Adriatic Sea geometry and bathymetry. The transect indicate the approximate limit of the Northern Adriatic Sea.

3.3.1 Sedimentation and accumulation rates

In the area between the Gulf of Trieste and the Middle Adriatic, three zones can be identified on the basis of recent data on sedimentation rates ($\frac{cm}{yr}$) and accumulation rates ($\frac{g}{cm\ \ yr}$), obtained using 137Cs and 210Pb (Table 5). For the southern Adriatic and in the western-central zone, Frignani and Langone (1991) and Frignani et al. (1996) report accumulation rates of 0.06 $\frac{g}{cm\ \ yr}$. The three northern and middle Adriatic areas are described below.

TABLE 32.5
Accumulation and sedimentation rates in the Adriatic Sea

Area	Accumulation rates (g/cm^{2}y)	Sedimentation rate (cm/y)	References
Trieste-Adige Delta			
Isonzo mouth		8.2	Frignani et al., 1990
Tagliamento mouth		2.1	Delfanti et al., 1994
Gulf of Venice (15–20 m deep)	0.45–1		Alvisi et al., 2000
Adige prodelta (15–20 m deep)	0.5->1.2		
Adige (20 m deep)	0.8		Frignani et al., 1990
Delta Po-Ancona			
Po di Levante-Po di Pila			
Depth: 7 m	0.86–1.86		Frignani et al., 1990; 1989
20 m	0.77		
27 m	0.22–0.48		
> 30 m	0.3–0.4		
Depth: 5 m	1.6		Alvisi et al., 2000
10 m	0.3		
Punta Maestra (20 m deep)		2	Frignani et al., 1990; 1989
Po di Goro			
Depth: 14 m	0.84		
18 m	1.8		
21 m	1.0		
23 m	0.37		
24 m	0.39–1.20		
Depth: 10 m	0.8–1.5		Alvisi et al., 2000
20 m	0.3–1.4		
<30 m	0.4–1		
	0.77		Frignani et al., 1996
Porto Garibaldi			
Depth: 13 m	0.26		Frignani et al., 1987
26 m			Guerzoni et al., 1984
	0.39		
Depth: 10–20 m	0.3		Alvisi et al., 2000
20–30 m	0.4		
Fiumi Uniti mouth (20 m deep)	0.27		Frignani et al., 1987 Guerzoni et al., 1984
Bevano mouth			
Depth: 14 m	0.4		
21 m	0.31		
Rubicone mouth			
Depth: 9 m	0.22		Giordani et al., 1992
12 m	0.24		
20 m	0.22		
24 m	0.26		
41 m	0.28		
Depth: 10 m	0.2–0.4		Alvisi et al., 2000
<20 m	0.2		
	0.3		Frignani and Langone 1991
Ancona-Gargano			
Chienti mouth			
Depth: 28 m	0.52–0.8		
62 m	0.17		
90 m	0.07–0.08		

TABLE 32.5
Accumulation and sedimentation rates in the Adriatic Sea

Area	Accumulation rates (g/cm^2y)	Sedimentation rate (cm/y)	References
Middle Adriatic deep			Frignani et al., 1987
Depth: 176 m			Sorgente et al., 1996
>200 m	0.05		
	0.04		
Central Adriatic	0.12		Frignani and Langone 1991; Frignani et al. 1996
Southern Adriatic Basin			
South Adriatic Pit	0.06		

3.4 Northern Adriatic: Trieste—Adige Delta

Near the mouth of the river Isonzo (Fig. 9), the sedimentation rate is seasonally high, in fact, short lived radionuclide dating shows a speed of 8.2 cm/year (Frignani et al., 1990). Near the Tagliamento (Fig. 9), there is a high seasonal sedimentation rate (2.1 $\frac{cm}{yr}$) (Alvisi et al., 2000; Delfanti et al., 1994) together with a low accumulation rate on century scale. It can be hypothesised that the difference is due to the erosive action of the currents. Off the mouth of the river Adige (Fig. 9), the accumulation rate amounts to 0.8 $\frac{g}{cm\ yr}$ (Frignani et al., 1990). Radiometric analysis indicates an absence of sedimentation in the sandy belt off the coast.

3.3.2 Po Delta—Ancona area

In the area off the Po Delta, there are high accumulation and sedimentation rates as the inflow of material from the hinterland is very high. Data in the literature (Frignani et al., 1989; 1990) confirm that the sedimentation rate decreases as the water becomes deeper. The sedimentation rate values drop rapidly moving from around the Po delta in the southward direction due to the current action that advects sediments along the coasts, away from the delta. South of the Po Delta, the values are lower and more or less constant (Frignani and Langone, 1991; Frignani et al., 1996). Off the coast of Emilia Romagna, the accumulation rate is below 0.5 $\frac{g}{cm\ yr}$ (Frignani et al., 1990; Guerzoni et al., 1984; Giordani et al., 1992). Recent studies of sediment resuspension and deposition due to the Po river delta runoff have established that the specific combination of wave and currents climate in the Adriatic determines the sediment deposition as a function of grain size along the western Adriatic coastlines (Wang and Pinardi, 2003).

3.3.3 Ancona-Gargano area

The area is characterised by extreme spatial and temporal variations in sedimentation. The greatest variability occurs in a land-sea direction and many parameters vary in bands parallel to the coast (Sorgente et al., 1996; Sorgente, 2002).

Off the mouth of the river Chienti, one of the rivers with the greatest solid flow in this section of coast (2.2x106 t/y), accumulation rates decrease gradually as water depth increases, with a high seasonal sedimentation rate at some points. The characteristics at stations sampled in the Middle Adriatic depressions are similar, with shaly sedimentation from currents arriving from the north transporting fine material from the Padana Plain. Sedimentation rates amount to about 0.05 $\frac{g}{cm\ yr}$ at a depth of 176 m and 0.04 $\frac{g}{cm\ yr}$ at a depth of more than 200 m (Frignani et al., 1987; Sorgente et al., 1996).

3.3.4 Organic carbon

Data gathered during recent surveys indicate that the highest organic carbon values are found near the mouth of the Po river and in the Middle Adriatic depressions (Giordani et al., 2002 - Table 6). The Adriatic Sea is characterised by a nutrient concentration and primary productivity trend decreasing in the land-sea and north-south directions, in accordance with inflow from the land and hydrodynamic patterns. The quantity of organic carbon reaching the water-sediment interface amounts to 42–56% of primary productivity in the shallower zones of the northern shelf and about 25% further out in the deeper sea. On the southern shelf, values vary between 26% and 3%, from zones nearest the coast to the deeper areas. The majority of the organic material characterising the Adriatic sediments is of land origin, but the autochthonous biogenic fraction increases proceeding southwards. The C/N ratio, whose southward decrease reflects these variations, can be used as an indicator of the degree of preservation of sediments (Faganeli et al., 1994).

TABLE 32.6
Primary productivity and organic carbon data for a number of stations in the Adriatic Sea (Giordani et al., 2002, modified).

Area	Station	Depth (m)	Primary productivity (gC/m^{2}y)	OrgC (%)	C/N (at. ratio)
North Adriatic	S3 (44°20.07'-12°39.95')	30	216	0.85	6.7
	S6 (43°42.20'-13°38.50')	41		0.93	7.9
Middle Adriatic Depressions	P1 (42°51.00'-14°45.00')	246	60	0.87	7.2
Southern Adriatic Pit	A1 (41°50.74'-17°44.71')	1196	97	0.65	7.9
Strait of Otranto	O2 (39°49.72'-18°57.48')	870	66	0.46	6.8
Ionian sea	I1 (38°29.10'-17°59.27')	2360	62	0.55	6.3

In areas with high sediment accumulation rates (>0.1 $\frac{g}{cm\ yr}$), more than 50% of carbon produced is not oxidised and accumulates in the sediments, while in areas with a lower rate, only about 1% is conserved (Canfield, 1993). The Middle

Adriatic depression is a special case as, in addition to normal accumulation in the water column, there is also strong lateral transport from the more productive northern shelf (Hopkins et al., 1999).

3.4 The Sicilian-Tunisian Platform

The Strait of Sicily is characterised by various depositional environments (Fig. 3.4, Argnani, 1992; Bowles et al., 1992). The southern Sicilian shelf is characterised by the inflow of terrigenous material from the Atlantic Ionian Stream-AIS branch (see section 4) that forms a wedge of well-stratified sands and silty shale varying in thickness from about 5–6 metres near the coasts to almost zero at the edge of the shelf (Colantoni et al., 1985). An exception is the Adventure Bank (Fig. 10) characterised by a virtually flat surface with a mean depth of about 80–90 metres. It is isolated from the inflow of terrigenous material by the strong currents and deposition is therefore authigenic: 1) heterometric carbonate sands consisting mainly of the remains of organisms (bryozoans, red algae, serpulidae, foraminiferida, gastropods and corals) living in the extensive eelgrass and seaweed meadows and 2) fragments of biogenic concretions (coraligen) (Colantoni et al., 1985).

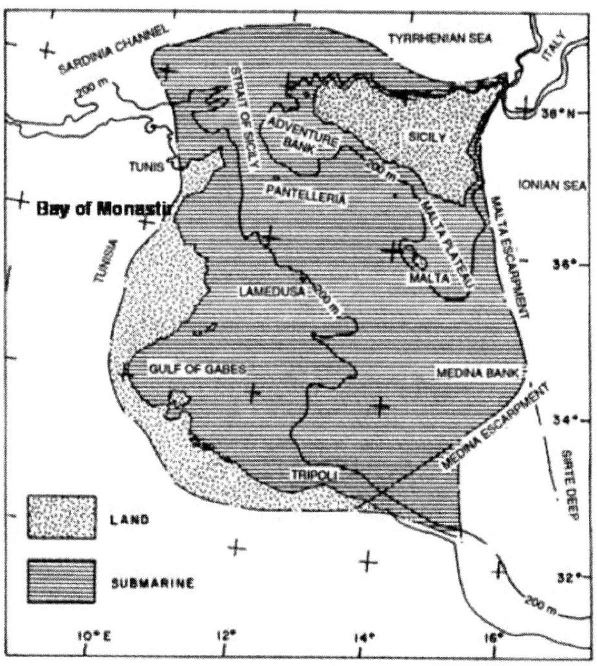

Figure 32.10 General map showing the location of the Sicilian-Tunisian Platform (from Max and Colantoni, 1992). Reprinted with permission

In the Lampedusa Bank the sediments are very similar to those of the Adventure Bank, but with a higher percentage of sandy fraction. The calcium carbonate content is very high (>90%) (Tonarelli et al., 1992).

In marked contrast to the surface deposits of the Adventure and Lampedusa Banks, the Malta Plateau sediments are finer and consist largely of shaly silt with low carbonate content (about half). The carbonate sediments, consisting of organogenic and algal debris, are limited to waters shallower than 75 metres. In deeper waters of the Malta plateau, pelagic, hemipelagic and terrigenous debris, probably eolian in origin, represent the main clastic depositional components (Tonarelli et al., 1992).

The Tunisian continental shelf is the most extended in the Mediterranean Sea as a whole. Recently the littoral between Tunis and the Gulf of Gabes has suffered from eutrophication and, as a result, build-up of algae has gradually lead to the deposit of sediments rich in organic material. Information about the whole shelf is difficult to find but data are available for the Bay of Monastir, which is south of Tunis (Fig. 11).

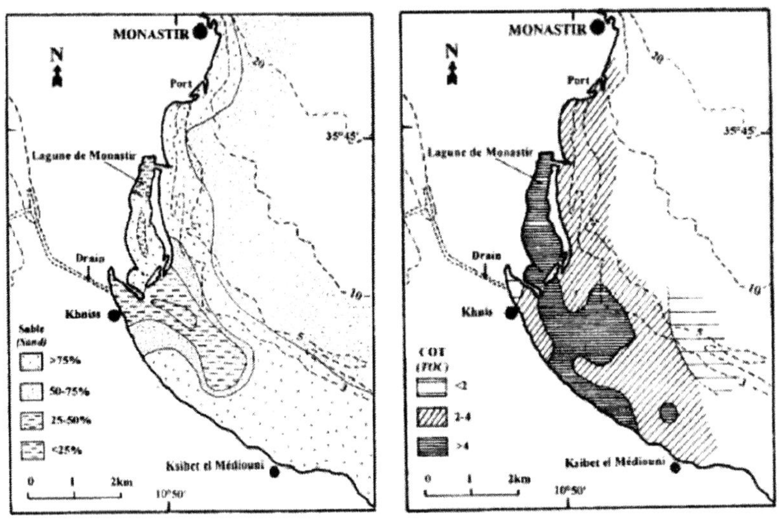

Figure 32.11 a Percentage of sand (φ>63μm) in the surface sediments (from Sassi et al., 1998). b Lateral distribution of organic carbon content (TOC) in the surface sediments (from Sassi et al., 1998). Reprinted with permission.

In this Bay, the sediments are largely sandy, while in the coastal zone of Khnis, the pelitic fraction predominates as a result of the high inflow of fine sediment from the drainage canal (Fig. 11a). The river runoff and waste-waters flow is directed into a drainage canal discharging in the Khnis zone. The whole area is characterised by currents of low amplitude (see section 4) due largely to the shallow water depth (Sassi et al., 1998).

There is a high organic carbon content (TOC) in the surface sediments, with values between 2–6%. The highest percentages (>4%) can be found in the Khnis area, confirming the high levels of organic pollution. In the rest of the coastal strip, they vary between 2–4%, attenuating beyond the 3m isobath (1% at a depth of greater than 5m). The nitrogen (N) content is also high (0.5–1%), reaching the highest levels in the zones richest in organic carbon. The C/N ratio is indicative of the nature of the organic material present in the recent sediments and the degree of mineralisation. In this zone, the ratio is less than 6 beyond the 3m isobath and reflects the silty marine origin (algae, eelgrass, plankton, etc) of the organic material, while the values of between 5 and 10 in the innermost areas indicate a composite origin, predominantly marine, but with a continental component. Locally (mouth of the drainage canal and Monastir lagoon) there are values of more than 10 (fig. 11b).

3.5 The Nile Delta zone

In the Nile Delta, the predominant sediments are terrigenous debris, shale and dark grey silt. The dark grey colour of the sediments is largely due to the high percentage of finely scattered organic debris and crystalline pyrite aggregates. There are few calcareous skeletal remains (foraminiferida, pteropods and heteropods) probably due to a dilution effect caused by the high percentage of clastic sediments. Mineralogical analysis indicates that the majority of the silts and sands are made up of mica and quartz (Herman, 1972). Construction of the Aswan Dam in 1964 drastically altered the water flow and the sedimentation pattern along the entire coast, considerably reducing the sedimentation rate.

In the absence of large quantities of sediment inflow from the Nile, wave and current action has caused heavy coastal erosion. However, in the Manzala lagoon (Fig. 12), accumulation rates are currently high (1.2 cm/year), greater than those occurring during the Holocene period (0.7 cm/year). This is due to the high input of water and sediment from the dense network of irrigation canals flowing into the lagoon (Benninger et al., 1998) and the high subsidence rates in the basin.

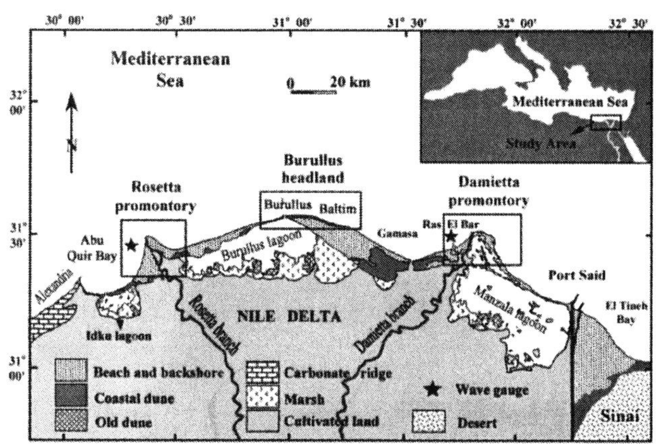

Figure 32.12 Map of the main promontories in the Nile delta area (Rosetta, Burullus and Damietta) and the geomorphologic units present (from Frihy et al., 2003). Reprinted with permission.

In the western side of the delta area (Figure 12), the recent building of protective structures (1990–2000) near the principal promontories (Rosetta, Burullus and

Damietta) has reduced and modified coastal erosion. Previously erosion was concentrated in the most projecting part of the promontories with regression values of up to 88 m/year (Rosetta promontory) and sediment was largely transported by currents running parallel to the coast in an easterly direction. Now erosion is distributed more uniformly along the coast with local areas of accumulation where the coast is growing by up to 37 m/year (Frihy et al., 2003).

3.6 The Israeli continental shelf

On the basis of current and wind direction measurements (Emery and Neev, 1960; Golik, 1997) and the characteristics of the sediment (Pomerancblum, 1966; Rossignol, 1969), numerous authors consider that the sandy sediments deposited along the coasts of Sinai and Israel derive largely from the Nile outflow. More recent studies carried out after construction of the Aswan dam (1964) which considerably reduced the contribution of sediment from the Nile, showed that the contribution of Israeli rivers is also significant in this area (Stanley et al., 1997; 1998).

The Israeli continental shelf extends for a maximum of 60 km in the Gaza area, shrinking to 25 km off the coast corresponding to the border with Lebanon. It consists largely of Pliocene-Quaternary sediments transported by the Nile and the sedimentation rates drop as the distance from the Nile increases (Ross and Uchupi, 1977; Almagor, 1993). Currents parallel to the coast are the main sediment transport agent in the South East Mediterranean from Sinai to the Bay of Haifa. North of Haifa, near the border with Lebanon, transport along the coast is less significant and the sandy fraction consists almost entirely of biogenic debris (Golik, 1997).

Marine sediments on the inner shelf south of Haifa consist of quartz, biogenic and detrital carbonates (5–10%), small percentages of heavy minerals and feldspars and shaly minerals. Sands are widespread in areas near the coast to a depth of 25 m, while shaly silt occurs at a depth of > 30 m. The shaly fraction represents more than 50% of the sediment at a depth of > 50 m.

Mineralogical study of the shaly fraction suggests that only 30–40% derives from inflow from the Nile. It is calculated that about 5–10% consists of grains transported by sandstorms and that the net contribution from Israeli rivers amounts to about 50% (Sandler and Herut, 2000).

4. The physical shelf regimes and the meteorological forcing

4.1 Hydrodynamic regimes classification at the large scales

The general circulation of the basin has been studied extensively in the past twenty years and it is now becoming very well mapped by observational data sets and reproduced by numerical simulations. One recent overview of the general circulation is given by Theocharis et al. (1998). The circulation can be subdivided into three main components: the large scale vertical circulation or thermohaline circulation, the sub-basin scale together with the Gibraltar-Atlantic water current system and the mesoscales.

As we have seen before, the Mediterranean is a concentration basin, i.e. water losses exceed water gains from precipitation and runoff. In addition, the net heat budget of the basin is negative (Bethoux, 1979), so that the vertical thermohaline circulation is negative or anti-estuarine, with waters exiting the Mediterranean at depths and entering from the Atlantic at the surface (Bryden and Kinder, 1991). The general characteristics of the thermohaline circulation are schematized in Fig. 13. This circulation is characterized by multi-decadal time scales and it is propelled by the water mass transformation processes that occur in the open ocean areas of the Northern Mediterranean. Both deep and intermediate waters form in the regions offshore the Gulf of Lions, the southern Adriatic and the northern Levantine basin, forced by intense heat losses during late winter (February-March) and influenced by the presence of large scale cyclonic circulations driven by wind stress curl. The new feature of such a conveyor belt is the introduction of the Aegean Sea as a source of deep waters for the Ionian Sea abyssal plains. This event has occurred at the end of the eighties and first half of the nineties and it has been documented as the Eastern Mediterranean Transient (EMT, Roether et al., 1996). Recently, Manca et al. (2003) have found that the Aegean has stopped the production of deep waters and that the Adriatic Sea has started to be again the unique site of production of the deep waters for the Eastern Mediterranean.

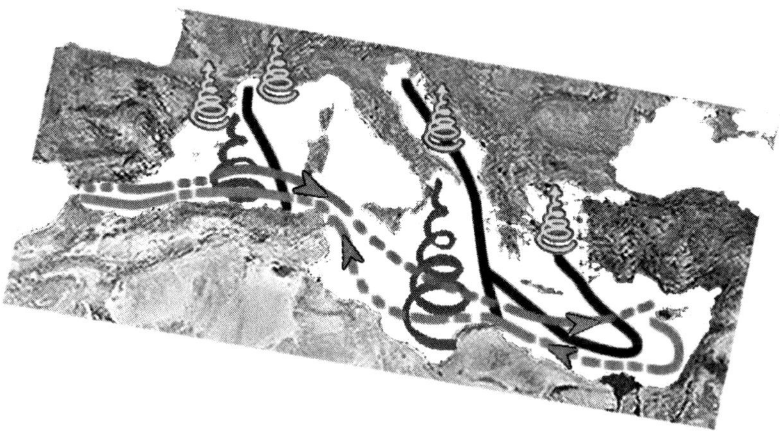

Figure 32.13 The three conveyor belts of the Mediterranean Sea. The first, dashed and in the zonal direction, is the surface-intermediate water masses circulation that is forced by Gibraltar and the Levantine Intermediate Water (LIW) formation processes occurring in the Northern Levantine basin. The second is the meridional vertical circulation in the Eastern Mediterranean, that is forced by deep water formation in the Southern Adriatic and Aegean Seas. The third is the meridional circulation in the western Mediterranean, forced by open ocean deep water convection in the Gulf of Lions. The spirals indicate the preferential site for strong heat losses from the ocean to the atmosphere during winter time.

The knowledge of the sub-basin scale circulation is relatively new (Robinson et al., 1991): it has several time scales and an important one is the steady state component. The latter consists of cyclonic and anticyclonic permanent gyres that are wind and thermal fluxes driven, superimposed to and interacting with the Gibraltar inflow system (Pinardi and Masetti, 2000). In Fig. 14 we show the time mean sea surface height (SSH) from a long term simulation of an ocean general circulation model devised for dynamical studies (Demirov and Pinardi, 2002): it shows the northern regions of the basin occupied by cyclonic gyres (the Gulf of Lions gyre, the Tyrrhenian gyre, the southern Adriatic gyre and the Rhodes gyre) and the southern regions interested by anticyclonic gyres and generally high values of SSH. This circulation bears a resemblance to the double gyre circulation of the middle-latitude oceanic regions with reduced space scales and amplitude, due to the limited meridional extension of the basin, the different wind stress strength and the peculiar forcing of Gibraltar.

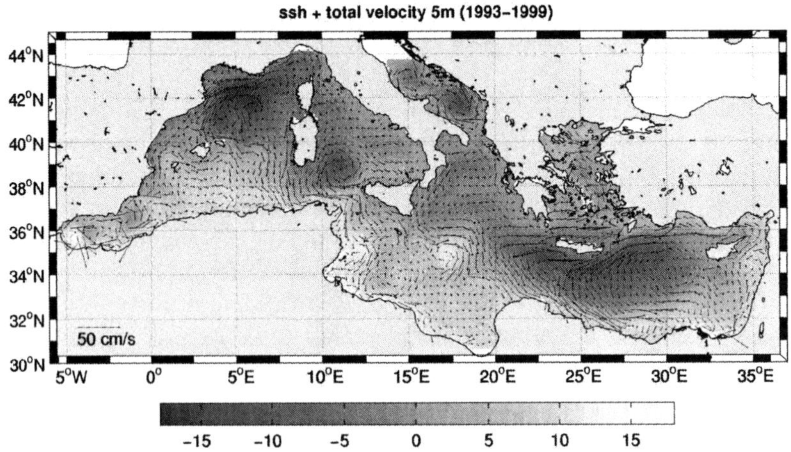

Figure 32.14 The mean sea surface height (ssh) from a 1993–1999 model simulation together with the surface velocity field. The grey bar is in cm and the basin average ssh is set equal to zero.

The time mean anticyclonic gyres of the circulation, while they are evident in theoretical studies of the wind driven circulation without Gibraltar inflow (Pinardi and Navarra, 1993, Molcard et al., 2001), are notably changed by interaction with the Gibraltar-Atlantic current system. The final result is that anticyclonic motion is either limited to a narrow band encompassing the continental slopes and shelves of the south-eastern Mediterranean (see for example the anticyclonic gyre centered at 28^0 E and 31.5^0 N or centered around 35^0 E and 35^0 N) or present in extended areas of the Algerian basin, the Sicily Strait and the southern Ionian Sea. These extended anticyclonic areas are normally occupied by mesoscales and non-permanent anticyclonic gyre structures so that the time mean circulation results in a weak downwelling area.

The sub-basin scale cyclonic (anticyclonic) gyres imply that at their centers upwelling (downwelling) motion prevails while at their borders the contrary occurs. This means that for cyclonic permanent gyres that impinge on the continental slope and sometimes on the shelf, there is an induced downwelling tendency on the shelf. In Fig. 15 we show the SSH again with the schematic of the upwelling/downwelling motion as deduced from the SSH slope. In general we see that downwelling motion prevails near the shelf areas, thus inducing the conclusion that the average sub-basin scale circulation induces open ocean upwelling and shelf areas downwelling. This has a profound influence on the shelf scale ecosystem dynamics, as we will discuss later. The only large shelf upwelling centers are associated with wind driven permanent or seasonal upwellings as in the case of the Sicilian and eastern Aegean coastal areas.

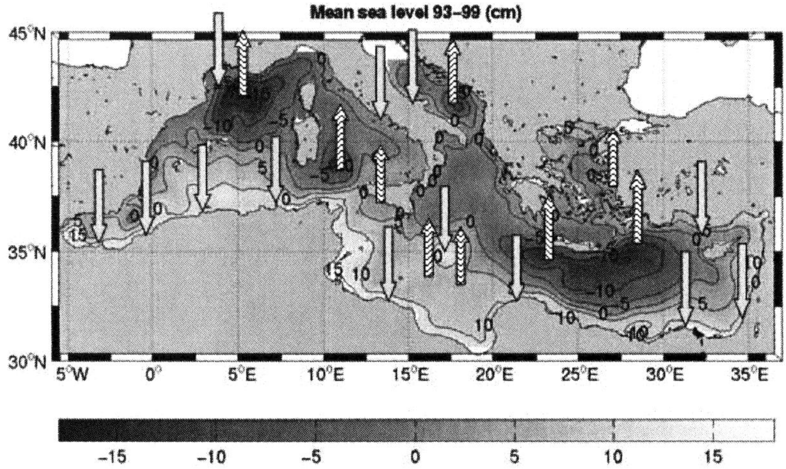

Figure 32.15 The time mean sea surface height of Fig. 32. 14 shown here only with amplitude contours. Arrows indicate the large scale vertical motion direction. The arrows start and point to the specific areas where the motion has the largest amplitude.

The sub-basin scale structures have amplitudes that are seasonally varying mainly due to the large wind stress variability. Coupled to such a seasonal variability there is also a large signal of interannual variability (Molcard et al., 2001, Korres et al., 2000). This implies that the general circulation structure can be quite different each year from the long term mean shown in Fig. 14 and 15. The more realistic schematic of the sub-basin scale structures, which considers interannual structures is presented in Fig. 16. The striking feature of this new picture is that the number of sub-basin scale features has increased and the size of the gyres has decreased.

The Gibraltar forced flow field is now decomposed in several sub-basin current systems, notably the Algerian current in the western Mediterranean, the Atlantic-Ionian Stream in the Ionian Sea and the Mid-Mediterranean jet in the Levantine basin. On the right side of the Gibraltar-Atlantic current system looking downstream (in the eastward direction), the prevailing motion is anticyclonic: we note the anticyclonic Algerian current eddies, the Mersa-Matruh and Shikmona gyres. The anticyclonic tendency in the southern Mediterranean is increased with respect to the steady state mean by the presence of sub-basin scale structures that vary at interannual time scales. Specifically, interannual atmospheric forcing helps to form anticyclonic gyres on the southern flank of the Gibraltar-Atlantic flow system (Korres et al., 2000) and in addition, flow instabilities would favor the formation of anticyclonic gyres on this side of the current system. This process will increase the amplitude of the open ocean downwelling areas thus increasing the north-south oligotrophic gradient suggested by Fig. 6.

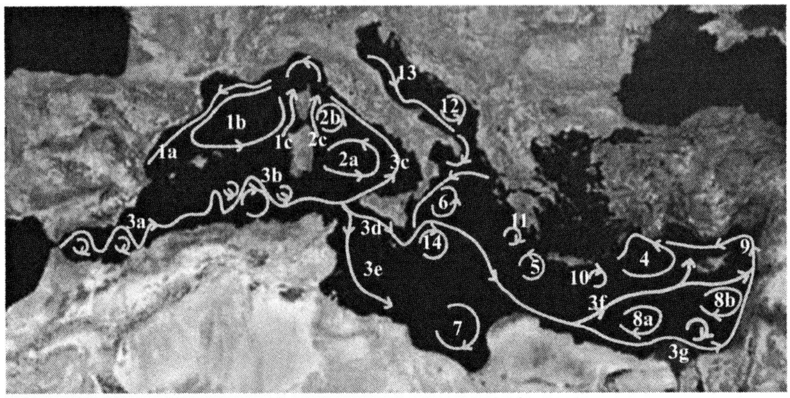

1a Liguro-Provencal-Catalan current (LPC)
1b Gulf of Lyon Gyre
1c Western Corsica Current
2 Northward Tyrrhenian current and gyres:
2a Northward current and Southern Tyrrhenian Gyre
2b Northern Tyrrhenian Gyre
2c Eastern Corsica Current
3 Gibraltar-Atlantic current system
3a Alboran basin Gyres and meanders
3b Algerian current gyres, eddies and meanders
3c Tyrrhenian bifurcation/current
3d Atlantic-Ionian Stream
3e African MAW (Modified Atlantic Water) Current
3f Mid-Mediterranean Jet
3g Southern Levantine current
4 Rhodes Gyre
5 Western Cretan cyclone
6 Western Ionian cyclonic Gyre
7 Syrte Gyre
8 Anticyclonic system of the South-eastern Levantine basin
8a Mersa-Matruh Gyre system
8b Shikmona Gyre system
9 Asia Minor current
10 Iera-Petra Gyre
11 Pelops Gyre
12 Southern Adriatic cyclonic Gyre
13 Western Adriatic Coastal Current
14 Western Ionian anticyclonic Gyre

Figure 32.16 Schematic of the surface circulation from recent observational data and model simulations. Names of structures and currents are listed.

The last circulation scale, the mesoscales, has by definition a shorter time scale than the one associated with the thermohaline circulation and the subbasin scale flow field structures. However current amplitudes are large and eddies are pervasive in the basin (Robinson et al., 1996). The mesoscales have been studied in the past in several subregions of the Mediterranean Sea (Robinson et al., 1987, Hecht et al., 1987, Millot, 1987, Paschini et al., 1993). The eddy-mean flow interaction mechanisms have not been studied yet but there is evidence from satellite altimetry of westward eddy propagation and a seasonal cycle in the eddy kinetic energy (Ayoub et al., 1998). In the Adriatic, evidence suggests that eddies are most frequent during spring and summer, when the atmospheric forcing relaxes and energy is converted from larger to smaller spatial scales (Artegiani et al., 1997).

4.2 Hydrodynamic regimes classification at the shelf scale

In this section we will overview the known circulation structures for the six study case areas: the Gulf of Lions, the Algerian shelf, the Sicily Strait, the Adriatic Sea, the Nile Delta and the Israeli shelf.

4.2.1 The circulation in the Gulf of Lions shelf

As introduced before, the northern shelves are, by and large, Regions Of Freshwater Influence (ROFI) that have been recently classified by Simpson (1997) and the Gulf of Lions is no exception to that.

The current system of this region is dominated on the slope boundary by the Liguro-Provencal-Catalan current (LPC) that flows along the Liguro Provençal and Catalan coast (see Fig. 16). The LPC is 30–50 km wide and it is formed upstream by the convergence of the Western Corsica Current (WCC in Fig. 16) and Eastern Corsica Current (ECC in Fig. 16). The current flows along the edge of the continental shelf, but it can form meanders penetrating the shelf (Auclair *et al.*, 2001; Flexas *et al.*, 2002, Echevin et al., 2003). Millot (1990a) estimated that in the Gulf of Lions the exchanged water with the open sea due to the LPC is 1000 times larger than the Rhône runoff (Table 2).

The LPC carries Modified Atlantic Waters (MAW, Astraldi and Gasparini, 1992) westward during summer that have a significant subsurface salinity minima. In the Gulf of Lions, Winter Intermediate Waters (WIW) can form that are very cold (T<12 °C) and then leave the shelf, forming lenses in the deep ocean at around 600 m depth. The seasonal water mass structure along transect 4 of Fig. 1a is shown in Fig. 17. The shelf cool and fresh waters in March contrast with the warmer and saltier water in the open sea, where the vertical homogenization and open ocean upwelling is evident down to the depth of 500 meters. It is interesting to notice the mixing between shelf and open ocean waters occurring between the depths of 100 and 500 meters, probably due to the WIW leaving the shelf. The summer picture shows a smaller contrast in temperature between shelf and open ocean waters but larger differences in salinity. The shelf, fresher waters now extend further in the open ocean and the subsurface Modified Atlantic Water signal is evident. During summer the LPC is clearly deeper, around 200 meters depth and hugging the escarpment (visible from the downwelling slope of the 38.2 isohaline in Fig. 17).

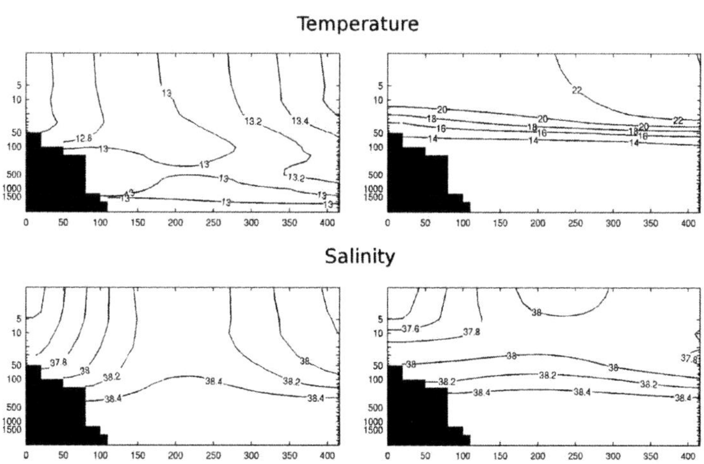

Figure 32.17 The climatological March (left panels) and August (right panels) temperature (upper panels) and salinity (lower panels) distributions along the transect 4 of Fig. 1a. The climatological data are taken from the Medar (2002) data set.

The dynamical nature of the LPC is still under debate but it is evident from Fig. 14 that it is the intensified border of the Gulf of Lions gyre. Its strength and position is influenced by the wind stress curl structure over the region (Herbaut et al., 1996, Korres et al., 2000) and by the frictional dissipation mechanisms against the continental escarpment (Pinardi and Navarra, 1993).

An important topographic feature of the shelf escarpment are the canyons (see section 3) and the so-called Rhône fan that has been indicated as a favorable site for the occurrence of open ocean deep convection (Madec et al., 1991). The LPC forms a deep meander over the Rhone fan (shifting inside the shelf) that produces an offshore cyclonic eddy that could be the preconditioning site for deep water homogenization to occur (Madec et al., 1991).

The westward transport of the LPC on the shelf around 5^0 E is about 2 Sv (Echevin et al., 2003) with a seasonal cycle of 0.2 Sv. Meanders of 30 km and 110 km wavelength form along the slope, propagating westward with 10 cm/s (see Fig. 18). The mesoscale variability has energy peaked at 3.5 and 7 days (Alberola et al., 1995). On the westward side of the Gulf of Lions shelf, relatively large anticyclonic eddies can form that trap waters on the shelf and then dissipate by moving westward (Echevin et al., 2003). The shelf is dominated by transient mesoscale variability induced both by the LPC meandering and directly by the wind.

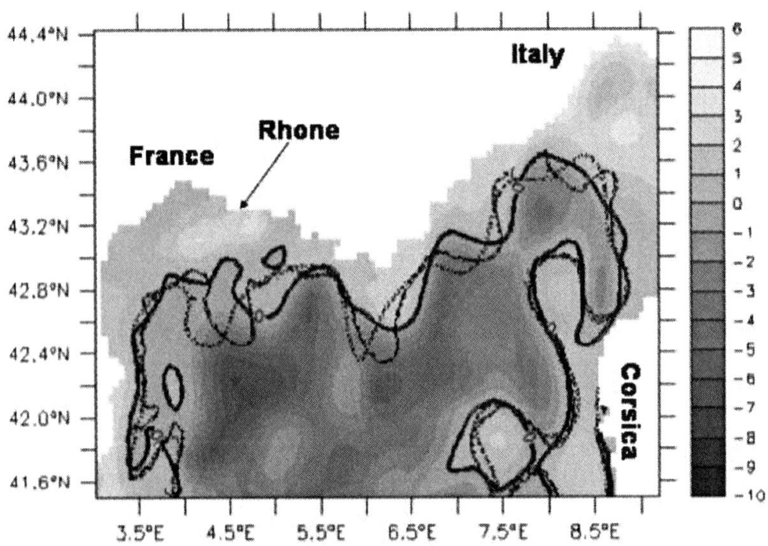

Figure 32.18 Meanders of the LPC depicted from a high resolution model of the north-western Mediterranean (reprinted with permission from Echevin et al, 2003).

The Rhone river plume moves cyclonically on the shelf and extends up to the westward side of the Gulf. During the winter, the shelf waters remain confined on the shelf while during the summer the Rhone plume and the shelf waters in general can extend offshore (Echevin et al., 2003).

4.2.2 The circulation near the Algerian shelf

The Algerian shelf current system is dominated by the dynamics of the Algerian current that is part of the Gibraltar-Atlantic water flow system described previously. Due to the shelf narrowness (see Fig. 2 and Table 1) the open ocean/slope currents intrude on the shelf. Here the slope current is formed by the Algerian current jets that show amplitudes of 50–60 cm/s. These are the largest current values in the Mediterranean basin and the critical shelf extension computed in (1) becomes larger, indicating that the shelf here is totally dominated by slope current intrusions up to the near-shore boundary layer.

The Algerian current forms large meanders and forms well known eddies due to mixed barotropic and baroclinic instabilities of the current. The largest and most persistent eddies are anticyclones (Fig. 19): they separate from the coast after their birth and follow a north-eastward path at the beginning of their life. After they get far enough from the coasts, they propagate backward, in the westward direction. The meandering Algerian current transports Atlantic Water (AW) into the Mediterranean Sea and the waters appear as filaments around the intense eddy field (not shown here).

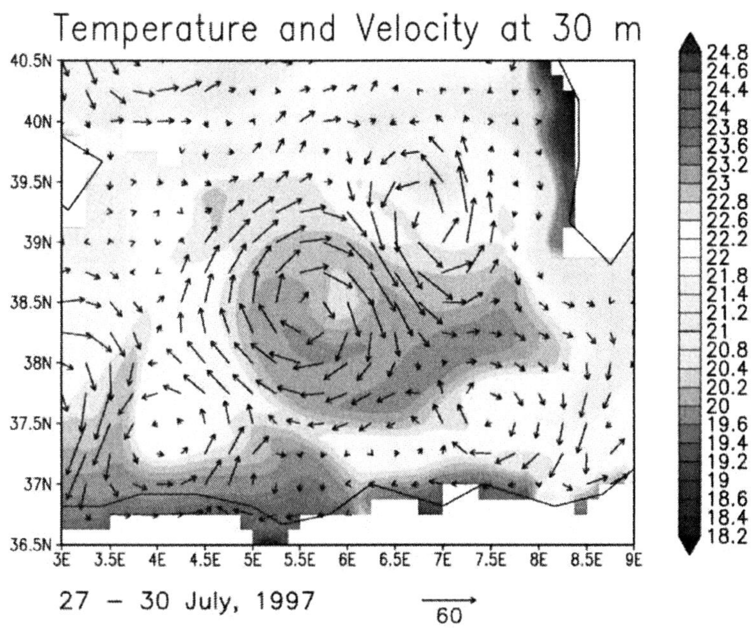

Figure 32.19 The 30 m currents and temperature field from a model simulation for the period 27–30 July 1997 in the Algerian basin.

Smaller scale cyclones are also found (Fig. 19) but they are more evanescent in time and it is difficult to observe them from satellite sea surface imagery. Thus their effects on the overall dynamics of the open ocean areas is quite unknown and in particular their effect on the shelf areas. Millot et al. (1990b, 1997) describes the dynamics of the mesoscale eddies of this region in great details.

This intense mesoscale field produces mushrooms of coastal waters that are pushed offshore by dipoles of cyclonic/anticyclonic eddies (Fig. 19 where warmer coastal waters are meandering outwards from the coasts). On the other hand, anticyclonic eddies, when they are still close to the coasts, might advect nutrient reach open ocean waters on the shelf, allowing for enhanced primary productivity (see also section 6). These blooms are episodic and relatively low in biomass. This process, due to the upwelling occurring at the anticyclones borders impinging on the shelf, has not been fully studied yet.

4.2.3 The circulation in the Sicily Strait shelves

The circulation in the Sicily Strait has been recently mapped by Robinson et al. (1999) and Sammari et al. (1999) and it has been simulated in details by Sorgente et al. (2003) and Drago et al. (2003). Recently an overview of the variability has been published by Lermusiaux and Robinson (2001). The Sicily Strait currents form a two layer system with 1–2 Sv transport in each layer and strong seasonal variability (Manzella et al., 1988, Astraldi et al., 1996). The surface flow field is directed eastward and the subsurface is westward, carrying Modified Levantine Intermediate Waters-MLIW. Here we will concentrate on the entrance of the Gibraltar-Atlantic current system in the Strait (transect 6 in Fig. 2), on the Tunisian shelf and the shelf between Sicily and Malta (part of transect 7 in Fig. 2).

Herbaut et al. (1996) have described for the first time the bifurcation of the Gibraltar-Atlantic flow system before entering the Sicilian Strait. Three branches form (Fig. 16), one going northeastward into the Tyrrhenian Sea, the other entering the central part of transect 6 and the third propagating close to the Tunisian coasts, on the shelf.

The seasonal climatology of transect 6 is shown in Fig. 20. The low salinity isolines mark the entrance of the Algerian current in the Strait from the surface down to 50 meters and on the Tunisian side of the transect. By contrast, a sharp halocline divides the surface waters from the intermediate, salty waters coming from the eastern basin. The wintertime temperature distribution shows interleaving layers of different temperature and the Italian side of the transect is warmer than the Tunisian. Below 100 m., the Italian side of the transect is occupied by a warm, salty water patch that might correspond to a vein of MLIW progressing westward. The summer picture is very different, the MAW layer is now subsurface, between 30 and 50 meters depth and waters have warmed up substantially (up to 10 ^{0}C at the surface) also down to 50–100 meters.

Fig. 21 shows the most recent simulation of the circulation in this area by Sorgente et al. (2003). This picture contains the salient structures of the circulation described in the literature. Robinson et al. (1999) called the segment of the Gibraltar-Atlantic current system entering the central Sicily Strait the Ionian-Atlantic Stream (AIS, Fig. 16) while the current on the Tunisian shelf area has been called by Sorgente et al. (2003) the African MAW current. The AIS flows near the Sicilian coasts, forms two large meanders (especially during the summer months, see Fig. 21b) and exits in the deep Ionian Sea between Malta and Sicily. On the northward side of the AIS upwelling centers form due to the upwelling favorable winds always present in this area. The AIS itself forces upwelling on its northern side and it does so especially during the summer time, when it reaches maximum amplitude. The African MAW current shows smaller wavelength meanders along the shelf slope and its amplitude is seasonal. In the Gulf of Gabes area (Fig. 21) the MAW mixes strongly with adjacent waters and increases its salinity from winter to summer. It is interesting to notice (not confirmed yet by observations) that the simulation shows a reversal of the along slope African MAW current in the region between 14^0 and 17^0 E, going from summer to winter, probably due to the extension in summer of the anticyclonic Syrte gyre (Fig. 16). Due to this summer reversal, upwelling centers may develop along the Libyan shelf.

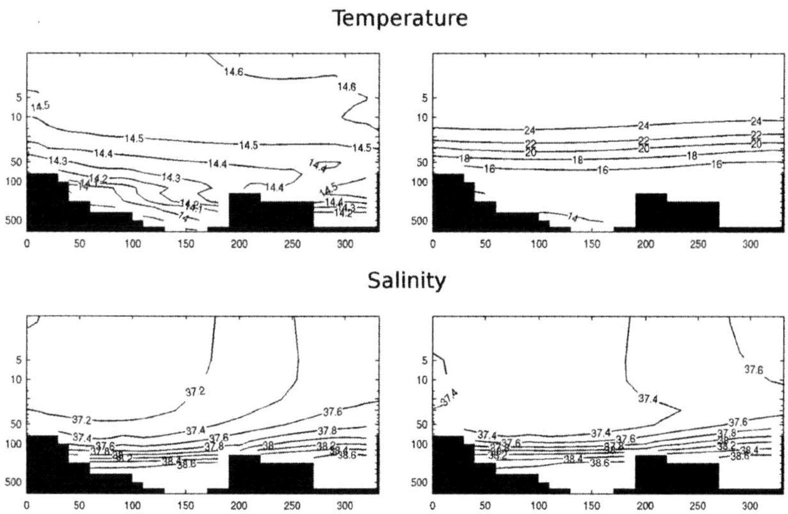

Figure 32.20 The climatological March (left panels) and August (right panels) temperature (upper panels) and salinity (lower panels) distributions along the transect 6 of Fig. 1a. The left side of the transect corresponds to the Tunisian shelf, the left to the Italian shelf. The climatological data are taken from the Medar (2002) data set.

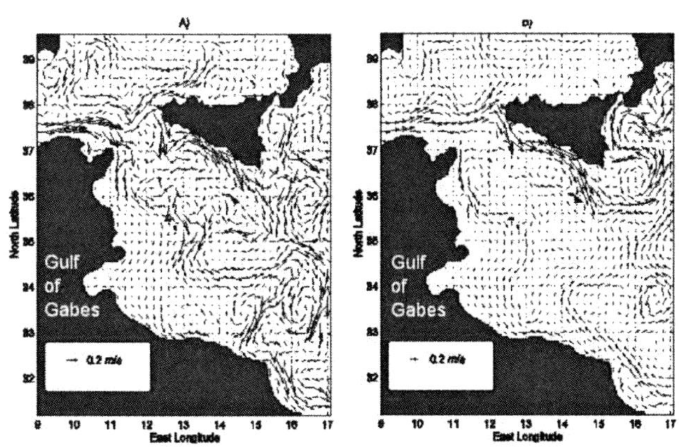

Figure 32.21 Surface circulation for (a) February and (b) August from Sorgente et al. (2003) simulations. Reprinted with permission.

As we said in section 2, Malta and Sicily are connected by an extended shelf (see Fig. 1a) called the Malta channel by Drago et al. (2003) where the Sicily upwelling center can extend 100 km offshore from the Sicily coast. The dynamics of this area is dominated by the latitudinal excursion and the amplitude changes of the AIS. This current structure determines the extreme productivity of these shelf waters where spawning and nursing grounds are found for most fish species of the area (see section 7).

4.2.4 The circulation in the Adriatic shelf areas

The circulation of the North and Middle Adriatic Sea has been described recently from observations and numerical simulations (Cushman-Roisin et al., 2001, Zavatarelli et al., 2002, Zavatarelli and Pinardi, 2003). The actual shelf break for the Adriatic Sea is commonly considered to be at the Pelagosa sill (see Fig. 9) but for this discussion we will concentrate only on the shallower part of the Adriatic Sea, down to Ancona (see Fig. 9).

The Adriatic Sea shelf area is controlled by air-sea fluxes and inflow of heat and salt at the shelf break. It is clearly a ROFI area influenced not only by the Po river but also by all the northern Adriatic rivers discharging along the Italian side of this area (see Fig. 9). The fluxes of heat and water combine in the so-called buoyancy budget that is very nearly zero averaged over the whole Adriatic Sea, due to the contrasting contribution of the heat and fresh water fluxes in this area. The buoyancy flux is written in fact:

$$B \quad \frac{{}_T g}{C_p} Q \qquad g \; S \; (E \quad P \quad \frac{R}{A}) \tag{4}$$

where Q is the net heat flux at the air sea interface (Maggiore et al., 1998), C_p is the specific heat, ${}_T$ is the coefficient of thermal expansion, is the coefficient of haline expansion, g is gravity, S is the surface salinity, E the evaporation, P the precipitation and R the runoff ($\frac{m^3}{s}$) divided by the cross-sectional area , A, at the river mouth.

In the Adriatic, Q is negative on a mean annual basis (22 $\frac{W}{m^2}$ from Artegiani et al., 1997) and also ($E \quad P \quad \frac{R}{A}$) is negative, mainly due to R. This gives rise to a balancing effect in (4). This means that the basin would be forced by the heat flux to work as an anti-estuarine marginal sea, forming deep waters, while the water budget would impose an estuarine circulation, as for a dilution basin. The dynamics of the Adriatic shelf is dominated by the balance between these two competing mechanisms in (4).

The overall water mass structure of the Northern and Middle Adriatic basin is dominated by the seasonal cycle in the heat fluxes and the runoff as it can be seen from Fig. 22. The water masses are totally renewed every year in the Northern Adriatic while the Middle Adriatic receives or renews locally its deep waters (between 200 and 250 meters depth) every few years only (Zavatarelli et al, 1998, Cushman-Roisin et al, 2001). The deep waters of the Northern Adriatic are the heaviest waters in the Mediterranean ($_{T}$ 29.5) and they slide southward, toward the Middle and the Southern Adriatic, entraining local waters and losing their signature. The signature of the cold waters is present in Fig. 22 where the isotherms bulge southward at the bottom of the transect. To be noted is the inverse vertical temperature gradient with waters cooler at the surface than at depths. This is allowed by the fresher surface waters that maintain a stable vertical density gradient. During summer, due to the spring discharge the water are fresher at the northern end of the Adriatic Sea while they maintain the same salinity in the middle Adriatic Sea. The large seasonal thermocline forms every year, between 10 and 30 meters depth, isolating the deep, cold waters present in the Middle Adriatic Pit (located at about 350 km along the transect of Fig. 22).

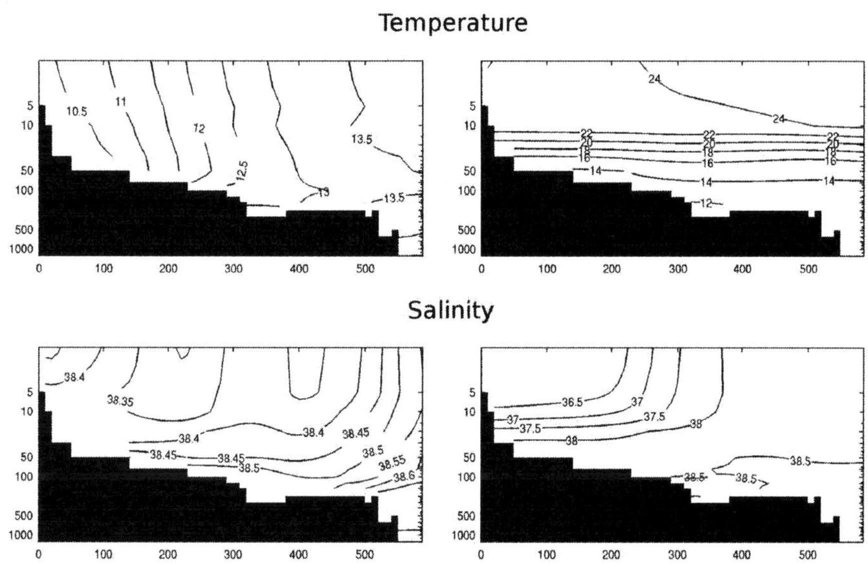

Figure 32.22 The climatological March (left panels) and August (right panels) temperature (upper panels) and salinity (lower panels) distributions along the transect 9 of Fig. 1a. The left side of the transect corresponds to the Northern Adriatic. The climatological data are taken from the Medar (2002) data set.

The other important driving mechanism for this area is given by the wind stress that has two very important regimes: the Bora or easterly wind and the Sirocco or southerly wind. The Bora is channeled through the Dinaric Alps and it forms intense jets that have positive and negative curls (Pullen et al., 2003). This induces strong changes in the vorticity sign of the circulation and sometimes reversal of the cyclonic circulation of the basin occurs in portions of the basin, especially the shallow, northern parts. The Scirocco, on the other hand, is forcing high sea level in the Northern Adriatic and could reverse the sign of the currents along the Italian coastlines (Poulain et al., 2003).

The circulation is generally cyclonic in the northern shelf areas and it is composed of a western boundary intensified current, called the Western Adriatic Coastal Current (WACC). The latest observational evidence of the Adriatic Sea circulation structure is given by Poulain (2001). In Fig. 23 we show the result of the simulation from Zavatarelli and Pinardi (2003), which synthesize the evidence for the changes in the seasonal circulation. During winter the circulation is dominated by the WACC and the overall scales are large. We note however, that the northward current on the eastern side of the shelf (along the Istrian coasts) tends to separate from the coasts, cutting the cyclonic circulation north of the Po River from the circulation south of it. We call this the cyclonic cut-off. During summer, the circulation is still cyclonic along the Italian coasts but a very large countercurrent (called the Istrian Coastal Counter-Current (ICCC), Supic et al., 2000) forms on the eastern side of the northern Adriatic. During summer, the cyclonic cut-off of the circulation has branched into two currents: one is southward of its winter position, toward the Middle Adriatic Sea, leaving a larger area dominated by mesoscales. The anticyclonic circulation north of the cyclonic cut-off can be also wind induced, if the Bora jet occurring normally above the Istrian peninsula is shifted southward or it is highly sheared. In conclusion we can say that the Adriatic shelf circulation is dominated by wind stress curl and thermohaline forcings, changing structure and direction of currents at least seasonally.

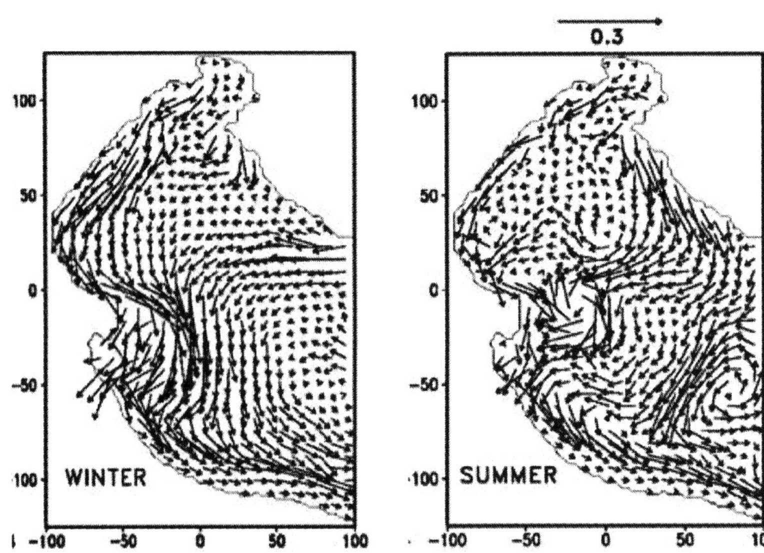

Figure 32.23 Northern Adriatic shelf circulation from model simulations (Zavatarelli and Pinardi, 2003). The reference arrow is in m/s.

4.2.5 The circulation in the south-eastern Levantine: the Nile Delta and Israeli shelf areas

The Egyptian and Israeli shelf areas are opposite in extension (see Table 1) but the latter is strongly interconnected by the open ocean flow field to the Egyptian slope current regime and then to the sediment and biochemical fluxes of that region. Thus they will be examined together in terms of the open ocean/slope currents.

For the near coast flow field the information on the Nile Delta coastal area is very scarce. It is evident that the Egyptian shelf is still a ROFI area (Hamza et al., 2003) but very different from pre-Aswan dam period. The damming completely changed the Nile runoff from 1968 onward, the annual total discharge averaging only one-tenth of the average value for the period prior to 1964 (see Table 2) and having the maximum discharge in winter. On the other hand, the Israeli shelf is very narrow both by morphology and by dynamical considerations. The runoff by the local rivers can account for a conspicuous part of the sedimentation rate of the Israeli shelf (see section 3) but nothing is known about the influence on the local hydrodynamics.

The climatological distribution of temperature and salinity for the two months of March and August is shown in Fig. 24. First of all, the salinity distribution shows that the Nile signal is present on the Egyptian shelf and it extends as far as several hundred kilometers from the coasts. The Levantine Intermediate Water layer is present both in winter and summer and hugs the shelf escarpment between 100 and 300 meters depth, thus influencing the shelf slope area. It is interesting to notice that during summer the Modified Atlantic Water is present on the shelf by a subsurface salinity minimum between 30 and 100 meters. The isotherms and isohalines are quite flat but a dominant downwelling structure can be recognized in the distributions near the shelf break.

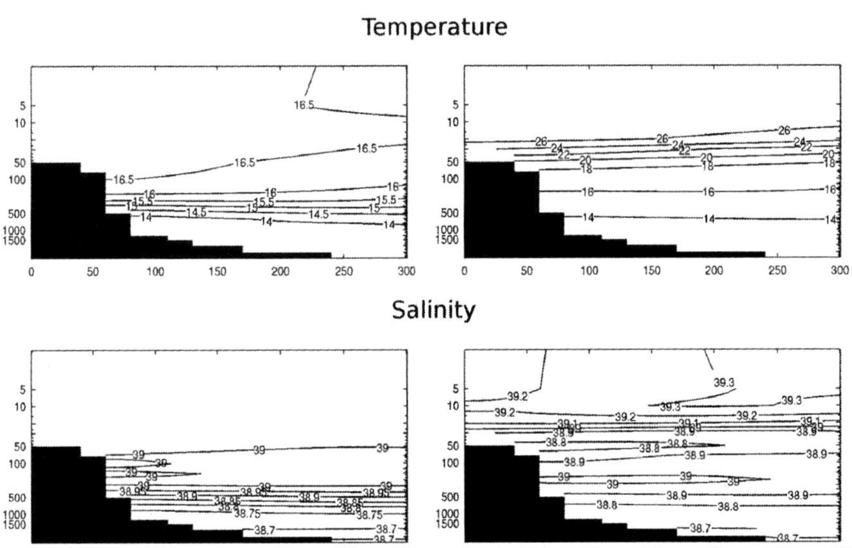

Figure 32.24 The climatological March (left panels) and August (right panels) temperature (upper panels) and salinity (lower panels) distributions along the transect 15 of Fig. 1a. The left side of the transect corresponds to the Egyptian coast. The climatological data are taken from the Medar (2002) data set.

The open ocean flow field in the southeastern Levantine has been mapped first by the MC cruises (Hecht et al., 1987) and then by the POEM experiment (the POEM group, 1992). It has been also recently simulated by a high numerical resolution seasonal model of the Levantine basin by Korres and Lascaratos (2003). These simulations offer a unique opportunity to see the possible current regimes on the Egyptian shelf areas and slope. The simulations are reproduced in Fig. 25 for the winter month of february. The slope current field is always in the eastward direction but intense meandering is occurring upstream of the Nile Delta shelf area, around the Mersa-Matruh anticyclonic gyre (see Fig.16). The Nile Delta slope current is formed by the convergence of a southward branch of the Mid-Mediterranean Jet (MMJ) and the slope eastward current, called the southern Levantine current in Fig. 16, always present along the African coasts. On the shelf, it seems that anticyclonic circulation is possible, in the near-shore area. From the circulation picture of Fig. 25 it is evident that the Nile Delta area behaves as an extended shelf area where the slope current and the near coast currents do not interact strongly.

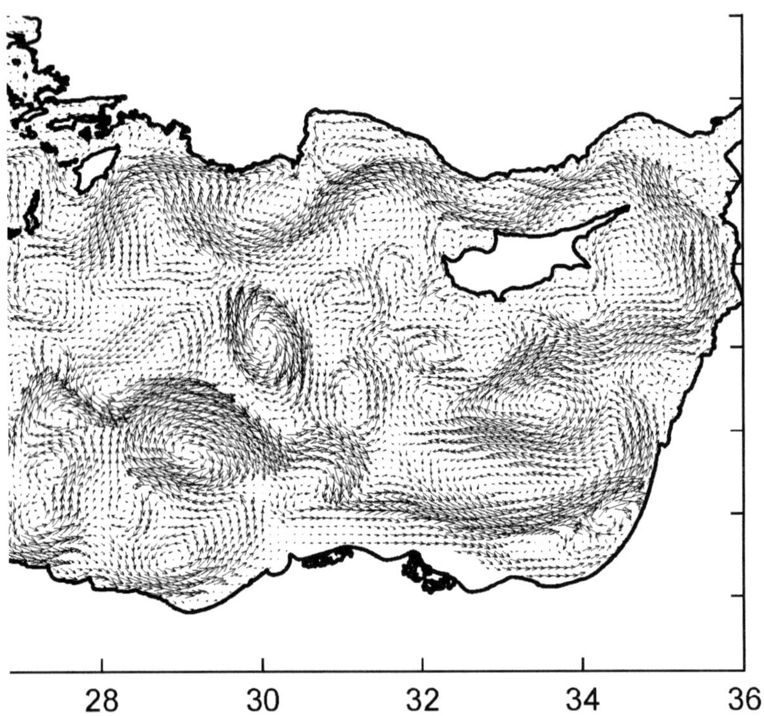

Figure 32.25 Circulation of the southeastern Levantine basin from a model simulation (from Korres and Lascaratos, 2003)

The Israeli slope currents are northward but large meandering occurs also there, together with eddy detachment (Brenner, 2003). The large anticyclonic eddy in the open ocean areas of Fig. 25, in front of the Israeli coasts, is the Shikmona gyre (see

Fig. 16) that has been documented by many authors (Hecht et al., 1988, Brenner, 1991). It is possible that, due to the nonlinear dynamics of such eddy formation processes, convergence and divergence areas would appear in near shore areas, with local upwelling and downwelling centers. During summer, the shelf areas are much warmer than the open sea and the mesoscale activity is at the minimum. Current meter data on the shelf (Brenner, 2003) show that the shelf currents are all the year long northward except for September where local reversals may occur, when the slope current is farthest from the continental shelf edge.

4.3 Meteorological forcing of the circulation

The meteorological forcing of the Mediterranean Sea is particularly complex and presents a strong seasonal variability. The atmosphere forces the ocean and its coastal areas by transferring momentum, heat and water across the sea surface. The heat and water fluxes combine in the buoyancy flux that has been written in equation (4)

The water flux has been discussed in section 2.2 and we know that on the long-term mean it is positive. The heat flux in the Mediterranean Sea is known to have a negative long-term mean equal approximately to $7\frac{W}{m^2}$ (Garrett, 1983, Castellari et al., 1998). The structure of the heat flux is consistent with large heat losses during the winter that overcome the heat gains during summer. The areas with largest heat losses are the Gulf of Lions gyre, the Rhodes gyre, the northern Aegean Sea and the Adriatic Sea, where deep and intermediate waters form (heat losses greater than $1000\frac{W}{m^2}$ have been recorded in the Gulf of Lions, Schott and Leaman, 1991).

Roussenov et al. (1995) and Korres et al. (2000) noted that the southern Mediterranean shelf areas are sites of intense air-sea interactions during winter, due to the relatively large temperature differences between ocean and atmosphere. During summer, the upwelling areas in the southern Sicily and Turkish coasts of the Aegean Sea are sites of intense heat gain due to the relatively low temperatures of the water.

In order to quantify the effect of shelf areas on the total heat budget of the basin, we calculated the heat flux with and without shelf areas (defined as the part of the basin with depth less than 200 m) from a 21-year simulation of the circulation with the model described by Pinardi et al. (2003). The results are presented in Table 7: they show that the shelf areas account for about 20% of the basin mean heat losses and their contribution is extremely variable with the years. The overall negative contribution to the mean heat budget due to the shelf areas reinforces the concept that the Mediterranean shelf areas are predominantly downwelling sites where the heat exchange with the atmosphere is large and it contributes to the negative heat budget of the basin. The model used in the computations of Table 7 does not contain the Northern Adriatic, thus this important contribution is neglected. However, this would make the results presented here even more consistent with the interpretation given above since the annual mean heat flux over the Northern Adriatic is negative.

TABLE 32.7

The basin mean heat flux over the basin in units of $\frac{W}{m^2}$.

The heat flux has been computer from a model simulation that does not include the northern Adriatic Sea shelf area (seen in fig. 14 by the absence of arrows above 43° N) oin the overall basin mean.

Year	Basin mean heat flux	Basin mean without shelf	Difference	Percentage of change
1979	9.6	10.2	-0.5	6.
1980	8.7	8.8	-0.1	1.
1981	0.5	0.6	-0.1	24.
1982	-8.2	-8.5	0.3	4.
1983	4.2	4.5	-0.3	8.
1984	2.4	2.2	0.2	8.
1985	2.5	2.3	0.2	7.
1986	-5.9	-6.2	0.3	6.
1987	-2.4	-3.3	0.9	39.
1988	-4.0	-4.6	0.6	16.
1989	9.8	9.9	-0.1	2.
1990	1.4	1.2	0.3	19.
1991	-9.9	-10.5	0.6	7.
1992	-5.0	-6.0	1.0	20.
1993	15.9	15.9	0.0	1.
1994	9.6	12.4	-2.9	31.
1995	4.5	7.4	-2.9	65.
1996	-2.4	-0.8	-1.6	66.
1997	4.7	6.1	-1.4	30.
1998	-4.8	-4.2	-0.7	14.
1999	-3.1	-3.7	0.6	21.
Total	1.4	1.7	-0.3	19

The momentum flux is given by the wind stress at the sea surface. In Fig. 26 we show one estimate of the wind stress for the period 1993–1999. The wind stress is dominated by the two intense structures of the Mistral in the Western Mediterranean and Etesian winds in the Eastern basin. These two structures are both present in the annual mean but they have larger amplitudes in winter for the Mistral and in summer for the Etesian. We should remark that this mean wind stress pattern is changing with a decadal time scale, following the NAO index phases. This change is documented in Demirov and Pinardi (2002) and Samuel et al. (1999) and it has profound influences on the circulation. Raicich et al. (2003) remark also that the Etesian winds are the lower branch of the African Hadley cell system that during summer extends its influence over the Mediterranean Sea.

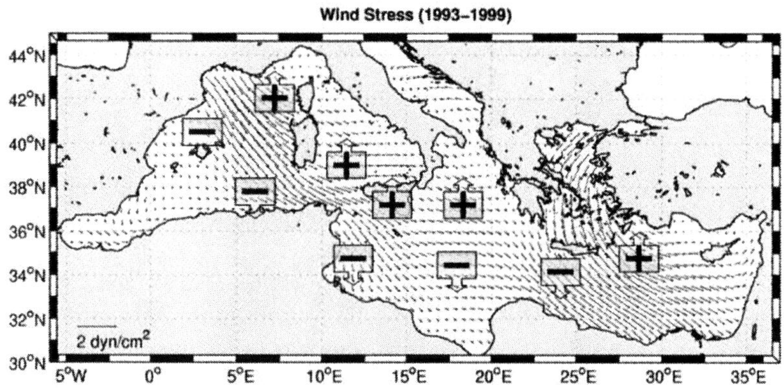

Figure 32.26 The average wind stress computed from four daily ECMWF operational surface wind analyses for the period 1993–1999. The boxes indicate the Ekman certical velocity sign. Positive means upwelling, negative downwelling.

The other important wind systems (Scirocco and Libeccio, Bolle, 2002) are northward but they are less frequent and intermittent than the Mistral and the Etesian and they do not appear on the average. However, they can start intense upwelling along the Italian coasts of the Adriatic Sea (Poulain et al., 2003) and/or along the western coasts of the Ionian Sea.

Another important forcing on the open ocean as well as the shelf areas of the Mediterranean Sea is the curl of the wind stress. The curl of the wind stress has been described in great details in many other papers (Molcard et al., 2002, Demirov and Pinardi, 2002, Josey et al., 2000) and it is formed by dipoles of positive and negative vorticity input on the right and left sides of the winds looking downward in their direction. These dipoles are depicted in Fig. 26 and are responsible for forcing many of the permanent and semi-permanent cyclonic and anticyclonic gyres described in section 4.1. Thus the wind stress curl is important and is related to the vertical velocity, w_E at the base of the surface Ekman layer:

$$w_E \quad \hat{k} \qquad \frac{\vec{\ }}{f} \qquad \frac{}{x}\,\frac{y}{f} \qquad \frac{}{y}\,\frac{x}{f}$$

where $\vec{\ }$ $\quad {}_{x}, {}_{y}$ is the wind stress with its components, is the density and f is the Coriolis parameter. In Fig. 27 we show the basin average wind stress curl over the Mediterranean Sea. The wind stress curl is generally positive over the basin, thus inducing a cyclonic vorticity input but it weakens and even reverses during summer, thus producing the enhancement of the anticyclonic circulation structures, permanent and semi-permanent. Note the very low cyclonic vorticity input in the 1989–1990–1991 years that is partly responsible for the changes in circulation that occurred in these years and that induced a strong anticyclonic vorticity in the flow field (Demirov and Pinardi, 2002). After 1993, the cyclonic input of vorticity in the basin almost tripled thus generating stronger slope currents such as the LPC and the Asia Minor Current (Fig. 16). This is clearly going to be important for the shelf area dynamics, which, especially for the narrow shelf areas, is profoundly affected by the slope current intensity.

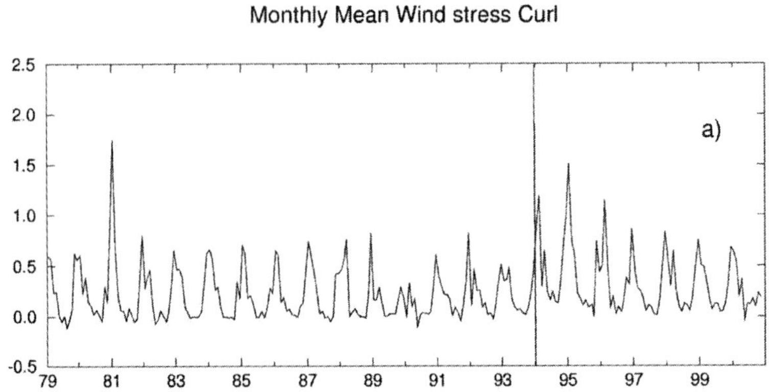

Figure 32.27 The area average wind stress curl for the 23 years from 1979 to 2001 calculated from ECMWF surface wind field. Units are dyne/cm^2.

5. Open ocean-shelf areas coupling

5.1 The Continental Shelf Pump

Runoff from rivers annually delivers ~0.8 Pg of carbon to the Mediterranean margins , 45% of which is organic, primarily in dissolved form (Liu et al., 2000), leading some authors to assert that the continental margins are net heterotrophic systems (Smith and Mackenzie, 1987). However a recent study suggests that the margins as a whole are a weak net CO_2 sink of about 0.1 Pg yr^{-1} (Liu et al., 2000). The ocean shelves can act as net carbon sinks if they transfer a substantial part of the primary production organic carbon to the oceanic interior through the combination of physical and biological processes known as Continental Shelf Pump (CSP) (Tsunogai et al., 1999). In the Mediterranean, even if the Adriatic shelf has been recognized as the main candidate as an effective CSP site (G. Civitarese, pers. comm.) proper quantification and its relative importance is still far from being fully assessed.

There is another important aspect of the Mediterranean shelves that should also be considered for its biogeochemical implications. Although the shelf surface is proportionally large compared with the World Ocean, the shelves are much narrower. What makes the Mediterranean Sea peculiar under this aspect is that almost all well developed shelves are located on the northern side of the basin and dense waters formed on the shelves are prone to replenish and spread into the ocean interiors from the northern side of the basin, contributing in creating north-south gradients in the system.

Moreover, Adriatic and Aegean Seas are elongated inlets where the exchanges are confined to a relatively small open boundary. This means that the shelf water has a mean residence time that can be long (time scales of months-years). This long lasting permanence allows the editing of the quality and quantity of organic matter and major chemical constituents in the water column along its seaward transit through production, decomposition, and other biotic and abiotic modifications. Biochemical (bacterial production and decomposition, photochemical modification), benthic-pelagic coupling and physically-mediated processes (adsorption, flocculation, settling) perturb the original biochemical signature of the land derived organic materials that arrive to their ultimate long term storing compartments heavily transformed.

As an example, the relevance of this complex suite of biogeochemical processes has been testified in a recent study carried out in the Ionian Sea (Seritti et al, 2003) where the slope in DOC/AOU profiles has been connected with the origin and age of the water masses (Adriatic vs Aegean). In general, the resulting composition of organic compounds and key elements in all their forms leaving the continental margins are dependent on a shelf-specific history and in turn, in the long run, affect the whole Mediterranean biogeochemical cycles.

5.2 The deep ocean-shelf exchanges

We have seen in section 2 and 4 that the mechanisms active in the deep ocean–coastal ocean interactions in Mediterranean are site-specific. However the common feature that allows some classification of the shelf dynamical regime can be found in the flow field that develops on the barrier constituted by the shelf slope. Along the shelf slope, the current field $\mathbf{v}$ (u, v) is in geostrophic balance, that means that :

$$\bar{\mathbf{k}} \quad \mathbf{v} \quad \frac{\mathbf{v}}{z} \quad 0$$

where $\bar{\mathbf{k}}$ is the unit vector in the vertical direction (Brink, 1998). As Sanchez-Arcilla and Simpson (2003) pointed out, the slope currents will interact with the shelf areas if they break the geostrophic constraint. The advection of momentum is the term that is responsible of the departure from geostrophy, more than other ageostrophic terms. The relative importance of nonlinearity is usually estimated by means of the Rossby number,

$$Ro \quad \frac{U}{fL}$$

already defined in section 2. Ro is O[10^{-3}] in Mediterranean if we take the open ocean flow scale variables, and thus the advection is small compared with the Coriolis term. However near the shelf break length scales become narrower and the geostrophic constraint can be broken by nonlinear advective processes. Several categories of non-linear processes control in the Mediterranean the open sea-shelf exchanges and they will be treated separately in the sections below.

5.2.1 Slope and coastal currents spill-outs over the topographic obstacles

The diverting of a general circulation current (also called slope current) into a shelf is commonly found in the Adriatic and Gulf of Lions areas. The pressure gradient induced by buoyancy fluxes in conjunction with the wind driven circulation (with variable intensity) explains the departure from geostrophy of the mean circulation. For sake of an example, we will discuss here the Adriatic case only.

Gacic et al. (1999) calculated from direct measurements that about 25% of the water entering Otranto Strait is able to pass over the Pelaguza sill (see Fig. 3.3) shelf break toward the Central Adriatic. These waters are carried by the large-scale northward Adriatic slope current that develops especially during autumn and winter (Artegiani et al., 1997), connected to the Otranto inflow. We call this an example of slope current spill out over the shelf break.

On the other hand, the WACC carries nutrients southward in a way that can be compared to a 'coastal current spill-out' out of the shelf break (located south of the Pelaguza sill). In fact the net seasonally integrated nutrient transport estimated at the Pelaguza transect is always directed seaward with a summer-to-winter variability that ranges from 7140–14780 Mmol DIN to 456–588 Mmol phosphate. Silicate seasonal variability closely follows DIN dynamics (11760–25780 Mmol)

These figure are subjected to changes according with the general circulation pattern of the region. The change of the thermohaline circulation in the Eastern Mediterranean (Roether et al, 1996), the EMT, has been found to have consequences on the Adriatic shelf dynamics. Due to the EMT, the Levantine Intermediate Water has been substituted with Cretan Intermediate Water and this has allowed the intrusion of alloctonous plankton species, typical of Levantine basin, up the northernmost part of the Adriatic (Gulf of Trieste).

Frontal instabilities Instabilities of the slope and coastal fronts are usually related to the strong baroclinic structure of these features and the quasi-geostrophic two-layer theory predicts that the scale of such instabilities is related to the wavelength of the fastest-growing meanders (Pedlosky, 1979) that is:

$$2\ R_m$$

where R_m is the geometric mean of internal Rossby deformation radii in each of the two layers.

Slope current instabilities produce eddies that may detach from the slope. Long-lasting anticyclonic northwestward-prograding eddies found in the Algerian basin (Arnone and LaViolette, 1986 and section 4) are generated by baroclinic instability of the Algerian Current and they impinge on the shelf areas modifying the nutrient budgets. In section 4 we have discussed their effects on the shelf areas that produce open ocean-shelf exchanges and in section 6 we will discuss their effects on biogeochemical fluxes.

Baroclinic instabilities of the Western Adriatic Coastal Current-WACC, that is the most noticeable coastal current front in the whole Mediterranean Sea, are commonly found in images in the visible band and, seasonally, in the infrared band, but only after the Adriatic shelf break, located at the Pelaguza sill, the frontal instabilities can be effective in transferring shelf waters offshore.

Cold filaments have been observed in the Adriatic Sea, (Borzelli et al., 1999) and have been explained as a baroclinic extension of the Rossby adjustment on a sloping bottom. Filaments-type instabilities were also reported along the Spanish sector of the LPC (see Fig.4.4, Font et al., 1995).

Another frontal instability occurring in the Mediterranean shelf areas is related to the upwelling centers. Mesoscale structures are found in the Sicily Strait, where wind driven coastal upwelling generates cold filaments that intrude in the vein of

MAW entering the EMed through the Sicily Strait (Buongiorno Nardelli et al., 1999).

Coastal and slope currents, upwelling front instabilities deserve more studies to assess the effectiveness of this mechanism in determining the resulting lateral fluxes, despite the large number of observations already available. Although mesoscale processes are frequently described in the slope areas of the Mediterranean shelves, a quantitative assessment of their importance is not yet available.

Wind driven coastal upwellings Upwelling processes determine exchange of water masses and properties at the shelf break and thus open ocean-shelf interactions. Alongshore windstress induces cross-shore Ekman transport. The long-wave approximation (time scales longer than inertial period, small frictional effects, alongshore scales >> cross-shore scales) allows finding a stationary solution of the depth integrated equations of motion, where surface and bottom Ekman transports balance the interior inviscid transport. This simplified view of the upwelling is modified by the space and time variability of the wind regime, which are prominent features of the Mediterranean area (if compared with the steady winds present in the open ocean, see Fig. 4.12). In this case spatial variability induces Ekman pumping while the short-term wind pulses can drive coastal upwelling also in the case of substantial cross-shore wind stress component (Brink, 1998).

Recurrent upwelling areas in the Mediterranean encompass the southern Sicily and Calabria coasts, Southern Spain and the eastern Adriatic and Aegean Sea coasts. In general, the Mediterranean wind driven coastal upwelling is believed to be not very efficient if compared with similar processes occurring in other parts of World Ocean and therefore its study in Mediterranean has been largely neglected. However, owing to the scarcity of macronutrients in the euphotic zone during most of the year, a reassessment of the upwelling areas importance for the nutrient basin budgets would be highly desirable.

5.2.2 Buoyancy forces: ROFI areas

Large Mediterranean ROFI areas where buoyancy input due to river runoff determines the coastal currents that eventually cross the shelf break are present in the Adriatic Sea (Po), Gulf of Lions (Rhone) and Catalan (Ebro) shelves. Typical transport can be calculated for the Mediterranean coastal currents assuming that buoyancy is balanced by inertial forces (and friction/mixing can be in first approximation disregarded because of small energy associated with tides). The transport in these wedge-like currents is:

$$Q \quad \frac{g' h_0^2}{2f}$$

where g' is the reduced gravity, and h_0 is the depth of the wedge (Hill, 1998). Entrainment in the current is an effective way to increase the transport/export of lower layer water and its properties.

The interactions with the wind show that in the case of downwelling favourable wind regime, the coastal current tend to adjust along the coast and maintain its integrity, while in the case of upwelling, the coastal current can be radically changed from its adjusted configuration.

5.2.3 Dense water formation downflows

Dense water formed in the Mediterranean shelves contributes to the ventilation of deep layers (Gulf of Lions and North Adriatic) of the Mediterranean. The down-sloping of dense water from the shelves offers another open-ocean shelf exchange mechanism that is extremely active in the Mediterranean Sea.

Generally, more attention has been paid to the open ocean dense water formation processes in the Mediterranean Sea but it is evident that the shelf deep waters contribute to the open ocean deep-water formation processes mainly because of their vicinity to the area of open ocean formation. In the Adriatic, the thermohaline characteristics of the dense water outflowing at the Otranto sill clearly require a contribution from the northern shelf waters.

Recently Aegean dense waters formed in the Cycladic Plateau have been found to be a major contributor to the Cretan Deep Water (CDW) masses that spilled over the Cretan sills into the abyssal plains, giving rise to the Eastern Mediterranean Transient-EMT.

Since the CDW is characterized by higher nutrient and lower oxygen concentrations than those normally measured at the same depths of the Cretan Sea, we can say that the deep ocean has been 'fertilized by the shelf areas'. The biogeochemical implications of this fertilization has been claimed as the putative origin of sporadic phytoplankton blooms reported in '90s in the Cretan Sea (A. Tselepides, pers. comm.). Sediment trap experiments put in evidence that the Aegean deep waters exported about 2 mg $m^{-2}d^{-1}$ of organic carbon. The POC content in settling material was relatively high (3.5%-12%) indicating a significant organic input to the benthic system; during the period of highest stratification (June-September), near-bottom mean organic carbon fluxes amounted to 1.4 mg $m^{-2}d^{-1}$ on the Aegean side and 8.7 mg m^{-2} d^{-1} on the Ionian side.

6. Pelagic ecosystem functioning in the Mediterranean Sea shelf areas

6.1 Background

Primary productivity estimates for the whole Mediterranean Sea range between 80 (Sournia, 1973) and 125 g C m^2 y^{-1} (Morel and Andrè, 1991; Antoine *et al.,* 1995). Estrada (1996) notes as these productivity estimates agree with the general view of the Mediterranean Sea as an oligotrophic ecosystem, this characteristic being maintained mainly (Hopkins, 1985), but not only (Crispi *et al.,* 2001), by the anti-estuarine thermohaline circulation of the basin and marked by a generalised phosphorus limitation (Berland *et al.,* 1980; Krom *et al.,* 1991; Thingstad and Rassoulzadegan, 1995; Thingstad *et al.,* 1998; Zohary and Robarts, 1998). The latter affects the structure and the functioning of the marine food web (Thingstad and Rassoulzadegan, 1999). In the Mediterranean pelagic system the food web is temporally and spatially dominated by the microbial food web where a large part of the carbon fluxes go through the small phytoplankton, protozoa and bacteria (Thingstad and Rassoulzadegan, 1999; Tanaka and Rassoulzadegan, 2002). The microbial food web is characterized also by phytoplankton-bacteria competition for inorganic nutrients (Thingstad and Rassoulzadegan, 1995; Thingstad *et al.,* 1997) and bacterial carbon production constitutes a significant proportion of the primary production (Turley *et al.,* 2000).

It is within this general ecosystem structure of oligotrophy, phosphorus limitation and microbial food web dominance that the coastal areas of the Mediterranean Sea are embedded.

It has been noted elsewhere in this chapter (see section 2), that in the Mediterranean Sea, two different kinds of coastal areas can be identified in dependence of the extension of the continental shelf (see fig. 1a and Table 1) and the dynamical control by the slope currents (eq. 1). Moreover, for the southern Mediterranean coastal areas, characterized mainly by narrow shelves and by absent (or non significative) river runoff contribution, the open ocean current regime (depicted in section 4) determines the shelf ecosystem productivity, since the upwelling of open ocean waters is the almost unique process providing nutrients to the euphotic zone. Therefore, the ecosystem functioning in the narrow shelf areas is expected to be mostly similar to the pelagic open ocean environment and characterized by a microbial food web dominance.

On the other hand, Mediterranean Sea coastal areas characterized by extended continental shelves have all significant runoff (Catalan shelf, Gulf of Lions area, Northern Adriatic), determining significant input of nutrient and organic matter. Thus they are less influenced by the open ocean pelagic ecosystem structure. The open ocean general circulation affects episodically, but still significantly (see section 5) such areas, in the form of current meanders or intrusions over the shelf. In particular, for extended shelf areas along the northern Mediterranean coastlines, the shelf dynamics is dominated by an overall downwelling vertical motion (see section 4), which does not allow for an internal (to the system) nutrient supply and therefore enhance the role of external nutrient inputs. It should be therefore expected the local ecosystem structure of the extended shelves to be dominated by the so-called "herbivorous" food web (Cushing, 1989), dominated by microphytoplankton (more specifically diatoms) and mesozooplankton.

Fig. 28 gives a schematic representation of the conditions leading to the development of the coastal and oceanic marine ecosystems. In dependence of the physical structure of the water column and of the prevailing limiting factor, the ecosystem structure is "rigidly" shaped into a coastal or open sea structure characterised respectively by the herbivorous and the microbial loop food chains. However, Legendre and Rassoulzadegan (1995) argue that the marine ecosystem might be shaped in a much wider spectrum of conditions (in dependence of changing environmental conditions) along a trophic continuum of which the two ecosystem structures mentioned above represent the two opposite extremes. We argue here that the opposite ends of the ecosystem structure are connected to the different seasons in the Mediterranean, so that the hypothesis of the trophic continuum can be achieved in a temporal sequence during the year. Vichi et al. (2003) shows this concept for the shelf areas of the Northern Adriatic (see section 6.2.3).

The trophic continuum can also be achieved spatially in extended shelf areas, due to the complex interplay of open-sea/coastal processes that take place there. In these shelves, there are sharp and highly variable (in space and time) ecological and biogeochemical horizontal and vertical gradients that could produce the continuum of ecosystem structures between the opposite ends of Fig. 28.

We shall try to demonstrate the above concept with examples from the Algerian coastal area, the Israeli coast, the Gulf of Lions and the northern Adriatic.

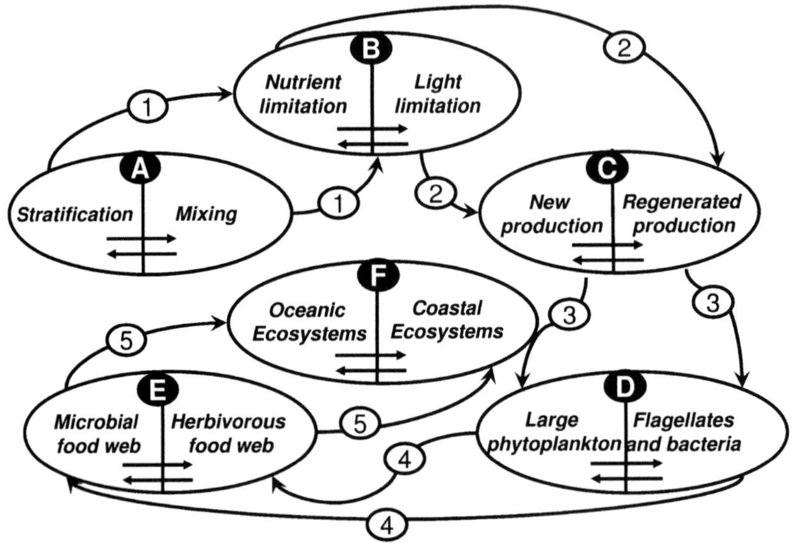

Figure 32.28 Representation of conditions leading to the establishment of a coastal or oceanic pelagic ecosystem. The sequence is a: the water column vertical structure; b: the factors limiting growth; c: the type of primary production; d: the type of organisms; e: the resulting food web; f: the type of ecosystem. Redrawn with modifications from Legendre and Rassoulzadegan (1995).

6.2 Study cases

6.2.1 The Gulf of Lions

Besides the wind, the two primary forcing elements for this region are the discharge of the Rhône River and the Liguro Provençal Catalan (LPC) Current described in 4.2.1. The Rhône annually averaged runoff and nutrient discharge is shown in Table 8 (after Moutin *et al.*, 1998):

The annually averaged vertical distribution of chlorophyll-a, phosphate and dissolved inorganic nitrogen concentration (ΣDIN = Nitrate+Nitrite concentration) along transect 4 of fig. 1a are shown in Fig. 29 in order to highlight differences between the coastal and the open sea domains. The Gulf of Lions appears as an area with a higher upper layers phytoplanktonic biomass (Fig. 29a), decreasing from the shelf to the offshore. The higher primary producers biomass is most probably due to the nutrient input from the Rhône ROFI area, which is particularly evident in the phosphate (Fig. 29b) and (to a lesser extent) in the dissolved inorganic nitrogen distribution (Fig. 29c). The Chlorophyll-a distribution also shows the signal of the summer subsurface chlorophyll maximum, a feature common to both the coastal and open sea areas. The slope of the dissolved nutrients iso-surfaces shows the open ocean upwelling, in the centre of the Lions Gyre, the downwelling at its borders, as expected.

TABLE 32.8
Rhône River annually averaged runoff and P, N and Si discharge into the Gulf of Lions. From Moutin et al. (1998)

Runoff m^3/s	Total P Kt/y	Total N Kt/y	Total Si Kt/y
1690	6.5–12.2	115–127	135–139

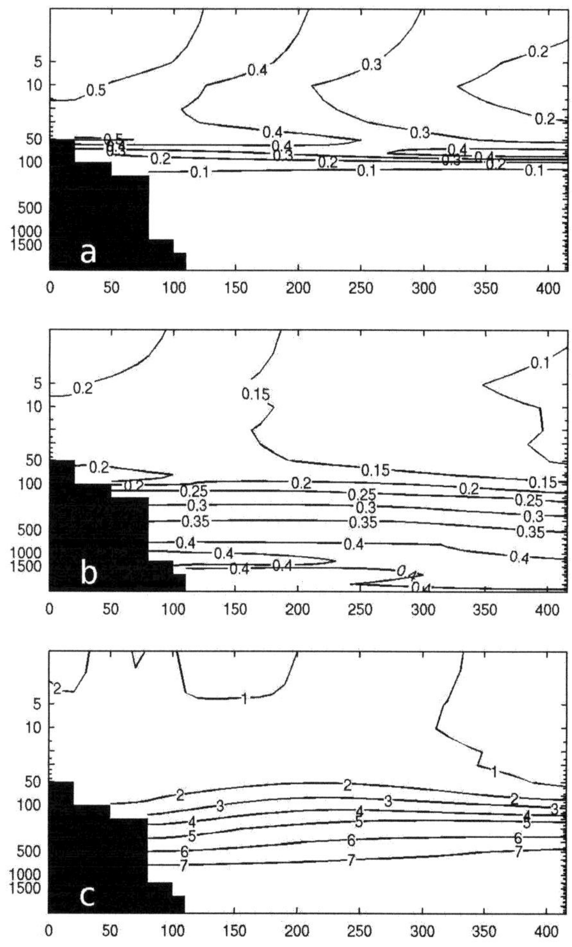

Figure 32.29 Annually averaged vertical distribution of: Chlorophyll-a (a), Phosphate (b) and ΣDIN (c) along section 4 of Fig. 32. 1a (Gulf of Lions). Units are mg m^{-3} for Chlorophyll-a and mmol m^{-3} for ΣDIN and Phosphate. Section starts from the Gulf of Lions (left) and extends into the Liguro-Provençal basin. The vertical scale is logarithmic and the horizontal scale is in km. Data are taken from the Medar (2002) database.

On the basis of primary production characteristics the Gulf of Lions can be divided into 4 distinct sub regions (Lefèvre et al., 1997): The Marseille Gulf, with low primary productivity rates (88 gC m^{-2} y^{-1}); the Rhône river plume with high primary productivity rates (300 to 1550 gC m^{-2} y^{-1}) a dilution zone with primary productivity rates ranging between 86 and 142 gC m^{-2} y^{-1} and the shelf break region where the LPC is located, with low primary productivity rates, but with local phenomena of relatively high production (up to 500 gC $m^{-2}y^{-1}$ in coincidence of frontal areas where the LPC and the dilution area waters meet). The seasonal production cycle has been investigated by Conan et al (1998a, b) and characterised by three distinct periods: 1) a winter period with variable production, 2) a late winter surface phytoplankton bloom, 3) a summer period with average productivity rates and with a clear subsurface chlorophyll maximum. The Rhône plume spatial and temporal variability is highly dependent on local meteorological conditions (wind) and on the magnitude of the river outflow (Broche *et al.*, 1998). In the Rhône plume area Chlorophyll-concentrations are constant at about 1 mg m^{-3} without any marked seasonal variability (Morel and André, 1991); however, primary production (Minas and Minas, 1991) and nutrient concentration (Bianchi *et al.*, 1999) decline from the mouth of the river to offshore (Minas and Minas, 1991).

The phytoplanktonic population is constituted in almost equal proportions by pico-nanoplankton and microplankton (Owens *et al., 1989;* Videau and Levanu, 1990; Woodward *et al.*, 1990). Moreover, in the transition zone between the plume and the offshore, a strong variation in the Bacterial-C/Phytoplankton-C ratio (Yoro et al., 1997) and high concentrations of microzooplankton are observed (Gaudy, 1990). These findings are calling for an ecosystem structure with a significant microbial component and this seems also confirmed by the results of the study on the Rhône river plume microbial communities carried out by Naudin *et al.* (2001) that are indicating the occurrence in the plume of competition for nutrients between phytoplankton and bacteria.

Another feature typical of the Mediterranean pelagic ecosystem that is also present in the Gulf of Lions is the occurrence of phosphorus limitation. Diaz *et al.* (2001) found very strong evidence (fig. 30) for phosphorus limitation (N:P ratios higher than the Redfield value and even higher than the value proposed by Mc Gill, 1969 for Mediterranean water) all over the Gulf, in the plume region and in the dilution area, intermittently affected by the river discharge and by the LPC. Although they did not carry out any measurement of the bacteria activity they could not rule out the possible role of bacteria competition for inorganic P in establishing the Phosphorus limitation condition.

In conclusions, the Gulf of Lyons extended shelf area is an example of heterogeneous environmental conditions, between open ocean and ROFI, where the ecosystem functioning is particularly homogeneous in spite of the large differences in primary production biomass and rates. In all parts of the shelf in fact, phosphorous limitation dominates and bacterial food web is an important component of the ecosystem, together with a mixture of micro and nanophytoplankton communities.

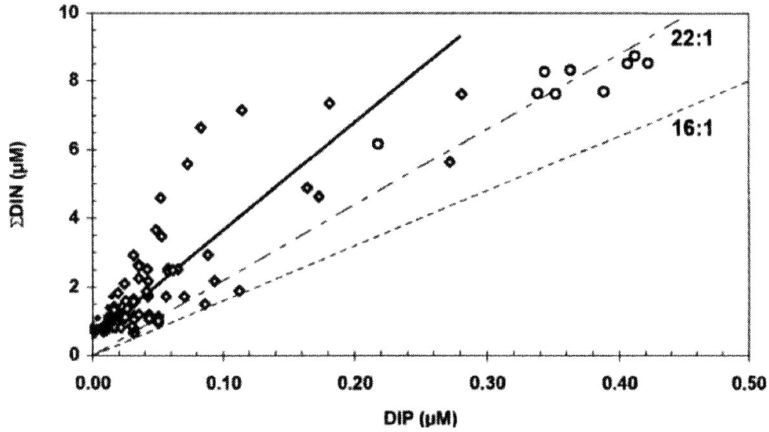

Figure 32.30 ΣDIN versus Phosphate (DIP) relationship. (◊) 0–150 m layer. (○) 200–400 m layer. Solid line represent the linear regression of data in the 0–150 m layer. Dashed and dashed-dotted lines indicate the 16:1 and 22:1 ratios respectively. From Diaz et al. (2001). Reprinted with permission.

6.2.2 The Algerian Coastal area

The Annually averaged vertical distribution of chlorophyll-a Phosphate and ΣDIN along transect 2 of Fig. 1a is shown in Fig. 31. The distribution of such properties confirm that in the case of narrow shelves, the open ocean nutrient and primary production characteristics extend over the shelf. In Fig. 31 the coastal areas as well as the open ocean have a noticeable subsurface Chlorophyll-a maximum (Fig. 31a). The largest difference between the near shelf areas and the open ocean can be noted in the intensity of the subsurface chlorophyll maximum (30–50 m depth) that is marked by higher concentrations in the vicinity of the Algerian coast.

The nutrient distribution (Figs 31b, c) is uniformly low between the shelf and the open ocean and the values are similar to the one present in the open ocean area in front of the Gulf of Lions. However, the ΣDIN concentrations are larger in the Algerian basin than in the Gulf of Lions area, hinting to a larger phosphorous limitation mechanism at work in this region.

As we said in section 4.2.2, the circulation feature strongly affecting this narrow shelf is the Algerian current (Millot, 1985, see also Fig. 16), which flows on the slope and the anticyclonic eddies that may induce coastal upwelling phenomena (Millot et al., 1990b). These processes are not shown in Fig. 31 since it is a climatological picture. A detailed study on the trophic characteristics of the Algerian current under all the physical features summarized above was carried out by Mòran *et al.* (2001). Their results point to a strong coupling of the marine ecosystem functioning (autotrophic and heterotrophic production) with the mesoscale dynamics. They show that the presence of mesoscales can induce nutrient increase in the upper layers and a consequent increase of primary producer biomass and production rates (Fig. 32). The primary production rates were found slightly higher (but comparable) to the values given for the open western Mediterranean waters. The bacterial biomass accounted for a significant source of carbon for the higher trophic levels. These two findings provide a further confirmation that in reduced extension shelf areas the coastal ecosystem functioning is strongly constrained by the ecological characteristics of the open sea waters.

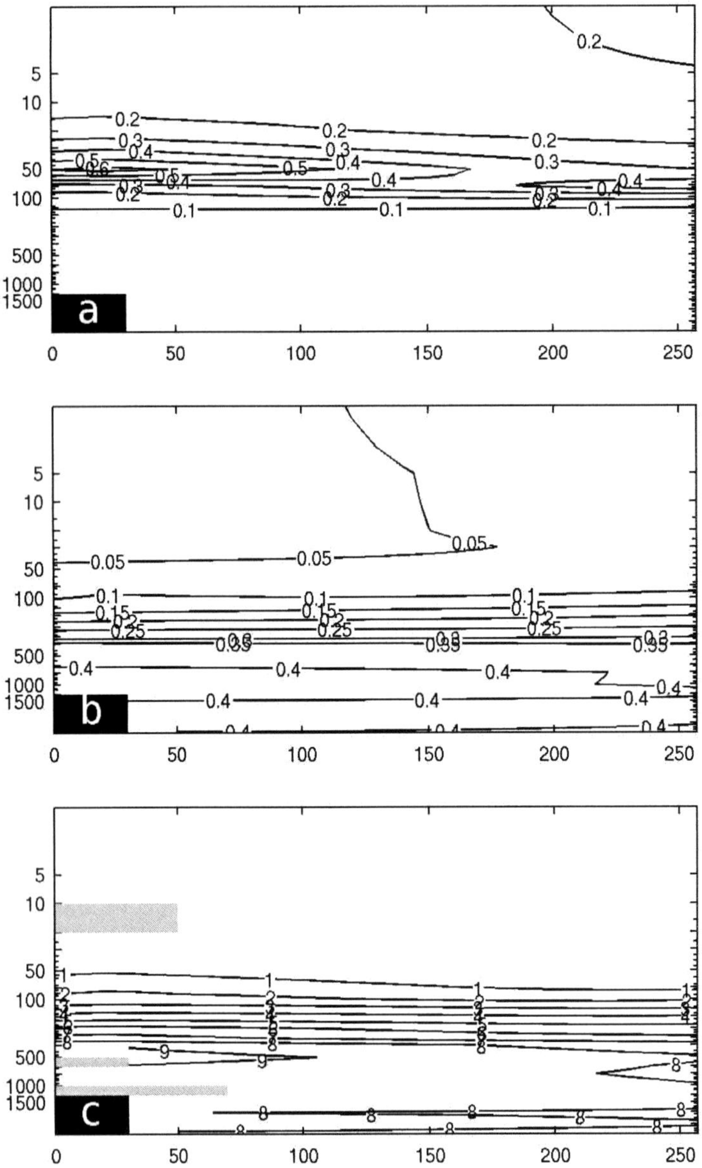

Figure 32.31 Annually averaged vertical distribution of: Chlorophyll-a (a), Phosphate (b) and ΣDIN (c) along section 2 of Fig. 32. 1a (Algerian current region). Units are mg m-3 for Chlorophyll-a and mmol m^{-3} for ΣDIN and Phosphate. Section starts from the Algerian coast (left) and extends into the Algerian basin. The vertical scale is logarithmic and the horizontal scale is in km. Data are taken from the Medar (2002) database. Grey shaded areas denote lack of data.

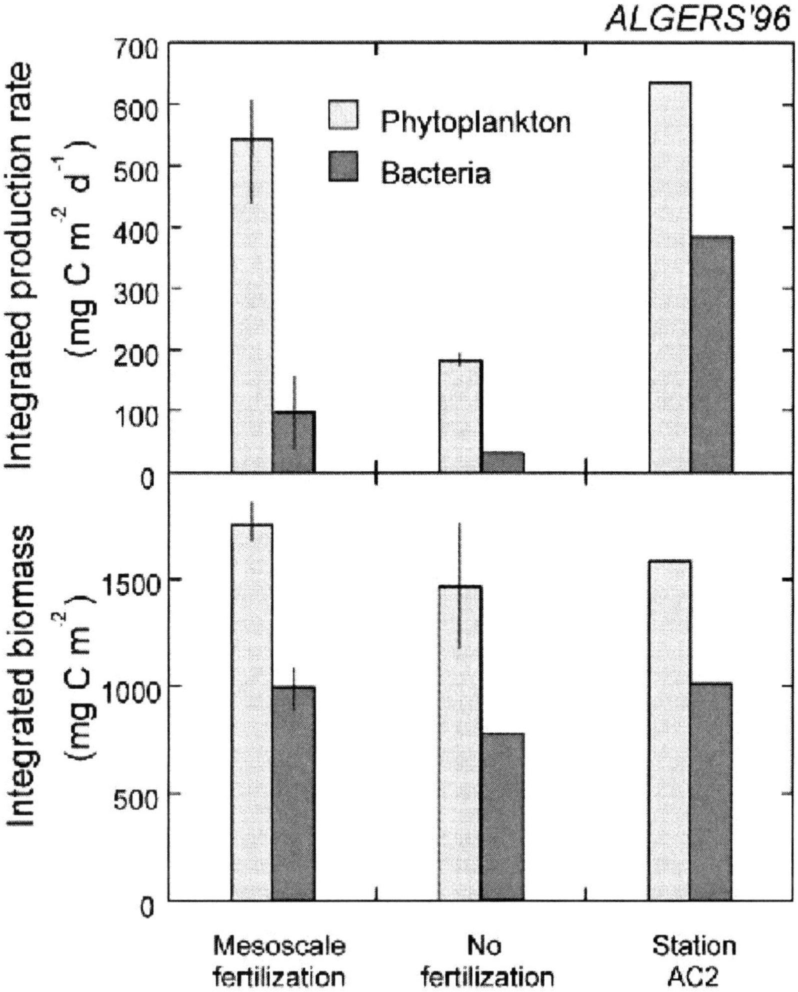

Figure 32.32 Averaged integrated values of phytoplanktonic and heterotrophic bacterial biomass and production under different hydrodynamic conditions in the Algerian basin: with and without mesoscale mechanism of fertilization. Mòran et al., 2000. Station AC2 is a fixed station in the Algerian basin. Reprinted with permission.

6.2.3 The Sicily Strait area

The productivity of this area is not very well known and is generally considered moderate to low, which is somehow in contrast with the development of an intensive fishery, as we will discuss in section 7. In Fig. 33 we show the distribution of the chlorophyll, phosphorous and total inorganic nitrogen along the two transects across the Strait, called 6 and 7 in Fig. 1a. Both transects are characterized by a subsurface chlorophyll maximum but the amplitude of the maximum is reduced between the western and eastern transect, following the general oligotrophic gradient of the basin. The subsurface chlorophyll maximum is generally spread across the transect 6 while along transect 7 the absolute maximum is present in the shelf region between Malta and Sicily (Fig. 33d). On the other hand, the chlorophyll abundance seems to spread toward the surface on the Tunisian side of the westernmost transect of the Strait bringing further evidence of the connection between Algerian current nutrient enrichment and local production.

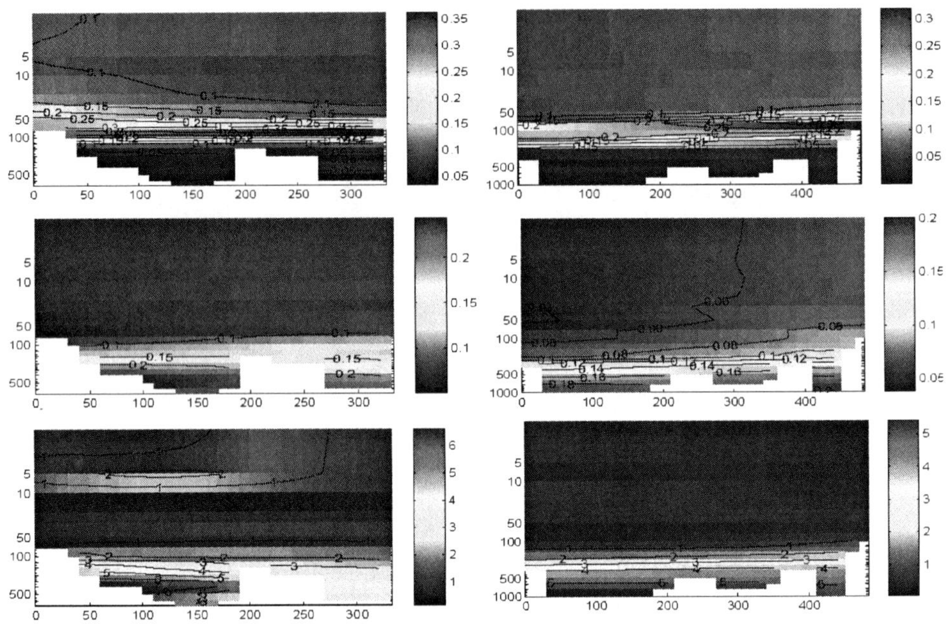

Figure 32.33 Annually averaged vertical distribution of: Chlorophyll-a (a,d), Phosphate (b,e) and ΣDIN (c,f) along transects (6,7) of Fig. 32. 1a (Strait of Sicily region). Units are mg m^{-3} for Chlorophyll-a and mmol m^{-3} for ΣDIN and Phosphate. Section starts from the Tunisan coast (left) and extends to the Siclian coasts. The vertical scale is logarithmic and the horizontal scale is in km. Data are taken from the Medar (2002) database.

The nutrient concentrations diminish of about 50% in the surface waters going across the Strait in the eastward direction and by 20% in the waters below 100 meters. However, the N:P ratio seems to remain constant in the Strait hinting to an internal regulatory effect on the nutrient ratios in accordance with the absence of river inputs in this area.

It is believed that the Modified Atlantic Waters arriving in this area could be one source of enrichment (Maurin, 1977) as well as upwelling and downwelling zones associated with the Atlantic Ionian Stream-AIS structures that dominate the current flow regime of this area. Agostini and Bakun (2002) describe the general upwelling and downwelling motion generated by the upwelling favourable winds in this area. The Sicilian side of the Strait is the site of powerful upwelling that should increase the productivity of this area. From our transects, it seems however that the subsurface maximum value changes only 10–20% going from the Tunisian to the Sicilian side of the Strait transects. Thus the AIS fertilisation effect may be much larger than the wind upwelling in this region.

6.2.4 The northern Adriatic

As we have seen in section 4.2.3 and 3.3, the northern Adriatic is the northernmost part of the Adriatic Sea strongly affected by riverine fresh water and nutrient inputs. The main river discharging in the northern basin is the Po but smaller rivers are giving a significant contribution of fresh water and nutrients (Degobbis and Gilmartin; 1990, Raicich, 1996. See Table 9).

TABLE 32.9
northern Adriatic. Annually averaged runoff and P, N and Si discharge. From Degobbis and Gilmartin (1990) and Raicich (1996).

	Runoff m^3/s	Total P Kt/y	Total N Kt/y	Total Si Kt/y
Po river	1585	15	162	168
All other rivers in the northern Adriatic	2728	28	272	234

The role played by the Po river runoff in forcing the marine ecosystem characteristics is highlighted in Fig. 34. Surface layers in the northern Adriatic are marked by higher concentrations of chlorophyll-a (Fig. 34a) and nutrient concentration (Fig. 34b, c). The signal of the subsurface chlorophyll maximum is particularly evident at depths of about 20–50 m. Note the very small concentration of phosphorous that is four times smaller than the one found in the Gulf of Lions shelf area (Fig. 29) and it is closer to the open ocean values (see Fig. 31). The total nitrogen is however only a factor of two less than in the open ocean bringing us to the conclusion that phosphorous limitation is less severe here than in the open waters. The nutrient isolines are downwelling from the shelf break toward the northern shelf making the process of enrichment from the open ocean not very plausible.

The surface phytoplankton seasonal cycle of the northern, middle and southern Adriatic Sea from satellite observations is shown in Fig. 35. The higher phytoplankton biomass in the northern Adriatic Sea is evident, while the seasonal cycle is characterised by three distinct periods of phytoplankton growth: winter, summer and late autumn, all connected to the Po runoff discharge cycle.

The Po river low salinity and nutrient rich waters are advected south along the WACC (see section 4). The offshore current extension is marked by a strong (salinity dependent) density front. Inflow of higher salinity, nutrient depleted water into the basin occurs along the eastern coast. The magnitude of the southerly inflow is modulated by the strength and intensity of the cyclonic gyres characterizing the circulation (Artegiani et al., 1997, Zavatarelli et al., 2002) as well as the presence of the counter current developing in summertime along the eastern coast of the Northern Adriatic (Supić *et al.*, 2000).

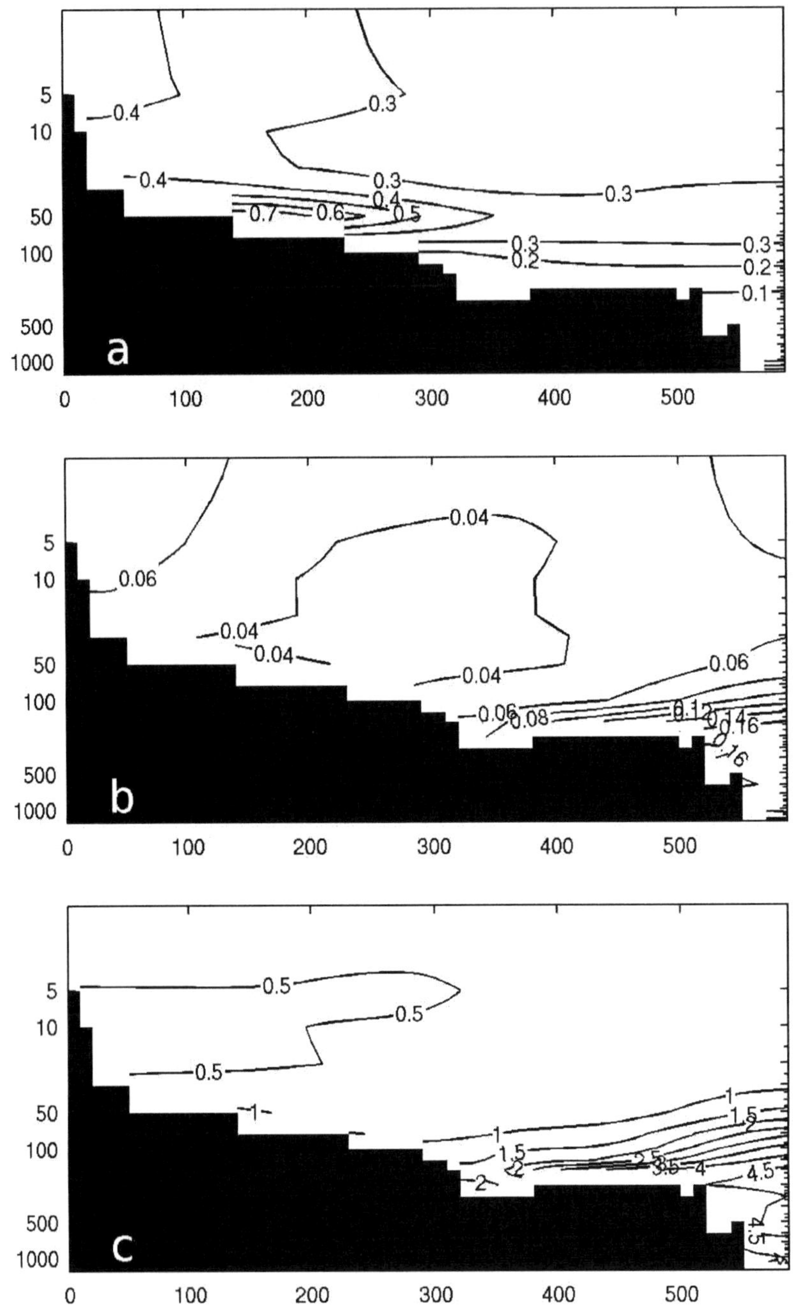

Figure 32.34 Annually averaged vertical distribution of: Chlorophyll-a (a), Phosphate (b) and ΣDIN (c) along section 9 of Fig. 32. 1a (Adriatic Sea). Units are mg m^{-3} for Chlorophyll-a and mmol m^{-3} for ΣDIN and Phosphate. Section starts from the northern Adriatic coast (left) and extends into the Middle Adriatic Sea. The vertical scale is logarithmic and the horizontal scale is in km. Data are taken from the Medar (2002) database.

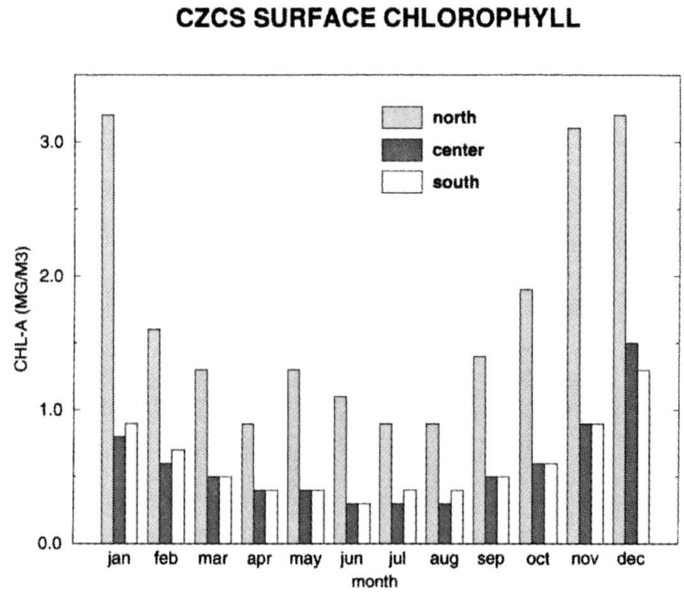

Figure 32.35 Monthly averaged surface chlorophyll-a concentration (mg m-3) for the northern (north), middle (centre) and southern (south) Adriatic Sea. From Zavatarelli et al. (2000). Reprinted with permission.

It is well known that the western coasts of the northern Adriatic basin suffer from eutrophication phenomena leading to the development of anoxic conditions during the summer season (Vollenweider *et al.*, 1992). The higher biomass production in the western coastal areas is well reflected by the Chlorophyll-a concentration and primary production data reported by Fonda Umani (1996) for the eastern and the western part of the basin (Table 10).

The average phytoplankton biomass in the western part of the basin is about three times higher than in the eastern part and this applies also to the primary production maximum values measured. On the basis of that Fonda Umani (1996) identified two areas: An "open sea" area with low biomass and a western coastal zone with higher biomass separated by a very sharp trophic gradient roughly coincident with the frontal system separating the western coastal current from the offshore water. However, it can be noticed from Table 10, that minimum values in both areas are comparable. This points to a strong variability of the trophic conditions (with shift from the meso-eutrophic to the oligotrophic state) in the area mostly affected by river nutrient input. In fact, potential for phosphorus limitation in the northern Adriatic western coastal area was reported by Zavatarelli *et al.* (1998) in their analysis of the climatological nutrient and biomass seasonal distribution in the Adriatic Sea.

TABLE 32.10

Chlorophyll-a mean values and maximum and minimum of Primary production (14C method) in the western and eastern North Adriatic. From Fonda Umani (1996).

		Western basin	Eastern basin
Chlorophyll-a mg/m^3		2.87	0.9
PP (mgC/(m^{3}h)	Max	30	10
	Min	<1	<1

The food web structure in the western coastal areas, according to Fonda Umani (1996), has nanophytoplankton as numerically prevailing primary producers, while microphytoplankton (mostly diatoms) has a temporal and spatial distribution highly dependent on the riverine input variability, but in general they have two period of strong abundance (spring and autumn). The nanophytoplankton biomass is mostly grazed by microzooplankton. However, also in this region the role of bacteria as secondary producers and as source of food for microzooplankton is quite important. In the Po delta region Puddu et al. (1998) measured bacterial carbon demand that indicates an organic carbon processing activity by the bacterial compartment comparable (or even larger, in dependence of the riverine dissolved organic carbon input) with the correspondent observed primary production. The Po delta dissolved organic input provides an important carbon source for the microbial food web, whose importance in the northern Adriatic coastal water is further highlighted by the results of the study on the dynamics of the microbial plankton communities carried out by Fonda Umani and Beran (2003) in the Gulf of Trieste. They found that most of the carbon flux in the food web was directed through the microbial compartment through grazing of microzooplankton on heterotrophic nanoflagellates and bacteria.

The occurrence of food web shift mediated by nutrient (phosphate) availability is further highlighted by the results of a one dimensional modelling exercise carried out by Vichi *et al.* (2003) at three different locations in the northern Adriatic Sea (Po river plume, centre of the northern basin, Gulf of Trieste) characterised by different hydrological and environmental conditions. In fig. 36a is shown the ratio between the modelled carbon flow due to herbivorous and to microbial grazers (this ratio has been proposed by Legendre and Rassoulzadegan, 1994, as an index of ecosystem functioning) for each of the three modelled sites. In winter spring the transfer of carbon occur mainly through the herbivorous food chain (ratio > 1), while during the summer season a microbial food web develop (ratio < 1). This shift in the food web structure is matched by a deterioration of the quality of the substrate (dissolved organic matter) for bacteria growth (Fig. 36b), inducing consequently the shift of the microbial community from nutrient remineralizers to competitors for inorganic resources (Fig. 36c).

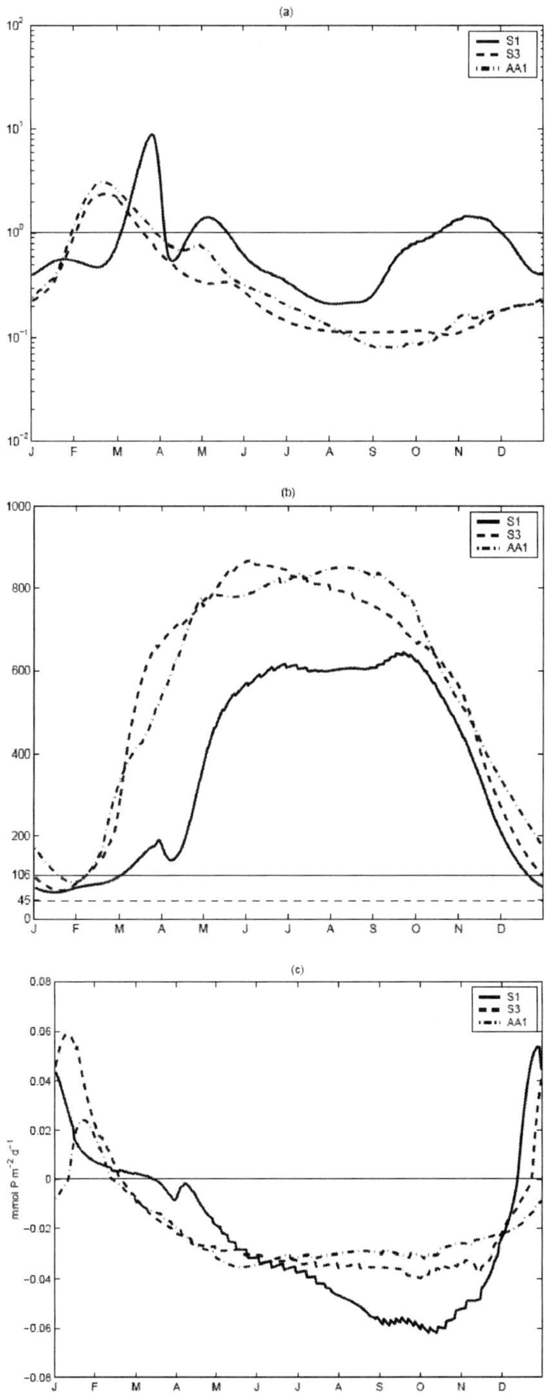

Figure 32.36 Indexes of ecosystem functioning and matter transfer pathways from the one-dimensional modelling of Vichi et al. (2003) at three different locations in the northern Adriatic Sea. S1: Po River plume, S3: Centre of the basin, AA1 Gulf of Trieste. A: ratio between the carbon flow due to herbivorous grazing and the one due to microbial grazing (in semi logarithmic scale). B: ratio between the C- and P-component of the dissolved organic matter. The optimal ratio for phytoplankton (106:1, Redfield et al., 1963) and bacteria (45:1, Goldman et al., 1987) are marked in the plot. C: phosphorus flux between bacteria and phosphate. Positive values indicate active phosphorus remineralization by bacteria. Negative values indicate bacterial phosphate consumption. Reprinted with permission.

6.2.5 The Nile Delta and Israeli shelf areas

Prior to the construction of the Aswan dam the marine ecosystem dynamics of this area was strongly conditioned by the Nile runoff imposing a major phytoplankton bloom in summer-autumn (Sharaf El Din, 1977, Dowidar, 1984) extending from the Egyptian to the Israeli coast (Herut *et al.,* 2000). After the damming the phytoplankton seasonality shifted to a winter bloom period (Berman *et. al.,* 1986; Azov, 1986) since the Nile is now regulated to discharge during winter.

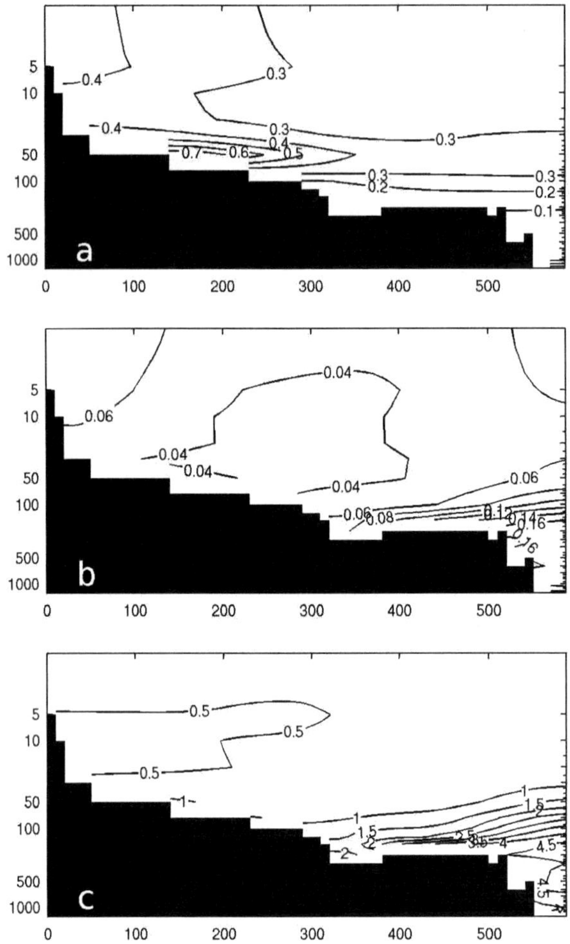

Figure 32.37 annually averaged vertical distribution of: Chlorophyll-a (a), Phosphate and ΣDIN along section 15 of Fig. 32. 1a (Nile delta region). Units are mg m^{-3} for Chlorophyll-a and mmol m^{-3} for ΣDIN and Phosphate. Section starts from the Egyptian coast (left) and extends into the eastern Levantine basin. The vertical scale is logarithmic and the horizontal scale is in km. Data are taken from the Medar (2002) database.

The Levantine basin is known to be a severely oligotrophic area (Berman *et al.,* 1984; Krom *et. al.,* 1991) controlled by phosphorus limitation. This characteristic appears to be extended also to the coastal domain (particularly after the Nile damming), as the chlorophyll-a measurements during blooms never exceeded 0.9 mg m^{-3} (Berman *et. al.,* 1986; Azov, 1986). This is further confirmed by the annually averaged vertical distributions of Chlorophyll-a, phosphate and ΣDIN along transect 15 of Fig. 1a that are shown in fig. 37, where almost no difference in concentration can be noticed between the coastal and the open sea domain of the transect. Notice also the low values of phosphate and ΣDIN with respect to the western basin transects, almost a factor of two in the deep waters. Notice also the relatively high surface water chlorophyll values extending from the coasts outward, probably connected to the Nile outflow extending far from the coasts. The largest vertical gradients in nutrients starts from the center of the LIW layer, as shown in Fig. 4.12, and located below 100 meters.

The temporal dynamics of the deep phytoplankton bloom is typical of open sea oligotrophic waters with a winter-spring bloom followed by the formation of a deep chlorophyll maximum. No information is available about the bacterial production and it is therefore difficult to assess the main carbon pathways through the trophic web. However the results of Herut *et al.* (2000) seem to indicate for the spring phytoplankton bloom a certain incidence of microphytoplankton (diatoms), but given the extremely oligotrophic conditions of the region it is likely that picophytoplankton and the microbial loop play a major role.

7. Exploitation of fisheries resources in the shelf areas

Through history, the coastal area of the Mediterranean Sea has always been very densely populated. It is not surprising that continental shelf areas were and are all subject to intense fishery exploitation. Total marine fishery catches for the Mediterranean are about 1,000,000 tonnes per year (average 1992–2001 GFCM FAO Fishery Statistics Database, (*www.fao.org/fi/statist*), equivalent to approximately 1% of the total world marine fish catch.

Three main kinds of fisheries resources are exploited in the Mediterranean: demersal and small and large pelagic resources. Small pelagic fish as sardine (*Sardina pilchardus*), anchovy (*Engraulis encrasicolus*), sardinella (*Sardinella* spp.) and mackerels (*Scomber* spp. *and Trachurus* spp.*)* live in midwater or near the surface in shelf areas and are usually the target of purse seiners and midwater trawlers. Their occurrence is not spread evenly throughout the Mediterranean and is determined by the ecological characteristics of each shelf area. Sardines are predominant in certain areas while in others sardinella or anchovy are more abundant.

Demersal species include fish, cephalopods, crustaceans and bivalves, all living in close proximity to the seabed shelf and slope areas, where they are exploited by bottom trawl fisheries. Up to about 100 demersal species are commercially exploited even if most of the catches comprise a pool of less than 20 species (Papacostantinou and Farrugio, 2000): red mullets (*Mullus barbatus* and *Mullus surmuletus*), sole (*Solea solea*), gurnards (*Trigla* spp.), poor cod (*Trisopterus minutus capelanus*), angler fish (*Lophius* spp.), shrimps (*Penaeus kerathurus* and *Parapenaeus longirostris*), spiny lobster (*Palinurus elephas*), cuttlefish, squids and octopuses (*Sepia* spp., *Loligo* spp., *Eledone* spp. and *Octopus vulgaris*). Some important demersal resources extend their range from the shelf to the slope, such as hake (*Merluccius merluccius*) and Norway lobster (*Nephrops norvegicus*).

Large pelagic fishes include tuna like fishes and pelagic sharks. Their share of the total catch is only about 4% (Lleonart and Recasens, 1996) but their economic importance is much higher. They are highly migratory open sea species, which cannot be considered strictly fishery resources of the shelf areas and thus will not be treated here.

Although the Mediterranean is an oligotrophic sea, it sustains some locally very important fisheries, suggesting that there are areas with substantial organic production (Bakun and Agostini, 2001). These areas are mainly shelf areas where river runoff is important, such as the Catalan Coast (Ebro river), the Gulf of Lions (Rhone river), the Adriatic Sea (Po river), the northern Aegean (Thracian rivers and Marmara Sea waters). These northern Mediterranean shelf areas have well developed fisheries, whereas on the southern side of the Mediterranean the situation appears to be quite different.

The main fishery area on the southern side is the Sicilian channel and the north-African shelf off Tunisia and Libya (Charbonnier and Garcia, 1985) where river run off is absent. The Eastern basin is very oligotrophic and the only important shelf fishery, located on the Egyptian shelf, has been heavily affected by the closure of the Nile outflow by the Aswan Dam (Wadie, 1982). Levi and Troadec (1974) pointed out that fish productivity tends to decrease from west to east and the rate of fishery exploitation is higher along the northern Mediterranean coasts and lower along the southern and eastern coasts. Biodiversity shows the same negative gradient eastwards (Garibaldi and Caddy, 1998).

Research on the links between oceanographic characteristics of the different shelf areas and fishery resources is not particularly advanced in the Mediterranean. Here we shall try to give an overview of the current status of knowledge analysing examples of fisheries exploiting shelf areas with different physical and oceanographic characteristics. Narrow shelf areas with nutrient inputs coming from upwelling and/or by advection from other areas like the Algerian shelf; medium-sized shelf areas where nutrients come from river runoff and upwelling phenomena like the Gulf of Lions; large shelf areas with (Adriatic) and without (Tunisian shelf and Sicilian channel) important river runoff; and finally a shelf area where nutrient input from river runoff has been artificially interrupted by man-induced changes (Egyptian shelf area).

We argue that higher fishery production in the Mediterranean is found in correspondence of substantial river runoff, although upwelling phenomena can be locally important (Bakun and Agostini, 2001). Each area will be briefly described in terms of the main fisheries taking place and the most important species captured. For this purpose, official landings statistics for the Mediterranean (GFCM Statistics database) were used. Specific process studies trying to link fish production to oceanographic features will be reviewed. These are mostly, if not exclusively centred, on small pelagic fish like anchovy (*Engraulis encrasicolus*) and sardine (*Sardina pilchardus*) two species which, due to their abundance and economic importance, have been the object of more detailed ecological research.

7.1 Processes and study cases

7.1.1 Gulf of Lions

The Gulf of Lions is one of the shelf areas with the highest productivity in the Mediterranean thanks to various mechanisms of fertilisation (see sections 4, 5 and 6) such as the permanent front which extends from the Ligurian Sea to Catalunia (associated with the Liguro-Provencal current), the river Rhone runoff, the process of large scale Ekman upwelling associated the prevailing winds and other more localised oceanographic phenomena (Caddy and Oliver, 1996; Bakun and Agostini 2001). This area is exploited mainly by Spanish and French trawlers but also by longliners, gillnetters and purse seiners, and there is a highly developed small scale fishery in inshore waters and lagoons (Papacostantinou and Farrugio, 2000). Average annual total landings in the period 1992–2001 were about 37,000 tonnes. The most important species, or groups of species, account for 80% of the total landings and are reported in Table 11.

TABLE 32.11
Average annual landings (1992–2001) in the Gulf of Lions divided by representative species or group of species (source GFCM Fishery Statistics database).

species	scientific name	average landings 92–01 tonnes	%
sardine	*Sardina pilchardus*	12216	33
anchovy	*Engraulis encrasicolus*	7251	20
hake	*Merluccius merluccius*	2608	7
mackerels	*Scomber* spp.; *Trachurus* spp.	2583	7
cephalopods	*Octopus* spp.; *Loligo sp. etc.*	1751	5
mullets	*Mugilidae*	817	2
poor cod	*Trisopterus minutus capelanus*	717	2
conger	*Conger conger*	601	2
angler	*Lophius spp.*	403	1
sole	*Solea spp.*	321	1
red mullets	*Mullus spp.*	275	1
total		29541	80

Small pelagic fishes like sardine and anchovy represent over than 50% of total landings and are caught both by purse seiners and trawlers. Anchovy ranks second in landings but it is by far most sought by fishermen, its price per unit landed being much higher than sardine. Sardine probably is still under exploited.

The peak spawning period for anchovy in this region is June-July (Garcia and Palomera, 1996) and eggs are mostly distributed along the shelf border from 10 to 50 miles offshore (fig. 38). Anchovy eggs are always mainly found in the first 20 m of depth, but larvae are found in day time from 20 down to 40–60 m depth (Olivar et al., 2001). This may be linked to feeding behaviour since during the summer nutrients at the surface are depleted because of the strong thermal stratification, and microzooplankton is concentrated close to the Deep Chlorophyll Maximum at 40–60 m depth (Estrada and Salat, 1989).

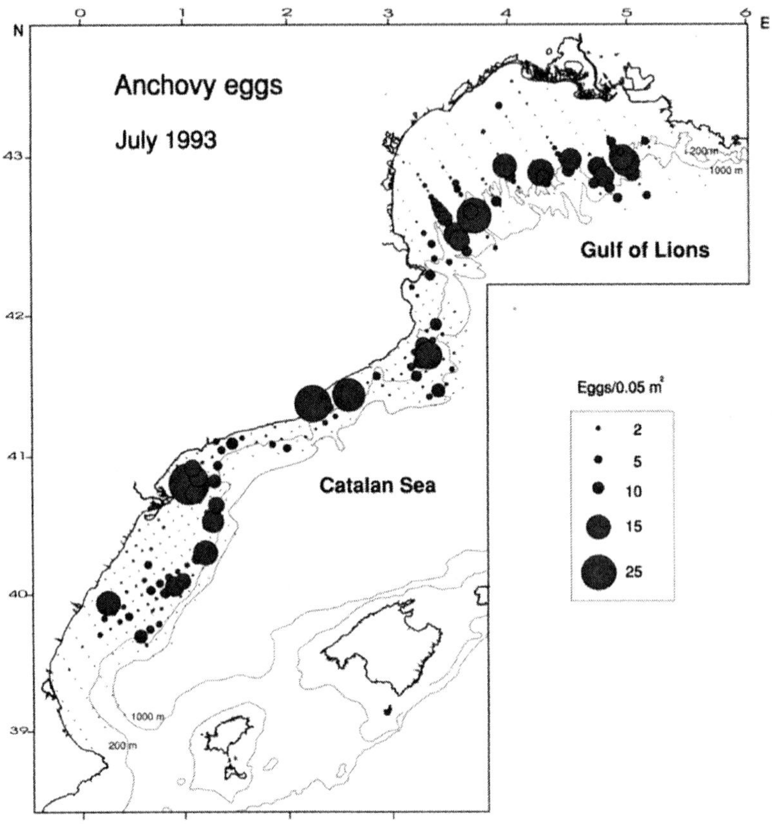

Figure 32.38 Spawning area of anchovy in the Gulf of Lions and Catalan sea (from Garcia and Palomera, 1996). Reprinted with permission.

The presence of the shelf slope front (Sabates 1990) and the corresponding distribution of eggs suggests that the enhanced productivity associated with the front may be exploited by anchovy (Garcia and Palomera, 1996). Agostini and Bakun (2001) proposed that the massive Rhone river outflow enriches the surface waters in the shelf slope front area, where eggs and larvae mostly live (Olivar et al., 2001). The shelf-slope front has here an effect of retention of the nutrients and thus of enhancement of the primary and secondary production which in turn supports larvae growth and survival. In addition, river flow enhances water column stability by stratification and the need for this stability is a general feature of anchovy spawning grounds throughout NW Mediterranean (Palomera, 1992).

Continental, relatively fresh surface waters formed in the Gulf of Lions (*CIW Continental Influence Water*) have higher nutrient and phytoplankton concentrations than adjacent oceanic waters (Cruzado and Velasquez, 1990), as well as higher densities of microzooplankton and copepods (Tudela and Palomera 1995). These waters offer a spawning ground for anchovy, larvae and following the southwestward direction of the Liguro-Provencal current, they could be exported from the Gulf of Lions to the Catalan coast, where another important fishery for anchovy takes place (Castellon et al., 1985; Sabates et al. 2001).

Sardine is a winter (October to May) spawner in the Mediterranean. In the Gulf of Lions, larvae in November tend to aggregate in the region near the Rhone river delta where water circulation is mainly wind driven (Rasoanarivo et al. 1991). Westerly winds in this season concentrate nutrients of fresh water origin in the coastal areas (Travers and Travers, 1972) promoting phytoplankton production, which is utilised by sardine larvae as a source of food.

In this area, anchovy is more influenced by river runoff in connection with the shelf-slope front in summer while sardine is reliant upon inshore water circulation of freshwater origin in winter. In both cases the nutrient enrichment effect of freshwater inflow seems to be crucial for sustaining these two important commercial fish populations. This has been recently confirmed for the adjacent Ebro river delta area by Lloret et al. (2004) by means of time series analysis of anchovy and sardine landings in relation to river run off and wind mixing.

7.1.2 Algerian shelf

This area of narrow shelf is exploited by purse seiners for small pelagic fish together with some mid-water trawlers fishing the same resources. Demersal species are fished by small scale, fixed gear fleet on rocky areas while trawlers exploit rather heavily the narrow trawlable areas (Caddy and Oliver, 1996). The constant inflow of Atlantic surface waters is an important source of nutrients and thus of fish productivity (Levi and Troadec, 1974). In addition, the Algerian coast is an area of coastal downwelling during the winter season (Bakun and Agostini, 2001). In summer coastal upwelling can take place locally near anticyclones boundaries (see section 4.2.2 and 6.2.2).

TABLE 32.12
Average annual landings (1992–2001) in the Algerian shelf area by representative species or group of species (source GFCM Fishery Statistics database).

species	scientific name	average landings 92–01 tonnes	%
sardine	*Sardina pilchardus*	58406	58
sardinellas	*Sardinella* spp.	11716	12
mackerels	*Scomber* spp.; *Trachurus* spp.	6322	6
seabreams	*Sparidae*	4754	5
marine fishes		3153	3
crustaceans		2890	3
bogue	*Boops boops*	2689	3
anchovy	*Engraulis encrasicolus*	2645	3
clupeoids	*Sardina, Sardinella and Engraulis*	2347	2
red mullets	*Mullus spp.*	1505	2
hake	*Merluccius merluccius*	1267	1
cephalopods	*Octopus sp.; Loligo sp. etc.*	998	1
total		98692	98

Average annual total landings in the period 1992–2001 were about 100,000 tonnes. The most important species, or groups of species, account for 98% of the total landings and are reported in Table 12.

The relative importance of the small pelagic fishery is evident; sardine, sardinellas, mackerels, anchovy and clupeoids together reach more than 80% of the total landings. Anchovy landings are rather low compared to sardine and sardinellas and this could be due to the lack of low salinity waters in this region, which has no substantial river outflows. The predominance of small pelagic fish catches is a general feature of the Mediterranean; here it is particularly enhanced by the narrow continental shelf, which limits the availability of suitable habitats for demersal species.

In this region there is a lack of studies on the link between environment and resources and, as pointed out by Caddy (1999), there is a difficulty in defining discrete fish stocks inhabiting narrow shelf areas, which can be a thousand kilometres long and only few kilometres wide. The systems of larval concentration and retention that can be observed or postulated on the basis of the water circulation in the Gulf of Lions or the Adriatic are not likely to be the same for the Algerian shelf and other narrow shelf areas in the Mediterranean. There is a possibility of there being some genetic discontinuity along the shelf, with separate fish stocks of the same species along the coast in correspondence of specific spawning grounds, but no data are available. Detailed research on mesoscale oceanography, larval dispersal, and population genetics is needed to improve the management of these stocks (Caddy, 1999).

7.1.3 Tunisian shelf and Sicily channel

The shelf areas of Tunisia and the Sicily channel are the largest and most productive fisheries of the Mediterranean after the Adriatic (Caddy and Oliver, 1996; see Table 13 below for annual landings). This area sustains an important trawl fishery mainly composed of industrial scale Italian vessels, together with a more coastal but equally intensive Tunisian trawl fishery on the southern part, mostly in the Gulf of Gabes (Papacostantinou and Farrugio, 2000).

As shown in Table 13, the pelagic fishery is not predominant in this area. Landings of sardine, sardinellas, anchovy and mackerels are about 27% of the total and stocks of these species are generally believed to be smaller in proportion to other areas (e.g. Adriatic and Gulf of Lions). Although in the past scientific investigations detected important biomasses of small pelagic fish, this was never reflected in the landings, and could be due to overestimation of biomass or lack of an adequate market for this resource (Papacostantinou and Farrugio, 2000).

The productivity of this area seems to be influenced by the modified Atlantic Waters (Maurin, 1997) as well as upwelling and downwelling areas associated with the Ioman Atlantic Stream (AIS). A recent study (Garcia Lafuente et al., 2002) has linked the general circulation pattern of the AIS to the reproductive strategy of anchovy in the Sicilian Channel (Fig. 39). The current is responsible of advection and retention mechanisms for eggs and larvae. Local, wind induced small-scale upwelling phenomena along the Sicilian coast contribute in terms of nutrient enrichment (Agostini and Bakun, 2002).

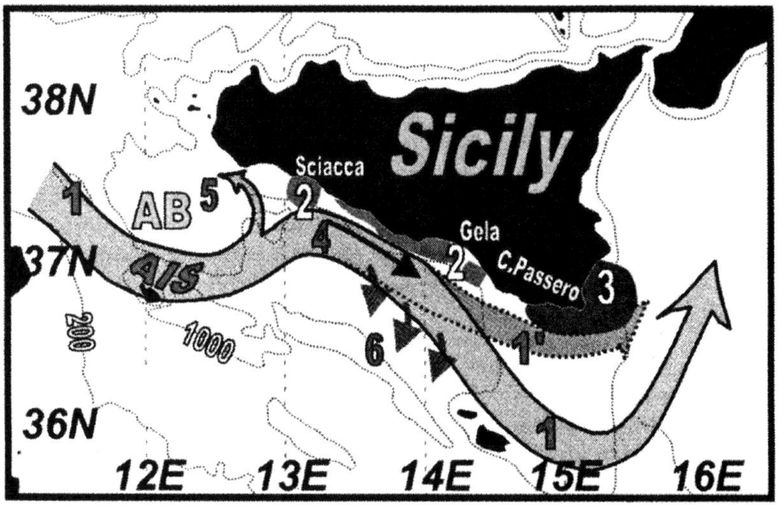

Figure 32.39 Schematic model of the spawning strategy of the Sicilian Channel anchovy (Garcia Lafuente et al., 2002). Main spawning grounds (2), Atlantic Ionian Stream (1and 1'), eggs and larvae transported downstream (4), main nursery ground of Capo Passero (3), possible offshore advection (5 and 6). Reprinted with permission.

TABLE 32.13
Average annual landings (1992–2001) in the Tunisian shelf and Sicily channel area divided by representative species or group of species (source GFCM Fishery Statistics database). Total landings are about 180,000 tonnes per year.

		average landings 92–01	
Species	scientific name	tonnes	%
other marine fishes		23736	13
cephalopods	*Octopus* sp.; *Loligo* sp. *etc*	21183	12
Sardine	*Sardina pilchardus*	19611	11
Hake	*Merluccius merluccius*	16354	9
crustaceans		15473	9
Mackerels	*Scomber* spp.; *Trachurus* spp.	12948	7
Seabreams	*Sparidae*	9667	5
Anchovy	*Engraulis encrasicolus*	8881	5
Sardinellas	*Sardinella spp.*	7855	4
red mullets	*Mullus spp.*	6808	4
sharks and rays	*Selachii*	6011	3
Bogue	*Boops boops*	4321	2
Mullets	*Mugilidae*	3975	2
Gurnards	*Trigla spp.*	2050	1
angler fish	*Lophius spp.*	1898	1
Total		160768	88

The southern area of the Gulf of Sirte and the Gulf of Gabes is rather poor in oceanographic data: high surface values of chlorophyll in the Gulf of Gabes detected by remote sensing have been interpreted as an artefact due to interference from the shallow sea floor (Agostini and Bakun, 2002). On the contrary, we argue that annual Tunisian landings are substantial on a Mediterranean scale (about 70–80,000 tonnes) and mostly come from the Gulf of Gabes and southern areas (Hachemi, 1996). Thus we think that the issue of productivity and nutrients along the Tunisian shelf calls for further research.

Demersal resources are predominant in landings and reflect both the wide continental shelf and the intense trawl fishery that takes place in this area. In Table 13 demersal species comprise 44% of the total, not including the item "other marine fishes" which is probably dominated by demersal species. The presence of sharks and rays in relevant quantity might be considered as an index of lighter exploitation compared to other Mediterranean areas where this group of species has nearly disappeared from landings (Myers and Worm 2003). It is also possible that high yields of demersal species are the result of a lower exploitation compared to other, generally overexploited, Mediterranean shelf areas (Charbonnier and Garcia, 1985). As mentioned before the mechanism sustaining fish productivity in this area is not clear both at pelagic and at demersal species level, and this should be a target for future research in fishery oceanography in the Mediterranean.

7.1.4 Adriatic Sea

The productive Northern and Central Adriatic Sea is characterised by an extended shelf area with muddy or sandy soft bottoms, which are well suited for trawl fishing. The high productivity in this area is a consequence of the strong nutrient outflow from rivers and of an additional input of Mediterranean waters (Papacostantinou and Farrugio, 2000) from the Otranto Strait. The Adriatic has a high productivity of molluscan shellfish (clams and mussels), and a variety of commercial invertebrates and fish (see Table 14). The area is exploited by bottom trawls for demersal resources, pelagic trawls and purse seines for small pelagic fish and an important dredge fishery for clams in the Italian inshore coastal waters.

The clam fishery is perhaps the most valuable monospecific fishery in the Mediterranean and it exploits the area up to one nautical mile offshore (up to a depth of 12–15m), along the Italian coast from Trieste to the Gargano promontory (Froglia, 1989). The small pelagic fishery is also very important in terms of value and landings, with anchovy being the most sought species. Traditionally anchovy is caught mainly along the more productive, less saline western Adriatic waters by the Italian fleet, and sardine is the main target of eastern Adriatic fishermen and it is generally more abundant in less productive and more saline eastern waters. Demersal resources are exploited all over the continental shelf and species distribution is influenced more by the change in the sediments (mud to sand eastward) than by the very gentle decrease in depth.

TABLE 32.14
Average annual landings (1992–2001) in the Adriatic divided by representative species or group of species (source GFCM Fishery Statistics database). Total landings are about 130,000 tonnes per year.

Species	scientific name	Average landings 92–01 tonnes	%
Sardine	*Sardina pilchardus*	30133	22
Clam	*Chamelea gallina*	28177	21
Anchovy	*Engraulis encrasicolus*	25910	19
cephalopods	*Octopus* sp.; *Loligo* sp. *etc*	10435	8
crustaceans		6656	5
other marine fishes		6008	4
Hake	*Merluccius merluccius*	4475	3
Mackerels	*Scomber* spp.; *Trachurus* spp	3474	3
red mullets	*Mullus spp*	2879	2
Mullets	*Mugilidae*	2194	2
Sole	*Solea solea.*	1395	1
Seabreams	*Sparidae*	1180	1
shark and rays	*Selachii*	912	1
Total		123829	91

The nutrient inputs in the Adriatic are dominated by the Po River and the other northern Italian rivers: these inputs are constrained by the Western Adriatic Coastal Current-WACC to flow along the Italian coast where downwelling motion prevails (see section 4.2.4). On the eastern side of the Adriatic Sea wind induced upwelling prevail due to the climatological wind conditions (Agostini and Bakun, 2001; Cushman-Roisin et al., 2001). These two oceanographic features are linked with the reproductive biology of the two most important small pelagic species, anchovy and sardine.

Anchovy mainly spawns on the western part of the Adriatic (Regner, 1996, fig. 40). This area covers the shallow northern Adriatic waters and the zone along the western coast, down to the Gargano peninsula and corresponds to the area of highest nutrient inputs and primary productivity. There are other spawning areas along the eastern Adriatic coasts but the intensity of spawning is substantially lower (Regner, 1996). Anchovy spawns from April to October with peaks between May and September and, in general, maximum egg production occurs earlier in open waters than in coastal areas. In general little is known about the shifting of spawning centres during the spawning season, the transport of larvae by the circulation of Adriatic surface waters and the various environmental factors, which could affect larval survival. Recently Coombs et al. (2003) studied anchovy spawning and larval distribution associated with the Po river plume, and described the region around the Po river mouth as favourable to survival of anchovy larvae partly because of the additional water column stability caused by the low salinity river outflow. Despite this, a strict relationship between enhanced water column stability and increased survival of anchovy larvae could not be proven. Eggs are released inshore and offshore but late post-larvae and early juveniles are found only along the Italian coast from September to January. The Western Adriatic Coastal Current (WACC, see paragraph 4.2.4) probably plays a major role in the transport of larvae and post-larvae from spawning to nursery ground but a detailed investigation is needed.

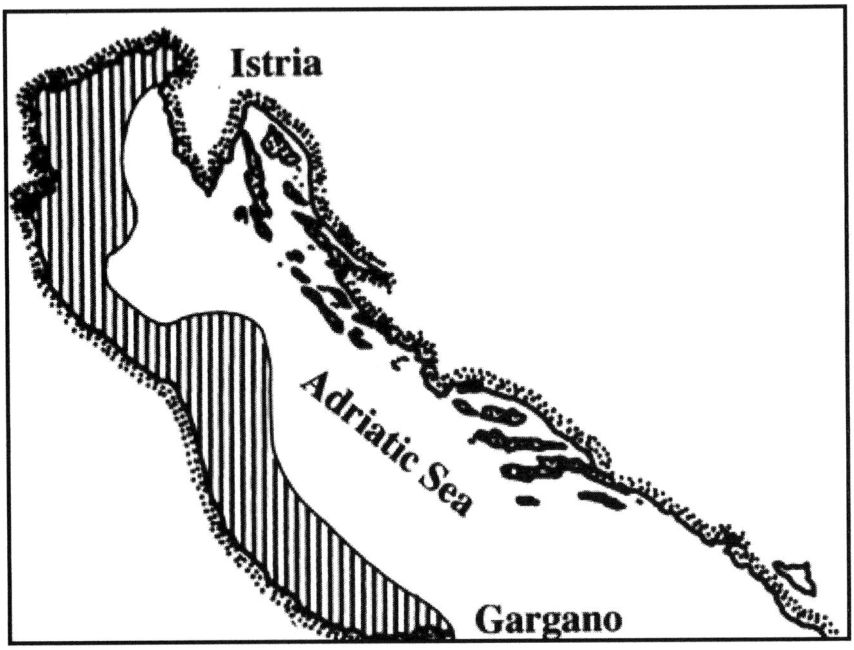

Figure 32.40 The main spawning areas of the Adriatic anchovy (Regner, 1996) shown by dashed areas. Reprinted with permission.

Sardine is a winter spawner in the Mediterranean and in the Adriatic it spawns from Autumn to Spring at temperatures ranging between 9 and 15 °C in the open water zone of the Adriatic (Gamulin and Hure, 1955). Two main spawning centres (see Fig.7.4) exist: a northern one between Ancona and Zadar and a southern one in the area of the Pelagosa Island (Fig. 9, Regner et al. 1987). These main spawning centres are located in oligotrophic areas and, as observed by Regner et al. (1988), temperature might act as a boundary, impeding sardine to spawn in the richer but colder western Adriatic waters in winter. Within the spawning season, the maximum of egg production is in Autumn; immediately along the frontal zone between the western cold waters and the warmer but poorer eastern waters. Sardine tends to spawn at the boundaries of these two zones until temperatures fall below 10 °C; Subsequently sardine moves to the less productive waters along the eastern coast where most intensive spawning has been observed in correspondence with local upwelling phenomena (Regner et al., 1987). On the basis of these observations, Regner et al. (1987) proposed a pattern of spawning and transport of sardine larvae along the eastern side of the Adriatic, which is reported in fig. 41.

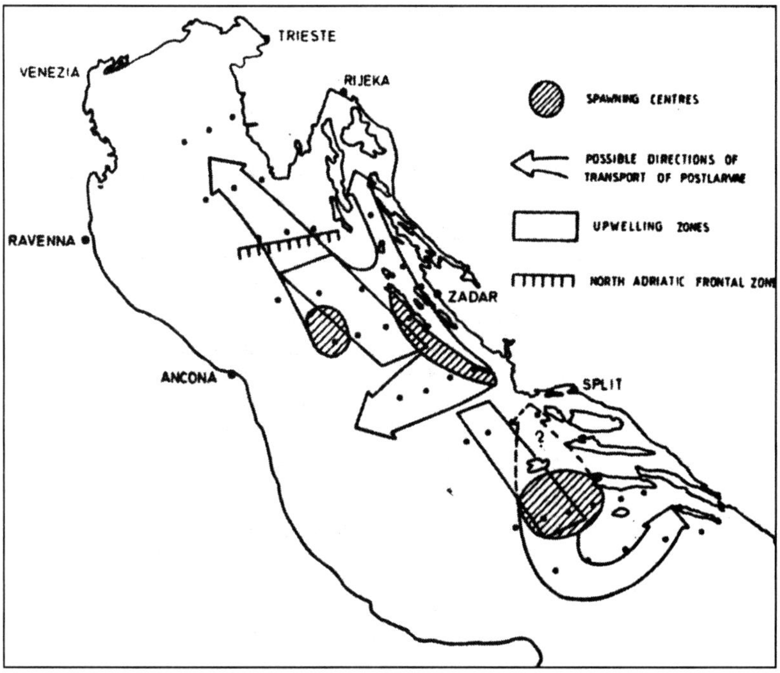

Figure 32.41 Supposed pattern of transport of sardine larvae and post-larvae (Regner et al., 1987). Reprinted with permission.

Spawning centres in fig. 41 should be considered as "main" spawning centres, because sardine eggs are found in most of the Adriatic. The hypothesis is that sardine, within its physiological range of water temperatures, concentrate to spawn in zones enriched either by the mixing of different water masses as frontal areas, or in zones enriched by local upwelling phenomena. Larvae are then transported by the main cyclonic Adriatic circulation or by local gyres that can carry the larvae over the western Adriatic side.

Demersal species are of high economic importance but in general their stock size and landings are not comparable with anchovy and sardine. The Adriatic bottom trawl fishery exploits a pool of species (Table 14) including cephalopods (octopus and cuttlefish), crustaceans (Norway lobster and mantis shrimps), hake, red mullets etc .. The high importance of cephalopods in the catches may be related to the intense exploitation of the demersal resources, with a shift of species composition from long lived (sharks and large ichthyophagous fishes) to fast growing and short lived predators like cephalopods, and planktivorous pelagic fishes (Pauly et al., 1998). For some of the demersal resources of the Adriatic (e.g. Norway lobster, hake and red mullet) general biological parameters, spawning and nursery areas are well known, as well as reproductive periods, although there is a lack (as for most of the Mediterranean) of detailed investigation aimed at clarifying the links between marine environment and population ecology.

7.1.5 Egyptian shelf

This area has a low level of biological productivity due to the low nutrient concentration. The fishing fleet is composed mainly of small-scale vessels, medium size trawlers and purse seiners (Caddy and Oliver, 1996). Before the construction of the Aswan dam the river Nile provided seasonal inflow of sediment and nutrients (see also sections 3.2.1 and 3.2.2). The dramatic reduction of Nile river flow drastically reduced nutrient inflow and had an immediate impact on primary production, small pelagic resources and also demersal resources as shrimps (Bebars and Laserre, 1983; Wadie and Abdel Razek, 1985).

The small pelagic fishery exploits sardinellas, sardine and round herring *(Dussumieria acuta)* a lessepsian migrant species, which has made its entrance from the Red Sea in the last decades. This fishery uses purse seines (Wassef et al., 1985.) and accounts for about one third of the total landings (Table 15). Shrimps of the family Penaeidae are second only to sardinella in the Egyptian Mediterranean fishery (Wadie and Abdel Razek, 1985).

TABLE 32.15

Average annual landings (1992–2001) in the Egyptian shelf divided by representative species, or groups of species (source GFCM Fishery Statistics database). Total landings are about 54,000 tonnes per year.

species	scientific name	Average landings 92–01 tonnes	%
sardinellas	*Sardinellas* spp.	15689	29
Marine fishes nei		5287	10
shrimps	*Penaeidae*	4267	8
silversides	*Atherinidae*	3254	6
seabreams	*Sparidae*	2537	5
mullets	*Mugilidae*	2452	5
Bogue	*Boops boops*	2205	4
red mullets	*Mullus* spp	1871	3
Marine crabs	*Portunidae*	1683	3
round herring	*Dussumeria acuta*	1403	3
Sharks and rays	*Selachii*	1323	2
cephalopods	*Octopus* sp.; *Loligo* sp. *etc*	1238	2
Total		43206	80

The Aswan damming is a good example of the fact that declines in nutrient-rich runoff will reduce the total catch and the proportion of small pelagic fish it contains (De Leiva Moreno et al. 2000). In the period of the Aswan damming catches of Clupeids dropped by 80% in a few years. The change was induced not only by the reduction in quantity but also by the sudden change in the seasonality of the outflow. A recent recovery of catches took place and seems to be associated to the nutrient outflow of drainage waters from the Nile river delta (Papacostantinou and Farrugio, 2000). The role played by nutrient rich runoff in fisheries production is well established for freshwater system and in some cases in marine ecosystems (e.g. Grimes, 2001). The Egyptian shelf case provided a large-scale experimental evidence of this fact and calls for a detailed scientific reassessment of the present situation in terms of oceanography and fishery.

8. Final remarks

In this chapter we have carried out a first attempt to classify the main characteristics of the Mediterranean shelf areas in terms of physical, sedimentological and ecosystem structures and variability.

First of all we have described the different shelf structures in terms of their morphological and dynamical shelf extension. From a general point of view, the shelf extension increases going from the southern to the northern shores with the noticeable exception of the Sicily Strait. Most of the Mediterranean shelf areas can be defined as 'narrow' (less than 20–30 km from the near shore area to the 200 m isobath). These narrow shelves are dominated by open ocean/shelf exchange processes. On the contrary, most extended shelf areas are dominated by river inputs that interconnect the shelf dynamics to the hydrological cycle of the Mediterranean basin. Shelf extension as well as precipitation regimes have a pronounced positive latitudinal gradient, which produces the first, large scale structural characterization of the Mediterranean shelves. Important exceptions to this north-south rule exist in the Strait of Sicily and the Nile Delta.

The sediment structure and composition of the Mediterranean is mainly connected to the river inflow in the extended shelf areas. The biogenic productivity of surface waters plays a major role both in the extended and narrow shelves. In narrow shelves the sediments arrive also from remote areas carried by the along slope currents. Atmospheric deposition is also very important along the southern shelves of the Mediterranean (Tunisian, Nile Delta and Israeli shelves).

The overall structure of the general circulation of the basin and its atmospheric forcing imposes a dominant downwelling character to the Mediterranean shelf areas, since the circulation is generally cyclonic in the open sea. However, exceptions to this rule occur in wind induced upwelling areas (Sicily Strait, eastern Aegean Sea and Adriatic Sea) and in eddy forced narrow shelves, such as the Algerian and Israeli coasts. The shelf areas account for 20% of the overall net heat flux at the air-sea interface (not including the northern Adriatic Sea): the shelves emerge as a negative heat flux engine, helping the overall basin to loose efficiently the heat stored, thus confirming the prevalent downwelling dynamics of the shelf areas.

The analysis has been carried out by considering examples of shelf regimes throughout the Mediterranean, both for narrow and extended shelves. We have examined the Gulf of Lions and the Algerian shelf in the western Mediterranean, the Adriatic Sea and the Sicily Strait in the Central Mediterranean, the Nile Delta and the Israeli shelves in the Eastern Mediterranean.

The northern extended shelves (Gulf of Lions and Adriatic) have large sedimentation rates, they are ROFI areas and they work as net CO_2 sinks. The ecosystem dynamics at the level of primary and secondary producers is partially dominated, even in these nutrient rich areas, by a microbial food web. The fish resources are large and they can be connected simply to the higher primary productivity of these areas.

The southern shelves are less productive at all levels of the ecosystem due to the absence of substantial river inputs. However, an exception exists for the Sicily Strait that can only be explained in terms of favourable physical conditions connected to the Atlantic water inflow along the shelf break.

It is possible to conclude that the Mediterranean Sea shelf structure and variability is a composite of site-specific dynamic scenarios, involving different open ocean/shelf exchange mechanisms and strong connections with land derived inputs. The possibility that such a system could be highly variable with the low frequency variability of atmospheric regimes and strongly dependent from the river loads makes it difficult to be modified and managed in the long term.

Acknowledgements

This work was partially funded by the EU Project 'Mediterranean Forecasting System Toward Environmental Predictions' and the Italian Ministry of Environment and Territory through the Project 'ADRIatic sea integrated COastal areaS and river basin Management system pilot project- ADRICOSM'. Two authors, Nadia Pinardi and Marco Zavatarelli, thank the skilful technical support of Dr. Luca Giacomelli and Dr. Claudia Fratianni. Another author, Mariangela Ravaioli, would like to thank Dr. Giovanna Orsini (Marine Science Department- Univ. of Urbino) for help in the bibliography search. All authors would like to express their greatest appreciation to Prof. A.R.Robinson for his suggestions and very helpful comments to an earlier version of the manuscript.

Bibliography

Agostini V.N. and A. Bakun, 2002. 'Ocean triads' in the Mediterranean Sea: physical mechanisms potentially structuring reproductive habitat suitability (with example application to European anchovy, *Engraulis encrasicolus*). Fish. Oceanogr. 11 (3): 129–142.

Albérola, C., Millot, C., and Font, J., 1995. On the seasonal and mesoscale variabilities of the northern current during the PRIMO-0 experiment in the western Mediterranean. Oceanologica Acta, 18, 2, 163–192.

Almagor, G., 1993. Continental slope processes of northern Israel and southernmost Lebanon and their relation to onshore tectonics. Marine Geology, 122, 151–169.

Alvisi, F., Ravaioli, M., Frignani, M. and Molducci, M., 2000. The study of climate related short timescale sedimentation pattern in prodelta areas of the northern Adriatic Sea. Bollettino Geofisico, a. XXIII, n° 3–4, luglio-dicembre 2000.

Anati, D. A. and J. R. Gat, 1989. Restricted marine basins and marginal sea environments. Handbook of Environmental Isotope Geochemistry, Chapter 2, The Marine Environment, A. Elsevier Science, vol. 3: 29–73.

Antoine, D., A. Morel and J. M. Andrè, 1995. Algal pigment distribution and primary production in the eastern Mediterranean as derived from Coastal Zone Colour Scanner observations. *J. Geophys. Res.*, 100(C8), 16193–16209.

Argnani, A., 1992. Neogene tectonics of the Strait of Sicily, 1992. In: Max, M.D. and Colantoni, P. (Eds.), Geological Development of the Sicilian-Tunisian Platform. UNESCO Reports in Marine Science, Proceedings of International Scientific Meeting, Urbino University, Italy, Nov. 1992., pp. 55–60.

Arnone, R., and P.E. La Violette, 1986: Satellite definition of the bio-optical and thermal variation of coastal eddies associated with the African current. *J. Geophys. Res.*, **91,** 2,351–2,364

Artegiani, A., D. Bregant, E. Paschini, N. Pinardi, F. Raicich and A. Russo, 1997a. The Adriatic Sea general circulation. Part I. Air-sea interactions and water mass structure. *J. Phys. Oceanogr.*,**27,** 1492–1514.

Artegiani, A., D. Bregant, E. Paschini, N. Pinardi, F. Raicich and A. Russo, 1997b. The Adriatic Sea general circulation. Part II. Baroclinic circulation structure. *J. Phys. Oceanogr.*,**27,** 1515–1532.

Arz, Helge W., F. Lamy, J. Pätzold, P. J. Müller, M. Prins, 2003. Mediterranean Moisture Source for an Early-Holocene Humid Period in the Northern Red Sea. *Science,* 300, 118–121.

Astraldi, M. and Gasparini, G.P., 1992. The seasonal characteristics of the circulation in the north Mediterranean basin and their relationships with atmospheric-climatic conditions, J. Geophys. Res., 97 (C6), 9531–9540.

Astraldi, M., Gasparini, G.P., Sparnocchia, S., Moretti, M., and Sansone, E., 1996. The characteristics of the water masses and the water transport in the Sicily Strait at long time scales the eastern and western Mediterranean through the Strait of Sicily, Bulletin de l'Institut Oceanographique, Monaco, CI-ESM Science Series n. 2, 95–115.

Auclair, F., P. Marsaleix and C. Estournel, 2001. The penetration of the northern current over the Gulf of Lions (Mediterranean Sea) as a downscaling problem. *Oceanol. Acta,* **24,** 529–544.

Ayoub, N., Le Traon, P.-Y. and De Mey, P., 1998. A description of the Mediterranean surface variable circulation fromcombined ERS-1 and TOPEX/POSEIDON altimetric data. J. Mar. Syst. 18:3–40.

Azov, Y., 1986 Seasonal patterns of phytoplankton productivity and abundance in nearshore oligotrophic waters of the Levant basin (Mediterranean). *J. Plankton Res.,* **8,** 41–53.

Bakun, A. and V. N. Agostini, 2001. Seasonal pattern of wind induced upwelling/downwelling in the Mediterranean Sea. Sci. Mar., 65, 243–257.

Bebars, M. I. and G. Laserre, 1983. Analyse des captures des pêcheries marines et lagunaires d'Egypte de 1962 à 1976, en liaison avec la construction du haut barrage d'Assouan achevé en 1969. *Oceanol. Acta,* **6,** 417–426.

Benninger, L.K., Suayah, I.B. and Stanley, D.J., 1998. Manzala lagoon, Nile delta, Egypt: modern sediment accumulation based on radioactive tracers. Environmental Geology, 34 (2/3), 183–193.

Berland, B. R., D. Bonin, and S. Y. Maestrini, 1980. Azote ou phosphore? Considérations sur le paradoxe nutritionel de la Mèditerranèe. *Oceanol. Acta,* **3,** 135–142.

Berman, T., D. W. Townsend, S. Z. El Sayed, C. C. Trees and Y. Azov, 1984. Optical transparency, chlorophyll and primary productivity in the eastern Mediterranean near the Israeli coast. *Oceanol. Acta,* 7, 367–371.

Berman, T., Y. Azov, A. Schellner, P. D. Walline and D. W. Townsend, 1986. Extent, transparency and phytoplankton distribution of the neritic waters overlying the Israeli coastal shelf. *Oceanol. Acta,* 9, 439–447.

Bethoux, J. P .. Budgets of the Mediterranean Sea: their dependence on the local climate and on the characteristics of the Atlantic waters. Oceanol. Acta, 2, 157–163, 1979.

Bethoux, J. P., L. Prieur, and J. H. Bong, 1988. Le Courante Ligure au large de Nice. *Oceanol. Acta,* 9, 59–67.

Bolle, H.-J., 2003. Mediterranean Climate. Variability and Trends. Springer, 372 pp.

Borzelli, G., G. Manzella, S. Marullo, R. Santoleri, 1999,Observation of coastal filaments in the Adriatic Sea. *J. Mar Sys.,* 20, 187–203.

Boukthir, M., and B. Barnier. Seasonal and inter-annual variations in the surface freshwater flux in the Mediterranean Sea from ECMWF re-analysis project. Jour. Marine Systems, 24, 343–354, 2000.

Bowles, F.A., Lambert, D.N. and Richardson, M.D., 1992. In: Max, M.D. and Colantoni, P. (Eds.), Geological Development of the Sicilian-Tunisian Platform. UNESCO Reports in Marine Science, Proceedings of International Scientific Meeting, Urbino University, Italy, Nov. 1992., pp. 129–134.

Brankart, J.M., N. Pinardi, 2001. Abrupt cooling of the Mediterranean Levantine Intermediate Water at the beginning of the eighties: observational evidence and model simulation. Journal of Physical Oceanography, Vol. 31, No 8, Part 2, pages 2307–2320.

Brenner, S., 2003. High-resolution nested model simulations of the climatological circulation in the southeastern Mediterranean Sea. Annales Geophysicae, 20: 267–280.

Brink, K.,N., 1998. Deep-Sea Forcing and exchange processes. In: *The Sea Vol.10,* Brink, K.,N. and A.,R. Robinson (eds.), pg. 151–167, Wiley.

Bryden, H. L., and T. H. Kinder, 1991. Recent progress in strait dynamics. Rev. Geophys. (Suppl.), 617–631.

Broche, P., J. L. Devenon, P. Forget, J. C. de Maistre, J. J. Naudin and G. Cauwet, 1998. Experimental study of the Rhone plume. Part I: physics and dynamics. Oceanologica Acta, 21(6), 725–738.

Brunetti, M., M.Maugeri, T.Nanni and A.Navarra, 2000. Droughts and estreme events in regional daily italian precipitation series. Int. Jour. Clim,.

Buongiorno Nardelli, B., R. Santoleri, S. Marullo, D. Iudicone, and S. Zoffoli. 1999. Altimetric signal and three dimensional structure of hte sea in hte Channel of Sicily. J. Geophys. Res., 104, C9, 20,585–20,603.

Buscail, R., Pocklington, R., Daumas, R. and Guidi, L., 1990. Fluxes and budget of organic matter in the benthic boundary layer over the northwestern Mediterranenan margin. Continental Shelf Research, 10, (9–11), 1089–1122.

Buscail, R., Pocklington, R. and Germain, C., 1995. Seasonal variability of the organic matter in a sedimentary coastal environment: sources, degradation and accumulation (continental shelf of the Gulf of Lyon – northwestern Mediterranean Sea). Continental Shelf Research, 15, n° 7, 843–869.

Buscail, R. and Germain, C., 1997. Present-day organic matter sedimentation on the NW Mediterranenan margin: importance of off-shelf export. Limnol. Oceanogr., 42, 217–229.

Buscail, R., Foudil-Bouras, A.E. and Pauc, H., 1999. Matière organique et pollution par les hydrocarbures dans les sédiments superficiels du golfe d'Arzew (mer Méditerrannée, Algérie). Oceanologica Acta, 22, 3, 303–317.

Caddy, J. F. and P. Oliver, 1996. Some future perspectives for assessment and management of Mediterranean fisheries for demersal and shellfish resources, and small pelagic fish. *Stud. Rev. GFCM,* 66, 19–60.

Caddy, J. F., 1999. Eléments pour l'aménagement des pêcheries méditerranéennes: unités géographiques et contrôle de l'effort. *Stud. Rev. GFCM,* 70, 1–30.

Canfield, D.E., 1993. Organic matter oxidation in marine sediments. In: Wollast, R., Mackenzie, F.T., Chou, L. (Eds.), Interactions of C, N, P and S Biogeochemical Cycles and Global Change. NATO ASI Ser., 14, 333–363.

Carter, T.G., Flanagan, J.P., Jones, C.R., Marchant, F.L., Murchison, R.R., Rebman, J.A., Sylvester, J.C. and Whitney, J.C., 1972. A new bathymetric chart and physiography of the Mediterranean Sea. In: The Mediterranean Sea, D.J. Stanley editor, Dowden, Hutchinson and Ross, Inc., Stroudsburg, Pennsylvania, 1–23.

Castellari, S., N. Pinardi, and K.Leaman, 1998. A model study of air-sea interactions in the Mediterranean Sea. Journal of Marine Systems, 18, 89–114.

Castellari, S., N. Pinardi, K. Leaman, 2000. Simulation of water mass formation processes in the Mediterranean Sea: influence of the time frequency of the atmospheric forcing. Journal of Geophysical Research, Vol. 105, C10, 24157–24181.

Castellon, A., J. Salat and M. Masó, 1985. Some observations on Rhone river fresh water plume in the Catalan coast. *Rapp. Comm. int. Mer Médit.,* **29,** 119–120.

Caulet, J.P., 1972. Recent biogenic calcareous sedimentation on the algerian continental shelf. In: The Mediterranean Sea, D.J. Stanley editor, Dowden, Hutchinson and Ross, Inc., Stroudsburg, Pennsylvania, 1–23.

Charbonnier, D. and S. Garcia, 1985. Atlas of the fisheries of the western and central Mediterranean. FAO, Rome.

Colantoni, P., Cremona, G., Ligi, M., Borsetti, A.M. and Cati, F., 1985. The Adventure Bank (off Southwestern Sicily): a present day example of carbonate shelf sedimentation. Giornale di Geologia, 47/1–2, 165–180.

Conan, P., M. Pujo-Pay, P. Raimbault and M. Leveau, 1988a. Variabilité hydrologique et biologique du golfe du Lion.I. Transport en Azote et productivité potentielle. Oceanologica Acta, 21 (6) 751–765.

Conan, P., M. Pujo-Pay, P. Raimbault and M. Leveau, 1988b. Variabilité hydrologique et biologique du golfe du Lion.II. Productivité sul le bord interne du courant. Oceanologica Acta, 21(6)767782.

Coombs, S. H., O. Giovanardi, N. C. Halliday, G. Franceschini, D. V. P. Conway, L. Manzueto, C. D. Barrett and I. R. B. Mc Fadzen, 2003. Wind mixing, food availability and mortality of anchovy larvae *Engraulis encrasicolus* in the northern Adriatic Sea. *Mar. Ecol. Prog. Ser.,* 248, 221–235.

Courp, T. and Monaco, A., 1990. Sediment dispersal and accumulation on the continental margin of the Gulf of Lyon. Sedimentary budget. Continental Shelf Research, 10, (9–11), 1063–1087.

Crispi, G. R. Mosetti, C. Solidoro, and A. Crise, 2001. Nutrient cycling in Mediterranean basino: the role of the biological pump in the trophic regime. *Ecological Modelling,* **138,** 101–114.

Cruzado, A. and Z. R. Velasquez, 1990. Nutrients and phytoplankton in the Gulf of Lyon, northwestern Mediterranean. *Cont. Shelf Res.,* 10, 931–942

Cushing, D. H., 1989. A difference in structure between ecosystems in strongly stratified waters and in those that are only weakly stratified. *J. Plankton, Res.,*11, 1–13.

Cushman-Roisin, B., M. Gacic, P. M. Poulain and A. Artegiani, 2001. Physical Oceanography of the Adriatic Sea. Kluwer Academic Publishers, Dordecht, The Netherlands, 304 pp.

Degobbis, D. and M. Gilmartin, 1990. Nitrogen, phosphorus and silicon budget for the northern Adriatic Sea. *Oceanol. Acta,*13, 31–45.

De Leiva, M., V. N. Agostini, J. F. Caddy and F. Carocci, 2000. Is the pelagic demersal ratio from fishery landings a useful proxy for nutrient availability? A preliminary data exploration for the semi-enclosed seas around Europe. *ICES J. Mar. Sci.,* 57, 1091–1102.

Delfanti, R., Frignani, M., Langone, L., Papucci, C. and Ravaioli, M., 1994. The role of the rivers in Chernobyl radiocesium delivery, distribution and accumulation in coastal sediments of northern Adriatic Sea. Seminar on Freshwater Radioecology, Lisbon, March 21–255, 1994.

Demirov, E. and N.Pinardi, 2002. The Simulation of the Mediterranean Sea circulation from 1979 to 1993. Part I: The interannual variability. *J. Marine* Systems, 33–34, pp. 23–50.

Diaz, F., P. Raimbault, B. Boudjellal, N. Garcia and T. Moulin, 2001. Early spring phosphorus limitation of primary productivity in a NW Mediterranean coastal zone (Gulf of Lions). *Marine Ecol. Progr. Ser.,* 211, 51–62.

Dowidar, N. M., 1984. Phytoplankton biomass and and primary productivity of the south-eastern Mediterranean. *Deep Sea Res.,* 31, 983–1000.

Drago, A. F., Sorgente, R., and Ribotti, A., 2003. A high resolution hydrodynamic 3-D model simulation of the malta shelf area. Annales Geophysicae, 21: 323–344.

Durrieu de Madron, X., Abassi, A., Heussner, S., Monaco, A., Aloisi, J.C., Radakovitch, O., Giresse, P., Buscail, R., and Kerherve, P., 2000. Particulate matter and organic carbon budgets for the Gulf of Lyon (NW Mediterranenan). Oceanologica Acta, 23, 6, 717–730.

Echevin, V., Cré pon, M., and Mortier, L. Simulation and analysis of the mesoscale circulation in the northwestern Mediterranean Sea. Annales Geophysicae, 21: 281–297, 2003.

Emery, K. O. and Neev, D., 1960. Mediterranean beaches of Israel. Isr. Geol. Surv. Bull., 26, 1–23.

Estrada, M. and J. Salat, 1989. Phytoplankton of deep and surface water layers in a Mediterranean frontal zone. *Sci. Mar.,* 52, 203–214.

Estrada, M., 1996. Primary production in the northwestern Mediterranean, *Scientia Marina,* 60 (suppl.), 55–64.

Fabricius, F. and Schmidt-Thomè, P., 1972. Contribution to recent sedimentation on the shelves of the Southern Adriatic, Ionian, and Syrtis Sea. In: The Mediterranean Sea, D.J. Stanley editor, Dowden, Hutchinson and Ross, Inc., Stroudsburg, Pennsylvania, 1–23.

Faganeli, J., Pedzic, J., Ogorolec, B., Misic, M. and Naialek, M., 1994. The origin of sedimentary organic matter in the Adriatic Sea. Continental Shelf Research, 14, 365–384.

Flexas, M., Garcia, M. A., Durrieu de Madron, X., Canals, M., and Arnau, P., 2002. Flow variability in the Gulf of Lions during the MATER HFF experiment (March-May 1997). J. Marine Systems, 33/34, 197–214.

Fonda Umani, S., 1996. Pelagic production and biomass in the Adriatic Sea. *Scientia Marina,* 60 (suppl.), 65–77.

Fonda Umani, S. and A. Beran, 2003. Seasonal variations in the dynamics of microbial plankton communities: first estimates from experiments in the Gulf of Trieste, northern Adriatic Sea. *Mar. Ecol. Progr. Ser.,* 247, 1–16.

Font, J.,E. Garcia-Ladona,E. Garcia-Gorritz, 1995. The seasonality of the mesoscale motion in the Northern Current of the western Mediterranean, Oceanologica Acta, 18, 207–219.

Fredj G., Bellan-Santini D., Meinardi M.,1992. Etat des connaissances sur la faune marine méditerranéenne. Bulletin de l'Istitut océanographique, Monaco, n° spécial 9:133–145.

Frignani, M., Langone, L., and Ravaioli, M., 1987. Profili e bilancio di massa di 137Cs e 210Pb in alcune carote dell'Adriatico centrale. IV Congresso Nazionale SIRR – S. Teresa (SP), 15–16 September 1987, 17.

Frignani, M., Langone, L., and Ravaioli, M., 1989. Interpretation of radionuclide activity-depth profiles in three sediment cores from the Middle Adriatic. Giorn. Geol., 51/2, 131–142.

Frignani, M., Langone, L., Ravaioli, M. and Cadonna, A., 1990. Sediment fluxes on a 100 yr time scale in different environments of the Adriatic Sea (Italy). XXXII Congress CIESM, Perpignan, 15–20 October, 1990.

Frignani, M. and Langone, L., 1991. Accumulation rates and 137Cs distribution in sediments off the Po River delta and the Emilia-Romagna coast (north-western Adriatic Sea, Italy). Continental Shelf Research, 11, 525–542.

Frignani, M., Langone, L., Beks, J. and Alvisi, F., 1996. Sediment accumulation rates from cores collected in the central and southern Adriatic Sea. In: Transfer pathways and fluxes of organic matter and related elements in water and sediments of the northern Adriatic Sea and their importance on the eastern Mediterranean Sea. Final report EUROMARGE-AS Project, EU/MAST2-CT93–0052 Commission of the European Union, 148–156.

Frihy, O.E., Debes, E.A. and El Sayed, W.R., 2003. Processes reshaping the Nile delta promontories of Egypt: pre- and post-protection. Geomorphology, 53, 263–279.

Froglia, C., 1989. Clam fisheries with hydraulic dredges in the Adriatic Sea. In *Marine invertebrate fisheries: their assessment and management,* J. F. Caddy ed., John Wiley & Sons., New York, pp 507–524.

Gacic, M., G. Civitarese, L. Ursella. 1999. Spatial and seasonal variability of water and biogeochemical fluxes in the Adriatic Sea. In: *The eastern Mediterranean as a Laboratory Basin for the Assessment of Contrasting Ecosystems.*335–357,Kluver.

Gamulin, T. and J. Hure, 1955. Contribution à la connaissance de l'écologie de la sardine (*Sardina pilchardus* Walb.) dans l'Adriatique. *Acta Adriat.,* **7,** 1–23.

Garcia, A. and I. Palomera, 1996. Anchovy early life history and its relation to its surrounding environment in the Western Mediterranean basin. *Sci. Mar.,* 60 (supl.2), 155–166.

Garcia Lafuente, J., A. Garcia, S. Mazzola, L. Quintanilla, J. Delgado, A. Cuttita and B. Patti, 2002. Hydrographic phenomena influencing early life stages of the Sicilian Channel anchovy. *Fish. Oceanogr.,* 11, 31–44.

Garibaldi, L. and J. F. Caddy, 1998. Biogeographic characterization of Mediterrannean and Black Sea faunal provinces using GIS procedures. *Ocean Coast. Manage.,* 39, 211–227.

Garrett, C., Outerbridge, R., Thompson, K., 1993. Interannual variability in Mediterranean heat and buoyancy fluxes. J. Climate 6, 900–910.

Gaudy, R. 1990. Zooplankton feeding on seston in the Rhône River plume area (NW Mediterranean Sea) in May 1988. *Hydrobiologia.,* **207,** 241–249.

Gensous, B., Williamson, D., and Tesson, M., 1993. Late Quaternary transgressive and highstand deposit of a shelf (Rhone delta, France). Spec. Pub. Ass. Sedim., 18, 197–211.

GFCM FAO Fishery Statistics Database. *www.fao.org/fi/statist/statist.asp*

Giordani, P., Hammond, D.E., Berelson, W.M., Montanari, G., Poletti, R., Milandri, A., Frignani, M., Langone, L., Ravaioli, M., Rovatti, G. and Rabbi, E., 1992. Benthic fluxes and nutrient budgets for sediments in the Northern Adriatic Sea: burial and recycling efficiencies. International Conference Marine Coastal Eutrophication, Bologna, 21–24 March. The Science of the Total Environment, Supplement 1992, 251–275.

Giordani, P., Helder, W., Koning, E., Miserocchi, S., Danovaro, R. and Malaguti, A., 2002. Gradients of benthic-pelagic coupling and carbon budgets in the Adriatic and Northern Ionian Sea. Journal of Marine System, 33–34, 365–387.

Golani D., 1996. The marine ichthyofauna of the eastern Levant-History, inventory, and characterization. Israel Journal of Zoology, 42 (1): 15–55.

Goldman D., J. Caron and M. Dennett, 1987. Regulation of gross growth efficiency and ammonium regeneration in bacteria by substrate C:N ratio. Limnol Oceanogr., 32, 1239–1252.

Golik, A., 1997. Dynamics and management of sand along the Israeli coastline. Bull. Instit. Oceanog., Monaco, n° special 18-Transformation and Evolution of the Mediterranean Coastline, 97–110.

Got, H. and Aloisi, J.C., 1990. The Holocene sedimentation on the Gulf of Lyon margin: a quantitative approach. Continental Shelf Research, 10, (9–11), 841–855.

Grimes, C. B, 2001. Fishery production and the Mississipi river discharge. *Fisheries,* **26,** 17–26.

Guerzoni, S., Ravaioli, M., Rovatti, G. and Suman, O.D., 1984. Comparison of 210Pb trace metals (Hg, Pb, Cu, Cr) profiles and river discharge in a core off the Po della Pila river mouth (Italy). VIIes Journèes Etud. Pollutions, Lucerne, C.I.E.S.M ..

Hachemi, M., 1996. Les ressources halieutiques de la Méditerranée sud centrale. *FAO Fish. Rep.,* **533** (suppl.), 191–229.

Hamza, W., P.Ennet, R.Tamsalu and V.Zalensky, 2003. The 3D physical-biological model study in the Egyptian Mediterranean coastal area. Aquatic Ecology, (37), 307–324.

Hecht, A., N.Pinardi, A.R.Robinson, 1988. Currents, Water Masses, Eddies and Jets in the Mediterranean Levantine Basin. Journal of Physical Oceanography, Vol. 18, No. 10, pp. 1320–1353.

Herbaut, C., Mortier, L., and Crépon, M., 1996. A sensitivity study of the general circulation of the western Mediterranean Sea, Part. I: the response to density forcing through the straits, Journal of Physical Oceanography 26: 65–84.

Herbaut, C., Codron, M., Crépon, M., 1998. Separation of a Kelvin front at the entrance of a strait. Example of the Strait of Sicily. Journal of Physical Oceanography 28 (7), 1346–1362.

Herman, Y., 1972. Quaternary Eastern Mediterranenan sediments: micropaleontology and climatic record. In: The Mediterranean Sea, D.J. Stanley editor, Dowden, Hutchinson and Ross, Inc., Stroudsburg, Pennsylvania, 1–23.

Herut, B., A. Almogi-Labin, N. Jannink and I. Gertman, 2000. The seasonal dynamics of nutrients and chlorophyll a concentrations on the SE Mediterranean shelf slope. *Oceanol. Acta,* **23,** 771–781.

Hill, A.,E.,1998. Buoyancy effects in coastal and shelf seas. In: *The Sea Vol.10,* Brink, K.,N. and A.,R. Robinson (eds.), pg. 21–62, Wiley.

Hopkins, T. S., 1985, Physics of the Sea. *In:* Western Mediterranean (R. Margalef Ed.), Pergamon Press, Oxford, 100–125.

Hopkins, T.S., Artegiani, A., Bignami, F. and Russo, A., 1999. Watermass modification in the Northern Adriatic. A preliminary assessment from the ELNA data set. In: Hopkins, T.S., Artegiani, A., Cauwet, G., degobbis, D., Malej, A. (Eds.), Ecosystem Res. Report n. 32. The Adriatic Sea (EUR 18834). European Commission Directorate-General for Research, Brussel, 3–23.

Ignatiades, L., S. Psarra, V. Zervakis, K. Pagou, E. Souvermezoglou, G. Assimakopoulou and O. Gotsis-Cretas, 2002. Phytoplankton size-based dynamics in the Aegean Sea (eastern Mediterranean). J. Mar. Sys., 36, 11–28.

Jago, C.F. and Barusseau, J.P.,1981. Sediment entrainment on a wave-graded shelf, Roussillon, France. Marine Geology, 42, 279–299.

Josey, Simon A., 2003. Changes in the heat and freshwater forcing of the eastern Mediterranean and their influence on deep water formation. Journal of Geophysical Research, Vol. 108, No. C7, 3237, doi:10.1029/2003JC001778.

Korres, G., N. Pinardi, A. Lascaratos, 2000. The ocean response to low frequency interannual atmospheric variability in the Mediterranean Sea. Part I: Sensitivity experiments and energy analysis. *Journal of Climate,* Vol. 13, No. 4, 705–731.

Korres, G., N. Pinardi, A. Lascaratos, 2000. The ocean response to low frequency interannual atmospheric variability in the Mediterranean sea. Part II:Empirical Orthogonal Functions Analysis. Journal of Climate, Vol. 13, No. 4, 732–745.

Korres, G., and A. Lascaratos, 2003. A one-way nested eddy resolving model of the Aegean and Levantine basins: implementation and climatological runs. Annales Geophysicae, 21: 205–220.

Krom, M. D., N. Kress, S. Brenner and L. I. Gordon, 1991. Phosphorus limitation of primary productivity in the eastern Mediterranean Sea. *Limnol. Oceanogr.,* **36,** 424–432.

Lefèvre, D., H. J. Minas, M. Minas, C. Robinson, P. J. Le B. Williams and E. M. S. Woodward., 1997. Review of gross community production, primary production, net community production and dark community respiration in the Gulf of Lions. *Deep Sea Res. II,* 44, 801–832.

Legendre, L. and F. Rassoulzadegan, 1995. Plankton and nutrient dynamics in marine waters. *Sarsia,* 41, 153–172.

Lermusiaux, P.F.J., A. R. Robinson, 2001. Features of diminat mesoscale variability, circulation patterns and dynamics in the Strait of Sicily. *Deep Sea Res.,* I, 48, 1953–1997.

Levi, D. and J. P. Troadec, 1974. The fish resources of the Mediterranean and the Black Sea. *Stud. Rev. GFCM,* 54, 29–52.

Liu, K.K., K. Iseki and S.Y. Chao, 2000. Continental margin carbon fluxes.In: *The Changing Ocean Carbon Cycle,* R.B. Hanson, H.W. Ducklow, & J.G. Field (Eds), pp. 187–239. Cambridge: Cambridge University Press.

Lleonart J. and Recasens L., 1996. Fisheries and the environment in the Mediterranean Sea. Studies and Reviews GFCM, 66: 5–18.

Lloret J., Lleonart J., Sole I., Fromentin J.,2001. Fluctuations of landings and environmental conditions in the north-western Mediterranean Sea. Fisheries Oceanography, 10 (1): 33–50.

Lleonart, J. and L. Recasens, 1996. Fisheries and the environment in the Mediterranean Sea. *Stud. Rev.* GFCM, 66, 5–18.

Lloret, J., I. Palomera, J. Salat and I. Sole, 2004. Impact of freshwater input and wind on landings of anchovy (*Engraulis encrasicolus*) and sardine (*Sardina pilchardus*) in shelf waters surrounding the Ebre (Ebro) River delta (north-western Mediterranean). *Fish. Oceanogr.,* 13, 1–9.

Madec, G., Chartier, M., and Crépon, M., 1991. The effect of thermohaline forcing variability on deep water formation in the western Mediterranean Sea: a high-resolution three-dimensional numerical study, Dynamics of the Atmosphere and Oceans, 301–332.

Manca B., Budillon g., Scarazzato P. and Orsella L., 2003 Evolution of dynamics in the Estern Mediterranean affecting water mass structures and properties in the Ionian and Adriatic Seas. Jour. Geophys. Res., 108, C9, 101029 - 101046.

Manzella, G. M. R., Gasparini, G. P., and Astraldi, M., 1988. Water exchange between the eastern and western Mediterranean through the Strait of Sicily, Deep-Sea Res. I, 35, 1021–1035.

Manzella, G., 1992.'La dinamica degli oceani', pg. 160. Nuova ERI.

Maouche, S., 1987. Mécanismes hydrosédimentaires en baie d'Alger (Algérie): approche sédimentologique, géochimique et traitement statistiques, thèse de 3e cycle, univ. Perpignan, 214 p.

Mariotti, A., M.V. Struglia, N. Zeng, K.-M. Lau, 2002. The Hydrological Cycle in the Mediterranean Region and Implications for the Water Budget of the Mediterranean Sea. American Meteorological Society, Journal of Climate, vol. 15: 1674–1690.

Maurin, C., 1977. Les problème halieutiques en Méditerranée et en Tunisie. *Bull. Off. natn. Pêch. Tunisie,* **1,** 9–16.

Max, M.D. and Colantoni, P., 1992. Introduction: geomorphological position and geology of the Sicilian_Tunisian Platform. In: Max, M.D. and Colantoni, P. (Eds.), Geological Development of the Sicilian-Tunisian Platform. UNESCO Reports in Marine Science, Proceedings of International Scientific Meeting, Urbino University, Italy, Nov. 1992., pp. 1–2.

Mc Gill, D. A., 1969. A budget for dissolved nutrient salts in the Mediterranean Sea. *Càh. Océanogr.,* 21, 543–554.

MEDAR Group, 2002. Mediterranean and Black Sea database of temperature, salinity and biochemical parameters and climatological atlas. 4 CD-ROM and www.ifremer.fr/sismer/program/medar/. European Commission Marine Science and Technology Programme (MAST). Ed. IFREMER.

Millot, C., 1985. Some features of the Algerian Current. *J. Geophys. Res.,* 90, 7169–7176.

Millot C., 1987. Circulation in the Western Mediterranean. Oceanol. Acta, 10, 2: 143–149.

Millot, C., 1990a. The Gulf of Lions Hydrodynamics. Continental Shelf Research, 10, 9/11, 885–894.

Millot, C., I. Taupier Letage and I. Benzohra. 1990b. The Algerian eddies. *Earth Sci. Rev.,* 27, 203–219.

Millot C., M. Benzhora, I. Taupier-Letage, 1997. Circulation in the Algerian Basin inferred from the MEDIPROD-5 current meters data. Deep-Sea Res., 44: 1467–1495.

Minas, M and H. J. Minas, 1991. Hydrological and chemical conditions in the Gulf of Lions and relationships to primary production encountered during CYBELE (First Leg, 12–29 April 1990). *Water Pollution Research Reports,* **28,** 131–144.

Molcard A., N. Pinardi, M. Iskandarani and D.B. Haidvogel, 2002. Wind driven general circulation of the Mediterranean Sea simulated with a Spectral Element Ocean Model. Dynamics of Atmospheres and Oceans, Vol. 35, pp. 97–130.

Monoley, C.L. and Field, J.G., 1991. Modelling carbon and nitrogen flows in a microbial plankton community. In: Protozoa and their role in marine processes, Springer, P.C. Reid et al editor, 443–474.

Moran, X. A. G., I. Taupier-Letage, E. Vázquez-Domìnguez, R. Ruiz, L. Arin, P. Raimbault and M. Estrada, 2001. Physical-Biological coupling in the Algerian Basin (SW Mediterranean) : Influence of mesoscale instabilities on the biomass and the production of phytoplankton and bacterioplankton. *Deep Sea Res. I,* **48,** 405–437.

Morel, A and J. M. André, 1991. Pigment distribution and primary production in the Western Mediterranean Sea as derived and modeled from coastal Zone color scanner observations.*J. Geophys. Res.,* 96(C7), 12685–12698.

Moutin, T., P. Raimbault, H. L. Golterman and B. Coste, 1998. The input of nutrients by the Rhône river into the Mediterranean Sea: recent observations abd comparison with earlier data. *Hydrobiologia,* 373/374, 237–246.

Müller, P.J. and Suess, E., 1979. Productivity, sedimentation rate and sedimentary carbon content in the oceans, organic carbon preservation. Deep-Sea Research, 26A, 1347–1362.

Myers, R. A. and B. Worm, 2003. Rapid worldwide depletion of predatory fish communities. *Nature,* 423, 280–283.

Naudin, J. J., G. Cauwet., C. Fajon, L. Oriol, S Terzić J. L. Devenon and P. Broche, 2001. Effect of mixing of microbial communities in the Rhône river Plume. *J. Mar. Sys.,* **28,** 203–227.

Nittrouer, C.A. and Wright, L.D., 1994. Transport of particles across continental shelves. Rev. Geophys., 32, 85–113.

Olivar, M. P., J. Salat and I. Palomera, 2001. Comparative study of spatial distribution patterns of the early stages of anchovy and pilchard in the NW Mediterranean Sea. *Mar. Ecol Prog. Ser.,* 217, 111–120.

Owens, N. J. P., A. P. Rees, E. M. S. Woodward and R. F. C. Mantoura, 1989. Size-fractionated primary production and nitrogen assimilation in the north-western Mediterranean Sea during January 1989. *Water Pollution Res. Rep.,* **13,** 126–135.

Palomera, I., 1992. Spawning of anchovy *Engraulis encrasicolus* in Northwestern Mediterranean relative to hydrographic feature in the region. *Mar. Ecol. Prog. Ser.,* 79, 215–223.

Papacostantinou C. and Farrugio H., 2000. Fisheries in the Mediterranean. Mediterranean Marine Science, 1/1: 5–18.

Paschini, E., A.Artegiani and N. Pinardi, 1993. The mesoscale eddy field of the middle Adriatic Sea. Deep Sea Research I, Vol. 40, No. 7 , pp 1365–1377.

Pauc, H., 1991. La nature minéralogique des apports en suspensions sur la marge algérienne et leur relation avec les sédiments. Actes du 3e Congrès français de sédimentologie, Brest 18–20 novembre 1991, 225–226.

Pauly, D., V. Christensen, J. Dalsgaard, R. Froese and Jr. F. Torres, 1998. Fishing Down Marine Food Webs. *Science,* 279, 860–863.

Pedlosky, J, 1987. *Geophysical Fluid Dynamics.* Springer-Verlag, pp. 724,.

Peixoto, J. P., M. De Almeida, R. D. Rosen, and D. A. Salstein, 1982: Atmospheric moisture transport and the water balance of the Mediterranean Sea. Water Resour. Res., 18, 83–90.

Phillips, N.A., 1954. Energy transformations and meridional circualtions associated with simple baroclinic waves in a two-level quasi-geostrophic model. *Tellus,* 6 273–286.

Pinardi, N., A. Navarra, 1993. Baroclinic wind adjustment processes in the Mediterranean Sea. *Deep Sea Research II,* Vol. 40, No. 6, 1299–1326.

Pinardi, N., and E. Masetti, 2000. Variability of the large scale general circulation of the Mediterranean Sea from observations and modelling: a review. Palaeogeography, palaeoclimatology, palaeoecology, 158 (2000) 153–173.

Pinardi, N., I.Allen , E. Demirov, P. De Mey, G.Korres, A.Lascaratos, P-Y. Le Traon, C.Maillard, G. Manzella, C. Tziavos. The Mediterranean ocean Forecasting System: first phase of implementation (1998–2001). Annales Geophysicae, 21: 3–20, 2003.

The POEM Group. General circulation of the Eastern Mediterranean. Earth Sciences Review, 32, pp 285–309, 1992.

Pomerancblum, M., 1966. The distribution of heavy minerals and hydraulic equivalents in sediments of the Mediterranean continental shelf of Israel. J. Sediment. Petrol., 36, 162–174.

Poulain, P.-M., 2001. Adriatic Sea surface currents as derived from drfiter data between 1990 and 1999. J.Mar. Syst.,29, 3–32.

Poulain, P.-M., E.Mauri and L.Ursella, 2003 Unusual upwelling event and current reversal off the Italian Adriatic coast in summer, submitted.

Puddu, A., R. La Ferla, A. Allegra, C. Bacci, M. Lopez, F. Oliva and C. Pienotti, 1998. Seasonal and spatial distribution of bacterial production and biomass along a salinity gradient (northern Adriatic Sea). *Hydrobiologia,* 363, 271–282.

Pullen, J.,J. D. Doyle, R. H. A. Ogston, J. W. Book, H. Perkins, and R. Signell. Coupled ocean-atmosphere nested modeling of the Adriatic Sea during winter and spring 2001. J. Geophys. Res., 108, C10, 3320, doi:10.1029/2003JC001780, 2003

Raicich, F., 1996. On the fresh water balance of the Adriatic Sea. *J. Mar. Sys.,* 9, 305–319.

Raicich, F., N. Pinardi, A. Navarra, 2003. Teleconnections between Indian Monsoon and Sahel rainfall and the Mediterranean. Int. Jour. of Clim., 23, 173–186.

Rasoanarivo, R., J. Folack, G. Champalbert and B. Becker, 1991. Relations entre les communautés phytoplactonique et l'alimentation des larves de *Sardina pilchardus* Walb. Dans le golfe de Fos (Méditerranée occidentale): influence de la lumière sur l'activité alimentaires des larves. *J. Exp. Mar. Biol. Ecol.,* 151, 83–92.

Redfield, A., B. Ketchum and F. Richards, 1963. The influence of organisms on the composition of sea water. In: The Sea (M. Hill Ed.), Vol 1, 26–77. Interscience New York

Regner, S., D. Regner, I. Marasović and F. Kršinić, 1987. Spawning of sardine, *Sardina pilchardus* (Walbaum, 1792), in the Adriatic under upwelling conditions. *Acta Adriat.,* 28,161–198.

Regner, S., G. Piccinetti-Manfrin and C. Piccinetti, 1988. The spawning of the sardine (*Sardina pilchardus* Walb.) in the Adriatic as related to the distribution of temperature. *FAO Fish. Rep.,* 394, 127–132.

Regner, S., 1996. Effects of environmental changes on early stages and reproduction of anchovy in the Adriatic sea. *Sci. Mar.,* 60 (Supl.2), 167–177.

Robinson, A.R., A.Hecht, N. Pinardi, Y.Bishop, W.G.Leslie, Z.Rosentroub, A.J.Mariano, S.Brenner, 1987. Small synoptic/mesoscale eddies: the energetic variability of the Eastern Levantine basin. Nature, Vol. 327, No. 6118, pp. 131–134.

Robinson, A.R., Golnaraghi, M., Leslie, W.G., Artegiani, A., Hecht, A., Lazzoni, E., Michelato, A., Sansone, E.,Theocharis, A., Unluata, G. The eastern Mediterranean general circulation: Features, structure and variability. Dynamics of Atmospheres and Oceans 15 (3–5), pp. 215–240, 1991.

Robinson, A.R .. Physical processes, field estimation and an approach to interdisciplinary ocean modelling. Earth Science Reviews, 40, 3–54, 1996.

Robinson, A.R., Sellschopp, J., Warn-Varnas, A., Anderson, L.A. and Lermusiaux, P.F.J.: The Atlantic Ionian Stream, J. Marine Systems, 20, 129–156, 1999.

Roether, W., B.B. Manca, B. Klein, D. Bregant, D. Georgopoulos, V. Beitzel, V. Kovacevic and A. Lucchetta.1996. Recent changes in eastern Mediterranean deep waters. *Science,* 271, 333–335.

Ross, D.A. and Uchupi, E., 1977. Structure and sedimentary history of southeastern Mediterranean Sea-Nile Cone area. AAPG Bull., 61, 872–902.

Rossignol, M., 1969. Sedimentation palynologique dans la domaine marine quaternaire de Palestine. Notes mem. Moyen-Orient., 10, 1–272.

Roussenov, V. E. Stanev, V. Artale and N. Pinardi, 1995. A seasonal model of the Mediterranean Sea. Journal of Geophysical Research, Vol. 100, C7, 13515–13538.

Sabates, A., 1990. Distribution pattern of larval fish populations in the Northwestern Mediterranean. *Mar. Ecol. Prog. Ser.,* 59, 75–82.

Sabates, A., J. Salat and M. P. Olivar, 2001. Advection of continental water as an export mechanism for anchovy, *Engraulis encrasicolus,* larvae. *Sci. Mar.,* 65 (Supl.1), 77–87.

Sammari, C., Millot, C., Taupier-Letage, I., Stefani, A., Brahim, M., 1999. Deep-Sea Research I 46: 1671–1703.

Sánchez-Arcilla, A., J.H. Simpson, 2002. The narrow shelf concept : couplings and fluxes, *Continental Shelf Research,* 22, 153–172.

Sandler, A. and Herut, B., 2000. Composition of clays along the continental shelf off Israel: contribution of the Nile versus local sources. Marine Geology, 167, 339–354.

Sassi, R., Souissi, F., Soussi, N., Boukaaba, M. and Belayouni, H., 1998. Organic matter geochemistry to analyse the degradation of the Monastir-Ksibet el Mediouni littoral (Eastern Tunisia) [French]. Comptes Rendus de I Academie des Sciences Serie II Fascicule A-Sciences de la Terre et des Plantes, 327, 303–308.

Seritti, A., B.B. Manca, C. Santinelli, E. Murru, A. Boldrin, L. Nannicini. 2003. Relationships between dissolved organic carbon(DOC) and water mmass structure in the Ionian Sea (winter 1999*). J. Geophys. Res.,* 108(C9), 8112,PBE13.1–13.

Sharaf El Din, S. H., 1977. Effect of the Aswan high dam on the Nile flood and on the Estuarine and coastal circulation pattern along the Mediterranean Egyptian coast. *Limnol. Oceanogr.,* 22, 194–207.

Simpson, J.H., 1997. Physical processes in the ROFI regime. J.Mar. Syst., 12, 3–15.

Smith, S.V. and F. T. Mackenzie,, 1987. The ocean as a net heterotrophic system: Implications from the carbon biogeochemical cycle, *Global Biogeochemical Cycles,* 1, 187–198.

Sorgente, D., Frignani, M., Langone, L. and Ravaioli, M., 1996. 210Pb e 137Cs in sedimenti dell'Adriatico centrale come indicatori dei processi attuali di sedimentazione e accumulo. Plinus-Supp. Italiano all'European Journal of mineralogy, 16, 200–201.

Sorgente D., Frignani, M., Albertazzi, S., Giuliani, S., Ravaioli, M., Langone, L. and Alvisi, F., 2002. 137Cs and 210Pb sediment accumulation in the central Adriatic Sea. Atti del Convegno "La radioattività nel contesto degli studi ambientali" Isola del Giglio, 9–10 luglio 2002, pp. 221–228.

Sorgente, R., Drago A.F., and Ribotti, A., 2003. Seasonal variabilità in the Central Mediterranean Sea circulation. Annales Geophysicae, 21: 299–322.

Sournia, A., 1973. La production primaire en Mèditerranèe: Essai de mise à jour. *Bulletin de l'ètude en commun de la Mèditerranèe,* 5.

Stanley, D.J., Mart, Y. and Nir, Y., 1997. Clay mineral distributions to interpret Nile Cell provenance and dispersal: II. Coastal plain from Nile delta to northern Israel. J. Coast. Res., 13, 506–533.

Stanley, Nir, Y. and Galili, E., 1998. Clay mineral distributions to interpret Nile Cell provenance and dispersal: III. Nile delta shelf to northern Israeli margin. J. Coast. Res., 14, 196–217.

Struglia, M.V., A.Mariotti, A. Filograsso, 2004. River discharge into the Mediterrabeab Sea: climatology and aspects of the observed variabilità. Submitted to J.Climate.

Supić, N., M. Orlić and D. Degobbis, 2000. Istrian coastal countercurrent and its year to year variability. *Est. Coast. Shelf Sci.,* 51, 385–397.

Tanaka, T. and F. Rassoulzadegan, 2002. Full depth profile of bacteria, heterotrophic nanoflagellates and ciliates in the NW Maediterranean Sea: Vertical partitioning of microbial trophic structures. *Deep Sea Res. II,*49, 2093–2107.

Theocharis, A., M.Gacic and H. Kontoyiannis, 1998. Physical and dynamical processes in the coastal and shelf areas of the Mediterranean. In: The Sea, Vol. 10, 863–888.

Thingstad, T. F. and F. Rassoulzadegan, 1995. Nutrient limitations, microbial food web and "biological C-pumps": suggested interactions in a P-limited Mediterranean. *Marine Ecology Progress Series,* 117, 299–306.

Thingstad, T. F., Å Hagström, and F. Rassoulzadegan, 1997. Accumulation of degradable DOC in surface waters: caused by a "malfunctioning" microbial loop?. *Limnol. Oceanogr.,* 42, 389–404.

Thingstad, T. F., U. L. Zweifel and F. Rassoulzadegan, 1998. P-limitation of both phytoplankton and heterotrophic bacteria in NW Mediterranean summer surface waters. *Limnol. Oceanogr.,* 43, 89–94.

Thingstad, T. F. and F. Rassoulzadegan, 1999. Conceptual models for the biogeochemical role of the photic zone microbial food web with particular reference to the Mediterranean Sea. *Progress in Oceanography,* 44, 271–286.

Tonarelli, B., Turgutcan, F., Max, M.D. and Akal, T., 1992. Shallow sediment composition at four localities on the Sicilian-Tunisian Platform. In: Max, M.D. and Colantoni, P. (Eds.), Geological Development of the Sicilian-Tunisian Platform. UNESCO Reports in Marine Science, Proceedings of International Scientific Meeting, Urbino University, Italy, Nov. 1992., pp. 123–128.

Travers, A. and M. Travers, 1972. Données sur quelque facteur de l'écologie du plancton dans la région de Marseille. 1: Les Vents. *Tethys,* 4, 3–26.

Tsunogai, S., S. Watanabe, and T. Sato, 1999. Is there a "continental shelf pump" for the absorption of atmospheric CO_2? *Tellus. Ser. B,* 701–712.

Tudela, S. and I. Palomera, 1995. Microzooplankton and feeding of anchovy larvae in the northwestern Mediterranean. *Rapp. Comm. Int. Mer Médit.,* 34, 259.

Turley, C. M., M. Bianchi, U. Christaki, P. Conan, J. R. W. Harris, S. Psarra, G. Ruddy, E. D. Stutt., A. Tselepides and F. Van Wambeke, 2000. Relationship between primary producers and bacteria in an oligotrophic sea – the Mediterranean and biogeochemical implications. Marine Ecology Progress *Series,*193, 11–18.

Vichi M., P. Oddo, M. Zavatarelli, A. Coluccelli, G. Coppini, M. Celio, S. Fonda Umani, N. Pinardi, 2003. Calibration and validation of a one-dimensional complex marine biogeochemical flux model in different areas of the northern Adriatic shelf. Annales Geophysicae, 21: 413–436.

Videau, C. and M. Leveau, 1990. Phytoplanktonic biomass and productivity in the Rhône River plume in the spring time. Compte Rendue Acad. Sci. Paris, 311(serie III), 219–224.

Vollenweider, R. A., A. Rinaldi and G. Montanari, 1992. Eutrophication, structure and dynamics of a marine coastal system: reults of a tem-year monitorino along the Emilia Romagna coast (Northwesten Adriatic Sea). Sci. Tot. Env., suppl., 63–106.

Yoro, S. C., R. Sempére, C. Turley, M. A. Unante, X. Durrieu de Madron and M. Bianchi, 1997. Cross-slope variations of Organic Carbon and Bacteria in the Gulf of Lions in relation to water dynamics (northwestern Mediterranean). Marine Ecology Progress Series, 161, 255–264.

Wadie, W. F., 1982. Observations on the catch of the most important fishes along the Egyptian continental shelf in the South Eastern part of the Mediterranean Sea. *Bull. Inst. Oceanogr. Fish.,* 8, 212–227.

Wadie, W. F. and F. A. Abdel Razek, 1985.The effect of damming on the shrimp population in the south-eastern part of the Mediterranean sea. *Fish. Res.,* **3,** 323–335.

Wang, X.H. and N. Pinardi, 2002. Modelling the Dynamics of Sediment Transport in the Northern Adriatic Sea. Journal of Geophysical Research, vol. 107, N0 C12, 3225 (pp. 23).

Wassef, E., A. Ezzat, T. Hashem and S. Faltas, 1985. Sardine fishery by purse-seine on the Egyptian Mediterranean coast. *Mar. Ecol. Prog. Ser.,* 26, 11–18.

Woodward, E, M. S., N. J. P. Owens, A. P. Rees and C. S. Law, 1990. A seasonal survey of nutrient cycling and primary productivity in the Gulf of Lions during 1988 and 1989. *Water Pollution Res. Rep.,* 13, 79–86.

Zavatarelli, M., F. Raicich, D. Bregant, A. Russo and A. Artegiani, 1998. Climatological biogeochemical characteristics of the Adriatic Sea. *J. Mar. Sys.,* 18, 227–263.

Zavatarelli, M.,Baretta, J.W, Baretta-Bekker, J.G, N. Pinardi, 2000. The Dynamics of the Adriatic Sea ecosystem. Part I: an idealised model study. Deep Sea Research, Vol.47, 937–970.

Zavatarelli, M., N. Pinardi, V.H. Kourafalou, A. Maggiore, 2002. Diagnostic and prognostic model studies of the Adriatic Sea general circulation. Part 1: The seasonal variability. Journal of Geophysical Research, vol. 107, (C1): art. n. 3004.

Zavatarelli M. and N. Pinardi, 2003. The Adriatic Sea modelling system: a nested approach. Annales Geophysicae, 21: 345–364.

Zohary, U. L., and R. D. Robarts, 1998. Experimental study of microbial P limitation in the eastern Mediterranean. *Limnol. Oceanogr.,* 43, 387–395.

Zuo, Z., Eisma, D., and Berger, G.W., 1991. Determination of sediment accumulation and mixing rates in the Gulf of Lyon, Mediterranenan Sea. Oceanologica Acta, 14, 253–262.

Zuo, Z., Eisma, D., Gieles, R. and Beks, J., 1997. Accumulation rates and sediment deposition in the northwestern Mediterranean. Deep-Sea Research II, 44, n° 3–4, 597–609.

Chapter 33. PYHSICAL AND BIOGEOCHEMICAL CHARACTERISTICS OF THE BLACK SEA (28,S)

TEMEL OGUZ, SULEYMAN TUGRUL, A. ERKAN KIDEYS, VEDAT EDIGER AND NILGUN KUBILAY

Middle East Technical University, Institute of Marine Sciences, Erdemli, Turkey

Contents

1. Introduction

The Black Sea, located approximately between latitudes of 41° to 46°N and longitudes of 28° to 41.5°E, is an elongated and nearly-enclosed basin connected with the Bosphorus Strait to the Mediterranean Sea. It has experienced one of the worst environmental degradations of the world oceans during the last three decades. The environmental crisis and subsequent dramatic changes in the ecosystem and its resources were a direct consequence of anthropogenic pollution due to an enormous increase in nutrients and pollutant load from rivers discharging into the northwestern region of the sea, uncontrolled industrial and municipal wastewater inputs around the periphery, dumping of wastes (including radioactive substances and solid wastes) into open parts of the sea, and accidental and operational releases of oil. Introduction of a jellyfish-like animal (*Mnemiopsis leidyii*), and overfishing have added further complications to the problem. At the beginning of the 1960s, total inorganic nitrogen, phosphate and silicate input from the Danube was 140 kt yr^{-1}, 12 kt yr^{-1}, and 790 kt yr^{-1}, respectively (Almazov, 1961). Three decades later, the Sulina branch of the Danube (one of its three main branches) alone discharged 800 kt yr^{-1} of total inorganic nitrogen, 32 kt yr^{-1} phosphate, and 1500 kt

The Sea, Volume 14, edited by Allan R. Robinson and Kenneth H. Brink
ISBN 0–674–01527–4

yr^{-1} silicate into the Black Sea (Cociasu et al., 1996). The total sediment load into the basin from the rivers around the periphery is about 145 million ton yr^{-1}, 65% of which enters into the northwestern shelf region (Hay, 1994). The Turkish rivers together contribute only 20% of the total sediment load (Hay, 1994). As a result, a major part of the Black Sea, particularly its northwestern shelf region, has become critically eutrophic and hypoxic (Zaitsev and Mamaev, 1997; Lancelot et al., 2002a). The exploitation of resources has been unsustainable during the last few decades as a result of dramatic reduction in fish stocks and falling recruitment.

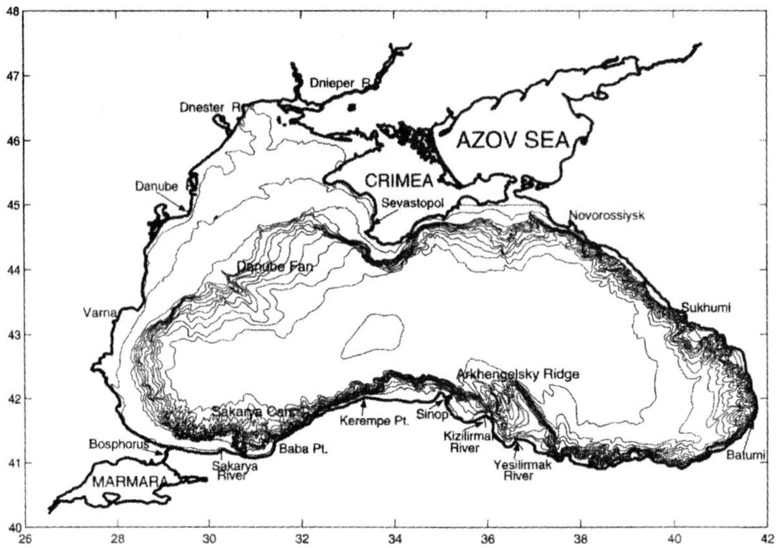

Figure 33.1 The Black Sea: geographic setting, main rivers and bathmetric features (from Besiktepe et al., 2001).

The last decade has introduced a new era in Black Sea oceanography because of collaborative research and management programs developed by the riparian states, and supported by various international organizations. They were aimed to address environmental and socio-economic issues facing the region, and to explore oceanographic characteristics of this relatively unknown, and in many respects, challenging sea. The present paper provides an overview of recent advances achieved within the framework of these efforts. Sections 2 and 3 outline the physical characteristics including the topography, water budget, and stratification, as well as major features of the upper layer horizontal circulation, respectively. This is followed in section 4 by description of the vertical biogeochemical characteristics of the upper layer water column up to the anoxic interface. This section also covers a brief overview of biogeochemical exchanges with the atmosphere. In Section 5, paleoceanographic characteristics are presented with special emphasis on connection of the Black Sea to the Aegean Sea, and sediment geochemical characteristics. Section 6 deals with major changes that took place in Black Sea ecosystem characteristics since 1970s. An overview of interdisciplinary modeling efforts is then provided in section 7. Conclusions are provided in section 8.

2. Physical Characteristics

The Black Sea, with a surface area of 423,000 km^2, is approximately one-fifth of the surface area of the Mediterranean. It has a total volume of 547,000 km^3, and a maximum depth of around 2200 m. It contains narrow shelves and very strong topographic variations around its periphery (Fig. 33.1). The northwestern shelf (NWS), occupying ~20% of the total area, is the only major shelf region with discharges from three of Europe's largest rivers: Danube, Dniepr and Dniestr. In the north, the sea is connected to the shallow Sea of Azov by the Kerch Strait. At its southwestern end, it communicates with the Aegean basin of the Mediterranean Sea through the Sea of Marmara and the Bosphorus and Dardanelles Straits. The Black Sea has always been a basin with a positive water balance. According to the data presented by Unluata et al. (1989) (see also Ozsoy and Unluata, 1997), the sum of fluxes due to precipitation (~300 $km^3 yr^{-1}$) and runoff (~350 $km^3 yr^{-1}$) exceeds that of evaporation (~350 $km^3 yr^{-1}$). The freshwater excess of 300 $km^3 yr^{-1}$ is balanced by the net outflow through the Bosphorus defined as the difference between the transports of its two layers. This particular net transport value agrees well with the estimates for 1980s and 1990s obtained by long-term hydrological-meteorological data (Peneva et al., 2001). It is also found to be consistent with the net Bosphorus transport computed independently by means of a two layer hydrodynamic Bosphorus model (Oguz et al., 1990a).

Today, the intermediate and deep water masses below a permanent halocline at depths of 100–150 m possess almost vertically uniform characteristics defined by T~9°C, S~22, σ_t~17.0 kg m^{-3} (Murray et al., 1991). The deepest part of the water column covering the entire abyssal plain of the sea, approximately below 1700m, involves a vertically homogeneous and horizontally uniform water mass formed during several thousands of years by convective mixing due to the geothermal heat flux of about 40 W m^{-2} from the bottom (Murray et al., 1991). The physical characteristics of the Mediterranean underflow, including its volume, velocity, temperature and salinity, are modified considerably by mixing with the upper layer waters as they cross the shelf (Oguz and Rozman, 1991; Latif et al., 1991; Murray et al., 1991; Ozsoy et al., 1993, 2001; Di Iorio and Yuce, 1998; Gregg and Ozsoy, 1999). The effluent issuing from the Bosphorus is first transported in a narrow channel, and then enters the shelf with typical values of T~12–13 °C and S~30. It follows persistently a north-northwestward track regulated by the small scale topographic variations in the shelf (Latif et al., 1991; Ozsoy et al., 2001), spreads out as a thin layer along the bottom of the shelf, and becomes highly diluted by entrainment of relatively colder and less saline ambient waters of the Cold Intermediate Layer (CIL) origin. The modified Mediterranean water is then injected through intrusions at intermediate depths in the form of multiple layers extending towards the interior from the continental slope (Oguz et al., 1990b; Murray et al., 1991; Ozsoy et al., 2002; Konovalov et al., 2003). Signature of the Mediterranean inflow within the interior parts of the basin is best monitored within the uppermost 500 m depth (Ozsoy et al., 1993), where the residence time of the sinking plume varies from ~10 years at 100 m depth to ~400 years at 500 m (Ivanov and Samodurov, 2001; Lee et al., 2002).

The density within the upper 100 m layer changes seasonally as much as σ_t~3–4 kg m^{-3}. In winter, the northwestern shelf and near-surface levels on top of the thermohaline domes of the cyclonic cell exhibit vertically uniform conditions in response to strong atmospheric cooling, evaporation and intensified wind mixing associated with a succession of strong, cold and dry continental wind events. The upper layer, homogenized up to ~50 m depth, is identified by T~5–6°C, S~18.5–18.8 and σ_t ~14.5 kg m^{-3} (Oguz et al., 1990b; Krivosheya et al., 2002). As the spring warming stratifies the surface water, the convectively generated cold water remains confined below the seasonal thermocline, and forms the CIL of the thermohaline structure. The summer mixed layer with depths less than 20 m has typical characteristics of T~25°C, S~18 and σ_t ~10.0–11.0 kg m^{-3}.

The water and salt budget calculations for the Bosphorus-Black Sea system, represented in the form of a two-layer box for the Bosphorus, two boxes for the interior basin and the western shelf in the upper layer, and two boxes representing the Bosphorus-Black Sea junction region and the interior basin in the lower layer (Fig. 33.2), suggest that a surface outflow of ~604 km^3 yr^{-1} exits from the basin through the Bosphorus. It comprises 54 km^3 yr^{-1} coastal flow of fresh water origin along the western shelf with salinity 16.5, and 550 km^3 yr^{-1} from the rest of the basin with an average salinity of 18. In return the denser Mediterranean water with the average salinity of 35.5 enters into the Bosphorus-Black Sea junction region as an underflow at a rate of 304 km^3 yr^{-1}. There, the Mediterranean plume with salinity of 26.5 entrains the upper layer flow of 426 km^3 yr^{-1} from the CIL to form a total transport of 730 km^3 yr^{-1} flowing into deeper parts of the lower layer of the basin interior having an average salinity of 22. This input is balanced by the difference of downward and upward fluxes of 639 and 1369 km^3 yr^{-1}, respectively, across the interface between the upper and lower layers.

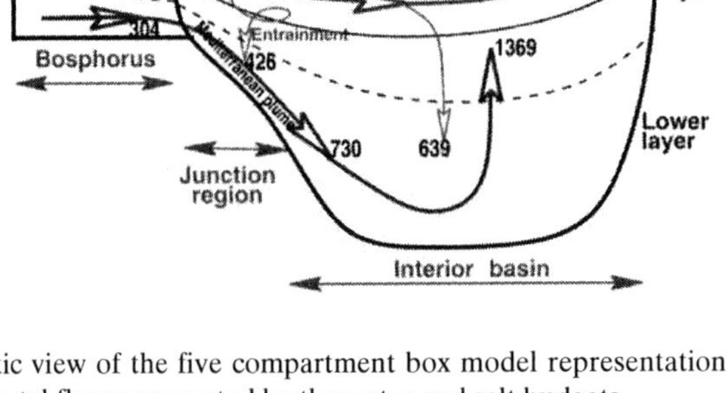

Figure 33.2 Schematic view of the five compartment box model representation of the Black Sea, and main intercompartmental fluxes computed by the water and salt budgets.

The upper layer water column was subjected to considerable climate-induced warming during the 1990s (Oguz et al., 2003). The basin-averaged winter-mean (December-March) sea surface temperatures (SSTs) derived from a 9 km monthly-mean AVHRR data set for regions deeper than 200 m suggested relatively uniform values of 8.1±0.3 °C from 1985 to 1991. They were followed by a strong cooling phase in 1992–1993 with the minimum SST value of 6.8 °C, and subsequently by the 1994–1996 strong winter warming phase with ~2 °C rise in the SST (Fig. 33.3a). The winter warming continued at a more gradual level after 1996. The warmer winter SSTs were related to weaker heat loss to the atmosphere, and weaker wind stress forcing exerted on the sea surface (Nezlin, 2001). The cooling-warming cycle seen in the AVHRR winter-mean SST data was supported by a similar cycle in the winter-mean air and SST temperatures (Fig. 33.3b) measured near Novorossik located along the northeastern coast of the Black Sea (Titov, 2000; Krivosheya et al., 2002). The annual-mean SST variations, also depicted in Fig. 33.3a, suggest warming of the Black Sea surface waters in the form of a linear trend with temperatures rising from ~14.2 °C in 1993 to ~16.4 °C in 2001. The upper layer stratification and circulation characteristics were modified subsequently by gradual depletion of the CIL (Staneva and Stanev, 2002; Krivosheya et al., 2002), rising of the mean winter sea level and accompanying weakening of the basinwide cyclonic circulation system (Stanev and Peneva, 2002), as well as ~10 m rise of the σ_t ~16.2 kg m^{-3} isopycnal surface characterizing the position of the anoxic interface (Yakushev et al., 2001).

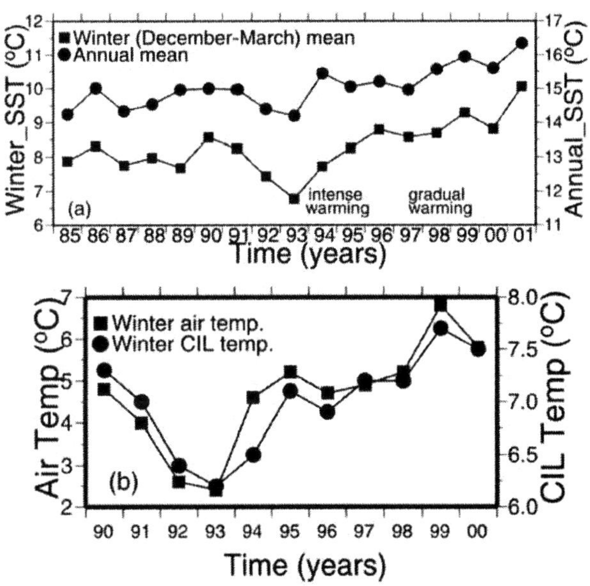

Figure 32.3 (a) The basin-averaged, winter (December-March)-mean (squares) and annual-mean (dots) AVHRR sea surface temperature distributions in the Black Sea from 1985 to 2001. The data are obtained by 9 km monthly-mean gridded AVHRR Oceans Pathfinder data set. The basin-averaging excludes the shelf areas shallower than 200 m (redrawn from Oguz et al., 2003), (b) the winter-mean air temperature (squares) and surface mixed layer temperature (dots) near Novorossik along the northeastern coast of the Black Sea.

3. Circulation characteristics

The upper layer waters of the Black Sea are characterized by a predominantly cyclonic, strongly time-dependent and spatially-structured basinwide circulation. Many details of the circulation system have been explored using recent hydrographic data (Oguz et al., 1993, 1994, 1998; Oguz and Besiktepe, 1999; Gawarkiewicz et al., 1999; Krivosheya et al., 2000), AVHRR data (Oguz et al., 1992; Sur et al., 1994, 1996; Sur and Ilyin, 1997; Ginsburg et al., 2000, 2002a; Afanasyev et al., 2002; Zatsepin et al., 2003), altimeter data (Korotaev et al., 2001 and 2003; Sokolova et al., 2001), and CZCS and SeaWIFS data (Ozsoy and Unluata, 1997; Oguz et al., 2002a; Ginsburg et al., 2002b). These analyses reveal a complex, eddy-dominated circulation with different types of structural organizations within the interior cyclonic cell, the Rim Current flowing along the abruptly varying continental slope and margin topography around the basin, and a series of anticyclonic eddies in the onshore side of the Rim Current. The interior circulation comprises several sub-basin scale gyres, each of them involving a series of cyclonic eddies. They evolve continuously by interactions among each other, as well as with meanders, and filaments of the Rim Current. The Rim Current structure is accompanied by coastal-trapped waves with an embedded train of eddies and meanders propogating cyclonically around the basin (Sur et al., 1994; Sur et al; 1996; Oguz and Besiktepe, 1999; Krivosheya et al., 2000; Ginsburg et al., 2002a,b). Over the annual time scale, westward propogating Rossby waves further contribute complexity to the basinwide circulation system (Stanev and Rachev, 1999). According to the Acoustic Doppler Current Profiler measurements (Oguz and Besiktepe, 1999), the Rim Current jet has a speed of 50–100 cm/s within the upper layer, and about 10–20 cm/s within the 150–300 m depth range. The mesoscale features evolving along the periphery of the basin as part of the Rim Current dynamic structure apparently link coastal biogeochemical processes to those beyond the continental margin, and thus provide a mechanism for two-way transports between nearshore and offshore regions. Taking the relatively narrow width of the basin into account, such mesoscale processes can easily give rise to meridional transports from one coast to another.

Apart from complex eddy-dominated features, larger scale characteristics of the upper layer circulation system possess a distinct seasonal cycle, as suggested by objectively analyzed, optimally interpolated and dynamically assimilated sea level anomaly data provided by the Topex-Poseidon and ERS-1/2 altimeters for the period from 1 January 1993 to 31 December 1998 (Korotaev et al., 2003). As shown by the model-derived circulation patterns (Fig. 33.4) for the middle of February, July and October, the interior cyclonic cell in winter months involves a two-gyre system surrounded by a rather strong and narrow peripheral jet without any appreciable lateral variations (Fig. 33.4a). This system transforms into a multi-centered composite cyclonic cell surrounded by a broader and weaker Rim Current zone in summer (Fig. 33.4b). The interior basin flow field further weakens and finally disintegrates into smaller scale cyclonic features in autumn (Fig. 33.4c). A composite peripheral current system is hardly noticeable in this season (Afanasyev et al., 2002). The turbulent flow field is, however, rapidly converted into a more intense and organized structure after November-December.

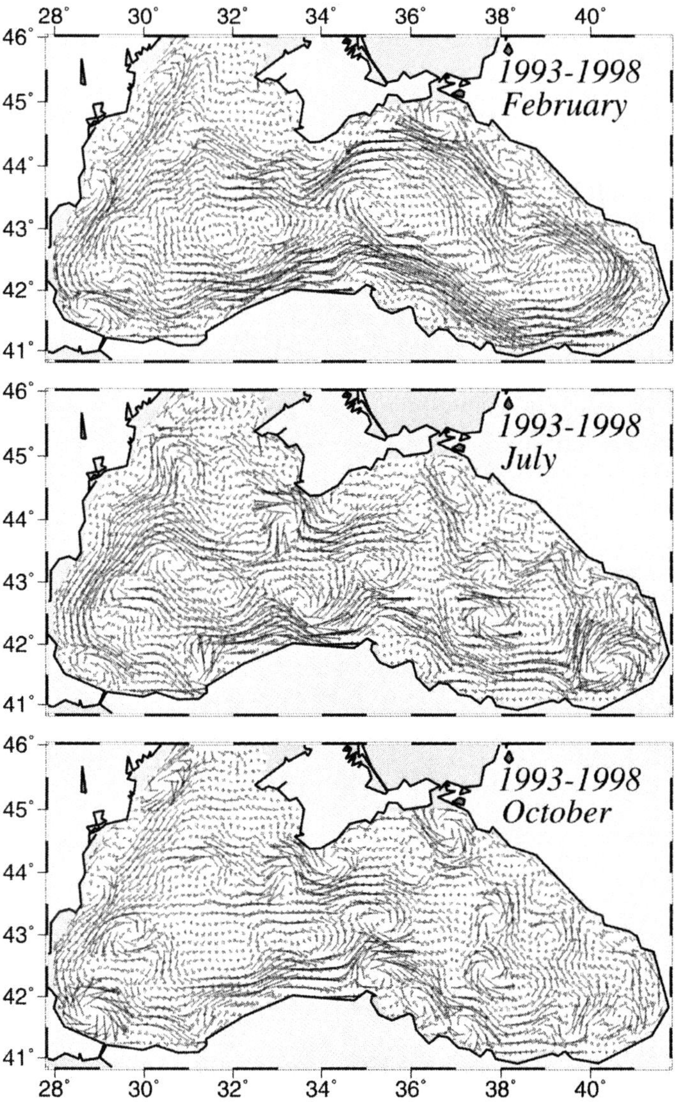

Figure 33.4 The upper layer circulation maps for (a) mid-February, (b) mid-July, (c) mid-October, constructed by the six year (1993–1998) averaging of the daily circulation fields computed by assimilating the Topex-Poseidon and ERS-I,II altimeter data into a 1.5 layer reduced gravity model described by Korotaev et al. (2003).

The most notable quasi-persistent and/or recurrent features of the circulation system, as schematically presented in Fig. 33.5, include (i) the meandering Rim Current system cyclonically encircling the basin, (ii) two cyclonic sub-basin scale gyres comprising four or more gyres within the interior, (iii) the Bosphorus, Sakarya, Sinop, Kizilirmak, Batumi, Sukhumi, Caucasus, Kerch, Crimea, Sevastopol, Danube, Constantsa, and Kaliakra anticyclonic eddies on the coastal side of the Rim Current zone, (iv) bifurcation of the Rim Current near the southern tip of the

Crimea; one brach flowing southwestward along the topographic slope zone, and the other branch deflecting first northwestward into the shelf and then contributing to the southerly inner shelf current system, (v) convergence of these two branches of the original Rim Current system near the southwestern coast, (vi) presence of a large anticyclonic eddy within the northern part of the northwestern shelf.

The basic mechanism which controls the flow structure in the surface layer of the northwestwern shelf is spreading of the Danube outflow. Wind stress, and Rim Current structure along the offshore side of the shelf are additional modifiers of this system. The freshwater discharge influences not only the circulation and mixing properties, but also the ecosystem of the entire shelf region along the western coast. The Danube plume generally forms an anticyclonic bulge confined within the upper 25 m layer. The leading edge of this plume protrudes southward (i.e downstream) as a thin baroclinic boundary current along the western coastline. The coastal jet is separated from the interior waters by a well defined front with salinity differences of more than 3.0 over an approximately 50 km zone along the coast. It is often unstable, exhibits meanders and spawns filaments, which extend across the wide topographic slope zone. The shelf and interior waters undergo cross-shelf exchanges as reported consistently in hydrographic surveys, satellite imagery, and altimeter data. An anticyclonic circulation system accompanying with small-scale structures over the northwestern shelf, shown in Fig. 33.5, have also been reproduced by modeling studies (e.g. Oguz et al., 1995; Staneva et al., 2001; Beckers, et al., 2002).

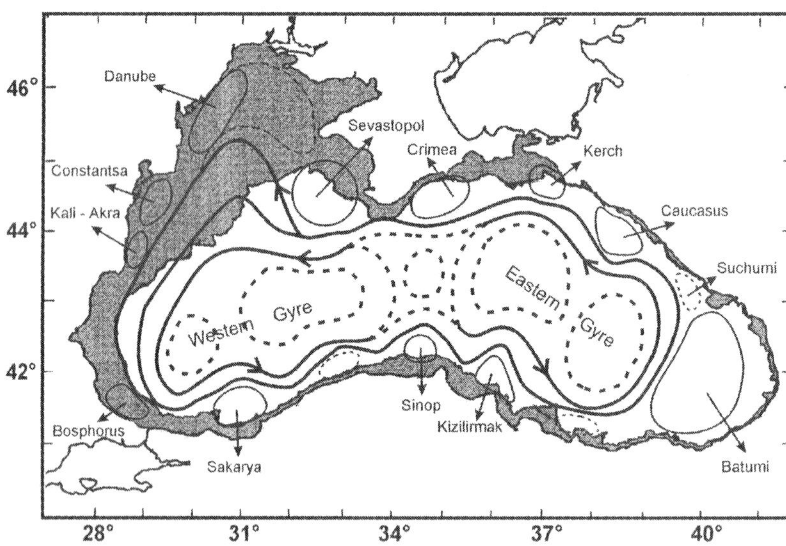

Figure 33.5 The schematic diagram showing major quasi-permanent/recurrent features of the upper layer circulation identified by synthesis of hydrographic studies and analysis of the Topex-Poseidon and ERS-I,II altimeter data.

4. Major features of the vertical biogeochemical structure

4.1 Nutrient and organic matter characteristics

The upper layer biogeochemical structure overlying the deep and lifeless anoxic pool (except anaerobic bacteria) involves four distinct layers. The uppermost part from the free surface to the depth of 1% light level is covered by a shallow euphotic zone with a maximum thickness of nearly 50 m. This is the layer of active planktonic processes (e.g uptake, grazing, mortality, microbial loop, etc.), and is characterized by high oxygen concentrations on the order of 300 μM as well as seasonally varying nutrient and organic material concentrations supplied laterally from rivers and vertically from sub-surface levels through vertical mixing. In the interior basin, the surface mixed layer waters are poor in nutrients for most of the year except for occasional incursions from coastal regions, and by wet precipitation. Below the seasonal thermocline and in the deeper part of the euphotic zone, nitrate concentrations increase due to their recycling as well as continuous supply from the nutricline. Nitrate accumulation in this light-shaded zone generally supports summer subsurface phytoplankton production (Oguz et al. 2000). In winter, nutrient stocks in the euphotic zone waters are renewed from the nutricline depths through upwelling, vertical diffusion and seasonal wind and buoyancy-induced entrainment processes, and depleted by biological utilization. About 90% of the sinking particles are remineralized inside the euphotic zone and the subsequent 20–30 m part of the oxygenated, aphotic zone (the so-called "upper nitracline" zone) where nitrate concentrations increase up to ~8 μM at ~70–80 m in cyclonic regions, and are re-supplied back to the surface waters to refuel the biological pump. Only a small fraction of particulate matter sinks to the deeper anoxic part of the sea (Lebedeva and Vostokov, 1984; Karl and Knauer, 1991), which occupies the water column below ~100 m depth within the interior parts and ~200 m in the onshore, anticyclonically-dominated side of the Rim Current. This loss is compensated by lateral nitrogen input mainly from the River Danube (Cociasu et al., 1996), by wet deposition and nitrogen fixation. The nutrient fluxes of anthropogenic origin are transported across the shelf and around the basin through the Rim Current system, and supplied ultimately to the interior basin, and some of which is lost in the form of Bosphorus surface flow in winter months (Polat and Tugrul, 1995). The river supply in the NWS gives rise to a high N/P ratio within the shallow water column, and further constitutes a major source of selectively nitrate-enriched CIL in winter. The river influence markedly weakens toward the south along the coast and offshore for most of the year due to photosynthetic consumption of dissolved inorganic nutrients. Nevertheless, below the seasonal thermocline, the thicker CIL in coastal regions contains measurable concentrations of nitrate but very low (<0.02 μM) phosphate values, yielding abnormally high N/P ratios (Codispoti et al., 1991; Basturk et al., 1998a).

When nitrate profiles are plotted against density, the position of the peak concentration coincides approximately with the σ_t ~15.5 kg m^{-3} level (Fig. 33.6a) (Tugrul et al., 1992; Saydam et al., 1993; Basturk et al.,1994, Yilmaz et al., 1998a), although some degree of variability in its position between σ_t ~15.3 and 15.6 kg m^{-3} isopycnal surfaces as well as in its maximum concentrations from 6 to 10 μM are observed in the data (Basturk et al., 1998a; Oguz et al., 2000). The nitrate structure is accompanied by occasional peaks of ammonium on the order of 0.5 μM near the base of the euphotic zone due to inputs from excretion and aerobic organic matter decomposition following subsurface plankton production. Ammonium concentrations are then subject to a linear trend of decrease towards trace concentrations at the anoxic interface (Basturk et al., 1998a). Within the oxygen deficient part of the water column below σ_t ~15.6 kg m^{-3}, organic matter decomposition proceeds via denitrification. This results in formation of the "lower nitracline" zone with sharp decrease of nitrate concentrations at a thickness of about 30–40 m from their peaks to their trace values around 100 m depth or σ_t ~16.0 kg m^{-3} isopycnal surface (Fig. 33.6a). Nitrate consumption due to oxidation of reduced manganese and ammonium may also contribute to reduction of nitrate concentrations within the lower part of the suboxic zone (Murray et al., 1995). Importance of the Anammox type reactions ($NO_2^- + NH_4^+ \rightarrow N_2$) in the Black Sea was recently shown by Kuypers et al. (2003). In these processes, bacteria utilize nitrate and nitrite ions to oxidize organic matter, reduced manganese and ammonium. The nitrate is then reduced to nitrogen gas with nitrite as an intermediate product. A nitrite peak with concentrations up to 0.5 μM is usually observed at σ_t ~15.85±0.05 kg m^{-3} located approximately 10 m (or, equivalently, σ_t ~0.1 kg m^{-3}) above the position of the zone of nitrate depletion (Fig. 33.6a). This coincides with the position of the phosphate minimum (Fig. 33.6b) (Codispoti et al., 1991). The thickness of the nitrite peak therefore marks the denitrification zone. The deep sulphide-bearing waters in the Black Sea contain no measurable nitrate, but constitute large pools of ammonium and dissolved organic nitrogen. Reduction in subsurface nitrate concentrations, however, does not appreciably limit primary production since the euphotic zone and extent of the winter mixed layer always lie above its maximum concentration zone.

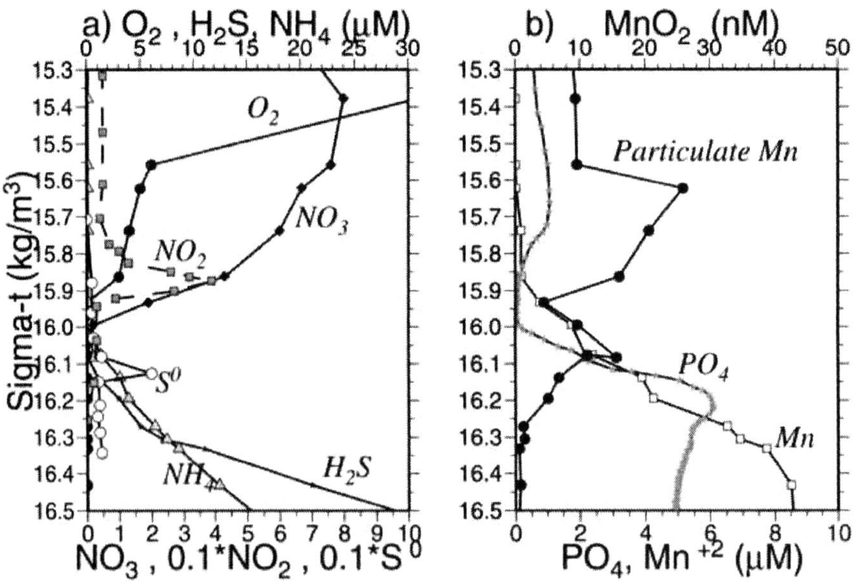

Figure 33.6 (a) Depth profile of density for the upper 150 m water column, (b) O_2, HS^-, NH_4^+, NO_3^-, NO_2^-, S^0 profiles, and (c) MnO^2, Mn, PO_4 profiles plotted versus density (sigma-t) at station 43°N, 34°E during RV Knorr survey of 13 June 1988 (from Oguz et al., 2001a).

The vertical structure of phosphate concentrations resembles that of nitrate in the upper layer but is quite complicated in the suboxic/anoxic layers (Fig. 33.6b). Phosphate concentrations increase gradually within the deeper part of the euphotic layer and the upper nitracline zone up to a maximum value of 1.0–1.5 μM around σ_t ~15.6 kg m^{-3}, and then decreases to minimum of about 0.05–0.1 μM in the suboxic zone around σ_t ~15.9±0.1 kg m^{-3} where nitrite locally displays a peak in the cyclonic basin (Codispoti et al., 1991; Tugrul et al., 1992; Murray et al., 1995). It then increases abruptly to peak values of 5.0–8.0 μM near σ_t ~16.2 kg m^{-3} coinciding with the first apperance of sulfide in the water column. The formation of this peak has been explained by dissolution of phosphate-associated iron and manganese oxides (Shaffer, 1986; Codispoti et al., 1991). Silicate possesses a relatively simpler vertical structure with a steady increase of concentrations below the euphotic layer up to about 70–75 μM at σ_t ~16.2 kg m^{-3} defining the anoxic boundary in the open sea, and then to about 150 μM at the depth of σ_t ~16.8 kg m^{-3} in the upper anoxic water.

A major contribution to total organic carbon (TOC) in the Black Sea comes from its dissolved form (DOC) (Sorokin, 2002; Tugrul, 1993). Both DOC and slowly sinking suspended particulate organic matter (SPOM) are especially abundant within the NWS waters due to large, river-borne organic matter and nutrient input. Their concentrations in the surface layer of the western central basin is 160–250 μM for DOC (Tugrul, 1993; Polat and Tugrul 1995), and from 60–80 μM to ~10 μM for POC (Burlakova et al., 1997; Coban-Yildiz et al., 2000a,b). The riverine supply alone contributes around 130 μM of the total DOC content within surface waters, and the rest represents contribution from local biological production. The DOC concentration decreases steadily with depth to background values of 110–130 μM within the oxic/anoxic interface and further below to the bottom (Tugrul,1993). Various references however have reported different DOC concentrations, such as from ~300 μM at the surface to ~265 μM at 100m and ~190 μM near the bottom (Sorokin, 2002), or from 100 μM at the surface to ~60 μM in the suboxic-anoxic interface zone (Karl and Knauer, 1991). SPOM concentrations also decrease towards lower levels of ~2–3 μM for particulate organic carbon (POC) in the suboxic zone. At the suboxic/anoxic transition zone, SPOM exhibits a small maximum (POC>10 μM) probably due to bacterially-mediated redox reactions (Coban-Yldz et al., 2000a,b, 2003). The peak is more distinguishable at the shelf-break regions of the southwestern Black Sea possibly due to additional supply of chemical energy through lateral injection from the Bosphorus plume. During productive periods, the C/N ratio of SPOM within the surface layer is similar to the Redfield ratio. The C/N ratio in the surface layer, however, increases during less productive periods (i.e. summer-autumn) due to selective decay of nitrogeneous compounds. The ratio then decreases to the Redfield value towards the suboxic/anoxic transition zone, in accordance with increasing SPOM concentrations. Despite distinct seasonal and regional variations in SPOM and chlorophyll concentrations, the isotopic compositions of carbon ($\delta^{13}C$) and nitrogen ($\delta^{15}N$) in SPOM stay nearly vertically uniform within the euphotic zone (Calvert and Fontugne, 1987; Fry et al., 1991; Kodina et al., 1996; Coban-Yildiz et al., 2003). At deeper levels, $\delta^{15}N$ increases as $\delta^{13}C$ decreases in the upper nitracline zone; $\delta^{15}N$ values initially increase from 3–4.5% to around 7.2 to 9.1 ‰ at the depth of the nitrate maximum then decrease to a minimum of 2.0 ‰ within the anoxic interface of the interior basin (Coban-Yildiz et al., 2003).

4.2. Suboxic layer and suboxic-anoxic interface zone characteristics

The euphotic layer oxygen concentration undergoes pronounced seasonal variations within a broad range of values from about 250 to 450 μM. The period from the beginning of January until mid-March exhibits vertically uniform mixed layer concentrations of ~300–350 μM, ventilating the upper ~50 m of the water column as a result of convective overturning. The rate of atmospheric oxygen input in the ventilation process is proportional to the excess of saturated oxygen concentration over the surface oxygen concentration. The maximum contribution of oxygen saturation is realized towards the end of February during the period of coolest mixed layer temperatures, coinciding with the maximum and deepest winter oxygen concentrations during the year. After March, initiation of the warming season is accompanied by oxygen loss to the atmosphere, thus reducing oxygen concentrations within the uppermost 10 m to 250 μM during the spring and summer months. A subsequent linear trend of increase across the seasonal thermocline links low near-surface oxygen concentrations to those of relatively higher sub-thermocline concentrations. Depending on the strength of summer phytoplankton productivity, the sub-thermocline concentrations exceed 350 μM in summer. Irrespective of the season, the oxygen concentration then decreases almost linearly within the upper nitracline zone to concentrations of about 100 μM at σ_t ~15.3 kg m^{-3} and about 10 μM at σ_t ~15.6 kg m^{-3} due to intense oxygen consumption during the decomposition process of organic matter. Oxygen concentrations vanish completely near the anoxic interface located at σ_t ~16.2 kg m^{-3} (Fig. 33.6a).

The oxygen deficient (O_2 < 10 μM), non-sulfidic layer having a thickness of 10-to-40 m coinciding with the lower nitracline zone is more commonly referred to as the "Suboxic Layer (SOL)". It has been identified for the first time by Murray et al. (1989, 1991). Earlier observations generally measured dissolved oxygen concentrations more than 10 μM inside the sulfidic layer (Sorokin, 1972; Faschuk, et al., 1990; Rozanov et al., 1998). Grashoff (1975) was the first to point out that co-existence of dissolved oxygen and H_2S was probably an artifact of atmospheric contamination during sampling. More recent observations (Tugrul et al., 1992; Saydam et al., 1993; Buesseler et al., 1994; Eremeev, 1996; Basturk et al., 1994, 1998a; Konovalov et al., 2003) have supported existence of the SOL along the density surfaces of σ_t~15.55±0.05 and 16.15±0.05 kg m^{-3}. Anayzing the data available since the 1960s, Tugrul et al. (1992), Buesseler et al. (1994) and Konovalov and Murray (2001) showed that the suboxic zone was in fact present earlier than its first observation in 1988, but it was masked because of low sampling resolution and contamination of water samples with atmospheric oxygen.

Below the suboxic zone is the deep anoxic layer containing high concentrations of hydrogen sulphide and ammonium. The boundary between the suboxic and anoxic layers is the site of a series of complicated redox processes (Murray et al., 1995; Rozanov, 1996). As dissolved oxygen and nitrate decrease towards zero concentrations at the suboxic-anoxic interface, dissolved manganese, ammonium and hydrogen sulfide begin to increase at the interface (Fig. 33.6a,b). Marked gradients of particulate manganese around this transition zone near σ_t ~16.0 kg m^{-3}

(Fig. 33.6b) reflect the role of manganese cycling as proposed by Spencer and Brewer (1971), Brewer and Spencer (1974), Kempe et al. (1991), Tebo (1991), Tebo et al. (1991), Lewis and Landing (1991), and Oguz et al. (2001a). The deep ammonium, sulfide and manganese pools have accumulated as a result of organic matter decomposition within the last 5000 years, after the Black Sea was converted into a two-layer stratified system. The ammonium concentrations, increasing sharply below σ_t ~16.0 kg m^{-3}, reach at values of 10 μM at 150 m (σ_t ~16.5 kg m^{-3}) and 20 μM at 200 m (σ_t ~16.8 kg m^{-3}). The gradient of ammonium profiles in the vicinity of the suboxic-anoxic interface implies that no ammonium is supplied to the photic zone from the anoxic region.

As pointed out above, major characteristic features of the Black Sea vertical biogeochemical structure are customarily expressed in terms of density used as the vertical coordinate. The assumption of isopycnal uniformity and independence of biogeochemical properties from the circulation features was first suggested by Vinogradov and Nalbandov (1990). It was then supported by Tugrul et al. (1992) and Saydam et al. (1993) when the oxygen, sulfide, and nutrient profiles from the 1988 Knorr and 1991 Bilim surveys were plotted all together against density. Because most of the data taken in these surveys covered the interior basin characterized by a similar type of cyclonic circulation system, this assertion was found to be reasonably valid. More recently, re-analysis of historical data together with model simulations (Oguz, 2002) noted some differences in different parts of the sea. This implies sensitivity of the vertical biogeochemical structure to local circulation characteristics and subsequently to the direction and intensity of vertical diffusive and advective fluxes within the water column. The vertical variations therefore can not be expressed in terms of density without taking into account physical charactristics of the water column. For example, as shown in Fig. 33.7, the upper boundary of the SOL in the eastern Black Sea defined by the position of 10 μM oxygen concentration markedly varies from σ_t ~15.55±0.1 kg m^{-3} in the cyclonic gyre (centered at 42.3°N, 38.5°E) to σ_t ~15.9±0.1 kg m^{-3} in the adjacent anticyclone (centered at 41.8°N, 40.2°E) during September 1991 (Oguz et al., 1994). The anticyclonic nature of the latter region implies a more pronounced downward oxygen transport resulting in a narrower SOL.

The anaerobic sulfide oxidation and nitrogen transformations coupled to the manganese and iron cycles have been considered one of the mechanisms to maintain stability of the interface structure between the suboxic and anoxic layers (Murray et al., 1995; 1999). The upward fluxes of sulfide and ammonium are oxidized by Mn(III, IV) and Fe(III) species, which are generated by Mn(II) and Fe(II) oxidation by reactions with nitrate. The upward flux of ammonium is also consumed by NO_3^-/NO_2^- via anammox reaction (Murray et al., 1995; Kuypers et al., 2003). These oxidation-reduction reactions are microbially catalyzed, but dissolved chemical reduction may also play a role in Mn(IV) reduction with sulfide. Modeling studies (Oguz et al., 2001a) demonstrate that this mechanism alone could provide the observed redox structure. Anaerobic photosynthesis is considered an additional mechanism contributing to the SOL formation. The reduced chemical species (HS^-, Mn^{2+}, Fe^{2+}) are oxidized by anaerobic phototrophic bacteria in association with phototrophic reduction of CO_2 to form organic matter. This mechanism was supported by the discovery of large quantities of bacteriochlorophyll pigments near the suboxic-anoxic boundary (Repeta et al., 1989; Repeta and Simpson, 1991; Jorgensen et al., 1991; Jannasch et al., 1991). A particular bacterium is capable of growth using reduced S (H_2S or S^0) at very low light levels (<<0.1% of the incident radiation at the surface). The third mechanism is oxidation of H_2S by oxygen and particulate manganese injected horizontally into the anoxic layer (Murray et al., 1989; Tebo et al., 1991; Basturk et al., 1998a). Konovalov and Murray (2001) show that more than 50% of the upward flux sulfide could be consumed by this pathway.

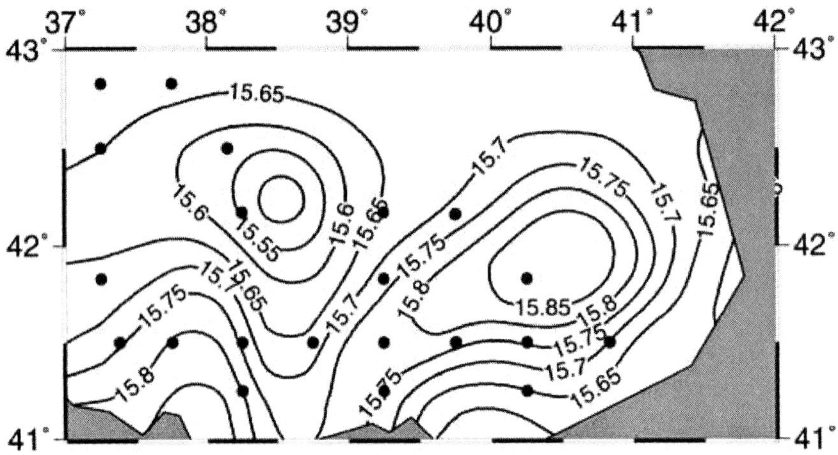

Figure 33.7 Horizontal distributions of σ_t (kg m^{-3}) at the depth of 10 μM oxygen concentration obtained from measurements of RV Bilim 1991 survey in the southeastern part of the Black Sea. Station locations are shown by dots. (redrawn from Oguz, 2002).

It is difficult to quantify the anoxygenic photosynthesis as a mechanism of basinwide relevance. Its contribution must be limited to cyclonic regions where the anoxic interface zone is shallow enough to be able to receive sufficient light to maintain bacterial photosynthetic activity. Presence of a persistent suboxic zone structure with its lower boundary located at depths of around 160–180 m within the quasi-permanent anticyclonic gyre of the eastern basin (Basturk, et al., 1998a) therefore suggests that anaerobic sulfide oxidation should control first-order dynamics of the redox structure in the Black Sea. Anaerobic photosynthesis is expected to have an additional contribution to the dissolved oxygen-hydrogen sulfide separation, and thus to formation of somewhat thicker SOL at shallower depths within cyclonic areas of the basin.

Because the sulfide layer is located only about 100 m below the surface in most parts of the Black Sea, the possible vertical extent of horizontal ventilation of the upper layer water column is a critically important issue. One possible ventilation mechanism is local convective overturning following intense winter atmospheric cooling episodes. There is, however, no data to support the ventilation of suboxic and anoxic layers through atmospheric oxygen supply. According to model simulations (Oguz et al., 2000; Oguz, 2002), episodic strong cooling events under realistic ranges of heat fluxes could only increase the density of the mixed layer up to σ_t

~15.0 kg m^{-3} which corresponds to deepening of the mixed layer to about 75 m within the cyclonic regions. Thus, it seems that the strong stability of the water column hardly allows any seasonal variation of oxygen below the oxycline due to atmospheric ventilation. Oxygen supplied by the Mediterranean waters of Bosphorus origin also contributes locally to sulfide oxidation (Konovalov and Murray, 2001; Neretin et al., 2001). The effects of ventilation by the Bosphorus plume can be dramatically seen in the vertical profiles of T, S, O_2 and H_2S in the southwestern region of the Black Sea (Konovalov et al., 2003). Basturk et al., (1998b) provided an example of the cross-shelf ventilation of the interior basin along the periphery of the Black Sea, induced by onshore-offshore exchanges of water masses due to meanders and complex eddy activities of the Rim Current system. Quasi-lateral injection of oxygen rich shelf waters offshore into a fine particle layer (Kempe et al., 1991; Rozanov et al., 1998) situated between isopycnal surfaces of σ_t ~15.9 kg m^{-3} and σ_t ~16.4 kg m^{-3} was identified by dissolved oxygen concentrations of about 20 μM within this layer.

4.3. Biogeochemical exchanges with atmosphere

Biologically produced gases in the surface ocean have a major impact on the global atmospheric cycling of elements such as sulfur, nitrogen and carbon, and can play an important role in the global climate system. Dimethyl sulfide (DMS) is the principal and most abundant biogenic organic sulfur compound entering the atmosphere, where it undergoes photo-oxidation and transformation to methanesulfonate (MSA) and SO_4 aerosol. In addition to anthropogenic and volcanic sources, it provides a biogenic contribution to non-sea-salt sulfate (nss-sulfate) in marine aerosols. The importance of nss-sulfate comes from its effect on the Earth's radiation budget by backscattering solar radiation to space (Charlson et al., 1991) and by controlling the formation of cloud condensation nuclei (Charlson et al., 1987). It has been suggested that its contribution to atmospheric cooling nearly compensates the estimated warming due to increased carbon dioxide and other greenhouse gases. The global oceanic contribution is about 40 % of the total sulfur burden of the atmosphere, comparable to manmade contributions (Simo, 2001). Recent studies have indicated the potential role of the Black Sea in the production of DMS (Cokacar et al., 2001; 2004), its transfer into the atmosphere, and subsequently aerosol transport processes over the region (Kubilay et al., 2002). Aerosol MSA and nss-sulfate concentrations measured in samples collected during January 1996-December 1999 at two coastal stations along the Mediterranean coast of Turkey near Erdemli and at the island Crete near Finokalia were found to be associated with (i) the presence of high level and almost continuous phytoplankton production over the entire Black Sea by means of DMS producing species (such as coccolithophorids, flagellates, etc) in response to intense eutrophication that developed within the last two decades, and (ii) the presence of a persistent northerly low level boundary layer atmospheric transport prevailing over the region.

Kubilay et al. (2002) observed a clear signature of seasonal variations of biogenically-derived nss-sulfate concentrations in the Eastern Mediterranean atmosphere from low winter to high summer values. Their summer concentrations measured at Erdemli were among the highest reported in the world. The period of their high summer concentrations correlated very well with basinwide blooming of coccolithophorids *E. Huxleyi* in the Black Sea during June and July every year, as suggested by the SeaWIFS mean normalized water-leaving radiance data (Cokacar et al., 2004). Low-level meridional atmospheric transport carried nss-sulfate aerosols over Anatolia into the marine atmosphere of the Eastern Mediterranean Sea roughly from the end of May to the end of September. The lateral aerosol supply from the Black Sea was found to terminate after September as the direction of low level air motions was shifted preferentially to northwesterlies during the autumn and winter months.

There is growing evidence that aeolian transported materials, particularly enriched in elements of ecological concern, such as iron, nitrogen, phosphorus, trigger phytoplankton production in the oceans (Duce et al., 1991). The Black Sea is under the particular influence of long-range aeolian transport from the Sahara, Middle East, Eastern Europe and Russian mainland in the northeast. A number of case studies confirmed transport from North Africa towards the Black Sea, particularly during spring and autumn months (Kubilay et al., 2000). On the basis of measurements performed during July 1992, dust deposition provided a total nitrogen supply of 44 kt yr^{-1}, which roughly corresponded to 13% of the total inorganic nitrogen input by the Danube outflow (Kubilay et al., 1995), and thus aerosol transport is important intermittently if not on the annual time scale.

5. Paleoceanographic characteristics

During the last glacial maximum before 12,000 BP, the Black Sea, the Sea of Marmara and Aegean Sea were about 120 m below their present levels, and were therefore decoupled from each other by shallow sills of the Bosphorus and Dardanelles. Freshwater conditions used to prevail both the Black Sea and Marmara Sea whereas the Aegean Sea reflected marine conditions. With the rise of sea level after the end of glaciation, the global sea level reached the 80 m sill depth of the Dardanelles around 12, 000 BP, and salty Mediterranean water started filling the Sea of Marmara, leading to the formation of a sapropel between 10,600 and 6400 BP (Cagatay et al., 2000). The date upon which saline Mediterranean water first entered the Black Sea during the Holocene is more uncertain and indeed a contraversial issue. The traditional view (Ross and Degens, 1974; Stanley and Blanpied, 1999), later elaborated by Aksu et al. (1999, 2002), suggested that the Black Sea during the period from 12,000 to 9,500 BP still remained as a freshwater lake, receiving a large freshwater inflow from the receding European ice sheet through the rivers around the periphery of the basin. These factors resulted in a substantial rise of the Black Sea level from its pre-flooding depth of -120m; by ~11,000–10,000 BP, the Black Sea has risen to the Bosphorus sill depth of -40m and subsequently began to spill large volume of waters first into the Marmara Sea across the Bosphorus, and later into the Aegeran Sea across the Dardanelles Strait. Evidence for this view includes the presence of a sapropel in the Sea of Marmara (Cagatay et al., 2000), studies of shelf sediments in the Black Sea (Gorur et al., 2001), westerly-oriented bedforms in the Sea of Marmara (Aksu et al., 1999), and the analysis of planktonic and benthic foraminifera (Yanko et al., 1999; Kaminski et al., 2002). Following incursions of saline Mediterranean water into the Marmara Sea after the rise of the Aegean Sea level above the Dardanelles sill depth during the period from 12,000 to 9,500 BP, the rising sea level enabled Mediterranean inflow into the

Black Sea across the Bosphorus Strait starting by 9000 BP. The saline wedge had possibly penetrated well into the strait 500–1000 yr later, and led to a gradual salination of the Black Sea, and development of a two-layer stratification and formation of anoxic conditions by perhaps 8000 BP.

On the basis of sedimentary data from the northwestern and northern shelves of the Black Sea, Ryan et al. (1997) offered an alternative and contrasting view for the timing and development of marine connection between the Black Sea and the Sea of Marmara following the last deglaciation. According to this so-called "flood hypothesis" the Mediterranean-Black Sea post-glacial connection occured as a result of refilling of the Mediterranean basin and then flooding catastrophically into the Black Sea in less than 2 years at a flow rate of more than 50 km^3 day^{-1} during 7150±100 yr BP. This hypothesis has largely been based on the rapid first appearance of euryhaline (Mediterranean) mollusks on the Black Sea shelves at ~7,500 BP. They further speculated that this flooding event was actually the reason for the migration of early Neolithic peoples from the region as mentioned in the biblical story of Noah's flood (Ryan and Pitman, 1999). Ballard et al. (2000) identified the location of ancient beach at 155 m water depth below the present day sea surface in the south-central Black Sea, and infered the marine flooding of the Black Sea between 7.46 and 6.82 ka by means of radiocarbon dating of mollusk shells. Ryan et al. (2003) later refined the hypothesis by considering large amounts of additional data. Aksu et al. (2002) believe that the euryhaline colonization of mollusks was not a consequence of catastrophic flooding but rather the outcome of a slow establishment of two-way flow in the Bosphorus and a time lag during which the fresher waters of the deep Black Sea were replaced by more saline inflow, eventually allowing marine organisms to colonize the Black Sea shelves. They suggested that mollusks arrived at the region about 7500 years ago when the level of salty Mediterranean water rose to the 100 m depths where mollusks thrive.

The Black Sea sediments deposited during the last 30,000 yr consist of three different environmental conditions of the Late-Quaternary history of the basin (Cagatay, 1999). At the bottom, *Unit 3* deposition during ~30,000–7,000 BP is a laminated clay with a low (~15%) carbonate content, and signifies the fresh water environmental conditions prior to inflow of the Mediterranean water through the Bosphorus. It also includes dark laminae that are formed by high concentrations of unstable iron monosulfides. *Unit 2* is ~ 40 cm thick sapropel deposition consisting of mainly gelatinous organic matter with some coccolith remains, clays, inorganically precipitated aragonite, iron monosulfides and pyrite. Sapropels are occasionally interrupted by turbidite layers of terrigenous origin. Sapropel unit was deposited during a period of high plankton productivity after the flooding of the lacustrine Black Sea basin by the Mediterranean waters via the Bosphorus Strait at around 7000 BP and terminated around 2000–1600 BP. Prior to termination of Unit 2 deposition, there is a period of time with intermittent coccolith invasions whose characteristics reveal some regional variability within the basin. *Unit 1* is ~30 cm thick coccolith mud, consisting of alternations of light- and dark-colored microlaminae. The light-colored laminae are composed mainly by calcareous coccolith remains deposited within the last 2000 yr after the invasion of the Black Sea coccolithophore *Emiliania huxleyi.* The dark laminae consist of clays and organic matter. The clay minerals include predominantly chlorite, smectite and illite with high chlorite/illite and smectite/illite ratios Three peak periods of coccolith deposition implied by maximum carbonate concentrations occurred around 450, 1050, and 1500 BP. They match closely with the transitions from high to low sea level changes that took place in the history of the Black Sea.

6. Changes in the ecosystem characteristics since the 1970s

The last three decades of the Black Sea have been characterized by profound changes in its pelagic ecosystem. These changes were first noted in the biomass, taxonomic structure and succession of the phytoplankton community, particularly in the northwestern shelf. The natural phytoplankton annual cycle with spring and autumn maxima in biomass has been replaced by a pattern characteristic of eutrophied waters identified by several exceptional maxima- the summer one being the most pronounced. In addition to an increase in the frequency of blooms and the number of blooming species, individual blooms have become more monospecific. Diatoms, which were the most abundant group of the annual phytoplankton community structure prior to the 1970s, were replaced by more predominant blooms of dinoflagellates and coccolithophores (Moncheva and Krastev, 1997; Mikaelyan, 1997; Uysal et al., 1998). This phenomenon may have been caused by changes in the silicon to nitrogen ratio due to eutrophication as well as a reduction in the dissolved silicate load of the River Danube (Moncheva and Krastev, 1997) as a result of dam construction in the early 1970s (Humborg et al., 1997) and/or increased eutrophication within the Danube itself (Garnier et al., 2002). Average phytoplankton biomass in the northwestern shelf area increased from 1 g m^{-2} in the 1960s to 19 g m^{-2} in the 1970s and 30 g m^{-2} in the 1980s (Zaitsev and Mamaev, 1997). A similar trend with lower intensity was also reported for other parts of the basin (cf. Fig. 33.8a, and Mikaelyan, 1997; Kovalev et al. 1998). The transparency (as revealed from Secchi Disk measurements shown in Fig. 33.8b) of even open waters was decreased during the 1970s and 1980s. At first the ecosystem responded favorably to increased primary production by producing higher mesozooplankton and fish stocks during the second half of the 1970s and early 1980s (cf. Fig. 33.8c,e, and Porumb, 1989). By the mid-1980s, a five-fold decrease in total mesozooplankton biomass was observed due to their consumption by opportunistic species such as *Noctiluca scintillans, Aurelia aurita, Pleurobrachia rhodopis* and *Mnemiopsis leidyi.* The total abundance of these new organisms reached 99% of the total zooplankton wet weight (Shushkina et al., 1998; Kovalev et al, 1998; Shiganova, 1998; Kideys and Romanova, 2001). As quantified by the Flow Network Analysis (Gucu, 2002), overfishing that took place during the early phase of the eutrophication (i.e. early 1980s) may have triggered destabilization of the ecosystem and the population explosion of gelatinous species.

Increase in the population of gelatinous carnivores apparently led to increases in particulate and dissolved organic matter content in the upper layer water column. This resulted in enhanced bacterial production and more active organic matter decomposition (Lancelot et al., 2002a), which then resulted in increased oxygen deficiency within the upper nitracline-oxycline zone of the water column and more denitrification and associated nitrate consumption within the suboxic zone. Two implications of such modifications in the biogeochemical structure were broadening of the suboxic zone from its position at σ_t~15.9 kg m^{-3} in the 1960s to σ_t~15.6 kg m^{-3} during the 1980s, and a change in the gradient of the subsurface nitrate structure as well as upward shifting of the nitrate peak by about 10 m (Konovalov and Murray, 2001). An increase in the value of the nitrate maximum from about 2.0 mmol m^{-3} to more than 6.0 mmol m^{-3} during 20 years of nitrate accumulation in the water column was roughly equivalent to the contribution from anthropogenic nutrient load after the 1960s. The greater contribution of POM export flux to the anoxic zone led to an increase in the rate of sulfate reduction by about 15%, and thus increased sulfide concentrations within the anoxic zone (Konovalov and Murray, 2001; Konovalov et al., 2001; Neretin et al., 2001).

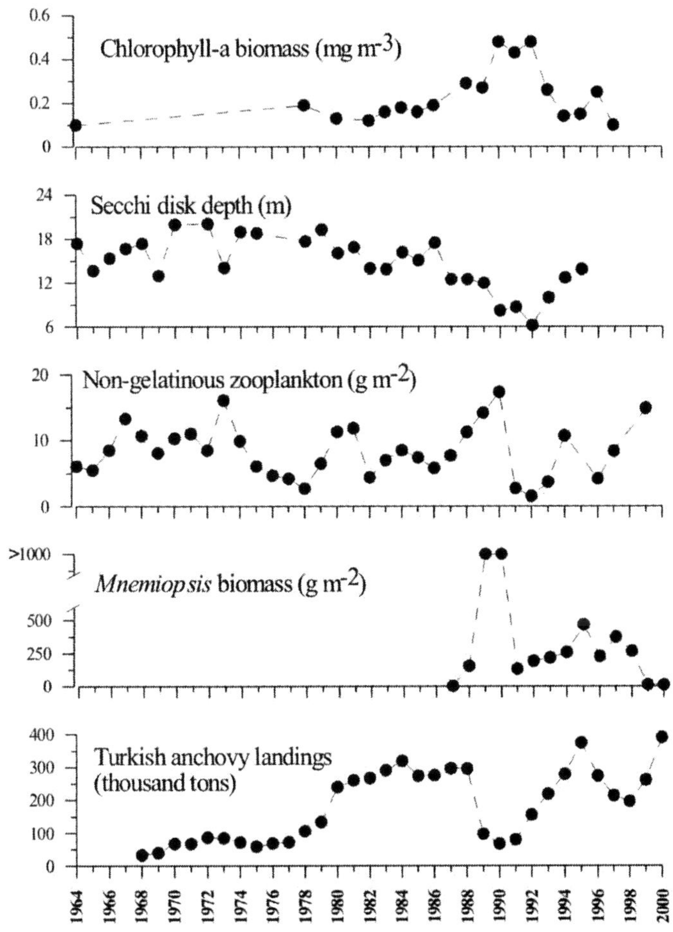

Figure 33.8 Long-term variability of (a) chlorophyll concentration (mg m-3), (b) secchi disk depth (m), (c) non-gelatinous mesozooplankton biomass (g m-2), (d) *Mnemiopsis* biomass (g m-2), (e) Turkish anchovy landings (ktons) (from Kideys 2002). Surface chlorophyll-a—mean values (of about 1000 samples) during May-September; Secchi disk depth (to show water transparency)—annual mean (of a total 1042 measurements during February-December period) for the central Black Sea; non-gelatinous zooplankton biomass—mean values of March, May, June, July and August (collected from several stations from the open waters at approx. biweekly intervals); Mnemiopsis biomass data—average of summer values (except the value of 1990 belonging to April) from around a total of 700 stations;. since the 1980s, Turkish catches comprise the bulk of anchovy.

The interannual and seasonal biomass variations of phytoplankton and zooplankton communities over the last 30 years may be classified mainly into four distinct phases depending on the major top predators controlling the lower trophic food web structure. The period from the mid-1970's to 1987 was dominated by the top predator jellyfish *Aurelia aurita*. Available data show that the impact of this species was considerable on the mesozooplankton. For example, very low mesozooplankton biomass was observed during the peak levels of this species in 1978 (Fig. 33.8c) (calculated to be around 400 million tons for the entire sea; Kideys and Romanova 2001). The phytoplankton community exhibited a major bloom during the late winter-early spring season (Fig. 33.9), following the period of active nutrient accumulation in the surface waters at the end of the winter mixing season and as soon as the water column receives sufficient solar radiation. The phytoplankton bloom was first followed by a mesozooplankton bloom of comparable intensity, which reduced the phytoplankton stock to a relatively low level and then by an *Aurelia* bloom that similarly grazed down the mesozooplankton. The phytoplankton recovered and produced a weaker late spring bloom, which triggered a steady increase in *Noctiluca* biomass during the mid-summer. As the *Aurelia* population decreased in August, first the mesozooplankton and then the phytoplankton and *Aurelia* gave rise to successive blooms during September-October. These blooms were followed by a secondary *Noctiluca* bloom in November. *Aurelia* biomass exhibited two seasonal peaks of about 2–3 gC m^{-2} during May and October, and attained minimum levels during the summer and winter seasons. The summer reduction in the *Aurelia* biomass seemed to be related to food competition as both *Aurelia* and small pelagic fishes (such as anchovy and horse mackerel) feed on the same trophic levels, and the abundance of small pelagic fishes was maximal during the summer period (Gucu, 2002). Bacteria and microzooplankton biomass stayed at much lower levels than the other groups, and did not show any appreciable variation throughout the year.

The years 1989–1991 constituted the second phase in which *Aurelia* blooms were almost totally replaced by those of *Mnemiopsis* (Fig. 33.8d). Following its accidental introduction into the Black Sea in ballast waters of tankers during the early 1980s, *Mnemiopsis* community quickly dominated the entire ecosystem, because it had no predators in the Black Sea. The sudden increase in the *Mnemiopsis* population caused further reduction in the biomass of the mesozooplankton community (Fig. 33.8c) as well as fish eggs and larvae during the late 1980s (Shushkina et al., 1998). This effect, together with overfishing, ultimately caused a collapse of commercial fish stocks (anchovy, sprat and horse-mackerel) during the early 1990s (Fig. 33.8e, and Rass, 1992). *Aurelia* biomass decreased below 1 gC m^{-2} throughout the year during this period. *Mnemiopsis* biomass, in contrast, which was never at measurable quantities before, reached about 1.0 gC m^{-2} in August 1988. During 1989, the *Mnemiopsis* maximum was about 2.0 gC m^{-2} in February-March, and was followed by a decreasing trend in June-July, and later by a second peak of about 3.0 gC m^{-2} during August-September. The next set of measurements, performed during late winter-early spring 1990, also recorded a strong *Mnemiopsis* peak on the order of 3.0 gC m^{-2}, whereas the *Aurelia* biomass remained only around 0.5 gC m^{-2}. *Mnemiopsis* biomass again decreased in the late spring-summer period.

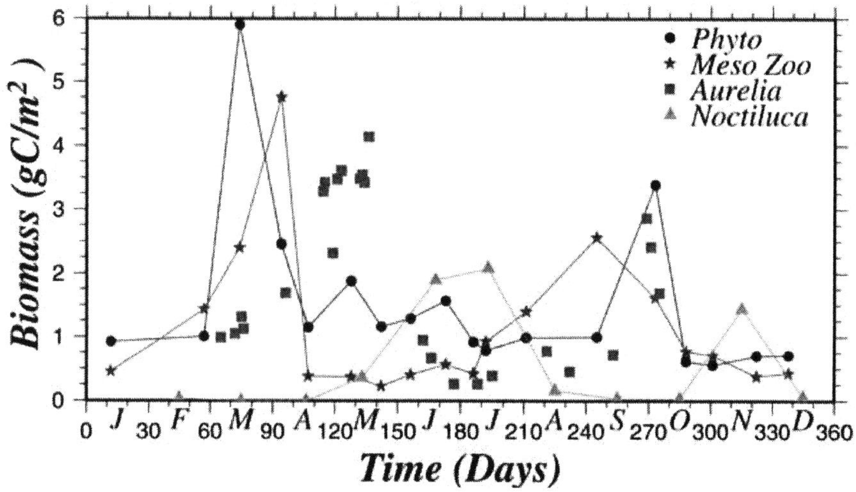

Figure 33.9 Annual distributions of total phytoplankton, mesozooplankton, *Noctiluca* and *Aurelia* biomass representing the conditions of pre-*Mnemiopsis* phase of the Black Sea ecosystem. The data for phytoplankton and mesozooplankton biomass are taken from measurements carried out at 2–4 weeks intervals during January-December 1978 at a station, off Gelendzhik along the Caucasian coast. The *Noctiluca* and *Aurelia* biomass data are taken from measurements on the Romanian shelf and the interior basin, respectively, during the late 1970s and early 1980s (from Oguz et al., 2001b).

Because the outbreak of the *Mnemiopsis* population has been limited to a relatively short period of time, its impact on all trophic levels could not be observed systematically. Measurements carried out within the interior Black Sea during February-April 1991 (Shushkina et al., 1998) indicated some shifts in timing of the phytoplankton and mesozooplankton blooms. Phytoplankton biomass showed an increasing trend starting in early January and reaching a value of ~6 gC m^{-2} toward the end of February. The data further indicated enhanced mesozooplankton stocks during March following the phytoplankton bloom. The surface chlorophyll-a data set formed by combining the Turkish, Russian, Ukrainian, Bulgarian and Romanian measurements for the period of 1990–1995 (Yilmaz et al., 1998b; Yunev et al., 2002) also exhibited peaks in winter (January-February), as well as in the spring-early summer period (May-June). Moreover, the composite data formed by measurements within the interior basin during the last decade suggested an order of magnitude decrease in *Noctiluca* biomass from the 1980s to the early 1990s (Kovalev and Piontkovski, 1998). Measurements from Sevastopol Bay during 1989 and 1990 revealed approximately a two-month shift in the *Noctiluca* biomass peaks from July to May and from November to September (Oguz et al., 2001b).

The period after 1991 represented a third stage in the ecosystem transformation in which *Mnemiopsis* stocks stabilized around 30% of their bloom level and became comparable with *Aurelia* stocks (around ~1.0 gC m^{-2}). The main factor for reduction in the *Mnemiopsis* biomass during the 1992–1993 period, (Fig. 33.8d) was the intense winter cooling and reduction of the sea surface temperatures down to 5°C, which were unfavorable for survival of *Mnemiopsis.* After 1993, the gelatinous macrozooplankton community no longer reached a level critically competing for food with pelagic fish groups (Kideys and Romanova, 2001). The 1993–1995 period of the ecosystem was therefore characterized by some positive sign of recovery such as an increase in fish stocks (Fig. 33.8e) and re-appearance of some zooplankton species (e.g. *Oithona nana, Sagitta setosa*).

The climate-induced warming during the second half of the 1990s (Fig. 33.3) caused loss or weakening of the early spring phytoplankton bloom as a result of weaker turbulent mixing and upwelling, stronger stratification and subsequently reduced upward supply of nutrients from subsurface levels. After 1995, the climate-induced bottom-up type nutrient limitation affected the entire food web. The new annual phytoplankton structure suggested by SeaWIFS chlorophyll data (Fig. 33.10) showed a major decrease of biomass during the late winter-early spring period without any major bloom signature (Oguz et al., 2003). Many higher trophic level species were characterized by reduced stocks (Fig. 33.8e).

Another peculiar feature of the ecosystem after 1995 was development of a strong early summer coccolithophore bloom over the entire basin from early-May until mid-July, contrast to their more limited abundance during the early 1980s. A comparision between the SeaWiFS and the historical coastal zone color scanner data (Cokacar et al., 2001, 2003), as well as some long-term continuous localized measurements (Moncheva and Krastev, 1997) indicates an increase in both coverage and abundance of coccolithophorids during the 1990s. The increasing contribution of coccolithophores to the early summer phytoplankton community structure during the last decade is consistent with the dramatic shifts in taxonomic composition from diatoms to coccolithhophores and flagellates, as a part of transformations that took place in the Black Sea biogeochemistry and ecosystem structure under changing anthropogenic and climate forcing during the 1980s and 1990s, respectively. The climate-induced changes in the physical and biogeochemical structure of the water column (e.g. enhanced stratification and light absorption characteristics, shallower mixed layer, decreased inorganic nutrient availability) should also support wider coverage of coccolithophore blooms in the Black Sea. Finally, the period after 1998 represents a new era in which the Mediterranean species *Beroe ovata* settled in the Black Sea and started consuming *Mnemiopsis,* especially in coastal and shelf waters (Vinogradov et al., 2000; Finenko et al., 2001: Shiganova et al., 2001; Kideys, 2002). Accordingly, *Mnemiopsis* biomass started decreasing (Fig. 33.8d) at the expense of increasing mesozooplankton biomass (Fig. 33.8c) due to their reduced top-down grazing pressure on the mesozooplankton community. The top-down induced improvement in the food web may be the major cause of an increase in the pelagic fish stocks as seen in the Turkish anchovy landing data towards the end of the 1990s (Fig. 33.8e).

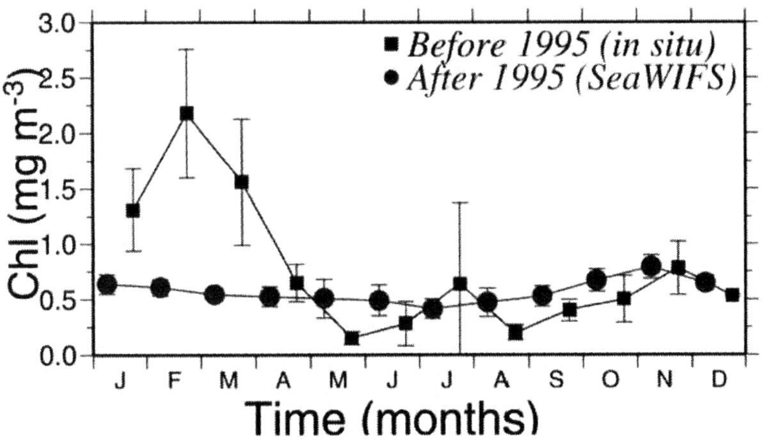

Figure 33.10 Basin-averaged, monthly-mean surface chlorophyll concentration distributions (in mg m^{-3}) before and after 1995. After 1995 distribution is obtained by averaging the OCTS and SeaWiFS ocean color data during 1997–2002 period for regions deeper than 200 m and are shown by dots. The pre-1995 data is composed from different sets of measurements carried out in deep parts of the sea and are shown by squares (from Oguz et al., 2003).

7. Interdisciplinary modeling studies

7.1. Circulation models

A series of three dimensional primitive equation-based circulation models has been used to study characteristcs of the Black Sea circulation system. While earlier models had somewhat coarser horizontal and vertical resolutions, and therefore were not able to properly resolve strong horizontal and vertical density gradients as well as the narrow and steep continental slope around the periphery of the basin (e.g. Stanev, 1990), more recent models acquired eddy-resolving character by introducing more than 20 levels in the vertical and a typical grid size less than 10 km in the horizontal with respect to the baroclinic radius of deformation of greater than 20 km (e.g. Oguz et al., 1995; Besiktepe et al., 2001; Staneva et al., 2001; Beckers et al., 2002). Implementation of five different types of models having comparable complexities are particularly noted among others; the Modular Ocean Model (MOM) by Stanev (1990), Princeton Ocean Model (POM) by Oguz et al. (1995), GHER model by Stanev and Beckers (1999), Harvard Open Ocean Prediction Systems (HOPS) by Besiktepe et al. (2001), and DieCAST model by Staneva et al. (2001). MOM and HOPS have similar characteristics using z-grid on the vertical, rigid-lid, and empirical-based functional representation of the vertical eddy diffusivity and viscosity, whereas POM and GHER have bottom-following σ-coordinate, free surface, and parameterize vertical mixing by means of fairly sophisticated turbulence closure models. DieCAST also resembles MOM but has very low dissipation and fourth-order accuracy. All these models explored various dynamical features of the sea concerned with water-mass formation processes, mesoscale dynamics related to eddy generation- dissipation processes and eddy-eddy, and eddy-mean flow interactions, relative contributions of different processes controlling the circulation system, coastal-open sea interactions and exchanges, wave characteristics, tracer distributions, forecast-related data-driven simulations, etc. They indicated that the overall basin circulation is controlled primarily by the wind stress forcing during the year, and further modulated by the seasonal evolution of the surface thermohaline fluxes, and mesoscale features arising from the basin's internal dynamics. Topography, together with the coastline configuration of the basin, is shown to exert important roles in controlling the pattern of the Rim Current system. The fresh water discharge from the Danube was also shown to generate a basin-wide buoyancy-driven circulation similar to those of the wind forcing.

Rossby waves were shown form spontaneously within the eastern basin and propagate to the west within the deep basin with a phase velocity of ~3 km day^{-1}, and eventually dissipate within the western basin (Rachev and Stanev, 1997; Stanev and Rachev, 1999). Different modes of Rossby waves were thus considered to contribute to the number of gyres forming the interior cyclonic cell. Moreover, they were shown to induce strong changes in the depth of the pycnocline, enhance the amplitude of the Rim Current and modulate exchange between cyclonic and anticyclonic eddies located on two different sides of the Rim Current frontal zone. These exchanges are further controlled on shorter time and spatial scales by mesoscale dynamics introduced by the baroclinic-barotropic (hence mixed) instability mechanism associated with strong vertical and horizontal shears associated with the Rim Current frontal structure. On the other hand, the models (Oguz et al., 1995; Stanev and Staneva, 2000; Besiktepe et al., 2001; Staneva et al., 2001; Beckers et al., 2002) provided clear examples of meander steepening, eddy detachment and growth and splitting, and therefore further contributed to variability of the circulation. A further complexity was shown to be introduced by coastal-trapped waves propagating cyclonically around the basin (coast to the right of the propagation direction) with phase speed of about 5–8 km day^{-1} (Staneva et al., 2001). The data-driven simulations by Besiktepe et al. (2001) attempted to predict a season-long transformation of the circulation during summer-autumn 1992 following initialization of the model by a particular hydrographic data set optimally interpolated to the model grid, and subsequently forwarded in time under wind stress and fresh water inflow forcings. The model identified distinct dynamical processes responsible for generation and evolution of important mesoscale circulation features in different parts of the basin. Some of the models (Oguz and Rizzoli, 1996; Stanev et al., 1997; Staneva and Stanev, 2002) were designed to explore cold intermediate water mass formation process, with particular emphasis on the relative contributions of the shelf and cyclonic interior to the formation process, and depth of convection depending on the intensity and number of cooling events as well as their interannual variability. In some studies, the circulation models were extended to study tracer distributions when coupled with appropriate tracer models. Staneva et al. (1999) used the distribution of the artificial radionuclide, ^{137}Cs, to investigate mixing and ventilation characteristics of the Black Sea surface and intermediate waters. The simulations suggested isopycnal sinking and spreading as the main mechanism of the tracer distribution. Diapycnal mixing had only secondary contribution in some particular regions during the time of CIW formation.

7.2. *Food web models*

Because fishery plays such an important role in the region's economy and because the biological community has been so heavily impacted over the past decades, much attention has been given to constructing food web models of the Black Sea ecosystem. The simplest approach for modeling the structure and functioning of the plankton community was to employ zero-dimensional biological models representing the vertically-averaged conditions within a prescribed surface layer, and introducing physical processes diagnostically from the available data (Lebedeva and Shushkina, 1994; Eeckhout and Lancelot, 1997; Lancelot et al., 2002b). The second approach was to utilize depth-dependent models providing finer vertical representation and physical-biogeochemical coupling through simulation of temperature and vertical diffusivity in the water column by means of a physical model (Oguz et al., 1996, 1999, 2000, 2001b; Gregoire et al., 1998, Gregoire and Lacroix, 2001). Models having an intermediate complexity in terms of their vertical resolution constituted the third approach (Oguz and Salihoglu, 2000; Oguz et al., 2001c). They involved multi-layer representation of the vertical structure, an entrainment parameterization, and coupling with a layered circulation model. They therefore provided a computationally more efficient tool particularly in three-dimensional applications. Majority of these models were applied for one-dimensional simulations (i.e. no horizontal dependence), except a few basin-wide, three-dimensional implementations (Gregoire et al., 1998; Gregoire and Lacroix, 2001; Oguz and Salihoglu, 2000).

Lebedeva and Shushkina (1994) explored ecosystem characteristics before and after the introduction of *Mnemiopsis* for the central Black Sea conditions. Their model included single groups of phytoplankton, bacteria, protozoa (i.e. microzooplankton), mesozooplankton, medusae and *Mnemiopsis* together with organic and inorganic forms of phosphate. The model was able to explain quantitatively how the shift in major gelatinous carnivore population from *Aurelia aurita* to *Mnemiopsis leidyi* resulted in changes in two isolated phytoplankton blooms taking place in autumn and early spring periods of the pre-*Mnemiopsis* phase to an extended autumn-to-winter bloom ended by a stronger early spring bloom. Their simulation of higher winter phytoplankton biomass was related to less efficient mesozooplankton grazing pressure on phytoplankton, and was found to be consistent with the observed winter bloom of 1991 within the interior part of the sea.

Eechout and Lancelot (1997) and Lancelot et al. (2002b) studied the role of nutrient enrichment on destabilization of the northwestern shelf ecosystem within the last three decades. Particular emphasis was given to establishing the link between changes in nutrients, phytoplankton composition and food web structures during the course of ecosystem evolution. Their ecosystem model configuration is the most sophisticated one used in the Black Sea, and incorporates carbon, nitrogen, phosphorus and silicon cycling of both planktonic and benthic systems. Phytoplankton were represented by three different groups (diatoms, flagellates, opportunistic species), microzooplankton, mesozooplankton, opportunistic herbivore *Noctiluca,* and gelatinous carnivore species *Aurelia, Mnemiopsis,* as well as two different biodegradability classes of particulate and dissolved organic matter. The simulations suggest that phosphorus is the major limiting nutrient for the northwestern shelf instead of nitrate or silicate. In the case of a well-balanced

N:P:Si nutrient enrichment scenario, the planktonic food web was characterized by a linear, diatom-copepod type food chain. The gelatinous carnivores were enhanced through their feeding on copepods. In the case of unbalanced nutrient inputs, such as nitrogen or phosphate deficiency, the food chain was dominated by microbial processes. Under these conditions, significant reduction in gelatinous organisms was predicted as observed in the 1990s. The major implication of the latter simulations was to relate the observed positive sign of recovery in the ecosystem to the reduction in anthropogenic nutrient supply, in particular phosphate.

According to the modeling studies by Oguz et al. (1996, 1998, 1999, 2001b), an early spring diatom bloom and a subsequent increase in mesozooplankton stocks are robust signatures of the annual plankton structure of the Black Sea ecosystem, and are seen in every data set irrespective of the type of top-down grazing control by top predators (e.g. Sorokin, 2002; Vedernikov and Demidov, 1997). The autotrophs and heterotrophs, however, have different responses during the rest of the year, depending on the nature of grazing pressure exerted by gelatinous predators. Oguz et al. (2001b) utilized a nitrogen-limited ecosystem model comprising two groups of phytoplankton (small and large fractions), microzooplankton, mesozooplankton, bacterioplankton, *Noctiluca, Aurelia, Mnemiopsis,* dissolved and particulate organic nitrogen, ammonium and nitrate compartments. They performed simulations to explore characteristic features of the *Aurelia*- and *Mnemiopsis*-dominated ecosystems similar to those studied by Lebedeva and Shushkina (1994). The *Aurelia*-dominated ecosystem model configuration was shown to quite realistically simulate the annual cycles of phytoplankton, mesozooplankton, *Noctiluca, Aurelia* (Fig. 33.11) consistent with the observations (Fig. 33.9). The top-down control by *Mnemiopsis* was then shown to alter the autotroph and heterotroph annual biomass structures. The annual phytoplankton structure consisted of three successive and intense bloom events during winter, spring and summer. The winter bloom may be regarded as a modified version of the late winter event seen in the pre-*Mnemiopsis* era. The other two blooms may also be interpreted as the two-month shifts in the late spring-early summer and the autumn events of the pre-Mnemiopsis case. The winter phytoplankton bloom was developed by a particular form of grazing pressure arising due to the specific annual *Mnemiopsis* biomass cycle. It caused an almost complete depletion of the microzooplankton, mesozooplankton and *Noctiluca* stocks towards the end of the autumn season. The lack of grazing on the phytoplankton community then promoted earlier growth starting by the beginning of January. In the previous case of *Aurelia* dominance, on the other hand, the zooplankton community developed following the autumn bloom event prevented early initiation of phytoplankton growth in winter months. The other interesting feature was the shift of *Noctiluca* peaks to two months earlier with respect to the pre-*Mnemiopsis* period, as reported by the observations.

Gregoire et al. (1988) and Gregoire and Lacroix (2001) attempted to model the three-dimensional structure of the Black Sea ecosystem using a sophisticated physical-biological model using 15 km horizontal grid resolution and 25 vertical levels. In Gregoire et al. (1988), the physical model providing three dimensional fields of currents, temperature and vertical eddy diffusivity was coupled with 13 compartment biological model representing the ecosystem by three functional groups of phytoplankton, two groups of herbivorous zooplankton, bacteria, ammonium, nitrate, phosphate, silicate, particulate and dissolved organic nitrogen and particulate silicate. In Gregoire and Lacroix (2001), the biological model was more simplified involving single groups of phytoplankton and zooplankton as well as a simplified nitrogen cycle, but oxygen dynamics were included. The results of the model simulations reproduced major horizontal and vertical structures of physical and biochemical properties such as the seasonal cycle of phytoplankton blooms within the interior basin and more extended blooms within the northwestern shelf. A similarly simple nitrogen-based model formed by two groups of phytoplankton and zooplankton, detritus, ammonium and nitrate using a three layer representation of the upper layer water column was used to assess impact of eddy-dominated horizontal circulation on the spatial and temporal variations of plankton biomass (Oguz and Salihoglu, 2000). All these models suggested combination of wind-driven coastal upwelling, eddy-pumping, entrainment due to mixed layer deepening, and vertical diffusion controlling the nutrient supply to the euphotic zone, and generating phytoplankton blooms of both new and regenerated origin during October-December and March-May periods. Eddy-induced lateral transports and asymmetries at the mixed- and intermediate-layer depths, as well as on nutrient fluxes, support complex patterns of biomass distributions.

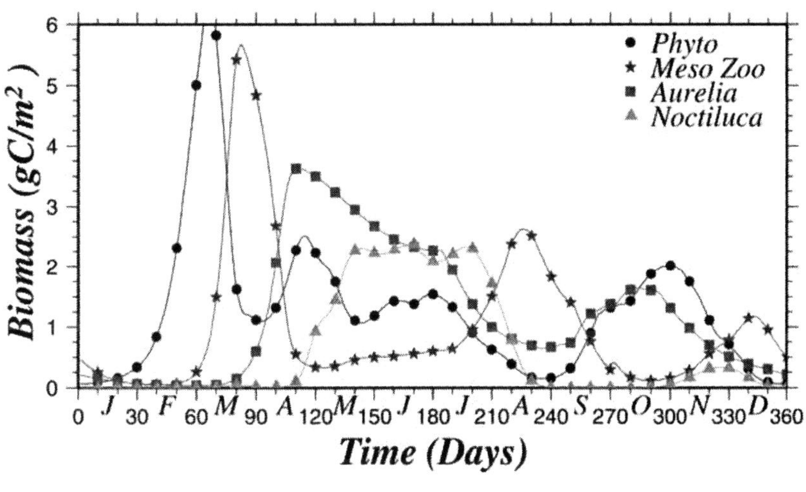

Figure 33.11 Annual distributions of total phytoplankton, mesozooplankton, *Noctiluca* and *Aurelia* biomass representing the conditions of pre-*Mnemiopsis* phase of the Black Sea ecosystem simulated by the model described by Oguz et al. (2001b).

7.3. Redox cycling models

Because of the relatively broad and stable character of the suboxic-anoxic transition zone, the Black Sea is an ideal site to model redox processes. The nitrogen and sulfur cycles were first modeled by Yakushev and Neretin (1997) constraining sulfide oxidation by oxygen alone. Sulfur cycling involved abiogenic oxidation to thiosulfate followed by its bacterial oxidation to sulfate by thiobasili. Because this bacterium requires oxygen, sulfur oxidation depends on the availability of oxygen. The lowest oxygen concentration requirement for oxidation was set to 2 μM, and its maximum efficiency was assumed to take place at oxygen concentrations greater than 9 μM. The Yakushev and Neretin model thus requires a continuous supply of oxygen to drive the sulfur cycle. This supply was provided by a downward diffusive oxygen flux using a vertical eddy diffusivity of $1x10^{-5}$ m^2 s^{-1}, which appears to be an order of magnitude higher than those estimated from microstructure measurements and hydrographic data for such strongly stratified conditions (Gregg and Ozsoy, 1999). Their simulations therefore show overlapping dissolved oxygen and sulfide concentrations. In Yakushev (1998) and Debolskaya and Yakushev (2002), this model was extended to incorporate simplified manganese cycling in which particulate manganese used for oxidizing hydrogen sulfide was produced by oxidation of dissolved manganese with oxygen as the data seem to indicate. A similar model based on the same concept of sulfide-oxygen interaction was given by Belyaev et al. (1997). This model was developed specifically for the northwestern shelf ecosystem, and a case study of its implementation was described by Lyubartseva and Lyubartsev (1998).

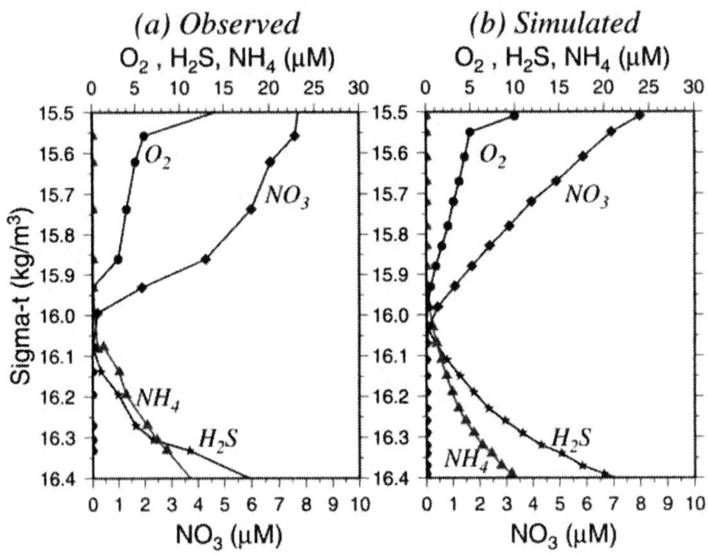

Figure 33.12 (a) Observed, and (b) simulated profiles of dissolved oxygen, hydrogen sulphide, nitrate and ammonium within the Suboxic Layer and onset of the sulfide zone (from Oguz et al., 2002b).

Oguz et al. (2001a) proposed a model with a somewhat different set of redox reactions catalyzed by the manganese cycle. The model was based on the hypothesis that the anaerobic sulfide oxidation and nitrogen transformations were the primary mechanisms controlling the interface structure between the suboxic and anoxic layers. Following Murray et al. (1995, 1999), it was proposed that the upward fluxes of sulfide and ammonium are oxidized by Mn(III, IV) and Fe(III) species, whereas the downward flux of nitrate may be reduced by dissolved manganese and ammonium. Mn(II) oxidation and Mn(IV) reduction are both microbially catalyzed (Tebo, 1991; Francis and Tebo, 1999), but dissolved abiotic, chemical reduction is also thought to play a role in Mn(IV) reduction by sulfide (e.g. Millero, 1991). In Oguz et al. (2001a), only manganese cycling was included and the additional contribution of iron cycling was neglected for simplicity. The anammox reaction, whose importance in the Black Sea was recently shown by Kuypers et al. (2003), was also not included. Dissolved manganese oxidized by nitrate was found to be responsible for the production and cycling of particulate manganese, which in turn oxidized hydrogen sulfide and ammonium transported upwards from deeper levels. Even with such a highly simplified representation of the redox processes, the model provided a realistic suboxic-anoxic interface zone structure, and was able to give quantitative evidence for the presence of an oxygen depleted and non-sulfidic suboxic zone (Fig. 33.12). This model pointed out the crucial role of the downward supply of nitrate from the overlying nitracline zone and for the upward transport of dissolved manganese from the anoxic pool below for maintenance of the suboxic layer.

The food web and the suboxic-anoxic interface redox cycle models provided by Oguz et al. (2001a,b) were then coupled with each other as well as a water column oxygen cycling model in order to provide a model capable of providing a unified description of the dynamically coupled oxic-suboxic-anoxic system (Oguz et al., 2000). The combined model (Fig. 33.13) was comprised of a nitrogen-based pelagic plankton system coupled through particulate and dissolved organic matter fluxes to the water column nitrogen cycle model including transformations among ammonium, nitrite and nitrate. This system was then coupled with a model of oxygen and redox dynamics describing oxygen variations within the euphotic zone, oxycline and near the suboxic-anoxic interface zone. These processes were finally linked to manganese and sulfur transformations. The model further allowed online coupling with the physical model providing vertical eddy diffusivity and temperature data to the biological model. The physical model, which was documented by Oguz et al. (1996), was based on the Princeton Ocean Model in which wind- and buoyancy-induced vertical mixing were parameterized by a level 2.5 turbulence closure model. The simulations indicated that oxygen consumption during remineralization and nitrification, together with the lack of ventilation of subsurface waters due to the presence of strong vertical stratification, are two main factors limiting aerobic biogeochemical activity to the upper 75 m of the water column, which approximately corresponds to the level of nitrate maximum. The position of the upper boundary and thus the thickness of the suboxic layer are controlled by upper layer biological processes. The quasi-permanent character of this layer and the stability of the suboxic-anoxic interface within the last several decades are maintained by a constant rate of nitrate supply from the nitrate maximum zone.

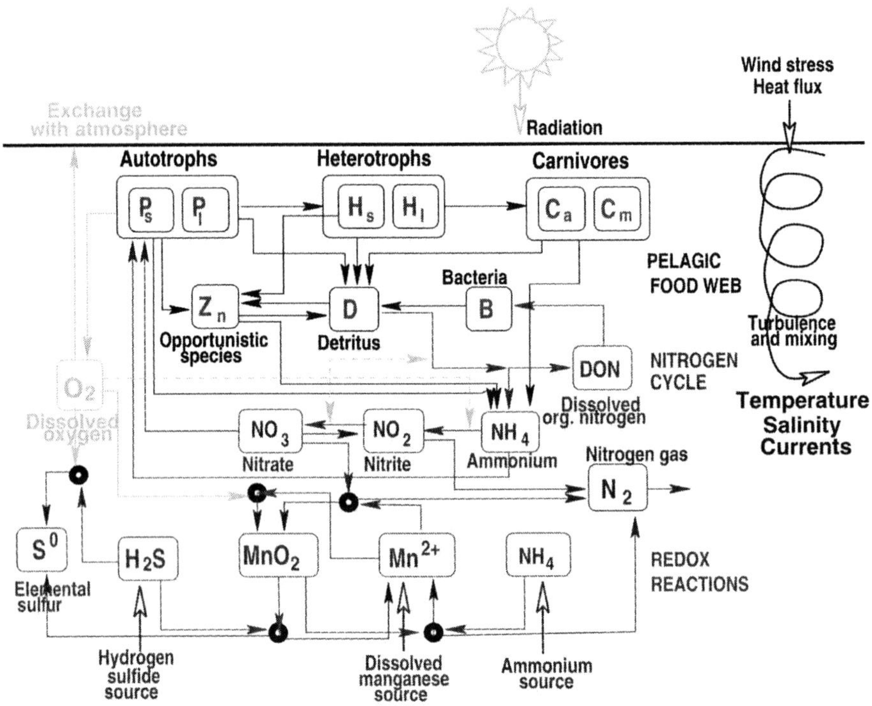

Figure 33.13 Schematic representation of the major processes and interactions between biogeochemical model compartments. P_l, P_s denote large and small phytoplankton groups, H_s is microzooplankton, H_l is mesozooplankton, C_a is jellyfish *Aurelia aurita,* C_m is ctenephore *Mnemiopsis leidyi,* Z_n is opportunistic dinoflagellate *Noctiluca scintillans,* MnO_2 is particulate manganese. Meaning of the other variables are as shown in the figure. The biogeochemical model is coupled to the physical model through vertical diffusivity and temperature (from Oguz et al., 2002b).

8. Conclusions

Implementation of various contemporary ocean research programs in the last decade have generated considerable progress toward understanding the Black Sea's basic physical and biogeochemical processes. A series of basin-wide, quasi-synoptic, multi-ship research cruises were accomplished, and complemented by analysis of satellite data and interdisciplinary modeling studies. Physical and biogeochemical data accumulated from these studies were able to resolve many features of the Black Sea circulation and ecosystem structure. The description and understanding of basin scale dynamical features has reached maturity. Modelling and data assimilation tools have been developed, and implemented but they still require further calibration and validation exercises by availability of new data sets. These efforts also led to substantial improvement in our understanding of biogenic element cycling, pelagic and benthic ecosystems carbon and energy flows, various distinct biogeochemical processes taking place across the anoxic interface, as well as their coupling to the circulation, and horizontal/vertical transport mechanisms. Knowledge and modeling of the coastal and shelf areas are still scarce and await implementation of a systematic observation and modeling program. More comprehensive future studies on the climate-driven ecosystem changes are expected to provide insights for sustainable use of marine living resources in the Black Sea, and effective management policies. More elaborate deterministic models will be particularly useful for the mechanistic understanding of the ecosystem as a whole, testing various hypotheses put forward for explaining observed features, assessing relative contributions of different mechanisms, predicting optimal near-real time estimates of large scale flow and biogeochemical fields and forecasting future state of the ecosystem.

Bibliography

Afanasyev, Y.D., A.G. Kostianoy, A.G. Zatsepin, and P.M. Poulain, 2002. Analysis of velocity field in the eastern Black Sea from satellite data during the Black Sea '99 experiment. *J. Geophys. Res.*, **107,** 10.1029/2000JC000578.

Aksu, A.E., R.N. Hiscott, and D. Yasar, 1999. Oscillating Quarternary water levels of the Marmara Sea and vigorous outflow into the Aegean Sea from the Marmara Sea-Black Sea drainage corridor. Marine Geology, **153,** 275–302.

Aksu, A.E., R.N. Hiscott, M.A. Kaminsky, P.J. Mudie, H. Gillespie, T. Abrajano, and D. Yasar, 2002. Last glacial-Halocene paleoceanography of the Black Sea and Marmara Sea: Stable isotopic, foraminiferal, and coccolith evidenced. Marine Geology.

Almazov, N. M., 1961. Stok ratverennykh soley I biogennykh veschetv kotorye vynoseatsya rekami USSR v Chernoe More. Naukovi Zapiski Odes. Biol. St. Kiev 3, pp. 99–107.

Ballard, R.D., D.F. Coleman, G.D. Rosenberg, 2000. Further evidence of abrupt Holocene drowning of the Black Sea shelf. *Mar. Geol.,* **170,** 253–261.

Basturk, O., C. Saydam, I. Salihoglu, L. V. Eremeeva, S. K. Konovalov, A. Stoyanov, A. Dimitrov, A. Cociasu, L. Dorogan, and M .Altabet, 1994. Vertical variations in the principle chemical properties of the Black Sea in the autumn of 1991. *J. Marine Chemistry,* **45,** 149–165.

Basturk, O., S.Tugrul, S.Konovalov, I., Salihoglu, 1998a. Effects of circulation on the saptial distributions of principle chemical properties and unexpected short- and long-term changes in the Black Sea. In: *Ecosystem modeling as a management tool for the Black Sea.* Ivanov, L.I. and T. Oguz (eds.), NATO ASI Series, Vol. 47, Kluwer Academic publishers, Netherlands, pp. 39–54.

Basturk, O., I. I. Volkov, S. Gokmen, H. Gungor, A.S Romanov, and E.V. Yakushev, 1998b. International expedition on board R.V. Bilim in July 1997 in the Black Sea. *Oceanology (English trans.),* **38,** 429–432.

Beckers, J.M., M.L. Gregoire, J.C.J. Nihoul, E. Stanev, J. Staneva and C. Lancelot, 2002. Hydrodynamical processes governing exchanges between the Danube, the North-Western continental shelf and the Black Sea's basin simulated by 3D and box models. *Estuarine, Coastal and Shelf Science,* **54,** 453–472.

Belyaev, V. I. , E.E. Sovga, and S.P. Lyubartseva, 1997. Modelling the hydrogen sulfide zone of the Black Sea. *Ecological Modelling,* **13,** 51–59.

Besiktepe, S.T., C.J., Lozano, and A.R., Robinson, 2001. On the summer mesoscale variability of the Black Sea. *J. Mar. Res.,* **59,** 475–515.

Brewer, P.G. and D.W. Spencer, 1974. Distribution of some trace elements in the Black Sea and their flux between dissolved and particulate phases. In: *The Black Sea- Geology, Chemistry, and Biology.* E.T. Degens and D.A. Ross, eds. The American Association of Petroleum Geologists, Memoir 20, pp. 137–143.

Buesseler, K.O., H.D. Livingston, L.Ivanov, and A. Romanov, 1994. Stability of the oxic-anoxic interface in the Black Sea. *Deep-Sea Res. I,* **41,** 283–296.

Burlakova, Z.P., L.V., Eremeeva, and S.K. Konovalov, 1997. Particulate organic matter of Black Sea euphotic zone. In: *Sensitivity to change: Black Sea, Baltic Sea and North Sea,* Özsoy, E. and A. Mikaelyan. (eds.), NATO ASI Series, Environment-Vol. 27, Kluwer Academic publishers, Netherlands, pp. 223–238.

Cagatay, M.N., 1999. Geochemistry of the late Pleistocene-Holocene sediments of the Black Sea basin. In: *Environmental degradation of the Black Sea: Challenges and Remedies.* S. Besiktepe, U. Unluata and A. Bologa, (eds). NATO-ASI Series, Environmental Security, Vol. 56, Kluwer Academic Publishers, Dordrecht, pp. 9–22.

Cagatay, M.N, N. Gorur, A. Algan, C.J. Eastoe, A. Tchapalyga, D. Ongan, T. Kuhn, I. Kuscu, 2000. Late Glacial_Holocene paleoceanography of the Sea of Marmara: timing of connections with the Mediterranean and the Black Seas. *Mar. Geol.* **167,** 191–206.

Calvert, S.E. and Fontugne, 1987. Stable carbon isotopic evidence for the marine origin of the organic matter in the Halocene Black Sea sapropel. *Chemical Geology* (Isotope Geoscience Section), **66,** 315–322.

Charlson, R. J., J. E. Lovelock, M. O. Andreae and S. G. Warren, 1987. Oceanic phytoplankton, atmospheric sulphur, cloud albedo and climate: A geophysiological feedback, *Nature,* **326,** 655–661.

Charlson, R. J., J. Longner, H. Rodhe, C. B. Leavy, and S. G. Warren,. 1991. Perturbation of the Northern hemisphere radiative balance by backscattering from anthropogenic sulphate aerosols, *Tellus,* **43A,** 152–163.

Coban-Yildiz, Y., G., Chiavari, D., Fabbri, A.F., Gaines, G., Galetti, S.Tugrul, 2000a. The abundance and elemental composition of Black Sea seston: confirmation by pyrolysis-GC/MS. *Marine Chemistry,* **69,** 55–67.

Coban-Yildiz, Y., S., Tugrul, D., Ediger, A., Ylmaz, S.C. Polat, 2000b. A comparative study on the abundance and elemental composition of POM in three interconnected basins: the Black, the Marmara and the Mediterranean Seas. *Mediterranean Marine Science,* **1,** 51–63.

Coban-Yildiz Y., M.A, Altabet, A., Ylmaz, S. Tugrul and I. Salihoglu, 2003. Carbon andnitrogen isotopic ratios of suspended particulate organic matter in the Black Sea water column. *Deep-Sea Research,* submitted manuscript.

Cociasu, A., L. Dorogan, C. Humborg and L. Popa, 1996. Long-term ecological changes in the Romanian coastal waters of the Black Sea. *Marine Pollution Bulletin,* **32,** 32–38.

Codispoti, L.A., G.E. Friederich, J.W. Murray, and C.M. Sakamoto, 1991. Chemical variability in the Black Sea: implications of continuous vertical profiles that penetrated the oxic/anoxic interface. *Deep-Sea Res. I,* **38,** Suppl.2, S691-S710.

Cokacar, T., N. Kubilay, and T. Oguz, 2001. Structure of *Emiliania huxleyi* blooms in the Black Sea surface waters as detected by SeaWIFS imagery. *Geophys. Res. Letters,* **28(24),** 4607–4610.

Cokacar, T., T. Oguz, and N. Kubilay, 2004. Interannual variability of the early summer coccolithophore blooms in the Black Sea: impacts of anthropogenic and climatic factors. To appear in *Deep-Sea Res. I.,* **51,** 1017–1031

Debolskaya, E.I., and E.V. Yakushev, 2002. The role of suspended manganese in hydrogen sulfide oxidation in the Black Sea redox-zone. *Water Resources,* **29,** 72–77.

Di Iorio, D. and H. Yuce, 1998. Observations of Mediterranean flow into the Black Sea. *J. Geophys. Res.,* **104,** 3091–3108.

Duce, R.A., Liss, P.S., Merill, J.T., Atlas, E.L., Buat-Menart, P., Hicks, B.B., Miller, J.M., Prospero, J.M., Arimoto, R., Church, T.M., Ellis, W., Galloway, J.N., Hansen, L., Jickells, T.D., Knap, A.H., Reinhardt, K.H., Schneider, B., Soudine, A., Tokos, J.J., Tsunogai, S., Wollast, R. and Zhou, M., 1991. The atmospheric input of trace species to the world ocean, *Global Biogeochem. Cycles,* **5,** 193–259.

Eeckhourt, D.V. and C. Lancelot, 1997. Modeling the functioning of the northwestern Black sea ecosystem from 1960 to present. In: *Sensitivity to change: Black Sea, Baltic Sea and North Sea,* NATO Sci. Partnership Sub-series 2, vol. 27, E. Ozsoy and A. Mikaelyan, eds. Kluwer Acad., Norwell, Mass., pp. 455–468.

Eremeev, V.N., 1996. Hydrochemistry and dynamics of the hydrogen-sulfide zone in the Black Sea. *Unesco Reports in Marine Science,* R.C. Griffiths, eds. Unesco 1996, 114 pp.

Faschuk, D.Ya., T.A. Ayzatullin, V.V. Dronov, T.M. Pankratova, and M.S. Finkelshteyn, 1990. Hydrochemical structure of the layer of co-existence of oxygen and hydrogen sulfide in the Black Sea and a possible mechanism of its generation. *Oceanology (English transl.),* **30,** 185–192.

Finenko, G.A., Z.A. Romanova, G.I. Abolmasova, B.E. Anninsky, L.S. Svetlichny, E.S. Hubareva, L. Bat, and A.E. Kdeys, 2003. Ingestion, growth and reproduction rates of the alien Beroe ovata and its impact on the plankton community in the Black Sea. *J. Plankton Res.,* **25,** 539–549.

Francis, C.A. and B.M. Tebo, 1999. Marine bacillus spores as catalyst for oxidative precipitation and sorption of metals. *J. Mol. Microbiol. Biotechnol.,* **1,** 71–78.

Fry, B., H.W., Jannasch, S.J., Molyneaux, C.O., Wirsen, J.A. Muramoto, and S., King, 1991. Stable isotope studies of the carbon, nitrogen and sulfur cycles in the Black Sea and the Cariaco Trench. *Deep-Sea Res.,* **38,** S1003-S1019.

Garnier, J., G. Billen, E. Hannon, S. Fonbonne, Y. Vdenina and M. Soulie, 2002. Modeling transfer and retention of nutrients in drainage network of the Danube River. *Estuarine, Coastal and Shelf Science,* **54,** 285–308.

Gawarkiewicz, G., G. Korotaev, S. Stanichny, L. Repetin, and D. Soloviev, 1999. Synoptic upwelling and cross-shelf transport processes along the Crimean coast of the Black Sea. *Cont. Shelf Res.,* **19,** 977–1005.

Ginsburg, A.I., A.G., Kostianoy, D.M., Soloviev, S.V., Stanichny, 2000. Remotely sensed coastal/deep-basin water exchange processes in the Black Sea surface layer. In Satellites, *Oceanography and Society,* D. Halperneds. Elsevier Oceanography Series, Amsterdam, 63, pp. 273–285.

Ginsburg, A.I., A.G., Kostianoy, N.P., Nezlin, D.M., Soloviev, S.V., Stanichny, 2002a. Anticyclonic eddies in the northwestern Black Sea. *J. Mar. Syst.,* **32,** 91–106.

Ginsburg, A.I., A.G., Kostianoy, V.G., Krivosheya, N.P., Nezlin, D.M., Soloviev, S.V., Stanichny, V.G. Yakubenko, 2002b. Mesoscale eddies and related processes in the northeastern Black Sea. *J. Mar. Syst.,* **32,** 71–90.

Gorur, N., M.N. Cagatay, O. Emre, B. Alpar, M. Sakinc, Y. Islamoglu, O. Algan, T. Erkal, M. Kecer, R. Akkok, G. Karlik, 2001. Is the abrupt drowning of the Black Sea shelf at 7150 yr BP a myth? *Mar. Geol.,* **176,** 65–73.

Grashoff, K., 1975. The hydrochemistry of landlocked basins and fjords. In: *Chemical Oceanography,* J.P. Riley and C. Skirrow, eds, Academic Press, New York, pp. 456–497.

Gregg, M.C. and E. Ozsoy, 1999. Mixing on the Black Sea shelf north of the Bosphorus. *Geophys. Res. Letters,* , **26,** 1869–1872.

Gregoire, M., J.M. Beckers, J.C.J. Nihoul and E. Stanev, 1998. Reconnaissance of the main Black Sea's ecohydrodynamics by means of a 3D interdisciplinary model. *J. Mar. Syst.,* **16,** 85–106.

Gregoire, M., and G. Lacroix, 2001. Study of oxygen budget of the Black Sea waters using a 3D coupled hydrodynamical-biochemical model. *J. Mar. Syst.,* **31,** 175–202.

Gücü, A.C., 2002. Can overfishing be responsible for the successful development of *Mnemiopsis? Estuarine, Coastal and Shelf Science,* **54,** 439–451.

Hay, B.J., 1994. Sediment and water discharge rates of Turkish Black Sea rivers before and after hydropower dam constructions. *Environmental Geology,* **23,** 276–283.

Humborg, C., V. Ittekkot, A. Cociasu, and B. Bodungen, 1997. Effect of Danube River dam on Black Sea biogeochemistry and ecosystem structure. *Nature,* **386,** 385–388.

Ivanov, L.I. and A.S. Samodurov, 2001. The role of lateral fluxes in ventilation of the Black Sea. *J. Mar. Syst.,* **31,** 159–174.

Jannasch, H.W., C.O. Wirsen, and S.J. Molyneaux, 1991. Chemoautotrophic sulfur-oxiding bacteria from the Black Sea. *Deep-Sea Res.,* **38,** Suppl.2A, S1105-S1120.

Jorgensen, B.B., H. Fossing, C.O. Wirsen, and H.W. Jannasch, 1991. Sulfide oxidation in the anoxic Black Sea chemocline. *Deep-Sea Res.,* **38,** Suppl.2A, S1083-S1104.

Kaminski, M.A:, A. Aksu, M. Box, R.N. Hiscott, S. Filipescu, M. Al-Salameen, 2002. Late Galacialto Holocene benthic foraminifera in the Sea of Marmara: implications for Black Sea-Mediterranean Sea connections following the last deglaciation. *Mar. Geol.,* **190,** 165–202.

Karl, D.M., and G.A. Knauer, 1991. Microbial production and particle flux in the upper 350 m of the Black Sea. *Deep Sea Res.,* **38,** Supp. 2A, S655-S661.

Kempe, S., A.R. Diercks, G. Liebezeit, and A. Prange, 1991. Geochemical and structural aspects of the pycnocline in the Black Sea (R/V Knorr 134–8 Leg 1, 1988). In: *Black Sea Oceanography,* NATO ASI Series C-Vol.351, E.Izdar and J.W. Murray, eds, Kluwer Academic Publishers, pp. 89–110.

Kideys, A.E., 2002. Fall and rise of the Black Sea ecosystem, *Science,* **297,** 1482–1484.

Kideys, E.A., and Z.A. Romanova, 2001. Distribution of gelatinous macrozooplankton in the southern Black Sea during 1996–1999, *Mar. Biol.,* **139,** 535–547.

Kodina, L.A., M.P., Bogacheva, S.V., Lyutsarev, 1996. Particulate organic carbon in the Black Sea: Isotopic composition and origin. *Geochemistry International,* **9,** 884–890.

Konovalov, S.K., and J.W. Murray, 2001. Variations in the chemistry of the Black Sea on a time scale of decades (1960–1995). *J. Mar. Syst.,* **31,** 217–243.

Konovalov, S.K., L.I. Ivanov, and A.S. Samodurov, 2001. Fluxes and budget of sulpfide and ammonia in the Black Sea anoxic layer. *J. Mar. Syst.,* **31,** 203–216.

Konovalov, S.K., G.W. Luther, III, G.E. Friederich, D.B. Nuzzio, B.M. Tebo, J.W. Murray, T. Oguz, B. Glazer, R.E. Trouwborst, B. Clement, K. W. Murray, A. S. Romanov, 2003. Lateral injection of oxygen with the Bosphorus plume-fingers of oxidizing potential in the Black Sea. *Limnol. Oceanogr.,* **48,** 2369–2376.

Korotaev, G.K., O.A., Saenko, C.J., Koblinsky, 2001. Satellite altimetry observations of the Black Sea level. *J. Geophys. Res.,* **106,** 917–933.

Korotaev, G.K., T. Oguz, A. Nikiforov, C. J. Koblinsky, 2003. Seasonal, interannual and mesoscale variability of the Black Sea upper layer circulation derived from altimeter data. *J. Geophys. Res.,* **108(C4),** 3122, doi:10.1029/2002JC001508.

Kovalev, A.V. and S.A. Piontkovski, 1998. Interannual changes in the biomass of the Black Sea gelatinous zooplankton. *J. Plankton Res.,* **20,** 1377–1385.

Kovalev, A.V., U., Neirman, V.V., Melnikov, V., Belokopytov, Z., Uysal, A.E., Kideys, M., Unsal, and D., Altukov, 1998. Long term changes in the Black Sea zooplankton: the role of natural and anthropogenic factors. In: *Ecosystem Modeling as a Management Tool for the Black Sea.* L. Ivanov and T. Oguz, eds. NATO ASI Series 2, Environment-Vol.47, Kluwer Academic Publishers, Vol. 1, pp. 221–234.

Krivosheya, V.G., V. B. Titov, I.M. Ovchinnikov, R.D. Kosyan, A.Yu. Skirta, 2000. The influence of circulation and eddies on the depth of the upper boundary of the hydrogen sulfide zone and ventilation of aerobic waters in the Black Sea. *Oceanology (Eng. Transl.),* **40,** 767–776.

Krivosheya, V.G., I.M. Ovchinnikov, and A. Yu. Skirta, 2002. Interannual variability of the cold intermediate layer renewal in the Black Sea. In: Multidisciplinary *investigations of the northeast part of the Black Sea,* A.G. Zatsepin and M.V. Flint, eds, Moscow, Nauka, pp. 27–39.

Kubilay, N. and A.C., Saydam, 1995. Trace elements in atmospheric particulates over the Eastern Mediterranean; concentrations, sources, and temporal variability. *Atmospheric Environment,* 29, 2289–2300.

Kubilay, N., S. Nickovic, C. Moulinc and F. Dulac (2000) An illustration of the transport and deposition of mineral dust onto the eastern Mediterranean. *Atmospheric Environment,* 34, 1293–1303

Kubilay, N., M. Kocak, T. Cokacar, T. Oguz, G. Kouvarakis, N. Mihalopoulos, 2002. The Influence of Black Sea and Local Biogenic Activity on the Seasonal Variation of Aerosol Sulfur Species in the Eastern Mediterranean Atmosphere. *Global Biogeochem. Cycles,* **16(4),** 1079, doi:10.1029/2002GB001880.

Kuypers, M.M.M., A.O. Sliekers, G. Lavik, M. Schmid, B.B. Jorgensen, J.G. Kuenen, D.J.S. Sinninghe, M. Strous, and M.S.M. Jetten, 2003. Anaerobic ammonium oxidation by anammox bacteria in the Black Sea. *Nature,* **422,** 608–611.

Lancelot, C., J. M. Martin, N. Panin, and Y. Zaitsev, 2002a. The North-western Black Sea: A pilot site to understand the complex interaction between human activities and the coastal environment. *Estuarine, Coastal and Shelf Science,* **54,** 279–283.

Lancelot, C., J. Staneva, D. Van Eeckhout, J. M. Beckers, and E. Stanev, 2002b. Modeling the impact of the human forcing on the ecological functioning of the northwestern Black Sea. *Estuarine, Coastal and Shelf Science,* **54,** 473–500.

Latif, M.A., E. Ozsoy, T. Oguz, and U. Unluata, 1991. Observation of the Mediterranean inflow into the Black Sea. *Deep-Sea Research,* **38,** Supl.2A, S711-S733.

Lebedeva, L.P. and S. V. Vostokov, 1984. Studies of detritus formation processes in the Black Sea. *Oceanology (Engl. Transl.),* **24(2),** 258–263.

Lebedeva, L.P. and E.A. Shushkina, 1994. Modelling the effect of *Mnemiopsis* on the Black Sea plankton community. *Oceanology (English transl.),* **34,** 72–80.

Lee, B-S., J.L.Bullister, J.W. Murray, and R.E. Sonnrup, 2002. Anthropogenic chlorofluorocarbons in the Black Sea and the Sea of Marmara. *Deep-Sea Res. I,* **49,** 895–913.

Lewis, B.L. and W.M. Landing, 1991. The biogeochemistry of manganese and iron in the Black Sea. *Deep-Sea Res.,* **38,** Suppl.2A, S773-S804.

Lyubartseva, S.P. and V. G. Lyubartsev, 1988. Modeling of the Black Sea anoxic zone processes. In: *Ecosystem Modeling as a Management Tool for the Black Sea.* L. Ivanov and T. Oguz, eds. NATO ASI Series, 2-Environmental Security-47, Kluwer Academic Publishers, Vol. 2, pp. 385–396.

Mikaelyan, A. S., 1997. Long-term variability of phytoplankton communities in open Black Sea in relation to environmental changes. In: *Sensitivity to change: Black Sea, Baltic Sea and North Sea,* NATO Sci. Partnership Sub-series 2, vol. 27, E. Ozsoy and A. Mikaelyan, eds., Kluwer Acad., Norwell, Mass., pp. 105–116.

Millero, F.J., 1991. The oxidation of H_2S in Black Sea waters. *Deep-Sea Research,* **38,** Supl.2A, S1139-S1150.

Moncheva, S., A., Krastev, 1997. Some aspects of phytoplankton long-term alterations off Bulgarian Black Sea Shelf. In: *Sensitivity to Change: Black Sea, Baltic Sea and North Sea,* E. Ozsoy, A. Mikaelyan, eds. NATO ASI Series, Vol. 27. Kluwer Academic Publishers, Dordrecht, pp. 79–94.

Murray, J.W., H. W. Jannash, S. Honjo, R. F. Anderson, W.S. Reeburgh, Z. Top, G.E. Friederich, L.A. Codispoti, and E. Izdar, 1989. Unexpected changes in the oxic/anoxic interface in the Black Sea. *Nature,* **338,** 411–413.

Murray, J. W., Z. Top, and E. Ozsoy, 1991. Hydrographic properties and ventilation of the Black Sea. *Deep-Sea Res.,* **38,** Suppl.2A, S663–690.

Murray, J.W., L.A. Codispoti, and G.E. Friederich, 1995. Oxidation-reduction environments: The suboxic zone in the Black Sea. In: *Aquatic chemistry:Interfacial and interspecies precosses.* C.P. Huang, C.R.O'Melia, and J.J. Morgan, eds. ACS Advances in Chemistry Series No.224. pp. 157–176.

Murray, J.W., B-S Lee, J. Bullister, and G. W. Luther III, 1999. The suboxic zone of the Black Sea. In: *Environmental Degradation of the Black Sea: Challenges and Remedies.* S. Besiktepe, U. Unluata and A. Bologa, eds. NATO ASI Series 2., 75–92.

Neretin, L.V., I.I. Volkov, M.E. Bottcher, V.A. Grinenko, 2001. A sulfur budget for the Black Sea anoxic zone. Deep-Sea Res.I, 48, 2569–2593.

Nezlin, N.P., 2001. Unusual phytoplankton bloom in the Black Sea during 1998–1999: Analysis of remotely-sensed data, *Oceanology, (English transl.),* **41,** 375–380.

Oguz, T., 2002. Role of physical processes controlling oxycline and suboxic layer structure in the Black Sea, *Global Biogeochem. Cycles,* **16,** 10.1029/2001GB001465.

Oguz,T. and L. Rozman, 1991. Characteristics of the Mediterranean underflow in the southwestern Black Sea continental shelf/slope region. *Oceanology Acta,* **14(5)**, 433.444.

Oguz, T. and P. Malanotte-Rizzoli, 1996. Seasonal variability of wind and thermohaline driven circulation in the Black Sea: Modeling studies. *J. Geophysical Research,* **101,** 16551–16569.

Oguz, T., S., Besiktepe, 1999. Observations on the Rim Current structure, CIW formation and transport in the western Black Sea. *Deep Sea Res.I,* **46,** 1733.1753.

Oguz, T. and B. Salihoglu, 2000. Simulation of eddy-driven phytoplankton production in the Black Sea. *Geophys. Res. Letters,* **27,** 2125–2128.

Oguz, T., E. Ozsoy, M.A. Latif, H.I. Sur and U. Unluata, 1990a. Modeling of ydrographically controlled exchange flow in the Bosphorus Strait. *J. Phys. Oceanogr.,* **20,** 945–965.

Oguz,T., M.A. Latif, H. I. Sur, E. Ozsoy, and U. Unluata, 1990b. On the dynamics of the southern Black Sea. In: *The Black Sea Oceanography,* J. Murry and E. Izdar, eds, NATO/ASI Series, Kluwer Academic Publishers, Dordrecht, pp. 43–64.

Oguz, T., P., La Violette, U., Unluata, 1992. Upper layer circulation of the southern Black Sea: Its variability as inferred from hydrographic and satellite observations. *J. Geophys. Res.,* **97,** 12569–12584.

Oguz, T., V.S., Latun, M.A., Latif, V.V., Vladimirov, H.I., Sur, A.A., Makarov, E., Ozsoy, B.B., Kotovshchikov, V.V., Eremeev, U., Unluata, 1993. Circulation in the surface and intermediate layers of the Black Sea. *Deep Sea Res. I,* **40,** 1597–1612.

Oguz, T., D.G., Aubrey, V.S., Latun, E., Demirov, L., Koveshnikov, H.I., Sur, V., Diacanu, S., Besiktepe, M., Duman, R., Limeburner, V.V., Eremeev, 1994. Mesoscale circulation and thermohaline structure of the Black sea observed during HydroBlack'91. *Deep Sea Research I,* **41,** 603–628.

Oguz, T., P., Malanotte-Rizzoli, D., Aubrey, 1995. Wind and thermohaline circulation of the Black Sea driven by yearly mean climatological forcing. *J. Geophys. Res.,* **100,** 6846–6865.

Oguz, T., H. Ducklow, P. Malanotte-Rizzoli, S. Tugrul, N. Nezlin, and U. Unluata, 1996. Simulation of annual plankton productivity cycle in the Black Sea by a one-dimensional physical-biological model. *J. Geophys. Res.,* **101,** 16585–16599.

Oguz. T, L.I. Ivanov, S. Besiktepe, 1998. Circulation and hydrographic characteristics of the Black Sea during July 1992. In: *Ecosystem Modeling as a Management Tool for the Black Sea,* L. Ivanov and T. Oguz (eds). NATO ASI Series, Environmental Security-Vol.47, Kluwer Academic Publishers, Vol. 2, pp. 69–92.

Oguz, T., H. Ducklow, P. Malanotte-Rizzoli, J.W. Murray, V.I. Vedernikov, and U. Unluata, 1999. A physical-biochemical model of plankton productivity and nitrogen cycling in the Black Sea. *Deep-Sea Res. I.* **46,** 597–636.

Oguz, T., H. W. Ducklow, and P. Malanotte-Rizzoli, 2000. Modeling distinct vertical biogeochemical structure of the Black Sea: Dynamical coupling of the Oxic, Suboxic and Anoxic layers. *Global Biogeochem. Cycles.,* **14,** 1331–1352.

Oguz, T., J. W. Murray, and A. E. Callahan, 2001a. Modeling redox cycling across the suboxic-anoxic interface zone in the Black Sea. *Deep-Sea Res. I.,* **48,** 761–787.

Oguz, T., H. W. Ducklow, J. E. Purcell, and P. Malanotte-Rizzoli, 2001b. Modeling the response of top-down control exerted by gelatinous carnivores on the Black Sea pelagic food web. *J. Geophys. Res.,* **106,** 4543–4564.

Oguz, T., P. Malanotte-Rizzoli, and H.W. Ducklow, 2001c. Simulations of phytoplankton seasonal cycle with multi-level and multi-layer physical-ecosystem models: The Black Sea example. *Ecological Modelling,* **144,** 295–314.

Oguz, T., A. G. Deshpande, and P. Malanotte-Rizzoli, 2002a. On the Role of Mesoscale Processes Controlling Biological Variability in the Black Sea: Inferrences From SeaWIFS-derived Surface Chlorophyll Field. *Continental Shelf Research,* **22,** 1477–1492.

Oguz, T., P. Malanotte-Rizzoli, H.W. Ducklow, J.W. Murray, 2002b. Interdisciplinary studies integrating the Black Sea biogeochemistry and circulation dynamics. *Oceanography,* **15(3),** 4–11.

Oguz, T., T. Cokacar, P. Malanotte-Rizzoli, and H. W. Ducklow, 2003. Climatic warming and accopanying changes in the ecological regime of the Black Sea during 1990s. *Global Biogeochem. Cycles,* **17(3),** 1088,doi:10.1029/2003.

Ozsoy, E. and U. Unluata, 1997. Oceanography of the Black Sea: a review of some recent results. *Earth Sci.Rev.,* **42,** 231–272.

Ozsoy, E., U. Unluata and Z. Top, 1993. The Mediterranean water evolution, material transport by double diffusive intrusions and interior mixing in the Black Sea. *Prog. Oceanogr.,* **31,** 275–320.

Ozsoy, E., D. Di Iorio, M. C. Gregg, and J.O. Backhaus, 2001. Mixing in the Bosphorus Strait and the Black Sea continental shelf: observations and a model of dense water outflow. *J. Mar. Syst.,* **31,** 99–135.

Ozsoy, E., D. Rank, and I. Salihoglu, 2002. Pycnocline and deep mixing in the Black Sea: stable isotope and transient tracer measurements. *Estuarine, Coastal and Shelf Science,* **54,** 621–629.

Peneva, E., E. Stanev, V. Belokopytov, and P.Y. Le Traon, 2001. Water transport in the Bosphorus Strait estimated from hydro-meteorological and altimeter data: seasonal and decadal variability. *J. Mar. Syst.,* **31,** 21–33.

Polat, S.Ç. and S., Tugrul, 1995. Nutrient and organic carbon exchanges between the Black and Marmara Seas through the Bosphorus Strait. *Continental Shelf Research,* **15,** 1115–1132.

Porumb F., 1989. The influence of eutrophication on zooplankton communities in the Black Sea waters. *Certerari Marine,* **22,** 233.246.

Rachev, N.H., E.V. Stanev, 1997. Eddy processes in semi-enclosed seas. A case study for the Black Sea. *J. Phys. Oceanogr.,* **27,** 1581–1601.

Rass, T.S., 1992. Changes in the fish resources of the Black Sea. *Oceanology (Eng. Trans.),* **32,** 197–153.

Repeta, D.J. and D.J. Simpson, 1991. The distribution and recycling of chlorophyll, bacteriochlorophyll and carotenoids in the Black Sea. *Deep-Sea Res.,* **38,** Suppl.2A, S969-S984.

Repeta, D.J., D.J. Simpson, B.B Jorgensen, and H.W. Jannash, 1989. Evidence for anoxic photosynthesis from distribution of bacteriochlorophylls in the Black Sea. *Nature,* **342,** 69–72.

Ross D.A. and E.T. Degens, 1974. Recent sediments of the Black Sea. In: *The Black Sea-Geology, Chemistry, and Biology,* Degens.E.T and Ross.D.A, eds, American Association of Petroleum Geologists Memoir Vol.20, Tulsa, Oklahoma, U.S.A., pp. 183–199.

Rozanov, A.G. 1996. Redox stratification in Black Sea waters. *Oceanology (Engl. Transl.),* **35,** 500–504.

Rozanov, A.G., L.N. Neretin, and I.I. Volkov, 1998. Redox Nepheloid Layer (RNL) of the Black Sea: Its location, composition and origin. In: *Ecosystem Modeling as a Management Tool for the Black Sea.* L. Ivanov and T. Oguz, eds. NATO ASI Series, 2-Environmental Security-47, Kluwer Academic Publishers, Vol. 1, pp. 77–92.

Ryan, W.B.F. and W.C. Pitman, 1999. Noah's Flood: The new scientific discoveries about the event that changed history. Simon and Schuster, New York, 319p.

Ryan, W.B.F., W.C. Pitman, C.O. Major, K. Shimkus, V. Moskalenko, G.A. Jones, P. Dimitrov, N. Gorur, M. Sakinc, and H. Yuce, 1997: An abrupt drowning of the Black Sea shelf. *Marine Geology,* **138,** 119–126.

Ryan, W.B.F., C.O. Major, G. Lericolais, S.L. Goldstein, 2003. Catastrophic flooding of the Black Sea. *Annu. Rev. Earth Planet. Sci.* **31,** 525–554.

Saydam, C., S. Tugrul, O. Basturk, and T. Oguz, 1993: Identification of the oxic/anoxic interface by isopycnal surfaces in the Black Sea. *Deep Sea Research,* **40,** 1405–1412.

Shaffer, G., 1986: Phosphate pumpsand shuttles in the Black Sea. *Nature,* **321,** 515–517.

Shiganova, T.A., 1998: Invasion of the Black Sea by the ctenophore Mnemiopsis leidyi and recent changes in pelagic community structure, *Fish. Oceanogr.,* **7,** 305–310.

Shiganova, T.A., Yu. V. Bulgakova, S.P. Volovik, Z.A. Mirzoyan, and S.I. Dudkin, 2001: The new invader *Beroe ovata* Mayer 1912 and its effect on the ecosystem in the northeastern Black Sea. *Hydrobiologia,* **451,** 187–197.

Shushkina, E.A., M.E. Vinogradov, L.P. Lebedeva, T. Oguz, N.P. Nezlin, V. Yu. Dyakonov, and L.L. Anokhina, 1998: Studies of structural parameters of planktonic communities of the open part of the Black Sea relevant to ecosystem modeling. In: *Ecosystem Modeling as a Management Tool for the Black Sea,* vol. 1, NATO Sci. Partnership Sub-ser., 2, vol. 47, L.I. Ivanov, and T. Oguz, eds. Kluwer Acad., Norwell, Mass., pp. 311–326.

Simo, R., 2001. Production of atmospheric sulfur by oceanic phytoplankton: biogeochemical, ecological and evolutionary links, *TRENDS in Ecology & Evolution,* **16,** 287–294.

Sokolova, E., E.V. Stanev, V. Yakubenko, I. Ovchinnikov, R. Kosyan, 2001. Synoptic variability in the Black Sea. Analysis of hydrographic survey and altimeter data. *J. Mar. Syst.,* **31,** 45–63.

Sorokin, Y.I., 1972. The bacterial population and the process of hydrogen sulfide oxidation in the Black Sea. *J. Conseil Int. pour l'Explorotion de la Mer,* **34,** 423–454.

Sorokin, Y.I., 2002. *The Black Sea ecology and oceanography.* Backhuys Publishers, Leiden, 875 pp.

Spencer, D.W. and P.G. Brewer, 1971. Vertical advection, diffusion and redox potentials as controls on the distribution of manganese and other trace metals dissolved in waters of the Black Sea. *J. Geophys. Res.,* **76,** 5877–5892.

Stanev, E.V., 1990. On the mechanisms of the Black sea circulation. *Earth-Sci. Rev.,* **28,** 285–319.

Stanev, E. V. and J.M. Beckers, 1999. Numerical simulations of seasonal and interannual variability of the Black Sea thermohaline circulation. *J. Mar. Syst.,* **22,** 241–267.

Stanev, E. V. and N.H. Rachev, 1999. Numerical study of the planetary Rossby modes in the Black Sea. *J. Mar. Syst.,* **21,** 283–306.

Stanev, E.V., J.V. Staneva and V.M. Roussenov, 1997. On the Black Sea water mass formation. Model sensitivity study to atmospheric forcing and parameterization of some physical processes. *J. Mar. Sys.,* **13,** 245–272.

Stanev, E.V. and J.V. Staneva, 2000. The impact of the baroclinic eddies and basin oscillations on the transition s between different quasi-stable states of the Black Sea circulation. *J. Mar. Syst.,* **24,** 3–26.

Stanev, E.V. and E. Peneva, 2002. Regional sea level response to global climatic change: Black Sea examples. *Global and Planetary Changes,* **32,** 33.47.

Staneva, J.V. and E.V. Stanev, 2002. Water mass formation in the Black Sea during 1991–1995. *J. Mar. Syst.,* **32,** 199–218.

Staneva, J.V., K.O. Buesseler, E.V. Stanev, and H.D. Livinston, 1999. The application of radiotracers to a study of Black Sea circulation: validation of numerical simulations against observed weapons testing and Chernobyl ^{137}Cs data. *J. Geophys. Res.,* **104,** 11099–11114.

Staneva, J.V., D.E., Dietrich, E.V., Stanev, M.J., Bowman, 2001. Rim Current and coastal eddy mechanisms in an eddy-resolving Black Sea general curculation model. *J. Mar. Systs.,* **31,** 137–157.

Stanley, D.J., and C. Blanpeid, 1999. Late Quaternary water exchange between the eastern Mediterranean and the Black Sea. *Nature,* **285,** 537–541.

Sur, H.I. and Y.P. Ilyin, 1997. Evolution of satellite derived mesoscale thermal patterns in the Black Sea. *Prog. Oceanogr.,* **39,** 109–151.

Sur, H.I., E., Ozsoy, U., Unluata, 1994. Boundary current instabilities, upwelling, shelf mixing and eutrophication processes in the Black Sea. *Progr. Oceanogr.,* **33,** 249–302.

Sur, H.I., E., Ozsoy, Y.P., Ilyin, U., Unluata, 1996. Coastal/deep ocean interactions in the Black Sea and their ecological/environmental impacts. *J. Marine Systems,* **7,** 293–320, 1996.

Tebo, B.M., 1991. Manganese(II) oxidation in the suboxic zone of the Black Sea. *Deep-Sea Res.,* **38,** Suppl.2A, S883-S906.

Tebo, B.M., R.A. Rosson, and K.H. Nealson, 1991. Potential for manganese (II) oxidation and manganese (IV) reduction to co-occur in the suboxic zone of the Black Sea. In: *Black Sea Oceanography,* E.Izdar and J.W. Murray, eds. NATO ASI Series C-Vol.351, Kluwer Academic Publishers, pp. 173–186.

Titov, V.B., 2000. Dependence of the formation of the winter hydrological structure in the Black Sea on the severity of winter conditions. *Oceanology (Engl. Transl.),* **40,** 777–783.

Tugrul, S. 1993. Comparison of TOC concentrations by persulphate-UV and HTCO techniques in the Marmara and Black Seas. *Marine Chemistry,* **41,** 265–270.

Tugrul, S., O. Basturk, C. Saydam, and A. Yilmaz, 1992. Changes in hydrochemistry of the Black Sea inferred from water density profiles. *Nature,* **359,** 137–139.

Unluata, U., T. Oguz, M.A. Latif and E. Ozsoy, 1989. On the physical oceanography of Turkish straits. In: *The physical oceanography of sea straits,* L.J. Pratt, eds. NATO ASI Series, Kluwer, The Netherlands.

Uysal, Z., A. E. Kideys, L. Senichkina, L. Georgieva, D. Altukhov, L. Kuzmenko, L. Manjos, and E. Eker, 1998. Phytoplankton patches formed along the southern Black Sea coast in spring and summer 1996. In: *Ecosystem Modeling as a Management Tool for the Black Sea,* L. I. Ivanov and T. Oguz, eds. vol 1, Kluwer Academic Publishers, Netherlands, pp. 151–162.

Vedernikov, V.I. and A.B. Demidov, 1997. Vertical distributions of primary production and chlorophyll during different seasons in deep part of the Black Sea, *Oceanology, (English trans.),* **37,** 376–384.

Vinogradov, M.E., and Yu.R. Nalbandov, 1990. Effect of changes in water density on the profiles of physicochemical and biological characteristics in the pelagic ecosystem of the Black Sea. *Oceanology (English trans.),* **30,** 567–573.

Vinogradov, M.E., E.A. Shushkina, L.L. Anokhina, S.V. Vostokov, N.V. Kucheruk, and T.A. Lukashova, 2000. Mass development of the ctenophore *Beroe ovata* Eschscoltz near the northeastern coast of the Black Sea. *Oceanolgy (Engl. Transl.),* **40,** 52–55.

Yakushev, E.V, 1998. Mathematical modeling modeling of oxygen, nitrogen, sulfur and manganese cycling in the Black Sea. In: *Ecosystem Modeling as a Management Tool for the Black Sea,* L. Ivanov and T. Oguz, eds. NATO ASI Series, 2-Environmental Security-47, Kluwer Academic Publishers, Vol. 2, pp. 373–384.

Yakushev,E.V. and L.N. Neretin, 1997. One dimensional modeling of nitrogen and sulfur cycles in the aphotic zone of the Black and Arabian Seas. *J. Global Biogeochemical Cycles,* **11,** 401–414.

Yakushev, E.V., D.E. Besedin, Yu. F. Lukashev, and V.K. Chasovnikov, 2001. On the rise of the upper boundary of the anoxic zone in the density field of the Black Sea. *Oceanology (Eng. Transl.),* **41,** 654–659.

Yanko, V., J. Kennett, H. Koral, J. Kronfield, 1999. Stable isotope evidence from the Holocene Sea of Marmara sediments for two-way watermass interchange between the Black Sea and the Mediterranean Sea. *S. Afr. J. Sci.,* **95,** 201–204.

Yilmaz, A., S. Tugrul, C. Polat, D. Ediger, Y. Coban, and E. Morkoc, 1998a. On the production, elemental composition (C,N,P) and distribution of photosynthetic organic matter in the Southern Black Sea. *Hydrobiologia,* **363,** 141–156.

Yilmaz, A., O.A. Yunev, V.I. Vedernikov, S. Moncheva, A.S. Bologa, A. Cociasu, and D. Ediger, 1998b. Unusual temporal variations in the spatial distribution of chlorophyll-a in the Black Sea during 1990–1996. In: *Ecosystem Modeling as a Management Tool for the Black Sea.* L. Ivanov and T. Oguz, eds., NATO ASI Series, 2, Environment-Vol.47, Kluwer Academic Publishers, Vol. 1, pp. 105–120.

Yunev, O.A., V.I. Vedernikov, O. Basturk, A. Yilmaz, A. E. Kideys, S, Moncheva, and S. Konovalov, 2002. Long-term variations of surface chlorophyll-a and primary production in the open Black Sea, *Mar. Ecol. Prog. Ser.,* **230,** 11–28.

Zaitsev, Yu. and V. Mamaev, 1997. *Marine Biological Diversity in the Black Sea: A Study of Change and Decline,* GEF Black Sea Environmental Programme, United Nations Publications, 208pp.

Zatsepin, A.G., A.I. Ginzburg, A.G. Kostianoy, V.V. Kremenitskiy, V.G. Krivosheya, S.V. Stanichny, P-M. Poulain, 2003. Observations of Black Sea mesoscale eddies and associated horizontal mixing. *J. Geophys. Res.,* **108,** 3246, doi:10.19/2002JC001390.

Chapter 34. SEAS OF THE ARABIAN REGION (29,S)

CLAUDIO RICHTER

Center for Tropical Marine Ecology, Bremen, Germany

AHMAD ABU-HILAL

Yarmouk University, Irbid, Jordan, and University of Bahrain

Contents

1. Introduction

The seas around the Arabian Peninsula as important repositories of marine biodiversity and non-living resources contribute significantly to the economic, social and cultural prosperity of the region (Gladstone et al., 1999; Sheppard et al., 1992). Their rich resources and services provide subsistence and commercial food supplies to the local communities, domestic and international tourist destinations, important international shipping routes, as well as a rich cultural heritage. Population growth rates in the bordering countries are high by global standards, and industrial and urban development are growing rapidly in many places. The northern Red Sea, for example, boasting some of the most spectacular coral reefs in the world, is subjected to sweeping rates of tourism putting growing pressures on the marine environment (Saleh, 1995). The anthropogenic pressures are particularly damaging, as the extreme environmental régime of temperature and salinity already imposes natural stresses which are close to the physiological limits of the organisms (Gladstone et al., 1999; Sheppard et al., 1992). The restricted water exchange in the almost land-locked Red Sea and Arabian (or Persian) Gulf, render these marginal seas particularly vulnerable.

The Sea, Volume 14, edited by Allan R. Robinson and Kenneth H. Brink
ISBN 0–674–01527–4

The unique features of the marginal seas, such as the warm deep waters of the Red Sea, and the geopolitical importance of the region has spurred intense research activities over the past years, encompassing all marine disciplines. Earlier work has been summarized i.a. by Edwards and Head (1987), Reiss and Hottinger (1984), Sheppard et al. (1992), and Sheppard and Dixon (1998). Complementary to these reviews, and Johns et al.'s (2000) recent report on the physical oceanography of the region, we here provide an updated overview of the marine ecology of the Arabian region, including recent results of interdisciplinary research programs on the physical, chemical and biological oceanography of the area.

2. Geography and Geology

2.1. General features

The Arabian Peninsula lies wedged between marginal seas of very different bathymetry and outline (Fig. 34.1). On the western side of the peninsula lies the elongated trench of the Red Sea extending north-south over 18° of latitude. Its deep central channel of over 2000 m depth is interrupted by a narrow and shallow entrance strait at Bab el Mandeb, which leads into the deep Gulf of Aden linking into the Indian Ocean. At the eastern end of the region, the Arabian Gulf differs notably in that it is quite shallow (less than 65 m deep) over most of its area. Its entrance strait (Strait of Hormuz) notably has no blocking sill. The Arabian Gulf/Hormuz complex also has a deep gulf (Gulf of Oman) linking it to the Indian Ocean which is very similar bathymetrically to the Red Sea's Gulf of Aden. The long littoral region of the southern Arabian Peninsula, the Arabian Margin, well known for intense upwelling filaments extending several hundred kilometers offshore, stretches along the Yemeni and Omani coasts and links the marginal sea complexes (Johns et al., 2000).

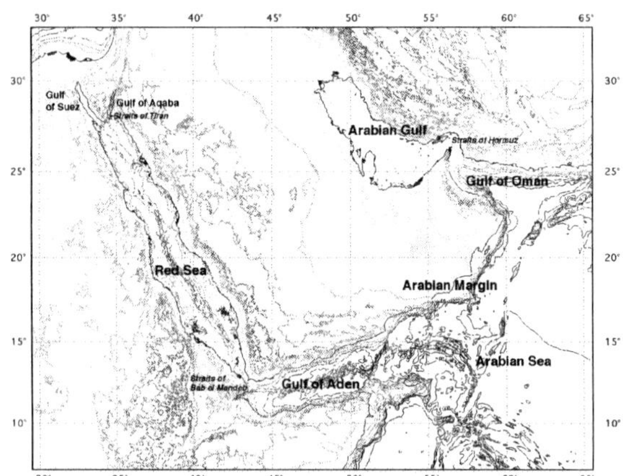

Figure 34.1 Map of the study area, showing the desert-enclosed Red Sea, Arabian/Persian Gulf and straits connecting to adjacent gulfs. (Source: PANGAEA, http://www.pangaea.de/Software/PanMap/)

2.2. *Arabian Gulf*

The Arabian Gulf is a semi-enclosed, relatively shallow body of water lying entirely within an arid zone north of the Tropic of Cancer. It is bordered by Iran and seven Arab countries including Iraq, Kuwait, Saudi Arabia, Bahrain, Qatar, United Arab Emirates and Oman. Its eastern coastline is marked by mountains and cliffs, whereas the western shore is often sandy (Fig. 34.2).

The Gulf is separated from the Gulf of Oman by the Strait of Hormuz, which represents the bottleneck of the Gulf with only 56 km width at its narrowest point. The trough reaches 100 m through the Strait and deepens to more than 2000 m within 200 km of the Gulf of Oman, outside the Strait. The length of the Gulf is about 1000 km while the maximum width is only 338 km between the United Arab Emirates and Iran. The depth of most of the Gulf is less than 65 m, and the mean depth is only 35 m, whilst the region between Kuwait and Qatar is much shallower. The surface area is about 240,000 km^2 and the average volume is 8,630 km^3.

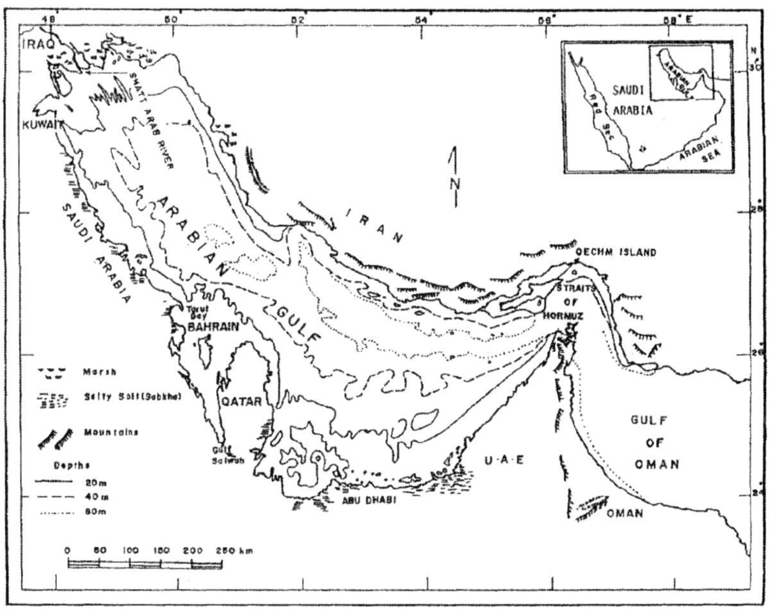

Figure 34.2 Topographic features of the Arabian Gulf. (Reprinted from Al-Muzaini and Jacob, 1996, © 1996, with permission from Elsevier)

The Gulf is relatively young geologically speaking, having been formed 3 or 4 million years ago. The Gulf basin was originally shaped by a Tertiary fold system, which caused its deepest depression to run along its northern side. The northwest – southeast axis of the Gulf separates a stable and gradually sloping Arabian shoreline from a geologically unstable and steeper Iranian shoreline (Purser and Seibold, 1973). Several islands in the Southern Gulf and in the Strait of Hormuz are the product of salt diapirism, due to thick (>2 km) salt deposits sandwiched between cenozoic to mesozoic sediments and the pre-Permian continental basement (Khattab, 1995).

During the Pleistocene of the Quaternary, global and extensive sea level fluctuations occurred causing sea level variations of 30 to 80 m below present sea level. During this period, the deeper parts of the Gulf were flooded and substantial reef building and development occurred, and strong biogenic limestone formation took place. Many of these reef areas were then uplifted and the relatively mobile salt deposits below the carbonate formations were forced to flow upwards by the younger and denser rocks, which had accumulated on top of them. The domes of the uplifted (over lied) carbonates became much closer to light, and less susceptible to sedimentation than most of the Gulf's sea floor. Consequently, coral communities were able to grow on these uplifted hard base limestone formations.

During the late Pleistocene, the Gulf dried out almost completely, but reconnected with the Indian Ocean during the Holocene transgression known as *Flandrian Transgression.* The lowest sea level (-120 m) occurred 20,000 years ago, when only a finger of water from the Gulf of Oman was passing the Strait of Hormuz along the Iranian shore. The Gulf proper was dry, just before the sea level rose again to its present levels during the Holocene transgression about 5,000 years ago (Kassler, 1973). Due to the gradual topography of the Gulf and of the conditions favouring biogenic carbonate production, the Gulf became a sedimentary basin, dominated mainly by soft biogenic sediments produced mainly by foraminifera and other microorganisms, and to a lesser extent by corals, calcareous algae and other limestone producing flora and fauna. However, terrigenous sediments from the Tigris, Euphrates and the Karun rivers are present at the northwestern portion of the Gulf near the Shatt Al-Arab, affecting wide areas off Kuwait and Iraq. The carbonaceous sediments predominate along the coasts of Saudi Arabia, Bahrain, Qatar and the United Arab Emirates (UAE). On the Iranian side of the Gulf, these sediments are mixed with substantial amounts of terrigenous sediments carried out by the wind and riverine inputs from the Zagros Mountains.

2.3. Red Sea

With a surface area of ~450,000 km^2, the Red Sea is almost twice as large as the Arabian Gulf, but much deeper along the axial rift reaching a maximum depth of 2,850 m. The Red Sea extends over a length of almost 2,000 km in NNW-SSE direction with a maximum width of 354 km in its southern section, between Eritrea and Saudi Arabia.

The Red Sea is separated from the Gulf of Aden by the Strait of Bab el Mandeb, stretching only 29 km across at its narrowest point. A shallow (130 m) sill 140 km north of the Strait restricts the circulation to the upper layers, effectively cutting the Red Sea off the cold deep waters of the world ocean.

In the southern part of the Red Sea the shelf is very wide, extending up to more than 150 km offshore (Fig. 34.3). North of Port Sudan, the shelf narrows considerably, restricting coral reef development to the coastal fringes. Away from the coast, the bottom drops off to about 500 m in the central trough of the Red Sea, sloping gradually to 1000 m near the center axis. The narrow central rift or axial trough is about 1500 m deep along most of its length, with a number of deep pits reaching over 2500 m depth.

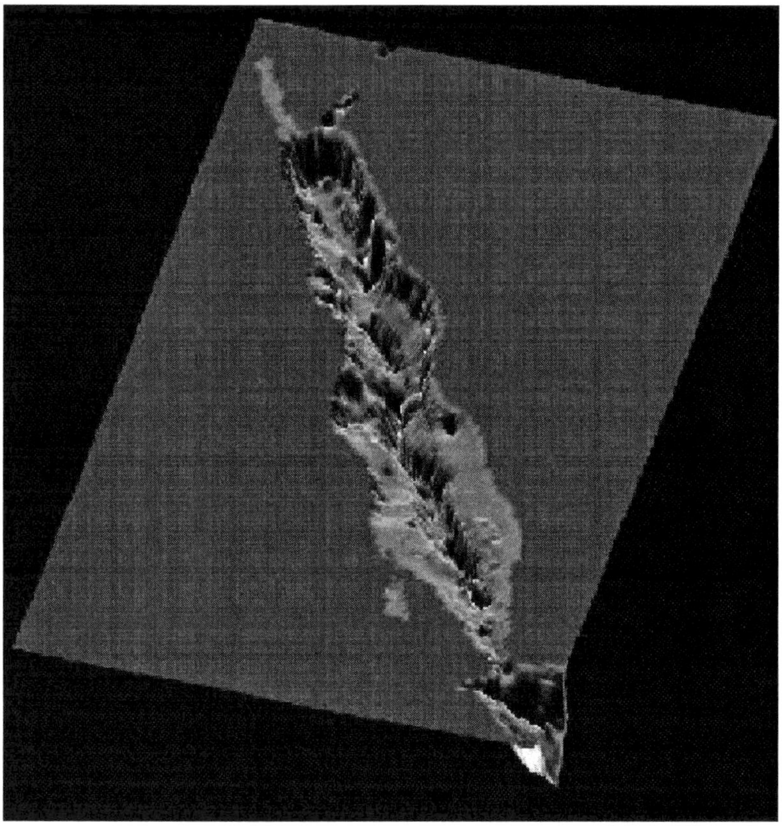

Figure 34.3 3-D rendering bathymetry of the Red Sea, showing the deep axial trough, broad southern and narrow northern shelves, the shallow Gulf of Suez and deep Gulf of Aqaba. (Courtesy S. E. Welsh)

The central rift extends northeast from the Red Sea into the Gulf of Aqaba. Separated from the Red Sea proper by a narrow and shallow sill in the Straits of Tiran (sill depth 265 m, Hall, 1975), the Gulf of Aqaba provides in itself a 1:10 model of the Red Sea. In spite of its less than 30 km width, the steep-sided gulf reaches more than 900 m throughout most of its 180 km length, resulting in a unique depth to width ratio (Hulings, 1989; Morcos, 1970) and a maximum depth of more than 1800 m. By contrast, the Gulf of Suez, forming the shallow northwestern extension of the Red Sea, is a shallow 55–70 m deep sedimentary basin, similar to the Arabian Gulf.

No rivers discharge into the Red Sea, and high evaporation and negligible rainfalls render the 250,000 km^3 of water one of the most saline waters in the ocean, the bulk of which features salinities in excess of 40.

The shape, bathymetry, and structure of the Red Sea resulted from the Cenozoic break-up of the Arabian-African plate, creating both, the deep rift in the zone of crustal expansion between the continental plates and the uplift of mountain ranges along the entire length of the western and eastern shores of the Red Sea.

The Red Sea forms part of a much larger rift system, extending from the Zagros-Taurus Mountains in Turkey through the Red Sea into the East African Rift Valley (Gregory, 1896), and from the Gulf of Aden into the Indian Ocean.

While the rift became established by the Oligocene, the area underwent periods of alternating emersion and flooding. In the late Miocene, the Red Sea was isolated from the Indian Ocean by a narrow land bridge. A shallow connection across the Isthmus of Suez provided only intermittent supply of Mediterranean water, so that entire Red Sea basin became a large evaporation pan with substantial precipitation of salt deposits. Tectonic activities during the Pliocene resulted in the formation of the deep axial trench, closure of the Isthmus of Suez and opening of Bab el Mandeb, connecting the Red Sea with the Indian Ocean.

Large sea-level changes during the Pleistocene Ice Ages with alternating periods of dry and wet climates (Arz et al., 2003; Naqvi and Fairbanks, 1996) resulted in strong oscillations between marine and extremely hypersaline conditions, the latter with salinities comparable to the ones in the Dead Sea. Present sea-levels were reached only 5,000 years ago, so that the Red Sea biota today are evolutionary very young.

3. Climate

3.1. General features

The dominant phenomenon affecting the oceanography and meteorology of the region is the summer monsoon, extending from the African coast, half way up the Red Sea, all along the Yemeni-Omani coast, across the entire Arabian (or Persian) Gulf eastward into India (Johns et al., 2000). The monsoon forcing is most intense over the coastal Arabian Sea and southern Red Sea, while the northern Red Sea and Arabian Gulf are influenced more strongly by continental patterns. The largest seasonal changes in wind forcing occur in the coastal Arabian Sea, where the strong SW winds of the summer monsoon alternate with the NE winter monsoon winds (Fig. 34.4).

The Somali coast and northern part of the Arabian Peninsula are subjected mainly to the SW monsoon, and the intensity of these winds diminishes greatly into the Gulf of Aden and Arabian Gulf. Conversely, the winter monsoon extends well into the Gulf of Aden and southern Red Sea, causing a seasonal reversal in the winds over this entire region.

Orographic influences on the atmospheric circulation are very significant, due to the high mountains bordering much of the coastline in this region (Fig. 34.1): The Ethiopian highlands and coastal mountain range of northern Somalia focus the SW monsoon winds offshore of Somalia and the Arabian Peninsula and shield much of the Gulf of Aden from strong SW monsoon forcing (Johns et al., 2000). In winter these mountains are responsible for steering the NE monsoon winds into the Gulf of Aden and southern Red Sea. High mountain ranges along the full length of the Red Sea cause the winds to be closely aligned along the axis of the Sea, except at a few locations in the central Red Sea where gaps in the mountains exist. The Zagros Mountains of Iran and the Jebel al Akhdar range in northern Oman play a similar role in focusing the winds over the Arabian Gulf and southern Gulf of Oman.

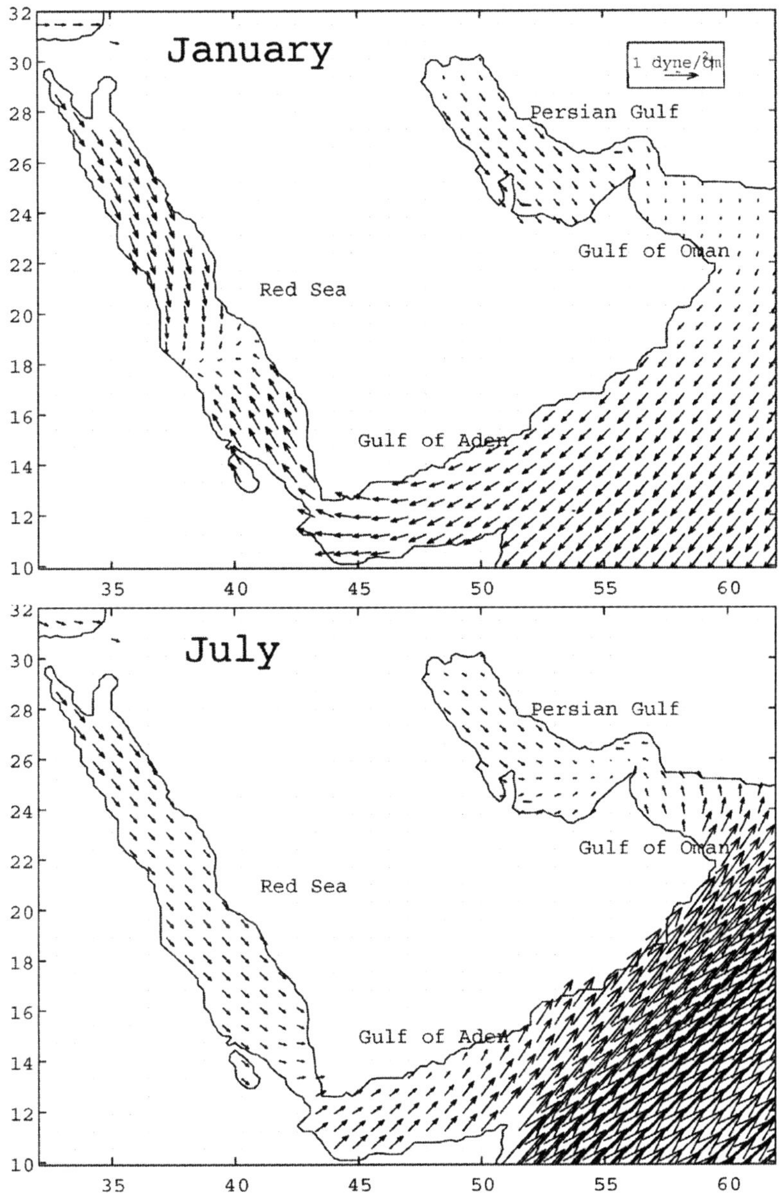

Figure 34.4 Climatological winds over the seas around the Arabian peninsula at the height of the NE and SW monsoon periods (January and July, respectively). (Johns et al., 2000)

The winds and the extreme aridity of the bordering lands enhance the moisture flux from the seas to the atmosphere. Extreme evaporation rates in the order of ~2 m yr^{-1} result in dense high salinity water masses driving the large scale circulation in the marginal seas, and filling the bulk of their interior volumes.

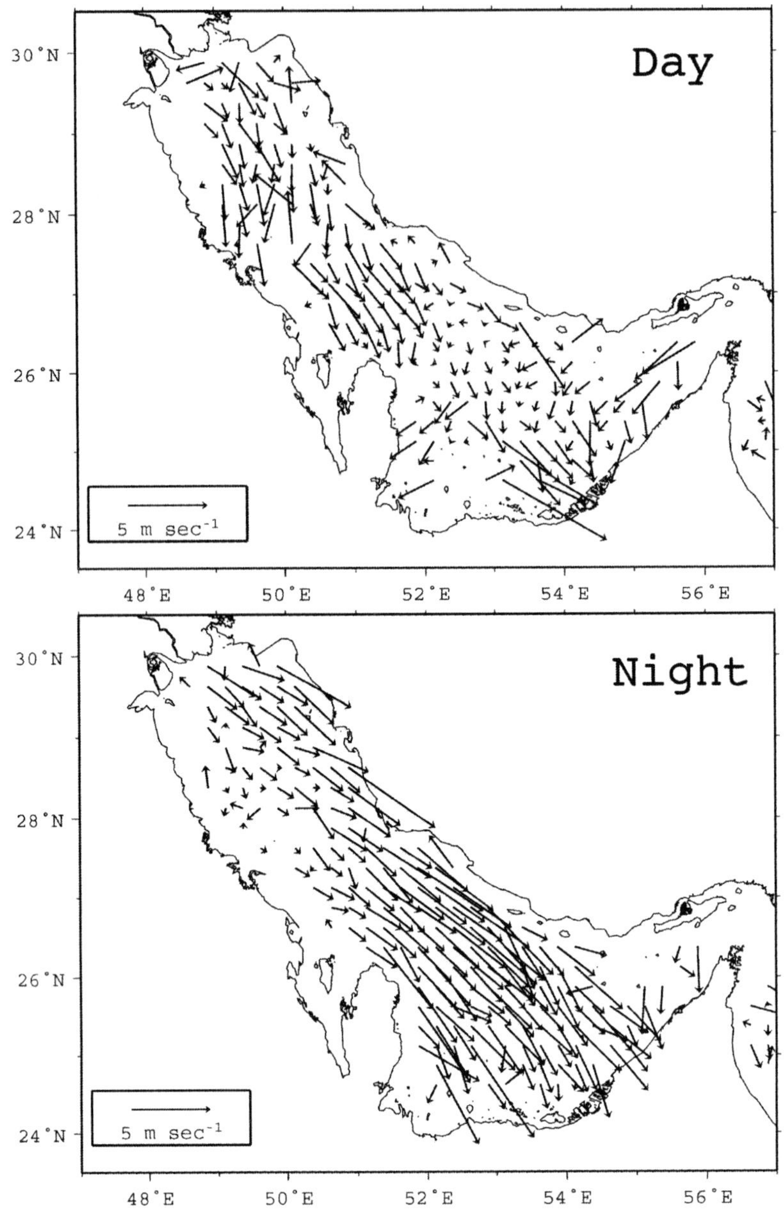

Figure 34.5 Average day (top) and night (bottom) winds over the Arabian Gulf for June 1997. (Johns et al., 2000)

3.2. Arabian Gulf

The Gulf is located between latitudes 24–30°N where most of the Earth's deserts are also found. The region consists of low-lying central areas enclosed by the Mountains of Taurus in southern Turkey, the Pontic Mountains in northeast Turkey, the Caucasus Mountains of Iran, and the Hajar and Hejaz Mountains of the

Arabian Peninsula. The cold air arrives in the Gulf area via Turkey, then across the eastern part of the Mediterranean Sea, then it travels either over or around the low mountains in Syria and Lebanon, and finally into the Tigris-Euphrates Valley. The surrounding orography also affects the airflow within the Arabian region; the sharply rising mountains to the north and east, and more gently rising to the west and southwest from a northwest – southeast axis that strongly influence the tracks of extra-tropical storms from northwest to southeast. This brings about the most interesting and well-known weather phenomenon in the Arabian Gulf Region, called *Shamal* which occurs all the year around. *Shamal* is an Arabic word meaning north. In the meteorological context it is referred to as the seasonal northwesterly winds that occur during winter as well as summer in the region. The summer *Shamal* is practically continuous from early June through July and is associated with the relative strength of the Indian and thermal lows. The winter *Shamal* occurs mainly during November to March and is associated with mid-latitude disturbances traveling from west to east. The cold and strong northwesterly winds mainly occur during December to February. The winter *Shamal* brings some of the strongest winds and highest seas of the season at most locations in the Arabian Gulf. This wind is related to synoptic weather systems to the northwest and can exceed 20 knots. Winds in the area, ahead of an approaching cold front, blow from the southeast and are called *Kaus* in Arabic and *Shakki* in Persian. These slowly increase in intensity as the front approaches. They may reach gale force before the passage of the cold front and with the onset of the *Shamal.* Due to the channeling effects of the low-level air flow by the Zagros Mountains, in western Iran, the strongest of these southerly winds occur on the eastern side of the Arabian Gulf. The *Shamal* occurs first in the northwest part of the Gulf and then spreads south and east behind the advancing cold front. Average wind speeds in the Gulf, range from 30 to 40 knots, with peak winds in excess of 50 knots in the long duration *Shamal,* compared to the speeds of the winds over the northern part of the Gulf which tend to be 5 to 15 knots less than the above values, on an average. The effect of local winds is considered more important to marine biota, especially in terms of local winds exerting stress, than the broad scale wind systems (Sheppard, 1993).

Along the entire coast of the Arabian Gulf, in summer, strong winds build-up in the afternoons from sea breeze effects, where the hot land heats overlying air, which rises due to convection, bringing surface flows and adds a landward component to all winds which some times reach tens of m s^{-1} compared to <5 m s^{-1} during most of the winter (Fig. 34.5). This land breeze can occur also at night, although less frequent, and have powerful cooling effects on the intertidal and coastal areas (Murty and El-Sabh, 1984; Reynolds, 1993; Sheppard, 1993).

Evaporation in the Gulf is much higher than both the river inflow and precipitation. Estimated evaporation ranges between 144 and 500 cm yr^{-1} (Privett, 1959). Meshal and Hassan (1986) estimated an evaporation rate of 200 cm yr^{-1}. Measurements show that most evaporation occurs in winter, and mainly due to the high wind speed. The river runoff from Tigris, Euphrates, and Karun is 1456 m^3 s^{-1} whereas other Iranian rivers provide an average of 2034 m^3 s^{-1}. The total runoff is 110 km^3 yr^{-1}, equivalent to 46 cm yr^{-1} (Reynolds, 1993), which is higher than 16 cm yr^{-1} values previously used by (Al Hajiri, 1990). The annual rainfall in the Gulf region is small, providing only about 7 cm yr^{-1} and thus making a negligible contribution to the fresh water budget of the Gulf.

3.3. Gulf of Oman and Arabian Margin

The climate in the Gulf of Oman is noticeably different from the climate in the Arabian Gulf. While the Gulf is affected mainly by the extra-tropical weather systems from the northwest, the Gulf of Oman is at the northern edge of the tropical weather systems in the Arabian Sea and the Indian Ocean (cf. Naqvi et al., this volume). This monsoon circulation produces southerly winds in summer and strong northerlies in winter. Thus the Strait of Hormuz regions form the boundary between the two systems.

3.4. Red Sea

The Red Sea climate is generally hot and arid. With an average annual air temperature of 34°C, Danakil Depression near the Eritrean side of the Red Sea, is the hottest place on earth (National Geographic Society, 2002). Air temperatures in the central Red Sea (Jeddah) reach their maximum during July-August (37/27°C, avg. monthly high/low) and their minimum during January-February (29/18°C). In the northern Red Sea and adjoining Gulfs of Suez and Aqaba, temperature extremes may lie up to 40°C apart. Because of the desert climate, diel variations in air temperature may also be very high there, often exceeding 15°C. Mean annual rainfall ranges from <4 cm yr^{-1} in Aqaba to >10 cm yr^{-1} near Bab el Mandeb. All rainfall occurs during the period October-May and 60% of the total occurs during December-February. The Red Sea receives no river runoff or any other major fresh water input, except for flash floods following heavy rains with only local effects.

Wind patterns are governed by the seasonal strengths and positions of the monsoonal trough and Mediterranean low pressure systems, and are strongly affected by the surrounding orography (Figs. 34.1 and 34.4). Mountain ranges on both sides of the Red Sea and Gulf of Aqaba channel air flow along the center axes of the basins. Winds blow consistently from the north to northwest from May to September along the entire length of the Red Sea. Wind speeds average >7 m s^{-1} in the northern Red Sea, and around 1 m s^{-1} in the south (Patzert, 1974a). During October to April, the seasonal advance of the low pressure trough towards the north causes a reversal of winds south of 20°N, blowing from the south to southeast during that period. SSE winds may reach 6 m s^{-1}, NNW winds are around 4 m s^{-1}.

Evaporation is estimated at 2.1±0.2 m yr^{-1}, i.e. orders of magnitude higher than precipitation, resulting in a volume loss of 0.03 Sv (1 Sverdrup = 10^6 m^3 s^{-1}) over the whole area of the Red Sea (Sofianos et al., 2002).

The heat flux into the Red Sea is positive in summer and negative in winter. In January, the net heat flux between 14–28° N was shown to vary between -20 and 240 W m^{-2}, where positive values denote heat losses to the atmosphere. In August, the net heat flux in the Red Sea ranged between -120 and -160 W m^{-2}. Insolation and wind contribute to higher heat fluxes in the southern portions of the Red Sea, both winter and summer. Recent estimates give an annual heat loss of 11±5 W m^{-2}, but in the light of the wide range of estimates in the literature (less than -80 W m^{-2} to more than 20 W m^{-2}) the issue seems not finally settled (Sofianos et al., 2002).

The seasonality in heat fluxes leads to a more or less pronounced seasonality in the vertical density structure of the water column, particularly in the northern Red

Sea, Gulf of Suez and Gulf of Aqaba, where stratified waters in May-November alternate with a mixed water column in between December and April (Genin et al., 1995; Wolf-Vecht et al., 1992).

4. Tides, Circulation and Mixing

4.1. General aspects

The previous chapters have highlighted important differences, e.g. in the atmospheric forcing and bathymetric constraints between the Red Sea and the Arabian Gulf complexes, affecting tides and circulation patterns (Johns et al., 2000): tidal currents are of first order importance inside the Arabian Gulf but of second order inside the Red Sea, and first order in both entrance straits. Hypersaline deep-water production appears to occur in totally different areas of the two marginal seas due largely to bathymetric contrasts.

4.2. Arabian Gulf

Due its shallow topography, the Arabian Gulf is influenced by both, wind driven and thermohaline forces. The details of the interior circulation of the Arabian Gulf, however, remain essentially unobserved, not only for the coastal currents but also in the broad interior.

Tides in the Gulf co-oscillate with those in the Strait of Hormuz, which opens into the deep Gulf of Oman, the tides of which then co-oscillate with those in the Arabian Sea. The tide range is large, throughout the Gulf, with values greater than 1m everywhere and exceeding 3 m at Shatt Al-Arab (Lehr, 1984). Off Kuwait, spring tidal ranges are 2 m in the south and up to 4 m in the north (Jones, 1986a) while off Bahrain the range is 2 m at extreme spring (Jones, 1986b).

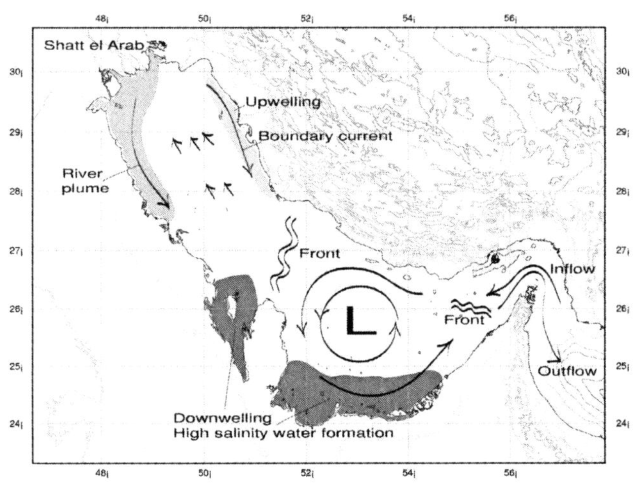

Figure 34.6 General circulation scheme of the Arabian Gulf. (Sources: PANGAEA, Johns et al., 2000)

The tides in the Gulf are complex standing waves and the dominant pattern varies from being primarily semi-diurnal to diurnal. The dimensions of the Gulf are such that resonance amplification of the tides can occur, and the result is that the semi-diurnal constituents have two amphidromic points, at the northwest and southeast ends, and the diurnal constituents have a single amphidromic point in the centre, near Bahrain (Lehr, 1984). Tides are important for stirring and mixing water vertically and on a horizontal scale of 10 km, but they do not make an important contribution to the residual circulation of the Gulf. Tidal currents averaging longer than a day have negligible residual energy, and as a result, basin-scale advection from tides is not considered by general circulation models. Tides are important on smaller scales (<10 km) of horizontal length and time (<24 h).

The persistent southward wind stress in the northern half of the Gulf sets up coastal current régimes along the Iranian (upwelling) and Saudi (downwelling) coasts (Fig. 34.6).

The latter, Saudi-Emirate coastal current, appears to be amplified by freshwater input from the Iraqi Shatt Al-Arab waterway. The overall circulation process inside the Gulf is counterclockwise. Surface water moves westwards along the Iranian coast in the north, and eastwards along the Emirates coast in the south. The surface water gains salt as it moves from the mouth (Strait of Hormuz) inwards (Wooster et al., 1967). A number of sources for the deep-water formation have been suggested in the shallow southern part of the Gulf (Johns et al., 2000). Sinking of saltier and denser water starts in the northwestern part of the Gulf. Dense saline water originating from the semi-enclosed lagoons in the south increases the salinity of this deep water (Linden et al., 1990) and exits below Indian Ocean surface water (IOSW) entering from the Gulf of Oman (Fig. 34.7). Based on true measurements and model analysis, Reynolds (1993) concluded that the inflow current along the Iranian coast is weakened by *Shamal* winds in winter, but in summer it strengthens and extends almost to the head of the Gulf. Cyclonic circulation gyre fills the southern Gulf and is driven by the inflowing surface water through the Strait of Hormuz. Runoff from Shatt Al-Arab in the northwest Gulf maintains a cyclonic circulation where otherwise an anti-cyclonic gyre would be expected. A southward coastal jet exists between the head of the Gulf and Qatar, and extends east of Qatar, depending on the wind. In the southern Gulf, the mean wintertime surface currents' flow is mainly density driven, with surface flow inward from the Strait of Hormuz and adjacent to the Iranian coast; whereas a southward coastal flow is present along the entire southern coast, which stagnates east of Qatar, where high evaporation and sinking forms a dense bottom flow to the northwest and out of the Strait of Hormuz. In the northern Gulf, circulation is predominantly wind driven, with surface flow along both coasts in a southerly direction. In summer, the surface inflow is recognizable as far north as 28°N, and the outflow from Shatt Al-Arab is carried by counter-clockwise circulation in a westerly direction and down the Kuwait and Saudi Arabian coast. A persistent thermal front across the Arabian Gulf about the latitude of Qatar appears related to the thermohaline exchange through the Hormuz Strait (Johns et al., 2000). Estimates of residence times vary between 2–5 years (Hughes and Hunter, 1979; Hunter, 1983).

Tides, winds and waves, and evaporation are the three major factors affecting mixing of the water column in the Arabian Gulf. The strong tidal currents create a turbulent friction layer near the bottom which may extend upwards into the water column, as evidenced by homogenous water properties in the hydrographic sections (Fig. 34.7). Wind friction creates a surface shear layer and also creates surface wave fields, both of which form a surface mixed layer. Evaporation of fresh water enhances mixing by increasing the salinity and density of surface water, and reducing the surface stability. In winter, combined cooling and evaporation increase the density sufficiency to overturn the water column, north of Qatar, to create a well-mixed water column.

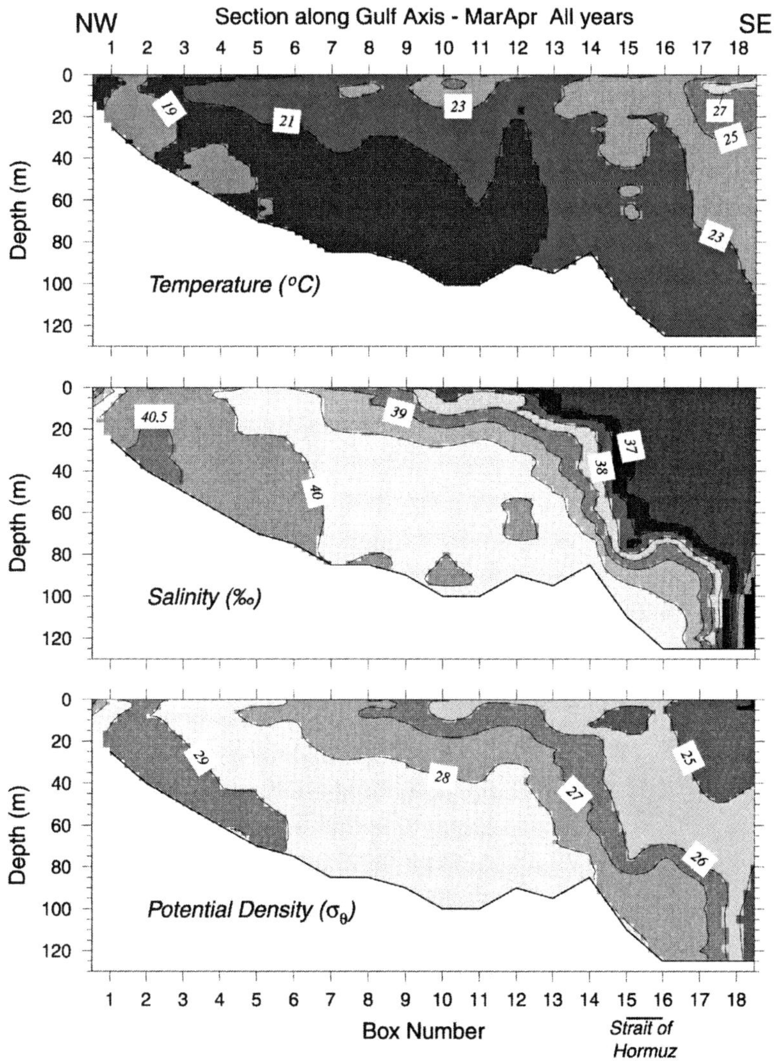

Figure 34.7 Axial section of the Arabian Gulf in winter (March–April) showing cold, saline water outcropping at the sea surface of the Tigris-Euphrates-Karun delta and warm, low salinity IOSW at the sea surface moving towards the head of the Gulf. Local density inversions are likely an artifact of binning irregularly sampled data. (Swift and Bower, 2003, reproduced by permission of American Geographical Union)

4.3. Strait of Hormuz and Gulf of Oman

The Strait of Hormuz differs from Bab el Mandeb in the absence of any sill to confine the flow such as in the Red Sea. The circulation in and out of the strait is not well understood. After transiting the Strait of Hormuz, the Arabian Gulf deep water (PDW) cascades downslope into the Gulf of Oman, forming sub-mesoscale eddies (*Peddies*) as it equilibrates and moves toward the Indian Ocean (Johns et al., 2000; Fig. 34.8). The surface inflow layer, rather than being laterally uniform, occurs preferentially on the eastern (Irani) side, while the outflow occurs preferentially on the western (Omani) side into an organized coastal current that transits along the Emirate and Omani coast to the major promontory of Ras al Hadd (Swift and Bower, 2003). A major front and accompanying jet form at the collision of this Gulf outflow and the eastward flowing coastal current along the southern Arabian margin (Boehm et al., 1999). The Ras al Hadd jet is highly variable, sometimes extending out to the east as shown in the figure and extending northeastward or southeastward at other times. During the SW monsoon the transport of the Ras al Hadd jet is believed to be at least 10 Sv (Elliot and Savidge, 1990). This feature is also referred to as the Ras al Hadd front because it forms the seasonal boundary between the northern Arabian Sea and the Gulf of Oman. The Ras al Hadd front appears to be the major physical-biological oceanographic feature in the region separating a cyclonic gyre in the southern Gulf of Oman from an anticyclonic gyre in the bounding northern Arabian Sea.

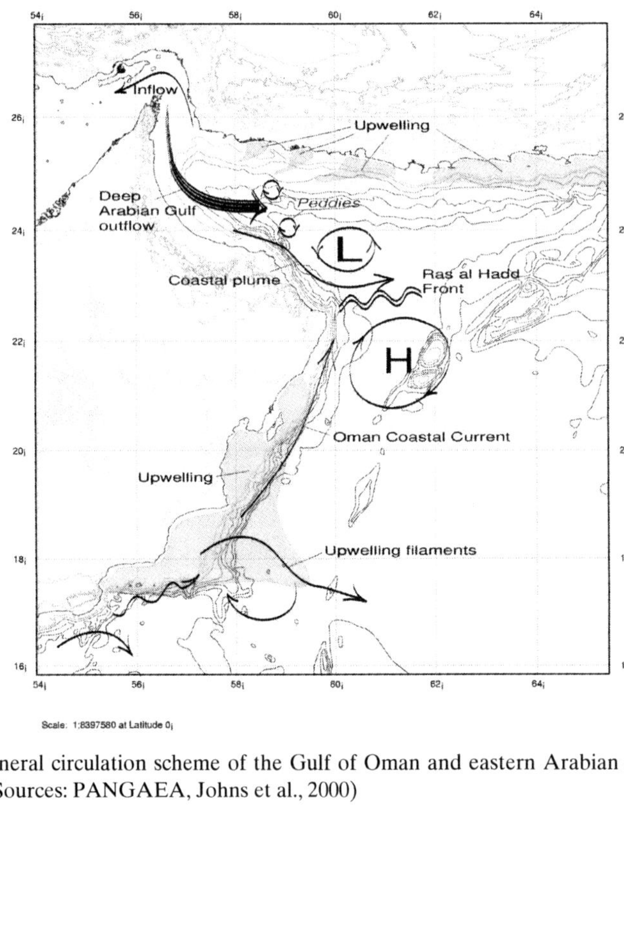

Figure 34.8 General circulation scheme of the Gulf of Oman and eastern Arabian margin during the SW monsoon. (Sources: PANGAEA, Johns et al., 2000)

Other important elements involved in the Gulf of Oman's circulation are the seasonal, but interannually very variable, upwelling along the coast of Iran to the north and the complicated mesoscale dynamics associated with the extension of the south coastal Oman upwelling system into filaments extending off Ras al Hadd (Johns et al., 2000). The latter is complicated by the shallow Murray Ridge that extends across the mouth of the Gulf (Quraishee, 1984).

4.4. Arabian Margin and Gulf of Aden

Upwelling favorable winds along the coasts of Somalia, Yemen and Oman begin with the onset of the SW monsoon in May, associated with a northeastward flowing Oman Coastal Current (OCC) whose speeds reach 40 cm s^{-1} and extend 200 km offshore (Fig. 34.8). The OCC current turns offshore to the east off Ras al Hadd, entraining filaments and jets capable of exporting cool and extremely productive upwelled waters hundreds of km offshore. The reversal to the NE or winter monsoon in November causes the reversal of the OCC, so that a southward current then extends all along the northeast coast of Oman, around Ras al Hadd, and continues south to about 20°N. Thereafter the current is entrained into offshore squirts and jets and disintegrates into southward propagating eddies (Johns et al., 2000). A detailed account of the oceanography of the Arabian margin is found in Naqvi et al. (this volume).

The oceanography of the Gulf of Aden is under the influence of both the Arabian Sea and the Red Sea. The latter (as the Arabian Gulf) acts as an inverted estuary, where dense salty water formed by evaporation and convection eventually flows out into the adjoining Gulf of Aden.

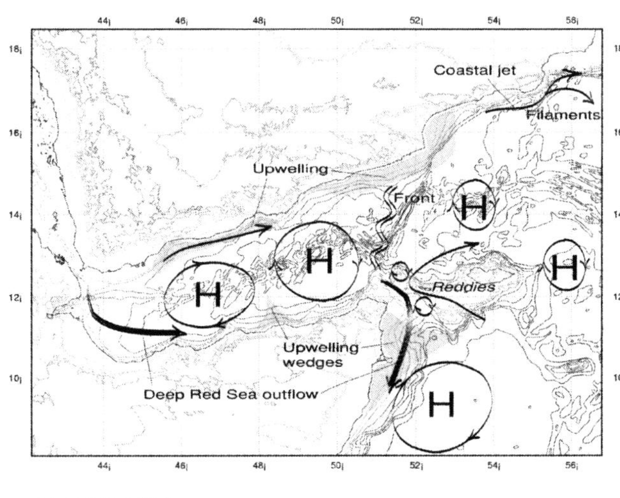

Figure 34.9 General circulation scheme of the Gulf of Aden. (Sources: PANGAEA, Johns et al., 2000)

The dense outflow spills down to about 600 m into the Gulf of Aden and is thought to evolve into mesoscale eddies (*Reddies* in Fig. 34.9; Johns et al., 2000).

The Gulf of Aden is seen in recent oceanographic observations to be 'choked up' with large, energetic, deep-reaching mesoscale eddies that fundamentally influence the spreading rates and pathways of intermediate-depth Red Sea Water (RSW; Bower et al., 2002).

These features appear to propagate westward from the mouth of the Gulf toward the Red Sea and, according to anomalous water properties in the center of an anticyclonic eddy, their origin may be linked to the propagation and decay of eddy features in the Somali Current. All eddies investigated reached nearly to the 1000–2000 m deep sea floor, with speeds as high as 0.2–0.3 m s^{-1}, extending through the depth range of RSW.

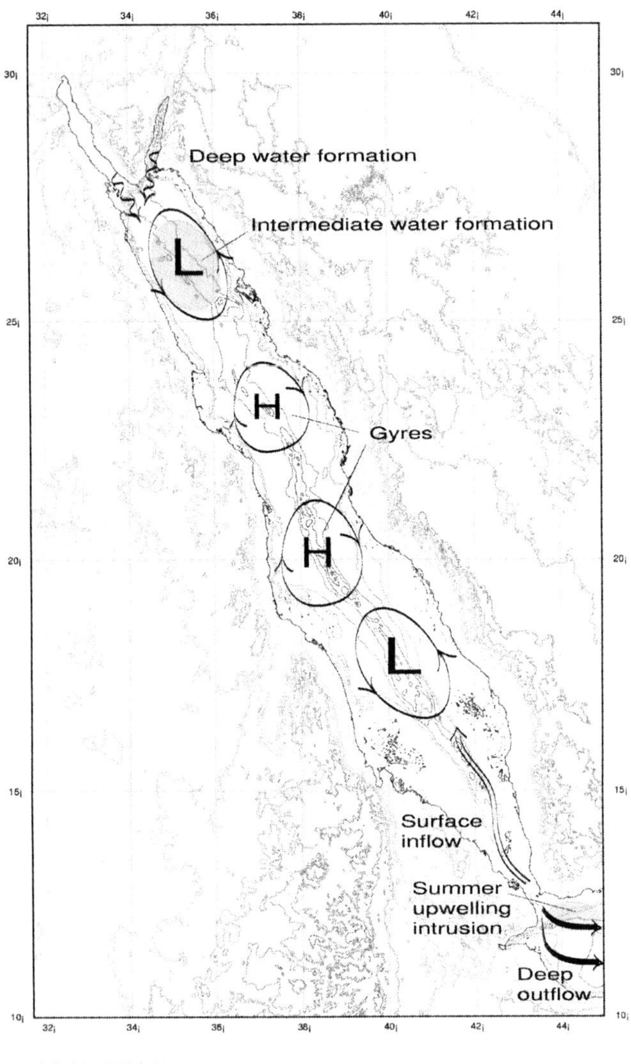

Figure 34.10 General circulation scheme of the Red Sea. (Sources: PANGAEA, Johns et al., 2000)

4.5. Red Sea

Because of the restricted connection through the Straits of Bab el Mandeb to the Gulf of Aden, tidal ranges in the Red Sea are rather small, with some amplification of tides on the shallow shelves in the southern Red Sea near the Straits of Bab el Mandeb and in the northern extremes of the gulfs of Suez and Aqaba (max. spring tides around 1 m; Genin and Paldor, 1998; Morcos, 1970). The tides are predominantly semidiurnal (period of 12.4 h), and account for up to 30% of the variance in the longshore circulation in the Gulf of Aqaba (Genin and Paldor, 1998).

Circulation in the Red Sea and adjoining gulfs is mainly governed by spatial and temporal gradients in winds, sea level and sea water density. The negative water balance due to excess evaporation drives an anti-estuarine circulation across Bab el Mandeb, with a surface inflow of comparatively fresh Gulf of Aden water (GASW) into the Red Sea, and an outflow of saline Red Sea water (RSW) over the bottom of the sill into the Gulf of Aden (Figs. 34.10 and 34.11).

RSW spreading into the Arabian Sea and southward along the African coast is considered one of the most important intermediate water masses in the Indian Ocean, whose characteristic temperature and salinity maximum has been observed as far south as the retroflection region of the Agulhas Current (Beal et al., 2000; Fine et al., 2000).

The anti-estuarine circulation across Bab el Mandeb is typical for the winter (Siedler, 1969), where the northeast monsoon piles up water in the Gulf of Aden, and both, the pressure gradient and SSE winds in the southern part of the Red Sea combine to drive a strong surface inflow into the Red Sea (Fig. 34.11).

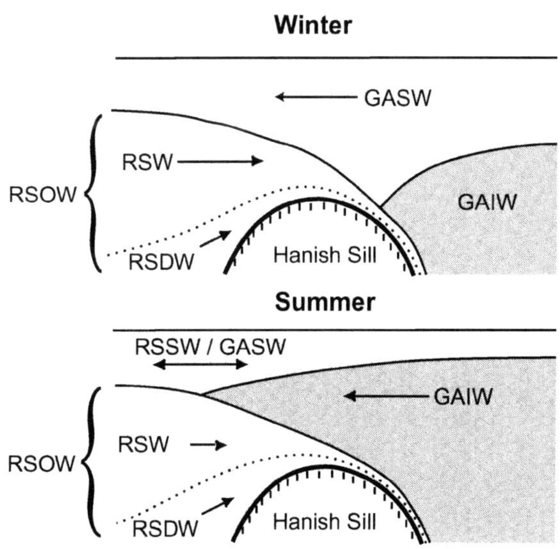

Figure 34.11 Summer and winter exchange across Bab el Mandeb. GASW: Gulf of Aden Surface Water, GAIW: Gulf of Aden Intermediate Water, RSSW: Red Sea Surface Water, RSW: Red Sea Water, RSDW: Red Sea Deep Water, RSOW: Red Sea Overflow Water (RSW+RSDW). (Reprinted from Siddall et al., 2002, © 2002, with permission from Elsevier)

Beyond 17–18°N the surface inflow meets opposing winds proceeding north at reduced velocities (Patzert, 1974a). In summer, the surface flow reverses under the combined effect of NNW winds, blowing along the entire length of the Red Sea, and a weak pressure gradient in response to both, lowered sea-level differences and an uplift of the denser deep waters in the Gulf of Aden due to upwelling in the Gulf of Aden. The outflow of RSW weakens significantly during that period and both, shallow surface and deep water outflows are compensated by an inflow of subsurface Gulf of Aden Intermediate Water (GAIW) into the Red Sea, giving rise to a three-layered flow pattern (Maillard and Soliman, 1986) (Figs. 34.11 and 34.12).

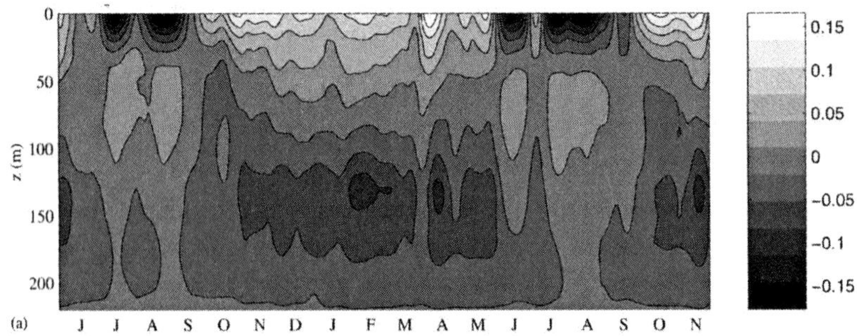

Figure 34.12 Seasonality in circulation across Bab el Mandeb (Perim narrows south of Hanish sill) showing a 3-layered pattern during the summer SW monsoon and a 2-layered pattern during the winter NE monsoon. Negative values (dark shades) denote Red Sea outflow, positive values (light shades) inflow of Gulf of Aden waters. (Reprinted from Sofianos et al., 2002, © 2002, with permission from Elseviers)

Indirect estimates of the transport of RSW through Bab el Mandeb Strait suggest an annual mean transport of 0.33 Sv (Siedler, 1969), varying from approximately 0.6 Sv in winter to nearly zero in late summer (Patzert, 1974b). These estimates are corroborated by recent measurements showing maximum outflow of RSW in February of more than 0.6 Sv and a minimum of 0.05 Sv in August, associated with an intrusion of GAIW of about 0.3 Sv (Murray and Johns, 1997; Sofianos et al., 2002). The estimated net inflow of 0.036 Sv agrees well with the yearly evaporation flux of 0.03 Sv.

A connection with the Mediterranean has been established with the construction of the Suez Canal, completed in 1869. It is insignificant in terms of the bulk volume flow, but non-negligible in terms of gene-flow of Red Sea biota into the Mediterranean, coined *Lessepsian migration* (Por, 1978).

The flow through the Suez Canal is determined by differences in sea level between the Red Sea and Mediterranean Sea, local winds, and the salinity of the canal waters due to the solution of the evaporite layers in the Bitter Lake. The sea level is higher in Suez (Red Sea) than in Port Said (Mediterranean), resulting in a surface flow from the Red Sea to the Mediterranean except in July-September, when pressure gradient and flow direction reverse (Wüst, 1934).

The horizontal circulation inside the Red Sea, based on drifters and model results, appears to be a number of both, transient and semi-permanent energetic gyres or eddies distributed along the length of the basin (Fig. 34.10; Quadfasel and

Baudner, 1993; Saad et al., 1999). A cyclonic gyre located in the northern Red Sea is believed to be associated with the formation of intermediate waters in winter (Clifford et al., 1997). In the central Red Sea, two anti-cyclonic eddies are locked topographically near 18–19°N and 23–24°N (Quadfasel and Baudner, 1993). Non-permanent cyclonic and anti-cyclonic eddies in the southern Red Sea, associated with seasonally changing wind fields in this area, may be involved in the formation of intermediate waters. With velocities of up to more than 0.5 $m\ s^{-1}$ these eddies override the ~0.1 $m\ s^{-1}$ surface flows associated with the large-scale thermohaline circulation of the Red Sea.

Seasonally reversing surface flows are associated with alternating downwelling and upwelling periods in the northern Red Sea during winter and summer, respectively.

Wyrtki (1974) proposed three different sources of RSW: overflow of Gulf of Aqaba water (GAW) across the Strait of Tiran, inflow from the Gulf of Suez (GSW), and open ocean convection in the northern Red Sea. While deep convection in the open northern Red Sea has never been directly observed, more recent work shows that both, overflow GAW and winter convection south of the Sinai peninsula are important for the deep water formation in the Red Sea (Cember, 1988), refuting earlier assumptions that the Strait of Tiran would be a barrier separating the Gulf of Aqaba from the circulation in the Red Sea (Neumann and McGill, 1962). Woelk and Quadfasel (1996) showed that plume convection from the Gulf of Suez may contribute significantly to RSW. Recent work combining CTD, chlorofluorocarbon (CFC) tracer and direct current measurements show that the deep water masses in the Red Sea are separated into two parts (Plaehn et al., 2002): The upper part is dominated by the inflow from the Gulf of Suez marked by a salinity maximum, spreading into the Red Sea at a core depth of about 900. The lower part is mainly formed by the dense outflow from the Gulf of Aqaba, marked by a CFC-12 anomaly at the bottom of the Red Sea, below the isopycnal σ_θ=28.6. With temperatures >21.3°C and salinities >40.5, the Red Sea features the warmest and saltiest bottom waters of all oceans.

Outflow of GAW across the Strait of Tiran is estimated at 0.03 Sv, and balanced by a inflow of Red Sea surface water into the Gulf of Aqaba (Murray et al., 1984).

The residence time of the deep waters (1.35×10^{14} m^3) have recently been estimated at 30–45 years (Plaehn et al., 2002), about half as long as previously assumed (Woelk and Quadfasel, 1996; Wyrtki, 1974).

A unique feature of the Red Sea are the hot brine pools in various sea-floor depressions along the axial trough due to solution and geothermal heating of mineral deposits through vents in the ocean crust (Degens and Ross, 1969). The resulting brine is dense enough to remain at the ocean floor even at very high temperatures. Values of nearly 70°C have been recorded (Hartmann et al., 1998a), together with brine salinities in excess of 300, differing however markedly in chemical composition from the true RSW salinity. The transition from high saline brine to normal seawater salinity extends over a depth of only a few decimeters or less, resulting in a staircase structure of the water column between convective layers of distinct temperature and salt properties (Fig. 34.13).

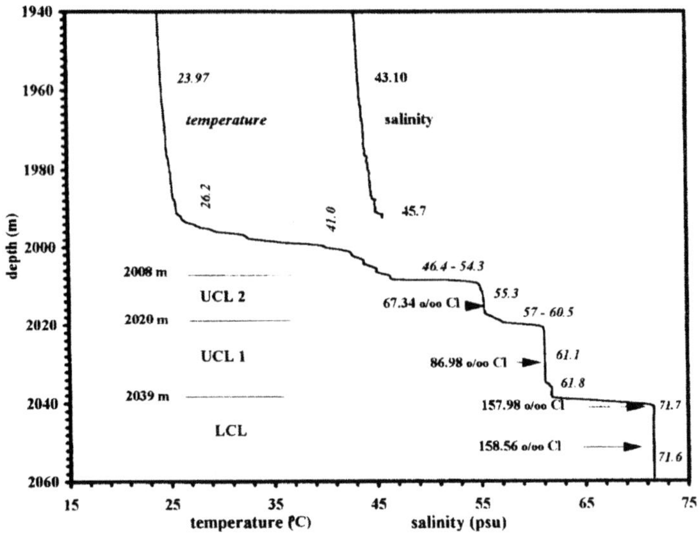

Figure 34.13 Temperature and salinity profiles in the Atlantis II Deep SW basin (Reprinted from Hartmann et al., 1998b, © 1998, with permission from Elseviers). Note that the curves are offset by about +4°C and 3 salinity units due to a calibration error (Hartmann et al., 1998a), but the general staircase shape of the profile remains valid.

Interannual variations in temperatures and brine levels suggest ongoing hydrothermal activities at in some of the Deeps (Fig. 34.14).

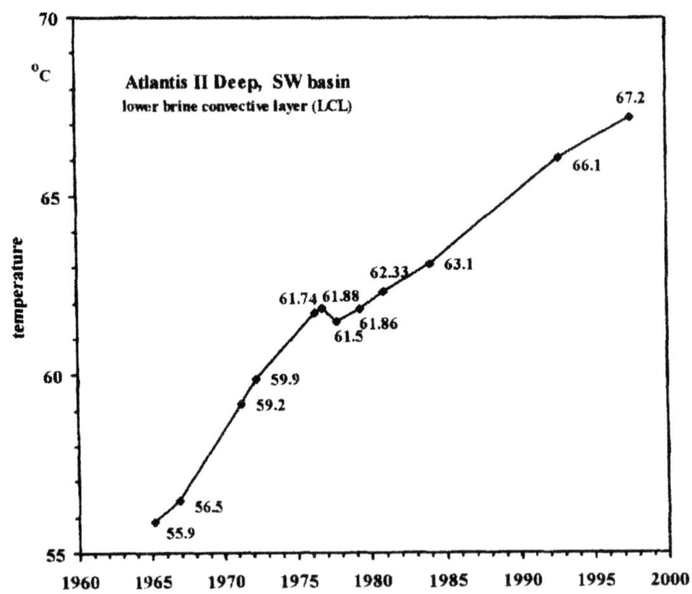

Figure 34.14 Temperature change in the Atlantis II Deep lower brine between 1965 and 1997. (Reprinted from Hartmann et al., 1998a, © 1998, with permission from Elseviers)

5. Productivity

5.1. General aspects

The productivity of the seas around the Arabian peninsula is largely governed by the geographical setting and physical forcing, notably the Arabian monsoon affecting primarily the coastal Arabian Sea and southern Red Sea, continental patterns affecting the northern Red Sea and Arabian Gulf, and the degree of closure of the marginal seas due to straits and sills. The external energies injecting nutrients into the productive surface layers thus vary significantly between the above régimes: the large-scale wind pattern forcing coastal upwelling in the Gulf of Aden, Arabian margin and Iranian coast of the Gulf of Oman; convective mixing in the northern and/or shallow areas of the Red Sea and Arabian Gulf; and oceanographic fronts and the presence and geometry of straits constraining the flow of nutrients between the basins. The occurrence of N^2-fixing cyanobacteria suggests that nitrogen is limiting productivity in stratified waters (Post et al., 2002), whereas the supply of dust-borne iron may be limiting productivity in the monsoonal upwelling areas (Tamburini et al., 2003).

5.2. Arabian Gulf

The Gulf is considered a moderately productive body of water, with primary production averaging 124 gC m^{-2} y^{-1} (Jones et al., 2002). However, considerably higher values can often be found in mixed central waters and shallow bays and creeks (arab. *Khores*). Primary production is particularly high in waters influenced by the Shatt Al-Arab, northwest of the Gulf, where it can reach up to 3 gC m^{-3} h^{-1} in eutrophied canals of the Shatt el Arab. The highest coastal values (up to 0.6 gC m^{-3} h^{-1}) were recorded during intense red tides (Rao et al., 1999).

Nutrient contents of the Arabian Gulf waters are lower than those of the Gulf of Oman due to the upwelling that occurs in the Gulf of Oman. A phosphate concentration of 0.47 mmol m^{-3} was reported for the Gulf of Oman compared to 0.36 mmol m^{-3} for the Strait of Hormuz and a range of 0.12–0.23 mmol m^{-3} for the Arabian Gulf between Kuwait and United Arab Emirates (El-Samra, 1988). The same trend was also reported for nitrate, which was 0.41 mmol m^{-3} for the Gulf of Oman waters compared to 0.30–0.26 mmol m^{-3} for the rest of the Gulf. There is evidence of nutrient limitation in some areas of the Gulf, even though most of the seabed lies within a strongly illuminated area (Jones, 1985). Productivity within the Arabian Gulf is limited by nitrogenous nutrients and not silicate (El-Samra, 1988).

Marked seasonal and regional variations were observed in the abundance and species distribution of the dominant phytoplankton (Dorgham, 1991). Species diversity is high, while standing crop is low. A total of 514 species were recorded, among which diatoms were represented by 284 taxa. Dinoflagellates, the second dominant group, were represented by 225 taxa. Densities of total phytoplankton in the Gulf water between Saudi Arabia and Karan Island, amounted to 100–150 $\times$ 10^6 cells m^{-3} in summer and 150–250 $\times$ 10^6 cells m^{-3} in winter (Basson et al., 1977). Much lower counts were reported in Qatari waters ranging between 2.01–115 $\times$ 10^6 cells m^{-3} (Dorgham, 1991). Chlorophyll *a* concentrations in the NW Arabian Gulf range between 0.17 mg m^{-3} at the height of the stratified period (October) and 4.05 mg m^{-3} during winter mixing (January) (Al-Saadi et al., 1976).

Zooplankton diversity is somewhat lower than in the Arabian Sea, featuring more than 100 genera and species (Michel et al., 1986a), predominantly copepods (40 species) and coelenterates (28 species). Zooplankton densities are similar to phytoplankton in showing pronounced temporal and regional variations. Abundances were highest in the Central Gulf with up to 3000 ind. m^{-3} (Sheppard, 1993). Biomass of the bongo net samples (wet weight) varied between 34 mg m^{-3} in autumn and 290 mg m^{-3} in summer. Most zooplankton groups reached maximum abundance in summer (June–August) and some forms produced secondary peaks in fall (October) (Michel et al., 1986b).

The distribution of foraminifera in the sediment of the Arabian Gulf suggests that the foraminiferal assemblages depend mainly on the supply of allochthonous materials: The high diversity on the Iranian side of the Gulf is attributed to appreciable amounts of nutrients from drainage in the topographically high hinterland compared to low faunal diversity on the Arabian side, as the arid Arabian hinterland is practically deprived of active drainage. Relatively low diversity occurs in the deep parts of the Gulf. Lower diversity is near the Strait of Hormuz and highest diversities are near the mouth of Shatt Al-Arab, which suggests that the amount of nutrients provided by drainage debouching landlocked basins is an important factor determining the constitution of the foraminifera of bottom-sediments (Cherif et al., 1997).

5.3. Red Sea

Due to its desert-enclosed setting and shallow sill, the Red Sea is nutrient-poor and rather unproductive, except for the coral reefs flourishing along its margins and on parts of its southern shelves (Edwards and Head, 1987; Reiss and Hottinger, 1984). In the absence of rivers, nutrient inputs into the Red Sea are limited to the advection of fertile Gulf of Aden water across the narrow and shallow straits of Bab el Mandeb and, to some extent, the deposition of wind-borne dust and biological fixation of N_2 (Naqvi et al., 1986).

During the SW monsoon, the nutrient-rich Gulf of Aden subsurface waters carry important amounts of nutrients into the Red Sea, and turbulent mixing due to current shear between the three-layered flow pattern enhances primary production at the depth of the Deep Chlorphyll *a* Maximum (DCM) layer (Hansen et al., 1992). The strongest inflow of extrinsic organic material, by contrast, occurs during the NE monsoon period, where plankton-rich Gulf of Aden surface waters enter the Red Sea (Halim, 1984). Large cyanobacteria and diatom blooms have been observed in the Gulf of Aden and southern Red Sea during this season (Veldhuis et al., 1997), with average primary production values (1.5 mg C m^{-2} d^{-1}) three times higher than during the SW monsoon (Wiebinga et al., 1997) and zooplankton stocks at their annual maximum (Van Couwelaar, 1997). Mixing with the much saltier and warmer Red Sea water results, however, in heavy mortalities in the expatriated plankton (Beckmann, 1984; Weikert, 1987) with important material losses from the surface layer due to the bulk fall-out of carcasses and other biogenic matter. The strong vertical density gradient in the area of fresh Gulf of Aden water overlaying Red Sea subsurface waters limits the supply of new nutrients due to mixing from below, so that the northbound surface waters become progressively oligotrophic. The general pattern of reduced fertility with increasing distance from the fertile Gulf of Aden is modulated by mesoscale features. Very low nutrient and productivity levels in the central Red Sea are associated with a topographically and geographically locked quasi-stationary anticyclonic eddy near 23°N (Quadfasel and Baudner, 1993; Quadfasel and Verch, 1987). By contrast, a cyclonic gyre near 27°N, as well as boundary mixing along the northern shelf margins and the overall weak vertical density stratification in the northern Red Sea explain the higher productivity in this area (Quadfasel and Baudner, 1993; Weikert, 1987).

The waters of the Gulf of Aqaba - often considered a 1:10 replica of the Red Sea proper, similarly separated from the adjoining deep waters by a shallow sill (Hulings, 1989; Reiss and Hottinger, 1984) - are particularly oligotrophic. Below the shallow wind-mixed surface layer the water column is nearly isothermal. The weak density stratification leads to a pronounced seasonality in the northern gulf, where deep convective mixing may extend up to more than 800 m depth in winter (Genin et al., 1995; Reiss and Hottinger, 1984; Wolf-Vecht et al., 1992). Winter mixing accounts for the very dense and well oxygenated deep waters of the Gulf of Aqaba (Klinker et al., 1978). The nutrients mixed into the surface layers set the stage for a marked annual cycle in pelagic production, featuring a more or less pronounced spring bloom between March and May (average 0.8 mg Chl *a* m^{-3}, Genin et al., 1995; Labiosa et al., 2003), low summer concentrations between June and September (<0.2 mg Chl *a* m^{-3}) and intermediate concentrations between 0.3–0.5 mg Chl *a* m^{-3} during the fall and winter months (Fig. 34.15). The timing, duration and intensity of the sping bloom appears to be directly related to the heat flux in the preceding winter, and may vary considerably between years. For example, an anomalously strong, delayed and prolonged bloom followed the exceptionally cold winter after the eruption of Mount Pinatubo, with peak concentrations of more than 3 mg Chl *a* m^{-3} (Genin et al., 1995).

At all times, more than 95% of the photosynthetic biomass is due to <8 µm ultraplankton (Lindell and Post, 1995; Sommer et al., 2002), amounting to 99.99% of total algal numbers (Sommer et al., 2002). Seasonal succession proceeds from small eukaryotic to minute prokaryotic algae (Lindell and Post, 1995): <8 µm Cryptophyte and Chlorophyte algae dominate the phytoplankton community during the winter months, <2 µm blue-green *Synechococcus* in spring, and <1 µm *Prochlorococcus* in summer (Fig. 34.16). The recurrence of this pattern between years contrasts marked short-term but also inter-annual variability in total integrated biomass, suggesting complex re-adjustments of the phytoplankton community to differential bottom-up (light, nutrients) and top-down (grazing, viral lysis) control.

Local and brief decoupling of nutrient-enhanced primary production from the double lock of grazing and viral lysis may help explain episodic massive blooms of diatoms, coccolithophores and also pteropods in the area (Kimor and Golandsky, 1977; Reiss and Hottinger, 1984; Winter et al., 1979), but the often great depths at which these blooms occur (300 m and 400 m in the case of diatoms and coccolithophorids, respectively; Kimor and Golandsky, 1977; Winter et al., 1979) has remained enigmatic.

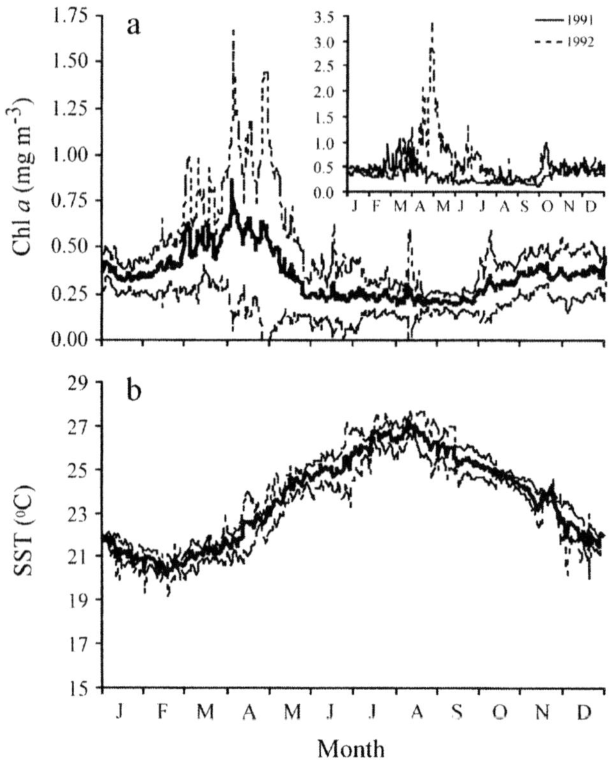

Figure 34.15 Seasonal changes in thermal stratification and convective mixing (shown in the seasonality of sea surface temperatures, SST) drive phytoplankton dynamics (Chlorophyll *a*) in the northern Gulf of Aqaba, Red Sea. Inset shows the strong phytoplankton bloom following the exceptionally cold winter of 1991. (Labiosa et al., 2003, © 2003 by the American Society of Limnology and Oceanography)

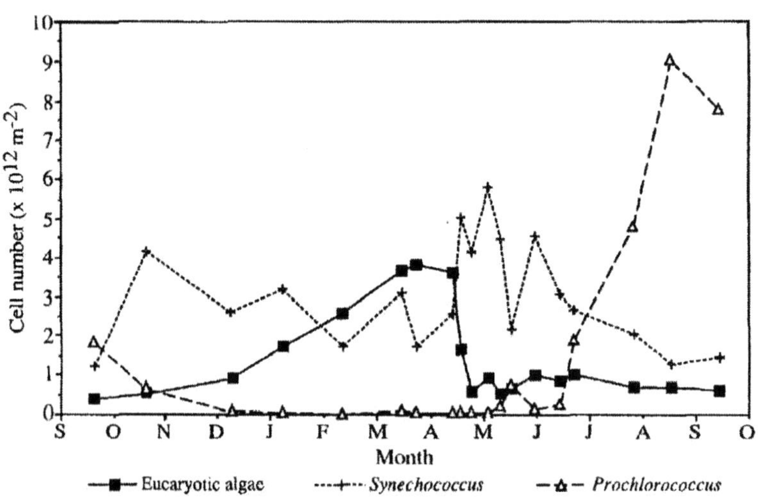

Figure 34.16 Seasonal abundances of ultraphytoplankton in the Gulf of Aqaba, integrated over a 600 m water column. (Lindell and Post, 1995, © 1995 by the American Society of Limnology and Oceanography)

Nutrient levels in the euphotic zone (~1 mmol N m^{-3} and 0.2–0.3 mmol P m^{-3}) are driven below detection limits in summer (Badran, 2001; Badran and Foster, 1998; Klinker et al., 1978). Nitrogen-limitation fosters the development of the diazotrophic filamentous blue-green *Trichodesmium* at the height of the stratification in summer and early fall (Post et al., 2002), as well as symbiotic associations of other N^2-fixing cyanobacteria with diatoms and dinoflagellates (Gordon et al., 1994; Kimor et al., 1992). Nitrogen-to-phosphorous (N:P) ratios at or below the Redfield ratio of 16 (Badran, 2001; Badran and Foster, 1998; Levanon-Spanier et al., 1979) suggest that the limiting nutrient in stratified waters is N. However, for the Red Sea as a whole, the N:P ratio appears to be higher than the Redfield ratio (Naqvi et al., 1986), and enhanced alkaline phosphatase activities of *Trichodesmium* spp. populations (Stihl et al., 2001) indicate that P-stress may ultimately govern the final successional stages of phytoplankton in the Red Sea.

The above seasonality finds its spatial equivalent in the spring time distribution of physical and biological properties between the well-mixed Gulf of Aqaba and the stratified Red Sea proper (Fig. 34.17).

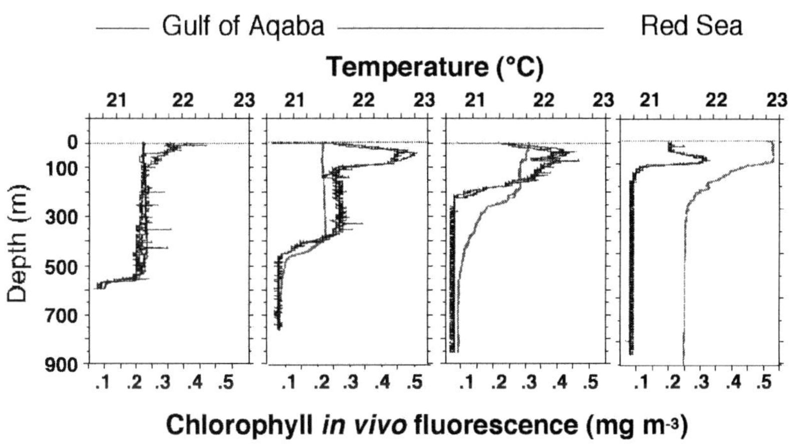

Figure 34.17 Spatial distribution of temperature (light lines) and phytoplankton chlorophyll *a* (dark ragged lines) between the Gulf of Aqaba (left) and the Red Sea proper (right) during early March 1999.

Although the section extends across less than 250 km, the northern Gulf of Aqaba features a typical winter situation with a deep mixed layer and moderately high phytoplankton stocks, while the Red Sea shows a typical summer situation with a DCM at the bottom of the mixed layer. Physical mixing and phytoplankton production explain the shallowing of the Chl *a* profiles along the transect, with highest integrated values in the re-stratifying waters of the central gulf, while post-bloom nutrient depletion in the Red Sea proper lead to the characteristic DCM of permanently stratified waters, with overall low productivities (around 0.1 g C m^{-2} d^{-1}, Thiel and Weikert, 1984) and picoplanktonic cyanobacteria contributing >75% of primary production (Gradinger et al., 1992). The DCM in the summerly Gulf of Aqaba and Red Sea anticyclonic gyres coincides with low surface concentrations of nutrients (<0.1 mmol $PO_4^{3-}m^{-3}$, <0.5 mmol $NO_3^{-}m^{-3}$, Klinker et al., 1978; Weikert, 1987) and a dynamic microbial loop, where the growth rates of picoplankton algae

(0.6 d^{-1}, Sommer et al., 2002) and heterotrophic bacteria (0.3–2.3 d^{-1}, Sommer et al., 2002; Weisse, 1989) are kept in check by protozoan grazing (0.2–2.6 d^{-1}), resulting in fairly constant levels of bacteria (around 5–9 × 10^5 cells ml^{-1}, Grossart and Simon, 2002; Sommer et al., 2002; Weisse, 1989). Mesozooplankton grazing rates are between 0.01–0.03 d^{-1}, i.e. two orders of magnitude lower than protozoan grazing rates (Sommer, 2000; Sommer et al., 2002) (Fig. 34.18).

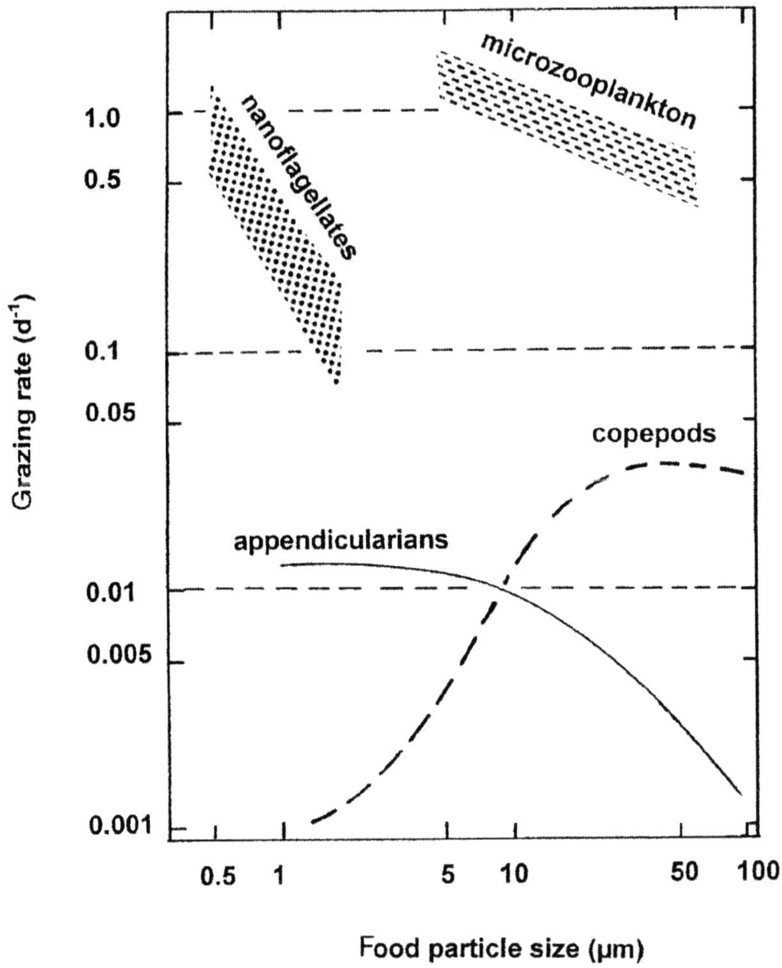

Figure 34.18 Scheme of the Red Sea food web with double-logarithmic plot of the grazing rates of various zooplankton functional groups on food particle size in the Red Sea. (Sommer et al., 2002, reproduced by permission of **[TK]**-Research)

Microphagous mesozooplankton show highest preferences for bacteria (Appendicularia, Ostracoda) and small algae (*Doliolum* spp.), while copepods (*Rhincalanus nasutus, Calanus* sp. and small calanoids) seem to prefer the larger (40–100 µm greatest axial linear dimension) algal size classes (Sommer et al., 2002). *Trichodesmium* spp. is fed upon selectively by the harpacticoid copepod *Macro-*

setella gracilis (Boettger-Schnack and Schnack, 1989). However, recent analyses of polyunsaturated fatty acid profiles in Red Sea zooplankton show that unselective uptake by *Salpa maxima* (but not by the tunicate *Doliolum denticulatum*) may be an even more important up to now overlooked trophic link controlling *Trichodesmium* spp. populations in the Red Sea (Post et al., 2002).

The small size of the phototrophs at the base of the food chain renders large part of the primary production unavailable to the classical food chain via copepods and pelagic fish, and the bulk of the energy is dissipated in the microbial loop (picoplankton-heterotrophic nanoflagellates-ciliates) before reaching higher trophic levels. As a result, pelagic fish stocks are low, and most of the catches occur in the shallow expanses of the southern shelves or the Gulf of Suez (Sanders and Kedidi, 1981).

The deep waters of the Red Sea are among the most nutrient-, biomass- and organic matter-impoverished in the world (Weikert, 1987). Microbial mineralisation of the organic material settling out of the productive layers is very rapid, enhanced by the high temperatures of the deep waters (Weikert, 1982; Grossart and Simon, 2002). Directly beneath the DCM at the base of the nutrient-depleted euphotic zone, successive increases in the concentrations of ammonia ($NH4^+$), nitrite (NO_2^-) and nitrate (NO_3^-), and concomitant decreases in the concentrations of particulate organic matter (POM) and dissolved oxygen, indicate intense oxidation of organic material at intermediate depths (Badran, 2001; Badran and Foster, 1998). These degradation processes result in the formation of a mesopelagic oxygen minimum zone at a core depth of 400–500 m with concentrations of 0.3–0.5 ml O_2 l^{-1} (Weikert, 1980b; Weikert, 1987), corresponding to 6–10% saturation. The reduction of POM below 100 m is associated with a strong decline in the abundance and diversity of mesozooplankton consumers, with concentrations as low as 1 ind. per 250 m^3 at 1500 m depth (Beckmann, 1984; Weikert, 1980a), concentrations found at much greater depths elsewhere in the world ocean (Weikert and Koppelmann, 1993). The planktocline is interrupted by a secondary abundance and diversity maximum coinciding with a deep scattering layer between 300–600 m (Weikert, 1980a; Weikert, 1980b). The intermediate maximum is due to a zone of overlap of the upper and lower mesopelagic biota around the oxygen minimum layer. It is dominated by the calanoid copepod *Pleuromamma indica* (30% of the mesozooplankton in this layer) carrying out large diel vertical migrations between the oxygen minimum layer (or below) and the surface, and a resting stock of the large (5 mm) *Rhincalanus nasutus* (Fig. 34.19). The latter appears to be dormant for most of the year, possibly taking advantage of the low O_2 levels to keep a low metabolism in spite of the high temperatures (Weikert, 1987), coming only to the surface for a very brief period to feed and reproduce (Farstey et al., 2002).

Similar to the deep-sea pelagic biota, the Red Sea deep-sea benthos is significantly impoverished relative to comparative depths in other ocean basins (Thiel, 1987). Throughout the area, the sediment is a pteropod ooze containing very low concentrations of organic matter (0.5% and 0.05% organic C and N, respectively, (Thiel and Weikert, 1984). Meiofauna (<1 mm) densities in deep-sea sediments were 600 $(10\ cm^2)^{-1}$, several orders of magnitude lower than meiofauna in shallow sediments of the Red Sea (Grelet, 1985), comparable to meiofauna densities in abyssal regions of the Pacific and Atlantic oceans (Thiel, 1979). Macro- and megafauna was similarly impoverished, with densities of less than 1000 m^{-2} and <20

$(100\ m^2)^{-1}$, respectively (Thiel, 1979; Thiel and Weikert, 1984). The low faunal biomass coincides with elevated metabolism in the deep sea biota. The high respiration and electron transport system activities measured in deep sediments are believed to be associated with the high metabolic demand due to the uniquely elevated deep sea temperatures in the Red Sea (Thiel et al., 1987; Thiel and Weikert, 1984). The extreme paucity of food, the high temperatures and the geologically short period since the establishment of present-day deep-sea conditions in the Red Sea appear to have converged into preventing the establishment of a true deep-sea fauna in the Red Sea (Thiel, 1987). Whereas many mesopelagic fish species are identical with those in the Indian Ocean, the bathypelagic, bathybenthopelagic and bathybenthic fauna appears to be completely different (Klausewitz, 1986), with many of the species being invaders from shallow waters (Thiel, 1987), the colonization of the deep-sea thus being an ongoing process of niche expansion (Uiblein, 1997).

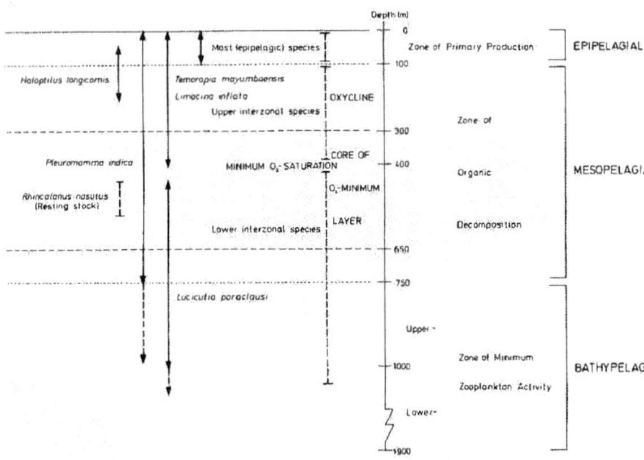

Figure 34.19 Scheme of the vertical distribution of mesozooplankton in the upper, middle and lower water column of the central Red Sea, showing depths of occurrence of characteristic species and ranges of diel vertical migration (arrows). (Weikert, 1982, reproduced by permission of **[TK]**-Research)

5.4. *Gulf of Aden, Arabian Margin and Gulf of Oman*

The seasonal productivity peak in the western Gulf of Aden is similar to the one in the southern Red Sea, but opposite to the one in the Arabian Sea (Johns et al., 2000). The NE monsoon pushing into the Gulf of Aden and southern Red Sea mixes the upper water column under the combined effect of wind and cooling, entraining subsurface nutrients into the euphotic layer, enhancing pelagic production during winter. Large phytoplankton blooms may occur on either side of Bab el Mandeb during this season, leading to a doubling of mesozooplankton grazer biomass relative to summer (Van Couwelaar, 1997). Although upwelling occurs along the coast of Yemen during the SW monsoon and some blooms due to this are observed near Bab el Mandeb, large parts of the western Gulf of Aden and the southern Red Sea remain protected from the winds, so that the strong insolation leads to a vernal stratification, with formation of a deep Chl *a* maximum.

Outside the orographic shielding by the Erithrean mountain range, the strong longshore winds during the SW monsoon generate intense upwelling in the eastern Gulf of Aden and along the Arabian Margin. The Arabian/Oman upwelling area, particularly the Ras al Hadd frontal region, supports one of the richest yet largely unexploited potential fisheries in the world. The paucity of data until very recently on the physical-biological oceanographic interactions in this dynamic region between the coastal circulation, upwelling filaments and meandering front, and the primary producers, their grazers and fish (Hitchcock et al., 2002) has triggered intense research in the frame of the Joint Global Ocean Flux Study (JGOFS), summarized by Naqvi et al. (this volume).

6. Coastal ecosystems

6.1. *General aspects*

As opposed to the large spatial and temporal differences in pelagic production in the seas around the Arabian peninsula, the diverse communities of the litoral fringe and shallow benthos are all very productive. The shallow relief of large parts of the Arabian Gulf, particularly along the northwestern-southwestern portion extending between Kuwait and the U.A.E. creates conditions suitable for the development of a sedimentary environment with soft substrate ecosystems, which may support extensive seagrass meadows and algal beds providing important feeding and breeding habitats for many types of marine organisms such as pearl oysters, shrimps and green turtles. Salt-marshes and mangroves are important intertidal habitats along the coasts of the Arabian peninsula, providing nurseries for marine fish and invertebrates and refuges and breeding areas for marine and terrestrial wildlife. High energy beaches and macroalgae dominated rocky shores are more common along the Arabian margin, and are important in supporting and sustaining many commercially important species. Carbonate platforms and coral reefs supporting high biological diversity occur on parts of the shallow Arabian Gulf and southern Red Sea shelves, but constitute a nearly continuous fringe along the entire length of the central and northern Red Sea and Gulf of Aqaba.

6.2. *Mangroves, tidal flats and salt marshes*

Salt-marshes and mangroves are dominant habitats of the intertidal zone along the coast of Saudi Arabia, United Arab Emirates and other Gulf states. These habitats are considered an essential part of the characteristic habitats found along the shore of these states. Their productivity is of great local and regional importance for marine ecosystems. They are important nesting sites and resting areas for migrating birds. Furthermore, mangroves and salt marshes are important feeding grounds for fish and crustaceans (Böer, 1996). In the Arabian Gulf *Avicennia marina* is the only mangrove tree that occurs naturally. This is due to the extremes of salinity (38–45) and water temperatures (12–35°C or more, in very shallow water) present in this region. The presence of only one species in the Gulf contrasts strongly with the higher diversities found on contiguous shorelines of northern India and the Red Sea where *Avicennia* occurs with three other species (*Rhizophora mucronata, Bruguiera gymnorhiza* and *Ceriops tagal* (Price et al., 1993). The northern limit of mangroves, both in the Arabian Gulf and the Red Sea (Gulf of Aqaba) is 27–28°N. In the Arabian Gulf, the mangroves are dwarfed and poorly developed trees, rarely exceeding a height of 2 m compared to those along the Gulf of Oman and southern Red Sea, which are well developed and reach 6 m and 7–8 m height, respectively (Price et al., 1987; Salm and Jensen, 1989). Whereas most mangroves grow on muddy bottom, the mangroves on the Sinai peninsula are rooted on hard fossil coral bottom (Por et al., 1977).

Mangrove production is low to moderate, relative to global standards, ranging from 0.5 kg C m^{-2} y^{-1} under dwarfed conditions to 5.1 kg C m^{-2} y^{-1} in well developed stands (Sheppard et al., 1992). Yet, it is orders of magnitude higher than the adjoining open waters. How this high productivity is maintained in the absence of estuarine inputs remains enigmatic. Large part of the mangrove productivity is believed to be based on regenerated nutrients (Por et al., 1977).

The mangroves in the region, like many other intertidal vegetation, are affected or threatened by grazing activities, land filling, dredging, oil spills and unplanned coastal development. Each of these activities has contributed to the reduced level of mangrove vegetation. In the Gulf area, the extent of mangroves has declined dramatically, with only about 125–130 km^2 remaining; 90 km^2 off Iran, 10 km^2 off Saudi Arabia and Bahrain and the remainder in the United Arab Emirates in Abu Dhabi and Umm-Al-Auwain. In Saudi Arabia, more than 40 percent of the Arabian Gulf coastline has been filled and 50 percent of the mangrove lost (Jameson et al., 1995; Price et al., 1993). The area of Bahrain has increased about 50 km^2 in the last 25 years by adopting the policy of land filling and reclamation which has affected the intertidal zone, and mangroves of the Tubli Bay to a great extent.

Cyanobacterial mats are considered a major contributor to primary productivity in the muddy tidal flats. These flats are also an important area for migrating birds (Zwarts et al., 1991). *Sabkha* is an extensive habitat in this region. It supports mats of cyanobacteria, bacteria and diatoms, which fix large amounts of nitrogen and contribute highly to the overall productivity. In addition, it hosts high populations of invertebrates and fish.

6.3. *Seagrass and macroalgal beds*

The dominant species of seagrasses in the Arabian Gulf sub-littoral waters are *Halodule uninervis,* a species tolerant of high salinity. Other species such as *Halophila ovalis* and *Halophila stipulacea* are also present. Seagrass beds occur mainly in shallow coastal waters. In the Arabian Gulf, they flourish in 2–5 m and are rarely found in depth of more than 10 m. At the Jubail Marine Life Sanctuary of Saudi Arabia *Haulodule uninervis* was dominant on the sand-bottom, less than 5 m deep, with *Halophila ovalis* and *Halophila stipulacea* also present. At the shallower intertidal fringe, seagrass beds were sparser with the best-developed beds at depths of 3–4 m. Below depths of 5 m seagrass was not found (Richmond, 1996). In Bahrain, however, they are more extensive, but generally they do not extend deeper than approx. 8 m (Price et al., 1993). Basson et al. (1977) reported that seagrass beds were host for more than 530 species of plants and animals. This number has been exceeded by now, as a result of recent research (Coles et al., 1990; Mccain, 1984; Richmond, 1996). These authors found that the infauna of seagrass beds was dominated by gastropods, bivalves and polychaetes. Richmond (1996) reported 233 infaunal species at the Jubail Marine Wildlife Sanctuary of Saudi Arabia and 95% of all infauna of seagrass beds belong to those three taxa.

The results of many phycological inventories in the Arabian Gulf were compiled in many reports between the late 19th century (Endlicher and Diesing, 1845) and the 20th Century (Basson et al., 1989). These results were summarized in the checklist of Basson (1992). However, the most ecologically oriented research was made by Basson et al. (1977), Mccain (1984), and Sheppard et al. (1992).

Macroalgae of the sub tidal zone are very important in supporting life of very important species in the Arabian Gulf. Ninety taxa were recorded at the Jubail Marine Wildlife Sanctuary in Saudi Arabia. Rhodophyta were the most dominant (44) followed by phaeophyta and chlorophyta, which were represented by 25 and 20 taxa respectively. Forty-two of these species were new in the Saudi Arabian Gulf and 15 were new in the Arabian Gulf. Macro algal beds are particularly important for the life of commercially important fish and shrimp *Penaeus semisulcatus* as well as the pearl oyster *Pinctada* spp.

6.4. Coral reefs

The most well-developed coral reefs in the Arabian region are found in the Red Sea. They form predominantly narrow fringing reefs (hence *contour* reefs; Fishelson, 1971), where the shelves are narrow or non-existent, such as along the northern and central margins of the Red Sea basin, and its northern extensions into the deep Gulf of Aqaba. Along the shallow Gulf of Suez and broad shelves of the Central and Southern Red Sea, also pinnacle reefs and reef barriers occur. Detailed accounts on the Red Sea coral reef communities are given i.a. by Fishelson (1971), Loya (1972), Mergner (1980), Mergner and Schuhmacher (1974), Mergner and Schuhmacher (1985), and Schuhmacher and Mergner, (1985) and the general features have been reviewed i.a. by Edwards and Head (1987) and Sheppard et al. (1992), and may thus not be repeated again.

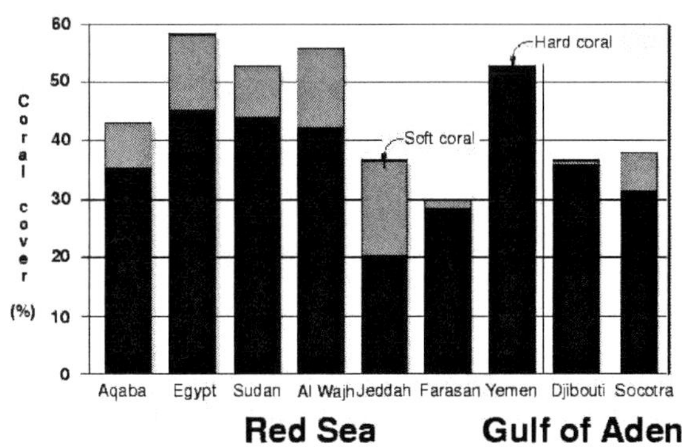

Figure 34.20 Hard and soft coral cover at 5m depth in 9 sites from the northern tip of the Gulf of Aqaba (left) along the length of the Red Sea to Djibouti (Gulf of Aden) and the island of Socotra (Arabian Sea). (After Hassan et al., 2002)

Recent census data show that live hard coral cover in the Red Sea is high, by global standards, averaging 20–50% (Hassan et al., 2002) (Fig. 34.20).

Highest coral cover cover and diversity occurs in the northern and central parts of the Red Sea, where water visibilities may exceed 40–50m (Pilcher and Abou Zaid, 2000). The more fertile and turbid waters of the southern Red Sea and Gulf of Aden result in poor coral development, particularly in deeper waters due to visibilities of sometimes less than 5 m (Hassan et al., 2002).

Where the shelf is narrow, steep slopes and rapid coral growth give rise to a fairly open unconsolidated framework with little sediment infilling (Barnes and Lazar, 1993). The resulting crevices, fissures and holes interlacing the reef framework are home to a diverse cryptic community (Wunsch and Richter, 1998; Richter and Wunsch, 1999). Suspension feeders, notably sponges, occupy large areas of the crevice walls and contribute significantly to the trophic balance of the coral reef ecosystem in the nutrient-impoverished waters (Richter et al., 2001). Oceanic nutrients harnessed by the cryptic filter-feeders may thus fuel up to 25% of the gross primary production of the overlying corals and algae, estimated at 4 g C m^{-2} d^{-1} (Erez, 1990; Richter et al., 2001). Zooplanktivorous fish, making up 60% of numbers of individuals and species (Khalaf and Kochzius, 2002a; 2002b), as well as suspension feeding invertebrates dwelling on the exposed reef surface (Fabricius et al., 1995; Yahel et al., 1998; Kappner et al., 2000) further contribute to pelagic-benthic coupling in Red Sea coral reefs.

In spite of the generally healthy state of Red Sea coral reefs, relative to global standards (Wilkinson, 2002), there is mounting evidence that live coral cover and fish populations have started to decline in many places (Jameson et al., 1997).

Several small outbreaks of crown-of-thorns starfish (COTS) have been recorded, some local bleaching events and an increase in bioeroding organisms such as the urchin *Diadema setosum* and the coral eating gastropods *Drupella* and *Coralliophila.* Threats appear to be associated with the increasing rate of coastal development and thus differ within the region, being particularly imminent in the northern sections of the Red Sea where the most spectacular reefs occur. The major threats are land filling, dredging, sedimentation, sewage discharge and effluents from intensive aquaculture and desalination plants. In major tourism areas, there is physical damage by tourists and boat anchors. Fishing pressure is constantly increasing throughout the region to satisfy demands of growing and more affluent populations.

In the Arabian Gulf, great variations in temperature, high salinity and high turbidity form an extreme environment for corals (Sheppard and Well, 1988). In this region, Coles et al. (1991) measured the lowest ever recorded temperature in coral reefs. Coral reefs in the Arabian Gulf occur mainly as patch reefs. Fringing reefs are usually found around offshore islands. These reefs have large reef flats and reef slopes connected at their deepest part to the soft substrate at a depth of 10 m or more. Many of these islands are found in the waters of Saudi Arabia, Kuwait, Bahrain and Qatar and are considered the best-developed and most diverse reefs in terms of species composition in this region. At the Jubail Marine Wildlife Sanctuary in Saudi Arabia it was found that offshore and inshore reefs varied greatly in terms of species composition and seasonality. Coral reefs in this protected area are restricted to a very limited depth range of only few meters. They rarely extended deeper than 10 m. However, single coral colonies may sometimes be found as deep as 20–25 m. Most of the offshore reefs showed similar species composition, and the scleractinian cover was poor in shallower parts of the reefs. At the inshore reefs, the poor live coral cover consisted mainly of *Porites compressa* and with increasing depth, corals of the genus *Acropora* occurred more frequently and dominated the deeper reef (Vogt, 1996).

Coral diversity (50–60 species) in the Arabian Gulf is low compared to the Red Sea (nearly 200 species; Jameson et al., 1995). Impacts from oil spills during the Gulf war of 1991 were much less than generally expected, indicating a high resilience among reef communities (Sheppard et al., 1992).

Bibliography

Al Hajiri, K. 1990. The circulation of the Arabian (Persian) Gulf: A model study of its dynamics. The Catholic University of America.

Al-Muzaini, S., and P. G. Jacob. 1996. Marine plants of the Arabian Gulf. *Environment International,* **22,** 369–376.

Al-Saadi, H. A., R. A. Hadi, and M. F. Hug. 1976. Preliminary studies on phytoplankton of the northwest Arab Gulf. *Bangl. J. Bot.,* **5,** 9–21.

Arz, H. W., F. Lamy, J. Paetzold, P. J. Mueller, and M. Prins. 2003. Mediterranean moisture source for an early-Holocene humid period in the northern Red Sea. *Science,* **300,** 118–121.

Badran, M. I. 2001. Dissolved oxygen, Chlorophyll a and nutrients: Seasonal cycles in waters of the Gulf of Aqaba, Red Sea. *Aquatic Ecosystem Health & Management [Aquat. Ecosyst. Health Manage.].* **4,** 139–150.

Badran, M. I., and P. Foster. 1998. Environmental quality of the Jordanian coastal waters of the Gulf of Aqaba, Red Sea. *Aquatic Ecosystem Health & Management [Aquat. Ecosyst. Health Manage.].* **1,** 75–89.

Barnes, D. J., and B. Lazar. 1993. Metabolic performance of a shallow patch reef near Eilat on the Red Sea. *J. Exp. Mar. Biol. Ecol.* **174,** 1–13

Basson, P. W. 1992. Checklist of marine algae of the Arabian Gulf. *Jr. Univ. Kuwait (Science),* **19,** 217 - 230.

Basson, P. W., J. E. Burchard, J. T. Hardy, and A. R. G. Price. 1977. Biotopes of the Western Arabian Gulf. ARAMCO, Dhahran, 284 pp.

Basson, P. W., S. A. Mohamed, and D. K. Arora. 1989. A survey of the benthic algae of Bahrain. *Bot. Mar.,* **32,** 27–40.

Beal, L. M., A. Field, and A. L. Gordon. 2000. Spreading of Red Sea overflow waters in the Indian Ocean. *Journal of Geophysical Research. C. Oceans [J. Geophys. Res.,* **105,** 8549–8564.

Beckmann, W. 1984. Mesozooplankton distribution on a transect from the Gulf of Aden to the Central Red Sea during the winter monsoon. *Oceanol. Acta,* **7,** 87–102.

Boehm, E., J. M. Morrison, V. Manghanani, H.-S. Kim, and C. N. Flagg. 1999. The Ras al Hadd Jet: Remotely sensed and acoustic Doppler current profiler observations 1994–1995. *Deep-Sea Research (Part II, Topical Studies in Oceanography) [Deep-Sea Res. (II Top. Stud. Oceanogr.)],* **46,** 1531–1549.

Böer, B. 1996. Trial planting of mangroves (Avicenia marina) and salt-marshes plant (Salicornia europaea) in oil-impacted soil in the Jubail area, Saudi Arabia, In *A Marine Wildlife Sanctuary for the Arabian Gulf. Environmental Research and Conservation Following the 1991 Gulf War Oil Spill.* F. Krupp, Abuzinada, A.H., and Nader, I.A. ed. NCWCD, Riyadh and Senckenberg Research Institute, Frankfurt a.M. p. 186–192.

Boettger-Schnack, R., and D. Schnack. 1989. Vertical distribution and population structure of *Macrosetella gracilis* (Copepoda: Harpacticoida) in the Red Sea in relation to the occurrence of *Oscillatoria*

(*Trichodesmium*) spp. (Cyanobacteria). *Marine Ecology Progress Series [Mar. Ecol. Prog. Ser.].* **52,** 17–31.

Bower, A. S., D. M. Fratantoni, W. E. Johns, and H. Peters. 2002. Gulf of Aden eddies and their impact on Red Sea Water. *Geophys. Res. Lett.,* **29,** 1–4.

Cember, R. P. 1988. On the sources, formation, and circulation of Red Sea deep water. *J. Geophys. Res. (C Oceans),* **93,** 8175–8191.

Cherif, O. H., Al-Ghadban, A.N., and Al-Rifaiy. 1997. Distribution of foraminifera in the Arabian Gulf. *Micropaleontology,* **13 A, 43,** 253 - 280.

Clifford, M., C. Horton, J. Schmitz, and L. H. Kantha. 1997. An oceanographic nowcast/forecast system for the Red Sea. *Journal of Geophysical Research. C. Oceans [J. GEOPHYS. RES.,* **102,** 101–125.

Coles, S. L., and Fadlallah, Y.H. 1991. Reef Coral survival and mortality at low temperature in the Arabian Gulf: New species - specific lower temperature limits. *Coral Reefs,* 231–237.

Coles, S. L., and McCain, J.C. 1990. Environmental factors affecting benthic infaunal communities of the Western Arabian Gulf. *Mar. Environ. Res,* **19,** 289–315.

Degens, E. T., and D. A. Ross. 1969. Hot brines and recent heavy metal deposits in the Red Sea. Springer. p. 600.

Dorgham, M. N. 1991. Temporal and spatial variations of the phytoplankton in Qatari Waters, Arab Gulf. *J.Fac. Sci. UAE (United Arab University),* **3,** 105–128.

Edwards, A. J., and S. M. Head. 1987. Key environments: Red Sea. Pergamon. p. 441.

El-Samra, M. I. 1988. Chemical Observation in the Arabian Gulf and the Gulf of Oman. *Arab Gulf J. Scient. Res., Math. Phys. Sci.,* **A6,** 205–215.

Elliot, A. J., and G. Savidge. 1990. Some features of the upwelling off Oman. *J. Mar. Res.,* **48,** 319–333.

Endlicher, S. L., and C. M. Diesing. 1845. Enumeratio algarum, quas ad oram insulae Karek, sinus Persici, legit Theodorus Kotschy. *Botanische Zeitung,* **3,** 268–269.

Erez, J. 1990. On the importance of food sources in coral reef ecosystems, In *Coral Reefs.* Z. Dubinsky ed. Elsevier, New York, p. 411–418

Fabricius, K. E., Y. Benayahu, and A. Genin. 1995. Herbivory in asymbiotic soft corals. *Science (Wash.),* **268,** 90–92.

Farstey, V., B. Lazar, and A. Genin. 2002. Expansion and homogeneity of the vertical distribution of zooplankton in a very deep mixed layer. *Marine Ecology Progress Series [Mar. Ecol. Prog. Ser.].* **238,** 91–100.

Fine, R. A., M. J. Warner, and R. F. Weiss. 2000. Water mass modification at the Agulhas Retroflection: chlorofluoromethane studies. *Deep-Sea Res.,* **35,** 311–332.

Fishelson, L. 1971. Ecology and distribution of the benthic fauna in the shallow water of the Red Sea. *Marine Biology,* **10,** 113–133.

Genin, A., B. Lazar, and S. Brenner. 1995. Vertical mixing and coral death in the Red Sea following the eruption of Mount Pinatubo. *Nature,* **377,** 507–510.

Genin, A., and N. Paldor. 1998. Changes in the circulation and current spectrum near the tip of the narrow, seasonally mixed Gulf of Elat. *Isr. J. Earth Sci.,* **47,** 87–92.

Gladstone, W., N. Tawfiq, D. Nasr, I. Andersen, C. Cheung, H. Drammeh, F. Krupp, S. Lintner. 1999. Sustainable use of renewable resources and conservation in the Red Sea and Gulf of Aden: issues, needs and strategic actions. *Ocean & Coastal Management [Ocean Coast. Manage.].* **42,** 671–697.

Gordon, N., D. L. Angel, A. Neori, N. Kress, and B. Kimor. 1994. Heterotrophic dinoflagellates with symbiotic cyanobacteria and nitrogen limitation in the Gulf of Aqaba. *Marine Ecology Progress Series [Mar. Ecol. Prog. Ser.].* **107,** 83–88.

Gradinger, R., T. Weisse, and T. Pillen. 1992. Significance of picocyanobacteria in the Red Sea and the Gulf of Aden. *Bot. Mar.,* **35,** 245–250.

Gregory, J. W. 1896. Great Rift Valley - Narrative of a Journey to Mount Kenya and Lake Baringo with Some Account of the Geology, Natural History, Anthropology and Future Prospect of British East Africa. Frank Cass Publishers.

Grelet, Y. 1985. Vertical distribution of meiobenthos and estimation of Nematode biomass from sediments of the Gulf of Aqaba (Jordan, Red Sea). M. Harmelin-Vivien, B. Salvat, C. La Croix, C. Gabrie and J. L. Toffart eds. Antenne Museum-EPHE. p. 251–256.

Grossart, H. P., and M. Simon. 2002. Bacterioplankton dynamics in the Gulf of Aqaba and the northern Red Sea in early spring. *Mar. Ecol. Prog. Ser.,* **239,** 263–276.

Halim, Y. 1984. Plankton of the Red Sea and the Arabian Gulf. *Deep-Sea Research [DEEP-SEA RES.],* **31,** 969–982.

Hall, J. K. 1975. Bathymetric chart of the Straits of Tiran. *Israel J. Earth Sci.,* **24,** 69–72.

Hansen, U., G. Paetzold, and R. Sachse. 1992. Deep chlorophyll a maxima in the Red Sea observed by in-vivo flash fluorometry. *Hydrobiologia,* **238,** 183–188.

Hartmann, M., J. C. Scholten, and P. Stoffers. 1998a. Hydrographic structure of brine-filled deeps in the Red Sea: Correction of Atlantis II Deep temperatures. *Marine Geology [MAR. GEOL.].* **144,** 331–332.

Hartmann, M., J. C. Scholten, P. Stoffers, and F. Wehner. 1998b. Hydrographic structure of brine-filled deeps in the Red Sea—new results from the Shaban, Kebrit, Atlantis II, and Discovery Deep. *Marine Geology [MAR. GEOL.].* **144,** 311–330.

Hassan, M., M. M. A. Kotb, and A. A. Al-Sofyani. 2002. Status of Coral Reefs in the Red Sea-Gulf of Aden, In *Status of coral reefs of the world: 2002.* C. R. Wilkinson ed. Australian Institute of Marine Science. p. 45–52.

Hitchcock, G. L., P. Lane, S. L. Smith, J. Luo, and P. B. Ortner. 2002. Zooplankton spatial distributions in coastal waters of the northern Arabian Sea, August, 1995. *Deep-Sea Research (Part II, Topical Studies in Oceanography) [Deep-Sea Res. (II Top. Stud. Oceanogr.)],* **49,** 2403–2423.

Hughes, P., and J. R. Hunter. 1979. A proposal for a physical oceanography program and numerical modeling of the KAP region. UNESCO, Division of Marine Science, Marine 27, 16 Oct. 1979.

Hulings, N. C. 1989. A Review of Marine Science research in the Gulf of Aqaba. *Publications of the Marine Science Station Aqaba, Jordan,* **6,** 267 pp.

Hunter, J. R. 1983. A review of the residual circulation and mixing process in the KAP region with reference to applicable modeling techniques. *UNESCO Reports in Marine Science,* **28,** 37–45

Jameson, S. C., McManus, J.W. and Spalding, M.D. 1995. State of the reefs, regional and global perspective. International Coral Reef Initiative Executive Secretariat, Background Paper. p. 1–32.

Jameson, S. C., H. M. Mostafa, and B. Riegl. 1997. Rapid ecological assessment of diving sites in the Egyptian Red Sea. Environmentally Sustainable Tourism project Report to USAID, Winrock International (Arlington VA).

Johns, W. E., G. A. Jacobs, J. C. Kindle, S. P. Murray, and M. Carron. 2000. Arabian marginal seas and gulfs. University of Miami RSMAS. Technical report 2000–01

Jones, D. A. 1985. The biological characteristics of the marine habitats found within the ROPME Sea Region, In *Proceeding of ROPME Symposium on Regional Marine Pollution Monitoring and Research Programmes (ROPME/GC-4/2).*

Jones, D. A. 1986a. A field guide to the sea shores of Kuwait and the Arabian Gulf. University of Kuwait and Blandford Press, Poole.

Jones, D. A. 1986b. The biological characteristics of the marine habitats found within the ROPME (Regional Organization for the Protection of the Marine Environment) Sea Area (30°30'N, 47°50'E, 16°39'N, 61°25'E). In: *Proceedings of ROPME Symposium on Regional Marine Pollution Monitoring and Research Programmes.* (ROPME/GC-4/2) ROPME, Kuwait. p. 71–89.

Jones, D. A., A. R. G. Price, F. Al-Yamani, and A. Al-Zaidan. 2002. Coastal and marine ecology, In *The Gulf Ecosystem - Health and Sustainability.* N. Y. Khan, M. Munawar and A. R. G. Price eds., Ecovision World Monograph Series. Backhuys.

Kappner, I., S. M. Al-Moghrabi, and C. Richter. 2000. Mucus-net feeding by the vermetid gastropod Dendropoma maxima in coral reefs. *Mar. Ecol. Prog. Ser.,* **204,** 309–313.

Kassler, P. 1973. The structure and geomorphic evolution of the Persian Gulf, In *The Persian Gulf.* B. H. Purser ed. Springer Verlag. p. 11–32.

Khalaf, M. A., and M. Kochzius. 2002a. Changes in trophic community structure of shore fishes at an industrial site in the Gulf of Aqaba, Red Sea. *Marine ecology progress series [Mar. Ecol. Prog. Ser.].* **239,** 287–299.

Khalaf, M. A., and M. Kochzius. 2002b. Community structure and biogeography of shore fishes in the Gulf of Aqaba, Red Sea. *Helgoland Marine Research [Helgol. Mar. Res.].* **55,** 252–284.

Khattab, M. M. 1995. Interpretation of magnetic and gravity surveys in the southern Arabian Gulf, the Strait of Hormuz, and the northwesternmost Gulf of Oman: Implications of pre-Permian basement tectonics. *Marine Geology [MAR. GEOL.].* **123,** 105–116.

Kimor, B., and B. Golandsky. 1977. Microplankton of the Gulf of Elat: Aspects of seasonal and bathymetric distribution. *Marine Biology,* **42,** 55–67.

Kimor, B., N. Gordon, and A. Neori. 1992. Symbiotic associations among the microplankton in oligotrophic marine environments, with special reference to the Gulf of Aqaba, Red Sea. *J. Plankton Res.,* **14,** 1217–1231.

Klausewitz, W. 1986. The characteristics of the deep sea ichthyofauna of the Red Sea, In *Proc. 2nd Int. Conf. on Indo-Pacific Fishes.* T. Uyeno, R. Arai, T. Taniuchi and K. Matsuura eds. Tokyo National Museum. p. 945.

Klinker, J., Z. Reiss, C. Kropach, I. Levanon, H. Harpaz, and Y. Shapiro. 1978. Nutrients and biomass distribution in the Gulf of Aqaba (Eilat), Red Sea. *Mar. Biol.,* **45,** 53–64.

Labiosa, R. G., K. R. Arrigo, A. Genin, S. G. Monismith, and G. van Dijken. 2003. The interplay between upwelling and deep convective mixing in determining the seasonal phytoplankton dynamics in the Gulf of Aqaba: Evidence from SeaWiFS and MODIS. *Limnology and Oceanography,* **48,** 2355–2368.

Lehr, W. J. 1984. A brief survey of oceanographic modeling and oil spill studies in the KAP region. UNESCO Reports in Marine Science 28.

Levanon-Spanier, I., E. Padan, and Z. Reiss. 1979. Primary production in a desert-enclosed sea. The Gulf of Eilat. (Aqaba), Red Sea. *Deep-Sea Res. (A Oceanogr. Res. Pap.),* **26,** 673–685.

Lindell, D., and A. F. Post. 1995. Ultraphytoplankton succession is triggered by deep winter mixing in the Gulf of Aqaba (Eilat), Red Sea. *Limnology and Oceanography,* **40,** 1130–1141.

Linden, O., Abdulraheem, M.Y., Gerges , M.A., Alam, I., Behbehani M., Borhan, M.A. and Al-Kassab, F. 1990. State of the marine environment in the ROPME Sea Area. UNEP Rep. p. 1, 34 pp.

Loya, Y. 1972. Community structure and species diversity of hermatypic corals at Eilat, Red Sea. *Marine Biology,* **13,** 100–123.

Maillard, C., and G. Soliman. 1986. Hydrography of the Red Sea and exchanges with the Indian Ocean in summer. *Oceanol. Acta,* **9,** 249–269.

Manasrah, R. 2002. The general circulation and water masses characteristics in the Gulf of Aqaba and northern Red Sea, In *Faculty of Mathematics and Natural Sciences.* Rostock University. 121 pp.

McCain, J. C. 1984. Marine ecology of Saudi Arabia. The nearshore soft bottom benthic communities of the northern area, Arabian Gulf, Saudi Arabia. *Fauna of Saudi Arabia,* **6,** 79 -97.

Mergner, H. 1980. Ecology of a fore-reef area near Aqaba (Red Sea), In *Proceedings of the Symposium on the Coastal and Marine Environment of the Red Sea, Gulf of Aden and Tropical Western Indian Ocean.* ALECSO Red Sea and Gulf of Aden Environmental Programme, Jeddah (Saudi Arabia). p. 39–76.

Mergner, H., and H. Schuhmacher. 1974. Morphologie, Ökologie und Zonierung von Korallenriffen bei Aqaba, (Golf von Aqaba, Rotes Meer). *Helgoländer wissenschaftliche Meeresuntersuchungen,* **26,** 238–358.

Mergner, H., and H. Schuhmacher. 1985. Quantitative analysis of coral communities of Sanganeb-Atoll (central Red Sea). 1. The community structure of outer and inner reefs exposed to different hydrodynamic regimes. *Helgol. Meeresunters.*, **39,** 375–417.

Meshal, A. H., and H. M. Hassan. 1986. Evaporation from the coastal waters of the central part of the Gulf. *Arab J. Sci. Res.*, **4,** 649 - 655.

Michel, H. B., Behbehani, M., and Herring, D. 1986a. Zooplankton of the Western Arabian Gulf south of Kuwait waters. *Kuwait Bull. Mar. Sci.*, **8,** 1 - 36.

Michel, H. B., Behbehani, M., Herring; D., Arar, M., Shoushani, M. and Brakoniecki, T. 1986b. Zooplankton diversity, distribution and abundance in Kuwait waters. *Kuwait Bull. Mar. Sci.*, **8,** 37 - 105.

Morcos, S. A. 1970. Physical and chemical oceanography of the Red Sea. *Oceanogr. Mar. Biol. A. Rev.*, **8,** 73–202.

Murray, S. P., A. Hecht, and A. Babcock. 1984. On the mean flow in the Tiran Strait in winter. *J. Ar. Res*, **42,** 265–287.

Murray, S. P., and W. Johns. 1997. Direct observations of seasonal exchange through the Bab el Mandab Strait. *Geophysical Research Letters [GEOPHYS. RES. LETT.].* **24,** 2557–2560.

Murty, T. S., and M. I. El-Sabh. 1984. Storm tracks, storm surges and sea state in the Arabian Gulf, Striat of Hormuz and the Gulf of Oman. In *Oceanographic modeling of the Kuwait Action Plan (KAP) region* M. I. El-Subh ed. UNESCO, *Marine Science* **28,** 12–36

Naqvi, S. W. A., H. P. Hansen, and T. W. Kureishy. 1986. Nutrient uptake and regeneration ratios in the Red Sea with reference to the nutrient budgets. *Oceanol. Acta,* **9,** 271–275.

Naqvi, W. A., and R. G. Fairbanks. 1996. A 27,000 year record of Red Sea outflow: Implication for timing of post-glacial monsoon intensification. *Geophys. Res. Lett.*, **23,** 1501–1504.

Naqvi, S.W.A., P.V. Narvekar and E. Desa. 2004. Coastal Biogeochemical Processes in the North Indian Ocean. In *The Sea.* A. R. Robinson and K. H. Brink eds. Harvard University Press, this volume.

National Geographic Society. 2002. National Geographic Family Reference Atlas of the World. National Geographic Society, Washington. 352 p.

Neumann, A. C., and D. A. McGill. 1962. Circulation of the Red Sea in early summer. *Deep Sea Research,* **8,** 223–235.

Patzert, W. C. 1974a. Seasonal reversal in Red Sea circulation, In *L'océanographie physique de la Mer Rouge.* CNEXO. p. 55–89.

Patzert, W. C. 1974b. Volume and heat transports between the Red Sea and Gulf of Aden, and notes on the Red Sea heat budget, In *L'océanographie physique de la Mer Rouge.* CNEXO. p. 191–201.

Pilcher, N., and M. M. Abou Zaid. 2000. The status of coral reefs in Egypt - 2000, In *Status of coral reefs of the world: 2000.* C. R. Wilkinson ed. Australian Institute of Marine Science. p. 1–17.

Plaehn, O., B. Baschek, T. H. Badewien, M. Walter, and M. Rhein. 2002. Importance of the Gulf of Aqaba for the formation of bottom water in the Red Sea. *J. Geophys. Res. (C Oceans),* **107,** 1–18.

Por, F. D. 1978. Lessepsian migration: the influx of Red Sea biota into the Mediterranean by way of the Suez Canal. Springer, New York, 228 p.

Por, F. D., I. Dor, and A. Amir. 1977. The mangal of Sinai: limits of an ecosystem. *Helgol. Wiss. Meeresunters.*, **30,** 295–314.

Post, A. F., Z. Dedej, R. Gottlieb, H. Li, D. N. Thomas, M. El-Absawi, T. El-Naggar, M. El-Gharabawi, U. Sommer. 2002. Spatial and temporal distribution of *Trichodesmium* spp. in the stratified Gulf of Aqaba, Red Sea. *Mar. Ecol. Prog. Ser.*, **239,** 241–250.

Price, A. R. G., P. A. H. Medley, R. J. McDowall, A. R. Dawson-Shepherd, P. J. Hogarth, and R. F. G. Ormond. 1987. Aspects of mangal ecology along the Red Sea coast of Saudi Arabia. *J. Nat. History,* **21,** 449–464.

Price, A. R. G., C. R. C. Sheppard, and C. M. Roberts. 1993. The Gulf: its biological setting. *Mar. Pollut. Bull.*, **27,** 9–15.

Privett, D. W. 1959. Monthly chart evaporation from the North Indian Ocean, including the Red Sea and the Persian Gulf. *Q.J.R. Meteor - Soc,* **85,** 424–428.

Purser, B. H., and E. T. Seibold. 1973. The principal environmental factors influencing Holocene sedimentation and diagenesis in the Persian Gulf, In *The Persian Gulf.* B. H. Purser ed. Springer, New York. p. 1–9.

Quadfasel, D., and H. Baudner. 1993. Gyre-scale circulation cells in the Red Sea. *Oceanol. Acta,* **16,** 221–229.

Quadfasel, D., and N. Verch. 1987. Seasonal variability of temperature in the Red Sea: XBT-sections from MCS Ubena in 1985 and 1986. *Inst. Meereskd. Tech. Rep. 1,* 25 p.

Quraishee, G. S. 1984. Circulation in the north Arabian Sea at Murray Ridge. *Deep-Sea Res. (a Oceanogr. Res. Pap.),* **31,** 651–664.

Rao, D. V. S., F. Al-Yamani, A. Lennox, Y. Pan, and T. F. O. Al-Said. 1999. Short communication. Biomass and production characteristics of the first red tide noticed in Kuwait Bay, Arabian Gulf. *Journal of Plankton Research [J. Plankton Res.].* **21,** 805–810.

Reiss, Z., and L. Hottinger. 1984. The Gulf of Aqaba (Eilat) - ecological micropaleontology. Springer, Berlin, 354 p.

Reynolds, R. M. 1993. Physical Oceanography of the Gulf, Strait of Hormuz, and the Gulf of Oman - Results from the Mt. Mitchell expedition. *Mar. Pollut. Bull.,* **27,** 35 - 59.

Richmond, M. D. 1996. Status of subtidal biotopes of the Jubail Marine Wildlife Sanctuary with special reference to self-substrata communities, In *A Marine Wildlife Sanctuary for the Arabian Gulf. Environmental Research and Conservation following the 1991 Gulf War Oil Spill.* F. Krupp, Abuzinada, A.H., and Nader, I.A. ed. NCWCD, Riyadh and Senckenberg research Institute, Frankfurt a.M. p. 159–176.

Richter, C., and M. Wunsch. 1999. Cavity-dwelling suspension feeders in coral reefs - a new link in reef trophodynamics. *Marine Ecology Progress Series [Mar. Ecol. Prog. Ser.].* **188,** 105–116.

Richter, C., M. Wunsch, M. Rasheed, I. Koetter, and M. I. Badran. 2001. Endoscopic exploration of Red Sea coral reefs reveals dense populations of cavity-dwelling sponges. *Nature [Nature],* **413,** 726–730.

Saad, N. N., E. E. Mohamed, and S. M. El-Nady. 1999. Seasonal variation of geostrophic current and water transport in the central part of the Red Sea. *Journal of King Abdulaziz University. Marine sciences. Jeddah [J. King Abdulaziz Univ.,* **10,** 17–38.

Saleh, M. A. 1995. Tourism development of the Red Sea coasts, In *IOC-PERSGA-ACOPS Workshop on Oceanographic Input to Integrated Coastal Zone Management in the Red Sea and Gulf of Aden.* Y. Halim and S. Morcos eds. UNESCO, Paris, Workshop Rep. IOC. p. 97–109.

Salm, R. V., and R. A. C. Jensen. 1989. Oman. Coastal zone management plan. Dhofar. IUCN, Gland.

Sanders, M. J., and S. M. Kedidi. 1981. Summary Review of Red Sea Commercial Fisheries Catches and Stock Assessments Including Maps of Actual and Potential Fishing Grounds. FAO, Rome/UNDP. RAB/777008/19, 49 p.

Schuhmacher, H., and H. Mergner. 1985. Quantitative analysis of coral communities of Sanganeb-Atoll (Central Red Sea). 2. Comparison with a reef area near Aqaba (northern Red Sea) at the northern margin of the Indo-Pacific reef belt. *Helgol. Meeresunters.,* **39,** 419–440.

Sheppard, C., A. Price, and C. Roberts. 1992. Marine ecology of the Arabian region. Academic Press, London, 359 p ..

Sheppard, C. R. C. 1993. Physcial Environment of the Gulf relevant to marine pollution : an overview. *Mar. Pollut. Bull.,* **27,** 3 - 8.

Sheppard, C. R. C., and S. Well. 1988. Directory of Coral Reefs of International Importance. Vol. 2 Indian Ocean Region. IUCN Gland and UNEP.

Sheppard, R. C., and D. J. Dixon. 1998. Seas of the Arabian region (29,S), In *The global coastal ocean.* K. H. Brink ed., The Sea. J. Wiley & Sons.

Siddall, M., D. A. Smeed, S. Matthiesen, and E. J. Rohling. 2002. Modelling the seasonal cycle ofthe exchange flow in Bab El Mandab (Red Sea). *Deep-Sea Research I,* **49,** 1551–1569.

Siedler, G. 1969. General circulation of the water masses in the Red Sea, In *Hot brines and recent heavy metal deposits in the Red Sea.* E. T. Degens and D. A. Ross eds. Springer. p. 131–137.

Sofianos, S. S., W. E. Johns, and S. P. Murray. 2002. Heat and freshwater budgets in the Red Sea from direct observations at Bab el Mandeb. *Deep Sea Research,* **49,** 7–8.

Sommer, U. 2000. Scarcity of medium-sized phytoplankton in the northern Red Sea explained by strong bottom-up and weak top-down control. *Mar. Ecol. Prog. Ser.,* **197,** 19–25.

Sommer, U., U. G. Berninger, R. Böttger-Schnack, A. Cornils, W. Hagen, T. Hansen, T. Al-Najjar, A. F. Post, S. B. Schnack-Schiel, H. Stibor, D. Stübing, S. Wickham. 2002. Grazing during early spring in the Gulf of Aqaba and the northern Red Sea. *Mar. Ecol. Prog. Ser.,* **239,** 251–261.

Stihl, A., U. Sommer, and A. F. Post. 2001. Alkaline phosphatase activities among populations of the colony-forming diazotrophic cyanobacterium Trichodesmium spp. (Cyanobacteria) in the Red Sea. *Journal of Phycology [J. Phycol.].* **37,** 310–317.

Swift, S. A., and A. S. Bower. 2003. Formation and circulation of dense water in the Persian/Arabian Gulf. *Journal of Geophysical Research - Oceans,* **108,** (DOI 10. 1029/2002JC1000 1360).

Tamburini, F., K. B. Foellmi, T. Adatte, S. M. Bernasconi, and P. Steinmann. 2003. Sedimentary phosphorus record from the Oman margin: New evidence of high productivity during glacial periods. *Paleoceanography [Paleoceanography],* **18,** np.

Thiel, H. 1979. First quantitative data on Red Sea deep benthos. *Mar. Ecol. (Prog. Ser.),* **1,** 347–350.

Thiel, H. 1987. Benthos of the deep Red Sea, In *Key environments: Red Sea.* A. J.Edwards and S. M. Head eds. Pergamon. p. 112–127.

Thiel, H., O. Pfannkuche, R. Theeg, and G. Schriever. 1987. Benthic metabolism and standing stock in the central and northern deep Red Sea. *Pubblicazioni della Stazione zoologica di Napoli I: Marine ecology,* **8,** 1–20.

Thiel, H., and H. Weikert. 1984. Biological oceanography of the Red Sea oceanic system. *Deep-Sea Research [DEEP-SEA RES.],* **31,** 829–831.

Uiblein, F. 1997. Faunal composition and depth distribution in warm oceans, the legacy of the Austro-Hungarian deep-sea expeditions, In *Conference One Hundred Years of Portuguese Oceanography. Symposium.* L. Saldanha and P. Re eds. Museu Bocage, Mus. Nac. de Historia Natural, Lisbon (Portugal). p. 235–244.

Van Couwelaar, M. 1997. Zooplankton and micronekton biomass off Somalia and in the southern Red Sea during the SW monsoon of 1992 and the NE monsoon of 1993. *Deep Sea Research,* **44,** 6–7.

Veldhuis, M. J. W.-., G. W. Kraay, J. D. L. Van Bleijswijk, and M. A. Baars. 1997. Seasonal and spatial variability in phytoplankton biomass, productivity and growth in the northwestern Indian Ocean: The southwest and northeast monsoon, 1992–1993. *Deep-Sea Research (Part I, Oceanographic Research Papers) [Deep-Sea Res. (I Oceanogr. Res. Pap.)],* **44,** 425–449.

Vogt, H. 1996. Investigations on coral reefs in the Jubail Wildlife Sanctuary using underwater video recordings and digital image analysis, In *A Marine Wildlife Sanctuary for the Arabian Gulf. Environmental Research and Conservation following the 1991 Gulf War Oil Spill.* F. Krupp, Abuzinada, A.H., and Nader, I.A. ed. NCWCD. Riyadh and Senckenberg research Institute, Frankfrut a.M. p. 302–326.

Weikert, H. 1980a. On the plankton of the central Red Sea. A first synopsis of results obtained from the cruises Meseda I and Meseda II, In *Proc. Symposium on Coastal and marine Environment of the Red Sea, Gulf of Aden and tropical western Indian Ocean.* p. 135–167.

Weikert, H. 1980b. The oxygen minimum layer in the Red Sea: Ecological implications of the zooplankton occurrence in the area of the Atlantis II Deep. *Meeresforsch./Rep. Mar. Res.,* **28,** 1–9.

Weikert, H. 1982. The vertical distribution of zooplankton in relation to habitat zones in the area of the Atlantis II Deep, Central Red Sea. *Mar. Ecol. (Prog. Ser.).* **8,** 129–143.

Weikert, H. 1987. Plankton and the pelagic environment, In *Key Environments: Red Sea.* A. J. Edwards and S. M. Head eds. Pergamon. p. 90–111.

Weikert, H., and R. Koppelmann. 1993. Vertical structural patterns of deep-living zooplankton in the NE Atlantic, the Levantine Sea and the Red Sea: A comparison. *Oceanol. Acta,* **16,** 163–177.

Weisse, T. 1989. The microbial loop in the Red Sea: Dynamics of pelagic bacteria and heterotrophic nanoflagellates. *Mar. Ecol. (Prog. Ser.).* **55,** 241–250.

Wiebinga, C. J., M. J. W. Veldhuis, and H. J. W. De Baar. 1997. Abundance and productivity of bacterioplankton in relation to seasonal upwelling in the Northwest Indian Ocean. *Deep-Sea Research (Part I, Oceanographic Research Papers) [Deep-Sea Res. (I Oceanogr. Res. Pap.)],* **44,** 451–476.

Wilkinson, C. R. 2002. Status of coral reefs of the world: 2002. Australian Institute of Marine Science, Townsville.

Winter, A., Z. Reiss, and B. Luz. 1979. Distribution of living coccolithophore assemblages in the Gulf of Elat (Aqaba). *Mar. Micropaleont.,* **4,** 197–223.

Woelk, S., and D. Quadfasel. 1996. Renewal of deep water in the Red Sea during 1982–1987. *J. Geophys. Res. (C Oceans),* **101,** 18155–18165.

Wolf-Vecht, A., N. Paldor, and S. Brenner. 1992. Hydrographic indications of advection/convection effects in the Gulf of Elat. *Deep-Sea Res. (A Oceanogr. Res. Pap.),* **39,** 1393–1401.

Wooster, W. S., Schaefer, M.B., Robinson, M.K. 1967. Atlas of the Arabian Sea for fishery oceanography. La Jolla, University of Calif. Inst. Of Marine Resources.

Wüst, G. 1934. Salzgehalt und Wasserbewegung im Suezkanal. *Naturwissenschaften,* **36,** 447–450.

Wunsch, M., and C. Richter. 1998. The CaveCam–an endoscopic underwater videosystem for the exploration of cryptic habitats. *Mar. Ecol. Prog. Ser.,* **169,** 277–282.

Wyrtki, K. 1974. On the deep circulation of the Red Sea. Conference Symposium de l'Association Internationale des Sciences Physiques de l'Ocean (I.A.P.S.O.), Paris (France), 9 Oct 1972 (Physical oceanography of the Red Sea). L'Oceanographie physique de la Mer Rouge. no. 2, [Publ.CNEXO (Actes Colloq.)]

Yahel, G., A.F. Post, K.E. Fabricius, D. Marie, D. Vaulot, and A. Genin. 1998. Phytoplankton distribution and grazing near coral reefs. *Limnol. Oceanogr.,* **43,** 551–563.

Zwarts, L., Felemban, H., and Price, A.R.G. 1991. Wader counts along the Saudi Arabian Gulf coast suggests that the Gulf harbours millions of waders. *Wader Stud. Group Bull.,* **63,** 25 - 32.

Chapter 35. INTERACTIONS BETWEEN PHYSICAL, CHEMICAL, BIOLOGICAL, AND SEDIMENTOLOGICAL PROCESSES IN AUSTRALIA'S SHELF SEAS (30,W-S)

SCOTT A. CONDIE

CSIRO Marine Research
GPO Box 1538, Hobart, Tasmania 7001, Australia
Email: scott.condie@csiro.au

PETER T. HARRIS

Geoscience Australia, Petroleum and Marine Division
GPO Box 378, Canberra ACT 2601, Australia
Email: Peter.Harris@ga.gov.au

Contents

1. Introduction

Australia is an island continent extending from tropical to midlatitude waters with an Exclusive Economic Zone of some 8.6 million square kilometres. Its regional seas are exposed to climatological conditions ranging from the westerly Roaring Forties winds in the south, to monsoon and tropical cyclone conditions in the north. It also encompasses regions of extreme biodiversity, with 80 percent of southern temperate species endemic to the region.

The focus of marine research in Australia is becoming more interdisciplinary in response to factors such as the national government's recent Oceans Policy, which emphasises sustainable development and ecological based management (Reichelt and McEwan, 1999). However, historically there have been relatively few major interdisciplinary studies in Australian waters. Initiatives, such as the Leeuwin Current Interdisciplinary Experiment (Smith et al., 1991) might be more accu-

The Sea, Volume 14, edited by Allan R. Robinson and Kenneth H. Brink
ISBN 0–674–01527–4

rately described as multidisciplinary, with very limited effort devoted to integration. Interdisciplinary work that has been undertaken has tended to be concentrated in a few regions. Most notably in the Great Barrier Reef where there has been a strong interest in terrestrial inputs and larval exchange between reefs, in the Tasman Sea adjacent to Sydney where the East Australian Current has a strong influence on upwelling and primary productivity, and off the west coast where larval dispersion impacts recruitment levels in high value fisheries.

Because an interdisciplinary review could in principle include most of the marine science undertaken in Australian waters, it has been important to limit its scope. The focus is therefore restricted to shelf scale processes, including major coastal and offshore influences, but excluding the surf zone, intertidal zone, estuaries and bays. This distinction is important since a large proportion of the interdisciplinary marine research conducted in Australian waters has been undertaken in these coastal environments. The other significant limitation is that the review is largely restricted to the mainstream scientific literature. In exceptional cases where reports have been cited, web access details have been provided where possible.

This review will begin in Section 2 with an overview of the broad-scale characteristics of Australia's shelf seas, considering physical, chemical, biological, and sedimentological aspects of both its tropical and temperate regions. Interdisciplinary studies will be reviewed on a regional basis in Section 3, along with recent attempts to synthesize information at the national scale. Research directions and emerging scientific issues will then be discussed in Section 4.

2. Characteristics of Australia's Shelf Environment

This section provides general descriptions of the geomorphological, sedimentological, physical, chemical, and biological properties of the shelf-slope seas around Australia. These descriptions will be based on bathymetric and sediment datasets, hydrographic climatologies, satellite imagery, biogeographic descriptions, and key in situ observations.

2.1 Geographical and Geomorphologic Setting

The shape of the Australian continent owes its origin to the rifting apart of the Gondwanaland super-continent. About 65 million years ago Australia was still a part of Gondwanaland, connected along its southern margin to what is now Antarctica. Rifting and seafloor spreading caused the splitting away of Australia and its last connection with Antarctica, at the southern edge of Tasmania, was severed about 50 million years ago. Australia moved northward until 10 to 15 million years ago when the northern margin collided with the Pacific Plate causing tectonic uplift, volcanic activity and eventually the creation of Papua New Guinea. Today, Australia is bounded by three oceans (Pacific, Indian and Southern) and four marginal seas (the Timor, Arafura, Coral, and Tasman Seas; Fig. 1).

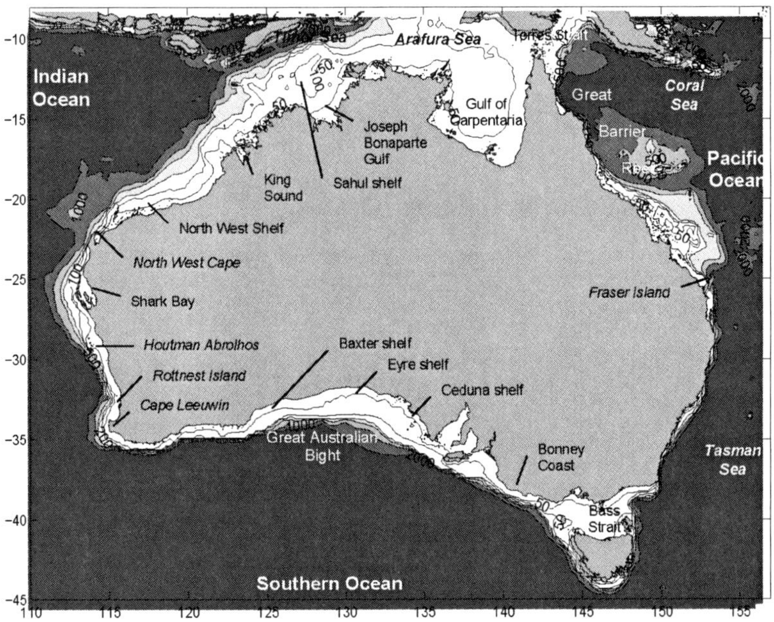

Figure 35.1 Bathymetry and location map for the Australian continent.

Australia's continental shelf is highly variable in terms of its width, depth and profile. The shelf break is located about 10km offshore from Fraser Island and North West Cape, but the Arafura Shelf is over 500km wide. In vertical profile, most of Australia's shelf is a relatively smooth and dips gently seawards (Fig. 1). In contrast, some parts are rimmed by shelf edge barrier reef systems, such as the Great Barrier Reef. During Pleistocene periods of lower relative sea level, the shelf was exposed and the Australian mainland was joined to several of the adjacent large islands such as Tasmania and Papua New Guinea. Such "land bridges" are considered to have facilitated the migration of animals and humans in the late Pleistocene ice age. The shallow seas comprising Bass Strait, the Sahul Shelf, Gulf of Carpentaria and Torres Strait were thus subjected to erosion and sedimentation by rivers and wind on several occasions during the past 150,000 years. Basins perched on the continental shelf in Bass Strait, the Gulf of Carpentaria and Bonaparte Gulf are considered to have been the sites of large fresh to brackish water lakes and lagoons during these periods of emergence (Blom and Alsop, 1988; Jones and Torgersen, 1988; Chivas et al., 2001).

2.2 *Sedimentological Environment*

As on all continental shelves, the nature and distribution of sediments around Australia is determined by a range of past and contemporary influences, including climate and sea level change, past and present shelf energy regimes, sediment supply, morphology, biology and chemistry. These factors are often interconnected in terms of their effect on shelf sediments, and significant effort has been directed towards understanding different processes and their sedimentological expression.

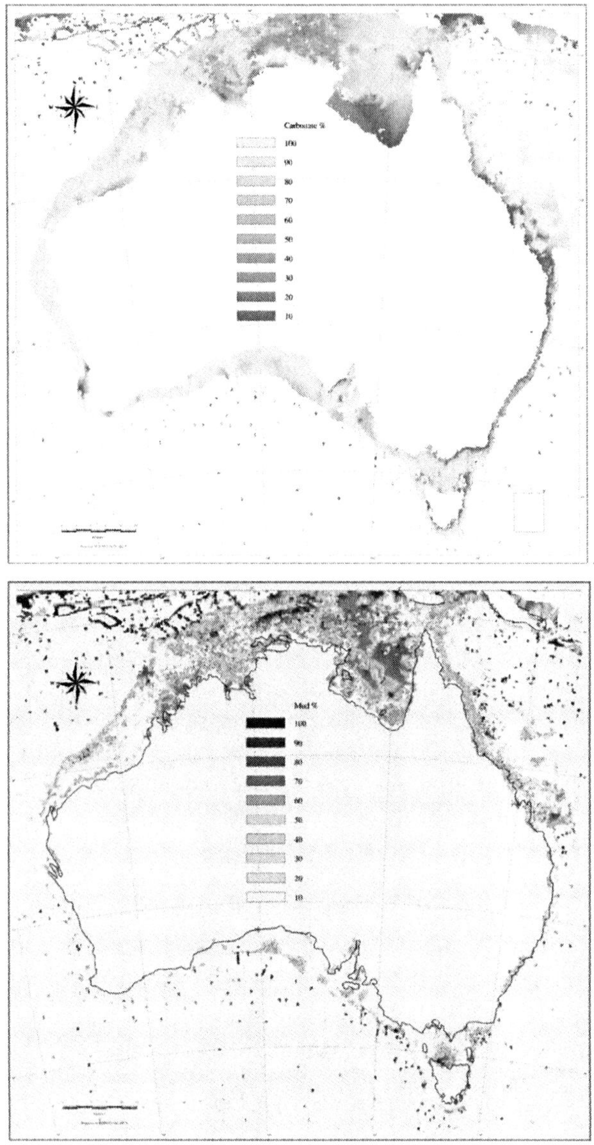

A

B

Figure 35.2 Maps of (A) calcium carbonate content and (B) mud content for surficial sediments on the Australian continental shelf (water depth less than 500m), based on data extracted from the AUSEA-BED database (Jenkins, 1997).

One of the most important parameters characterizing shelf surficial sediment composition and type is the percentage of calcium carbonate content. Australia is distinguished from other continents by comparatively high carbonate content over most of its shelf, exceeding 80% over large parts of the southwest, west and northwest (Fig. 2). These distributions are largely explained by the small riverine sediment loads around most of the continent, supplying less than 150 million tonnes per annum to the entire coastline. Significant regions of non-carbonate sediments of terrigenous origin are only found in the southern Gulf of Carpentaria (Jones 1987), southern Great Barrier Reef (Maxwell, 1968), and southeastern Australian inner shelf.

Another key parameter characterizing shelf sediments is the mud content. On the Australian shelf, enhanced mud content can be attributed to fluvial inputs in a few regions, such the northern Great Barrier Reef (Belperio and Searle, 1988) and eastern Australia (Short, 1979). However, significant mud content is found more commonly in the tide-dominated regions of the north (Jones, 1987), southern Great Barrier Reef (Maxwell, 1968), and Bass Strait (Fig. 2). It has been proposed that continuous reworking of soft calcareous sediment by tidal currents causes grains to fracture and gradually disintegrate, thus generating silt-sized carbonate detritus (Harris, 1994). In this way, carbonate mud is generated and deposited on tide-dominated parts of the shelf, whilst the mean carbonate content of the shelf deposits remains relatively constant (Fig. 2). The lower mud content associated with wave-dominated regions can be attributed to the lower energy available to transport carbonate sand as bedload and the winnowing of fine particles and their export to deeper waters.

2.3 Physical Environment

The Australian continent spans a number of climatic zones, ranging from wet tropics in the northeast and dry tropics in the northwest to temperate conditions in the south. The large-scale wind patterns show significant seasonal variability (Fig. 3). They are influenced by the tropical monsoon in the north, where persistent southeasterly trade winds are replaced by northwesterlies during summer. The trade winds extend over the subtropics and are then replaced by the westerly Roaring Forties, which reach southern Australia during winter and spring.

Oceanic temperatures, salinities, and current patterns also exhibit strong seasonality and major contrasts between tropical and temperate systems (see review by Church and Craig, 1998; Ridgway et al., 2002). Currents throughout the shallow tropical shelf seas from North West Cape to the southern Great Barrier Reef on the east coast are dominated by mainly semidiurnal tidal flows. The peak tidal range occurs at King Sound (>10 m), with strong tidal currents also common at bathymetric constrictions such as Torres Strait and parts of the Great Barrier Reef. An M_2 amphidrome off the southwest of the continent results in diurnal tides locally, with semidiurnal and mixed tides in other subtropical and temperate regions. With the exception of semi-enclosed regions, such as Bass Strait and the South Australian Gulfs, tidal elevations and currents are relatively small compared to those in the tropical regions.

Wind-driven circulation is significant over much of shelf and is known to generate coastal-trapped ways around much of the coastline (e.g., Church et al., 1986). With the possible exception of southwestern Australia, significant wind-driven upwelling events tend to be infrequent and localized (Schahinger, 1987; Gersbach et al., 1999; Pearce and Pattiaratchi, 1999). On the other hand, stronger winds (Fig. 3) and surface cooling during the winter months often results in complete vertical mixing of shallow seas such as the Gulf of Carpentaria and Bass Strait.

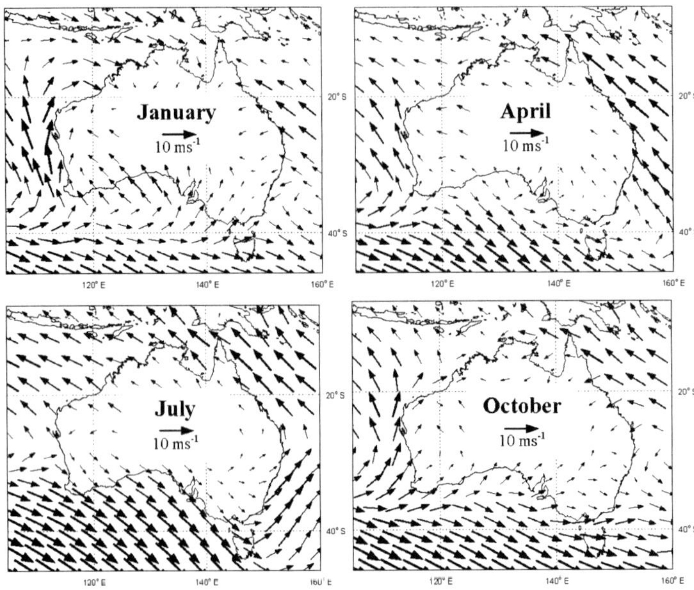

Figure 35.3 Large-scale seasonal wind patterns illustrated using wind vectors at 10 m from the NCEP-NCAR Reanalysis product (Kalnay et al.1996).

Regional current systems have a major impact on Australia's shelf and slope seas, particularly off the east and west coast (Fig. 4). The major western boundary current in the South Pacific is the East Australian Current. This system is fed from the east by a complex pattern of flows forming the South Equatorial Current (Ridgway, 2003), which split into northward and southward branches following the continental slope offshore of the Great Barrier Reef (Brinkman et al., 2001). To the south, the East Australian Current develops as a series of intense anticyclonic eddies, carrying warm low-nutrient water southward (Figs. 5 and 6). It usually separates from the upper slope at around 33°S to form the southern boundary of the South Pacific subtropical gyre. East Australian Current eddies extend down the east coast of Tasmania during summer, but retreat to the north as the system weakens during winter (Fig. 4). Variability in the East Australian Current is believed to be closely related to upwelling events observed along the east coast.

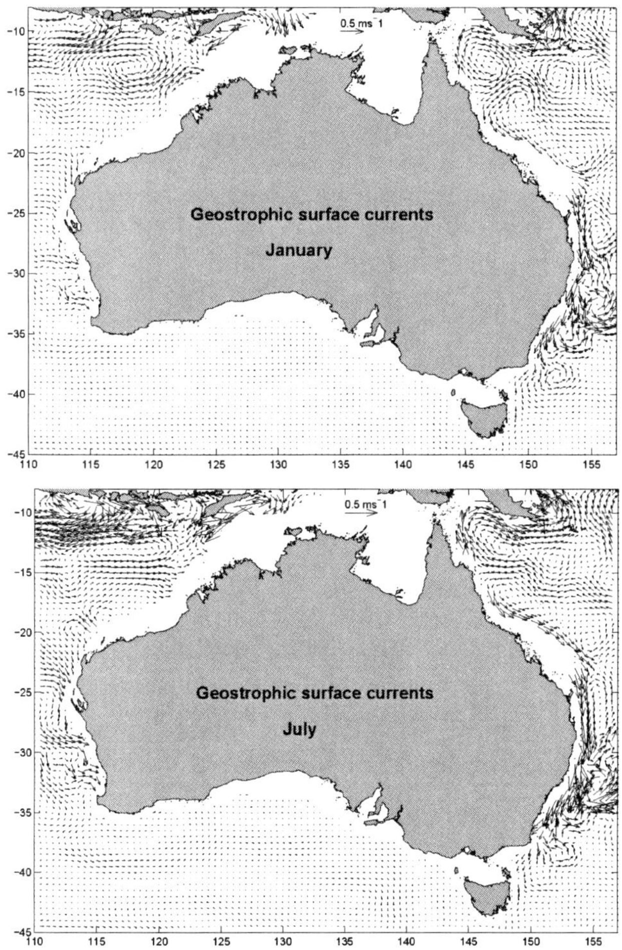

Figure 35.4 Geostrophic surface currents for summer (January – upper) and winter (July – lower) relative to a reference depth of 2000 m. These estimates were made from dynamics height fields based on the CSIRO Atlas of Region Seas (CARS) temperature and salinity fields (Fig. 5) for waters deeper than 200 m.

On the west coast, the poleward flowing Leeuwin Current intensifies along the upper continental slope (Fig. 4) and carries warm, low-salinity water southward (Fig. 5; Godfrey and Ridgway, 1985). It opposes the prevailing southerly winds (Fig. 3) and effectively prevents the formation of a major upwelling system like those observed off the west coasts of other continents. The system is weak during summer, allowing northward flow to develop over the shelf including localized upwelling associated with the Capes Current in the south (Pearce and Pattiaratchi, 1999). The Leeuwin Current then develops rapidly in autumn (Fig. 4), eventually wrapping around the southwest corner of the continent and continuing eastward across the Great Australian Bight to form what is arguably the longest continual coastal current in the world (Ridgway and Condie, 2004).

Another major current system coincides with the Subtropical Front, south of Australia (e.g. Rintoul and Trull, 2001). This feature separates the high nutrient waters of the Subantarctic Zone from more depleted subtropical waters (Fig. 6). It is also evident in the mixed layer depth distribution, but tends to be smeared out in the seasonal temperature and salinity fields due to variability in the location of the frontal zone (Fig. 5). During most of the year the Subtropical Front is south of the Australian continent, but tends to encroach onto the Tasmanian continental slope during the winter months.

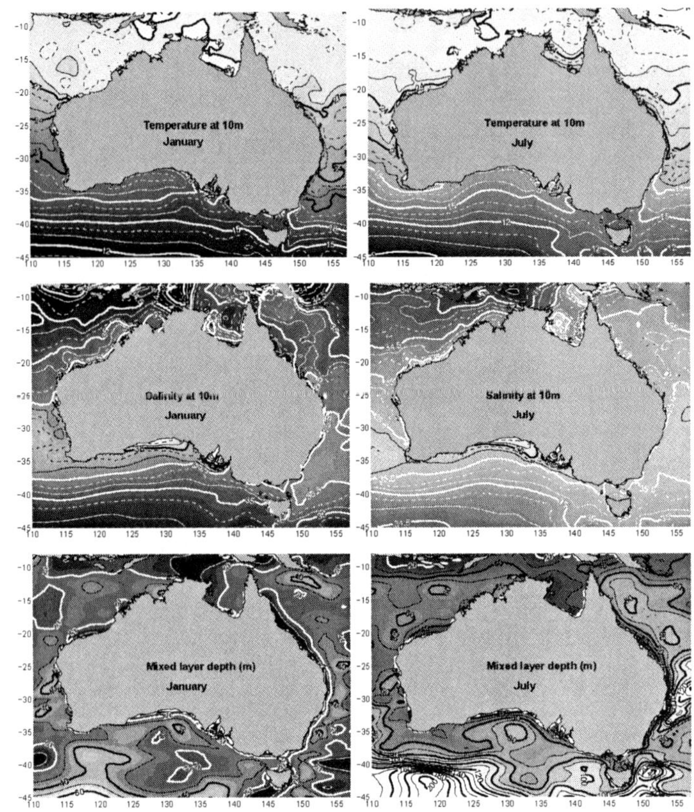

Figure 35.5 Temperature in °C (upper) and salinity in PSU (middle) at a depth of 10 m from the CSIRO Atlas of Regional Seas (CARS, Dunn and Ridgway 2002, Ridgway and Dunn 2002), and an associated estimate of mixed layer depth in metres (lower, Condie and Dunn 2004). Fields are for summer (January – left) and winter (July – right). The base of the mixed layer is defined as the depth where the temperature falls more than 0.04° below the value at 10 m depth or the salinity rises more than 0.03 PSU above the value at 10 m depth.

2.4 *Chemical Environment*

Australia's shelf seas tend to be strongly oligotrophic (Rochford, 1984; Condie and Dunn, 2004). Relatively arid conditions across most of the continent restrict the input of nutrients from riverine sources. As in most tropical systems, nutrients in the surface waters around northern Australia are rapidly consumed and recycled, so that the only significant replenishment is from below the mixed layer. In subtropical waters, both the East Australian Current and Leeuwin Current tend to envelope the continent in low nutrient subtropical water (nitrate < 0.5 μM, Fig. 6). The influence of these systems is seasonal in the southeast, where nitrate and phosphate are moderately enhanced during winter by the withdrawal of the East Australian Current and northward migration of the Subtropical Front (Rochford, 1984).

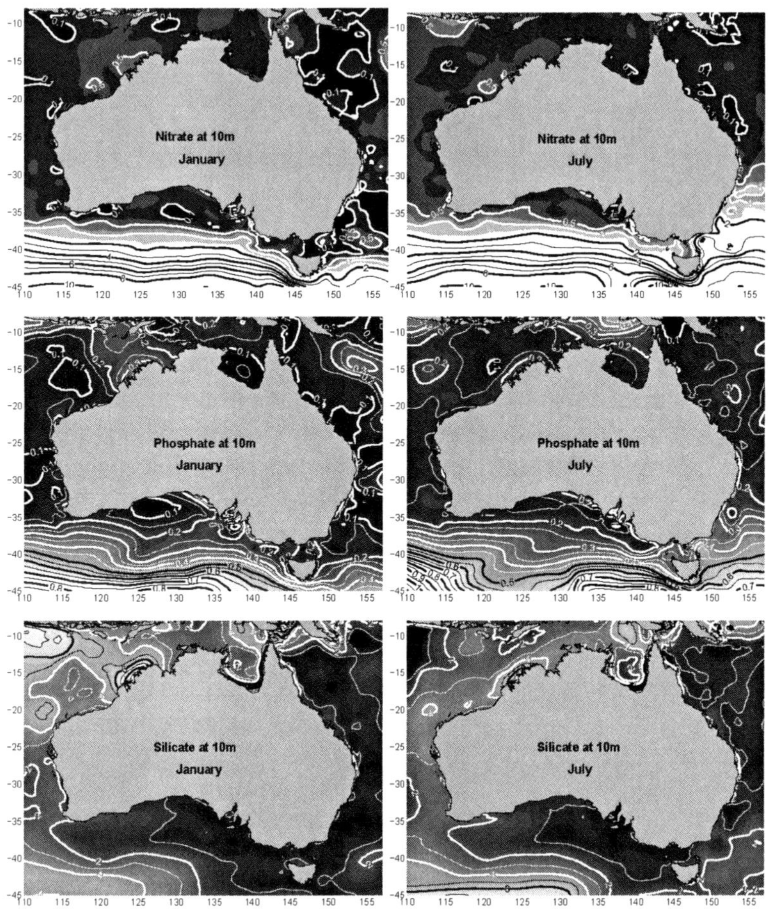

Figure 35.6 Nitrate (upper), phosphate (middle), and silicate (lower) at a depth of 10 m from the CSIRO Atlas of Regional Seas (CARS, Condie and Dunn 2004). Fields are for summer (January – left) and winter (July – right) and are in micromolar units (μM).

Wind-driven upwelling events around Australia, such as occur on the Bonney Coast and Cape Leeuwin areas, tend to be infrequent and localized (Lewis, 1981). Any enhancement in nutrient levels is therefore short-lived and often restricted to subsurface waters, leaving little evidence of upwelling activities in the seasonal fields (Figs. 5 and 6). However, wintertime mixing associated with stronger winds (Fig. 3) and surface cooling in shallow seas, such as the Gulf of Carpentaria and Bass Strait, tend to produce a longer-term enhancement in near-surface nutrient levels (Gibbs et al., 1986).

While eutrophication and other significant chemical contamination have been documented in a number of estuaries and embayments around Australia, there is little evidence of major impacts on the continental shelf. Areas of potential concern include oil, produced formation water, and drilling fluids from oil and gas production on the North West Shelf (Furnas and Mitchell, 1999; Holdway and Heggie, 2000), deep-water sewage outfalls off Sydney (Gray, 1996b; Middleton et al, 1997), and the exposure of the Great Barrier Reef Lagoon to riverine inputs of nutrients and contaminants such as organochlorine compounds, hydrocarbons, and heavy metals (Haynes and Johnson, 2000; Haynes and Michalek-Wagner, 2000).

2.5 *Biological Characteristics*

Information on biological characteristics tends to be concentrated in localized regions with particular conservation or fisheries significance. In the past, significant effort was directed at national collations of species level phytoplankton information (Jeffrey and Hallegraeff, 1990) and zooplankton biomasses (Tranter, 1962). However, more recently collected data, including species distributions, biomass and productivity measurements, and satellite ocean-color data, are still to be incorporated into these descriptions. Other biological distributions have been summarized at various spatial scales in terms of bioregionalisations, which were derived using a diverse range of data inputs and expert opinion.

The phytoplankton biogeography around Australia was described by Jeffrey and Hallegraeff (1990). While nanoplankton species (2–20 μm) are remarkably similar around the continent, significant regional differences are evident in diatom and dinoflagellate communities (30–100 μm) (Fig. 7). Australia's tropical shelf seas contain diatoms, dinoflagellates, and cyanobacteria (*trichodesmium*), but with a high proportion of nanoplankton (70–95% of total chlorophyll). The west and southwest coast and much of the east coast include diatoms, cyanobacteria, and a great diversity of dinoflagellates. Cell concentrations are particularly low off the northeast (Coral Sea), where nanoplankton and picoplankton form a high percentage of total chlorophyll (70–95%). The southeastern region around Tasmania includes a wide variety of diatoms and dinoflagellates, with nanoplankton contributing 50–80% of chlorophyll, except during diatom blooms when this range falls to 10–40%. During summer the southeastern plankton community contracts to the south along the NSW coast to be replaced by Coral Sea species carried by the East Australia Current. Detailed species level information for the three regions can be found in Jeffrey and Hallegraeff (1990) and the earlier compilations of Wood (1954), Crosby and Wood (1958), and Revelante et al. (1982).

Figure 35.7 Regionalisation based on phytoplankton community distributions around the Australian continent (adapted from Jeffrey and Hallegraeff 1990). The three distinct communities reside in the tropical northern Australian shelf seas including part of the North West Shelf (solid grey); the Eastern Indian Ocean and Coral Sea regions (horizontal hatching); and the southeast (diagonal hatching). During summer the southeastern community contracts to the south to be replaced by Coral Sea species (indicated by dashed lines).

Attempts to calibrate satellite ocean-color in Australia's shelf waters have only commenced recently and reliable algorithms are not yet available. This is particularly problematic in the turbid waters off northern Australia, where large suspended sediment concentrations introduce a significant non-biological contribution to the ocean color signal. Under these circumstances the absolute chlorophyll levels are not always reliable, although the temporal and spatial trends are still informative. Seasonal chlorophyll estimates over the shelf typically fall within the range of 0.3 to 0.9 mg m^{-3}, with peak values usually occurring in the cooler months (Fig. 8). For example, the afore' mentioned wintertime mixing and entrainment of nutrients in the Gulf of Carpentaria and Bass Strait, correspond to almost a doubling in estimated chlorophyll levels according to both ocean color and in situ estimates (Fig. 8 and Table 1). The main exception to winter chlorophyll enhancement is around southern Tasmania, where shallow summer mixed layers (~ 30 m, Fig. 5) and moderately enhanced nutrient levels (nitrate ~ 1 μM, Fig. 6) appear to support quite high chlorophyll concentrations extending offshore to the east (> 0.5 mg m^{-3}, Fig. 8). While localized upwelling events are clearly visible on individual ocean-color images, they have little impact on the seasonal fields due to their intermittent nature. Measurements of primary production are extremely sparse around the Australian shelf, but published rates range up to 3 g C m^{-2} d^{-1} in both tropical and temperate waters (Table 1).

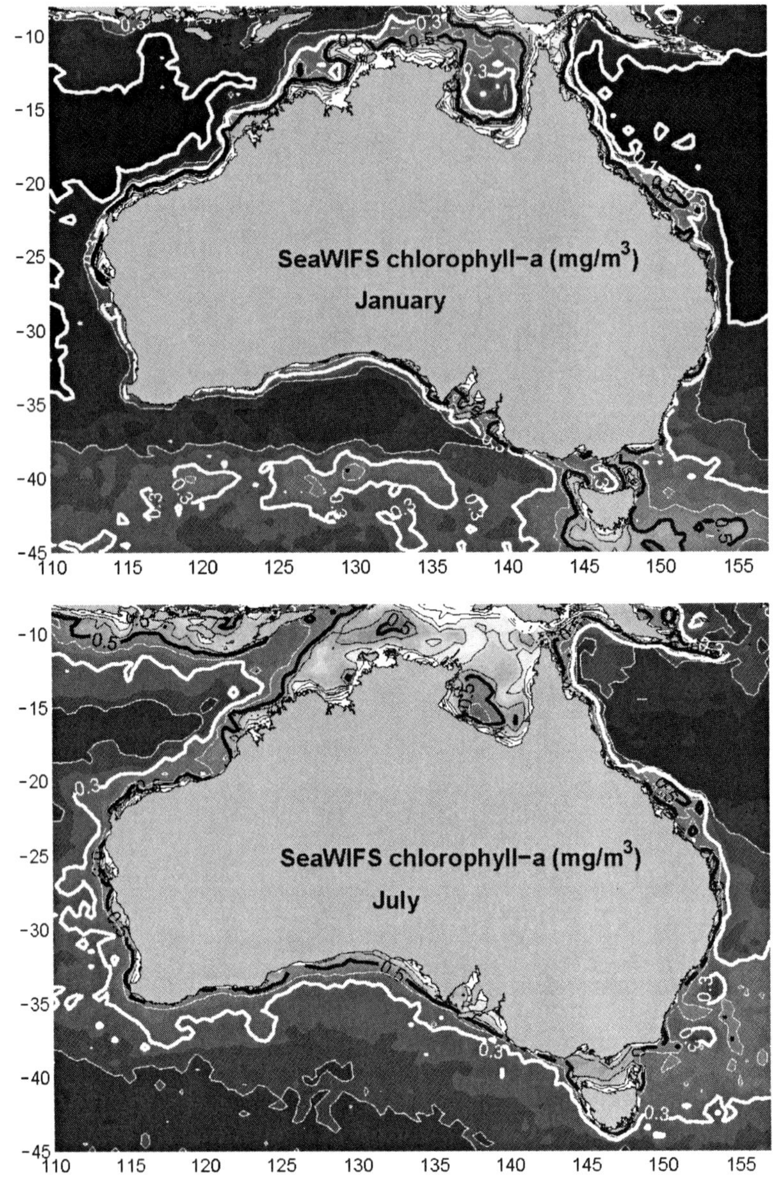

Figure 35.8 Surface chlorophyll concentration (mg m^{-3}) for summer (January – upper) and winter (July – lower) based on SeaWiFS data. These fields are seasonal composites derived using the spatial and temporal mapping methods adopted for the CSIRO Atlas of Region Seas (CARS).

TABLE 1
Estimated chlorophyll concentration and light saturated primary production of phytoplankton in Australian shelf waters. This is an updated version of a table presented by Jeffrey and Hallegraeff (1990), but restricted to shelf waters.

Area	Chlorophyll ($mg\ m^{-3}$)	Productivity ($g\ C\ m^{-2}\ d^{-1}$)	Season	Reference
Central Great Barrier Reef	0.3 – 1.0	0.55 ± 0.23	Summer	Furnas and Mitchell 1987
(mid-shelf)	0.2 – 0.5	0.39 ± 0.18	Winter	Furnas and Mitchell 1987
Central Great Barrier Reef	0.2 – 0.5	0.41 ± 0.17	Summer	Furnas and Mitchell 1987
(shelf-break)	0.3 – 0.5	0.39 ± 0.25	Winter	Furnas and Mitchell 1987
Bass Strait and Tasmanian east coast	1.5 – 2.4	1.3 – 2.9	Spring	Harris et al. 1987
Tasmanian west coast		0.3 – 0.5	Winter	Harris et al. 1987
Great Australian Bight	0.1 – 0.4	0.44	Summer	Motoda et al. 1978
Exmouth Gulf	0.2 – 0.4	0.15 – 0.25	Spring	Ayukai and Miller 1998
North West Shelf (Barrow–Monte Bello Is.)	0.25 – 0.26	0.45 – 2.5	Spring	Furnas and Mitchell 1999
Gulf of Carpentaria -	1 – 2	1.43 ± 0.40	Summer	Rothlisberg et al. 1994
coastal	0.29 ± 0.02	0.95 ± 0.32	Summer	Burford and Rothlisberg 1999
	0.24 ± 0.03	0.75 ± 0.38	Winter	Burford and Rothlisberg 1999
Gulf of Carpentaria -	0.1– 1.6	3.2	Summer	Motoda et al. 1978
central	0.1– 0.2	0.66 ± 0.11	Summer	Rothlisberg et al. 1994
	0.34 ± 0.02	0.96 ± 0.13	Summer	Burford and Rothlisberg 1999
	0.66 ± 0.03	0.56 ± 0.35	Winter	Burford and Rothlisberg 1999

Information on secondary production of zooplankton and higher trophic levels is extremely patchy over the Australian shelf. Tranter (1962) collated zooplankton data from around Australia and reported that biomass levels on the shelf varied from 82 to 213 mg m^{-3} with an average of around 100 mg m^{-3}, which is relatively low compared to many other parts of the world. While significant quantities of new data have been gathered within the intervening 40 years, there has been surprisingly little progress in updating Tranter's analysis or expanding its scope to include more species level information.

Higher trophic levels have been studied mainly in a fisheries context. The most extensive datasets have been collected for the large multi-species fishery operating in southeastern Australia, where dietary data from commercial species has allowed food-webs to be constructed and trophic exchanges to be quantified (Young et al., 1997; Bulman et al., 2001). The focus for seabed habitat studies has followed similar trends, motivated largely by concerns about the impacts of trawling (e.g., Bax et al., 1999; Bax and Williams, 2001). However, substantial effort has also been devoted to maximizing the use of limited habitat data so as to develop preliminary characterizations of the distributions of ecosystems in Australian waters at a range of spatial scales (Fig. 9, Interim Marine and Coastal Regionalisation for Australia Technical Group, 1998; The Marine Science and Technology Plan Working Group, 1999).

TABLE 2:

Studies categorized according to province and dynamical process.

	Great Barrier Reef	East Coast	Southeast Coast	South-west & Great Australian Bight	West Coast	North West Shelf	Northern Australian Seas
Transport of nutrients and sediments from nearshore	Birkeland 1982, Alongi 1990, Mitchell et al. 1997, Larcombe & Woolfe 1999, Orpin et al. 1999, Lambeck & Woolfe 2000, Woolfe et al. 2000, Cappo & Kelly 2001, Lourey et al. 2001, Brunskill et al. 2002, Neil et al. 2002						Wolanski et al. 1999, Hemer et al. 2003
Transport of nutrients from offshore	Wolanski et al. 1988, Furnas and Mitchell 1996	Newell 1966		Lewis 1981, Schahinger 1987		Holloway et al. 1985	Harris 1989, 1991
Primary productivity and plankton blooms	Revelante et al. 1982, Revelante & Gilmartin 1982, Furnas and Mitchell 1986, Furnas and Mitchell 1987	Tranter et al. 1986, Hallegraeff & Reid 1986, Hallegraeff & Jeffrey 1993, Gibbs 2000	Harris et al. 1987, Clementson et al. 1989, Gibbs et al. 1986, 1991, Bax et al. 2001	Motoda et al. 1978	Kimmerer et al. 1985	Tranter & Leech 1987, Ayukai & Miller 1998, Furnas & Mitchell 1999	Motoda et al. 1978, Rothlisberg et al. 1994, Burford et al. 1995, Burford & Rothlisberg 1999
Larval transport and recruitment	Alldredge & Hamner 1980, Doherty et al. 1985, Sammarco & Andrews 1988, 1989, Doherty et al. 1988, Dight et al. 1990a,b, James et al. 1990, Black et al. 1991, Keesing & Halford 1992, James & Scandol 1992, Scandol and James 1992, Brodie 1992, Oliver et al. 1992, Black et al. 1995, Jones et al. 1999	Rothlisberg et al. 1995, Gray 1996, Smith 2000	Harris et al. 1988, Gunn et al. 1989, Bruce et al. 2001a,b,c	Fletcher et al. 1994, Maxwell & Cresswell 1981	Simpson 1991, Caputi & Brown 1993, Hutchins & Pearce 1994, Caputi et al. 1996, Lenanton et al. 1996, Caputi et al. 2001, Griffin et al. 2001	Condie et al. 2003	Rothlisberg et al. 1983, Rothlisberg et al. 1996, Condie et al. 1999

TABLE 2 (*continued*)

Biophysical distributions	Williams & Hatcher 1983, Doherty 1987, Done 1992, Moltschaniwskyj & Doherty 1995, Newman et al. 1997,	Gray 1996a, Dempster et al. 1997, Smith et al. 1999, Smith & Suthers 1999, Gray & Mi-skiewicz 2000, Smith 2000, Young et al. 2001	Harris et al. 1988, Jordan et al. 1995, Bax & Williams 2001, Williams and Bax 2001	Bone & James 1996, Griffin et al. 1997, Li & McGowran 1998, James et al. 2001	Collins et al. 1996, James et al. 1999	McLoughlin & Young 1985, Penn & Caputi 1986, Wilson et al. 2003	Somers 1987, 1994, Wolanski & Ridd 1990

3. Regional and National Interdisciplinary Studies

The Large Marine Ecosystems (LMEs, Fig. 9) represent major marine provinces based on physical characteristics, such as the geomorphology, sediment distributions, and physical and chemical oceanography (as described above), with further refinements based on selected biological data such as fish distributions (The Marine Science and Technology Plan Working Group, 1999). Studies with a significant focus on interactions between the physical, chemical, biological, or sedimentological components of the Australian shelf system will now be described by considering each of these provinces in turn. An attempt has also been made in Table 2 to categorize these studies in terms of the dynamical processes under investigation.

3.1 Great Barrier Reef

The Great Barrier Reef system is made up of almost 3000 individual coral reefs extending 2600 km along Australia's east coast from the Tropic of Capricorn in the south to the coast of Papua New Guinea in the north (Fig. 9). The region is exposed to southeast trade winds throughout most of the year and there is a relatively high incidence of tropical cyclones impacting the reefs (Done, 1992). The reefs cover about 5% of the shelf, while the remainder is characterised by inter-reef sediments, ranging from nearshore terrigenous material to outer shelf carbonate facies (Woolfe et al., 2000). This offshore zonation is also reflected in the long-term carbon burial rates, which consist almost entirely of organic carbon on the inner shelf and carbonate on the mid- and outer-shelf (Brunskill et al., 2002). It also appears to have a marked influence the distributions of various fish communities (Williams and Hatcher, 1983; Doherty, 1987; Moltschaniwskyj and Doherty, 1995; Newman et al., 1997).

Riverine inputs of terrigenous sediment are supplied to the Great Barrier Reef shelf at an annual rate of about 10 tonnes per metre of coastline (Belperio and Searle, 1988). They have high mud content, which is reflected in the nearshore facies (Fig. 2). One exception is the southern Capricorn Channel, where tidal current processes contribute to the production of carbonate mud as described previously (Harris, 1994). High carbonate facies are restricted to areas of reef growth and adjoining areas where reef derived sediments are dispersed by waves and currents. In the high tidal energy zone of the southern Great Barrier Reef, the area of carbonate dispersal is up to 20km distant from the nearest reef, whereas further north it rarely extends more than 2km beyond reef margins (Maxwell, 1968).

There is significant potential for riverine inputs of terrigenous sediment, carbon, nutrients, and other contaminants to impact biota within the Great Barrier Reef lagoon (Isdale, 1984; Haynes and Johnson, 2000; Wolanski, 2001; Cappo and Kelley, 2001; Suzuki et al., 2001). While sediment loads have been increased approximately four-fold by changes in land use (Neil et al., 2002), the impact on coral communities located over the mid- and outer-shelf is still unclear. For example, Larcombe and Woolfe (1999) argue that sediment accumulation and turbidity are limited by wave and current dynamics rather than sediment supply. Sediments, resuspended mainly by wave action on the inner-shelf, migrate northward with the wind-driven currents and may be carried offshore in the lower water column by downwelling flows (Orpin et al., 1999; Lambeck and Woolfe, 2000).

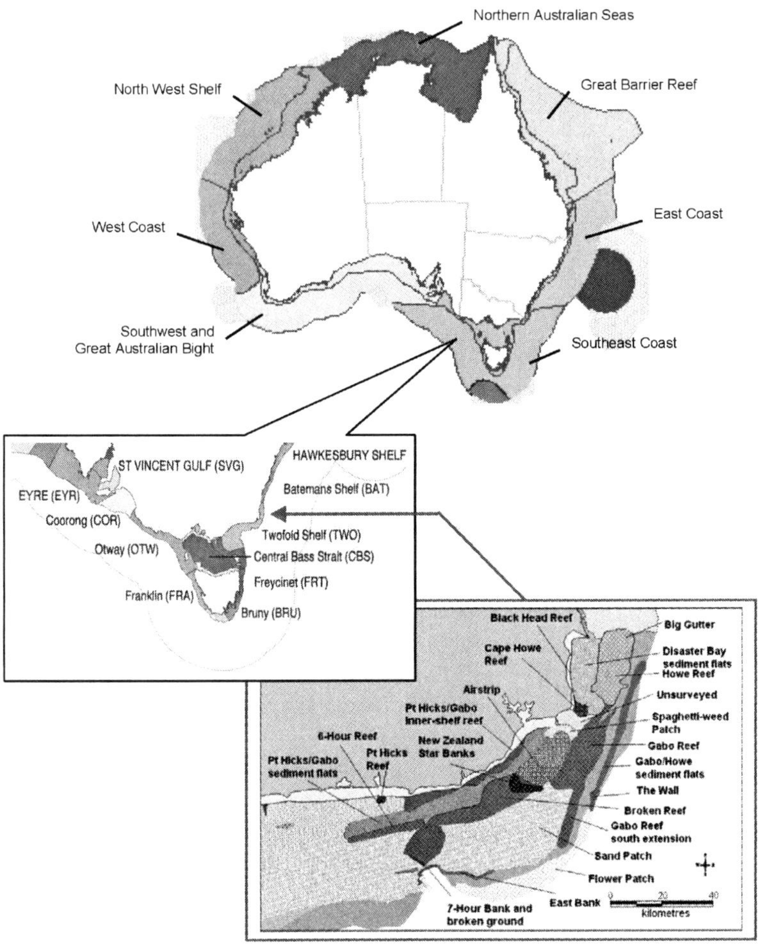

Figure 35.9 A Large Marine Ecosystem (LME) regionalisation for the seas surrounding the Australian continent (upper, The Marine Science and Technology Plan Working Group 1999); the southeast sector of a meso-scale regionalisation referred to as the Interim Marine and Coastal Regionalisation of Australia (IMCRA) for the continental shelf (center left, Interim Marine and Coastal Regionalisation for Australia Technical Group 1998); and a still smaller scale mapping of seabed habitats over part of the south-eastern continental shelf (lower right, adapted from Bax and Williams 2001).

Riverine nutrient inputs are highly variable with strong peaks during floods induced by tropical cyclones (Birkeland, 1982; Mitchell et al., 1997; Cappo and Kelley, 2001). Significant inputs have also been associated with tidal exchange of detritus from coastal mangroves (Alongi, 1990) and regeneration in bottom sediments (Lourey et al., 2001). However, riverine inputs are largely responsible for the seasonality in phytoplankton, particularly the inner-shelf diatom community, which peaks during periods of high rainfall (Revelante et al., 1982; Revelante and Gilmartin, 1982). Nutrient supply to the outer reef from offshore has been documented through summer upwelling events (Furnas and Mitchell, 1986; 1996) and exchanges associated with internal tides (Thompson and Golding, 1981). Wolanski et al. (1988) estimated that the tidal mechanism could easily provide the entire nitrogen requirements of the *Halimeda* banks found inside the outer ribbon reef. However, the waters are much more oligotrophic than those on the inner-shelf, so that dinoflagellates appear to out compete diatoms for available nutrients (Revelante et al., 1982) and when upwelling is absent nanoplankton and picoplankton form most of the biomass (Furnas and Mitchell, 1986). Consequently, total phytoplankton productivity is relatively low and there is little evidence of a regular seasonal cycle (Furnas and Mitchell, 1987; Table 1).

There have been a large number of larval transport studies in the Great Barrier Reef region, applied to recruitment of coral larvae and coral fish larvae, and aggregations of other zooplankton (Doherty et al., 1985; Doherty and Williams, 1988; Williams and English, 1992). These have focused on either retention of larvae around individual reefs (Jones et al., 1999; Carleton et al., 2001), or inter-reef exchange (Dight et al., 1990a,b). Retention or aggregation of larvae adjacent to reefs and coastal features have been associated in this region with physical features such as wake eddies (Alldredge and Hamner, 1980; Sammarco and Andrews, 1988; 1989), tidal fronts, and Langmuir cells (Kingsford, 1990; Kingsford et al., 1991). Even with very modest vertical migratory behavior, larvae can be trapped in small-scale features for extended periods. This may increase food availability and increase the chance of recruitment back to the spawning reef. However, the small length scales of these features and patchiness of larval distributions provides a major challenge for dispersion models (Black et al., 1991; Oliver et al., 1992).

Outbreaks of the coral-eating crown-of-thorns starfish (*Acanthaster planci*) have had major ecological impacts on the Great Barrier Reef, with recovery time-scales of 20 years or more (Johnson, 1992). Understanding of the outbreaks has focused on the chain of reproduction, dispersal, and recruitment. While elevated nutrient levels may enhance larvae survival (Birkland, 1982; Brodie, 1992), physical factors appear to be more important (Keesing and Halford, 1992) with outbreaks favored by weak residual currents and high local retention of larvae (Black et al., 1995). Models of circulation and larval dispersion suggest that once initiated, outbreaks propagate as a wave of secondary outbreaks (Dight et al., 1990a,b; James and Scandol, 1992; Scandol and James, 1992). The spatial extent of the outbreak is restricted by larval behavior and the time-scales over which they are competent to settle (Johnson, 1992).

3.2 East Coast

The east coast province extends from south of the Great Barrier Reef to the southeastern corner of mainland Australia (Fig. 9). The shelf is relatively narrow (< 50 km) and sediments range from clean quartzose sand with low carbonate and mud contents on the inner shelf, to calcareous gravelly sands with high carbonate content on the outer shelf (Short, 1979; Jones and Davies, 1979; Fig. 2). The carbonate material is derived from bryozoans and red algae occurring as rhodoliths off Fraser Island and commonly form cemented hardgrounds (Harris et al., 1994).

The presence of shallow water fossil species is also suggestive of a relict origin (e.g. Roy and Thom, 1981).

The east coast is the strip of shelf most strongly influenced by the East Australian Current and its energetic eddy field. Newell (1966) was the first to propose that nutrient enrichment observed in this area is caused by upwelling of slope water, and subsequent studies have identified a range of current, wind, and topographic interactions that may drive the upwelling (Boland, 1979; Tranter et al., 1986; McClean-Padman and Padman, 1991; Condie, 1995; Gibbs et al., 1997, 1998, 2000; Marchesiello et al., 2000; Oke and Middleton, 2000, 2001). These flows also influence the transport of primary treated sewage plumes from offshore outfalls near Sydney. Salinity fronts associated with these plumes appear to attract large numbers of juvenile fish, although any consequences for contamination of the local food web are still unclear (Gray, 1996b; Middleton et al., 1997).

Results from a long-term monitoring station off Sydney (100 m isobath) have identified variability in the plankton community associated with the East Australian Current (Hallegraeff and Reid, 1986), as well as short-term succession of diatoms and dinoflagellates in spring and summer (Grant and Kerr, 1970), and more persistent groups such as the nanoplankton (Hallegraeff, 1981). Subsequent observations confirm that the phytoplankton response extends over most of the east coast region (Tranter et al., 1986; Hallegraeff and Jeffrey, 1993; Gibbs, 2000) and pelagic species such as yellowfin tuna appear to aggregate in response (Young et al., 2001). While the modeling studies mentioned above have successfully simulated a number of physical upwelling events, plankton responses and other biogeochemical processes are yet to be explicitly represented in these models.

The cross-shelf circulation on the east coast appears to have a significant influence on the vertical distribution of larval fish. One consequence is that there is no simple correlation between species distributions and water masses (Miskiewicz et al., 1996; Dempster et al., 1997; Gray and Miskiewicz, 2000). However, consideration of offshore transport during upwelling favorable winds has allowed larval distributions to be tracked over timescales of days to weeks (Smith et al., 1999; Smith and Suthers, 1999; Smith, 2000). For other species living in such a dynamic environment, larval behavior may be the dominant influence (Gray, 1996a,b). For example, the vertical migratory behavior of prawn larvae has been shown to assist recruitment to east coast estuaries through tidal advection processes (Rothlisberg et al., 1995).

3.3 Southeast Coast

The southeast coast region encompasses all the waters surrounding the island of Tasmania, including Bass Strait (Fig. 9). The area covers a temperate, carbonate-shelf environment, where the lack of large rivers results in a characteristically low level of terrigenous material being deposited (Rao, 1986; Blom and Alsop, 1988; Lavering, 1994; Harris, 1994). Sediments are reworked and dispersed by tidal and ocean currents and storm waves, with sediment grain sizes and composition often being closely related to the controlling physical processes (Malikides et al., 1988, 1989; Harris, 1994). Fine sediments (muds and silty sands) occur only in pockets along the east coast of Tasmania and in the central Bass Strait where carbonate content is also very high (Fig. 2). A notable plume of terrigenous (low-carbonate) sediments, located off the west coast of Tasmania, probably reflects the influx of fluvial sediments from Macquarie Harbour.

Winter cooling of the shallow waters of Bass Strait generates a strong frontal region where they meet the warmer waters of the Tasman Sea. This feature feeds into the Bass Strait Cascade, which carries nutrient rich water north along the mainland east coast and supports enhanced phytoplankton and zooplankton levels (Gibbs et al., 1986, 1991; Bax et al., 2001). Further south, the Tasmanian shelf is influenced by the Subtropical Front separating the low nutrient subtropical waters of the East Australian Current from more productive subantarctic waters. Long-term monitoring off eastern Tasmania (~ 40 years) reveals high interannual variability in the timing and duration of the spring phytoplankton bloom (Harris et al., 1987; Clementson et al., 1989, Table 1). This variability is driven by a combination of seasonal and episodic events related to ENSO and the location of the westerly wind belt, which influence environmental factors such as the positioning of East Australian Current eddies and local mixed layer depths. It also appears to have a significant influence on spawning times and recruitment levels in local fisheries (Harris et al., 1988; Jordan et al., 1995).

The shelf-break current system known as the Zeehan Current travels anticlockwise around western and southern Tasmania (Baines et al., 1983). Studies using aged larval data suggest that the recruitment of finfish spawning over the Tasmanian slope may be strongly influenced by advection patterns within this system (Gunn et al., 1989; Bruce et al., 2001b,c). This conclusion has been further supported by recent modeling of dispersion patterns in the southeast region suggesting that larvae sampled off the southeast mainland were spawned off eastern Bass Strait slope, rather than the known spawning grounds off the west coast of Tasmania (Bruce et al. 2001a).

3.4 The Southwest and Great Australian Bight

This region extends from the southwest corner of mainland Australia eastward across the Great Australian Bight to the gulfs of South Australia. It comprises the largest cool-water carbonate province in the modern world (James et al., 2001). The facies are characterised by abundant bryozoans and are contrasted from tropical carbonates by the lack of corals and Halimeda (Bone and James, 1993). In the Great Australian Bight biogenic sediment production is balanced by surface-wave erosion processes to yield zero net sedimentation (James et al., 1994). Sediments found in less than 100 m water-depth contain mainly coarse-grained sand and gravel consistent with wave reworking and slow accumulation rates (James et al., 2001). Sediment facies exhibit contrasts in upwelling, downwelling and sedimentation patterns between the central Bight (Eyre shelf) and adjacent shelves to the west (Baxter shelf) and east (Ceduna shelf), where periodic upwelling occurs (James et al., 2001). These gradations are also reflected in the distributions of planktonic foraminifera (Li and McGowran, 1998). The inner Baxter shelf experiences warm (22°C) oceanic conditions due to the southern arm of the Leeuwin Current, which supports luxuriant algal growth and the deposition of abundant rhodoliths, while sediments on the middle shelf and upper slope are dominated by molluscs and bryozoans.

Seasonal variations in the southern arm of the Leeuwin Current also impact the dispersion of eggs and larvae along the southwest coast from both small pelagics such as pilchards (Fletcher et al. 1994) and more tropical species (Maxwell and Cresswell, 1981). Because the Leeuwin Current is low in nutrients and there are few terrestrial inputs, productivity tends to be low over most of the Great Australian Bight (Motoda et al. 1978, Table 1). However, seasonal winds off Kangaroo Island can lead to episodic upwelling of higher nutrient slope water (Lewis 1981, Schahinger 1987) and localized phytoplankton blooms. Griffin et al. (1997) considered the potential role of these processes in initiating a mass mortality of adult pilchards in 1995, but concluded that environmental conditions were not abnormal for the region at the time of the outbreak.

3.5 *West Coast*

The west coast shelf is relatively broad in the north where the main feature is Shark Bay, then gradually narrows to the south (Fig. 1). Sediments deposited on the shelf along the southwest margin of Western Australia reflect the transition from sub-tropical to temperate climate regime that characterizes this region. West-coast rivers carry little sediment and most of this is trapped in estuaries. Thus terrigenous sediment is typically a minor constituent in shelf deposits, which are mostly algal remains, shell fragments, foraminifera, and bryozoa. The southern Rottnest Shelf sediments consist of mobile medium-to-coarse sands, transitioning to finer silts and clays beyond the 90 m isobath (Collins, 1988). In contrast, the northern Rottnest shelf is characterised by luxuriant seagrasses growing on coralline-encrusted hardgrounds (James et al., 1999). The Houtman Abrolhos reef platforms support widespread coral growth with surrounding sediments dominated by bryozoans and coralline red algae (Collins et al., 1996).

Adjacent to Shark Bay, the Dirk-Hartog shelf exhibits mostly relict sediment of sorted planktonic foraminiferal sands under strong bottom currents, or spiculitic mud. James et al. (1999) suggested that relict sediments are due to arrested carbonate sedimentation caused by downwelling and episodic outflows of saline waters from Shark-bay onto the outer shelf and upper slope. Within Shark Bay the plankton community is dominated by diatoms and small copepods (Kimmerer et al., 1985), and algae trap and bind sediment in columnar structures referred to as stromatolites (Logan et al., 1964; Playford, 1979). These structures develop preferentially in the hypersaline parts of Shark Bay, where grazing pressure by metazoans such as gastropods is low. Here extreme nutrient limitation also favors a plankton community consisting mostly of dinoflagellates and demersal forms of zooplankton (Kimmerer et al., 1985).

Biophysical studies along Australia's west coast have focused on the role of the Leeuwin Current on larval advection and recruitment of finfish and invertebrates such as the western rock lobster (Caputi et al., 1996). For example, the main spawning of tailor on the inner shelf coincides with the presence of wind-driven northward coastal currents capable of transporting eggs and larvae towards coastal nursery areas (Lenanton et al., 1996). Rock lobsters have a much longer planktonic larval phase (9–11 months) and the correlation between Leeuwin Current strength and settlement is well established (Pearce and Phillips, 1988; Caputi and Brown, 1993; Caputi et al., 2001). However, sophisticated models of larval movements have not been able to explain the observed interannual variability in settlement (Griffin et al., 2001).

The Leeuwin is also responsible for the recruitment of tropical reef fishes as far south as Rottnest Island off Perth (Hutchins and Pearce, 1994), and influences the recruitment of other species such as scallop and pilchards (Caputi et al., 1996), and possibly corals (Simpson, 1991; Simpson et al., 1993). The presence of this poleward flow also suppresses upwelling along the west coast and the resulting low primary productivity levels appear to strongly limit fish densities (Williams et al., 2001).

3.6 North West Shelf

Australia's North West Shelf encompasses the tropical waters from Northwest Cape northeast almost to Joseph Bonaparte Gulf. It is characterised by large tidal ranges, including internal tides along the outer shelf, and a high incidence of tropical cyclones (Lough, 1998). Each of these processes contributes to the movement of sediments across the North West Shelf (Ribbe and Holloway, 2001). The composition of the sediments is predominantly calcareous sands and gravels derived from relict skeletal debris (Jones, 1973; McLoughlin and Young, 1985). Calcareous pellets (pisoliths) and oolites comprise more than 50% of the sediments over much of the middle and outer shelf (Jones, 1973), indicating that most of the shelf is currently starved of sediments. Finer silts and clays appear to be restricted to coastal environments, such as Exmouth Gulf (Brunskill et al., 2001), where water column turbidity is also high.

With minimal terrestrial inputs (Lough, 1998; Burns et al., 2003), nutrients are supplied to the shelf through a combination of barotropic and internal tidal motions, summertime wind-driven upwelling, and episodic cyclone events (Holloway et al., 1985). Because upwelling does not extend to the surface, nutrient levels in surface waters remain very low. As a consequence, phytoplankton biomass is also low, although rapid recycling of nutrients supports high primary productivity rates (Table 1, Furnas and Mitchell, 1999; Burns et al., 2003) and higher than average zooplankton abundance (Tranter, 1962; Wilson et al., 2003). While the biomass levels are similar in Exmouth Gulf, lower primary productivity and high grazing pressure have been reported in this system (Table 1, Ayukai and Miller, 1998). Over the mid- and outer-shelf phytoplankton biomass tends to be concentrated below the thermocline or in the bottom mixed layer, where benthic production is also high (Tranter and Leech, 1987).

Recent modeling studies indicate that primary production is closely coupled to the spring-neap tidal cycle, with only relative weak seasonal variability (NWSJEMS). However, El Nino conditions also affect the region by lowering the incidence of tropical cyclones, weakening the Indonesian Throughflow, and allowing increased upwelling. Recent studies suggest these conditions support higher chlorophyll and zooplankton abundance on the shelf (Wilson et al., 2003). A relationship has also been demonstrated between tropical cyclone incidence and recruit survival for tiger prawns in Exmouth Gulf (Penn and Caputi, 1986). The influences of seasonal and interannual variability on larval dispersion are only now being studied for coral spawning events, commercial pearl oysters, and other applications (Condie et al., 2003).

3.7 Northern Australian Seas

The tropical coastal waters of northern Australia are characterised by broad shallow shelf seas, such as the Timor Sea (Sahul Shelf), the Arafura Sea, the Gulf of Carpentaria, and Torres Strait. Currents in the Timor Sea are influenced by the eastern arm of the Indonesian Throughflow (Godfrey, 1996), as well as seasonal winds driving currents to the northeast during the summer monsoon and to the southwest during winter (Cresswell, 1993). However, most of the shelf is dominated by tidal motions, which are particularly strong in Joseph Bonaparte Gulf and Torres Strait.

The surficial sediment composition varies significantly across the region. On the Sahul Shelf grain size is negatively correlated with calcium carbonate content (Fig. 2), with coarse-grained carbonate sediments contrasting with fine-grained terrigenous mud (Van Andel and Veevers, 1967). This relationship breaks down in Joseph Bonaparte Gulf where sediments are enriched with quartzose terrigenous sand deposited as ebb tidal deltas (Lees, 1992). On the Arafura Shelf carbonate content diminishes toward the coast as fluvial inputs become more significant (Fig. 2).

In the Gulf of Carpentaria, bottom sediments and suspended sediments generally trend parallel with the coastline (Jones, 1987; Wolanski and Ridd, 1990). They include sandy near-shore facies characterised by relatively high sedimentation rates and mud-sand mixtures in deeper areas characterised by low sedimentation rates. These patterns have been shown to influence the distribution of commercial prawns, with some species preferring muddy sediments and others sandy sediments (Somers, 1987, 1994). Tidal sand banks and subtidal dunes can also be found around the major island groups. The latter have been shown to reverse their orientation in Torres Strait in response to seasonal reversals in wind direction (Harris, 1988, 1991). Sedimentation rates in northern Torres Strait may also be enhanced by inputs from the Fly River, with potential impacts on coral and seagrass communities (Wolanski et al., 1999; Hemer et al., 2003).

Turbidity tends to be high in the northern seas, particularly within the coastal zone (0 – 20 m), where phytoplankton production has been found to be light limited (Table 1, Rothlisberg et al., 1994; Burford and Rothlisberg, 1999). While nitrate and phosphate levels are relatively low, the availability of silicate allows diatoms to dominate the phytoplankton community with chlorophyll levels peaking over the wet season (Burford et al., 1995). Further offshore thermal stratification develops during summer and surface production becomes nutrient limited. The net effect is limited seasonal variability in chlorophyll without any obvious species succession of diatoms (Burford et al., 1995).

Because of their commercial importance, a number of modelling studies have addressed the advection and recruitment of prawn larvae to mangrove and seagrass nursery beds (Rothlisberg et al., 1983, 1996; Condie et al., 1999). These studies have demonstrated how larvae can move rapidly towards the coastline and estuaries by coinciding periods of pelagic swimming with flood tides, and periods of benthic habitation with ebb tides. However, the results of these models are sensitive to details of the larval behavior, which are yet to be fully quantified.

3.8 National integration

Section 2 described various attempts to develop geographically integrated descriptions of the physical, chemical, biological, or sedimentological environment around the Australian continent. A small number of studies have progressed to exploring interdisciplinary links at the national scale. For example, Porter-Smith et al. (2004) have used oceanographic conditions and sediment characteristics from around the entire continental shelf to identify where bottom sediments are likely to be mobilized and by what processes. Their results indicate that sediments are mobile over most of the mid- and inner-shelf, with the North West Shelf, Northern Australia (excluding the southern Gulf of Carpentaria), southern Great Barrier Reef, and parts of Bass Strait dominated by tidally induced resuspension, and the remainder by wave induced resuspension (Fig. 10).

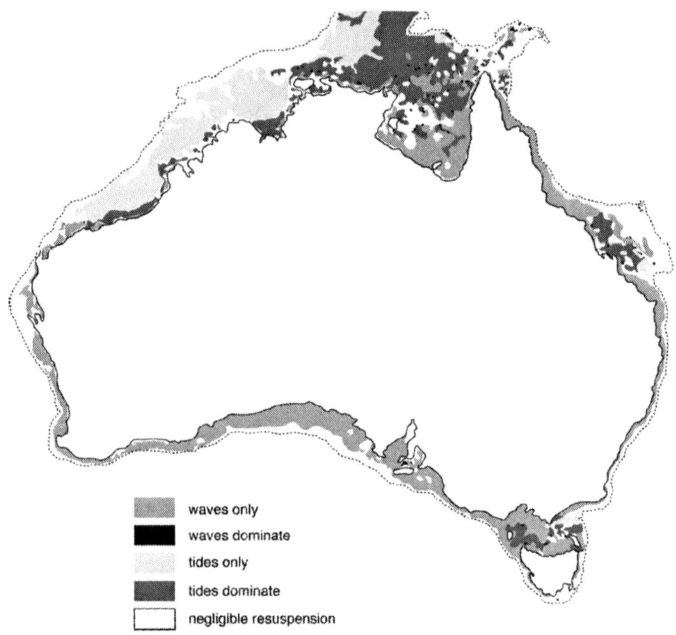

Figure 35.10 Regions of the Australian shelf in which sediment resuspension occurs and whether it is caused by surface waves or tidal currents (adapted from Porter-Smith et al. 2004).

The national biogeographic description used to structure this review (Fig. 9) was developed mainly on the basis of benthic and epibenthic information. However, pelagic descriptions have also been constructed on the basis of seasonal patterns of phytoplankton distributions (Fig. 8) and the physical and chemical conditions limiting primary production (Figs. 5 and 6). While the global biogeographic analysis of Longhurst (1998) does not adequately resolve features on the Australian shelf, finer-scale analyses are consistent with nutrient limitation of surface layer primary productivity over nearly all the shelf (Condie and Dunn, 2004). The main exceptions are the turbid tropical coastal waters of the North West Shelf and Northern Australia, where light limitation is a significant factor.

Approaching the topic of larval advection at a national scale is difficult because of potential dependencies on larval behavior and small-scale circulation features. However, by focusing on the large-scale transport of larvae residing in the upper water column, Condie et al. (2004) have recently developed a national analysis of marine connectivity patterns for advection timescales of days to months. While not replacing the finer-scale larval advection studies needed in complex coastal environments, it provides a generalized description within an easily accessible web-based user-interface (*http://www.per.marine.csiro.au/aus-connie*).

4. Research Directions and Issues

While particular regions or topics have received special attention due to their conservation significance (e.g. nutrient inputs and larval transport on the Great Barrier Reef) or relevance to major fisheries (e.g. larval transport on the West Coast), the number of studies that have focused on interdisciplinary processes in the Australian marine environment is still relatively small (Table 2). However, this situation is changing rapidly as the increasing focus on ecosystem-based management prompts a range of new initiatives. For example, a diverse range of characterization and process studies are progressing on the Great Barrier Reef and providing a basis for the development of a more integrated understanding of the system. While there is significantly less data available for the North West Shelf, highly integrated models have recently been developed for this ecosystem and its interactions with human activities (Section 3.6). Other major ecosystem studies have also been initiated on the west coast, where the focus is on biophysical interactions of the offshore Leeuwin Current system with the shelf and coastal environment, and in the shallow tropical waters of Torres Strait where the focus is on seabed habitats and their response to oceanographic and sedimentary processes. A range of fisheries and ecologically related studies in the Southeast are also beginning to address issues such as the relationship between ocean circulation, primary production, recruitment, and trophic interactions.

4.1 Major information gaps

A comprehensive assessment of information gaps is beyond the scope of this review. However, on such a large continental shelf surrounding a sparsely populated continent, there are clearly large data gaps across all disciplines. Even a seemingly large dataset, such as historic hydrography, is insufficient to characterize the seasonal cycle in many areas of the Australian shelf (Ridgway et al., 2002). Biological distributions are typically very patchy, but for key variables such as primary productivity there are estimates available from only a few localities (Table 1). There are other important ecosystem components that have received almost no attention. For example, studies in other parts of the world suggest that microphytobenthos play a much larger role in nutrient cycling on continental shelves than previously suspected. Information on zooplankton distributions and secondary production is also quite limited and limited progress has been made in collating this data and interpreting it within an ecosystem context. While obtaining the necessary field data to address these issues is major challenge, they represent critical sources of uncertainty in both biogeochemical models and tropic models linking primary production to fisheries production.

4.2 Scale related issues

A recurring issue when considering interdisciplinary interactions in Australian shelf waters (and elsewhere) is that temporal and/or spatial scales are often not directly compatible. This may relate to the scales of the field sampling, model resolution, or the dynamical scales of underlying processes. For example, while it has been clearly established that El Nino cycles effect Australia's marine environment, local ecological impacts of climate variability and change are largely unexplored. Scale mismatches can be particularly problematic where water column and benthic processes are coupled, the latter usually being associated with much smaller length scales. Scales are also important when considering the influence of larval behavior on transport patterns. While most behaviors have limited ability to effect long pelagic phases (e.g. inter-reef exchange), they may be critical during the settlement phase (e.g. retention around reefs). Improved strategies are clearly required for re-scaling information and coupling processes operating at different scales.

4.3 Whole-of-system approaches

This review has described a diverse range of studies exploring specific interdisciplinary links. However, the ultimate goal is to expand this approach into a whole-of-system understanding through the development of conceptual, statistical, or dynamical integrated models. While many of the necessary components for such models are available for intensively studied regions such as the Great Barrier Reef, it has so far only been seriously attempted on the North West Shelf (North West Shelf Joint Environmental Modelling Study, 2002; Condie et al., 2003). The outcomes of this study are centered around coupled ecosystem models incorporating ocean circulation, waves, sediment dynamics, nutrient cycling, primary and secondary production, and higher trophic interactions. This approach has revealed strong interactions between the spring-neap tidal cycle, suspended sediments, primary productivity, and habitat and fish distributions. These ecosystem models have also been dynamically linked to models describing the cumulative impacts of human activities (Fig. 11). This approach is supporting management of the region by allowing a wide range of management strategies to be evaluated in terms of the sometimes-conflicting needs of multiple human uses and the natural ecosystem.

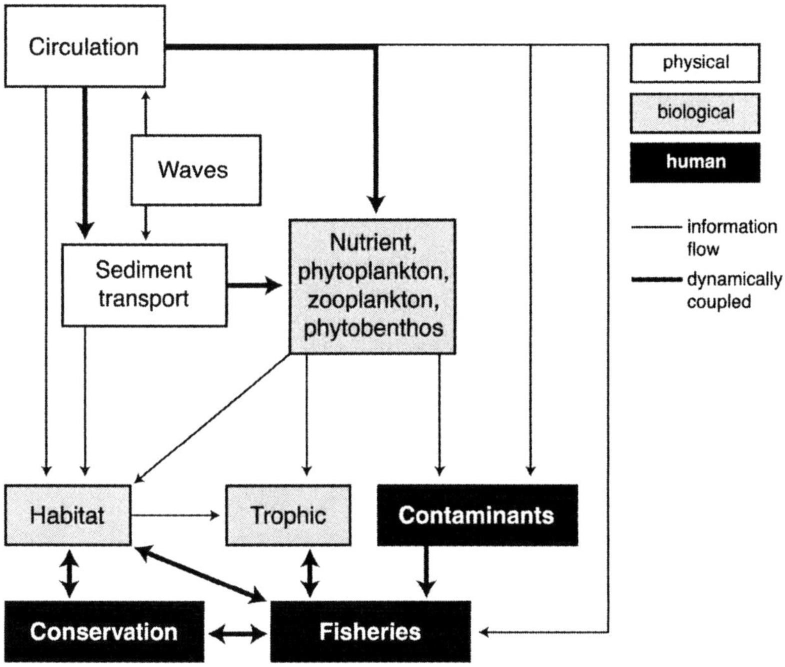

Figure 35.11 Major components of the integrated model applied to the North West Shelf ecosystem and human activities.

The North West Shelf study represents a first attempt at a whole-of-system model for a small part of Australia's continental shelf. It has provided a framework for identifying the processes most critical in addressing particular issues, as well as helping to prioritize needs for further observations (including ongoing monitoring) and improvements in model components and parameterizations. It should also provide a template for regional ecosystem studies in other shelf environments. Focusing interdisciplinary effort on a small number of geographical areas appears to be the most viable strategy for progressing understanding on the Australian continental shelf ecosystem.

Acknowledgements

Our thanks to Jeff Dunn, Madeleine Cahill, Alan Williams, Bruce Barker, and Lea Croswell for the production of figures.

References

Alldredge, A. L. and W. M. Hamner, 1980. Recurring aggregation of zooplankton by a tidal current. *Estuarine and Coastal Marine Science,* 10, 31–37.

Alongi, D. M., 1990. Effect of mangrove detrital outwelling on nutrient regeneration and oxygen fluxes in coastal sediments of the Central Great Barrier Reef Lagoon. *Estuarine, Coastal and Shelf Science,* 31, 581–598.

Ayukai, T. and D. Miller 1998. Phytoplankton biomass, production and grazing mortality in Exmouth Gulf, a shallow embayment on the arid, tropical coast of Western Australia. *Journal of Experimental Marine Biology and Ecology,* 225, 239–251.

Baines, P. G., R. J. Edwards and C. B. Fandry, 1983. Observations of a new baroclinic current along the western continental slope of Bass Strait. *Australian Journal of Marine and Freshwater Research,* 34, 155–157.

Bax, N. J., M. A. Burford, L. A. Clementson and S. Davenport, 2001. Phytoplankton blooms and production sources on the south-east Australian continental shelf. *Marine and Freshwater Research,* 52, 451–462.

Bax, N. J., R. Kloser, A. Williams, K. Gowlett-Holmes and T. Ryan, 1999. Seafloor habitat definition for spatial management in fisheries: a case study on the continental shelf of southeast Australia. *Oceanologica Acta,* 22, 705–720.

Bax, N. J. and A. Williams, 2001. Seabed habitat on the south-eastern Australia continental shelf: context, vulnerability and monitoring. *Marine and Freshwater Research,* 52, 491–512.

Belperio, A. P. and D. E. Searle, 1988. Terrigenous and carbonate sedimentation in the Great Barrier Reef province. In *Carbonate - Clastic Transitions* L. J. Doyle and H. H. Roberts, ed. Amsterdam, pp. 143–174.

Birkeland, C., 1982. Terrestrial runoff as a cause of outbreaks of *Acanthaster planci* (Echinodermata: Asteroidea) . *Marine Biology,* 69, 175–185.

Black, K. P., P. J. Moran, D. Burrage and G. De'ath, 1995. Association of low-frequency currents and crown-of-thorns starfish outbreaks. *Marine Ecology Progress Series,* 125, 185–194.

Black, K. P., P. J. Moran and L. S. Hammond, 1991. Numerical models show coral reefs can be self-seeding. *Marine Ecology Progress Series,* 74, 1–11.

Blom, W. M. and D. B. Alsop, 1988. Carbonate mud sedimentation on a temperate shelf: Bass Basin, southeast Australia. *Sedimentary Geology,* 60, 269–280.

Boland, F. M., 1979. A time series of expendable bathythermograph sections across the East Australian Current. *Australian Journal of Marine and Freshwater Research,* 30, 303–313.

Bone, Y. and N. P. James, 1993. Bryozoans as carbonate sediment producers on the cool-water Lacepede Shelf, southern Australia. *Sedimentary Geology,* 86, 247–271.

Brinkman, R., E. Wolanski, E. Deleersnijder, F. McAllister and W. Skirving, 2001. Oceanic inflow from the Coral Sea into the Great Barrier Reef. *Estuarine, Coastal and Shelf Science,* 54, 655–668.

Brodie, J. E., 1992. Enhancement of larval and juvenile survival and recruitment in Acanthaster planci from the effects of terrestrial runoff: a review . *Australian Journal of Marine and Freshwater Research,* 43, 539–554.

Bruce, B. D., S. A. Condie and C. A. Sutton, 2001a. Larval distribution of blue grenadier (Macruronus novaezelandiae Hector) in south-eastern Australia: further evidence for a second spawning area. *Marine and Freshwater Research,* 52, 603–610.

Bruce, B. D., K. Evans, C. A. Sutton, J. W. Young and D. M. Furlani, 2001b. Influence of mesoscale oceanographic processes on larval distribution and stock structure in jackass morwong (Nemadactylus macropterus: Cheilodactylidae). *ICES Journal of Marine Science,* 58, 1072–1080.

Bruce, B. D., F. J. Neira and R. W. Bradford, 2001c. Larval distribution and abundance of blue and spotted warehous (Seriolella brama and S. punctata: Centrolophidae) in south-eastern Australia. *Marine and Freshwater Research,* 52, 531–636.

Brunskill, G. J., A. R. Orpin, I. Zagorskis, K. J. Woolfe and J. Ellison, 2001. Geochemistry and particle size of surface sediments of Exmouth Gulf, Northwest Shelf, Australia. *Continental Shelf Research,* 21, 157–201.

Brunskill, G. J., I. Zagorskis and J. Pfitzner, 2002. Carbon burial rates in sediments and a carbon mass balance for the Herbert River region of the Great Barrier Reef continetal shelf, north Queensland, Australia. *Estuarine, Coastal and Shelf Science,* 54, 677–700.

Bulman, C., F. Althaus, X. He, N. J. Bax and A. Williams, 2001. Diets and trophic guilds of demersal fishes of the south-eastern Australian shelf. *Marine and Freshwater Research,* 52, 537–548.

Burford, M. A. and P. C. Rothlisberg, 1999. Factors limiting phytoplankton production in a tropical continental shelf ecosystem. *Estuarine, Coastal and Shelf Science,* 48, 541–549.

Burford, M. A., P. C. Rothlisberg and Y. G. Wang, 1995. Spatial and temporal distribution of tropical phytoplankton species and biomass in the Gulf of Carpentaria, Australia. *Marine Ecology Progress Series,* 118, 255–266.

Burns, K. A., J. K. Volkman, J. Cavanagh and D. Brinkman, 2003. Lipids as biomarkers for carbon cycling on the Northwest Shelf of Australia: results from a sediment trap study. *Marine Chemistry,* 80, 103–128.

Cappo, M. and R. Kelley, 2001. Connectivity in the Great Barrier Reef World Heritage Area - an overview of pathways and processes . In *Oceanographic Processes of Coral Reefs: Physical and Biological Links in the Great Barrier Reef* E. Wolanski, ed. Boca Raton, Florida, pp. 161–186.

Caputi, N. and R. S. Brown, 1993. The effect of environment on puerulus settlement of the western rock lobster (Panulirus cygnus) in Western Australia. *Fisheries Oceanography,* 2, 1–10.

Caputi, N., C. F. Chubb and A. F. Pearce, 2001. Environmental effects on recruitment of the western rock lobster, Panulirus cygnus. *Marine and Freshwater Research,* 52, 1167–1174.

Caputi, N., W. J. Fletcher, A. F. Pearce and C. F. Chubb, 1996. Effect of the Leeuwin Current on the recruitment of fish and invertebrates along the Western Australian coast . *Marine and Freshwater Research,* 47, 147–155.

Carleton, J. H., R. Brinkman and P. J. Doherty, 2001. The effects of water flow around coral reefs on the distribution of pre-settlement fish (Great Barrier Reef, Australia). In *Oceanographic Processes of Coral Reefs: Physical and Biological Links in the Great Barrier Reef* E. Wolanski, ed. Boca Raton, Florida, pp. 209–230.

Chivas, A. R., A. Garcia, S. van der Kaars, M. J. J. Couapel, S. Holt, J. M. Reeves, D. J. Wheeler, A. D. Switzer, C. V. Murray-Wallace, D. Banerjee, D. M. Price, S. X. Wang, G. Pearson, N. T. Edgar, L. Beaufort, P. De Deckker, E. Lawson and C. B. Cecil, 2001. Sea-level and environmental changes since the last interglacial in the Gulf of Carpentaria, Australia: an overview. *Quaternary International,* 83–85, 19–46.

Church, J. A. and P. D. Craig, 1998. Australia's shelf seas: diversity and complexity - coastal segment (30,W-S). In *The Sea* , Vol. 11, A. R. Robinson and K. H. Brink, ed. New York, pp. 933–964.

Church, J. A., H. J. Freeland and R. L. Smith, 1986. Coastal trapped waves on the east Australian continental shelf. I. Propogation of wave modes. *Journal of Physical Oceanography,* 16, 1929–1943.

Clementson, L. A., G. P. Harris, F. B. Griffiths and D. W. Rimmer, 1989. Seasonal and inter-annual variability in chemical and biological parameters in Storm Bay, Tasmania. I. Physics, chemistry and the biomass of components of the food chain. *Australian Journal of Marine and Freshwater Research,* 40, 25–38.

Collins, L. B., 1988. Sediments and history of the Rottnest Shelf, southwest Australia: a swell-dominated, non-tropical carbonate margin. *Sedimentary Geology,* 60, 15–49.

Collins, L. B., Z. R. Zhu and K. H. Wyrwoll, 1996. The structure of the Easter Platform, Houtman Abrolhos Reefs: Pleistocene foundations and Holocene reef growth. *Marine Geology,* 135, 1–13.

Condie, S. A., 1995. Interactions between western boundary currents and shelf waters: a mechanism for coastal upwelling . *Journal of Geophysical Research,* 100, 24,811–24,818.

Condie, S. A. and J. R. Dunn, 2004. Seasonal characteristics of the surface mixed layer in the Australasian region: Implications for primary productivity and biogeography. *Marine and Freshwater Research,* (submitted).

Condie, S. A., C. Fandry, D. McDonald, J. Parslow and K. Sainsbury, 2003. Linking ocean models to coastal management on Australia's North West Shelf. *Eos, Transactions, American Geophysical Union,* 84, 49–53.

Condie, S. A., N. R. Loneragan and D. J. Die, 1999. Modelling the recruitment of tiger prawns *Penaeus esculentus* and *P. semisulcatus* to nursery grounds in the Gulf of Carpentaria, northern Australia: implications for assessing stock-recruitment relationships . *Marine Ecology Progress Series,* 178, 55–68.

Condie, S. A., J. Waring, J. Mansbridge and M. Cahill, 2004. Ocean connectivity patterns around the Australian continent. *Environmental Modelling and Software,* (in press).

Cresswell, G., A. Frische, J. Peterson and D. Quadfasel, 1993. Circulation in the Timor Sea. *Journal of Geophysical Research,* 98, 14379–14389.

Crosby, L. H. and E. J. F. Wood, 1958. Studies of Australian and New Zealand diatoms. I. Planktonic and allied species. *Transactions of the Royal Society of New Zealand.* 85, 483–530.

Dempster, T., M. T. Gibbs, D. Rissik and I. M. Suthers, 1997. Beyond hydrography: daily ichthyoplankton variability and short term oceanographic events on the Sydney continental shelf. *Continental Shelf Research,* 17, 1461–1481.

Dight, I. J., L. Bode and M. K. James, 1990a. Modelling the larval dispersal of *Acanthaster planci.* I. Large scale hydrodynamics . *Coral Reefs,* 9, 115–123.

Dight, I. J., M. K. James and L. Bode, 1990b. Modelling the larval dispersal of *Acanthaster planci.* II. Patterns of reef connectivity. *Coral Reefs,* 9, 125–134.

Doherty, P. J., 1987. Light-traps: selective but useful devices for quantifying the distributions and abundances of larval fishes. *Bulletin of Marine Science,* 41, 423–431.

Doherty, P. J. and A. Williams, 1988. The replenishment of coral reef fish populations. *Oceanography and Marine Biology: an Annual Review,* 26, 487–551.

Doherty, P. J., D. M. Williams and P. F. Sale, 1985. The adaptive significance of larval dispersion in coral reef fishes. *Environmental Biology of Fishes,* 12, 81–90.

Done, T. J., 1992. Effects of tropical cyclone waves on ecological and geomorphological structures on the Great Barrier Reef. *Continental Shelf Research,* 12, 859–872.

Dunn, J. R. and K. R. Ridgway, 2002. Mapping ocean properties in regions of complex topography. *Deep-Sea Research I,* 49, 591–604.

Fletcher, W. J., R. J. Tregonning and G. J. Sant, 1994. Interseasonal variation in the transport of pilchard eggs and larvae off southern Western Australia . *Marine Ecology Progress Series,* 111, 209–224.

Furnas, M. J. and A. W. Mitchell, 1986. Phytoplankton dynamics in the central Great Barrier Reef- I. Seasonal changes in biomass and community structure and their relation to intrusive activity. *Continental Shelf Research,* 6, 363–384.

Furnas, M. J. and A. W. Mitchell, 1987. Phytoplankton dynamics in the central Great Barrier Reef- II. Primary production. *Continental Shelf Research,* 7, 1049–1062.

Furnas, M. J. and A. W. Mitchell, 1996. Nutrient inputs into the Great Barrier Reef (Australia) from subsurface intrusions of Coral Sea Waters: a two-dimensional displacement model . *Continental Shelf Research,* 16, 1127–1148.

Furnas, M. J. and A. W. Mitchell, 1999. Wintertime carbon and nitrogen fluxes on Australia's North-west Shelf. *Estuarine, Coastal and Shelf Science,* 49, 165–175.

Gersbach, G. H., C. B. Pattiaratchi, G. N. Ivey and G. R. Cresswell, 1999. Upwelling on the south-west coast of Australia - source of the Capes Current . *Continental Shelf Research,* 19, 363–400.

Gibbs, C. F., G. H. Arnott, A. R. Longmore and J. W. Marchant, 1991. Nutrient and plankton distribution near a shelf break front in the region of Bass Strait Cascade . *Australian Journal of Marine and Freshwater Research,* 42, 201–217.

Gibbs, C. F., M. Tomczak Jr. and A. R. Longmore, 1986. The nutrient regime of Bass Strait . *Australian Journal of Marine and Freshwater Research,* 37, 451–466.

Gibbs, M. T., 2000. Elevated chlorophyll a concentrations associated with a transient shelfbreak front in a western boundary current at Sydney, south-eastern Australia. *Marine and Freshwater Research,* 51, 733–737.

Gibbs, M. T., P. Marchesiello and J. H. Middleton, 1997. Nutrient enrichment of Jervis Bay, Australia, during the massive 1992 coccolithophorid bloom . *Marine and Freshwater Research,* 48, 473–478.

Gibbs, M. T.; P., Marchesiello, and J.H. Middleton, 2000. Observations and simulations of a transient shelfbreak front over the narrow shelf at Sydney, southeastern Australia. *Continental Shelf Research.* 20, 763–784.

Gibbs, M. T., J. H. Middleton and P. Marchesiello, 1998. Baroclinic response of Sydney shelf waters to local wind and deep ocean forcing. *Journal of Physical Oceanography,* 28, 178–190.

Godfrey, J. S., 1996. The effect of the Indonesian throughflow on ocean circulation and heat exchange with the atmosphere: A review. *Journal of Geophysical Research,* 101, 12217–12237.

Godfrey, J. S. and K. R. Ridgway, 1985. The large-scale environment of the poleward-flowing Leeuwin Current, Western Australia: longshore steric height gradients, wind stresses and geostrophic flow. *Journal of Physical Oceanography,* 15, 418–495.

Grant, B. R. and J. D. Kerr, 1970. Phytoplanktonnumbers and species at Port Hacking station and their relationship to the physical environment. *Australian Journal of Marine and Freshwater Research,* 21, 35–45.

Gray, C. A., 1996a. Do thermoclines explain the vertical distributions of larval fishes in the dynamic coastal waters of south-eastern Australia? *Marine and Freshwater Research,* 47, 183–190.

Gray, C. A., 1996b. Intrusions of surface sewage plumes into continental shelf waters: interactions with larval and presettlement juvenile fishes . *Marine Ecology Progress Series,* 139, 31–45.

Gray, C. A. and A. G. Miskiewicz, 2000. Larval fish assemblages in south-east Australian coastal waters: seasonal and spatial structure. *Estuarine, Coastal and Shelf Science,* 50, 549–570.

Griffin, D. A., P. A. Thompson, N. J. Bax, R. W. Bradford and G. M. Hallegraeff, 1997. The 1995 mass mortality of pilchard: no role found for physical or biological oceanographic factors in Australia. *Marine and Freshwater Research,* 48, 27–42.

Griffin, D. A., J. L. Wilkin, C. F. Chubb, A. F. Pearce and N. Caputi, 2001. Ocean currents and the larval phase of Australian western rock lobster, Panulirus cygnus. *Marine and Freshwater Research,* 52, 1187–1199.

Gunn, J. S., B. D. Bruce, D. M. Furlani, R. E. Thresher and S. J. M. Blaber, 1989. Timing and location of spawing of blue grenadier, Macruronus novaezelandiae (Teleostei: Merlucciidae), in Australian coastal waters. *Australian Journal of Marine and Freshwater Research* , 40, 97–112.

Hallegraeff, G. M., 1981. Seasonal study of phytoplankton pigments and species at a coastal study off Sydney: importance of diatoms and nanoplankton. *Marine Biology,* 61, 107–118.

Hallegraeff, G. M. and S. W. Jeffrey, 1993. Annually recurrent diatom blooms in spring along the New South Wales coast of Australia. *Australian Journal of Marine and Freshwater Research,* 44, 325–334.

Hallegraeff, G. M. and D. D. Reid, 1986. Plankton species successions and their hydological environment at a coastal station off Sydney. *Australian Journal of Marine and Freshwater Research.* 37, 361–377.

Harris, G. P., P. Davies, M. Nunez and G. Meyers, 1988. Interannual variability in climate and fisheries in Tasmania. *Nature,* 333, 754–757.

Harris, G. P., C. Nilsson, L. A. Clementson and D. Thomas, 1987. The water masses of the east coast of Tasmania: seasonal and interannual variability and the influence on phytoplankton biomass and productivity . *Australian Journal of Marine and Freshwater Research,* 38, 569–590.

Harris, P. T., 1994. Comparison of tropical, carbonate and temperate, siliciclastic tidally dominated sedimentary deposits: examples from the Australian continental shelf. *Australian Journal of Earth Science,* 41, 241–254.

Harris, P. T., 1991. Reversal of subtidal dune asymmetries caused by seasonally reversing wind-driven currents in Torres Strait, northeastern Australia. *Continental Shelf Research,* 11, 655–662.

Harris, P. T., 1988. Sediments, bedforms and bedload transport pathways on the continental shelf adjacent to Torres Strait, Australia - Papua New Guinea. *Continental Shelf Research,* 8, 979–1003.

Haynes, D. and J. E. Johnson, 2000. Organochlorine, heavy metal and polyaromatic hydrocarbon pollutant concentrations in the Great Barrier Reef (Australia) Environment: a Review. *Marine Pollution Bulletin,* 41, 267–278.

Haynes, D. and K. Michalek-Wagner 2000. Water quality in the Great Barrier Reef World Heritage Area: Past perspectives, current issues and new research directions. *Marine Pollution Bulletin,* 41, 428–434.

Hemer, M. A., P. T. Harris, R. Coleman and J. Hunter, 2003. Sediment mobility due to currents and waves in the Torres Strait - Gulf of Papua region. *Continental Shelf Research,* 24, 2297–2316.

Holdway, D. and D. T. Heggie, 2000. Direct hydrocarbon detection of produced formation water discharge on the Northwest Shelf, Australia. *Estuarine, Coastal and Shelf Science,* 50, 387–402.

Holloway, P. E., S. E. Humphries, M. Atkinson and J. Imberger, 1985. Mechanisms for nitrogen supply to the Australian North West Shelf . *Australian Journal of Marine and Freshwater Research,* 36, 753–764.

Hutchins, J. B. and A. F. Pearce, 1994. Influence of the Leeuwin Current on recruitment of tropical reef fishes at Rottnest Island, Western Australia. *Bulletin of Marine Science,* 54, 245–255.

Interim Marine and Coastal Regionalisation for Australia Technical Group, 1998. Interim Marine and Coastal Regionalisation for Australia: an ecosystem-based classification for marine and coastal environments. Version 3.3. Environment Australia, Commonwealth Department of the Environment. Canberra, 104pp. <*http://neptune.oceans.gov.au/* >

Isdale, P., 1984. Fluorescent bands in massive corals record centuries of coastal rainfall. *Nature,* 310, 578–579.

James, M. K. and J. P. Scandol, 1992. Larval dispersion simulations: Correlation with the crown-of-thorns starfish outbreaks database . *Australian Journal of Marine and Freshwater Research,* 43, 569–582.

James, N. P., Y. Bone, L. B. Collins and T. K. Kyser, 2001. Surficial sediments of the Great Australian Bight: facies dynamics and oceanography on a vast cool-water carbonate shelf. *Journal of Sedimentary Research,* 71, 549–567.

James, N. P., T. D. Boreen, Y. Bone and D. A. Feary, 1994. Holocene carbonate sedimentation on the west Eucla Shelf, Great Australian Bight: A shaved shelf. *Sedimentary Geology,* 90, 161–177.

James, N. P., L. B. Collins, Y. Bone and P. Hallock, 1999. Subtropical carbonates in a temperate realm: Modern sediments on the southwest Australian shelf. *Journal of Sedimentary Research,* 69, 1297–1321.

Jeffrey, S. W. and G. M. Hallegraeff, 1990. Phytoplankton ecology of Australasian waters. In: Clayton, M. N. and King, R. J., Editors. *Biology of Marine Plants.* Melbourne: Longman Cheshire, pp. 311–348.

Johnson, C., 1992. Reproduction, recruitment and hydrodynamics in the crown-of-thorns phenomenon on the Great Barrier Reef: introduction and synthesis. *Australian Journal of Marine and Freshwater Research,* 43, 517–523.

Jones, G. P., M. J. Milicich, M. J. Emslie and C. Lunow, 1999. Self-recruitment in a coral reef fish population. *Nature,* 402, 802–804.

Jones, H. A., 1973. Marine Geology of the Northwest Australian Continental Shelf. *Bureau of Mineral Resources Bulletin,* 136, Canberra.

Jones, H. A. and P. J. Davies, 1979 . Preliminary studies of offshore placer deposits, eastern Australia. *Marine Geology,* 30, 243–268.

Jones, M. R., 1987. Surficial sediments of the western Gulf of Carpentaria. *Australian Journal of Marine and Freshwater Research,* 38, 151–167.

Jones, M. R. and T. Torgersen, 1988 . Late Quaternary evolution of Lake Carpentaria on the Australia - New Guinea continental shelf. *Australian Journal of Earth Science,* 35, 313–324.

Jordan, A., G. Pullen, J.-A. Marshall and H. Williams, 1995. Temporal and spatial patterns of spawning in Jack Mackerel, Tranchurus declivis (Pisces: Carangidae), during 1988–91 in eastern Tasmanian waters. *Marine and Freshwater Research,* 46, 831–842.

Kalnay, E., M. Kanamitsu, R. Kistler, W. Collins, D. Deaven, L. Gandin, M. Iredell, S. Saha, G. White, J. Woollen, Y. Zhu, M. Chelliah, W. Ebisuzaki, W. Higgins, J. Janowiak, K.C. Mo, C. Ropelewski, J. Wang, A. Leetmaa, R. Reynolds, R. Jenne and D. Joseph, 1996. The NCEP/NCAR 40-year reanalysis project. *Bulletin of the American Meteorological Society,* 77, 437–471.

Keesing, J. K. and A. R. Halford, 1992. Importance of postsettlement processes for the population dynamics of *Acanthaster planci* (L.). *Marine and Freshwater Research,* 43, 635–651.

Kimmerer, W. J., A. D. McKinnon, M. J. Atkinson and J. A. Kessell, 1985. Spatial distributions of plankton in Shark Bay, Western Australia. *Australian Journal of Marine and Freshwater Research.* 36, 421–432.

Kingsford, M. J., 1990. Linear oceanographic features: a focus for research on recruitment processes. *Australian Journal of Ecology,* 15, 391–401.

Kingsford, M. J., E. Wolanski and J. H. Choat, 1991. Influence of tidally induced fronts and Langmuir circulations on distribution and movements of presettlement fishes around a coral reef . *Marine Biology,* 109, 167–180.

Lambeck, A. and K. J. Woolfe, 2000. Composition and textural variability along the 10 m isobath, Great Barrier Reef: evidence for pervasive northward transport. *Australian Journal of Earth Science,* 47, 327–335.

Larcombe, P. and K. J. Woolfe, 1999. Increased sediment supply to the Great Barrier Reef will not increase sediment accumulation at most coral reefs. *Coral Reefs,* 18, 163–169.

Lavering, I. H., 1994. Marine environments of southeast Australia (Gippsland shelf and Bass Strait) and the impact of offshore petroleum exploration and production activity. *Marine Georesources and Geotechnology,* 12, 201–226.

Lees, B. G., 1992. Recent terrigenous sedimentation in Joseph Bonaparte Gulf, northwestern Australia. *Marine Geology,* 103, 199–213.

Lenanton, R. C., S. G. Ayvazian, A. F. Pearce, R. A. Steckis and G. C. Young, 1996. Tailor (*Pomatomus saltatrix*) off Western Australia: where does it spawn and how are the larvae distributed. *Marine and Freshwater Research,* 47, 337–346.

Lewis, R. K., 1981. Seasonal upwelling along the south-eastern coastline of south Australia . *Australian Journal of Marine and Freshwater Research,* 32, 843–854.

Li, Q. and B. McGowran, 1998. Oceanographic implications of recent planktonic foraminifera along the southern Australian margin . *Marine and Freshwater Research,* 49, 439–445.

Logan, B. W., R. Rezak and R. N. Ginsburg, 1964. Classification and environmental significance of algal stromatolites. *Journal of Geology,* 72, 68–83.

Longhurst, A., 1998. *Ecological Geography of the Sea,* Academic Press, San Diego.

Lough, J. M., 1998. Coastal climate of northwest Australia and comparisons with the Great Barrier Reef: 1960 to 1992. *Coral Reefs,* 17, 351–367.

Lourey, M. J., D. M. Alongi, D. A. J. Ryan and M. J. Devlin, 2001. Variability of nutrient regeneration rates and nutrient concentrations in surface sediments of the northern Great Barrier Reef. *Continental Shelf Research,* 21, 145–155.

Malikides, M., P. T. Harris, C. J. Jenkins and J. B. Keene, 1988. Carbonate sandwaves in Bass Strait. *Australian Journal of Earth Science,* 35, 303–311.

Malikides, M., P. T. Harris and P. M. Tate, 1989. Sediment transport and flow over sandwaves in a non-rectilinear tidal environment: Bass Strait, Australia. *Continental Shelf Research,* 9, 203–221.

Marchesiello, P.; Gibbs, M. T., and Middleton, J. H. Simulations of coastal upwelling on the Sydney continental shelf. Marine and Freshwater Research. 2000; 51577–588.

Maxwell, J. G. R. and G.R. Cresswell, 1981. Dispersal of tropical marine larvae to the Great Australian Bight by the Leeuwin Current. *Australian Journal of Marine and Freshwater Research,* 32, 493–500.

Maxwell, W. G. H., 1968. *Atlas of the Great Barrier Reef,* Elsevier,. Amsterdam,

McClean-Padman, J. and L. Padman, 1991. Summer upwelling on the inner continental shelf: the relative roles of local wind forcing and mesoscale eddy encroachment. *Continental Shelf Research,* 11, 321–345.

McLoughlin, R. J. and P. C. Young, 1985. Sedimentray provinces of the fishing grounds of the North West Shelf of Australia: grain-size frequency analysis of surficial sediments. *Australian Journal of Marine and Freshwater Research,* 36, 671–681.

Middleton, J. H., D. Cox and P. Tate, 1997. The oceanography of the Sydney region. *Marine Pollution Bulletin,* 33, 124–131.

Miskiewicz, A. G., B. D. Bruce and P. Dixon, 1996. Distribution of tailor (Pomatomus saltatrix) larvae along the coast of New South Wales, Australia. *Marine and Freshwater Research,* 47, 331–336.

Mitchell, A. W., R. G. V. Bramley and A. K. L. Johnson, 1997. Export of nutrients and suspended sediment during a cyclone-mediated flood event in the Herbert River catchment, Australia . *Marine and Freshwater Research,* 48, 79–88.

Moltschaniwskyj, N. A. and P. J. Doherty, 1995. Cross-shelf distribution patterns of tropical juvenile cephalopods sampled with light-traps. *Marine and Freshwater Research,* 46, 707–714.

Motoda, S., T. Kawamurra and A. Taniguchi, 1978. Differences in productivities between the Great Australian Bight and the Gulf of Carpentaria, Australia, in Summer. *Marine Biology,* 46, 93–99.

Murray, A. G. and J. S. Parslow, 1999. Modelling of nutrient impacts in Port Phillip Bay - a semi-enclosed marine Australian ecosystem. *Marine and Freshwater Research,* 50, 597–611.

Neil, D. T., A. R. Orpin, P. V. Ridd and B. Yu, 2002. Sediment yield and impacts from river catchments to the Great Barrier Reef lagoon. *Marine and Freshwater Research,* 53, 733–752.

Newell, B. S., 1966. Seasonal changes in the hydrological and biological environments off Port Hacking, Sydney . *Australian Journal of Marine and Freshwater Research,* 17, 77–91.

Newman, S. J., D. M. Williams and G. R. Russ, 1997. Patterns of zonation of assemblages of the Lutjanidae, Lethrinidae and Serranidae (Epinephelinae) within and among mid-shelf and outer-shelf reefs in the central Great Barrier Reef. *Marine and Freshwater Research,* 48, 119–128.

North West Shelf Joint Environmental Modelling Study, 2002. North West Shelf Joint Environmental Modelling Study Interim Report, 140pp. (*http://www.marine.csiro.au/nwsjems/reports/index.html*).

Oke, P. R., 2000. Topographically induced upwelling off Eastern Australia. *Journal of Physical Oceanography,* 30, 512–531.

Oke, P. R. and J. H. Middleton, 2001. Nutrient enrichment off Port Stephens: the role of the East Australian Current . *Continental Shelf Research,* 21, 587–606.

Oliver, J. K., B. A. King, B. L. Willis, R. C. Babcock and E. Wolanski, 1992. Dispersal of coral larvae from a lagoonal reef - II. Comparisons between model predictions and observed concentrations. *Continental Shelf Research,* 12, 873–889.

Orpin, A. R., P. V. Ridd and L. K. Stewart, 1999. Assessment of the relative importance of major sediment-transport mechanisms in the central Great Barrier Reef lagoon. *Australian Journal of Earth Science,* 46, 883–896.

Pearce, A. F. and C. B. Pattiaratchi, 1999. The Capes Current: a summer countercurrent flowing past Cape Leeuwin and Cape Naturaliste, Western Australia. *Continental Shelf Research,* 19, 401–420.

Pearce, A. F. and B. F. Phillips, 1988. ENSO events, the Leeuwin Current, and larval recruitment of the western rock lobster. *Journal du Conseil, Conseil International pour l'Exploration de la Mer,* 45, 13–21.

Penn, J. W. and N. Caputi, 1986. Spawning stock-recruitment relationships and environmental influences on the tiger prawn (Penaeus esculentus) fishery in Exmouth Gulf, Western Australia . *Australian Journal of Marine and Freshwater Research,* 37, 491–505.

Playford, P. E., 1979. Stromatolite research in Western Australia. *Journal of the Royal Society of Western Australia*, 62, 13–20.

Porter-Smith, R., P. T. Harris, O. Anderson, R. Coleman, D. Greenslade and C. J. Jenkins, 2004. Classification of the Australian continentalshelf based on predicted sediment threshold exceedances from tidal currents and swell waves. *Marine Geology,* 211, 1–20.

Rao, C. P., 1986. Geochemistry of temperate-water carbonates, Tasmania, Australia. *Marine Geology,* 71, 363–370.

Reichelt, R. E. and A. D. McEwan, 1999. Australia's marine science and technology plan: an action plan for Australia's Oceans Policy. *Marine and Freshwater Research,* 50, 711–716.

Revelante, N. and M. Gilmartin, 1982. Dynamics of phytoplankton in the Great Barrier Reef Lagoon. *Journal of Plankton Research,* 4, 47–76.

Revelante, N., W. T. Williams and J. S. Bunt, 1982. Temporal and spatial distributions of diatoms, dinoflagellates and trichodesmium in waters of the Great Barrier Reef. *Journal of Experimental Marine Biology and Ecology,* 63, 27–45.

Ribbe, J. and P. E. Holloway, 2001. A model of suspended sediment transport by internal tides. *Continental Shelf Research,* 21, 395–422.

Ridgway, K. R. and S. A. Condie, 2004. The 5500-km long boundary flow off western and southern Australia. *Journal of Geophysical Research,* 109, CO4017, doi: 10.1029/2003JC001921.

Ridgway, K. R. and J. R. Dunn, 2003. Mesoscale structure of the mean East Australian Current System and its relationship with topography. *Progress in Oceanography,* 56, 189–222.

Ridgway, K. R., J. R. Dunn and J. L. Wilkin, 2002. Ocean interpolation by four-dimensional weighted least squares – application to the waters around Australia. *Journal of Atmospheric and Oceanic Technology,* 19, 1357–1375.

Rintoul, S. R. and T. W. Trull, 2001. Seasonal evolution of the mixed layer in the Subantarctic Zone south of Australia. *Journal of Geophysical Research,* 106, 31447–31462.

Rochford, D. J., 1984. Nitrates in Eastern Australia coastal waters. *Australian Journal of Marine and Freshwater Research,* 35, 385–397.

Rothlisberg, P. C., J. A. Church and C. B. Fandry, 1995. A mechanism for near-shore concentration and estuarine recruitment of post-larval Penaeus plebejus Hess (Decapoda, Penaeidae). *Estuarine, Coastal and Shelf Science,* 40, 115–138.

Rothlisberg, P. C., J. A. Church and A. M. G. Forbes, 1983. Modelling the advection of vertically migrating shrimp larvae. *Journal of Marine Research,* 41, 511–538.

Rothlisberg, P. C., P. D. Craig and J. R. Andrewartha, 1996. Modelling penaeid prawn larval advection in Albatross Bay, Australia: defining the effective spawning population. *Marine and Freshwater Research,* 47, 157–168.

Rothlisberg, P. C., P. C. Pollard, P. D. Nichols, D. J. W. Moriarty, A. M. G. Forbes, C. J. Jackson and D. Vaudrey, 1994. Phytoplankton community structure and productivity in relation to the hydrological regime of the Gulf of Carpentaria, Australia, in summer. *Australian Journal of Marine and Freshwater Research,* 45, 265–282.

Roy, P. S. and B. G. Thom, 1981. Late Quaternary marine deposition in New South Wales and southern Queensland: an evolutionary model. *Journal of the Geological Society of Australia,* 28, 471–489.

Sammarco, P. W. and J. C. Andrews, 1989. The Helix experiment: differential localized dispersal and recruitment patterns in Great Barrier Reef corals. *Limnology and Oceanography,* 34, 896–912.

Sammarco, P. W. and J. C. Andrews, 1988. Localized dispersal and recruitment in Great Barrier Reef reef corals: the Helix experiment. *Science,* 239, 1422–1424.

Scandol, J. P. and M. K. James, 1992. Hydrodynamics and larval dispersal: a population model of *Acanthaster planci* on the Great Barrier Reef. *Australian Journal of Marine and Freshwater Research,* 43, 583–596.

Schahinger, R. B., 1987. Structure of coastal upwelling events observed off the south-east coast of South Australia during February 1983 - April 1984. *Australian Journal of Marine and Freshwater Research,* 38, 439–459.

Short, A. D., 1979. Three dimensional beach stage model. *Journal of Geology,* 87, 553–571.

Simpson, C. J., 1991. Mass spawning of corals on Western Australian reefs and comparisons with the Great Barrier Reef. *Journal of the Royal Society of Western Australia* , 74, 85–91.

Simpson, C. J., J. L. Cary and R. J. Masini, 1993. Destruction of corals and other reef animals by coral spawn slicks on Ningaloo Reef, Western Australia. *Coral Reefs,* 12, 185–191.

Smith, K. A., 2000. Active and passive dispersal of Centroberyx affinis (Berycidae) and Gonorynchus greyi (Gonorynchidae) larvae on the Sydney shelf. *Marine and Freshwater Research,* 51, 229–234.

Smith, K. A., M. T. Gibbs, J. H. Middleton and I. M. Suthers, 1999. Short term variability in larval fish assemblages of the Sydney shelf: tracers of hydrographic variability . *Marine Ecology Progress Series,* 178, 1–15.

Smith, K. A. and I. M. Suthers, 1999. Displacement of diverse ichthyoplankton assemblages by a coastal upwelling event on the Sydney shelf. *Marine Ecology Progress Series,* 176, 49–62.

Smith, R. L., A. Huyer, J. S. Godfrey and J. A. Church, 1991. The Leeuwin Current off Western Australia. *Journal of Physical Oceanography,* 21, 323–345.

Somers, I. F., 1987. Sediment type as a factor in the distribution of commercial prawn species in the western Gulf of Carpentaria, Australia. *Australian Journal of Marine and Freshwater Research,* 38, 133–149.

Somers, I. F., 1994. Species composition and distribution of commercial penaeid prawn catches in the Gulf of Carpentaria, Australia, in relation to depth and sediment type. *Australian Journal of Marine and Freshwater Research,* 45, 317–335.

Suzuki, A. and H. Kawahata, 2001. The oceanic CO_2 system and carbon budget in the Great Barrier Reef, Australia. *Geophysical Research Letters,* 28, 1243–1246.

The Marine Science and Technology Plan Working Group, 1999. Australia's Marine Science and Technology Plan. Department of Industry, Science and Resources, Commonwealth of Australia, pp146.

Thompson, R. O. R. Y. and T. J. Golding, 1981. Tidally induced "upwelling" by the Great Barrier Reef. *Journal of Geophysical Research,* 86, 6517–6521.

Tranter, D. J., D. J. Carpenter and G. S. Leech, 1986. The coastal enrichment effect of the East Australian Current eddy field . *Deep-Sea Research (Part A),* 33A, 1705–1728.

Tranter, D. J. and G. S. Leech, 1987. Factors influencing the standing crop of phytoplankton on the Australian Northwest Shelf seaward of the 40 m isobath. *Continental Shelf Research,* 7, 115–133.

Van Andel, T. H. and J. J. Veevers, 1967. Morphology and sediments of the Timor Sea. *Bureau of Mineral Resources Bulletin,* 83,

Williams, A., J. A. Koslow and P. R. Last, 2001. Diversity, density and community structure of the demersal fish fauna of the continental slope off Western Australia (20 to 35° S) . *Marine Ecology Progress Series,* 212, 247–263.

Williams, D. M. and S. English, 1992. Distribution of fish larvae around a coral reef: direct detection of a meso-scale, multi-specific patch? *Continental Shelf Research* , 12, 923–937.

Williams, D. M. and A. I. Hatcher, 1983. Structure of fish communities on outer slopes of inshore, mid-shelf and outer shelf reefs of the Great Barrier Reef. *Marine Ecology Progress Series,* 10 , 239–250.

Wilson, S. G., J. H. Carleton and M. G. Meekan, 2003. Spatial and temporal patterns in the distribution and abundance of macrozooplankton on the southern North West Shelf, Western Australia. *Estuarine, Coastal and Shelf Science,* 56, 897–908.

Wolanski, E., 2001. Physics-biology links in the Great Barrier Reef. In *Oceanographic Processes of Coral Reefs: Physical and Biological Links in the Great Barrier Reef* E. Wolanski, ed. Boca Raton, Florida , pp. 7–18.

Wolanski, E., E. Drew, K. M. Abel and J. O'Brien, 1988. Tidal jets, nutrient upwelling and their influence on the productivity of the alga *Halimeda* in the Ribbon Reefs, Great Barrier Reef. *Estuarine, Coastal and Shelf Science,* 26, 169–201.

Wolanski, E. and P. V. Ridd, 1990. Mixing and trapping in Australian tropical coastal waters. In *Residual currents and long-term transport* R. T. Cheng, ed. New York, pp. 165–183.

Wolanski, E., S. Spagnol, B. King and T. Ayukai, 1999. Patchiness in the Fly River plume in Torres Strait. *Journal of Marine Systems,* 18, 369–381.

Wood, E. J. F., 1954. Dinoflagellates in the Australian region. *Australian Journal of Marine and Freshwater Research,* 5, 171–347.

Woolfe, K. J., P. Larcombe and L. K. Stewart, 2000. Shelf sediments adjacent to the Herbert River delta, Great Barrier Reef, Australia. *Australian Journal of Earth Science,* 47, 301–315.

Young, J. W., R. W. Bradford, T. D. Lamb, L. A. Clementson, R. Kloser and H. Galea, 2001. Yellowfin tuna (thunnus albacares) aggregations along the shelf break off south-eastern Australia: links between inshore and offshore processes. *Marine and Freshwater Research,* 52, 463–474.

Young, J. W., T. D. Lamb, D. Le, R. W. Bradford and A. W. Whitelaw, 1997. Feeding ecology and interannual variations in diet of southern bluefin tuna, Thunnus maccoyii, in relation to coastal and oceanic waters off eastern Tasmania, Australia. *Environmental Biology of Fishes,* 50, 275–291.

Chapter 36. NEW ZEALAND SHELF REGION (31,S)

JANET BRADFORD-GRIEVE

National Institute of Water and Atmospheric Research, New Zealand

KEITH PROBERT

Department of Marine Science, University of Otago, Dunedin, New Zealand

KEITH LEWIS, PHILIP SUTTON, JOHN ZELDIS, ALAN ORPIN

National Institute of Water and Atmospheric Research, New Zealand

Contents

1. Introduction

The islands of New Zealand rise from a submerged continental block stretching from 18°S to 56°S (Fig. 36.1). The total shelf area is about 300,000 km^2 ranging from a width of less than 1 km wide off Fiordland to a 250 km wide central western shelf (Sharples, 1998). There are three regions of extensive continental shelf (Greater Cook Strait, Pegasus Bay and the region south of South Island) while in other places the continental shelf is almost nonexistent (Fig. 36.2). New Zealand tides are characterized by a coastally-trapped Kelvin wave that travels anticlockwise around the shelf forming a degenerate amphidrome in the centre of New Zealand. The country also sits astride the *roaring forties* on the pole-ward boundary of the South Pacific gyre and has climatic features that produce high precipitation on the western slopes of the Southern Alps and very dry conditions in central Otago (Tomlinson, 1992). This set of unique circumstances produces a range of nearshore conditions, upwelling, tidal mixing and upper-ocean mixing that determines the continental shelf environment. For the location of geographic features mentioned in the text, refer to figures 36.2b, 36.7, 36.11 and 36.15.

The Sea, Volume 14, edited by Allan R. Robinson and Kenneth H. Brink
ISBN 0–674–01527–4

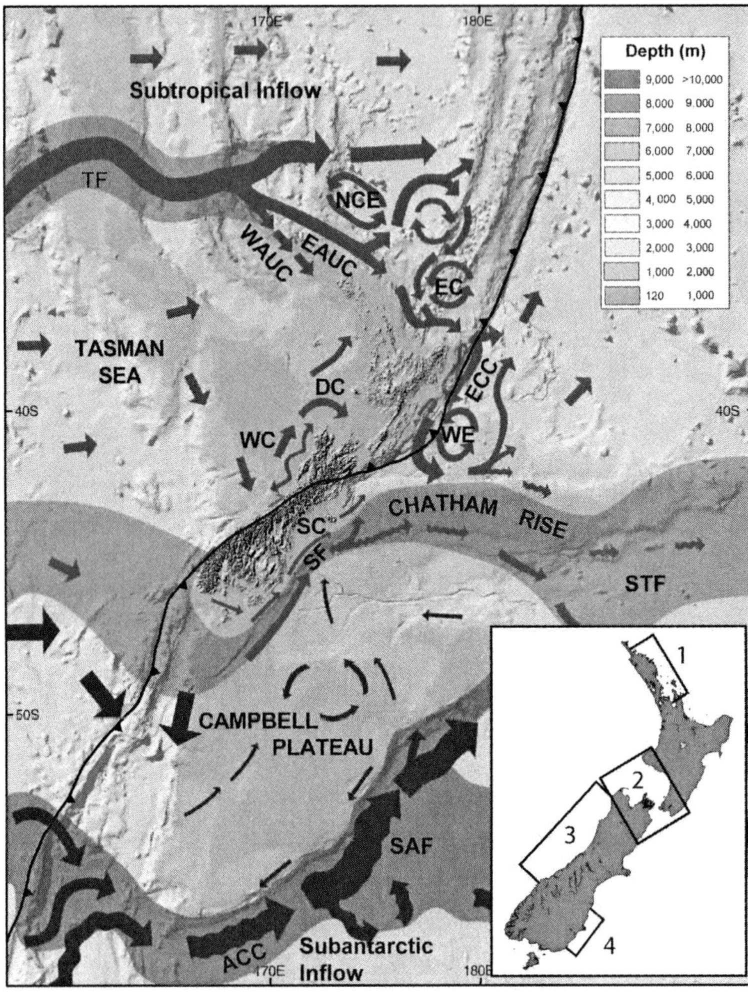

Figure 36.1 New Zealand region bathymetry, major currents, and the location of the major plate boundary (after Carter et al. 1998). ACC = Antarctic Circumpolar Current, DC = D'Urville Current, EAUC = East Auckland Current, EC = East Cape Eddy, ECC = East Cape Current, NCE = North Cape Eddy, SAF = Subantarctic Front, SF = Southland Front, STF = Subtropical Front, TF = Tasman Front, WE = Wairarapa Eddy, WAUC = West Auckland Current. Inset showing the four case studies mentioned in section 2. 1 = Northeast shelf, 2 = Greater Cook Strait, 3 = Western South Island shelf, 4 = Otago shelf. See color insert.

1.1 Geographical and geomorphologic settings

The New Zealand continental block is split by a plate boundary that runs NE – SW (Walcott, 1987) (Fig. 36.1). Along most of this boundary, the Australian and Pacific plates push past and into one another. It is the increase in the component of convergence over the last few million years that has made New Zealand mountainous. In turn, these mountains supply a massive 209 Mt y^{-1} of sediment (New Zealand contributes about 1% of the mud supply to the world's oceans) to mainly narrow continental shelves (Zeldis et al., submitted; Hicks and Shankar, 2003) (Fig. 36.2a). The largest rates of supply of sediment to the New Zealand coast (> 60 t km^{-2} y^{-1}) off the East Cape region and South Westland are amongst the highest in the world (e.g. Milliman and Meade, 1983; Orpin et al. 2002). During Cyclone Bola in 1988 flooding of the Waipaoa River discharged sediment to the coast as a sea-bed hugging hyperpycnal flow which generated a fluid mud layer that smothered benthic organisms (Foster and Carter, 1997).

The nature of shelf sediments can be directly affected by subduction processes. When the subducting Pacific Plate reaches 80 – 100 km deep, it gives off volatiles that rise to form a line of volcanoes and a huge caldera that has filled to form Lake Taupo (Cole, 1979, 1990). Volcanic ash derived directly from airfall and washed in by rivers is a major component of the sediments of the Bay of Plenty and the shelf to the east and to a lesser extent the west (Lewis and Pantin, 1984).

East of North Island, the dense floor of the oceanic Pacific Plate is subducting at a rate of 40–50 mm y^{-1} (Lewis, 1980). Sand and mud scraped off the subducting plate is pressed against and under eastern North Island producing a rising margin of young, soft, and easily eroded, muddy coastal hills (Lewis and Pettinga, 1993). The three major rivers that drain the Raukumara Ranges carry over 30% of total sediment input to the New Zealand coast (Hicks and Shankar, 2003) (Fig. 36.2a). Thick, soft mud now blankets the narrow shelf (Foster and Carter, 1977). Only at Hawke Bay, where the sea has broached the coastal hills, is the shelf wider (Lewis, 1971).

At the southern end of South Island, the plate boundary is almost a mirror-image of that in North Island. The oceanic Australian Plate is being subducted, at rates of 30–40 mm y^{-1}, obliquely beneath the Fiordland edge of the Pacific Plate (Walcott, 1998). The mirror-image systems are linked by the Alpine Fault, a *dextral transpressive transform* fault, which has sheared the western side of South Island and simultaneously pushed it beneath the eastern side. Continental crust of the Australian plate is colliding with the Pacific Plate and results in the formation of the Southern Alps mountain chain. Short, steep, often glacier-fed rivers draining the metamorphic rocks of the high alps carry mineral-rich sands and gravels to the western and eastern continental shelf (Carter, 1975). Immediately east of the main divide, much of the modern input is presently trapped in freshwater glacial lakes, although glaciers once carried huge amounts of debris to the very narrow shelf or straight into canyons (Barnes et al., 2001).

In relatively stable parts of New Zealand away from the plate boundary, including northern and western North Island, western Cook Strait, and southern and southeastern South Island, rivers sourced from the mountains have had time to build relatively wide shelves over many cycles of low and high sea level. Because of their width, the modern sand and mud prism is generally confined to the inner shelves, the outer shelves being covered either by sands, gravels or shell debris that are mainly relict near-shore deposits of glacially lowered sea-level (e.g. Carter, 1975).

Around much of New Zealand, the continental shelf edge is incised by canyons, which, like the shelves themselves, are presumed to be a product of glacial lowerings of sea-level. (Carter, 1975; Carter and Carter, 1993). Where the canyon heads are close to shore, gravel, sand and mud is trapped in them and when shaken by rupture of plate boundary faults may avalanche down canyons and, in a few cases, travel over 2000 km as high-velocity turbid flows along deep-ocean channels. In places, shelf-edge instability is enhanced by seeps of methane-rich fluids that support extensive chemosynthetic faunas and carbonate concretions (Lewis and Marshall, 1996; Orpin, 1997). Massive slope failures have occurred and probably generated tsunamis large enough to stir large areas of the shelf (e.g. Collett et al., 2001).

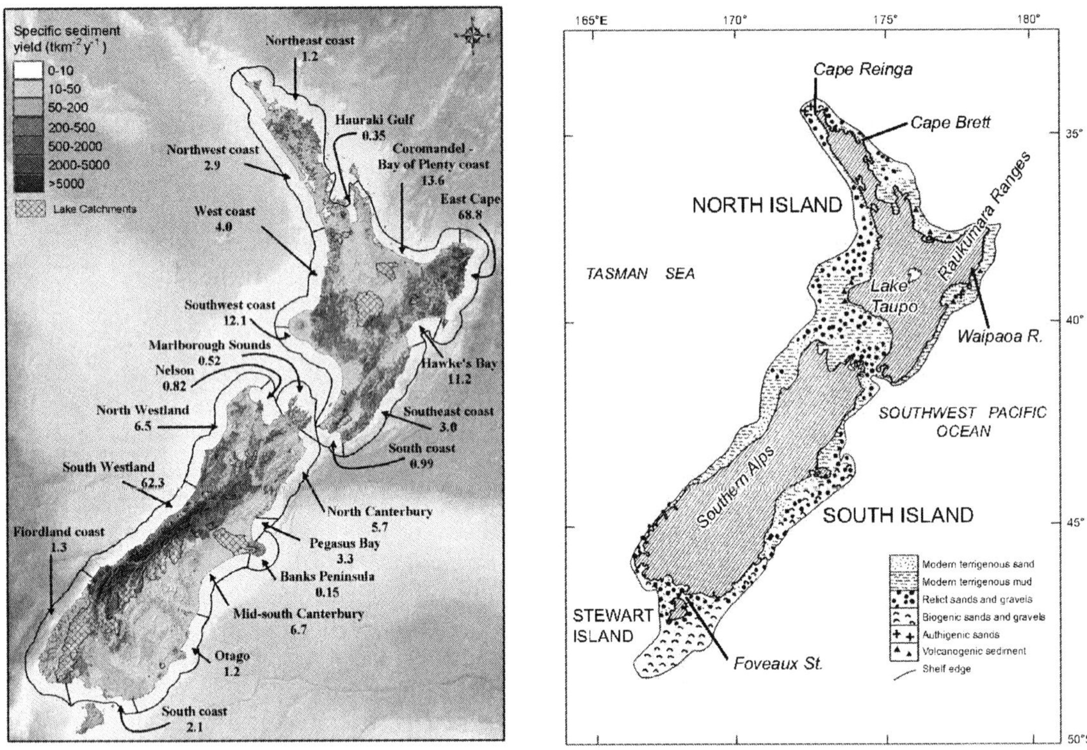

Figure 36.2 a, Source map of river suspended sediment ($t.km^{-2}y^{-1}$), with sediment yields to the coast totalled by region ($Mt.y^{-1}$). Hatched areas define catchments of large natural lakes (adapted from Hicks and Shankar, 2003); b, Distribution of main sediment types on the New Zealand continental shelf (from Carter and Heath, 1975) and location of some geographic features mentioned in the text.

The geomorphologic setting, tidal characteristics and wave climate result in a shelf mantled mainly by modern and relict terrigenous and biogenic sediments with a wide range of grain sizes (Carter and Heath, 1975) (Fig. 36.2b). They conclude that sediments suspended during storms are re-distributed by tidal currents and to some extent by the mean circulation.

1.2 Physical environment

New Zealand is a relatively long, narrow archipelago that lies athwart the West Wind Drift and forms the western boundary to the South Pacific Ocean south of 34°S. This results in shelf edge currents and oceanic eddies that interact with coastal waters over much of the narrow continental shelf and maintain an intimate contact of oceanic waters with the coastal zone (Sharples, 1998). Offshore currents are particularly important in maintaining shelf-break fronts, e.g. the East Auckland Current and the Southland Current/STF.

The South Pacific western boundary current, the East Australian Current, separates from the coast of Australia and a flow of water crosses the Tasman Sea forming the Tasman Front. A portion of this warm, salty Subtropical Water (STW) lies adjacent to the New Zealand landmass to form the East Auckland Current, which transports between –5 and 24 Sv of water south along the northeast continental shelf break with a mean transport of 9 Sv (Stanton et al., 1997, Stanton and Sutton, 2003). Most of this flow deflects south around East Cape as the East Cape Current before forming the northern side of the Subtropical Front (Tilburg et al., 2001; Carter et al., 1998; Heath, 1985). Three permanent eddies lie offshore of this boundary current: the North Cape, East Cape and Wairarapa Eddies (Roemmich and Sutton, 1998) (Fig. 36.1). These large-scale offshore features may impact coastal ecosystems, for example, through, for example, the retention of the larvae of coastal organisms (Chiswell and Booth, 1999) or by entraining coastal water and taking it offshore (Bradford and Chapman, 1988b).

New Zealand intersects the circumpolar Subtropical Front (STF), which separates warm, salty STW from cold, fresh Subantarctic Water (SAW). The STF passes south of Australia and Tasmania and extends towards New Zealand at around 45°S off Fiordland. The front deviates south along the continental margin before following the shelf break northwards along the east coast of South Island where it is locally known as the Southland Front and has an associated current called the Southland Current. The Southland Current advects roughly 8 Sv (with measurements ranging between 3 and 13 Sv) of mainly SAW with peak surface speeds of 20–30 cm s^{-1} (Chiswell, 1996; Sutton, 2003). The STF turns east along the crest of Chatham Rise at 43.5°S where it is constrained by the shallow bathymetry to a limited depth of 300–350 m and a narrow width of ~100km (Sutton, 2001). Although the southern limits of South Island are at latitudes associated with SAW, in fact the entire coastal region is bathed in water of STW origin, with the transition to SAW (i.e. the STF) occurring at the continental shelf break around the southern extreme of South Island.

Coastal seas around New Zealand are strongly influenced by atmospheric forcing. The local climate is strongly influenced at all times of year by the passage of transient anticyclones and depressions in the prevailing westerlies (Renwick et al., 1998 and references therein). The region is dominated by fronts resulting from the interaction of subtropical and polar air masses. The topography has dramatic effects on wind speeds, especially in Cook Strait, and the west coast of South Island is the wettest region in New Zealand due to the interaction of the Southern Alps with the persistent moist on-shore flow (Sturman and Tapper, 1996).

The oceanographic conditions around New Zealand are highly variable in both time and space. The temperature variability in STW is strongly correlated with ENSO, with near-surface temperatures typically being cooler during El Niños (Gordon, 1986; Greig et al., 1988; Stanton et al., 1997; Goring and Bell, 1999; Sutton and Roemmich, 2001). There are also thought to be correlations with the Interdecadal Pacific Oscillation (IPO), although temperature time-series records are too short to confirm this.

New Zealand's semidiurnal tides (M_2 and N_2) have a complete 360° range of phase around New Zealand (Walters et al., 2001). The semidiurnal tides have been characterized as a coastally-trapped Kelvin wave traveling anticlockwise around the shelf. Tidal elevations increase towards the coast with a degenerate amphidrome situated in the centre of New Zealand (Heath, 1985). A by-product of this geometry is that the tides are always 180° out of phase through Cook Strait, resulting in very high tidal velocities through the strait. High tidal velocities also occur north of Cape Reinga and in Foveaux Strait. These areas of strong tides are associated with tidal mixing (e.g. Bowman et al., 1980).

River flow regimes are influenced by a wide range of combinations of physical and climatic features (Duncan, 1992). The wettest areas of the country are on the western slopes of the Southern Alps where average rainfall is greater than 6400 mm y^{-1}, and the driest are in central Otago (Tomlinson, 1992). This results in mean flows ranging from 563 $m^3 s^{-1}$ from the Clutha River south (Duncan, 1992) down to very small flows from minor rivers, such as along the northeastern coast of North Island.

The seasonal pattern of mixed-layer depth varies from north to south, with water mass and with large-scale circulation (Longhurst, 1998). The seasonal pattern of mixed layer depth in coastal waters interacts with the nutrient characteristics of the water masses, fresh water inflow and the light regime (mainly through the depth of the photic zone) to determine the seasonal patterns of primary production and nutrient depletion in coastal surface waters (e.g. Bradford and Chang, 1987; Zeldis, in press).

1.3 Nutrient and chemical environment

New Zealand is located on the pole-ward boundary of the South Pacific gyre in the southwest Pacific Ocean. For this reason shelf edge (at 200 m) nitrate concentrations (Fig. 36.3a) are modest (5–15 mmol m^{-3}) in comparison with many regions of the world (Conkright et al., 2002) and are similar in range to the northeastern Atlantic Ocean. The absolute concentrations of dissolved inorganic nutrients in oceanic water around New Zealand depend on the water mass involved. Coastal water is mainly of subtropical origin, although SAW water lies adjacent to the southeastern South Island slope. These two water masses have different nutrient characteristics that result in a north to south gradient in mean nitrate at 200 m at the shelf break (Ridgway et al., 2002) (Fig. 36.3a). The distribution of mean nitrate at 200 m off the east coast shelf break of New Zealand ranges from 10 mmol m^{-3} in the northeast to 16 mmol m^{-3} in the southeast. The distribution on the west coast ranges from 10 mmol m^{-3} in the northwest to 12 mmol m^{-3} to the southwest, with a minimum off the central west coast of < 6 mmol m^{-3}.

Subtropical, South Pacific Central Water has a typical mix of nutrients (Tomczak and Godfrey, 1994) and nitrate (NO_3) and dissolved reactive silica (DRSi) are depleted more or less together (Zentara and Kamykowski, 1981) (Fig. 36.3b). SAW, as well as being low in Fe and Cu (Sedwick et al., 1997; Croot and Hunter, 1998), has excess NO_3 relative to DRSi (Zentara and Kamykowski, 1981). Atmospheric transport of iron from arid and semi-arid parts of Australia may be a source of iron to surface seawater in this region (Kieber et al., 2001; Boyd et al., 2004).

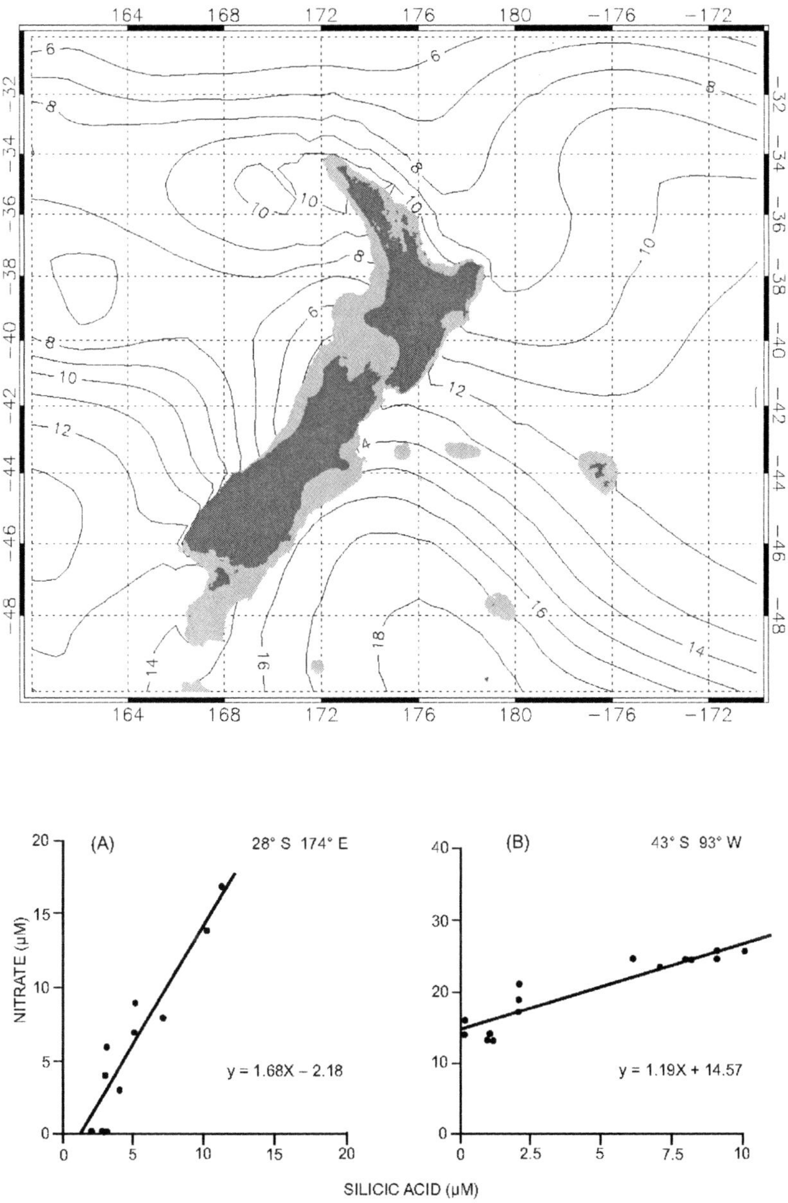

Figure 36.3 Upper panel: Mean nitrate (mmol m^{-3}) at 200 m derived from a four-dimensional ocean interpolation system based on locally weighted least squares fitting (Ridgway et al, 2002). Lower panel: Scatter diagrams and regression lines of silicic acid vs nitrate concentration showing A, a near simultaneous decline of both silicic acid and nitrate to near zero values in South Pacific Central Water; B, the depletion of silicic acid when about 15 mmol m^{-3} nitrate remains in Subantarctic Surface Water (adapted from Zentara and Kamykowski, 1981).

The contribution of freshwater inflow to dissolved nutrients in the nearshore can be estimated from river water concentrations. Nutrient concentrations in New Zealand's rivers depends on run-off, and are proportional to the area of catchment composed of soft sedimentary rock, intensive pasture use (which is negatively correlated with steep slopes and high altitudes), exotic forest, and volcanically-derived rocks (Close and Davies-Colley, 1990). Rivers cluster into eleven types. For example, Cluster 1 rivers tend to be the high altitude or less-developed catchments draining major mountain ranges; this type is particularly common in the South Island and fairly low in DRP, Si and NO_3 (0.13, 391, 3.93 mmol m^{-3}) respectively. It is possible that freshwater input of nutrients is not significant on the open coast, given the degree of dilution that is achieved (e.g. Hawke and Hunter, 1992). Deep winter mixing and upwelling of deep waters are more likely to be dominant in enhancing nutrient flux in New Zealand waters (Zeldis et al. 2004.).

In addition to dissolved inorganic nutrients, river waters may carry other substances. Land derived dissolved / colloidal organic material (yellow substance), as indicated by the extinction coefficient at 440 nm (*g*440), reaches a high of 1.8 m^{-1} in the southwestern fiords, 0.10 m^{-1} in Foveaux Strait, and 0.04–0.09 m^{-1} off the west coast of South Island (Davies-Colley, 1992). Even though some rivers, e.g. Cluster 5 and 11 of Close and Davies-Colley (1990) have high *g*440 (2.2, 4.0), it is probable that measures of this parameter in coastal waters are not primarily due to freshwater run-off (Davies-Colley, 1992 and references therein). Concentrations of yellow substance explain nearly 50% of total variation in light absorption in New Zealand waters and optical properties mainly conform to Jerlov optical types II and III (Davies-Colley, 1992).

1.4 Ecosystem characteristics

Ecosystem characteristics of New Zealand coastal waters are known from only a few regions. Information has been contributed by a small number of workers (e.g. Phytoplankton: Cassie, 1966; Chang, 1988. Zooplankton: Bradford, 1972; Jillett, 1971; Murdoch, 1989). From this work we can make the following generalizations.

The New Zealand shelf region has a biota similar to that of other temperate regions (e.g. Colebrook et al., 1961). The seasonal succession of phytoplankton appears to be typical of temperate waters (Parson et al., 1977). Early spring bloom populations on the northeastern continental shelf are dominated by chain-forming diatoms such as *Lauderia annulata* and *Cerataulina pelagica,* followed by smaller *Chaetoceros* spp. in late spring (Chang et al., 2003). As nutrients become depleted by early and late summer, diatoms are succeeded by a summer community consisting of dinoflagellates, small nanoflagellates and picophytoplankton, with the proportion of non-photosynthetic dinoflagellates increasing (Chang et al., 2003). Species diversity of diatoms, thecate dinoflagellates, and nanoflagellates decreases in summer although non-thecate dinoflagellate diversity increases.

Biomass in terms of chlorophyll *a* may be > 2 mg m^{-3} (Chang et al., 2003) in spring, but drops back to < 1 mg m^{-3} in late spring. The mean oceanic phytoplankton biomass (as measured by SeaWiFS), adjacent to New Zealand, is not remarkable in a global context (e.g. Doney et al., 2003). This observation may be a natural consequence of the modest potential nutrient supply at 200 m.

Coastal zooplankton populations around New Zealand reflect the degree of riverine input, latitude and longitude, and the proximity of offshore populations. Neritic forms such as ctenophores (*Pleurobrachia pileus*), cadocerans (*Evadne nordmanni, Penilia avirostris, Pleopsis polyphemoides*) and copepods (Fig. 36.4) are typical of New Zealand coastal waters. Over the shelf, species of *Acartia, Calanus, Centropages, Clausocalanus, Corycaeus, Paracalanus, Temora,* and the euphausiid *Nyctiphanes australis* are common (Jillett, 1971; Bradford, 1985) (Fig. 36.4). These species are either endemic to New Zealand, are confined to New Zealand and southeastern Australian waters and in one case (*Calanus australis*) also to South American coastal waters, or are widespread tropical / subtropical species. Some oceanic species are commonly encountered on the shelf depending on the fauna of the adjacent oceanic waters and the time of year (e.g. Jillett, 1971; Bradford, 1972, 1985) including chaetognaths (*Serratosagitta tasmanica*), salps (*Thalia democratica, Salpa fusiformis*), amphipods (*Themisto gaudichaudii*) and copepods (Fig. 36.4).

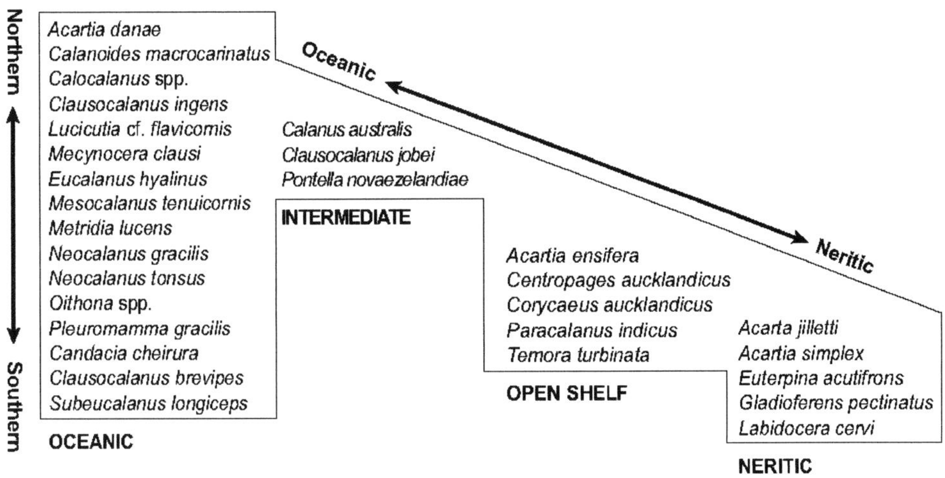

Figure 36.4 Onshore / offshore distribution of species of copepods in New Zealand coastal waters and adjacent oceans.

The sedimentary and hydraulic regime governs the broad pattern of benthic populations on the New Zealand shelf. The major infaunal benthic assemblages have been classified on the basis of the contained bivalve, echinoderm, and polychaete faunas and the sediment substrate (McKnight, 1969). Suspension-feeder dominated benthos is found on modern terrigenous sands (often as a nearshore prism) and on relict terrigenous or biogenic sediments. These are the *Venus* communities of McKnight (1969). In contrast, deposit-feeder dominated assemblages are found on shelves with high deposition of modern fine-grained sediment and are represented by the *Nemocardium* communities of McKnight (1969). Coarse relict, biogenic sediments also provide a substrate for an epifauna of sessile suspension feeders (Probert et al., 1979; Cranfield et al., 1999).

The seasonal cycle of phytoplankton biomass in New Zealand shelf waters is known only from a few studies. It appears that a spring bloom occurs from late August to early September in phytoplankton and zooplankton, with lower, sometimes variable biomass the rest of the year (e.g. Jillett, 1971; Bradford, 1972).

2.0 Coastal biogeochemistry – four case studies

The following four case studies have been chosen as they represent the best-known sections of the New Zealand continental shelf, from a multidisciplinary / interdisciplinary perspective (see Fig. 36.1 inset for locations). In the absence of a completed typology of New Zealand coastal environments (Snelder et al., 2001) we are not able to say how typical these regions are of the whole coast. We have a general appreciation of the occurrence of each type of phenomenon that drives coastal ecosystems over most of the New Zealand region (see Section 1). Nevertheless, very little is known about the northwest coast of North Island.

Multidisciplinary / interdisciplinary studies of the New Zealand shelf region have been driven by various priorities that have been either scientist driven or more lately driven by the priorities of the funding agencies. Studies in region 1, the northeast shelf of North Island, are mainly physical – biogeochemical and interdisciplinary in focus and are driven mainly by priorities relating to aquaculture and wild fisheries in the region. In region 2, greater Cook Strait, the multidisciplinary studies were first prompted by the need for an Environmental Impact Statement relating to the installation of an offshore gas platform. This work was aimed mainly at identifying the physical phenomena impacting biological spatial variability in summer. In region 3, western South Island shelf, studies were aimed at understanding the breeding environment of hoki (*Macruronus novaezelandiae*) that is New Zealand's largest fisheries by tonnage. This multidisciplinary study was focused mainly on ecosystem dynamics. In region 4, the Otago shelf, studies were not contemporaneous and represent an incremental approach to gaining an understanding of the functioning of the coastal ecosystem of this region by staff and students of the University of Otago.

2.1 Northeast shelf of North Island

The oceanography of northern New Zealand is dominated by a general west to east input of oceanic water originating from the East Australian Current that has a transport volume 15 Sv (Tomczak and Godfrey, 1994). Part of this flow attaches to the northeastern New Zealand continental platform forming the East Auckland Current (EAUC) (Stanton et al., 1997; Roemmich and Sutton, 1998; Zeldis et al., 2004). Much of the northeast shelf is < 40 km wide and the proximity of slope waters has significant implications for the inner shelf and coastal zone physics and nutrient supply (Sharples and Greig, 1998).

There is a wind-forced component of the circulation on the northeastern shelf of North Island. Strong seasonal upwelling and downwelling results in this region being one of the most productive shelf regions of New Zealand (Zeldis et al., 2004 press and references therein). There is also low freshwater and sediment input to this coast (Carter, 1975; Duncan, 1987, 1992).

Analysis of the non-tidal circulation over the shelf, slope and Hauraki Gulf separated wind-forced upwelling and downwelling (Fig. 36.5) from the outer shelf and slope dynamics that are dominated by long-period variability in the EAUC. Late summer downwelling is accompanied by more stratified conditions on the shelf and in Hauraki Gulf (Zeldis et al., 2004). Forcing of the shelf circulation arises largely from local winds. The EAUC serves to maintain the shelf-break front and near-bed onshore flows at the shelf edge, while the formation of a strong seasonal thermocline in late summer enables offshore surface water to intrude onto the shelf. In addition, the strong internal tide present on this shelf in summer (Sharples and Greig, 1998; Sharples et al., 2001) has the capacity to mix nutrients across the pycnocline. They calculate a flux of about 12 mmol N m^{-2} d^{-1} and suggest that internal tidal mixing could potentially drive annual new production of about 100 g C m^{-2}.

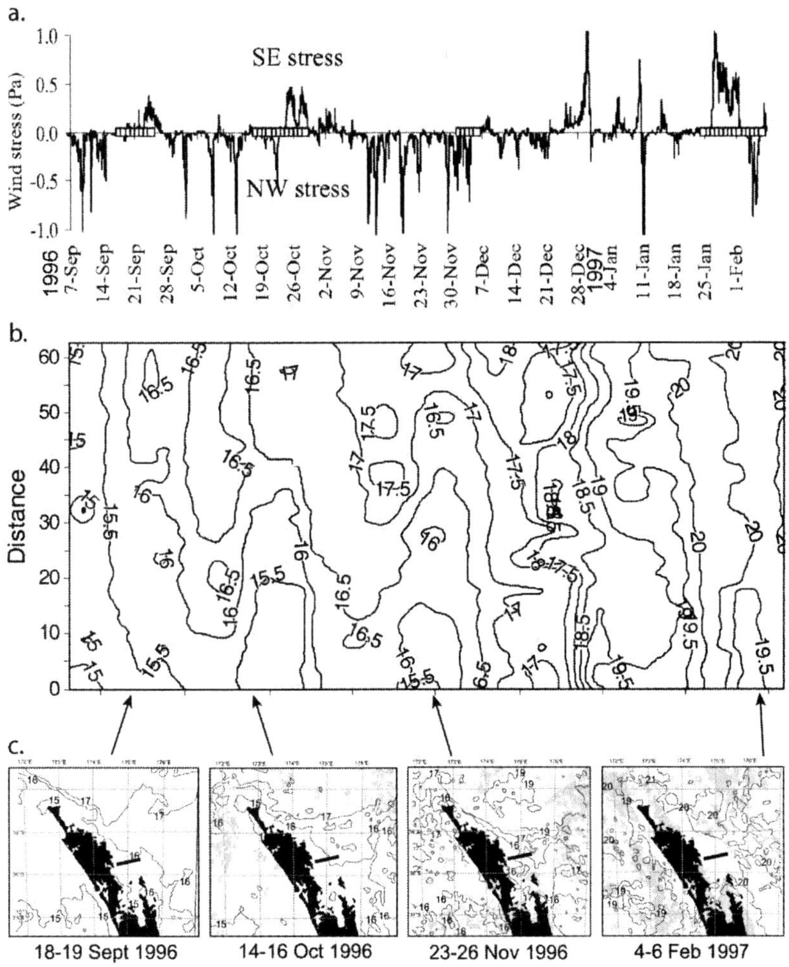

Figure 36.5 Wind stress, distance / time diagram of surface temperature along the indicated transect. a, Wind stress from the NW has negative stress, and from the SE positive stress; b, AVHRR sea-surface temperature data along section (indicated in panel c) contoured in distance offshore with time. Four voyage periods, illustrated in c below, are shown by the boxes on the x-axis; c, Spatial AVHRR images are shown representing each voyage period. Arrows connecting the voyages with their dates (from Zeldis et al. 2004). See color insert.

Strong seasonal variation in ecosystem functioning has been observed between the upwelling-to-downwelling seasonal transition (Zeldis et al., 1995; Zeldis, 2004). Winter mixing and spring upwelling cause extensive nutrient enrichment and elevated phytoplankton growth within the water column. The dominance of autotrophic processes at this time is indicated by high ratios of new-to-regenerated nutrients (*f*-ratio 0.7), high chlorophyll concentrations (> 3 mg m^{-3}), low phaeopigment concentrations, and significant upper water-column oxygen evolution. Diatoms dominate the phytoplankton at this time (Chang et al., 2003). During the summer downwelling phase, decreasing *f*-ratio (< 0.6), low chlorophyll concentrations (0.5 mg m^{-3}), increased phaeopigments, and increases in apparent oxygen utilisation (AOU) indicate the increasing importance of heterotrophic (respiratory) processes (Fig. 36.6). During this phase, assemblages are composed of dinoflagellates, diatoms, nanoflagellates and picophytoplankton (Chang et al., 2003).

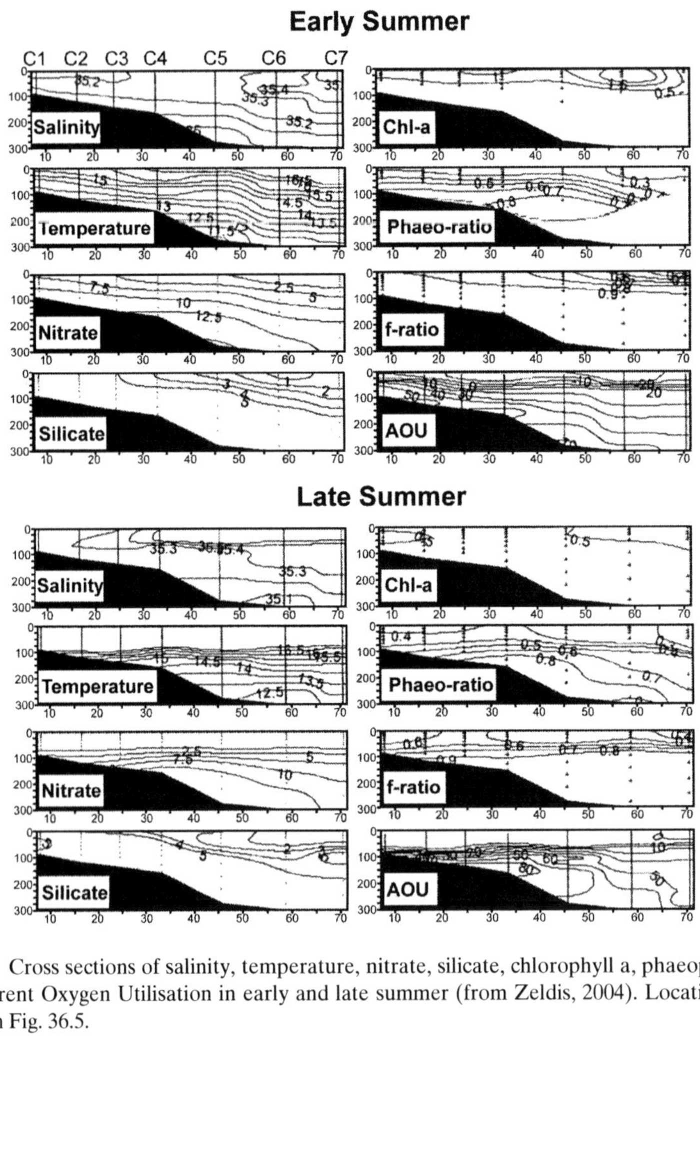

Figure 36.6 Cross sections of salinity, temperature, nitrate, silicate, chlorophyll a, phaeophytin ratio, f-ratio, Apparent Oxygen Utilisation in early and late summer (from Zeldis, 2004). Location of transect indicated on Fig. 36.5.

Dissolved inorganic nitrate concentrations in spring at the shelf edge at 200 m depth were about 10 mmol m^{-3} and increased in early summer to about 13 mmol m^{3}. At the surface inshore nitrate concentrations were about 2.5 mmol m^{-3} in spring and increased to 7.5 mmol m^{-3} in early summer during upwelling, before decreasing during the downwelling event to < 2.5 mmol m^{-3} by late summer.

As the system evolves from upwelling to downwelling, it appears that remineralised nutrients become more important relative to upwelled new nutrients. A mixing model was used to test this hypothesis and estimate the relative proportions of 'new' dissolved inorganic nitrogen (DIN) supplied by upwelling, *vs* 'secondarily remineralised' DIN supplied locally by benthic nitrification within the inner shelf system (Zeldis, 2004). In spring and early summer, the advected (deep ocean) DIN supply was found to dominate the remineralised (shelf) DIN supply. It was also found that this import of nutrient in spring was partially balanced in early summer by lateral export of organic matter to beyond the shelf break by surface advection associated with strong upwelling. It appears however, that most of the inner shelf production during the spring bloom was lost through the sedimentation of large diatoms, during a 'bloom collapse' in late spring (Chang et al., 2003). In late summer, during the downwelling phase, strong stratification caused near-bottom, remineralised DIN to accumulate but, on a water-column integrated basis, remineralised nutrient concentration changed little between spring and summer. However, because the supply had decreased greatly with the cessation of upwelling, the steady supply of remineralised DIN between the seasons probably sustained shelf phytoplankton in summer, when the upwelled supply was low. There was also less surface transport of biologically-fixed material seaward across the shelf break in summer, because of the downwelling conditions.

The mixing model (Zeldis, 2004) also showed that remineralising activity was related to current speed and sediment type at different locations along the shelf. In both spring and summer, there was less DIN remineralised immediately south of Cape Brett than there was at greater distances to the south where the shelf is wider. Similar results were obtained for spatial distributions of remineralised silica and oxygen utilisation. At Cape Brett the shelf is narrow and strong currents allow only coarse sediment deposition (Eade 1974, Zeldis et al., 2004). Further south, the shelf is wider, the currents are weaker, and sediments are finer grained (Carter and Eade, 1980; Zeldis et al., 2004). It was also found that the ratio between O_2 and remineralised DIN was above the value expected when organic matter is oxidised, suggesting that denitrification may also be a significant process in the region (Zeldis and Smith, 1999; Giles, 2001).

The nutrient concentrations and stratification of the water column on this coast result in several clearly defined stages in the development of phytoplankton populations. The inner shelf had a relatively stable assemblage of small autotrophs from early spring to late summer, overlaid by a highly variable assemblage of large autotrophs in early spring with chlorophyll *a* concentrations of about 1.2 mg m^{-3}. In contrast, the outer shelf, early spring assemblage was composed of small cells with much lower chlorophyll *a* biomass (0.3 mg m^{-3}) and therefore a lesser capacity to sink (Chang et al., 2003).

Early spring bloom populations on the inner shelf, where the euphotic zone was shallower than the depth of mixing, were dominated by chain-forming diatoms such as *Lauderia annulata* and *Cerataulina pelagica* (Chang et al., 2003). These populations are supported by upwelling of nutrients and also, presumably, deep winter mixing. The collapse of this bloom was captured by a moored fluorometer and sediment trap records in late September and was driven by depletion of DIN and DRSi. The large diatom assemblage was followed by smaller diatoms such as *Chaetoceros* spp. in late spring and small-celled *Nitzchia* in early summer.

Strong upwelling in spring was associated inshore with lower salinities, higher nutrients in shelf surface waters and the offshore transport of diatom populations to the shelf break (Chang et al., 2003; Zeldis et al. 2004). These populations were replaced inshore by a low-biomass, dinoflagellate-dominated assemblage similar to the deep assemblage present at the outer shelf. The entrainment of deep water during upwelling in early summer was faster than the growth rate of the phytoplankton, preventing biomass accumulation at inner shelf depths. Another consequence of cross-shelf transport in summer during downwelling, combined with favourable conditions for dinoflagellate growth in Hauraki Gulf, is the likely establishment of populations of toxic dinoflagellate species of offshore origin (Sharples, 1997; Chang et al., 1995b).

At mid-shelf to upper-slope depths, the mean ratio of the depth of the euphotic zone to the depth of mixing was low. Algal growth was limited here in spring due to the depth of mixing and the impact of internal waves (see above). Therefore, nanoflagellates, dinoflagellates and picophytoplankton were the main groups present (Chang et al., 2003). Species diversity of diatoms, thecate dinoflagellates, and nanoflagellates decreased in summer, although non-thecate dinoflagellate diversity increased. A small number of nitrogen-fixing *Trichodesmium erythraeum* were found in late summer after the advection of oligotrophic, subtropical water onto the outer shelf.

The maximum in spring phytoplankton biomass was not observed because of a mismatch between the field work and the timing of the spring maximum (Zeldis, 2004). An inner shelf deployment of time-incremental sediment traps recorded a peak in particulate organic carbon flux at the end of September, although this flux was not substantial compared with shelf regions globally (Dr S. Nodder, NIWA, pers comm.). This inner shelf organic loading promoted high respiration in the sediments (Zeldis, 2004). As well as being exported downwards, the production on the inner shelf is also, on occasion, exported laterally beyond the shelf break by the upwelling circulation (Chang et al., 2003). This export occurred well after the decline of the spring bloom, so on that occasion it was a minor component of shelf production. Were the upwelling to occur closer to the peak spring biomass period, potentially large amounts of production could be exported offshore, rather than be remineralised *in situ* on this narrow shelf region.

Meso- and microzooplankton composition has not been described for this coast. Nearshore mesozooplankton populations are likely to be similar to those of the nearby Hauraki Gulf with copepods dominant (Jillett, 1971). Other groups included in the Gulf fauna are cladocerans, appendicularians, salps, euphausiids, medusae, ctenophores, with the addition of chaetognaths, pteropods and heteropods in the outer Gulf. Salps (*Thalia democratica*) can dominate the assemblage in spring if shelf water exchange with the Gulf has been high (Zeldis et al., 1995). In the inner Gulf, common copepod genera (e.g. *Acartia, Paracalanus, Centropages, Temora*) tend not to have a strong seasonal signal in their abundance, whereas in the outer Gulf there was a distinct spring increase in abundance.

Upwelling onto the shelf and into Hauraki Gulf bottom waters depends on the local winds, therefore new nutrient supply should vary with climatological variation in winds. Climate-scale variation in the wind field is expected to induce interannual variation in upwelling and downwelling. Preliminary results suggest the El Niño phase of the Southern Oscillation favours upwelling whereas the La Niña phase favours downwelling (Zeldis et al., 2000). These differences are predicted to cause a cascade of biological effects which impact fisheries recruitment and aquaculture productivity in the region.

2.2 *Greater Cook Strait*

Greater Cook Strait, the region between North and South Islands of New Zealand, is composed of the broad, flat, South Taranaki Bight and Cook Strait Narrows through which Cook Strait Canyon extends (Fig. 36.7). The region is exposed to strong westerly winds (Harris, 1990 and references therein), with orographic influences funnelling prevailing winds through Cook Strait Narrows (winds up to 240 km h^{-1} have been measured), tidal currents are strong (up to 3.5 m s^{-1}) (Heath, 1974), there are large spatial changes in tidal energy dissipation (Bowman et al. 1980), and there is moderate freshwater input (Duncan, 1992). Winds and tides drive most of the variability in sediment distribution and the physical, chemical and biological properties of the region.

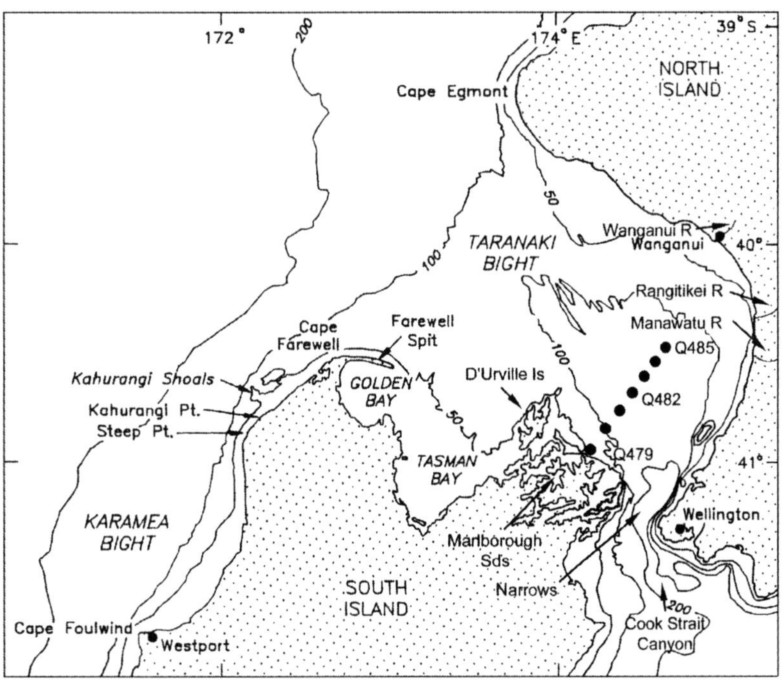

Figure 36.7 Bathymetry and locality diagram for greater Cook Strait region.

The sediments of the region are predominantly mud thought to originate from discharges from the Manawatu and Wanganui Rivers (van der Linden, 1969; Lewis and Eade, 1974; McDougall, 1975; Carter, 1983). A broad band of sand lies southwest / northeast, dividing the mud region, near the western entrance. This sand has probably been carried up the west coast of South Island. Gravels occur in the Narrows, off Cape Egmont, and west of Cape Farewell. Bowman et al. (1980) note the relationship between high mean bottom frictional stress during a tidal cycle and the distribution of coarse sediments. Sediment type has a strong bearing on the type of benthic communities found in the region (McKnight, 1969). Macrobenthic biomass of western Cook Strait lies between 200–650 mg m^{-2} although these figures are based on sparse data (Probert and Anderson, 1986).

The most conspicuous feature of greater western Cook Strait in summer, the Kahurangi upwelling plume (Fig. 36.8), results from the interaction of the wind-driven Westland Current with the local bathymetry (e.g. Shirtcliffe et al., 1990). Winds induce a fall in sea level near Cape Farewell, and the resulting favourable sea surface slope accelerates flows of deep water over the bathymetric rise inshore of Kahurangi Shoals and also raises sea level at Cape Egmont. The hydraulic response of the thermocline, coupled with the coastal convergence of the bottom Ekman flow, produce a strong upwelling source near Kahurangi Point. When the westerly wind drops, warm water sometimes moves southwards round Cape Farewell inshore of the upwelling plume.

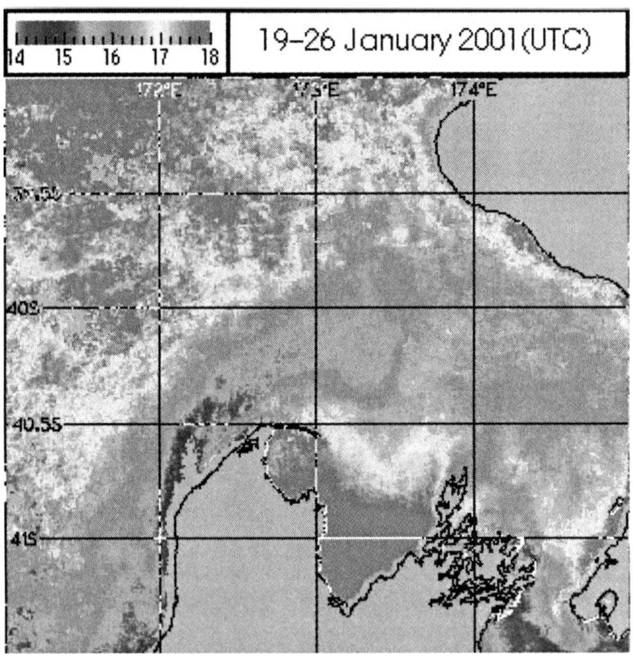

Figure 36.8 AVHRR of sea surface temperature showing the Kahurangi plume from the NIWA SST archive (Uddstrom and Oien, 1999). See color insert.

This Kahurangi Point upwelling results in a north-eastwards shedding of cold, nutrient-rich (> 2 mmol m^{-3}) water associated with greater concentrations of chlorophyll *a* (> 2 mg m^{-3}) (Fig. 36.9) and primary production than the surrounding waters (e.g. Bradford et al., 1986). Cross-sections through upwelling features show that the maximum chlorophyll concentrations are usually located adjacent to the coldest, most nutrient-rich water (Kibblewhite et al., 1982). Surface primary productivity follows approximately the same pattern (Bradford et al., 1986).

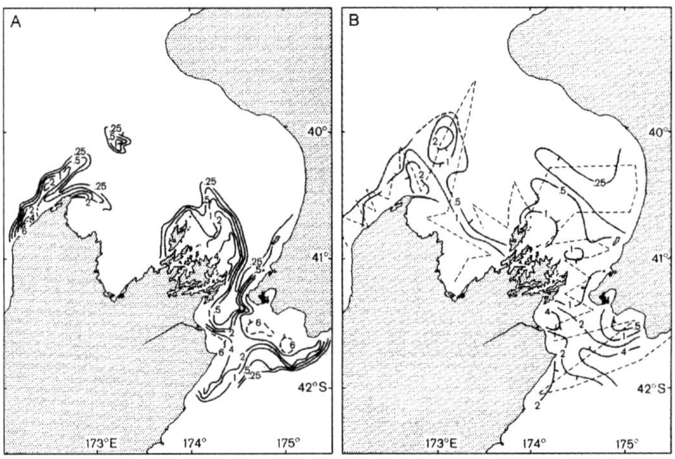

Figure 36.9 Surface properties during January – February 1980: A, Nitrate nitrogen (mmol m^{-3}); B, Chlorophyll *a* (mg m^{-3}) (adapted from Bradford et al., 1986 and references therein).

Potential primary production (measured continuously over 24 hours in an on-deck incubator and corrected for diel rhythms in photosynthetic capacity) as great as 12 mg C m^{-3} h^{-1} was recorded in greater Cook Strait (Bradford et al., 1986). Nevertheless, Viner (1990) showed that non-photosynthetic ^{14}C uptake in the plume region can be very high (up to 88%). The most elevated rates of dark ^{14}C fixation were associated with the warmer, more mature parts of the plume and with elevated ambient NH_4 concentrations and NH_4 uptake. Viner concluded that the dark uptake resulted mainly from high bacterial nitrification rates because nitrification uses inorganic carbon as a source of carbon and NH_4 as an energy source. Observed nitrification rates may have supplied about 40% of the algal requirements for NO_3 (Priscu and Downes, 1985). The f-ratio ranged from 0.06 to 0.87 and the highest ratios imply that new nitrate-nitrogen was a more important source of N than NH_4 in the upwelling plume itself (Viner and Wilkinson, 1988). Nitrate uptake varied from 1–8.5 μg (mg N)$^{-1}$ h^{-1} and was strongly inhibited in the presence of high levels of ambient NH_4 (Viner and Wilkinson, 1988). Upwelling also was associated with the lowest proportion of ^{14}C incorporation into protein compared with flow into storage products (Priscu and Priscu, 1984).

The spatial distribution of summer zooplankton populations and their biomass are formed by advective as well as *in situ* processes (James and Wilkinson, 1988; Bradford-Grieve et al., 1993). Zooplankton within the plume divide into five geographically coherent assemblages, generally reflecting temporal changes in the zooplankton community as the upwelled water advected from west to east (Bradford-Grieve et al., 1993). Near the source of the plume, inshore plankton biomass decreased as did the numbers of several coastal species, and oceanic species were introduced into nearshore waters. The capacity of copepods and the euphausiid *Nyctiphanes australis* to reproduce was reduced (Bradford and Chapman, 1988a), and the diversity indices and proportion of herbivorous copepods decreased, relative to water south of the plume source. Downstream in the plume, many zooplankton species were distributed in a manner reflecting the physical characteristics of the plume, whereas oceanic forms were entrained along the offshore border of the plume. In the eastern plume the proportion of omnivorous copepods was reduced and the capacity of herbivorous copepods and *N. australis* to reproduce increased. The abundance of common coastal copepods was apparently related to their vulnerability to offshore transport on an upwelling coast. Zooplankton biomass (as wet weight) is about 100 - 300 mg m^{-3} in autumn but can be as high as 8000 mg m^{-3} in summer when dominated by the salp *Thalia democratica* (Kibblewhite et al., 1982).

Small copepods, such as *Acartia ensifera, Oithona similis,* and *Paracalanus indicus,* had the largest influence over the patterns of distribution of metabolic parameters. Their carbon ingestion and NH_4 excretion is more highly weight specific compared with *Nyctiphanes australis* (James and Wilkinson, 1988). Calculations of the carbon requirements of zooplankton suggests that at the upwelling source their grazing may have overtaken phytoplankton production, although it is more likely that microzooplankton were an important part of their diet.

Bradford et al. (1993) compared the Kahurangi upwelling plume with the Mauritanian plume, West Africa, in the eastern subtropical North Atlantic Ocean. The mauritanian upwelling has much greater NO_3, chlorophyll *a* concentrations and primary productivity than the Kahurangi system (Table 36.1), while zooplankton biomass attained much higher biomass near the Kahurangi plume. This suggests that the Kahurangi plume is either converting primary production into zooplankton biomass more efficiently or the structure of the food web differs. We do not know if the benthos benefits from the vertical export of the overlying production in the western part of greater Cook Strait.

TABLE 36.1
Comparison of the Mauritanian and greater Cook Strait upwelling plumes (from Bradford-Grieve et al., 1993).

	MAURITANIA	GREATER COOK STRAIT
Latitude	19°N	40°S
Surface NO_3 maximum	> 20 mmol m-3	> 7.4 mmol m-3
Surface chlorophyll *a* maximum	> 20 mg m^{-3}	> 4–5 mg m^{-3}
Surface primary production maximum	> 60 mgC m^{-3} h^{-1}	> 9 mgC m^{-3} h^{-1}
Maximum zooplankton dry weight	1800 mg m^{-2}	5926 mg m^{-2}
Principle herbivorous copepods	*Calanoides carinatus*	*Clausocalanus jobei* *Paracalanaus indicus*
Maximum percentage herbivorous copepods	78%	68%

Tidal mixing has been demonstrated theoretically (Bowman et al., 1980) and confirmed from field observations (Kibblewhite et al., 1982) to be a source of nutrient renewal in summer in Cook Strait Narrows and off D'Urville Island and Marlborough Sounds; tidal mixing also occurs in Taranaki Bight but the shallow depth of the region ensures that nutrients are depleted throughout the water column (Bradford et al., 1986). Vertical sections off Marlborough Sounds show profiles of a number of parameters typical of tidal mixing. For example, near-surface isotherms bend towards the surface and deeper isotherms bend towards the seafloor adjacent to a region of completely mixed water inshore (Fig. 36.10). Nevertheless, the concentrations of chlorophyll *a* were low (<2 mg m^{-3}) (Bradford et al., 1986) compared with those found in the western approaches to the English Channel by Pingree et al. (1978), even though the levels of nutrients were similar. The Marlborough Sounds tidal mixing region is a deepwater front (Pingree et al., 1978) where inshore, cold water remains nutrient-rich because photosynthesis is light-limited there. Thus, higher chlorophyll *a* levels are found only in areas of weakly stratified water adjacent to the well-mixed water (Bradford et al. 1986).

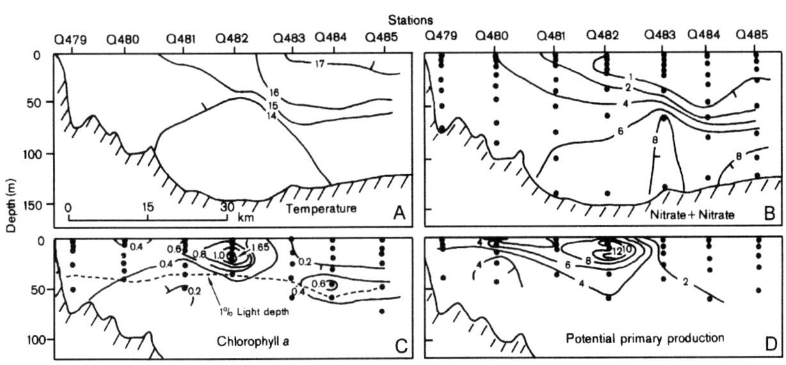

Figure 36.10 Vertical profiles along section off Marlborough Sounds (see Fig. 36.7): A, Temperature (°C); B, Nitrate nitrogen (mmol m^{-3}); C, Chlorophyll *a* (mg m^{-3}); D, Potential primary production (mg C m^{-3} h^{-1}) (from Bradford et al., 1986).

Freshwater discharges into greater Cook Strait are greatest along the Manawatu coast of North Island where the Wanganui, Rangitikei, and Manawatu rivers open onto the coast (Harris, 1990 and references therein). These rivers are part of Cluster 8 on the east coast of North Island (Close and Davies-Colley, 1990) and have average silica content (356 mmol m^{-3}) but low NO_3 and DRP (1.1, 0.13 mmol m^{-3} respectively). This freshwater is responsible for shoaling of the surface mixed layer and has been implicated in elevated reactive silicate, chlorophyll *a,* and primary production levels off the Manawatu River (Bradford et al., 1986). The reasons for, and extent of, enhanced primary productivity near rivers have not yet been clearly established. Freshening also occurs in Tasman and Golden Bays but not to the same extent. In Tasman Bay, freshwater runoff from agricultural land has been implicated in local nutrient and phytoplankton enrichment (MacKenzie and Gillespie, 1986). This nutrient enrichment may be responsible for enhanced microphytobenthic production and denitrification that is known to occur in shallow regions of Tasman Bay (Christensen et al., 2003). They calculate that denitrification accounted for 11–20% of the total carbon remineralised – one of the highest numbers reported for coastal sediments.

The Greater Cook Strait area is highly complex. We have not integrated our understanding of the interactions between the physical, chemical, and biological processes and cannot yet be predictive. Bowman et al. (1980) noted an association between the squid fishing fleet and the Kahurangi upwelling plume. Also, there is a moderate jack mackerel fishery in the region but we have not quantified the supply of organic material to the sea floor nor its impact on the benthic food web. In spite of the apparently biologically rich nature of at least the pelagic ecosystem, Greater Cook Strait is only a moderate focus for fisheries activity.

2.3 Western South Island shelf

The western South Island coast is flanked by a nearby, high mountain range. The continental shelf ranges from <25 to 100 km wide (Fig. 36.11), has high freshwater input, very high riverine sediment discharges, and is exposed to prevailing westerly weather systems resulting in marked cloudiness, windiness, upwelling, and variability in mixed layer depth. A model driven by observed meteorological data, produced a realistic seasonal cycle in sea surface temperature, mixed layer depth and surface nitrate concentration over the continental slope (Hadfield and Sharples, 1996). The continental slope off this region is an important breeding ground of hoki (*Macruronus novaezelandiae*) that forms New Zealand's largest fishery and whose numbers increase many-fold during winter spawning migrations (Patchell, 1982).

New Zealand's bathymetric platform constrains the west to east oceanic flow to the south along the southern flank of Challenger Plateau, although northwards flowing currents are observed close inshore (Heath, 1982). The general circulation affects the distribution of properties near shore with high surface and subsurface temperatures and salinities have been observed to the left of the southwestwards flow along the coast (Stanton, 1976) thus aligning isotherms and isohalines parallel to the shore. These water properties and the mixed-layer depth (Bradford and Chang, 1987) attain maxima near the continental slope, because of the juxtaposition with cold, less saline coastal water.

Coastal currents are highly variable with a weak mean alongshore flow of about 4 cm s^{-1} (Cahill, et al., 1991). The bulk of the alongshore flow variance within 50–100 km of the coast is explained by a combination of low mode and wind-forced coastally trapped waves (CTW) (Stanton and Greig, 1991; Cahill et al., 1991; Stanton, 1995). At subtidal frequencies the CTW field is the dominant signal in the shelf currents and is generated by broad-band wind forcing over Cook Strait. Variable flow through Cook Strait is considered to be the major forcing mechanism for CTWs on the Westland shelf (Stanton and Greig, 1991). It is also possible that CTWs are involved with shelf-to-offshore exchange of water and biological material through events known as *coastal squirts* (Moore and Murdoch, 1993).

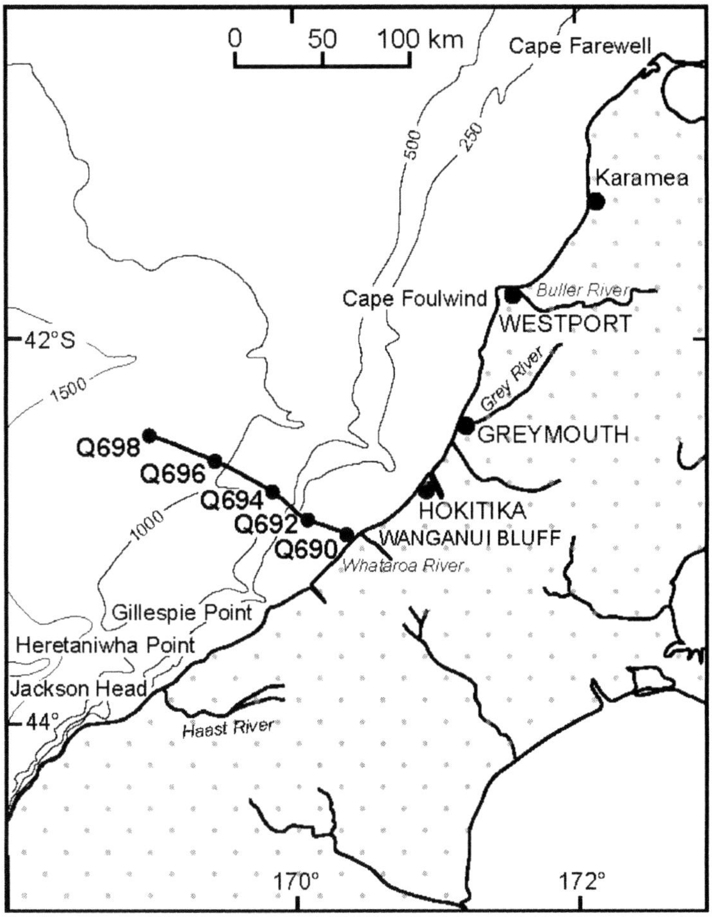

Figure 36.11 Bathymetry and locality diagram for the west coast of South Island.

Upwelling occurs in sporadic events lasting 2–12 days and rarely reaches a discernible steady state (Stanton and Moore, 1992). Upwelling-favourable winds occur 40–50% of the time, but calculations place the Westland coast in the lower part of the upwelling range compared with the California/Oregon, Peru, and Northwest African coasts where upwelling seasons can last for months (Stanton and Moore, 1992). It appears that southwesterly winds are not the only forcers of upwelling. On some occasions upwelling has been observed when winds were not upwelling-favourable (Chang et al., 1992; Moore and Murdoch, 1993). It is possible that CTWs are important in the spin-up of upwelling along this coast, although this hypothesis has not yet been investigated (Moore and Murdoch, 1993).

The impact of upwelling on near-surface nutrient concentrations is variable (set in a background of relatively low nutrient at 200 m) (Fig. 36.3a) on the western, upper panel South Island coast) and sometimes does not have surface expression because a cap of dilute river water water overlies the upwelled water (Bradford, 1983; Bradford and Chang, 1987). Field measurements of dissolved inorganic nutrients show that both upwelling and river inflow impact coastal nutrient levels, although the importance of these two sources to the biological production system has not yet been quantitatively assessed.

Rivers contribute a relatively large proportion of the shelf water volume (Stanton, 1976), modify the salinity distribution, and add dissolved inorganic nutrients and yellow substance (Cole and Davies-Colley, 1990). Sometimes the influence of river water may be very large and produces strong, non-seasonal temperature and salinity variations (Heath and Ridgeway, 1985). There is also 62 Mt y^{-1} input of terriginous sediment (Fig. 36.2a) (Griffiths, 1979; Hicks and Shankar, 2003) resulting in a shore-attached sediment wedge composed of sand and mud (Carter, 1980; Probert and Swanson, 1985), and attenuation of the penetration of light into the water column (Chang and Bradford, 1985).

Cloud cover on this coast is highly variable and can cause summer incident solar radiation to be as low as in winter (Chang and Bradford, 1985; Bradford and Chang, 1987). A large proportion of variability in production per unit chlorophyll *a* integrated over the euphotic zone for both winter and summer is explained by incident solar radiation received (Fig. 36.12) (Bradford and Chang, 1987). The attenuation of light is consistently greatest in near surface waters towards the coast (e.g. Bradford and Chang, 1987) with a 1% light level of 21–77 m. Chang and Bradford (1985) noted that the attenuation of photosynthetically active radiation brought about by non-photosynthetic particles was significantly negatively-correlated with salinity and therefore likely to be associated with river input. Although some of this light attenuation may be due to terrigenous sediment, concentrations of yellow substance explain nearly 50% of the variation in light absorption in southern New Zealand waters (Davies-Colley, 1992; Howard-Williams et al., 1991, 1995).

All of these physical processes impact the biological production system in ways not yet fully understood. The highest concentrations of chlorophyll *a,* primary production, and zooplankton biomass (Table 36.2) are usually found in a narrow band close inshore in both winter and summer (Bradford, 1985; Chang and Bradford, 1985; Bradford and Chang 1987). In inshore waters the 2–20 μm phytoplankton size fraction is the more dominant (Chang, 1988), but very small autotrophic cells (< 5 μm) comprise a large proportion (40–90%) of the total chlorophyll *a* and production offshore. Winter neritic populations were dominated by diatoms, with phytoflagellates concentrated on the outer shelf and dinoflagellates most dominating away from the coast (Chang, 1983). In summer, dinoflagellates dominated populations with *Prorocentrum micans* particularly abundant and diatoms were most numerous close inshore (Chang, 1988). Although upwelling appears to enhance biological production (Fig. 36.13) (Chang et al., 1992), if it is too strong it can alter light attenuation by re-suspending sediment from the sea floor and introducing phytoplankton with low light-saturated photosynthetic capacity into near surface waters (Chang and Bradford, 1985). Under these conditions, upwelling did not have an appreciable effect on phytoplankton biomass.

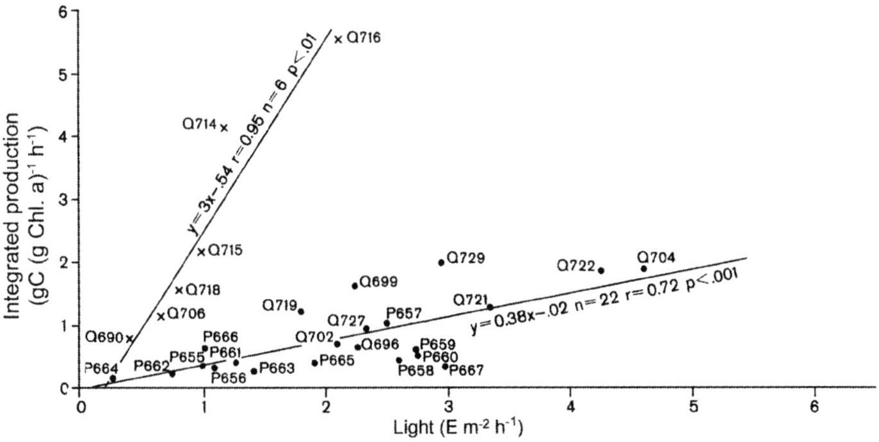

Figure 36.12 Production per unit chlorophyll *a* (P^B) integrated over the euphotic zone (g C (g Chl *a*)$^{-1}$ h^{-1}) plotted against incident solar radiation (I) (E m^{-2} h^{-1}). • = stations not dominated by *Prorocentrum micans;* x = stations dominated by *Prorocentrum micans* (from Bradford and Chang, 1987).

Studies of the utilisation of nitrogen by phytoplankton showed that both *new* and *regenerated* forms of production occur on this coast. Nitrate was the most important source of nitrogen in winter (Chang et al., 1989, 1992). Nevertheless, all size classes of phytoplankton were shown to have a strong preference for NH_4. The f-ratio for the < 2 μm fraction fell below 0.5 in winter and 0.3 in summer suggesting the picophytoplankton-sized organisms are more efficient utilisers of NH_4 than their larger counterparts. f-ratios were generally high inshore both in summer and winter (Chang et al, 1989, 1992, 1995a). Mean bacterial biomass was 869 and 928 mg C m^{-2} in winter and spring, respectively (Smith and Hall, 1997). This biomass represents 65 and 37% of the total plankton biomass in winter and spring, respectively, and indicates the importance of recycling processes on this coast.

Limited work has been carried out on the role of micro- and mesozooplankton as grazers and recyclers of nutrients. Microzooplankton herbivory was 80–194% of primary production in winter and 20–64% in spring and grazing on bacterial production was 92–154% and 79–250% respectively (James and Hall, 1998). These grazing rates indicate the microzooplankton grazing can control phytoplankton and bacterial biomass. Neritic mesozooplankton are found in large concentration close to shore (Bradford, 1985). Here, their filtering rates were greatest in populations dominated by *Acartia ensifera* (James, 1989) and were sometimes at levels that had a negative impact on the biomass of phytoplankton. On other occasions, only 1–4% of primary production was ingested by mesozooplankton implying that other sources of food (microzooplankton) are necessary if they are to obtain sufficient food for growth (Bradford-Grieve et al., 1998). James (1989) also established that the regeneration of nitrogen was related to zooplankton biomass and that up to 24% of the NH_4 uptake by phytoplankton is potentially regenerated by zooplankton > 55 μm.

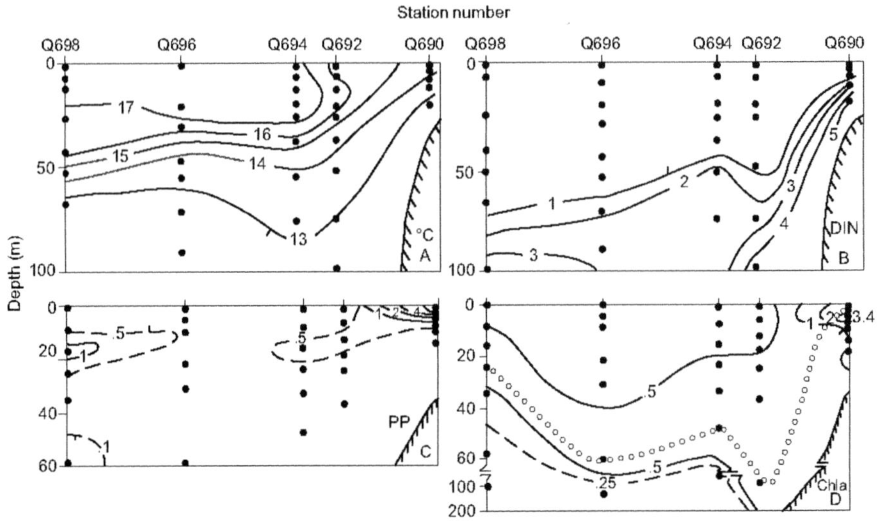

Figure 36.13 Vertical profile of water properties along a transect of Wanganui Bluff, June 1979. A, temperature (°C); B, nitrate-nitrogen (mmol m^{-3}); C, primary production (mg C m^{-3} h^{-1}); D, chlorophyll *a* (mg m^{-3});. Ticks directed towards regions of minimum values. For location of transect see Fig. 36.11 (from Bradford and Chang, 1987).

TABLE 36.2.

Phytoplankton biomass and production and zooplankton biomass on the continental shelf off the western coast of South Island. Chl *a* = chlorophyll *a*, PP = primary production, WW = wet weight (from Chang and Bradford, 1985; Bradford and Chang, 1987).

Season	Location	Chl *a* mg m^{-2}	Chl *a* surface mg m^{-3}	PP mgC m^{-2} d^{-1}	Zooplankton WW 0–200 m mg m^{-3}
winter	inshore	22–41	0.75–1.42	47–147	25 - >1000
	offshore	29–52	0.64–0.99	55–152	<25 - 100
summer	inshore	15–84	0.37–8.76	175–4885	>100 - >400
	offshore	110–401	0.26–0.45	110–401	25 - >100

Certain characteristics of slope waters appear to be important to the survival of young hoki larvae. Hoki aggregate to breed off the western South Island coast from the Campbell Plateau east of New Zealand and other regions, from late June through July and August (Murdoch, 1992 and references therein). Hoki lay their eggs in the water column above the shelf edge (Zeldis et al., 1998), although in some years very few eggs and larvae have been found. Older stage hoki larvae have been recorded further inshore than eggs and newly hatched larvae, suggesting shoreward transport. The diet of hoki larvae consists primarily of copepod adults and copepodites (Murdoch, 1990) and they actively select copepods of the genera *Calocalanus* and *Paracalanus* and the tintinnid *Dyctyocysta* (Fig. 36.14) (Murdoch and Quigley, 1992).

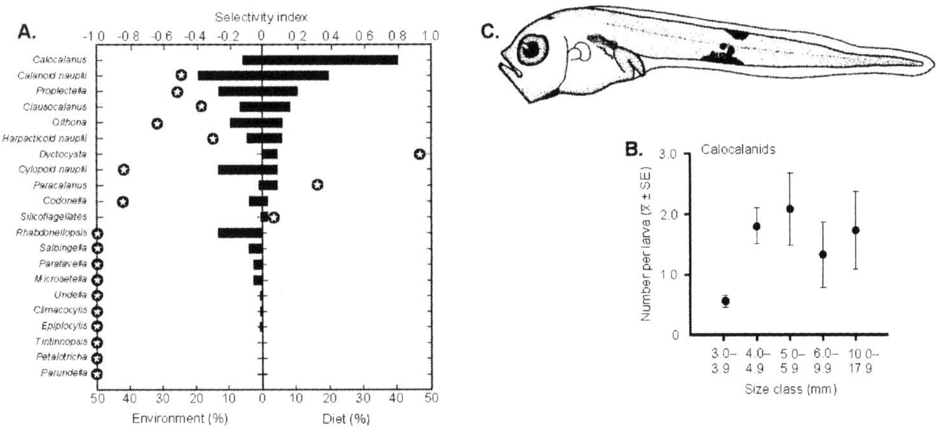

Figure 36.14 Diet of hoki (*Macruronus novaezelandiae*) larvae. A, Percentages of planktonic species in the environment (left) and diet (right) of larvae. Value of alpha selectivity index denoted by star: 0 indicates random feeding, values > 0 positive selection, values < 0 negative selection (thecate dinoflagellates and diatoms excluded); B, Mean number of selected prey types per larvae size classes (see Murdoch, 1992). C, larva standard length 5.2 mm (Patchell et al., 1987).

Attempts have been made to identify the optimal physical conditions for the survival of hoki larvae, but at the time such a study was carried out there were no hoki larvae present in the samples so the resulting hypotheses have not been tested (Bradford-Grieve et al., 1996). Bradford-Grieve et al. (1996) note that in early winter the depth of mixing appears to have a strong impact on the distribution of small zooplankton species that are actively selected for or are important in the diet hoki. For example, *Calocalanus* spp. and *Oithona similis* were distributed, not only in highest concentrations, but also over a much greater depth range in surface waters above the continental slope when the depth of mixing was deepening (158 m), nitrate-nitrogen was increasing (3.8 mmol m^{-3}) and the percentage of chlorophyll *a* was in the > 20 μm size fraction was increasing (32%). Hindcasts of the depth of mixing (Hadfield and Sharples, 1996) for 1987 and 1988 which contributed strong year classes to the adult hoki fishery indicate that the early and gradual onset of winter mixing was associated with the replenishment of nitrate in surface waters to 1 mmol m^{-3} by the end of June (Bradford-Grieve et al., 1996). Whereas, in 1990 when the hoki year class was weak, the hindcasted winter mixing began later, nitrate replenishment reached 1 mmol m^{-3} a month later, at the end of July, and mixing progressed rapidly to depths > 200 m (Bradford-Grieve et al., 1996) that would have inhibited net primary production.

High sedimentation and episodic upwelling promote a deposit-feeder dominated macrobenthos with opportunistic taxa seemingly well represented (Probert and Grove 1998; Probert et al., 2001). However, the amount of primary production that reaches the seafloor and is utilised there is not clear. Macrofaunal benthic biomass ranges from 24–143 g wet weight m^{-2} with biomass and faunal density is inversely related to depth (Probert and Anderson, 1986). The predominant source of energy to the benthos is planktonic production (calculated to be 182 g C m^{-2} y^{-1}), but calculated benthic production (39, 3.6, and 1.8 g C m^{-2} y^{-1} for bacteria, meiofauna, and macrofauna, respectively) seems low relative to the overlying primary production (Probert, 1986). Rivers would be expected to provide a large input of carbon to the shelf, based on annual flow rates, but given that the catchments are forested, much of the particulate organic carbon is likely to be structural plant tissue not readily assimilable by the benthos. Probert suggests that efficient bacterial remineralisation of phytodetritus, its export off the shelf, and/or burial could all be explanations for apparently low benthic production.

Gillespie et al. (unpubl. data) examined the relative microbial activity of sediments. Both the rate of ^{14}C-glucose mineralisation (V_{max}) and ATP concentration declined by more than an order of magnitude from shelf to upper slope sites. Ratios of sulphate reduction to oxygen consumption indicated that the relative importance of anaerobic mineralisation of organic C decreased dramatically with distance offshore. High ratios observed at nearshore stations suggested that a significant or even major fraction of total C mineralisation within the sediments may occur through the anaerobic pathway. Kaspar et al. (1985) recorded a higher denitrification capacity of sediments at shelf sites than on the upper slope. At an inner shelf (50 m) site, 75% of net mineralised N was denitrified. By markedly reducing the flux of NH_4 and NO_3 from some sediments, microbial denitrification may have an indirect regulating effect on primary production on the shelf.

2.4 *Otago shelf*

The South Otago continental shelf averages 30 km in width between Molyneux Bay at the mouth of the Clutha River and Blueskin Bay north of Otago Peninsula, and is constricted to only about 10 km in width off the out-jutting volcanic complex of the Otago Peninsula (Fig. 36.15). The shelf break occurs at water depths of 125–150 m. The shelf edge is incised by several submarine canyons that connect with tributary channels of the Bounty Trough (Carter et al., 1985).

Modern sediment input to the South Otago shelf is predominantly from the Clutha River, which now contributes an estimated 0.39 Mt y-1 (reduced from 2.3 Mt y-1 before damming of the river that commenced in the 1950s) (Hicks and Shankar, 2003) (Fig. 36.2a). Much of the bedload (sand and gravel) reaching the shelf is stored within a large nearshore sand wedge off the Clutha mouth, whilst bedload that escapes storage is transported north-eastwards and deposited on littoral and inner shelf environments to north of Otago Peninsula (Carter, 1986; Carter and Carter, 1986) (Fig. 36.15). Bedload transport to the north-east is initiated by tides and storm events, with mean flow (Southland Current) playing a minor role, but reinforced by southerly storm events, which may stir sediments to mid-shelf depths (Carter and Heath, 1975). Modern terrigenous sediments are thus confined to an inshore sand wedge (< 40 m water depth with increasing silt content > 20 m). The inner shelf sand benthos is similar to faunas of well-sorted terrigenous sands at 20–40 m at open sea locations worldwide (in terms of, for example, amphipod and polychaete taxa), though with a conspicuous suspension-feeding trochid gastropod component (Pérès, 1982; Probert and Wilson, 1984).

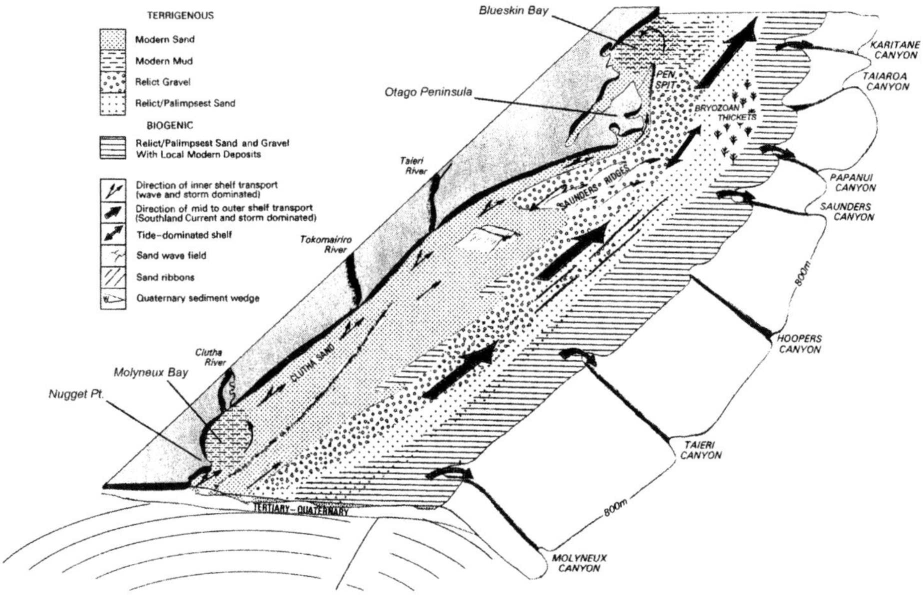

Figure 36.15 Hydraulic and sedimentary regimes of the Otago coast (from Carter et al., 1985).

Mean flow over the south-east New Zealand shelf is to the northeast associated with the Southland Current (Heath, 1972) (Fig. 36.1). The presence of this current is thought to be due to remote forcing from strong topographically constrained abyssal currents located approximately 12° east of New Zealand (Tilburg et al., 2002) although variability in the current speed has been shown to be correlated with local winds (Chiswell, 1996). Mean current speeds in the Southland Current at the 100 m isobath off Oamaru (North Otago) and Nugget Point (South Otago) were 14 and 24 cm s^{-1} respectively (Chiswell, 1996). Southland Current water on the shelf derives from the region of the STF west of New Zealand and consists mainly of STW with some SAW (Heath, 1975), while off the shelf the Southland Current advects SAW horizontally (Sutton, 2003). To the east, shelf water is bounded by a narrow (mean width 8 km) well-defined front (temperature differential of 1.5–2.0° C), the Southland Front, which represents the sector of the STF off south-east New Zealand separating modified subtropical shelf water from offshore SAW (temperature differential of c. 1.8° C). The mean position of the Southland Front follows the 500 m isobath (Shaw and Vennell, 2001).

The Southland Current on the continental shelf is recognisable as a relatively warm, high salinity flow (typically > 9.5° C and > 34.4 in winter; > 12.0° C and > 34.6 in summer) (Jillett, 1969; Hawke, 1989) (Fig. 36.16). As a consequence of this strong along-shelf flow, a meso scale lee eddy is generated in the embayment north of Otago Peninsula. This eddy entrains plankton (Robertson, 1980; Zeldis, 1985; Murdoch, 1989; Murdoch et al., 1990). Nearshore, riverine input forms neritic water with temperature and salinity of <10° C and <34.4 in winter, and >12° C and < 34.6 in summer (Jillett, 1969; Hawke, 1989; Vincent et al., 1991; Chiswell, 1996). The main freshwater source is the Clutha River (mean flow 563 m^3s^{-1}), 100 km

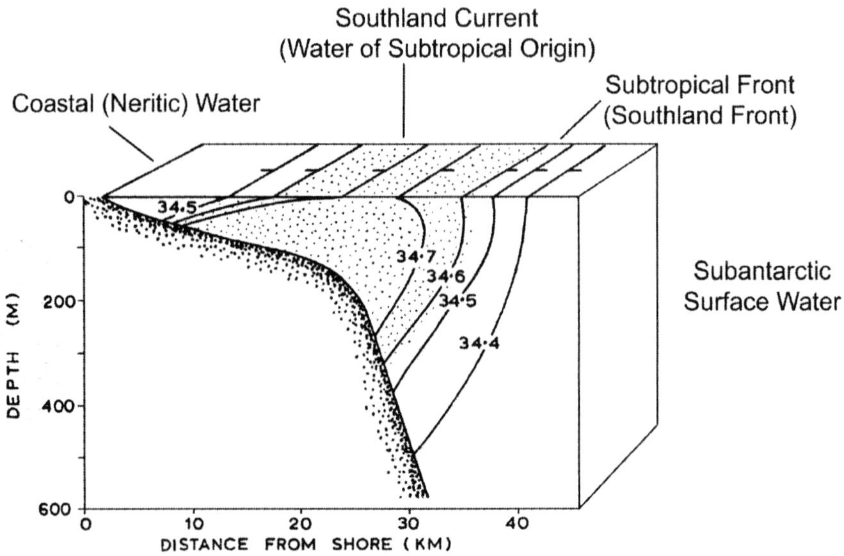

Figure 36.16 Vertical section of salinity distribution across the Otago coast (from Jillett, 1969).

south of Otago Peninsula. Periods of high Clutha flow produce northward-propagating pulses of lower salinity water with higher suspended sediment (Murdoch et al. 1990) (Figs. 36.17 and 36.18). El Niño events produce a marked cooling of sea surface temperature (Greig et al., 1988), e.g., by < 2.9° C and by < 3.4° C for Southland Current and neritic water, respectively, between 1990 and mid-1992 (Shaw et al., 1999).

Reactive phosphorus levels are depleted by primary production processes over the shelf in summer relative to winter (0.12 and 0.78 mmol m^{-3}, respectively) (Hawke, 1989; Kirchlechner, 1999), corresponding to minima in the partial pressure of CO_2 in the surface waters (Currie and Hunter, 1999). Reactive phosphate concentrations are greatest in SAW offshore (see also Fig. 36.3, top panel) although the vertical density gradient associated with the Southland Front probably inhibits transfer to inshore surface waters (Hawke and Hunter, 1992; Croot and Hunter, 1998). There is also evidence of nutrient renewal processes off the Otago Peninsula and nutrient depletion in the lee eddy north of the peninsula. Riverine input appears to be the main source of reactive silicate (Butler et al., 1992), with high concentrations occurring after Clutha flood events (Vincent et al., 1991; Hawke, 1995). Resuspension of shelf sediments is suggested to be the main source of dissolved iron although iron is unlikely to be transported offshore because the Southland Front is an impediment to onshore / offshore transport (Croot and Hunter, 1998).

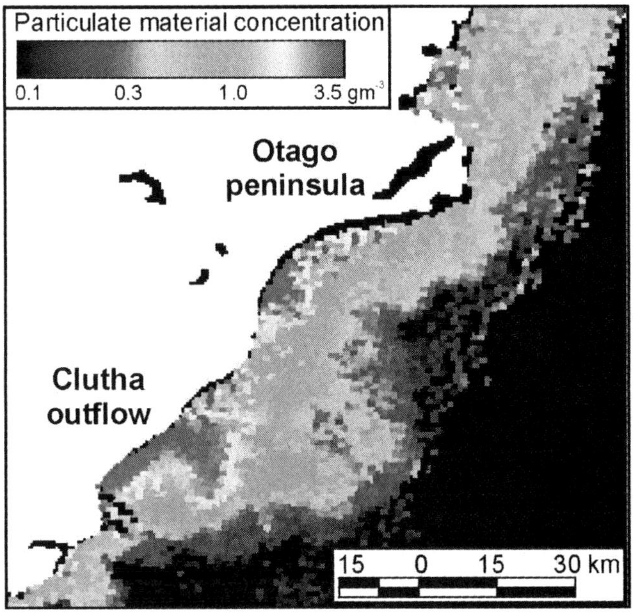

Figure 36.17 Satellite image of suspended particulate material concentration in the Otago region on 5 July 2000, showing sediment-rich waters being discharged by the Clutha River. Turbid waters are shown in a red colour, and clearer waters are shown in a purple colour. Data were received by the SeaWiFS ocean colour satellite (used courtesy of NASA SeaWiFS project and Orbimage Inc) and processed at the National Institute for Water and Atmospheric Research, Wellington, New Zealand. See color insert.

Primary production is estimated at 150–200 g C m^{-2} y^{-1} for the neritic to neritic/subtropical zone (60–120 m) and 115 g C m^{-2} y^{-1} for STW (300 m). Diatoms dominate the neritic zone phytoplankton, but the contribution of nanoflagellates, particularly prymnesiophytes, increases in Southland Current and STF waters (Kirchlechner, 1999). Compared to STW, primary production in the vicinity of the Southland Front appears to be somewhat elevated (~ 200 g C m^{-2} y^{-1}). Enhancement of primary production at the STF is also indicated by CO_2 draw-down (Currie and Hunter, 1998).

Neritic zooplankton are characterised by an abundance of meroplanktonic larvae of inshore species (e.g., stomatopod larvae, barnacle nauplii, porcellanid zoeae) and the harpacticoid *Euterpina acutifrons.* Among species typical of shelf waters in general are the euphausiid *Nyctiphanes australis, Munida gregaria* post-larvae, the salp *Ihlea magalhanica,* and the copepod *Calanus australis* (Jillett, 1976; Murdoch, 1989). Ichthyoplankton similarly indicate fish spawning distributions associated with water masses (e.g., *Sprattus antipodum* in neritic water; *Seriolella brama* in Southland Current) (Robertson, 1980). With the STF off Otago there is a major biogeographic boundary adjacent to the shelf (Robertson et al., 1979; Jackson et al., 2000). Proximity of SAW facilitates the incursion of subantarctic species into shelf waters, such as the copepod *Neocalanus tonsus,* which is seasonally abundant in the Southland Current and probably introduced each spring from deeper SAW (Jillett, 1968; Ohman, 1987). Its high lipid content may be of significance to the shelf food web (Ohman et al., 1989).

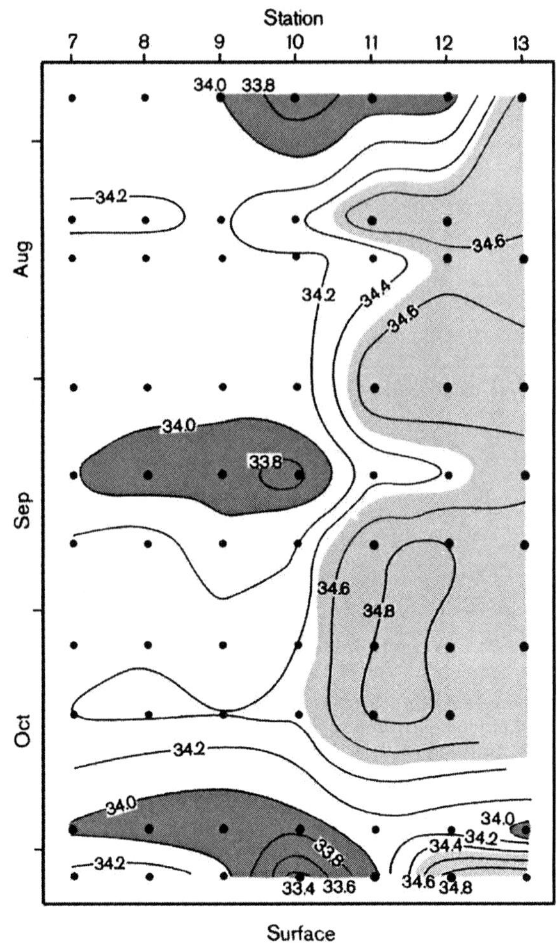

Figure 36.18 Temporal variability in surface salinities on the Otago shelf showing the pulses of riverine input to the neritic zone (from Murdoch et al., 1990).

Particularly distinctive and seasonally important to shelf trophic structure are surface swarms of *Nyctiphanes australis* and *Munida gregaria.* The galatheid *Munida gregaria* is a subantarctic species important to both pelagic and benthic systems (Williams, 1980; Zeldis, 1985). Adults are benthic, but the postlarvae are pelagic and in the warmer months (November-March) form large aggregations (> 100 m across) (Fig. 36.19). These aggregations are controlled by hydrodynamic processes, such as internal waves that may provide a mechanism for their onshore transport (Zeldis and Jillett, 1982; Jillett and Zeldis 1985). Aggregations of *Munida* postlarvae (< 1.5 kg m^{-3}) (Zeldis 1985) and of *Nyctiphanes* are important prey for planktivorous fish such as barracouta (*Thyrsites atun*), jack mackerel (*Trachurus murphyi*) and slender tuna (*Allothunus fallai*) (O'Driscoll, 1998a; O'Driscoll and McClatchie, 1998), as well as for seabirds (*Larus scopulinus, L. bulleri, L. dominicanuss, Puffinus griseus*) (McClatchie et al., 1989; O'Driscoll, 1998b; Cruz et al., 2001).

Figure 36.19 Aerial photo of *Munida gregaria* aggregation at a headland front south east of Taiaroa Head, taken from 1100 m altitude. (from Zeldis and Jillett, 1982). See color insert.

The benthos of the middle and outer shelf reflects the fact that the region is starved of modern sediment input and is characterised by coarse relict (from last glacial maximum) and biogenic sediments (Andrews, 1973; Carter et al., 1985). Suspension feeders are again well represented in the infauna (McKnight, 1969; Probert and Wilson, 1984). The coarse sediments support a well-developed sessile epifauna, most notably mid-shelf bryozoan thickets associated with relict gravels off the Otago Peninsula where the inshore flow accelerates as it is constricted between the Peninsula and the SAW offshore (Probert et al., 1979) (Fig. 36.15). Several bryozoan species (e.g., *Cinctipora elegans, Hippomenella vellicata, Celleporina* spp.) form thicket-like structures, mostly at 75–110 m water depth, that significantly enhance benthic habitat complexity and associated biodiversity (Batson and Probert, 2000). Bryozoan-dominated habitat may be important in the life history of commercially important fish and shellfish (Vooren, 1975; Bradstock and Gordon, 1983; Cranfield et al., 1999). The mid to outer shelf epibenthos also supports a diverse and abundant mobile fauna (e.g., echinoderms, gastropods, decapod crustacea) (Probert et al., 1979). Benthic recruitment of *Munida gregaria* occurs to these middle and outer shelf sediments (Zeldis, 1985). Also distinctive of the sediment-starved outer shelf is the scallop *Zygochlamys delicatula* (Powell, 1950; Orpin et al., 1998).

Thus, the narrow Otago continental shelf is a highly advective system dominated by along-shelf flow of subtropical origin. Its unusual close proximity to the STF, located over the upper slope, facilitates the introduction of distinctive subantarctic elements to the biota. Modern sedimentation is low and confined to an inner shelf sand wedge. Relict and biogenic sediments dominate the middle and outer shelf and provide for a well-developed sessile epifauna.

3 Prospects: Research directions and issues

The requirement for an evidence-based approach to the management of human activities on the continental shelf is driving the research agenda currently in New Zealand. The aim is to ensure that human activities are sustainable and do not limit the options of future generations. No matter how closely research is targeted to current management issues, this targeted research will not achieve its aims if it is not underpinned by basic knowledge of ecosystem functioning. Optimal results are always obtained if the work is carried out in an integrated, interdisciplinary manner. This requires that the physical oceanographers first identify the processes that create spatial and temporal variability in order that the biological studies can be properly planned so that the impacts of the physical phenomena can be identified. Unfortunately, it has not always been possible to carry out this first step and studies have proceeded simultaneously.

The current level of integrated knowledge of physical, biological, chemical, and sedimentological processes in New Zealand coastal seas is reflected in the way studies were planned and carried out. On the southeastern shelf of South Island, studies progressed separately to some extent and information has been added incrementally, whereas, in the greater Cook Strait region, and the western shelf of South Island there was an attempt to conduct the research in a coordinated, multidisciplinary manner, although with limited resources. This project came to an end just as significant physical phenomena (squirts and coastally trapped waves) were identified, therefore, their impact on the patterns of productivity in the region were not able to be identified at that time. The most recent, and ongoing, study off the northeastern shelf of North Island and Hauraki Gulf has the best interdisciplinary integration of information, with the gaining of knowledge of physical processes partly preceding the biological studies. This study is also tackling the issue of interannual biological variability and its drivers. This work is an essential component of the information necessary for estimating the suitability of coastal regions for aquaculture or for the harvest of wild marine organisms. The main drivers of coastal biological variability are likely to differ in various sections of New Zealand's coast because of the New Zealand's complex physical characteristics.

For its size, the New Zealand shelf encompasses a considerable range of characteristics relating to geomorphology, sedimentary regime, hydrography, and patterns of biological productivity. The whole New Zealand shelf is bathed in water of mainly subtropical origin and modest potential nutrient status, but there are marked regional differences in the importance of riverine input, advection, upwelling / downwelling, and inshore / offshore exchange, all of which have important implications for nutrient dynamics, patterns of primary productivity, pelagic-benthic coupling and interannual variability. The particular mix of characteristics and drivers results in a number of already identified issues.

From a benthic point of view it is striking that New Zealand has shelf areas that have very different modern sediment input. Some regions receive very high modern sediment input (on a global basis), while others are relatively starved of sediment. These differences apparently result in strong contrasts in benthic structure and function (e.g., between western and southeastern South Island shelves). One of the impediments to creating a marine classification (Snelder et al., 2001) is the absence of appropriate information on the contribution that river water quantity and quality makes to the biogeochemistry of the coastal zone. Therefore, some priority should be given to estimating average sediment input to coastal waters, derived from satellite sea surface colour data, and to devising methods of mapping the average freshwater fraction.

Ongoing work on the northeast shelf of North Island includes studying the fluxes of nutrient from benthos in sustaining production; the role that bottom stresses, associated with the passage of internal waves, play in the resuspension and transport of sediments across the shelf break; and the relative roles of benthic and pelagic nutrient forcing. There are good possibilities to understand climatological predictors of production at the El Niño / Southern Oscillation and longer time scales up to the Interdecadal Pacific Oscillation. The processes driving the benthic deposition of organic matter and the importance of shelf areas as sources of greenhouse gases are also subjects that could be advanced in New Zealand shelf regions (especially NE coast).

In the greater Cook Strait region knowledge of physical-biological-chemical-sedimentological interactions in the system have not yet been well integrated. The most obvious gaps in our knowledge are: the impact of water column processes on benthic biomass and benthic productive potential, the impacts on organisms higher in the food web, and the impact of interannual variability on ecosystem functioning.

Knowledge of the physical processes operating on the western shelf of South Island were not well known when the research in that area began. Therefore the chemical, biological and sedimentological studies were not planned in such a way that the impacts of river inflow, upwelling, CTWs, offshore eddies, and mixed layer depth maxima over the continental slope could be separately identified. When and if the prediction of the survival of hoki larvae on this coast becomes an issue again for New Zealand fisheries management, we have an excellent base from which to launch work to refine and integrate our understanding and predictive capability.

A great deal is known about certain aspects of the Otago shelf, eastern South Island. Nevertheless, the role of winter mixing, recycling from shelf sediments, and advection of dissolved inorganic nutrients needs attention, as well as the relative impact of all these processes on the level of primary production. A possibly unique feature of this coastal region is the presence of a major ocean water mass front situated over the continental slope. It is possible that there is little inshore / offshore exchange across this front.

The codification of the current state of knowledge into tools such as a classification system facilitates marine environment management in the presence of sparse data and little prospect of acquiring better data quickly. In this context, the further development of a typology and ecological geography of the New Zealand shelf is timely now we appreciate some of the main drivers.

Low cost proxies for estimating the health of ecosystems are also in demand so that managers can be alerted to problems before they become unmanageable. Unfortunately, much of the work that is necessary to verify that potential indices (e.g. diversity measures or other proxies for ecosystem state, natural or otherwise) actually measure what we think they are measuring in New Zealand coastal waters, has not yet been done. There is a need to review a range of specific indices that are possible candidates in relation to particular environmental attributes or human activities and to carry out the work to tailor these indices to the use for which they are intended. Such work would need to be placed in a setting where there is a good knowledge of seasonal, geographic and interannual variability.

Acknowledgements

This work was supported by the institutions of the authors and New Zealand Foundation for Research Science and Technology, NSOF C01X0223. We thank Dr Vanessa Sherlock for SST images, Dr Matthew Pinkerton for preparing figure 36.17, Dr Mark Hadfield for preparing figure 36.3, upper panel, Dr John Jillett for discussions on the Otago shelf, Jonathan Sharples for his constructive criticism and valuable input from two referees.

Bibliography

Andrews, P. B. 1973. Late Quaternary continental shelf sediments off Otago Peninsula, New Zealand. *N. Z. J. Geol. Geophys.*, **16,** 793–830.

Barnes P. M., R. Sutherland, B. Davy and J. Delteil, 2001. Rapid creation and destruction of sedimentary basins on mature strike-slip faults: an example from the offshore Alpine Fault, New Zealand. *J. Struct. Geol.*, **23,** 1727–1739.

Batson, P. B. and P. K. Probert, 2000. Bryozoan thickets off Otago Peninsula. *N. Z. Fish. Assessment Rep.*, **2000/46,** 1–31.

Bowman, M. J., A. C. Kibblewhite and D. E. Ash, 1980. M_2 tidal effects in Greater Cook Strait, New Zealand. *J. Geophys. Res.*, **85(C5),** 2928–2742.

Boyd, P. W., G. McTainsh, V. Sherlock, K. Richardson, S. Nichol, M. Ellwood and R.D. Frew, 2004. Episodic enhancement of phytoplankton stocks in New Zealand subantarctic waters: Contribution of atmospheric and oceanic iron supply. *Global Biogeochem. Cycles,* **18,** GB1029.

Bradford, J. M. 1972. Systematics and Ecology of New Zealand central east coast plankton sampled at Kaikoura. *N. Z. Oceanogr. Inst. Mem.*, **54,** 1–87.

Bradford, J. M., 1983. Physical and chemical oceanographic observations off Westland, New Zealand, June 1979. *N. Z. J. Mar. Freshwat. Res.*, **17,** 71–81.

Bradford, J. M. 1985. Distribution of zooplankton off Westland, New Zealand, June 1979 and February 1982. *N. Z. J. Mar. Freshwat. Res.*, **19,** 311–326.

Bradford, J. M. and F. H. Chang, 1987. Standing stocks and productivity of phytoplankton off Westland, New Zealand, February 1982. *N. Z. J. Mar. Freshwat. Res.*, **21,** 71–90.

Bradford, J. M. and B. Chapman, 1988a. *Nyctiphanes australis* (Euphausiacea) and an upelling plume in western Cook Strait. *NZ. J. Mar. Freshwat. Res.*, **22,** 237–247.

Bradford, J. M. and B. E. Chapman, 1988b. Epipelagic zooplankton assemblages and a warm-core eddy off East Cape, New Zealand. *J. Plankt. Res.* **10,** 601–619.

Bradford-Grieve, J. M., R. C. Murdoch and B. E. Chapman, 1993. Composition of macrozooplankton assemblages associated with the formation and decay of pulses within an upwelling plume in greater Cook Strait, New Zealand. *N. Z. J. Mar. Freshw. Res.*, **27,** 1–22.

Bradford, J. M., P. P. Lapennas, R. A. Murtagh, F. H. Chang and V. Wilkinson, 1986. Factors controlling summer phytoplankton production in greater Cook Strait, New Zealand. *N. Z. J. Mar. Freshwat. Res.*, **20,** 253–279.

Bradford-Grieve, J. M., R. C. Murdoch and B. E. Chapman, 1993. Composition of macrozooplankton assemblages with the formation and decay of pulses within an upwelling plume in greater Cook Strait, New Zealand. *N. Z. J. Mar. Freshwat. Res.*, **27,** 1–22.

Bradford-Grieve, J. M., R. C. Murdoch, M. R. James, M. Oliver and J. Hall, 1996. Vertical distribution of zooplankton > 39 μm in relation to the physical environment off the west coast of South Island, New Zealand. *N. Z. J. Mar. Freshwat. Res.*, **30,** 285–300.

Bradford-Grieve, J. M., R. C. Murdoch, M. R. James, M. Oliver, J. McLeod, 1998. Mesozooplankton biomass, composition and potential grazing pressure on phytoplankton during austral winter and spring 1993 in the Subtropical Convergence region near New Zealand. *Deep-Sea Res. I,* **45,** 1709–1737.

Bradstock, M. and D. P. Gordon, 1983. Coral-like bryozoan growths in Tasman Bay, and their protection to conserve commercial fish stocks. *N. Z. J. Mar. Freshwat. Res.*, **17,** 159–163.

Butler, E. C. V., J. A.Butt, E. J. Lindstrom, P. C. Tildesley, S. Pickmere and W. F. Vincent, 1992. Oceanography of the Subtropical Convergence Zone around southern New Zealand. *N. Z. J. Mar. Freshwat. Res.*, **26,** 131–154.

Cahill, M. L., J. H. Middleton and B. R. Stanton, 1991. Coastal-trapped waves on the west coastal of south Island, New Zealand. *J. Phys. Ocean.*, **21,** 541–557.

Carter, L., 1975. Sedimentation on the continental terrace around New Zealand: a review. *Mar. Geol.*, **19,** 209–237.

Carter, L., 1980. Ironsand in continental shelf sediments off western New Zealand – a synopsis. *N. Z. J. Geol Geophys.*, **23,** 455–468.

Carter, L., 1983. Sediments, shallow seismic stratigraphy, and structure in the vicinity of the submarine power cables, Cook Strait, New Zealand. *N. Z. O. I. Oceanogr. Fld Rep.*, **19,** 1–14.

Carter, L. 1986. A budget for modern-Holocene sediment on the South Otago continental shelf. *N. Z. J. Mar. Freshwat. Res.*, **20,** 665–676.

Carter, L. and R. M. Carter, 1986. Holocene evolution of the nearshore sand wedge, South Otago continental shelf, New Zealand. *N. Z. J. Geol. Geophys.* **29,** 413–424.

Carter, R. M. and L. Carter, 1993. Sedimentary evolution of the Bounty Trough : A Cretaceous rift basin, Southwestern Pacific Ocean. In: Balance, P.F. (ed.) Sedimentary Basins of the World. Elsevier, Amsterdam, Pp 52–67.

Carter, R. M., L. Carter, J. J. Williams and C. A. Landis, 1985. Modern and relict sedimentation on the South Otago continental shelf, New Zealand. *N. Z. Oceanogr. Inst. Mem.*, **93,** 1–43.

Carter, L and J. V. Eade, 1980. Hauraki sediments. *N. Z. Oceanogr. Inst. Chart, Misc. Ser.* 1-200,000.

Carter, L.; R. D. Garlick, P. Sutton, S. Chiswell, N. A. Oien, B. R. Stanton, 1998. Ocean Circulation New Zealand. *NIWA Chart Misc. Ser.*, **76.**

Carter, L. and R. A. Heath, 1975. Role of mean circulation, tides, and waves in the transport of bottom sediment on the New Zealand continental shelf. *N. Z. J. Mar. Freshwat. Res.*, **9,** 423–448.

Cassie, V. 1966. Diatoms, dinoflagellates and hydrology in the Hauraki Gulf 1964–65. *N. Z.J. Sci.* **9,** 569–585.

Chang, F. H. 1983. Winter phytoplankton and microzooplankton populations off the coast of Westland, New Zealand, 1979. *N. Z. J. Mar. Freshwat. Res.*, **17,** 279–304.

Chang, F. H., 1988. Distribution, abundance, and size composition of phytoplankton off Westland, New Zealand, February 1982. *N. Z. J. Mar. Freshwat. Res.*, **22,** 345–367.

Chang, F. H. and J. M. Bradford, 1985. Standing stocks and productivity of phytoplankton off Westland, New Zealand, June 1979. *N. Z. J. Mar. Freshwat. Res.*, **19,** 193–211.

Chang, F. H., J. M. Bradford-Grieve, W. F. Vincent and P. H. Woods, 1995a. Nitrogen uptake by the summer size-fractionated phytoplankton assemblages in the Westland, New Zealand upwelling system. *N. Z. J. Mar. Freshwat. Res.*, **29,** 147–161.

Chang, F. H., L. MacKenzie, D. Till, D. Hannah and L. Rhodes, 1995b. The first toxic shellfish outbreaks and the associated phytoplankton blooms in early 1993, in New Zealand. In: Lassus, P., G. Arul, E. Erad, P. Gentien,, P. C. Marcallou (eds). Harmful marine algal blooms. Paris Lavoisier Interceot Ltd., pp. 145–150.

Chang, F. H., W. F. Vincent and P. H. Woods, 1989. Nitrogen assimilation by three size fractions of the winter phytoplankton off Westland, New Zealand. *N. Z. J. Mar. Freshwat. Res.,* **23,** 491–505.

Chang, F. H., W. F. Vincent and P. H. Woods, 1992. Nitrogen utilisation by size-fractionated phytoplankton assemblages associated with an upwelling event off Westland, New Zealand. *N. Z. J. Mar. Freshwat. Res.,* **26,** 287–301.

Chang, F. H., J. Zeldis, M. Gall and J. Hall, 2003. Seasonal and spatial variation of phytoplankton functional groups on the northeastern New Zealand continental shelf and in Hauraki Gulf. *J. Plankt. Res.,* **25,** 737–758.

Chiswell, S. M. 1996. Variability in the Southland Current, New Zealand. *N. Z. J. Mar. Freshwat. Res.,* **30,** 1–17.

Chiswell, S. M. and J. D. Booth, 1999. Rock lobster *Jasus edwardsii* larval retention by the Wairarapa Eddy off New Zealand. *Mar. Ecol Prog. Ser.,* **183,** 227–240.

Christensen, P.B., Glud, R.N., Dalsgaard, R.N. and P. Gillespie, 2003. Impacts of longline mussel farming on oxygen and nitrogen dynamics and biological communities of coastal sediments. *Aquaculture,* **218,** 567–588.

Cole, J. W., 1979. Structure, petrology and genesis of Cenozoic volcanism, Taupo Volcanic Zone, New Zealand - a review. *N. Z. J. Geol. Geophys.,* **22,** 631–657.

Cole, J. W., 1990. Structural control and origin of volcanism in the Taupo volcanic zone, New Zealand. Bull. *Volcanol,* **52,** 445–459.

Colebrook, J. M., R. S. Glover and G. A. Robinson, 1961. Continuous plankton records: contributions towards a plankton atlas of the north-eastern Atlantic and North Sea. *Bull. Mar. Ecol.,* **5,** 67–80.

Collet, J.-Y., K. Lewis, G. Lamarche and S. Lallemand, 2001. The giant Ruatoria debris avalanche on the northern Hikurangi margin, New Zealand: Result of oblique seamount subduction. *J. Geophys. Res.,* **106(B9),** 19271–19297.

Conkright, M.E., Locarnini, T.A., Garcia, H.E., O'Brien, T.D., Boyer, T.P., Stephens, C. and J.I. Antonov, 2002. World Ocean Atlas 2001: objectives analyses, data statistics, and figures CD-ROM documentation. National Oceanographic Data Center Internal Reports 17, 17 pp. http://www.nodc.noaa.gov/OC5/WOA01F/nusearch.html

Cranfield, H. J., K. P. Michael and I. J. Doonan, 1999. Changes in the distribution of epifaunal reefs and oysters during 130 years of dredging for oysters in Foveaux Strait, southern New Zealand. *Aquat. Conserv.,* **9,** 461–483.

Croot, P. L. and K. A. Hunter, 1998. Trace metal distributions across the continental shelf near Otago Peninsula, New Zealand. *Mar. Chem.,* **62,** 185–201.

Cruz, J. B., C. Lalas, J. B. Jillett, J. C. Kitson, P. O. Lyver, M. Imber, J. E. Newman and H. Moller, 2001. Prey spectrum of breeding sooty shearwaters (*Puffinus griseus*) in New Zealand. *N. Z. J. Mar. Freshwat. Res.,* **35,** 817–829.

Currie, K. I. and K. A. Hunter, 1998. Surface water carbon dioxide in the waters associated with the subtropical convergence, east of New Zealand. *Deep-Sea Res. I,* **45,** 1765–1777.

Currie, K. I. and K. A. Hunter, 1999. Seasonal variation of surface water CO_2 partial pressure in the Southland Current, east of New Zealand. *Mar. Freshwat. Res.,* **50,** 375–382.

Davies-Colley, R. J., 1992. Yellow substance in coastal and marine water round the South Island, New Zealand. *N. Z. J. Mar. Freshwat. Res.,* **26,** 311–322.

Doney, S.C., D.M. Glover, S.J. McCue and M. Fuentes, 2003. Mesoscale variability of Sea-viewing Wide Field-of-view Sensor (SeaWiFS) satellite ocean color: Global patterns and spatial scales. J. Geophys. Res. (C2. Oceans). **108,** Article 6, 15 pp.

Duncan, M. J. 1987. River hydrology and sediment transport. In: Viner, A. B. ed. Inland waters of New Zealand. *Dept Sci. Ind. Res. Bull., Wellington,* **241,** 113–137.

Duncan, M. J. 1992. Flow regimes of New Zealand Rivers. In: Mosley, M. P. (ed.) Waters of New Zealand. New Zealand Hydrological Society, Caxton Press, Christchurch, pp. 13–28.

Eade, J. V. 1974. Poor Knights Sediments. *N. Z. Oceanogr. Inst. Chart, Coast. Ser.* 1:200,000.

Foster, G. and L. Carter, 1997. Mud sedimentation on the continental shelf at an accretionary margin - Poverty Bay, New Zealand. *N. Z. J. Geol. Geophys.*, **40,** 157 - 173.

Giles, H. 2001. Denitrification in continental shelf sediments. Unpublished MSc Thesis held in the University of Waikato Library, Hamilton, New Zealand.

Gordon, N. D. 1986. The Southern Oscillation and New Zealand weather. *Monthly Weather Rev.*, **114,** 371–387.

Goring, D. G., R. G. Bell, 1999. El Niño effects on sea-level variability in northern New Zealand: a wavelet analysis. *N. Z. J. Mar. Freshwat. Res.*, **33,** 587–598.

Greig, M. J., N. M. Ridgway, and B. S. Shakespeare, 1988. Sea surface temperature variations at coastal sites around New Zealand. *N. Z. J. Mar. Freshwat. Res.*, **22,** 391–400.

Griffiths, G. A. 1979. High sediment yields from major rivers of the western Southern Alps. *Nature, Lond.*, **282,** 61–63.

Hadfield, M. G. and J. Sharples, 1996. Modelling mixed layer depth and plankton biomass off the west coast of South Island, New Zealand. *J. Mar. Syst.*, **8,** 1–29.

Harris, T. F. W. 1990. Greater Cook Strait form and flow. DSIR Marine and Freshwater, Wellington. 212 pp.

Hawke, D. J. 1989. Hydrology and near-surface nutrient distribution along the South Otago continental shelf, New Zealand, in summer and winter 1986. *N. Z. J. Mar. Freshwat. Res.*, **23,** 411–420.

Hawke, D. J. 1995. Reactive silicate in shelf waters near Otago Peninsula, New Zealand. *Mar. Freshwat. Res.*, **46,** 427–433.

Hawke, D. J. and K. A. Hunter, 1992. Reactive P distribution near Otago Peninsula, New Zealand: an advection-dominated shelf system containing a headland eddy. *Estuar. Coast. Shelf Sc.*, **34,** 141–155.

Heath, R. A. 1972. The Southland Current. *N. Z. J. Mar. Freshwat. Res.*, **6,** 497–533.

Heath, R. A. 1974. The lunar semi-diurnal tide in Cook Strait. *Dt. Hydrogr. Z.*, **27,** 214–224.

Heath, R. A. 1975. Oceanic circulation and hydrology off the southern half of South Island, New Zealand. *N. Z. Oceanogr. Inst. Mem.*, **72,** 1–36.

Heath, R. A. 1982. What drives the mean circulation on the New Zealand west coast continental shelf? *N. Z. J. Mar. Freshwat. Res.*, **16,** 215–226.

Heath, R. A. 1985. A review of the physical oceanography of the seas around New Zealand - 1982. *N. Z. J. Mar. Freshwat. Res.*, **19,** 79–124.

Heath, R. A. and N. M. Ridgeway, 1985. Variability of the oceanic temperature and salinity fileds on the west coast continental shelf South Island, New Zealand. *N. Z. J. Mar. Freshwat. Res.*, **19,** 233–245.

Hicks, D.M. and U. Shankar, 2003. Sediment from New Zealand rivers. *NIWA Chart Misc. Ser.*, **79.**

Howard-Williams, C., R. Davies-Colley, W. Vincent, S. Pickmere, A. Schwarz and S. Falconer, 1991. Optical properties of the coastal and offshore waters of the South Island in May 1989: Report on the data set for Cruise 2027. *Taupo Res. Lab. Rep.*, **118.**

Howard-Williams, C., R. Davies-Colley and W. F. Vincent, 1995. Optical properties of the coastal and oceanic waters off the South Island, New Zealand: Regional variation. *N. Z. J. Mar. Freshwat. Res.*, **29,** 589–602.

Jackson, G. D., A.G. P. Shaw and C. Lalas, 2000. Distribution and biomass of two squid species off southern New Zealand: *Nototodarus sloanii* and *Moroteuthis ingens. Polar Biol.*, **23,** 699–705.

James, M. R., 1989. The role of zooplankton in the nitrogen cycle off the west coast of the South Island, New Zealand, winter 1987. *N. Z. J. Mar. Freshwat. Res.*, **23,** 507–518.

James, M. R. and J. Hall, 1998. Microzooplankton grazing in different water masses associated with the Subtropical Convergence round the South Island, New Zealand. *Deep-Sea Res. I*, **45,** 1689–1707.

James, M. and V. Wilkinson, 1988. Biomass, carbon ingestion, and ammonia excretion by zooplankton associated with an upwelling plume in western Cook strait, New Zealand. *N. Z. J. Mar. Freshwat. Res.*, **22,** 249–257.

Jillett, J. B. 1968. *Calanus tonsus* (Copepoda, Calanoida) in southern New Zealand waters with notes on the male. *Aust. J. Mar. Freshwat. Res.*, **19,** 19–30.

Jillett, J. B. 1969. Seasonal hydrology of waters off the Otago Peninsula, south-eastern New Zealand. New Zealand. *N. Z. J. Mar. Freshwat. Res.*, **3,** 349–375.

Jillett, J. B. 1971. Zooplankton and hydrology of Hauraki Gulf, New Zealand. *N. Z. Oceanogr. Inst. Mem.*, **53,** 1–103.

Jillett, J. B. 1976. Zooplankton associations off Otago Peninsula, south-eastern New Zealand, related to different water masses. *N. Z. J. Mar. Freshwat. Res.*, **10,** 543–557.

Jillett, J. B. and Zeldis, J. R. 1985. Aerial observations of surface patchiness of a planktonic crustacean. *Bull. Mar. Sci.*, **37,** 609–619.

Kaspar, H. F., R. A. Asher and I. C. Boyer, 1985. Microbial nitrogen transformations in sediments and inorganic nitrogen fluxes across the sediment/water interface on the South Island west coast, New Zealand. *Estuar. Coast. Shelf Sci.* **21,** 245–255.

Kieber, R. J., K. Williams, J. D. Willey, S. Skrabal and G. B. Avery, 2001. Iron speciation in coastal rainwater: concentration and deposition to seawater. *Mar. Chem.*, **73,** 83–95.

Kibblewhite, A. C., P. R. Bergquist, B. A. Foster, M. R. Gregory and M. C. Miller, 1982. Maui development environmental study. Shell and BP and Todd Oil Services Ltd., Auckland. 174 pp.

Kirchlechner, T. M. 1999. Biogeochemical aspects of the New Zealand sector of the Southern Ocean. PhD thesis, University of Otago, Dunedin, New Zealand.

Lewis, K. B., 1971. Growth rate of folds using tilted wave-planed surfaces: coast and continental shelf, Hawke's Bay, New Zealand. *R. Soc. N. Z., Bulletin,* **9,** 225–231.

Lewis, K. B., 1980. Quaternary sedimentation on the Hikurangi oblique-subduction and transform margin, New Zealand. In: P.F. Ballance and H.G. Reading (Editors), Sedimentation in oblique-slip mobile zones. *Spec. Publ. Int. Assoc. Sedimentol.,* pp. 171–189.

Lewis K. B. and J. V. Eade, 1974. Sedimentation in the vicinity of the Maui Gasfield. *N. Z. O. I. Oceanogr. Summ.*, **6,** 1–6.

Lewis, K. B. and B. A. Marshall, 1996. Seep faunas and other indicators of methane-rich dewatering on New Zealand convergent margins. *N. Z. J. Geol. Geophys.*, **39,** 181 - 200.

Lewis, K.B. and H. M. Pantin, 1984. Intersection of a marginal basin with a continent: Structure and sediments of the Bay of Plenty, New Zealand. In: B.P. Kokelaar, M.F. Howells (Eds), Marginal Basin Geology. *Spec. Publ. Geol. Soc. (Lond)*, pp. 121 - 135.

Lewis, K. B. and J. R. Pettinga, 1993. The emerging, imbricate frontal wedge of the Hikurangi Margin. In: P.F. Ballance (Ed.), South Pacific Sedimentary Basins. Sedimentary Basins of the World, 3, Elsevier, pp. 225 - 250.

Longhurst, A., 1998. Ecological geography of the sea. Academic Press, London, 398 pp.

Mackenzie, A. L. and P. A. Gillespie, 1986. Plankton ecology and productivity, nutrient chemistry and hydrography of Tasman Bay, New Zealand, 1982–1984, *N. Z. J. Mar. Rreshwat. Res.*, **20,** 365–395.

McClatchie, S., D. Hutchinson and K. Nordin, 1989. Aggregation of avian predators and zooplankton prey in Otago shelf waters, New Zealand. *J. Plankt. Res.*, **11,** 361–374.

McDougall, J. C., 1975. Cook sediments. *N. Z. Oceanogr. Inst. Chart, Oceanic Ser.* 1:1,000,000

McKnight, D. G., 1969. Infaunal benthic communities of the New Zealand continental shelf. *N. Z. J. Mar. Freshwat. Res.*, **3,** 409–444.

Milliman, J.D. and R.H. Meade, 1983. World-wide delivery of river sediment to the oceans. *J. Geol.* **91,** 1–21.

Moore, M. I. and R. C. Murdoch., 1993. Physical and biological observations of coastal squirts under nonupwelling conditions. *J. Geophys. Res.*, **98 (C11),** 20,043–20,061.

Murdoch, R. C. 1989. The effects of a headland eddy on surface macro-zooplankton assemblages north of Otago Peninsula, New Zealand. *Estuar. Coast. Shelf Sci.,* **29,** 361–383.

Murdoch, R. C., 1990. Diet of hoki larvae (*Macruronus novaezelandiae*) off Westland, New Zealand. *N. Z. J. Mar. Freshwat. Res.,* **24,** 519–527.

Murdoch, R. C., 1992. A review of the ecology of hoki *Macruronus novaezelandiae* (Hector), larvae in New Zealand waters. *Proc. Bur. Rural Resour.,* **15,** 3–16.

Murdoch, R. C., R. Proctor, J. B. Jillett and J. R. Zeldis, 1990. Evidence for an eddy over the continental shelf in the downstream lee of Otago Peninsula, New Zealand. *Estuar. Coast. Shelf Sci.,* **30,** 489–507.

Murdoch, R. C. and B. Quigley 1992. A patch study of mortality, growth and feeding of the larvae of the southern gadoid *Macruronus novaezealandiae. Mar. Biol.,* **121,** 23–33.

O'Driscoll, R. L. 1998a. Feeding and schooling behaviour of barracouta (*Thyrsites atun*) off Otago, New Zealand. *Mar. Freshwat. Res.,* **49,** 19–24.

O'Driscoll, R. L. 1998b. Description of spatial pattern in seabird distributions along line transects using neighbour *K* statistics. *Mar. Ecol. Progr. Ser.,* **165,** 81–94.

O'Driscoll, R. L. and S. McClatchie, 1998. Spatial distribution of planktivorous fish schools in relation to krill abundance and local hydrography off Otago, New Zealand. *Deep-Sea Res. II,* **45,** 1295–1325.

Ohman, M. D. 1987. Energy sources for recruitment of the subantarctic copepod *Neocalanus tonsus. Limnol. Oceanogr.,* **32,** 1317–1330.

Ohman, M. D., J. M. Bradford, and J. B. Jillett, 1989. Seasonal growth and lipid storage of the circumglobal, subantarctic copepod, *Neocalanus tonsus* (Brady). *Deep-Sea Res.,* **36,** 1309–1326.

Orpin, A. R., 1997. Dolomite chimneys as possible evidence of coastal fluid expulsion, uppermost Otago continental slope, southern New Zealand. *Mar. Geol.* **138,** 51 - 67.

Orpin, A.R., Carter, L., Kuehl, S.A., Trustrum, N.A., Lewis, K.B., Alexander, C.R. and B. Gomes, 2002. Deposition from very high sediment yield New Zealand Rivers is captured in upper margin basins. *Margins Newsletter,* **9,** 1–4.

Orpin, A. R., P. R. Gammon, T. R. Naish and R. M. Carter, 1998. Modern and ancient *Zygochlamys delicatula* shellbeds in New Zealand, and their sequence stratigraphic implications. *Sediment. Geol.,* **122,** 267–284.

Parson, T.R., M. Takahashi and B. Hargrave, 1977. Biological Oceanographic Processes. Pergamon Press, Oxford, 332 p.

Patchell, G. J. 1982. The New Zealand hoki fisheries 1972–82. *Occas. Publ. Fish. Res. Div. Minist. Agric. Fish. (N. Z.),* **38,** 1–23.

Patchell, G. J., M. S. Allen and D. J. Dreadon, 1987. Egg and larval development of the New Zealand hoki *Macruronus novaezelandiae. N. Z. J Mar. Freshwat. Res.,* **21,** 301–313.

Pérès, J. M. 1982. Major benthic assemblages. In Marine Ecology vol. V, Ocean Management, part 1 (O. Kinne ed.). John Wiley, Chichester, UK, 373–522.

Pingree, R. D., P. M. Holligan and G. T. Mardell, 1978. The effects of vertical stability on phytoplankton distributions in the summer on the northwest European shelf. *Deep-Sea Res.,* **25,** 1011–1028.

Powell, A. W. B. 1950. Mollusca from the continental shelf, eastern Otago. *Rec. Auckland Instit. Mus.,* **4,** 73–81.

Priscu, J. C. and M. T. Downes, 1985. Nitrogen uptake, ammonium oxidation and nitrous oxide (N_2O) levels in the coastal waters of western Cook Strait, New Zealand. *Estuar. Coast. Shelf Sci.,* **20,** 529–542.

Priscu, J. C. and L. R. Priscu, 1984. Photosynthate partitioning by phytoplankton in a New Zealand coastal upwelling system. *Mar. Biol.,* **81,** 31–40.

Probert, P.K., 1986. Energy transfer through the shelf benthos off the west coast of South Island, New Zealand. *N. Z. J. Mar. Freshwat. Res.*, **20**, 407–417.

Probert, P. K. and P. W. Anderson 1986. Quantitative distribution of benthic macrofauna off New Zealand, with particular reference to the west coast of the South Island. *N. Z. J. Mar. Freshwat. Res.*, **20**, 281–290.

Probert, P. K., E. J. Batham, and J. B. Wilson, 1979. Epibenthic macrofauna off southeastern New Zealand and mid-shelf bryozoan dominance. *N. Z. J. Mar. Freshwat. Res.*, **13**, 379–392.

Probert, P. K. and S. L. Grove, 1998. Macrobenthic assemblages of the continental shelf and upper slope off the west coast of South Island, New Zealand. *J. R.. Soc. N. Z.* **28**, 259–280.

Probert, P. K., G. B. Read, S. L. Grove and A. A. Rowden, 2001. Macrobenthic polychaete assemblages of the continental shelf and upper slope off the west coast of the South Island, New Zealand. *N. Z. J. Mar. Freshwat. Res.*, **35**, 971–984.

Probert, P. K. and K. M. Swanson, 1985. Sediment texture of the continental shelf and upper slope off the west coast of the South Island, New Zealand. *N. Z. J. Mar. Freshwat. Res.*, **19**, 563–573.

Probert, P.K. and Wilson, J.B. 1984. Continental shelf benthos off Otago Peninsula, New Zealand. Estuar. *Coast, Shelf Sci.*, **19**, 373–391.

Renwick, J.A., Hurst, R.J. and J.K. Kidson. 1998. Climatic influences on the survival of southern gemfish (*Rexea solandri*, Gempylidae) in New Zealand waters. *Int. J. Climat.*, **B18**, 1655–1667.

Ridgway, K. R., Dunn, J. R. and J.L. Wilkin. 2002. Ocean Interpolation by Four-Dimensional — Weighted Least Squares: Application to the Waters around Australasia. *J. Atmos. Ocean. Technol.*, **19**, 1357–1375.

Robertson, D. A. 1980. Hydrology and the quantitative distribution of planktonic eggs of some marine fishes of the Otago coast, south-eastern New Zealand. *N. Z. Fish. Res. Bull.*, **21**, 1–69.

Robertson, D.A., P. E. Roberts, and J. B. Wilson, 1979. Zoogeographical significance of fronts and convergences around New Zealand. Proceedings of the International Symposium on Marine Biogeography and Evolution in the Southern Hemisphere. *N. Z. Dep. Sci. Ind. Res, Inform. Ser.*, **137**, 525–538.

Roemmich, D. and P. Sutton, 1998. The mean and variability of ocean circulation past northern New Zealand: determining the representativeness of hydrographic climatologies. *J. Geophys. Res.*, **103(C6)**, 13041–13054.

Sedwick, P. N., P. R. Edwards, D. J. Mackey, F. B. Griffiths and J. S. Parslow. 1997. Iron and manganese in surface water of the Australian subantarctic region. *Deep-Sea Res.*, **44**, 1239–1253.

Sharples, J., 1997. Cross-shelf intrusion of subtropical water into the coastal zone or northeast New Zealand. *Con. Shelf Res.*, **17**, 835–857.

Sharples, J., 1998. Chapter 34. Physical processes on the New Zealand shelf and the rest of the world's islands. In: Robinson, A. R., K. H. Brink (eds). The global coastal ocean: regional studies and syntheses. *The Sea*, **11**, 965–996. John Wiley and Sons, Inc, New York.

Sharples, J. and M. J. N. Greig, 1998. Tidal currents, mean flows, and upwelling on the northeast coast of New Zealand. *N. Z. J. Mar. Freshwat. Res.*, **32**, 215–231.

Sharples, J., M. C. Moore and E. R. Abraham, 2001. Internal tide dissipation, mixing, and vertical flux at the shelf edge of NE New Zealand. *J. Geophys. Res.*, **106**, 14069–14081.

Shaw, A.G. P., L. Kavalieris, and R. Vennell, 1999. Seasonal and inter-annual variability of SST off the east coast of the South Island, New Zealand. *Geocarto Int.*, **14**, 27–32.

Shaw, A.G. P. and R. Vennell, 2001. Measurements of an oceanic front using a front-following algorithm for AVHRR SST imagery. *Remote Sens. Environ.*, **75**, 47–62.

Shirtcliffe, T. G. L., M. I. Moore, A. G. Cole, A. B. Viner, R. Baldwin and B. Chapman, 1990. Dynamics of the Cape Farewell upwelling plume, New Zealand. *N. Z. J. Mar. Freshwat. Res.*, **24**, 555–568.

Smith, R. and J. Hall, 1997. Bacterial abundance and production in different water masses around South Island, New Zealand. *N. Z. J. Mar. Freshwat. Res.*, **31**, 515–524.

Snelder, T., J. Grieve, T. Hume and J. Zeldis, 2001. Draft design for a classification system for New Zealand's marine environment. Report prepared for the New Zealand Ministry for the Environment, Wellington, 69 pp.

Stanton, B. R., 1976. Circulation and hydrology off the west coast of the South Island, New Zealand. *N. Z. J. Mar. Freshwat. Res.,* **10,** 445–467.

Stanton, B. R., 1995. Intraseasonal sea level variability on the west coast of New Zealand. *N. Z. J Mar. Freshwat. Res.,* **29,** 213–222.

Stanton, B. R. and M. J. N. Greig, 1991. Hindcasting currents on the west coast shelf of South Island, New Zealand. Pp 523–527. *In:* Healy T.R. (ed.) *Coastal Engineering – Climate for Change.* Proceedings of 10th Australasian Conference on Coastal and Ocean Engineering, Auckland, New Zealand, 2–6. December 1991. Water Quality Centre Publication **21,** 572 p.

Stanton, B. R. and M. I. Moore, 1992. Hydrographic observations during the Tasman Boundary Experiment off the west coast of South Island, New Zealand. *N. Z. J. Mar. Freshwat. Res.,* **26,** 339–358.

Stanton, B. R. and P. J. H. Sutton, 2003. Velocity measurement in the East Auckland Current north-east of North Cape, New Zealand. *N. Z. J. Mar. Freshwat. Res.,* **37,** 195–204.

Stanton, B. R., P. J. H. Sutton and S. M. Chiswell, 1997. The East Auckland Current, 1994–95. *N. Z. J. Mar. Freshwat. Res.,* **31,** 537–549.

Sturman, A. and N. Tapper, 1996. The weather and climate of Australia and New Zealand. Oxford University Press, Melbourne, 476 pp.

Sutton, P. 2001. Detailed structure of the Subtropical Front over Chatham Rise, east of New Zealand. *J. Geophys. Res.,* **106 (C12),** 31,045–31,056.

Sutton, P. J. H. 2003. The Southland Current: A subantarctic current. *N. Z. J. Mar. Freshwat. Res.,* **37,** 645–652.

Sutton, P.J.H. and D. Roemmich. 2001. Ocean temperature climate off north-east New Zealand. *N. Z. J. Mar. Freshwat. Res.,* **35,** 553–565.

Tilburg, C. E., H. E. Hurlburt, J. J. O'Brien and J. F. Shriver, 2001. The dynamics of the East Auckland Current System: The Tasman Front, the East Auckland Current, and the East Cape Current. *J. Phys, Oceanogr.,* **31,** 2917–2943.

Tilburg, C. E., H. E. Hurlburt, J. J. O'Brien and J. F. Shriver. 2002. Remote topographic forcing of a baroclinic Western Boundary Current: An explanation for the Southland Current and the pathway of the Subtropical Front east of New Zealand. *J. Phys. Oceanogr.,* **32,** 3216–3232.

Tomlinson, A. I. 1992. Precipitation and the Atmosphere. In: Mosley, M. P. (ed.) Waters of New Zealand. New Zealand Hydrological Society, Caxton Press, Christchurch, pp. 63–74.

Tomczak, M. and J. S. Godfrey, 1994. Regional oceanography: an introduction. Pergammon, London, 422p.

Uddstrom, M. J. and N. A. Oien, 1999. On the use of high-resolution satellite data to describe the spatial and temporal variability of sea surface temperatures in the New Zealand region. *J. Geophys. Res.,* **104(C9),** 20729–20751.

van der Linden, W. J. M., 1969. Off-shore sediments, north-west Nelson, South Island, New Zealand. *N. Z. J. Geol. Geophys.,* **12,** 172–207.

Vincent, W.F., C. Howard-Williams, P. Tildesley and E. Butler, 1991. Distribution and biological properties of oceanic water masses around the South Island, New Zealand. *N. Z. J. Mar. Freshwat. Res.,* **25,** 21–42.

Viner, A. B. 1990. Dark ^{14}C uptake, and its relationship to nitrification and primary production estimates in a New Zealand upwelling region. *N. Z. J. Mar. Freshwat. Res.,* **24,** 221–228.

Viner A. B. and V. H. Wilkinson, 1988. Uptake of nitrate and ammonium, and distribution of related variables, in the upwelled plume of western Cook Strait/Taranaki Bight, New Zealand. *N. Z. J. Mar. Freshwat. Res.,* **22,** 565–576.

Vooren, C. M. 1975. Nursery grounds of tarakihi (Teleostei: Cheilodactylidae) around New Zealand. *N. Z. J. Mar. Freshwat. Res.*, **9,** 121–158.

Walcott, R. I., 1987. Geodetic strain and the deformation history of the North Island of New Zealand during the late Cenozoic. *Phil. Trans. Roy. Soc. Lond., Ser. A,* **321,** 163–181.

Walcott, R. I., 1998. Modes of oblique compression: late Cenozoic tectonics of the South Island of New Zealand. *Rev. Geophys.,* **36,** 1 - 26.

Walters, R. A., D. G. Goring, R. G. Bell, 2001. Ocean tides around New Zealand. *N. Z. J. Mar. Freshwat. Res.,* **35,** 567–579.

Williams, B. G. 1980. The pelagic and benthic phases of post-metamorphic *Munida gregaria* (Fabricius) (Decapoda, Anomura). *J. Exp. Mar. Biol. Ecol.,* **42,** 125–141.

Zeldis J.R. 1985. Ecology of *Munida gregaria* (Decapoda, Anomura): distribution and abundance, population dynamics and fisheries. *Mar. Ecol. Progr. Ser.,* **22,** 77–99.

Zeldis, J. R. 2004. New and remineralised nutrient supply and ecosystem metabolism on the northeastern New Zealand continental shelf. *Cont. Shelf Res.,* **24,** 563–581.

Zeldis, J. R., C. S. Davis, M. R. James, S. L. Ballara, W. E. Booth and F. H. Chang, 1995. Salp grazing: effects on phytoplankton abundance, vertical distribution and taxonomic composition in a coastal habitat. *Mar. Ecol. Prog. Ser.,* **126,** 267–283.

Zeldis, J., M. Gall, M. Uddstrom and M. Greig, 2000. La Niña shuts down upwelling in northeastern New Zealand. *Water and Atmosphere,* **8,** 15–18.

Zeldis, J. R. and J. B. Jillett, 1982. Aggregation of pelagic *Munida gregaria* (Fabricius) (Decapoda, Anomura) by coastal fronts and internal waves. *J. Plankt. Res.,* **4,** 839–857.

Zeldis, J. R., R. C. Murdoch, P. L. Cordue and M. J. Page, 1998. Distribution of hoki (Macruronus novaezelandiae) eggs, larvae, and adults off Westland, New Zealand, and the design of an egg production survey to estimate hoki biomass. *Can. J. Fish. Aquat. Sci.,* **55,** 1682–1694.

Zeldis, J. R. and S. V. Smith, 1999. Water, salt and nutrient budgets for Hauraki Gulf, New Zealand. In: S. V. Smith (ed.). Australasian Estuarine Systems: Carbon, Nitrogen and Phosphorus Fluxes. *LOICZ Reports and Studies,* **12,** ii + 182 pp. LOICZ IPO, Texel, The Netherlands.

Zeldis, J. R., R. A. Walters, M. J. N. Greig and K. Image, 2004. Circulation over the northeastern New Zealand continental slope, shelf and adjacent Hauraki Gulf, from spring to summer. *Cont. Shelf Res.* **24,** 543–561.

Zeldis, J., M. Hicks, N. Trustrum, A. Orpin, U. Shankar, K. Currie, S. Nodder and K. Probert, submitted. New Zealand continental margin. In: L. Atkinson, K.K. Liu, R. Quiñones, and L. Talaue-McManus (eds), *Carbon and Nutrient Fluxes in Global Continental Margins.* IGBP.

Zentara, S. J. and D. Kamykowski, 1981. Geographic variation in the relationship between silicic acid and nitrate in the South Pacific Ocean. *Deep-Sea Res.,* **28A,** 455–465.

Chapter 37. OCEANOGRAPHIC INFLUENCES ON ANTARCTIC ECOSYSTEMS: OBSERVATIONS AND INSIGHTS FROM EAST ANTARCTICA (0° TO 150°E) (32,P)

STEPHEN NICOL, ANTHONY P. WORBY

Australian Antarctic Division and Antarctic Climate and Ecosystems Cooperative Research Centre

PETER G. STRUTTON

College of Oceanic and Atmospheric Sciences, Oregon State University

THOMAS W. TRULL

Antarctic Climate and Ecosystems Cooperative Research Centre

Contents

1. Introduction

The Antarctic coastal zone is characterized by the following features, which distinguish it from other coastal regions around the globe.

- The effective location of the coast changes dramatically on an annual basis, because sea-ice extends from hundreds to thousands of kilometers from the land in winter, before melting back close to the shore in summer. This annual expansion of sea-ice covers ~19 million km^2 in winter but only ~4 million km^2 remains in summer (Gloersen et al., 1992). The presence of sea ice is such a dominant aspect of the waters surrounding Antarctica that any assessment of

The Sea, Volume 14, edited by Allan R. Robinson and Kenneth H. Brink
ISBN 0–674–01527–4

processes in the coastal zone must include the entire region annually covered in sea ice.

- The coastal shelf surrounding Antarctica is narrow and deep, effectively separating pelagic and benthic processes, and permitting free exchange with deep ocean water masses.
- There is a distinct lack of chemical or biological inputs from the continent to the coastal zone – rivers are restricted to ephemeral summer melt-streams, mineral aerosol deposits are minimized by ice cover, and the ice-shelf and ice-bergs provide only very pure freshwater.
- The prevailing surface circulation patterns are circum-continental, and at relatively constant latitude. Therefore, while processes that occur 'upstream' certainly influence local physics, chemistry and biology, the circumpolar currents do not pass through significantly different biogeographic zones, as is the case for, say, the Leeuwin Current, California Current and the Gulf Stream.

As the sea ice zone effectively defines the Antarctic coastal zone, in this review we will concentrate on what has been defined as the marginal ice zone (MIZ; Arrigo et al., 1998a), both on and off the Antarctic shelf. In particular, we will examine the links between sea-ice and the structure of Antarctic ecosystems, including both spatial aspects such as the distribution of phytoplankton and zooplankton in relation to sea-ice, and temporal aspects such as the seasonal cycles of biological activity in relation to ice-melting and water column stratification in spring. We focus on East Antarctica for several reasons:

- Relatively little attention has been paid to this region in comparison to the intensively studied marine ecosystems of the Antarctic Peninsula and the Weddell and Ross Seas;
- It offers interesting contrasts and comparisons with these other regions. For example, the East Antarctic coast effectively runs east-west with little geographic variability in the form of bays or peninsulas. Therefore it is dominated by circumpolar circulation rather than the large gyres which influence the Weddell and Ross Seas. It is also south of the milder climate which influences the Antarctic Peninsula;
- Recent studies of the Indian Ocean sector provide a broad overview of biological, physical and chemical properties, which can be compared with other sectors; and
- Volume 11 of "The Sea" described the oceanography of the Atlantic and Pacific sectors of the Southern Ocean (Hofmann and Klink, 1998). In addressing the remaining sector here (from 0°E eastwards to 150°E), we complete the survey of the Antarctic coastal zone.

We first briefly summarize the physical oceanography of the region from 0° to 150°E, then we discuss the associated chemical and biological properties. Finally we summarize the similarities and differences between this sector and other regions of the Antarctic coast. In particular we examine whether it is appropriate to view the circumpolar sea-ice zone as a region of relatively homogenous ecosystem properties.

A number of recent reviews have addressed aspects of the Antarctic marine environment. In addition to Hofmann and Klink's (1998) review of the shelf zone from 150°E to the Greenwich Meridian, the Antarctic Circumpolar Current System has been extensively reviewed (Jacobs and Weiss, 1998; Rintoul et al., 2001), and the location of Southern Ocean circumpolar fronts has been described (Orsi et al., 1995). The ecology of the Antarctic pack ice has been discussed (Lizotte and Arrigo, 1998; Brierley and Thomas, 2002) as have the spatial and temporal variability of the sea ice and its physical properties (Jeffries, 1998; Zwally et al., 2002). The biogeography of the Southern Ocean has been addressed in general terms (Longhurst, 1998) and with particular attention to biogeochemical cycles based on shipboard observations (e.g. Treguer and Jacques, 1992; Le Fevre et al., 1998; Trull et al., 2001; Nelson et al., 2002), satellite remote sensing (e.g. Comiso et al., 1993; Arrigo et al., 1998a; Moore and Abbott, 2000; Moore et al., 2000) and paleoceanographic studies focusing on the influence of biogeochemistry on glacial-interglacial cycles (e.g. Kumar et al., 1995; Francois et al., 1997; Anderson et al., 2002). There have also been efforts to describe circumpolar biological and physical interactions in relation to fisheries management (Constable et al., 2003).

The gross physical features of the Southern Ocean appear on initial inspection to be relatively simple. Far offshore the waters of the ACC (known in earlier literature as the West Wind Drift) circulate from west to east, while inshore the Antarctic Coastal Current (known in earlier literature as the East Wind Drift, and defined here as bounded by the coast and the southern boundary of the ACC, or SBACC; see Figure 1) circulates from east to west. A series of frontal zones separate the different currents and water masses (Orsi et al., 1995), and overlain on this general circulation pattern is the annual north-south advance and retreat of the sea ice. Closer inspection reveals a much higher degree of complexity. The Antarctic Coastal Current consists of a number of interlinked gyres rather than being a coherent zonal current, and is often defined by topographic or bathymetric features as shown in Figure 3. This, in turn, leads to considerable meridional variation in physical properties such as the formation of bottom water or the distribution of sea ice (Bindoff et al., 2000), and to regional variability in the ecosystems along a coastline that is relatively uniform compared to the Ross and Weddell Seas and the Antarctic Peninsula (Nicol et al., 2000a).

The ecosystems of the Antarctic coastal zone vary most obviously in a latitudinal gradient. There are distinct biological communities associated with the coastal zone, the shelf break, offshore waters and the shifting pack ice (Hosie 1994; Hosie et al. 2000). The offshore waters of the ACC are characterised as high nitrate/nutrient, low chlorophyll (HNLC) because they contain high levels of the macro-nutrients phosphate, nitrate and silicate, but support only modest levels of phytoplankton growth and accumulation, primarily because of a lack of the essential micro-nutrient iron (Martin et al., 1990; de Baar et al., 1995; Boyd et al., 2000). In contrast, the waters of the Coastal Current, which are seasonally covered in pack ice, can be highly productive. Elevated productivity appears to be associated with areas where the Coastal Current is widest and seasonal pack ice is at its most extensive. Conversely, depressed productivity occurs where the Coastal Current is narrowest, the pack ice is least extensive and where the oligotrophic waters of the ACC come closest to the coast (Nicol et al., 2000a). The physical environment, particularly the width of the Coastal Current and the extent of winter pack ice, varies considerably and systematically across the Indian Ocean sector. Therefore, this region is especially well-suited to examining hypotheses that have been developed in other regions concerning the role of physical processes in driving seasonal and interannual variations in biological productivity (Loeb et al., 1997; Tynan, 1998).

Studies of the Indian Ocean sector of the Southern Ocean have generally been regional in their focus. Earlier work includes the Biological Investigations of Marine Antarctic Systems and Stocks (BIOMASS) Program in Prydz Bay (70° – - 80°E) and to the west (0° –, -70°E; Hosie, 1994; Miller, 1985; Miller and Monteiro, 1988; Smith and Treguer, 1994; Smith et al., 1984; Terazaki and Wada, 1986). More recently Nunes Vaz and Lennon (1996) and Wong et al. (1998) have addressed the physical oceanography of Prydz Bay. The 80–150°E region has been more intensively studied of late with a single large-scale biological-physical survey (Nicol et al., 2000b) and six repeat transects along World Ocean Circulation Experiment (WOCE) line SR3 (140°E between Tasmania and Antarctica) between 1991 and 1996 (Rintoul and Bullister, 1999). Smaller scale efforts include studies of the full annual cycle of water column physical, chemical and biological properties close to shore in the vicinity of Antarctic stations (Gibson and Trull, 1999; Gibson et al., 1999) and extensive investigations of sea ice properties and processes (e.g., Heil and Allison, 1999; Worby et al., 1998). Long-term studies of the population biology of the land-based vertebrates (seals and seabirds) are now being linked to conceptual models of variability in secondary production (Barbraud and Weimerskirch, 2001a and b; Wilson et al., 2001). The distribution of completely pelagic vertebrates (cetaceans, fish, pack ice seals) is more difficult to quantify, but attempts have been made to link their abundance to oceanographic patterns (Tynan, 1997) and to wider ecological studies (Nicol et al., 2000a; Murase et al., 2001; Thiele et al., 2000). Historical fishing records are scarce but modern fisheries investigations have successfully determined links between biological features and regional oceanography (Ichii, 1990) while providing considerable information on the distribution, if not necessarily abundance, of commercial species.

A number of papers have recently attempted to synthesise observations relating the biological oceanography of the Southern Ocean to physical features such as sea ice variability (Loeb et al., 1997), frontal systems (Tynan, 1998), the intrusion of Circumpolar Deep Water (Prézelin et al., 2000), surface water mass distribution (Nicol et al., 2000a), and other variables including wind stress and the rate of sea ice retreat (Constable et al., 2003). Attempts to transplant models developed for one region to another have met with limited success and there is evidence that the considerable physical differences observed around the Antarctic continent are reflected in regional biological variability (Constable et al., 2003). Thus, it makes sense to examine the Antarctic coastal zone as a series of interlinked regions rather than as a single entity.

2. Overview of major physical features and circulation of the Antarctic coastal zone

The Antarctic is unique oceanographically because of the current system that rings the continent. The climatic and ecological effects of these circulation patterns have been widely written about and will not be discussed in detail here (Deacon, 1982;

Knox, 1994; Rintoul et al., 2001). Despite the zonal linkages that the large-scale circulation pattern confers both physically and biologically, there is still considerable regional variation around the continent. We have defined the coastal zone as being contained within the marginal ice zone (MIZ; Longhurst, 1998; Arrigo et al., 1998a), and this region will be the focus of this chapter. But first we examine the regional variability in the physical and biogeographic zones that make up the Southern Ocean.

The Polar Front (PF) defines the Southern Ocean (Figure 1), and only near South Georgia does it come close to the coastal zone. The northern limit of winter ice extent defines the southernmost limits of the Permanently Open Ocean Zone (POOZ). The Southern Boundary of the Antarctic Circumpolar Current (SBACC) lies within the seasonal sea ice zone and reflects the transition from the eastward flowing ACC to the westward flowing Antarctic Coastal Current. The SBACC is defined as the southernmost edge of the Upper Circumpolar Deep Water (UCDW) signal, and is generally south of the Southern ACC Front (SACCF) – the southernmost deep-reaching front in the Southern Ocean. This transition is also referred to as the Antarctic Divergence (AD; Deacon 1937) and corresponds to the point of maximum wind stress curl between the westerlies to the north and the easterlies to the south (Bindoff et al., 2000).

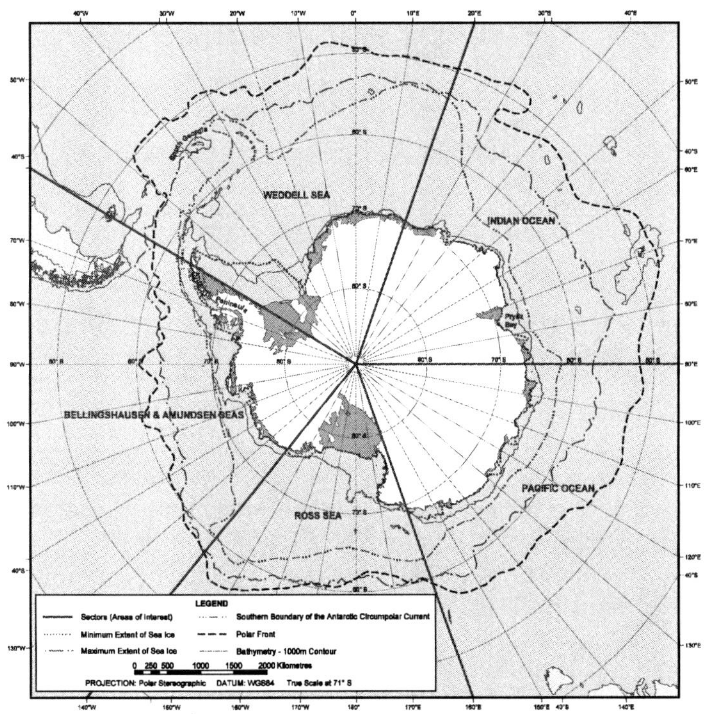

Figure 37.1 The major physical features of the Southern Ocean. The maximum and minimum sea ice extents were derived from passive microwave data compiled at the Antarctic Cooperative Research Centre, Australia between 1989 and 1999. The locations of the Polar Front and the Southern Boundary of the Antarctic Circumpolar Current are from Orsi et al. (1995). The 1000 m bathymetric contours were sourced from the GEBCO Digital Atlas 1994, published by the British Oceanographic Data Centre on behalf of the IOC and IHO. Map produced under the guidance of D. Smith, Australian Antarctic Data Centre.

TABLE 37.1
The bounding and typical values for density, temperature and salinity that define each of the water masses found in the 80°-150°E region (Bindoff et al., 2000) and in the Prydz Bay region (Smith and Treguer, 1994) compared to the water mass properties found in the other regions of the Antarctica descorbed by H & C (Hoffmann and Klinck, 1998)

Region (west to east)	Water mass	Temperature(°C)	Salinity (ppt)
80–140°E	Antarctic Surface Water	-1.84 – 2.0	>34
	Circumpolar Deep Water	>1.8	~34.7
	Modified Circumpolar Deep Water	<1.8	<34.7
	Antarctic Bottom Water	-1.7 - < 0	34.65 - < 34.72
	Ross Sea Bottom Water	~ -0.4	>34.68
	Adélie Land Bottom Water	<0.5	34.66 - < 34.68
	Shelf Water	<-1.7	<34.72
Prydz Bay	Winter Water	-1.87	34.34
(70–80°E)	Circumpolar Deep Water	0.61	34.68
	Continental Shelf Water	-1.85	34.56
	Ice Shelf Water	-2.00	34.5
	Antarctic Bottom Water	-0.51	34.66
150°E–0°E	Antarctic Surface Water	0.0 – 1.5	34.0 – 34.4
	Winter Water	<-1.5	34.0 – 34.4
	Circumpolar Deep Water	>0	34.6 – 34.73
	Modified Circumpolar Deep Water	1.0 – 1.4	34.6 – 34.7
	Ice Shelf Water (Ross and Weddell Sea)	<-1.8	34.36 – 34.42
	Shallow	<-2.0	34.62
	Deep		
	Bottom Water (Ross Sea)		
	Low salinity	-0.1	34.65
	High salinity	0.5	34.71

The water masses of the Antarctic have been well described in earlier publications (Hofmann and Klinck, 1998; Orsi et al., 1995; Bindoff et al., 2000) and comparative properties of these water masses between East and Western Antarctica are presented in Table 1. Temperature-salinity diagrams for the various regions can be found in the source publications. Apart from the highly regional water masses, such as Weddell Deep Water (Hofmann and Klinck, 1998), most of these water mass types are found in all regions around the Antarctic. Of particular interest in East Antarctica is the presence of regional water masses, including Ross Sea Bottom Water, Adélie Land Bottom Water (Bindoff et al., 2000) and Ice Shelf Water. The latter emerges from the vicinity of the Amery Ice Shelf in Prydz Bay (Smith and Treguer, 1994). Overlain on the gross and relatively conservative water mass structure is the summer surface water, which can exhibit a wide range of temperatures and salinities resulting from the seasonal melting of sea ice and ice shelves, cooling events and surface mixing (Smith and Treguer, 1994). This surface layer can vary considerably in depth and therefore have a considerable effect on biological productivity (see Section 3).

The remaining physical features of importance include the edge of the continental shelf and the Antarctic coastline itself. The areas encompassed by each of these physical boundaries have been calculated (Table 2) and there is considerable regional variability. Presentation of these regions as areas is subject to bias because the area encompassed per degree of latitude increases as latitude decreases, i.e., regions in the north of each 10° sector become larger, which tends to distort the analysis. Never the less, there is some value in comparing the relative proportions of sectors occupied by the various zones – these relationships are summarized in Table 1, Figure 2 and below:

- When the SBACC is further offshore (i.e. the coastal current is widest) the width of the sea ice zone is greatest
- When the coastline is further south there is a greater annual sea ice extent
- The widths of the oceanic zones are greatly affected by physical features, which affect the circulation. For example, the Antarctic Peninsula/Drake Passage in the South Atlantic and the Kerguelen Plateau in the South Indian Ocean.
- There are striking regional differences in the relative areas of the different zones, and some relationships between features that are apparent in some sectors are absent in others.

There are noticeable effects associated with the major geographic features. Variability in the width of the continental shelf is clearly associated with the dominant geographical features – the Antarctic Peninsula, the Weddell Sea, the Ross Sea and to a lesser extent Prydz Bay (Figure 2a). The continental shelf in Antarctic waters is unusually deep (average depth >500m) compared to other continents, because of the effect of the ice sheet mass on the continent. Some sectors, such as the Pacific, have a coastline that is entirely south of 70°S. Although the coastline of East Antarctica is relatively uniform, there is a significant gradient in latitude from approximately 70°S in the west to 62°S in the east. Consequently the area covered annually by sea ice in the 0° – 10°E sector (1.02 million km^2) is more than three times that covered in the 130° – 140°E sector (0.31 million km^2). The coastline near 130° E is the farthest from the pole (with the exception of the Antarctic Peninsula) and was reported by Gloersen et al. (1992) to be climatologically warmer than the other regions around Antarctica. The area between the coastline and the SBACC varies considerably, as does the relationship between the SBACC and the SACCF —from being nearly coincident in the 140° – 160°E region to being widely separated in the 40°E region (Figure 1 and Orsi et al, 1995).

The POOZ, MIZ, ACC and Coastal Current (Figure 1) vary in their spatial extent between the major Antarctic sectors (Figure 2 b). in the following ways:

- In the South West Indian Ocean there is a substantial decrease in the area occupied by the Coastal Current moving from the west to the east associated with the Weddell Gyre. This change is mirrored by a decrease in area occupied by the MIZ and an increase in the width of the ACC.
- There is an increase in the area of the POOZ, ACC and MIZ in the Prydz Bay/Kerguelen Plateau region. All three decrease in area across the Indian/Pacific and into the Ross Sea whilst the decrease in the width of the Coastal Current is less dramatic.
- In the Ross Sea, the area of the POOZ and the width of the ACC decline as the Southern Boundary and the MIZ extend northwards.

- In the Amundsen-Bellingsahusen Seas there is little change in the relative areas of the four regions, but
- In the Antarctic Peninsula region, all four appear to increase simultaneously. To some extent this increase is due to the distortion of area with latitude as described above. The north-south alignment of the coastline is also influencing the calculation of sector areas here. For the rest of the Antarctic coastline, the distance to the ice edge and fronts is essentially perpendicular to the coast. For the peninsula this is not the case, and so the MIZ is at a minimum.

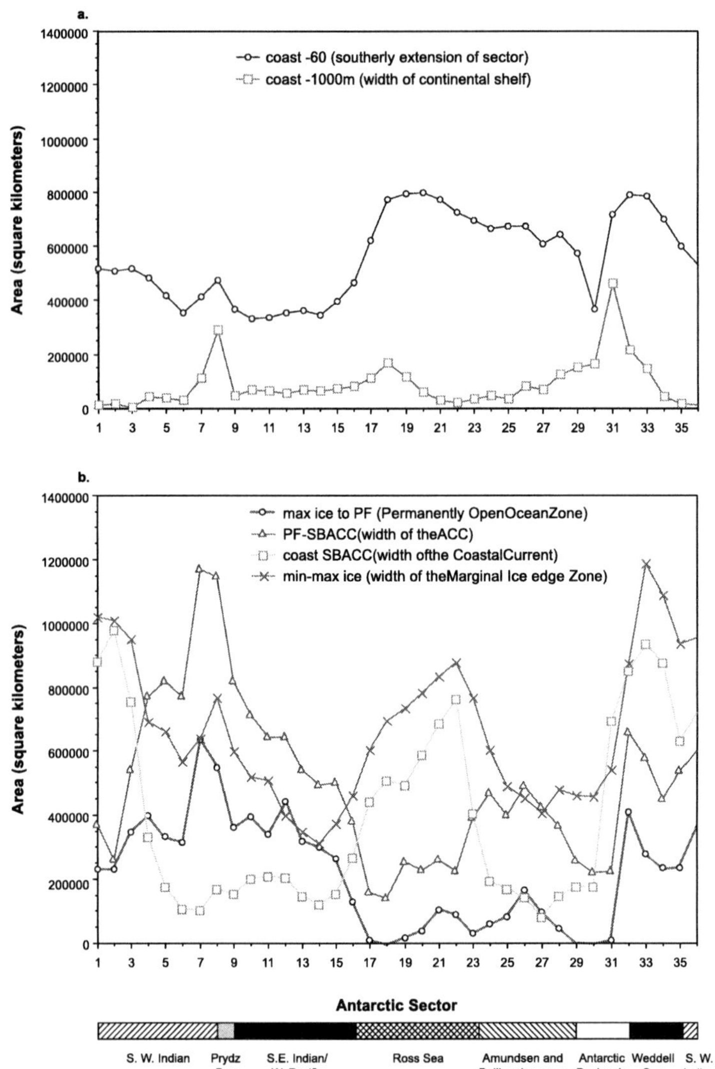

Figure 37.2 Graphical representation of the areas between physical boundaries by sector. (a.) The areas between the coast and 60°S and between the coast and the shelf break (defined as the 1000m isobath). (b.) The sectoral width of each of the major biogeographical zones defined as the areas between the various oceanographic boundaries.

TABLE 37.2 PART I

The area within geographic zones in 10° sectors around the Antarctic. 10° sectors were chosen to resolve regional differences and to provide a sufficient large sample size (36) for statistical purposes. The areas were calculated using a Lambert Azimuthal Equal Area map projection and ArcGISm8.2 software (ESRI, USA). Notes: The first of the two min. ice to max. ice columns includes ice only to the PF, while second column includes ice beyond the PF.

	Sector	Coast-60°S	Coast polar Front (km^2)	Coast-South Boundary Acc (km^2)	Polar Front southern Boundary Acc (km^2)	Coast-1000m Bathy Contour (km^2)	Min Ice Extent-Max Ice Extent (km^2)	Min Ice Extent -Max Ice Extent (km^2)	Max Ice Extent-Polar Front(km^2)
	0°–10°E	519973	1252077	881244	370883	14641	1020008	1020008	232119
	10°–20°E	509515	1242787	980130	262658	16840	1008780	1008780	234007
	20°–30°E	517071	1300560	757889	542671	6327	951944	951944	348617
Western Indian Ocean	30°–40°E	483930	1106568	334456	772112	44971	693448	693448	401926
	40°–50°E	420496	997849	178301	819548	41009	663474	663474	334626
	50°–60°E	353539	883904	109522	774382	32128	568225	568225	315843
	60°–70°E	413512	1273066	106936	1171932	113265	641111	641111	637770
Prydz Bay	70°–80°E	474435	1318056	170525	1149486	289009	769457	769457	550607
	80°–90°E	367632	974892	155329	819563	48697	600787	600787	363986
	90°–100°E	332971	920336	204499	715838	70953	522136	522136	398410
	100°–110°E	335722	855252	211253	643999	64982	510053	510053	341956
Eastern Indian Ocean	110°–120°E	352744	853786	207979	645807	57744	399860	399860	445073
	120°–130°E	364419	692514	150062	542452	68041	349276	349276	321696
	130°–140°E	345228	617117	122563	494555	65416	313298	313298	303821
	140°–150°E	397452	657204	155016	502188	74622	374807	374807	267108
	150°–160°E	466892	649744	268613	381131	82344	464628	464628	131079
	160°–170°E	621823	602230	443028	159202	111268	586938	605872	9826
	170°–180°	775607	652933	509480	143453	168859	652866	697589	73
Ross Sea	170°–180°W	794390	751816	495421	256394	116391	732522	736616	19395
	160°–170°W	798726	821973	591209	230764	61328	780982	782699	40991
	150°–160°W	772837	951613	690841	260773	31974	834223	834223	104509

TABLE 37.2 PART II

The area within geographic zones in 10° sectors around the Antarctic. 10° sectors were chosen to resolve regional differences and to provide a sufficient large sample size (36) for statistical purposes. The areas were calculated using a Lambert Azimuthal Equal Area map projection and ArcGISm8.2 software (ESRI, USA). Notes: The first of the two min. ice to max. ice columns includes ice only to the PF, while second column includes ice beyond the PF.

	Sector	Coast-60°S	Coast polar Front (km^2)	Coast-South Boundary Acc (km^2)	Polar Front southern Boundary Acc (km^2)	Coast-1000m Bathy Contour (km^2)	Min Ice Extent-Max Ice Extent (km^2)	Min Ice Extent -Max Ice Extent (km^2)	Max Ice Extent-Polar Front(km^2)
	140°–150°W	724443	988380	763973	224407	23062	880296	880296	91805
	130°–140°W	694673	802362	409701	392662	33732	768877	768993	31814
	120°–130°W	666161	667140	197469	469671	45420	606211	606258	60935
Bellingshausen and	110°–120°W	673852	575477	172915	402562	34193	492228	492228	82896
Amundsen Seas	100°–110°W	675696	635531	144036	491495	83323	457307	457307	169497
	90°–100°W	608728	507606	82825	424942	67074	407600	407600	100166
	80°–90°W	644575	516215	149428	366787	126456	467824	482539	48391
Antarctic Peninsula	70°–80°W	572956	436733	178767	257966	149205	409472	463255	400
	60°–70°W	368567	400780	178380	222400	163447	359640	459602	0
	50°–60°W	719140	923725	696488	227237	464288	514989	543593	9808
	40°–50°W	790140	1512220	854106	658115	217992	873697	873697	413689
Weddell Sea	30°–40°W	784523	1514019	937342	580342	145308	1190214	1190214	280241
	20°–30°W	701764	1329256	878469	450788	43012	1090808	1090808	238764
	10°–20°W	601739	1174207	634114	540094	16394	938773	938773	235434
	0°–10°W	533187	1323183	720845	602338	11437	954930	954930	368254

These physically delimited zones have biogeographic counterparts, which result in differences around the continent in phytoplankton productivity (section 3) and higher trophic levels (section 4). The factors that determine the annual extent of sea ice in a sector (Section 2.4) include the circulation patterns, latitude, bathymetry and the interactions of these factors with each other. In many sectors the area of the sea ice zone is a reflection of the area of the coastal current in that sector (eg. the extreme west of the southwest Indian Ocean, the eastern portion of the southeast Indian, the west of the Ross Sea and the western Antarctic Peninsula region). Given the influence of sea ice on productivity, generalisations on the relationship between physical and biological variability are likely to hold true only for particular regions. Attempts to apply conceptual models derived in areas such as the Antarctic Peninsula to regions with a considerably different physical environment such as the Ross Sea should be viewed with caution.

2.1. East Antarctica

The eastern end of the Weddell gyre reaches approximately 20°E (Deacon, 1937) at which point the ACC turns south toward the coast. The ACC influences the winter ice boundary such that the ice edge runs essentially north-south in this region. The Weddell gyre is characterized by chlorophyll concentrations ~1 mg m^{-3}, particularly in December and January, while chlorophyll between the coast and the maximum ice extent east of 20°E is ~0.1 mg m^{-3}. The movement of CDW toward the continental margin at the eastern end of the gyre is important for recirculating water back into the Weddell gyre and eventually forming Antarctic Bottom Water (AABW) in the Weddell Sea. Whitworth et al. (1998) claimed that this only occurs at two places around Antarctica, the second being in the Ross Sea, where bottom water is also formed. However recent results from Bindoff et al. (2000) and Rintoul (1998) show that this is also important near 140°E. As discussed below, the Adélie Land region is now acknowledged as a significant source of AABW formation.

Prydz Bay, between 70° and 80°E, represents the only substantial embayment along the coastline between 0° and 150°E. Numerous hydrographic surveys have been undertaken in the region and reported by Grigor'yev (1967), Smith et al. (1984) and Smith and Treguer (1994). Prydz Bay is bounded to the south by the Amery Ice Shelf and is characterized by depths near 600 m, with Fram Bank (to the west) and Four Ladies Bank (to the east) providing a partial barrier to exchange with the deep ocean. The general circulation pattern between 60° and-90°E, discussed by Smith and Treguer (1994), is characterized by a clockwise circulating gyre south of the Antarctic Divergence, fed by a broad inflow from the NE and a stronger, coastally-confined outflow toward the west, connecting the continental shelf waters with the offshore flow of the ACC (Figure 3). Northward outflows occur near the West Ice Shelf (around 90°E) and MacRobertson Land near 60°E. Geostrophic calculations indicate that substantial amounts of cold water may be carried into Prydz Bay from near the West Ice Shelf, and this may have a bearing on the water mass balance within the bay. The flow within Prydz Bay and the outflow past Fram Bank toward the west are at least partly controlled by bathymetry and, in the case of the outflow, are thought to influence the distribution of biological communities (Smith et al., 1984). Data from drifting buoys deployed in the ice cover of Prydz Bay reveal a split in the flow north of Cape Darnley, with some buoys heading west parallel with the coast, while others drift almost due north along 70°E before turning eastward in the flow of the ACC (Allison, 1989). Heil and Allison (1999) observed that this northward extension of the Prydz Bay gyre interrupts the eastward flow of ice north of the Antarctic Divergence. The depression between Four Ladies Bank and Fram Bank permits only limited exchange with waters outside the bay and is seen as a key element in the deep circulation with Prydz Bay. It is the lack of a mechanism for mixing the warm, salty Circumpolar Deep Water with the cold shelf water that is seen as a major obstacle to bottom water formation in this region (Smith et al., 1984). Smith and Treguer (1994) comment that the presence of the West Ice Shelf near 85°E is significant for determining the circulation and characteristics of the shelf waters in this region. The coldest water in the region has been observed west of the ice shelf (Smith et al., 1984).

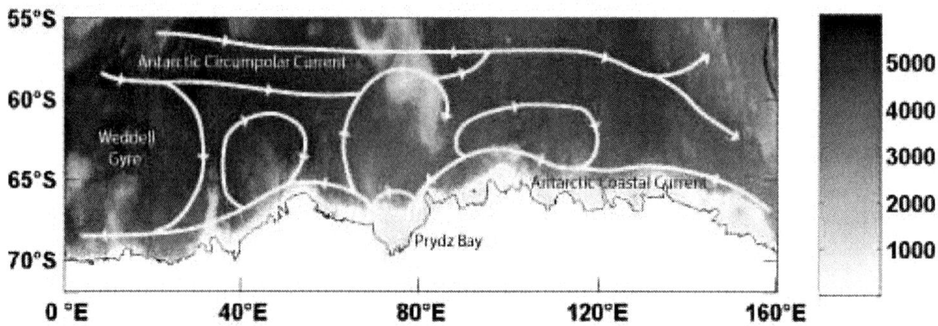

Figure 37.3 The general surface circulation pattern off east Antarctica. Diagram produced from figures in Mackintosh (1972), Bindoff et al. (2000), and Pakhomov (1995a). The Kerguelen Plateau is to the North of Prydz Bay separated from the Antarctic Continental Shelf by the Princess Elizabeth Trough. Bathymetry is shown in grey scale.

Prydz Bay is ice covered for much of the year, with the exception of a short period in January-February (Worby et al., 1998). Within Prydz Bay, surface water conditions are typically highly variable and closely associated with ice conditions (Smith et al., 1984). Over the continental shelf, sea surface temperatures have been observed to be higher in regions where the pack ice retreats earliest. A recurring feature is an area of anomalous warm water north of the Amery Ice Shelf in the central part of the bay, that has been linked to polynyas in the region (e.g., Knapp, 1969; Leont'yev, 1973). Shelf water salinities can vary over a large range depending upon brine rejection during sea ice formation, convection and mixing with subsurface water and horizontal advection and interleaving of water masses. Low salinity shelf water originating from the east near the West Ice Shelf can be differentiated from high salinity continental shelf water which is formed by freezing and convection within the bay (Smith et al., 1984). While this is not a region of AABW formation, deep mixing is driven by sea ice formation, transporting water with surface properties into the deep ocean. During the ice melt season the water column is re-stratified, adding a fresh surface layer that is readily heated by solar radiation. An intense spring/summer bloom of >5 mg m^{-3} chlorophyll occurs within the bay (70° – 80°E), reaching its peak in January – SeaWiFS imagery shows that it is bounded to the north by the shelf break. Between the shelf break and the maximum extent of winter ice (55° – 56°S) surface chlorophyll is ~0.1 mg m^{-3} in the western part of this sector, and 0.5 mg m^{-3} or slightly higher in the east. North of the winter ice boundary chlorophyll is ~0.5 mg m^{-3} in the west and ~2.0 mg m^{-3} in the east over the Kerguelen plateau. This topographical feature is probably impacting surface chlorophyll concentrations in two ways. First, the shoaling bathymetry likely interacts with the Polar Front to cause enhanced vertical mixing and meandering of the front, both of which will bring nutrients into the euphotic zone (Moore et al., 1999). Second, the land mass of Kerguelen Island and its shelf (admittedly small) act as an iron source to the waters 'downstream' (to the east) in the same way as the South Georgia Rise in the South Atlantic and the Galapagos and Marquesas Islands in the equatorial Pacific.

Ice Shelf Water (ISW) has also been identified in Prydz Bay, in the vicinity of the Amery Ice Shelf. Wong et al. (1998) showed that there was significant cooling and freshening of the water that flowed under the Amery Ice Shelf and derived a net melt rate from the bottom of the ice shelf between 10.7 and 21.9 Gt $year^{-1}$ (Gt = gigaton = 10^{12} kg). This region is presently the focus of an intensive study called AMISOR (Amery Ice Shelf Ocean Research) to examine the interactions between the ice shelf and the ocean (Allison, 2003). Preliminary data show strong outflows of ISW between 150 and 400 m, and a total overturning circulation of approximately 1.3 Sv. Williams et al. (2001) used an ocean circulation model to simulate the ocean cavity beneath the Amery Ice Shelf and showed that the sub-shelf circulation is predominantly barotropic, steered by the cavity topography, and is strongly influenced by melting and freezing processes at the ice-ocean interface. The model results showed basal freezing of several Gt $year^{-1}$, sufficient to provide substantial layers of accreted ice.

The coastal circulation patterns between 80° and 150°E have been described recently by Bindoff et al. (2000). They reported the results of an experiment comprising eight meridional CTD sections between the coast and the SBACC, and one zonal section along approximately 63°S, during February and March 1996. The distribution of water masses along each of the meridional sections suggest that the continental shelf, break and slope regions have complex water-mass and frontal structures that vary along the entire survey area. On most of the sections, gradients in the ocean interior suggest the presence of cyclonic eddies or meanders. Near 120°E, direct observation of the intrusion of Modified Circumpolar Deep Water (MCDW) onto the shelf has important implications for the salinity of the shelf waters and for the exchange of heat from offshore onto the shelf. The MCDW in this region is relatively cold and fresh compared to that found in the Adélie Depression (Rintoul, 1998) and is therefore not associated with the formation of AABW. Wong et al. (1998) showed that it is the more salty MCDW that is important for this process. The bottom water properties at the western end of the region are warmer and fresher, in comparison to the eastern end between 140°E and 150°E, where Adélie Land Bottom Water is formed. The distribution of tracer data suggests that the bottom waters spread westward along the continental rise (Bindoff et al., 2000).

The surface flow implied from hydrographic data shows the strongest part of the ACC passing north of the Kerguelen Plateau, with a small branch passing south into the Antarctic Basin on the northern side of the Princess Elizabeth Trough

(PET) (Figure 3). A branch of the ACC passes eastward through the PET and follows the Kerguelen Plateau to the north, forming a western boundary current (Speer and Forbes, 1994). On the eastern side of the Kerguelen Plateau these two branches coalesce into a single current system that heads southeast towards the Antarctic continental shelf near 145°E (Orsi et al., 1995). These eastward-moving water masses are characterized by a shallow oxygen minimum layer (Lower Circumpolar Deep Water). The southern-most edge of this water mass has been used to define the location of the southern boundary of the eastward flowing ACC (Orsi et al., 1995) and in the 80° – 150°E region corresponds quite closely to the location of the Antarctic Divergence (Bindoff et al., 2000). It appears likely that the strong currents along the slope and the broad eastward flow at the western end of the region form a partially closed gyre (Bindoff et al., 2000). The flow field and water mass properties suggest that this gyre is widest in the western part of the basin near 100°E, and narrowest near 140°E, consistent with the transport of sea ice as observed by drifter data (Heil and Allison, 1999).

South of the Antarctic Divergence information on the surface ocean currents comes from buoy drift data (from sea-ice and drogued buoys; Bindoff et al., 1997, 2000; Heil and Allison, 1999) and from iceberg drift data (Tchernia and Jeannin, 1984). These data show a strong westward circulation of the surface waters along the Antarctic continental shelf from 150°E to near the Princess Elizabeth Trough, strongest near the Antarctic Slope Front (Whitworth III et al., 1998, Bindoff et al., 2000) and weaker offshore towards the Divergence. The tracks of icebergs and drifting buoys are strongly affected by bottom topography and follow isobaths for long periods of time. Bottom topography also appears to play a strong role in creating large quasi-stationary eddies (Wakatsuchi et al., 1994; Bindoff et al., 2000), seen in both the drifting buoy tracks and hydrographic data at depth.

Drifting buoys and icebergs east of 80°E do not escape from the Australian-Antarctic Basin and are deflected north and then east at the Princess Elizabeth Trough and return south of the Divergence. Water mass properties (Rodman and Gordon, 1982; Heywood et al., 1999; Bindoff et al., 2000) indicate that the bottom waters in this basin are distinct from those found further west in the Weddell-Enderby Basin, suggesting that relatively little water mass exchange occurs between the Weddell-Enderby Basin and Australian Antarctic Basin through the Princess Elizabeth Trough (Rintoul, 1998). At the sea-ice maximum (September) the sea-ice extent is greatest near the Kerguelen Plateau, and least near 140°E (Gloersen et al., 1992). Sea-ice is transported with the westward surface currents (and winds) around the gyre between 80° and 155°E, melting during the summer months. This distribution of sea-ice leads and surface currents means that the Antarctic Surface Waters (Whitworth III et al., 1998) characterized by their low salinity (and including the temperature minimum layer) also have the same spatial pattern as the gyre.

Figure 4 (after Heil and Allison, 1999) shows the meridional and zonal ice velocities from a compilation of data from 39 satellite-tracked drifting buoys off East Antarctica between 1985 and 1996. The buoys were deployed in newly forming ice or on ice floes, mostly during the March-May period to maximise their time in the pack ice. Individual buoy tracks show that the pack ice drift is highly variable in space and time, however the large-scale circulation features described in Figure 3 can clearly be seen in the ice drift climatology. The zonal flow presented in Figure

4 (top panel) shows that the region of westward ice transport, which is adjacent to the coast or fast ice, is only 2 – 3° of latitude wide in the western Pacific sector (82.5° – 150°E), while in the Indian ocean sector (20° – 65°E) the current is up to 5° of latitude wide. The zonal velocity component has an average daily value of 0.23 m s^{-1} in the Indian Ocean sector, compared with 0.17 m s^{-1} for the western Pacific and 0.11 m s^{-1} for the Prydz Bay region between 65° and 82.5° E. Features such as the Shackleton Ice Shelf near 95°E locally reduce the westward speed of the coastal drift, and other coastal protruberances in the Pacific sector slow the westward drift in some regions (Heil and Allison, 1999). Within the shear zone associated with the Antarctic Divergence, zonal ice speed is low. North of the Divergence drift speeds are generally slower and to the east. There is also less variability in this region suggesting that the ice drift is more influenced by steady oceanic forcing in the ACC rather than highly variable atmospheric forcing. The meridional ice drift is shown in the bottom panel of Figure 3. There are a number of areas of net northward drift where sea ice is transported from the westward flowing coastal current into the ACC. These occur in the Prydz Bay region at 67°E and 77°E, and also at 85°E, 98°E, 110°E, 125°E and 135°E (Heil and Allison, 1999). The average meridional motion south of the AD is southward but only -0.02 m s^{-1}. North of the Divergence the drift is variable between -0.1 to 0.42 m s^{-1}.

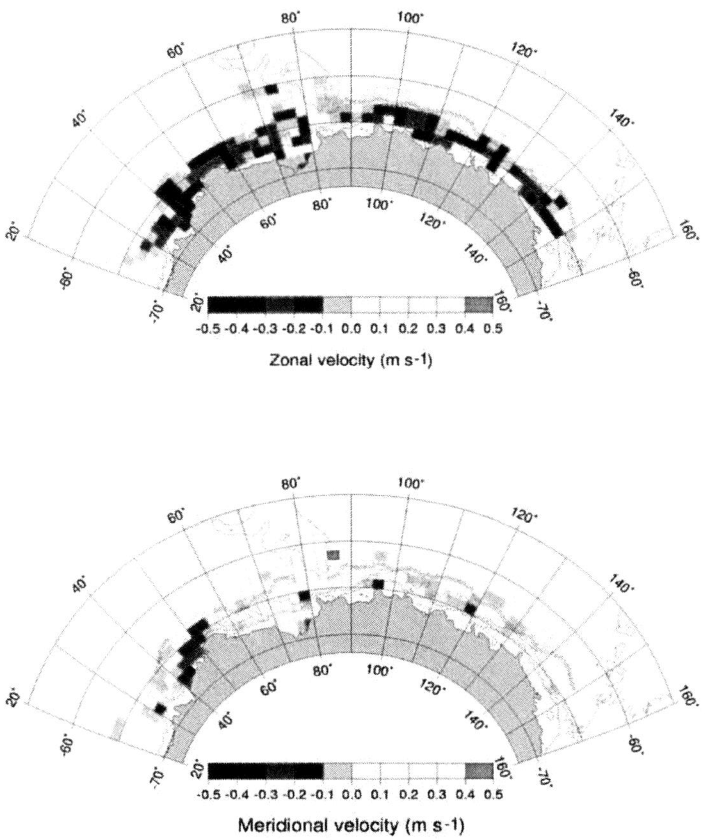

Figure 37.4: Sea ice drift velocities off East Antarctica. Figure 37. From Heil and Allison (1999).

2.2. Sea ice

Antarctic sea ice undergoes an extreme seasonal change from a minimum area of approximately 4 million km^2 in February to a maximum of approximately 19 million km^2 in September (Gloersen et al., 1992). This represents one of the greatest seasonal changes in physical properties anywhere on earth. Significant regional variability in sea ice thickness occurs around Antarctica but the average thickness of the pack ice is approximately 0.5 - 1.0 m, while fast ice and multi-year ice reach average thicknesses between 1.0 and 3.0 m. Numerous studies from different regions of the Antarctic pack have reported that frazil ice growth is the dominant mechanism for ice formation (Jeffries et al., 1994; Lange and Eicken, 1991), and that ridging, rather than thermodynamic growth, is primarily responsible for thickening the ice beyond about 0.5 m (Worby et al., 1998).

The sea ice in different sectors of the Antarctic pack may exhibit significantly different characteristics. The large embayments of the Weddell and Ross Seas contain cyclonic gyres that influence the drift and distribution of the ice. The western Weddell Sea contains the thickest ice and also accounts for 80% of the multi-year ice around Antarctica (Gloersen et al., 1992). The other region of persistent multi-year ice is along the Bellingshausen-Amundsen coast where minimum extent typically occurs in March, unlike other regions where it occurs in February (Gloersen et al., 1992). The Ross Sea region and the East Antarctic coast are typically ice free during the summer months, except for relatively small, isolated embayments where sea ice may become trapped.

The East Antarctic sea ice cover is contained in a relatively narrow and highly mobile zone, which near 140°E, extends only several hundred kilometers offshore at maximum extent. There is significant interannual variability in the rate of ice edge advance and also in maximum ice extent, consistent with other regions of the pack. The East Antarctic region has been the focus of numerous studies in the past decade that have focused on the structure, properties and drift of the ice. Worby et al. (1998) presented a first look at the ice thickness distribution based on observational data collected from icebreakers. This data set has subsequently been expanded under the auspices of the Scientific Committee on Antarctic Research ASPeCt (Antarctic Sea Ice Processes and Climate) program and now includes data from more than 80 voyages to all regions of Antarctica between 1980 and 2004. The seasonal ice thickness distributions for the Indian (20°-90°E) and Pacific Ocean (90°-160°E) sectors are shown in Figure 5. The average sea ice thickness in the Indian sector is 0.60 m compared with an average of 0.75 m in the Pacific sector. In all seasons the "tail" of the distribution curves contain more ice in the Pacific sector, which represents the thicker deformed ice. At the thinner end of the curves, the seasonal evolution is quite similar in the two sectors. The highest open water fraction and lowest concentrations of thin ice types occur in the summer (December-January-February, DJF) when the ice has retreated toward the coast and much of the remaining ice is multi-year. In the autumn (March-April-May, MAM) there is a much higher fraction of each ice type up to 1 m thick as air temperatures fall and ice forms rapidly and expands northward. There is some evidence in the Pacific sector of a bimodal distribution in the winter (June-July-August, JJA) and spring (September-October-November, SON) periods which may be a result of deformation. This is consistent with the higher average ice thickness in this sector. These data, which are obtained from vessels underway, provide an excellent first order estimate of sea ice thickness around Antarctica. Errors in ice thickness are estimated to be ±20%, but may be greater in areas of deformed ice where a simple model of the undeformed ice thickness, average sail height and fractional area ridged are used to determine ice thickness. The observational techniques and thickness calculations are described in detail in Worby and Allison (1999). Errors in ice concentration are estimated to be ±10%.

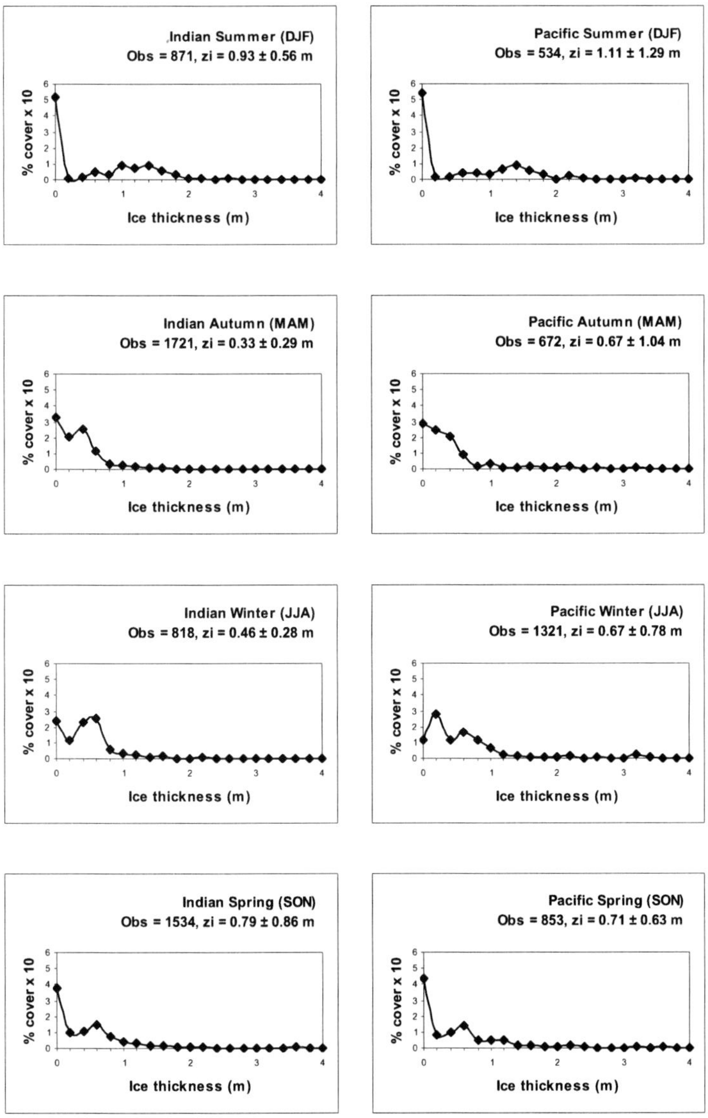

Figure 37.5: Monthly distribution of sea ice and snow thickness on sea ice off East Antarctica. 37. From Worby et al. (1998)

There has been considerable research into trends in the distribution, abundance and characteristics of the sea ice that forms annually around Antarctica. Any persistent changes in the amount of annual sea ice would have significant implications for the ecosystems of the region and for global climate. Unfortunately, the extent of sea ice is best measured using satellite remote sensing and the passive microwave satellite records only extend back to 1973. Considerable ingenuity has been expended in finding proxies of past ice edges to compare to modern satellite records, and long-term variability in the location of the sea ice edge has been examined using various combinations of ship data from both historical and contemporary records, satellite data and proxies of sea ice extent present in sedimentary records and in ice cores. Parkinson (1990) compared historical data from explorers' logs of the early 19th century with the first three years of passive microwave data (1973 – 76) and found no evidence of a substantial change. In contrast, de la Mare (1997) reported a southward shift in the summer ice edge (October – April) of 2.8° latitude between the mid 1950s and early 1970s, based on a comparison of whaling data and satellite passive microwave data. Ackley et al. (2003) however, claim that the variability observed by de la Mare (1997) is within the present day variability once the satellite data are corrected for a summer bias. Worby and Comiso (2001) reported the location of the ice edge observed by research vessels to be routinely 1 – 2° latitude north of the passive microwave ice edge in summer. This is because summer ice is typically warm, low concentration and saturated with sea water and therefore has a passive microwave signature more like that of open water. Furthermore, an examination of ice edge trends using different algorithms to interpret the passive microwave data show variable results, suggesting that the current time series from remote sensing data is insufficient to draw firm conclusions (Cavalieri et al., 1997; Bjørgo et al., 1997; Comiso, 2003). Another potential proxy for Antarctic sea ice extent, the concentration of methyl sulphonic acid (MSA) in continental ice sheet cores, suggest a decrease in sea ice extent in the 80 – 140°E region of up to 25% since the 1950s (Curran et al., 2003). There is, however, considerable seasonality and variability in the MSA record and this proxy for sea ice has not been validated elsewhere in Antarctica.

2.3. Polynyas

The East Antarctic region in winter is characterized by numerous recurring areas of open water in the sea ice, known as polynyas. Adjacent to the coast these are known as "latent heat" polynyas because they are predominantly formed and maintained by cold density driven (katabatic) winds that drain off the ice sheet. These winds may reach 80 knots at the coast where, once in contact with the ocean, they rapidly form sea ice and advect it northward away from the shore. The heat required to balance loss to the atmosphere and hence maintain the open water is provided by the latent heat of fusion of the continually forming ice (Smith et al., 1990). Consequently these are regions of thin ice cover but high ice production, and they play an important role in vertical convection and water mass modification due to the high input of brine from sea ice formation. Massom et al. (1998) identified 28 recurrent coastal polynyas around the East Antarctic coast and these are shown in Figure 6. The streamlines of near-surface gravity drainage winds that cause katabatics are also shown and discussed further by Parish and Bromwich (1987).

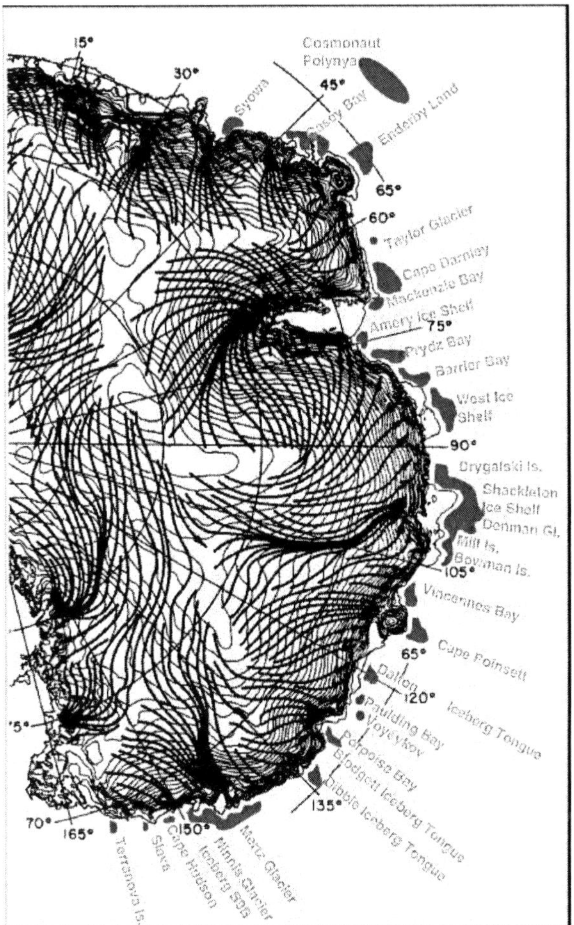

Figure 37.6: Predicted streamlines of near-surface gravity drainage winds over Antarctica. Note that the Mertz Glacier polynya (23,000 km^2) is in a region of intensely cold and persistent katabatic winds and high ice production, linked to the formation of Adélie Land Bottom Water. Figure 37. Adapted from Parish and Bromwich (1987) and Massom et al. (1998).

Sensible heat fluxes may also play a part in maintaining coastal polynyas as suggested by Jacobs (1989), by upwelling relatively warm water from below. This may occur in regions where intrusions of Modified Circumpolar Deep Water (MCDW) come onto the shelf, bringing sufficient heat to maintain ice-free conditions. This could in fact lead to increased ice production near the coast (latent heat region) by removing or reducing the amount of ice to the north and enhancing the offshore removal of new ice by winds. Such conditions occur in the Mertz Glacier polynya region and provide part of the mechanism by which Antarctic Bottom Water (AABW) is formed there, as described in the next section. A number of studies (e.g., Adolphs and Wendler, 1995; Lytle et al., 2001) also show that synoptic winds may play a significant role in ice removal and that many polynyas are formed and maintained by a combination of processes. Additionally, coastal protuberances such as ice shelves or shoals of grounded bergs may contribute to the maintenance of polynyas by reducing the flow of sea ice into the polynya regions.

Polynyas may also be important regions for biological activity, and Massom et al. (1998) commented that although the relationship between regional productivity and biomass at higher trophic levels in polynya regions is largely unknown, polynyas may be a critical component in the ecology of certain species. Cetaceans are known to utilise polynyas (Thiele and Gill, 1999), and it is likely that emperor penguins depend on polynyas to access the ocean close to shore during late winter and early spring (Wienecke and Robertson, 1997). However, limited results indicate that polynyas do not sustain high productivity in winter, presumably due to light limitation. Arrigo and van Dijken (2003) carefully examined the persistence and productivity of 37 polynyas. They showed that several large polynyas (in the Ross Sea, Ronne Ice Shelf, Prydz Bay and Amundsen Sea) account for >75% of total polynya primary productivity. The significance of polynyas for the survival of Adélie penguins is greatest in east Antarctica, where 91% of all colonies are associated with polynyas, and polynya primary productivity is related to the size of the penguin colonies.

As well as near the coast, polynyas may occur in the open ocean if they are driven purely by sensible heat fluxes (known as "sensible heat" polynyas). They occur in regions where there is sufficient oceanic heat at the surface to prevent sea ice forming. There is only one recurrent sensible heat polynya in the East Antarctic region, centred near 65°S, 52°E, in the Cosmonaut Sea and initially observed in satellite imagery (Comiso and Gordon, 1987). Although it can be as large as 114,000 km^2 and has occurred in satellite imagery every year since 1972, it rarely occurs for longer than one month at a time. It is not generally characterized by elevated productivity, particularly in comparison with the Shackleton and Mertz polynyas. Comiso and Gordon (1996) showed that the centroid of the polynya varies only slightly with each formation, both during a given year and interannually. They suggest that its formation may not involve convective overturning, but rather locally enhanced upwelling of warm deep water induced through local topography and circulation interactions. Similar circulation anomalies were invoked by Wakatsuchi et al. (1994) to explain notable decreases in ice thickness and concentration with upwelling and eddy formation, associated with north-south ocean ridges that intersect the ACC between 50°E and 130°E.

Cavalieri and Martin (1985) used passive microwave data and atmospheric model analysis combined with bulk-transfer calculations, to estimate sea ice production in polynya regions to be 8.8 – 11.1 m year^{-1}, compared with approximately 1 m annual growth for adjacent fast ice. Given an ice growth season of approximately 100 days, this is an average of 8.8 – 11.1 cm day^{-1}, although their study does show large variability during the winter. Recent studies have shown that ice production rates may vary significantly between polynyas depending on the strength and direction of local winds, air and water temperatures, and any effect from geographic barriers such as ice shelves or grounded icebergs. The two largest and most frequently recurring polynyas along the East Antarctic coast are the Shackleton Ice Shelf polynya (~30,000 km^2) at 95°E and the Mertz Glacier polynya (~23,000

km^2) at 145°E (Massom et al., 1998). The Shackleton polynya, despite its larger size, is not associated with AABW production (Bindoff et al., 2001), while the Mertz polynya, which is associated with some of the world's strongest and most persistent winds (Ball, 1957; Adolphs and Wendler, 1995), plays an important role in AABW formation due to the high ice production. Both polynyas are associated with high chlorophyll concentrations (> 2 mg m^{-3}) during summer, the Shackleton perhaps more so on average. Sambrotto et al. (2003) observed a dense *Phaeocystis* bloom with chlorophyll concentrations >10 mg m^{-3} in the Mertz polynya in the summer of 2000–2001.

2.4. *Bottom water*

Ice production in katabatic affected coastal regions can be intense and this is an important process that drives vertical convection and leads to the formation of Antarctic Bottom Water (AABW). The production of such dense water masses at high latitudes has a critical effect on global ocean circulation and on world climate. Turbulent heat fluxes from the ocean in the range 400 – 500 W m^{-2} are common (e.g. Fahrbach et al., 1994) resulting in high sea ice growth rates such as the 4.9 to 8 cm day^{-1} reported in winter in the Mertz Glacier polynya (67°S, 145°E) as cited above.

The dense water formed as a result of sea ice production may accumulate in depressions on the continental shelf and eventually spill over the sill at the shelf break or find paths down canyons to the deep sea (Baines and Condie, 1998). In the process it entrains and mixes with the warmer CDW over the slope and if subsequently dense enough, will sink to the bottom of the slope constituting part of the Antarctic Bottom Water. If not, the downflow will mix with the Southern Ocean at shallower depths.

East Antarctica has been identified as a site for the formation of AABW (Gordon and Tchernia, 1972), but until recently the sources of AABW have been considered relatively weak in regions other than the Weddell and Ross Seas. Bottom water off the Terra Adélie coastline (140°E) for example, was thought to be the product of mixing between Ross Sea bottom water (RSBW) and Weddell Sea bottom water (WSBW; Carmack, 1977). Recent hydrographic observations however have shown a local maximum in bottom CFC-11 concentrations and a minimum in bottom temperatures (Bindoff et al., 1997; Rintoul and Bullister, 1999) that cannot be explained by mixing of these two water masses. This has led to the identification of a local source of bottom water on the continental shelf in the Adélie Depression near the Mertz Glacier that is estimated at three times the volume of RSBW (Rintoul, 1998), or 24% of the total AABW production. This contradicts much earlier results by Carmack (1977) who suggested that less than 0.5% of AABW production could be attributed to this source. The Adélie Land region is unique in that it does not have the broad continental shelves or large ice shelves that are dominant features of the Weddell and Ross Seas, and have long been considered to dominate AABW formation. The two dominant factors leading to the formation of AABW in this region are the large amounts of brine produced in the Mertz Glacier polynya as a result of sea ice production, and the onshore flow of Modified Circumpolar Deep Water (MCDW), which increases shelf water salinity, and potentially supplies sufficient heat to help maintain the polynya (Rintoul, 1998).

Bindoff et al. (2001) developed a model of the circulation in the Mertz Glacier polynya region for estimating the ice production and heat flux for the polynya/glacier system. They suggested two scenarios by which the Highly Modified Circumpolar Deep Water (HMCDW) exchanges with the Ice Shelf Water. The first, based on observational data collected during an experiment in July-August 1999, is that HMCDW is converted to Winter Water (WW) and then to Ice Shelf Water (ISW). The pycnocline separating WW from the underlying High Salinity Shelf Water (HSSW) suggests that these water masses were not mixing during the experiment. The second flow pattern is suggested for the period prior to July, when there is active formation of HSSW, which interacts with the Mertz Glacier to form ISW. This flow pattern represents the process that would create the greatest brine rejection from sea ice formation. The ice production rate and heat flux out of the polynya for the latter flow pattern were estimated at 7.7 cm day^{-1} and 310.7 W m^{-2} respectively, while for the period of the experiment these values were 4.9 cm day^{-1} and 196.3 W m^{-2}. These values compare well with ice formation rates from direct observations of sea ice conditions in the polynya. Lytle et al. (2001) reported an average value of 8 cm day^{-1}, approximately half of which could be attributed to thermodynamics and half of which was the result of deformation. Roberts et al. (2001) reported that close to the coast where the most extreme conditions of the polynya are prevalent, ice growth rates as high as 25 cm day^{-1} are possible.

3. Phytoplankton communities

The unique characteristics of the Antarctic coastal zone significantly impact phytoplankton primary production. The strong high-latitude seasonality in light availability is amplified further by the shading effects of sea-ice, because it persists well beyond the spring equinox over most of the East Antarctic shelf, and often lingers past the summer solstice along the coast (e.g. Worby et al., 1998). Sea-ice further retards production by buffering seasonal warming of surface waters and maintaining temperatures close to freezing, until it is gone. It then enhances production by inducing stratification as it melts, which acts to keep phytoplankton in well-lit surface waters. Sea-ice is itself colonized by algae, both within brine channels and attached to its bottom surface. These algae appear to extend the overall season of primary production, by beginning earlier in spring than phytoplankton production beneath the ice, and acting as a seed population for the spring-summer bloom (e.g. Gibson and Trull, 1999). As noted in the introduction, the ecology of sea-ice algal communities has been recently reviewed. Here we focus on open ocean phytoplankton communities, which account for most of the total annual production. Production models for sea-ice (Arrigo et al., 1998b) and an annual study of export production in Prydz Bay (Gibson et al., 1999) attribute only about 20% of total annual production to sea-ice algae.

The annual sea-ice cycle also contributes to the abundance of phytoplankton macro-nutrients (phosphate, nitrate, and silicate) observed in all Antarctic waters. These nutrients are brought close to the surface by upwelling Circumpolar Deep

Water as part of the broad scale meridional overturning circulation that affects all the Southern Ocean south of the Polar Front. Further mixing of these waters into the surface layer is driven by the rejection of brine during the formation of sea-ice each autumn. This upwelling provides one of the dominant features of Antarctic biogeochemistry – phytoplankton macro-nutrients are abundant everywhere, and virtually never limit primary production. This differs from the seasonal cycles of most coasts, which often include a period of macro-nutrient limitation in summer.

The abundance of macro-nutrients is counter-balanced by limitation of phytoplankton production by the micro-nutrient iron, at least in off-shore waters. The importance of iron has been clearly demonstrated by shipboard incubations, water column measurements, and open ocean iron fertilization experiments (e.g. Martin et al., 1990; de Baar et al., 1995; Boyd et al., 2000; Coale et al., 2004). Much less is known about the importance of iron limitation in Antarctic coastal waters, although shipboard experiments suggest iron limitation does occur within the Ross Sea marginal ice zone (Sedwick and DiTullio, 1997) and low levels of iron have been observed in east Antarctic coastal waters in Adélie Land (Sambrotto et al., 2003). Sources of iron to coastal waters include 1) upwelling of deep waters, and their penetration onto the shelf, which is enhanced by the depth of the shelf; 2) shelf-sediments, which despite the deep shelf can supply surface waters because of the deep winter-time convection and furrowing by ice-berg keels (Barnes and Lien, 1988; Beaman and Harris, 2003); 3) coastal sediments from the break-out of fast ice, and 4) aerosols from distant sources that have accumulated in snow on melting sea-ice, although initial measurements suggest levels are low (Edwards and Sedwick; 2001).

In combination, all these aspects of the interaction of sea-ice with production – light levels, ocean temperatures, ocean stratification, iron supply, provision of 'seed' populations and/or nutrients derived from their breakdown – have fuelled the dominant paradigm regarding summer primary productivity in Antarctic waters: Production follows the retreat of the marginal ice zone southward. Given the complexity of these interactions, it is perhaps not surprising that there is some uncertainty regarding the development of ice-edge blooms. For example, Bathmann et al. (1997) observed no real increase in chlorophyll concentrations in the marginal ice zone (MIZ) of the Weddell Sea in the wake of ice retreat in 1992. For approximately the same time period, near the Antarctic Peninsula, Savidge et al. (1995) also observed no ice-edge bloom and attributed its absence to particularly rapid ice retreat.

Using remotely sensed chlorophyll and ice cover data, Constable et al. (2003) investigated the relationship between ice dynamics and regional productivity over a period of 4 years, 1997–2001. No consistent relationship was found between ice retreat rates and chlorophyll concentrations. Enhanced chlorophyll was found associated with rapid ice retreat in some areas (Ross Sea, Weddell Sea and Prydz Bay) with slower ice retreat in others (Bellingshausen Sea, consistent with Savidge et al. (1995)) and in some areas there was no correlation at all. There was also no simple relationship between productivity in an area and the average annual coverage of sea ice, which might be predicted from an extension of hypotheses developed at single locations (Loeb et al., 1997). Coastal regions are typically the most productive, but movement of this production offshore could occur in gyres. There was also evidence of localised upwelling associated with bathymetric features causing increased open ocean productivity. This phenomenon has previously been observed, notably by Bathmann et al. (1997), Savidge et al. (1995) and Moore and Abbott (2000).

The areas of greatest primary production are generally the coastal waters to the south of the SBACC (Nicol et al., 2000a, Constable et al., 2003). The reduced wind stress, presence of sea ice (and consequent ice melt) and nutrient-rich water provide for high rates of primary and secondary production. The latitudinal range of production appears to be related to sea ice extent but may instead be driven by the northward extension of the Coastal Current. Localized northward extensions of this production occur at the western sides of the major gyral systems as a result of currents and winds. This production may then be retained, transported eastward in the ACC, and gradually depleted. It is in these areas where the relationship between biomass and ice extent breaks down. This process can be enhanced in areas without gyres as a result of a combination of the east-west Coastal Current, wind stress and ice formation causing an enhanced northward movement of ice, thereby extending the latitudinal range of the MIZ, such as in eastern Antarctica (Nicol et al., 2000a).

For the region considered by this chapter, in contrast to other regions such as the Ross, Weddell, Bellingshausen and Scotia Seas, there are few studies of the large-scale distribution of phytoplankton biomass or productivity based on *in situ* sampling. Therefore, in order to understand the relationship between physics and primary productivity, we rely here on satellite chlorophyll to describe large-scale patterns of phytoplankton biomass as a proxy for productivity. Underway measurements have been made during the voyages of the Japanese icebreaker 'Shirase' that each year travels along the coast of East Antarctica between 40° and 150°E (Suzuki and Fukuchi, 1997). In particular, records of surface chlorophyll, salinity and temperature are available for coastal waters spanning 80° – 150°E for seven years. This provides an indication of longitudinal variations in summer conditions across the area surveyed. Although the latitudinal position of the cruise track varies according to the degree of seasonal ice cover, the general feature most noticeable from these data is a peak in chlorophyll a concentration between 80° and 110°E which is evident in five of the seven records presented. This is consistent with the work of Strutton et al. (2000) and Nicol et al. (2000a) who observed enhanced productivity in this region compared to the area between 120° and 150°E.

3.1. Relationships between physical features and primary productivity

The zonal gradient in chlorophyll and productivity observed by Strutton et al. (2000) and Nicol et al. (2000a) in 1996 is also observed in mean summer SeaWiFS chlorophyll images spanning 1997–2003 (Figure 7). Nicol et al. (2000a, b) proposed that elevated productivity was confined to the south of the SBACC. Between 80° and 150°E the SBACC corresponds with the northwards extent of the partially closed gyre (Figure 3) described by Bindoff et al. (2000). It is this physical feature that gives rise to the observed west to east decrease in biomass and productivity, since the SBACC extends several degrees further north between 80° and 120°E, compared to the region between 120° and 150°E. In 1996, this elevated productivity was observed across trophic levels from phytoplankton (Strutton et al., 2000) to krill (Pauly et al., 2000), whales (Theile et al., 2000) and seabirds. A transition in trophic structure was also observed from west to east. While the western portion of the study area was dominated by diatoms and krill, salps (Hosie et al., 2000) and picoplankton (Wright and van den Enden, 2000) dominated in the east, where the nearshore physical conditions more closely resembled the oceanic waters to the north. There was no evidence of macronutrient limitation of photosynthesis anywhere in the study area, but enhanced silicate uptake rates and data obtained from a fast repetition rate fluorometer (FRRF) indicate that the phytoplankton community in the eastern transects may have been iron-deficient (Strutton et al., 2000). Iron deficiency in this general area was confirmed by the SOIREE experiment (Boyd et al., 2000), which took place at 61°S, 140°E in February 1999 (location labelled 'S' in Figure 7).

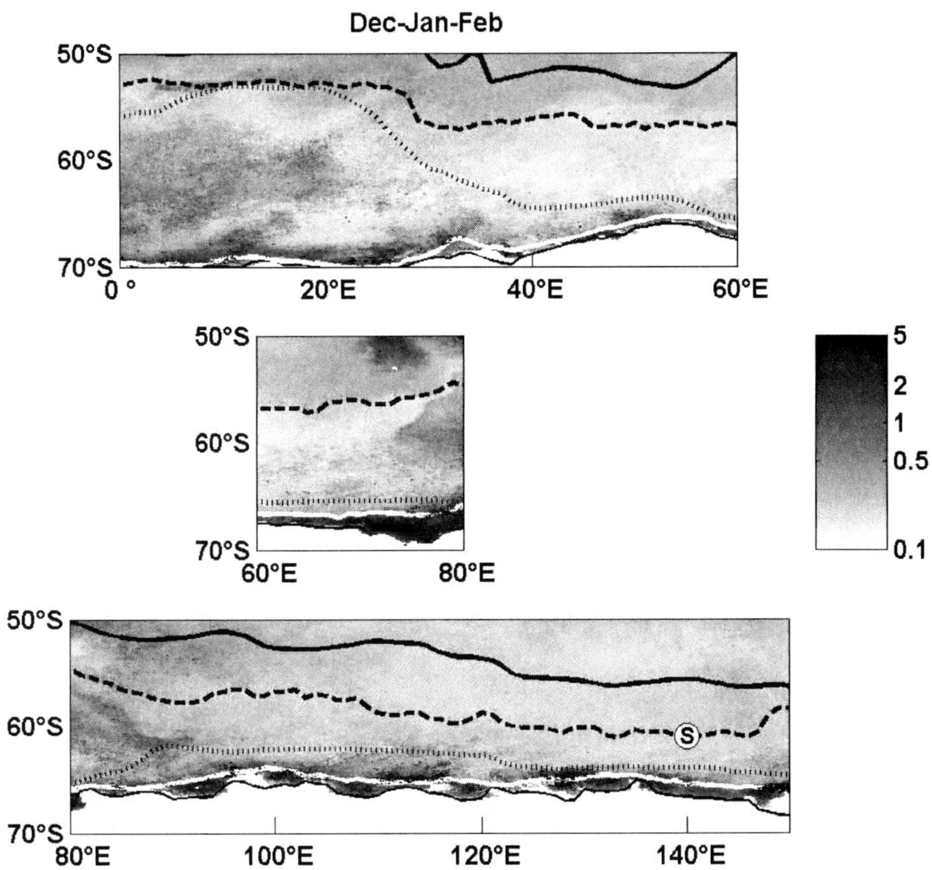

Figure 37.7: Mean summer chlorophyll (Dec-Jan-Feb, SeaWiFS data spanning Dec-1997 to Feb-2003) for the region of interest, split into three subsections: The SE Atlantic/SW Indian sector (top), Prydz Bay (middle) and the region south of Australia (bottom). Also plotted are the 1000m isobath (solid white line), the southern boundary of the ACC (SBACC; dotted line), the maximum extent of winter sea ice (dashed line) and the polar front (solid line). The location of the SOIREE experiment is indicated by 'S'.

As described above, Constable et al. (2003) used time series of ice coverage data from the Special Sensor Microwave/Imager (SSM/I) in conjunction with SeaWiFS chlorophyll to calculate ice retreat rates at weekly time scales. The ice retreat data were grouped into 10° longitude bins around the Antarctic coastline, and compared with mean chlorophyll in the corresponding region. The time period studied was the Austral summers of 1997–98 through 2000–01. Here we take a climatological approach by comparing the latitudinal extent (N-S distance) over which ice advance/retreat occurs (based on SSM/I data from the period 1989 to 1999) with the mean chlorophyll in that area as quantified by SeaWiFS satellite data from the Austral summers of 1997–98 to 2002–03. The data are divided into 1° as opposed to 10° (Constable et al., 2003) longitude bins.

Given that ice retreat at any longitude around the Antarctic coastline takes place over approximately the same period of time (October through February), the north-south extent of sea ice should be an indication of the rate of the retreat. In regions where the north-south extent is large, the retreat should be rapid, and the converse should also be true. Figure 8(a) shows the relationship between the N-S extent of the MIZ, and the mean chlorophyll encompassed by that area. Consistent with Constable et al. (2003), it appears that the relationship between the rate of ice retreat and primary productivity varies depending on location. For the entire coastline, the correlation is positive and significant ($r = 0.178$, $n = 360$), indicating that rapid ice retreat favours enhanced chlorophyll. On closer inspection it seems that this relationship is mainly driven by the strong positive correlation between ice extent and MIZ chlorophyll in the regions we have identified as 'bays and peninsula': Prydz Bay (70°E to 80°E), the Ross Sea (170°E to 150°W) and the Antarctic Peninsula/Weddell Sea (70°W to 0°). Elsewhere the correlation is weak.

Given the importance of iron to phytoplankton production, and the probability that the continental shelf may represent a source of iron to offshore waters through sediment transport, a larger continental shelf area may represent a larger source of iron. Here we investigate this mechanism by comparing the productivity of a region to the extent of the continental shelf. Figure 8(b) shows the correlation between mean chlorophyll south of the SBACC (referred to as coastal chlorophyll) and the N-S extent of the shelf. For the entire Antarctic coast the relationship is significant ($r = 0.294$, $n = 360$), and even more so for the region 80° – 170°E ($r = 0.550$, $n = 90$). These relationships also hold for mean chlorophyll south of the PF (data not shown) and mean chlorophyll in the MIZ (Figure 8(c)). This suggests that a broader continental shelf is associated with greater productivity both on the shelf and further offshore. Chlorophyll concentration at the shelf break (defined as ±2° north and south of the 1000m isobath) is positively correlated with the N-S extent of the shelf for the region 80°–170°E ($r = 0.523$, $n = 90$) but not for the Antarctic coast as a whole (Figure 8(d)). This relationship for east Antarctica probably occurs because, with the exception of the Antarctic Peninsula, this is the region where the SBACC corresponds most closely with the shelf break.

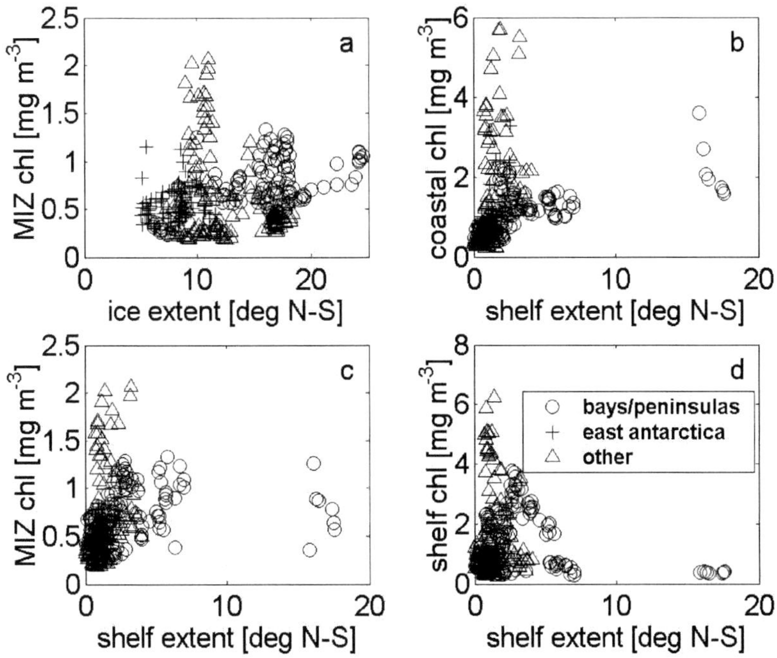

Figure 37.8: Relationship between the area bounded by a given physical feature and the mean chlorophyll within that area. Mean chlorophyll data depicted are the Dec-Jan-Feb average for December 1997 through February 2003 (as for Figure 7). Each data point represents a 1° bin (i.e. n = 360). The data are presented by area, with sub-regions defined at (1) Bays and Peninsulas (Prydz Bay, Ross Sea, Weddell Sea and the Antarctic peninsula, plotted as open circles), (2) East Antarctica (80°E to 170°E, plotted as + symbols) and (3) the remainder of the coast (plotted as open triangles). The four panels depict: (a) Relationship between mean chlorophyll within the MIZ and the north-south extent of the MIZ. (b) Relationship between mean chlorophyll south of the SBACC and the extent of the continental shelf. (c) Relationship between mean chlorophyll within the MIZ and the extent of the continental shelf, and (d) Relationship between mean chlorophyll at the shelf break and the extent of the continental shelf.

4. Distribution, abundance and production of ecological communities

Our understanding of higher trophic levels in the pelagic zone off east Antarctica is improving, but there are a number of areas which remain unsampled and there is a dearth of longer-term studies. Descriptions of the pelagic communities in the South Indian Ocean occurred in the 'Discovery Reports' and because of the geographic scope of these observations they still provide some of the best information on the distribution of species such as salps (Foxton, 1966) and krill (Mackintosh, 1972; Marr, 1962). Analyses of pelagic communities in the Prydz Bay region, using multivariate analytical techniques, was undertaken as part of the BIOMASS (Biological Investigations of Marine Antarctic Systems and Stocks) program (Miller, 1985; Hosie, 1994) and wider studies have extended the information base (Hosie and Cochran, 1994; Hosie et al., 2000; Pakhomov, 1995a, b, 2000). These studies can be divided into those that (a) describe the distribution and relative abundance of species and groups in focused (Miquel, 1991, Hosie, 1994), or wider areas (Hosie et al., 2000); (b) examine regional ecological relationships (Nicol et al., 2000a; Pakhomov, 2000) or long-term variability (Pakhomov, 2000); and (c) those that examine production, or proxies of production such as recruitment or growth (Pakhomov, 1995a, b; Nicol et al., 2000c). Although Antarctic krill has been the focus of many studies there have been sufficient reports on other species such as the pelagic tunicate, *Salpa Thomsoni* (Pakhomov et al., 2002), and of overall community structure (Hosie et al., 2000), to allow more general observations on ecosystem structure to be made. There are now sufficient data that some comparisons can be made between the East Antarctic and the better studied South Atlantic.

Voronina (1968) described the general distribution of total zooplankton biomass in the Southern Ocean, which later studies have confirmed. In the Prydz Bay region four pelagic communities have been identified (Hosie, 1994; Hosie and Cochran, 1994) which correspond to the general latitudinal ocean regimes. These consist of a neritic community located in the southern part of the Prydz Bay continental shelf, dominated by the coastal krill species *Euphausia crystallorophias,* gammarid amphipods and larvae of the fish *Pleuragramma antarcticum.* A second community, located mainly along the continental shelf edge, was characterized by the dominance of the Antarctic krill *E.superba* and also by the paucity of zooplankton. The Oceanic community was prevalent in the region north of the shelf edge to approximately 62° – 63°S, with the major components being copepods, chaetognaths, siphonophores and the euphausiid *Thysanoessa macrura.* To the north of the coastal zone a northern oceanic community was identified and comprised of species generally confined to the Antarctic Circumpolar Current, *Salpa thompsoni,* small euphausiids and hyperiid amphipods. Pakhomov (1993) did not include copepods in his analyses but otherwise described similar species assemblages and distribution patterns in the Prydz Bay region. The pelagic community in the coastal zone in the 80° – 150°E region was structured in a remarkably similar way to that observed in the Prydz Bay region (Hosie et al., 2000) although there were onshore-offshore differences in the extent of the communities which were related to regional environmental differences (Nicol et al., 2000b). An examination of the spatio-temporal variability of the community structure in the 80° – -160°E region further confirmed the relative distribution of the species groups but indicated that their geographic location can vary with physical forcing (Chiba et al., 2001). Water temperature explains a large amount of variation in the community patterns for 80 – 150°E region (Hosie et al., 2000) and in Prydz Bay (Hosie, 1994

Other studies have mainly focussed on the distribution, abundance and, occasionally, the life history of individual species, e.g. salps (Casareto et al., 1986), euphausiids (Terazaki and Wada, 1986; Nicol et al., 2000c) and chaetognaths (Terazaki, 1989). Since krill and salps are the species in this region that have been most intensively studied we will focus on these to examine the distribution and abundance of secondary production. The 'Discovery Reports' provide some of the few historical observations on krill abundance in the South East Indian Ocean (Marr, 1962). Krill in this sector were concentrated between 90° and 120°E and distinctly absent in the 120° – 150°E region. Mackintosh (1973) defined a distinct Kergulen-Gaussberg stock between 85° – 100°E where krill were described as 'plentiful' but in the 120° – 150°E region, he described krill as being confined close to the coast in the East Wind drift which 'here forms a narrower belt near the coast than anywhere else'. He pointed out that krill may be plentiful in this narrow band but the overall quantity in this region may not be great because of its restricted latitudinal distribution.

There have been a number of later studies in restricted parts of this region. Japanese acoustic studies have investigated the distribution of Antarctic krill (*Euphausia superba*) between 115° and 124°E and between 145° and 151°E (Inagake et al., 1985). The distribution of euphausiids between 110° and 150°E was investigated as part of the BIOMASS Program (Terazaki and Wada, 1986) using both nets and acoustics. A single large-scale acoustic survey described krill distribution and abundance between 0° and 150°E (Pauly et al., 2000). The results of this survey agreed with the general distribution pattern suggested by the Discovery Investigations with krill being confined mainly in the Coastal Current which was wider in the 80° – 115° E sector and narrower in the 115° –150°E region.

The relationship between the distribution of krill and the physical environment has been investigated in the Antarctic coastal zone between 15° and 30°E (Miller and Montiero, 1988; Weber and El Sayed, 1985; Hampton, 1985). There was a lack of correlation between krill abundance and chlorophyll *a,* dissolved oxygen, latitude, density, stability, dissolved organics and nitrate, and krill abundance was negatively correlated with phytoplankton biomass. Krill were found to concentrate near the top of the thermocline and were mostly confined to the top 100 m of the water column. Krill densities in this area were considered to be low, however, the survey design and the target strength used in these studies means that it is difficult to compare them to more recent surveys in the 10° – 150°E region. Net-derived estimates of krill density in the 52° – 64°E region (5.2g m^{-2}) are at the lower end of densities obtained by nets from other regions of the South Atlantic (Miller, 1985) though were higher than those estimated for the 0° – 150°E region (0.158 g m^{-2}; Nicol et al., 2000c). In this region, as in areas further to the East, krill were most abundant in the southern strata of the survey, associated with the waters of the Antarctic Coastal Current (Miller, 1986).

Fishery-related studies have described where the Japanese and Soviet krill fishing fleets encountered fishable concentrations of krill and some of the biological characteristics of the krill (Dolzhenkov and Timonin, 1990; Ichii, 1990). The krill fishery in the Indian Ocean Sector has been a summer fishery with all catches being reported between November and May and a peak in mid-January (Nicol and Endo, 1997). The fishery operated along the continental shelf and shelf break throughout the 0° – 150°E region and targeted fishable concentrations which extend a few tens of kilometres in horizontal extent and consist of swarms denser than 100g m^{-2} (Ichii 1990). The Japanese fishery has concentrated in the 120° – 140°E region because the retreat of ice there in December is most rapid. Krill trawling, presumably reflecting the presence of dense krill concentrations, has focussed on the shelf-slope frontal zone. Vertical sections clearly indicate increased trawling activity at 141°E and 147°30'E and in sections from other areas of the Antarctic (Ichii 1990). The fishery is an indicator of the distribution of fishable aggregations of krill but is probably not a particularly useful indicator of the overall distribution of krill because it tends to concentrate where catches are predictable and where the ice conditions are not difficult. Fishing fleets are also not concerned with the overall distribution and abundance of krill, but concentrate on local abundance, which is enough to sustain catches, but may not be representative of the gross abundance over wide geographic areas.

Indications of large-scale salp (*Salpa thomsonii*) abundance in the South East Indian Ocean come from the 'Discovery reports' (Foxton, 1966) and from Japanese studies (Casareto et al. 1986). Despite the paucity of samples for this area, there are indications that salps are more abundant close to the coast in the 120° – 150°E region than in the 80° – 120°E region. Between 60° and 90°E there are indications that salps are to be found mainly north of 62°S and are indicative of offshore, oceanic water in the region (Hosie 1994).

5. Interannual variability

The region between 30° and 120°E was extensively studied as part of Soviet fisheries investigations between 1977 and 1990 (Pakhomov, 2000). The life cycle of krill in this region was similar to that observed in other studies off East Antarctica (Miquel, 1991; Hosie, 1994; Nicol et al., 2000c) and in the Atlantic sector (Siegel, 1986; Siegel et al., 1990), with spawning occurring offshore and spatial segregation of adults, juveniles and larvae. The long time series of studies allowed an examination of interannual variability in krill stocks and their relationship to the biological and physical environment. Previously identified associations between krill variability and either oceanographic or atmospheric processes were tested using a correlation analysis and no meaningful correlations were identified between krill biomass and any other biological or environmental parameter examined. There was, however, a positive cross correlation between krill biomass and the influx of warm deep water into the coastal area during the previous year, which the author attributed to a passive concentration of the juvenile year classes (Pakhomov, 2000). Years with high levels of warm deepwater input from the north to the shelf areas led to large numbers of salps occurring north of the Antarctic Divergence and successful krill spawning. These years were also characterised by low ice coverage and were subsequently followed by intense ice accumulation the following winter promoting successful krill recruitment. These processes occurring in this area were similar to those occurring off the Antarctic Peninsula (Siegel and Loeb, 1995; Siegel et al., 1997).

In the South Atlantic, the abundance of krill has been linked to the annual extent of sea ice (Loeb et al., 1997). This relationship to sea ice extent has also been observed regionally off East Antarctica; sectors with greater sea ice extent harbouring greater concentrations of krill than areas with less extensive annual sea ice (Nicol et al., 2000a). The explanations put forward for these two sets of observations differ. In the South Atlantic the sea ice is viewed as an overwintering habitat for krill and therefore the more sea ice is present, the greater the habitat and the greater the success of the krill population. Off East Antarctica the amount of sea ice and the regional abundance of krill were related to the latitudinal extent of the cooler water, which is the habitat of krill. Neither set of relationships is likely to hold on a circumpolar scale (Constable et al., 2003). Although regional sea ice extent is positively related to the width of the cool Antarctic Coastal Current, the abundance of krill on a circum-Antarctic scale (where data exist) does not follow this general trend (Nicol et al., 2000a). Krill are most abundant in the Scotia Arc and this is the area in which annual sea ice is at its minimum. Although the areas with the greatest extent of annual sea ice are difficult to survey it is unlikely that they are regions with high krill biomass. Interannual variation in production may be driven by variation in the strength of coastal gyres, wind stress and the location of the ACC. Variation in these conditions promotes spatial changes in the relative distributions of the different pelagic communities.

6. Vertebrates

Although vertebrates are undoubtedly key elements of the coastal ecosystem of Antarctica there are few long-term studies in the East Antarctic region, which are able to provide more general ecological insights. We will concentrate on the information from (1) Béchervaise Island (67°35'S, 62°48'E) which has been a part of the Ecosystem Monitoring Program conducted under the auspices of the body that regulates the fisheries of the Southern Ocean, the Commission for the Conservation of Antarctic Marine Living Resources (CCAMLR) (Agnew, 1997) since 1990, and (2) from studies conducted at the French station Dumont d'Urville (66.7°S, 140.0°E). At Béchervaise Island, annual changes in breeding success and other biological parameters of Adélie penguins have been observed. Sometimes these changes have been catastrophic and have amounted to almost complete breeding failure (Irvine et al., 2000). These changes in breeding success have been attributed to changes in food availability, but this does not seem simply linked to sea ice extent.

There do appear to be direct linkages between the movement of sea ice in East Antarctica and the winter migrations of Adélie penguins from Béchervaise Island and other colonies in the Prydz Bay region (Clarke et al., 2003). The gyral ocean circulation between 0° and 80°E moves the sea ice westwards in the inshore areas with northward components in the west, which direct ice out into the eastward flow further offshore. This circulation pattern appears to aid penguins dispersing westwards during early winter when sea ice is less extensive and returning to their colonies during spring when sea ice is at its maximum extent. This study indicates that winter sea ice has effects, which are more complex than those that can be examined merely by reference to the north-south extent.

The effect of sea ice on the biology of individual species is a result of a complex interaction that also involves regional ocean circulation patterns. Consequently regional, seasonal and interannual differences are to be expected. Studies on the effect of sea ice cover on the breeding success of snow petrels near Dumont d'Urville show indications that seasonal sea ice can have both positive and negative effects, highlighting the complex relationship between physical variability and biological response. High levels of winter sea ice resulted in lower numbers of breeding birds but extensive spring sea ice could result in enhanced breeding success and improved fledgling weight (Barbraud and Weimerskirch, 2001a). Emperor penguins at a nearby site showed similarly ambiguous responses to sea ice changes (Barbraud and Weimerskirch, 2001b). Mortality rates of emperor penguins increased when annual sea ice cover was reduced, a phenomenon associated with increased sea surface temperatures. In contrast, emperor penguins hatched fewer eggs when sea ice was extensive. Studies on other species at the same site indicated long-term trends that ranged from significant increases (Adélie Penguins: +3.5% over 14 years), through those registering slight increases (Southern Fulmars: +0.4%) to those indicating significant decreases (Giant Petrels: -3.9%; Micol and Jouventin, 2001).

These results reflect the multiple ecological effects of sea ice; it can impede access to marine food resources but at the same time there appears to be some linkages between biological productivity and the extent of sea ice. Sea ice may also act as a co-variable with some biological measures of production rather than being the causative factor behind changes in productivity which may be oceanographically determined (Nicol et al., 2000a). The complex responses of organisms to sea ice requires that studies on ecological processes in relation to sea ice changes take full account of seasonal effects as well as the underlying causes for observed changes in sea ice cover. Additionally, there are likely to be regional effects, which may determine whether the sea ice cover has a positive or a negative impact on a particular species. Finally, there are certain to be inter-specific differences in the effect that sea ice cover has on survival and breeding success – long term changes in winter sea ice cover may favour some species over others.

Cetacean research in this area has largely been restricted to studies related to commercial whaling. There is a considerable database on the distribution of sightings and catches of whales in IWC Areas IV and V which cover 70° – 130°E and 130°E – 170°W respectively (Mizroch et al., 1985). Whale catches in the 0° – 150°E region reveal that there were certain areas from which above average numbers of whales were taken. For fin, blue and humpback whales these areas were concentrated around the 30°E and 90°E lines of longitude. Records of baleen whales from summer sighting surveys in this area have provided evidence of uneven distribution along this coastline (Kasamatsu et al., 1996). Of the abundant species, humpbacks had a latitudinal peak of abundance between 62° and 66°S, and a longitudinal peak in this area between 80° and 120°E, whereas minkes were distributed much further south (60° – 80°S) and had much less pronounced variation in longitudinal abundance. Other studies of whales in the east Antarctic have related interannual changes in body fat condition of minke whales to krill availability and to sea ice conditions (Ichii et al., 1998). Sea ice extent was thought to have an effect on whale feeding because extensive sea ice excluded minke whales from entering the areas where krill aggregated. Large-scale interdisciplinary research in the 80 – 150°E region indicated a greater abundance of whales in the area with the greatest sea ice extent but that this area coincided with increased primary production, higher krill abundance and with a wider Antarctic Coastal Current (Nicol et al., 2000a). Whales, however, do not utilise the available habitats evenly. Some species such as minkes are found closer to the ice edge whereas species such as humpbacks use the offshore and continental shelf zones, even though they depend on the same species of prey, Antarctic krill (Thiele et al., 2000; Murase et al., 2001).

7. Conclusions

There has been a tendency in the past to underestimate the complexity of the physical and biological systems of the Antarctic. Although this is not unique to the Antarctic region (Kachel et al., 2002) it remains a prevalent viewpoint because of the geographic nature of Antarctica, ringed by concentric current and ice systems, which confer an impression of continental uniformity. Concepts such as a single simple Antarctic ecosystem have given way to the idea of a series of complex interrelated communities, often delineated by oceanographic boundaries or related to particular water masses. East Antarctica has not received the attention that has been focussed on features such at the Antarctic Peninsula, the Ross and Weddell Seas, or some of the subantarctic islands, but recent studies in this region have improved our understanding of both physical and biological processes. The geographic variation in the physical features that we have highlighted in this review is a useful tool for further investigating physical-biological interactions.

The annual advance and retreat of sea ice around Antarctica sets it apart from other coastal zones around the globe. Although the North Pacific can experience extensive winter sea ice that is comparable to many areas of the Antarctic (Schumacher and Stabeno, 1998), generally the extreme seasonal variability observed in the Antarctic is not present in the Arctic. The sea ice around Antarctica has a profound influence on other physical, biological and chemical processes in the Southern Ocean through a complicated series of positive and negative feedback mechanisms. In this review we have highlighted the significance of sea ice for Antarctic Bottom Water (AABW) formation and the importance of this for ventilating the deep ocean and linking the global ocean basins. Persistent coastal polynyas, where sea ice production is greatest, are an integral component of AABW formation, but bathymetry and ocean circulation patterns also play an important role. Recently, Rintoul (1998) identified the Mertz Glacier polynya off the Adélie Land coast as being the second largest producer of AABW around Antarctica. Thus the East Antarctic coastline plays a critical role in global ocean circulation and any changes in the sea ice environment may have a disproportionately large effect on other systems.

The Antarctic sea ice zone is also highly dynamic and this affects the thickness distribution of the ice, through the formation of pressure ridges and open water leads. The distribution of sea ice in turn affects the interaction between the atmosphere and ocean, and also determines the usefulness of sea ice as a habitat for vertebrates. We have discussed contrasting examples that illustrate the potential impacts of sea ice variability on the feeding, survival and breeding success of cetaceans and birds. Sea ice acts as a habitat for some vertebrates, as a barrier to others and as a substrate for many invertebrates and microorganisms. The annual formation and melting of sea ice also govern the vertical stability of the water column and affect the overall productivity of the system. Changes in sea ice extent, concentration and type thus have a wide range of effects on the ecosystems of the Antarctic. There is no doubt that the populations of a number of well-studied species in East Antarctica have experienced considerable change over the last 50 years. Some have increased in number, some have decreased and some may have changed their distributional range. For most species there is insufficient information to even speculate on their recent history.

The East Antarctic coastline can be described as relatively 'linear', reaching gradually further northward from around 70° to 62°S between 0° and 150°E. This shift in the latitude of the coastline is reflected in the extent of sea ice from the coast, which is greatest to the west (0°) and only several hundred kilometres near 150°E. This has implications for primary productivity, since sea ice cover and ocean circulation have a significant impact on phytoplankton community dynamics. This has been particularly well illustrated by the large-scale observational programs we have discussed (Suzuki and Fukuchi, 1997; Strutton et al., 2000; Nicol et al., 2000a). Our simple sectoral examination of the relationships between the physical features of the Southern Ocean and biological productivity indicates that the relationships between these variables are not simple and that there are considerable regional variations. There are regions of the coastal zone around Antarctica that can be treated as relatively homogeneous, however, inter-regional comparisons should be treated with caution and, in particular, comparisons between the major embayments and the Antarctic Peninsula and other regions should be avoided.

There is some evidence that there have been fundamental changes in the distribution of water masses in the Southern Ocean and this may be the result of general oceanic warming (Wong et al., 1999). Other water mass changes are less well described. There is presently no firm evidence to indicate whether the extent, concentration or thickness of Antarctic sea ice is changing at time scales of the order of decades. A number of studies based on satellite passive microwave data are contradictory in their assessment of ice edge trends, and there is also disagreement in the literature with regard to recent studies based on records from whaling ships and from other proxies. There is however, agreement on the important role sea ice plays in the climate system and the need to monitor the Antarctic sea ice zone for change. The extreme natural seasonal variability in sea ice growth and decay makes this a particularly difficult task, however satellite data time series are approaching the necessary lengths for reliable trend analyses. The same is true for the expanding *in situ* and remotely sensed databases for physical, chemical and biological oceanographic parameters. These data, in conjunction with modelling studies, will undoubtedly continue to advance our understanding of both regional processes and long-term trends.

Bibliography

Ackley, S. F., P. Wadhams, J. C. Comiso and A. P. Worby, 2003. Decadal decrease of Antarctic sea ice extent inferred from whaling records revisited on the basis of historical and modern sea ice records. *Polar Research,* **22**(1), 19–25.

Adolphs, U. and G. Wendler, 1995. A pilot study on the interactions between katabatic winds and polynyas at the Adélie coast, eastern Antarctica. *Antarctic Science,* **7**(3), 307–314.

Agnew, D. J., 1997. Review: the CCAMLR ecosystem monitoring programme. *Antarctic Science,* **9**(3), 235–242.

Allison, I., 1989. The East Antarctic sea ice zone: Ice characteristics and drift. *GeoJournal,* **18**(1), 103–115.

Allison, I., 2003. The AMISOR project: ice shelf dynamics and ice-ocean interaction of the Amery Ice Shelf. *FRISP Report,* **14,** 9 pp.

Anderson, R.F., Z. Chase, M.Q. Fleischer, and J. Sachs, 2002. The Southern Ocean's biological pump during the Last Glacial Maximum, *Deep-Sea Res. II,* **49,** 1909–1938.

Arrigo, K. R., D. Worthen, A. Schnell and M. P. Lizotte, 1998a. Primary production in Southern Ocean waters. *J Geophys. Res.,* **103**(C8), 15,587–15,600.

Arrigo, K., D.L. Worthen, P. Dixon, and M.P. Lizotte, 1998b. Primary productivity of near surface communities within Antarctic pack ice. In *Antarctic Sea Ice Biological Processes, interactions, and variability,* M. P. Lizotte, and K. R. Arrigo, eds. Antarctic Research Series, Vol. 73. p. 2–American Geophysical Union, Washington, D.C., pp. 23–43.

Arrigo, K. R. and van Dijken, G. L., 2003. Phytoplankton dynamics within 37 Antarctic coastal polynya systems, *J. Geophys. Res.,* **108**(C8), 27–1–27–18.

Ball, F. K., 1957. The katabatic winds of Adélie Land and King George V Land. *Tellus,* **9,** 201–208.

Baines, P. G. and S. Condie, 1998. Observations and modelling of Antarctic downslope flows, 1998. In *Ocean, Ice, and Atmosphere: Interactions at the Antarctic Continental Margin,* S. S. Jacobs and R. F.

Weiss, eds. Antarctic Research Series, Vol. 75. American Geophysical Union, Washington, D.C., pp. 29–50.

Barbraud, C. and H. Weimerskirch, 2001a. Contrasting effects of the extent of sea-ice on the breeding performance of an Antarctic top predator, the snow petrel *Pagodroma nivea*. *J. Avian Biology* **32,** 297–302.

Barbraud, C. and H. Weimerskirch, 2001b. Emperor penguins and climate change. *Nature,* **411,** 183–186.

Barnes, P. W., and R. Lien, 1988. Icebergs rework shelf sediments to 500 m off Antarctica, *Geology,* **16,** 1130–1133.

Bathmann, U. V., R. Scharek, C. Klaas, C. D. Dubischar, and V. Smetacek, 1997. Spring development of phytoplankton biomass and composition in major water masses of the Atlantic sector of the Southern Ocean, *Deep-Sea Res. II,* **44**(1–2), 51–67.

Beaman, R. J., and P. T. Harris, 2003. Seafloor morphology and acoustic facies of the George V Land shelf, *Deep-Sea Res. II,* **50,** 1343–1356.

Bindoff, N. L., M. J. Warner and S. Nicol, 1997. The Antarctic margin experiment, 80–150°E. *International World Ocean Circulation Experiment (WOCE) Newsletter,* **26,** 36–38.

Bindoff, N. L., M. A. Rosenberg and M. J. Warner, 2000. On the circulation of the waters over the Antarctic continental slope and rise between 80 to150°E. *Deep-Sea Res. II,* **47**(12–13), 2299–2326.

Bindoff, N. L., G. D. Williams and I. Allison, 2001. Sea-ice growth and water-mass modification in the Mertz Glacier Polynya, East Antarctica, during winter. *Ann. Glaciol.,* **33,** 399–406.

Boyd, P. W., Watson, A. J., Law, C. S., Abraham, E. R., Trull, T., Murdoch, R., Bakker, D. C. E., Bowie, A. R., Buesseler, K. O., Chang, H., Charette, M., Croot, P., Downing, K., Frew, R., Gall, M., Hadfield, M., Hall, J., Harvey, M., Jameson, G., LaRoche, J., Liddicoat, M., Ling, R., Maldonado, M. T., Michael McKay, R. M., Nodder, S., Pickmere, S., Pridmore, R., Rintoul, S., Safi, K., Sutton, P., Strzepek, R., Tanneberger, K., Turner, S., Waite, A. and Zeldis, J., 2000. A mesoscale phytoplankton bloom in the polar Southern Ocean stimulated by iron fertilization. *Nature,* **407,** 695–702.

Brierley, A. S. and D. N. Thomas, 2002. Ecology of Southern Ocean pack ice. *Advances in Marine Biology,* **43,** 171–276.

Bjørgo, E., O. M. Johannessen, and M. W. Miles, 1997. Analysis of merged SSMR-SSMI time series of Arctic and Antarctic sea ice parameters 1978–1995, *Geophys. Res. Lett.,* **24**(4), 413–416.

Carmack, E. C., 1977. Water characteristics of the Southern Ocean south of the Polar Front. In *A Voyage of Discovery,* M. Angel, ed. G. Deacon 70th Anniversary Volume, *Deep Sea Res.,* (Suppl.). Pergamon Press, Elmsford, N.Y. pp. 15–42.

Casareto, B. E., T. Nemoto and T. Hoshiai, 1986. Salps of the Southern Ocean (Australian Sector) during the 1983–84 summer, with special reference to the species *Salpa thompsoni,* Foxton 1961. *Proceedings of the seventh symposium on polar biology - Memoirs Special Issue* (40), T. Nemoto and Y. Naito, eds. Tokyo, National Institute of Polar Research, pp. 221–239.

Cavalieri, D. J., P. Gloersen, C. L. Parkinson, J. C. Comiso, and H. J. Zwally, 1997. Observed hemispheric asymmetry in global sea ice changes, *Science,* **278,** 1104–1106.

Cavalieri, D. J., and S. Martin, 1985. A passive microwave study of polynyas along the Antarctic Wilkes Land coast. In *Oceanology of the Antarctic continental shelf,* S. S. Jacobs, ed. Antarctic Research Series, Vol. 43. American Geophysical Union, Washington, D.C., pp. 227–252.

Chiba, S., T. Ishimaru G. W. Hosie and M. Fukuchi, 2001. Spatio-temporal variability of zooplankton community structure off east Antarctica (90 to 160°E). *Mar. Ecol. Prog. Ser.,* **216,** 95–108.

Clarke, J., K. Kerry, C. Fowler, R. Lawless, R. Eberhard and R. Murphy, 2003. Post fledging and winter migration of Adélie penguins *Pygoscelis adeliae* in the Mawson region of east Antarctica. *Mar. Ecol. Prog. Ser.,* **248,** 267–278.

Coale, K. H., Johnson, K.S., Chavez, F. P., Buesseler, K. O., Barber, R. T., Brzezinski, M. A., Cochlan, W. P., Millero, F. J., Falkowski, P. G., Bauer, J. E., Wanninkhof, R. H., Kudela, R. M., Altabet, M. A., Hales, B. E., Takahashi, T., Landry, M. R., Bidigare, R. R., Wang, X., Chase, Z., Strutton, P. G.,

Friederich, G. E., Gorbunov, M. Y., Lance, V. P., Hilting, A. K., Hiscock, M. R., Demarest, M., Hiscock, W. T., Sullivan, K. F., Tanner, S. J., Gordon, R. M., Hunter, C. N., Elrod, V. A., Fitzwater, S. E., Jones, J. L., Tozzi, S., Koblizek, M., Roberts, A. E., Herndon, J., Brewster, J., Ladizinsky, N., Smith, G., Cooper, D., Timothy, D., Brown, S. L., Selph, K. E., Sheridan, C. C., Twining, B. S. and Johnson, Z. I., 2004. Southern Ocean iron enrichment experiment: Carbon cycling in high- and low-Si waters. *Science,* **304,** 408–414.

Comiso, J.C., 2003. Large scale characteristics and variability of the global sea ice cover. In *Sea Ice - an introduction to its physics, biology, chemistry, and geology,* Thomas, D. and G. Dieckmann, eds. Blackwell Scientific Ltd., Oxford, UK, pp. 112–142.

Comiso, J. C. and A. L. Gordon, 1987. Recurring polynyas over the Cosmonaut Sea and the Maud Rise. *J. Geophys. Res.,* **92**(C3), 2819–2833.

Comiso, J. C. and A. L. Gordon, 1996. Cosmonaut polynya in the Southern Ocean: structure and variability. *J. Geophys. Res.,* **101**(C8), 18,297–18,313.

Comiso, J. C., C. R. McClain, C. W. Sullivan, J. P. Ryan, and C. L. Leonard, 1993. Coastal zone color scanner pigment concentrations in the Southern Ocean and relationships to geophysical surface features, *J. Geophys. Res.,* **98,** 2419–2451.

Constable, A. J., S. Nicol and P. G. Strutton, 2003. Southern Ocean productivity in relation to spatial and temporal variation in the physical environment. *J. Geophys. Res.,* **108**(C4), 6–1 - 6–21.

Curran, M. A. J., T. D. van Omen, V. I. Morgan, K. L. Philips and A. S. Palmer, 2003. Ice core evidence for Antarctic sea ice decline since the 1950s. *Science,* **302,** 1203– - 1206.

de la Mare, W. K., 1997. Abrupt mid-twentieth century decline in Antarctic sea ice extent from whaling records. *Nature,* **389,** 57–60.

Deacon, G. E. R., 1937. The hydrology of the Southern Ocean. *Discovery Rep.* **15,** 3–152.

Deacon, G. E. R., 1982. Physical and biological zonation in the Southern Ocean. *Deep-Sea Res.,* **29**(1A), 1–15.

de Baar, H. J. W., J. T. M. de Jong, D. C. E. Bakker, B. M. Loscher, U. V. Bathmann, and V. Smetacek, 1995. Importance of iron for phytoplankton blooms and carbon dioxide drawdown in the Southern Ocean, *Nature,* **373,** 412–415.

Dolzhenkov, V. N. and V. P. Timonin, 1990. Krill (*Euphausia superba Dana*) resources and distribution in the Wilkes Land area in the seasons 1986 to 1989. *CCAMLR Selected Scientific Papers, 1990* (SC-CAMLR-SSP/7). CCAMLR, Hobart, Australia, 149–162.

Edwards R. L., Sedwick P. N., 2001. Iron in East Antarctic snow: Implications for atmospheric iron deposition and algal production in Antarctic waters. *Geophys. Res. Lett.,* **28,** 3907–3910.

Fahrbach. E., R. G. Peterson, G. Rohardt, P. Schlosser and R. Bayer, 1994. Suppression of bottom water formation in the southeastern Weddell Sea, *Deep-Sea Res.,* **41,** 389–411.

Francois, R. F., M. A. Altabet, E.-F. Yu, D. M. Sigman, M. P. Bacon, M. Franck, G. Bohrman, G. Bareille, and L. D. Labeyrie, 1997. Water column stratification in the Southern Ocean contributed to the lowering of glacial atmospheric CO_2, *Nature,* **289,** 929–935.

Foxton, P., 1966. The distribution and life history of *Salpa thomsoni* Foxton with observations on a related species, *Salpa gerlachei. Discovery Rep.,* **34,** 1–116.

Gibson, J. A. E. and T. W. Trull, 1999. Annual cycle of $f\,CO_2$ under sea-ice and in open water in Prydz Bay, East Antarctica. *Marine Chemistry,* **66,** 187–200.

Gibson, J. A. E., T. W. Trull, P. D. Nichols, R. E. Summons and A. McMinn, 1999. Sedimentation of 13C-rich organic matter from Antarctic sea-ice algae: A potential indicator of past sea-ice extent. *Geology,* **27**(4), 331–334.

Gloersen, P., W. J. Campbell, D. J. Cavalieri, J. C. Comiso, C. L. Parkinson and H. J. Zwally, 1992. *Arctic and Antarctic sea ice 1978–1987: Satellite passive-microwave observations and analysis,* NASA SP-511, National Aeronautics and Space Administration, Washington, D.C., USA, 290 pp.

Gordon, A. L., and P. Tchernia, 1972. Waters of the continental margin off Adélie coast, Antarctica. In *Antarctic Oceanology II: The Australian-New Zealand Sector,* D. E. Hayes, ed. Antarctic Research Series, Vol. 19. American Geophysical Union, Washington, D.C., pp. 59–69.

Grigor'yev, Y. A., 1967. Circulation of the surface waters in Prydz Bay. *Soviet Antarctic Expeditions,* **7,** 74–76.

Hampton, I., 1985. Abundance, distribution and behaviour of *Euphausia superba* in the Southern Ocean between 15 and 30°E during FIBEX. In *Antarctic Nutrient Cycles and Food Webs,* W. R. Siegfried, P. R. Condy and R. M. Laws, eds. Berlin Heidelberg, Springer-Verlag, pp. 294–303.

Heil, P. and I. Allison, 1999. The pattern and variability of Antarctic sea-ice drift in the Indian Ocean and western Pacific sectors. *J. Geophys. Res.,* **104**(C7), 15,789–15,802.

Heywood, K. J., M. D. Sparrow, J. Brown and R. R. Dickson, 1999. Frontal structure and Antarctic bottom water flow through the Princess Elizabeth Trough, Antarctica. *Deep-Sea Res.,* **46,** 1181–1200.

Hofmann, E. and J. M. Klinck, 1998. Hydrography and circulation of the Antarctic Continental Shelf: 150°E to the Greenwich Meridian. Coastal segment (32,P). In *The Sea,* Vol. 11, A. R. Robinson and K. H. Brink, eds. John Wiley & Sons Inc., New York, N.Y., pp. 997–1042.

Hosie, G. W., 1994. The macrozooplankton communities of the Prydz Bay region, Antarctica. In *Southern Ocean Ecology, the BIOMASS perspective,* S. Z. El-Sayed, ed. Cambridge University Press, pp. 93–123.

Hosie, G. W. and T. G. Cochran, 1994. Mesoscale distribution patterns of macrozooplankton communities in Prydz Bay Antarctica - January to February 1991. *Mar. Ecol. Prog. Ser.,* **106,** 21–39.

Hosie, G. W., M. B. Schultz, J. A. Kitchener, T. G. Cochran and K. Richards, 2000. Macrozooplankton community structure off East Antarctic (80–150°E) during the austral summer of 1995/96. *Deep Sea Res. II,* **47,** 2437–2463.

Ichii, T., 1990. Distribution of Antarctic krill concentrations exploited by Japanese krill trawlers and Minke whales. *Proceedings of the National Institute of Polar Research Symposium on Polar Biology,* **3,** 36–56.

Ichii, T., N. Shinohara,Y. Fujise, S.Nishiwaki and K. Matsuoka, 1998. Interannual changes in body fat condition index of minke whales in the Antarctic. *Mar. Ecol. Prog. Ser,* **175,** 1–12.

Inagake, D., N. Matsuura and Y. Kurita, 1985. Stock and quantitative distribution of the Antarctic krill (*Euphausia superba* Dana) in the Antarctic Ocean South of Australia in January and February 1984. *Transactions of the Tokyo University of Fisheries,* **6,** 139–147.

Irvine, L., J. R. Clarke and K. R. Kerry, 2000. Poor breeding success of the Adélie penguin at Béchervaise Island in the 1998/99 season. *CCAMLR Science,* **7,** 151–167.

Jacobs, S. S., 1989. Marine controls on modern sedimentation on the Antarctic continental shelf. *Marine Geology,* **85,** 121–153.

Jacobs, S. S. and R. F. Weiss, 1998. *Ocean Ice and Atmosphere.* Antarctic Research Series, Vol. 75, American Geophysical Union, Washington D.C., 380 pp.

Jeffries, M. O., R. A. Shaw, K. Morris, A. L. Veazy, and H. R. Krouse, 1994. Crystal structure, stable isotopes ($\delta^{18}O$), and development of sea ice in the Ross, Amundsen, and Bellingshausen seas, Antarctica, *J. Geophys. Res.,* **99**(C1), 985–995.

Jeffries, M., 1998. Antarctic Sea Ice: Physical Processes, Interactions and Variability. Antarctic Research Series, Vol. 74. American Geophysical Union, Washington D.C., 407 pp.

Kasamatsu, F., G. G. Joyce, P. Ensor and J. Mermoz, 1996. Current occurrence of baleen whales in Antarctic waters. *Reports of the International Whaling Commission,* **46,** 293–304.

Kachel, N. B., Hunt, G., S. A. Salo, J. D. Schumacher, P. J. Stabeno and T. E. Whitledge, 2002. Characteristics of the Inner Front of the Southeastern Bering Sea. *Deep Sea Res. II, Topical Studies in Oceanogr.,* **49,** 5889–-5909.

Knapp, W. W., 1969. A satellite study of large stationary polynyas in Antarctic coastal waters. *PhD thesis,* University of Wisconsin, 118 pp.

Knox, G.A., 1994. *The Biology of the Southern Ocean.* Cambridge University Press. 444 pp.

Kumar, N., R. F. Anderson, R. A. Mortlock, P. N. Froelich, P. Kubik, B. Dittrich-Hannen, and M. Suter, 1995. Increased biological productivity and export production in the glacial Southern Ocean, *Nature,* **378,** 675–680.

Le Fevre, J., L. Legendre, and R. B. Rivkin, 1998. Fluxes of biogenic carbon in the Southern Ocean: roles of large microphagous zooplankton, *J. Marine Systems,* **17,** 325–345.

Lange, M. A. and H. Eicken, 1991. Textural characteristics of sea ice and the major mechanisms of ice growth in the Weddell Sea, *Ann. Glaciol.,* **15,** 210–215.

Leont'yev, Y. B., 1973. Ice conditions in Prydz Bay (December 1971–February 1972). *Soviet Antarctic Expeditions,* **8,** 557–558.

Lizotte, M. P. and K. R. Arrigo, 1998. Antarctic sea ice: Biological processes, interactions and variability. Antarctic Research Series, Vol. 73. American Geophysical Union, Washington D.C., 198 pp.

Loeb, V., V. Siegel, O. Holm-Hansen, R. Hewitt, W. Fraser, W. Trivelpiece and S. Trivelpiece. 1997. Effects of sea-ice extent and salp or krill dominance on the Antarctic food web. *Nature,* **387,** 897–900.

Longhurst, A. R., 1998. *Ecological geography of the sea.* London Academic Press, 398 pp.

Lytle, V. I., A. P Worby, R. Massom, M. J. Paget, I. Allison, X. Wu and A. Roberts, 2001. Ice formation in the Mertz Glacier polynya, East Antarctica, during winter. *Ann. Glaciol.,* **33,** 368–372.

Mackintosh, N. A., 1972. Life cycle of Antarctic krill in relation to ice and water conditions. *Discovery Rep.,* **36,** 1–94.

Mackintosh, N. A. 1973. Distribution of post-larval krill in the Antarctic. *Discovery Rep.,* **36,** 95–156.

Marr, J. W. S., 1962. The natural history and geography of the Antarctic krill (*Euphausia superba* Dana), *Discovery Rep.,* **32,** 33–464.

Martin, J. H., S. E. Fitzwater, and R. M. Gordon, 1990. Iron deficiency limits growth in Antarctic waters, *Global Biogeochemical Cycles,* **4,** 5–12.

Massom, R. A., P. T. Harris, K. Michael, and M. J. Potter, 1998. The distribution and formative processes of latent heat polynyas in East Antarctica. *Ann. Glaciol.,* **27,** 420–426.

Micol, T. and P. Jouventin, 2001. Long-term population trends in seven Antarctic seabirds at Pointe Géologie (Terre Adélie). Human impact compared with environmental change. *Polar Biol.,* **24,** 175–185.

Miller, D. G. M., 1985. The South African SIBEX-I Cruise to Prydz Bay region, 1984. *South African J. Antarct. Res.,* **15,** 33–41.

Miller, D. G. M., and P. M. S. Monteiro, 1988. Variability in the physical and biotic environment of the Antarctic krill (*Euphausia superba* Dana), south of Africa: Some results and a conceptual appraisal of important interactions. In *Antarctic Ocean and Resources Variability,* D. Sahrhage, ed. Springer-Verlag, Berlin Heidelberg, pp. 245–257.

Miquel, J. C., 1991. Distribution and abundance of post-larval krill (*Euphausia superba* Dana) near Prydz Bay in summer with reference to environmental conditions. *Antarctic Science,* **3**(3), 279–292.

Mizroch, S. A., D. W. Rice, J. L. Bengtson, J. L., S. W. Larson, 1985. Preliminary atlas of Balaenopterid whale distribution in the Southern Ocean based on pelagic catch data. *CCAMLR Selected Scientific Papers,* **2,** 113–193.

Moore, J. K., M. R. Abbott, J. G. Richman, W. O. Smith, T. J. Cowles, K. H. Coale, W. D. Gardner, and R. T. Barber, 1999. SeaWiFS satellite ocean color data from the Southern Ocean. *Geophys. Res. Lett.,* **26,** 1465–1468.

Moore, J. K. and Abbott, M. R., 2000. Phytoplankton chlorophyll distributions and primary production in the Southern Ocean. *J. Geophys. Res.,***105,** 28,709–28,722.

Moore, J. K., M. R. Abbott, J. G. Richman, and D. M. Nelson, 2000. The Southern Ocean at the last glacial maximum: A strong sink for atmospheric carbon dioxide, *Global Biogeochemical Cycles,* **14**(1), 455–476.

Murase, H., K. Matsuoka, T. Ichii and S. Nishiwaki, 2001. Relationship between the distribution of euphausiids and baleen whales in the Antarctic (35°E - 145°W). *Polar Biology,* **25,** 135–145.

Nelson, D. M., R. F. Anderson, R. T. Barber, M. A. Brzezinski, K. O. Buesseler, Z. Chase, R. W. Collier, M.-L. Dickson, R. Francois, M. R. Hiscock, S. Honjo, J. Marra, W. R. Martin, R. N. Sambrotto, F. L. Sayles, and D. E. Sigmon, 2002. Vertical budgets for organic carbon and biogenic silica in the Pacific sector of the Southern Ocean, 1996–1998, *Deep-Sea Res. II,* **49,** 1645–1674.

Nicol, S. and Y. Endo, 1997. *Krill fisheries of the world.* Rome, FAO Fisheries Technical Paper 287, 100 pp.

Nicol, S., T. Pauly, N. Bindoff, S. Wright, D. Thiele, G. Hosie, P. Strutton and E. Woehler, 2000a. Ocean circulation off East Antarctica affects ecosystem structure and sea-ice extent. *Nature,* **406,** 504–507.

Nicol, S., T. Pauly, N. L. Bindoff and P. G. Strutton, 2000b. BROKE: A biological/ oceanographic survey of the waters off East Antarctica (80–150°E) carried out in January-March 1996. *Deep-Sea Res. II,* **47,** 2281–2298.

Nicol, S., J. Kitchener, R. King, G. W. Hosie and W. K. de la Mare, 2000c. Population structure and condition of Antarctic krill (*Euphausia superba*) off East Antarctica (80–150°E) during the Austral summer of 1995/1996. *Deep Sea Res. II,* **47,** 2489–2517.

Nunes Vaz, R., and G. Lennon. 1996. Physical oceanography of the Prydz Bay region. *Deep-Sea Res.,* **43,** 603–641.

Orsi, A. H., T. Whitworth III, and W. D. J. Nowlin, 1995. On the meridional extent and fronts of the Antarctic Circumpolar Current. *Deep-Sea Res.,* **42,** 641–673.

Pakhomov, E. A., 1993. The faunistic complexes of macroplankton in the Cooperation Sea (Antarctica). *Antarctica,* **32,** 94–110 (in Russian).

Pakhomov, E. A., 1995a. Demographic studies of Antarctic krill (*Euphausia superba*) in the Cooperation and Cosmonaut Seas (Indian sector of the Southern Ocean). *Mar. Ecol. Prog. Ser.,* **119**(1–3), 45–61.

Pakhomov, E. A., 1995b. Natural age-dependent mortality rates of Antarctic krill, *Euphausia superba* Dana, in the Indian sector of the Southern Ocean. *Polar Biol.,* **15,** 69–71.

Pakhomov, E. A., 2000. Demography and life cycle of Antarctic krill, *Euphausia superba* in the Indian sector of the Southern Ocean: long-term comparison between coastal and open-ocean regions. *Can. J. Fish. Aq. Sci.,* **57,** (Supplement S3), 68–90.

Pakhomov, E. A., P. W. Froneman and R. Perissinotto, 2002. Salp/krill interactions in the Southern Ocean: Spatial segregation and implications for the carbon flux. *Deep Sea Res. II,* **49,** 1881–1907.

Parish, T. R., and D. H. Bromwich, 1987. The surface windfield over the Antarctic ice sheets. *Nature,* **328**(6125), 51–54.

Parkinson, C. L., 1990. Search for the little ice age in Southern Ocean sea ice records. *Ann. Glaciol.,* **14,** 221–225.

Pauly, T., S. Nicol, and I. Higginbottom, 2000. Distribution and abundance of Antarctic krill (*Euphausia superba*) off East Antarctica (80–150°E) during the Austral summer of 1996, *Deep-Sea Res. II,* **47**(12–13), 2465–2488.

Prézelin, B. B., E. E. Hofmann, C. Mengelt, and J. M. Klinck, 2000. "The linkage between Upper Circumpolar Deep Water (UCDW) and phytoplankton assemblages on the west Antarctic Peninsula continental shelf. *J. Marine Res.,* **58,** 165--202.

Rintoul, S. R., 1998. On the origin and influence of Adélie Land bottom water. In *Ocean, Ice, and Atmosphere: Interactions at the Antarctic Continental Margin,* S. Jacobs and R. Weiss, eds. Antarctic Research Series, Vol. 75. American Geophysical Union, Washington, D.C., pp. 151–171.

Rintoul. S. R. and J. L. Bullister, 1999. A late winter hydrographic section from Tasmania to Antarctica. *Deep-Sea Res.,* **46,** 1417–1454.

Rintoul, S. R, C. Hughes, D. Olbers, 2001. The Antarctic Circumpolar Current System, In *Oceans, Circulation and Climate,* G. Siedler, J. Church and J. Gould, eds. Academic Press, San Diego, pp. 271–310.

Rodman, M. and A. Gordon, 1982. Southern Ocean bottom water of the Australian–New Zealand sector. *J. Geophys. Res.,* **87,** 5771–5778.

Roberts, A., I. Allison and V. I. Lytle, 2001. Sensible- and latent-heat-flux estimates over the Mertz Glacier polynya, East Antarctica, from in-flight measurements, *Ann. Glaciol.,* **33,** 377–384.

Sambrotto, R. N., A. Matsuda, R. Vaillancourt, M. Brown, C. Langdon, S. S. Jacobs, and C. Measures, 2003. Summer plankton production and nutrient consumption patterns in the Mertz Glacier Region of East Antarctica, *Deep-Sea Res. II,* **50,** 1393–1414.

Savidge, G., D. Harbour, L. C. Gilpin and P. W. Boyd, 1995. Phytoplankton distributions and production in the Bellinghausen Sea, Austral spring 1992, *Deep-Sea Res. II,* **42**(4–5), 1201–1224.

Schumacher, J. D., and P. J. Stabeno,1998. Continental shelf of the Bering Sea. In *The Sea, Vol. 11* John Wiley & Sons, Inc., New York, N.Y., pp. 789–822.

Sedwick P. N., and G. R. DiTullio, 1997. Regulation of algal blooms in Antarctic shelf waters by the release of iron from melting sea ice. *Geophys. Res. Lett.,* **24,** 2515–2518.

Siegel, V., 1986. Structure and composition of the Antarctic krill stock in the Bransfield Strait (Antarctic Peninsula) during the Second International BIOMASS Experiment (SIBEX). *Archiv fur Fischereiwiss,* **37**(Beih 1), 51–72.

Siegel, V. and V. Loeb, 1995. Recruitment of Antarctic krill *Euphausia superba* and possible causes for its variability. *Mar. Ecol. Prog. Ser.,* **123,** 45–56.

Siegel, V., B. Bergström, J. O. Strömberg and P. O. Schalk, 1990. Distribution, size frequencies and maturity stages of krill, *Euphausia superba,* in relation to the sea-ice in the northern Weddell Sea. *Polar Biol.,* **10,** 549–557.

Siegel, V., W. K. de la Mare, and V. Loeb, 1997. Long-term monitoring of krill recruitment and abundance indices in the Elephant Island area (Antarctic Peninsula). *CCAMLR Science,* **4,** 19–35.

Smith, S. D., R. D. Muench, and C. H. Pease, 1990. Polynyas and leads: an overview of physical processes and environment. *J. Geophys. Res.,* **95**(C6), 9461--9479.

Smith, N., and P. Treguer, 1994. Physical and chemical oceanography in the vicinity of Prydz Bay, Antarctica. In *Southern Ocean Ecology: the BIOMASS Perspective,* S. Z. El Sayed, ed. Cambridge University Press, pp. 25–43.

Smith, N. R., D. Zhao Qian, K. R. Kerry, and S. Wright, 1984. Water masses and circulation in the region of Prydz Bay, Antarctica. *Deep-Sea Res.,* **31,** 1121–1147.

Speer, K. and A. Forbes, 1994. A deep western boundary current in the South Indian Basin. *Deep-Sea Res.,* **41**(9), 1289–1303.

Strutton, P. G., B. F. Griffiths, R. L .Waters, S. W. Wright and N. L. Bindoff, 2000. Primary productivity in the waters off East Antarctica (80–150°E) January to March 1996. *Deep-Sea Res. II,* **47**(12–13), 2327–2362.

Suzuki, T. and M. Fukuchi, 1997. Chlorophyll *a* concentration measured with a continuous water monitoring system during the cruise to Syowa Station, Antarctica, JARE-27 (1985/86) to JARE-35 (1993/94). Center for Antarctic Environment Monitoring, National Institute of Polar Research, Tokyo, Japan, 1–60.

Tchernia, P. and P. F. Jeannin, 1984. Circulation in Antarctic waters as revealed by iceberg tracks, 1972–1983. *Polar Rec.,* **22,** 263–269.

Terazaki, W., 1989. Distribution of chaetognaths in the Australian sector of the Southern Ocean during the BIOMASS SIBEX cruise (KH-83–4). *Proceedings of the NIPR Symposium on Polar Biology,* **2,** 51–60.

Terazaki, M. and M. Wada, 1986. Euphausiids collected from the Australian Sector of the Southern Ocean during the BIOMASS SIBEX cruise (KH-83–4). *Memoirs of the National Institute of Polar Research, Spec. Issue 40,* 97–109.

Thiele, D., E. T. Chester and P. Gill, 2000. Cetacean distribution off eastern Antarctica (80–150°E) during the Austral summer of 1995/96. *Deep-Sea Res., II,* **47,** 2543–2572.

Thiele, D. and P. C. Gill, 1999. Cetacean observations during a winter voyage into Antarctic sea ice south of Australia. *Antarctic Science,* **11**(1), 48–53.

Treguer, P., and G. Jacques, 1992. Dynamics of nutrients and phytoplankton, and fluxes of carbon, nitrogen, and silicon in the Antarctic ocean, *Polar Biol.,* **12,** 149–162.

Trull, T. W., S. R. Rintoul, M. Hadfield, and E. R. Abraham, 2001. Circulation and seasonal evolution of polar waters south of Australia: Implications for iron fertilisation of the Southern Ocean, *Deep-Sea Res., II,* **48** (11/12), 2439–2466.

Tynan, C. T., 1997. Cetacean distributions and oceanographic features near the Kerguelen Plateau. *Geophys. Res. Lett.,* **24**(22), 2793–2796.

Tynan, C. T., 1998. Ecological importance of the Southern Boundary of the Antarctic circumpolar current. *Nature,* **392,** 708–710.

Voronina, N. M., 1968. The distribution of zooplankton in the Southern Ocean and its dependence on the circulation of water. *Sarsia,* **34,** 277–284.

Wakatsuchi, M., K. I. Ohshima, M. Hishida and M. Naganobu, 1994. Observations of a street of cyclonic eddies in the Indian Ocean sector of the Antarctic Divergence. *J. Geophys. Res.,* **99**(C10), 20,417–20,426.

Weber, L. H. and S. Z. El-Sayed, 1985. Spatial variability of phytoplankton and the distribution and abundance of krill in the Indian Ocean sector of the Southern Ocean. In *Antarctic Nutrient Cycles and Food Webs,* W. R. Siegfried, P. R. Condy and R. M. Laws, eds. Berlin, Heidelberg, Springer-Verlag, pp. 284–293.

Whitworth III, T., A. H. Orsi, S. -J. Kim and W. D. Nowlin Jnr., 1998. Water masses and mixing near the Antarctic Slope Front. In *Ocean, Ice, and Atmsophere: Interactions at the Antarctic Continental Margin,* S. S. Jacobs and R. F. Weiss, eds. Antarctic Research Series, Vol. 75. American Geophysical Union, Washington, D.C., pp. 1–27.

Wienecke, B. C. and G. Robertson, 1997. Foraging space of Emperor Penguins *Aptenodytes forsteri* in Antarctic Shelf waters in winter. *Mar. Ecol. Prog. Ser.,* **159,** 249–263.

Williams, M. J. M., K. Grosfeld, R. C. Warner, R. Gerdes and J. Determann, 2001. Ocean circulation and ice-ocean interaction beneath the Amery Ice Shelf, Antarctica. *J. Geophys. Res.,* **106**(22), C10, 383–399.

Wilson, P. R., D. G. Ainley, N. Nur, S. S. Jacobs, K. J. Barton, G. Ballard and J. C. Comiso, 2001. Adélie penguin population change in the pacific sector of Antarctica: relation to sea-ice extent and the Antarctic Circumpolar Current. *Mar. Ecol. Prog. Ser.,* **213,** 301–309.

Wong, P. S., N. L. Bindoff, and A. Forbes. 1998. Ocean-ice shelf interaction and possible bottom water formation in Prydz Bay, Antarctica, In *Ocean, Ice, and Atmosphere: Interactions at the Antarctic Continental Margin,* S. S. Jacobs and R. F. Weiss, eds. Antarctic Research Series, Vol. 75. American Geophysical Union, Washington, D.C., pp. 173–187.

Wong, A. P. S., Bindoff, N. L., Church, J. A. 1999. Coherent large-scale freshening of intermediate waters in the Pacific and Indian Oceans. *Nature,* **400,** 440–443

Worby, A. P. and I. Allison, 1999. A technique for making ship-based observations of Antarctic sea ice thickness and characteristics. *Antarctic Cooperative Research Centre Research Report,* **14,** Hobart, Australia, 63 pp.

Worby, A. P. and J. C. Comiso. Studies of the Antarctic Sea ice edge and ice extent from satellite and ship observations, 2004. Remote Sensing of Environment, **92,** 98–111.

Worby, A. P., R. A. Massom, I. Allison, V. I. Lytle and P. Heil, 1998. East Antarctic sea ice: A review of its structure, properties and drift. In *Antarctic Sea ice: Physical processes, interactions and variability,* M. O. Jeffries, ed. Antarctic Research Series, Vol. 74. American Geophysical Union, Washington, D.C., pp. 41–67.

Wright, S. W. and R. L. van den Enden, 2000. Phytoplankton community structure and stocks in the East Antarctic marginal ice zone (BROKE survey, January-March 1996) determined by CHEMTAX analysis of HPLC pigment signatures. *Deep-Sea Res. II,* **47,** 2363–2400.

Zwally, H. J., J. C. Comiso, C. L. Parkinson and D. J. Cavalieri, 2002. Variability of Antarctic sea ice 1979–1998. *J. Geophys. Res.,* **107,** 9–1 – 9–19.

INDEX

Zooplankton, 18-19, 21, 25, 26, 53, 64, 66, 73, 239, 275, 288, 358, 370, 418, 420-421, 422, 431-432, 494, 501, 509, 511, 520-527, 536, 656, 786, 793, 794, 810, 823, 904, 985; gelatinous, 423, 424